1

Fatal Army Air Forces Aviation Accidents in the United States, 1941–1945

Fatal Army Air Forces Aviation Accidents in the United States 1941–1945

ANTHONY J. MIRELES

Volume 1 : Introduction, January 1941–June 1943

McFarland & Company, Inc., Publishers
Jefferson, North Carolina, and London

Volume 1

LIBRARY OF CONGRESS CATALOGUING-IN-PUBLICATION DATA

Mireles, Anthony J., 1963–
Fatal Army Air Forces aviation accidents in the
United States, 1941–1945 / Anthony J. Mireles.
p. cm.
Includes bibliographical references and index.

ISBN-13: 978-0-7864-2106-0
3 volume set : softcover : 50# alkaline paper ♾

1. Aircraft accidents — United States — History — 20th century.
2. United States. Army Air Forces — History.
I. Title.
TL553.52.M57 2006 358.4'134 — dc22 2006008342

British Library cataloguing data are available

On the front cover: A Stearman PT-13B (AAF serial number 40-1643), Cope, South Carolina, March 6, 1945, United States Army Air Forces *(courtesy Craig Fuller, Aviation Archeology Investigation and Research);* fire and smoke ©2006 PhotoSpin

Manufactured in the United States of America

McFarland & Company, Inc., Publishers
Box 611, Jefferson, North Carolina 28640
www.mcfarlandpub.com

To the pilots and crewmembers
of the U.S. Army Air Forces
who flew and died in the United States
during the Second World War

Acknowledgments

The author would like to acknowledge the following individuals: Chris Baird, Ben Bilski, William J. Borah, Helen Borberly, Trey Brandt, Gary Bretz, Dennis Case (AFHRA), David Casiano, Joe Caver (AFHRA), Archie DiFante (AFHRA), Mike Ferrantelli, Craig Fuller (AAIR), Ed Furfano, Mary Grantner, Charles Hale, Wes Henry (AFM), Dale Housley (NPS), Steven Jensen, Norby King, Raymond J. Kowal Jr., Anita J. Kunke, Doug Lantry (AFM), Sara Legler, Suzanne Long, John Lopez Jr., Dave Maraville, Marie Maraville, Louis Marcus, Patrick McGraw, Terry Mireles, Eileen Moore, Tony Moore (X-Hunter), Richard Morrisroe, Kerri Nelson (NPS), Brian Padjen, Judy Pallay, Brenda Pietrowski (AFHRA), Herb Poppe, Chris Poulos, Dave Quinn, Stan "the Big Heat" Ridgway, Essie Roberts (AFHRA), Dee Dee Schmitt, Paul A. Schweizer, Thomas A. Starcevich, SSgt. Amy Stewart (USAF), Edward Stibili Ph.D., Steve Stimac, Nick Veronico, Ann Webb (AFHRA), Steve Wilson, Francis H. Wojciechowski, John B. Wojciechowski, Paul Wolframski, Joseph Yacoub Ph.D., Edward Zivich Ph.D., all my friends and family and Caroline Jerles. Thank you to all of you.

The author would also like to acknowledge the following firms and organizations: Aviation Archaeological Investigation and Research; Bulb Man, Reno, Nevada; Calumet College of St. Joseph, Whiting, Indiana; Census Microfilm, Salt Lake City, Utah; DeLaney Printing, Hammond, Indiana; Ferrantronics, Hammond, Indiana; Gray Supply Company, East Chicago, Indiana; The Hammond Public Library, Hammond, Indiana; Micro Maintenance Imaging Inc., South Holland, Illinois; Micrographic Equipment Services, Lansing, Illinois; National Park Service; Raytheon Corp., Wichita, Kansas; RFCU, Hammond, Indiana; Top Bulb, East Chicago, Indiana; USAF Historical Research Agency, Maxwell AFB, Alabama; USAF Museum, Wright-Patterson AFB, Dayton, Ohio; www.warbirdcrash.com. Thank you.

Contents

Volume 1

Volume 2

Volume 3

Introduction

As a young history student I read the book *Mission with LeMay* by U.S. Air Force General Curtis E. LeMay (with MacKinlay Kantor) in which the general stated that during the summer of 1945 it was actually safer to fly a combat mission over a defended Japanese city in a Boeing B-29 than it was to fly a B-29 training mission back in the United States. The fatality rate in the B-29 training program late in the war exceeded the fatality rate in combat, said LeMay.

As time marched on I never forgot those amazing remarks by one of the Army Air Forces' most effective wartime combat commanders. And a few years later while I was studying 1940s era newspapers while doing research for a college paper on World War II Hollywood, I kept coming across curious little newspaper articles describing military airplane crashes that were occurring in the continental United States during that time. The articles would read something like this:

> **TEN FLIERS KILLED IN FATAL BOMBER CRASH**
>
> ROSWELL, N.M.—(News Service)—Ten Army fliers were killed last night when their four-engine bomber crashed 17 miles southeast of Roswell. Army authorities did not immediately announce the names of the dead airmen.

Articles similar to this fictional example appeared in American newspapers every day during 1943 and 1944. Sometimes there were two or three similar air crash articles buried in the back pages, crowded out by the big war stories of the day.

These anonymous articles raise inevitable questions: Why did this airplane crash? Who were these unidentified airmen? Yet these vague articles likely raised little real concern among the general population, unless of course a loved one or friend was one of the unidentified airmen. And as the war went on, fatal stateside Army Air Forces (AAF) aircraft accidents began to reach epidemic proportions.

Hollywood acknowledged the phenomenon of stateside aviation accidents in its own way in the poignant U.S. home front–themed motion picture *Since You Went Away* (David O. Selznick Productions/United Artists 1944). In the film, Claudette Colbert and Joseph Cotten attend a servicemen's dance at the local airfield, where they encounter a local boy wearing flight officer insignia and pilot's wings. He greets them and tells them that he is looking for one of his gunners in preparation for a "routine flight." My grim imagination immediately went into action with the words "routine flight," and I was not surprised that some minutes later air officers alerted Joseph Cotten and Claudette Colbert that the young pilot's bomber had crashed in flames just outside of town with the loss of all on board. It was pretty close to reality. By the end of 1943 AAF aircraft accidents were occurring with an appalling frequency, alarming and angering some AAF pilots and commanders.

In his famous book, *Yeager—An Autobiography,* U.S. Air Force General Chuck Yeager (who had the rank of Flying Officer in 1943) addresses the AAF aviation accident problem in a brutally frank manner, referring to the unfortunate pilots as "weak sisters" who were inevitably and cruelly weeded out before they could cause problems in combat. Yeager was angered by the high amount of accidents and the loss of life. His unit experienced the loss of 13 pilots during six months of intense training at Tonopah, Nevada, in late 1943. Yeager—and LeMay even more so in his command capacity—was especially angered by the loss of expensive airplanes by careless pilots. But not all AAF airplane crashes were caused by pilot carelessness or aviation horseplay. Some accidents were caused by poor aircraft and component design, inadequate training, the carelessness of others, bad weather, improper maintenance, and sometimes just plain bad luck.

The incredible number of AAF aircraft accidents that occurred in the United States during World War II was the inevitable precipitate of the enormous AAF expansion and reorganization that occurred after 20 June 1941, when the Air Corps evolved into the Army Air Forces. In June of 1941 war was looming large in the future of the United States. The armies of the Third Reich had defeated Poland in 1939 and by June of 1940 had defeated France and her continental allies. They had kicked the British off of the European continent and were bombing English cities nightly. Two days after the AAF reorganization, the Wehrmacht rolled east into Soviet Russia and within a few months had smashed their way to the gates of Moscow. Relations between the United States and Imperial Japan, already rocky, deteriorated further after aggressive Japanese moves in Indochina, and it was just a matter of time before the United States became involved in the globe-spanning conflict.

In response to the gathering war clouds, forward-looking leaders like AAF General H.H. Arnold and President Franklin D. Roosevelt made the expansion of the Army Air Forces a national priority. President Roosevelt had called for an aircraft production program capable of producing "50,000 planes a year." At the time, in early 1940, even industry insiders thought these numbers unreachable. The country was deeply isolationist prior to the Japanese attack on Pearl Harbor, and most people wanted the United States to stay out of the war. In 1941, necessary legislation for national defense passed Congress by a one-vote margin and the AAF began getting the resources it needed to expand. The AAF added nearly 4,000 aircraft of all types to its inventory by the end of June 1941, with a total of 6,777 aircraft of all types on hand, up from 2,966 the previous year. U.S. aircraft factories produced 19,433 airplanes for all of 1941 and the AAF received 15,861 of these aircraft. This number includes aircraft accepted by the AAF under the Lend-Lease Act on behalf of Great Britain. Aircraft production more than doubled in the year after the United States entered the war — U.S. aircraft factories produced more than 48,000 airplanes for all of 1942. This number was nearly doubled in 1943 with 85,898 military aircraft being produced. For the year 1944 U.S. aircraft factories produced an astounding 96,318 aircraft, the majority going to the AAF. AAF factory acceptance for 1944 exceeded 70,000 military aircraft and trainers. Production was cut back as the war began to wind down. More than 48,000 airplanes were produced by the United States during 1945, but by 1946 AAF factory acceptances had dropped to almost nothing.

The expansion of the AAF aircraft inventory demanded a corresponding increase in pilots, crew members and ground crew. The entry requirements for Aviation Cadet training were stringent, but the AAF had no shortage of would-be flyers. The AAF required all cadet applicants to be at least 20 years of age and in perfect health, with at least two years of college. A series of examinations determined an applicant's suitability for flight training. The flight training was demanding, washing out one-third of the cadets in each of the initial three phases (Primary, Basic and Advanced). After the successful completion of ground school training and 200 hours of flying time, the cadet was commissioned as a second lieutenant or flight officer in the Army Air Forces and assigned to transition training. For multi-engine pilots this training consisted of a ten-week course and 105 hours of tactical training in first-line combat aircraft; single-engine pilots were required to take a five-week transition course.

By late 1942 the AAF was training thousands of pilots and crew to fly the thousands of newly produced airplanes, and the fatal accident rate went up accordingly. In 1941 fatal AAF aircraft accidents numbered 233, resulting in 417 fatalities, 86 injuries and 259 aircraft damaged or destroyed. During 1942 the number of fatal aircraft accidents had increased to 1,108, resulting in 2,395 deaths, 277 injuries and the loss of over 1,200 aircraft. The large increase in fatal AAF flying accidents was alarming and some spectacular accidents began to attract government attention. A fatal head-on runway collision between a Douglas B-18B and a civilian-operated Douglas C-49 at 36th Street Airport in Miami, Florida, on 22 September 1942 and a fatal mid-air collision between a Lockheed-Vega B-34 and an American Airlines Douglas DC-3 over the San Gorgonio Pass near Palm Springs, California, on 23 October 1942 triggered two separate congressional investigations. Despite this official attention, the number of fatal AAF aircraft accidents increased unabated.

In 1942 AAF commanders addressed the fatal accident problem internally, producing a classified memo titled "Our Safety Problem." Apparently the self-analysis did not help — there were 2,268 fatal AAF aircraft accidents in 1943, averaging to an amazing six fatal accidents per day. Those accidents resulted in over 5,600 fatalities, over 1,000 injuries and 2,536 aircraft damaged or destroyed. The number of fatal AAF accidents in the United States dropped as more and more pilots and airplanes were sent overseas. There were 1,948 fatal AAF aircraft accidents in 1944, a 14 percent drop from 1943. Despite the drop in fatal accidents, another 5,000 airmen were killed. By early 1945 the AAF had begun to cut back on flight crew training and the fatal accident rate began to drop even further. In the final year of the war there were 793 fatal AAF accidents, resulting in over 2,100 deaths and the destruction of nearly 900 aircraft. There were over 6,350 fatal AAF aviation accidents in the continental United States during World War II, resulting in the loss of 7,114 airplanes and the deaths of over 15,530 pilots, crew members and persons on the ground. And no one was immune — the leading U.S. fighter ace, Maj. Richard I. Bong, was killed in a crash 6 August 1945 while testing the jet-powered Lockheed P-80 Shooting Star at Burbank, California.

During the Second World War, 439 AAF personnel were killed while undergoing primary flight training; 1,175 AAF personnel were killed during basic flight training; 1,888 AAF personnel were killed during advanced training. The AAF aircraft accident rate for its flying schools during World War II (both fatal and non-fatal accidents) was as follows: Primary training phase, 48 accidents per 100,000 flight hours; Basic training phase, 27 accidents per 100,000 flight hours; Advanced training phase, 55 accidents per 100,000 flight hours.

During the Second World War, thousands of AAF fliers were killed and thousands of AAF aircraft were destroyed in what essentially amounted to a third air front over the continental United States. It is interesting to point out that the AAF lost over 4,500 aircraft in actual combat against Japanese army and naval air forces, at the same time losing over 7,100 aircraft in the states while moving airplanes around and teaching people how to fly. The reader should keep in mind that there were also thousands of non-fatal accidents and incidents involving AAF aircraft operating in the United States, injuring thousands of personnel and destroying or damaging a large number of aircraft. It should also be noted that a large number of fatal U.S. Navy, Marine Corps, Coast Guard, Civil Air Patrol and civilian aircraft accidents occurred in the United States during those same war years. The thousands of largely forgotten AAF personnel that were killed while flying in the United States were unfortunate victims of the cataclysmic struggle that was World War II just as if they had been shot dead by a German or Japanese soldier on a battlefield. Their ultimate sacrifice and their tragic stories should be remembered.

The author would like to advise the reader that to avoid confusion all aircraft are referred to by their primary designer/manufacturer. For example, a B-17F manufactured by Douglas or Lockheed-Vega will always be referred to as a Boeing B-17F in the text. The author has also done away with most hyphenated aircraft manufacturer names such as Consolidated-Vultee or Boeing-Stearman. Although technically correct, they are cumbersome for the author and confusing to the reader who is unaware of the vagaries of U.S. aircraft manufacturers of the 1940s. Because of incomplete information on the original AAF Form # 14 Aircraft Accident Reports, there will be no distinction made of variants and sub-variants of a particular aircraft model or series. For example, an aircraft will be referred to in the text as a Bell P-39D rather than as a Bell P-39D-20-1.

The vast majority of the accident summaries in this book are based on information from the original AAF Form # 14 Aircraft Accident Reports. The author is merely reporting facts and is not finding fault with the flying ability or the judgment of any named or unnamed individuals, living or dead. The author does not intend to reach any definitive conclusions concerning the performance of any aircraft, component or aircraft accessory. In order to make subsequent editions of this work as complete and accurate as possible, the author encourages readers to submit corrections or other pertinent information via the publishers.

Anthony J. Mireles
Calumet City, Illinois

Abbreviations

1Lt.	First Lieutenant
2Lt.	Second Lieutenant
AAB	Army Air Base
AAC	Army Air Corps
AAF	Army Air Forces
A/C	Aviation Cadet
AFCE	Automatic Flight Control Equipment
agl	above ground level
A/S	Aviation Student
Aspt.	Aspirant
CAA	Civil Aeronautic Administration
CAP	Civil Air Patrol
Capt.	Captain
Col.	Colonel
Contact	Visual flying conditions
Cpl.	Corporal
CWT	Central War Time
Ens.	Ensign
EWT	Eastern War Time
F/C	Flying Cadet
F/O	Flying Officer
GMT	Greenwich Mean Time
Hg	mercury (in inches)
IFR	Instrument Flight Rules
Lt.	Lieutenant
Lt. Col.	Lieutenant Colonel
Lt. (j.g.)	Lieutenant (junior grade)
Maj.	Major
MSgt.	Master Sergeant
msl	mean sea level
MWT	Mountain War Time
Pfc.	Private First Class
Pvt.	Private
PWT	Pacific War Time
RAF	Royal Air Force
RCAF	Royal Canadian Air Force
Sgt.	Sergeant
SSgt.	Staff Sergeant
Sub-Lt.	Sub-Lieutenant
TSgt.	Technical Sergeant
USMA	U.S. Military Academy
VFR	Visual Flight Rules
WASP	Women's Air Force Service Pilots
W/O	Warrant Officer
Zulu Time	Greenwich Mean Time

A Note about Dates, Times, and Entry Numbers

All times are local unless otherwise indicated and are based on World War II era time zones. Beginning February 9, 1942, most regions in the continental United States moved clocks one hour ahead to "War Time," which ended in October of 1945.

Dates serve as entry numbers and are listed in the Month/Day/Year format. Example: the entry 1-2-41 represents January 2, 1941. When more than one accident occurred on a given day, alphabetical indicators have been attached to the date entries to show the order in which the accidents were recorded and to help the reader distinguish among them. Example: the entry 7-19-43D represents the fourth fatal accident recorded for July 19, 1943.

1941

January

1-2-41. Laredo, Texas. At 1040, a North American BC-1I attempting a landing stalled and spun into a golf course east of Laredo Municipal Airport, Laredo, Texas, killing pilot Capt. Eugene C. Tips and passenger TSgt. Hector E. Chessi. The aircraft was in the pattern for the airport at about 600 feet agl when witnesses heard a dramatic change in propeller pitch. The airplane then began a 30-degree banked left turn from base to final when it went into a nose high attitude and a steep right bank before spinning to the right and into the ground. There was no fire but the aircraft was totally destroyed, the fuselage breaking in a shearing fashion just aft of the firewall. The Accident Classification Committee stated that the pilot was new to the aircraft and had "... erred in his selection of levers, moving the flaps to the [full] down position instead of [lowering] the wheels as he intended, and the propeller [pitch] control to full high instead of full low position. That the effect of these mistakes combined with a probable reduced flying speed caused the airplane to execute the characteristic vertical reversement [sic] and resultant spin at an altitude too low to allow recovery from the spin."

1-3-41. Kelly Field, Texas. At 0040, a North American AT-6A on a night training mission stalled and spun into Kelly Field, San Antonio, Texas, from an altitude of 50 feet agl during a landing attempt, killing A/C Harry K. Hemingsen.

1-6-41. Barksdale Field, Louisiana. At 1735, two Seversky AT-12 airplanes (similar to the P-35 type) collided on the runway at Barksdale Field, Shreveport, Louisiana, killing 2Lt. Henry G. Strachan, Jr. (AT-12 #41-17536) and seriously injuring 2Lt. Edward A. Sanders (AT-12 #41-17538). Lt. Sanders started a take-off run and collided with the aircraft of Lt. Strachan, who was momentarily stopped on the runway in front of him. Lt. Sanders had landed behind Lt. Strachan and apparently failed to see that Lt. Strachan was still on the runway when he began his take-off. The Accident Classification Committee noted that Lt. Sanders' sunglasses, the approach pattern, the lack of radios in the subject airplanes, poor visibility in AT-12 type airplanes and pilot error as causative factors in the accident.

1-7-41. Langley Field, Virginia. At 1505, a flight engineer was killed when he was crushed between a tow cable and a Douglas B-18A bomber that he and his crew were trying to free from mud at Langley Field, Hampton, Virginia. Sgt. Robert C. Homan, 25, Center County, Pennsylvania, was killed when the B-18, using its engines and a light tow tractor, suddenly lurched out of the mud and over ran the cable. The resultant slack allowed the cable to be fouled by the aircraft landing gear. When the tow tractor maneuvered to get out of the way of the moving aircraft, the cable slack was taken up with Sgt. Homan caught between the cable and the aircraft, fracturing his skull and killing him instantly.

1-16-41. Morton, Washington. At 1050, a Douglas B-18A collided with Deschutes Peak seven miles NNW of Morton, Washington, killing seven crewmembers. The B-18 took off from McChord Field, Tacoma, Washington, and was not seen again until its wreckage was found on 2-3-41 by two civilian woodsmen, Harry Studhalter and Tom Harper. The airplane had been flying on instruments in heavy overcast prior to the accident. The seven fliers killed were: 1Lt. Robert M. Krummes, pilot; 2Lt. Charles T. Neilsen, co-pilot; 2Lt. John F. Geis, navigator; TSgt. Hearn A. Davis, engineer; SSgt. Leo H. Neitling, radioman; Sgt. Paul L. Mass, bombardier; 1Lt. Lewis E. Mackey, passenger. Lt. Krummes had attempted the same flight the day before but had been turned back by bad weather. The Accident Classification Committee stated, "From the appearance of the wreckage, it appears that both engines were operating at the time the crash occurred and the airplane was in near level flight. The airplane was on a heading of true north and struck the face of the peak at almost a right angle [25 feet from the top of the 5,692-foot peak]. A possible cause for the crash may have been an erroneous reading by the pilot of his altimeter, believing himself to be at an altitude greater than that at which he was actually flying. The crash appeared to have occurred at approximately 1050 as evidenced by one wristwatch and a face of an airplane clock."

1-29-41A. Tucson, Arizona. At 0850, a North American BT-14 collided with a large cactus while maneuvering at low altitude near Tucson, Arizona, killing pilot 2Lt. Joseph A. Neil, Chattanooga, Tennessee, and seriously injuring passenger A/C Leroy T. Tempest, Fairfield, Washington. The Accident Classification Committee stated, "[The] pilot flew low at the request of passenger in order that the latter might take photos of giant cactus. Photos secured (8mm movie) disclosed pilot to have flown at 10 to 15 feet altitude over terrain covered with many giant cactus. The cactus range in height from 20 to 30 feet. Last frame of movie show plane banking sharply to right to avoid cactus with left wing, at which time plane must have hit cactus with right wing. From this point on pilot lost control of plane — it progressed approximately [one quarter of a] mile further, hit another cactus and catapulted onto its back with wing sheared off, engine and forward stations of fuselage broken off and various parts scattered over at least [one] acre of ground."

1-29-41B. Shafter, California. At 0805, a Vultee BT-13 on a routine training mission dove into the ground at full power near Shafter, California, killing Aviation Cadet Robert E. Baylis, 25, Wenatchee, Washington. The Accident Classification Committee stated, "The student either tried to dive through a hole to get under the fog patches and struck the ground with full power, or was doing stalls or lazy eights just above the fog and stalled and struck the ground with full power while trying to regain flying speed."

1-31-41. Maxwell Field, Alabama. At 1030, a Vultee BT-13 stalled and spun into the ground ten miles southeast of Maxwell Field, Montgomery, Alabama, killing flight instructor 2Lt. Frederick H. Morris and student Aviation Cadet Woolford B. Baker. The BT-13 struck the ground out of control and with a high rate of vertical descent.

February

✓ **2-4-41A. Centerville, Illinois.** At 0930, a Fairchild PT-19 suffered a catastrophic structural failure of the starboard wing and crashed near Centerville, Illinois, killing civilian flight instructor S.L. Harris and student A/C E.E. Hersman. Investigation revealed that one of the propeller blades failed at the hub and flew off, striking the starboard wing. The resultant excess vibration of the damaged propeller caused the starboard wing to fail and break off about one foot beyond the center section, the airplane plummeting to the ground and bursting into flames upon impact. The instructor and the student were not able to parachute to safety.

2-4-41B. Northport, Long Island, New York. At 1440, a Curtiss P-40 involved in high altitude tests crashed after suffering a catastrophic failure of the port wing while in level flight over Northport, Long Island, New York, killing pilot 2Lt. S. E. Denny. The P-40 was seen to go into a steep dive sometime before the accident. Eyewitness accounts indicate that the airplane had traveled straight and level for some minutes before the wing came off.

2-6-41A. Wyandanch, Long Island, New York. At 1035, two Curtiss P-40 aircraft collided in mid-air while in simulated combat over Wyandanch, Long Island, New York, killing pilot 1Lt. John H. Eakin, 25, aboard P-40 #39-214. 2Lt. Joseph T. Laycock, 23, was seriously injured parachuting from P-40 #39-202 (See 5-3-42B). Lt. Laycock stated, "My airplane had slowed up until it was now making about the same forward speed as his. As I watched the other airplane, it seemed to come sideways into mine and they crashed wing to wing with considerable force. I had no time turn away from the other airplane to avoid the collision. A moment after the collision, as the other airplane passed over the tail section of mine, some part of it struck the tail or canopy of my airplane and caused the canopy to shatter. Glass particles flew into my face. I was unable to keep my airplane in a flying attitude and it fell off on the left wing, which was crumpled, and went into a dive. After the collision, which occurred at an altitude approximately 9,000 feet, I did not again see the other airplane."

2-6-41B. Lovelock, Nevada. At 0940 PST, a Boeing B-17B flying in instrument conditions suffered a catastrophic structural failure and crashed onto the west slope of Ragged Top Mountain in the Trinity Range 19 miles west of Lovelock, Nevada, killing eight fliers. The B-17B, which was based at Ladd Field, Fairbanks, Alaska, and was specially equipped as a cold weather laboratory, took off from McClellan Field, Sacramento, California, on a flight to Wright Field, Dayton, Ohio, via Denver, Colorado. The airplane crashed over an hour after taking off. Civilian Conservation Corps camp foreman Lloyd M. Carpenter witnessed the crash, which scorched over 100 yards of mountainside. The airplane was flying at about 12,000 feet when it entered instrument conditions. The B-17 encountered very turbulent conditions and apparently went out of control, entering a violent spin. The B-17 was observed spinning out of the overcast, which was at 8,500 feet msl, or about 1,500 feet agl, shedding pieces. The engine cowlings failed and separated as the airplane plunged to earth. The pilot apparently stressed the airplane beyond its design limitations in an attempt to re-establish level flight. The outboard wings failed and separated, the pieces colliding with the tail section and causing its progressive failure. The airplane broke up and smashed to the ground, exploding into flames upon impact. The crew was unable to bail out. Wreckage was scattered over an area of two miles on the west slope of Ragged Top Mountain. The wings, vertical fin and horizontal stabilizers were found over two miles from the main wreckage. Investigation revealed that the airplane was improperly loaded and the center of gravity was aft of the allowable limit, rendering the aircraft longitudinally unstable. The ship encountered very turbulent air and icing conditions after it entered the clouds. It was speculated that severe icing and turbulent conditions, aggravated by the loading error, caused the pilot to lose control of the airplane while flying in instrument conditions. It was speculated that the windscreen was covered with ice at the time of the accident. Killed in the crash were: Capt. Richard S. Freeman Winnamac, Indiana, pilot; 1Lt. E.W. Ketcham, co-pilot; TSgt. E.H. Gilreath, Clinchport, Virginia, engineer; Sgt. Joseph P. Davies, Dayton, Ohio, passenger; Sgt. Frank C. Whidden, Hanes City, Florida, passenger; Sgt. Elmer S. Trainer, San Jose, California, passenger; Sgt. Everette Crabb, passenger; Pfc. Charles E. Applegate, Canon City, Colorado, passenger. Capt. Freeman, who was the commander of Ladd Field and a pioneering Army Air Mail Service pilot, was a Mackay Trophy winner and had been awarded the Distinguished Flying Cross for his service in the Air Corps.

2-11-41. San Angelo, Texas. At 1620, a North American BT-14 crashed into a riverbank three and one-half miles from the Air Corps Basic Flying School at San Angelo, Texas, killing pilot 1Lt. Don Cascio and passenger TSgt. Thomas C. Cobb. 1Lt. William H. Turner stated to the Accident Classification Committee, "Lt. Cascio decided to leave the formation, not because he was in trouble or formation was not going well. I saw him pull violently up and turn to the left. The airplane half-rolled to the left and was approximately upside down when he started dive out. When he was in an inverted dive position, he apparently realized he did not have room to complete recovery from the dive and attempted to roll left, which he completed. Before he could level off he contacted the ground. I saw his left wing hit the riverbank and the airplane fall into the river."

2-19-41. El Paso, Texas. A Lockheed Hudson (A-29), one of ten in a shipment to Great Britain, crashed into a radio tower two miles from the municipal airport at El Paso, Texas, killing two civilian ferry pilots. The Hudson had just taken off from El Paso and was en route from the Lockheed factory in Burbank, California, to Canada. The cause of the crash was determined to be the pilot's failure to see the radio tower. The two civilian ferry pilots killed were pilot Marino Guglielmetti, 42, Petaluma, California, and co-pilot Robert McKee, 30, Spokane, Washington.

2-21-41. Amesville, Ohio. At about 2110 EST, a Northrop A-17A flying in a severe snowstorm struck a large oak tree and crashed near Amesville, Ohio, killing two fliers. It is believed that the pilot was trying to get under the ceiling while on a flight to Wright Field, Dayton, Ohio. The aircraft struck a large oak tree at the top of a ridge, losing its port wing before smashing into the ground some 200 yards later and coming to a rest against a 100-gallon fuel oil tank on the farm of Jenny Calentine. Pilot Capt. Lawrence J. Eyler, 47, Xenia, Ohio, and passenger Capt. Russell L. Montgomery, Stocksburg, Indiana, were killed in the crash.

2-26-41. Moffett Field, California. At 1525, a Vultee BT-13 on an instrument training flight collided with a power line and crashed into a valley 20 miles ENE of Moffett Field, Sunnyvale, California, killing pilot 2Lt. W. S. Rutherford and seriously injuring pilot-rated passenger 2Lt. H. G. Macdonald. The Accident Classification Committee stated, "Lieutenant Rutherford was flying quite low up the valley and failed to see a power line hanging about 200 feet above the floor of the valley. Examination of the area revealed that about eight feet of the right wing was cut by the power lines and lying near where he struck it. The line that was broken was dragged about 600 feet up the valley by the plane. The plane itself crashed about 1,500 feet up the valley from where the line was hit."

2-27-41. Maxwell Field, Alabama. A North American BC-1A on a routine training flight crashed under unknown circumstances at 1405 just north of the Alabama River ten miles west of Maxwell Field, Montgomery, Alabama, killing Aviation Cadet J. Raymond King, 23, Winnamac, Indiana. Cadet King had lost visual contact with the leader of his formation in the middle of a large cloud system and had been missing for over 24 hours before his aircraft was located on 2-28-41. The Accident Classification Committee stated, "The airplane struck the ground in an almost vertical position. Marks on the ground indicated that the airplane was not spinning at the time of impact and that the airspeed was so high that the airplane was practically buried in hard clay to a depth of over four feet. The body of Cadet King was found at the edge of the hole made by the airplane just to the rear of the tail surfaces, which were extending above the ground. While it is impossible to exactly determine the sequence of events from the time Cadet King left his formation to the time he crashed, the committee has noted the fact that he was approximately in the center of a layer of clouds about 2,000 feet thick."

2-28-41A. San Diego County, California. At 1430, a Ryan PT-20A stalled and spun into the ground on Kearney Mesa in San Diego County, California, killing Aviation Cadet J. E. Spillman and slightly injuring civilian flight instructor William E. Clark, who parachuted to safety. Instructor Clark stated to accident investigators, "I had been instructing my students in stalls and spins. The airplane was functioning normally. At [1400] of this date I began the flight with Cadet Spillman, which ended in the fatal accident at about [1430]. Cadet Spillman had a total of about five hours of dual instruction. At about 3,500 feet altitude and after practicing stalls, the student entered into what was to be a two-turn spin to the left. When the two turns were completed, I told him to come on with the right rudder, and because of his slow action, I started to apply right rudder, in doing this, I felt a sudden snap which seemed to be in the rudder. The airplane did not recover from the spin and I continued to hold on right rudder and tried to use the throttle, which did not help. Realizing the critical situation, I told the student to jump. I continued to hold right rudder and tried using flaps, but nothing helped. I continued to tell the student to jump, but he only stood up in the cockpit. I climbed onto the right wing as the ground neared and pushed away from the airplane. I struck the ground before realizing it. I ran over to the ship, which was badly wrecked with the student lying close by. He showed no signs of life."

2-28-41B. Randolph Field, Texas. At 2025, a North American BT-14 on a scheduled night training flight stalled and spun into the ground one-quarter of a mile west of Randolph Field, San Antonio, Texas, killing student Aviation Cadet Robert R. Reed. Investigation revealed that the airplane entered the spin at an altitude too low to allow recovery. The airplane remained in the spin until it smashed to the ground. The airplane burst into flames upon impact.

March

3-1-41. Kingman, Arizona. A North American Harvard (AT-6 type aircraft) en route to Canada via Salt Lake City, Utah, crashed at an unknown time 15 miles from Kingman, Arizona, shortly after take-off from the airport there, killing ferry pilot Joseph S. Martinek, Los Angeles, California. The airplane and pilot had been missing for five days before being spotted from the air on 3-6-41 by Paul Penrose, a pilot for North American Aviation Company (see 8-31-41B). The aircraft was one of a flight of seven that had departed the North American factory in California for delivery to the Royal Canadian Air Force. The aircraft were not equipped with radios and one was forced to land in Las Vegas, Nevada, because of poor weather.

3-3-41. Bakersfield, California. At 1305, a Vultee BT-13 stalled and spun into the ground one-quarter of a mile east of Rosedale Station near Bakersfield, California, killing flight instructor 1Lt. W. J. Alsop and student Aviation Cadet Donald J. Stanford. The Accident Classification Committee concluded, "The airplane was doing lazy eights at an altitude of approximately 3,000 feet. From this maneuver the plane rolled over on its back and dived out in an approximate half roll. Upon recovery from the dive the plane turned to the right and began the final spin from an altitude of 1,000 feet from which there was no apparent effort to recover. The committee had no evidence upon which to base an opinion that structural failure caused the failure to come out of the spin and the crash, and is forced to assume that the pilot failed to start recovery in time to stop the spin before reaching the ground. The airplane was so badly wrecked that any conclusions drawn from examination of the wreckage would be without basis and inconclusive."

3-5-41. Randolph Field, Texas. At 0945, a North American BT-9B stalled and spun into the ground eight miles northwest of Randolph Field, San Antonio, Texas, killing pilot 2Lt. William A. Sally and seriously injuring passenger 2Lt. Paul H. Snow. Investigators stated, "This accident was caused by the pilot's allowing the airplane to stall close to the ground while practicing forced landings; he attempted to recover, but allowed the airplane to stall the second time, to such an extent that it broke onto its back at such a low altitude that the pilot was unable to recover before the airplane struck the ground."

3-6-41. Langley Field, Virginia. At 1353, a Martin B-10BM crashed into the back river area of Langley Field, Hampton, Virginia, after experiencing carburetor icing and loss of power in the starboard engine, killing pilot 2Lt. Bernard P. Smith. The B-10 was attempting to land at the time of the accident when it went into a climbing turn to the right before the pilot cut the throttles causing an abrupt and violent stall. The airplane spun to the right and into the ground from about 75 feet.

3-8-41A. Lake Salvador, Louisiana. At 1140, a Martin B-10BM flying in poor weather was abandoned and crashed 10 miles south of Lake Salvador, Louisiana, killing one crewman. The airplane was lost and running low on fuel. The B-10 took off from Orlando Army Air Base, Orlando, Florida, at 0750 EST on a navigation training flight. The airplane began experiencing radio trouble near Mobile, Alabama, at about 1007. Private Winston T. Gant was killed when, for reasons unknown, he delayed his jump and struck the ground before his parachute opened. His body was found in close proximity to the wreckage, indicating that Pvt. Gant left the aircraft at a very low altitude. Pilot 2Lt. Donald W. Lang and 2Lt. Wilson Moore both parachuted to safety.

3-8-41B. Shafter, California. At 1455, a St. Louis PT-15 stalled and spun into the ground near Shafter, California, killing pilot 2Lt. Paul W. Bunch.

3-9-41. Coos Bay, Oregon. At 1925, a Northrop A-17, lost due to poor weather, ran out of fuel and ditched close to shore two miles north of the north jetty at Coos Bay, Oregon, killing pilot Major Roy P. MacDonald, 48. Passenger 1Lt. John J. Trauernicht, 34, was only slightly injured and swam to safety after trying to free Major MacDonald, who was stuck in the cockpit after the aircraft overturned in the ditching attempt. This was the fourth aircraft accident that Lt. Trauernicht had survived.

3-10-41. Hemet, California. At 1410, a civilian flight instructor and his student were killed when their Ryan PT-16A collided in mid-air with a Stearman PT-13 at Hemet, California. The pilot of the PT-13, Aviation Cadet Charles W. Ryckman, parachuted to safety. He later stated that at about 700 feet altitude, one half mile southeast of Hemet Airport, he felt a crash and saw his right wing fold up. He saw no other aircraft at this time. After some difficulty releasing his safety belt, Cadet Ryckman bailed out. Killed in the collision were the student, Aviation Cadet Lynn E. Rodgers and civilian flying instructor Gordon F. Rosenstock.

3-14-41. Moffett Field, California. At 1300, a Vultee BT-13 stalled and spun into the ground near Moffett Field, Sunnyvale, California, killing flight instructor 2Lt. Mortimer D. Gager and student, Aviation Cadet T. D. Rose. The airplane was apparently attempting a turn without power while performing a simulated forced landing. Lt. Gager had been a flight instructor for about a month.

3-15-41. Fontana, California. At 1315, a Stearman PT-13B crashed out of control after entering a power-on spin while practicing aerobatics near Fontana, California, killing Aviation Cadet Warren C. Dunaway. The Accident Classification Committee stated that "Flying Cadet Dunaway was on an authorized flight and in the practice of acrobatic maneuvers, attempted a series of snap rolls, the last of which began at stalling speed, thereby, precipitating a power spin resulting in a fatal crash, the airplane burning after impact with the ground."

3-22-41. San Marcos, Texas. At an unknown time, North American BC-1 crashed five miles northeast of the Army Air Field at San Marcos, Texas, killing Aviation Cadet Cecil F. Porter. Investigators determined by examination of the airplane and the surrounding area that "the airplane was descending from the southwest, that it first struck the tops of small trees and continued in a straight line, disintegrating along a distance of about 100 feet, with the main part of the fuselage being found at the final point of travel. Fire occurred at impact, and the plane was almost completely demolished. The pilot had been burned severely and thrown from the plane. The almost complete [destruction of the aircraft] indicated very high speed at the time of impact. In view of these circumstances, it is the opinion of the board, that: The airplane had been in a spin from which the pilot was unable to accomplish a recovery in time to prevent the accident."

3-25-41. Coalinga, California. At 0725, two Curtiss P-40 airplanes collided head on while flying at 2,000 feet agl directly over Coalinga Airport, Coalinga, California, killing 2Lt. George K. Williams (P-40 #40-294) and 2Lt. Charles T. Engeman (P-40 #39-242). Investigators determined that the cause of the collision was the failure of both pilots to see each other and avoid the collision.

3-26-41A. Mt. Meigs, Alabama. Sometime after 1800, a training aircraft that had taken off from Maxwell Field, Montgomery, Alabama, was abandoned between Mt. Meigs and Waugh, Alabama, after the aircraft began experiencing engine trouble. The pilot, 2Lt. Charles D. Bird, Bellefonte, Ohio, was killed attempting to parachute to safety. An unidentified flier parachuted to safety from this aircraft.

3-26-41B. Tallapoosa, Georgia. At 2055, a North American BC-1A on a night navigation training flight crashed out of control near Tallapoosa, Georgia, after encountering very heavy rain and reduced visibility, killing Aviation Cadet William F. Peak. The Accident Classification Committee stated, "Evidence indicates that Cadet Peak descended to an altitude of approximately 500 feet, lowered his landing gear and his flaps and approached the field at Tallapoosa from the northwest. After passing over the field he turned to his right toward the southwest and apparently stalled his airplane and started to spin to the left over a hill approximately 100 feet higher than the level of the field. It appears that at the time the spin started he was not over 300 or 400 feet above the level of the airport. Evidence indicates that the airplane struck the ground almost vertically and continued spinning to the left until it was headed approximately northeast." Poor weather was given as the official cause and was attributed as the cause of five non-fatal accidents involving this same night navigation training flight. Those accidents occurred at Tallapoosa, Carrollton; LaGrange, Georgia; Greenville, South Carolina; and Anniston, Alabama.

3-30-41. New Braunfels, Texas. At an unknown time, a North American BC-1I on a night navigation training flight crashed at the Circle Dot Ranch 14 miles north of New Braunfels, Texas, killing Aviation Cadet E. L. Hastings. The flight began at 2020 CST. Accident investigators stated, "Flying Cadet Hastings was on a routine night triangular cross country — Kelly Field, Texas, to Buchanan Dam, Texas, to Temple, Texas, and return via the Waco-Austin-San Antonio beacon line to Kelly Field at an altitude of 4,000 feet. He was checked in and out of Temple by radio with the ground control airplane and started toward San Antonio. He was overdue at Kelly Field and aerial searchers did not discover the wreckage until late the following afternoon. The crash occurred in an isolated spot, several miles from the nearest habitation or road and no witnesses or evidence was available as to the cause of the crash. The Committee, therefore, finds the cause of the accident 100% undetermined." See 4-1-41.

April

4-1-41. New Braunfels, Texas. At 0940, a North American AT-6 crashed 15 miles north of New Braunfels, Texas, killing two fliers. The airplane crashed while the pilot was directing ground parties to the wreckage of a North American BC-1I that had crashed 14 miles north of New Braunfels, Texas, on 3-30-41 (see 3-30-41). Pilot Capt. David N. Crickette and passenger TSgt. B. J. Williams were killed when the AT-6 stalled out of a low altitude turn and spun to the ground on the Wegner Ranch.

4-5-41. Banning, California. At 1045, a Stearman PT-13 crashed at the Auxiliary Airfield at Banning, California, killing A/C Lewis S. Craswell. The airplane entered a spin at too low an altitude to allow the student to recover before striking the ground.

4-10-41. Wright Field, Ohio. At approximately 1330, a Wright Field test pilot was killed when the Curtiss P-40 he was flying suffered a catastrophic structural failure six miles south of Wright Field, Dayton, Ohio. 1Lt. John N. Carlton, 27, St. Peter, Minnesota, was killed when the P-40 that he was piloting disintegrated after a high-speed dive and an apparent abrupt pull-up. Lt. Carlton had been testing P-40 cockpit enclosures and canopy crank handle operation during high-speed dives just prior to the crash. The pilot's parachute did not open and he apparently bailed out at less than 500 feet above ground level.

4-11-41. Camp Beauregard, Louisiana. At 0920, a North American O-47B crashed on take-off from Camp Beauregard, Louisiana, killing the pilot and seriously injuring two crewmembers. The Accident Classification Committee stated, "It was a normal take-off to the south with the motor functioning properly. The aircraft had reached a point about 1,500 feet south of the south edge of runway at an altitude of about 125 feet when the motor failed. The pilot continued southerly, turning about ten degrees in an attempt to set down in a fire lane in the woods. The airplane lost flying speed at a point where a stream flowed east and west striking the [south] bank of the stream, which was about 20 feet high." The pilot, 2Lt. Wilmer Esler, 26, was killed and 1Lt. Paul R. Smith, 40, and Pvt. Charles Yeasolones, 26, were both seriously injured. All were from Detroit, Michigan. Investigators could not determine the cause of the engine failure.

4-14-41A. San Angelo, Texas. At 1105, a Vultee BT-15 stalled and spun into the ground at the Campbell Ranch after an engine failure 25 miles east of the Air Corps Basic Flying School at San Angelo, Texas, killing Aviation Cadet Raymond L. Stephenson, 22, Sweetwater, Texas. The Accident Classification Committee stated, "The pilot apparently attempted to make a rapid 180 degree turn to the south (into the wind) and toward the field when the engine failed; in doing so the plane stalled and crashed. The possibility that the gas supply had become exhausted on one tank was considered by the committee."

4-14-41B. Veribest, Texas. At 1415, a Vultee BT-15 practicing forced landings crashed on a farm three miles south of Veribest, Texas, killing instructor 2Lt. Eliam B. Burns, Jr., 22, Ecru, Mississippi, and student A/C Frank E. Pogue, 25, Duluth, Minnesota. Farmer Henry Mazur, who witnessed the crash, stated that the aircraft appeared to be "idling" and crashed from a low altitude. The crash occurred on Pogue's 25th birthday.

4-21-41A. Coalinga, California. At 1336, a North American AT-6A on a cross-country training flight collided with a 4,000-foot mountain ridge 17 miles northwest of Coalinga, California, killing Aviation Cadet V. H. Duncan. The weather was clear, with unlimited visibility and light winds. The aircraft appeared to have struck the ridge in a climbing attitude, crashing violently into the ground 300 yards from the point of impact.

4-21-41B. Baton Rouge, Louisiana. At 1600, a North American AT-6A flying in poor weather crashed into a densely wooded area while attempting a forced landing eight miles northwest of Baton Rouge, Louisiana, killing two students. The aircraft was one of a flight of seven that had flown to the Florida coast for gunnery practice. On the way back to Barksdale Field, Shreveport, Louisiana, the flight broke up due to a 500-foot ceiling and thunderstorms in the area of the field. Three aircraft of this flight made forced landings within a few miles of the crash site. Killed in the crash were A/C Edward J. Gibbons, 26, Jersey City, New Jersey, and A/C Albert J. Fisher, 24, Hammond, Illinois. According to published accounts of the crash, both cadets were to receive their wings and commissions upon graduation on 4-25-41, and Cadet Gibbons was to marry Miss Virginia Richards, Bangor, Maine, that same day.

4-25-41. Gunter Field, Alabama. At about 1500, a Vultee BT-13 crashed after apparently trying to avoid a mid-air collision in the traffic pattern for Gunter Field, Montgomery, Alabama. Aviation Cadet William A. Tippet was killed in the crash, which occurred about one and a half miles northwest of the field. The Accident Classification Committee stated, "Cadet Tippet was returning to Gunter Field from a student training flight and was maneuvering to join south bound traffic for a landing to the north. The pilot of the other airplane was also joining Gunter Field traffic. [Cadet Tippet] did not see the other airplane soon enough to avoid a mid-air collision except by over controlling his airplane. An attempt was made to find the pilot of the other airplane in question but to no avail. The pilot of the other airplane either did not see the accident or did not recognize the emergency."

4-26-41. Trumbull Field, Connecticut. At 0730, a Curtiss P-40 crashed after suffering an engine failure while attempting to take off from Trumbull Field, Groton, Connecticut, killing pilot 1Lt. William A. Webber, Westbury, Long Island, New York. Investigators discovered that a rag that had been inadvertently left in the engine intake manifold screen was the direct cause of the engine failure. The P-40, which was part of the 33rd Pursuit Squadron, 8th Pursuit Group, crashed from an altitude of about 400 feet, narrowly missing a highly populated area.

4-27-41. United States. *Note: Most regions in the continental United States move clocks one hour ahead to pre-war daylight savings time.*

May

5-5-41. Wichita, Kansas. At 1100, an "experimental" Beech AT-10 crashed on take-off from Wichita Municipal Airport, Wichita, Kansas, killing pilot Maj. George P. Moody. After an extensive investigation, the Accident Classification Committee could not determine why the aircraft crashed out of control while entering a high performance climb out.

5-6-41. Arroyo Grande, California. A Stearman PT-13A experiencing engine trouble crashed into a moving van while trying to make a forced landing on U.S. Highway 101 near Arroyo Grande, California, fatally injuring the pilot and injuring one of the truck's three occupants. The PT-13 was part of a flight of eight airplanes being ferried from the Army Air Corps Depot at Sacramento, California, to the Hancock College Training Detachment, Hancock Field, California. Witnesses stated that the aircraft glided toward a wide part of the highway and collided with the truck at the top of a hill. The left wing struck the cab of the truck just above the windshield. The airplane plunged over the highway guardrail and, careening across the ground at the foot of a small hillside, crashed through a barbed wire fence before coming to rest half way up a hillside on the north side of the highway. The aircraft engine came to rest 20 feet further up the hill. The crash occurred at about 1407 and the pilot, 2Lt. Archibald Campbell, Jr., died about an hour later at Santa Maria Hospital. He was the son of Reverend Archibald Campbell, Sr., Princeton, New Jersey. Truck occupant Jack Burt was slightly injured.

5-9-41A. Sikeston, Missouri. At 0920, two Stearman PT-18 airplanes collided on the runway during take-off at an Army Auxiliary Airfield eight miles north of Sikeston, Missouri, killing one cadet and seriously injuring another. Aviation Cadet Dean W. Cooper, 21, Lafayette, Indiana, was killed when his PT-18 (40-1972) was struck from behind by a PT-18 (40-1996) flown by Aviation Cadet Reed C. Owens, Praise, Kentucky, after an idle problem caused Cooper's engine to die momentarily while on the runway. Cadet Owens had landed behind Cadet Cooper but failed to see that Cooper's airplane had not cleared the runway before attempting his take-off.

5-9-41B. New Orleans, Louisiana. At 1900, a North American AT-6A that had landed on an inactive runway collided with a U.S. Government truck operating at New Orleans Municipal Airport, New Orleans, Louisiana, seriously injuring the driver and fatally injuring the pilot. Aviation Cadet Seeburn E. Paine, 26, Meridian, Mississippi, whose airplane was part of a 38-airplane formation training flight from Maxwell Field, Alabama, fell forty minutes behind his flight and had to attempt his landing as daylight was fading away. The runway that the cadet landed on was not in service at the time of the accident and the U.S. Department of Engineers truck, driven by Casmir Mellerine, 26, Violet, Louisiana, did not have its headlights illuminated while operating on the inactive runway. Mellerine suffered a broken leg, broken arm, and cuts and contusions to the head. It is not known why Cadet Paine fell forty minutes behind his flight.

5-12-41. Hicks Field, Texas. At 1536, a Fairchild PT-19A crashed 200 yards west of Hicks Field, Fort Worth, Texas, killing Aviation Cadet Robert A. Whitford, Memphis, Tennessee. The Accident Classification Committee stated, "While on 360 [degree] overhead stage, the student made the first of two turns of 135 [degrees] each and turned into the last 90 [degree] turn. The airplane spun out of the turn to the right, executed two complete turns, and hit slightly nose first. According to the testimony witnesses, the student tried to bring the ship out of the spin, as he used throttle in the attempted recovery. There was insufficient altitude to effect a recovery."

5-13-41. Oxford, Michigan. At 1430, a Stearman PT-17 crashed out of control while performing aerobatics about one mile from Oxford, Michigan, killing pilot 2Lt. Jack S. Slade, 23, El Dorado, Arkansas, and mechanic/passenger Sgt. George Dellich, Anvil Location, Michigan. The airplane stalled and fell out of control after pulling up into a near vertical maneuver and going into a full wing stall with inadequate altitude for recovery. The airplane crashed onto the estate of M.E. Coyle, president of General Motors' Chevrolet Division.

5-16-41A. Carbon Hill, Ohio. At 2035 EST, a Beech AT-7 flying in a severe thunderstorm crashed into a hillside two miles north of Carbon Hill, Ohio, killing the crew of five. The AT-7, bound for Pittsburgh, Pennsylvania, took off from Louisville, Kentucky, at 1909 and encountered severe thunderstorms in the vicinity of Carbon Hill at 2020. High winds, severe lightning, heavy rain and hail accompanied the storms. Investigators determined that the airplane had circled the Carbon Hill area for about 15 minutes before it dove into a hillside and roadway at a 45-degree angle. The aircraft slammed into the hill with terrific force and wreckage was scattered over a wide area. Killed in the crash were: 2Lt. Robert Sonenfield, pilot; 2Lt. William Wiandt, co-pilot; 2Lt. James Criswell, passenger; 2Lt. R. L. Brown, passenger; Sgt. John H. Davis, radioman. The crew was based out of Barksdale Field, Shreveport, Louisiana.

5-16-41B. Martinsville, Ohio. At 2045, a North American AT-6 flying in severe thunderstorms collided with a hillside about one and a half miles east of Martinsville, Ohio, killing Capt. Ralph Von Derau and Capt. John C. Stanley. The AT-6 took off from Bolling Field, Washington, D.C., at 1756 on a navigation training flight and encountered thunderstorms that contained heavy rain, severe winds, lightning and hail. The aircraft contacted the ground with such force that the engine burrowed several feet into the ground and pieces of the airplane blocked a state road for a short time.

5-19-41A. Randolph Field, Texas. At 0955, a North American BT-14 stalled and spun out of control while practicing formation flying one mile northwest of Randolph Field, San Antonio, Texas, killing Aviation Cadet L. A. Dial. Investigators determined that Cadet Dial overran his leader and in his attempt to recover stalled the airplane.

5-19-41B. Rains, South Carolina. At 1150, two Douglas A-20A bombers collided in flight near Rains, South Carolina, killing all four crewmembers. The accident occurred during a formation training flight when aircraft #40-102 descended into aircraft #40-99, the lead aircraft of the element of three. Aircraft #102 severed aircraft #99's starboard wing with its port propeller. The pilot of aircraft #40-99, 2Lt. Allan J. Moore, was killed when he bailed out too close to the ground. Gunner Sgt. David L. Boyd was found in the wreckage that crashed to the ground 20 feet from the house of James N. Ludlam, burning down the house. 2Lt. Frank B. Ralston, pilot, and Cpl. Herbert O. Pruitt, gunner, were killed when Lt. Ralston attempted a crash landing of A-20 #102.

5-22-41A. Milpitas, California. At an unknown time, a Vultee BT-13 practicing a simulated forced landing crashed three miles south of Milpitas, California, killing instructor 2Lt. Warren P. Rece and student Aviation Cadet R. V. Nipper. The aircraft crashed into a wooded area and burst into flames upon impact.

5-22-41B. Poquoson River, York County, Virginia. At about 1600, a Stearman PT-17 that had taken off from Langley Field, Hampton, Virginia, crashed out of control into Back Creek, a branch of the Poquoson River, York County, Virginia, striking a small boat and killing its fisherman occupant. The aircraft had been buzzing small boats in the area just prior to the crash. Killed in the accident was Linwood Saunders, 38, Poquoson, Virginia. Pilot 2Lt. William H. Howard was slightly injured and his passenger 2Lt. Paul M. Lindsey was uninjured.

5-22-41C. Fontana, California. At 1510, a Stearman PT-13B crashed at Fontana, California, killing civilian flight instructor B. W. Ells and seriously injuring student Aviation Cadet R. B. Whitney. The airplane stalled and spun into the ground from an altitude of 200 feet during a simulated forced landing exercise.

5-23-41. Midvale, Utah. At 1020, a Stearman PT-13 crashed while performing aerobatics eight miles east of Midvale, Utah, killing pilot 2Lt. Ralph M. Tidwell and passenger Pvt. Guy V. Richards. Investigators speculated that Lt. Tidwell was practicing intentional spins when the crash occurred.

5-27-41. Kelly Field, Texas. At 1435, two North American AT-6A aircraft collided while attempting to take-off from Kelly Field, San Antonio, Texas, killing one pilot and seriously injuring another. The airplanes were part of an 18-ship student formation. Aviation Cadet Jack O. Glaze was killed when he lost control of AT-6A #41-674 and collided with AT-6A #41-681, piloted by 2Lt. Oscar G. Johnson, during the take-off roll. Lt. Johnson later stated, "His airplane suddenly swerved into mine from the left and out of my line of vision. I caught a fleeting glimpse of Cadet Glaze in his airplane just prior to the crash. Cadet Glaze was watching me and my airplane intently. The two airplanes were interlocked and skidded around to the right and then separated a few feet. I jumped out of the right hand side of my airplane and rushed to the aid of Cadet Glaze and tried to get him out of the cockpit. He was limp and slumped over the left edge of the cockpit and the canopy was open. The heat was too intense for me to extricate Cadet Glaze from his airplane."

5-28-41. Los Altos, California. At 1030, a Stearman PT-13 crashed out of control into a peach orchard after colliding with a pine tree at Los Altos, California, killing pilot 2Lt. Theodore C. Barrett, 23, Wichita, Kansas, and passenger Sgt. Albert Newmeyer, 23, Alhambra, California. Witnesses reported that the airplane had been buzzing the area prior to the crash.

5-30-41. Pacific Ocean, California. At 1120, a Curtiss P-40 crashed into the Pacific Ocean off Stewarts Point, California, after the pilot became separated from his wingman in poor weather, killing pilot 2Lt. Paul R. Rowe, 23, Fargo, North Dakota. Flying with Lt. Rowe was 2Lt. Lacy F. Mangleburg who stated to the Accident Classification Committee, "We had been instructed to make a formation flight, unit combat assembly. We encountered heavy broken clouds at about 1,000 feet to 1,200 feet northwest of the field, and crossed the mountains to the west and flew just off the beach where the ceiling was higher, about 1,500 feet. After flying in the vicinity just north of Stewarts Point for about 30 minutes the weather began getting worse and we started back. The clouds closed in on the mountains, so we started south, down the coast. It began raining ahead of us so I turned out toward the sea and began climbing through. At 8,000 feet I told Lt. Rowe to use oxygen and watched until placed the tube in his mouth. We came out at about 15,000 feet and I turned to head for Hamilton Field [California]. I attempted for 20 to 30 minutes to get Lt. Rowe to assemble on my wing and fly back but he continued to circle where he was. I used the radio and the assembly signal but to no avail. Lt. Rowe finally assembled, but as I turned to fly to Hamilton Field, he veered off to the left and disappeared into the overcast. I called for him to turn right as he veered to the left but he continued straight ahead. As he went into the overcast I called and told him to jump if necessary. Lt. Rowe tried to answer almost everything I radioed to him but I could seldom understand his transmission. Lt. Rowe never returned to Hamilton Field."

June

6-2-41A. Abilene, Texas. At 1440, three civilian Works Progress Administration workers were killed when the automobile that they were in was struck by a North American BT-14 aircraft that had inadvertently landed on a runway that was under construction at Abilene Municipal Airport, Abilene, Texas. It was not known why the privately owned automobile was parked on the closed runway. The accident occurred during a rainstorm and very poor visibility, slightly injuring pilot 2Lt. A. J. Richard. George E. Kniffen. Charles E. Belvin and Albert H. Walker were killed when the propeller of the BT-14 sheared the automobile's body to shreds.

6-2-41B. San Diego, California. A Consolidated LB-30 being test flown prior to delivery to the Royal Air Force was destroyed and four crewmembers killed when the bomber crashed and burned shortly after taking off from the Consolidated factory airfield at San Diego, California. Investigators speculated that a loose bolt had jammed the elevator controls. Among those killed in the crash was Consolidated Aircraft Company test pilot William Wheatly.

6-8-41A. Banning, California. At 0300, a North American BT-9 crashed into Hathaway Canyon five miles northeast of Banning, California, seriously injuring pilot 2Lt. Kent R. Granzow and killing passenger 2Lt. J. R. Brosheer, Medical Corps Physician. The flight originated at Tucson, Arizona, and encountered heavy clouds and fog at Banning Pass. The BT-9 crashed into the canyon while executing a 45-degree turn to the right to avoid entering the pass at too low an altitude. Despite serious head and leg injuries, Lt. Granzow walked for several miles to the Morongo Indian Reservation to summon help.

6-8-41B. Lyman, Wyoming. At 1425, a Douglas B-18 flew into rising terrain 15 miles east of Lyman, Wyoming, killing six crewmembers. The B-18 took off from Salt Lake City, Utah, en route to Chicago, Illinois, on a day navigation flight. Apparently the pilot lost visual contact with the ground and misjudged his location and altitude. Investigators determined that the aircraft had been in a steep descent at the time of the crash. Killed in the crash were: Capt. David N. Motherwell, 35, Gordonsville, Virginia, pilot; 2Lt. Lyman K. Harvey, Lakewood, Ohio, co-pilot; 1Lt. Theodore F. Holsteen, Burlington, Ohio, navigator; 2Lt. John V. Gallagher, 27, Bayonne, New Jersey, observer; SSgt. Terrence McGurgan, 26, Kenosha, Wisconsin, radioman. Close relatives of Capt. Motherwell alleged that unknown persons removed cash and a West Point graduation ring from the pilot's body. After an investigation, authorities could not find evidence that theft had occurred. The bodies were badly mutilated and the aircraft destroyed completely. The aircraft wreckage had been located by search planes on 6-10-41. Investigators discovered that civilian juveniles had visited the crash scene before Army authorities could reach the site. Published newspaper accounts reported that civilians had turned in the two propellers and other items from the wreck to the local scrap drive.

6-11-41. Yukon, Oklahoma. At 1555, a Fairchild PT-19A spun into the ground after entering an intentional spin five miles north of Cimarron Field, Yukon, Oklahoma. Killed in the crash was Aviation Cadet Robert K. Graham, 22, Chicago, Illinois.

6-12-41A. Selfridge Field, Michigan. At 1516, a Northrop A-17 crashed on take-off after suffering an engine failure due to fuel starvation on the southeast portion of Selfridge Field, Mt. Clemens, Michigan, killing pilot 2Lt. B. E. Pearson and Air Corps meteorologist Aviation Cadet Morris Hallett. Investigators stated, "On take-off, at about 100 feet altitude, the engine stopped. The pilot attempted a turn back to the field, lost flying speed, and crashed. On examination of the wreckage, the left gasoline selector valve was found in the 'off' position; the right selector valve was turned on the right front tank, which was known to be empty. This tank is normally kept empty for local flying to improve the flying characteristics of the airplane."

6-12-41B. Chattanooga, Tennessee. At 2205, a Douglas A-20A crashed during a forced landing at Chattanooga Municipal Airport, Chattanooga, Tennessee, killing one crewman instantly and fatally injuring another. A third crewman suffered minor injuries. The Accident Classification Committee found that the number 12 cylinder head on the port engine was driven off by impact with the piston of that cylinder, indicating a failure of the crankshaft and connecting rod linkage in flight. Bombardier Cpl. L. W. Henley, Columbus, Georgia, was killed instantly in the crash. The pilot, 1Lt. Henry P. Bacot, Lamar, South Carolina, died of his injuries at a Chattanooga hospital on 6-15-41. Pvt. Joseph Buturia, gunner was slightly injured.

6-18-41. Tulsa, Oklahoma. At 1340, a Fairchild PT-19A stalled and spun into the ground 12 miles north of Tulsa, Oklahoma, killing Aviation Cadet Joseph E. Quigley, 23, Vandalia, Ohio. The Accident Classification Committee stated, "Cadet Quigley, just prior to his fatal accident was engaged in a 180-degree overhead accuracy stage at Auxiliary Field No. 5, twelve miles north of Tulsa Municipal Airport. On his last pattern while negotiating his second power-off turn he allowed the airplane to stall and spin into the ground. From statements of witnesses it was ascertained that at the time of the stall and subsequent spin, he was approximately 800 feet high."

6-19-41. Oscoda, Michigan. At 0800, a Curtiss P-40 experiencing an engine failure stalled and spun into the ground after taking off from Camp Skeel Gunnery Range, Oscoda, Michigan, killing pilot 2Lt. Claude L. Van Zant, Jr. The pilot, who was scheduled to fly a gunnery mission, had made a normal take-off and climbed to 700 feet to wait for the rest of the flight to take off when he began to experience engine trouble. Lt. Van Zant radioed the field that the engine had quit and that he was going to attemt a dead stick landing. He attempted to make a complete circle of the field, lowering the landing gear after about 160 degrees of turn and then started his remaining 200 degrees of turn. After lowering the flaps he spun in from an altitude of approximately 250 feet.

✓ **6-20-41. Tacoma, Washington.** At 0815, a Stearman PT-17 collided with terrain while maneuvering at low altitude near Tacoma, Washington, killing pilot 2Lt. John R. Stapleton and pilot-rated passenger 2Lt. Arthur K. Patterson. The pilot apparently had performed a loop below 1,500 feet and was unable to pull out in time to avoid colliding with the ground. The airplane smashed into the ground at the bottom of the loop.

6-22-41. Mountville, Georgia. At 1440, a Douglas A-20A collided with terrain after attempting a 360-degree turn in a severe thunderstorm two miles northeast of Mountville, Georgia, killing three crewmembers. Investigators stated, "It is believed that the pilot heading north, entered a heavy rain storm before realizing its intensity. It appears as if the pilot decided to turn around [360 degrees] and get under the storm. He throttled the engines, lost altitude in this turn and struck trees, headed north, at downward angle of about 30 degrees with wings level. It is believed that when the wings were torn loose the airplane was approximately 30 feet off the ground. It is apparent from close examination of the wreckage that the gasoline tanks exploded violently, completely destroying the airplane and occupants. There was very little fire before the airplane first hit the ground. The airplane was so completely demolished that examination of the propellers, controls, instruments, etc. indicated no reason for the accident." Killed in the crash were: 2Lt. Stratford Lee Morton, Jr., Clayton, Missouri, pilot; Capt. Henry W. Herlong, Jacksonville, Florida, ordinance specialist; Pvt. Thomas Simms, 19, Huttig, Arkansas.

6-23-41A. Rubio Canyon, California. At 0730, a North American AT-6A crashed into the East Fork of Rubio Canyon near Mount Lowe, California, after colliding with telephone lines while trying to fly through the saddle of two peaks, killing pilot 1Lt. Jesse Jackson and passenger SSgt. Fred L. Pankey. Investigators stated, "Apparently the flight path was directed [through] the saddle in the higher mountain ridge across which the telephone lines were strung at a height from 80 to 100 feet above the ground. The airplane came in contact with the wires, which ripped away several feet of the leading edge of the left wing and the wing tip on the same side, which were found a short distance away from the broken lines. It appears that the airplane went immediately out of control, falling into the adjacent canyon, the walls of which are nearly perpendicular."

6-23-41B. Atlanta, Michigan. At 1600, a Lockheed YP-38 dove into the ground near Atlanta, Michigan, killing pilot 2Lt. Guy Putman. Investigators stated, "This accident occurred when pilot was on routine training flight. Witnesses claim to have seen the airplane dive at about 60 to 70 degrees straight down from a considerable altitude. No obvious attempt to recover from the dive was noticed. Airplane practically disintegrated, pilot mangled beyond recognition. It is the belief of the board that the pilot either blacked out or became unconscious by some unknown reason which would account for this accident." It is likely that the pilot put the airplane in a diving maneuver from which he could not recover because of the tendency of the P-38 type airplane to suffer from severe control surface buffeting caused by compressibility.

6-25-41. Randolph Field, Texas. At 2335, a North American BT-14 descended into the ground while trying to let down through low hanging clouds 18 miles west of Randolph Field, San Antonio, Texas, killing pilot 2Lt. John W. Crews and passenger Pvt. D.A. Eckelberg.

6-26-41. San Jose, California. At 1130, a Vultee, BT-13A stalled and spun into the ground while maneuvering near San Jose Airport, San Jose, California, killing flight instructor 2Lt. William C. Senner and student Aviation Cadet Waldo E. McIntosh, Jr.

6-27-41. Steedman, South Carolina. At 2205, a Vultee BT-13A collided with terrain while trying to fly visual contact with the ground in poor weather near Steedman, South Carolina, killing pilot 2Lt. Norman E. Priester, Jr., Columbia, South Carolina, and passenger 1Lt. Patrick W. Hodges, Macon, Georgia. Lt. Priester's parents were waiting at the Municipal Airport at Columbia, South Carolina, for their son's landing there. The BT-13 was not located until 0200, 6-28-41.

6-30-41. Gunter Field, Alabama. At 0915, a Vultee BT-13 crashed into the ground after suffering a catastrophic structural failure of the tail section while performing aerobatics five miles northeast of Gunter Field, Montgomery, Alabama, killing flight instructor 2Lt. Murro A. McCracken. Aviation Cadet Walter W. Griffin, Jr. parachuted to safety. A Gunter Field flight instructor stated to investigators, "I took off with my student and we were just leaving traffic and were climbing to an altitude of 2,000 feet when I observed a plane in a spin. I looked at my altimeter again to see if I was correct. I wondered why a plane was doing a spin at that low altitude, because it was the vertical type and the spin was very slow. I followed the plane all the way down until I saw it flatten out and then I thought that he had recovered. He flattened out at approximately 100 feet. He was spinning to the right when he flattened out and he continued to turn into the right slowly, and crashed into the trees. The airplane seemed to burst into flames immediately upon the crash."

July

7-3-41. San Pablo Bay, California. At 0830, a Curtiss P-40 crashed into San Pablo Bay three and one half miles NNE of Hamilton Field, San Rafael, California, killing pilot 2Lt. Gerald W. Martin, 25, Santa Anna, California. Investigators were unable to determine what caused the airplane to dive into the bay.

7-8-41. Hayes Store, Virginia. At 1730, a Northrop A-17A crashed out of control after entering severe thunderstorms in the vicinity of Hayes Store, Virginia, killing pilot 2Lt. George Samec, Jamaica, New York, and passenger Sgt. Harry S. Walker, Atlanta, Georgia. Investigators discovered that the aircraft descended into the ground out of control at an angle of 45 degrees or greater with full power. There were no witnesses to the crash, which was to be an individual proficiency mission from Langley Field, Hampton, Virginia, to Bolling Field, Washington, D.C.

7-10-41. Kathwood, South Carolina. At 1323, two Vultee BT-15 aircraft collided in flight near Kathwood, South Carolina, killing three civilian flight instructors. Aviation Cadet Vincent K. Relyea parachuted to safety from BT-15 #41-10008. Killed in the crash of BT-15 #41-10095 were J.W. Dionne, K. A. Yawn, and D.C. Bush, all civilians. Investigation revealed that instructor Yawn was practicing pylon eights near where Cadet Relyea and instructor Dionne were practicing a simulated forced landing. Instructor Yawn's aircraft severed the tail of Cadet Relyea's BT-15. Cadet Relyea stated, "I felt the impact of the other ship striking ours, and in an instant I knew our tail assembly was gone, because there was no control possible. As the ship started down, I opened the hatch entirely, unloosed my safety belt and fell out."

7-15-41A. Cochran Field, Georgia. At 1015, a Vultee BT-13A stalled and spun into the ground after an extreme maneuver to avoid an apparent mid-air collision during a formation training flight five miles west of Cochran Field, Macon, Georgia. Killed in the crash were Aviation Cadet Frank E. During, 22, Hollywood, California, and instructor 2Lt. Fred H. Barrett, 23, Warren, Maine. An instructor who witnessed the accident stated, "When I first noticed the ship it was just raising above the lead ship going into a steep loop. When the plane went over on its back it was lost to my view, and then I broke formation circled to the left and saw the ship on the ground the instant it burst into flames. The ship was crowding in too close to the lead ship and was jerked away and above from the lead ship."

7-15-41B. Evergreen, Alabama. At 2110, a North American AT-6A collided with terrain while flying in heavy thunderstorms seven miles north of Evergreen, Alabama, killing Aviation Cadet Winston S. Porter. The Accident Classification Committee stated, "The evidence at the scene of the accident shows that that the left wing of the airplane struck some small trees at the crest of a slight rise, at a high rate of speed and in a left bank. The airplane cartwheeled down a hill into the swamp, the nose of the airplane striking a large tree. The center section and the landing gear were torn free here and came to rest in the top of a large oak tree 50 yards further on in a direct line with the path of the airplane through the woods. The airplane was completely demolished. It is the opinion of the committee that Cadet Porter was seeking either the lighted landing field at Evergreen or an open field in the vicinity of Evergreen in which to make a landing during a heavy thunder shower." Two aircraft that were three minutes ahead and three minutes behind Cadet Porter made it around the thunderstorms.

7-16-41. Fontainebleau, Mississippi. At 1530, a Douglas O-38E crashed into a bayou after striking a power line three miles southwest of Fontainebleau Emergency Field, Fontainebleau, Mississippi, killing pilot 2Lt. James E. Baggett, Jr. and aerial observer SSgt. Emmett A. Cambre.

7-18-41. Eastport, Long Island, New York. At 1505, a Curtiss P-40 crashed into a wooded area one and a half miles from Eastport, Long Island, New York, decapitating the pilot, 2Lt. Robert E. Hetrick, 24, Dimondale, Michigan. The P-40 was caught in dense fog that enveloped the majority of Long Island and forced four other P-40 airplanes to land at alternate fields. The airplane struck a large tree and smashed into the ground at high speed at an angle exceeding 45 degrees, scattering wreckage over a quarter mile.

7-19-41. Seale, Alabama. At 0650 (eastern time), a North American AT-6A crashed under unknown circumstances near Seale, Alabama, killing the two pilots. The airplane disappeared soon after taking off from Lawson Field, Fort Benning, Georgia, on a flight to Kelly Field, San Antonio, Texas. The airplane was declared missing and an extensive search conducted. The airplane was finally located on 7-23-41 at 1330, 15 miles east of Lawson Field. An erroneous position report issued by Maxwell Field (Montgomery, Alabama) personnel hampered the initial search. Killed in the crash were Maj. Fred S. Stocks, 33, Honolulu, Hawaii, Executive Officer, Lawson Field, Fort Benning, Georgia, and pilot-rated passenger Capt. William S. Hargan, Pine Grove, Kentucky. Investigators determined that the aircraft struck the ground with a high rate of speed. Weather was reported to be scattered clouds at 900 to 1,000 feet with visibility at seven miles at Lawson Field. Two other aircraft had taken off with Maj. Stocks but returned to Lawson Field due to bad weather.

7-23-41. Alpena, Michigan. At 0905, a Lockheed YP-38 crashed near Alpena, Michigan, fatally injuring pilot 2Lt. James R. Taylor. Lt. Taylor stated to investigators before dying on 7-26-41, "When I took off everything was normal. Shortly after take-off, the right engine started to make a loud clanking noise followed by complete engine failure, propeller stopped still, had about 150 feet of altitude. Pushed left throttle practically wide open and attempted to correct for torque by use of rudder tabs, was unable to maintain full control of ship which turned towards the dead engine. Attempted to return to the field and came within 300 yards of making it ... lowered the gear ... ship more difficult to control. Ship finally turned into the ground and crashed."

7-24-41A. Ashley, Ohio. At 1542, a Royal Air Force Lockheed Hudson being ferried from Burbank, California, to Montreal, Canada, crashed and burned four and one-half miles from Ashley, Ohio, killing pilot 1Lt. R.F. Rush and co-pilot 2Lt. N.L. Warner (both AAF). The airplane had taken off from Patterson Field, Ohio, at 1500. Investigators stated, "Witnesses stated that the airplane was losing altitude with motors sputtering.... Examination of the [number-two engine] revealed that the lower edge of three (3) cylinder walls inside the crankcase had been broken on either side by being struck by the articulating rods, and also the end seal on the main bearing was broken. It is the opinion of the Board that the engine failure was the primary cause of the accident."

7-24-41B. Wasco, California. At 1430, a Vultee BT-13A crashed after striking a power line 12 miles west of Wasco, California, killing an instructor and his student while they were practicing simulated forced landings. The crash was attributed to the failure of the instructor and the student to see the power line. Killed in the crash were instructor 2Lt. C.M. Cox and student Aviation Cadet Byron Sylvestor.

7-24-41C. Corning, California. At 2100, a North American AT-6A crashed under unknown circumstances six miles south of Corning, California, killing Aviation Cadet Lowell C. Redfield, 23, Ogden, Utah. The aircraft took off from Stockton Field, California, at 2010 on a night navigation mission and was reported missing when it failed to return. The aircraft was found 7-25-41 at about 1630. Investigators described the aircraft as have hitting the ground in a slight dive and at very high speed with wings level. Weather at the time of the accident was reported as clear with unlimited visibility.

7-25-41. Albany, Georgia. At 0835, an RAF Flying Cadet was killed when RAF Flying Cadet R. Light crashed Stearman PT-17 #41-995 into a PT-17 parked on the flight line after over shooting his landing at Darr Aero Tech, Albany Municipal Airport, Albany, Georgia. Two civilian mechanics were also injured, one seriously. RAF Flying Cadet D. H. Wilson was killed in the accident. C. Holiey was seriously injured and L. D. Robinson sustained minor injuries.

7-29-41A. Fort Sam Houston, Texas. At 0915, 2Lt. Jules F. Rosenthal was killed when he walked into the rotating propeller of his Vultee-Stinson O-49 after landing on the polo grounds at Fort Sam Houston, Texas, with another O-49. The pilot of the other O-49, 2Lt. J. H. Soules, stated to investigators, "Lt. Rosenthal and I had landed two O-49 airplanes on the Sam Houston polo field. He taxied along side of my plane. He said something to me but not understanding what he said I shook my head. He parked the airplane with the engine idling and got out and started to walk around the right side of the engine and the propeller blade hit him in the head. He was thrown in front of the plane with his head almost torn off."

7-29-41B. Waskom, Texas. AT 0930, a flying instructor and his student were killed when their Cessna AT-8 was seen to go into a dive and continue straight into the ground four miles south of Waskom, Texas. Investigators could find no evidence to indicate structural or mechanical failure of any of the aircraft's systems. The aircraft was completely destroyed and was seen by witnesses to have entered a steep dive from a reasonably high altitude with no apparent effort to recover by the pilots. Weather was not a factor in the flight, which originated at Barksdale Field, Shreveport, Louisiana, at 0715. Killed in the crash were instructor 2Lt. R.J. Redd, 21, El Paso, Texas, and student A/C David W. Hancock, 25, Morocco, Indiana.

7-29-41C. Brooks Field, Texas. At 1145, a North American BC-1I crashed after suffering a structural failure of the port wing while apparently doing spins seven miles ESE of Brooks Field, San Antonio, Texas, killing instructor pilot 2Lt. Wallace B. Brown and student Aviation Cadet George N. Weber. Aviation Cadet Maynard E. White stated to investigators, "I was flying along at 3,000 feet when suddenly I noticed a plane in a spin at my left and I would say approximately 2,000 feet above me.... I immediately thought that the plane was in trouble and watched it close. At about my level I saw the wing tear from the rest of the plane ... the plane appeared to be in a very close, fast spin before the wing came off. Afterwards, it spun violently in a straight downward position and crashed, immediately bursting into flames and the wing falling close by."

7-31-41. Warrior, Alabama. At 2300, a North American AT-6A flying in poor weather collided with terrain near Warrior, Alabama, killing Aviation Cadet V. W. Ginaquitti. A witness on the ground stated that the aircraft dove into the ground in a shallow dive. Visibility in the area was reported as very poor. Cadet Ginaquitti was on a night navigation flight and only days from graduating from the Advanced Flying School at Maxwell Field, Montgomery, Alabama.

August

8-1-41. Marina, California. At 0915, one flier parachuted to safety and three others were killed when two Douglas O-38E type aircraft collided in mid-air three miles north of Marina Beach, Marina, California. The aircraft were part of a three-airplane formation training flight that had originated at Salinas Army

Air Base, Salinas, California. Killed in the crash of O-38E #33-06 were pilot 2Lt. John C. Rounsaville, 21, Alta, Texas, and passenger, Pvt. Marcus W. Owen, 22, Los Angeles, California. Their O-38 crashed into the sea. 2Lt. Burt Franklin Myers, 22, Rollin, Wyoming, parachuted to safety from his O-38E, which crashed onto the beach. His passenger, Pvt. Earl E. Potter, 20, Westville, Oklahoma, was killed in his bail out attempt when he was struck in the head by the horizontal stabilizer. His parachute was not deployed. The pilot of the third ship, 2Lt. Myers L. McCrory, stated to investigators, "Lt. Meyers and Lt. Rounsaville and I took off at about 0840. Lt. Meyers led for about 15 minutes; then Lt. Rounsaville led for about 15 minutes; then I took over the lead. At 0915, we were at about 2,000 feet near Marina, California, [and] I signaled for echelon right, which put Lt. Meyers behind me, followed by Lt. Rounsaville. Then I signaled a Lufbery circle then went down through a hole in the fog (ceiling estimated to be at 800 feet). I flew along the beach for about a mile then I went back up through the fog, when I got on top I waited for Lt. Meyers and Lt. Rounsaville. When they failed to appear I went back down through and after not finding them I went back to my home field and landed."

8-8-41. Gulf of Mexico, Florida. At 1015, a Seversky P-35 practicing gunnery on water targets crashed into the Gulf of Mexico eight miles south of Panacea, Florida, killing pilot 2Lt. Harold R. Levinson. Investigators determined that the P-35 pulled out too late from too steep an approach and caught the starboard wing tip in the water, causing the aircraft to cartwheel into the sea. The airplane originated its flight at Dale Mabry Field, Tallahassee, Florida, at 0910.

8-11-41. Atlantic Ocean, Virginia. At 2245, a North American O-47A flew into the Atlantic Ocean near Virginia Beach, Virginia, killing three fliers. The O-47 had taken off from Langley Field, Hampton, Virginia, at 2045 on a searchlight mission. Killed in the crash were pilot 2Lt. Adrian I. Wilcombe, radio operator Sgt. Allen McRogers and observer Pvt. Phillip R. Wheeler. 2Lt. C. H. Ferris (AAF) witnessed the accident and later stated to investigators, "At approximately [2230], a plane approached the Virginia Beach section from the north, flying south along the edge of the beach at an estimated 800 to 1,000 feet. It flew down the shore from my observation point at the *Surf Club* until its movements were unobserved. About five minutes later, the plane reappeared from the south, flying to the north, this time about one half a mile off the beach at an estimated 500 feet. As it came opposite the Surf Club, it made a ninety-degree turn to the left, flying west directly towards the club. This time the plane appeared to be in a glide, losing altitude. As it approached the club the landing lights came on and remained on until it passed over. Over the club the plane appeared to have about 300 feet of altitude and the engine sounded as if it were at about cruising throttle. No engine trouble was observed. After passing over the club the plane flew straight for only a few seconds and executed an extremely steep turn to the right and continued on around 180 degrees out to sea. In the execution of this turn it appeared to be at a lower altitude than when it flew over the club. Although my vantage point at that time was not good, the ship appeared from my angle to level out and fly level for about five seconds before disappearing into the sea." It was later speculated that the steep turn caused the airplane to lose altitude, causing the pilot to inadvertently fly the airplane into the sea.

8-12-41. Starkville, Mississippi. At 1201, a North American BC-1A crashed after an apparent engine malfunction or failure on take-off from Starkville Municipal Airport, Starkville, Mississippi, killing Aviation Cadet E. V. Brazeau. A civilian witness stated to investigators "[Cadet Brazeau] started his take-off to the southwest. About the middle of the field, it seemed he throttled back for just a second and then throttled up again. He then left the ground and appeared to be having some trouble clearing the power line at the south end of the field. Approximately over the power line a cloud of smoke came from his engine and then the nose of the plane began to drop and the port wing started down. From my position it appeared that the port wing hit the ground first."

8-13-41. Coleman, Texas. At 1150, a Fairchild PT-19A crashed seven miles southeast of Coleman, Texas, killing civilian flight instructors J.B. Ottzen, Tulsa, Oklahoma, and E.E. Driscoll, 23, Rochester, New York. The instructors were performing aerobatics at low altitude. The airplane stalled out of a maneuver and smashed to the ground on the farm of Frank McIntosh. The Accident Classification Committee stated, "This board is of the opinion that the accident was the result of an unintentional spin resulting from a hurried recovery from a preceding spin, the recovery of which was made dangerously low." According to published accounts of the crash, instructor Driscoll was to marry Miss Mary Wingfield of Norton, Kansas on 8-16-41.

8-14-41. East Wareham, Massachusetts. At 1252, a Douglas O-46A crashed after striking high-tension power lines while on a message dropping mission near East Wareham, Massachusetts, seriously injuring pilot 2Lt. Malcolm F. Nash and killing observer 2Lt. Alden C. Cole. Lt. Cole died of electrocution when he came in contact with the live wire during the collision. The aircraft took off Otis Field, Camp Edwards, Massachusetts at 1210. Lt. Nash stated to the Accident Classification Committee, "I was engaged in a message dropping mission during division maneuvers in the vicinity of the [command post] at the Miles Standish Reservation. I located the message dropping ground, which was difficult to find at low altitude after you flew away; and I made a pass to drop the first message from 200 feet. My observer, Lt. Cole, said I was too high so another pass was made at 150 feet when Lt. Cole threw out the message bag. My course was approximately from north to south with right hand turns and passing over a high-tension line at the end of the course. The first message was not observed to land so a copy of the message was prepared for the next try. This time we flew across the dropping ground in a glide with plenty of speed but quite low, and almost immediately after dropping the message I had just started to bank when I felt a sudden jolt and felt the ship dive. I knew nothing else until I came to in the ambulance."

8-19-41. Patterson Field, Ohio. At 1115, a Lockheed P-38D stalled and spun into Patterson Field, Ohio, killing pilot 2Lt. Perry H. Van Patten. The aircraft suffered the failure of the port engine just after take-off on a maintenance test flight. The aircraft had been flown to the field on 8-17-41 after experiencing problems with the port engine. The Accident Classification Committee stated that the cause of the engine failure was a fire of undetermined cause in the port engine intake manifold.

8-21-41A. Windsor Locks, Connecticut. At 0955, a Curtiss P-40C crashed out of control while engaging another P-40 in simulated combat over the Army Air Field at Windsor Locks, Connecticut, killing pilot 2Lt. Eugene M. Bradley. The other P-40 landed safely. The Accident Classification Committee stated, "It is the opinion of this committee that insufficient evidence exists to permit an exact [cause] of this accident. The accident followed a short period of individual combat, which ended when the pilot spun out of a violent left turn at about 5,000 feet. The airplane then followed an indefinite turning descent to about 800 feet where it ceased to turn and dove into the ground at a slight angle." The committee speculated that the pilot might have blacked out during the tight left turn or rapid descent causing a loss of control.

8-21-41B. Luke Field, Arizona. At 2330, a North American AT-6A on a local night training mission flew into terrain while in the pattern for Luke Field, Phoenix, Arizona, killing Aviation Cadet Virgil T. Sander, 24, Hayes, Kansas. Cadet Sander was a second lieutenant in the infantry reserves before entering aviation training as a cadet. The airplane was part of a group of several AT-6 type airplanes practicing night landings. 2Lt. Otis L. Elliot witnessed the crash and later stated to investigators, "[Cadet Sander] was circling at 2,300 feet clockwise. He appeared from my position on the ground to be in an approximate bank of 45 degrees and losing altitude at the rate of 1,500 to 2,000 feet per minute. When his position was approximately 500 feet above the ground, I became alarmed and called him on the radio to level out. Other ships in the air heard this call but there was no answer from [Cadet Sander]. His rate of descent and degree of bank did not appear to change from the time he started to lose altitude until he crashed at 2330. The plane crashed approximately one-half mile south of the southwest corner of Luke Field."

√**8-29-41. March Field, California.** At 1445, two Stearman PT-17 type aircraft collided in mid-air while attempting aerobatics in an unauthorized three-airplane formation eight miles east of March Field, Riverside, California, killing four fliers. The aircraft, based out of March Field, were scheduled for individual navigation training flights. Killed in the crash of PT-17 #41-8080 were pilot 2Lt. John J. Mitchell, Pauhuska, Oklahoma, and passenger TSgt. Richard E. Stafford, Nashville, Tennessee. Killed in the crash of PT-17 #41-8081 were pilot 2Lt. Edgar A. Hose, Grass Valley, California, and passenger Pvt. James D. Glenn, Fort Worth, Texas. The pilot of the third PT-17, 2Lt. Arthur P. Hustead, stated to investigators, "I left March Field on a cross-country mission to Palm Springs, California. Arriving at Palm Springs at [1345], I was joined by two more PT-17 airplanes, piloted by Lt. Mitchell and Lt. Hose respectively. We decided to fly back in formation to March Field as we had to make the flight back at approximately the same times. We left Palm Springs at [1400] in a string formation. After crossing the mountains we changed to a second position formation. At approximately [1445], the signal for loop and spread was given by Lt. Mitchell, [who] was acting as formation leader. In attempting this loop, the lead ship stalled at the top of the loop and fell into the left wing ship, striking it squarely. The ships stuck together for several seconds and then fell apart and fell straight to the ground. I would estimate the altitude at approximately 3,500 feet. The ships took 20 to 30 seconds to reach the ground after colliding and upon hitting the ground the second ship burst into flames, which shortly engulfed the first plane when it struck the ground. There was no visible move made on the part of any of the occupants of either plane to jump or free themselves from the plane as it fell. I circled the scene of the crash and then landed beside the airplanes. When it was apparent that I could be of no assistance whatsoever, I immediately left for March Field, arriving there at [1530] where I filed a report of the accident with the proper authorities."

8-30-41A. Mobile, Alabama. At 0045, two North American AT-6A type airplanes on individual night navigation training flights collided in mid-air 12 miles northeast of Mobile, Alabama, killing Aviation Cadet Albert Horsfall, Exeter, New Hampshire, aboard AT-6A #41-366. The pilot of AT-6A #41-610, Aviation Cadet Malcolm D. Holt, Lynchburg, Virginia, parachuted to safety. Investigation revealed that Aviation Cadet Horsfall inadvertently ascended into Cadet Holt's airplane.

8-30-41B. Los Angeles, California. A North American B-25 crashed 100 yards from the boundary of the municipal airport at Los Angeles, California, while attempting a turn from base to final during an emergency landing attempt just minutes after taking off on a factory test flight. The B-25 burst into flames when it impacted the ground, scattering wreckage over a large area. Killed in the crash was crew chief/radio operator James Knight, 24, Manhattan Beach, California. Pilot Wilcox B. Wild, 37, Los Angeles, California, and co-pilot Paul Penrose, 23, Dallas, Texas (see 3-1-41), were both seriously injured in the crash. All three were employees of North American Aviation Company.

September

9-2-41A. Location Unknown. At 1000, a Stearman PT-17 took off from the AAF Training Detachment airfield at Lakeland, Florida, and vanished with its RAF student pilot soon after. Thunderstorms were reported about 10 miles north of the Lakeland training area. An extensive search of central Florida and the Gulf of Mexico by numerous aircraft failed to locate Flying Cadet R.A. Cleave or his airplane.

9-2-41B. Hempstead, New York. At 1345, a Bell P-39D that had been abandoned by its pilot after an engine failure and an in-flight fire crashed into the southwest corner of Florence and Jerusalem avenues at Hempstead, Long Island, New York, killing three Hempstead children on the ground. P-39 pilot 2Lt. Roy W. Scott parachuted to safety. He was cut down from a tree by three youths at Rutland Road. He suffered very minor injuries and walked to the accident site before being whisked away to Mitchel Field, Long Island, New York. The aircraft struck the curb of a vacant lot next to the Hempstead Reform Church at high speed in a nose down attitude. Wreckage and flaming gasoline were scattered and sprayed down the street, hitting three children who were playing in front of the Giannone Barber Shop at 409 Jerusalem Avenue. Fatally burned in the accident were Georgene Kramer, 3, 407 Jerusalem Ave., Pauline Cuccio, 7, and Jasper Cuccio, 5, 414 Jerusalem. All three died just hours after the accident at Meadowbrook Hospital, Hempstead. Arthur Weston, 47, Uniondale, Long Island, Peter Cuccio (father of Pauline and Jasper), Dr. Arthur Kramer, 32, and Mrs. Evelyn Kramer, 28 (father and mother of Georgene), all received minor burns to their arms and hands after ripping off the flaming clothing of the three children and beating out the flames. The crash shattered the windows of several shops and started a small fire at a drug store at 406 Jerusalem. Children had been around the front of the Giannone Barber Shop throughout the day, getting haircuts for the beginning of the new school year that was commencing the next day. Lt. Scott later stated to investigators, "I took off ... and headed for the Whitestone Bridge. Just before I got there, the motor cut out and started up again immediately. I switched gas tanks and started back for the field, approximately 5,000 and 6,000 feet. As I came over the northwest corner of the field, the motor cut out again and started immediately. I called the tower and told them I was coming in for an emergency landing. I started down from 6,000 feet at the southeast corner of the field and the propeller ran away. When I reached the northwest corner of the field at approximately 3,000 feet, the motor began to get very rough. Then the cockpit began to fill up with smoke and I called in and told them to prepare for a sudden landing. Just as I finished this call, the cockpit completely filled up with fire. I pulled the left hand door emergency release and bailed out at approximately 400 feet. The airplane crashed and burned." Investigators noted that Mrs. Kramer had telephoned Mitchel Field several times and repeatedly threatened the life of the pilot.

9-4-41. Bay Minette, Alabama. At 2210, a North American AT-6A crashed near Bay Minette, Alabama, killing Aviation Cadet A.J. Trehanon. Investigation revealed that the student attempted to bail out at an altitude of less than 100 feet agl. Investigators speculated that the student might have lost engine power due to his failure to switch fuel tanks when the port tank ran empty.

9-5-41A. Stamford, Texas. At 0903, a Stearman PT-17 stalled and spun into the ground from a very low altitude one mile south of Stamford, Texas, killing civilian flight instructor Louie C. Cearley and student Aviation Cadet Stanley A. Siegel.

9-5-41B. Bakersfield, Texas. At 1415, a Vultee BT-13 stalled and spun into the Hat A Ranch, 17 miles south of Bakersfield, Texas, killing pilot 2Lt. Allen R. Bailey and passenger 1Lt. Austin T. Heffelman. The BT-13 had been buzzing the house of Mr. Arthur Harrel just prior to the accident. Mr. Harrel stated to investigators, "At about [1415] we were aware of a plane flying very close to our house and flying very low. We went outside of our ranch house and were told by friends of ours, who had been outside, that the airplane had just come from our son's house, about six miles below us, where it had been circling very low for two or three times. When the airplane circled our house, it was about 100 feet high and made two or three turns around our house. The pilot in front was waving at us. Some of the pilots from San Angelo [Texas] had visited at our ranch and knew where my son lived. We thought that he had gone to my son's house and finding no one home, had come up to look at the main ranch house. Lt. Bailey was a good friend of my son and had visited here at the ranch. When he circled the house, the motor seemed normal, but when he pulled up abruptly, the plane turned over and fell."

9-5-41C. Winslow, Arizona. At 1820, a Douglas C-39 suffering the failure of the starboard engine just after take-off crashed into houses one-eighth of a mile from the edge of the runway of the municipal airport at Winslow, Arizona, killing a civilian woman and injuring her husband. Two crewmembers aboard the C-39, Pfc. Douglas D. Melton and Sgt. Raymond W. Hedstrom, were also injured. Mrs. Jacoba Garcia, 63, was fatally injured when the C-39's starboard wing smashed through her kitchen, burying her under hundreds of pounds of debris and pinning her against a hot stove. Her husband, Nazario Garcia, received minor injuries. The airplane smashed into two other homes but no occupants were injured. Pilot 2Lt. Lorrill A. Palm and co-pilot 2Lt. George W. Vaughn were uninjured. Lt. Palm had reported that his aircraft had been running very sluggish on the flights prior to the accident flight. He tried to abort the take-off when the engine failed about two-thirds of the way down the runway, but was unable to stop before going through the airport boundary fence and into the houses. The cargo flight, which had originated in South Bend, Indiana, was to deliver materials to the North American Aviation Company factory at Inglewood, California, by way of St. Louis, Missouri, Tulsa, Oklahoma, and Winslow.

9-6-41A. Shreveport, Louisiana. At 1159, a Douglas O-38B collided with power lines and crashed into the Red River eight miles northwest of Barksdale Field, Shreveport, Louisiana, killing pilot 2Lt. William H. Binns and injuring passenger 2Lt. Ira A. Hassler. Lt. Hassler stated, "We were returning to [Barksdale Field] roughly following the course of the Red River when I believe Lt. Binns' attention was attracted by something in the river or nearby. He then flew lower to get a better view. Suddenly we struck some wires [that] were strung across the river about one-half mile north of [Barksdale Field]. I am sure that neither of us ever saw the wires. The plane immediately went into a violent, skidding dive and crashed in mid-stream. Swimming from the submerged plane, only parts of which were visible, I shed my parachute and clothing as I floated downstream. I then attempted to swim upstream to the plane so as to get Lt. Binns out. Against the current I got nowhere so I swam ashore, ran along the bank to a point above the plane and entered the water again. The current carried me past the plane before I could reach it. I was resting, nearly exhausted, on the bank when help arrived a short time later. They got Lt. Binns out right away, but he had died instantly in the crash."

9-6-41B. Victoria, Texas. At 1240, a North American BT-14 dove into the ground while performing low altitude aerobatics just after take-off from the municipal airport at Victoria, Texas, killing pilot 2Lt. James Joseph Meehan, Jr. and seriously injuring passenger SSgt. W.V. Bray. In 1940, Lt. Meehan resigned a regular commission as a lieutenant in the U.S. Marine Corps to enter the Air Corps as a cadet. Investigation revealed that Lt. Meehan thought of himself as "a daredevil pilot" and exhibited "delusions of grandeur as to his flying ability." 1Lt. Sidney A. Price witnessed the accident and stated to investigators, "Lt. Meehan took off on the North-South runway flying south. He immediately lowered his left wing tip until it was not visible because of the [high] grass. About where the two runways cross he had approximately 20 feet altitude. He circled to the left [in a tight climbing turn to an altitude of about 50 feet] and went down to about two-thirds the length of the field and circled [to the left] back toward the little office building, leveling out and going into a slight dive. Just before reaching the office building he pulled up to the left, immediately banking over steeply to the right in a high steep climbing turn, which seemed to be a cross between a wingover and a split-S turn. At the top of the climb, I would judge he reached an altitude of approximately 300 feet and was above and slightly northeast of the farmer's house, and in this position it was possible to see part of [Meehan's] lap and his left leg and the inner surface of the left side of the fuselage. Continuing the circling maneuver, he dived towards the ground a little to the east of the gas pumps and seemed to be leveling off normally but suddenly the wheels began dragging the ground. A large cloud of dust was thrown up around the ship." The Accident Classification Committee stated, "Lt. Meehan immediately after take-off made a sharp turn at low altitude to the left. While still over the airdrome, he pulled up steeply to the left, immediately banking over into a steep right climbing turn. At the top of this turn he was in a partial stall. He recovered in a dive back at the ground but did not have sufficient altitude or else used poor judgment in delaying to level off longitudinally, then striking the ground in a very tail high position, causing the landing gear to give way. The airplane burst into flames after impact and skidded along the ground for about 300 feet. It is the opinion of the committee that the accident was caused by the pilot failing to use normal flying caution, by using poor judgment in regard to safe altitude in his desire to demonstrate to himself and others that he was a superior pilot and that the pilot used excessive bottom rudder on the top of the climbing turn, causing the airplane to turn more abruptly and more steeply toward the ground than he anticipated."

9-6-41C. Houston, Texas. At 1243, a North American O-47B performing practice landings drifted off the runway and collided with a dump truck performing maintenance at the municipal airport at Houston, Texas, fatally injuring civilian dump truck driver Charles R. Tollis. The pilot, 2Lt. John R. Dyas, was undergoing a transition check flight when the crash occurred. He was not injured.

9-7-41. Greencastle, Indiana. At 0435, a Northrop A-17 flying in severe thunderstorms crashed into a heavily wooded area near Greencastle, Indiana, killing pilot Maj. Fred H. Murchison, 43, and passenger Lt. Col. Rufus B. Davidson. The aircraft smashed into the ground out of control at a high rate of speed, bursting into flames upon impact.

9-9-41. Mt. Constance, Washington. A Douglas B-18A flew into Mt. Constance in the Olympic Range, 50 miles southwest of Seattle, Washington, killing the crew of six. The bomber collided with a sheer cliff at the 6,000-foot level while on a night navigation training flight. Investigators speculated that the bomber entered instrument conditions at approximately 0430 and collided with the mountain soon after. The wreck was not discovered until 9-25-41. Killed in the crash were: 2Lt. E. H. Valorz, pilot; 2Lt. V. W. Winship, co-pilot; Sgt. V. W. Page, engineer; SSgt. R. D. Dexter, radio operator; Pvt. J. J. Huppert, crewmember; Pvt. M. Knight.

9-10-41. Barksdale Field, Louisiana. At 1640, a Lockheed A-29 collided with a water truck while attempting to take-off from Barksdale Field, Shreveport, Louisiana, fatally injuring two army aviation cadets and killing a civilian truck driver instantly. Pilot Aviation Cadet Nelson R. Slayton, 24, Memphis, Tennessee, and co-pilot Aviation Cadet Benjamin L. Sykes, 24, Jackson, Mississippi, were making their first team solo flight in the Lockheed Hudson type airplane. The pilots lost control of the airplane on the take-off roll, veering towards a civilian work detail at the north end of the field. In an effort to avoid a collision, the pilots attempted to get airborne prematurely, stalling the airplane. The left wing struck the ground first and the airplane careened into the tank truck, bursting into flames and killing the driver Charlie Cheeks, 38, Haughton, Louisiana. The aircraft continued on, smashing into a cement-mixing truck and lightly injuring L. P. Thomas and Allen S. Minter, both of Shreveport. Minter suffered a mild concussion when, in an attempt to escape the bomber's path, he was struck on the back of the head by the wing. The construction crews were installing new taxiways adjacent to the runway when the crash occurred.

9-15-41A. Coushatta, Louisiana. At 0730, a Curtiss P-40 collided with a U.S. Navy Douglas SBD Dauntless while engaging in simulated combat over the Red River near Coushatta, Louisiana, killing the P-40 pilot 2Lt. John W. Brobroski. U.S. Navy pilot Lt. (j.g.) W. Woodhall and gunner Radioman 3C Albert W. Garlow were both able to parachute to safety, suffering only minor injuries. Five P-40 type aircraft from Barksdale Field, Shreveport, Louisiana were sent to intercept a group of navy airplanes that were "attacking" a bridge near Coushatta. The navy airplanes made three simulated attacks on the bridge with the army airplanes in pursuit. The P-40 pilot apparently misjudged his dive during his third interception and collided with the SBD bomber. The P-40 came to rest on the riverbank about 100 yards from the bridge.

9-15-41B. Brady, Texas. At 0905, a Vultee BT-13 crashed while buzzing a house on the W. J. White Ranch, 12 miles SSW of Curtis Field, Brady, Texas, killing civilian flight instructor Daniel R. Conway, 24, Shaker Heights, Ohio, and student A/C Dale N. Smith, 22, Anamosa, Iowa. Mr. Conway was a former army flight instructor. The airplane crashed when the starboard wing struck some oak brush when the pilots made a vertically bank turn at extremely low altitude. The aircraft cartwheeled into the ground and burst into flames, careening across the terrain for about 100 feet before coming to a rest about 300 yards from the W.J. White Ranch house.

9-16-41A. San Antonio, Texas. At 1200, two North American AT-6A aircraft collided in mid-air during a formation training flight two and one-half miles east of Kelly Field Auxiliary Field #2, San Antonio, Texas, killing two fliers and lightly injuring a third. All three had taken to their parachutes. Flight instructor 2Lt. Barry C. Francks, Jr. and Aviation Cadet Joseph C. Foley were killed when they bailed out of AT-6A #41-584 at too low an altitude to allow their parachutes to open. Both struck the ground near the wreckage of their plane. 2Lt. Robert E. Halley, Kelly Field flight instructor, suffered minor injuries parachuting to safety from AT-6A #41-672. Aviation Cadet John J. Fisher stated to investigators, "The first squadron took off to the north and we took off to the west. There was sufficient spacing between the first and second elements of H-flight on the take-off, which was normal in all respects. While in the first turn off the field at 700 feet, the second element of H-flight overran the first element. I knew we had overrun because I saw the flash of Aviation Cadet Baron's wing below at a safe distance. We continued to turn to the left in a gentle bank of 15 [degrees] or less and to climb at a normal rate. After the brief view of Cadet Baron's airplane I saw none of the other airplanes in our squadron until No. 35 appeared below No. 40. Lt. Francks [No. 35] gestured to Aviation Cadet Bausser with his left hand to climb up and simultaneously pulled up into the fuselage of No. 40 [Halley]. The canopy of No. 35 cracked and the two airplanes whipped over backwards. I pulled out of position in a climbing turn to the left and banking to the right I saw a plane burning on the ground, with one parachute open swinging down and one parachute three-fourths open on the ground close to the burning airplane." AT-6A No. 35 (AAF serial number 41-584) can be seen in the motion picture *I Wanted Wings* (Paramount Pictures 1941).

9-16-41B. Maxwell Field, Alabama. At 2110, a North American BC-1A struck power lines two miles west of Maxwell Field, Montgomery, Alabama, killing A/C Calvin Richard Shaw, 24, Stewartstown, York County, Pennsylvania. Cadet Shaw was returning to Maxwell Field after completing a flight to Birmingham, Alabama, with three other airplanes. The crew of a Mobile and Ohio Railroad switching locomotive found the wreckage of the downed BC-1.

9-19-41A. Natchitoches, Louisiana. At 1030, a Lockheed P-38D and a Curtiss P-40 collided in mid-air while engaging in simulated combat 10 miles northwest of Natchitoches, Louisiana, killing P-40 pilot 2Lt. Leland A. Viar, 23, Dunlap, Kansas, and P-38 pilot 2Lt. R. H. Keith, 23, Durham, North Carolina, formerly of Menomonie, Wisconsin. Lt. Keith was struck by pieces of the exploding aircraft during his bail out attempt, severing both of his legs and destroying his parachute. Lt. Viar's P-40 struck the ground with tremendous force, out of control and inverted, auguring the wreckage several feet into the ground. Investigators stated, "Two flights of P-40s took off from Natchitoches airport on a patrol mission in that area. Lt. Salisbury was leading A-flight and Lt. Viar was leading B-flight. Several miles west of Natchitoches, the P-40s sighted a flight of nine P-38s and decided to attack. As the attack was started the flight of P-38s broke formation and scattered. On signal, A-flight attacked a single P-38, which peeled off to the right, and B-flight attacked a P-38 to the left. After one pass, Lt. Viar called his flight to break off the attack and assemble. The flight commander of A-flight also ordered his flight to assemble. While Lt. Viar was circling to the left so his flight could assemble, the P-38s combined to dive individually on the assembling flight of P-40s. During one such individual diving attack, the P-38 in question [Lt. Keith] came down on Lt. Viar's airplane. The collision was from the right and slightly below. It is doubtful if Lt. Viar ever saw the P-38 since he was making a left turn and his right wing obscured his vision to his lower flank."

9-19-41B. Parksley, Virginia. At 1530, a North American BT-14 crashed out of control at Parksley Field, Parksley, Virginia, killing pilot 2Lt. Carlyle C. Hoch, Newark, New Jersey, and passenger 2Lt. Angus J. MacDonald, Los Angeles, California. The aircraft had just taken off and was conducting a series of dives and climbing turns low over the field when it stalled and spun in from less than 100 feet. Winds were reported to be northeast at 25 mph, gusting to 30. Henry West, a civilian who witnessed the crash, stated to investigators, "We were standing in front of the hangar and had been talking to Lt. Hoch, who was on the ground at Parksley Field waiting to be picked up by one of the pilots who was practicing landings. When the airplane came in at about 1530 to pick him up, Lt. Hoch climbed into the front seat and the other pilot got into the rear seat. The airplane taxied down to the southeast end of the runway and took off toward the northwest. After taking off, it made a right turn and flew over the field at low altitude in a southwest direction. It then made a left turn and passed over the field again to the northeast. Then it made a right turn and started a dive toward us, still turning to the right. Just as it was over the runway, the airplane seemed to go out of control and suddenly turned halfway around and hit the ground nose first. It immediately burst into flames."

✓**9-19-41C. Boyd, Texas.** At 1545, a Fairchild PT-19 suffered a catastrophic failure of the starboard wing after performing aerobatics three miles south of Boyd, Texas, killing civilian flight instructor J. C. Foster, 22, Fort Worth, Texas, and student Aviation Cadet Donald B. Bierer, 21, Brooklyn, New York. Witnesses on the ground stated that the aircraft had been doing a series of loops just prior to the wing coming off.

9-19-41D. Marshall, Texas. At 1630, a North American O-47A experiencing engine trouble clipped trees and crashed into a field during a forced landing at Marshall Airport, Marshall, Texas, killing two crewmembers and seriously injuring the pilot. The pilot, 2Lt. Raymond M. Baldwin, Evansville, Indiana, was rescued by civilians, intense flames thwarting the rescue of the other two crewmembers. Killed in the crash were observer 1Lt. Walter H. Boyd, Long Beach, California, and radio operator Cpl. Artie M. Howard, Salyersville, Kentucky. Lt. Baldwin, suffering from serious burns, stated to investigators, "I took off with 10 degrees of flaps and 35 [inches] Hg [mercury]. After getting off I pumped up the flaps and started climbing on my course, pulling the throttle back to 30 inches Hg and prop pitch back to 1,850 [rpm]. Then manifold pressure dropped to 20 inches [for] no apparent reason. Then I headed back to the field, thinking the carburetor was icing. I applied a little carburetor heat but this had no effect so I pushed the lever to the off position. I motioned to Boyd to tell Howard to get out of the belly of the ship while going in for a landing. Approaching the field with wheels and flaps (45) down the manifold pressure dropped to 15 inches while still 500 feet in the air. I pulled the throttle all the way back but the motor kept on turning over at a speed in excess of idling speed. I hit the field at 110 mph and I saw it was impossible to land without running into the trees at the other end of the field, so I pushed the throttle forward and pulled the wheels up, leaving the flaps down, intending to make a wheels up landing in a clearing on the other side of the road. The plane mushed across the rest of the field and hit the top of a tree at the other end of the field."

9-21-41. Lubbock, Texas. At 1405, an RAF Douglas DB-7 crashed while maneuvering at low altitude at the municipal airport at Lubbock, Texas, killing pilot 2Lt. Shelby Harper, 24, Lubbock, Texas. The crash occurred in front of a crowd of nearly 500 hometown spectators, including Lt. Harper's mother, father and brother. Lt. Harper and his brother, 2Lt. Bryan B. Harper, had performed a formation take-off at about 1350 and then began circling and diving on the field. The pilots had picked up the bombers in Long Beach, California, on 9-19-41 to deliver them to the RAF on the east coast and had stopped over in Lubbock to see their parents and receive birthday congratulations for Shelby, whose twenty-fourth birthday was 9-20-41. Lt. Shelby was killed when he stalled the DB-7 in a steep, climbing left turn that he was executing in an attempt to close the distance between the two ships. The DB-7 spun in from an altitude of about 500 feet above ground level and burst into flames when it struck the ground nose down with the left wing nearly vertical.

9-22-41. San Pablo Bay, California. At 0805, two Curtiss P-40 aircraft collided in mid-air over San Pablo Bay, California. The airplanes were part of a flight of six P-40s that had just taken off from Hamilton Field, San Rafael, California. One pilot was killed and the other was injured parachuting to safety. The six airplanes had taken off at 0800 and were attempting formation assembly over San Pablo Bay when one P-40 crossed under and ascended into his wingman. Killed in the crash of P-40 #39-239 was 2Lt. William E. Scott, 22, Newkirk, Oklahoma. 2Lt. James B. Morehead, 25, also of Oklahoma, received only slight cuts and bruises when he parachuted to safety from P-40 #39-263. Three months prior to this accident, Lt. Morehead successfully bailed out of a burning airplane near Petaluma, California, again surviving with only minor injuries. 2Lt. Paul C. May, who witnessed the crash, stated to investigators, "[A] routine formation flight of six airplanes were scheduled to take-off, led by 2Lt. Donald Stoebe (see 10-9-41C). We took off in two-plane elements, the first element being led by Lt. Stoebe, the second by Lt. Morehead and the third by myself. In order to overtake the first element, the second and third elements have to cut the pattern short. The three elements were just coming up into a single formation when the accident occurred. In order to get directly in line with the lead element upon joining it, a steep turn is necessary. When Lt. Morehead made this turn, Lt. Scott, not seeing the turn, hit Lt. Morehead's ship. He first hit the tail assembly and then cartwheeled around Lt. Morehead's ship and came down on top of him, the ships locking. I led my element up above them and then watched the proceedings. I called the tower immediately and told them of the accident. The ships finally broke apart and hurtled earthward. Lt. Morehead left the ship just when they broke apart. He seemed to come out with his chute open. Lt. Scott's plane dived directly down at about 30 degrees on its back while Lt. Morehead's plane made about half a turn before striking the earth. They both disappeared from sight, striking in soft mud at the edge of San Pablo Bay one mile southwest of Black Point. They hit about 200 yards apart. I followed the planes and Lt. Morehead down until Lt. Morehead, who had also landed in the mud got up and waved to me."

9-23-41. Waskom, Texas. At 1015, a Martin B-12AM crashed 10 miles north of Waskom, Texas, killing A/C Donald E. Curtis. The aircraft had been part of a mixed aircraft formation flight that had taken off from Barksdale Field, Shreveport, Louisiana, at 0900. The pilot had trouble getting an engine started just prior to the flight and during the flight had trouble keeping up with the formation, gradually falling back and eventually losing contact. The flight leader, 2Lt. R. L. Jackson, made several efforts to make contact with the B-12, visually and by radio, without success. Clouds were scattered at about 3,000 feet but were not considered as threatening. Investigators could not determine the cause of the accident.

9-26-41. Chico, California. At 1325, a North American AT-6A suffered a catastrophic failure of its port wing 10 miles NNE of Chico, California, killing Aviation Cadet John L. Driscoll and injuring instructor 2Lt. Joseph H. Carter, who parachuted to safety. Cadet Driscoll's mangled headset and shattered pieces of the canopy were found some distance away from the crash site. Investigators speculated that the wing might have struck A/C Driscoll, killing him or preventing him from attempting a bailout.

9-28-41. Lawton, Oklahoma. At 2005, a North American O-47A dove into the ground after entering unexpected overcast near Lawton, Oklahoma, killing the crew of three. Investigators stated, "After thorough investigation and detailed inspection of available instruments and parts from the airplane, [it was determined] that the airplane struck the ground in excess of 200 miles per hour, causing instantaneous death to the three occupants. That the pilot was flying contact five minutes prior to his estimated time of arrival at destination, and he flew into the overcast [that] he did not anticipate since the ceiling had lowered two to four hundred feet. In the transition between contact and instrument flying, it is the opinion of the committee that the pilot experienced vertigo and lost control of the airplane resulting in the crash." Pilot 2Lt. Ralph Altee McNabb, observer Capt. Oliver C. Appleton and radio operator Pvt. James D. Smith were killed in the crash.

October

10-1-41. Turner Field, Georgia. At 1800, three North American AT-6A type airplanes collided in midair near Turner Field, Albany, Georgia, killing one cadet and fatally burning another. The airplanes were part of a 12-airplane formation flight that was preparing to land at the field. Instructor 2Lt. Paul J. Hydo was uninjured and able to safely land damaged AT-6A #41-531 at Turner Field after the collision. A/C Albert E. Kaufman, 21, Norfolk, Virginia, was killed in the crash of AT-6A #41-487 and A/C Charles G. Kadel, 21, Memphis, Tennessee, was severely burned when his clothing caught fire during his attempt to parachute to safety from AT-6A #41-258. He descended all the way to the ground in his parachute with his flight suit on fire. He died 10-3-41 at Phoebe Putney Memorial Hospital, Albany, Georgia. Investigators determined that Kaufman's AT-6 pulled up into A/C Kadel's airplane, causing A/C Kadel to collide with 2Lt. Hydo's airplane. The twelve airplanes were in an echelon formation to the right with landing gear lowered. The leader of the first flight of six, 2Lt. W.D. Ceely, gave a hand signal to peel off for landing at six-second intervals and then peeled off. The pilot of the second airplane hesitated on his peel-off and the formation began to loosen up and gain altitude. Lt. Hydo later stated to investigators, "As I observed the airplanes beginning to gain altitude, I began to skid out of position to my right to take my formation away and form my own flight. As I did this it seemed as if number five airplane [Kaufman] had to pull up suddenly to avoid hitting airplane number four, and this maneuver caused him to collide with number six airplane that in turn hit my left wing with his antenna and right wing. I saw flames and then heard a slight explosion. After my airplane was hit it began to fall off to the right and was out of control for a few seconds. After violent movement of the stick, it began to react to my actions and I righted the airplane.... When my airplane was hit I could feel the heat from the fire and saw the pilot [Kadel] in the airplane in position number six of the echelon throw up his hands and turn his face toward my airplane. Then airplane number five and number six seemed to melt together and they began spinning down to the left.... I called by radio to my flight and gave instructions for them to follow me and be careful of other airplanes flying around loose, and for them to land individually, as my plane was not flying as it should. It began to drop to the left and was very wing heavy. I called Turner Field tower and came in and landed."

10-2-41. Flagstaff, Arizona. At a time after 2030 Pacific Time, a Douglas B-18A crashed into the 10,500-foot level of 12,340-foot Agazzi Peak in the San Francisco Mountains ten miles north of Flagstaff, Arizona, killing six fliers. The airplane had been reported missing and was not discovered until 10-6-41. The B-18 took off at 1726 Pacific Time from Sacramento Air Depot, Sacramento, California, on a material ferry flight to Albuquerque, New Mexico, and was last heard from at about 2030 near Needles, California. Investigators noted that the pilot was off course and that on the day of the accident severe snowstorms were reported in the area of the crash between 2030 and 2400. Investigation revealed that the airplane had been in a slight climbing attitude and flying at approximately 120 miles per hour when it struck the mountain. The airplane's port wing topped six trees, cutting off the top ten feet of them before smashing into mountain. The airplane did not strike the mountain nose first but pancaked into the terrain. The time of the accident could not be determined with any accuracy. Killed in the crash were: 1Lt. Arnold T. King, Laramie, Wyoming, pilot; 2Lt. R. H. Conway, Wilbur, West Virginia, co-pilot; Cpl. Leo E. Gillem, Spokane, Washington, engineer; Pvt. Lawrence T. Morfeld, Springerville, Arizona, radio operator; 2Lt. W. R. Grim, Pendleton, Oregon, passenger; 2Lt. D.C. Boyd, Pendleton, Oregon, passenger.

10-8-41. Riverdale, Georgia. At 1900, a Douglas DB-7B suffering an in-flight fire in the starboard engine crashed near Riverdale, Georgia, killing the pilot. 2Lt. Kenneth E. Elliot was killed when his parachute fouled on the airplane and was damaged during his bail out attempt 16 miles south of Chandler Field, Atlanta, Georgia. Investigators speculated that the pilot might have opened the parachute prematurely. The airplane was one of a flight of three DB-7 airplanes being delivered to the Royal Air Force.

10-9-41A. Crestview, Florida. At 0945 Central Time, a Vultee BT-13 that had apparently ran out of fuel crashed while attempting to make a forced landing at Crestview, Florida, killing Aviation Cadet William H. Snyder. Investigators stated, "This airplane took off at approximately 0800 Central Time for a navigation flight to Marianna, Florida, via Crestview, Florida, and return [to Gunter Field, Montgomery, Alabama]. It had flown approximately one hour and forty-five minutes, all of which was on the right fuel tank (this amount of flying time would lower the level of gasoline in the right tank to about the level of the reserve stand pipe), with the gasoline selector valve set on the right tank. Cadet Snyder overshot the field on his first attempt for a landing at Crestview, advanced the throttle, and was making a second flight around the field. The engine cut out after each of his first two turns. This was no doubt caused by the fact that there was not enough gasoline in the right tank, with the gasoline selector valve set on the right tank instead of reserve, for the engine to continue to run."

10-9-41B. Monongahela River, Pennsylvania. At 1645, a Bell P-39D flying in poor weather crashed into the Monongahela River at Courtney, Pennsylvania, killing pilot 2Lt. John R. Lyons, 27, Savannah, Georgia. The airplane took off from Patterson Field, Ohio, on a ferry mission to Mitchel Field, Long Island, New York. The pilot was originally assigned to fly P-39D #41-6799 but was given a replacement because he had inadvertently pulled the emergency door release, causing the starboard door to fall off of the airplane while he was taxiing out for take-off. He was then assigned P-39D #41-6846 and accidentally released the door of this airplane as well. He was delayed while mechanics reattached the door. The pilot finally took off at 1502 and reported over Columbus, Ohio, at 1515. At about 1630, Lt. Lyons entered the smoke-filled and very foggy Monongahela River Valley, lost, low on fuel and looking for an airport. Pittsburgh Airport reported visibility of less than three miles. The P-39 circled Finleyville, Pennsylvania, for a few minutes while the pilot apparently contemplated a parachute jump. The P-39's door was released from about 2,000 feet over the town. The surrounding terrain was very hilly and did not lend itself well to a crash landing. The pilot was trying to follow the river and get his bearings when the airplane narrowly missed colliding with a steamboat before plunging into the river at high speed in a 90-degree left bank. Investigators speculated that Lt. Lyons was probably low on fuel following all of the additional taxiing that was done when the airplane had to go back to the hangar to get the door reattached and he apparently was not able to make his refueling stop at Harrisburg, Pennsylvania. Civilian riverboat pilot Lee E. Underwood, Pittsburgh, Pennsylvania, witnessed the crash and later stated to investigators, "I was standing watch as pilot on the steamboat *Donora*. I was coming up the Monongahela River with a tow of six empty barges, and at the head of Cincinnati Bend, which is at Courtney, Pennsylvania, I saw a plane coming quartering across the river, about the middle of the river the plane banked to the left and headed due west, which at this point is straight down the river. The plane was headed directly at the boat, and just as I thought that the plane would strike the boat the plane went into a full left bank, putting the wings in a vertical position and missing the boat by inches. The plane failed to straighten out from this position and struck the water and never came to the surface. I then immediately stopped the boat and reversed the engines, and ordered a yawl put into the river and the yawlsmen picked up the body of one flyer which was turned over to the coroner."

10-9-41C. Pacific Ocean, California. At 1930, a Curtiss P-40 crashed into the Pacific Ocean about 500 yards northwest of Tennessee Point, California, killing pilot 2Lt. Donald R. Stoebe (see 9-22-41). The P-40 was part of a flight that was conducting a searchlight/harbor defense mission with ground units deployed near Bonita Point, California. 1Lt. John P. Hines, an anti-aircraft battery commander, stated, "[I] was standing by for harbor defense drill when a low flying plane came over Rodeo Beach from the southeast going northwest. It proceeded over Tennessee Point where searchlights [one] and [two] were playing on the water and then appeared to go into a dive. Just before the plane came over Tennessee Point it appeared to be dodging the beam of the searchlights that were located in the Rodeo Beach and Bonita Point areas. It never came out of the dive and it finally crashed into the water at about a 45-degree angle. The impact was so great that it sounded like an explosion as the plane hit the water with a large splash. I immediately called the Bonita Point Coast Guard Station and informed them of the accident location. At about 2000, the Coast Guard boat arrived on the scene. At the same time the Coast Guard was called, Townsley radio operator, Private Zink, called Harbor Defense Radio Station 7 and informed them of the accident and asked them if they could contact [the Hamilton Field rescue boat] so that it could proceed to the scene of the accident."

10-11-41. Duncan Field, Texas. At 1046, a Boeing B-17C crashed while attempting a take-off at Duncan Field, San Antonio, Texas, killing the two pilots and seriously injuring two passengers. Five enlisted men on the ground received minor injuries. Eleven passengers and crew were uninjured. The airplane was taking off on a flight to Albuquerque, New Mexico, when it began to swerve down the runway. The out of control B-17, having tried to abort its take-off, careened 1,500 feet across the field, smashing through the base security-operations building and an automobile, bursting into flames and coming to rest in front of the base hospital, which it barely missed colliding with. Investigators speculated that the pilots were nullifying each other's attempt to gain control of the B-17, which never gained sufficient airspeed for take-off. The airplane lurched into the air momentarily and then came down hard on its port main wheel. The brakes were applied (indicated by skid marks on the pavement) and directional control was lost. The B-17 then skidded across the field and collided with the building and the automobile, exploding into flames. The B-17 had flown in from Washington state and had been delayed in San Antonio because of a fuel pressure problem that was discovered during the flight. 1Lt. Maurice A. Morgan, San Antonio, Texas, rated B-17 first pilot and instructor (flying in the co-pilot seat), and pilot 2Lt. Robert Reichstadt, Omaha, Nebraska, were killed. It was noted by investigators that Lt. Reichstadt was at the controls in the left seat during the take-off and had very little flying time in B-17 type airplanes and that the airplane was trimmed in a tail-heavy attitude. Two passengers aboard the B-17, 2Lt. T.L. McKnight and SSgt. J.J. Jones, were seriously injured. The driver of the automobile, Cpl. Hubert O. Criss, received minor injuries. Pvt. Sam Miller, Pvt. Martin D. Paseo, Pvt. Albert Ranzau, and Pvt. Andrew McKnutt, who were in the building, received minor injuries. Uninjured in the crash were: 2Lt. W. F. Meenagh, navigator; TSgt. J.R. Broyles, engineer; Pvt. M. M. Newcomer, radio operator; Sgt. J.J. Bruce; 2Lt. H.R. Taylor; 2Lt. T.C. Parkinson; Aviation Cadet J.A. Crockett; Aviation Cadet C.E. Bergdoll; Maj. T.R. Stander; 2Lt. P.R. MacPhee; 1Lt. D.K. Farnsworth.

10-12-41. Beaumont, California. At 1430, a Douglas B-23 flying in poor weather crashed into the north side of a mountain in the San Gorgonio Pass two miles northeast of Beaumont, California, killing seven crewmembers. Residents of the area reported hearing the airplane flying low, east to west, in low-laying clouds near the mountains and a moment later heard the airplane strike the mountain. Weather in the area was reported to be very poor with rain, fog, low clouds and high winds. The airplane had taken off at 1210 MST from Albuquerque, New Mexico, for a flight to March Field, Riverside, California. Killed in the crash were: 2Lt. John E. Overstreet, Riverside, California, pilot; 2Lt. J.S. Park, Sunnymead, California, co-pilot; 1Lt. J.W. Templeton, Wichita, Kansas, passenger; SSgt. B.G. Bell, engineer; SSgt. R.J. Thomas, radio operator; SSgt. R.J. Farlow, passenger; Cpl. J.E. Smith, passenger.

10-13-41. Ash Canyon, New Mexico. At a time unknown, a North American O-47A collided with terrain and crashed into rugged and desolate Ash Canyon in the San Andreas Mountains ten miles west of Lake Lucero, New Mexico, killing the crew of three. The O-47 was reported missing at 1115 and was not located until the afternoon of 10-18-41. The O-47, which was based out of Biggs Field, El Paso, Texas, had been participating in combined ground operations with an anti-aircraft battery of the 260th Coastal Artillery and had failed to arrive for its 1115 mission. Investigators determined by careful examination of marks on the ground that the airplane struck a small hillside with the port wing while making a turn in the canyon and crashed down a ravine, bursting into flames at the bottom of the canyon. Killed in the crash were: 2Lt. Willis B. Hunt, 24, Blytheville, Arkansas, pilot; 2Lt. Caesar S. Kaiser, 26, Louisville, Kentucky, observer; SSgt. Richard B. Lauck, 28, Denver, Colorado, gunner. Weather in the area at the time of the crash was reported as clear with ceiling and visibility unlimited.

10-14-41. Youngwood, Pennsylvania. At a time between 1806 and 1812, a North American AT-6 flying in severe thunderstorms crashed to the ground in a violent spin near Youngwood, Pennsylvania, killing pilot 1Lt. Marshall E. Stewart and passenger 1Lt. J. Schleuger, Medical Corps. The AT-6, which was based out of Kelly Field, San Antonio, Texas, was following instructions from Pittsburgh air traffic control on an extended cross-country navigation flight. The aircraft had taken off from Philadelphia, Pennsylvania, at 1520 and was to land at Pittsburgh, Pennsylvania.

10-15-41. Richmond, Michigan. At 0920, a Lockheed P-38D crashed out of control near Richmond, Michigan, killing 2Lt. Alfred Voss, Elgin, Illinois, who plunged to his death when he was ripped out or fell out of his parachute harness after bailing out of the twin-engine fighter and deploying his parachute. Michigan State Police reported that the empty parachute fell into the St. Claire River. Investigators speculated that the pilot or the parachute became fouled on either the rear canopy enclosure or the tail section resulting in the tearing or cutting of the parachute from the body of the pilot. Lt. Voss' body was found about 300 yards from the wreckage of the P-38. 2Lt. Robert E. Williams, flying a P-38 at 12,000 feet, observed Lt. Voss flying at 10,000 feet, making a steep left turn and then disappearing from his sight. Lt. Williams made a right turn but could not see the other P-38 upon its completion. Observing a parachute floating several thousand feet below, he flew down to investigate and found the parachute to be empty. Investigators could not determine why the pilot had to leave the airplane.

10-16-41A. Shafter, California. At 1340, a Vultee BT-13A stalled and spun into the ground near Shafter, California, killing instructor 2Lt. Thomas V. Bright and student Aviation Cadet Albert F. Owens. 2Lt. Edward P. Denton stated to investigators, "Lt. Bright and his student were engaged in cross under and echelon turns. On a cross under back to V-formation, Lt. Bright and Aviation Cadet Owens' ship fell off into a flat spiral to the left. At approximately 1,000 [feet] their ship assumed flight straight ahead except for the fact that the wings were wobbling in what appeared to be a controlled stall. Immediately thereafter, [the airplane] fell off to the right in the same type of flat spiral. In this condition it continued to fall to the earth."

10-16-41B. Beaufort, South Carolina. At 2330, a Lockheed P-38D crashed while attempting an emergency forced landing near Beaufort, South Carolina, killing pilot 2Lt. John S. Costello. The P-38 had taken off from the Army Air Base at Savannah, Georgia, on a night searchlight mission at 1930. The pilot apparently became lost and tried to land the airplane in a swampy area 1,000 feet from an emergency airfield located about 40 miles from the Savannah Air Base. Lt. Costello attempted to land the airplane with the landing gear in the extended position, causing the airplane to flip over on its back. Lt. Costello was knocked unconscious during the forced landing and drowned when he was unable to escape from the airplane.

10-17-41. Jackson, Georgia. At 0610, a Seversky P-35 descending through heavy overcast suffered the catastrophic failure of its port wing and dove into the ground at Jackson, Georgia, killing pilot 2Lt. Carlyle O. Loverud. The airplane was one of a flight of three P-35 airplanes on a night cross-country navigation flight from MacDill Field, Tampa, Florida, to Atlanta, Georgia. The other two airplanes landed safely at Atlanta. Investigators speculated that the pilot might have overstressed the airplane while maneuvering or descending through the overcast, causing the wing failure.

10-20-41A. Cunningham Field, North Carolina. At about 1250, a Curtiss P-40E that was attempting a forced landing on a road crashed near Cunningham Field, North Carolina, killing pilot 2Lt. Walter E. Houston. The airplane was part of a four-airplane formation that had become lost and low on fuel. The three other P-40 airplanes made emergency forced landings with landing gear retracted at Cunningham Field, which was still under construction. 2Lt. Chauncey S. Irvine stated to the Accident Classification Committee, "We then flew to the coast and flew a course of 85 degrees along the coast for about 25 minutes. I buzzed a town and identified it as Newport, North Carolina. We did not have enough gas to get back to Wilmington [North Carolina] or any other marked airport. I notified the flight that we would have to make a forced landing and that we would land wheels up. The field we picked was a runway under construction at Cunningham Field, North Carolina. I landed first followed by 2Lt. Allison Embry and 2Lt. Herbert E. Smith; all three landings were wheels up with no injury to personal. Lt. Houston, after making several passes at the other fields and roads in the vicinity attempted a wheels down landing on a road. A truckload of workmen coming down the road forced him to take off again after his wheels had already touched the ground. He pulled up to 200 feet, made a 180-degree turn and then spun into the ground. Workers in the vicinity pulled him free of the burning ship."

10-20-41B. Brady, Texas. At 2140, a Vultee BT-15 suffering an apparent engine failure stalled and spun into the ground one-quarter of a mile east of Curtis Field, Brady, Texas, killing Aviation Cadet Jack H. Fletcher. The Accident Classification Committee stated, "Cadet Fletcher had taken off and then made a wheel landing near the end of the field, and finally, with motor apparently functioning normally, [the airplane] had cleared the field and the Brownwood Highway. When he had climbed to about 100 feet altitude and was about 150 yards east of the highway the motor failed completely. Cadet Fletcher turned on his landing lights and started a turn to the left. Near the completion of the turn, the airplane stalled and started to spin to the left, striking the ground in a nearly vertical attitude."

10-22-41A. San Antonio, Texas. At 0205, a North American BC-1 and a North American AT-6A flew into the ground in formation while letting down through clouds on a night cross-country 12 miles north of Kelly Field, San Antonio, Texas, killing BC-1 pilot 2Lt. Harold H. Jessen and seriously injuring student A/C George D. Meador. Killed in the crash of the AT-6 was A/C Andrew B. O'Brien. The two aircraft had taken off at 0100 from Austin Municipal Airport, Austin, Texas, on a flight to Kelly Field. The BC-1 was serving as a control ship, helping the students fly back to their base at Kelly Field. Maj. William B. Wimer stated to the Aircraft Accident Classification Committee, "I received a telephone call from Post Operations to the effect that we had a crash involving two airplanes which occurred on the Fredericksburg Road about eight or ten miles north of San Antonio. In company with Maj. J.H. Bundy and Capt. W.F. Chapman, we went to the scene of the crash. Upon arriving at the scene of the accident we found the state highway patrolmen and members of the Sheriff's force in charge of the scene. From the statements of a farmer living adjacent to the scene of the crash, and from one of the Sheriff's Deputies, the time of this crash is fixed at between [0200] and [0205]. This time is further fixed by the fact that a watch belonging to one of the three men involved in this accident was stopped a 2:05 A.M. The Sheriff's Deputy who was the first one at the scene of the accident was only about a mile away at the time of the crash; his car was equipped with a two-way radio and he called Kelly Field immediately informing us of the crash. Nothing had been disturbed prior to our arrival. We found the fuselage of BC-1 airplane, A.C. No. 38-419, on the west side of Fredericksburg Road headed north. The motor of this airplane was in the center of the highway. The wings and other portions of this airplane were in a field to the east of Fredericksburg Road and considerably south of the point where the fuselage came to rest. Lt. Jessen's body was lying on the side of the highway just off of the shoulder. Cadet Meador, who was Lt. Jessen's passenger, had already been removed from the wreckage and sent to the hospital by the Sheriff's Deputies. The wreckage of Cadet O'Brien's airplane, AT-6A, A.C. No. 41-670, was in a field to the west of Fredericksburg Road and had come to rest at a point about 200 yards north of the wreckage of Lt. Jessen's airplane. Portions of this wreckage had caught fire but the fire had been extinguished by the Sheriff's Deputies who had arrived at the scene of the crash shortly after it occurred. Our first information was to the effect that these two airplanes had collided prior to crashing. Subsequent investigation, however, showed that this could hardly be the case because in company with one of the State Highway Patrolmen who had a powerful portable spotlight with him, I climbed the embankment and fence of the field to the east of Fredericksburg Road and retraced the path of Lt. Jessen's airplane. Indications showed that the first contact made by Lt. Jessen's airplane was made in tall weeds, which had been cut down [by the airplane] plowing a path through the weeds then making three distinct contacts with the ground which tore off the wings and part of the tail surfaces. The fuselage went through a fence and down an embankment onto the Fredericksburg Road at which time the motor was knocked out of the fuselage and the fuselage then skidded across the road and came to rest against the embankment on the opposite or west side of the road facing in a northerly direction. The entire path of Lt. Jessen's airplane was in the general direction from south to north. Prior to the accident it was apparent that Lt. Jessen was leading Cadet O'Brien and the airplanes were echeloned to the left and rear of Lt. Jessen's airplane, and it was in this general formation that the two airplanes struck the ground. Examination of Cadet O'Brien's airplane showed that it was in a field on the opposite side of Fredericksburg Road which would be the west side facing north, and it had flown into the ground and had overshot the leader of the formation, which would be the natural position of a formation of this kind suddenly coming to rest. In reconstructing the events immediately preceding the accident it was evident that Lt. Jessen was leading Cadet O'Brien home following the light line from Austin to San Antonio. By remaining on the right side of the light line it would place Lt. Jessen and Cadet O'Brien in approximately the location where the crash occurred without some other maneuver. In all probability, Lt. Jessen came to the overcast sky condition and attempted his let-down. He must have made visual contact with automobile headlights on Fredericksburg Road and had in mind coming in under the overcast, maintaining contact until he reached Kelly Field. If this were the case, he made a wide turn in order to come back on Fredericksburg Road and this would account for the fact that the airplanes were headed in a northerly direction instead of a southerly direction which would have pointed them toward Kelly Field."

10-22-41B. Kelly Field, Texas. At 0205, a North American AT-6A (41-670) collided with terrain 12 miles north of Kelly Field, San Antonio, Texas, killing Aviation Cadet Andrew B. O'Brien (see 10-22-41A).

10-24-41A. Twin Sisters Mountain, California. At about 0930, a Douglas B-18A collided with Twin Sisters Mountain 15 miles northwest of Fairfield, California, killing the five crewmembers. The B-18 had taken off from Hamilton Field, San Rafael, California, at 0915 on a flight to the bomber's home station at Fort Douglas, Salt Lake City, Utah. The B-18 had inadvertently flown into instrument conditions and the pilots were unaware of their position relative to the peak, which was shrouded in thick, low hanging clouds. Witnesses on the ground reported hearing the bomber in the clouds and then hearing the impact and explosion moments later. A short time later the clouds cleared away and the burning airplane could be seen on the mountainside. Killed in the crash were: Capt. Franklin S. Nelson, 34, Salt Lake City, Utah, pilot; 2Lt. E.F. Sell, 24, Lander, Wyoming, co-pilot; TSgt. Andrew Zeik, 33, Salt Lake City, Utah, engineer; TSgt. Robert M. Kinney, 32, Salt Lake City, Utah, radio operator; Pfc. John W. Phillips, 23, Salt Lake City, Utah. Capt. Nelson was an airline pilot prior to joining the Air Corps.

10-24-41B. Kings Canyon, California. Five Curtiss P-40 type airplanes crashed or were abandoned at Kings Canyon National Park, California, after becoming separated from their formation in heavy overcast on a flight from March Field, Riverside, California, to McClelland Field, Sacramento, California. Two pilots were killed in crashes and three pilots parachuted to safety, two being rescued after enduring eight days trapped in the mountains. The flight consisted of 19 P-40 airplanes of the 57th Pursuit Group, which had spent several days flying from their station at Windsor Locks, Connecticut, to participate in simulated war maneuvers on the west coast. Investigation revealed that the flight, which had taken off from March Field between 1100 and 1130, broke up a short time later over the southern end of the Sierra Nevada Mountains, SSW of Mt. Whitney. According to the commander of the flight, Maj. Clayton E. Hughes, 57th Pursuit Group (see 11-2-41C), not all of the airplanes had functional radios, only flight leaders had full map sets, and many of the airplanes had dysfunctional oxygen systems. Maj. Hughes reported coming out on top of the overcast at about 17,000 feet, staying at this altitude for only a few minutes before leading seven airplanes back down through and to a safe landing at Smith, Nevada. One P-40 was damaged in a taxiing incident at Smith. 1Lt. Frank H. Mears led five other P-40 airplanes to the intended destination at Sacramento. 1Lt. William Henry Birrell was found dead in P-40 #39-200, which had smashed into 5,500-foot Gray's Mountain near Bass Lake. The actual time of the accident could not be determined. 1Lt. Richard N. Long was killed when P-40 #39-287 smashed into the side of a mountain at the 11,200-foot level west of South Guard Lake in the Cunningham Creek area (see 10-24-41C). 2Lt. John Pease bailed out of P-40 #39-213 north of Lake Isabelle after his engine suffered a complete loss of oil pressure. He was rescued almost immediately after parachuting to safety, his airplane crashing to earth at Kennedy Meadows, California. 1Lt. Leonard C. Lydon, 27, bailed out of P-40 #39-194 near Barton Peak at about 1245. At about the same time, and also near Barton Peak, 2Lt. Jack C. West, 22, safely parachuted from P-40 #39-285. He immediately found a mountain cabin stocked with canned food and survival gear, including matches, blankets, firewood, fishing poles, and a tent. The next day Lt. Lydon arrived at the cabin looking for help only to find Lt. West. Both flyers made a few attempts to walk to safety, once coming within a quarter mile of a ranger station before turning back for their camp. After eight days they were spotted by a B-18 bomber and rescued on 11-1-41. As of this writing, the wreckage of Lt. Lydon's early model P-40 (39-194) has not been found and is believed to be in the vicinity of Mt. Brewer.

10-24-41C. Kings Canyon, California. At an unknown time, a Curtiss P-40 flying in instrument conditions collided with rising terrain at an elevation of 11,200-foot msl west of South Guard Lake in the Cunningham Creek area of Kings Canyon National Park, California, killing pilot 1Lt. Richard N. Long. The airplane was part of a 19-ship flight that took off from March Field, Riverside, California, on a flight to McClelland Field, Sacramento, California. This airplane and pilot were discovered about five miles southeast of the Roaring River Ranger Station in the summer of 1959. See 10-24-41B.

10-26-41. United States. *Note: Most regions in the continental United States move clocks one hour back to standard time.*

10-28-41A. Lon, New Mexico. At 0115, a Beech AT-7 smashed into the ground after the failure of its port engine five miles west of Lon, New Mexico, killing the crew of five. The aircraft had taken off from Albuquerque, New Mexico, at 0005 on night cross-country navigation flight. The AT-7 circled the Lon area three times in a steady rain and on one engine before plunging to the ground and exploding into flames, killing the crew instantly. Killed in the crash were: 2Lt. W.H. Carpenter, pilot; 2Lt. G.W. Jones, co-pilot; Aviation Cadet E.B. LaRoache, student navigator; Aviation Cadet K.C. Manchee, student navigator; Aviation Cadet M. Marantz, student navigator.

10-28-41B. Tulare, California. At 1135, a Stearman PT-17 stalled and spun into the ground following an engine failure five miles south of Tulare, California, killing instructor 2Lt. John I. Lothrop, 26, San Francisco, California, and student Aviation Cadet Edward L. Riggs, 21, Rupert, Idaho.

10-29-41. Phoenix, Arizona. At 1130, a North American AT-6A crashed out of control after suffering a catastrophic structural failure of the starboard wing near Phoenix, Arizona, killing Aviation Cadet John F. Pinette, 22, Tacoma, Washington, and Aviation Cadet

Jack H. Saunders, 26, Sacramento, California. The airplane was on an instrument training flight with one student under the hood and the other acting as safety pilot. The aircraft was reported to have entered a tight spin with the pilots slow to recover. Such an extreme maneuver was needed to stop the spin that structural failure occurred. The starboard wing folded up and struck the cockpit canopy, shattering it. The aircraft began to roll in a clockwise direction, scattering debris along both sides of the flight path. The fuselage struck the ground inverted and immediately burst into flames, rolling end over end for over 400 feet from the initial point of impact with the ground.

✓ **10-30-41A. Albany, Georgia.** At 1220, two Stearman PT-17 airplanes collided in mid-air during a formation take-off at Darr Aero Tech Flying School, Albany Municipal Airport, Albany, Georgia, killing civilian flight instructor Joe Frank Machamer, Knoxville, Tennessee, and RAF Flying Cadet Eric N. Furze, Plymouth, England, aboard PT-17 #40-1820. The RAF student aboard PT-17 #41-953, Flying Cadet Kenneth Archibald, Southall, Middlesex, England, was seriously injured. The Accident Classification Committee stated, "After take-off the ships apparently drifted toward each other, the occupants of neither airplane seeing the other. Whether ship 41 (Archibald) drifted toward ship 59 (Furze), or vice versa or whether both drifted toward each other cannot be determined. When the collision occurred, at approximately 200 feet altitude, ship 41 was slightly lower, to the left of and angling toward ship 59. Ship 41's propeller cut the left aileron of ship 59 in numerous places, and 41's right wings tore off the rudder and left elevator of ship 59. Ship 41 then ballooned over and to the right of ship 59 and spun to the ground, no fire followed the crash. Ship 59 appeared to immediately plunge downward, straight ahead, bursting into flames immediately after striking the ground." Cadet Archibald was too severely injured to give a statement to the committee.

10-30-41B. Dungeness Point, Washington. At 1116, a Curtiss P-36A dove into the Straights of Juan De Fuca off of Dungeness Point, Washington, after simulating attacks on two U.S. Navy Consolidated PBY-5 Flying Boats, killing P-36 pilot 2Lt. Francis M. Cox. The PBY-5s began a descent after a series of simulated attacks and on the final attack the P-36 pilot misjudged his altitude and dove his airplane into the sea. Ensign Beecher Snipes, pilot of PBY 42-P-2, stated to investigators, "About 1050, 42-P-1 and 42-P-2 were dived on by an Army pursuit plane. My altimeter read 3,000 feet at this time. The attacking plane made dives on us at about two minute intervals until about 1100, when 42-P-1 and 42-P-2 joined up with the second section [of PBY-5 airplanes] and took station about 600 yards astern. Our altitude was decreased to about 1,500 feet at the time we joined the second section. The attacking plane made a dive from our port hand to attack the lead section. He pulled out under us and gained altitude on our starboard bow. The entire squadron was losing altitude at about 300 feet per minute. We were now at about 1,000 feet and flying slightly above the lead section. At about 1108, the attacking plane started another dive from our starboard bow, passed under the lead section, appeared on my port side, very close to the water, and crashed. The plane hit the water with his starboard wing, flipped over and exploded immediately. The plane appeared to tumble about two times on the water, covered with bright flames and smoke, and then sank immediately. I broke off from my section leader, made a 270 degree turn and landed alongside the floating wreckage."

November

11-1-41. Mt. Cory, Ohio. At 1315, a North American B-25B dove into the ground three miles north of Mt. Cory, Ohio, killing three AAF fliers and two civilian mechanics. The B-25 took off from Wright Field, Dayton, Ohio, on an accelerated service test flight of the automatic flight control equipment in preparation for a maximum range mission. The airplane was seen to dive out of the overcast at high speed, pull up abruptly at approximately 1,000 feet agl, level out for about a quarter of a mile then plunge to the ground at about a 45-degree angle. Weather at the time of the accident was reported as windy with misting rains with 1,500 to 2,000 foot ceiling and was not considered a factor in the accident. Investigators speculated that the aircraft got into an unusual attitude while flying in instrument conditions and in the abrupt pull-up close to the ground, the elevator surfaces failed. They further speculated that the pilots might have caged and un-caged the gyro instruments while the aircraft was not in straight and level flight. The B-25 struck the ground with terrific force, hurling wreckage and bodies over a six-acre area on the farm of A. H. Clymer. Killed in the crash were: 1Lt. Thomas W. Bafford, Fallon, Nevada, pilot; 2Lt. George W. Smith, Asheville, North Carolina, co-pilot; SSgt. Lee D. Chambers, 24, Hinsdale, New York, engineer; George Hebeman, Dayton, Ohio, civilian mechanic; John Southard, 20, Springfield, Ohio, civilian mechanic.

11-2-41A. Niles, California. At 0945, a Vultee BT-15 on a navigation flight collided with terrain near Niles, California, killing pilot 2Lt. Robert E. Agnew, Donalds, South Carolina, and passenger Pvt. Dan Fisk. The aircraft, missing for three days, reportedly had radio trouble while in instrument conditions. The wrecked airplane was found on 11-5-41. Investigators stated, "The elevation of the hill on which the crash occurred is approximately 1,900 feet. Plane hit the ground in a right turn with approximately [twenty-five] degrees of bank. The estimated speed upon impact was 180 miles per hour. From the disturbed ground, it appeared as though the pilot endeavored to climb just prior to impact. The motor cowling and propeller stopped approximately 100 feet from the point of impact; the right wing approximately 200 feet; the left wing approximately 400 feet. The remarks on Form 1A indicate that the radio was not functioning. It is the opinion of the board that there are two possible findings to be made; first, that the pilot was letting down out of the clouds just prior to impact, or second, that the pilot was endeavoring to fly contact with a very low ceiling, misjudged the proper valley and hit the ground while trying to make a course to Moffett Field [Sunnyvale, California]."

11-2-41B. Georgetown, California. At 1229, a Boeing B-17C suffered a catastrophic structural failure of the tail section and elevators while flying in severe winds and turbulence, crashing near 6,000-foot Tells Peak in the Sierra Nevada Mountains 25 miles east of Georgetown, California. Pilot 1Lt. Leo M.H. Walker, Salt Lake City, Utah, was killed in an unsuccessful parachute jump. An abrupt maneuver might have contributed to the in-flight break up of the airplane, which rained down in many pieces over a three-mile area. Eight crewmembers were able to parachute to safety, all receiving minor injuries. The pilot was missing after bailing out and was not found until several days after the accident. Heavy ground fog hampered the initial search for Lt. Walker. Forest Ranger J. L. Ogden found most of the crew at a nearby ranch. Two additional crewmembers were found at another ranch in the area. Parachuting to safety were 2Lt. John R. Mode, Fort Worth, Texas, co-pilot; SSgt. Eugene M. Clemens, Chamberlain, South Dakota, engineer; Cpl. Sterling H. Isom, hometown unknown, engineer; Pfc. Harold W. Salisbury, Bangor, Maine, radio operator; Cpl. Walter Glantz, Salt Lake City, Utah, passenger; Pfc. Alden H. Stookey, Corning, California, passenger; Pfc. Fred E. Pekuri, Portland, Oregon, passenger; Pfc. Robert V. Moseley, hometown unknown, passenger.

11-2-41C. San Anselmo, California. At 1730, two Curtiss P-40C airplanes flying in formation in instrument conditions collided with Bald Peak, Marin County, near San Anselmo, California, killing both pilots instantly. Heavy cloud cover was prevalent in the area. A third pilot parachuted to safety from his P-40C in Lucas Valley, two miles from Hamilton Field, San Rafael, California. The three P-40s were part of a 13-airplane flight commanded by Maj. Clayton E. Hughes, 57th Pursuit Group (see 10-24-41B). The airplanes had taken off from Medford, Oregon, at about 1530 on a flight to Fresno, California. Weather was reported to be a 1,500 to 2,000 feet ceiling at Fresno, with good visibility underneath. A major front extended all the way up the coast and well inland. Maj. Hughes planned to go "over the top" and let down through the overcast at Fresno. The flight let-down somewhere near Sacramento, well north of Fresno. The formation broke up into four separate flights during the descent. Maj. Hughes and two airplanes landed at Oakland, California. 1Lt. Frank Mears led Lt. Aiken and Lt. Clark to Stockton following the Sacramento beam. Capt. McGoldrick, leading Lt. Fairland, Lt. Thomas, and Lt. Carey, also followed this beam to Stockton. 2Lt. Thomas L. Truax (Madison, Wisconsin) was leading 2Lt. Russell Speckman (Ottawa, Illinois) and 2Lt. Walter V. Radovitch north of Hamilton Field, San Rafael, California. Lt. Radovitch lost contact with Lt. Truax and Lt. Speckman in the clouds and decided to bail out while he could. Lt. Truax and Lt. Speckman, flying in formation, slammed into Bald Peak, Marin County, while trying to orient themselves. Both were killed instantly. Mrs. Gerda E. Chamberlain, San Anselmo, stated to investigators, "I was standing on my door step and saw two airplanes pass over the house. In about two or three seconds they hit the mountain about a quarter mile away. I heard the crash and immediately saw an illuminated area at the point of the crash."

11-2-41D. San Anselmo, California. At 1730, a Curtiss P-40C (41-13454) flying in formation with another P-40C in instrument conditions slammed into Bald Peak, Marin County, near San Anselmo, California, killing pilot 2Lt. Russell L. Speckman, Ottawa, Illinois (see 11-2-41C).

11-4-41A. Bennettsville, South Carolina. At about 1355, a North American O-47A crashed shortly after take-off five miles southwest of Bennettsville, South Carolina, killing pilot 1Lt. Robert B. Womble, Jr. and observer 2Lt. Joel A. White. Pvt. Henry R. Williams was seriously injured. The pilot could not attain the necessary airspeed to keep the O-47 in the air after an extended take-off roll on a very soft field. Pvt. Williams stated to investigators, "We taxied over to the take-off position. I strapped myself in tighter than I usually do. The ship felt like it was going terribly fast and then seemed to be jerked up. As the wheels left the ground, we seemed to go into a slow right turn. It felt as though we hit a down draft, which seemed to take the ship out of his control. The ship felt like it was acting funny. It felt like a sloppy turn to the right. As we were running along the ground, it felt like he couldn't get the front of the ship up. I think we had enough altitude to clear trees, but the ship was banking slowly to the right and then bang, the right wing hit the ground. Motor sounded like it was running OK to me. I do not think that it stopped. I kept thinking to myself as to why the pilot was turning to the right, as this was not his normal way of taking off. I remember the plane hitting the ground and felt myself still strapped in my seat. I remember undoing the safety belt and walking away from the ship. I felt my hair on fire and remember walking away from the ship and looking around to see where the pilot and observer were. I could not see them, as I wanted to go back and try to pull them out. I remember seeing that the plane was on fire but after that I don't remember much as I seemed to be burning all over."

11-4-41B. Glendale, California. At 1147, a Lockheed YP-38 suffering a catastrophic structural failure of the tail booms dove into a back yard at 1147 Elm

Street, Glendale, California, killing Lockheed Aircraft Company test pilot Ralph B. Virden. Investigators noted that the AAF had lent this aircraft back to Lockheed so it could be modified with a view to solving the P-38's compressibility problem, which was just then being identified. This aircraft had two additional trim tabs installed on the elevator and was then sent out on a series of diving tests. The airplane took off from the Lockheed Air Terminal at 1035. Investigators stated, "Apparently Mr. Virden also had instructions to pass over the Lockheed Factory in sight of the speaker's stand at about 1145 if his tests were complete. Three dives and recoveries were successfully completed. A fourth dive was made. The altitude at which the fourth dive was started is unknown. It may have started at high altitude or it might possibly have been the start of a dive for the purpose of passing the speaker's stand at exceptionally high speed. In either case the airplane was observed in a sustained dive of about 15 degrees just prior to the structural failure.... [It] is believed that that considerable power was used as some altitude above 5,000 feet and that the airplane as it approached the 5,000-foot level was traveling at a speed greater than normal, but not at an excessively high speed. At approximately 5,000 feet it was observed that the tail surfaces started to leave the airplane in increments with great rapidity, the bare booms only remaining. The airplane continued on course momentarily then pitched forward completing the first 180 degrees of an outside loop with very little loss in altitude. From this point on the airplane was observed in a series of inverted flat spin gyrations until it crashed in a flat inverted position." Apparently, stresses were incurred that compromised the structural integrity of the tail section, causing its failure and resulting in a loss of control and the airplane diving into the ground. Numerous qualified observers, including Lockheed employees, had seen the airplane enter its fourth high-speed dive just prior to the crash. Lockheed engineers and Army investigators determined that the tail section underwent severe buffeting and tore away where the tail section meets the booms. The damaged YP-38 struck the ground with terrific force, driving the engines and pieces of the airframe several feet into the ground. Published reports of the accident reported that "thousands" of spectators flocked to the crash site, causing traffic problems for civilian emergency personnel and Army investigators. Army authorities, police, and employees of Curtiss-Wright and Lockheed confiscated the film from several cameras at the scene. No one on the ground was injured, but the rear bedroom of resident Jack Jensen's home at 1147 Elm Street was burned and severely damaged. For a short time, neighboring homes on either side of the Jensen home were in danger of catching on fire. Firefighters quickly extinguished the fire. Water damage to the Jensen home was considerable.

11-4-41C. Selfridge Field, Michigan. At 1648, a Bell P-39D crashed on the edge of Selfridge Field, Mt. Clemens, Michigan, after overshooting the runway during a landing attempt, killing pilot 2Lt. Donald J. Stone. 2Lt. R. E. Keyes stated to investigators, "I was sitting in my P-39 airplane at the east end of the field when Lt. Stone came in to land. He came in directly over my head, using power and a rate of speed [that] appeared to me excessive. I saw him land about the middle of the field and then run off the edge of the field. To the best of my knowledge and belief, his flaps were up."

11-6-41. Daniel Field, Georgia. At 1230, a North American B-25B suffering the failure of the port engine just after take-off crashed while attempting an emergency forced landing at Daniel Field, Augusta, Georgia, killing the crew of four. The B-25 was taking off on a simulated bombing mission. Killed in the crash were: 2Lt. Jack H. Butler, pilot; 2Lt. Harold C. Jepson, co-pilot; SSgt. Harold E. Brown, engineer; Pvt. Earl L. Willburn. The Accident Classification Committee stated, "Ship was last of six airplanes taking off on a mission in co-operation with IV Corps. Pilot ran up the engines and tested switches prior to take-off. Ship took a longer run [than the] other ships but this is not thought to be significant. After reaching about 300 feet in altitude, left motor cut out completely and pilot pulled back throttle, propeller control and mixture control on that side. Flaps were still down 15 degrees and it was impossible to tell if left propeller was feathered. Propeller control on right engine was pushed forward or was still in take-off position. Pilot started a gradual turn to the right evidently intending to circle field for a landing. For some reason the pilot was unable to maintain flight and decided to make a belly landing in a small clearing. Approach was made over trees about 50 feet in height. Pilot evidently lost too much speed and dropped vertically in normal flying attitude from about 75 feet just after passing the trees. Ship stopped at point of impact on a rising of about 15 degrees. Ship burst into flames at impact or soon thereafter and was completely destroyed. All occupants were killed and partially burned." Another B-25 suffered a similar (but non-fatal) engine failure on take-off three days later. Investigators determined a collapsed fuel line caused the engine failure in both airplanes.

11-10-41. Raleigh, North Carolina. At 2013, a Douglas A-24 attempting a landing at the municipal airport at Raleigh, North Carolina, crashed into an infantry bivouac area after striking trees, killing pilot 2Lt. Harrison P. Taylor and seriously injuring passenger Pvt. Norman L. Toussaint. Two army enlisted men on the ground were also seriously injured. They were Army Cpl. N. Bartley and U.S. Navy Seaman James Fetton. The Accident Classification Committee stated, "It is apparent that the pilot circled the field at too low an altitude and allowed himself to lose sufficient altitude to put him in the trees about at the completion of his last turn before landing. He struck a group of trees about 600 yards short of the point he crashed and it is believed that in striking these trees the propeller was damaged to such an extent as to render it incapable of keeping the airplane in level flight. The crash occurred when he struck the next group of trees [that] in which the campsite was located. At this point it was seen and heard that the pilot was making every effort to gain altitude by shoving the throttle to the open position and pulling the stick to the rear. The ship struck the second group of trees, cut through about 100 yards of pines ranging in diameter from five inches to fourteen inches and crashed. The airplane burned for three hours before the fire could be extinguished."

11-11-41. Muskogee, Oklahoma. At 1315, a Fairchild PT-19 spun to the ground approximately 15 miles west of Muskogee, Oklahoma, killing student pilot Pvt. Robert E. Fritsch, Jr. A civilian witness, Julian S. Huff, stated that the aircraft spun down from quite a ways up and that the pilot added power and stopped spinning just prior to striking the ground. According to Huff, the aircraft "flew all to pieces" when it hit the ground.

11-12-41A. Mohawk, Arizona. At 1816, a Douglas DB-7B (AL-882) being ferried to the east coast for delivery to the Royal Air Force encountered very poor weather and flew into terrain at Mohawk, Arizona, killing pilot 2Lt. Ivan J. May. The Aircraft Accident Classification Committee stated, "Airplane was cleared from Long Beach, California, as wing airplane in two airplane formation led by DB-7B AL-880 [piloted by 2Lt. Lawrence A. Barrett, Jr., 22, Gary, Indiana]. The weather sequences showed contact flight ceilings and visibilities with some rain. The take-off was made at 1440 PST, or 1540 MST. Sundown at Tucson, Arizona, the destination, on November 12, 1941, was 1729 MST. Estimated time en route was 2 hours and 15 minutes, giving E.T.A. as 1755 MST. Even in good weather, the flight could not have reached Tucson before dark, and with reported weather should not have been cleared for daylight flight. Lt. May lost contact with [Lt. Barrett] at some unknown time, evidently prior to 1728 MST, as he communicated with Phoenix [Arizona] at that time and reported himself as lost. The Committee interviewed Mr. Roy Kelland of Ralph's Mill, a service station near Wellton, Arizona, 20 miles southwest of Mohawk, Arizona. His remarks were to the effect that at about 1730 [MST], a two-motored airplane circled for about ten minutes a mile or two east of his station, and then flew off to the southeast. (This period of circling was evidently the time when Lt. May was in communication with Phoenix.) Mr. Kelland said that it was raining and getting dark at the time. The Committee obtained the statements of three witnesses and decided that Lt. May then started to follow the highway, was attracted by the lights at Mohawk, saw the hills too late to complete an attempted turn-away from them, and crashed in an almost flat position on a heading approximately 120 degrees to the left of his route up the highway. The airplane had its wheels up at the time of the crash. The empennage section was the only part not completely demolished by the impact. Fire completed the destruction." A former Army Air Service Lieutenant and WWI veteran, Edward S. Waterman, Mohawk Lodge, Mohawk, stated to investigators, "At about [1700], a heavily powered airplane passed over the rear of the building at Mohawk Lodge very close to the ground. I was inside the building at the time. Shortly afterwards I went outside to look at the weather conditions. It was raining very hard at the time and had been raining for several hours. The clouds were hanging down the sides of the peaks near us. These nearby peaks are not high. The ceiling was very low. I went back into the building, and had looked at my watch at [1750]. A few minutes later another heavily powered airplane passed very close to our building. It was dark and all of our lights were on. The airplane motors were running satisfactorily. It was raining hard. My wife and our employees ran out front. The airplane had hit the ground just behind a hill (which is about 200 feet high) and is just across the road from us. The sky was immediately brightly illuminated." Lt. May was killed instantly. Lt. Barrett, who was leading Lt. May, was reported as missing and he and his airplane were not located until 12-19-42, 22 miles south of Wellton, Arizona. See 11-12-41B.

11-12-41B. Wellton, Arizona. At an unknown time, a Douglas DB-7B (AL-880) being ferried to the east coast for delivery to the Royal Air Force encountered very poor weather and collided with terrain 22 miles south of Wellton, Arizona, killing pilot 2Lt. Lawrence A. Barrett, Jr., 22, Gary, Indiana. The airplane took off from Long Beach, California, at 1440 PST for a flight to Tucson, Arizona. Lt. Barrett had taken off with and was leading a DB-7B (AL-882) piloted by 2Lt. Ivan J. May (who was killed when his airplane collided with terrain at Mohawk, Arizona, at 1816 MST). Lt. Barrett and Lt. May encountered poor weather and darkness and soon became separated. It was speculated that Lt. Barrett passed over Mohawk, Arizona, at about 1750 MST and flew away to the south. Apparently the pilot began to circle in an attempt to determine his location or get back to the highway and flew into terrain a short time later. Weather in the area at the time was reported as heavy rain and low hanging clouds. Lt. Barrett and his airplane were reported as missing and were not located until 12-19-42. War Department engineers Lester P. Heacock and Edward L. Stephenson found the wreckage in the early evening while they were working in the area. See 11-12-41A.

11-13-41A. Maxwell Field, Alabama. At 1020, a North American AT-6A and a North American BC-1A collided in mid-air while preparing to land at Maxwell Field, Montgomery, Alabama, killing RAF Flying Cadet Richard N. Moss, 20, Preston, Lancaster, England (BC-1A), and RAF Flying Cadet Frank V.

Marhoff, 29, Boxmoor, Herts, England (AT-6A). When Cadet Moss was on his downwind leg, an Army instructor leading another flight had noticed that Cadet Moss had not lowered his landing gear and had called him on the radio to alert him. He received no response from the cadet. When Cadet Moss was on his base leg, the instructor again called the cadet to warn him about the landing gear and again received no response. The flight leader, 2Lt. M.A. Miller, stated to investigators, "We came in and made a 90-degree turn directly into the landing leg. After this turn I was ready to cut my throttle and I saw this ship off to my left above me with his wheels up. As I saw him, he was over us, since my students were not in formation, I knew the men were behind me. I stayed about where I was because he was above us. Then as I watched him he dropped his wheels. By that time he was probably straight out from me to my left. Then as I continued on I looked back, thinking maybe he would come in. I couldn't tell whether he was dropping or not. Then suddenly the two ships collided. I circled on around the field and came in and landed."

11-13-41B. Palm Springs, California. At 0930, a Lockheed P-38D suffering a control problem due to compressibility dove into the ground near Palm Springs, California, killing pilot 2Lt. Elery G. Gross. Lt. Gross had been following his leader 1Lt. Charles W. Dunning through a series of diving and climbing turns. The Accident Classification Committee stated, "At some point in [the] dive, the airplane apparently reached the critical speed at which the characters of the P-38 type airplane cause it to become uncontrollable and Lt. Gross was unable to bring it out of the dive."

11-14-41. Long Field, Arkansas. At 1020, a Fairchild PT-19A stalled and spun into the ground one-half mile from Long Field, Arkansas, eight miles southeast of Grider Field, Pine Bluff, Arkansas, killing Aviation Cadet Marion N. Cahill. Cadet Cahill stalled in the second turn of a 360-degree overhead approach and spun in, the airplane bursting into flames upon impact.

11-15-41. Springfield, Maine. At 1932, a Douglas B-18A collided with terrain ten miles south of Springfield, Maine, killing the crew of four. The B-18 took off at 1643 from Westover Field, Chicopee, Massachusetts, on a flight to Bangor, Maine, when it became lost letting down through the overcast. The airplane was seen about 1920 approaching Springfield from the south, flying very low under the clouds. At 1930, three hunters saw the B-18 fly over the Carry Farm at a very low altitude, heading toward the southwest. At about 1932, they heard the airplane crash and then seen the flames of the wreckage out in the distance on the 1,000 acre bog in which the crash occurred. Dense fog was reported in the area at the time of the crash. Killed in the crash were: 2Lt. Peyton W. Beckman, 28, Houston, Texas, pilot; 2Lt. Wyman O. Thompson, 21, Underwood, South Dakota, co-pilot; Cpl. Jacob L. Parson, 28, Holbrook, Pennsylvania, engineer; Pfc. Lee E. Rothermel, 30, Valley View, Pennsylvania, radio operator.

11-16-41. Patterson Field, Ohio. At 0938, a Bell P-39D crashed at Patterson Field, Ohio, after suffering an engine failure just after take-off, killing pilot 1Lt. Eiler C. Sundorph. Maj. Ralph O. Brownfield, who witnessed the crash, stated to investigators, "Immediately upon his motor failure, the pilot went into a steep bank to the left and dropped his nose at a very sharp angle, and made a sharp diving turn into the ground. There was no spin in this case, as he had excessive speed up until the time he struck the ground, and it is my belief that the pilot banked to the left, then looked down into the cockpit, probably with the intention of lowering his wheels, and when he did, the nose dropped unknown to him and struck the ground before he realized it." Investigators discovered that the intake valve for the port number three cylinder was distorted to an extent, which permitted fire or hot gases to pass back into the induction system, blanketing the remaining cylinders and causing the engine to stop. A safe crash landing could have been made straight ahead but the pilot chose to execute a steep turn at very low altitude.

11-17-41A. Park City, Utah. At 0015, a Douglas B-18 suffering a malfunctioning airspeed indicator was abandoned by its crew and crashed four miles west of Park City, Utah, killing two crewmembers. Five members of the crew were able to parachute to safety. The B-18 had departed Lowry Field, Denver, Colorado, at 2030 (11-16-41) on flight to Salt Lake City, Utah. The flight encountered instrument conditions near Park City, picking up moderate to minimal ice while flying through frozen sleet. All de-icing equipment was functioning. The pilot noticed a steady drop in airspeed and pushed the nose down to arrest the apparent loss, getting no indication of an increase in airspeed on his instrument. After a few moments the pilot ordered the crew to abandon ship. After the crew bailed out they watched the B-18 enter a gentle turn to the left and circle the area. It was during this turn that the pilot-less aircraft struck and fouled the fully deployed parachute of passenger Maj. R.E.L. Pirtle, Commanding Officer 88th Reconnaissance Squadron, Fort Douglas, Utah, causing him to plummet to his death. Sgt. Jack D. Anderson failed to leave the airplane and was killed when it crashed. The crewmembers parachuting to safety were: 1Lt. William E. Basye, pilot; 2Lt. Mabry Simmons, co-pilot; 2Lt. C.A. Smith, bombardier; SSgt. E.V. Bynum, engineer; Pfc. Raymond L. Togerson, radio operator. Investigation revealed that a short circuit in the pitot heater caused the airspeed indicator to malfunction.

11-17-41B. Florence, South Carolina. At 0830, a Bell P-39D crashed while attempting a wheels-down emergency forced landing on a plowed field after running out of fuel two miles north of Florence, South Carolina, killing pilot 2Lt. William E. Howard. The pilot lost control of the airplane as soon as it touched down, collapsing the nose landing gear and causing the P-39 to flip over on its back.

11-17-41C. Turner Field, Georgia. At 1108, a North American AT-6A stalled and spun into the ground five miles northeast of Turner Field, Albany, Georgia, killing RAF students Flying Cadet Robert S. Wilson and Flying Cadet Jack Hartley. The airplane was on an instrument training flight. There were no witnesses to the crash.

11-17-41D. Bear Canyon, Sandia Mountains, New Mexico. At 1157, a Northrop A-17 crashed at the head of Bear Canyon at an elevation of 8,000 feet on the west slope of a peak in the Sandia Mountains, New Mexico, killing pilot 2Lt. Seldon T. Miller and passenger SSgt. Howard L. Edwards. Investigation revealed that this airplane had done a considerable amount of buzzing in the area of the Estancia Valley prior to the accident. Strong winds were listed as a contributing factor in the accident, which occurred 20 miles north of Albuquerque. It was determined that the airplane was banking to the left when it struck trees, causing the pilot to lose control of the airplane and crash.

11-18-41A. McBean, Georgia. At 0736, a Vultee BT-15 stalled and spun into the ground four miles west of McBean, Georgia, killing Aviation Cadet Louis C. Selby and seriously injuring civilian instructor Harold S. Andrews. Andrews stated to investigators, "While riding in the rear cockpit of a BT-15 airplane with Aviation Cadet Selby as pilot we approached a field on our left for a simulated forced landing at approximately 1,300 feet. Mr. Selby, seeing that the base leg was too close, attempted to S-turn to the right, but in making the turn he stalled and the airplane began to spin. At that time I took the controls and attempted to recover but did not have enough altitude. We hit a fence post, then a small barn, and then the ground. The airplane apparently went end over end. I got out and tried to help Mr. Selby out, who was obviously injured badly. Some people nearby came to our aid and took us to the hospital."

11-18-41B. Patterson Field, Ohio. At 1330, a Schweizer XTG-2 two-place glider stalled and spun into a field adjoining Patterson Field, Ohio, killing 2Lt. Claude L. Luke, Salt Lake City, Utah, and seriously injuring 1Lt. Floyd J. Sweet. Lt. Sweet stated to investigators, "I felt the ship stall and the nose drop. I remember Lt. Luke saying we were going to crash, and I could see the ground coming up at us as I looked over his shoulder. I have no memory of our hitting the ground. We crashed from an altitude of about 150 feet and I believe that it was caused from a stall in downwind turn." Lt. Sweet also reported that Lt. Luke had been watching and commenting on another glider in the pattern just prior to the accident.

11-18-41C. North Hampton, Ohio. At 1400, a Martin B-26 suffered a catastrophic structural failure of the tail section and dove into the ground on the Miranda Farm at North Hampton, Ohio, killing the crew of three. The Accident Classification Committee stated, "This accident took place while on an official Accelerated Service Test Mission in a B-26 type airplane. From the testimony of witnesses and subsequent examination of the airplane it appears that a structural failure of the vertical fin and rudder occurred at a high but undetermined altitude. As a result the plane fell into an uncontrolled spin, crashed and burned, killing the crew of three and demolishing the airplane. The vertical fin and rudder were found scattered over an area approximately [one-half] mile in radius from the point of impact, both the rudder and the fin having broken into three separate pieces. The left side of the pilot's escape hatch cover and the right nose wheel door were also found at some distance from the point of impact. It is impossible to definitely determine the cause of the structural failure, but the following possibility is offered on the basis of testimony of officers familiar with the B-26 type airplane. The left half of the pilot's escape hatch cover found approximately [three-eighths] of a mile from the crash and had been cut by the propeller blade. The emergency release on this hatch had not been operated. Further there was dent in the left side of the leading edge of the vertical fin approximately one foot from the top, which may have been caused by the escape hatch cover striking the vertical fin in flight. This is based on the thought that the escape hatch cover may have come open accidentally during flight and had been torn from its hinges by the air stream. It is the testimony of officers familiar with this type of airplane that difficulty is occasionally experienced in properly securing the hatch cover and that they have been known to come open during flight. Progressive failure of the vertical fin and rudder could have resulted from such an impact." Killed in the crash were: 1Lt. Thomas P. Bacon, Bradford, Virginia, pilot; 2Lt. James G. Lynch, West Haven, Connecticut, co-pilot; SSgt. Edgar E. Johnson, Jackson, Mississippi, engineer. The well-known early B-26 phenomenon of high wing loading notwithstanding, this is the first of a series of fatal B-26 accidents in the United States and overseas in which one of the pilot's escape hatches separates from the airplane and collides with the tail section, causing its progressive failure.

11-19-41A. Myrtle Beach, South Carolina. At 1100, a Bell P-39D suffering an engine failure crashed into a heavily wooded area about three miles WNW of the municipal airport at Myrtle Beach, South Carolina, killing pilot 2Lt. Leonard Charles Lee. The pilot was attempting to make a forced landing in a cornfield. The Accident Classification Committee stated, "A few seconds prior to the crash the pilot reported to the airport control tower a dead engine at 400 feet. Pine and hardwood timber surrounding the cornfield averaged approximately 100 feet high. The approach was made lengthwise with the landing gear retracted. Plane entered the woods at 20 feet, the left wing hit a tree, shearing it off at the wing guns, driving one .30 caliber gun eight inches into the wood before the tree split off. The right [wing] was sheared off at the outer edge of the fuel tanks, uprooting and cutting into a tree ten inches in diameter. The nose section and fuselage hit a large tree, throwing the engine clear and breaking the fuselage, which rolled to an upside down position. The empennage was thrown on top of the pile by the impact. The accident was caused by complete engine failure, cause undetermined."

11-19-41B. Luke Field, Arizona. At a time between 1430 and 1730 MST, a North American P-64 crashed 20 miles northwest of Luke Field, Phoenix, Arizona, killing Aviation Cadet Charles C. Ball. The P-64 hit the ground in a high-speed vertical spin from an unknown altitude in a wings level attitude. The pilot tried to parachute to safety but bailed out about 200 feet from the ground, too low to allow his parachute to fully deploy. The Accident Classification Committee stated, "The airplane was descending vertically in excess of 200 miles per hour as evidenced by the two foot hole made by the engine in the ground. The engine was turning over in excess of 1,800 rpm as evidenced by the straight propeller blades. The blades rotated into the ground rather than by impact. The third blade, which evidently rotated into the ground first, assumed a depth of four or five feet at the tip. The airplane was rotating as though in a violent vertical spin. Evidence of this is that the wings bounced from the trough made by the first impact and turned about five degrees before the inertia of the spin was stopped."

11-25-41. Kelly Field, Texas. At 1130, a North American BC-1 dove into the ground at high speed at the Gallagher Ranch 30 miles northwest of Kelly Field, San Antonio, Texas, killing pilot 2Lt. John R. Harshey. Witnesses stated that the pilot was performing aerobatics at low altitude prior to the accident. The BC-1 exploded violently into flames upon striking the ground.

11-29-41. Gunter Field, Alabama. At 0930, a Vultee BT-13 crashed out of control ten miles south of Gunter Field, Montgomery, Alabama, killing pilot 2Lt. James H. Butler, Baltimore, Maryland, and passenger Pvt. Neil Sullivan, Braidwood, Illinois. The cause of the accident could not be determined.

December

12-1-41A. Post Field, Oklahoma. At 1430, a Piper O-59 crashed while attempting a landing during a short field landing exercise at Post Field, Fort Sill, Oklahoma, killing pilot 2Lt. Charles W. Maxwell and seriously injuring passenger SSgt. Charles E. Byrd. SSgt. Byrd stated to the Accident Classification Committee, "We were following Lt. Moore and landing in small fields on the reservation. Lt. Moore landed in a field and we flew low over it to look it over. Lt. Maxwell started a steep turn to land in the field, our left wing struck a tree and we crashed. I remember Lt. Maxwell throttling back the engine as he started the turn and to the best of my knowledge and belief the engine was idling and did not stop."

12-1-41B. Poteet, Texas. At 1705, a North American BC-1 collided with terrain after suffering an apparent control surface failure after an extreme spin six miles north of Poteet, Texas, killing A/C Eugene C. McCaffery and injuring instructor 2Lt. Victor Milner, Jr. Lt. Milner reported blacking out during an extreme maneuver made by the student to avoid flying into clouds. The BC-1 apparently entered a very tight spin at about 13,000 feet. The instructor regained consciousness and partial control of the airplane very close to the ground before striking a stand of black jack oak trees in a near flying position. The airplane bounced off the treetops and descended into a small field with forward momentum and in a near flying attitude.

12-2-41A. Kingsbury, Texas. At 1130, two North American BT-9B aircraft collided while flying a formation training flight near Kingsbury, Texas, killing pilot 2Lt. Carleton H. Hibbard aboard BT-9B #37-193. A/C Howell P. McCorkle landed lightly damaged BT-9A #36-107 without incident at Randolph Field, San Antonio, Texas. Cadet McCorkle stated to investigators, "I was flying in No. 2 position, with Lt. Hibbard leading the formation. [Aviation Cadet Joseph A. Nischan (see 8-20-42A)] was flying No. 3 position. We made several turns and approached Marion Field where we made one landing, after which we took off again and climbed to approximately 4,000 feet. After we had reached that altitude, we began making our turns steeper. We made a climbing turn and a steep turn to the right and started diving. In the process of this turn, Mr. Nischan, flying in No. 3 position, was lost from the formation and fell behind. He was on the outside of the turn. After diving, I lost sight of the leader immediately. I then thought that the most sensible thing to do was to try to clear myself and I pulled out of the dive into a climb. I slowed the speed of the plane down to approximately 90 mph and was in the process of leveling out when I felt my plane jar and I was thrown slightly to the right side of the cockpit. Immediately after realizing that I had been hit, I glanced to the right and saw the plane flown by Lt. Hibbard was the plane that collided with the plane that I was flying. I do not know at what altitude we were when we collided. I could see that his plane was damaged; the left end of his wing was bent. I also noticed that my wing was damaged, observing a small dent on the top of my right wing. After that I kept my eye on Lt. Hibbard's plane as it made from two to four spirals to the left and as it neared the ground I would say that it went into a spin. Immediately after seeing the plane crash, I turned my attention to my plane and proceeded to return to Randolph Field. The only part of Lt. Hibbard's plane that I could see was damaged was the wing. I could see Lt. Hibbard sitting upright in the cockpit and the canopy did not look as if it had been crushed."

12-2-41B. King City, California. At 1505, a Ryan PT-22 stalled and spun into the ground while practicing forced landings seven miles WNW of the King City Airport, King City, California, killing civilian instructor Arthur J. Hook and student SSgt. William A. Cashin. The Accident Classification Committee determined that the crash resulted from a downwind turn made at low speed, causing the airplane to stall and crash into the ground from an altitude of about 200 feet.

12-3-41A. Placerville, California. At 1415, a North American AT-6A crashed in the Sierra Nevada Mountains 35 miles east of Placerville, California, killing Aviation Cadet Stephen Grundy, 24, San Bruno, California. The aircraft was part of a ten-ship formation flight from Stockton Field, Stockton, California, to Lake Tahoe, Nevada, and had been missing for several months.

12-3-41B. Dozier, Alabama. At 1830, a North American BC-1A flying in poor weather collided with terrain eight miles west of Dozier, Alabama, killing pilot 2Lt. Charles M. Jose, Montgomery, Alabama, and passenger Pvt. Thomas J. Stevens. The aircraft had been returning from gunnery practice at Eglin Field, Valparaiso, Florida. Investigators noted that Lt. Jose was the son of Maj. E.H. Jose, U.S. Army, Greenville, Mississippi, and the brother of AAF pilot Lt. E.H. Jose, Jr., Barksdale Field, Shreveport, Louisiana.

12-5-41. Camas, Washington. At 1235, a Douglas DB-7B flying in heavy fog collided with Prune Hill near Camas, Washington, killing pilot 2Lt. Harry O'Neal. The aircraft was part of a flight of 18 DB-7B aircraft being ferried to Canada for delivery to the Royal Air Force. The aircraft barely missed colliding with three houses in the immediate area after it and two other DB-7B airplanes became separated from the formation while letting down through thick clouds and fog that had suddenly formed after the flight's departure from Portland Army Air Base, Portland, Oregon. The wreckage and pilot were strewn over a ten-acre area.

12-6-41. Rockaway Beach, Long Island, New York. At 1430, a Curtiss P-40 crashed into the Atlantic Ocean at Rockaway Beach, Long Island, New York, killing Aviation Cadet Charles B. Davenport, 24, Silsbee, Texas. The student was attempting a forced landing because of an engine failure of an undetermined nature. The airplane stalled at an altitude too low to allow recovery. The airplane spun into the surf, killing the student instantly.

12-7-41. Pinelawn, Long Island, New York. At 1140, a Curtiss P-40 suffering an engine failure collided with two automobiles while attempting an emergency forced landing at Montefiore Cemetery, Pinelawn, Long Island, New York, killing civilian bystander Hyman Wiener, 34, Brooklyn, New York. Five civilians and the Army pilot, A/C Joseph F. Rybek, were injured in the accident. Seriously injured were: Charles Zucker, 41, Brooklyn, New York; Soloman Schtierman, 69, Brooklyn, New York; Morris Leff, 33, Brooklyn, New York; Elmer Rhodes, Lynbrook, New York. The pilot and Mrs. Jacob Schtierman, 33, Brooklyn, New York, were lightly injured. The airplane, which had taken off from Mitchel Field, Long Island, about 20 minutes earlier, suffered a sudden loss of power and the pilot was unable to make his auxiliary field for an emergency landing. The student picked the nearest field he could spot and set up his approach. The student then shot the landing and felt his airplane strike an object and cause him to lose control. The P-40 had collided with a 1939 Plymouth Station Wagon owned by the cemetery and a 1930 Packard Sedan owned by Dr. Morris Zucker, Brooklyn, New York. Mr. Wiener was standing near the automobiles and was killed instantly when he was struck by the airplane.

12-8-41. Coahuila, Mexico. At an unknown time after 2100 CST, a Seversky AT-12 (similar to the P-35 type fighter) collided with rising terrain 20 miles northwest of the El Fortin Ranch, Coahuila, Mexico, killing pilot 2Lt. James E. Snyder. The airplane had taken off at 1830 CST from Biggs Field, El Paso, Texas, and was reported missing on a ferry flight to Duncan Field, San Antonio, Texas. Investigators stated, "After an extended air and ground search the crash of the airplane and the body of the pilot were located on 12-21-41 approximately 20 miles northwest of the El Fortin Ranch, state of Coahuila, Mexico, slightly above 7,000 foot level on the northwest side of [an] unnamed [9,000-foot] mountain in that vicinity, in an extremely inaccessible location. The pilot was overheard by one of his comrades, at the time of preparing his flight maps, to remark, 'The heading back is 110 [degrees].' This heading, however, is at least 15 [degrees] off the proper heading. An airplane was reported, by telegraph the morning of 12-9-41, in response to radio broadcasts for information, as having been observed passing over the town of Terlingua, Texas, about 2000 the previous night. Terlingua is almost directly on line between El Paso and the scene of the accident. In view of the above opinion of the committee is that Lt. Snyder met his death when his aircraft struck the mountains, which are higher in that vicinity than the map indicates, and which he undoubtedly did not see due to the darkness of the moonless night. Causes of the accident were the inexperience of the pilot in flight preparation and execution and the darkness of the night."

12-9-41A. Biggs Field, Texas. At 0900, a Martin B-26 suffering an engine failure crashed while attempting an emergency landing at Biggs Field, El Paso, Texas, killing three crewmembers and seriously injuring four others. The Accident Classification Committee stated, "Pilot made one run of about [two-thirds of a] mile without being able to take-off, then returned for a second run without stopping for any check of the engines. Took off after second run of about [three-quarters of a] mile, barely clearing a fence at end of one-mile runway, with wheels retracting. Ship never gained more than 100 feet of altitude. After a wide circle to the left, during which he passed El Paso Municipal Airport about one mile south of Biggs Field, pilot apparently attempted a wheels up landing about three miles east of Biggs Field but the left wing struck a large sand knoll, whirled the ship almost 180 degrees to the left and caught fire. Investigation indicated that the left engine was delivering less power than the right, but however, the propeller settings were determined to be in the take-off position." Killed in the crash were: Maj. Mark K. Lewis, pilot; Maj. D. P. Laubach, co-pilot; SSgt. M. A. Miullo, engineer. Seriously injured were Maj. Stuart G. McLennan, Cpl. F. J. Rizzi, Cpl. Leland H. Whitman, and Pvt. W. S. Tracz.

✓ **12-9-41B. Skidaway Island, Georgia.** At 1415, a Stearman PT-17 performing spins and aerobatic maneuvers crashed out of control on Skidaway Island six miles southeast of Savannah, Georgia, fatally injuring pilot 2Lt. John P. Robbins and passenger Pfc. Wayne E. Bradley. Both died a short time after the accident at a civilian hospital in Savannah. Investigation revealed that an unsecured toolbox in the baggage compartment might have fouled the control cables within the fuselage. Civilian Hurley Cook, Savannah, Georgia, stated to investigators, "The first time that I noticed the plane, he was flying at what appeared to be about 1,500 feet. The pilot began stunting the airplane, looping and rolling the airplane. He went into a spin, and made three or four turns and came down to about 500 feet above the ground and began a series of stunts, including one loop and then another tail spin, coming out of the second spin at what I would judge to be about 100 feet. After coming out of the second spin the airplane went into a normal glide, and proceeded straight ahead until it struck the ground, about 50 feet from the bank of the Skidaway River. All of this took place immediately above the plantation home of R.C. Roebling, located on Skidaway Island. I was at work at the Roebling place at the time, about 100 yards from where the plane struck the ground. I assisted in removing the injured men from the airplane, and placing them on a boat for transport to Savannah. The injured men were both unconscious and were not able to make any statements as to the cause of their accident."

✓ **12-10-41A. Chickasha, Oklahoma.** At 0855, a Fairchild PT-19 practicing spins and spin recovery spun into the ground out of control four miles northeast of Chickasha Field, Chickasha, Oklahoma, killing Aviation Cadet John A. Coons. Cadet Coons' instructor, John A. Myers, who was flying with another student, witnessed the crash. He stated to investigators, "While flying at an altitude of 1,500 feet above the terrain, I saw another trainer at approximately 3,000 feet in a right hand spin about two miles north and slightly east of my position. I saw this airplane continue four or five turns in a normal spin and suddenly flatten out and continue [in a] flat spin four or five more turns into a small wooded gully, located about five or six miles northeast of the airport. I immediately flew to the wreckage, circled and noted that the airplane number was 110, and from all indications the pilot had been instantly killed."

12-10-41B. Taft, California. At 1645, an RAF lend-lease Douglas DB-7B flying in poor weather on a ferry flight collided with a small hill on the P.A. Klipstein Ranch 15 miles west of Taft, California, killing pilot 2Lt. William C. Daniel III and passenger Sgt. John K. Copely. The Accident Classification Committee stated, "Pilot Daniel was en route from the Sacramento Air Depot [California] to Lerdo [California], Air Corps Basic Flying School, the pilot, in attempting to stay under low clouds and rain, hit a small hill while turning. The ship was completely wrecked and partially burned."

12-10-41C. Boulder City, Nevada. At 2025, a North American AT-6A participating in a blackout drill crashed out of control in an inverted position two miles north of Boulder City, Nevada, killing pilot 2Lt. John G. Kelso and passenger 2Lt. George E. Turner. The Accident Classification Committee stated, "The fact that a blackout was in progress is a small contributing factor inasmuch as the flight was for the purpose for a check on the blackout effectiveness. As the blackout became more and more complete due to the pilot's efforts, he finally found himself with no point of reference other than his instruments, and was unable to attain complete control in time to recover from the attitude of inverted flight attained inadvertently following the extinguishing of the last light. The airplane landed in an inverted position with the fuselage at an angle of approximately 45 degrees with the horizontal, completely demolishing the plane and killing both occupants."

12-10-41D. Gainesville, Alabama. At 2230, a North American AT-6A crashed while attempting a landing at Gainesville Airport, Gainesville, Alabama, killing RAF Flying Cadet Peter Green and RAF Flying Cadet Stanley Holden. The two cadets became lost on a night cross-country flight from Maxwell Field, Montgomery, Alabama, to Maxwell Field via Birmingham and Gainesville, Alabama. An investigator stated to the Accident Classification Committee, "From all indications on the field, it appeared that the pilot was attempting a wheels up landing into the northwest. He first hit the ground about [one-third of the way] down the field, at which point the air scoop was found buried in the field, and a series of marks of the propeller were found digging in about 14 inches; then for about ten yards the engine seemed to have scooped up the ground to a maximum depth of about one foot. At this point the ship seemed to have left the ground again, since there were no marks for a space of about 300 feet down the line of flight. At that point the ground was plowed up again for about 20 feet to where the airplane came to rest right side up facing southwest. The engine had bent forward on the airplane at an angle of about 30 degrees. The airplane appeared to be intact with the exception of the position of the engine. There was no collapse of cockpit structure, and the apparent cause of death was through striking their heads, in the case of the man in the rear cockpit, against the instrument panel, which was shattered and bent forward about three inches; and in the case of the man in the front cockpit, by striking the head against the cowling inside the cockpit. The front windshield was also shattered, indicating he may have struck that. Although both safety belts were still latched and intact, the seats in both cockpits had broken away from their supports in places. It is recommended that the improved shoulder type safety belt be installed on all training planes. It is believed that had this type belt been worn, these two fatalities would not have occurred."

12-11-41. Groton, Connecticut. At 1300, a Curtiss P-40C crashed into a wooded ravine at Groton, Connecticut, killing pilot 2Lt. Hudson Gow Dunks, 26, Monroe, Michigan. The P-40 had been seen executing a series of tight turns at low altitude just prior to the accident. The P-40 plunged to the ground and exploded into flames after suffering a catastrophic structural failure of the port wing while making a vertically banked left turn at low altitude. In the middle of the turn the port wing separated and struck the tail section, causing it to separate. The rest of the airplane continued on out of control through a stand of trees and ultimately struck the ledge on the side of a wooded ravine.

12-12-41. Bishop, California. At an unknown time, a Douglas B-18 flying in a snowstorm collided with rising terrain at Thumb Lake (also known as Kidd Lake) 15 miles southwest of Bishop, California, killing eight passengers and crew. The B-18 had been reported missing on a flight from Phoenix, Arizona, to Hamilton Field, San Rafael, California, and was last reported two miles south of Palmdale, California, at 1955. Despite an extensive search by civilian and military authorities, the airplane remained missing and was ultimately found two miles northwest of Birch Mountain on 5-7-42 by Roy G.B. Burns, Spokane, Washington, father of the co-pilot. The aircraft and bodies were strewn over a 25 acre area that was covered with up to five feet of snow. All of the bodies had not been removed until the end of July 1942 because of the inaccessibility of the 11,000-foot crash site. Killed in the crash were: Capt. James G. Leavitt, pilot; 1Lt. Homer C. Burns, co-pilot; SSgt. Stephen W. Hoffman, engineer; Pvt. Samuel J. Van Hamm, radio operator; Major General Herbert A. Dargue, Commander of the First Air Force, Mitchel Field, New York, passenger; Lt. Col. Charles W. Bundy, passenger; Lt. Col. George W. Rickey, passenger; Maj. Hugh F. McCaffery, passenger.

12-13-41. Richmond, Indiana. At 2120, a Northrop A-17 flying in a snowstorm spun into the ground 12 miles east of Richmond, Indiana, killing passenger Richard Lee Finch, 32, civilian engineer at Wright Field, Dayton, Ohio. Pilot 1Lt. William A. Filer received minor injuries parachuting to safety. Investigation revealed that the airplane was suffering from faulty or inoperative instruments.

12-14-41A. Paine Field, Washington. At 1028, a Republic P-43A crashed on take-off from Paine Field, Everett, Washington, killing pilot 2Lt. Karl F. Leabo. The airplane was taking off in a squadron formation and failed to reach flying speed. The airplane never exceeded an altitude of 20 feet above ground level. Investigators speculated that the P-43, taking off uphill with a slight quartering tailwind, might have been caught in the propeller vortex of the preceding airplane. The subject airplane went out of control and smashed into the runway.

12-14-41B. Florence, Alabama. At 1405, an RAF Lockheed Hudson (A-29) on a ferry flight crashed three miles east of Florence, Alabama, killing pilot 2Lt. Norris W. Browne, Hastings, Nebraska, and co-pilot 2Lt. James W. Riddell. The airplane was one of a flight of four that had taken off at 1300 from Nashville Municipal Airport, Nashville, Tennessee, on a flight to New Orleans, Louisiana. Witnesses reported that the airplane had flown over Muscle Shoals Airport, Alabama, at a very low altitude and was flying at about 30 feet above the terrain before it began a steep left turn close to the ground. The port wing tip struck the terrain, the airplane cartwheeling into the ground and exploding into flames. Due to conflicting witness statements, the Accident Classification Committee could not determine the exact cause of the accident.

12-16-41A. Gardner Field, California. At 1055, a Vultee BT-13A stalled and spun into the ground while turning from base to final at Gardner Field, Taft, California, killing student Aviation Cadet Milton J. Vakos. The student had been practicing landings at the field when the accident occurred. Investigators speculated that the cadet did not use the flaps correctly.

12-16-41B. Gunter Field, Alabama. At 1830, a Vultee BT-13A crashed 24 miles southeast of Gunter Field, Montgomery, Alabama, killing pilot 2Lt. George H. Olson, 24, Chicago, Illinois, and passenger 1Lt. Charles J. Young, 34, Cincinnati, Ohio, Army Medical Corps. The BT-13 struck the ground with great speed in a vertical position and exploded into flames. The airplane had taken off from Gunter Field at 1720 to check the area weather for student night flying. The pilot was last heard from at 1800. Lt. Young, in the rear cockpit, was very interested in aviation medicine and two pilots who had flown with Lt. Young previously reported to investigators that they thought that Lt. Young sometimes braced himself on the flight controls in the rear cockpit. They reportedly felt the control stick and rudder pedals hitting Lt. Young while he was conducting his experiments in aviation medicine in flight. A ground crewman on the flight line testified to the Accident Classification Committee, "I was standing by the ship when two officers came out to fly. When they got in the airplane I noticed this officer in the rear cockpit having trouble; he was too tall to get the canopy closed. I got on the wing to help him. The officer in the front cockpit asked me to see if he fit in all right. I said he did, and he told me to plug in the headset and fasten the safety belt. He had on a flying suit and the [seat] straps were too short, so I was trying to lengthen the safety belt out. In the meantime the officer in front had already started the motor. I had one side of the safety belt lengthened and was trying to lengthen out the other side when he started to taxi out. I was hanging there on the edge of the cockpit, so I jumped off. He was taxiing pretty fast; I was almost caught by the tail of the ship. [The pilot] was holding my flashlight when he was getting in the airplane. After I jumped off the plane and was running alongside yelling for him to stop, he gave me my flashlight." The Accident Classification Committee stated, "From the evidence available, the Committee has been unable to determine the cause of this accident. Probable causes of the accident are that the controls may have been jammed by the loose safety belt; the pilot may have been performing some kind of aerobatics and the passenger, possibly not fastened in the seat properly with the safety belt, may have interfered with or frozen the controls; or there may have been engine failure."

✓ **12-17-41A. Yukon, Oklahoma.** At 0840, a Fairchild PT-19A crashed into a plowed field after a delayed recovery from an intentional spin one-half mile from Cimarron Field, Yukon, Oklahoma, killing Aviation Cadet Anthony F. Incorvaia, 24, Cleveland, Ohio.

12-17-41B. Geiger Field, Washington. At 0943, a Boeing B-17B attempting an emergency landing crashed one and one-half miles northeast of Geiger Field, Spokane, Washington, killing five members of the crew and seriously injuring three others. The airplane, carrying a load of bombs for a coastal defense mission, was flying with the number-one engine feathered and was losing altitude rapidly as it approached the field. Investigators could not determine the reason for the failure of the number-one engine. The B-17 was totally destroyed when the bomb load detonated soon after the crash. 1Lt. John A. Way, who was flying a B-18A airplane in the vicinity, witnessed the crash. He later stated to investigators, "We were on a routine local training flight in a B-18A. We asked for landing instructions and were told to stand by for a ship making an emergency landing. We saw the ship, a B-17 type, approaching from the southwest. He was at an altitude of approximately 2,500 feet when four miles from the field. We maneuvered so as to be behind and above him to see what was the trouble. We saw that his number-one engine was feathered and stopped. The pilot continued towards Spokane losing altitude and when in the vicinity of Hangman's Creek, over southwest Spokane, turned very slowly to the left and headed west. He was very low, probably 250 feet above the ground after finishing his turn. He continued due west on the north side of the highway. He made no attempt to approach Runway 20 on which he had been directed to land. Suddenly, when almost opposite the east edge of the field, the left wing dropped as if the pilot had suddenly saw the runway and was going to turn toward it. The ship banked up to approximately seventy degrees, the nose dropped and ship hit a small clump of trees. As it banked over I noticed what seemed an excessive amount of black smoke coming from the number-two engine. The ship hit on the left wing and nose and wallowed around so that the fuselage was pointed northeast and a small fire started in the vicinity on number-two engine nacelle. We circled the wreck approximately 1,500 feet twice trying to see if anybody was getting out. We didn't see anybody get out. The fire grew larger and there was a big explosion. The [smoke] billowed up to about 1,000 feet and the concussion gave our ship a definite jolt. The explosion tore up the ship and scattered parts of the airplane and flames about the area. The explosion followed the crash by two or three minutes. I could not see whether the flaps were down at the time of the crash." Killed in the crash were: Capt. Eugene D. Zadentseff, 45, St. Louis, Missouri, pilot; 1Lt. Elbert D. Hoffman, 26, Drexel Hill, Pennsylvania, co-pilot; Aviation Cadet Edwin B. Perkins, 23, Cincinnati, Ohio, navigator; SSgt. Ester Brannon, 30, Clayton, North Carolina, engineer; Cpl. Clarence A. Cunningham, 30, New Castle, Pennsylvania, bombardier. Seriously injured in the crash were: Sgt. Donald W. Shutt, radio operator; Cpl. Clarence Roman, gunner; Pfc. Ernest L. Pugh, gunner.

12-18-41A. Muroc, California. At 0615, a Martin B-26 crashed on take-off from the Muroc Bombing and Gunnery Range, Muroc, California, killing the crew of six. The airplane was taking off with another B-26 airplane on an armed coastal patrol mission. 1Lt. James W. Anderson Jr. witnessed the crash and later stated to investigators, "The craft seemed to become airborne under great difficulty, though not uncommon for overloaded ships. The initial altitude gained during the first ten seconds was approximately one hundred feet. At the time the ship passed over the edge of the lake, his power was heard to take a distinct reduction. This was probably due to a lowering of rpm and manifold pressure. As the aircraft passed beyond my forward view, and also as the left wing dropped, there was visible upon the wing and tail surfaces a definite deposit of frost. From this time until the ship crashed, there seemed to be a gradual, but definite change in the ship's flying attitude. It seemed to take on a tail-heavy, nose-up position, which, I believe, resulted in the wing over to the left and the nose downward plunge into the earth. At the time of the impact there was a large, enveloping flash and fire, which was followed in about twenty seconds by the explosions of the bomb load aboard." Killed in the crash were: 2Lt. John H. Work, 25, New York, pilot; 2Lt. Theodore M. Richards, 23, New York, co-pilot; Pfc. Frank Serao, 20, New York, bombardier; Pfc. Gerald J. Lucian, 20, New York, radio operator; Pfc. Thomas J. Kennedy, 21, New York, gunner; Pfc. Alfred G. Leganhauser, 22, New Jersey, engineer.

12-18-41B. Buda, Texas. At 1138, a North American BC-1 flying in instrument conditions smashed into the ground on the Wilson Ranch at Buda, Texas, killing A/C Arthur R. Oliver. The BC-1 descended into the ground at a steep angle in a flying attitude.

12-18-41C. LaGrange, Texas. At approximately 2210, a North American BC-1I flew into the ground and exploded while letting down through fog near La Grange, Texas, killing A/C Eugene A. Poletto. The pilot was on a night navigation training flight from Brooks Field, San Antonio, Texas, to Brooks Field via Temple and Yoakum, Texas.

12-19-41. Tallahassee, Florida. At 1045, a North American BT-14 flew into the ground after performing a series aerobatics eight miles southeast of Dale Mabry Field, Tallahassee, Florida, killing pilot 2Lt. George Reed Matthews and passenger Pfc. Michael G. Malinario. Investigators could not determine the exact chain of events that led to the accident.

12-20-41. Lambert Field, Missouri. At 0803, a Curtiss P-40E crashed after taking off with five other P-40 airplanes from Lambert Field, St. Louis, Missouri, killing pilot 2Lt. Howard D. Cory, 25, Clayton, Missouri. The Accident Classification Committee stated, "It is believed that this pilot, in an attempt to join his formation quickly, banked steeply and, in this position, was caught in the turbulent propeller wash of preceding airplanes and thrown out of control at an altitude which was not sufficient to recover control prior to striking the ground. The airplane struck the ground about two and one-half miles southeast of Lambert Field in a nearly normal attitude with the landing gear up. The auxiliary fuel tank exploded on contact, setting fire to the airplane. The pilot was thrown clear after the airplane first struck the ground." Lt. Cory's parents, Mr. and Mrs. Walter T. Cory, along with Lt. Cory's wife, the former Miss Betty Jehle, were at the airport to witness his departure.

12-21-41A. Pacific Ocean, California. At 1105, two Lockheed P-38D airplanes collided in flight and fell into the Pacific Ocean 15 miles west of North Island Naval Air Station, San Diego, California, killing both pilots. 2Lt. Foy Roberson, Jr. was killed aboard P-38D #40-770; 2Lt. William L. Northern, Jr. was killed aboard P-38D #40-805. The airplanes were part of a six-airplane flight on a coastal patrol mission and were attempting to execute a crossover turn from left to right when the collision occurred. 2Lt. William B. Peterson, flying in the formation, witnessed the collision. He later stated to investigators, "It appeared that Lt. Roberson had lost sight of his flight and was attempting to locate them when the collision occurred. The planes' wings hit first and the planes fell into the water together, the one lying on top of the other. Fire broke out simultaneously and the wreckage burned for only a few minutes before disappearing into 100 fathoms of water."

12-21-41B. Jackson, Mississippi. At 1330, a Martin B-26 crashed into a railroad switchyard shortly after take-off from the Army Air Base at Jackson, Mississippi, killing the crew of five. The Accident Classification Committee stated, "Subject airplane took off headed south-southeast. [A witness standing] near the south end of the runway noticed a high-pitched whine, which suggested a propeller running away. He watched the plane until it was some 300 feet high. He did not notice whether or not the landing gear was retracted after take-off. A few seconds later, he notice the plane east of the field with landing gear down. It veered suddenly, possibly 15 degrees, and then started losing altitude gradually, continuing to turn from an east heading to a northerly one. The plane was mushing along. It gradually lost altitude and disappeared from view of personnel on the base behind a line of trees. A few seconds later a tremendous burst of flame told of the plane's crash. In those last few seconds the plane, while at very low altitude, must have made a comparatively sharp turn of about ninety degrees. Just prior to impact it must have stalled and flipped over on its back. Evidence of the impact marks supports this theory. Pieces of the right motor propeller hub extension were imbedded in the middle of a cross-tie of the railroad [track] and were at the left edge of the plane trace. Glass from the windshield was found just in front of the railroad track. Both motors were scuffed from front to rear near their tops. Bodies of four were found headfirst in the mud, facing forward, but upside down. Seats were upside down facing forward and one body was still strapped in the seat. Position of the tail section of the fuselage would indicate a spin flip from right to left. AFCE [automatic flight control equipment] was dumped out fairly intact immediately beyond the pilot and co-pilot. The right motor broke the rail of the railroad [track] and did not roll far. The left motor rolled nearly a hundred feet beyond the railroad. The Curtiss propeller man found the left propeller to be in full high position. Propeller of the right engine was in normal position. Number one and two cylinder heads were broken off the left engine but there is no direct evidence to support a theory of these heads having failed in flight prior to impact. No smoke was observed nor was any unusual engine noise heard by reliable witnesses. The plane was out of sight of reliable pilot-observers for only a few seconds before impact. Statements of civilian witnesses are vague and cannot be relied upon. It is the opinion of the board that the left propeller ran away on or shortly after take-off." Killed in the crash were: 1Lt. John J. Dorr, Poughkeepsie, New York, pilot; 2Lt. Claude A. Knight, Ocala, Florida, co-pilot; 2Lt. Robert A. Saner, Fremont, Ohio, navigator; 2Lt. Leslie M. Shimeall, Chicago, Illinois; Sgt. Harold A. Wissinger, Indiana, Pennsylvania.

12-22-41. Cuero, Texas. At 1540, a Stearman PT-17 stalled and spun into the ground six miles northeast of the AAF station at Cuero, Texas, killing student Pvt. James M. Campbell. Witnesses first sighted the airplane in a spin at approximately 3,000 feet. The PT-17 spun down to about 500 feet when it straightened out into a vertical dive and then dove into the ground, bursting into flames upon impact.

12-23-41. Cuero, Texas. At 1330, a Stearman PT-17 stalled and spun into the ground six miles northeast of the AAF station at Cuero, Texas, killing student Sgt. William C. Bruenig. Investigators stated, "The pilot was flying below the minimum altitude permissible by local flying regulations. He made numerous turns and dives. Pulling up from one dive in a steep climbing turn to the left he barely avoided another airplane and continued the turn to a stalling point. At this point, approximately 900 feet above the ground, the airplane went into a spin from which the pilot never recovered."

12-24-41. Floresville, Texas. At 2005, a North American BC-1 stalled and spun into the ground on the Doc Marshall Ranch six miles west of Floresville, Texas, killing Aviation Cadet David B. Anderson, Jr. Investigators speculated that the engine quit when the pilot tried to switch fuel tanks. A civilian named Ace Fuller witnessed the crash while setting off fireworks with his family. He reported that the engine had quit and then restarted, followed by the roar of the engine apparently running at full throttle. Maj. William B. Wimer reported to the Accident Classification Committee, "The motor was running apparently full throttle up to the moment of impact. Examination of the wreckage and the ground immediately surrounding it showed without doubt that the airplane did fall in a spin and struck the ground in an almost vertical dive as the motor had dug a huge hole in the ground, and one wing in its turning motion had swept the ground just a wing's length away from the motor. The airplane was utterly demolished and the pilot's body was underneath the wreckage."

12-26-41A. Conococheague Mountain, Pennsylvania. At 1140, a Northrop A-17 flying in poor weather crashed into the 1,700-foot level of 1,950-foot Conococheague Mountain near Blaine, Pennsylvania, killing pilot 2Lt. William E. Luetzow, 22, Milwaukee, Wisconsin. The A-17 had departed Bolling Field, Washington, D.C., at 0945 on a flight to the AAF Air Depot at Middletown, Pennsylvania. The top 500 feet of the mountain was enveloped in clouds when the aircraft plowed through a stand of trees four to ten inches in diameter for over 100 feet. Both wings were sheared off before the fuselage came to rest against a ridge. Wreckage was scattered over a wide area and the engine came to rest some distance from the fuselage. The pilot was found strapped in his seat in the cockpit. Investigators estimated that the airplane struck the southeastern slope at over 140 miles per hour.

12-26-41B. Skipperton, Georgia. At 1435, a Vultee BT-13A suffering an engine failure crashed one and one-half miles west of Skipperton, Georgia, killing instructor 2Lt. John E. Loehrke, 26, Maywood, Wisconsin, and RAF student Flying Cadet John K. Briers, 21, Derby, England.

12-27-41A. Craig Field, Alabama. At 0305, a North American AT-6A crashed when it ran off the runway while taking off at Craig Field, Selma, Alabama, killing RAF student Flying Cadet Douglas F. Leman and seriously injuring instructor 2Lt. Walter W. Griffin. The aircraft was on a night training flight. The Accident Classification Committee stated, "UK Cadet Leman was flying the airplane. The airplane left the runway immediately after it had started rolling on take-off. Lt. Griffin was undoubtedly observing for other airplanes and was not aware the airplane had left the runway as it continued to swerve farther from the runway. The instructor saw the runway lights on the right side of the runway and thinking them to be lights on the left side of the runway pulled the ship still farther off the runway in an attempt to straighten it out. Due to the large number of obstacle lights on the field, it is easily understood that the instructor could have been confused as to which line of lights he was looking at. The airplane hit a large pilot of dirt about 50 yards off the runway resulting in the fatal injury of the pilot and the serious injury of the instructor and the complete wrecking of the airplane."

12-27-41B. Wittman, Arizona. At 0755, a North American AT-6A collided with terrain two miles west of Wittman, Arizona, killing A/C Joseph M. Brownback, Sarasota, Florida, and A/C Winston P. Bruun, Salt Lake City, Utah. The airplane had taken off from Luke Field, Phoenix, Arizona, on a routine training mission. A locomotive engineer for the Santa Fe Railroad witnessed the airplane crash, which occurred approximately 15 miles northwest of Luke Field. Investigators stated, "The plane came in from the east in a vertical dive and leveled off just before striking the ground. The right wing seemed to have taken the brunt of the impact, as the bottom cylinders of the motor were the only parts of the engine severely damaged. The plane seemed then to have rolled, shearing off both the right and left wings, coming to rest in a south-north position, with the nose of the fuselage facing south. The ship rolled approximately three hundred feet after first striking the ground. The debris was slightly scattered. From all indications it would seem that the wheels of the airplane were in the up position. The body of Cadet Brownback, who was in the front cockpit, was found a few feet to the right of the fuselage, his parachute still intact. Cadet Bruun, who was in the rear cockpit, was found a few feet to the left of the fuselage, his parachute also intact. It appeared that Cadet Bruun was under the hood in the rear cockpit at the time of the crash. The fact that blood and particles of flesh were found on the hood substantiates the theory." Cadet Brownback was the brother-in-law of Cincinnati Reds pitcher Paul Derringer.

12-27-41C. Masonboro, North Carolina. At 1450, a Curtiss O-52 collided with the surface of the Atlantic Ocean and crashed onto the beach at Masonboro, North Carolina, killing the pilot and seriously injuring the passenger. The airplane had been maneuvering on a gunnery-training mission one mile off shore. The pilot had misjudged his altitude while attempting a steep turn over calm water. The aircraft struck the water with the starboard wingtip then bounded 50 feet into the air before hitting a sand dune on the shoreline, rolling several times after impact. Pilot 2Lt. Roy Robertson, 27, Spartanburg, South Carolina, was killed and passenger Cpl. Louis A. Walker, Meridian, Mississippi, was seriously injured.

12-28-41. Pacific Ocean, California. At 0720, a Republic P-43 on a gunnery training exercise dove into the Pacific Ocean near North Island Naval Air Station, San Diego, California, killing pilot 2Lt. Allan B. Scales. 2Lt. John L. Moutier, Jr., who was flying an airplane in the exercise, stated to the Accident Classification Committee, "We dove down on the water to fire our guns, the formation was spread wide with B-flight on the right and C-flight on the left. After the completion of our dive, A-flight, in which I was flying, made a gradual climbing turn to the right. When we completed about 180 degrees of our turn we were at approximately 1,500 feet with the other two flights at 1,000 feet. I was leading the second element of A-flight and in our turn I glanced back over my left shoulder to check my position with the third element. In doing so, I saw Lt. Scales at an altitude of about 200 feet. His ship was in a vertical dive with its belly facing A-flight. He continued to dive and I could see no effort to recover. The ship hit the water nose first, there was a large splash and nothing but oil appeared on the surface."

12-29-41A. Sikeston, Missouri. At 0755, a Stearman PT-18 stalled and spun into the ground while turning from the base leg to final approach for the AAF auxiliary airfield near Sikeston, Missouri, killing Aviation Cadet Richard A. Reid, 24, White Plains, New York. The Accident Classification Committee stated, "Aviation Cadet Reid was practicing landings in a scheduled 180 degree overhead accuracy stage. He had made several landings. The student established his glide and turned to the base leg, on which he made his glide dangerously slow. His extremely slow glide apparently reached the stalling point as he applied controls to make a right turn off of the base leg causing the plane to enter a right spin. The altitude at that moment was approximately 175 feet [above ground level]. The plane completed one turn of a spin before it struck the ground approximately 700 feet from the boundary of the field. The student was fatally injured by the impact of his head against the windshield and the top of the instrument panel. The airplane was totally demolished."

12-29-41B. Sumner, Georgia. At 1030, two North American AT-6A airplanes engaging in simulated combat collided in mid-air four miles south of Sumner, Georgia, killing RAF Flying Cadet Edward S. Headington, London, England, aboard AT-6A #41-244. AAF instructor 2Lt. Preston C. Newton was able to parachute to safety from AT-6A #41-278. The airplanes collided as they made a head-on pass at one another at about 10,000 feet. J.D. Duckworth, a civilian who witnessed the collision, stated to investigators, "The lead airplane made a steep bank to the right and collided with the second airplane. When [the two airplanes] collided, the wings of the second airplane were knocked off and I saw the man in the second airplane leave the airplane and his parachute opened. As soon as the man in the second airplane left the airplane, it began spinning and crashed to the ground. The other airplane seemed to float around and finally settled to the ground. I did not see anyone leave this plane."

12-30-41A. Oakland, California. At 1151, a Curtiss P-40E dove into the ground near the municipal airport at Oakland, California, killing pilot 2Lt. John A. Finkin. The airplane was part of a formation flight of five P-40 airplanes. 1Lt. Bruce F. Parcell, who was flying behind Lt. Finkin, stated to the Accident Classification Committee, "We were flying a string formation and I was third in a string behind Lt. Finkin. After leveling out from a dive, I saw the left wing of Lt. Finkin's plane drop suddenly, as though it was in propeller wash from the ship ahead. Then the nose suddenly dropped and the plane whipped over to the right. I immediately pulled away to the right, losing sight of Lt. Finkin's plane. My first thought was that he had caught propeller wash, and then his plane whipped over. I expected him to pull back up into formation and I continued looking for him for a minute or two. I then called [1Lt. James] Ferguson, the flight leader, and told him that Lt. Finkin had dropped out of the string. I do not know the altitude at the time this happened, but estimate it to have been between 3,000 and 4,000 feet. The flight reassembled and landed, and while circling I saw Lt. Finkin's burning plane on the ground."

12-30-41B. Keller Peak, California. At 1642, a Martin B-26 flying in poor weather crashed into 7,882-foot Keller Peak, San Bernardino Mountains, California, killing nine fliers. The airplane had taken off from Muroc Bombing and Gunnery Range, California, with eight other B-26 type airplanes on a positioning flight to March Field, Riverside, California. The B-26 failed to arrive at March Field and was reported missing. After an extensive search effort, army search planes located the wreckage on 1-14-42. Severe winds had forced the B-26 formation into instrument conditions. 1Lt. Hubert J. Konopacki stated to investigators, "I was leading nine B-26 airplanes on a flight from Muroc to March Field, California. Departure from Muroc was approximately 1620. When over Cajon Pass, en route to March Field, at 8,500 feet, I encountered a severe downdraft and was forced to go on instruments for about two minutes. Instructions were given for a loose formation prior to take-off. At the time instrument flight was undertaken there was distance of approximately ten miles between first and the last elements. The sudden violence of the downdraft and subsequent dropping into the clouds did not warrant sufficient time to issue further instructions. Approximately two minutes after going into the clouds, I broke out at 9,000 feet riding an updraft. The lowest altitude I dropped to was 7,500 feet. When I broke out of the clouds, I was approximately three miles southeast of March Field. A quick glance around showed me that I had two wingmen with me and the other flights were coming through. It was not until after I had landed that I found a ship was missing. The ship was number 40-1475 ... flying on my left wing." Killed in the crash were: 2Lt. Frank A. Kobal, Queens, New York, pilot; 2Lt. Joseph B. Maloney, Waterbury, Connecticut, co-pilot; TSgt. Waldo Jensen, Langley Field, Virginia; Sgt. Roger F. Organ, Springfield, Ohio; Pvt. William R. Chinn, Ironton, Pennsylvania, bombardier; Pvt. Vernon H. Englebrecht, Fort Wayne, Indiana, engineer; Pvt. George C. May, Foxworth, Mississippi, radio operator; Pvt. Jack C. Shirley, Lavonia, Georgia, gunner; Pvt. Robert N. Enyeart, Columbus, Ohio.

12-30-41C. Delano, California. At 1930, a Vultee BT-13A on a night navigation flight inadvertently entered instrument conditions and crashed ten miles north of Delano, California, killing Aviation Cadet George E. Werner, Jr. The Accident Classification Committee stated, "The student flew into a rain cloud, became confused and crashed. This accident occurred during a student individual night navigation flight of twenty-five airplanes flying a triangular course from Lerdo, California, to Porterville, California, Lost Hills and return. Inasmuch as other airplanes preceded and followed this student pilot on this flight without encountering difficulties with the weather, it is believed that the pilot lost control of the airplane while attempting to fly through a rain shower."

1942

January

1-1-42A. New Hyde Park, New York. At 0725, a North American B-25A suffering the failure of its starboard engine crashed just after take-off from Mitchel Field, Long Island, New York, killing the crew of five. The B-25, which was carrying live bombs for a combat patrol, narrowly missed crashing into homes in Garden City and Minneola before smashing into a sand pit at New Hyde Park. The airplane had trouble gaining altitude and eventually stalled and spun into the ground from an altitude of approximately 500 feet. The bomb load detonated in the crash, breaking windows on several structures in the vicinity. The Accident Classification Committee noted that the pilot had been flying combat patrols nearly everyday since the Japanese attack on 12-7-41. The cause of the engine failure could not be determined and eyewitness accounts varied widely. Killed in the crash were: 2Lt. Charles W. Van Eeuwen, 23, Allendale, Michigan, pilot; 2Lt. James J. Orr, Oak Park, Illinois, co-pilot; Aviation Cadet Earl W. Ray, 21, Kings Mills, Ohio, navigator; Pfc. Edwin A. Onufrowicz, 21, Ipswich, Massachusetts, engineer; Pfc. Joseph W. Gallick, 22, Pittsburgh, Pennsylvania, radio operator.

1-1-42B. Langley Field, Virginia. At 0905, a Curtiss P-40B went out of control while attempting a take-off from Langley Field, Hampton, Virginia, and collided with a group of Works Progress Administration workers repairing box drains adjacent to the runway, killing three and injuring a fourth. The pilot, 2Lt. George R. Daugherty, who was on a transition training flight, was not injured. The aircraft was slightly damaged, suffering a broken port main gear strut, a damaged port wing and propeller. The Accident Classification Committee stated that the pilot applied full throttle for take-off and was not able to check the propeller torque with the rudder, causing a loss of control and a ground loop to the left. The three civilian workers killed in the accident were Frank Mollicot, Cossie Clayton, and Floyd Edwards. Hildra Ebron received minor injuries.

1-2-42A. Gunter Field, Alabama. At 1500, a Vultee BT-13A smashed into the ground inverted while attempting to perform low altitude aerobatics eight miles south of Gunter Field, Montgomery, Alabama, killing pilot 2Lt. George C. Rodge, Fords, New Jersey, and passenger Pfc. James V. Dougherty, Norfolk, Virginia.

1-2-42B. Alvarado, California. At 1540, a Curtiss P-40E crashed two and one-half miles southwest of Alvarado, California, killing pilot 2Lt. John M. Gentry. Civilian witnesses on the ground stated that the aircraft had been maneuvering in the vicinity before crashing to the ground in a vertical dive, exploding violently into flames upon impact. The pilot, who was undergoing transition training, was killed instantly.

1-3-42A. Godman Field, Kentucky. At 0605, a Curtiss O-52 crashed 1,000 yards from the runway while turning from base to final for Godman Field, Fort Knox, Kentucky, killing pilot 2Lt. Roy L. Drew, 22, Rothschild, Wisconsin, and fatally injuring observer 1Lt. Robert W. Harner, 24, Newton, Mississippi, who died at 1700 at the Fort Knox Post Hospital. Witnesses stated that the airplane appeared to slip out of a left turn, dropping to the left and into power lines and trees. The airplane hit the ground in a 70-degree bank and immediately burst into flames upon impact (see 1-3-42B).

1-3-42B. Godman Field, Kentucky. At 0655, a Curtiss O-52 stalled and spun into the ground 600 yards from Godman Field, Fort Knox, Kentucky, killing pilot 2Lt. Wilbur Camp, 26, Lafayette, Wisconsin, and observer 1Lt. Addie J. Hogan, 25, Ruston, Louisiana. The crash occurred while the airplane was circling the wreckage of an O-52 that had crashed fifty minutes earlier (see 1-3-42A). The airplane struck the ground 400 yards from the earlier accident. 1Lt. Daniel F. Tatum (AAF) stated to investigators, "I looked up and saw an airplane at low altitude apparently starting a turn to the right. His right wing kept going down, the nose dropped, and the airplane went straight into the ground in almost a vertical position, but slightly on its back. The aircraft exploded when it hit."

1-6-42. Ashland, Oregon. At about 1100, a Beech F-2 (photo-reconnaissance version of the AT-7 type airplane) collided with the sheer face of 6,094-foot Pilot Rock 15 miles southeast of Ashland, Oregon, killing the crew of three. Weather in the area at the time of the crash was reported as heavy fog, freezing rain, snow, and extremely limited visibility. The aircraft was overdue at its station and had been reported missing. Eliza Miller, a state of Oregon employee, found the airplane on 6-8-42. Between 1-9-42 and 2-18-42, several searches for the airplane and crew were conducted without success. Poor weather hampered the initial search efforts. The aircraft had taken off at 0856 from Gray Field, Washington, and was last heard from at 1049 by the Medford, Oregon, range station. Killed in the crash were pilot 1Lt. Raymond A. Stockwell, and passengers TSgt. Paul W. Stone and TSgt. Randolph Jones.

1-7-42A. Eglin Field, Florida. At 1024 CWT, a Martin B-12AM, suffering the failure of the starboard engine just after take-off, crashed and exploded into flames in a densely wooded area north of Eglin Field, Florida, killing pilot 2Lt. Garland O. Peel.

1-7-42B. Ozark, Alabama. At 1625, a North American AT-6A crashed into a lake six miles southwest of Ozark, Alabama, killing RAF Flying Cadet Kenneth Barlass and seriously injuring RAF Flying Cadet Lawrence S. Bell. Cadet Bell stated to investigators, "I was riding in the rear seat, and Cadet Barlass was in the front as the safety pilot. I went under the hood and flew about 30 minutes after which I came out for the usual rest period. When Cadet Barlass took over the controls we were at 2,000 feet. On taking over he made the remark, 'I'll show you what a poor pilot I am.' We dived down over the lake at Ozark and came across the dam and flew out over the water. In about the center of the lake we hit the water with the prop. Cadet Barlass pulled up to about 50 feet, but could not hold flying speed and it was apparent we were going to stall back into the water. I then unfastened my safety belt, and after we hit the next thing I remember is that I was swimming toward the shore."

1-7-42C. James River, Virginia. At about 2035, a Douglas B-18A ditched into the James River one mile north of the James River Bridge near Brandon Heights, Virginia, killing three crewmembers and injuring four others. The B-18 had run out of fuel after flying a combat patrol of the Atlantic Ocean. After completing its patrol mission and returning to the mainland in deteriorating weather conditions, the crew had trouble finding Langley Field, Hampton, Virginia, first sighting land near what they thought was Newport News, Virginia. Officers at Langley Field ordered the crew to fly out to sea and jettison their bombs and then return to the field for a landing. The B-18 flew out to sea and jettisoned its bombs, again becoming lost over the mainland. After flying around for some time to get their bearings, the crew found the James River and began to fly it back to Langley Field. The starboard engine stopped because of fuel exhaustion and then the port engine lost power. The pilot put the airplane into a landing glide and descended into the river. Killed in the accident were: Pfc. James B. Shannon, radio operator; Cpl. Harold E. Decker, engineer; Pfc. Dusan J. Craven, gunner. Pilot Maj. Ephraim M. Hampton was seriously injured. Co-pilot Capt. Carl R. Russell (who was piloting the airplane from the right seat during the ditching), navigator Aviation Cadet Robert W. Wert, and bombardier Pvt. William McKinley received minor injuries.

1-8-42A. Cochran Field, Georgia. At 1545, a Vultee BT-13A stalled and spun into Cochran Field, Macon, Georgia, while turning from base to final, killing instructor pilot 2Lt. Dalton R. Hardy, 22, Syracuse, New York, and RAF Flying Cadet Kenneth N. Thompson, 27, Edinburgh, Scotland. The BT-13 spun in from approximately 150 feet.

1-8-42B. Duncan Field, Texas. At 1720, a Bell P-400 (export version of the P-39 type airplane) crashed out of control after inadvertently entering instrument conditions about one mile north of Duncan Field, San Antonio, Texas, killing pilot 2Lt. Lewis H. Miller. Investigators stated that the pilot became disoriented after inadvertently entering overcast while in a high-performance climb. The airplane was seen by witnesses on the ground to fall out of the overcast and flutter to the ground.

1-8-42C. Hannah, Wyoming. At 2200, a Consolidated LB-30 was abandoned by its eight-man crew ten miles south of Hannah, Wyoming, after experiencing severe carburetor icing while flying through thick clouds. Navigator 2Lt. Phillip T. Wolking was killed during the bail out. Because the body was badly mutilated, investigators speculated that the airplane probably struck Lt. Wolking as it circled unmanned through the crew as they were descending in their parachutes. The pilot had climbed to 16,000 feet to get on top of the overcast and was unable to hold flying speed because of a severe loss of power. Pilot 1Lt. Frank W. Potter, co-pilot 2Lt. Harold W. Edmonds, radio operator Pfc. William M Edwards and engineer SSgt. Wayne T. Doyle received minor injuries parachuting to safety; Sgt. James A. Carson, SSgt. Harry B. Shields and pilot rated passenger 2Lt. Jerral W. Derryberry (see 6-12-43F) received serious injuries.

1-8-42D. Luke Field, Arizona. At 2245, a North American AT-6A struck and killed a U.S. Army guard 50 feet from the end of the runway at Luke Field, Phoenix, Arizona. Aviation Cadet Robert G. McCormick struck Pvt. William K. Prue with the starboard wing of his airplane after landing with excessive speed and a tail wind, over-running the end of the runway. Pvt. Prue had been guarding aircraft at the south end of the field when the accident occurred.

1-9-42. Brighton, Ohio. At 2340 EST, a Beech C-45 flying through a snowstorm crashed into a farm field at Brighton, Ohio, killing the crew of three. The aircraft struck the ground with its starboard wing first, cartwheeling across the field for more than 150 yards until coming to rest against a large hickory tree. Pieces of the airplane were strewn over a large area of a farm owned by Joseph Olinger. Pilot 2Lt. Harold W. Wolfe, co-pilot 2Lt. Clayton L. Head, and engineer Pvt. George M. Hopkins were killed instantly in the crash.

1-10-42. Galveston Bay, Texas. At 0900, a North American AT-6A suffered a catastrophic structural failure of the starboard wing after executing an extreme maneuver, falling into Galveston Bay, Texas, three miles southeast of Red Fish Reef Lighthouse, and killing pilot 2Lt. Randall R. Schamp and passenger 2Lt. Jack Clark. Witnesses reported hearing three loud reports, after which the airplane came diving out of the overcast minus at least one wing. The wing floated down out of the clouds and struck the water some 20 seconds after the fuselage hit.

1-12-42A. Madison, Mississippi. At 0905, a Stearman PT-17 stalled and spun into the ground at Madison, Mississippi, killing civilian flight instructor William H. Brown Jr., Jackson, Mississippi, and student Aviation Cadet James F. Wood, 22, Valdosta, Georgia. The pilots were practicing simulated forced landings when the accident occurred.

1-12-42B. Phenix City, Alabama. At 1130, an Aeronca L-3A crashed into the ground nose first with power near Phenix City, Alabama, two miles northwest of Lawson Field, Fort Benning, Georgia, killing pilot 2Lt. Lucius D. Edwards, 22, Yazoo City, Mississippi, and passenger Pvt. Clarence A. Puder, 23, Dover, New Jersey. Investigators stated, "The accident occurred due to the maneuvering of the airplane at low altitude, not allowing the pilot sufficient time to make necessary corrections to resume normal flight. A possible cause of the accident is the possibility of the rear control stick having become caught in the bracing of the radio."

1-12-42C. Dale Mabry Field, Florida. At 1725, a Bell P-39D stalled and spun into the ground while turning from base to final in the traffic pattern for Dale Mabry Field, Tallahassee, Florida, killing pilot 2Lt. Henry T. Horton, Kershaw, South Carolina. Maj. William G. Lee, Jr., AAF, observed the airplane at an altitude of 500 feet with its gear down and flying just above stalling speed. Maj. Lee stated to investigators, "There was Lockheed Hudson ahead of the P-39, and the Hudson was apparently cleared to the field as the number one to land. The Hudson slightly overshot his last turn and found it necessary to [S-turn] on his final approach. The P-39 apparently had overrun the Hudson and it appeared to me that the pilot was attempting to avert the necessity to circle the field by slowing up and taking distance. When the P-39 airplane was in position to take its final turn and began to take the turn, the airplane apparently lost flying speed and slid [to the left] into the ground."

1-13-42. Craig Field, Alabama. At 2345, a North American AT-6A flew into the ground three and one-half miles from Craig Field, Selma, Alabama. The airplane was returning from a night cross-country flight to Birmingham, Alabama, via Montgomery, Alabama. RAF Flying Cadet John F. Rimer was killed when he descended into the ground under control while letting down through low hanging overcast while attempting a landing at the field. Investigation revealed that the engine was producing power at the time of the crash. The airplane smashed itself to pieces upon impact.

1-14-42A. Harlingen, Texas. At 0835, a Bell P-39D crashed near Harlingen, Texas, killing pilot 2Lt. Woodrow W. Christian, Oklahoma City, Oklahoma. The Accident Classification Committee stated, "On final approach for landing the pilot made a nose-high slow turn with power off and flaps down at approximately 300 feet altitude. Aircraft stalled, striking the ground about one-half mile north of the North-South runway and resulting in death to the pilot and almost complete destruction of the aircraft."

1-14-42B. Gowen Field, Idaho. At 1422, a Martin B-26B crashed in brush-covered hills near Black Creek Reservoir nine miles southeast of Gowen Field, Boise, Idaho, killing the crew of nine. The airplane was part of a formation of five B-26 airplanes on a positioning flight to McChord Field, Tacoma, Washington, and had taken off third from Gowen Field at 1415, apparently crashing soon after. A road grader operator working in the area stated that he had seen the airplane in a steep bank just prior to the crash. The airplane was headed in a southeast direction and banked sharply to the right apparently to avoid a fog bank. The airplane half-rolled to the right and went in a nose-down inverted position, heading southwest. The airplane struck the ground inverted and at a steep angle, exploding violently into flames and scattering wreckage over a wide area. Ammunition detonated in the fire following the accident. The pilot of the B-26 that had taken off just after the accident airplane noticed the column of smoke but did not associate it with an air crash. A Union Pacific Railroad section crew that had been working in the area discovered the wreckage in a small clearing covered by about one foot of snow. Investigators could not determine the cause of the accident but speculated that the white snow on the ground and the fog might have caused the pilot to lose the horizon and over-control the airplane. Wristwatches found in the wreckage and on the bodies established the time of impact. Killed in the crash were: 2Lt. Elmer M. Munn, Fort Worth, Texas, pilot; 2Lt. Darrell S. Wing, Phoenix, Arizona, co-pilot; SSgt. Carmen R. Gismondi, Oliver, Pennsylvania, radio operator; SSgt. Carl Funkhouser (hometown unknown), engineer; Pvt. Leland Saunders, Odessa, Missouri, bombardier; Pfc. Andre A. Chisholm, Woodside, New York, crewmember; Pfc. Stanley Sansenbach, Parkersburg, Iowa, crewmember; Pvt. Robert F. Adams, Vale, Oregon, gunner; Pvt. Lloyd Knight, Thompsonville, Illinois, gunner.

1-14-42C. Craig Field, Alabama. At 1830, a North American AT-6A crashed one mile north of Craig Field, Selma, Alabama, killing RAF Flying Cadet George B. Whigham. "It is the opinion of the Accident Classification Committee that the student became confused while night flying. It was a dark night with very little horizon, and as there are very few lights on the ground in this vicinity it is believed that the student flew into the ground while letting down from his assigned zone and altitude."

1-14-42D. Atlantic Ocean, Virginia. A Vultee BT-13A crashed into the Atlantic Ocean off the Virginia coast, killing passenger 2Lt. Rush H. Willard, Bay City, Michigan. Pilot 2Lt. James A. Fannin was missing and presumed lost at sea. The airplane had taken off at 1830 CST from Knoxville, Tennessee, on a pilot proficiency flight to Bolling Field, Washington D.C. When the airplane failed to arrive at Bolling Field, it was reported missing. On 1-15-42, the body of Lt. Willard and two wheels from a BT-13 type airplane were recovered from the sea 45 miles east of Cape Charles, Virginia. Despite a thorough search effort, the body of Lt. Fannin has never been recovered. No other wreckage was recovered and the cause of the accident remains undetermined.

1-14-42E. Mt. Moosilauke, New Hampshire. At 2004, a Douglas B-18A returning from an anti-submarine patrol mission over the Atlantic Ocean crashed into Mt. Moosilauke, New Hampshire, killing two crewmembers and seriously injuring five others. The aircraft, attempting to get back to its base at Westover Field, Chicopee, Massachusetts, was lost and off course due to the navigator's inability to compute the drift because of unexpected winds that were strong and adverse. The pilot and the navigator had calculated that they would be at their base at approximately 2000. The pilot stated that at about 2000, the airplane was flying in instrument conditions in the overcast at an altitude of about 4,000 feet. The night was very dark and the visibility poor. The pilot was flying the airplane on instruments and the co-pilot was watching for visual contact conditions. A few minutes later, the co-pilot shouted that there was a mountain ahead and grabbed the control wheel. The pilot looked up and saw the mountain rapidly approaching. They attempted to make a steep, climbing turn to the right when the airplane struck the mountain and the pilot lost consciousness. The co-pilot, who never lost consciousness during the accident, stated that the airplane was in a skidding turn to the right and that he had gave the airplane some left rudder and then pulled back on the control column and the airplane mushed in. The aircraft tore itself apart as it collided with tall trees. The shattered airplane came to rest and a few small fires were burning among the wreckage. The five surviving crewmembers were able to escape the airplane before it erupted in flames and the ordinance exploded. The crew got away from the airplane by rolling down the hill and soon three large explosions were heard. They struggled down the hill through deep snow until rescued by civilians a short time later. Killed in the crash were the radio operator Pfc. N. W. Philips and the bombardier Pvt. F. J. Lawrence. Pilot 1Lt. Anthony Benvenutto, co-pilot W. A. Kantner, navigator 2Lt. F. M. Craig, photographer Pfc. R. P. Pickard, and engineer Pfc. R. G. Chubb were seriously injured.

1-15-42. Nixon, Texas. At 2305, a North American BC-1I on a night navigation training flight crashed three miles east of Nixon, Texas, killing Aviation Cadet John C. Blade. The airplane had taken off at 2140 from Brooks Field, San Antonio, Texas. Investigators stated, "An examination of the scene of the crash indicated that the airplane struck the ground at a high rate of speed at approximate angle of 30 degrees. The airplane was on course, on schedule, and heading in the proper direction. At the time of the accident, the airplane had been in the air one hour and twenty-five minutes. The cause of the accident remains undetermined. It is probable, however, that the gas had been exhausted in one tank. The pilot experienced difficulty in switching to the other tank and dove the airplane in order to keep the motor turning. The motor caught for a few moments before the airplane struck the ground."

1-16-42A. Dupo, Illinois. At 1353, a Fairchild PT-19 performing aerobatics entered an inverted spin and crashed four miles west of Dupo, Illinois, killing Aviation Cadet Robert H. Hornbeck, 24, Huron, South Dakota. The student had been training at Parks Air College, East St. Louis, Illinois. Investigation revealed that Cadet Hornbeck had received some civilian flight training before reporting for duty as an aviation cadet.

1-16-42B. Las Vegas, Nevada. At 1923 PST, a Douglas DC-3 Skyclub Transport operated by Transcontinental & Western Airlines as TWA Flight #3 collided with the 7,769 foot level of 8,304-foot Potosi Peak 33 miles southwest of Las Vegas, Nevada, killing the civilian crew of three and nineteen passengers. Four of the passengers were civilians and fifteen were officers and enlisted men serving with the Army Air Forces Ferry Command, Long Beach, California. The DC-3 had taken off from New York, New York, at 1040 EST on a continental flight to Los Angeles, California. In the afternoon the aircraft took off from Indianapolis, Indiana, and flew to Lambert Field, St. Louis, Missouri, where it was delayed on the ground for two hours because of weather. The weather cleared and the aircraft took off, flying to its scheduled stops at Wichita, Kansas, and Albuquerque, New Mexico. The DC-3 took on a new flight crew and the 15 AAF passengers (and all of their flying equipment and parachutes) at Albuquerque. The aircraft took off but soon encountered strong head winds, arriving over Boulder, Nevada, after dark. The crew had planned on stopping at Boulder for fuel, but since Boulder could not handle night arrivals because of a lack of runway lighting the transport diverted to the Western Airlines Airport at Las Vegas, Nevada, landing at 1830 to fuel up for the last leg of the flight to Los Angeles. After a short time on the ground the fully fueled DC-3 took off, becoming airborne at 1907 local time. The pilots had been advised to fly a 208-degree course out of Las Vegas and climb to 10,000 feet. The pilot flew a 218-degree course and began a slow climb to 8,000 feet at a speed of about 150 mph. At 1923, the DC-3 slammed into Potosi Peak (also known as Double or Nothing Mountain, Table Rock Mountain, or Double-U Mountain) and exploded into flames upon impact, setting fire to a stand of pine trees and melting the snow in the vicinity of the crash. The bodies of the 22 victims were badly mutilated and several burned beyond recognition. Investigation revealed that the collision with terrain was a result of the pilots being ten degrees off course, flying at an altitude lower than known obstacles in the area, and their failure to see the mountain on a dark, moonless night. Weather at the time of the accident was reported as CAVU and not considered a factor in the accident. Investigation of the Captain's flying history revealed that he had been discharged from TWA in August of 1933 for "insubordination and repeated non-compliance with rules and regulations," including flying non-regulation routes at night. The National Labor Board reinstated him in October of 1933. The Captain, a former Army Air Corps aviator, had had no serious accidents until the fatal flight. Killed in the crash were: Capt. Wayne C. Williams, pilot, TWA; First Officer Morgan A. Gillette, co-pilot, TWA; Alice Getz, stewardess, TWA; Cpl. Milton B. Affrine, AAF, Philadelphia, Pennsylvania; 2Lt. James C. Barham, AAF, Long Beach, California; 2Lt. Hal E. Brown Jr., AAF, Long Beach, California; Sgt. Albert M. Belejchak, AAF, Braddock, Pennsylvania; Sgt. Frederick P. Cook, AAF, Reidsville, North Carolina; 1Lt. Robert E. Crouch, AAF, Long Beach, California; 2Lt. Frederick J. Dittman, AAF, Long Beach, California; 2Lt. Kenneth T. Donahue, AAF, Long Beach, California; 2Lt. Charles D. Nelson, AAF, Long Beach, California; 1Lt. Robert F. Nygren, AAF, Dunbar, Pennsylvania; Sgt. Edgar A. Nygren, AAF, Dunbar, Pennsylvania; 2Lt. Stuart L. Swenson, AAF, Long Beach, California; Pvt. Martin W. Tellkamp, AAF, La Moille, Illinois; Sgt. David Tilghman, AAF, Snow Hill, Maryland; Pvt. Nicholas Varsamine, AAF, Bronx, New York; Carole Lombard, 33, Los Angeles, California, screen actress for Metro-Goldwyn-Mayer Studios and wife of screen actor Clark Gable; Elizabeth K. Peters, Los Angeles, California, mother of Carole Lombard; Otto Winkler, Hollywood, California, Press Agent for Metro-Goldwyn-Mayer Studios; L. Hamilton, Detroit, Michigan, passenger. Carole Lombard was returning to California from a government sponsored war bond rally in Indianapolis during which she sold $2 million worth of war bonds. Because of her successful war bond sales, her group was allowed to remain on the government-chartered flight at Albuquerque. Reportedly, a coin flip decided whether the Lombard entourage would return to Los Angeles by rail or by air.

1-17-42A. Curtis Field, Texas. At 0745, a Vultee BT-15 stalled and spun to the ground while attempting to land at Curtis Field, Brady, Texas, killing Aviation Student Houston W. Savage. The student, being graded on 180-degree approaches, had undershot a prior attempt to land and he gunned the engine and went around for another attempt. The pilot was observed to make a left turn from the downwind leg to the base leg. About halfway through the turn the airplane whipped over to the right and spun to the ground. The spin occurred at too low an altitude to allow recovery.

1-17-42B. Pendleton Field, Oregon. At 1130, a Boeing B-17E crashed immediately after take-off from Pendleton Field, Oregon, killing the crew of eight. The B-17 traveled about 1,200 feet on its take-off run and was seen to suddenly lift off and enter a steep climb to an altitude of about 150 feet before entering a stall. The airplane struggled to remain airborne, mushed along for a short distance and was then seen to fall off side ways to the right before plunging nose down into the ground two and one-half miles northeast of the field. Witnesses reported that the landing gear had not been retracted and the airplane never appeared to be under control. Investigators speculated that the white clouds and the white snow on the ground might have confused the pilots, causing them to lose the horizon. The brand new B-17, which was being ferried to MacDill Field, Tampa, Florida, had been flown to Pendleton Field from the Boeing factory airfield at Seattle, Washington, on 1-1-42 and had a total of 23 hours of flying time. Investigation revealed that the pilot had only five hours of first-pilot/co-pilot time in B-17 type equipment. Killed in the crash were: 2Lt. Andrew J. Francisco, pilot; 2Lt. Ruel C. Shows, co-pilot; 2Lt. Leroy E. Grindle, navigator; SSgt. Allen D. Spiers, engineer; Sgt. Donald Clark, radio operator; Cpl. Vernon A. Learman, bombardier; Pvt. George Vrable, assistant engineer; Pvt. Lewis Fagan, assistant radio operator.

1-17-42C. Riverside, California. At 1620, a Curtiss P-40 crashed five miles west of Riverside, California, killing pilot 2Lt. Herbert F. Soest. Investigation revealed that the pilot entered a vertical power dive at a dangerously low altitude, estimated to be about 5,000 feet above ground level, and was unable to pull out before striking the ground and exploding into flames. Wreckage was scattered for 300 feet along the bank of a small creek. Two Vultee O-49 airplanes (similar to the L-1 type) landed near the wreckage soon after the crash. The P-40 had been part of a two-airplane formation that had been maneuvering in the area.

1-18-42A. Lake Apopka, Florida. At 0930, a North American B-25B threw a propeller blade and crashed into shallow water at Lake Apopka, Florida, killing the crew of three. Killed in the crash were: 2Lt.

Walter C. Isely, 24, Sioux Falls, South Dakota, pilot; 2Lt. William T. Gardner, 24, Detroit, Michigan, co-pilot; Pfc. John R. Parrish, Charleston, South Carolina, engineer. Witnesses on the ground and aboard boats on the lake observed the airplane entering a steep turn to the right over the lake at an altitude of about 500 feet. The port engine shook itself loose and fell from the airplane and then the port wing folded up and struck the fuselage. The airplane dove into the lake in water about ten feet deep and buried itself in the mud up to fifteen feet. Gasoline and wreckage covered about one acre of the lake. The body of the engineer was discovered intact but only the torso of the co-pilot was found. The body was so mangled that investigators doubted that the crash impact alone could have caused such damage. A large intestine was found and it was assumed to be the pilot's as the other bodies contained the large intestine. Investigators speculated that while airplane was flying at about 1,200 feet above ground level, the propeller blade was flung from its hub, sliced through the fuselage and entered the cockpit, killing the pilots. The airplane then entered a steep bank to the right and the resultant vibration caused by the damaged propeller caused the progressive structural failure of the port engine and wing before the B-25 plunged into the lake. Investigators could not determine what caused the propeller failure because the port engine and propeller could not be found and examined.

1-18-42B. Rayville, Louisiana. At 1130, a Cessna AT-8 crashed one mile west of Rayville, Louisiana, killing pilot 2Lt. Donn W. Piatt and co-pilot 2Lt. Frank G. J. Micieli. Investigation revealed that the pilots were lost and the airplane out of fuel when an emergency landing was attempted. The pilot was unable to reach the field he had chosen and the airplane struck a large cypress tree, crashing to the ground and killing the pilots instantly. Weather was not considered a factor in the accident.

1-19-42A. Dill, Oklahoma. At 1030, a Vultee BT-15 on an engineering test flight stalled and spun into the ground nine miles west of Dill, Oklahoma, killing pilot 2Lt. Paul W. Anthony, 25, Winnebago, Minnesota, and passenger SSgt. John H. Hinkle, 23, Canute, Oklahoma. Civilians had observed the airplane maneuvering at very low altitude prior to the crash. The airplane was seen to dive at the ground and pull up. The airplane entered a stall in a nose-high attitude and fell off into a spin, spinning to the ground and bursting into flames upon impact. The spin occurred at too low an altitude to allow the pilot to recover before striking the ground. Weather was not considered a factor in the accident and there was no evidence of structural or mechanical failure.

1-19-42B. Selma, Alabama. At 1245, a Vultee BT-13A crashed while attempting a formation take-off at the municipal airport at Selma, Alabama, killing pilot 2Lt. Melvin L. Rake. The pilot took off in formation on a narrow, partially completed runway without the pilot of the lead ship being aware that he was taking off in formation with him. Immediately after take-off, the pilot of the lead ship turned toward the flight path of the accident airplane. The accident airplane was on the inside of the turn at an altitude of about 200 feet and attempted to maneuver out of the way of the lead airplane. The airplane stalled, entered a snap-roll and pancaked into the ground, killing the pilot instantly. The airplanes had landed at Selma because they were lost on a cross-country flight from Meridian, Mississippi to Gunter Field, Montgomery, Alabama.

1-19-42C. McChord Field, Washington. At 1440, a Martin B-26 crashed while attempting a take-off at McChord Field, Tacoma, Washington, killing the crew of three. The airplane suffered an apparent failure of the starboard engine while attempting the take-off. The B-26 was barely able to get airborne and was skidding and mushing along above the treetops. The starboard propeller and wing struck the trees, shearing off about six feet of the wing from the wingtip. The pilot jettisoned his bombs and fully loaded bomb bay fuel tank. The airplane then veered to the right 45 degrees, entered a half-roll to the right and struck the ground inverted. The airplane bounded 30 feet into the air and over a small tree before smashing into the ground again 50 feet further on and bursting into flames. Killed in the crash were: 2Lt. Jesse P. Ottosen, Salt Lake City, Utah, pilot; 2Lt. Connor G. Hopkins, Parsons, Kansas, co-pilot; Sgt. Lee C. Osborn, Los Angeles, California, engineer.

1-20-42A. Verbena, Alabama. At 0350, a North American AT-6A on a night navigation training flight crashed near Verbena, Alabama, killing RAF Flying Cadet Edgar Hopkin. The student became lost attempting to return to Napier Field, Dothan, Alabama, and attempted a forced landing when his fuel supply became exhausted. He had last checked in on time at 0300.

1-20-42B. Craig Field, Alabama. At 0840, a North American AT-6A crashed two miles southeast of Craig Field, Selma, Alabama, killing RAF Flying Cadet Alfred B. Kinnear. Investigators speculated that the pilot, in the act of raising the landing gear after an unusually long take-off roll, put his head down in the cockpit and inadvertently flew the airplane into trees 2,000 feet from the end of the runway. The airplane bounded back into the air for a few moments before descending at high speed into heavily wooded terrain. The airplane broke off trees and was torn to pieces as it smashed to the ground.

1-20-42C. Richmond, California. At 1610, a Curtiss P-40E stalled and spun to the ground two miles east of Richmond, California, killing pilot 2Lt. Richard J. Sander. The P-40 was part of a four-airplane formation flight that had taken off from Hamilton Field, San Rafael, California, at 1545. The formation climbed to 9,000 feet and in the vicinity of Richmond the leader of the formation initiated a left turn toward Sausalito, California. During the maneuver the P-40 was seen to fall out of the turn in a spin. The pilot stopped and re-entered the spin several times before the P-40 struck the ground and exploded into flames. The flight leader radioed the pilot to ease forward on the stick when he observed him struggling with the recovery. When the P-40 spun to about 4,000 feet, the flight leader repeatedly radioed the pilot to bail out but to no avail.

1-22-42. Mahomet, Illinois. At 0945, a Bell YFM-1A suffering the failure of its starboard engine was abandoned and crashed five miles northeast of Mahomet, Illinois, killing pilot 1Lt. James O. Reed. Engineer SSgt. Chester I. Smith was uninjured when he parachuted to safety at an altitude of 3,500 feet. Lt. Reed's body was found about 150 yards from the wreck. His parachute had not been deployed. The in-service experimental twin-engine pusher-propeller fighter had taken off from Chanute Field, Rantoul, Illinois, at 0927 on a ferry flight to Keesler Field, Biloxi, Mississippi. The airplane spun to the ground and exploded into flames upon impact. The Accident Classification Committee stated, "It is believed that the accident was caused by the failure of a connecting rod in the right engine. This failure caused excessive vibration, leakage of oil, and finally fire which necessitated abandoning the airplane. It is believed that the pilot failed to clear the airplane in time to successfully use his parachute. A similar failure, with the exception of the resulting fire, had been experienced on another airplane of this type at this station. Inherent defects in this design caused constant maintenance difficulties and the flying of this type had been very limited both at this station and at their prior stations. [It is recommended that] all YFM-1 type airplanes be grounded and placed in [a non-airworthy classification]."

1-23-42A. Luke Field, Arizona. At 1130, a North American AT-6A crashed 17 miles NNE of Luke Field, Phoenix, Arizona, killing Aviation Cadet Jack L. Whitfield, Virginia, Illinois. "It is the opinion of this Committee that the immediate cause of this accident was the impact of the airplane with the ground. Based on the following facts: A/C Whitfield was wearing the oxygen mask at the time of the impact, and it was found wedged in the vibrator inverter box; The cadet's head was cut off at the forehead by the instrument panel; it is the opinion of this Committee that the underlying cause of this accident was: A/C Whitfield passed out at high altitude and from the statements of witnesses, it is believed that A/C Whitfield regained consciousness momentarily and tried to pull the airplane out of the dive, blacked out, lost consciousness, slumped over the stick and then the airplane dived to the ground. The oxygen regulator was so badly damaged that it could not be determined whether the valve was on or off."

1-23-42B. Shaw Field, South Carolina. At 1402, a Vultee BT-15 crashed five miles north of Shaw Field, Sumter, South Carolina, killing Aviation Cadet William Offard Herndon. Investigators speculated that the student ran a tank dry and attempted a forced landing after the engine quit.

1-24-42. Maxwell Field, Alabama. At 1000, two Fairchild PT-19A airplanes collided in mid-air nine miles west of Maxwell Field, Montgomery, Alabama, killing civilian ferry pilot Blake W. Docker, who plummeted to his death in an unsuccessful parachute jump from PT-19A #41-20362. 2Lt. John J. Daunt, Jr. parachuted to safety from PT-19A #41-20370. Mr. Docker made a right turn and collided with Lt. Daunt's airplane, which was part of another formation. Investigators speculated that Mr. Docker had become separated from his formation and in his effort to locate his formation collided with Lt. Daunt.

1-25-42A. Marshall Ford Lake, Texas. At 1100, a Curtiss O-52 on a navigation training flight crashed into Marshall Ford Lake, Texas, killing 2Lt. Owen R. S. Graham. Passenger 2Lt. Clay U. Bullis was uninjured. The airplane had been maneuvering at low altitude prior to the accident. "It is the opinion of the Accident Classification Committee that the accident was caused by the pilot's attempt to fly too close to the surface of the water. First the propeller touched the water then apparently the whole airplane was pulled into the water and sank. Both occupants succeeded in getting out of the airplane. The passenger swam about 200 yards to shore and was saved. The pilot swam a short distance and then drowned."

1-25-42B. Santa Rosa, Arizona. At 1250, two Stearman PT-17 airplanes collided in mid-air and one crashed to earth on the Santa Rosa Ranch, Santa Rosa, Arizona, 65 miles west of Tucson, killing civilian ferry pilot Thomas Reed Hollandsworth aboard PT-17 #41-25380. Civilian ferry pilot Wendell Pendleton was uninjured and able to crash land heavily damaged PT-17 #41-25381. His PT-17 struck a clump of trees and flipped over. The airplanes were part of a five-airplane ferry flight from Wichita, Kansas, to King City, California, and had taken off from Tucson at 1224 for Yuma, Arizona. Pendleton was flying west at an altitude of 800 feet. Hollandsworth came up from the right and behind Pendleton and flew very close. The starboard wings of Pendleton's airplane contacted the port wings of Hollandsworth's airplane. The airplanes remained in contact briefly, separated and then collided again. Pendleton jerked his airplane up and away and portions Hollandsworth's port wings were severed at the tips. Hollandsworth made a right gliding turn and attempted to hold the damaged wing up. The airplane changed direction and started into a left spin. The airplane flipped over to an inverted attitude and then crashed to earth. Two of the three other airplanes landed nearby to render assistance. The third airplane flew back to Davis-Monthan Field, Tucson.

1-26-42A. Martindale Field, Texas. At 1450, a North American BT-9B crashed one-half mile northwest of Martindale Field, Texas, killing pilot 2Lt. Foster L. Walker and Aviation Cadet James H. Cousins. The Accident Classification Committee stated, "It is the opinion of the Committee that the pilot, while leading a formation of three [BT-9B] airplanes taking off from Martindale Auxiliary Field (auxiliary to Randolph Field, San Antonio, Texas), in executing a turn to the left [at an altitude of approximately 400 feet above ground level], apparently saw a civilian airplane (a yellow Piper Cub #NC 26081) directly in his path. He signaled to his wingmen to break up the formation and then pulled up into a steep climbing turn to the left to avoid the oncoming civilian airplane. In so doing, he lost flying speed and stalled his airplane, causing it to dive into the ground out of control. It has been determined by the Committee that this civilian airplane was being operated from the East End Airport approximately one and one-half miles due west of Martindale Field. Due to the enlarged student-training program at Randolph Field, the student traffic at Martindale Field is very heavy. Airplanes operating out of the East End Airport continually fly in the danger zone surrounding Martindale Field, getting in the traffic pattern of Martindale Field at dangerously low altitudes."

1-26-42B. Altheimer, Arkansas. At 1500, a Fairchild PT-19A crashed one mile south of Altheimer, Arkansas, killing Aviation Cadet Donald R. McCarnes. Investigators stated, "Inspection of the wreckage at the scene of the accident indicated that the airplane spun in from an undetermined altitude, estimated to be at least 2,000 feet. The airplane struck on the nose, landing gear, and left wing. The airplane was completely demolished but did not burn."

1-27-42. Cuero, Texas. At 1015, a North American AT-6A attempting an emergency forced landing crashed four miles southeast of Cuero, Texas, killing pilot 2Lt. Gordon O. Kibbe. Witnesses on the ground reported that the airplane was having engine trouble prior to the accident. The pilot had lowered the landing gear and was attempting a landing when he struck a dead tree with the port wing, causing the airplane to whip over to the left and strike the ground in a vertical attitude. Investigation revealed that the starboard fuel tank was empty and the port fuel tank contained about 30 gallons of gasoline. The tank selector lever was found in between the left and right positions. Investigators speculated that the pilot ran the starboard wing tank dry, causing the engine to quit.

1-28-42A. Wayne County, Michigan. At 1205, a Bell P-39F crashed near Merriam and Henry Ruff Roads, two miles northwest of the Wayne County Airport, Wayne County, Michigan, killing pilot 2Lt. Donald P. Baker, 24, Seattle, Washington. The airplane had taken off from Wayne County Airport at 1157 with three other P-39F airplanes, flying in the number-four position on the ferry flight to Selfridge Field, Mt. Clemens, Michigan. At the boundary of the field the engine was heard to misfire three times and it was observed that the pilot was slow in retracting the landing gear. The airplane attempted a left turn back toward the airport about three miles out and while in this turn the airplane entered a stall and spun to the ground. The airplane exploded into flames upon impact and the pilot was killed instantly. Published accounts of the accident reported that the pilot's wife had just relocated to the area to be near her husband.

1-28-42B. Camp Bullis, Texas. At 1600, a North American BT-14 crashed at Camp Bullis, Texas, killing Aviation Cadet Jack F. Whitright. U.S. Army troops observed the airplane maneuvering at an extremely low altitude prior to the accident. The witnesses on the ground reported that the airplane dove on the field that they were in from an altitude of about 700 feet, pulling up just above ground level and then buzzing a command car and a jeep, clearing them by three feet before pulling up. The airplane flew away at low altitude, estimated to be about twenty feet above ground level. The airplane flew up and over a small hill before the pilot attempted a steep left turn at low altitude, striking a tree and crashing to the ground.

1-29-42. Baltimore, Maryland. At 1549, a Martin B-26A rolled out of control and crashed in Snyder's Woods near Breezy Point, five miles southwest of the Glenn L. Martin Aircraft Company Airfield, Baltimore, Maryland, killing the civilian crew of three. The airplane had taken off from the factory field at 1542 and the pilot reported at 1547 that he was at 2,000 feet. The crew had been practicing take-offs and landings prior to the crash. Investigators speculated that the airplane might have suffered an engine failure and that the unequal distribution of power caused the airplane to roll out of control and enter a spin to the right, spinning to the ground before the pilots could recover the airplane. The B-26 was seen to spin four or five times from an altitude of approximately 1,200 to 1,500 feet before striking the ground and exploding into flames. The airplane struck the ground on its nose and port wing, burying the nose machine gun in the ground vertically up to the handles and the port propeller neatly dissecting a tree in its path. Killed in the crash were: Paul F. Diehl, 33, Baltimore, Maryland, pilot; George L. Cowell, 30, Baltimore, Maryland, co-pilot; Alden C. Bentley, New Hope, Pennsylvania, observer. All three were civilian members of the AAF Northeast Ferry Command.

1-30-42A. Gulf of Mexico, Louisiana. At 0930, a North American BC-1A suffering an engine failure while on an over-water patrol mission crashed into the Gulf of Mexico, three and one-half miles SSE of the Chandeleur Islands, Louisiana. The BC-1 had been flying the patrol with a North American O-47. On the inbound course, the BC-1 began experiencing engine trouble. The O-47, flying in the lead, set a course for Chandeleur Island. The BC-1 ditched and the crew of the O-47 observed both airmen alive in the water. They dropped a life raft, which landed about 30 yards from the men. Despite a massive search effort, pilot 2Lt. Arthur F. Davies and observer 1Lt. Walter F. Gardner could not be found and were presumed lost at sea. The airplane was later found submerged in 30 feet of water. The Accident Classification Committee recommended that single-engine airplanes such as the BC-1 not be used for over-water patrol flights.

1-30-42B. Brookwood, Alabama. At 1310, a Stearman PT-17 spun to the ground after entering a deliberate spin near Brookwood, Alabama, killing civilian instructor Carl E. Knauer and RAF student Cpl. Jack Leslie Ford. Investigation revealed that the instructor entered a deliberate spin at an altitude of 2,000 feet, 1,600 feet above ground level. The instructor recovered from the spin at an extremely low altitude and, in an attempt to avoid striking the ground, abruptly pulled back on the stick. The airplane lost flying speed, stalled and entered another spin. The PT-17 spun to the ground, falling into a grove of trees. The airplane did not burn but was completely destroyed.

1-30-42C. Randolph Field, Texas. At 1405, a North American BT-9B crashed seven miles northwest of Randolph Field, San Antonio, Texas, killing instructor pilot 2Lt. James R. Davidson and student Aviation Cadet Louis G. Hood. The Accident Classification Committee stated, "It is the opinion of the Committee that this student pilot was practicing elementary-eights at a low altitude, with a strong, gusty wind condition. It is believed that the pilot instructor allowed the student to make a steep climbing turn down wind and the airplane reached a stalling position without his applying corrective action. After stalling, the airplane fell off and struck the ground with the resultant damage to the airplane and injury to personnel."

1-31-42. Mitchel Field, New York. At 1131, a North American B-25 returning from a patrol mission suffered a collapsed nose landing gear upon landing at Mitchel Field, Long Island, New York, fatally injuring bombardier Pvt. Joe P. Hollingsworth. Five other crewmembers were uninjured. The airplane made an approach and touched down at the intersection and rolled for about 700 feet. The pilot opened the throttle and took off, going around for another approach. On the second approach, the pilot made a three-point landing. The airplane rolled the length of the runway and could not use its brakes effectively because of slippery conditions on the runway. When the pilot realized that he was going to run off the end of the runway, he executed a gradual turn to the left, crossing a taxiway. Twenty feet beyond the taxiway, the nose wheel struck an 18 inch graded embankment that was frozen solid. The nose wheel strut collapsed and the nose of the airplane smashed into the ground, crushing the bombardier. Uninjured in the crash were: 1Lt. Thompson F. Dow, Jr., pilot; 2Lt. Channing B. Emberson, co-pilot; Pvt. Kermit C. Buckner, engineer; Pvt. Richard A. Peterson, radio operator; Cpl. Ben J. Moore, photographer.

February

2-1-42. Long Beach, California. At 1630, a Vultee P-66 crashed while attempting an emergency forced landing just after take-off from Long Beach Municipal Airport, Long Beach, California, killing pilot 2Lt. Rudd V. Mann. The airplane suffered an apparent engine failure just after becoming airborne. Witnesses on the ground reported that the engine was "sputtering" and observed the airplane, with flaps in the extended position, in a dive at an altitude of about 300 feet above ground level. When the P-66 reached an altitude of 100 feet, a witness observed that the landing gear was being extended. The airplane entered a steep turn to the left before diving into the ground at a steep angle and exploding into flames. The pilot's hat, wristwatch and the airplane windscreen were found about 30 feet beyond the wreckage. The P-66 was on a ferry flight back to the Vultee factory at Downey, California.

2-2-42. Duncan Field, Texas. At 1045, a Douglas C-39 descended onto a North American BC-1 while both were attempting a landing at Duncan Field, San Antonio, Texas, killing BC-1 pilot 2Lt. Harris A. Stuart. C-39 pilot 1Lt. Felder W. Cullum, co-pilot 2Lt. William J. Bausser and engineer Pfc. M. A. Grissinger were uninjured. Both pilots entered long final approaches and were apparently unaware of the fact that they were very close to each other. There was a considerable amount of air traffic in the vicinity because of the proximity of several busy AAF airfields in the San Antonio area. The C-39 was on a high approach and the BC-1 was on a lower, shallower approach. Pilots on the ground and in the air reported that the tower had given the BC-1 number-one position to land. Soon after the landing instruction had been given, many qualified witnesses noticed that the airplanes were getting very close to each other and a collision seemed imminent. The BC-1 remained in the blind spot under the wings of the C-39 during the entire approach. The C-39 continued to descend and collided with the BC-1 at an altitude less than 20 feet above the runway. The C-39's starboard propeller struck the pilot and cockpit canopy, decapitating the pilot and killing him instantly. The C-39's starboard landing gear contacted the tail of the trainer and the BC-1 fell to the right, hitting the runway on its wheels and skidding to the right, the starboard landing gear collapsing a short distance later. The BC-1 came to rest at the runway intersection and the C-39 landed without further incident, the pilots feeling only a slight vibration from the damaged propeller and still not really aware that they had collided with another aircraft. When the co-pilot looked back he could see the

BC-1 coming to rest behind them with the crash truck and ambulance speeding onto the field. Investigators faulted air traffic control personnel at the field for allowing the airplanes to enter a simultaneous approach and both pilots for failing to see and avoid each other during the approach.

2-3-42A. Lake Tahoe, Nevada. At 1100, a North American AT-6A struck a tree with its port wing and then crashed into Lake Tahoe, Nevada, killing Aviation Cadet John E. Martin and Aviation Cadet Albert Mason. Civilians had observed the airplane maneuvering at low altitude for several minutes in the Lake Tahoe area. The airplane went out of view of witnesses on the ground and was then heard to strike a tree. The airplane came into view minus part of the port wing and in a ninety-degree bank to the left. The airplane then dove into the lake. Investigators noted that the airplane, based out of Mather Field, Sacramento, California, was more than 50 miles from the area where it was authorized to fly an instrument-training mission.

2-3-42B. Brooks Field, Texas. At 1335, a Curtiss O-52 stalled and spun to the ground at Brooks Field, San Antonio, Texas, killing pilot 2Lt. Charles Steen Jr. and observer1Lt. William H. Sills. The airplane had taken off from Brooks Field at 1328 and reached an altitude of 700 feet in its initial climb. Witnesses observed that the attitude of the climb was steeper than normal. The airplane, with landing gear extended, then began a climbing 45 to 35 degree turn to the right. The airplane lost flying speed and began to mush to the ground. The wings were observed to "wobble" and the airplane suddenly dropped off to the right. The airplane spun about one-half a turn to the right before diving straight toward the ground and exploding into flames upon impact.

2-3-42C. Rehoboth, Alabama. At 1515, a North American AT-6A collided with terrain in full flight near Rehoboth, Alabama, killing RAF Flying Cadet Paul D. Underwood and RAF Flying Cadet David W. Turner. "It is the opinion of the [Accident Classification] Committee that the accident occurred when the pilot flew very close to the ground to permit his passenger to take photographs. This belief is supported by the fact that three commercial cameras were found in the wreckage. Film from the only undamaged camera was printed, showing that aerial pictures had been taken. The testimony from the only available witness further showed that the aircraft had been operated very close to the ground immediately prior to the accident. Examination of the wreckage by members of the Committee at the scene of the accident showed that the right wing tip had been sheared from the aircraft when it struck a tree in almost a level flight position. The wing tip was found at the base of the tree at the extreme right of the path cleared by the airplane as it flew into the ground. This, it is believed, indicates that the airplane was in level flight at the time of impact with the tree. The extent of the damage to the aircraft and the size of the area over which the wreckage was strewn, indicated that the aircraft struck the trees at a speed in excess of normal landing speed. This, and the fact that open spaces were immediately adjacent to the scene of the accident, leads the Committee to believe further, that a forced landing was not being attempted."

2-3-42D. Trinity Center, California. At a time unknown, a Martin B-26A flying in poor weather crashed in the Trinity Alps northwest of Trinity Center, California, killing the crew of six. No other information is available.

2-4-42A. Wheeler County, Oregon. At an unknown time after 1545 PST, a Douglas B-18A crashed into Wolfe Mountain, Wheeler County, Oregon, killing the crew of four. The B-18 took off from McClelland Field, Sacramento, California, at 1415 and never arrived at Pendleton Field, Oregon, the airplane's intended destination on its trip back to its base at Elmendorf, Alaska. The B-18 last reported at 1545, 9,000 feet over Redding, California, and was reported as missing at 0000, when investigators estimated that the airplane would be out of fuel. Despite a massive search effort, the bomber was not located until 8-13-42, when it was discovered by two sheepherders. Killed in the crash were: 2Lt. Richard J. Heiderstadt, 22, Geneva, Nebraska, pilot; 2Lt. Walter V. McShane, 24, Troy, New York, co-pilot; TSgt. Michael T. Bittner, 43, Memphis, Tennessee, engineer; SSgt. Donald Kirtland, 33, Hagerman, Idaho, radio operator.

2-4-42B. Luke Field, Arizona. At 2130, a North American AT-6A crashed two miles southwest of Luke Field, Phoenix, Arizona, killing A/C Homer W. Goode, 26, Moody, Texas. The student was scheduled for night take-offs and landings at Luke Field. The student had overshot a landing attempt and elected to go around for another try. He apparently lost his horizon while in a turn to the right, flying the airplane into the ground. The airplane burst into flames upon impact and wreckage was scattered over an area of 400 feet.

2-5-42A. Santee, California. At 0948, a Ryan PT-20A struck power lines while executing a simulated forced landing five miles west of Santee, California, killing instructor 2Lt. Carl E. Danner, Jr., 24, Marion, Ohio. Aviation Cadet Allyn F.B. Taylor, 20, Selma, California, received minor injuries. The airplane smashed to the ground on Mission Valley Road, ten feet behind a sedan in which an elderly couple was riding, barely missing their vehicle. Published accounts of the crash reported that the couple was "shaken" after the accident. Electric power in the area was disrupted for a short time and automobile traffic on the road was blocked for an hour until the wreckage could be removed.

2-5-42B. Tyler, Alabama. At 1500, two North American AT-6A type airplanes collided in mid-air near Tyler, Alabama, killing RAF Flying Cadet George E. Rowley, 19, Bentley Yorkshire, England, who plummeted to his death in an unsuccessful parachute jump from AT-6A #41-745. RAF Flying Cadet John T. Rhys, London, England, was able to safely land AT-6A #41-750 at Craig Field, Selma, Alabama. Investigators stated, "The accident occurred during a formation training flight. The flight consisted of two elements of two ships each; F/C Rhys was the leader of the second element. F/C Rowley was the wingman of the second element. The accident occurred during a one needle-width turn to the left. There was no violation of instructions or flying regulations involved. It is the opinion of the Committee that the accident occurred as the result of the failure of F/C Rowley to maintain his proper position in the formation. As a result, he lost sight of his element leader, F/C Rhys, and collided with him in mid-air. Members of the Committee checked Cadet Rowley's parachute and found it in perfect working condition. The accident occurred at sufficient altitude for him to operate his parachute safely. It is believed that Cadet Rowley attempted a forced landing but changed his mind and abandoned the aircraft at too low an altitude to operate his parachute successfully. The parachute ripcord was found approximately 10 feet from the body, further indicating that the release was pulled at low altitude."

2-6-42A. Gloucester, Virginia. At 1545, two Curtiss P-40 aircraft collided in mid-air eight miles west of Gloucester, Virginia, killing 2Lt. Gail T. Updegraff (P-40 #39-279) and seriously injuring 2Lt. Harry Koral P-40 #39-218), who parachuted to safety. Lt. Updegraff was flying the number-three position and Lt. Koral was flying the number-four position in a four-airplane flight of P-40 type airplanes. The formation had been in a Lufbery circle for a short time before breaking up and joining a "rat race." Lt. Koral thought that the formation was reforming and took up a position on Lt. Updegraff's right wing, slightly behind and below. Lt. Koral then attempted a crossover maneuver behind Lt. Updegraff in an attempt to position himself on Lt. Updegraff's left wing. As Lt. Koral crossed over, he reported that Lt. Updegraff entered a nose down attitude. Lt. Koral stated that he gave his airplane power and pulled up but was unable to avoid the collision. The airplanes collided and it is likely that Lt. Updegraff was killed or rendered unconscious during the collision. He was found in the wreckage of his airplane. Lt. Koral parachuted from his spinning airplane. The airplanes fell to earth one-half mile from each other, each exploding into flames.

2-6-42B. Mount Shasta, California. At a time unknown, a Douglas DB-7B crashed into Mount Shasta, California, two miles southeast of the Horse Ridge Lookout Station, killing pilot 2Lt. Charles Samuel Riley. The airplane had taken off from Eugene, Oregon, at 1130 on a ferry flight and had been unreported, being declared missing at 1530. Despite a massive search effort, the airplane and pilot were not found until 9-22-43. Very poor weather and heavy snows hampered the initial search efforts and the search was discontinued on 2-18-42. The airplane was part of a three-airplane ferry flight to Oakland, California, with the other two pilots encountering poor weather en route. One airplane was wrecked and its pilot lightly injured when it made a forced landing at Ruth, California. The other DB-7 landed safely at Red Bluff, California. Investigators could not determine the cause of the accident, but it was speculated that the pilot had probably encountered poor weather.

2-6-42C. Merced, California. At 2030, a North American AT-6A crashed 20 miles south of Merced, California, killing A/C Dennis R. Francis. Investigators stated, "No witness was able to give a detailed account of the accident. Investigation at the scene of the crash indicated that the airplane was upside down at the time of impact. It is the belief of this board that the pilot lost his horizon, due perhaps to the weather or looking at a map, and crashed because of inability to go directly on instruments."

2-7-42. Tallahassee, Florida. At 2005, a North American AT-6A crashed near Tallahassee, Florida, killing RAF Flying Cadet Ronald G. Robbins. Cadet Robbins had become lost on a day navigation flight earlier and eventually made it to his original destination of Tallahassee. That evening, Cadet Robbins took off with two other AT-6A airplanes piloted by two instructors for a night flight back to the student's base at Dothan, Alabama. After a short time in the air, Cadet Robbins dropped out of the formation and turned back toward Tallahassee. The instructors radioed him to turn 180 degrees and they would pick him by turning 360 degrees with their landing lights on. Cadet Robbins did not turn back and continued back toward Tallahassee. When the airplane reached the field at Tallahassee, the control ship radioed the correct course to the cadet and instructed the student to return to Dothan. The cadet then attempted the turn back to the course to Dothan at an altitude of 3,000 feet and was seen to enter a tight diving spiral to the ground, the airplane exploding into flames upon impact and killing the student instantly. Investigators could not determine the cause of the accident.

2-9-42. United States. *Note: Most regions in the continental United States move clocks one hour ahead to War Time, which is to be in effect until October 1945.*

2-9-42A. Sikeston, Missouri. At 1100, a Fairchild PT-19A crashed nine miles north of Sikeston, Missouri, killing Aviation Cadet Jack Roe Denton, 23, Hornell, New York. The Accident Classification Committee stated, "From the position of the airplane on the ground relative to the tall surrounding trees, which averaged 60 feet in height, with one tree 25 feet high being next to the fuselage and eight feet behind the trailing edge of the wing, it can be determined that the airplane stalled in, falling vertically for at least the last 25 feet. The airplane could not have stalled in from any great height as it had gathered no forward speed and did not strike the ground with great force. It did not hit any of the surrounding trees. The airplane was nosed up on the ground, with the tail 45 degrees in the air immediately after the accident. There was no evidence on the ground of forward speed. The ignition switch and the gas valve were turned off. No power was on at the time of the crash as the wooden propeller was not shattered."

2-9-42B. Gunter Field, Alabama. At 1430, a Vultee BT-13 stalled and spun to the ground six miles east of Gunter Field, Montgomery, Alabama, killing AAF instructor 2Lt. Willard T. DeGolyer and RAF student Cpl. Reginald A. Price. Investigators stated, "Apparently the student had been given a simulated forced landing by the instructor (as shown by the position of all controls after the accident), and the student or instructor allowed the airplane to assume a steep and or slow gliding turn which precipitated a spiral or incipient spin from which recovery was not made."

2-9-42C. Hermanas, New Mexico. At 2130, a North American AT-6A crashed seven miles east of Hermanas, New Mexico, killing A/C Walter S. Beamer. The airplane was part of a four-airplane flight that had taken off from Luke Field, Phoenix, Arizona, on a cross-country flight to Biggs Field, El Paso, Texas, via Tucson, Arizona, and return. The flight encountered an unexpected violent rainstorm and soon Cadet Beamer was seen to leave the formation. His airplane flew into the ground at flying speed on the Matthews Ranch, exploding into flames upon impact and scattering wreckage over a considerable area. The pilot was decapitated and eviscerated. Several flights from Luke Field on the same navigation problem encountered the unexpected storms. Four of those AT-6A airplanes crashed, killing three of the pilots (see 2-9-42D, E, F).

2-9-42D. Hachita, New Mexico. At 2140, a North American AT-6A crashed six miles east of Hachita, New Mexico, killing A/C Earl B. Fritz. The airplane was part of a three-airplane flight that had taken off from Luke Field, Arizona, on a cross-country flight to Biggs Field, El Paso, Texas, via Tucson, Arizona, and return. The flight encountered an unexpected violent rainstorm and soon Cadet Fritz became separated from the formation. His airplane slammed to the ground at high speed one mile north of State Highway 3 and the Southern Pacific Railroad line in Grant County, exploding into flames and scattering wreckage over a wide area. Several flights from Luke Field on the same navigation problem encountered the unexpected storms. Four of those AT-6A airplanes crashed, killing three of the pilots (see 2-9-42C, E, F).

2-9-42E. Hachita, New Mexico. At 2145, a North American AT-6A crashed eight miles east of Hachita, New Mexico, killing A/C Fay E. Book. The airplane was part of a three-airplane flight that had taken off from Luke Field, Phoenix, Arizona, on a cross-country flight to Biggs Field, El Paso, Texas, via Tucson, Arizona, and return. The flight encountered an unexpected violent rainstorm and soon Cadet Book became separated from the flight. His airplane flew into the ground at the Jack Missal Ranch, 200 feet south of State Highway 3 in Luna County. The airplane struck the ground at flying speed and exploded into flames upon impact, scattering wreckage over a wide area. The pilot was decapitated, badly mutilated, and severely burned. Aviation Cadet Thomas A. Bohr, who was flying in the same flight as Cadet Book, parachuted from his AT-6A eight miles east of Hachita, New Mexico, at 2145, receiving minor injuries. Several flights from Luke Field on the same navigation problem encountered the unexpected storms. Four of those AT-6A airplanes crashed, killing three of the pilots (see 2-9-42C, D, F).

2-9-42F. Hachita, New Mexico. At 2200, a North American AT-6A crashed 12 miles northeast of Hachita, New Mexico, killing A/C Charles B. Coats. The airplane was part of a three-airplane flight that had taken off from Luke Field, Phoenix, Arizona, on a cross-country flight to Biggs Field, El Paso, Texas, via Tucson, Arizona, and return. The flight encountered an unexpected violent rainstorm and soon Cadet Coats became separated from the flight. His airplane slammed into the ground at high speed on the Gitson Ranch in the Flying W Mountains in Luna County. The pilot was badly mangled and decapitated. Several flights from Luke Field on the same navigation problem encountered the unexpected storms. Four of those AT-6A airplanes crashed, killing three of the pilots (see 2-9-42C, D, E).

2-9-42G. Gunter Field, Alabama. At 2330, a Vultee BT-13A crashed 12 miles southeast of Gunter Field, Montgomery, Alabama, killing RAF Flying Cadet Charles W. Wadkinson. The student had just taken off from Taylor Field, made a left turn and flew parallel and just beyond the downwind leg of the pattern, the airplane crashing soon after. Pilots flying that night testified to the fact that the horizon was impossible to see because of overcast conditions and that it was a very dark night with few illuminated landmarks. Investigators stated, "The probable cause of the accident was the inability of the student to visually orientate himself, and his failure to check the flight instruments. As a result, he apparently went into a steep spiral from which he did not recover.... The path of the airplane through the woods on the edge of the field on which it crashed, the furrow ploughed in the ground where the plane first hit, the location of the pitot tube, and the wide dispersion of parts, point quite clearly to the fact that the plane was in a steep spiral just before it crashed, that it hit the ground in an inverted position, and that the gas tank exploded on the first impact with the ground."

2-10-42A. Blue Springs, Missouri. At 1632, a North American BC-1A crashed into an open field 100 yards north of the Main Street business district in Blue Springs, Missouri, killing pilot 2Lt. Boyd V. Mann, 29, Pittsburgh, Kansas, and passenger 2Lt.

Norman R. Meeks, 27, Wichita, Kansas. The airplane had taken off from Sherman Field, Fort Leavenworth, Kansas, at 1615 on a reconnaissance training flight. The airplane struck the ground at a steep angle but did not catch fire. The pilots were pulled from the wreckage immediately but were found to be dead.

2-10-42B. McChord field, Washington. At 1450, a Republic P-43D suffering apparent engine trouble crashed while attempting an emergency landing at McChord Field, Tacoma, Washington, killing pilot 2Lt. Gordon E. Houston. The Accident Classification Committee stated, "[The] pilot took off as an integral component of a formation on the NE/SW runway in a southwesterly direction. After making a left hand turn, while still in the formation, with wheels up at or about 500 feet, he apparently encountered motor trouble and started to follow Lt. Dusard (who was making an emergency forced landing) into the field on the N/S runway. Apparently believing himself too high he attempted to continue a complete 360-degree turn. Evidence presented indicates that he tightened his turn eventually falling into a spin. The airplane struck on its right side, shearing the right landing gear and pant. It then struck the left side, tearing the left gear loose from the body. The motor struck the ground, tearing it loose from the firewall and immediately enveloping the aircraft in flames. The motor hurtled forward 150 feet, struck a tree, and veered at a 45-degree angle for approximately 50 feet. The remainder of the airplane continued forward on its belly for 225 feet. Upon immediate examination of the cockpit after the removal of the pilot's body, the controls showed the wheels and flaps locked down and the propeller in full high pitch. It is recommended that all pilots be continuously cautioned that in the case of emergency landing, wheels will be kept in an up position in this type of airplane."

2-11-42A. Gardner Field, California. At 0850, a Vultee BT-13A stalled and spun to the ground at Gardner Field, Taft, California, killing instructor 2Lt. Gordon D. McKenney and student Aviation Cadet Miller Draughan. Investigation revealed that the student had stalled the airplane while turning from the base leg to final approach at an altitude of approximately 300 feet. The airplane spun in and smashed to the ground, killing the occupants instantly.

2-11-42B. East Granby, Connecticut. At 1550, a Lockheed A-29 suffering the failure of the starboard engine crashed at East Granby, Connecticut, killing the crew of six. A member of the crew bailed out of the airplane and was struck by the horizontal stabilizer, causing the progressive failure of the tail section. Killed in the accident were: 1Lt. Melvin W. Schoephoester, Windsor Locks, Connecticut, pilot; 2Lt. Walter C. Boyle, La Cross, Wisconsin, co-pilot; SSgt. Michael M. Kaufman, photographer; Sgt. Gordon G. Johnson, Renovo, Pennsylvania, photographer; Sgt. Thomas F. Quinn, Upper Darby, Pennsylvania, photographer; Pfc. John T. Howey, New York, New York, photographer. The Accident Classification Committee stated, "It is the opinion of the committee that the loss of one engine at high altitude [12,000 feet] was followed by a spin; that after the spin had started, some member of the crew jumped with his parachute harness insecurely fastened; that the parachute, still containing the man, struck the leading edge of the stabilizer and burst the parachute open. After opening, the parachute became entangled in the tail surfaces, pulled them away from the ship and threw the occupant out of the harness. The committee believes that the member of the crew who jumped was Sgt. Quinn for the reason that his arm, hand and sleeve were found some distance from the rest of his body and the wreck. It is believed that these members were severed as the tail gave way. Since all of the emergency exits, except the one over the pilot station, were found in the wreckage, it is believed that he left through this opening. It is probable that after the ship went into a spin, the crew was ordered to abandon ship and that Sgt. Quinn, who is believed to be in the nose of the ship, was the first to leave. At this time, it is believed that Lt. Boyle, whose body was found in the forward portion of the wreckage, was at the controls. If this is true, it is probable that Lt. Schoephoester would have been standing just behind him and that as the spin became more violent after the tail left the ship, Lt. Schoephoester was thrown out through the escape hatch without a parachute. Five parachutes were found in the wreckage. One of these was under the body of Lt. Boyle; the other four were in the aft section of the wreck. It is the opinion of this committee that this accident was primarily caused by the fact that the tail surfaces were torn away by a parachute. A contributing cause is an unknown condition which caused one member of the crew to leave the ship prior to the loss of the tail surfaces."

2-11-42C. Lemoore, California. At 1520, a Vultee BT-13A stalled and spun to earth 14 miles northwest of Lemoore Army Air Field, Lemoore, California, killing Aviation Cadet Weldon W. Woods, 25, Oklahoma City, Oklahoma. Investigation revealed that the airplane entered a gliding turn at an altitude of approximately 2,000 feet when it stalled and fell off into a spin to the right, describing four or five turns before diving into the ground. Investigators speculated that the pilot recovered from the spin at too low an altitude to prevent the airplane from diving into the ground.

2-11-42D. Fall City, Washington. At 2350, a Martin B-10BM crashed five miles north of Fall City, Washington, killing the crew of four. The airplane had been observed two miles north of Boeing Field, Washington, at an altitude of 3,000 feet. The B-10 flew into the terrain a short time later. The airplane struck the ground in a diving turn to the right, cartwheeling and exploding into flames upon impact, scattering wreckage and bodies over a wide area. Killed in the crash were: 2Lt. Harry A. Klein, pilot; 1Lt. H. Hosfelt, Battery A, 210th Coastal Artillery, observer; Cpl. Walter

W. Bently, radio operator; Pfc. Walter M. Dings, gunner.

2-12-42A. Portland, Arkansas. At 1512, a Douglas DB-7B on a low-altitude training mission collided with a tree and crashed near Portland, Arkansas, killing two crewmembers and seriously injuring pilot 1Lt. Edward L. Larner. Lt. Larner later stated to investigators, "I was leading a scheduled low altitude assembly mission. We took off at approximately [1400]. On our final leg to the target, I reached down to reset my periodic compass so that I might head out on my new course immediately after passing over the target. The sun was more or less in my eyes and it was considerably darker in the cockpit than outside. I think that this caused me to keep my head down longer than I should have because I could not read the compass until my eyes became accustomed to the change in light. Although flying at some 50 to 75 feet over and above all obstacles, I must have inadvertently pushed forward on the control column as I leaned down to adjust the compass. When I straightened up, I was confronted by a large tree which I struck before I could pull up over it." Crewmembers killed in the crash were: Sgt. J. C. Badura, Omaha, Nebraska, and Cpl. C. E. Moeller, Columbus, Ohio.

2-12-42B. Llano, Texas. At 2100, a North American BC-1 on a night navigation training flight crashed 19 miles south of Llano, Texas, killing instructor 2Lt. Clarence R. Simmons and student Aviation Cadet John M. Hemphill. Investigation revealed that the airplane struck the ground in a vertical bank to the right, cartwheeling along the ground before bounding back into the air and coming to rest facing opposite of the line of flight. The airplane had taken off at 2020 from Kelly Field, San Antonio, Texas. Witnesses on the ground stated that the airplane had circled Llano two or three times before it was heard to crash. Investigators speculated that the pilots might have been looking at a map or had lost the horizon, inadvertently flying the airplane into the ground.

✓ **2-13-42. Sikeston, Missouri.** At 0840, a Fairchild PT-19A crashed nine miles north of Sikeston, Missouri, killing Aviation Cadet William P. Klaus, 21, Beckley, West Virginia. Investigators speculated that Cadet Klaus was practicing aerobatics at low altitude and stalled the airplane in an inverted position. The cadet allowed the airplane to gather excessive speed in a vertical dive and was unable to regain control of the airplane before striking the ground and exploding on the farm of Jess Wheeler.

2-15-42A. Stratford, Connecticut. At 1049, a Bell P-39C crashed at Bridgeport Municipal Airport, Stratford, Connecticut, killing pilot 2Lt. Harry L. Matthews, 24, Gates, North Carolina. The Accident Classification Committee stated, "The pilot took off at 1045 EWT in a P-39C on an aerobatic mission. He circled the field once at approximately 300 feet altitude at which time he called for landing instructions. He was instructed to land on the E-W runway from east to west. After circling approximately a quarter circumference more with wheels down and, it is believed from examination of plane, flaps down, the pilot stalled and spun out of a tight bank from about 200 feet. The plane struck at 1049 EWT approximately [one and one-quarter] miles SSW of the Bridgeport Airport."

2-15-42B. Location Unknown. At a time between 0910 and 1230 EWT, a North American O-47B on an anti-submarine patrol mission disappeared over the Gulf of Mexico in the vicinity of Dry Tortugas and Key West, Florida. The crew of three was missing and presumed lost at sea. The O-47, based out of Miami, had taken off from Key West, Florida, at 0800 and never returned. Despite a thorough search effort, no trace of the crew or aircraft was found. Missing and presumed lost at sea were pilot 2Lt. Marsh W. Hemmenway, observer 1Lt. Erskine R. Kelly and gunner Cpl. Percy M. Stewart.

2-15-42C. Elk Mountain, Wyoming. At an unknown time after 1800 MWT, a Douglas DB-7B crashed at Elk Mountain near Hannah, Wyoming, killing pilot 2Lt. Robert S. Gleghorn, Asheville, North Carolina. The airplane had taken off from Ogden, Utah, at 1600 MWT with a formation of other DB-7B airplanes on a ferry flight from California to Canada via Salt Lake City, Utah, and Cheyenne, Wyoming. The flight leader reported seeing Lt. Gleghorn at about 1800 in the vicinity of Elk Mountain. The airplane was not found until July of 1942.

2-16-42. Ajo, Arizona. At 1430 MWT, a North American AT-6A on a gunnery training mission collided with the target tow cable and crashed 15 miles north of Ajo, Arizona, killing Aviation Cadet Arnold E. Sindlinger. The port wing was severed seven feet from the tip when the airplane collided with the cable (which was attached to a sleeve target and an AT-6A), cutting the cable and causing the student's AT-6A to spiral to the ground from an altitude of 4,800 feet above sea level. The airplane hit the ground in a vertical dive and exploded into flames upon impact, killing the student instantly. The other AT-6 was able to land safely at its base.

2-18-42A. McCutchenville, Ohio. At 1350 EWT, a Bell P-39F flying in poor weather crashed three miles southwest of McCutchenville, Ohio, killing pilot 2Lt. Thomas F. Almon. Investigators stated, "The pilot took off from Buffalo, New York, to ferry this P-39F to Wayne County, Michigan. He evidently flew to the south in order to avoid Lake Erie. He was known to have been following the railroad south of Tiffin, Ohio, just prior to running into a snow squall. From witness statements it appears that he circled back to get out of the snow squall but struck a tall tree in a small grove and plunged to the ground. There is no evidence of his motor missing or failing and the unusual loudness of the motor may have been because of the pilot's using low pitch under the emergency conditions."

2-18-42B. Brandon, Mississippi. At 1250 CWT, two Curtiss P-40 airplanes apparently collided in mid-air six miles northeast of Brandon, Mississippi, killing 2Lt. Harry H. Woodford, Pasadena, California (P-40 #40-306), and 2Lt. William P. Brady, Columbia, South Carolina (P-40 #40-484). It was speculated that the airplanes might have locked wings while flying in formation in instrument conditions. The Accident Classification Committee stated, "These two airplanes struck the ground at an angle of almost 90 degrees, and the points of impact were separated by [200] feet. Both appeared to have struck the ground heading approximately north, while their course should have been approximately east. These facts seem to indicate one of the following causes: Collision in flight, with resultant loss of control by both pilots, while flying at low altitude; Loss of orientation by formation leader while flying on instruments, resulting in a wingover or half-roll and straight dive into the ground. A front had passed Jackson [Mississippi] one hour and thirty minutes before these airplanes took off, and a resident of the vicinity of the crash reports that ceiling and visibility were very low at the time of the crash, with 'heavy mist.'" The airplanes struck the ground with terrific force, burying their engines at least eight feet in the ground and both creating a crater about ten feet in diameter. Wreckage was scattered over a considerable area. The airplanes had taken off from Jackson, Mississippi, at 1241 CWT on a ferry flight to an unknown location.

✓**2-18-42C. Millstadt, Illinois.** At 1630, a Fairchild PT-19A spun to the ground eight miles west of Millstadt, Illinois, killing Aviation Cadet Edward F. Crandall, 22, Minneapolis, Minnesota. Investigators stated, "The Board finds that Cadet Crandall crashed from a prolonged spin to the left. The airplane, when first sighted by witnesses, was well above 1,000 feet and descending in a spin. The Board has no evidence to indicate whether or not the spin was intentional, and no evidence to indicate the reasons for the cadet's inability to recover. It is the opinion of the Board that the spin was probably a normal one that the pilot was unable to recover from due to inexperience and excitement. However, there is some reason to believe that the spin was flat since the impact was not as great as can usually be expected from a normal spin." The student, being trained by Parks Air College, took off from East St. Louis, Illinois, at 1555.

2-18-42D. Goodfellow Field, Texas. At 2337, a Vultee BT-13A stalled and spun to the ground one-half mile east of Goodfellow Field, San Angelo, Texas, killing Aviation Cadet James Robert Lewis, 22, Independence, Kansas. The student had been practicing night flying for three hours before coming in to Goodfellow Field to practice night landings. Control tower personal observed the airplane in a diving turn to the right. The student was attempting his second landing and apparently stalled out of a turn from base to final or flew into the ground because of the lack of a horizon.

2-19-42A. Olinger, Virginia. At 1630, a Bell P-39D crashed three miles west of Big Stone Gap near Olinger, Virginia, killing pilot 2Lt. Edward J. Wilczewski, 26, Chicago, Illinois. The airplane slammed into a hill and exploded into flames while attempting to locate a suitable forced landing site in the Powell River Valley. The airplane was part of a six-ship flight (5th Pursuit Squadron, 52nd Pursuit Group) that had taken off from Bowman Field, Louisville, Kentucky, at 1525 on a flight to Knoxville, Tennessee. The flight became lost and separated from the flight leader. None of the pilots were able to orient themselves. The flight leader had been buzzing towns along the Tennessee River but was also unable to ascertain his position. Three other airplanes in the formation made emergency forced landings or were abandoned because of fuel exhaustion. 2Lt. Walter A. Kari was uninjured when he bailed out of his P-39 at 1630 near Wise, Virginia. 2Lt. N.L. McDonald and 2Lt. R.D. McGuire crash-landed their airplanes at 1715 near Fort Blackmore, Virginia, and both were uninjured. 2Lt. Ray R. Waynick was able to safely land his airplane at Tri-Cities Airport, Bristol, Tennessee, and the flight leader, 1Lt. George B. Deaton, was able to land his airplane in a field near the Tennessee River.

2-19-42B. Mather Field, California. At 1700, two Curtiss AT-9 type airplanes collided in mid-air near Mather Field, Sacramento, California, killing passenger Pfc. Joseph C. Daas, who fell to earth in an unsuccessful parachute jump from AT-9 #41-5857. He apparently bailed out at too low an altitude to allow his parachute to fully deploy. Pilot 2Lt. Lloyd D. Collins parachuted to safety and was uninjured. Lt. Collins had become entangled in the airplane and had to struggle to free his foot from the area between the rudder pedals and the sidewall. The pilot of AT-9 #41-5803, 2Lt. William G. Erck, parachuted to safety but received serious injuries. His passenger, Sgt. Laverne H. Kreutner, parachuted to safety and was uninjured. Lt. Collins had been flying on the right wing of another AT-9 when, unknowingly to the two pilots, they were joined by a third AT-9, piloted by Lt. Erck. The airplanes flew along in this formation at an altitude of 4,000 feet for several minutes before the flight leader gave Lt. Collins a signal to cross over to the left wing position. When Lt. Collins executed the maneuver he collided with the AT-9 flown by Lt. Erck. The propeller of Lt. Erck's airplane severed the tail of Lt. Collins' airplane and both airplanes fell out of control, the crews soon parachuting out. Both airplanes exploded into flames when they slammed into the ground.

2-20-42A. Hoquiam, Washington. At 0725, a Vultee O-49 crashed into Grays Harbor just after take-off from Moon Island Airport, Hoquiam, Washington, killing pilot 2Lt. Charles P. Clark and observer

1Lt. Rudy J. Binder. The airplane was about to go on a coastal patrol mission. Investigators stated, "Ship took off and turned right, entering ground fog at low altitude just at daybreak and pilot flew into the water before he could properly orient himself on instruments."

2-20-42B. Perrin Field, Texas. At 1140, a Vultee BT-13A crashed ten miles south of Perrin Field, Sherman, Texas, killing instructor 2Lt. Cyril Thomas Van Valkenburg and student Aviation Cadet Quinton Oswell Perkins. Investigation revealed that prior to the accident the airplane had executed a series of loops, commencing at an altitude of about 5,000 feet and working down to about 500 feet. The airplane allegedly performed a loop, stalled at an altitude of 500 feet above ground level, entered a spin and smashed into the ground while in a diving turn to the left. The pilot was just regaining control when the airplane slammed into the ground and exploded in flames, scattering wreckage over an area of 200 feet.

2-20-42C. Fort Lewis, Washington. At 0945, a Lockheed A-29 crashed near Fort Lewis, Washington, killing two crewmembers and seriously injuring three others. Radio operator Sgt. Kenneth W. Holmstrom and bombardier Cpl. Forrest W. Barlow were killed; pilot 2Lt. Charles R. Solberg, co-pilot Vincent E. Duffy, and gunner Cpl. Arnold R. Bell were seriously injured. Investigators stated, "Lt. Solberg departed McChord Field [Tacoma, Washington] 0855 for local training flight. The weather in this area was ceiling 1,000 feet overcast, lower broken clouds with visibility 4 to 7 miles, light to moderate icing in clouds and precipitation above 3,000 feet M.S.L. Evidence presented indicates that while cruising at an altitude of approximately 5,000 feet, clouds were encountered (icing conditions known to exist from an altitude of 3,000 feet up) resulting in carburetor trouble (ice) which apparently resulted in causing the left engine to cease functioning. It is believed that together with engine trouble and being required to fly on instruments momentarily, the plane immediately entered a flat spin. Further evidence presented also indicates that the pilot and co-pilot being unable to recover from the flat spin, moved to the rear of the airplane for the probable purpose of bailing out. It is believed this increased weight in the rear compartment forced the airplane out of the spin; as it righted itself, approximately 100 feet from the ground, the airplane started to climb at an angle of approximately 45 degrees. Due to this weight being in the rear of the airplane, it then stalled and mushed into the trees."

2-20-42D. Los Angeles, California. At 1230, a Lockheed A-29 on an AAF acceptance flight crashed at Canoga Park, Los Angeles, California, killing Lockheed Aircraft Company inspector pilot Harry S. Roberts, 32, Los Angeles, California, and Lt. Frank S. Dace, 33, Van Nuys, California, U.S. Navy aviator. The airplane slammed into the ground in a grove behind 20327 Stagg Street, narrowly missing a home belonging Mrs. W. L. Odell. The Accident Classification Committee stated, "That the airplane was flying in a northwesterly direction at approximately 1,500 feet above the ground [with everything apparently] operating normally. At this time, a rather rough, severe vertical bank was made to the left. The reason for this is unknown and it appears at this time that some trouble developed in the controls. The airplane assumed a diving attitude of about 70 degrees and maintained this to approximately 200 feet from the ground, at which point it climbed to about 350 feet, stalled, fell off on the left wing, and crashed and burned, killing both occupants." The cause of the accident could not be determined.

2-20-42E. Bolling Field, Washington, D.C. At 1700, a Douglas DB-7B suffering the failure of its port engine stalled and spun into the Potomac River near Bolling Field, Washington, D.C., killing civilian ferry pilot Jack H. Marine. The airplane had just taken off from Bolling Field on a ferry mission to an undisclosed location and had reached an altitude of about 200 feet. The airplane was then seen to make a sharp turn to the left (into the dead engine), plunging nose first into the Potomac River 500 yards from the edge of the field. The pilot was killed instantly. The spin occurred at too low an altitude to allow the pilot to recover before hitting the water.

2-20-42F. Woodville, Texas. At 2015, a Cessna AT-17 on an interception mission crashed six miles east of Woodville, Texas, killing A/C Arthur W. Clothier and A/C Joseph A. Corcoran. The airplane crashed into a heavily wooded area and was not found until the next day. Investigators could not determine the cause of the accident.

2-21-42. Westover Field, Massachusetts. At 0847, a Lockheed A-29 crashed while attempting a take-off at Westover Field, Chicopee, Massachusetts, fatally injuring co-pilot 2Lt. Gordon C. McArthur. The Accident Classification Committee stated, "Investigation revealed that the pilot took off with his aircraft trimmed slightly tail heavy. It is believed that this condition, coupled with strong gusty winds, caused the pilot to get off the ground and start climbing at a speed too slow for safety in this type of airplane with its wingtip stall characteristics. The pilot was unable to recognize the loss of aileron control as a stalling condition and hence followed his impulse to land as soon as possible because he felt that something was wrong with the aileron control system. Physical inspection of the plane after landing failed to show any evidence that the control system was fouled. Apparently in the absence of a footrest, the co-pilot leaned forward to brace himself against the impending shock of landing and in the swerving motion of the airplane to the right, he evidently was thrown into the control pedestal wherein occurred injuries which later resulted in his death." Pilot 1Lt. George C. Salvo, engineer Pvt. L. Helton, radio operator Pvt. G. W. Oakes, bombardier 2Lt. J.

R. Owen, bombardier Cpl. D.C. Appleman, and bombardier R. F. Knight were uninjured in the aborted take-off and forced landing at the end of the runway.

2-24-42. Atlantic Ocean, Massachusetts. At 1327, a Curtiss P-40E plunged into the Atlantic Ocean two miles east of the U.S. Coast Guard Radio Station at Humarock, Massachusetts, killing pilot 2Lt. Dennis J. Dowling, 22, Long Island, New York. The airplane was part of a three-airplane combat training flight that had taken off from Boston at 1310. The other pilots reported that Lt. Dowling had apparently fallen out of a left turn at an altitude of about 3,500 feet. After noticing that Lt. Dowling was not in the formation, the pilots observed wake rings and an oil slick on the surface of the sea. Investigators were not able to determine the cause of the accident.

2-25-42A. Chilili, New Mexico. At 1100, a Beech AT-11 on a bombardier training flight was abandoned by its four crewmembers eight miles north of Chilili, New Mexico, killing student bombardier A/C William E. West, 26, Hazel Park, Michigan, who was struck by the tail section of the airplane in an unsuccessful parachute jump. Pilot 1Lt. Charles M. Opiel, Cleveland, Ohio, parachuted to safety and was uninjured. Bombardier instructor 2Lt. Austin L. Miller, Madrid, Iowa, and student bombardier A/C Frederick L. Lopez, 26, Carpinteria, California, parachuted to safety and received minor injuries. The airplane had been flying at 17,500 feet in instrument conditions and began to pick up a heavy amount of ice. The airplane lost over 4,000 feet of altitude very rapidly and the pilot lost control of the airplane. He soon ordered the crew to bail out and while attempting to regain control of the airplane he felt a thud, which later proved to be the cadet striking the tail section. The pilot was unable to regain control of the airplane and he soon bailed out.

2-25-42B. Ravenna, Texas. At 1725, a Fairchild PT-19A crashed four miles west of Ravenna, Texas, killing Aviation Cadet Noel A. Brown. The Accident Classification Committee stated, "While practicing spins on a routine solo training flight the student lost control of the airplane and the ship spun down to within 100 feet of the ground. An examination of the wreckage showed that the majority of the impact was suffered by the right wing, landing gear and tail wheel indicating that the cadet had regained partial control just previous to the crash but without sufficient altitude to make a safe recovery."

2-26-42A. Key Field, Mississippi. At 1139, a Curtiss P-40E flying in poor weather crashed nine miles northeast of Key Field, Meridian, Mississippi, killing pilot 2Lt. William J. McGowan, who fell to his death in an unsuccessful parachute jump. Investigation revealed that the pilot had become lost in a snowstorm and when he became low on fuel elected to bail out. He apparently stood up in the cockpit and pulled his ripcord, being pulled over the tail section and fracturing his skull. He was killed instantly and his body descended to earth in the damaged parachute.

2-26-42B. Fort Dix, New Jersey. At 1555, a Curtiss O-52, stalled and spun to earth on the artillery range at the Fort Dix Military Reservation, New Jersey, killing pilot 2Lt. Donald L. Chase, 29, Milwaukee, Wisconsin, and observer 2Lt. Thomas M. Lewis, 27, Garden City, New York. The airplane, which had taken off from Fort Dix at 1544 on a message-dropping mission, was in a slow speed turn at an altitude of less than 200 feet when it fell abruptly toward the ground. At an altitude of about 50 feet, the airplane entered a spin and dove toward the ground. The airplane burst into flames upon impact.

2-26-42C. Skullton, Pennsylvania. At 1630, a Curtiss P-40E on a ferry flight crashed on the farm of Grant Johnson at Skullton Mountain, Skullton, Pennsylvania, killing pilot 2Lt. Robert O. Jessup. The airplane had taken off from Wayne County Airport, Wayne County, Michigan, at 1504. It was speculated that the pilot lost control of the airplane when he entered instrument conditions. A civilian witness on the ground reported that the airplane had flipped over on its back and then entered a steeply banked diving turn to the right and then collided with the ground. The airplane exploded into flames upon impact. Weather at the accident scene was reported as foggy with freezing rain and snow flurries.

2-28-42A. Curtis Field, Texas. At 0100, a Vultee BT-13A suffering an engine failure was abandoned in flight and crashed nine miles east of Curtis Field, Brady, Texas, killing civilian pilot Richard W. Clark, 29, Toledo, Ohio, who plummeted to his death in an unsuccessful parachute attempt. Civilian student instructor Russ Chitwood, 23, Conway Springs, Kansas, parachuted to safety and was uninjured. When the engine quit, Clark ordered Chitwood to leave the airplane. Chitwood hesitated and after repeated orders by Clark, Chitwood finally left the airplane. Clark left the airplane just as it stalled and entered a spin. His parachute was in working order but he had left the airplane at an altitude too low to allow it to fully deploy. The airplane smashed to earth on the Richards Ranch. Investigators speculated that carburetor icing was the cause of the engine failure.

2-28-42B. Marion, Virginia. At 1330, a Vultee L-1A flying in instrument conditions crashed at Glade Mountain near Marion, Virginia, killing civilian ferry pilot Andrew E. Majni, Nashville, Tennessee. The L-1 was part of a four-airplane ferry flight. The flight had been flying on top of the overcast when the flight leader dove through a hole in the clouds to establish visual contact with the ground. The other airplanes followed and Majni, flying in the number four position, apparently dove through the low overcast. When Majni broke out of the bottom, he was unable to level out and the airplane slammed into the mountain.

✓ **2-28-42C. Ontario, California.** At 1227, a Stearman PT-13B crashed at Ontario, California, killing civilian instructor J.H. Leedy and student A/C Chase Andrews. There were no known witnesses and investigators could not determine what caused the airplane to dive into the ground. The airplane struck the ground at a near 90-degree angle near the Cal Aero Corporation Airfield. Both pilots were fatally injured and died immediately after the accident. It was speculated that the PT-13 entered a spin at too low an altitude to allow the pilot to recover before striking the ground.

2-28-42D. Waukomis, Oklahoma. At 1445, a Vultee BT-13A crashed two miles west of Waukomis, Oklahoma, fatally injuring student Aviation Cadet Carroll W. Jones, 21, Okemah, Oklahoma. He died at 1740 at the AAF station hospital at Enid, Oklahoma. Instructor 2Lt. Vic Alberty, 26, Hollywood, California, received serious injuries, including a broken left leg. Investigators stated that the accident was likely caused by the student "momentarily freezing on the controls" during a simulated forced landing. At an altitude of about 200 feet above ground level, the airplane stalled, fell off on a wing and dove toward the ground. After struggling against the student on the controls, the instructor was able to recover partial control of the airplane but was not able to level off before the airplane smashed into the ground. The BT-13 crashed on the farm of Frank Hromas. Hromas and his family witnessed the crash and they pulled both airmen from the wreckage immediately after the accident.

2-28-42E. Marion, Texas. At 1535, a North American BT-9B spun to the ground three miles west of Marion, Texas, killing instructor pilot 2Lt. Ben H. Sherman and student Aviation Cadet W.H. Dahling. Randolph Field, Texas, instructor pilot 2Lt. J.S. Bagby witnessed the accident and stated to investigators, "I had just gotten up to my altitude of approximately 1,100 feet and leveled off and had just passed the town of Cibolo [Texas] when I saw an airplane, which I thought to be a BT-9, in a slight climb. The climb got steeper and steeper until it reached a vertical climb. It fell off on one wing — I believe the left wing — and started to spin. I pointed to the airplane, showing it to my left wingman [2Lt. James N. Hall]. We both watched it go down. When it was about 200–300 feet above the ground, it seemed to come out just for an instant — it did not get all the way out — and then went right back into a spin and crashed. When I first saw this airplane in the climb it was anywhere from 500 to 1,000 feet above me."

March

3-1-42. Atlantic Ocean, South Carolina. At 1045, a Curtiss P-40 dove into the Atlantic Ocean 200 yards off shore of Myrtle Beach, South Carolina, killing pilot 2Lt. Werner C. Van Berckefeldt. Investigators could not determine why the aircraft dove into the ocean.

3-2-42A. Barksdale Field, Louisiana. At 1130, a North American BT-14 crashed after it struck trees while maneuvering at low altitude 12 miles north of Barksdale Field, Shreveport, Louisiana, killing pilot 2Lt. James S. Carithers, 21, Leland, Mississippi, and safety pilot 2Lt. DeWitt H. Kelly, 24, Eldorado, Oklahoma. The BT-14, which was on an instrument training flight, was seen buzzing the area before it struck trees with the starboard wing. The airplane crashed out of control and burst into flames upon impact.

3-2-42B. San Francisco Bay, California. At 0934, a Vultee P-66 crashed while attempting to ditch into San Francisco Bay about 100 yards off shore of Treasure Island, California, killing 2Lt. Roy M. Crothers. The aircraft, one of a flight of six, was attempting to ditch after leaving formation for unknown reasons. Other pilots of the flight witnessed the crash and stated that the airplane ditched smoothly at first and then was violently whipped downward, the tail going into a vertical position. The pilot was killed when he was violently ejected from the aircraft, being hurled through the air for 30 yards before striking the water. The pilot's body and the aircraft were determined to be unrecoverable after a seven-day search.

✓ **3-3-42A. Coleman, Texas.** At 1030, a Fairchild PT-19A stalled and spun into the ground nine miles south of Coleman, Texas, killing civilian instructor pilot Fred W. Halbert and Aviation Cadet Stanley Howe. Witnesses observed the airplane spinning to the right from an altitude of about 400 feet before it struck the ground.

3-3-42B. Bithlo, Florida. At about 1245, a Curtiss O-52 was seen to spin to the ground from an altitude of about 300 feet two miles southeast of Bithlo, Florida, killing pilot 2Lt. John F. Stevenson and photographer Sgt. Arthur J. Seerig. The airplane struck the ground spinning to the left with little forward speed. Investigators noted that the landing gear was extended and the engine was producing power when the crash occurred.

3-3-42C. Cucamonga, California. At 1610, a Stearman PT-13B stalled and spun into the ground near Cucamonga, California, killing Aviation Cadet B. J. Bradley. The aircraft spun in from about 500 feet and burst into flames upon hitting the ground. Investigators could not determine what caused the stall and the fatal spin.

3-4-42A. Eglin Field, Florida. At 1346 CWT, a North American AT-6A crashed out of control after losing its port wing two miles north of Auxiliary Field #4, Eglin Field, Valparaiso, Florida, killing pilot 2Lt. Richard E. Baldsiefen and passenger 1Lt. John W. Smith. The airplane had peeled off from its flight just prior to the accident and pieces of the wing struck other aircraft in the flight. Investigators stated, "After diving from an altitude of 8,000 feet to approximately

3,000 feet, the left aileron tore away from the left wing and the outer left wing section tore away from the center section. The plane continued in a spinning dive to the earth, making it impossible for pilot and passenger to use their parachutes. Upon examination of the wreckage, a small government issue screwdriver was found in the center section of the left wing, adjacent to left center section joint, and near a control cable pulley. The pulley was chipped, but it is the opinion of the committee that the screwdriver had no bearing on the cause of the accident."

3-4-42B. Elgin County, Ontario. At 1635, a Bell P-39F suffering an apparent engine failure crashed in Malahide Township, Elgin County, Ontario, killing pilot 2Lt. Charles A. Kinzie, 24, Quinter, Kansas. The Accident Classification Committee stated, "A flight of five [P-39F airplanes] departed Niagara Falls, New York, for Wayne County Airport, Romulus, Michigan, on an authorized ferry flight. Approximately [50] minutes after departure Lt. Kinzie left the formation. A ground witness noticed an object drop from his ship, which later proved to be the left door. The ship continued on its westerly course at an estimated altitude of 1,200 feet for approximately two miles where it turned abruptly to the north, with motor misfiring. Almost immediately after the turn it went into a spin [that] continued to the ground where explosion and fire followed impact. During the spin, black smoke was noted issuing from the engine. The pilot was thrown clear with his open parachute over him. The ripcord was not found. The terrain in the immediate vicinity of the crash was not suitable for an emergency landing."

3-4-42C. Lemoore Field, California. At 1553, a Vultee BT-13A stalled and spun into the ground 12 miles west of Lemoore Field, Lemoore, California, killing instructor 2Lt. Jack W. Pounds, 25, Tyler, Texas, and seriously injuring Aviation Cadet Augustus F. Mangani, 25, Jackson Heights, New York. The airplane was making a turn from base to final for Fox Field, California, at about 500 feet when it stalled. A partial recovery was made and a second stall occurred. The aircraft hit the ground spinning to the right and skidded along the ground for about 100 feet. Witnesses stated that the engine was running at full throttle.

3-5-42A. Mather Field, California. At 0845, a Curtiss AT-9 was abandoned in flight after the student pilot damaged the port landing gear in a hard, bouncing landing attempt at Mather Field, Sacramento, California. Aviation Cadet Howard Corbin, Jr., 23, Mason City, Illinois, fell to his death when he failed to deploy his parachute after being struck by the airplane during his bailout attempt. Instructor pilot 2Lt. John B. Coalson Ranger, Texas, parachuted to safety and was uninjured.

3-5-42B. Barnegat Light, New Jersey. At 1325, a Douglas DB-7B struck the Atlantic Ocean with its wing tip while flying close to the surface 10 miles south of Barnegat Light, New Jersey. The pilot, 2Lt. David A. Southard, and bombardier, Pfc. Jack E. Maxey, were killed in the crash. Gunner Pvt. George F. Oswerk was rescued by the freighter *SS West Gotomska* but died of his injuries 3-7-42. The body of Lt. Southard was never recovered. The aircraft, part of three-plane flight on coastal patrol, left formation for an unknown reason and was seen by the Captain of the *SS West Gotomska*, Mitchel Leger, circling over the water at a low altitude. The Captain stated that he observed the port wingtip strike a swell and then the aircraft crashed into the sea.

3-5-42C. St. Lucie River, Florida. At 1850, a Douglas C-39 crashed into the St. Lucie River at Port Sewall, Florida, killing eight fliers. Investigators speculated that the pilot lost visual contact after entering an extremely severe rainstorm. The aircraft struck the river with its starboard wing tip while in a right turn. The pilot was apparently trying to orient himself in the poor visibility. Killed in the crash were: 1Lt. John R. Evans, Auburndale, Massachusetts, pilot; 1Lt. Harry Bullock, Lexington, Kentucky, co-pilot; Sgt. Ercel Stallard, Lathard, Kentucky, crew chief; Cpl. Clarence F. Ayers, Huntington, West Virginia, radio operator; 2Lt. Raymond D. Clement; 2Lt. Newton M. Simpson; Sgt. John A. Rutko. Witnesses on the ground reported that it was raining extremely hard at the time of the crash. The C-39 had taken off from Patterson Field, Ohio, on a cargo flight to Morrison Field, West Palm Beach, Florida.

3-5-42D. Randolph Field, Texas. At 2025, a North American BT-9B crashed one and one-half miles north of Randolph Field, San Antonio, Texas, killing Aviation Cadet Arthur P. Prouse. The aircraft was practicing night landings and take-offs when the accident occurred. Investigators could not determine why the airplane crashed.

3-5-42E. Location Unknown. At an unknown time after 1100 EWT, a Boeing Y1B-17 disappeared and was presumed to have crashed into the Atlantic Ocean with eight crewmembers. The airplane failed to return to base following a combat patrol over the Atlantic Ocean along the East Coast. The crew of eight was declared missing and presumed lost at sea. The Y1B-17 had taken off at 0730 EWT from Langley Field, Hampton, Virginia, and had 14 hours of fuel aboard. Its last reported position was off of Cape Hatteras, North Carolina, at 1100 EWT. An extensive search of the sea and inland areas was conducted without success. Missing and presumed lost at sea were: Lt. Col. Edward Flanick, pilot; 2Lt. James H. Miller, co-pilot; 2Lt. Robert L. Garrison, navigator; Sgt. Ellsworth W. Aitken, engineer; Pfc. Robert R. McIntyre, assistant engineer; Sgt. Steven J. Chohrach, radio operator; Sgt. Arthur L. Becker, assistant radio operator; Cpl. Draper Hoyt, bombardier.

3-6-42. Minter Field, California. AT 0820, A Vultee BT-13A stalled and spun into the ground five miles northwest of Minter Field, Bakersfield, California, killing 2Lt. Robert Borthick and Aviation Cadet Leonard G. Gard. The aircraft had been on a navigation check ride and then participated a simulated aerial combat sequence with another BT-13 that also was on a navigation check ride. The subject airplane apparently stalled out of an aerobatic maneuver and then dove into the ground at a 70-degree angle from an altitude of about 500 feet above ground level.

3-7-42A. Atlantic Ocean, Delaware. At 1050, a Curtiss P-40E crashed into the Atlantic Ocean 15 miles east of Dover, Delaware, killing pilot 2Lt. Henry G. Jaehning, Jr. The P-40 had been engaged in simulated combat with another P-40 at the time of the crash. Investigators speculated that the pilot had been suffering from a loss of oxygen or from carbon monoxide poisoning. The P-40 struck the water at an estimated 500 mph in a very steep dive from 14,000 feet. 2Lt. Carmon B. Boone later stated that Lt. Jaehning was flying erratically and had failed to notice that he was trying to engage him in their combat exercise. Lt. Jaehning entered a dive that approached 450 mph with Lt. Boone following. Lt. Boone pulled out of the dive with difficulty and lost sight of Lt. Jaehning. After 15 minutes of radio calls and visual searching, Lt. Boone could not locate the other P-40 and returned to base.

3-7-42B. Mooresville, Alabama. At 1345, two Stearman PT-17 airplanes collided in mid-air two miles north of Mooresville, Alabama, killing two aviation cadets and a civilian instructor. A/C Joseph H. Hopkins, 21, Arlington, Virginia, was killed in the crash of PT-17 #41-8278; A/C Wilfrid J. Mulligan, 22, Brooklyn, New York, and civilian instructor Seth R. Mapes, Jr., 23, Tuscumbia, Alabama, were killed in the crash of PT-17 41-8384. Investigators determined that Cadet Mulligan's airplane approached Cadet Hopkins' airplane from the rear. Cadet Hopkins then attempted to execute a chandelle type maneuver to the left, colliding with the starboard landing gear of Cadet Mulligan's airplane. The collision ripped open the fuel tank of Cadet Hopkins' airplane, setting it on fire. Both airplanes then plummeted to the ground with no effort of any of the pilots to use their parachutes.

3-8-42. Shaw Field, South Carolina. At 1035, a Vultee BT-13A stalled and spun into the ground five miles south of Shaw Field, Sumter, South Carolina, killing Aviation Cadet John W. E. Raygan and seriously injuring instructor 2Lt. Milton W. Minarcin. The aircraft had been practicing a simulated forced landing. Lt. Minarcin later stated to investigators, "I gave the student a forced landing but he did very poorly and nearly overshot. I took the controls and brought the ship back to the same spot to show him how it should have been done. I closed the throttle at the same spot and established my glide at approximately 700 feet. Just as I started to turn I got a glimpse of another ship ahead and below. At that point I steepened my bank and dropped the nose at the same time. The bank was a little too sudden and the ship stalled at approximately 625 feet. I opened the throttle, pushed the nose down, and broke the stall. I held the nose down to be sure that I had sufficient flying speed. While I was holding the stick forward, I felt a pressure opposing my effort, something similar to tension on the stabilizer. We were then at about 300 feet and were going to start pulling out of the dive. I relaxed my forward pressure on the stick and at that time it came back very suddenly. We had speed well over 100 when the ship stalled. I could not break the stall and we struck the trees. I had my canopy open; I always fly with my canopy open at low altitudes. I was in somewhat of daze and don't know just how long I had been on the ground before I opened my eyes and realized that there were flames. I unfastened my safety belt and felt all right at the time. I then climbed out of the left side of the cockpit. I tried to get up to the front cockpit where the student was but couldn't. Then I attempted to take off my parachute and was unable to do this because of my wrist giving way. About that time my foot slipped and I fell on the wing and then into the swamp. I saw Cadet Raygan still in the cockpit but was unable to get up because of the bog."

3-9-42A. Lemoore Field, California. At 0740, a Vultee BT-13A stalled and spun into the northwest corner of Lemoore Field, Lemoore, California, while turning from base to final, killing instructor 2Lt. Treman Barber and Aviation Cadet Charles O. Golson. The BT-13 developed engine trouble about four miles out from Lemoore and the instructor radioed he was heading back to the field. Witnesses at the field reported that the airplane was flying very slowly and in a nose-high attitude when it began its turn and stalled at about 500 feet, spinning one full turn before striking the ground.

3-9-42B. Victorville, California. At 1045, a Curtiss AT-9 suffering the loss of its starboard wing while on an instrument training flight crashed on U.S. Highway 395 five miles south of Victorville Army Air Field, California, killing safety observer Sgt. W. C. Blessing, Farmington, Illinois. Pilot 2Lt. Knox Parker, La Grande, Oregon, parachuted to safety. Lt. Parker later stated, "I was practicing basic instruments under the hood. I had made a few turns to the right when I started to lose altitude. The needle and ball were centered and I increased backpressure on the stick. The ship continued to dive and I could not get the nose up. At 4,000, I started to take the hood off and resume contact flying when the right wing came off, knocking off the left door. I unhooked my safety belt and bailed out."

3-10-42A. Key Field, Mississippi. At 0845, a Curtiss P-40E crashed 20 miles west of Key Field, Meridian, Mississippi, killing 2Lt. Frank H. Pulley. Investigators could not determine why the P-40 spun in from an altitude of about 5,000 feet with no effort by the pilot to recover or bail out. The P-40 was on a test flight to check for propeller vibration.

3-10-42B. Tonopah, Nevada. At 0950, a North American AT-6A crashed out of control two miles south of Tonopah Gunnery Range, Nevada, killing pilot 2Lt. James I. White, Brownsville, Texas, and seriously injuring passenger Pvt. Oliver W. Kruger (hometown unknown). The aircraft was on a gunnery mission and was circling a downed target sleeve when it suffered a wingtip stall. The pilot recovered but stalled the airplane again. The airplane struck the ground in a nose-down, left vertical bank position. Investigators noted that the landing gear was in the extended position at the time of the crash.

3-11-42. Hope, Texas. At 2105, a North American B-25A cartwheeled into the ground after striking terrain with its starboard wingtip while making a low altitude vertically banked turn one mile northeast of Hope, Texas, killing six fliers. The B-25 took off from Duncan Field, San Antonio, Texas, at 2019 on a night navigation mission. Investigation revealed that both engines were operating at cruising speed. Upon impact the fuel tanks ruptured causing an explosion so terrific that flaming wreckage was hurled 1,200 feet along the prolongation of the flight path. Investigators could not determine why the airplane was flying at very low altitude. Killed in the crash were: 2Lt. Dana W. Bradford, 24, Los Angeles, California, pilot; 2Lt. Ronald C. Hocking, 22, Absarka, North Dakota, co-pilot; 2Lt. Angus M. Johnston, 23, Taft, California, engineer; MSgt. C. G. Frazier, 37, San Rafael, California, observer; Sgt. Stanley A. Smith, Bison, South Dakota, radio operator; TSgt. Earl F. Ricks, 42, Bridgeport, Indiana, observer.

3-12-42. Long Beach, California. At 2210, a Lockheed P-38D crashed while trying to land at Long Beach Airport, Long Beach, California, killing 2Lt. James W. Lauck, 23, Toledo, Ohio. The P-38 was one of a flight of six that was echeloned to the left and peeling off for landing. The aircraft struck the ground with its port engine and wing with the landing gear down, exploding into flames upon impact. Investigators speculated that the aircraft might have suddenly lost the port engine due to fuel starvation but this could not be concluded with any degree of certainty. Pilots in the same flight later complained of being blinded by the searchlights surrounding the field. The P-38 crashed into a field off of Bellflower Boulevard near Carson Avenue in the vicinity of the Douglas Aircraft Company factory.

3-13-42A. Scranton, South Carolina. At 1615, a Bell P-39D crashed while maneuvering two miles east of Scranton, South Carolina, killing 2Lt. Robert G. Kasper, Lowry, Minnesota. The pilot was practicing individual attacks when he pulled out of a shallow dive and into a steep wingover from which he could not recover in time to avoid flying into the ground. Investigators speculated that the P-39 might have stalled in the wingover, which was attempted at low altitude, or a foreign body might have wedged in the elevator hinge preventing a pullout.

3-13-42B. San Diego Bay, California. At 1325, a Lockheed P-38D suffering the failure of the port engine crashed into San Diego Bay on take-off from North Island, San Diego, California, killing 2Lt. George E. Bennett. The P-38 had cleared the field and banked to the left, turning for a short time before sliding off to the left and diving into the bay.

3-13-42C. Fellows, California. At 2050, a Beech AT-7 collided with terrain while trying to execute a 180-degree turn after inadvertently entering a severe rainstorm near Fellows, California, killing four fliers. The aircraft was on night navigation training flight. Killed in the crash were pilot 2Lt. John S. Fennell, Aviation Cadet Robert L. Black, Aviation Cadet John J. Clifford and Aviation Cadet Leonard E. Sasheltz.

3-14-42. Brookley Field, Alabama. At 1745, a North American BC-1A crashed near Brookley Field, Mobile, Alabama, killing pilot Capt. Robert W. Rulkoetter, San Antonio, Texas, and passenger 1Lt. Edwin B. Woodward, Port Huron, Michigan. The Accident Classification Committee noted that a sudden thick fog enveloped the airfield and surrounding area just before the crash. Four other aircraft were forced down in the Mobile area because of the fog with no fatalities.

3-15-42A. Redwood City, California. At 0730, a Lockheed P-38 crashed into Redwood City Slough, Redwood City, California, killing 2Lt. Mario L. Biava. The aircraft was part of a six-ship flight on a routine coastal patrol mission. Only four of the six P-38 airplanes actually became airborne. Civilian George R. Fleming later stated, "I saw the plane dive and then level off. He tried to bank the ship but went into a dive from which he crashed into the ground. The engines were both running when I saw him. [He] was approximately 1,000 feet when I first saw him. The plane seemed to crash with both engines running full."

3-15-42B. Pendleton Field, Oregon. At 2355, a Boeing B-17E collided with a mountain 20 miles southwest of Pendleton Field, Pendleton, Oregon, killing the crew of ten. Light rain and poor visibility were reported in the area at the time of the crash. The last known position of the B-17, which had taken off at 2227 from Gowen Field, Boise, Idaho, was over Baker, Oregon. The aircraft was flying straight and level with all four engines producing power when it collided with the mountain. Investigation revealed the B-17 was off course at the time of the accident. Killed in the crash were: 2Lt. Lawler C. Neighbors, Goodwater, Alabama, pilot; 2Lt. John L. Bailey, Evansville, Indiana, co-pilot; 2Lt. Glen R. Metsker, St. Helens, Oregon, co-pilot; Sgt. Weldon G. Wilson, Montoursville, Pennsylvania, engineer; Sgt. Frank D. Gillis, Beverly, Massachusetts, gunner; Sgt. Raymond J. Paveglio, Concord, New Hampshire, gunner; Sgt. Charles C. Parsons, Duquoin, Illinois, gunner; Cpl. George R. Hall, Aldren Bridge, Louisiana, gunner; Cpl. James T. Mitchler, Falls City, Texas, radio operator; Cpl. Frank H. Spooner, New Bedford, Massachusetts, gunner.

3-16-42A. Gowen Field, Idaho. At 0220, a Boeing B-17E collided with a ridge four miles southwest of Gowan Field, Boise, Idaho, killing four crewmen and seriously injuring two others. The B-17 was preparing to land at the field after a seven-hour night navigation-training mission. The pilot had missed his approach to the field by choosing the wrong runway and was forced to go around after realizing his mistake. The airplane struck the ridge in a shallow right bank with its starboard main wheel and wing. Wreckage was scattered for about 1,000 yards. Investigators noted that the pilot was probably fatigued, having flown a long mission earlier in the day. The surviving airmen stated that the pilot had gotten ill on the flight, having vomited into his oxygen mask. He appeared "normal" when they arrived over Boise. Killed in the crash were: 2Lt. Charles F. Hosford, Butler, Pennsylvania, pilot; 2Lt. Duane T. Crosthwaite, Pleasant Ridge, Michigan, co-pilot; Pvt. Harold Gillian, Everett, Washington, gunner; Pvt. Arthur L. Schipper, gunner. Cpl. George A. Gerlach, Danbury, Connecticut, engineer, and Sgt. Morton L. Howard, Danville, Pennsylvania, radio operator were seriously injured.

3-16-42B. Winslow, Arizona. At 1830, a Vultee BT-13A suffering an apparent engine failure stalled and spun into the ground while attempting an emergency landing 40 miles southwest of Winslow, Arizona, killing pilot 1Lt. Thomas S. Bither and passenger TSgt. Thomas A. Dybicz. Investigation revealed that improper management of the fuel system led to the engine failure. The pilot tried to make a small clearing for an emergency landing, increased his bank with little speed and stalled the airplane. The airplane entered a spin to the right and dove into the ground from an altitude of about 300 feet.

3-17-42. Myrtle Beach, South Carolina. At 1045, a Bell P-39F stalled and spun into the ground while turning from base to final for the airport at Myrtle Beach, South Carolina, killing pilot 2Lt. Thomas D. Dillant. 2Lt. Clifton Shaw later stated, "I was following Lt. Dillant around the traffic pattern approximately 1,000 feet behind him at an altitude of about 1,500 feet. I saw him overrun the runway on his base leg. He banked the plane up in a steep bank and it appeared to me that he stalled in his turn at about 700 feet. The plane snapped over on the left wing and he pushed the nose down. He regained control of his plane but did not have enough altitude to pull out of the dive. He hit at an angle of approximately 30 degrees."

3-18-42A. Lima, Ohio. At 1145, a Bell P-39F leading a four-airplane flight flew the formation into terrain six and one-half miles east of Lima, Ohio, while trying to maintain visual contact with the ground in a blinding snowstorm. The formation leader, 2Lt. Edward H. Saunders, was killed in the accident as were the three other pilots. The flight had taken off from Wayne County Airport, Romulus, Michigan, at 1044 on a ferry flight to New Orleans, Louisiana, via Louisville, Kentucky. Investigators believe that the windscreen may have been frosted up and that a last second effort to pull up and avoid terrain might have been attempted (see 3-18-42B, C, D).

3-18-42B. Lima, Ohio. At 1145, a Bell P-39F that was part of a four-airplane flight collided with terrain in formation six and one-half miles east of Lima, Ohio, while trying to fly in a blinding snowstorm. 2Lt. Earl A. Houser was killed while trying to maintain formation with flight leader 2Lt. Edward H. Saunders. The flight had taken off from Wayne County Airport, Romulus, Michigan, at 1044 on a ferry flight to New Orleans, Louisiana, via Louisville, Kentucky. Investigators believe that a last second effort to pull up might have been attempted (see 3-18-42A, C, D).

3-18-42C. Lima, Ohio. At 1145, a Bell P-39F that was part of a four-airplane flight collided with terrain in formation six and one-half miles east of Lima, Ohio, while trying to fly in a blinding snowstorm. 2Lt. Armel J. Kennedy was killed while trying to maintain formation with flight leader 2Lt. Edward H. Saunders. The flight had taken off from Wayne County Airport, Romulus, Michigan, at 1044 on a ferry flight to New Orleans, Louisiana, via Louisville, Kentucky. Investigators believe that a last second effort to pull up might have been attempted (see 3-18-42A, B, D).

3-18-42D. Lima, Ohio. At 1145, a Bell P-39F that was part of a four-airplane flight collided with terrain in formation six and one-half miles east of Lima, Ohio, while trying to fly in a blinding snowstorm. 2Lt. Eugene H. Anderson was killed while trying to maintain formation with flight leader 2Lt. Edward H. Saunders. The flight had taken off from Wayne County Airport, Romulus, Michigan, at 1044 on a ferry flight to New Orleans, Louisiana, via Louisville, Kentucky. Investigators believe that a last second effort to pull up might have been attempted (see 3-18-42A, B, C).

3-18-42E. Durbin, West Virginia. At 1300, a Douglas A-20C crashed into a cloud covered mountain valley 40 miles south of Elkins, West Virginia, in the vicinity of Durbin, killing pilot, 2Lt. Kenneth E. Conley, Fairmont, West Virginia. Investigators speculated that the pilot attempted to turn around in the valley after fearing he would be boxed in or collide with rising terrain. It was reported that the pilot flew over his hometown of Fairmont just prior to the crash. The A-20 had taken off from Charlotte, North Carolina, at 1204 on a ferry flight to Bolling Field, Washington, D.C.

3-18-42F. Cochran Field, Georgia. At 2240, two Vultee BT-13 airplanes collided while on final approach to land at Cochran Field, Macon, Georgia, killing RAF Flying Cadet John F. Latta, Ayrshire, Scotland, aboard BT-13A #41-10820. RAF Flying Cadet Donald A. Macinnes landed BT-13A #41-1350 without injury. Cadet Latta was making an approach for a landing when an airplane piloted by Cadet Macinnes struck his airplane. The port wing of Cadet Macinnes' airplane sheared off the vertical stabilizer and rudder of Cadet Latta's airplane. Cadet Latta's BT-13 spun into the ground from an altitude of 400 feet. 2Lt. C.H. Gibson, Jr. stated to the Accident Classification Committee, "At the time of the crash I was serving as Control Ship Officer. The tower called Zone 3 (Cadet Latta) in for a landing. After an interval of time Zone 4 (Cadet Macinnes) was called in. Zone 3 was a bit slow flying to his base leg, resulting in two ships flying the base leg at the same time. When Zone 4 turned on his base leg, there was sufficient interval between the ships to avoid collision but not enough for landing. Zone 4 was ordered to execute a 360-degree turn and space himself behind the other ship. This order was given at least twice by the tower but Zone 4 did not execute this order. Zone 4 was traveling much faster than Zone 3 and quickly closed the distance between them. Zone 3 cut his throttle for his landing. I ordered him not to land but to give it the throttle. This order was repeated but Zone 3 continued a gliding turn into the field. Zone 4 overtook him in the turn. The two ships collided and Zone 3 started spinning. I ordered the Crash Truck to the scene immediately. Zone 4 maintained his approach but I had to order him around the field again to avoid his colliding with the Crash Truck. He continued around the field and landed safely."

3-18-42G. Castaic Lake, California. At 2045, a Vultee BT-13A crashed at Castaic Lake two miles east of Lebec, California, after inadvertently entering instrument flying conditions, killing A/C Eugene W. Crites. The aircraft was one of a flight of five (E-flight, 326th School Squadron, Minter Field, Bakersfield, California) that took off from Daggett Field, California, at 1935 on a night cross-country navigation training flight. This BT-13 and pilot had been reported missing and were not discovered until 4-8-42 (see 3-18-42H, J, K, L).

3-18-42H. Quail Lake, California. At 2046, a Vultee BT-13A crashed five miles northeast of Quail Lake, California, after inadvertently entering instrument flying conditions, killing A/C Donald Carne, 20, San Francisco, California. The aircraft was one of a flight of five (E-flight, 326th School Squadron, Minter Field, Bakersfield, California) that took off from Daggett Field, California, at 1935 on a night cross-country navigation training flight (see 3-18-42G, J, K, L).

3-18-42I. Moore Field, Texas. At 2255, a North American AT-6A crashed just after take-off one mile east of Moore Field, Mission, Texas, killing A/C Cecil Richardson. Witnesses stated that the airplane was climbing slower than what would be considered normal. The aircraft crashed into the ground in a 30-degree bank to the left with a descent angle of about 20 degrees and at about cruising speed from an altitude of approximately 500 feet. The airplane was on a night navigation flight.

3-18-42J. Rosamond, California. At 2105, a Vultee BT-13A crashed 20 miles west of Rosamond, California, after inadvertently entering instrument flying conditions, killing A/C Paul R. Everette. The aircraft was one of a flight of five (E-flight, 326th School Squadron, Minter Field, Bakersfield, California) that took off from Daggett Field, California, at 1935 on a night cross-country navigation training flight. This BT-13 and pilot were reported missing and were not discovered until 3-27-42 (see 3-18-42G, H, K, L).

3-18-42K. Tehachapi, California. At 2120, a Vultee BT-13A crashed 10 miles south of Tehachapi, California, after inadvertently entering instrument flying conditions, killing A/C Paul A. Davis, 23, Denver, Colorado. The aircraft was one of a flight of five (E-flight, 326th School Squadron, Minter Field, Bakersfield, California) that took off from Daggett Field, California, at 1935 on a night cross-country navigation training flight (see 3-18-42G, H, J, L).

3-18-42L. Tehachapi, California. At a time unknown, a Vultee BT-13A crashed in the vicinity of Oak Creek, 10 miles southwest of Tehachapi, California, after inadvertently entering instrument flying conditions, killing A/C Paul S. Day. The aircraft was one of a flight of five (E-flight, 326th School Squadron, Minter Field, Bakersfield, California) that took off from Daggett Field, California, at 1935 on a night cross-country navigation training flight. This BT-13 and pilot had been reported missing and were not discovered until 6-18-48 (see 3-18-42G, H, J, K).

3-19-42A. Atlantic Ocean, South Carolina. At 1615, a Bell P-39D crashed into the Atlantic Ocean two miles off shore of Myrtle Beach, South Carolina, killing 2Lt. Keith H. Fedde. Witnesses in the air observed the P-39 performing steep banking maneuvers at about 800 feet. The airplane entered another steep banking turn prior to flipping to almost inverted. The airplane stalled and then slammed into the ocean in a near vertical dive. Investigators stated, "The pilot stalled or spun the airplane, intentionally or unintentionally, while on a [formation] training mission and was unable to recover complete control of the airplane before crashing. The airplane had been successfully performing acrobatic maneuvers with no malfunctioning of airplane or engine."

3-19-42B Heflin, Alabama. At 2130, a Vultee BT-13 crashed three miles west of Heflin, Alabama, killing pilot 2Lt. Maurice E. Thompson, Dunn Center, North Dakota, and passenger Sgt. James M. Daley, Buffalo, New York. The aircraft had taken off from Gunter Field, Montgomery, Alabama, at 1935 on a night navigation mission. Investigators could not determine why the aircraft crashed.

3-19-42C. Las Vegas, Nevada. At 2220, a North American AT-6A flew into the ground under control three miles northeast of the Army Air Field at Las Vegas, Nevada, killing pilot 2Lt. Sidney Getzovitz,

Oakland, California, and fatally injuring passenger Sgt. Paul Harshaw, Dallas, Texas, who died 03-23-42. Investigators stated, "The pilot had made two landings in a normal manner and had taken off on his third flight. Upon being called in for his third landing, no answer from the pilot was received. A search of the area located the wreckage two hours later. Investigation indicated that the airplane had struck the ground at a very flat angle at high speed, at a point 3½ miles northeast of the field with the airplane heading directly towards the airport. The elevation of the scene of the accident was 80 feet above the level of the airport. At the point of the first contact, five marks made by the propeller tips, and a shallow groove made by the left aileron center hinge bracket were visible on the surface of the ground. The airplane ricocheted about 200 feet, striking the ground with considerable force and losing the left wing at that point. The fuselage was found 200 feet further on and the engine continued rolling 400 feet beyond the fuselage. The pilot was instantly killed and the passenger was fatally injured. The airplane was found to have struck the ground intact with power plant apparently operating normally. The cause of the accident is unknown."

3-20-42. Greenfield, Indiana. At 2045, a Martin B-26 crashed into the ground and exploded after striking a large tree four miles north of Greenfield, Indiana, killing four fliers. The aircraft became lost while on a transition flight. Witnesses on the ground stated that the airplane had circled the town for several minutes before striking a large hickory tree on the farm of Charles Gant. Wreckage was scattered over a large area. There were reports that this aircraft may have been flying near Anderson, Indiana, looking for a place to land earlier in the evening. Civilian law enforcement personnel attempted to illuminate the runway at Anderson with automobile headlights and to get the attention of the B-26 crew but without success. Investigators could not determine the cause of the crash and weather was not considered a factor. Killed in the crash were: 2Lt. Paul F. Hawkins, Ponca City, Oklahoma, pilot; 2Lt. James P. Van Story, Lincolnton, North Carolina, co-pilot; 2Lt. Lawrence J. Rux, Henderson, North Carolina, navigator; Sgt. Robert W. Morgan, Uniontown, Pennsylvania, engineer.

3-21-42. Memphis, Tennessee. At 0753, a Martin B-26 crashed on take-off three-quarters of a mile northwest of Memphis Municipal Airport, Memphis, Tennessee, killing six crewmembers and seriously injuring a seventh. The B-26 was being ferried from Wayne County, Michigan, to Jackson, Mississippi. The aircraft took off to the northwest; the pilot held the aircraft close to the ground after rotation then pulled up abruptly at the end of the runway. He leveled off at about 400 feet above ground level then started a shallow turn to the left. After completing approximately 90 degrees of the turn, the nose dropped and then the airplane whipped to the left and into the ground, exploding upon impact. The surviving passenger, 2Lt. Edward C. Huggs, Arkadelphia, Arkansas, reported that there had been some trouble fully retracting the left flap on the leg from Patterson Field, Ohio, made the day before. The paperwork that pertained to this matter was destroyed in the crash. Killed in the accident were: 2Lt. Joseph F. Simpson, Greenville, Mississippi, pilot; 2Lt. John P. Treadway, Miles, Texas, co-pilot; 2Lt. Frank E. Budenholzer, Bolan, New Mexico, navigator; TSgt. V. A. Costlow, Jackson, Mississippi, engineer; 2Lt. R. C. Jones, Jackson, Mississippi, passenger; TSgt. Richard F. Gemein, New York, passenger (died of injuries 03-23-42).

3-22-42. Condor Field, California. A Schweizer TG-2 glider collided with telephone lines one-half mile east of Condor Field, Twenty-nine Palms, California, at 1900, killing pilot 2Lt. Adam A. Wojcik. The glider collided with the telephone lines about 30 feet above the ground, which resulted in the glider being flipped over and the canopy being crushed. The glider then crashed into the ground in an inverted attitude.

3-23-42A. Midland, Texas. At 1115, two Beech AT-11 airplanes collided in flight while on a simulated bombing mission 40 miles north of Midland, Texas, killing four fliers aboard AT-11 #41-9535, which crashed to the ground and exploded. AT-11 #41-9572, piloted by 1Lt. Richard A. Sedgwick, made a safe landing at Midland Army Flying School. Lt. Sedgwick later stated, "I had just completed a run on target 4 and looking behind saw no one in the immediate vicinity. Up to the time of the crash I had seen no airplane close to my ship. At no time from the time I joined traffic until the collision occurred, did any airplane appear between me and the airplane I was following. The only warning before the crash occurred was just the flash of the other ship as it passed under my left wing. An immediate attempt was made to pull my ship out of danger but the warning was not quick enough. Immediately after the crash I saw the other ship ahead of mine and to the right as the tail peeled off. At this point the tower was contacted and information on the crash and its location were given. As the ship was vibrating excessively, no attempt was made to turn for some time as I wished to determine if possible the condition of my ship and the possibilities of bringing it home safely. All crewmembers were ordered to put their parachutes on and tighten straps. After this was done several turns were attempted and a landing at high altitude simulated in order to further determine the condition of the airplane. At this point I considered the airplane relatively safe and asked each individual crewmember to determine for himself whether to jump or to stay with the ship; each decided to stay. The tower was contacted by radio and told the exact conditions that existed at the time and that I was to attempt a landing. The ship was brought back to Midland without further damage. No member of the crew suffered any injury." Killed in the crash of AT-11 #41-9535 were: 2Lt. Robert W. Ingle, 25, Matoon, Illinois, pilot; 2Lt. William M. Bolton, 24, Griffin, Georgia, bombardier instructor; Aviation Cadet Oliver R. Harja, 27, Ironwood, Michigan, bombardier student; Aviation Cadet William T. Hallstead, 22, Lake City, Florida, bombardier student.

3-23-42B. Kaye Field, Mississippi. At 2113, a Cessna AT-8 crashed while performing touch and go landings at Kaye Field, Columbus, Mississippi, killing instructor pilot 2Lt. John C. Clifton, 26, San Bernardino, California, and A/C Harold K. Persinger, 22, Muncie, Indiana. The airplane had overshot slightly on its landing, its wheels touching the runway before the pilot initiated a take-off. The AT-8 became airborne and then plunged to the ground. Investigators speculated that the autopilot might have been inadvertently activated by one of the pilots causing the fatal plunge.

3-24-42A. Foster Field, Texas. At 0815, a North American AT-6A suffering engine trouble crashed just after taking off from Foster Field, Victoria, Texas, killing A/C Malcolm A. Rousseau, 22, Myrtle Beach, South Carolina. Lt. Col. H. H. Van Auken, who had witnessed the crash, stated to investigators, "I was riding in a government truck near the south end of the flying line, I heard an airplane engine stop running and then start again. I looked out of the truck and saw the airplane climbing at a very steep angle after take-off on the runway. The pilot started a turn to the right with the nose of the plane very high, apparently in a partial stall. After turning about 60 [degrees] the engine failed completely and the airplane stalled and fell off to the right, completing the 180 [degree] turn and striking the ground at about a 60 [degree] angle. The maximum altitude in this turn appeared to me to be about 200 feet. From the sound of the engine it appeared that it was not getting gasoline. Had the pilot gone straight ahead instead of attempting to turn back to the field, I believe that he could have landed with no injury to himself and minor damage to the airplane."

3-24-42B. Brinkley, Arkansas. At 0930, a Lockheed F-4 (photo-reconnaissance version of the P-38E) crashed after apparent engine trouble one-half mile east of Brinkley Municipal Airport, Brinkley, Arkansas, killing pilot 2Lt. James C. Cunningham, 25, Morrilton, Arkansas. The airplane had taken off from Little Rock, Arkansas, just 13 minutes prior to the crash. The F-4 was being ferried from Long Beach, California, to Bradley Field, Windsor Locks, Connecticut. Lt. Cunningham had taken off from Long Beach on 3-22-42 with 2Lt. Hugh D. Wild, who was piloting another Lockheed F-4, and arrived in Little Rock via Tucson, Arizona, and Dallas, Texas, on 3-23-42. Lt. Cunningham paid an overnight visit to his mother's home in Morrilton, Arkansas, and on the day of the accident, just one hour before he was to take off for the east coast, Lt. Cunningham married Miss Reba Pierce, 29, Morrilton dress shop owner, in downtown Little Rock. His mother, grandmother and his new bride were at the airport to bid him farewell and watch him take off on his leg to Memphis, Tennessee. Near Brinkley, Lt. Cunningham began losing power and having trouble maintaining altitude. He tried to make the airport near Brinkley when the P-38 crashed into the ground and exploded into flames. Investigators could not determine the cause of the engine failure.

3-24-42C. San Antonio, Texas. At 1113, a Bell P-39F crashed into the ground in a flat spin in a near vertical attitude 30 miles northwest of San Antonio, Texas, killing pilot 2Lt. Robert B. Brown, 23, Cincinnati, Ohio. The aircraft was on an accelerated test flight when it inadvertently entered instrument flight conditions. Investigators believe that the pilot exhausted his fuel trying to find his way down after being trapped on top of the overcast and probably became disoriented in the clouds resulting in the fatal spin.

3-24-42D. Morris Field, North Carolina. At 1516, a Douglas A-20C crashed one mile southwest of Morris Field, Charlotte, North Carolina, after the failure of the starboard engine, killing 2Lt. James A. Grant, Finley, Tennessee. The A-20 was attempting an emergency landing and was at approximately 50 feet when the pilot lowered the landing gear. The resultant drag and the torque of the port engine forced the starboard wing down. The aircraft lost flying speed and began sliding off to the right. The pilot gave the port engine full throttle and began a staggering climb to about 200 feet when the aircraft lost flying speed and crashed to the ground in a low, flat glide.

3-25-42. Goodsprings, Nevada. At 2040, a North American AT-6A suffered a catastrophic structural failure and crashed three miles northwest of Goodsprings, Nevada, killing pilot 2Lt. Willis D. Crippen. The airplane was on a routine night navigation training flight from Las Vegas, Nevada, to Silver Lake, California, and return. The right wing, both ailerons, right stabilizer, right elevator and upper half of the rudder were found scattered a considerable distance from the main wreckage. The pilot was killed when his parachute was either fouled by the aircraft or deployed at too low an altitude. Weather for the area at the time of the accident had been reported as clear with unlimited visibility with some clouds reported in the Mt. Charleston region. Investigators could not determine with certainty the reason for the in-flight structural failure or the pilot's inability to parachute to safety.

3-26-42A. Stratford, Connecticut. At 1030, a Bell YP-39 spun into St. Michaels Cemetery, Stratford, Connecticut, killing 2Lt. Edward G. Armstrong, 24. The airplane was observed performing a series of rolls prior to the accident. 2Lt. Herbert Schrauer (AAF) stated to investigators, "I saw Lt. Armstrong circling the field at about 3,000 feet. North of the field he executed a slow roll. He followed with another slow roll, but this time at about two-thirds of the roll he went into a spin. Attempting to pull out of the spin, he leveled the plane for a short while, but having no power the plane went into another spin. At about 500 feet the plane was leveled again, and then the nose dipped and the plane crashed nose down."

3-26-42B. Perrin Field, Texas. At 1355, a Vultee BT-13A stalled and spun into the ground while attempting to position itself during a formation landing one mile south of Perrin Field, Sherman, Texas, killing instructor 2Lt. Bernard F. Gallagher and student Aviation Cadet Theodore W. Dimke, Jr. Investigators stated, "Lt. Gallagher, when attempting to get into proper position in the formation that was preparing to land, allowed or caused his airplane to approach too fast to the lead airplane. When dangerously close, he stalled the airplane by an abrupt, violent use of the controls. The airplane apparently snapped to an inverted position and the pilot endeavored to recover by rolling it back to the right. Because of insufficient altitude and speed he was unable to complete the recovery and collided with the ground. From an investigation at the scene of the crash it has been determined that the airplane dove straight into the ground at a slow speed at an angle of about 60 degrees killing both student and instructor. The plane burned immediately after the crash."

3-27-42A. Thunderbird Field, Arizona. At 0835 MWT, a Stearman PT-17 dove into the ground nine miles south of Thunderbird Field, Glendale, Arizona, killing Aviation Cadet Eugene Black. Investigators determined that Cadet Black had deliberately flown the airplane inverted at 300 feet above ground level and struck the ground in a dive while attempting to reestablish level flight.

3-27-42B. Gulf of Mexico, Florida. At 1600 EWT, a Northrop A-17 crashed into the Gulf of Mexico six miles south of Sarasota, Florida, killing 2Lt. Thad P. Medlin, 21, Memphis, Tennessee, and passenger 2Lt. Joseph E. Anderson, 21, Henderson, Tennessee. Witnesses stated that the A-17 was performing aerobatics at a very low altitude just prior to the accident and that the airplane remained on the surface for some minutes after the crash before finally sinking below the waves. Investigators revealed that Lt. Medlin was not checked off on A-17 airplanes and that the flight was undertaken without official permission. The aircraft and the bodies of the airmen were not recovered until 3-29-42.

3-27-42C. Lordsburg, New Mexico. At an unknown time, a Royal Air Force Avro Lancaster crashed near Lordsburg, New Mexico, killing pilot Flt. Lt. N.K. Floyd, RAF, Kelowna, British Columbia, and seriously injuring three unnamed crewmembers. The RAF crew was touring the United States on a War Bond Drive when the airplane crashed and burned near Animas, New Mexico. No other information is available.

3-28-42A. Kelly Field, Texas. At 0955, a North American BC-1 stalled and spun into the ground 18 miles NNW of Kelly Field, San Antonio, Texas, killing the Aviation Cadet John J. Lamb, 24, Chelsea, Michigan. The crash occurred when Cadet Lamb was preparing to land at Auxiliary Field #5, 20 miles northwest of Kelly Field. Aviation Cadet Raymond P. Gerrity witnessed the crash and later stated to investigators, "I was coming in for a landing at Kelly Field No. 5 [when] I saw Cadet Lamb shooting a 180-degree side approach. He had completed at least a 180-degree turn but was heading for the center of the field. He continued to turn past 180 degrees. The right wing of his airplane dropped and it looked like he peeled off and dove into the ground, the airplane making approximately a one-half turn of a spin to the right. The airplane struck the ground approximately head on."

3-28-42B. Hudsonville, Pennsylvania. At 1130, a Curtiss P-36A circling a fire tower collided with Broad Mountain at about the 1,260-foot level near Hudsondale, Pennsylvania, killing pilot 2Lt. Fred J. Knauf, 22, Bark River, Michigan. Low ceilings were reported in the area at the time of the crash. The aircraft had buzzed the airport at Leighton five minutes prior to the crash, evidently to get its bearings after descending through the overcast, breaking out near Mauch Chunk. The P-36 was in a very steep bank to the right when its right wing contacted a tree, causing the airplane to cartwheel into the ground. The aircraft exploded upon impact with the ground and wreckage was scattered over a wide area, starting a series of small fires. Investigators speculated that Lt. Knauf may have been looking at his map and trying to orient himself when his airplane struck the tree. Contrary to published newspaper accounts of the crash, there were no eyewitnesses. The aircraft had taken off at 1005 from Stewart Field, Newburgh, New York, on a navigation training flight and was 40 miles off course for its intended destination of Reading, Pennsylvania.

3-28-42C. Uvalde, Texas. At 1043, a Fairchild PT-19A stalled and spun into the ground 12 miles northeast of Uvalde, Texas, killing pilot E. J. Sobrino, a Chilean civilian undergoing training. Farmer Charlie Muecke stated that the airplane was in a left turn when it spun to the left and crashed into the ground. Another Army PT-19 landed nearby to lend assistance.

3-28-42D. Chickasha, Oklahoma. At 1108, a Fairchild PT-19A stalled and spun into the ground 11 miles north of Chickasha, Oklahoma, killing Aviation Cadet Joseph D. Potenza, 26, Brooklyn, New York. The airplane had been practicing S-turns and steep turns prior to the accident. The airplane entered a flat spin to the right at about 500 feet above ground level. Witnesses stated that the airplane seemed to "fall" to the ground. The aircraft did not catch fire but was severely crushed.

3-28-42E. Lemoore Field, California. At 1325, a Vultee BT-13A stalled and spun into the ground 14 miles southwest of Lemoore Army Flying School, Lemoore, California, killing Aviation Cadet Charles F. Hufford, 22, Uhrichsville, Ohio. The airplane was turning from base to final for its fourth practice landing at Indian Field when it stalled and fell off into a two-turn spin from approximately 600 feet agl.

3-28-42F. Mayfield, Georgia. At 1700, two Vultee BT-13A airplanes collided in mid-air ten miles northwest of Mayfield, Georgia, killing two RAF students. Flying Cadet T. H. Hedger, 20, Sunderland, England (BT-13A #41-1278), was flying an unauthorized formation at low altitude with Flying Cadet R. E. Davies, 23, Kenilworth, Warwickshire, England (BT-13A #41-11230), when Cadet Davies' airplane struck Cadet Hedger's airplane from behind and slightly below, cutting off its tail. The aircraft collided at about 500 feet altitude and fell within 200 yards of each other into a heavily wooded area. Each pilot was cleared for an individual navigation training flight from Cochran Field, Macon, Georgia, to Siloam, Georgia.

3-29-42. Pacific Ocean, California. At 0745, a Lockheed P-38D crashed into the Pacific Ocean five miles northwest of Point Vicente, California, killing pilot 2Lt. Lawrence R. Schlotzhauer, 28, Long Beach, California. He had been leading a flight of P-38 airplanes on a coastal patrol. The P-38 suffered a failure of the port engine and peeled away from the formation as they were heading for Point Vicente. The pilot reported by radio that he had successfully feathered the propeller and the other pilots did not see anything wrong as Lt. Schlotzhauer fell behind. No pilots in the flight witnessed the crash and investigators could not determine with certainty the cause of the engine failure or why the airplane crashed. The lack of two generators on P-38 airplanes was noted as a possible cause of a total electric failure in the case of losing the engine that ran the single generator. The Accident Classification Committee recommended that all P-38 type airplanes be equipped with a generator for each engine. The pilot and aircraft have not been recovered and are considered lost at sea. Lt. Schlotzhauer escaped injury in a P-38 accident on 2-17-42 when his nose landing gear strut collapsed after he made a hard landing at North Island Naval Air Station, San Diego, California.

3-30-42. Tampico, Mexico. At about 1200, a Lockheed C-40D crashed at Villa Aldams 20 miles north of Tampico, Mexico, killing three fliers. The airplane had taken off from Brownsville, Texas, on a flight to Albrook Field, Panama. Investigators could not determine why the airplane crashed. Killed in the crash were: 2Lt. Theodore Corcoran, pilot, Assistant Military Attaché for Air Service, Columbia; Sgt. William W. Horgan, crewman; Sgt. Charles H. Carroll, crewman.

3-31-42. Turner Field, Georgia. At 1605, a Curtiss AT-9 suffered a catastrophic failure of its starboard wing five miles south of Turner Field, Albany, Georgia, killing instructor pilot 2Lt. John W. Wright, 23, Pleasant Garden, North Carolina, and student Aviation Cadet P. W. Kingston, 22, New York, New York. Darr Aero Tech Flight Instructor Robert E. Lawler witnessed the crash. He later stated to investigators: "I was watching an AT-9 in flight. It seemed as though the airplane was coming down, starting a fast spiral turn to the left. It looked as though he was trying a steep turn. When the airplane reached what appeared to be the vertical point and started to recover, just at that moment, as they recovered, the right wing snapped off. A short time after the wing snapped off it sounded as if the motor was wide open. It was a full roar and then it stopped. As [the wing] snapped off, there was a large puff of smoke. The airplane was flying at approximately 2,500 to 3,000 feet."

April

4-1-42A. Thomasville, Georgia. At 1430, a North American AT-6B collided with terrain seven miles northeast of Thomasville, Georgia, killing Aviation Cadet William F. Bridgeport, Savannah, Georgia, and Aviation Cadet Patrick E. Brennan, Yazoo City, Mississippi. The students had been scheduled for an instrument-training mission and were to be practicing at 3,000 feet. Witnesses on the ground reported that the airplane had been flying low in the area and appeared to operating normally before it attempted a right turn at low altitude, the starboard wing striking the ground. The airplane exploded into flames upon impact and wreckage was scattered for over 200 yards. Investigators speculated that the fliers may have been "hedgehopping."

4-1-42B. Minter Field, California. A Vultee BT-13A stalled and spun while making a turn from base to final at Minter Field, Bakersfield, California, at 2350, killing Aviation Cadet Leonard P. James, Denver, Colorado. The crash occurred one-half mile from the field. 2Lt. Jackson S. Wallace, who was the control officer for this flight, witnessed the crash from his airplane as he sat on the east end of the field. He later stated to investigators, "As the pilot banked the ship to turn he apparently did a partial snap roll from which he did not recover. The ship plunged to the ground and burst into flames."

4-2-42A. Paine Field, Washington. At 0649, a Lockheed P-38D attempting a take-off crashed into a mess hall at Paine Field, Everett, Washington, killing two enlisted men on the ground and seriously injuring pilot 2Lt. Oscar L. Morse. The P-38 struck the mess hall in a 90-degree left bank after suffering a runaway port propeller and striking the runway with the propeller and the semi-retracted landing gear. Investigators later stated, "This was the first airplane off in an alert take-off at 20 second intervals. It is evident that the pilot retracted the landing gear immediately after wheels left the runway. It is believed that the windshield then fogged up, which is a common difficulty with this airplane on morning take-offs, and the pilot let the airplane settle back to the runway striking heavily on the left propeller and partially retracted gear. This was done as he made an adjustment on propellers as one had run away. The shock of propeller striking the runway damaged the left engine causing a fire and loss of power. The pilot applied power to the right engine causing the airplane to turn left in an increasingly steep bank and then to strike the mess hall." Killed on the ground were Cpl. Henry H. Robertson, Elk City, Oklahoma, and Pfc. James R. Henly, Little Rock, Arkansas.

4-2-42B. Pearson, Georgia. At 1130, a Douglas DB-7B collided with trees while maneuvering five miles south of Pearson, Georgia, killing three fliers. The Accident Classification Committee speculated that the aircraft had been buzzing the area when the crash occurred. The pilot, 1Lt. Edmund J. Walker, and the gunner, Pvt. Frank L. Barber, were killed instantly in the crash. Bombardier/gunner Cpl. Emanuel A. Vollweiler died of his injuries at a hospital shortly after the crash.

4-2-42C. Helena, Arkansas. At 1135, a Stearman PT-17 dove into the ground three miles north of Thompson-Robbins Field, Helena, Arkansas, killing Aviation Cadet Emery H. Johnson, Jr., 20, Covington, Kentucky. Aviation Cadet Phillip E. Colman witnessed the crash and later stated to investigators, "We were flying at an altitude of 1,000 feet heading toward the Main Airdrome to the south of us. After checking off to our left to see if any airplanes were near I looked to our right and in front of us and noticed below our right wing, a plane in a vertical dive at approximately 500 feet. I watched the plane until it struck the ground at an angle of approximately 45 degrees. While in the dive the pilot of the plane did not leave the ship. We circled the wreckage once and returned to the Main Airdrome and landed. After the instructor reported the accident, we took off again and flew over the area and circled it twice and then returned."

4-2-42D. Goodfellow Field, Texas. At 1600, a Vultee BT-13A dove into the ground on the Nichols Ranch 12 miles south of Goodfellow Field, San Angelo, Texas, killing instructor pilot 2Lt. Phillip W. Larsen, 22, Westego, Louisiana, and student Aviation Cadet Burdette H. Hetzler, 22, Houston, Texas. Witnesses stated to investigators that the airplane began a moderate nose-down descent from about 4,000 feet above ground level with engine power reduced. As the dive became steeper, the aircraft produced a whistling sound and the attitude of the aircraft changed from a moderate nose-down attitude to an extreme nose-down attitude, striking the ground evenly along the leading edges of the wings. Cadet Hetzler left the aircraft at about 150 feet but he failed to deploy his parachute before hitting the ground. He was killed instantly. Investigators speculated that the instructor might have lost consciousness and that the student may have realized too late that the instructor was not properly on the controls.

4-2-42E. Planada, California. At 1630, a Vultee BT-13A stalled and spun into the ground three and one-half miles south of Planada, California, killing Aviation Cadet Donald J. Gann and Aviation Cadet Luther W. Jones. The airplane was executing a climbing turn to the left at approximately 500 feet when it stalled and plunged to the ground. The aircraft struck the ground nose first in an almost vertical attitude and exploded into flames upon impact.

4-2-42F. Leary, Georgia. At 2045, a Curtiss AT-9 on a night navigation training flight crashed after suffering a catastrophic failure of its starboard wing five miles south of Leary, Georgia, killing instructor pilot 2Lt. Hinton S. McLeod, 25, Fayetteville, North Carolina, and student Aviation Cadet Emerson W. Kaler, Crestline, Ohio. Civilian witnesses on the ground stated that the airplane had executed a series of turns prior to the accident. Investigators could not determine what caused the wing failure and noted a similar AT-9 accident that occurred on 3-31-42 at Turner Field, Albany, Georgia, which also involved the starboard wing (see 3-31-42).

4-3-42A. West Greenwich, Rhode Island. At about 0630, a North American B-25A crashed at West Greenwich, Rhode Island, killing the crew of five. The airplane slammed into the ground after its port engine malfunctioned shortly after take-off on a submarine patrol mission from Westover Field, Chicopee, Massachusetts. Investigators speculated that the port engine had either failed or quit completely causing the pilot to return to Westover Field. The pilot did not deem the situation serious enough to jettison the depth charges the airplane was carrying. On the way back to the field the starboard engine apparently quit, forcing the pilot to execute a steep descending glide to a clearing for an attempted crash landing. The pilot was able to maintain flying speed and make it to the clearing, but when he attempted to flare for the landing he stalled the airplane and it mushed into the ground in a near flying attitude. The aircraft was consumed by fire, which detonated the depth charges, the explosion hurling three bodies from the forward section of the airplane 100 yards. Killed in the crash were: 2Lt. George L. Dover, 24, Shelby, North Carolina, pilot; 2Lt. Neal W. Frame, 24, Porterville, California, co-pilot; SSgt. R. H. Trammell, 21, Brunswick, Georgia, radio operator; Pvt. R. R. Meredith, 21, Memphis, Tennessee, bombardier; Pvt. T. J. Rush, Philadelphia, Pennsylvania, gunner.

4-3-42B. Shaw Field, South Carolina. At 0850, two Vultee BT-15 airplanes collided while in a three-ship formation 10 miles NNE of Shaw Field, Sumter, South Carolina, killing A/C Glen E. Hedges aboard BT-15 #41-10235. A/C Hugh J. Haslett landed badly damaged BT-15 #41-10298 safely at Shaw Field. Investigators stated that the accident was a result of the two students getting out of position and then closing into position on the leader too rapidly, causing them to overrun their positions and collide. A/C Hedges attempted to parachute to safety but fouled his parachute on the ship when he opened it prematurely. It was later determined that A/C Hedges was struck across the back just below the shoulder blades by part of the aircraft before he plummeted to the ground in the damaged parachute. He died at 1935 at Tuomey Hospital, Sumter, South Carolina.

4-3-42C. Columbia, South Carolina. At 0900, a Douglas A-20A crashed shortly after take-off from Columbia Army Air Base, Columbia, South Carolina, killing the crew of three. The port engine had been problematic on previous flights and is thought to have failed, forcing the pilot to return to the field shortly after take-off. Capt. M.P. Howard witnessed the crash and later stated to the Accident Classification Committee, "He was taking off on the southwest runway and when about 1,500 feet above the ground I thought he was going to do a chandelle off the field. I later noticed that he seemed to be in trouble and that his left wing seemed to dip and carried him over the rise of the ground where he disappeared from view and it looked from where I was standing that he would crash but just a few seconds later, after making a 90-degree turn, he came over the rise and cleared the trees. The left wing was still low and I noticed at this time that his landing gear was still down. It seemed that he leveled off and I thought for a minute that he would probably make the field. Then his left wing again dipped and forced him immediately to the ground. It seemed to me at one time when he was leveling out both motors were working and that he had momentarily gained control of his ship but just as he was ready to level off to approach the field either his left motor gave out or he lost control of his ship, forcing him to make a steep turn to the left, then crashing to the ground, the ship immediately burst into flames." Pilot 2Lt. George F. Hayman, Jr., observer 1Lt. A. C. Darden, and gunner Pvt. C. D. Fischer were killed in the crashed.

4-3-42D. Espiritu Santo Bay, Texas. At 1350, a North American AT-6B dove into Espiritu Santo Bay, Texas, after failing to recover from a simulated diving attack on a small boat that was being executed by his six-ship flight. A/C George M. Duvall was killed in the crash. 2Lt. Clarence T. Baker, instructor and the formation leader, later stated to the Accident Classification Committee, "I took my formation off on a gunnery mission to Matagorda Island [Texas]. On arriving at this island another formation was using the range and we flew around at about an altitude of 1,000 feet, waiting for them to finish. As I was flying west along the right side of Matagorda Island I noticed a boat crossing the bay. I dived on this boat, pulling out at an altitude of not less than 200 feet, expecting the rest of the formation to do likewise. Evidently number five (5) man in this formation misjudged his altitude and crashed into the bay."

4-3-42E. Barksdale Field, Louisiana. At 1515, a North American B-25C crashed while attempting to take-off from Barksdale Field, Shreveport, Louisiana, killing one crewman. Investigation revealed that the airplane contacted the runway after the co-pilot retracted the landing gear before the airplane became fully airborne. The landing gear retracted fully and the airplane structure and propellers contacted the runway, causing the pilot to lose control. The airplane bounced along the runway until coming to a rest and bursting into flames, killing the engineer Sgt. Edward E. Hughey. Pilot 2Lt. John C. Townsend received minor injuries. Co-pilot 2Lt. Lloyd E. Moore, who was seriously injured, stated to investigators, "After taking off, when the airplane got into the air about 5 or 10 feet, upon the signal from the pilot, I pulled the wheels up. As the wheels were going up the plane began to settle back down and the right engine didn't seem to be giving out full power. As the plane settled further down the pilot pushed the throttles all the way forward but there was no action. The throttles remained forward but for a second and then were pulled completely back. At that point we hit the ground and slid. A second after the plane stopped it burst into flames. We cut the switches immediately after making contact with the ground. The pilot and I climbed out of the hatch."

4-3-42F. Elizabeth City, North Carolina. At 2040, a Curtiss P-40E collided with terrain while maneuvering six miles north of Elizabeth City, North Carolina, killing pilot 2Lt. Ralph D. Davis. The aircraft had been on a night navigation mission and witnesses reported that the aircraft had been flying low and at a high rate of speed. The P-40 attempted a steep turn to the right when the nose dropped and the airplane flew into the ground. The airplane crashed into a remote swampy region and the pilot's body was not found and recovered until 4-8-42.

4-3-42G. Strevell, Idaho. At 2055, a Boeing B-17E broke up in flight and crashed six miles north of Strevell, Idaho, killing eight fliers. Investigators stated, "The plane was cleared from Hill Field, Ogden, Utah, [to Gowen Field, Boise, Idaho] and took off at 2018 MWT. The pilot was cleared by the Weather Officer in questionable weather; and by the Clearing Officer at an altitude lower than the highest peaks along the route. At the time of clearance, contact conditions did exist on the airway. However, shortly after take-off the weather changed to lower ceilings and turbulence. Lt. Walker apparently climbed to 10,000 feet, the altitude at which he was cleared west, and apparently ran into the wind shift line into an area of great turbulence. In trying to control the bucking of the plane, either he or the co-pilot, or both pushed forward too violently on the wheel, causing the tail surfaces to fail, or an exceedingly violent updraft threw the plane on its back from which position it could not be righted, but started to disintegrate. The plane completely disintegrated in the air. The crew was unable to bail out due to the fact that none were wearing parachutes, and the buffeting of the plane made it impossible for them to get parachutes on and get out of the ship." Killed in the crash were: 2Lt. James R. Walker, Seattle, Washington, pilot; 2Lt. Albert J. Rich, Hemet, California, co-pilot; 2Lt. Maurice V. Bradley, Hydro, Oklahoma, navigator; Sgt. Hugh P. Jennings, New Bedford, Massachusetts, engineer; Pvt. Kenneth S. Biddinger, Indianapolis, Indiana, engineer; Sgt. H. O. Williams Jr., Belleville, Arkansas, radio operator; Pfc. Charles J. Gneiding, Richfield, New Jersey, gunner; Pvt. Joseph W. Koury, Kansas City, Missouri, gunner.

4-4-42A. Romulus, Michigan. At 1547, a Lockheed A-29 Hudson crashed one mile south of Wayne County Airport, Romulus, Michigan, shortly after take-off, killing pilot 2Lt. Eldon E. Powell and co-pilot Joe G. Frost, civilian ferry pilot. The aircraft was one of six A-29 type airplanes being ferried from the Lockheed factory in California. The Accident Classification Committee stated, "Lt. Powell took off in the fourth position in a flight of Hudsons on an authorized routine ferry flight. The take-off and flight proceeded in a normal manner to a point about one mile off the southeast runway. At this point, with an estimated altitude of 800 feet, a steep bank was made to the left. The nose dropped and a spin followed which continued to the ground resulting in the complete destruction of the ship and the death of the pilot and co-pilot. It is the opinion of the Committee that the pilot in his effort to overtake the flight pulled the turn too tight causing an unintentional stall with the resulting spin."

4-4-42B. Puget Sound, Washington. At 1706, a Lockheed P-38E dove into Puget Sound six miles southwest of Paine Field, Everett, Washington, killing pilot 2Lt. Larry E. Pinson. Investigators could not determine the cause of the crash but speculated that the pilot put the airplane in a diving maneuver from which he could not recover because of the tendency of the P-38 type airplane to suffer from severe control surface buffeting caused by compressibility. The resultant force probably caused an in-flight break-up of the airplane. The aircraft had been flying with another P-38E for high altitude training, but the other pilot reported that Lt. Pinson had been flying behind him and did not appear to be in trouble. He did not see the crash.

4-4-42C. Olympia, Washington. At 1730, a Lockheed P-38E dove into the house of Mrs. Arthur J. Thompson on East 22nd Street, Olympia, Washington, killing pilot 1Lt. Charles E. Heitz. The Thompson home was completely destroyed by the fire that followed the accident. Mrs. Thompson received very minor injuries. Mr. Thompson was not at home at the time of the accident. The Accident Classification Committee stated, "Apparently Lt. Heitz pulled out of a shallow dive at approximately 4,000 feet and at this time momentarily blacked-out. The airplane then started a very steep climb until it stalled off (whipstall). When the ship performed this whipstall, it threw the pilot (still unconscious) forward striking his head on the bulletproof windshield. Lt. Heitz, being rendered unconscious by the black out as well as by striking his head on the windshield gives reasonable explanation where there was no apparent attempt on the part of the pilot to cut the throttle or pull the airplane out of the fatal dive. It is estimated that the airplane hit the ground at approximately 500 miles per hour."

4-5-42A. Ellinger, Texas. At 0910, a Douglas DB-7B crashed while attempting a forced landing at Ellinger, Texas, killing the crew of three. Because of the complete destruction of the airplane and the vague and conflicting testimony of witnesses, investigators were not able to determine the cause of the crash with any amount of certainty. Killed in the crash were the pilot, 2Lt. C. W. Richardson, Del Cardin, Oklahoma, gunner Cpl. J. W. Caddell, Alva, Kentucky, and engineer SSgt. L. R. Attison, Galveston, Texas.

4-5-42B. Sherman Field, Kansas. At 1545, a Douglas O-38E crashed while maneuvering two miles southeast of Sherman Field, Fort Leavenworth, Kansas, killing pilot 2Lt. Harold C. Chandler and observer Pfc. Forrest Wilson. The pilot apparently was engaging in simulated combat with two L-59A type airplanes, entering a Lufbery Circle. The maneuver was begun at an altitude of 1,000 feet with the two L-59 airplanes pulling out at about 500 feet. The O-38 peeled off and dove down in an attempt to gain speed and continued diving into the ground. The aircraft burst into flames upon impact.

4-5-42C. Mines Field, California. At 1355, a Lockheed P-38E dove into the ground after performing a series of aerobatic maneuvers one mile north of Mines Field, Inglewood, California, killing 1Lt. Benjamin H. Bridges, Jr. 1Lt. Andrew Johnson witnessed the crash. He later stated to investigators, "I was attracted by the sound of a P-38 apparently in a dive. I looked up and saw a plane on its back and heading in a northeast direction and in a power dive. I would estimate the altitude as being from 4,000 to 5,000 feet at this time. The angle of the dive was increasing constantly when suddenly from the underside of the ship I saw a parachute snap out and appear to become fouled in the tail assembly. The pilot could be seen trailing behind the ship which by this time was traveling almost straight towards the ground with accelerating speed and both engines opened up. Suddenly the pilot and parachute broke away from the plane and fell clear. The parachute did not open more than a few feet at any time although I would estimate the altitude at which the parachute broke free from the plane at approximately 500-1,000 feet. The plane was almost in a vertical dive and struck the ground several seconds before the pilot. The instant the plane struck, it exploded in a sheet of flames and black smoke." It was speculated that the pilot put the airplane in a diving maneuver from which he could not recover because of the tendency of the P-38 type airplane to suffer from severe control surface buffeting caused by compressibility.

4-5-42D. Gunter Field, Alabama. At 2135, a Vultee BT-13 stalled and spun into the ground near Gunter Field, Montgomery, Alabama, killing RAF Flying Cadet Reginald A. J. Shotbolt. Investigation revealed that the airplane crashed when the student attempted to stretch a glide after suffering an engine failure just after take-off. His instructor, RAF Pilot Officer Nigel H. Cain, survived the crash without injury. He later stated to investigators, "On my first take-off the engine ran good the entire time I was in the air. We had been flying in zone three, in cruising pitch when we came in to land. We put down about 60 degrees of flaps, made a normal landing, gave it the gun, and began to roll flaps up slowly. At about 250 feet, the engine started to miss. I decided to make a left hand turn. The student turned up all flaps. Just about that time the engine cut out altogether. I called to the student to leave 20 degrees of flaps down. We got just under 300 feet and turned back into the field. I did not have enough height and had to tighten the turn. The ship stalled out from under me at about 300 feet. From where I crashed it shows I could not have gone into the field unless I had tightened the turn. When I stalled the left wing dropped. The throttle was back continuously, all the way around, and the landing lights were on as we came down. The last thing I said to the student was to put the landing lights on."

4-6-42A. Somersville, Connecticut. At 1130 EWT, a Lockheed P-322 collided with trees and smashed into the ground on the farm of William Meecham near Somersville, Connecticut, killing pilot 2Lt. Raymond A. Keeney, 24, Somersville, Connecticut. The pilot had been buzzing the Somersville home of his sister, Mrs. Willard F. Pinney, and of his father, Col. R. Leland Keeney, for almost an hour prior to the crash. The pilot had collided with trees at about 50 feet agl in his sister's back yard while performing an impromptu air show on his 24th birthday. The airplane severed the top ten feet of the tree and sections of the port wing and tail section were found on the line of flight from the tree to the crash site. The aircraft went into a steep climbing turn to the left and half rolled into an inverted attitude before diving into the ground about a quarter of a mile away. The aircraft burst into flames upon impact, rolling end over end and scattering wreckage over a large area. The pilot's body was thrown clear of the main wreckage. A civilian had shot motion pictures of the low flying that occurred prior to the crash, which was witnessed by numerous people in the vicinity. Published accounts of the crash reported that Lt. Keeney's wings were restored to him on 4-4-42 after he had been reprimanded for damaging an airplane and unsafe flying that occurred during his previous visit to his home town. It was reported that on the previous visit, Lt. Keeney's airplane had collided with a smokestack and water tower while he was buzzing Somersville Manufacturing Company, where his father was then vice-president and treasurer.

4-6-42B. Tucson, Arizona. At 1515 MWT, a Boeing B-17B suffering the failure of the number-one engine and the failure of the propeller feathering mechanism crashed 14 miles southeast of Davis-Monthan Field, Tucson, Arizona, killing five crewmembers. The co-pilot, 2Lt. Sidney L. Fouts, Santa Rosa, California, and aerial engineer Sgt. William T. Regan, Dunmore, Pennsylvania, both suffered minor injuries parachuting to safety from the stricken airplane. Investigators stated, "Number-one engine failed [for] reason undetermined resulting in excessive vibration. All attempts to feather the propeller failed due to the feathering switches having been burned out while practice feathering at high altitudes. An attempt to freeze the motor by starting resulted in the motor bursting into flames. The violent vibration and the fire caused pilot to decide to abandon ship. The Aerial Engineer and Co-pilot abandoned the ship on pilot's orders. The five remaining crewmembers did not abandon the ship for undetermined reasons although available evidence indicates all were wearing parachutes. The ship crashed from a steep spiral and [burst into flames upon impact]." Lt. Fouts had found himself caught on the horizontal stabilizer during the bailout attempt and freed himself by pulling the ripcord and deploying the parachute. Investigators speculated that a crewmember in the aft part of the ship might have deployed his parachute in the aircraft, preventing some of the crew from exiting. Killed in the crash were: 1Lt. Donald W. Johnson, Dunning, Nebraska, pilot; Sgt. Laurel D. Larsen, Mink Creek, Idaho, radio operator; Pvt. Herman W. Dunn, Miffletown, Pennsylvania, assistant engineer; Pvt. Emerson L. Wallace, Phillipsburg, Pennsylvania, assistant radio operator; Pvt. Leo W. Thomas, Lemoore, California, gunner.

4-8-42A. Bradley Field, Connecticut. At 0853, a Lockheed P-322 crashed just after take-off from Bradley Field, Windsor Locks, Connecticut, killing 2Lt. Philip R. McKevitt, Vinton, Iowa. Bradley Field control tower personnel reported that the airplane had been experiencing trouble with the starboard engine prior to an aborted take-off that resulted in the aircraft running off the right side of the runway. The pilot taxied back to the run-up area, and after making a determination that he could safely fly the airplane he attempted a second take-off. The airplane took off and gained about 150 feet of altitude before the starboard engine cut out altogether. The pilot then successfully executed two 90-degree right turns into the dead engine under gusty wind conditions. On his third turn to make the field, the aircraft nosed down, slipped to the right and crashed into the ground, bursting into flames upon impact.

4-8-42B. Dunlap Field, California. At 1430, a Vultee BT-13A stalled and spun into the ground while turning from base to final at Dunlap Auxiliary Army

Air Field, four miles east of Delano, California, killing instructor 2Lt. William B. Raabe, Seattle, Washington, and student Aviation Cadet Iril Woodward Crowe, Santa Clara, California. Witnesses stated that the airplane went into a spin to the right after entering a shallow flat turn at approximately 500 feet agl. The pilot appeared to have recovered the airplane from the initial stall but then entered a second stall and spin. The airplane dropped into the ground low and flat on the starboard wing and burst into flames upon impact. Investigators speculated that the student was on the controls during the turn from base to final.

4-9-42A. Troy, Alabama. At 0030, a North American AT-6A crashed 10 miles northeast of Troy, Alabama, after losing contact with the leader of its flight while descending through a cloud layer, killing RAF Flying Cadet G. L. Simpson. Investigators determined that the airplane crashed in a diving spiral to the left. The airplane was part of an 80-airplane night formation training flight.

4-9-42B. Randolph Field, Texas. At 1055, a North American BT-9 stalled and spun into the ground six miles northeast of Randolph Field, San Antonio, Texas, killing instructor 2Lt. Max G. Lewis and lightly injuring student Aviation Cadet J.L. Van de Venter, who parachuted to safety. The instructor was demonstrating various power off stalls and deliberately used too much bottom rudder. The airplane went into a spin to the left at an altitude of approximately 4,500 feet from which the instructor could not recover after repeated attempts. He ordered the student to jump but was unable to do so himself. Cadet Van de Venter stated to investigators, "He yelled at me to jump and I immediately pulled the hatch open and unhooked my safety belt and tried to climb out. I could not do this so I pushed myself out with my feet and hands and leaned over the side as much as possible. When I was halfway out, the wind caught me and took me the rest of the way. Something hit me on the back of the head as I left the plane and I immediately pulled my ripcord. When it opened my head was caught in between the four twisted lines and I could not look around. When the weight of my body untwisted the lines I looked around and the plane was already on the ground. I looked around for the instructor but could not see him. When I hit the ground I landed sideways and it knocked the wind out of me and twisted my right knee. The parachute dragged me through a soft plowed field for about 150 feet before I could stop it."

4-9-42C. Bagdad Auxiliary Airfield, California. At 2040, a Beech AT-7 crashed in the Mojave Desert while preparing to land at the Bagdad Auxiliary Airfield, 60 miles east of Barstow, California, killing the five-man crew. The airplane was on a night navigation training flight. Investigators estimated that the airplane entered a spin to the left at about 1,000 feet above ground level. Witnesses reported seeing the red and green wingtip lights spinning around, indicating a spin. The airplane, with landing gear extended, struck left wing tip down with the wings in an almost vertical position and cartwheeled upon impact approximately one and a quarter mile southwest of the field. The bodies were badly mangled and wreckage was scattered over an area of 200 yards by 500 yards. The AT-7 was one of a flight of seven that were navigating from Mather Field, Sacramento, California, to Tucson, Arizona. Killed in the crash were: 1Lt. Richard W. Ford, 25, San Anselmo, California, pilot; 2Lt. William W. Hovey, 24, Challis, Idaho, co-pilot; 2Lt. Byron F. Vandenburg, 22, Sacramento, California, pilot-rated navigator instructor; Aviation Cadet Robert A. Baker, 24, Binghamton, New York, navigation student; Aviation Cadet Leonard H. Ballif, Jr., 21, Portland, Oregon, navigation student.

4-10-42A. McLemore Auxiliary Airfield, Alabama. At 1430, a Vultee BT-13 crashed in an orchard near the edge of McLemore Auxiliary Airfield, Alabama, killing instructor 2Lt. William S. Mudd, Jr. and RAF Flying Cadet Frederick Nash. The student had failed an earlier check ride because he had done poorly in forced landing procedure and chandelle maneuvers and was undergoing additional training with his instructor when the crash occurred. Investigators determined through the testimony of witnesses that the pilots appeared to be practicing forced landing technique but could not determine what caused the fatal spin.

4-10-42B. Condor Field, California. At 1425, two Schweizer TG-2 gliders collided in mid-air five miles NNW of Condor Field, Twenty-Nine Palms, California, killing pilots 2Lt. Earl W. Helm, Jr. (TG-2 #42-8715) and 2Lt. Alexander Konchis (TG-2 #42-8716). The gliders were two of many that were ridge soaring near the field. The two gliders were ridge soaring in close proximity to each other and apparently unaware of each other's presence. The two gliders turned into each other almost head on and contacted each other where the wing joins the fuselage. The two inside wings folded over and the gliders stuck together with both outside wings in the vertical position. The gliders spun approximately five times locked together before striking the ground.

4-10-42C. Lowry Field, Colorado. At 1942, a Curtiss P-40E struck a small grass-cutting tractor while rolling out upon landing at Lowry Field, Denver, Colorado, killing the tractor operator SSgt. James E. Ryan. Pilot 2Lt. Albert H. Little was not injured in the crash. Lt. Little had been experiencing trouble getting his landing gear down prior to landing, and as a result he came in too fast on his landing, causing the P-40 to leave the runway while rolling out. The lawn tractor was 125 feet from the right side of the runway and going in the same direction as that of the airplane. The grass-mower turned left and into the path of the careening P-40, resulting in fatal injuries to the operator.

4-10-42D. Perote, Alabama. At 2230, a Vultee BT-13A spiraled into the ground while leading a formation on a night navigation flight two miles west of Perote, Alabama, killing RAF Flying Cadet Thomas W. Atkin, Middlesborough, England. The flight had taken off from Gunter Field, Alabama, at 2045 and was to fly back to Gunter via Evergreen, Dothan, and Ozark, Alabama. Cadet Atkin had left the formation four times, each time returning and apologizing to RAF instructor Pilot Officer Robert H. Wilson, who was supervising the flight. Instructor Wilson asked the cadet what was wrong but Cadet Atkin just continued to apologize. He broke formation to the right and Instructor Wilson went after him but lost him in the clouds somewhere near Troy, Alabama. Instructor Wilson tried to contact Cadet Atkin several times by radio but was unsuccessful. Investigators speculated that the pilot was suffering from vertigo.

4-11-42A. Jefferson, Georgia. At 0100, a Vultee BT-13A crashed seven miles south of Jefferson, Georgia, killing RAF Flying Cadet J.W. Hawes, 29, Essex, England. The aircraft took off at 2245 from Cochran Field, Macon, Georgia, on a night navigation flight to Waycross, Georgia, and return. Investigators stated, "Inspection of the wreckage failed to show any possible cause for the accident. The weather at that time was good. The visibility was excellent, and the pilot should have been able to see the lights of Cochran Field."

4-11-42B. Tarversville, Georgia. At 2230, a Vultee BT-13A crashed near Tarversville, Georgia, killing RAF Flying Cadet P. W. Winter, 20, Goverton Swansea, Wales. The aircraft had taken off from Cochran Field, Georgia, at 2157 on a night navigation flight to Waycross, Georgia, and return to Cochran Field. The Accident Classification Committee stated, "Inspection of the wreckage revealed absolutely no cause for the accident. The weather was good and the pilot must have been able to see the lights of Cochran Field. There was no possibility that his gas supply was exhausted and nothing to show that there might have been engine or material failure. The committee, therefore, believes that it is impossible to determine the cause of this accident."

4-12-42. Midland, Texas. At 2327, a Beech AT-11 flew into terrain five miles south of Midland Army Air Field, Midland, Texas, killing three fliers. The aircraft was preparing to land after completing a bombardier-training mission. Two AT-11 pilots on the same mission reported seeing the aircraft in a power-on descent while approaching seven miles south of the field. The AT-11 passed beneath the two airplanes and descended into the ground with no apparent attempt to pull up. The landing lights came on momentarily before the aircraft struck the ground and burst into flames. Investigators could not determine why the aircraft descended into the ground. Killed in the crash were: 2Lt. Frank C. Hamilton, 23, St. Paul, Minnesota, pilot; A/C Thomas L. Schade, bombardier cadet; Pfc. Lloyd Le Blanc, Church Point, Louisiana, crewmember. Published accounts later reported that Lt. Hamilton's widow, the former Miss Dorothy Murphy, St. Paul, Minnesota, married Lt. Hamilton's former St. Thomas College classmate, U.S. Navy flying instructor Ensign Francis T. Mahoney, who himself was later killed in a flying accident at Pasco, Washington, on 9-22-42 while giving dual instruction to a student.

4-14-42. Wewahitchka, Florida. At 1005, a Bell P-39F stalled and spun into the ground while trying to make an emergency landing following an apparent engine failure two and one-half miles west of Wewahitchka, Florida, killing Chinese Air Force Sub-Lieutenant Wu Kang. The aircraft struck the ground in a vertical attitude, tumbling end over end until coming to rest inverted and bursting into flames. Florida Park Service personnel quickly extinguished the fire. Investigators determined that the landing gear was in the up position and the flaps partially down.

4-15-42A. Location Unknown. A Boeing B-17E and six army fliers were reported missing and believed lost at sea after their airplane failed to return from a navigation-training mission in the vicinity of the Gulf of Mexico. The B-17 had taken off with five other B-17 aircraft from Hendricks Field, Sebring, Florida, at 0328 EWT on a graduation navigation flight to Brownsville, Texas, and return. The B-17 had been having radio transmission problems and was last reported at 27N/83W at 0534 and carried full survival gear and enough fuel to keep it airborne until 2045. On 4-19-42, a fishing schooner operating 60 miles SSW of Edgemont Key found a B-17E type bomb bay fuel tank, and on 4-20-42 a B-17E type life raft was found 30 miles off shore of Tampa, Florida. The Accident Classification Committee could not definitely determine with certainty that this equipment came from the missing B-17, but did note that information indicated that a bomb bay fuel tank was lost by a B-17E assigned to Sarasota, Florida, and that a life raft was lost by a Hendricks Field B-17E on 4-17-42 while searching for the lost B-17 crew. The AAF, the U.S. Navy, the U.S. Coast Guard, the Mexican Air Force, American Airways and the Civil Air Patrol participated in a search that lasted ten days and covered the Gulf of Mexico, Texas, Mexico, Florida, and the Atlantic Ocean, but no trace of the aircraft or crew was found. Weather was not considered a factor in the disappearance and the five other B-17s landed at Hendricks Field without incident. Missing and presumed killed were: Capt. Thomas C. Kennington, B-17 instructor pilot; 2Lt. Robert G. Gross, pilot; 2Lt. Earl W. Clifton, co-pilot; 2Lt. James R. Rizzolo, navigator; Sgt. Daniel J. Maloney, engineer; SSgt. Phillip B. Wenk, radio operator.

4-15-42B. Nash Field, Oklahoma. At 1940, a Vultee BT-13A stalled and spun into the ground four miles southwest of Nash, Oklahoma, killing instructor

2Lt. Dean B. Lewis, 24, Lafayette, Indiana, and student Aviation Cadet Robert C. Tracy, 23, Seattle, Washington. The airplane was on a night training mission. 1Lt. P. B. Smyth stated to the Accident Classification Committee, "I was out about 4 or 5 miles [from Enid Army Flying School] when, glancing back, I noticed a ship quite some distance back, coming up on my left. Not wanting him to catch me I swung off to the right. In about 10 minutes he had caught me and was on my right wing. We flew formation for a few minutes. Then I cut my gun completely, letting him shoot past me. I then pulled in about 1,000 yards behind and about 200 to 300 feet below him. We stayed that way for about five minutes, between 800 and 1,000 feet. Suddenly, he pulled up. I pulled up to about 1,300 feet, still about 300 or 400 feet below him. We flew that relative position for about five minutes. He made a couple of shallow turns, apparently looking for me. Then he started into a shallow dive; I would say about a 50-degree angle dive. I also started a dive, about the same angle. When my altimeter showed about 700 feet, he steepened his dive abruptly; went straight down. At that time, I reduced my speed to see what he intended doing. He made an abrupt pull up, straight in front of me. I then cut my gun completely. He came up over the top of me. It looked like he stalled at the top. He made about a three quarter turn of a spin before going out of sight behind my horizontal stabilizer. I didn't actually see the plane strike the ground. Apparently he stalled while on his back. I would say he was attempting a loop or an Immelmann."

4-15-42C. Charlotte, Texas. At 2300, a North American BC-1 dove into the ground 10 miles west of Charlotte, Texas, killing A/C Elzie L. Mize. Investigators speculated that the student might have become disoriented after losing the horizon. 2Lt. John C. Hall, who was leading the element of three that Cadet Mize was flying with, stated to investigators, "We were at 3,500 feet flying straight and level, and Cadet Mize pulled up to an altitude of about 200 or 300 feet above us, crossed over to the right, and came down barely missing my number-two man, and continued in a dive until he hit the ground."

4-16-42A. Potomac River, Virginia. At 1118, a Douglas DB-7B crashed into the Potomac River at its mouth six miles north of Burgess Store, Virginia, killing three fliers. Investigation revealed the crash occurred when the subject airplane attempted an Immelmann type maneuver while in simulated combat with three U.S. Navy Douglas SBD type airplanes. The subject airplane was part of a flight of three DB-7B airplanes that were on a familiarization flight. The flight leader decided to engage in simulated combat with the navy airplanes. Both formations began circling with each other at an altitude of about 1,500 feet with the army airplanes on the outside of the circle. The number-three DB-7 pulled up in a half loop, trying to roll out at the top. The subject airplane stalled during the maneuver and spun to the left and into the water. The airplane and the bodies of the crew could not be located and have not been recovered. Killed in the crash were: 2Lt. Ralph C. Graves, pilot; TSgt. Norman G. Simmons, radio operator; SSgt. Julius Young, bombardier.

4-16-42B. MacDill Field, Florida. At 1212, a Lockheed A-29B crashed while trying to take off from MacDill Field, Tampa, Florida, killing the crew of seven. The aircraft, piloted by the commanding officer of the 3rd Mapping Squadron, Maj. James H. C. Houston, was one of a flight of three A-29s taking off on a secret mission overseas. The airplane became airborne about half way down the runway in a tail-heavy attitude. Witnesses reported that the engines were going "full blast" and that the tail wheel did not leave the ground until the rest of the airplane was in the air. The airplane then climbed steeply until it stalled about 150 feet above ground level, falling off to the right about 90 degrees. The airplane recovered momentarily before stalling again and falling off to the left, turning about 270 degrees before striking the ground nose first and bursting into flames. Investigation revealed that all of the airplanes in the flight were at least 600 pounds overweight and that the center of gravity of the two remaining airplanes was two feet aft of the allowable limit. All of the airplanes were fitted with an unauthorized wooden rack in the dorsal turret that held four large non-airplane type oxygen cylinders that Maj. Houston insisted that he would need at his destination. Additionally, each airplane carried two large non-airplane type oxygen cylinders in the aft gun tub. Engineering officers at MacDill Field had refused Maj. Houston's request to install the oxygen cylinders and racks in the airplanes, citing the dangerous tail-heavy weight distribution and the opinion that the non-regulation oxygen cylinders posed a safety risk. Maj. Houston declared that he would install the racks himself and that he did not need the engineering staff's help to complete his mission. Data does not indicate the squadron's destination. Killed in the crash were: Maj. James H.C. Houston, pilot; 1Lt. Floyd Messer, co-pilot; TSgt. Sam F. Riegger, engineer; Sgt. Edgar F. Matthews, radio operator; TSgt. Malcolm R. Mathis, passenger; TSgt. Charles Mull, passenger; Cpl. Stephen F. Wysocki, passenger. The two other A-29s, which did not take-off, were to be piloted by 1Lt. Robert H. Payne and 1Lt. Albert M. Welsh. The two pilots were ordered to abort their scheduled mission after the crash.

4-17-42A. Lemoore Field, California. At 0750, a Vultee BT-13A dove into the ground eight miles southwest of Lemoore Army Flying School, Lemoore, California, killing Aviation Cadet Sherman H. Smith. Investigators stated, "The airplane was first seen at an altitude of approximately 600 feet and was diving at about a 30 degree angle with the motor full on. Neither the angle of the dive nor the heading of the airplane changed until it crashed with terrific force."

4-17-42B. Moody Field, Georgia. At 1115, two Cessna AT-17 airplanes collided in mid-air four and one-half miles southeast of Moody Field, Valdosta, Georgia, killing the four crewmembers aboard both airplanes. A/C Howard C. Berger and A/C Sammy S. Burrell were killed in the crash of AT-17 #42-159; A/C Lloyd B. Davis and A/C Richard Dexter, Jr. were killed in the crash of AT-17 #42-140. The aircraft were part of the first of two flights of seven-airplanes on a formation flying training mission. The flights were flying in two V-formations at an altitude of about 3,000 feet in light rain showers when the accident occurred. The leader of the second flight, 2Lt. Philip R. Burling, stated to investigators, "The airplane in #2 position [A/C Berger] apparently lost its position and started dropping gradually, slowly downward and back, when it became involved in the accident with the airplane in #3 position [A/C Davis]. Just as the #3 airplane was about to collide it seemed to have an upward swing to it and it was below the #2 airplane. It appears that the pilot had pulled back on the control column. The #3 airplane hit the bottom of the fuselage of the #2 airplane, apparently with the left propeller. The #3 airplane evidently thought the #2 airplane would pull up into position but he failed to do so. I saw the tail severed and the #2 airplane was in a climbing position with the tail cut off when I pulled out of the formation and lost sight of both airplanes. We were in the clear [of rain] for approximately two minutes before the accident occurred."

4-17-42C. Ballinger, Texas. At 1358, a Fairchild PT-19A stalled and spun into the ground while turning from base to final at the AAF Auxiliary Airfield seven miles northwest of Ballinger, Texas, killing Aviation Cadet William W. Pearce.

4-17-42D. Victorville, California. At 1705, a Vultee BT-15 suddenly plunged to the ground near El Mirage Dry Lake, 12 miles northwest of Victorville, California, killing pilot 1Lt. Albert B. Edwards, Jr., 29, Albuquerque, New Mexico, and passenger 2Lt. Curtis E. Bjorklund, 24, Salt Lake City, Utah. Witnesses reported seeing the airplane flying straight and level at an approximate altitude of 600 feet above ground level when the airplane suddenly rolled to the left and on its back, completed a partial recovery before entering a second stall and slamming into the ground with the nose and left wing. The aircraft burst into flames immediately upon impact.

4-18-42A. Tallahassee, Florida. At 1425, a Bell P-39F spun into the ground four miles north of Tallahassee, Florida, killing pilot Chinese Army Sub-Lt. Chian Chung Liang. The sole witness to the accident, 2Lt. A. F. Hennings, reported to investigators that while driving his automobile he saw the P-39 spin in from an altitude of approximately 2,000 feet in a nose up attitude. The pilot evidently made no visible effort to recover the airplane and made no attempt to use his parachute. Lt. Liang was undergoing training under the direction of the AAF.

4-18-42B. Agate, Colorado. At 1745, a Cessna AT-17 collided with terrain in poor weather conditions 16 miles WSW of Agate, Colorado, killing pilot 2Lt. Richard G. Rosebush and co-pilot 2Lt. John L. Rencher. The AT-17, which was on a navigation training flight, lost the leader in the clouds and was trying to let down through the overcast for a landing at Lowry Field, Denver, Colorado, when the collision occurred. The other AT-17, piloted by Capt. E. W. Best, landed safely at Lowry Field.

4-18-42C. San Diego, California. At 1715, two Lockheed P-38F airplanes collided in mid-air over North Island, San Diego, California, killing 1Lt. John F. Dixon (41-2333) and 2Lt. Richard D. Meer (41-2324). The aircraft had taken off from North Island Naval Air Station at 1700 for a high altitude-training mission. Witnesses on the ground reported that one P-38 ascended into and collided with the tail of the other. The aircraft were seen to fall from an altitude of approximately 7,000 feet by numerous U.S. Marines who were training in the area. Lt. Dixon's tail-less aircraft slammed into the ground inverted at an angle of about 45 degrees one-quarter of a mile east of Camp Kearney. The P-38 that was piloted by Lt. Meer straightened out momentarily before slamming into the ground at high speed and bursting into flames five miles north of Camp Kearney.

4-20-42A. Phoenix, Louisiana. At an unknown time after 1407 CWT, a Bell P-39F on an aerobatic practice flight crashed into a swamp 10 miles northeast of Phoenix, Louisiana, killing pilot 2Lt. Sidney L. Sims, Jr. The P-39 took off from New Orleans Army Air Base, Louisiana, at 1407 and was seen flying at about 8,000 feet just prior to the accident. The aircraft hit the swamp in a vertical dive with tremendous force. The airplane had been missing and had enough fuel to keep it airborne until approximately 1610 CWT. After that time, an extensive aerial search was launched before the crash site had been tentatively located in an extremely remote and inaccessible location at 1630 on 4-23-42. The aircraft wreckage and the pilot's remains were buried and submerged in about 25 feet of inaccessible swamp and were not recovered. A piece of human skull was found in the vicinity of the crash site leading investigators to speculate that the pilot had remained with the airplane. Investigators speculated that the pilot lost control of the airplane while performing an aerobatic maneuver and crashed into the swamp.

4-20-42B. Passmore, Alabama. At 2240, a North American AT-6A flew into the ground while trying to land at the AAF Auxiliary Airfield at Passmore, Alabama, killing the RAF Flying Cadet R.A. Parry. The cadet, attempting to land with excessive airspeed, had overshot the field and was instructed by the tower to go around. In his attempt to make his go-around, he made a right hand turn against traffic and then flew the airplane into the ground. Investigators determined that there had been no engine failure or mechanical problems that could account for the crash and they speculated that the cadet might have become disoriented and allowed the airplane to fly into the ground.

4-21-42A. Spence Field, Georgia. At 0035, a North American AT-6A dove into the ground two and one-half miles north of Spence Field, Moultrie, Georgia, killing Aviation Cadet Allen J. Perkins, Jr. The airplane was involved in tactical formation flying with several other planes. Investigators stated, "It is the opinion of this board that the student lost the horizon when he pulled out of the formation and dove straight into the ground. The airplane fell in dense woods and did not touch any of the trees around. There was no evidence of a spin. There were no witnesses who saw the crash. The plane had to be searched for and Air Corps officials were the first to the scene. The plane was damaged so bad that no information could be gathered from the scene."

4-21-42B. Pope Field, North Carolina. At 0930, a Douglas C-50A crashed six miles NNW of Pope Field, Fort Bragg, North Carolina, killing the two pilots and seriously injuring four crewmembers. The Accident Classification Committee stated, "It is the opinion of the Committee that failure of the left engine occurred while the pilot was making left circles of small radius at low altitude, approximately 800 feet, at the direction of the officers of the Airborne Command who were conducting air-to-air and ground-to-air radio communication equipment tests. These maneuvers were viewed from Pope Field, and the ship was observed to make approximately one complete left turn with the wing down before disappearing from sight. Statement of civilian witnesses were to the effect that noise from both engines stopped before the crash, which indicated that right engine was throttled in effort to regain control. Main ignition switch was found to be jammed in the open position, indicating that the pilot cut main switch before hitting the ground." The co-pilot, 2Lt. Frank B. Bragg, was killed instantly and the pilot, 2Lt. Barton B. Hewitt, Euclaire, Wisconsin, died of his injuries 4-22-42. Seriously injured in the crash were: Sgt. A. Monhof, crew chief; Maj. L. A. Walsh, observer; SSgt. H. A. Weidman, radio operator; Sgt. P. Adams, radio operator.

4-21-42C. Lake Wales, Florida. At approximately 1610, a Curtiss P-40E crashed 10 miles northeast of Lake Wales, Florida, killing pilot 2Lt. Irvin B. Hillman. Witnesses indicated that the airplane had spun at least once with the pilot apparently recovering before entering into a second spin and crashing into the ground. Despite and extensive aerial and land search, the aircraft and pilot were missing for eight days until they were discovered 4-29-42 by three civilians, Red Zipprer, W. C. Zipprer, and Oscar Mims. Investigators could not determine the cause of the accident.

4-21-42D. Fort Devens, Massachusetts. At 1955, a Curtiss 0-52 stalled and spun to the ground while preparing to land at Fort Devens Army Air Field, Massachusetts, killing pilot 2Lt. Robert W. Booker and aerial observer 1Lt. Gerald P. Kennedy. Investigators stated, "Pilot upon returning from a local reconnaissance mission lowered the landing gear and then made a complete circle of the landing field to the left. He momentarily leveled off and then started an abrupt second turn to the left. His altitude at this time is estimated to have been 500 feet above the ground. Observer's stated that the nose dropped from this turn and the ship started a spin to the left. About a quarter turn was completed before crashing into four feet of water at the edge of a small pond."

4-22-42A. Harding Field, Louisiana. At 1415, two Curtiss P-40 airplanes collided in mid-air while participating in combat training exercises five miles west of Harding Field, Baton Rouge, Louisiana, killing 2Lt. Ferdinand B. Paris, 24, Houston, Texas, aboard P-40 #39-254. 2Lt. Wesley F. Mink, Boise, Idaho, parachuted to safety from P-40 #40-304. Investigators speculated that the topside of Lt. Paris' fuselage and his right wing struck the bottom of Lt. Mink's airplane, probably knocking Lt. Paris unconscious or killing him. After the accident Lt. Mink stated to the Accident Classification Committee, "I was leading the third element of a six-ship flight. We were in a slight turn to the left when we were simulating attacking the squadron commander from above and head on. The first and third elements turned to the left, as we had decided upon beforehand. The second element turned to the right. The attacking plane was on the tail of the first element and my element was on his tail. The second element, Lt. Paris leading, climbed to the right, crossed over and dove on the attacking plane. He either did not see my element or misjudged his distance. There was quite an impact as the two ships came together. My ship went out of control momentarily and when I righted it, level flight could be maintained only with the stick clear back and considerable right rudder. I stayed with it for several minutes and decided it was definitely not airworthy. I bailed out, without injuring myself."

4-22-42B. Cimarron, New Mexico. At 2029, a Consolidated B-24D collided with terrain at the 10,100-foot level of 10,500-foot Trail Peak 20 miles southwest of Cimarron, New Mexico, killing the crew of seven. The B-24 had been missing while flying a navigation training flight from Kansas City, Missouri, to Albuquerque, New Mexico. The wreckage and bodies were not located until 5-2-42. A severe storm with 100 mile per hour winds and dangerous turbulence prevailed at the time of the B-24's disappearance. Investigators speculated that the number-four engine was not operating at the time of impact. The pilot had radioed Albuquerque that he couldn't get under the storm and that he was going on instruments at about 2011. Investigators stated, "After passing Las Vegas, New Mexico, the pilot encountered heavy weather over the Sangre De Cristo Mountains east of Santa Fe, New Mexico. The committee inspected the wreckage at the scene of the crash and it was evident that only three motors were operating under power at the point of impact. The airplane struck [the mountain] in almost level flight with a possibility of the left wing slightly low at the time of impact, flying at a true altitude of abut 10,100 feet on a true heading of about 80 degrees. At the time of impact the gear was up and the speed of the plane about 300 feet per second. The slope of Trail Peak was estimated at approximately thirty-five percent grade. The distance from the trees, which were broken off at the first point of impact to the terrain on the hill itself was approximately 60 feet. At this point there were five indentations on the ground, one for each motor and the nose. The plane commenced disintegrating at this spot and wreckage was strewn up to the point where it finally came to rest, approximately 250 feet further up the slope. The main structure of the airplane came to rest pointed approximately true north with the center section broken off where the left wing joins the fuselage and the entire nose section entirely demolished to the rear of the flight deck. The rear fuselage and the empennage were in approximately the correct position behind the main structure but were upside down. Two engines were approximately 100 feet ahead of the main plane, one was on the very brow of Trail Peak about 70 feet beyond the first two and one was completely disintegrated and parts scattered across the top of the peak and down the slope away from the wreck. Oxygen masks were found scattered through the wreckage and the general vicinity of the bodies in such a way that it is believed the crewmembers were using oxygen. All members of the crew had parachutes on, adjusted and fastened. The bomb bay doors were open and it is believed that preparations were being made to abandon the plane. One smashed Arco wristwatch was found in the wreckage of the flight deck [that] had all hands straight and had stopped at [2029]." Killed in the crash were civilian employees of TWA being used to help train Army Air Forces personnel at the four-engine school at Kirtland Army Air Base, Albuquerque, New Mexico. Killed in the crash were: Robert Redding, pilot and TWA multi-engine instructor, Minatare, Nebraska; Jonas G. Ruff, TWA co-pilot, San Jose, California; 2Lt. Charles O. Reynard, observer pilot, Hiram, Ohio; 2Lt. Roland L. Jefferies, observer pilot, Kansas City, Missouri; Cpl. Philip E. MacComber, crewmember, Pendleton Field, Oregon; Cpl. Duane Peterson, crewmember, Salt Lake City, Utah; George E. Van Hoozer, TWA aerial engineer instructor, Kansas City, Missouri.

4-23-42A. Blanchard, Oklahoma. At 0955, a Douglas A-20C stalled and spun into the ground four miles west of Blanchard, Oklahoma, killing 2Lt. Sherman W. Long, 21, Patterson, California, one day short of his 22nd birthday. Investigators determined that the airplane spun in from an altitude of about 2,500 feet while attempting a steep turn to the right. 2Lt. William J. Taylor witnessed the crash and stated to investigators, "I sighted Lt. Long and flew past him. He cut in behind me and followed. I attempted to lose him. After several minutes, he was in the lead. I looked at the clock and saw that it was time for me to be back at the field; so, I cut away from him. As I turned, I lost sight of him. I leveled off and looked back just as he went into the spin. He spun down to the ground and burst into flames. I don't know how he went into the spin, unless he made a steep bank in an attempt to follow me back to the field."

4-23-42B. Bakersfield, California. At 0814, a Vultee BT-13A stalled and spun into the ground 10 miles west of Bakersfield, California, killing students A/C Robert D. Murphy, Los Angeles, California, and A/C Edward Ortenburger, Bay City, Michigan. Aviation Cadet John N. Week, who witnessed the crash, stated to investigators that he saw the airplane about 300 feet above him in a spin at an altitude of about 2,600 feet and watched the airplane until it struck the ground.

4-23-42C. Spokane, Washington. At 1012, a Lockheed F-4 (the photo-reconnaissance version of the P-38) crashed at Hogan Street near Fairfield Avenue, Spokane, Washington, killing pilot 2Lt. Milton E. Connelly, Chicago, Illinois. The airplane had just taken off from Felts Field, Spokane, Washington, on a routine training mission. Investigators stated, "Statements indicate that this airplane had been given a daily inspection but the engines had not been warmed up. Lt. Connelly was informed of this and he, in turn, informed the crew chief that he would make the pre-flight. The pilot had difficulty in starting both engines. The pilot then obtained taxi clearance at 1005 from the control tower and proceeded to the east end of the field, whereupon he received take-off instructions at 1009. According to available testimony of witnesses, the pilot failed to run up the engines or check magnetos or hold the airplane in accordance with approved method of take-off. The pilot then took off and, after a long roll, gained an altitude of approximately 75 feet at the west end of the airport. The flight then continued westward for a distance of approximately two and a half miles, at which time a right turn was made, after which the accident occurred. The time of flight was approximately three minutes and was performed at a low altitude. At no time during the flight was it apparent that the aircraft was performing at rated power output. The landing gear was retracted soon after take-off. However, available testimony indicates that the wheels were lowered just prior to [the crash], indicating that the pilot was attempting to make an emergency forced landing. Just prior to contact with the ground, the right wing of the aircraft came in contact with a telephone pole. The aircraft then struck a small wooden shed and continued through another telephone pole, traveled approximately 50 feet further and crashed [into the garage of John Hirsch, N3004 Hogan] immediately bursting into flames. Two garages, a 1936 Chevrolet sedan and a boat were completely destroyed."

4-23-42D. Moore Field, Texas. At 1600, a North American AT-6A and an AT-6B collided in mid-air eight miles southeast of Moore Field, Mission, Texas, killing Aviation Cadet Douglas E. McCassey aboard AT-6A #41-698. Aviation Cadet Philip O. Works, Jr. parachuted to safety from AT-6B #41-17133. The airplanes were conducting simulated attacks under the supervision of flying instructors with several other airplanes in their class when Cadet Works and Cadet McCassey collided. Investigators stated, "Cadet McCassey was making his attack on Cadet Work's airplane and the right wing of Cadet McCassey's airplane struck the rear part of the fuselage and part of the empennage of Cadet Work's airplane, tearing off all affected parts. Cadet McCassey's airplane went into a cloud immediately and no one saw him make any attempt to jump. Cadet Works' airplane went out of control, he nosed over, and when he unfastened the safety belt he was thrown clear. He opened his parachute and landed without injury. Apparently Cadet McCassey pulled the rip cord of his parachute prior to getting out of the airplane and that the resistance of his body dragged the parachute out of the cockpit at a fairly slow rate. This is shown by the canopy railing grease marks on the shroud lines and on the silk of the parachute near the rim, and also by a piece of silk which tore loose and remained in the cockpit. The parachute also entangled with the tail wheel and a piece of silk was torn off and lodged there. Two whole panels of the parachute were torn loose and never found. The pilot chute was torn loose and fell free. It is believed that Cadet McCassey jumped at an altitude of more than 2,000 feet, but fell fairly close behind the rudder of the airplane, being held there by the parachute, which was dragging out of the cockpit. His body struck [the ground] 30 feet from the airplane."

4-23-42E. Burleson, Texas. At 1620, a North American AT-6B slammed into the ground in a vertical attitude after inadvertently entering a severe thunderstorm four miles southeast of Burleson, Texas, killing pilot 2Lt. Benjamin F. Slack and passenger Sgt. F.T. McLaughlin. Investigators could not determine exactly why the airplane crashed. It was noted that Lt. Slack was a low time pilot and not experienced enough to fly on instruments in poor weather.

4-23-42F. Leesburg, Georgia. At 1820, a North American AT-6A crashed while performing low altitude aerobatics over Leesburg Auxiliary Airfield, Leesburg, Georgia, killing pilot 2Lt. Martin N. Martino, who died of his injuries about 20 minutes after the accident. Lt. Martino had flown to Leesburg on an instrument training flight with 2Lt. Douglas Michel who stated to investigators, "At approximately 1720 we departed from Turner Field [Albany, Georgia] and climbed to an altitude of 5,000 feet. At this altitude Lt. Martino flew instruments under the hood. He flew on instruments for 55 minutes and then we landed at Leesburg Auxiliary Field. We both got out of the airplane and Lt. Martino said he wanted to go up alone. He took off at 1820, circled and came flying across the field at about 400 feet altitude, and at a speed of approximately 160 to 180 MPH. He pulled the nose up and proceeded to do a slow roll, whereupon, the nose of the airplane dropped and instead of rolling out smoothly the airplane barrel rolled and crashed flat, right side up, into the ground."

4-25-42A. Moore Field, Texas. At 0945, two Fairchild PT-19A airplanes collided in mid-air while participating in supervised individual aerial combat training four miles west of Moore Field, Mission, Texas, killing A/C John H. Simmons, 23, Fulton, South Dakota, aboard PT-19A #41-20484. A/C Richard J. Shipway, Jr. parachuted to safety from PT-19 #41-20483. Although these students were flying primary trainers, they were advanced students very close to graduation. Investigators stated, "2Lt. Edward S. Bippus was supervising individual combat practice between two of his students at an altitude of about 4,800 feet, about four miles west of Moore Field. The students had had between two and three hours of combat practice previously. This combat practice was being carefully supervised. After a breakaway from a head-on pass Cadet Shipway was in a chandelle to the right, and Cadet Simmons completed a turn and flew so as to cause his right wing to collide with the tail surfaces of Cadet Shipway's airplane. All affected parts were knocked off and fell free. Both airplanes went out of control, Cadet Simmons' airplane started to roll, and Cadet Shipway's airplane went into an inverted spin. Cadet Shipway used the parachute successfully. Cadet Simmons either made no attempt to jump, or he jumped too late for the parachute to function."

4-25-42B. Tehachapi, California. At 1045, a Vultee BT-13A collided with rising terrain in deteriorating weather conditions four miles west of Tehachapi, California, killing pilot 2Lt. Donald C. Graybill and passenger 2Lt. Harry J. Brown. Investigators speculated that the pilot tried to maintain visual contact with the ground while trying to fly through Tehachapi Pass. The airplane was at an altitude and on a course other than that for which it had clearance. The airplane had taken off from Lemoore Army Air Field, California, at 0929 on a flight described as "Official business of Finance Officer at Santa Ana."

4-26-42A. Ocklockonee Bay, Florida. At 1105, a Bell P-39D dove into Ocklockonee Bay, Florida, after failing to pull out of a simulated diving attack on a North American BT-14 type airplane, killing pilot 2Lt. Walter J. Suckling. 2Lt. James R. Willis stated to investigators, "We were flying formation at about 9,000 feet for about 15 minutes. Lt. Suckling was leading and I was flying on his right wing. After doing some general formation flying we peeled off and dropped down to about 2,000 feet. At a lower altitude of about 300 feet there was a BT-14, which we dove past. At the time we passed it I was in trail at about 100 feet behind him. After passing the BT-14 we chandelled to about 1,000 feet. At this point I fell behind about a mile. While climbing to about 1,000 feet we were turning to the left for about 180 degrees. From there Lt. Suckling's airplane was in a shallow dive with wings level when contact was made with the water."

4-26-42B. Kansas City, Missouri. At 1105, a North American B-25D crashed into a railroad switchyard after suffering the failure of its port engine on take-off from Kansas City Municipal Airport, Kansas City, Missouri, killing the civilian crew of five. The aircraft cartwheeled across the yard, bursting into flames immediately upon impact with the ground, scattering wreckage over a wide area and barely missing a line of boxcars. A North American Aviation Company flight crew was testing the B-25 for the AAF. Investigators stated, "Immediately after take-off, the pilot encountered difficulties, which were evidenced by the fact that the landing gear was not retracted and a flat turn to left was executed with very little altitude. At no time did the airplane attain more than 200 feet altitude and, in general, it was mushing or in a semi-stall condition. After crossing the Missouri River and the northern boundaries of the Kansas City Municipal Airport, a left turn resulting in approximately 80 degree of bank was made and immediately thereafter the airplane crashed on the railroad tracks of the Chicago, Burlington, and Quincy Railroad." The North American Aviation Company flight crew killed in the crash were: Raymond Quick, 44, pilot; Oran B. Sims, 38, co-pilot; Joseph R. Peacher, 26, engineer; Edward R. Krakora, 21, engineer; Charles F. Frederick, engineer; all of Kansas City, Missouri. Investigators for the Army Air Forces and North American Aviation Company could not determine with certainty the cause of the engine failure.

4-26-42C. Pacific Ocean, California. A Lockheed P-38E on a gunnery training mission with three other P-38 airplanes crashed into the Pacific Ocean five miles north of Point Reyes, California, killing pilot 2Lt. Wilbert C. Moss. None of the other three pilots actually saw Lt. Moss crash. An unidentified witness reported the accident to have occurred at 1836. A thorough search of the area revealed no trace of the airplane or the pilot and they were never recovered.

4-28-42A. Whitesboro, Texas. At 0900, a Vultee BT-13A flew into terrain at Whitesboro, Texas, after entering very poor weather and losing sight of the formation it was flying with, killing instructor 2Lt. Peter E. Richards and student Aviation Cadet Raymond E. Miller, Jr. The aircraft was reported to have tried to climb above the weather. There were no witnesses to the crash.

4-28-42B. Ocala, Florida. At 1720, a Stearman PT-17 spun into the ground 12 miles southeast of Taylor Field, Ocala, Florida, killing student pilot Aviation Cadet R. N. Cottrell. The student had been practicing spins in the area prior to the crash. There were no witnesses to the crash.

4-28-42C. Lawson Field, Georgia. At 1740, a Republic P-43D performing aerobatics spun into the ground four miles northwest of Lawson Field, Fort Benning, Georgia, killing 2Lt. Walter J. Schmidt. The P-43 was on a scheduled aerobatic practice flight. Trees surrounding the crash site were not damaged or shorn indicating that the aircraft slammed into the ground in a vertical attitude. The pilot had lowered the flaps and landing gear in an attempt to recover the airplane. The pilot attempted to parachute from the airplane at an altitude of approximately 70 feet above ground level, striking the ground about 40 feet from the airplane.

4-29-42A. Smyer, Texas. At 0310, a North American AT-6A on a night formation flight with two other AT-6A airplanes crashed three and one-half miles west of Smyer, Texas, killing Aviation Cadet Haynes G. Huggins, 23, Lockhart, Alabama. The student had been flying the number-two position in the formation when he suddenly left the formation and passed beneath and in front of his wingmen. The airplane appeared to be under control when it disappeared from sight. The Accident Classification Committee could not determine exactly why the airplane crashed or why the pilot left the formation.

4-29-42B. Orlando, Florida. At 1040, a Boeing B-17E and a Curtiss P-40E collided in flight nine miles west of Orlando, Florida, while participating in simulated combat, killing six crewmembers aboard the B-17 and the pilot of the P-40, 2Lt. Robert S. Boyce, 23, Chillicothe, Ohio. The B-17 was "attacked" by two P-40 airplanes piloted by Lt. Boyce and 2Lt. Charles J. Rosenblatt. Lt. Rosenblatt stated to the Accident Classification Committee, "[Lt. Boyce and I] took off about 1000 together in formation, climbed to 10,000 feet, flying in formation all the time. We spotted a B-17 about 9,000 feet about 1020. We then attacked them and in a few minutes 2Lt. Richard A. DeEulis and 2Lt. McCauley Clark came up and attacked also. When the accident occurred we were at 10,000 feet. Lt. Boyce and myself were making a head-on attack. I was about one-quarter of a mile behind him making a slight bank. He hit the right outboard engine and went over the top of the wing and burst into flames immediately. The B-17 pitched up and immediately went into a spin. I followed the P-40 down a few thousand feet. Afterwards I flew over to where the B-17 had crashed. I came back to the field and landed at about 1100." The B-17 fell into an orange grove at the Agriculture Sub-Tropical Fruit Research Station just off of Hiawassee Road and the P-40 smashed into the ground near Hoequist Field about two miles away. Killed in the crash of the B-17 were: 2Lt. Ernest W. Robertson, 25, Eugene, Oregon, pilot; 2Lt. Joseph S. Smith, 24, Yoakum, Texas, co-pilot; 2Lt. R. R. Carnevale, Columbus, Ohio, observer; Pvt. Leon A. Gerhart, Wernersville, Pennsylvania, engineer; Pvt. Eugene Wimp, Holland, Michigan, radio-operator; Pvt. Marvin C. Brown, Detroit, Michigan, gunner. The Accident Classification Committee noted in its official report, "Attacks on B-17s by P-40s have been daily occurrences since the middle of April. That a percentage of mistakes in judgment apparently must occur, with fatal results incident to the hazards of aviation training."

4-30-42. Santa Catalina, California. At 0710, a Lockheed P-38F suffering an apparent failure of the port engine crashed while attempting an emergency landing about one mile west of the Santa Catalina Island Air Field, California, killing pilot 2Lt. Walter F. Lichtenberger, Decatur, Illinois. The P-38 was one of a flight of five that was attempting to land at the field. The other four P-38 airplanes passed over the field at low altitude with the fifth falling behind. U.S. Navy Chief Boatswains Mate W. K. Stevenson witnessed the crash and stated to investigators, "A flight of five P-38 airplanes flew over the field. I noticed that one of the planes was considerably behind the rest of them. It finally dropped very low and went out of sight. The airplane attempted a landing on a hill and I went over to the edge of the field and saw the ship on the hill. A small fire had started and the airplane had come to a complete stop. I immediately went to the scene of the crash, arriving in about three minutes. The pilot was dead when I had arrived."

May

5-1-42A. Trumbull Field, Connecticut. At 1120, a Curtiss P-40E crashed while engaged in simulated combat with two other P-40 airplanes two miles north of Trumbull Field, Groton, Connecticut, killing 2Lt. Kenneth R. MacQuarrie, 25, New York, New York. The crash set fire to approximately six acres of woodland. Investigators stated, "The committee saw the accident occur. The pilot in attempting to shake off pursuers in individual combat executed a chandelle from an altitude of about 100 feet. On the top of the maneuver the plane was pulled through until the nose was well down and the plane 10 to 20 degrees past a vertical bank at not over 150 mph. The pilot, instead of rolling out toward the direction of least inclination, attempted to pull the plane through in a partial half-roll. On realizing his altitude was too low, he attempted to apply more elevator control with visible dropping of the left wing just above stalling with the plane reaching a horizontal position on the bottom of the attempted maneuver but mushing on down to strike the ground in a timbered area exploding as it hit."

5-1-42B. Fresno, California. At 1430, a Cessna AT-17 on an instrument training flight crashed three miles east of Fresno, Texas, killing instructor 2Lt. Hays Brantley, Jr., 22, Memphis, Tennessee, and student Aviation Cadet Willard Fairchild, 22, New York, New York. The AT-17 was part of a two-ship formation that entered a dense cloud layer at about 3,000 feet. The pilot of the other AT-17 did not see the airplane again after they entered the clouds. Witnesses on the ground saw the AT-17 spinning to earth with no obvious attempt by the pilots to recover the airplane.

5-2-42A. Wichita, Kansas. At approximately 0850, a Cessna AT-17 on an Army acceptance flight crashed into a house and garage at 907 Litchfield Street, Wichita, Kansas, killing pilot 2Lt. Finley K. Steele, San Angelo, Texas. The airplane had taken off from the Cessna factory airfield at 0835 and according to witnesses on the ground had executed a series of steep turns at low altitude just prior to the accident. The house belonging to Mr. R.E. Watson was slightly damaged but his garage containing a 1935 Ford V-8 and a 1929 Ford Town Sedan was completely destroyed.

5-2-42B. Brownsville, Texas. At 1055, a Lockheed A-29B crashed while taking off from Brownsville Municipal Airport, Brownsville, Texas, killing four fliers. Killed in the crash were: 1Lt. John H. Dougherty, pilot; 2Lt. Warren R. Gray, co-pilot; Sgt. William M. Allen, radio operator; TSgt. Louis J. Schwab, engineer. The aircraft was one of two airplanes of the 4th Mapping Squadron taking off for their home field at Laredo, Texas. 1Lt. Frank N. Graves, who was leading the flight, stated to investigators, "After take-off, I made a gentle climbing turn to the left, to allow Lt. Dougherty to take-off and join me in the air. At the time I first saw Lt. Dougherty's airplane, he was approximately one-third of the way down the runway making his take-off. I glanced at my instruments and upon looking back, saw that Lt. Dougherty's airplane was approximately 50 feet in the air, banking steeply to the left. Lt. Dougherty's airplane continued in an easterly direction with the left wing extremely low and was losing altitude at this time. I saw the left wing tip hit the ground and the airplane seemed to make a cartwheel on the ground, the rear section of the fuselage breaking off just behind the wing section. The airplane at this time burst into flames."

5-2-42C. Garwood, Texas. At 2110, a Curtiss O-52 stalled and spun into the ground eight miles west of Garwood, Texas, killing pilot 2Lt. William L. Bloomgren and observer 1Lt. William H. Boyce. The aircraft was on a night navigation training flight and was seen to tumble out of the sky and crash to the ground by Mr. E. E. Treybig, Garwood, Texas. Mr. Treybig discovered the wreck and the bodies the next morning on the property of Mrs. Charlie Leher. The two fliers had been killed instantly. The Accident Classification Committee speculated that the pilot might have suffered vertigo after losing the horizon while making a turn around "Light No. 11," which was the objective of the navigation mission. The pilot then pulled the airplane up into a stall and was unable to recover the airplane because of vertigo and spatial disorientation. The airplane had taken off from Brooks Field, Texas, at 2020.

5-2-42D. Seguin, Texas. At 2115, a North American AT-6A crashed in a remote area 13 miles south of Seguin, Texas, killing A/C Frank C. Rosendorf. The airplane had taken off from Brooks Field, Texas, at 2030 on a night navigation flight. Investigators speculated that the student might have flown into some low-hanging clouds that had been reported in the area, causing him to lose the horizon and suffer spatial disorientation. The student apparently inadvertently flew the airplane into the ground in a power-on spiral.

5-3-42A. Aberdeen, Mississippi. At 1500, a Cessna AT-17 crashed after suffering an in-flight failure of the leading edge of its port wing five mile south of Aberdeen, Mississippi, killing pilot 2Lt. Gerald L. Gray, Tonkawa, Oklahoma, and co-pilot 2Lt. Leroy Chapman, Kilgore, Texas. Pieces of fabric, ribs and plywood from the leading edge of the wing were found approximately two miles from the scene of the main wreckage. The airplane did not burn but was smashed to bits upon impact.

5-3-42B. Meacham, Oregon. At 1305, a Boeing B-17E collided with rising terrain while trying to let down through clouds and snowstorms near Meacham, Oregon, killing seven fliers and seriously injuring TSgt. Wesley E. Wallace, Philadelphia, Pennsylvania, who miraculously survived. Although seriously injured, Sgt. Wallace struggled overland for three miles through the woods of Blue Mountain to get help. Sgt. John H. Starr, Littleton, Massachusetts, also survived the crash only to die of his injuries a short time later. The B-17 was preparing to land at Pendleton Field, Pendleton, Oregon, 24 miles to the northwest. Investigators noted that the pilot had begun his let down over a particularly dangerous area and that the weather had closed in on the area a lot faster than originally forecasted. Investigation revealed a radio problem plaguing these early model B-17s. Investigators stated, "Army aircraft are seriously handicapped in instrument flying, particularly in making an instrument let-down due to the fact that the radio is almost invariably lost due to static as soon as the airplane goes on instruments. The instrument let-down procedures are based on the proper functioning of the airplane's radio. The airplane cannot work a let-down procedure without the use of the radio; yet the present type receivers fail due to static at the very time they are most needed. Since it was snowing at the scene at the time of the accident, it is very likely that the pilot in this fatal accident did not have full use of his radio receiver." Killed in the crash were: 1Lt. Joseph T. H. Laycock, 24, Baton Rouge, Louisiana, pilot (see 2-6-41A); 2Lt. James M. Brown, Pleasantville, Louisiana, co-pilot; SSgt. Harry G. Cameron, Medford, Massachusetts, engineer; Sgt. William A. Forrest, Salisbury, North Carolina, radio operator; SSgt. Richard A. Carmichael, Winewood, Pennsylvania; Cpl. Chester A. Nowak, Alpena, Michigan.

5-3-42C. Pleasant Valley, Oregon. At 1520, a Lockheed A-29B collided with terrain near Lindsey Peak, one and a half miles from Pleasant Valley, Oregon, killing the crew of six. The aircraft, which had taken off from Gowen Field, Boise, Idaho, at 1505 on a ferry mission, had encountered a heavy snowstorm mixed with fog at the time of the accident. Numerous civilians in the area heard the aircraft impact the mountain and reported that visibility was very poor with snow falling at the time. Investigators speculated that the pilot might have tried to turn around after encountering the poor weather. The aircraft exploded into flaming pieces upon impact, scattering wreckage and badly mangled bodies over an area of a quarter of a mile by 150 yards. The aircraft struck the ridge about 100 feet from the top. Among the wreckage was a copy of *Through the Overcast, The Weather and the Art of Instrument Flying*, a book on instrument flying by Assen Jordanoff. The aircraft was missing until 5-5-42, when it was discovered by two Union Pacific Railroad track foremen working in the area. Killed in the crash were: Capt. Harold E. Hammers, Spokane, Washington, pilot; 1Lt. Wilbert P. Matthews, Floydada, Texas, co-pilot; TSgt. Emmett C. Wheaton, Los Angeles, California, engineer; SSgt. R. O. Massingill, South Bend, Indiana, engineer; Sgt. John J. Carpenter, El Paso, Texas, radio operator; Cpl. Carroll B. Rettstatt, Rawlins, Wyoming, assistant radio operator.

5-3-42D. McChord Field, Washington. At 1630, a Lockheed A-29 crashed on take-off at McChord Field, Tacoma, Washington, killing four fliers. The pilot was on a check ride for first pilot on A-29 type airplanes. Investigators stated, "The airplane started on its take-off run and approximately half-way down the runway the ship veered to the left about 45 degrees. The pilot then applied full throttle and full right rudder and the airplane swung back to the right and took to the air with the right wing down. The airplane reached an altitude of about 50 feet, stalled, and crashed to the ground. The airplane hurtled along for approximately 200 feet scattering debris along its path. About 30 seconds later Sgt. [Randolph T.] Correll stumbled from the flaming wreckage with his clothing on fire." Sgt. Correll, Perryton, Texas, later died of his injuries. Killed in the crash were: 1Lt. William E. Comber, Frankfort, Pennsylvania, pilot; 2Lt. Adolph T. Frerichs, Fairfax, Missouri, co-pilot; Cpl. Frank W. Erwin, San Benito, Texas, radio operator. Investigators noted that the installation of the large dorsal turret might be creating turbulent airflow around the tail control surfaces and speculated that the physical shortness of the pilot might have prevented him from regaining control of the airplane.

5-4-42. Las Vegas, Nevada. At 1650, a Douglas A-33 (similar to the A-24 type dive bomber) dove into the ground 10 miles northwest of Las Vegas, Nevada, killing pilot 2Lt. Ares A. Smith and passenger 2Lt. George A. Ragas. Pilot 2Lt. George H. Johnston, who had witnessed the crash, stated to investigators, "I glanced out of the left side of my airplane and saw the A-33 heading almost straight down for the ground at an estimated altitude of 250 to 300 feet. The airplane crashed a moment later and burst into flames."

5-5-42. Mount Calm, Texas. At 1400, a Vultee BT-13A crashed while practicing stall-spin maneuvers three and one-half miles west of Mount Calm, Texas, killing instructor pilot 2Lt. William J. Fraley, Jr., High Point, North Carolina. Aviation Cadet George L. Kittle, Ferndale, Michigan, parachuted to safety, suffering minor injuries. Cadet Kittle stated to investigators, "[Lt. Fraley] demonstrated various types of stalls, namely, power on and off, characteristic and stick back. After completing the stalls he explained that we were to do a two turn spin to the right. He stalled the ship, kicked right rudder and the ship fell into a normal entry. He started the recovery at one and one-half turns kicking full left rudder and popped the stick forward. All this time I was following through on the controls. The airplane did not respond at all in my opinion and we kept spinning to the right. Soon the stick started moving back and fourth in the cockpit and it seemed as if Lt. Fraley was trying another recovery. The controls were moving so fast that I lost track and began to realize we were in trouble. About this time I noticed the altimeter reading. It was about 2,000 feet. Right after that Lt. Fraley shouted, 'Get out of here' over the interphone system and I immediately opened the canopy and tried to get out of the airplane. Because of the motion of the airplane and the slipstream, I was thrown back and forth in the cockpit and made two or three attempts to leave before I finally got out. As close as I can remember I was either half over the side of the cockpit or on the wing when the chute opened. The main canopy caught the leading edge and tip of the elevator and I hung there for moment before being thrown free. I then looked down and the ground was directly below me. I landed in a field directly adjacent to the crash and about 1,000 feet distant. I immediately unbuckled my harness, leaving my chute in the field and ran over to the ship. I looked in and saw Lt. Fraley and realized that there wasn't anything I could do for him."

5-6-42A. Lawson Field, Georgia. At 0910, Army paratrooper Sgt. Bryson Cash was killed instantly when he was struck by the rotating propeller of a Vultee L-1A that was taxiing off the flight line at Lawson Field, Fort Benning, Georgia. Apparently the trooper had fallen behind and was trying to catch up with his stick when he crossed the path of the moving aircraft. Investigators noted that a small dummy parachutist that Sgt. Cash was carrying might have blocked his vision to the side, preventing him from seeing the airplane. There was a Douglas A-20 and several other airplanes in the vicinity that were warming up their engines, probably making it impossible for Sgt. Cash to hear the L-1, which was manned by pilot 2Lt. Louis R. Burke and observer 1Lt. Daniel F. Hynes.

5-6-42B. Byron, Georgia. At 1415, a Vultee BT-13A stalled and spun into the ground one mile northeast of Byron, Georgia, killing instructor 2Lt. John B. Hoffman, 23, Columbus, Ohio, and RAF Flying Cadet Geoffrey Holmes, 20, Wakefield, England. The aircraft, which earlier had been flying formation with two other BT-13s, was reported to be doing steep climbing turns at an altitude of approximately 1,500 feet when it entered a stall and spun into the ground, bursting into flames upon impact.

5-6-42C. Muroc, California. At 1430, a Lockheed P-38E was abandoned by its pilot near Muroc, California, when he could not recover the airplane from a steep dive. SSgt. Rudolph C. Dear was killed when he bailed out at about 2,500 feet and his parachute failed to open completely. The pilot had been maneuvering in the area with other sergeant pilots flying P-38 airplanes. Witnesses on the ground observed the airplane at about 5,000 feet in a power dive of about 80 degrees from which it did not recover.

5-6-42D. Hope, Texas. At 2245, a North American AT-6B crashed five miles south of Hope, Texas, killing Aviation Cadet Richard E. Jacques. Investigators speculated that the pilot entered a bank of clouds and because of vertigo was unable to maintain control of the airplane, striking the ground in a spiral. The pilot might have seen the ground just before impact and applied full throttle to attempt a recovery.

5-6-42E. Location Unknown. At an unknown time after 1914 EWT, a Boeing B-17E disappeared and was presumed to have crashed in the Gulf of Mexico with its crew of ten. The B-17 had not been seen or heard from since it had taken off from MacDill Field, Tampa, Florida, at 1914 EWT on a local navigation training flight. The B-17 had enough fuel to allow it to remain airborne until 0330. Numerous search efforts were made between 5-7-42 and 5-11-42, covering the Gulf of Mexico, the Florida peninsula, the coast of Mexico and the western coasts of Cuba with no trace of aircraft or crew being found. Missing and presumed lost at sea were: 2Lt. Hugh W. Millis, 23, Houston, Texas, pilot; 2Lt. Charles E. Thomas III, 23, Macon, Georgia, co-pilot; SSgt. Wilbert George, 23, Tampa, Florida, engineer; Pvt. Elmer E. Mathauser, 23, Burwell, Nebraska, asst. engineer; Pvt. Charles R. Phillips, 20, Waukesha, Wisconsin, radio operator; Pvt. Raymond S. Lynch, 20, Lumpkin, Georgia, asst. engineer; Pvt. Robert I. Pigmon, 22, Berea, Kentucky, asst. engineer; Pvt. Melton L. Teal, 21, Dunkie, Louisiana, assistant radio operator; Pvt. Gordon J. Howard, 21, Green Bay, Wisconsin, crewman; Sgt. Eugene L. Kasmer, 22, Chicago, Illinois, crewman.

5-7-42A. Gunter Field, Alabama. At 0020, a Vultee BT-13A crashed into trees after attempting a take-off in the wrong direction at Gunter Field, Montgomery, Alabama, killing RAF Flying Cadet Oswald H. McDonald. The Gunter Field control officer had called the airplane in to the field so that it could be fueled up. After the fuel stop, the student was given clearance to take-off so he could complete his air work. A witness on the ground stated that the student taxied to the end of the runway, ran up the engine, and then took off straight into the trees.

5-7-42B. Gunter Field, Alabama. At 0140, a Vultee BT-13 crashed while maneuvering five miles southeast of Gunter Field, Montgomery, Alabama, killing RAF Pilot Officer George A. Butler and seriously injuring RAF Flying Cadet Lawton L. Morgan. Investigators determined that the plane was engaging in simulated aerial combat with another BT-13 when it went into a diving attitude and continued in this attitude until the it struck the ground. This dog fighting was reported to have occurred at an altitude of approximately 3,000 to 1,000 feet.

5-7-42C. Esler Field, Louisiana. At 1505, a North American B-25A crashed one and a half miles northeast of Esler Field, Alexandria, Louisiana, killing the crew of six instantly. The airplane was seen to be at an altitude of approximately 1,000 feet before it began a spin to the right and smashed into the ground, bursting into flames immediately upon impact. Investigators could not determine exactly what caused the airplane to crash. Killed in the crash were: 2Lt. Merton W. Ortmann, 23, Clinton, Wisconsin, pilot; 2Lt. Edward N. Walters, 23, Odebolt, Iowa, co-pilot; 2Lt. George S. Edwards, 25, Crawford, Texas, observer; 2Lt. Thomas S. Gilliam, 27, Appomattox County, Virginia, observer; 2Lt. Charles T. Willock, 25, Winthrop, Massachusetts, bombardier; SSgt. Thomas R. Numbers, 26, Glendale, California, crewman.

5-7-42D. Perrin Field, Texas. At 2325, two Vultee BT-13A aircraft collided in flight while in the traffic pattern for an attempted landing at Perrin Field, Sherman, Texas, killing student pilot Pvt. Gus A. Woynofsky aboard BT-13A #41-10553. Pfc. Kenneth M. Silverman, who stated that he had never seen Pvt. Woynofsky's airplane, landed safely in BT-13A #41-10836 at Perrin Field. He was uninjured. Investigators noted that the weather was clear, with ceiling and visibility unlimited and that there were three airplanes on the base leg of the pattern just before the accident.

5-8-42A. Phoenix, Arizona. At 0800, a civilian Howard type airplane (NC-22400) and a North American AT-6A collided in mid-air at Sky Harbor Air Park, Phoenix, Arizona, killing two pilots while a third pilot parachuted to safety. The AT-6, which had taken off from Luke Field on an instrument practice flight, struck the ground at 18th Avenue and Mojave Street, immediately bursting into flames upon impact. The Howard aircraft, which was en route to McChord Field, Washington, and was attached to the AAF Ferry Command, slammed to earth at 12th Avenue and Henshaw Road. Civilian ferry pilot Wray B. Smith, Atlanta, Georgia, was killed when he began a climbing turn to the left and collided with the AT-6. 2Lt. John K. Austin was unable to parachute from the AT-6A and was killed in the crash. 2Lt. Jack O. Scott parachuted to safety from the AT-6A. Investigators could not determine why Lt. Austin was unable to leave the airplane. The collision occurred at an altitude of approximately 5,000 feet and hundreds of Phoenix residents observed the airplanes breaking up and falling to earth.

5-8-42B. Randolph Field, Texas. At 1345, a North American BT-9B stalled and spun while turning from base to final at Randolph Field, San Antonio, Texas, killing Aviation Cadet Nicolas E. LeCompte. According to investigators, the student pulled the nose of the airplane up, lost flying speed and attempted a turn that resulted in a spin to the right. The student was unable to recover the airplane in time to prevent the crash.

✓ **5-8-42C. Santa Maria, California.** At 1431, a Stearman PT-17 airplane stalled and spun into the ground five miles southeast of Santa Maria, California, killing student pilot Cpl. Gordon M. Short. Investigators speculated that the student stalled out of a maneuver while in inverted flight, fell into an inverted spin and was unable to recover the airplane before colliding with the ground.

5-8-42D. Puget Sound, Washington. At 1529, a Douglas A-20A stalled and spun into Puget Sound about 1,000 yards off shore of the south end of Ketron Island, Washington, killing three fliers. The A-20 had taken off from McChord Field, Tacoma, Washington, at 1500 on a local flight. AAF pilot 2Lt. Jack W. Oliver witnessed the crash and stated to investigators, "I was cruising along about 2,000 feet on a local transition flight in a P-40 type pursuit airplane. My attention was attracted by a plane, which was about 4 miles away in a nose-down spin. The plane, when I first sighted it, was about 3,000 feet altitude and already in a spin. Apparently no attempt was made by the pilot to recover from the spin, as the attitude of the plane did not change from the time that I saw it until it struck the Bay about one and a half miles south of Ketron Island." Killed in the crash were: Capt. Angus C. B. MacPhee, pilot; Lt. Col. Perry O. Huff, passenger; Capt. Raymond C. Alexander, passenger.

5-9-42A. Will Rogers Field, Oklahoma. At 0254, a Douglas A-20C stalled and spun into the ground one mile north of Will Rogers Field, Oklahoma City, Oklahoma, killing pilot 2Lt. Raymond E. Miller, Jr., 23, Temple, Texas. The airplane was on a local night training flight when the pilot encountered unknown difficulties while preparing to land at the field. The pilot's attempt to inform the tower of his troubles was unsuccessful because of radio transmission problems. A witness in the control tower stated that the pilot had overshot on his turn from base leg to final approach. This necessitated that the pilot S-turn the airplane to bring it back in line with the runway.

5-9-42B. Craig Field, Alabama. At 0710, a North American AT-6A landed on top of another AT-6A that had inadvertently taxied onto the runway at Craig Field, Selma, Alabama, killing RAF Flying Cadet Ernest G. Gulliver. RAF Flying Cadet Leslie B. Cameron received minor injuries and instructor 2Lt, Joseph E. Couerette was uninjured in the accident.

Cadet Cameron and his instructor were coming in for a landing in AT-6A #41-207, and both testified that the runway was clear when they made their turn from base to final. At that time the tower alerted all of the airplanes waiting to take-off to watch out for inbound traffic and that no take-offs were permitted until Cadet Cameron's AT-6 landed. At about this time, Cadet Gulliver taxied AT-6A #41-16631 out onto the runway. Cadet Cameron's airplane then landed on top of Cadet Gulliver's AT-6 in a perfect three-point touch down, the propeller striking the cockpit area. Both airplanes stuck together as they careened down the runway for approximately 100 yards before coming to rest. There was no fire. It was unknown why the cadet taxied out onto the runway against orders.

5-9-42C. Augusta, Georgia. At 1348, a Vultee BT-15 crashed while maneuvering 15 miles south of Augusta, Georgia, killing civilian flight instructor Henry J. Dohm and student Cpl. Lewis E. Ward. Dr. J.R. Wainwright, Augusta, Georgia, witnessed the crash and stated to investigators, "[We were] driving south on U.S. Highway 25. While crossing Spirit Creek and going up the hill we saw an airplane flying toward us at an angle, or in a direction about northeast, and at about five or six hundred foot altitude. The plane banked to the right and started a slow right turn, as if to fly south over the road we were on. After about 180 degrees of the turn had been made, without appreciable loss of altitude, the right wing dipped sharply and the ship slipped or rolled to the right past the vertical and dived sharply. It came out of the right roll and went at once into a left roll, still diving steeply, as if trying to pick up speed to come out of it. Just as the wings seemed to be getting a grip on the air, the angle was lessened, the ship struck into the ground at about a 45 degree angle and burst instantly into flames. We parked [at the scene of the crash] in less than half a minute. The ship was resting in about a 3 to 4 foot crater, wings level, as if it struck that way. The body in the front cockpit was resting in a position forward and left, while the one in the rear was forward and right and high, as if the top housing had stopped it. Both were badly burned and charred. The ship struck with such force that it was obvious they must have been killed on impact and instantly."

5-9-42D. Moody Field, Georgia. At 1950, two Cessna AT-17 type aircraft collided in mid-air seven miles northwest of Moody Field, Valdosta, Georgia, killing four fliers. Pilot 2Lt. John E. Harris and co-pilot 2Lt. Gresham P. Redden were killed in the crash of AT-17 #42-108; pilot 2Lt. James H. Harper and co-pilot 2Lt. Jack Prinz, Jr. were killed in the crash of AT-17 #42-223. The airplanes were part of a three-ship formation training flight. The aircraft had completed the training mission and were seen maneuvering in the area just prior to the crash. A witness stated that the airplanes were chasing each other around when one airplane ran into the tail of the other. The AT-17 piloted by Lt. Harris and Lt. Redden immediately slammed into the ground in an open field 75 feet north of Battle Creek Road, three-quarters of a mile from the Valdosta/Nashville cutoff. The AT-17 piloted by Lt. Harper and Lt. Prinz leveled off and momentarily held a northwesterly course before spinning to the ground and exploding into flames upon impact. Investigators noted that the design of AT-17 airplanes does not allow for the quick exit of the crew in the event of trouble.

5-10-42A. Atlantic Ocean, New York. A Lockheed A-29 on a coastal patrol mission of the Atlantic Ocean failed to return to base and was reported missing and presumed lost at sea with its crew of five. The A-29 had taken off from Mitchel Field, Long Island, New York, at 1115 and was thought to have crashed soon after. Weather at the time of take-off was reported as generally fair with a fog bank along the Long Island coast extending up to 500 feet altitude. Visibility was reported to be three to five miles in haze with a winds west at five knots. A search of the sea in the Long Island area on 5-10-42 revealed an oil slick and debris evidence of an airplane crash about 40 miles east of Long Island. Parachutes, seat cushions and unopened life rafts of the type used on A-29 airplanes were observed on the surface by rescue aircraft. Missing and presumed lost at sea were: 2Lt. Kenneth A. Woodle, pilot; 2Lt. Jay W. Myers, co-pilot; 2Lt. Adam B. Kalmanowicz, navigator; Pfc. William P. Rutherford, radio operator; Pvt. Curtis Bausher, gunner.

5-10-42B. Morris Field, North Carolina. At 1605, a Curtiss P-40F stalled and spun into Morris Field, Charlotte, North Carolina, killing 2Lt. Harold D. Hays. Investigators stated, "The pilot overshot the field on his first landing attempt. Loss of full engine power caused the pilot to attempt an emergency landing from an impossible position after overshooting. In attempting to execute a 270-degree left turn back into the field, wheels and flaps extended, from an altitude of approximately 200 feet, the airplane stalled out of the last tight turn.... [Inexperience] caused the pilot to overshoot the field on the first attempt and an error in judgment [the turn back to the field] caused the fatal crash." Witnesses on the ground observed a thin streak of white smoke flash by the tail section when the pilot gave the P-40 full throttle for the go-around. The airplane did not respond to full throttle and when the pilot attempted to tighten his turn, the P-40 fell into a half-turn spin to the left, striking the ground in a vertical attitude, bounding over a fence and bursting into flames.

5-10-42C. Dale Mabry Field, Florida. At 1605, a Bell P-39D crashed into a wooded area three miles WSW of Dale Mabry Field, Tallahassee, Florida, while attempting a go around after overshooting the field, killing Chinese Army Sub-Lt. Chen Yen-Chien. Investigation revealed that the aircraft had run out of fuel. Chinese Army pilot Sub-Lt. Chang Chin-Lo, who was flying the same training mission, reported that the P-39 had taken a wrong turn at Madison, Florida, but had arrived at Dale Mabry Field ahead of him anyways. Investigators speculated that the pilot had used excessive gasoline in his attempt to get back to the field after making the wrong turn. The airplane sheared off the tops of several trees before striking the ground nose first, coming to rest upside down. The pilot's body had been thrown clear and was found lying about 20 feet beyond the main wreckage still strapped to the seat.

5-11-42A. Chef Menteur, Louisiana. At 1220, a Bell P-39D crashed into a swamp after performing a series of aerobatics four miles southeast of Chef Menteur, Louisiana, killing 2Lt. George D. Marine. Investigators stated, "After careful investigation the board is unable to determine the cause of the accident. The probable cause was a stall in acrobatic flight from too low an altitude to effect recovery."

5-11-42B. Shaw Field, South Carolina. At 1405, a Vultee BT-13A spun into the ground 10 miles southwest of Shaw Field, Sumter, South Carolina, killing A/C Leroy J. Franke. The student had been practicing spins and apparently attempted to parachute to safety after being unable to recover the airplane from an intentional spin. He attempted his jump from approximately 125 feet, striking the ground about 150 feet from where the airplane hit. The parachute did not open in the air but burst open upon impact with the ground. A/C Thomas E. Smith witnessed the crash and stated to investigators, "I was clearing myself for a stall when I noticed this other plane on my left. I was making a left turn at the time. I glanced back and noticed that he was in a spin. I stopped my turn and started watching him spin. He spun about four times and I thought he was coming out of it and then he continued to spin. He spun down about 500 or 1,000 feet of the ground. The spin stopped and the plane dove into the ground. It burst into flames immediately. He was about 5,000 feet from the ground when [the spin] started."

5-11-42C. Hunter Field, Georgia. At 1717, a Douglas DB-7B crashed while attempting to take-off from Hunter Field, Savannah, Georgia, killing 2Lt. Monte L. Crawford. Investigators stated, "Pilot received adequate pre-flight instruction and was properly supervised. He had made one take-off and one landing, but on his second take-off run, it is believed he [inadvertently] applied brake pressure throughout his take-off run. As a result of this continued application of brakes, the airplane failed to take-off and struck soft sand at the end of the runway, bounced into the air, stalled, and crashed. The ship never left the ground until it hit dirt at the end of the runway. It spun up on its side, rose up approximately 30 feet in a very sharp bank. It sank quickly and tore one wing off. When the fuselage hit the ground it immediately burst into flames and rolled over. The plane remained a flaming wreck for 40 minutes."

5-11-42D. Waukomis, Oklahoma. A Vultee BT-13A crashed under unknown circumstances at a time unknown six miles ESE of Waukomis, Oklahoma, killing Aviation Cadet Robert F. Shoup, 26, Butler, Pennsylvania. The airplane had taken off from the Enid Army Flying School, Enid, Oklahoma, at 2330 on a night navigation flight and was missing until 0800 the next morning when it was found crashed in a wheat field owned by Mr. V. B. Beard. Investigators could not determine the cause or nature of the accident. Weather was reported as contact and not considered a factor in the accident.

5-12-42A. Breckenridge, Oklahoma. At 0845, a Vultee BT-13A stalled and spun into the ground while making a turn from base to final at Breckenridge Auxiliary Airfield, Breckenridge, Oklahoma, killing AAF instructor 2Lt. Odie B. Fox, Jr., 25, Purcell, Oklahoma, and A/C William E. Halstead, 24, Denver, Colorado. Witnesses stated that the airplane was banked steeply before going into a one and a half turn spin.

5-12-42B. Blanchard, Oklahoma. At 1035, two Douglas A-20C aircraft collided in mid-air five miles east of Blanchard, Oklahoma, killing four fliers and injuring two others. The airplanes had taken off from Will Rogers Field, Oklahoma City, Oklahoma, at 1020 and were on a three-airplane formation-training mission. Killed in the number-two A-20C (41-19189) were 2Lt. Chester P. Toler, 22, Grants Pass, Oregon, pilot; Pvt. James D. Davis, 18, East Bend, North Carolina, gunner; Pvt. Samuel W. Dye, 32, Ogden, Utah, gunner. Killed in the number-three A-20C (41-19223) was Pvt. John R. Pellish, Jr., 19, Minesville, Pennsylvania, gunner; pilot 2Lt. Victor E. Gualtiere, 23, Atlantic City, New Jersey, and gunner Cpl. Louis N. Killeen, 21, Phoenix, Arizona, were lightly injured parachuting to safety. Cpl. Killeen stated to investigators, "The flight started out in a V-formation but changed to echelon formation with Lt. Gualtiere's plane in the number-three position, Lt. Toler's plane in number-two position, and [1Lt. A. M.] Thompson's plane in front. The change to the echelon formation had been made and the planes were flying in that position for several minutes. Lt. Toler's plane did not get in its proper position but remained behind, although there appeared to be plenty of room between the first and last positions. Lt. Thompson's plane was far ahead. First, Lt. Toler's plane approached ours from above and it seemed to me that the planes would collide. However, they drew apart again. Then Lt. Toler's plane came down toward us a second time, from behind. I started to inform the pilot by means of the interphone, as I did not believe he could see the other plane, but never finished this. First the rudder of our plane cut into the wing of the other. Then the propellers of Lt. Toler's plane cut into our tail. Apparently something from the plane hit me in the head and I do not remember exactly what happened for a time. The next thing I knew there was nothing left behind me of the tail and our plane was falling. I stepped to the edge and jumped. There was no trouble in getting out.

I have been told that my safety belt was broken but I remember unfastening it. After jumping, I saw the planes below me and it seemed there were other parachutes in the air. After a short time I landed in a tree, remaining there until help came."

5-12-42C. Paine Field, Washington. At 1545, a Lockheed C-40A crashed while attempting a landing at Paine Field, Everett, Washington, killing five fliers. The airplane crashed one-quarter of a mile west of the field. Killed in the crash were: 2Lt. Bailey B. Smalling, Altus, Oklahoma, pilot; 2Lt. Howard F. Porter, Indiana, Pennsylvania, co-pilot; 2Lt. John P. Lynch, Carlinville, Illinois, pilot observer; 2Lt. William H. Robinson, Jr., Coral Gables, Florida, pilot observer; Cpl. John L. Zurcher, Everett, Washington, engineer. Investigators stated, "Lt. Smalling was flying the plane from the right side and giving transition training to Lt. Porter in the pilot's seat. Upon contact with the runway, the left landing gear gave way allowing the left propeller to strike the runway. The airplane then veered to the left and the pilot took off in a left turn and continued to climb until at an altitude of 400 to 500 feet. Just west of the field the airplane fell off on the left wing making about one-half turn of a spin. It is believed that when the propeller struck the runway, the left engine was made ineffective and with this reduced amount of thrust, too steep a climb was attempted with a spin resulting." The pilot had made five landings and was attempting a sixth when the crash occurred. The airplane hit the ground in a near vertical dive with flaps down at 45 degrees and landing gear in the down position.

5-13-42A. Sacation, Arizona. At 0100, a North American AT-6A collided with terrain three miles southwest of Sacaton, Arizona, killing A/C Garland A. Dishman, 22, Okmulgee, Oklahoma. The airplane apparently flew into the ground under control. Investigators speculated that the student had lost the horizon or had fallen asleep, causing him to fly the airplane into the ground. Wreckage was scattered over a large area and the cadet's body was dismembered and badly mangled. The airplane was on a night navigation flight from Luke Field, Phoenix, Arizona, to Tucson, Arizona, and return. Weather was reported as clear with unlimited visibility and not considered a factor in the accident. Eighty-six other student training airplanes that made the same flight that evening encountered no trouble and safely completed the mission.

5-13-42B. Camp Shelby, Mississippi. At 1115, a Piper L-4A stalled and spun into the ground at Camp Shelby, Mississippi, seriously injuring Liaison Pilot TSgt. Nick J. Vlass and killing observer/passenger Capt. Harlan Moore. The airplane was on an artillery reconnaissance training flight. The Accident Classification Committee stated, "The pilot was flying at an altitude lower than 100 feet and in attempting to climb and make a turn to the left, apparently stalled the ship. The plane, with power on, fell off on the left side and struck the ground, slight left wing tip first. Switches in safety belt in rear cockpit came loose, resulting in death of Capt. Moore, who was thrown out by the force of the impact." The pilot maintains that the Captain's leaning forward in his seat had interfered with the controls and that the Captain unintentionally retarded the throttle when he moved his left elbow upon leaning back to a normal position.

5-13-42C. Eglin Field, Target Range P4, Florida. At 1145 CWT, a North American AT-6B on a gunnery training mission flew into the ground at Eglin Field Target Range P4, Florida, killing instructor pilot 2Lt. Jones R. Mulford and seriously injuring student A/C C.C. Tisdale, Jr. Investigators speculated that Lt. Mulford had misjudged his distance to the target and altitude while demonstrating gunnery to his student and became fixated on the target and consequently flew the airplane directly into the target.

5-13-42D. Bolling Field, Washington, D.C. At 2020, a Curtiss P-40F stalled and spun into the ground after suffering an apparent engine failure near Bolling Field, Washington, D.C., killing 2Lt. James Kriss, 25, Minneapolis, Minnesota. Maj. Norman L. Peterson witnessed the crash and stated to investigators, "I heard a P-40 taking off with no indication of any lack of power. Suddenly, I heard the engine quit completely. There was no sputtering of the engine before or after the sudden stoppage. I immediately looked out of the window and saw the airplane about 100 feet high and passing over Westover Avenue just south of Quarters No. 74. At that time, the airplane was banked slightly to the left. The pilot ceased to turn to the left and continued with the nose slightly high in the same attitude of flight for about 300 feet. By this time, the airplane had perceptibly lost speed. Suddenly, dropping the nose slightly, a fairly short turn to the right was started. The airplane had turned about 90 degrees, under control, when the nose dropped sharply, and the airplane made one full turn of a spin before striking the ground."

5-14-42A. Cimarron Field, Oklahoma. At 0925, a Fairchild PT-19A crashed in an inverted spin five miles west of Cimarron Field, Oklahoma, killing Aviation Cadet Ignatius A. Pelletere, 22, New Orleans, Louisiana. Civilian flying instructor L. L. McMillon witnessed the crash and later stated to investigators, "I was practicing pylon eights with [my student] and just as we rolled out to level flying position I saw [Pelletere's] airplane in an inverted position at a maximum of 100 feet altitude. The nose looked to be down at about a 15 degree or 20 degree angle, and the ship was rolling and turning as it hit the ground inverted."

5-14-42B. Poteau, Oklahoma. At 1030, a Douglas A-20C collided with rising terrain after encountering instrument flying conditions while on a pilotage navigation training flight 15 miles west of Poteau, Oklahoma, killing the crew of three. The airplane was seen by witnesses on the ground to be flying just under the clouds at an altitude of about 700 feet above ground level. There were no witnesses to the actual crash of the airplane. Investigators speculated that the airplane entered instrument conditions and either the pilot suffered disorientation or attempted a left turn at low altitude to get out of the weather, causing the airplane to fly into the ground. Killed in the crash were: 2Lt. Harry E. Wells, 24, Raymond, Washington, pilot; Cpl. Fred E. Richardson, 26, Crab Orchard, Nebraska, crewmember; Cpl. David S. Spargo, 23, Hot Springs, Arkansas, crewmember.

5-14-42C. Lubbock, Texas. At 1630, two Curtiss AT-9 airplanes collided in flight while flying in a three-ship formation training flight 11 miles west of Lubbock, Texas. A/C James W. Baer, 24, Oxford, Ohio, and A/C Hardin J. Barton, 23, Memphis, Tennessee, were killed when their AT-9 (41-12105) slammed to the ground one-half mile east of Lubbock Auxiliary Airfield #1, bursting into flames upon impact. Instructor 2Lt. Raymond A. Gilbert received minor injuries and student A/C Edward L. Dorr was uninjured in the collision and were able to land the damaged AT-9 (41-12131) safely at Lubbock Auxiliary Airfield #1.

5-15-42A. Pacific Ocean, California. At 1215, a Lockheed A-29 was seen to dive into the Pacific Ocean at Table Bluff, near Eureka, California, killing the seven Army and Navy crewmembers on board. Aircraft spotter Mrs. Susan Kegler and A.W. Mareley, Master of the civilian owned tanker *M/S Standard Service*, witnessed the crash. Master Mareley stated that the A-29 circled the ship a couple times before flying off in a westerly direction. Moments later the man saw the airplane dive into the sea. The depth charges carried aboard the A-29 detonated as the airplane sank. The crew of the tanker gathered pieces of wreckage and equipment but were not able to recover any of the bodies of the A-29 crew. Investigators could not determine why the airplane crashed. Killed in the crash were: 2Lt. William Z. Froehlingsdorf, Virginia, Minnesota, pilot; Ensign A. C. Maher, USN, New Orleans, Louisiana, co-pilot; Sgt. Wade E. Andrews, Putnam, Texas, engineer; TSgt. Wilfred J. Caissie, Chelsea, Massachusetts; bombardier; Cpl. Stanley E. Shaw, Richmond, Virginia, gunner; Sgt. Lawrence M. Christian, Jr., Shawnee, Oklahoma, radio operator; Seaman 2nd class J. E. Jones, USN, Whitney, Nevada, observer.

5-15-42B. Wendover Field, Utah. At 1430, a Boeing B-17E was abandoned in flight 10 miles northwest of Wendover Field, Wendover, Utah, killing the pilot and injuring two other crewmembers. The airplane went out of control and crashed after it encountered deteriorating weather that included strong turbulence. The B-17 went out of control three times with the pilot recovering it twice. The pilot ordered the crew to bail out when he could not regain control of the airplane. Three crewmembers parachuted to safety. Pilot 1Lt. James H. Payne, East St. Louis, Illinois, was killed when he bailed out of the B-17 without his parachute. No parachute was found near the pilot's body, which was not located until 5-31-42 about eight miles northwest of Wendover Field, Utah. The parachute was later found among the wreckage. Co-pilot 2Lt. Richard H. Drake and engineer Pfc. Robert E. Johnson received minor injuries and radio operator Pfc. Robert S. Evans received serious injuries parachuting to safety.

5-15-42C. Ignacio, California. At 1857, a Lockheed P-38E crashed at Ignacio, California, killing 2Lt. Edward E. Cahill, 27, Newark, New Jersey. The airplane was attempting to land at Hamilton Field, San Rafael, California. The P-38 was seen by numerous witnesses on the ground and in the air to go into a vertical type spin to the left from an altitude of approximately 500 feet, striking the ground nose first and bursting into flames immediately upon impact. One witness reported seeing the airplane take-off and then observed it maneuver in a left wing down attitude just prior to the crash. It was later speculated that the port engine had failed.

5-15-42D. Mason, Texas. At 2315, a North American BC-1 flew into terrain 19 miles southeast of Mason, Texas, killing A/C Andrew C. Smith. The BC-1 had taken off from Kelly Field, San Antonio, Texas, at 2200 on a night cross-country navigation flight to Brady, Texas, to Austin, Texas, and return to Kelly Field. The BC-1 was determined to be about 21 miles off course on the leg from Brady to Austin. Investigators could not determine why the student was off course or why the airplane crashed, speculating that he had either suffered vertigo or was looking at a map, causing him to fly the airplane into the ground. The wreckage and the pilot's body were not found until 1000 on 5-16-42.

5-17-42A. Gastonia, North Carolina. At 1045, a Curtiss P-40F struck the ground in a high speed vertical dive seven miles west of Gastonia, North Carolina, killing 2Lt. Edward M. Goode. Witnesses stated to investigators that the airplane stalled and entered a spin after performing a snap roll at an altitude of approximately 5,000 feet agl. The airplane spun at least twice before diving straight into the ground, exploding violently into flames upon impact. Witnesses observed what was thought to be a small amount of debris stream from the airplane just before it slammed into the ground. The pilot made no apparent attempt to leave the airplane.

5-17-42B. Turk's Head Mountain, Colorado. At a time between 0900 and 1200 MWT, a Lockheed F-4 (photo-reconnaissance version of the P-38) collided with Turk's Head Mountain, 35 miles southwest of Lowry Field, Denver, Colorado, killing 1Lt. Constant S. Owen. The airplane had failed to return to base and was reported missing. The pilot had encountered low ceilings and limited visibility and was attempting to turn his way toward better weather. Investigators speculated that the pilot had tried to maintain visual contact with the ground in instrument conditions while trying to execute this turn. The aircraft had struck the mountain on its western slope about 400 yards from the top. At approximately 1700 on 5-18-42, the Devil's Head Forest Ranger Station reported a small fire on Turk's Head Mountain, which was later investigated by a B-17E from Lowry Field. The B-17 crew reported to Lowry Field at 1900 that the fire was probably the wreckage of the F-4 that had been reported as missing the day before. A rescue party, guided by an unnamed local civilian, reached the crash site at about 0300 on 5-19-42. Wreckage and body pieces were strewn over an area of over 150 yards.

5-17-42C. Morris Field, North Carolina. At 1625, a North American O-47A crashed on take-off at Morris Field, Charlotte, North Carolina, killing pilot 2Lt. John O. Gaskin. Cpl. Kenneth Loomis, engineer, and Cpl. Harold E. Clark, gunner, received minor injuries in the accident. The O-47 had been at the south end of the North/South runway waiting with other aircraft to take-off when it began to taxi out onto the runway. The tower warned all aircraft to hold for two P-40s coming in to land and announced that no take-offs were permitted until the P-40s had landed. The O-47 continued to taxi out and began its take-off roll. The tower tried to contact the O-47 to prevent the attempted take-off and also warn the pilot of the P-40 of the O-47 runway incursion. The P-40 pilot on final approach simultaneously observed the O-47 on the runway and aborted his landing attempt. He gave the airplane full throttle and passed over the O-47, which was just becoming airborne. As the O-47 began climbing through the propeller turbulence created by the passing P-40 at approximately 100 feet above the runway, it suddenly entered a nearly 90-degree bank to the left. The aircraft lost flying speed, stalled and fell off on the port wing. The O-47 impacted the ground on its port wing and nose in between two hangars, bounding into the air and tumbling for a short distance. The pilot, still strapped to his seat, was thrown from the cockpit and hurled through the roof of a hangar, killing him instantly. There was no fire as a result of the crash. Control tower personnel stated that traffic at the field and in the vicinity was "heavy" at the time of the accident. Investigators could not determine why the O-47 commenced a take-off against orders from the tower and concluded that the propeller turbulence of the passing P-40 caused the pilot to lose control of and stall the O-47 at too low an altitude to allow recovery.

5-18-42A. Barksdale Field, Louisiana. At 0525, a Consolidated B-24D crashed while attempting a take-off from Barksdale Field, Shreveport, Louisiana, killing nine fliers. Weather at the time of the crash was a ceiling of 3,000 feet, visibility 5 miles with moderate rain and thunderstorms. Published newspaper accounts of the crash reported that the thunderstorms were sometimes "severe." The Aircraft Accident Classification Committee stated, "The pilot had departed from Barksdale Field on a patrol mission over the Gulf of Mexico, had returned almost immediately thereafter, and accomplished a landing. After covering about three-fourths of the length of the runway he applied power for an attempted take-off and crashed one and one-half mile south of the end of the runway. Investigation indicates that the pilot, with his airplane heavily loaded including full gas tanks and 1,800 pounds of bombs, made a fast landing and was overshooting the runway. At about three-fourths of the distance down the runway he applied power and attempted to take off again to attempt another landing. Wheel marks in the mud some 600 yards to the south of the end of the runway indicate that he once squashed into the ground. He then gained sufficient altitude to clear some 60-foot trees but failed to clear a higher tree about a mile and a half south of the field at which point the right aileron was torn off and his airplane crashed into the ground several hundred yards ahead." Killed in the crash were: 2Lt. Herbert W. Frawley, 23, Fort Monmouth, New Jersey, pilot; 2Lt. James R. Everhart, 26, Uniontown, Pennsylvania, co-pilot; 2Lt. Augustus H. Tate, 24, Greenwood, Mississippi, navigator; Sgt. Mansfield Crabtree, 27, Meadowview, Virginia, bombardier; Sgt. Lewis J. Helper, 22, Valley View, Pennsylvania, engineer; Pvt. Rudolph McJunkins, 22, Saratoga, Arkansas, assistant engineer; Sgt. Arlo V. Werley, 23, Shillington, Pennsylvania, radio operator; Pvt. Herman R. Sanneman, 28, Creston, Iowa, assistant radio operator; Cpl. Stanley C. Andrews, 19, Elba, New York, gunner.

5-18-42B. Denham Springs, Louisiana. At 1140, a Curtiss P-40 collided with trees two miles south of Denham Springs, Louisiana, killing pilot 2Lt. Rhyne C. Doggett. Investigators stated, "A formation of three ships in open two-ship formation dived at a cloud at 4,000 feet, continuing down in a turn to the right and making a gradual recovery at 300 to 500 feet, completing in this dive, 90 degrees of turn to the right from their direction as they passed the cloud. Lt. Doggett, flying on the wing of the leader [2Lt. Laune C. Erickson], continued in a right turn after the leader straightened out until he crashed at approximately 90 degrees from the path of the formation. The flight leader is considered to have led his flight rather close to the ground at that speed, but it seems improbable that this was the cause of the crash. The other pilots both noticed the normal amount of torque for that speed, but experienced little difficulty in controlling their airplanes. There was no tendency to blackout." The P-40 mushed a 150-yard path through the trees before coming to rest and catching fire.

5-18-42C. San Pablo Bay, California. At 1650, a Lockheed P-38D attempting an emergency landing after experiencing port engine trouble crashed into San Pablo Bay 500 yards from Hamilton Field, California, killing 2Lt. Mike T. Tangorra, Rockford, Illinois. Witnesses on the ground and in the air observed the port propeller was windmilling and had not been feathered. 2Lt. William J. Hoelle witnessed the crash and stated to investigators, "I was sitting at the head of runway 30, awaiting take-off instructions when I heard [Lt. Tangorra] call in for emergency landing instructions to the Hamilton Tower. He came directly southeast, about 2 miles from the field on his down wind leg. His altitude then was approximately 1,000 feet, and he was gradually descending. My guess as to his airspeed then would be about 150 mph. The plane made the turn into base leg satisfactorily, but his pattern was entirely too small for a successful landing. About the time he started turning into his landing leg, the pilot dropped his wheels. This sudden drag pulled the nose directly down, and as he attempted a recovery with back stick and throttle, the pull of but one motor flipped the ship over on its back and directly into the water. The plane hit about 500 yards southeast of runway 30. Just before the time of his turn toward the field, I attempted to instruct him to put down flaps and correct his position over the radio. However, the pilot was tuned to the tower frequency and did not hear my call."

5-18-42D. Arlington, California. At 1724, a Vultee BT-15 stalled and spun into the ground one and a half mile west of Arlington Auxiliary Airfield, Arlington, California, killing Aviation Cadet Joseph Morgan. The airplane was part of a three-airplane formation training flight that had just taken off from Arlington airfield, four miles southwest of Riverside. The BT-15 peeled off of the formation at about 600 feet altitude and continued in a steep left turn entering a steep spiral to the left. The aircraft momentarily leveled out at approximately 50 feet above ground level before it stalled and spun to the left, smashing into the ground at an angle of about 30 degrees. The airplane initially struck some high trees and then hit the ground near a row of small buildings, careening through a row of small eucalyptus trees before coming to rest in an upright position. Investigators determined that the engine had failed at some point during the flight but could not determine the cause of the engine failure.

5-20-42. Harding Field, Louisiana. At 1845, a Bell P-39F stalled and spun into the ground eight miles southwest of Harding Field, Baton Rouge, Louisiana, killing pilot 2Lt. Charles J. Borden. The testimony of witnesses indicates that the P-39 spun, partially recovered, spun again, and partially recovered again before diving into the ground. The airplane exploded into flames when it struck nose first in a bean field.

5-21-42A. Atmore, Alabama. Between 0025 CWT and 0120 CWT, seven Vultee BT-13 airplanes crashed in the vicinity of Atmore, Alabama, killing seven RAF student pilots. The seven airplanes were part of a 35-ship flight that took off from Crestview, Florida, on a night cross-country flight back to its home station at Gunter Field, Montgomery, Alabama. All of the students on the flight were RAF cadets and they were supervised by nine AAF instructors. Also near Atmore, two students from this same flight parachuted to safety after they abandoned their two BT-13 airplanes when both were trapped by poor weather and darkness, suffering minor injuries. Two other BT-13 airplanes crash-landed in the vicinity of the airport at Evergreen, Alabama, resulting in minor injuries to the students and destroying the two aircraft. Another BT-13 of this flight was damaged seriously when it crash-landed at an airfield at Mobile, Alabama. The student was reported as uninjured. This ill-fated flight began at 1545 on 5-20-42 when 35 BT-13 type airplanes took off from Gunter Field, Montgomery, Alabama, for Crestview, Florida, on a day navigation flight. The return trip was to be a night navigation flight from Crestview to Mobile, Alabama, and then returning to Gunter Field. The initial take-off had been delayed by poor weather. All of the airplanes landed safely at Crestview by 1813. The 2000 take-off was delayed because of deteriorating weather conditions. At 2115, officers at Gunter Field ordered the flight to remain on the ground unless the ceiling was 3,000 feet or better. At 2230, weather reports and pilot reports indicated that the weather along the route from Crestview to Mobile and Montgomery was clearing up enough to allow visual contact flying. At 2320, the 35 BT-13 aircraft took off from Crestview in eight, four-ship flights and one three-ship flight, each flight including one instructor. The students and instructors were ordered to turn back for Crestview if the ceiling deteriorated to 1,500 feet agl. All of the airplanes arrived at Mobile, Alabama, as scheduled where they turned on their course for Gunter Field. Beginning at approximately 0000 (5-21-42), the weather along the route began to rapidly deteriorate. After about 20 miles on the course for Gunter Field extremely poor visibility was encountered. The formations broke up and the individual student pilots were left on their own. Of the 35 aircraft on the flight, five landed safely at Gunter Field, Alabama; two returned to Crestview, Florida; three landed at Mobile, Alabama (one aircraft severely damaged in crash landing); one landed safely at Napier Field, Alabama; one landed at Atmore, Alabama; five landed safely at Evergreen, Alabama; one crash landed one mile south of Evergreen and another crash landed five miles northeast of Evergreen; five landed safely at Greenville, Alabama; two landed at the U.S. Navy Air Station at Pensacola, Florida; RAF Flying Cadets Robert M. Pinkerton and Richard Lewis bailed out of their airplanes near Atmore, Alabama. The first fatal accident occurred at 0025; RAF Flying Cadet Ronald E. Randall, Birmingham, England, was killed when his BT-13 crashed nine miles northeast of Atmore, Alabama. At 0030, RAF Flying Cadet Patrick G. M. Overton, Portcawl, South Wales, was killed when his BT-13 crashed 20 miles northeast of Atmore, Alabama. At 0035, RAF Flying Cadet David S. Peattie was killed when his BT-13 crashed 17 miles east of Atmore, Alabama. This airplane and student had been missing and were discovered 5-23-42. At 0100, RAF Flying Cadet Victor W. Lear, Rapstock, North Bath, Somerset, England, was killed when his BT-13 crashed seven miles northwest of Atmore, Alabama. At 0100, RAF Flying Cadet Michael E. Peachell was killed when his BT-13A crashed 14 miles north of Atmore, Alabama. This airplane and student had been missing and were discovered 5-23-42. At 0100, RAF Flying Cadet James E. Maddick, Dundee, Scotland, was killed when his BT-13 crashed ten miles southwest of Atmore, Alabama. At 0120, RAF Flying Cadet Arthur V. Lowe was killed when his BT-13A crashed 14 miles north of Evergreen, Alabama. This airplane and student had been missing and were discovered 5-27-42 (See 5-21-42B, C, D, E, F, G).

5-21-42B. Atmore, Alabama. At 0030, RAF Flying Cadet Patrick G. M. Overton, Portcawl, South Wales, England, was killed when his Vultee BT-13 crashed in poor visibility 20 miles northwest of Atmore, Alabama. The airplane was part of a 35-airplane night navigation flight (see 5-21-42A).

5-21-42C. Atmore, Alabama. At 0035, RAF Flying Cadet David S. Peattie was killed when his Vultee BT-13 crashed in poor visibility 17 miles east of Atmore, Alabama. This airplane and student had been missing and were discovered 5-23-42. The airplane was part of a 35-airplane night navigation flight (see 5-21-42A).

5-21-42D. Atmore, Alabama. At 0100, RAF Flying Cadet Victor W. Lear, Rapstock, North Bath, Somerset, England, was killed when his Vultee BT-13 crashed in poor visibility seven miles northwest of Atmore, Alabama. The airplane was part of a 35-airplane night navigation flight (see 5-21-42A).

5-21-42E. Atmore, Alabama. At 0100, RAF Flying Cadet Michael E. Peachell was killed when his Vultee BT-13A crashed in poor visibility 14 miles north of Atmore, Alabama. This airplane and student had been missing and were discovered 5-23-42. The airplane was part of a 35-airplane night navigation flight (see 5-21-42A).

5-21-42F. Atmore, Alabama. At 0100, RAF Flying Cadet James E. Maddick, Dundee, Scotland, was killed when his Vultee BT-13 crashed in poor visibility 10 miles southwest of Atmore, Alabama. The airplane was part of a 35-airplane night navigation flight (see 5-21-42A).

5-21-42G. Evergreen, Alabama. At 0120, RAF Flying Cadet Arthur V. Lowe was killed when his Vultee BT-13A crashed in poor visibility 14 miles north of Evergreen, Alabama. This airplane and student had been missing and despite a large search effort were not discovered until 5-27-42. The airplane was part of a 35-airplane night navigation flight (see 5-21-42A).

5-21-42H. Boyle, Mississippi. At 0340, a Vultee BT-13A crashed one mile east of Boyle, Mississippi, killing Aviation Cadet Joseph D. Molloy, 24, Ansonia, Connecticut. The airplane had been part of a three-airplane night navigation formation flight. Investigation revealed that the student had lost his leader and became confused. The student was panicked and became "semi-hysterical" while trying to follow Greenville tower air traffic controller's verbal radio directions to the Greenwood (Mississippi) radio range. The aircraft entered a steep spiral to the left that the student could not recover from and he attempted to parachute to safety. Investigators stated, "Cadet Molloy apparently pulled his ripcord immediately upon leaving the cockpit. The parachute opened and its canopy went to the rear, across the fuselage, to the right of the vertical fin. Cadet Molloy apparently fell to the left rear, passing below the left horizontal stabilizer. When the shroud lines pulled taut, Cadet Molloy apparently was thrown up and over the left horizontal stabilizer so firmly entangling his parachute that it could not be freed."

5-21-42I. Pixley, California. At 1235, a Stearman PT-17 dove into the ground six miles northeast of Pixley, California, killing Aviation Cadet Carl E. Larson, 24, Los Angeles, California. Investigators stated, "The airplane dove into the ground at an angle of 60 degrees. The aircraft did not spin, but flew straight into the ground and started burning immediately [upon impact]. The impact killed the pilot instantly. A thorough examination of the wreckage revealed no structural failure or sabotage. The cause of this accident is undetermined, however, there is reason to believe that it was caused by pilot error due to the straight path in which the aircraft descended in the dive, with no attempt being made by the pilot to either recover or use his parachute."

5-21-42J. Randolph Field, Texas. At 1510, a Vultee BT-13A crashed five miles south of Randolph Field, San Antonio, Texas, killing A/C John D. Finn, who fell to his death when his parachute failed to open. A/C David B. Eastham parachuted to safety. The aircraft was attempting a steep turn to the left when the airplane stalled and began to spin. The student regained control of the airplane momentarily. The airplane then stalled again and fell off to the right. The student struggled to get the airplane under control. Cadet Eastham (sitting in the rear cockpit) noticed Cadet Finn shouting something back to him. Cadet Eastham attempted to unbuckle his seat belt, but was thrown violently by the airplane. Cadet Finn, who appeared to be leaving the airplane, again shouted back to Cadet Eastham who then left the airplane at about 500 feet above ground level. Cadet Finn managed to leave the airplane but his parachute failed to fully deploy and he was killed instantly when he struck the ground.

5-22-42. Bonham, Texas. At 1400, a Fairchild PT-19A crashed into a field eight miles northwest of Jones Field, Bonham, Texas, while simulating a forced landing, killing A/C Thompson L. Gaut and fatally injuring civilian instructor Douglas Chandler, who died at approximately 1540. Investigators stated, "When first seen the airplane was at an altitude of from 400 to 500 feet going due north in a glide. From this point a gentle turn of 220 degrees was executed, the airplane continuing its turn until it struck the ground in the center of a field large enough to have been used for a simulated forced landing. Throttle was used at an estimated altitude of 30 feet. When the throttle was used for recovery, neither pilot was at the controls; each thinking the other would fly the airplane out. The instructor thinking so because the student had instructions to fly the airplane at all times. The student thinking so because the instructor had been on the controls enough to cause him to believe that the instructor was actually flying the airplane."

5-23-42A. Millville, New Jersey. At 1200, a Curtiss P-40F ran off the runway upon landing and collided with two parked pick-up trucks belonging to a construction crew working on an airfield drainage ditch at Millville Airport, Millville, New Jersey, killing civilian workman Joseph Guiseppi Ricciuto, 45, Millville. The aircraft was piloted by 2Lt. Edward L. Jett, who was not injured in the accident. Lt. Jett had attempted a landing at the airport after completing a formation training flight with two other P-40s. Landing number two behind his leader, Lt. Jett undershot his approach and was ordered to go around. Lt. Jett again undershot his approach on his second landing attempt and bounced upon landing. He made a good recovery and got the airplane down on the runway. As he rolled out he allowed the airplane to veer to the right, rolling off of the runway and toward a drainage ditch. The P-40 hit the ditch, careened over it and then collided with Mr. Ricciuto, a 1929 Ford Pick-up truck owned by Nick Hefcheck, and a 1936 Chevrolet Pick-up truck owned by Frank Minkwitz. Mr. Hefcheck and Mr. Minkwitz escaped injury. Mr. Ricciuto was struck by the airplane's starboard wing and was pronounced dead at the scene. This was the third landing accident at the airport in seven days.

5-23-42B. McChord Field, Washington. At 1040, a North American B-25 crashed while attempting to take-off from McChord Field, Tacoma, Washington, killing the five crewmembers on board. Witnesses on the ground observed an explosion in the nose landing gear wheel-well or on the right front of the aircraft just before the crash. Investigators speculated that the explosion in the nose gear wheel-well hurled pieces of the nose gear door into the starboard propeller causing its failure and the crash. The B-25 went into a steep right bank, leveling off momentarily before crashing into the ground nose first and exploding into flames. Pieces of an exploded carbon dioxide bottle were found along the line of flight of the airplane. The carbon dioxide bottle might have been inadvertently placed in the nose wheel-well or on the nose landing gear, exploding when it was compressed by the retracting landing gear. Investigation was inconclusive and the matter was handed over to U.S. Army Intelligence (G-2). Killed in the crash were: Lt. Col. Eugene Nall, 30, Atmore, Alabama, pilot; 1Lt. Charles E. Daily, 25, Tacoma, Washington, co-pilot; MSgt. Delana A. Shepard, 56, Tacoma, Washington, engineer; Cpl. Guinn N. Murdock, 19, Denton, Texas, radio operator; Pfc. Walter F. Rudesill, 20, Hot Springs, Arkansas, assistant engineer.

5-23-42C. Russell, Kentucky. At 1530, a Lockheed P-38F experiencing trouble with its starboard engine crashed while attempting a forced emergency landing 16 miles west of Russell, Kentucky, killing pilot 2Lt. Virgil E. Holman. The airplane was part of a flight of five P-38s that had taken off from Barksdale Field, Shreveport, Louisiana, on a ferry flight to Patterson Field, Ohio. There were no witnesses to the crash.

5-24-42. Howe Brook, Maine. At 1120, a Lockheed C-40D collided with terrain five miles east of Howe Brook, Maine, killing six fliers. The aircraft was on a personnel transport flight from Bolling Field, Washington, D.C., to Montreal, Canada, and return to Bolling Field via Houlton and Presque Isle, Maine. The transport took off from Montreal at 0934 for the return leg to Bolling Field. Weather in the Howe Brook vicinity at the time of the crash was reported as ceiling 600 feet with visibility approximately two miles. The aircraft crashed into the lowest area of the immediate terrain with no obstructions in its flight path. Broken trees near the crash site indicated that the airplane struck the ground at an angle of about 75 degrees in a north-northwesterly direction. The airplane exploded into bits upon impact with the ground, scattering wreckage and bodies for over 400 yards. Investigators could not determine why the airplane flew into ground, speculating that the deteriorating weather conditions could have been a factor. The airplane was so completely demolished that evidence of mechanical failure could not be determined with any certainty. Killed in the crash were: 1Lt. Clarence A. Wright, pilot; Capt. John D. Franciscus, passenger; Lt. Col. Louis S. Gimbel, passenger; 1Lt. Gilbert M. Herbach, passenger; 2Lt. E. R. Wilkerson, passenger; SSgt. Frederick J. Taylor, engineer.

5-25-42A. Offutt Field, Nebraska. At 1233 CWT, a Martin B-26A dove into the ground at Offutt Field, Omaha, Nebraska, killing two Glenn L. Martin Company test pilots. Investigation revealed that the airplane crashed after suffering a structural failure of the tail surfaces when they were struck in flight by the pilot's escape hatch, which had separated from the airplane during a high-speed pull-up following a diving maneuver on Offutt Field. Pilot Carl W. Hartley, 30, Omaha, and co-pilot Henry K. Meyers, 24, Omaha, were killed in the crash. The aircraft had taken off at 1228 and was on an authorized test flight following a major modification by the manufacturer. Investigators stated, "The take-off was to the southeast and the airplane continued in a straight line of flight to approximately 900 feet altitude. The airplane then banked to the left through east to north and continued gaining altitude. The airplane was then seen to bank to the left through north and west. An altitude of approximately 2,500 feet was reached at this point. The airplane again banked to the left to southeast and was seen diving to the northwest and southeast runway. The airplane approached the runway approximately 30 feet above the ground level with an estimated speed in excess of 300 miles per hour. After continuing approximately three-fourths the length of the runway, it was seen to zoom at an angle of approximately 50 degrees where it disappeared into a large cloud in the broken overcast. A competent witness who was watching with field glasses saw the pilot's escape hatch fly from the airplane previous to its entrance into the overcast. Investigation definitely indicates that the pilot's half of the hatch cover left the airplane during the high-speed pull up, hitting the vertical stabilizer and badly damaging the same. This damage undoubtedly caused the upper portion of the vertical stabilizer and the mating portion of the rudder to carry away. The lower portion of the rudder apparently followed, the lower portion of the fin remaining with the airplane and being torn away upon impact. The canopy fell from the airplane onto the airport runway. The airplane was seen to emerge from the overcast, obviously out of control, in a flat spiral to the left. At that instant, the nose dropped to approximately a 45-degree angle with the horizontal and continued in an unstable line of flight until it struck the ground. It is evident from the position of the body of pilot Hartley and from the testimony of witnesses that he was successful in abandoning the airplane at an altitude of approximately 250 feet before striking the ground. There is also sound evidence that the pilot struck some part of the tail surfaces when he left the airplane. An analysis of the airplane's flight path from the time of take-off to the time of the crash considers the facts that the airplane remained under a broken deck of clouds at approximately 2,000 feet and within very close proximity to the field indicates that neither altitude nor distance was available in which to exceed the design limitations of the airplane although exhibition flying was being indulged in." All B-26 type airplanes were grounded for a short time following this accident. This is one of several fatal B-26 accidents in the U.S. and overseas where the pilot's escape hatch separates from the airplane and strikes the tail section, causing the progressive failure of the empennage.

5-25-42B. San Francisco, California. At 1811, a Lockheed P-38D suffering the failure of the port engine crashed just after take-off from San Francisco Airport, San Francisco, California, killing pilot 2Lt. Robert G. Hampton, 25, Charlotte, North Carolina. The P-38 took off behind four others at 1808 and climbed to about 1,000 feet above ground level when it was observed by witnesses on the ground to trail smoke. The pilot attempted a left turn and stalled the airplane. The airplane fell off to the left and spun into an electric transmission line tower of the Pacific Gas and Electric Company. Electric power to the airport and the surrounding area was disrupted by the crash. The aircraft exploded into flames upon impact with the ground.

5-26-42. Moscow, Arkansas. At 1604, a Fairchild PT-19A stalled and spun into the ground one mile west of Moscow, Arkansas, killing civilian flight instructor Albert L. Smith, 28, Springfield, Missouri, and Aviation Cadet Victor S. Harrison, 25, Sardinia, New York. The aircraft was seen by a witness on the ground to be flying at slow speed in a southwesterly direction in a nose high wings level attitude at about 500 feet above ground level. The aircraft was then seen to stall and spin, striking the ground nose first. The PT-19 was totally destroyed but did not catch fire in the crash.

5-27-42A. Morris Field, North Carolina. At 1024, a Curtiss P-40F crashed while attempting to go around after overshooting a landing attempt at Morris Field, Charlotte, North Carolina, killing pilot Maj. Charles L. Robbins, Elkhart, Indiana. Maj. Robbins was leading a two-airplane flight from Norfolk, Virginia. Maj. Robbins overshot the landing, touching down at the runway intersection. He gave the airplane full power for the go-around and retracted the landing gear. The wingman did the same and climbed out. Maj. Robbins was seen to climb out to about 100 feet above ground level. At this point the aircraft ceased to climb but remained in a nose up attitude. The airplane quickly lost flying speed, stalled and spun off the end of the runway, exploding into flames upon impact with the ground.

5-27-42B. Tonopah, Nevada. At 0950, a North American AT-6A was abandoned and crashed into the side of an unnamed mountain at the AAF Gunnery Range at Tonopah, Nevada, after the student gunner shot the airplane's tail off. Gunnery student Pvt. John Kelleman, Jr., 26, East Chicago, Indiana, was killed while attempting to use his parachute. Pilot SSgt. Jack J. Chaney parachuted to safety. The Accident Classification Committee stated, "This accident occurred during a normal gunnery mission wherein the gunner accidentally fired an unknown number of rounds from the gun while in the stowed position causing sufficient damage to render the tail assembly controls inoperative. The plane went into an uncontrolled dive and crashed into the side of a mountain. The pilot realizing that he had no further control over the aircraft ordered his passenger to jump and prepared to jump himself. The pilot states that prior to jumping he noticed that his passenger was on the outside of the fuselage preparing to jump and then jumped himself. The pilot estimated that he left the ship at 8,500 feet with an airspeed of about 160 miles per hour. The estimated altitude at the point of his landing was 5,000 feet [above sea level]. The attempt of the passenger to jump was unsuccessful and the canopy of his parachute was found to be entangled in the tail section of the aircraft upon examination of the wreckage. The parachute ripcord was pulled prematurely either by the gunner himself or by accidentally fouling on the aircraft structure, causing the canopy to become entangled in the tail surfaces of the airplane. The gunner's body was found in such a position as to preclude the possibility of the gunner being in the rear cockpit at the instant of the impact."

5-27-42C. March Field, California. At 1015, a Lockheed P-38F crashed five miles northwest of March Field, Riverside, California, killing 2Lt. Richard O. Whyman, Jr. The P-38 was attempting a peel-off to the left at an altitude of approximately 2,500 feet above ground level while trying to follow his leader to the traffic pattern for a landing attempt at March Field. The aircraft was observed in a flat spin before it crashed in a near vertical attitude on top of a knoll. The aircraft exploded into flames upon impact with the ground, scattering wreckage over an area of 200 yards. Investigators could not determine the cause of the crash.

5-27-42D. Location Unknown. At an unknown time after 1130 PWT, a Lockheed P-38F disappeared with pilot SSgt. Eugene W. Babb and presumed lost at sea in the vicinity of San Clemente Island, California. The aircraft was part of a two-ship flight that had taken off at 1010 from Mines Field, Los Angeles, California, on a sea patrol of the San Clemente Island area and was last seen by the other pilot at approximately 1130.

5-27-42E. Hempstead, New York. At 1555, a Lockheed P-38E crashed into a vacant lot between Rhodes Avenue and Fairview Boulevard, Hempstead, Long Island, while trying to make an emergency landing at Mitchel Field, Long Island, New York, killing pilot 2Lt. Raymond V. Wetzel, 21, Gilman, Wisconsin. The airplane was part of a flight of P-38s on a ground gunnery-training mission. Investigators stated, "Lt. Wetzel notified the flight leader that his oil pressure was dropping and that he was going to return to Mitchel. He returned to Mitchel Field and came in with the left engine feathered at about 1,500 feet. He signaled to the tower by rocking his wings of his intentions to land. He made left hand turns (turning into his dead engine) went around the field, overshot, pulled his wheels up, and went around again. On his second try, he circled the field (still turning into his dead engine), made his base leg too close to the field and overshot again. Wheels remained down this time. As he started to go around this time, his right engine caught fire just after he passed the field, probably due to excessive manifold pressure and low rpm. The fire was no doubt caused by a rod failure on his right engine. When his good engine failed, the airplane rolled to the left and hit the ground upside down."

5-27-42F. Buffalo, New York. At 1720, a Curtiss P-40F crashed after suffering a structural failure while demonstrating high-speed dives and pullouts four miles north of Buffalo Municipal Airport, Buffalo, New York, killing Curtiss-Wright Corporation test pilot George H. Dorsey. The purpose of the flight was to obtain 8.0 G at 496 miles per hour calibrated airspeed. The canopy, elevators and the rudder failed in flight during one of the maneuvers. Investigators could not determine in what sequence the structures failed or what caused them to fail and although Army investigators and Curtiss-Wright engineers had numerous theories, a consensus could not be reached.

5-27-42G. Vicksburg, Mississippi. At 1644, a Vultee BT-13A crashed while performing aerobatics over Vicksburg Airport, Vicksburg, Mississippi, killing pilot 1Lt. J.A. Butner and passenger Sergeant Pilot W. Schalk. The aircraft had landed at the airport and remained on the ground for approximately 30 minutes. The BT-13 took off and circled the field at about 1,000 feet. The airplane performed a snap roll then went into a steep climb, at the top of which the airplane stalled. The airplane went into a spin to the left and crashed in the middle of the airport.

5-27-42H. Lakeland, Florida. At 2301, a Consolidated B-24D attempting an emergency forced landing after suffering the failure of the number-three and number-four engines crashed two miles ENE of the Army Air Base at Lakeland, Florida, killing six crewmembers and injuring two others, one seriously. Investigators stated, "The accident was caused by exhausted fuel supply in the right tanks. Both right [engines number-three and number-four] propellers were found feathered. The crew chief was attempting to transfer fuel out of the left tanks into the right tanks. The committee believes that the crew chief may have transferred all fuel from the left tanks thus starving the remaining two engines [engines number-one and number-two] at a time when the pilot was beginning his approach for landing. The fuel otherwise could not have been exhausted. The landing gear was down." Killed in the accident were: 2Lt. William A. Sims, pilot; 2Lt. Lawrence R. Abrams, co-pilot; 2Lt. Preston R. King, navigator; SSgt. Russell C. Tomlinson, engineer; Pvt. Henry O'Donnell, gunner; Pvt. Martin Goldstein, radio operator. Pvt. Charles G. Brock, gunner, was seriously injured and Pvt. James P. Buchanan, gunner, was slightly injured.

5-27-42I. Boise, Idaho. An AAF navigator was killed when he and two other crewmembers inadvertently parachuted from a Boeing B-17B airplane on a navigation mission 42 miles southeast of Gowen Field, Boise, Idaho. 2Lt. Vernon Lee Hall, Parsons, Kansas, plummeted to his death when he was dragged in his parachute and swept over a cliff after initially landing successfully on the extremely rugged terrain. The pilot, 2Lt. Gilbert J. Kuhn, landed the B-17 safely at Salt Lake City, Utah, and the other two crewmembers parachuted to safety 42 miles southeast of Boise. Lt. Hall's body was found on 5-28-42 at 1630.

5-28-42A. Gunter Field, Alabama. At 0825, a Vultee BT-13A crashed while maneuvering on an instrument training flight ten miles southeast of Gunter Field, Montgomery, Alabama, killing instructor 1Lt. Berton P. Marcley and RAF Flying Cadet Frank Rogers. The crash was witnessed by 1Lt. Benjamin S. Preston who stated to investigators, "When I first observed the airplane it was above and directly in front of me and was at approximately 1,800 to 2,000 feet, and was in a vertical dive. He pulled up from this dive at approximately 500 feet in a steep climbing turn to the right with about 45 degree of bank. As his speed fell off, the plane again started a steep diving turn to the right, turning about 270 degrees and stalled at high speed. The plane did about three-quarters of a turn of a snap roll out of this vertical right turn, coming out at approximately 400 feet in a steep dive straight ahead, and was pulling out slowly, straight ahead, right side up, when it crashed at about a 45 degree angle."

5-28-42B. Wright Field, Ohio. At 0946, a Hawker Hurricane stalled and dove into the ground near Wright Field, Dayton, Ohio, killing Wright Field test pilot 2Lt. William B. Exelby, Jr., 25, Cleveland, Ohio. The engine failure occurred just after take-off from Wright Field. After the engine failure, the pilot made a 90-degree right turn in an attempt to land back at the field, allowing the Hurricane to stall and enter a spin with too little altitude to effect a recovery. The aircraft fell off on its starboard wing and dove into the ground from an altitude of about 250 feet, bursting into flames upon impact.

5-28-42C. Brewton, Alabama. At 1415, a Douglas DB-7B crashed 25 miles southeast of Brewton, Alabama, after suffering an in-flight fire in the port engine nacelle, killing pilot 2Lt. Alan W. Detrick and bombardier Sgt. Samuel Cordell. Crew Chief Sgt. Joseph Braun parachuted to safety. The crash was witnessed by 2Lt. Raymond T. Schlanser, who later stated to investigators, "We were flying in a seven-plane formation led by Capt D. R. Dick, and Lt. Detrick was flying in position number two on the right wing of Capt. Dick. A stream of smoke appeared in the left engine of the plane piloted by Lt. Detrick, which gradually grew larger, becoming a white mask of smoke. After that a flame appeared to come out of the exhaust. Lt. Detrick dropped back out of formation, making a wide circle to the left and I lost sight of the plane. The plane was approximately 500 feet above the ground when I saw it again. The left engine was on fire. At about 100 feet from the ground the left engine seemed to explode. A parachute appeared approximately the same time as the explosion. I was circling at about 1,000 feet altitude, one-half mile from the scene of the crash, when the ship crashed to the ground."

5-29-42A. Peckham, Colorado. At 0954, a Lockheed A-29 collided with a power line three miles southeast of La Salle, Colorado, and then crashed at Peckham, Colorado, killing six fliers. After the aircraft struck the power line, it passed underneath another power line 300 yards further along the line of flight at an altitude of about 50 feet above the ground. The airplane continued to fly low and attempted a left turn at low altitude, slamming into the ground and exploding into flames at Peckham, three miles from the initial point of impact with the power line. Wreckage and bodies were scattered over a wide area. Killed in the crash were: 2Lt. Joe DaGrade, pilot; 2Lt. Carlos R. Kibler, co-pilot; Cpl. James H. Florney, engineer; Sgt. Alec Chararria, passenger; Sgt. Louis H. Garcia, passenger; Edward Paul Kelly, civilian engineer, San Antonio, Texas.

5-29-42B. Roswell, New Mexico. At approximately 1000, a Cessna AT-17 crashed 20 miles northeast of Roswell, New Mexico, killing pilot 1Lt. Lawrence N. Johndrow, Enid, Oklahoma, and co-pilot 1Lt. Charles E. Flanigan, Denton, Texas. The aircraft struck the ground in a vertical position as if hitting the ground as the result of a flat spin. There were three civilian witnesses to the crash but they were not reached and interviewed until 6-1-42. They reported that the airplane had been maneuvering and diving in the area prior to the crash. Their testimony was contradictory and not considered useful to the investigators in determining the cause of the crash. The airplane had taken off from Roswell at 0800 and was overdue at 1000. The aircraft and fliers were reported missing and after an extensive aerial search were finally located at about 1509 on 5-30-42.

5-30-42A. Lincolnton, North Carolina. At 1130, a Curtiss P-40 crashed into the ground out of control seven miles south of Lincolnton, North Carolina, killing pilot 2Lt. Leon M. Zele. Civilian witnesses on the ground reported that the P-40 had been flying at low altitude in the area and appeared to have been experiencing some engine problems as evidenced by black smoke emitting from the engine. The engine was seen to explode in flight. The airplane then slammed into the ground out of control, exploding into flames upon impact. The wreckage was scattered over an area in excess of 100 yards.

5-30-42B. McChord Field, Washington. At 0845, a Lockheed P-38E crashed six miles south of McChord Field, Tacoma, Washington, killing 2Lt. Peter A. Trick. Investigators stated, "[The] P-38 took off from the northeast to the southwest on Runway 21, McChord Field, Washington. Smoke was first observed coming from the airplane as it crossed the North/South runway during the take-off run. As the airplane continued its take-off it was observed that black smoke was issuing from the vicinity of the right engine. The airplane continued to climb, veering only slightly from left to right. Approximately three miles from the field on the projected extension of Runway 21 at approximately 500 feet altitude a bright glare emanated from the vicinity of the right engine resembling a magnesium fire. This glare was observed intermittently until it reached a position approximately 6 miles south where it suddenly veered to the right, nosed downward, and crashed. Examination of the wrecked right engine indicated that the piston of number 6 cylinder left bank had failed and that after this failure the connecting rod, being loose, knocked out the crankcase walls below cylinders 6 right and left bank."

5-31-42. Felts Mills, New York. At 1530, a Curtiss O-52 crashed one mile east of Felts Mills, New York, killing pilot 2Lt. Jack Lyons, 29, Portland, Oregon, and observer TSgt. Ernest W. Urban, 26, Stamford, Connecticut. The O-52, which was on a local reconnaissance flight, narrowly missed telegraph wires before it struck the ground at a gully in between a railroad track embankment and a small sand hill, bursting into flames upon impact.

June

6-1-42A. Billingsley, Maryland. At 1230, a Republic P-43A flying in instrument conditions crashed into a wooded area at Billingsley, Maryland, killing pilot Capt. Donald E. Wilburn, 25, Fountain City, Tennessee. Investigators noted that the pilot was instrument rated and had an approved instrument flight plan for the flight. It was not known why the pilot was trying to fly contact in the poor weather.

6-1-42B. Pine Grove, Pennsylvania. At 1415, a Vultee BT-13A on a cross-country navigation training flight descended into rising terrain while flying in instrument conditions near Pine Grove, Pennsylvania, killing pilot 1Lt. John C. Famenella and seriously injuring 2Lt. Henry A. Salm. Lt. Salm, who regained consciousness in the wreck on the morning of 6-2-42 and was found soon after, could not remember exactly how the crash occurred. Lt. Famenella, an instrument instructor at the time of the crash, probably misjudged his position relative to the high terrain when he began his let down.

6-2-42A. Fort Mill, South Carolina. A Curtiss P-40F suffering an engine failure stalled and spun into the ground while trying to make an emergency landing at a private airfield three-quarters of a mile NNE of Fort Mill, South Carolina, killing pilot 2Lt. Emerson R. Gross. The accident occurred at approximately 1105 at the grass airfield owned by Elliot White Springs. Investigators stated, "At a point about 2 miles north of Fort Mill the aircraft was observed to make a gentle turn to the left, power off, wheels down, and approach the private airport of Mr. Springs. The approach was made downwind from a westerly direction to the northeast of the southern boundary of the northeast-southwest landing strip at an altitude of about 100 feet. The aircraft was observed to make a sharp turn to the left, at which time the nose dropped and it struck the ground on the nose and wing. The pilot was killed instantly and the airplane demolished. Despite repeated warnings and instructions to the [pilots] to land with wheels retracted and flaps extended in the case of forced landings away from the base, nevertheless, the pilot lowered his wheels and attempted to extend his glide for a circuit of the [private airfield]."

6-2-42B. Fort Valley, Georgia. At 2150, a Vultee BT-13A collided with a 1937 Chevrolet Sedan after aborting a landing attempt at Harris Field, Fort Valley, Georgia, killing two civilians, seriously injuring another, and injuring two children. RAF instructor Pilot Officer J.J. Mulrooney, Jr. and RAF Flying Cadet Bruce Dix were injured. Pilot Officer Mulrooney overshot the landing and touched down about halfway down the runway. When he attempted to take-off, the engine would not perform and he applied the brakes. The engine then kicked in and he applied full throttle for the take-off. Approximately five yards south of the south boundary of the field, the BT-13 struck the unlighted parked automobile, which had been parked there so that the occupants could observe the traffic at the field. The airplane had struck the automobile on its right side with the starboard landing gear, knocking the Chevrolet over onto its left side. The airplane then nosed over and landed on its back in a peach orchard about five yards south of the car. The overturned aircraft had to be raised so that Cadet Dix could be extracted from the wreckage. Killed instantly in the crash was Mrs. Avis McDonald, 35, Fort Valley. The owner of the automobile, Robert Maurice Jones, Fort Valley, died of his injuries 6-6-42. Mrs. McDonald's niece, Miss Mary Lynn Morrow, Lineville, Alabama, was seriously injured. Mrs. McDonald's children, Margie, 9, and Carroll, 7, were lightly injured. Investigators determined that the car had been parked on a private road directly in the flight path of the active runway for about 30 minutes and that the airplane probably would have executed a successful take-off had the automobile not been parked in its path.

6-3-42A. Mobile, Alabama. At 1045, a North American B-25D crashed in a wooded area eight miles northwest of Mobile, Alabama, killing the crew of seven. Witnesses on the ground observed the airplane flying at approximately 500 feet above ground level in a north-easterly direction in a gentle turn to the right. The B-25 leveled out and made a steep turn to the left, leveled out again and then made a steep turn to the right. The airplane rolled through inverted, righting itself momentarily before diving into the ground and exploding violently into flames. Investigators could not determine the cause of the crash. Investigators stated, "The violent maneuvers of the airplane could have been caused by switching on the Automatic Flight Control Equipment [auto-pilot]. Several pilots in this group have experienced violent changes in attitude when switching on flight controls or AFCE. One other probable cause is that the pilot, when seeing himself lose altitude while flying so low, pulled up too sharply thereby stalling the airplane." Killed in the crash were: 2Lt. Emmett F. Hogue, Jackson, Mississippi, pilot; 2Lt. James D. Thompson, Marshfield, Missouri, co-pilot; SSgt. John L. Poythress, Farmville, North Carolina, radio operator; Pvt. Donald L. Delay, Gardner, Massachusetts, engineer; Pvt. William A. Ochs, Newton, Illinois, engineer; Pvt. Charles G. Hill, Jr., Memphis, Tennessee, gunner; Cpl. Richard C. Delay, Gardner, Massachusetts, passenger and brother of the above mentioned Pvt. Donald L. Delay.

6-3-42B. Spreckles, California. At 1115, a Taylorcraft L-2A stalled and spun into the ground following an apparent engine failure one-half mile north of Spreckles, California, killing pilot 2Lt. David B. Henderson, 21, Norwood, North Carolina. The airplane was seen by witnesses on the ground to complete a loop at about 3,000 feet and then to enter a second loop from which the airplane stalled out of and fell off to the left in a spin. The airplane spun several times before striking the ground. Witnesses reported that the engine was not running when the airplane fell out of the second loop. Investigators noted that engine failure had been known to occur to L-2 type airplanes when the throttle is suddenly closed and that pilots be warned of the limited performance capabilities of small observation type airplanes. The crash occurred on the ranch of Louis Nissen, two miles south of Salinas, California.

6-3-42C. Fort Valley, Georgia. At 1630, a Vultee BT-13A spun into the ground near Harris Field, Fort Valley, Georgia, killing RAF Flying Cadet Arthur W. Wakely. Investigators determined that the aircraft struck the ground while in a spin and in an almost vertical attitude. It could not be determined at what altitude the spin started. The aircraft exploded into flames upon impact with the ground.

6-3-42D. Williams Field, Arizona. At 2355, a Curtiss AT-9 crashed five miles northeast of Williams Field, Chandler, Arizona, killing Aviation Cadet John C. Eustice and Aviation Cadet Irving C. Frank. Investigators could not determine why the airplane crashed and there were no known witnesses. Weather in the area at the time of the crash was reported as ceiling and visibility unlimited.

6-4-42A. Blanco, Texas. At 2135, a North American BC-1 on a night navigation mission crashed 10 miles south of Blanco, Texas, killing Aviation Cadet Thomas J. Schmitt. The aircraft struck the ground in a vertical attitude and at a high rate of speed. Investigators speculated that the student succumbed to oxygen starvation, causing him to inadvertently stall and spin the airplane at an altitude of approximately 17,000 feet. The extreme forces of the spin caused the starboard wing to separate and tore some pieces from the airplane before it struck the ground. The airplane struck the ground with such force that much of the wreckage was buried by impact. The starboard wing was found almost two miles north of the main wreckage. The port wing, still attached to the airplane, had telescoped itself into a bundle along the leading edge. The pilot's badly mangled body had to be cut from the smashed fuselage. The airplane and pilot were found 6-5-42 on the property of Christian Schmitt, no relation to the dead pilot.

6-4-42B. Hamilton Field, California. At 2245, a Consolidated LB-30 collided with a hill three miles WNW of Hamilton Field, California, killing 14 passengers and crewmembers. The aircraft, crewed by employees of the Consolidated Aircraft Company under contract to the Army Air Forces, had taken off at 2239 from Hamilton Field carrying AAF personnel and Martin B-26 parts for a flight to Hickam Field, Oahu, Hawaii. The pilot had been experiencing trouble with the number-two engine, forcing him to feather the propeller and to call the Hamilton Field control tower for a return to the field and emergency landing instructions. The pilot, apparently losing track of the hills surrounding the field, began to turn the airplane back toward the field in an attempt to land to the southeast, the same direction that he had taken off from. Personnel at the field began preparing for the emergency landing and tower personnel cleared the LB-30 for landing. To tower personnel, it appeared that the pilot had the situation under control. The tower, which had been handling two B-17 aircraft awaiting take-off, had lost sight of the bomber and called the LB-30 twice on the radio but received no response. At about the same time, an explosion and flames were spotted in the hills to the northwest. Investigators could not determine with certainty the specific cause of the engine failure. The LB-30 was descending for the approach and struck a tree on top of the hill with the tail, causing it to be torn away. The aircraft crashed and burned in a ravine down the southeast side of the hill. The civilian flight crew killed in the accident were: Alfred O. Fisher, pilot; George W. Bowen, co-pilot; George D. Hilton, flight engineer; Herbert C. Valley, radio operator; Edwin Channing, navigator. The Army Air Forces passengers killed were: TSgt. Alex Mazeikas; SSgt. John Pilarick; SSgt. Joseph E. Kulis; Cpl. Walter E. Haynes; TSgt. Clyde E. Tweedy; TSgt. Loren VanKirk; TSgt. Floyd W. Gerald; Pvt. Travis Polk; Pvt. Alexander Constantopoulos. The passengers were the crew chiefs and technicians for the 70th Bombardment Squadron. The B-26 parts had been stripped from aircraft that were being ferried to the Pacific Theater of Operations via Hickam Field.

6-5-42A. Bradley Field, Connecticut. At 0741, a Curtiss P-40F crashed while maneuvering three miles south of Bradley Field, Windsor Locks, Connecticut, killing 2Lt. Richard M. Stafford, Summit, New Jersey. The pilot was practicing 360-degree overhead approaches at the field. Capt. David C. Shilling, who was grading the landings and approaches, stated to investigators, "He had just completed the first [130-degree] turn and had established the down wind leg when the ship yawed to the left and rolled over on its back. The ship's nose dropped and it was believed by the undersigned that he was going to attempt a roll and recover. The pilot pulled through and dove into the ground, striking it at an angle of approximately 30 degrees. I was standing at the intersection [of the] north/south and southeast/northwest runway, grading the landings. Lt. Stafford had completed four hours of the transition school and displayed better than average ability."

6-5-42B. Greenville, Mississippi. At 1430, two Vultee BT-13A aircraft collided in mid-air one-half mile southeast of the Army Air Field at Greenville, Mississippi, killing two instructors and two students. Killed in the crash of BT-13A #41-10506 were instructor 2Lt. Marvin Taylor and student Aviation Cadet Charles A. Rigsby. Killed in the crash of BT-13A #41-10472 were instructor 2Lt. Eric P. Nordeen and student Aviation Cadet Charles T. Rieder. Investigators stated, "The Committee finds that the two airplanes involved in this air collision made temporary contact at an altitude of approximately 700 feet in the traffic pattern. Lt. Taylor was climbing steeply and was overtaken from the rear by Lt. Nordeen. The left wing of [Lt. Nordeen's BT-13] struck and tore off the right wing of [Lt. Taylor's BT-13], causing this ship to catch fire and fall to the ground. The overtaking airplane, piloted by Lt. Nordeen, fell to the ground in a flat spin, landing on its back. Lt. Nordeen bailed out, but the altitude was insufficient for his parachute to open completely."

6-5-42C. Quonset Point, Rhode Island. At 1602, a Curtiss P-40E crashed while trying to make an emergency landing at the Naval Air Station at Quonset Point, Rhode Island, killing pilot 2Lt. Martin Taub. The P-40 was on a formation training flight with two other P-40 aircraft when it began to experience electrical trouble. Lt. Taub alerted his flight leader through hand signals that he was having trouble and needed to land. The aircraft was soon cleared by Navy personnel for an emergency landing at the station on runway 1. The pilot turned right for runway 1 and the electrical problem necessitated that the landing gear be pumped down manually. The pilot overshot his landing and at an altitude of about 100 feet made a steep left turn parallel with runway 28 with flaps and landing gear down. At the intersection of runway 28 and runway 16, the pilot made a steep left turn heading for the hangar line. The P-40 nearly struck Land Plane Hangar No. 4, clearing it by a few feet. The engine of the P-40 then stopped at an altitude of about 25 feet. The pilot attempted a vertical bank left turn for runway 5 but stalled and spun the airplane into the ground in the middle of runway 5, impacting on the nose and landing gear and collapsing the landing gear. The P-40 then bounced into the air and impacted the runway a second time, exploding into flames as it skidded east, coming to rest near the anti-aircraft machine gun pit. Navy rescue personnel quickly extinguished the flames, but investigation indicated that the pilot died from the violent impact of the airplane with the ground.

6-5-42D. Ropesville, Texas. At 1600, a North American AT-6A flying a formation training flight with another AT-6 airplane stalled and spun into the ground one and a half miles south of Ropesville, Texas, killing instructor 2Lt. Thomas L. Thrasher and student Pfc. James P. Nobles. The student in the other airplane stated that the two aircraft were in a Lufbery circle maneuver approximately 17 miles south of the Lubbock Army Airfield when Lt. Thrasher attempted to pull up the nose to tighten the circle and stalled, spinning to the left about two turns and smashing into the ground.

6-6-42. Ray City, Georgia. At 0230, a Cessna AT-17 on a night navigation mission collided with terrain while letting down through clouds one and a half miles east of Ray City, Georgia, killing Aviation Cadet Jordan T. Luhr and seriously injuring Aviation Cadet James D. Longanecker. The aircraft struck the ground at a shallow angle as a result of the student letting down too rapidly through scattered clouds over extremely dark terrain. Cadet Longanecker stated to investigators, "We took off from Moody Field [Valdosta, Georgia] at approximately 0140 EWT. We were just getting through some wispy clouds when the Control Tower called all airplanes to return to Moody Field. I called the pilot's attention to the altimeter which read 500 feet above sea level and said, 'We had better get some altitude.' That's all I remember."

6-7-42A. Coolidge, Arizona. At 1105, two North American AT-6A aircraft collided in mid-air seven miles SSW of Coolidge Field, Coolidge, Arizona. One flier was killed and three others parachuted to safety. Three AT-6 type aircraft took off on a local flight from Coolidge Field at approximately 0945 in order that one of the officers could take pictures of the aircraft flying in formation. After flying various formations for about 30 minutes, the airplanes began diving and chasing each other. The collision occurred during a diving left turn when the two pilots apparently lost sight of each other. 2Lt. Elbert R. D. Donham and passenger SSgt. Joseph R. Ferg parachuted to safety from AT-6A #41-16396 and were uninjured. Cpl. Lewis Betke, who was flying with 2Lt. Marcus Lee Hill aboard AT-6A #41-16454, was killed when he attempted his bail out at too low an altitude. He was found lying next to the airplane with the parachute draped over him. Lt. Hill received minor injuries in a successful parachute jump. Cpl. Betke, dressed only in swimming trunks and shoes, had temporarily traded places with Pvt. Rottersman, who became ill on the flight to Coolidge Field from Williams Field, Chandler, Arizona.

6-7-42B. Lovington, New Mexico. At 1230, a North American BT-14 dove into the ground on the R.E. Hilburn Ranch 10 miles west of Lovington, New

Mexico, killing pilot 2Lt. Herbert H. Ballew and passenger Cpl. Jack D. Zambrovitz. Witnesses on the ground and in the air observed the aircraft flying at extremely low altitude just prior to the accident. The aircraft had been diving on and buzzing the Hilburn ranch home when the pilot attempted a tight 270-degree left turn and lost control of the airplane, which slammed into the ground. There was no fire.

6-8-42A. Fort Benning, Georgia. At 0815, a Douglas A-20A crashed and three fliers were killed when a bomb they attempted to drop on the bombing range at Fort Benning, Georgia, during a VIP bombing demonstration inadvertently exploded, destroying the aircraft in mid-air and causing it to crash out of control on the range. Numerous officers and enlisted men who were observing the demonstration witnessed the accident, including Gen. George C. Marshall, Lord Louis Mountbatten, and Sir John Dill. A lengthy investigation generated several theories speculating on how and why the bomb exploded. Investigators, after reviewing a great amount of expert testimony, speculated that the bomb became armed on board the aircraft when the bomb doors were opened and a bomb-arming wire somehow fell away prematurely, allowing the fuse vane to begin spinning. The armed bomb fell about eight feet and then detonated under the aircraft, disabling it in mid-air and probably killing the crew instantly. The A-20, still relatively intact but shedding pieces, went out of control and crashed onto the bomb range in a nose down attitude from an altitude of approximately 5,000 feet, exploding into flames upon impact. Wreckage and dismembered bodies were scattered over an area of 500 yards. Killed in the accident were: Capt. Morris Pelham, pilot; 2Lt. Raymond J. Manley, bombardier; Cpl. William C. Rowland, gunner.

6-8-42B. Milstead, Alabama. At 0900, a North American AT-6A flying at low altitude over the Tallapoosa River struck a power line and crashed into the river one mile north of Milstead, Alabama, killing Aviation Cadet Robert A. Dawson and lightly injuring Aviation Cadet Walter I. Lawson. Cadet Dawson, who was previously a civilian flying instructor, had over 1,000 flight hours before enlisting in the Army.

6-8-42C. Prairie, Mississippi. At 1644, a Lockheed Hudson III suffering the failure of its port engine slammed into the ground on a farm one and a half miles north of Prairie, Mississippi, killing four fliers. Witnesses stated that the airplane was at an altitude of approximately 3,000 feet when the port engine stopped and the airplane entered a flat glide. The pilot began maneuvering in an attempt to make an emergency landing when smoke was seen to issue from the port engine nacelle. The pilot added full power to the starboard engine just as the port wing dipped. The aircraft entered a half-roll and spun to the left, smashing into the ground and exploding violently, scattering wreckage over an area of 300 yards. At an altitude of approximately 700 feet the crewmen opened the emergency escape hatches. By the time they were ready to jump they were too low to bail out. An open parachute was found in a tree 100 feet from the crash but it apparently left the airplane when the emergency hatches were opened. Killed in the crash were: 2Lt. Jasper K. Larkin, 21, Palatka, Florida, instructor; Aviation Cadet William K. Anderson, 27, Kent, Ohio, student; Aviation Cadet Harry B. Boulware, 27, Orlando, Florida, student; Aviation Cadet Frank H. Britton, 23, Richmond, Virginia, student.

6-9-42A. Famosa, California. At 1055, a Vultee BT-13A stalled and crashed two and one-half miles northwest of Famosa, California, killing instructor 2Lt. Ralph A. Robdeau and student Aviation Cadet Glenn J. McKean. The instructor and student were apparently practicing simulated forced landings when they let the airplane stall and mush into the ground from an altitude of approximately 400 feet. A witness in the air stated that the airplane struck the ground in a near flying attitude in a wobbly flat descent. The landing gear collapsed when the airplane impacted the ground. There was no fire and the fliers were killed by the impact with the ground.

6-9-42B. Brayton Field, Texas. At 1355, a Fairchild PT-19A collided with trees and crashed seven miles south of Brayton Field, Cuero, Texas, killing civilian flying instructor Frank R. Alden, Shawnee, Kansas, and student Cpl. William Glenn Burr, 21, Des Moines, Iowa. Both were killed instantly. The airplane was practicing a simulated forced landing and apparently lost power while attempting to climb out, colliding with the trees. The airplane smashed through two large trees before it came to rest in a wooded area near the small field. The airplane did not burn when it impacted the ground, but was destroyed utterly. There were no witnesses to the crash and no structural or mechanical cause was evident. Investigators speculated that the pilot might have allowed the engine to "load up," causing diminished performance when power was needed most.

6-9-42C. Half Moon Bay, California. At 1400, two Curtiss P-40F aircraft collided in mid-air over Half Moon Bay, California. Pilots 1Lt. George A. Reeve, Jr., Ashland, Kentucky (P-40F #41-13770), and 2Lt. Walter E. Rumbaugh, Hackensack, New Jersey (P-40F #41-13789), successfully parachuted from their aircraft only to drown while awaiting rescue. The P-40s were part of a four-airplane formation that was patrolling the San Francisco Bay area. The two airplanes executed a slow roll together and then leveled out. The leader then performed a second slow roll but the other P-40 did not follow the leader and flew straight ahead. The lead P-40 rolled up and into the second P-40. The Accident Classification Committee stated, "Investigation revealed that Lt. George A. Reeve's airplane rolled into Lt. Walter E. Rumbaugh's airplane while performing slow rolls in two-ship element formation; and that these acrobatics were being performed with belly tanks installed. This was in violation of existing orders." Lt. Reeve's body was never recovered from the bay.

6-11-42A. Waco, Texas. At 0040, a Vultee BT-13A attempting a landing crashed one-half mile southeast of Waco Army Flying School, Waco, Texas, killing Aviation Cadet J. C. McCrary, 21, Waco, Texas. The airplane was returning from a routine night navigation flight. The Accident Classification Committee stated, "The airplane went into a steep diving turn from approximately 500 feet from which it failed to recover before crashing into the ground in an almost vertical position. Apparently the student pilot was preparing to enter traffic in a normal manner for a night landing terminating his student night cross-country. It is believed that the student pilot lost his horizon while checking something in the cockpit and failed to orientate himself or realize the attitude of the airplane before the airplane crashed into the ground. There is no evidence to show that weather was a contributing factor."

6-11-42B. Portsmouth, Rhode Island. At 1345, a Curtiss P-40E attempting an emergency forced landing on a beach crashed into the Sakonnet River off of Seapowet Point one mile northeast of Portsmouth, Rhode Island, killing 2Lt. William K. Liggett. The P-40 was on a formation training flight with two other P-40 aircraft when it began suffering engine trouble. Witnesses on the ground stated that smoke was issuing from the aircraft as it dropped out of formation and attempted a forced landing on a beach. The pilot noticed that there were people on the beach in his landing path and abruptly pulled up, stalling the airplane and causing it to spin into the water.

6-11-42C. Randolph Field, Texas. At 1420, a Vultee BT-13A stalled and spun into the ground while attempting a simulated forced landing six miles northwest of Randolph Field, San Antonio, Texas, killing instructor 2Lt. Franklin J. Wilgus and student Aviation Cadet Stancell B. Sellers. Investigators determined that the student had shot the simulated forced landing, got too low and made an almost vertically banked turn at low altitude to avoid trees, causing the aircraft to stall and spin into the ground before the instructor could regain control.

6-12-42A. Sarasota Bay, Florida. At 0333, a Boeing B-17E on a celestial navigation training flight to Jackson, Mississippi, crashed into Sarasota Bay, Florida, after striking trees just after take-off from Sarasota Army Air Base, killing eight crewmen, seriously injuring one and slightly injuring another. The B-17 took off to the west on a 4,000-foot East-West runway with no wind and with a heavy coating of dew covering the airplane. One-quarter of a mile west of the end of the runway the B-17 struck two pine trees at approximately 40 feet above the ground. Six feet of the starboard wing and the ball turret were sheared off when the airplane hit the trees, causing it to crash into shallow water 200 feet off shore. The B-17 broke up and caught fire upon impact with the water. Assistant engineer Pfc. John Stratton was thrown clear through the broken fuselage and was lightly injured. He swam to the wreckage and rescued seriously injured assistant navigator 2Lt. Jack E. Hansma and, with witness Maj. C.L. Vandiver, recovered the body of Cpl. James Gee before flames drove them back to shore. Investigators stated, "Due to the conditions under which this take-off was made (comparatively short runway with no wind) the pilot probably had a tendency to 'pull' the airplane off at minimum flying speed. In this case, the airplane would tend to lose altitude slightly soon after leaving the ground. Since the take-off was made into total darkness, any instrument lag or momentary preoccupation of the pilot while adjusting throttles, etc., would have allowed the airplane to strike the trees which reached a height of about 60 feet approximately one-quarter mile west of the airbase. The theory that heavy accumulation dew on the airplane may have added enough drag to cause an abnormally long take-off was advanced by 2Lt. A.E. Hermance. This pilot took off on the same runway in a B-17E airplane approximately two minutes before the time of the accident. He stated that his airplane, covered with dew at the time, felt unusually [sluggish] and that his landing gear actually brushed the tops of the trees later hit by the airplane involved in the accident." Soon after the crash, all of the trees at the west end of the runway were cut down. Killed in the crash were: 2Lt. James H. Butler, pilot; 2Lt. Hugh Lee Kelly, co-pilot; 2Lt. Alfred J. Wilkerson, navigator; Sgt. Rosser Dilley, engineer; Cpl. James Gee, bombardier; Pfc. Edwin Clark, radio operator; Cpl. Jessie Greer, gunner; Sgt. Robert Fulton, passenger.

6-12-42B. Greenville, Mississippi. At 1615, a Vultee BT-13A stalled and spun into the ground five miles south of the Army Basic Flying School at Greenville, Mississippi, killing A/C William G. Allendorph. Investigators stated that Cadet Allendorph "pulled his turn into the traffic pattern so tight that he caused his airplane to stall, resulting in a right hand spin from which he did not have sufficient altitude to recover."

6-12-42C. Burleson, Texas. At 2350, a Vultee BT-13A on a night navigation training flight crashed three miles west of Burleson, Texas, killing A/C Thomas Kelly, 21, Flint, Michigan. A/C Kelly became lost and because of mental confusion disregarded verbal radio commands to fly either 20 miles north to Fort Worth or 20 miles south to Parker and attempted to find a field at Burleson. He misjudged his altitude and flew the airplane into the ground.

6-13-42A. Olympia, Washington. At 0955, a Lockheed P-38E suffering the failure of the port engine collided with a hospital and a power line and crashed at West 4th Street and Sherman Street, Olympia, Washington, killing pilot 2Lt. Ralph M. Edwards. The airplane had just taken off from Olympia Airport, Tumwater, Washington, on a routine training flight. Investigator stated, "The left motor cut out directly after leaving the ground on take-off. The pilot feathered the propeller but was unable to maintain altitude. From the course of the plane, an effort was evidently made to reach the bay approximately 500 yards to the right of the direction of take-off. The rapid loss of altitude forced the pilot to attempt an emergency landing on a city street. The right wing struck the top of St. Peter's Hospital, causing the ship to [strike a power line and] crash into the street." Investigation revealed that a small piece of a neoprene grommet that was replaced during maintenance had found its way into the carburetor nozzle and restricted the fuel flow to the port engine causing its failure. A Sheriff's Deputy suffered a minor injury when he was struck by exploding ammunition. See 7-3-42.

6-13-42B. McChord Field, Washington. At 1155, a Lockheed P-38E suffering the failure of its port engine crashed while attempting an emergency forced landing one-half mile south of the McChord Field, Tacoma, Washington, killing 2Lt. Roy D. Stone. The P-38 was part of a formation training flight and was attempting a landing with its leader when it overshot the landing and went around. Investigators believe that the pilot allowed the engines to "load up" on the initial approach. The P-38 smashed head-on into a large tree and exploded into flames when it slammed into the ground, killing the pilot instantly and completely destroying the airplane. 2Lt. Donald L. Perry witnessed the crash and stated to investigators, "I was sitting in my plane servicing at the pits on the ramp when I noticed Lt. Stone wasn't lined up with the runway for his landing. As he made a turn back to get lined up with the runway, I noticed a puff of white smoke come from his left engine. He was too close to the runway and too high to make a landing. The pilot gave the engine throttle, pulled up his wheels and went around. At the time he had passed over the field, his left engine was sputtering and puffs of white smoke, which turned black, were trailing the left engine. About the time he reached the end of the runway, a clanging noise was coming from the left engine. I turned on my radio and tried to contact him and tell him to feather the propeller on the dead engine and not to turn into the dead engine. He started a turn to the left into the dead engine, so he evidently did not hear me. At the time he first went over the field he had an altitude of about 100 feet. He then had gained approximately 200 feet more before the engine cut out. He leveled off and flew in level flight about 150 degrees around the field. He kept turning to the left and was slowly losing speed. Suddenly his left wing went down and he spun almost straight in."

6-13-42C. Deadman Dry Lake, California. At 1436, a Schweizer TG-2 glider crashed in a launching accident at Deadman Dry Lake, California, ten miles north of Twenty-Nine Palms, killing student glider pilot Pvt. Alexander E. Campbell. The Accident Classification Committee stated, "Accident [was the] result of the pilot using poor technique on a winch launching resulting in the porpoising of launching wire. Pilot released under great tension on tow wire, and with nose high had to dive to get up speed. In doing so, he dove under the tow wire being held in the air by a small parachute. The tow wire wrapped around the glider wing and fuselage, fouling the controls. The glider spun into the ground from approximately 700 feet altitude. Student was seen trying to get out of the glider but apparently the canopy release would not function."

6-13-42D. Mt. Meigs, Alabama. At 2115, two Vultee BT-13 aircraft collided in mid-air three miles east of Mt. Meigs, Alabama, killing RAF Flying Cadet Walter E. Bowden (BT-13 #40-1030) and Flying Cadet James A. Barnes (BT-13 #40-1062). Cadet Bowden's aircraft slammed into a barn one-quarter of a mile north of the Atlanta Highway. Cadet Bowden attempted to parachute to safety but bailed out too low for his parachute to fully deploy. Cadet Barnes' aircraft fell vertically in an inverted attitude into a wooded swamp one-half mile west of Cadet Bowden's aircraft. Only the tree branches directly above the aircraft wreckage were damaged and there was no evidence that the pilot attempted to use his parachute as his safety belt was still fastened. The aircraft were in the pattern to land at Mt. Meigs Airfield when Cadet Bowden missed his spacing turn and his airplane collided with Cadet Barnes' airplane.

6-15-42A. San Francisco, California. At 1130, a Lockheed P-38D suffering the failure of its port engine crashed into an Army Air Forces hangar while attempting an emergency landing at San Francisco Airport, California, killing pilot 2Lt. James H. Mitchell and AAF ground crewman Pvt. Levertt B. Thomas, who was working at the hangar and could not escape the path of the stricken aircraft. Investigators could not determine the cause of the engine failure. 2Lt. John H. Green, who witnessed the crash, stated to investigators, "I was sitting on the edge of the concrete apron starting another plane. Plane came in on extraordinary high approach. Pilot seemed to be having trouble holding ship in line with runway. At about 50 feet above runway, after having traveled about 200 yards down runway, the ship started a gentle turn to the left as if the right engine was pulling too much. The turn increased and ship slipped until the left wing touched the ground near the center of the concrete apron. The ship immediately came back to a level position, with enough force to buckle the landing gear. Ship caught fire and then crashed into the hangar #3 entrance."

6-15-42B. Turner Field, Georgia. At 1535, a Cessna AT-17 suffering the failure of its port engine crashed while trying to make an emergency landing at Turner Field, Albany, Georgia, killing instructor 2Lt. Ellery M. Christianson, 22, Albert Lea, Minnesota, and

RAF Flying Cadet Dennis Hope, 21, Cloverdale Gardens, Newcastle-on-Tyne, England. The aircraft had just taken off from Turner Field on a training flight. Investigation revealed that the port engine oil line failed. The pilot shut down the engine and feathered the propeller but failed to set the fuel selector switch in the "Right On" position, instead setting it in between "Both On" and "Right On," causing a restricted fuel flow to the starboard engine. The engine stopped and the airplane stalled and spun north of the field from an altitude of about 200 feet, striking trees and exploding into flames when it slammed into the ground.

6-16-42A. Dale Mabry Field, Florida. At 0800, a Bell P-39D on a formation training flight was abandoned by its pilot after entering a spin along its pitch axis four miles southeast of Dale Mabry Field, Tallahassee, Florida. Chinese Air Force pilot Sub-Lt. Chi-Chia Li attempted to parachute to safety but was struck by the starboard horizontal stabilizer and killed, his body floating back to earth in the fully deployed parachute. The aircraft had been maneuvering with other aircraft in the flight when the pilot stalled out of a steep, climbing left turn. The aircraft entered a normal type spin to the left and then pitched up and entered the end over end tumbling spin. The pilot decided to bail out. After Lt. Li left the aircraft, the P-39 went into a conventional spin that developed into a flat spin, striking the ground vertically in an upright wings level attitude with no forward speed. Investigators noted that P-39 type fighter airplanes had a tendency to enter end-over-end spins along the pitch axis under certain undeterminable circumstances.

6-16-42B. Glendale, California. At 1105, a Lockheed P-38F suffering a loss of power in its starboard engine struck and killed a civilian airport employee at Grand Central Terminal while making an emergency-landing attempt at Glendale, California. Civilian Alfred A. Machen, 65, Whittier, California, was killed instantly when pilot SSgt. Jess O. Yaryan ground looped the airplane to avoid leaving the runway. The pilot was uninjured. SSgt. Yaryan stated to investigators, "I took off from Glendale [California] Airport at 1045 for an interception mission. After climbing through overcast, my right engine began to cut out and miss resulting in a loss of power and rpm. I switched both propellers into manual with rpms reading 2,500 and circled back for a landing. After making a normal approach, I discovered that my landing gear was not completely down and the hydraulic pressure was zero. I attempted to put my gear up without result. So I pumped the wheels down using the emergency system. On my next approach my engine was missing much worse and I landed without flaps. In attempting to stop before running off the runway, I ground looped the airplane, washing out the right main wheel, the nose wheel, the right wing and the right boom, and rudder assembly. In the course of the landing, I struck a civilian with the ship."

6-16-42C. Sequoia Field, California. At 1428, two Ryan PT-22 type airplanes collided in mid-air while in the pattern to land at Sequoia Field, Visalia, California, killing civilian flying instructor Frank S. Boggs, Jr. aboard PT-22 #41-15509. Instructor Boggs' student, 1Lt. Hamden L. Forkner, and the pilot of PT-22 #41-20657, A/C Henry J. Rodgers, miraculously escaped. Lt. Forkner, under Instructor Bogg's supervision, turned from base to final and then descended into the airplane of Cadet Rodgers at an altitude of about 50 feet agl. Both airplanes were locked together in the collision and spun one-quarter of a turn before striking the ground.

6-17-42A. Prattville, Alabama. At 0050, a North American AT-6A on a night navigation mission dove into the ground with power 10 miles northeast of Prattville, Alabama, killing RAF Flying Cadet Charles D. Eaton. A hunter witnessed the aircraft enter a large cloud and a few moments later observed the aircraft diving out the cloud in a steep spiral. The airplane struck the ground at high speed at a steep angle, exploding violently into flames upon impact.

6-17-42B. Hunter Field, Georgia. At 1335, a Douglas DB-7B crashed while taking off from Hunter Field, Savannah, Georgia, killing 2Lt. James R. Barrow, Jr. The DB-7 was one of a flight of three taking off on a transition training flight. Investigators stated, "After the lead aircraft left the runway, Lt. Barrow started his take-off. The ground run appeared normal. The nose wheel came off the runway after the aircraft had traveled approximately 1,000 feet. Shortly thereafter, the aircraft became airborne. The landing gear wheels started retracting immediately and were in the retracted position when the aircraft was not more than 10 or 15 feet above the surface of the runway. At this point the aircraft appeared to tip its wings sharply to the right and the left. The nose came up in a steep climb and the aircraft assumed a constant attitude of a slow roll to the left about 50 or 60 feet above the surface of the runway. As the degree of bank became steeper, the rotating motion became increased. The aircraft rolled over and, in the inverted position with the right wing slightly lower than the left, struck the hard surface of the new East/West runway on its back. It slid in this position in an easterly direction for a distance of 450 feet from the point of initial impact. At about the time it came to rest, it exploded and burst into flames." Investigation revealed that when the control column and control wheel of this type of airplane were vigorously moved fore and aft and the control wheel was being rotated left to right and back, aileron lock can occur.

6-18-42A. Shafter, California. At 0830, two Vultee BT-13A type airplanes attempting a forbidden cross-over maneuver in formation collided in mid-air two miles north of Shafter, California, killing instructor 2Lt. Walter M. Hill aboard BT-13A #41-1432. A/C William M. Martin, Lt. Hill's student, parachuted to safety, as did the pilot of BT-13A #41-1366, A/C Magnus G. Bolken. The landing gear of Cadet Bolken's BT-13 was impacted by the vertical stabilizer of Lt. Hill's BT-13. It was not known why Lt. Hill was unable to bail out.

6-18-42B. Anaheim, California. At 1500, two Lockheed P-38F airplanes collided in mid-air five miles northwest of Anaheim, California, killing pilots SSgt. Victor E. Thacker (P-38F #41-7502) and SSgt. Willard E. Rath (P-38F #41-7516). The airplanes were part of a three-element, five-ship combat patrol formation flight. The aircraft began executing a series of diving and climbing turns in formation about one mile east of Midway, California, when the collision occurred. SSgt. Jack R. Kingery, sergeant pilot, witnessed the crash and stated to investigators, "A diving turn to the right was executed, followed by a climbing turn to the left. The crash was caused in part by too quick a pull up by SSgt. Rath, wingman of the first element, and partially by SSgt. Thacker, leader of second element, coming into position too fast, overrunning Rath and not being able to pull up and around as much as Rath. Our speed was approximately 300 mph and Thacker mushed in directly on top of Rath. I, being Thacker's wingman, avoided the entangled planes by a few feet and could feel my plane mush and shudder." The two aircraft locked together and made a slow, rotating flat spin from approximately 4,000 feet until they struck the ground and exploded into flames. Neither pilot was able to leave the aircraft.

6-19-42A. Seguin, Texas. At 0055, a North American BT-9B crashed under unknown circumstances two miles east of Seguin, Texas, killing Aviation Cadet Leon C. Harer. The airplane hit the ground at high speed and was completely demolished by the impact. There were no known eyewitnesses to the crash, but residents in the area heard the impact of the airplane with the ground.

6-19-42B. Copperas Cove, Texas. At 1625, a Piper L-4A on a simulated bombing mission crashed 10 miles north of Copperas Cove, Texas, killing pilot 2Lt. Herman C. Lamkin and observer 1Lt. John W. Jones, Jr. Investigators stated, "The pilot and observer flying at an altitude of approximately 200 feet dropped to an altitude of 50 feet to drop sacks of flour on a ground target in simulation of bombing. After passing over the target he banked steeply in a climbing turn for another approach on the target. But before the turn was completed the aircraft stalled, the nose dropped and the plane struck the ground before recovery could be effected." The aircraft burst into flames about 15 seconds after hitting the ground. The pilot lived for about two minutes after being pulled from the burning wreckage. Investigators noted that pilots should be warned about the limited performance capabilities of small observation type aircraft.

6-19-42C. San Antonio, Texas. At 1715, two North American BC-1 type aircraft collided in mid-air 15 miles south of Brooks Field, San Antonio, Texas, killing A/C Lewis Roberts (BC-1 #37-662) and A/C Eugene C. Scott (BC-1 #37-659). The aircraft were part of a six-airplane formation flight that was attempting a landing at Yturri Auxiliary Airfield because one of the BC-1s was experiencing engine roughness. The leader and the other three airplanes (including the BC-1 with the engine trouble) landed safely at Yturri. None of the pilots witnessed the collision, which was believed to have occurred at about 5,000 feet. Civilian witnesses on the ground at first believed that the airplanes tumbling down to earth were performing aerobatics. The engines of both airplanes detached during the collision and fell to earth separately from the main wreckage. The airplanes fell within 200 feet of each other. Neither student attempted to leave the airplane, probably being killed or knocked unconscious by the initial impact.

6-19-42D. Pamlico Sound, North Carolina. At 2135, a Douglas DB-7B on an anti-submarine patrol mission crashed into Pamlico Sound three miles northeast of Ocracoke Light, North Carolina, killing 2Lt. George D. Holmes. The pilot was unsure of his position and the aircraft low on fuel when he ordered the crew to bail out. Bombardier Pfc. Claude M. Rand, radio operator Pfc. Gerald J. Ridnour and passenger Cpl. Robert N. J. Morin parachuted to safety. Cpl. Morin landed in the sea and was quickly rescued by U.S. Coast Guard vessels that were operating in the area. The other two crewmembers landed on the beach uninjured. Although the wreckage of the DB-7 and the body of the pilot were recovered from the sound, investigators could not determine why the pilot did not leave the airplane.

6-20-42A. Hunter Field, Georgia. At 1111, a Douglas DB-7B crashed on approach to Hunter Field, Savannah, Georgia, killing pilot 2Lt. Jesse E. Burall. The subject airplane, which was one of flight of three that took off at 1011 from Hunter Field on a formation training flight, was attempting to land with its flight when it undershot the approach. The left tank ran dry on this low and slow approach, causing the port engine to stop. The aircraft began to lose altitude, prompting the pilot to give the starboard engine full power. The aircraft banked sharply to the left. The pilot attempted to recover but the aircraft skidded to the left and fell from 50 feet into a heavily wooded area near the field. A small fire erupted but was quickly extinguished. Investigators determined fuel mismanagement caused the engine failure. The port outboard fuel tank was completely empty and starboard tank contained about five gallons. Both inboard fuel tanks contained ample quantities of fuel. Investigation revealed that the pilot used the two outboard fuel tanks on the flight. One hour is about the flying time available using these two tanks.

6-20-42B. Grenier Field, New Hampshire. At 1404, a Bell P-39D suffering engine trouble crashed while making a landing approach one mile south of Grenier Field, Manchester, New Hampshire, killing pilot 2Lt. Clevio R. Rogo. The pilot had been scheduled for a two-hour transition training flight but returned to the field after about 25 minutes. The airplane entered the pattern at a reduced flying speed. Witnesses on the ground reported that the engine sounded like it was malfunctioning. The pilot attempted a steep left turn from base to final, stalling the airplane. The aircraft entered a one-turn spin to the left and dove into the ground from an altitude of approximately 350 feet.

✓ **6-20-42C. Victoria, Texas.** At 1315, a Fairchild PT-19A stalled and spun into the ground near Wagener Auxiliary Field, Victoria, Texas, killing Aviation Cadet Jay F. Chappell. The aircraft crashed at the corner of the base leg and downwind leg of the pattern. There were no witnesses to the crash and investigators could not find evidence of mechanical or structural failure.

6-21-42. Hemet, California. At 1905, a Ryan PT-22 maneuvering with two other PT-22 aircraft crashed into the ground two and one-half miles northeast of the airfield at Hemet, California, killing civilian flight instructor Fred C. Kopeinig. The aircraft were seen by witnesses on the ground to be diving at a road in pairs or singularly, pulling up within 50 feet of the ground. During a series of these maneuvers, the pilot apparently blacked out during an extreme pull up and crashed. The pilot had told another pilot that he had blacked out while performing similar maneuvers and had set the elevator trim tab in the nose up position to keep the aircraft in a positive climb attitude in the event that he blacked out again.

6-22-42A. Albuquerque, New Mexico. At 0815, a student bombardier plummeted to his death after he fell out of a Beech AT-11 without a parachute two miles southeast of the Army Air Field at Albuquerque, New Mexico. Aviation Cadet Herbert M. Barag was moving about the aircraft and apparently fell onto the starboard bomb bay door, causing it to give way and allowing him to fall from the airplane. The pilot, 1Lt. Donald E. Guthrie, speculated that the cadet fell into the bomb bay door when he was reaching for his cap, which had apparently fell off his head and onto the bomb bay catwalk. No one in the airplane actually saw the cadet fall into the door but bombardier student Aviation Cadet Winthrop T. Frazee saw the cadet's feet sticking straight up as he fell out of the airplane (heels facing Frazee), an indication that Cadet Barag probably fell onto the door on his back. Bombardier instructor 2Lt. James M. Fiumara, who was sitting in the co-pilot seat, did not see the cadet fall. The pilot also stated to investigators that he did not see the cadet fall. The crew left the cap in place on the catwalk. Investigators could find no evidence that the cadet was despondent and speculated that the cap probably fell off the cadet's head as he fell backward onto the bomb bay door.

6-22-42B. Seguin, Texas. At 1100, a North American BT-14 on an instrument training flight collided with trees while maneuvering at low altitude six miles northeast of Seguin, Texas, killing A/C Marcellus D. Bryant and injuring A/C Felix G. Brenner. Witnesses on the ground reported that the airplane had been buzzing the area and attempted a steep right turn at about 50 feet altitude when the pilot leveled out and then descended into a patch of mesquite trees. The aircraft plowed through the trees for about 150 feet, shearing off the wings and stabilizers and decapitating the pilot before coming to rest.

6-22-42C. San Francisco Bay, California. At 1445, a Lockheed P-38D dove into San Francisco Bay, California, one mile off the east shore and three miles north of the Dumbarton Bridge, killing 2Lt. Jules B. Allain. The P-38 was performing authorized aerobatics at an altitude of 10,000 feet with another P-38 just prior to the accident. The other P-38 pilot lost track of Lt. Allain during the maneuvers. A passenger in a U.S. Navy Vought-Sikorsky OS2U-3 observed the P-38 in a steep dive just prior to it hitting the water. Another P-38 pilot flying in the area observed the subject P-38 just as it impacted the surface. Investigators could not determine the cause of the crash. It is likely that the pilot put the airplane in a diving maneuver from which he could not recover because of the tendency of the P-38 type airplane to suffer from severe control surface buffeting caused by compressibility.

6-23-42A. Lordsburg, New Mexico. At 1600, a Consolidated B-24D crashed after suffering a fire in the number-four engine about 15 miles west of Lordsburg, New Mexico, killing co-pilot Maj. William A. Van Dusen, Lemon Grove, California, and civilian passenger Carl V. Gragg, hometown unknown. Pilot 1Lt. B. F. Jones parachuted to safety. Lt. Jones later stated to investigators, "[We] took off from Lordsburg airport en route to Tucson [Arizona] Municipal airport. After a duration of ten minutes flying, I felt a surge of power from the two left engines. I looked to the right and saw no. 4 engine was on fire. I immediately tried to feather no. 4 propeller but it would not feather. Next I instructed Maj. Van Dusen to turn engine fire extinguisher selector valve to no. 4 and pull handle to release fire extinguisher. No favorable results occurred. I gave orders for the occupants to proceed to the bomb bay and bail out. I sat there and trimmed the ship for three engine flying. After trimming the ship it went into long shallow turn to the right. The ship seemed to be flying fairly well and was slowly losing altitude. By this time the flames were quite large and pieces of materiel was falling from no. 4 engine. I kept yelling for the Major and Mr. Gragg to jump. Both were very hesitant about leaving the ship. I went back on the catwalk and begged them to jump. By this time we were well under a thousand feet from the ground. So I yelled 'Let's go!' and bailed out."

6-23-42B. Lubbock, Texas. At 2255, a Cessna AT-17 dove into the ground after encountering severe thunderstorms one mile east of the Army Air Field at Lubbock, Texas, killing A/C Charles A. Whittington, 25, San Gabriel, California, and A/C Donald C. Upham,

22, Chicago, Illinois. The aircraft was trying to land at the field during a severe rainstorm.

6-24-42A. Hillsgrove, Rhode Island. At 1900, a Curtiss P-40E crashed just after take-off from State Airport, Hillsgrove, Rhode Island, fatally injuring pilot 2Lt. Herbert C. Chamberlain. Witnesses on the ground stated that the engine was missing badly when the pilot warmed it up prior to take-off. Black smoke was observed issuing from the engine on the take-off roll. The aircraft became airborne and gained about 40 feet altitude over the runway intersection when the engine lost considerable rpm. The pilot began a shallow turn to the left and the airplane started to lose altitude until it descended into the ground on a golf course driving range adjacent to the west boundary of the field. The aircraft careened over a fence, across Post Road and burst into flames as it came to rest about 100 feet west of the road. Investigators determined that the pilot's shoulder harness was not fastened, causing him to slam into the instrument panel and gun sight when the airplane hit the ground. He died several hours after the accident.

6-24-42B. Camp Williams, Wisconsin. At 1810 CWT, a Douglas C-49F and a Douglas C-48B collided in mid-air three miles west of Camp Williams, Wisconsin, killing one crew of three. The aircraft were part of a six-ship flight and were attempting a gentle peel off to the left to enter the traffic pattern for Camp Williams Army Air Field when the accident occurred. The C-49 pilot, flying in the number three position, did not wait for the disbanding interval and peeled off in a steep left bank immediately after the C-48 pilot in the number two position began his gentle left turn, causing the collision with the C-48 at an altitude of approximately 3,000 feet. 2Lt. Theodore P. Tatum, the co-pilot of the C-48, stated to investigators, "We were flying a six-ship echelon to the right. I was in the co-pilot seat of the second ship. The pilots agreed on a 15 second disbanding interval and the signal was to be a wiggle of the wings. As we approached the field (Camp Williams) the lead ship wiggled his wings and we passed the signal on down the line. The lead ship then broke out of formation with a slow let down and gentle turn to the left. We counted 15 seconds and let down in a gentle turn to the left as the lead ship had done. I looked back to count off the 15 seconds for the third ship and it was right off our wing tip in a vertical climbing turn and headed to us at about a 90 degree angle to our line of flight. The trailing edge of the left wing of the third ship struck the leading edge of our right wing, starting at the tip and sliding down to the motor nacelle. The forward momentum of our airplane tended to steepen his bank and threw his airplane over our fuselage in a half roll. This was the last I saw of him." The left aileron and wing tip came off the C-49 during its contact with the C-48, the aileron wedging into the leading edge of the C-48 wing and the C-49 wing tip passing underneath the wing of the C-48. After the C-49 passed over the top of the C-48 and fell off into a half roll, the C-49 leveled out momentarily before it crashed to the ground in a vertical dive and exploded in flames. Killed when the C-49 crashed were: 2Lt. Eugene Clark Beadles, pilot; 2Lt. Hulon W. Williams, co-pilot; 2Lt. Samuel Meacham, Jr., passenger. C-48 pilot 1Lt. Maurice W. Wiley, co-pilot 2Lt. Theodore P. Tatum, and engineer SSgt. Claude W. Patterson landed safely at the Camp Williams Airfield. The C-49 (militarized DC-3) had been temporarily acquired by the Army Air Forces from Transcontinental & Western Airlines and the C-48 (militarized DC-3) had been temporarily acquired from United Airlines.

6-25-42A. Augustine Field, Mississippi. At 1545, a Stearman PT-17 stalled and spun into the ground six miles west of Augustine Field, Jackson, Mississippi, killing student pilot Pfc. Harold E. Morelli. The student stalled the airplane while attempting a steep climb at too low altitude to allow recovery. The aircraft struck the ground in a wooded area in an inverted nose down attitude with full power, bursting into flames upon impact.

6-25-42B. Bakersfield, California. At 1630, a Vultee BT-13A performing aerobatics on a maintenance test flight suffered a structural failure of its port wing and crashed 11 miles west of Bakersfield, California, killing pilot 2Lt. Clifford D. Reeder and civilian passenger George E. Peters, a government employee. Investigators determined that the pilot had been "maneuvering the aircraft in such a manner as to impose stresses in excess of those for which the airplane was designed." Further investigation revealed that maintenance personnel had been attaching wing bolts and nuts with excessive torque and this may have contributed to the in-flight wing separation.

6-27-42A. Dale Mabry Field, Florida. At 1310, a Bell P-39D suffering a backfiring engine crashed just after take-off one mile east of Dale Mabry Field, Tallahassee, Florida, killing Chinese Air Force pilot Sub-Lt. Hsun Li. Investigation revealed that pieces of the backfire screen had entered the combustion cylinders, causing the engine to fail. The pilot landed straight ahead when the engine failed but was killed when he struck the gun sight with his head. The aircraft hit the ground hard, careening through a fence and a garden before coming to rest. The airplane did not catch fire.

6-27-42B. Brooksville, Mississippi. At 1755, a North American AT-6B stalled and spun into the ground while maneuvering near Brooksville, Mississippi, killing pilot Capt. Albert A. Norrod, Washington, D.C., and passenger SSgt. Robert S. Landreth, Sparta, North Carolina. The AT-6 was seen by witnesses on the ground to make a steep right turn, stall and then spin to the right and into the ground. The aircraft was on a flight to Barksdale Field, Shreveport, Louisiana.

6-28-42A. Boston, Massachusetts. At 0915, a Curtiss P-40E stalled and spun into the ground while turning from base to final at Boston Airport, Boston Massachusetts, killing 2Lt. Albert J. Wiebe. The P-40 was part of a three-ship formation training flight. At 1,500 feet above the northwest area of Boston the pilot radioed his flight leader 2Lt. Ernest J. Trexler and reported that he was experiencing engine trouble and that he needed to land. The airplane broke off formation and performed a normal 360-degree overhead approach to the field. On the turn to final approach the P-40 banked sharply, lost flying speed and then dove into the ground from an altitude of approximately 100 feet.

6-28-42B. Quonset Point, Rhode Island. At 0934, a Curtiss P-40E engaging in authorized simulated combat with another P-40 suffered a catastrophic structural failure of the port wing while pulling out of a split-S maneuver east of the Quonset Point Naval Air Station, Rhode Island, killing pilot 2Lt. Robert M. Flanders. The two P-40s were maneuvering at 15,000 feet when Lt. Flanders rolled inverted, dove down and attempted to pull out of the split-S at approximately 8,000 feet when the left wing "exploded" off of the aircraft. The wingman, 2Lt. David H. Brown, reported that the airspeed was approximately 400 mph at the bottom of the maneuver.

6-28-42C. Lebanon, Ohio. At approximately 1630, a Lockheed A-29 crashed near Lebanon, Ohio, 25 miles south of Dayton, after suffering a port engine failure, killing the crew of four. The airplane had taken off from Patterson Field, Ohio, on a flight to Orlando Army Air Base, Orlando, Florida. The aircraft circled the Lebanon area for a couple minutes on one engine before stalling and diving into the ground, exploding into flames upon impact. The crew was killed instantly. Investigators determined that the aircraft was heavily loaded and that the pilot had violated several air traffic regulations at the airport in the course of his take-off. Killed in the crash were: 2Lt. William K. Van Zandt, pilot; 2Lt. Ralph A. Oehman, co-pilot; SSgt. Earle J. Corder, passenger; TSgt. William K. Elder, passenger.

6-28-42D. Key Field, Mississippi. At 1715, a Bell P-39D on a transition training flight crashed into a wooded area one mile southwest of Key Field, Meridian, Mississippi, killing pilot 2Lt. Call M. Norwood. The pilot took off at 1702 and soon returned to the field and asked the control tower to take a look at the landing gear. The pilot was instructed by the tower to fly past the tower from west to east at an altitude of 1,000 feet. A witness on the ground observed the right hand cockpit door leave the aircraft at an altitude of approximately 1,000 feet. The tower operator observed the aircraft one mile southwest of the field at an altitude of about 1,000 feet when it executed a sharp left bank and dove straight into the ground and burst into flames.

6-29-42A. Columbus, Mississippi. At 1000, a Curtiss AT-9 and a Cessna AT-8 collided in mid-air eight miles north of the Army Air Field at Columbus, Mississippi, killing A/C Forrest J. Johnson, Jr., North Hollywood, California, and A/C Raeburn F. Gordon, Farmville, Virginia in the AT-9. The pilots of the AT-8, A/C Desmond N. Fairbairne, Easton, Pennsylvania, and A/C Hoover Edwards (hometown unknown) parachuted to safety. The two aircraft were part of three-airplane formation training flight flying at an altitude of approximately 2,000 feet and attempting a cross over to a V-formation from echelon right. When moving from echelon to V-formation in a three-airplane flight, the third airplane moves over and the second airplane maintains position. The AT-9, which was flying in the second position, moved over to the left when the flight leader gave a signal to change the formation. The pilot, realizing his mistake, began to move back and collided with the AT-8, which was moving into the empty number two position. The tail of the AT-9 came in contact with the bottom of the AT-8 fuselage and the starboard propeller of the AT-8 contacted the fuselage of the AT-9. The AT-9 fell to the ground out of control and immediately burst into flames upon impact. The AT-8 leveled out momentarily but the pilots discovered that they had no elevator control and bailed out. The instructor, 2Lt. Richard N. Azar, thought that the AT-8 pilot might have missed two formation change signals, prompting the AT-9 to move over against procedure.

6-29-42B. Ellaville, Georgia. At 1530, a Stearman PT-17 crashed while maneuvering three miles east of Ellaville, Georgia, killing civilian flight instructor Warren E. McGlann, 24, Shellman, Georgia, and RAF Flying Cadet Morris Hislop, 20, Birmingham, England. The aircraft was seen performing a series of loops just prior to the accident. Witnesses on the ground reported seeing Mr. McGlann leave the airplane at an altitude of approximately 100 feet. His parachute did not fully deploy.

6-30-42. Seymour, Connecticut. At approximately 1535, a Republic P-47B flying in a three-airplane formation suffered a catastrophic structural failure of its tail section and crashed on East Hill at Seymour, Connecticut, killing pilot 2Lt. Henry A. Plahetka, 25, Chicago, Illinois. The aircraft crashed into a wooded area near Lake Zoar on the property of the Ansonia Water Company. Parts of the tail section were found within two miles of the crash site. The formation had been maneuvering for a short time when the flight leader and the flight dove down in a turning dive to the left from an altitude of 10,000 feet, pulling out at approximately 2,500 feet at a speed of approximately 375 mph. The entire tail empennage separated in pieces from Lt. Plahetka's P-47 during this pull up. The aircraft pitched up and to the left before it entered a flat spin to the left and struck the ground in a level, upright attitude, smashing trees in a sixty-foot radius. The pilot was ejected from the aircraft upon impact, his body being found 15 feet in front of the left side of the wreckage. The aircraft burst into flames ten minutes after hitting the ground. The pilot was to leave the next day on a five-day furlough to Chicago and planned to marry his fiancé on 7-4-42.

July

7-1-42A. March Field, California. At 0525, a Consolidated LB-30 crashed just after take-off from March Field, Riverside, California, killing the crew of nine. The aircraft was on a tactical patrol mission and crashed two and one-half miles west of the field. The airplane passed over the Camp Haan stockade at very low altitude, struggling to gain altitude in moderate to light foggy conditions. The LB-30 never gained more than 75 or 100 feet in altitude and cleared the 30-foot high Quartermaster's building by about 45 feet. Witnesses on the ground observed the airplane clear a hill and descend beyond it before it disappeared into the light fog, which was reported to be patchy and about 50 feet thick at the time of the accident. Visibility was reported to be four miles with ceiling unlimited. The aircraft then descended into the ground in a flying attitude at near flying speed. The airplane broke up and exploded into flames, careening across the terrain for several hundred yards. About a minute after the plane hit the ground and began burning the bomb load detonated, hurling wreckage and bodies over a wide area. The aircraft was carrying four 500 lb. bombs. Investigators noted that the area where the crash occurred was relatively flat and suitable for a straight ahead crash landing and speculated that the pilots could not sufficiently see the ground to make a safe crash landing or were confused by the fog and flew the airplane into the ground. Witnesses on the ground reported that the engines did not sound like they were producing full power and it appeared that the airplane was laboring to remain in the air. Killed in the crash were: 1Lt. Robert K. Murphy, 25, Twin Falls, Idaho, pilot; 2Lt. Jack W. Green, 24, Kansas City, Kansas, co-pilot; SSgt. Norman S. Anderson, 25, Enderlin, North Dakota, engineer; SSgt. Charles H. Davis, 27, New Boston, Ohio, bombardier; Sgt. John D. Rice, 25, Higbee, Missouri, radio operator; Cpl. Harry Scuffos, Jr., 19, Spartanburg, South Carolina, gunner; Pfc. James P. Stephens, 22, Remington, Virginia, gunner; Pvt. Jack B. Hearrelli, 29, Eldorado, Kansas, gunner; Pvt. William Rosser, 27, Anson, Texas, gunner.

7-1-42B. Premier, Virginia. At 1200, a Douglas C-49E on a personnel transport mission broke up in flight and crashed three-quarters of a mile southwest of Premier, West Virginia, killing four crewmembers and 17 passengers. The heavily loaded aircraft had taken off from Kellogg Field, Battle Creek, Michigan, at 0815 on a flight to Florence, South Carolina, with a stop en route at Lunken Field, Cincinnati, Ohio. Sometime after 1000, the pilot made four unsuccessful landing attempts at Lunken Field. On the pilot's first attempt to land the C-49, he approached with landing gear and flaps down but was misaligned with the runway. On the second and third landing attempts, the airplane approached with the gear and flaps down but the pilot overshot the field both times and went around. On the fourth attempt the airplane approached with excessive speed with gear down and no flaps. The pilot appeared to be forcing the airplane down and the C-49 hit hard on its main wheels and bounced about 20 feet into the air. The airplane descended to the ground and hit the runway hard again, apparently damaging the right main landing gear and compromising the structural integrity of the starboard wing. The airplane then bounced 30 feet into the air and was seen slightly slipping to the left. The pilot applied full power and disappeared to the east with no further radio contact with the Lunken control tower personnel. Investigators noted that the pilot should have landed immediately and inspected the aircraft for damage before continuing on the flight. Nearly two hours later, witnesses on the ground in the vicinity of Welch, West Virginia, observed the aircraft circle the area in and out of low-lying clouds and thunderstorms for about 30 minutes. At 1200, the C-49 was heard and observed to be approaching Premier, West Virginia, from the northwest. Witnesses reported that the engines were surging and backfiring and as the aircraft neared the western edge of Premier it was observed to plunge out of a very dark and dense cloud in a near vertical dive, pulling out at about 800 feet above ground level. During this pull out, the port and starboard wing tips, the outside portions of both ailerons, the starboard wing, both elevators and horizontal stabilizers, the starboard engine and landing gear, progressively separated from the aircraft. The starboard wing detached at the point just outside where the wing joins the engine nacelle. The aircraft then rolled to the right and entered a flat spin with the port engine producing power. The C-49 plunged to the ground, striking trees on the top of a ridge nose first then careening down a wooded ravine sloped about 50 degrees, coming to rest inverted at the base of a mountain on the opposite side of the ravine about 350 feet from the point of initial impact. The airplane immediately caught fire. Weather at the time of the accident was described by local civilians as consisting of low clouds, misting rain and very poor visibility. Wreckage was found scattered for about three-quarters of a mile to the northwest along the ridge that forms the east side of Bad Branch between where the right wing was found at Bad Branch Hollow and the main wreckage. Investigators determined that carburetor icing caused the engine surging. The C-49 was equipped with carburetors that could produce dangerous and unrecoverable carburetor icing conditions if mismanaged by the flight crew. This engine surging could have caused severe vibration of the wing and nacelle. Investigators speculated that if the engine had been operated for an appreciable time with surging, cutting out and backfiring at full power, it could "cause the rear counterweight to shift, creating a running misalignment of the crank shaft resulting in a very rough engine and serious vibration which, if continued, could cause eventual disintegration of the aircraft." Investigators listed the probable cause of the accident as the "Action of the pilot in subjecting the heavily loaded aircraft to abnormal severe shock and progressive torsional loads beyond those for which it is designed to withstand resulting in a structural failure of the right wing in flight." Newspaper accounts of the crash reported that "thousands" of local residents arrived in the area to view the crash scene. Killed in the crash were: 2Lt. Walter B. Faught, Sacramento, California, pilot; 2Lt. Edwin A. Harness, La Porte, Indiana, co-pilot; 2Lt. Harry Ellsworth Haney (hometown unknown), passenger and assistant co-pilot; Cpl. Claudius Haywood III, Ambler, Pennsylvania, engineer; TSgt. Salvatore T. Barone, Mt. Morris, New York; Pvt. Elmer J. Campbell, Luray, Virginia; Pvt. Edward F. X. Carr, Philadelphia, Pennsylvania; SSgt. Russell A. Carter, Meriden, Kansas; SSgt. Archie W. Champlin, Leavenworth, Kansas; SSgt. Leo T. Colburn, Tuscaloosa, Alabama; Pvt. Elmo Edmonds, Payson, Illinois; Pfc. Lester S. Erickson, South Range, Wisconsin; SSgt. George W. Eversull, Dexter, Iowa; Pfc. Denver W. Koeppe, Ney, Ohio; Cpl. Kermit S. Masters, Greenview, Missouri; SSgt. Leon A. Olin, Big Timber, Montana; Pvt. Chester F. Peterowski, Auburn, New York; SSgt. Pilot Anthony W. Belcher; SSgt. Joseph S. Cholewa; SSgt. Horace M. Dickson; Cpl. Bartola P. Purpura. On 7-4-42, a mass service with military escorts and a rifle salute was held for the victims at the Norfolk and Western Railroad Station at Welch, West Virginia, before the bodies were removed to their respective hometowns. The C-49E (militarized DC-3) had been temporarily acquired by the Army Air Forces from American Airlines.

7-1-42C. Thomasville, Georgia. At 2000, a Bell P-39F being ferried from Savannah, Georgia, to Tallahassee, Florida, crashed five miles southeast of Thomasville, Georgia, killing pilot 1Lt. F. Sharples. 1Lt. James P. Davis, who was accompanying Lt. Sharples on the ferry flight, witnessed the crash. He stated to investigators, "I was ferrying a P-39F from Savannah to Tallahassee leading a loose formation consisting of Lt. Sharples and myself. Lt. Sharples was flight commander but at the time I was leading and navigating. We were at 3,000 feet in clear weather east of Thomasville, Georgia, when Lt. Sharples pulled up on my right wing and motioned back with both hands with his palms facing him. He also pointed back or at himself with his right thumb. I retarded my throttle and he passed me. When about 200 yards in front of me and to my right he peeled off to the left going into a vertical bank and then a shallow dive. I noticed condensation trails at each wing tip. He pulled out but his controls apparently were not all intact for the ship kept slipping from side to side in what appeared to be a 'fishtail.' He crashed in a wooded area and flames immediately came from the wreckage. The plane had not at any time gone into a spin. I tried to contact the tower at Tallahassee but was unable to do so. I landed at Tallahassee twenty minutes later and reported the accident." Investigators could not determine the cause of the accident.

7-1-42D. Long Beach, California. At 1845, a Lockheed P-38F taking off from the municipal airport at Long Beach, California, collided with a Ryan PT-22 that had taxied onto the active runway (25L), killing civilian flight instructor Robert A. Turner, Prescott, Arkansas. The P-38 pilot, SSgt. Pincus P. Taback, was not injured. The P-38, leading a flight of four on a tactical mission, began his take-off roll with his wingman on his left. After the two aircraft reached a speed of about 70 mph and covered about one-third of the runway, the PT-22 attempted to taxi across the runway from right to left, taxiing directly in the path of the P-38s. Sergeant pilot Taback hauled back on the control column in an effort to clear the Ryan trainer but struck the cockpit area with his starboard tail boom and the PT-22 propeller with the nose wheel. The P-38 then landed straight ahead, the nose wheel collapsed and the propellers contacted the runway surface. The airplane ran off the end of the runway and veered to the right before coming to rest. The P-38 flight had been warned by the tower that four trainers would be crossing the runway. Three PT-22s crossed successfully and the P-38 flight apparently thought that all the trainers had crossed and proceeded with the take-off. Investigators faulted both pilots for not closely monitoring traffic and avoiding the collision.

7-2-42A. Helendale, California. At 1615, a Cessna AT-17 on an instrument training flight crashed after suffering a structural failure of its wing two miles northeast of Helendale, California, killing Aviation Cadet Wayne Kiger and Aviation Cadet Harvey B. Jensen. The Accident Classification Committee stated, "Accident is believed to be caused by too abrupt pull-out in recovering from stall. Evidence indicates plane was stalled steeply shortly after student went under hood to fly instruments, with resulting failure of the wing. Examination of fabric and plywood pulled from the plane in the air showed no evidence of failure from any cause other than overstress of members. Safety pilot considered partially responsible for permitting pilot to get plane in position where overstressing was necessary to recover."

7-2-42B. Roswell, New Mexico. At 2340, a North American AT-6A crashed while performing touch-and-go landings at the municipal airport at Roswell, New Mexico, killing Aviation Cadet Norman A. Pepper and seriously injuring flight instructor 2Lt. Winston J. Bradford. The airplane flew into the ground one-quarter of a mile northwest of the Civilian Conservation Corps (CCC) camp based near there. The

CCC workers rendered immediate first aid to the surviving pilot. Investigators stated, "Accident may have been caused by overshooting the runway on a follow-thru landing and a take-off at reduced speed and propeller pitch with a possibility of flaps being raised after take-off. It is considered possible that a stall resulted or when flaps were raised the ship lost altitude and in recovering the instructor had insufficient altitude and made contact with the ground in a normal flying position. The nature of the terrain was such that the fuselage was sheared from the wings and catapulted through the air."

7-3-42. Olympia, Washington. At 0930, a Lockheed P-38E suffering a power loss in the starboard engine crashed while attempting an emergency landing at the municipal airport at Olympia, Washington, killing pilot 2Lt. Walter M. Jordan. The P-38, which had taken off at 0740 from Olympia with three other P-38 airplanes on a patrol mission, was turning from base to final when it began to slip to the right and slammed into the ground, bursting into flames upon impact. Investigators discovered small pieces of a neoprene grommet in the carburetor nozzle of the starboard engine, restricting fuel flow to that engine. Lt. Jordan's friend, Lt. Ralph M. Edwards, was killed in an identical P-38 accident the month before (see 6-13-42A). Lt. Jordan had returned to Olympia three days earlier after accompanying Lt. Edwards' body east for a military funeral. Lt. Jordan, originally from Houston, Texas, resided in Olympia with his wife.

7-4-42A. Tyndall Field, Florida. At 0915, a Vultee BT-13A spun into a swamp ten miles north of Tyndall Field, Panama City, Florida, killing SSgt. Karl R. Berg, 22, Long Beach, California, and SSgt. John H. Liles, 21, Bailey, North Carolina. The Accident Classification Committee stated, "The airplane struck the ground in a vertical position with such force as to bury the engine approximately seven feet in the soft earth. The fuselage from the rear cockpit back was broken off and came to rest on the vertical stabilizer; both pilot and passenger were in their respective cockpits and below the surface of the ground. The immediate cause of the accident was an uncontrollable tailspin. At an altitude of 2,500 feet, the airplane was in a rather tight spin and that so far as could be seen, no motion of the airplane indicated any visible effort to recover from the spin or [effort of the pilots] to leave the airplane by parachute."

7-4-42B. Selma, Alabama. At 1229, a Seversky P-35 flew into the ground one mile north of Furniss Field, Selma, Alabama, killing pilot Maj. Ross T. Hopkins, Adairsville, Georgia. Investigators speculated that the pilot was attempting a landing at the field and struck the ground in a wooded area in a nose down, left wing low attitude. They could not determine why the airplane crashed.

7-4-42C. Hamilton Field, California. At 1145, a Lockheed P-38F crashed five miles northwest of Hamilton Field, California, killing pilot 2Lt. Ernest M. Trice. The airplane was one of a flight of five on a local formation flight and stalled out of a steep, climbing right turn in an attempt to rejoin the formation. 2Lt. Richard F. Kenny witnessed the accident and stated later, "I was on Lt. Trice's wing, when I dropped behind him because my wheels wouldn't come up. Trice called me and said 'Your wheels are down.' I called back and said, 'I know it. I can't get them up.' Then I retracted the wheels and called back saying, 'My wheels are up.' He then started turning more to the right to join the formation. He was very low and I was gaining on him. After this I saw him drop off to the right and do about a turn and one-half spiral. As his altitude gave out, he reversed the spiral and started in a steep dive with slight bank to the left and hit the side of a hill. The plane burst into flames about two seconds after striking and the flames covered about a fifty foot area."

7-5-42. Ambridge, Pennsylvania. At 1242, a Vultee BT-13A crashed while performing aerobatics at low altitude over Conway Airport, Ambridge, Pennsylvania, killing pilot 2Lt. Michael M. Majcher, 26, Ambridge, Pennsylvania. The Accident Classification Committee stated, "On July 3, 1942, 2Lt. Michael Majcher, on extended cross country flight landed at Conway Airport a short distance from his home in Ambridge. After about half an hour he flew to the Allegheny Airport, Pittsburgh, Pennsylvania, to leave the airplane at the Army hangar. Before leaving Pittsburgh on July 5, 1942, he apparently told his family and friends that he would pass over the Conway Airport en route back to his home station. On this date at approximately 1240, he approached the Conway Airport from the south. He did a left vertical turn around the field, at an estimated altitude of 300 feet, straightened out, climbed to 500 feet and snap rolled to the right. From the account of reputable witnesses, he failed to gain control after the roll, which resulted in an unexpected power-stall and spin from which he attempted to recover, but dove into the ground."

7-6-42A. Americus, Georgia. At 0102, a Beech AT-7 flew into terrain with power three miles west of Souther Field, Americus, Georgia, killing four fliers and seriously injuring navigation student Aviation Cadet Embry G. Scott. The aircraft was on a night navigation training flight when it encountered poor weather. The pilot evidently lost track of his altitude and flew the airplane into the ground. The altimeter found by investigators read 870 feet and the crash site was at an elevation of 470 feet. Rain was reported in the area at the time of the crash. Investigators could not determine why the pilot was flying low in the area and the surviving cadet could not offer an explanation. The aircraft encountered rain near Alma, Georgia, and was right on course up until that time. The aircraft was then slightly left of course for a time but this was corrected by the time the plane crashed. Killed in the crash were: 1Lt. Peter Mikkelson, Hartford, Connecticut, pilot; 2Lt. G. W. Eggeling, San Antonio, Texas, navigation instructor; Aviation Cadet W. M. Seligman, New York, New York, navigation student; Aviation Cadet Wilson C. Yeager, Albany, Georgia, navigation student.

7-6-42B. Lake Charles, Louisiana. At 1145, a North American AT-6A on a navigation training flight descended into terrain three and one-half miles southeast of Lake Charles, Louisiana, killing Aviation Cadet Jacob H. Neff. The aircraft impacted the ground at a high rate of speed and at a flat angle estimated by investigators to be approximately 30 degrees. Wreckage was scattered for over a hundred yards in a wooded area and a flooded rice paddy.

7-6-42C. Santa Anna, Texas. At 1315, a Fairchild PT-19A stalled and spun into the ground two miles east of Santa Anna, Texas, killing Aviation Cadet James R. DeVita. The Accident Classification Committee stated, "Aviation Cadet DeVita had been practicing spin recoveries on this training flight. He entered into a spin to the left at approximately 3,000 feet. In attempting to recover he failed to get the stick forward, which resulted in a continued spin. Since the airplane did not recover as soon as the pilot anticipated he probably became frantic and excited. The maneuver developed into a flat spin and continued that way until [the airplane] struck the ground."

7-6-42D. Hunter Field, Georgia. At 1546, a Vultee V-72 (similar to the A-31 type) suffering an engine failure due to fuel starvation crashed while attempting an emergency forced landing approximately three miles east of Hunter Field, Savannah, Georgia, killing pilot 1Lt. Douglas K. Strope. The descending aircraft struck an 85-foot jack pine tree approximately 18 feet from the top, careening forward for a hundred yards before coming to rest inverted in an open field. The pilot was thrown clear of the cockpit. Lt. Strope, who was on limited flying status, was warned that the aircraft was not fully serviced and that a flight of any appreciable time should not be attempted. Lt. Strope was airborne for approximately one hour and 52 minutes.

7-6-42E. Spartanburg, South Carolina. At 1720, a Bell P-39D stalled and crashed out of control into the ground while on approach to the airfield at Spartanburg, South Carolina, killing pilot 2Lt. Theodore F. Galloway. The aircraft was observed by other pilots in the pattern to be flying the approach at about 50 feet above ground level and at a slow speed. The port wing dropped and the airplane smashed into the ground, bursting into flames upon impact.

7-7-42A. Phoenix, Arizona. At 1050, a North American AT-6C stalled and spun into the ground while turning from base to final at Auxiliary Field #2, ten miles NNW of Luke Field, Phoenix, Arizona, killing A/C Kenneth D. Weir, 22, Denver, Colorado.

7-7-42B. Trenton, Georgia. At 1340, a Consolidated B-24D flying in a severe thunderstorm suffered a catastrophic failure of the port elevator and horizontal stabilizer in flight, causing the airplane to crash into a mountain one mile north of Trenton, Georgia, killing five crewmen. Killed in the crash were: 2Lt. Earl M. Hobson, 28, San Carlos, Arizona, pilot; 2Lt. Earl B. Wood, 27, McComb, Mississippi, co-pilot; SSgt. Warren E. McMurtray, 25, Beaumont, Texas, engineer; Sgt. Roland F. Carrigan, 22, St. Johnsbury, Vermont, radio operator; Cpl. Thomas R. Thurman, Jr., 19, Palestine, Texas, assistant engineer. The aircraft took off at 1202 CWT from Barksdale Field, Shreveport, Louisiana, on a personnel ferry flight to Lovell Field, Chattanooga, Tennessee. The Accident Classification Committee stated, "A section of the left elevator was found about 400 yards from the main body of the wreckage and about 250 yards from the left horizontal stabilizer and rudder, indicating that these parts of the airplane came off prior to the crash of the airplane. There were no trees broken off or no limbs of trees broken off except those actually hit by the airplane as it plunged into the ground at an angle estimated to be 80 to 90 degrees. The pilot was flying a northerly heading below the overcast, saw a rainstorm ahead and turned left. He then saw the mountain and increased his turn until he was headed nearly south. There was a cliff in front of him there, so he increased his turn and it was then that the tail come off."

7-7-42C. Fort Valley, Georgia. At 1445, a Vultee BT-13A stalled and spun into the ground while performing low altitude aerobatics at Harris Field, Fort Valley, Georgia, killing pilot 2Lt. John T. Offutt, 30, New Castle, Pennsylvania, and passenger 2Lt. Robert E. Adams, 22, Huron, North Dakota. The airplane had flown to Fort Valley so that Lt. Adams could ferry an aircraft back to Cochran Field, Georgia. When the aircraft arrived over the field, Lt. Offutt dove across the field and flew above the runway at very low altitude. As he reached the end of the runway he pulled up in a slight climb and executed a climbing slow roll. At the completion of the maneuver the aircraft stalled and spun in from about 200 feet. The aircraft struck the ground on the nose and port wing at an angle of approximately 45 degrees, exploding into flames upon impact.

7-7-42D. Wendover Field, Utah. At 2235, a Boeing B-17E collided with terrain 10 miles SSE of Wendover Field, Utah, killing the crew of seven. The B-17 had taken off from Wendover Field at 2015 and was on a local night instrument and night landing proficiency check flight. There were no eyewitnesses to the crash. The B-17 exploded violently into flames and wreckage was scattered over a wide area. Killed in the crash were: 1Lt. Arthur L. Birleffi, Spokane, Washington, pilot; 2Lt. Van Vanderbie, Los Angeles, California, co-pilot; 2Lt. Floyd R. Reno, St. Joseph, Missouri, assistant co-pilot; Pvt. Eugene P. Holsey, Tulsa, Oklahoma, engineer; Pvt. Jessie C. Swann, Albany, Georgia, radio operator; Pvt. Homer R. Enswiler, Columbus, Ohio, gunner; Pvt. Kenneth Gillies, Jackson, Michigan, gunner. "The opinion of the Accident

Classification Committee is that the pilot turned away from the field, lost flight horizon, and the airplane went into a right wing low diving turn, striking [the ground with] the tip of the right wing and nose simultaneously at approximately a 45 degree or better angle, burying the engines to an undetermined depth since only one engine could be extricated from the earth. The crash was followed immediately by a violent explosion and fire. No bombs were aboard the aircraft. [It was a] dark night with no horizon over smooth and level terrain [and] nearest hill or mountainous terrain some 15 miles distant. Terrain was desert, absolutely flat and substantial and an attempted forced landing could have been accomplished in any direction from the spot of impact. At 2230, the pilot reported to squadron operations, giving his position as seven miles NE of the station, course 180 degrees, altitude 9,000 feet. Approximately five minutes after the routine position report, flames were observed from the control tower at Wendover Field disclosing the scene of the crash, 10 miles SSE of the field."

7-8-42A. Luke Field, Arizona. At 0200, a North American AT-6B crashed into the ground and exploded into flames while turning from base to final one mile south of Luke Field, Phoenix, Arizona, killing Aviation Cadet Edgar B. Wilson, 21, Shreveport, Louisiana. Investigators speculated that the pilot lost the horizon while attempting a right turn to final approach over the dark terrain, was unable to orient himself, suffered vertigo and flew the airplane into the ground. One witness reported that the airplane "slipped off on the inside of the turn. It did not spin, but slipped all the way in."

7-8-42B. Spartanburg, South Carolina. At 1350, a Bell P-39D stalled and spun to the ground out of control 10 miles southwest of Spartanburg, South Carolina, killing pilot 2Lt. Jonathon A. Crites. The pilot bailed out of the airplane at an altitude too low for his parachute to fully deploy. The Accident Classification Committee stated, "It is the opinion of this committee that Lt. Crites while flying on specified Group Engineering Training Mission No. 6, which calls for tight vertical turns to the point of stalling and steep lazy-eights, either fell into a tumbling spin from lazy eight or snapped in a tight turn. It has been found in the past that inexperienced pilots rush recovery from spins without giving ship sufficient time or speed to properly recover. It was evident that the airplane was in a flat spin from attitude at which it hit the ground."

7-8-42C. Davis-Monthan Field, Arizona. At 1200, a Consolidated B-24D crashed while attempting an emergency landing after running out of fuel six miles southeast of Davis-Monthan Field, Tucson, Arizona, killing B-24 student 2Lt. Robert R. Barton, Jr., Charleston, West Virginia. Four crewmembers were seriously injured and three others suffered minor injuries when they were thrown clear of the aircraft when it broke apart. Four rated pilots were aboard for B-24 transition training, each alternating in the pilot's seat to perform take-offs and landings. The B-24 was approaching the field from the southeast when the number-three and number-four engines quit due to lack of fuel. Lt. Barton was flying in the left seat at the time of the crash and instructor 1Lt. Robert J. Colleron, La Crosse, Wisconsin, who suffered minor injuries, was in the right hand seat. Investigators stated, "The airplane had 1,200 gallons of fuel aboard at take-off about 0815; that the direct cause of the crash was complete consumption of the fuel in tanks number-three and number-four; the running out on the two tanks occurring within an interval of one minute; that no fuel was lost in flight; that tanks number-one and number-two contained more than the estimated 200 gallons at time of crash; that the unequal distribution of remaining fuel in the airplane at time of crash was caused by transfer of fuel in the air by the crew; and that a contributing factor was the pilot's delay in cutting engines number-one and number-two after the first indication power failure on number-four engine. Observation at the scene revealed that the aircraft struck the ground in flat brush covered desert terrain on the right wing tip with landing gear retracted at an assumed airspeed of 90 mph. Following the impact, the aircraft bounced into the air and continued approximately 200 feet before second contact at which time the fuselage broke off at the rear of the bomb-bay. All engines were broken from their mounts and the nose section was demolished. The main wing section came to rest in an inverted position. All occupants, except Lt. Barton, were thrown clear of the wreck. The aircraft was completely demolished, and wreckage strewn for a total distance of 150 yards along a southwesterly heading." The aircraft did not catch fire. 2Lt. E. L. Bauder, Aliquippa, Pennsylvania, co-pilot, and Sgt. R. E. Avery, Levelland, Texas, engineer, suffered minor injuries. Suffering serious injuries were: 2Lt. J. W. Davidson, Huntington, West Virginia, co-pilot; 2Lt. Ora T. Dawson, Stillwell, Oklahoma, co-pilot; Pvt. S. J. Hunter, Cleveland, Ohio, assistant engineer; Sgt. W. F. Miller, Absecon, New Jersey, radio operator.

7-8-42D. Sioux City, Iowa. At 1352, a Beech F-2 (photo-reconnaissance version of the AT-7) struck power lines while maneuvering to land five miles north of the Sioux City Army Air Base, Sioux City, Iowa, killing the crew of four. Three crewmembers attempted to parachute to safety but bailed out at an altitude too low to allow their parachutes to deploy. The aircraft had taken off from Huron, South Dakota, at 0918 CWT and arrived over Sioux City at 1255 CWT to take photographs. After the photographic mission the pilot was going to attempt to land at Rickenbacker Field but deemed the field too small to land his aircraft safely. The pilot then headed toward the Sioux City Army Air Base, flew an approach and then flew off. The pilot then returned to the field with gear down, no flaps and was in what appeared to be a normal landing glide attitude. The pilot let the aircraft descend very close to the ground near the runway threshold at a level below the temporary high voltage line strung near the northwest end of the field. The pilot spotted the power line and attempted to pull up and over it. The aircraft made contact with the top line with the port wing, taking 90 feet of cable with it. The pilot circled the field and asked the tower to take a look at the landing gear for damage. At about five miles north of the airbase, the cable wound its way around the port engine propeller shaft, causing the port engine to fail. The pilot was unable to maintain flying speed and the aircraft stalled and spun in from an altitude of approximately 500 feet. The pilot 2Lt. P. E. Everette, engineer SSgt. R. W. Shores, and crewmember Sgt. J. R. Jones were killed when they bailed out at too low an altitude. Crewmember Cpl. J. H. Plew went down with the airplane.

7-8-42E. Wilson Mills, North Carolina. At 1655, a North American B-25B flying in poor weather crashed at Wilson Mills, North Carolina, killing the crew of seven. The airplane broke up while executing an extreme pull-up maneuver to avoid rising terrain, crashing into a field and exploding into flames upon impact. Weather at the time of the accident was reported as thunderstorms with heavy rain and lightning. Investigators stated, "This pilot flew off his course to avoid storms which developed in his path after take-off and that finally in his endeavor to get back to destination he flew into the storm which Raleigh [North Carolina air traffic control] had reported at 1610. [The aircraft] developed excessive airspeed, came [diving out of the overcast] in the clear and being close to the ground pulled up, and in the pull-up put such stress on the wings as to cause failure. Break up of the left wing followed by loss of right wing tip and part of tail surfaces and was attended by fire in the left [fuel] tank area. Wing wreckage passing back over fuselage demolished the plexi-glass covering the upper turret, parts of which were located over half a mile from the scene of the crash. Exploding of remaining tanks at moment of impact scattered wreckage, two bodies and burning fuel to some distance from wreck." Killed in the crash were: Capt. James A. Plant, pilot; 2Lt. Howard E. Henthorn, co-pilot; Sgt. Arlyn A. Gustafson, engineer; SSgt. Louis F. Turner, radio operator; Pvt. Samuel A. McDonald, passenger; Pvt. Frank E. Sailer, passenger; Pvt. Joseph A. Schmidt, passenger.

7-8-42F. Needles, California. At 2215, a Cessna AT-17 flew into the ground two miles south of the municipal airport at Needles, California, killing A/C D.R. Collins and A/C R.J. Kirkpatrick. Witnesses on the ground observed the airplane fly over the field parallel to the runway and then begin a turn on approach to land. The airplane then dove into the ground and burst into flames. Investigators speculated that the pilots probably lost the horizon in the turn over the dark terrain.

7-9-42A. Maxwell Field, Alabama. At 0310, a North American AT-6A on a formation flight with two other AT-6 airplanes left the formation and flew into the ground 12 miles northwest of Maxwell Field, Montgomery, Alabama, killing A/C Leroy A. Wilkins.

7-9-42B. Fort Morgan, Colorado. At 0926, a Taylorcraft L-2A crashed nine miles northeast of the airfield at Fort Morgan, Colorado, killing civilian instructor Harold C. Moore, 32, Gillette, Wyoming, and student A/C James W. Scheurich, 26, Imperial, Nebraska. The Accident Classification Committee stated, "The airplane was flying at an altitude of approximately 500 feet. The ship went into a spin out of a shallow [left] turn, did one and one-half turns and recovered from the spin, but had insufficient altitude to recover from the dive. [The aircraft] crashed into the ground with both wings level."

7-9-42C. Gardner Field, California. At 0850, a Vultee BT-13A spun into the ground 10 miles east of Gardner Field, Taft, California, killing Aviation Cadet John A. Moran. The pilot had been practicing spins. The airplane was observed to spin down from about 5,000 feet, describing about five turns of spin before striking the ground in a nose down attitude and bursting into flames.

7-9-42D. Columbus, Mississippi. At 2230, a Curtiss AT-9 on a night formation flight crashed four miles north of the Columbus Army Flying School, Columbus, Mississippi, killing instructor 1Lt. Donald V. Thompson, 23, Fort Worth, Texas, and student Aviation Cadet Charles J. Dimmling, 26, Baltimore, Maryland. The aircraft had peeled off of the formation and had entered the pattern for the field when witnesses on the ground observed the airplane fly into the ground. The aircraft exploded into flames upon impact. The aircraft appeared to be functioning normally and it is likely that the pilots lost the horizon on the dark terrain while making a turn from the downwind leg to the base leg of the pattern. Lt. Thompson, a West Point graduate, was the son of Army Air Forces flying officer Col. B.S. Thompson, Fort Worth.

7-10-42A. Moore Field, Texas. At 1145, a North American AT-6A suffered a catastrophic failure of the port wing while maneuvering 10 miles northeast of Moore Field, Mission, Texas, slamming into the ground with high velocity and killing instructor 2Lt. Charles L. Barbour and student A/C Clifford R. Crow. The student was scheduled to do hood work on an instrument training flight. Witnesses on the ground observed the airplane at an altitude of approximately 500 feet above ground level when one third of the port wing separated at the root of the aileron well and fell one third of a mile from where the airplane crashed into the ground in a large ditch. Body parts and wreckage were scattered over an area of 300 feet. The airplane hit the ground with such force that the wreckage was severely crushed into the mud and the badly dismembered and mangled bodies had to be removed from the ground with the help of a hoist cable fastened to the parachute harness and seat backs. The propeller and engine were buried several feet into the ground. Investigators could not determine what caused the wing failure or if the pilot had executed an extreme maneuver and had caused an overstress of the wing.

7-10-42B. Mitchel Field, New York. At 1428, two Republic P-47B aircraft collided in mid-air near Mitchel Field, Long Island, New York, killing 2Lt. Lyman F. Rhodes, IV aboard P-47B #41-5934. Lt. Rhodes ascended into the leader of the three-airplane flight, contacting the lead airplane's propeller with the aft fuselage of his airplane. The pilot of the other airplane, Capt. Phillip E. Tukey, landed P-47B #41-5907 safely at Mitchel Field. Capt. Tukey later stated to investigators, "The accident occurred during the performance of a routine formation training mission. There were three P-47Bs in the flight. The mission proceeded normally with practice consisting of three-ship V and three-ship string formation. The breakdown signal was given and a mild 'rat race' followed. After a few minutes of steep climbing and diving turns, the assembly signal was given which should have brought the formation back to its original three-ship V. The circle was to the right. The no. 2 man assembled promptly. Lt. Rhodes was experiencing some difficulty in assembling and I was coaching him over the radio. In his first attempt he overshot, passed beneath me, and ended up to my left and somewhat behind. This was probably due to the fact that he had reduced power in an attempt to bring himself back to position. I directed him to cut the circle again. This time he appeared to be making a satisfactory assembly except that he was somewhat below my level. I sighted him to my right and watched him disappear below and slightly behind me. I glanced to my left several times, expecting to see him in formation but never sighting him until the instant of impact. The airplane appeared directly in front of me simultaneously with the crash. I saw his propeller and a portion of the forward part of his airplane, but due to a film of oil on the windshield I cannot definitely establish with what part of his airplane my propeller made contact. At the time of impact, Rhodes' ship was pointing up in at least a 45-degree climbing position. Directly after the impact, I caught a glimpse of a falling object, which I believe was the empennage of the other ship. It is possible that in attempting to get into position he misjudged or over-controlled and pulled himself up into my airplane rather than the no. 3 position."

7-10-42C. Augusta, Georgia. At 2310, a Vultee BT-15 stalled and spun into the ground after suffring an engine failure two miles northeast of Bush Field, Augusta, Georgia, killing Aviation Cadet William R. Case. The airplane spun in from an altitude of 800 feet while turning from the downwind leg to the base leg of the pattern.

7-11-42A. Oriental, North Carolina. At 1140, a Douglas DB-7B returning from an anti-submarine patrol crashed in the Meuse River near Oriental, North Carolina, while attempting a landing at Cherry Point Marine Corps Air Station, killing two fliers and injuring two others, one seriously. While the pilot was circling the Cherry Point area at an altitude of approximately 1,000 feet in preparation to enter the traffic pattern, the starboard engine quit. The pilot jettisoned the four depth charges over the river, made a turn to the left toward the general area of the field. The airplane apparently was losing altitude in the turn so the pilot abruptly leveled out the airplane. The pilot then got the engine started but the airplane continued losing altitude and the tail made contact with the river surface, dragging on it for about a half a mile before the nose finally slammed into the water. The aircraft flipped over and came to rest inverted. Engineer MSgt. Steven J. Shultz, who was seriously injured, and radioman SSgt. James L. Elrod, who was lightly injured, both were able to exit the aircraft. They tried to swim to shore but the river's current was too strong. After floating around for an hour and a half, the two were picked up by a fisherman in a rowboat. Investigation revealed that pilot 2Lt. Eugene P. Ulak, Newark, New Jersey, and bombardier SSgt. Everett C. Jackson, Corinth, Mississippi, were killed instantly when the nose slammed into the water and the airplane flipped over. Sgt. Elrod and Sgt. Schultz attempted to rescue the two men but were unable to get to them before the airplane went under.

7-11-42B. Key Field, Mississippi. At 1520, a ground crewman was killed when he walked into the rotating propeller of a taxiing North American AT-6A at Key Field, Meridian, Mississippi. The AT-6, piloted by 1Lt. Samuel Grisby Porterfield, was taxiing between two rows of parked airplanes with three other AT-6 airplanes in order to position themselves to take-off on a ferry flight after refueling at the field. The pilot was taxiing slow and in an S-pattern when he spotted the crewman 150 feet ahead directing the pilot of one of the parked aircraft. The pilot continued to taxi but upon seeing the crewman in close proximity to his airplane slowed to a halt. The ground crewman, Sgt. Charles G. Lewis, was facing away from Lt. Porterfield's airplane and motioning to an airplane on the line with upraised hands when he suddenly walked backward about thirty feet and into the moving propeller of the AT-6. At the time of the accident there were about 12 to 15 airplanes with engines running on the flight line and apparently Sgt. Lewis was unable to hear the airplane directly behind him.

7-12-42A. Clinton, Wisconsin. At 0900, a Taylorcraft L-2A dove into a field at Clinton, Wisconsin, killing civilian instructor Robert E. Ellis and student Sgt. Richard P. Leonard. The aircraft was seen to be buzzing a cornfield at an altitude of about 50 feet and made a couple of low level turns prior to diving into the ground. Testimony of eyewitnesses on the ground was vague and contradictory. Investigators could not determine the cause of the accident but speculated that the airplane lost flying speed in a low level turn.

7-12-42B. Whitestone, South Carolina. At 1545, a Republic P-43A stalled and spun to the ground at Whitestone, South Carolina, killing pilot 2Lt. Wayne V. Mosher. The aircraft was seen by another pilot in a diving, rolling spin at an altitude of approximately 1,500 feet. The aircraft hit the ground with great force and immediately burst into flames upon impact. The pilot was on an authorized aerobatic practice flight.

7-13-42A. Cimarron Field, Oklahoma. At 1020, two Fairchild PT-19A airplanes collided in midair three miles south of Cimarron Field, Oklahoma, killing civilian flight instructor Mervyn L. Davis and Aviation Cadet James F. Gillis aboard PT-19A #42-2524. Civilian flight instructor Gerald L. Bottorff and student Aviation Cadet Lloyd L. Gravely, Jr. parachuted to safety from PT-19A #42-2520. Instructor Bottorff and his student executed a series of climbing turns when he spotted the other PT-19 at an altitude of about 1,000 feet making a left turn into their flight path on an angle of 45 degrees. Instructor Bottorff pulled the airplane up avoiding a head on collision and managed a slight glancing blow with his tail section. After instructor Bottorff determined that he could not control the airplane, he ordered Gravely to bail out. Both received minor injuries. Both airplanes fell to the ground in close proximity to each other. Instructor Davis and Cadet Gillis were found in the wreckage of their airplane.

7-13-42B. Bridgeport, Connecticut. At 1130, a Republic P-47B crashed while turning from base to final near the municipal airport at Bridgeport, Connecticut, killing pilot 2Lt. Burdette L. Wertman. Investigators speculated that the P-47 made the approach with only the port flap extended. When the flap system fails and only one flap extends, the P-47 type aircraft exhibits a severe and virtually uncontrollable tendency to yaw toward the extended flap. The aircraft was seen to make the turn from base to final at a slow speed at about 150 feet. The port wing was seen to go down and the pilot apparently over controlled in an attempt to correct the airplane's attitude by banking right and adding power. The aircraft leveled out momentarily before the starboard wing was seen to drop. The P-47 fell to the right and hit the ground right wing down, cartwheeling across the ground and scattering wreckage for several hundred yards.

7-14-42A. Ellington Field, Texas. At 0030, a Curtiss AT-9 crash-landed four miles northeast of Ellington Field, Houston, Texas, after running out of fuel on a night navigation flight, killing Aviation Cadet Francis C. Wertz, 23, Chadron, Nebraska, and seriously injuring Aviation Cadet John B. Weiss, 20, Minneapolis, Minnesota.

7-14-42B. Location Unknown. At an unknown time after 0915 CWT, a Lockheed A-29 disappeared and is presumed to have crashed into the Gulf of Mexico with its crew of five. The airplane had taken off at 0850 from the Army Air Base at New Orleans, Louisiana, on an anti-submarine patrol. The aircraft was due back at base at 1400 and was last heard from at 0915 by a U.S. Navy ship operating in the area. In its last radio transmission, the aircraft reported being 60 miles south of New Orleans. Investigators believe that the A-29 probably crashed in the Gulf of Mexico although no evidence of the airplane was found. Missing and presumed lost at sea were: 1Lt. Rondell L. Cox, Cuyahoga Falls, Ohio, pilot; 2Lt. Joe P. Stanford, Mantee, Mississippi, co-pilot; TSgt. Guy C. Payne, New Orleans, Louisiana, bombardier; Sgt. Robert J. Albers, Dayton, Ohio, gunner; SSgt. Eual K. Martin, hometown unknown, radio operator.

7-14-42C. Verden, Oklahoma. At 1555, a flying student plummeted to the ground and was killed when he was flung from a Fairchild PT-19A during spin recovery practice three and one-half miles northeast of Verden, Oklahoma. The civilian flight instructor, William B. Naugle, landed the badly damaged aircraft safely in a nearby field soon after the incident. The Accident Classification Committee stated, "Aviation Cadet James R. Adams failed to fasten and secure safety belt before violent maneuvers, the result being he was thrown from the airplane upon a spin recovery and crashed into the vertical fin and fell to the ground from approximately 2,700 feet without pulling rip cord. Cadet Adams was dead when found."

7-14-42D. Salem, Florida. At 2300, a Cessna AT-17 flying in poor weather crashed four miles east of Salem, Florida, killing RAF flying cadets Ronald B. Pinsent and Sidney S. Platt. The AT-17 was part of a 39-airplane flight that took off at 0850 in three minute intervals from Moody Field, Valdosta, Georgia, on a night navigation mission to Dothan, Alabama, to Marianna, Florida, to Tallahassee, Florida, to Cross City, Florida, and back to Moody Field. Investigators speculated that the pilots flew the airplane into the ground while trying to fly under the heavy thunderstorms in the area (See 7-14-42E, F, G).

7-14-42E. Cherry Lake, Florida. At 2305, a Cessna AT-17 flying in poor weather crashed north of Cherry Lake, Florida, killing RAF flying cadets Philip W. Longbottom and William G. Lamont. The AT-17 was part of a 39-airplane flight that took off at 0850 in three minute intervals from Moody Field, Valdosta, Georgia, on a night navigation mission to Dothan, Alabama, to Marianna, Florida, to Tallahassee, Florida, to Cross City, Florida, and back to Moody Field. Investigators speculated that the pilots flew the airplane into the ground while trying to fly under the heavy thunderstorms in the area (See 7-14-42D, F, G).

7-14-42F. Madison, Florida. At 2320, a Cessna AT-17 flying in poor weather crashed four and one-half miles northwest of Madison, Florida, killing AAF flying instructor 2Lt. Samuel T. King and RAF Flying Cadet Kenward F. Stevens. The AT-17 was part of a

39-airplane flight that took off at 0850 in three minute intervals from Moody Field, Valdosta, Georgia, on a night navigation mission to Dothan, Alabama, to Marianna, Florida, to Tallahassee, Florida, to Cross City, Florida, and back to Moody Field. Investigators speculated that the pilots flew the airplane into the ground while trying to fly under the heavy thunderstorms in the area (See 7-14-42D, E, G).

7-14-42G. Lamont, Florida. At 2330, a Cessna AT-17 flying in poor weather crashed seven miles southeast of Lamont, Florida, killing RAF flying cadets Rowland W. Holmes and Bernard W. Howcraft. The AT-17 was part of a 39-airplane flight that took off at 0850 in three minute intervals from Moody Field, Valdosta, Georgia, on night navigation mission to Dothan, Alabama, to Marianna, Florida, to Tallahassee, Florida, to Cross City, Florida, and back to Moody Field. Investigators speculated that the pilots flew the airplane into the ground while trying to fly under the heavy thunderstorms in the area (See 7-14-42D, E, F).

7-15-42A. Walla Walla, Washington. At 0131, a Boeing B-17E collided with rising terrain and plunged down a ravine near Pike's Peak Road, seven and one-half miles southeast of the Army Air Base at Walla Walla, Washington, killing the crew of five. The B-17 had taken off from Walla Walla at 0122 on a local night flight and soon encountered instrument conditions southeast of the field. Investigation revealed that the airplane was in a gentle left bank when it initially struck 11 pine trees, snapping off the tops and shearing off the wing tips before sliding across a gully for about 150 feet. The B-17 then bounced gently off the ground for a short distance before striking the ground hard and careening 180 feet, striking two large trees, uprooting one of 27 inches diameter and snapping off the other 10 feet above the ground. The horizontal stabilizers and elevators were ripped from the fuselage and fire broke out on the starboard wing. The fuselage broke apart, ejecting two crewmembers and shooting the nose section down the ravine, the shattered tail section tumbling behind. The flight deck and mid-section exploded into flames; the three other crewmembers were found in the charred wreckage. The aircraft was fully serviced with 1,700 gallons of gasoline, which scorched the hillside for over one hundred yards when the fuel tanks ruptured and the gasoline ignited. Investigators noted that this was the pilot's first night flight in the vicinity and that had the B-17 cleared this ridge it would have certainly hit one of greater elevation on the other side of the gully. The mountains and hills in the area south and southeast of Walla Walla were obscured with clouds at the time of the accident and visibility was limited. Killed in the crash were: 2Lt. Richard Gordon Hill, Mart, Texas, pilot; 2Lt. Earl O. Stevens, Ravenna, Ohio, co-pilot; Pvt. Joseph Mallick, Corry, Pennsylvania, radio operator; Pvt. Kenneth White, Vale, Oregon, engineer; Pvt. Freddie E. Vanover, hometown unknown, assistant engineer.

7-15-42B. Clovis, New Mexico. At 0429, two Piper L-4A observation aircraft collided in mid-air at Clovis, New Mexico, killing two AAF glider pilots and a civilian instructor. Civilian flight instructor Fred R. Gibbs, Hot Springs, Arkansas, and student Sgt. John S. Russell, flying L-4A #42-36759, collided head on with L-4A #42-38398, which was piloted by Sgt. Thomas R. Moxley. Sgt. Russell was in the pattern for landing and Sgt. Moxley was trying to climb out of the area. Because of extremely high winds in the area at the time, Sgt. Russell allowed his aircraft to drift into the airspace adjacent to the traffic pattern and collided with Sgt. Moxley's airplane at an altitude of 1,000 feet above ground level. The aircraft stuck together and plummeted to the ground near the western boundary of the field, bursting into flames upon impact.

7-15-42C. Shaw Field, South Carolina. At 1030, a Vultee BT-13A stalled and spun into the ground from an altitude of approximately 500 feet while turning from base to final at Shaw Field, Sumter, South Carolina, killing A/C Clyde H. Wilson and A/C Charles B. Young. Although Cadet Wilson was on a scheduled instrument training flight with Cadet Young as safety observer, it was Cadet Young who was piloting the aircraft (from the front cockpit) at the time of the accident. The airplane spun 180 degrees before the pilot stopped the spin, leveled out momentarily and then dove into the ground. The aircraft burst into flames immediately upon striking the ground.

7-15-42D. English Field, Texas. At 1525, a Schweizer TG-2 crashed at English Field, Amarillo, Texas, after the tow cable broke just after take off, killing student glider instructor SSgt. Lloyd Smith and lightly injuring student glider instructor SSgt. Bernard C. Szmanski. When the tow cable broke at an altitude of 100 feet above ground level, Sgt. Smith attempted a 180-degree turn to the right, down wind, and went into a tight spiral to the right, crashing into the ground nose first. Wind velocity at the field was estimated at 20 mph at the time of the accident.

7-15-42E. Galveston, Texas. A North American O-47A crashed into Galveston Bay, Texas, while maneuvering on a maintenance test flight at 1635, killing two crewmembers and lightly injuring a third. Investigators stated that the pilot was seen to be flying at a very low altitude and that the surface of the water was extremely smooth. When the pilot made a turn toward the sun, he apparently misjudged his altitude above the surface and flew the airplane into the water. It was noted that the turn and bank indicator had been recently repaired and reinstalled and the pilot was reportedly checking the turn and bank indicator periodically during the flight. Investigators speculated that the pilot might have become distracted in the cockpit or misjudged his position relative to the surface. The pilot, Capt. Berle E. Sampson, 28, Galveston, Texas, and passenger TSgt. Ralph Bottolfson, 36, Des Moines, Iowa, were killed in the accident and

Sgt. John W. Knight, 24, Des Moines, Iowa received minor injuries.

7-16-42A. Spokane, Washington. At 0300, a Consolidated B-24D flying in a rainstorm and poor visibility crashed into Signal Peak 15 miles east of Geiger Field, Spokane, Washington, killing the crew of five. Killed in the crash were: 1Lt. William P. Marsh, Jr., Spokane, Washington, pilot; 2Lt. James H. Briggs, Gary, Indiana, co-pilot; SSgt. Fred W. Herring, Camillus, New York, engineer; Sgt. Robert J. O'Brian, Buffalo, New York, radio operator; Pvt. Raymond J. Luettgen, Milwaukee, Wisconsin, assistant engineer. The Accident Classification Committee stated, "Lt. Marsh was engaged in a routine night training flight. After making several landings, Lt. Marsh proceeded on a local night flight. He apparently proceeded eastward and flew in the vicinity of the foothills about 15 miles east of Spokane. He entered a heavy rain shower, which possibly blinded him. He flew low over the home of Mr. Dean D. Triggs, turning southeastward away from the valley in which lies the road to Coeur d'Alene. He next passed over the home of Mrs. J. J. Wagner on the slope of Signal Peak, proceeding at a dangerously low altitude on a course of southeast and climbing. A short time later he crashed on the side of Signal Peak going in a diagonal toward the top of it. The force of striking the ground and subsequent gas tank explosion had blown the bodies and the plane to pieces. Only two of the bodies were wearing parachutes. The co-pilot was still strapped in the bucket seat, which had been thrown clear and was wearing a parachute. The pilot was still strapped in the bucket seat, sitting on his parachute but not wearing it. The engineer was found in the approximate vicinity of the bomb bay, the normal exit from the plane, wearing a parachute. The ripcord had been pulled and lay in front of him. The parachute was open and spread out on the ground close to the body." The Army Air Forces had accepted the B-24 from the manufacturer only five days earlier.

7-16-42B. Bradley Field, Connecticut. At 0945, a Lockheed P-38F crashed into a fuel truck and trailer on the Southeast/Northwest runway at Bradley Field, Windsor Locks, Connecticut, killing truck driver Pfc. Norman E. Salters and fatally injuring pilot 2Lt. Albert R. Dawson. Ground crewman Pfc. Leon H. Terry received minor injuries. The Accident Classification Committee stated, "Two P-38s took off at 0945 on the SE/NW runway into the northwest. The leader was Lt. Bolling and the flight was echeloned to the right with Lt. Dawson as his right wing man. After the take-off had covered approximately 900 feet, a gasoline fuel unit came out from the right hand side of the taxiway, started onto the runway without waiting for the green light from the tower. The fuel unit upon sighting the approaching aircraft attempted to make a turn away from the path of the airplanes but the airplane flown by Lt. Dawson struck the fuel unit with the tip of his right wing and the contact was made at such a speed that the airplane immediately wheeled into the fuel trailer of the fuel unit so as to cause an explosion which threw the driver from the tractor and left his body on the runway. The airplane and the gasoline unit, being interlocked, swept on through the force of impact for approximately 200 yards. After the combined wreckage had come to a halt, the pilot, Lt. Dawson, released his hatch and removed himself from the plane and managed to walk [through the flames surrounding the wreck, walking] about halfway across the runway before being reached and taken to the hospital. Pfc. Terry was standing on the [right] running board and apparently did not see the approaching aircraft until the truck had advanced 20 feet onto the runway at which time he warned the driver to turn back. The fuel unit [turned and] was in a position parallel to the runway at the time of impact. Pfc. Terry jumped and was only slightly injured." Lt. Dawson died at 0130 on 7-17-42.

7-16-42C. Drew Field, Florida. At 1122, a Bell P-39D crashed into the ground out of control from an altitude of 50 feet one-quarter of a mile south of Drew Field, Tampa, Florida, killing pilot 2Lt. Richard K. Babbitt. The Accident Classification Committee stated, "It appears that the pilot approached Drew Field for landing with wheels up and flaps either up or nearly up, in a steep glide which resulted in excessive airspeed. When the airplane would have normally touched its wheels in the course of the landing, the propeller struck the runway in as much as the wheels were retracted. As the contact with the runway occurred, the pilot apparently decided to pull up and circle for a second approach. It is further believed that due to the damage to the propeller, the airplane failed to climb after the pull up from the runway and crashed out of control."

7-16-42D. Fredericktown, Maryland. At 1145, two Curtiss P-40F airplanes collided in mid-air and crashed at Fredericktown, Maryland, killing 2Lt. Lewis R. Contardi, 27, Iron River, Michigan, aboard P-40F #41-13679. The pilot of P-40F #41-14048, 2Lt. Robert J. Palenscar, parachuted to safety and was uninjured. He later stated to investigators, "I was flying in number-four position on number-three man's wing in a four-ship formation. Our flight had just taken part in an aerial encounter with an alert flight from Philadelphia [Pennsylvania] and after the first pass at the supposed enemy, Lt. Contardi dropped out of formation while I remained in position. After the encounter, Lt. Maneely the flight leader called the flight into formation and we proceeded back to base, during this time Lt. Contardi was to the rear of the second element and not yet in position. At the time of the collision my eyes were on my element leader so I never saw what actually hit my aircraft. When I looked out, my right wing was gone and I immediately fell off into a spin to the left. Seeing that I had no control, I proceeded to extract myself forcibly from the airplane."

7-16-42E. Stamford, Texas. At 1145, a Fairchild PT-19A dove into the ground two miles south of Arledge Field, Stamford, Texas, killing student pilot

Sgt. Mahlon A. Heinrich. The Accident Classification Committee stated, "Careful examination of the wreckage indicates that, at the time of the impact, the airplane was in a straight flight going almost straight down. It appears that, whatever the cause of the crash, the student made an attempt, apparently late, to get out of the airplane. The ripcord was outside the airplane on the ground and the safety belt in the student's cockpit was unfastened and intact. The airplane apparently came in from an altitude of at least 2,000 feet, but the reason is undetermined due to lack of competent testimony and difference of opinion of available witnesses."

7-16-42F. Barksdale Field, Louisiana. At 1600, a Martin B-26B crashed into a pecan tree grove while attempting to take-off from Barksdale Field, Shreveport, Louisiana, killing one flier and lightly injuring six others. Investigation revealed that the starboard engine was not producing full power on the take-off because of a carburetor problem. The aircraft never gained more than 25 feet of altitude before descending into the pecan tree grove. The aircraft broke apart and MSgt. Harold J. Courtney, 37, Indianapolis, Indiana, was killed. The other crewmembers were able to escape the wreckage before the aircraft burst into flames. Lightly injured in the accident were: 1Lt. Raymond N. Lipscomb, pilot; 2Lt. Edwin J. Soniat, co-pilot; TSgt. Harris K. Camp, engineer; 2Lt. Richard D. Ebeling, passenger; Joel Allen, civilian passenger; Lawrence Albert, civilian passenger. Investigators stated that the pilot recognizing the loss of engine power and aborting the take-off would have prevented the accident. The pilot stated that the airplane began pulling to the right and he initially thought the rudder was out of trim.

7-17-42A. Pacific Ocean, California. At 0115, a Vultee BT-13A lost on a night navigation training mission crashed into the Pacific Ocean two miles off shore of Hollywood by the Sea, California, killing student A/C Richard G. Parr. The student was to fly from Gardner Field, Taft, California, to Fresno, then back to Gardner via Avenal, California. The student missed Avenal and became lost. A flying officer witnessed the aircraft flying over Oxnard, California, at an altitude of approximately 4,000 feet descending with the landing lights on heading south. The aircraft was seen to turn west and fly out to sea, then to turn south and then east again before it descended into the sea. Investigators speculated that the aircraft ran out of fuel and that the student probably missed the airfield lights at Oxnard because of his illuminated landing lights. After an extensive search, the student and the airplane could not be found but the U.S. Coast Guard recovered a parachute belonging to the cadet on 7-24-42 near Hermosa Beach. Investigators could not determine why the student did not parachute to safety before the aircraft ran out of fuel or use his radio to attempt to confirm his position.

7-17-42B. Concord, North Carolina. At 1036, a Curtiss P-40F suffering engine trouble crashed nine miles southwest of Concord, North Carolina, killing pilot 2Lt. George E. Hutton, 22. Witnesses on the ground observed the airplane at an altitude of about 200 feet agl with white smoke issuing from the engine. The airplane, steadily losing altitude, was seen coming from the southwest before turning south and crashing into the ground in a near flying attitude with landing gear retracted. The aircraft bounded into the air for about 200 yards and then hit the ground again, striking a road. The P-40 bounced into the air again, careening through a cotton field before coming to rest inverted and bursting into flames. Investigation of the engine revealed that a camshaft retainer assembly failed, allowing the camshaft drive to become disengaged and allowing valves in the left cylinder bank to become out of time.

7-17-42C. Lutz, Florida. At 1140, a Bell P-39D crashed into the ground two and one-half miles west of Lutz, Florida, killing pilot 1Lt. Robert S. Cole. The pilot was leading a flight of three P-39 airplanes in a string formation at about 8,000 feet when he began a 45-degree dive. At about 2,500 feet the airplane made a sharp turn to the right, turning about 145 degrees to the right of its course before it began a rolling plunge straight down. The aircraft leveled out momentarily, rolled out of control and then slammed into the ground inverted in a shallow dive and at high speed. The aircraft burst into flames immediately upon impact and wreckage was scattered over an area of 150 yards. The two other P-39 aircraft pulled out of the dive with no difficulty.

7-17-42D. Gulf of Mexico, Florida. At approximately 2250, a Vultee BT-13A crashed into the Gulf of Mexico at Dead Man's Bay, 15 miles west of Cross City, Florida, killing pilot 2Lt. Francis J. Minor and passenger Capt. John P. Hyman, Jr. The Accident Classification Committee stated, "The committee finds that while on a routine navigation training flight from Lakeland, Florida, to Napier Field, [Dothan] Alabama, the airplane crashed into the Gulf of Mexico, killing both occupants. The committee in its investigation was unable to determine the true cause of the accident and believes that the extreme darkness of that night and the scattered thunder showers reported in that vicinity may have been a contributing factor to the accident." Lt. Minor's body, still strapped to his seat, was recovered from the sea five miles off shore of Dead Man's Bay on 7-21-42 and Capt. Hyman's body was recovered 7-23-42 in the wreckage of the airplane.

7-18-42A. Magma, Arizona. At 0730, a Beech AT-10 on a practice-bombing mission dove into the ground three miles east of Magma, Arizona (near Coolidge), killing A/C Joseph C. Zalesky and A/C Harry M. Wright. 2Lt. Robert McAtee witnessed the accident and later stated to investigators, "I was flying a bombing mission at an altitude of 1,000 feet on the back leg on target E-3 when I caught a glimpse of the AT-10 about six miles away at an altitude of about 500 feet in a nearly vertical dive at a very high speed. An instant later the airplane hit the ground and exploded [and was] still at a very steep attitude." Because the landing gear was discovered in the extended position, investigators speculated that the student might have been preparing for an emergency forced landing prior to being observed in the fatal dive. It was noted that all students be warned not to extend the landing gear when attempting off-field forced landings.

7-18-42B. Goodfellow Field, Texas. At 0900, a Vutltee BT-13 practicing forced landings stalled and spun into the ground near Goodfellow Field, San Angelo, Texas, killing students Aviation Cadet Richard L. Adler and Capt. Emmett F. Blakemore, Jr. The Accident Classification Committee stated, "It was determined that the airplane struck the ground in approximately a vertical angle and from a relatively low altitude, inasmuch as the tail group was not particularly damaged and the shock of the impact was absorbed by the engine and the leading edge of the wings, folding the landing gear straight back into the wings and blowing out only one tire."

7-18-42C. Bruce Field, Texas. At 1058, a Fairchild PT-19A stalled and spun into the ground while maneuvering 10 miles east of Bruce Field, Ballinger, Texas, killing Aviation Cadet Edward H. Insinger, Maplewood, New Jersey. Investigators determined that the aircraft entered a spin at an altitude of approximately 2,000 feet from which the student was unable to recover. Witnesses in the air observed the airplane spinning very rapidly. At approximately 100 feet above ground level the student attempted to use his parachute, but was too low for it to deploy and he struck the ground near the aircraft.

7-18-42D. North Reading, Massachusetts. At 1520, a Boeing B-17B flying in instrument conditions suffered a structural failure of the port wing and crashed into the Harold Parker State Forest near North Reading, Massachusetts, killing the crew of ten. Weather in the area at the time was reported as "Overcast, ceiling about 500 feet and drizzle. A low, heavy scud extended down to within 50 feet of the ground, creating a zero ceiling with visibility of less than one-half mile. The top was between 4,000 and 3,500 feet." The port wing separated from the aircraft and was found impaled on a tree a considerable distance from the main wreckage. The aircraft, rolling to the left, plunged to the ground in a near vertical dive on the slope of a wooded hill at a speed estimated to be around 350 mph, shattering large trees as it careened for 100 yards, exploding into flames and scattering wreckage and body parts over a wide area. The Accident Classification Committee stated, "A probable conclusion is that the pilot was making a let down through the overcast, developing excessive airspeed. A wing flutter started as he broke out of the overcast and in trying to pull out, the wing section let go. Evidence from debris picked up in the flight path indicate wing vibration due to wing flutter, losing first the aileron, second, the wing from the outboard motor to the tip. The altitude must have been over 500 feet." Killed in the crash were: 1Lt. Marion R. Klice, pilot; 1Lt. Don H. Johnson, co-pilot; 1Lt. James N. Phillips, bombardier; 2Lt. O. D. Andrews, navigator; Sgt. William E. Perkins, engineer; Cpl. William Bilocur, radio operator; Cpl. C. H. Torrance, assistant radio operator; Pfc. F. S. Koltun, assistant engineer; MSgt. Archie R. Jester, passenger; SSgt. Robert J. Aulsbury, passenger. The B-17 was en route to the AAF Depot at Middletown, Pennsylvania, from its duty station at Gander, Newfoundland, for an ordered equipment update.

7-20-42A. Highlands, Texas. At 0120, a North American AT-6A on a night cross-country navigation flight crashed into the ground in a vertical attitude three miles east of Highlands, Texas, killing Aviation Cadet J. D. Gilliam, 21, Corsicana, Texas. The aircraft hit the ground in an area covered with about two inches of water at high speed and with considerable force, burying the engine several feet under the surface. Investigators could not determine the cause of the crash.

7-20-42B. Tucson, Arizona. At 0440, a Consolidated B-24D flew into terrain three and one-half miles ESE of Davis-Monthan Field, Tucson, Arizona, killing eight fliers. The B-24 had been practicing night landings and the pilot had overshot on the previous two attempts. The B-24 began a go-around and had made a left turn away from the field when it flew into the ground, striking on its left wing and left front fuselage. The field elevation at the end of the southeast runway is 2,565 feet and the elevation at the initial impact point was 2,775 feet. Investigators stated, "The pilot, after overshooting the field, and while turning for his next approach and apparently losing the horizon or failing to go on instruments in sufficient time inadvertently contacted the ground with the left wing of his airplane, causing the resultant crash. From the appearance of the wreckage on the ground it is evident that the left wing tip struck a small tree and the airplane struck the ground, tearing loose two propellers at the point of impact. The B-24 then apparently cartwheeled through the air, catching fire and breaking off various parts of the aircraft until the right wing and fuselage came to rest, at which point the gasoline tanks exploded and the aircraft was destroyed by fire." The wreckage and bodies were scattered over a distance of 1,000 feet from the point of initial impact. Killed in the crash were: 1Lt. Blair K. Blacker, Corvallis, Oregon, pilot; 2Lt. Earl W. Howard, Cottonwood Falls, Kansas, co-pilot; 2Lt. Bishop M. Toms, assistant co-pilot; Cpl. William J. Brady, engineer; Pvt. Howard A. Peterson, assistant engineer; Cpl. Richard J. Anderson, radio operator; Pvt. Thomas B. Fish, assistant radio operator; Pvt. Thomas P. McGrath, gunner.

7-20-42C. Visalia, California. At 0833, a North American B-25A stalled and dove into the ground five miles northwest of the municipal airport at Visalia, California, killing the crew of five. The port engine apparently failed. Witnesses on the ground reported that the port engine was issuing puffs of black smoke and was heard backfiring. The aircraft was then seen to pitch up and then go into a diving spin to the left and into the ground from an altitude of approximately 700 feet above ground level. The B-25 burst into flames upon impact with the ground, scattering wreckage and bodies over a wide area. Killed in the crash were: 2Lt. Omer D. Curnutte, pilot; 2Lt. Dwight N. Harrelson, co-pilot; Pfc. James A. Dietrich, engineer; Pvt. Paul G. Wright, radio operator; Cpl. Clois E. Hawkins, assistant engineer.

7-20-42D. Foster Field, Texas. At 1130, a North American AT-6A on a student instrument team ride suffered the failure of its port wing and crashed to the ground out of control eight miles east of Foster Field, Victoria, Texas, killing Aviation Cadet Patrick L. Welsh and Aviation Cadet Curtis W. Torrens. At 1025, a report had come in that this aircraft had been buzzing areas near Yoakum. The aircraft landed at Yoakum and the students changed seats for their respective instrument practice. Investigators could not determine why the wing panel and aileron failed but speculated that either of the students might have stressed the aircraft with an extreme maneuver, compromising its structural integrity and causing it to fail. The wing panel and aileron were found over 200 yards away from the main wreckage. No evidence that the low flying aircraft struck an object on the ground or in the air could be found. The airplane struck the ground inverted and at an angle of about 45 degrees.

7-21-42A. Leesburg, Georgia. At 0150, a Curtiss AT-9 crashed while attempting a take-off at the Army Auxiliary Airfield at Leesburg, Georgia, killing RAF Flying Cadet Thomas Mosely, 22, Middlesborough, Yorkshire, England, and seriously injuring RAF Flying Cadet William J. Reynolds, 28, hometown unknown. Investigation revealed that the students did not advance the throttles to the full power take off position. The aircraft never gained flying speed, went off the end of the runway and smashed into a frame house belonging to an unnamed civilian. Four unnamed civilians were slightly injured.

7-21-42B. Chandler, Arizona. At 0940, a Beech AT-10 crashed seven miles south of Chandler, Arizona, after suffering the structural failure of the port wing, killing pilot 2Lt. Robert E. Thomas, 23, Sacramento, California, and passenger Pvt. John H. Keil, 22, Stevensville, Michigan. A witness on the ground reported that the airplane spun into the ground and exploded into a "thousand pieces." The airplane hit the ground with great force and came to rest in an inverted position. Both pilots were killed instantly. Investigation revealed that the wing spar bolts "pulled through" the main wooden spar.

7-21-42C. Tampa Bay, Florida. At 1518, a Martin B-26B crashed into Tampa Bay just after taking off from MacDill Field, Tampa, Florida, killing the crew of six. The aircraft struck the water approximately 200 feet south of the Gandy Bridge. A pilot who flew the aircraft on 7-17-42 testified that the aircraft exhibited indications that the left engine was not providing full power or that the aircraft was seriously out of trim when normal left turns were attempted. He reported that the port wing had a tendency to "drop" in the turn. After some examination, mechanics found a minor rudder cable problem that was easily repaired. A subsequent test flight revealed that the tendency of the left wing to drop continued. Upon arriving at MacDill Field, the problem seemed to correct itself. Engine trouble was ruled out because the aircraft could climb normally, but the control anomaly was never determined and apparently only exhibited itself when the aircraft attempted a left turn. Another pilot also testified to the aircraft's tendency to "yaw violently to the left." The B-26 was grounded for a short time but after an extensive examination no problem could be found. "The [Accident Classification] Committee finds that the airplane took off, reached an altitude of approximately 1,000 feet, then went into a flat spin to the left and crashed into the Tampa Bay. Because of a lack of evidence, the committee is unable to determine whether the accident was caused by failure of the left engine, locked controls or by the pilot exceeding the operational limits of the airplane. It is apparent that one of the above causes, or a combination of the three contributed to the accident." Killed in the accident were: 1Lt. Joseph J. Schabinger, 23, Evanston, Illinois, pilot; 2Lt. William P. Cawthon, hometown unknown, co-pilot; TSgt. William West, Jackson, Mississippi, engineer; SSgt. James D. Keaton, Henderson, Texas, engineer; 2Lt. Paul P. Cecil, Norfolk, Virginia, passenger; 2Lt. Phillip C. Cone, Little Falls, New York, passenger.

7-22-42A. Rosamond, California. At 0925, a Bell P-39D stalled and spun into the ground four miles north of Rosamond, California, killing pilot 2Lt. Robert W. Hamlin. The aircraft was part of a four-ship formation conducting training operations at Muroc Dry Lake, California. 2Lt. Clarence M. Caldcleugh witnessed the accident and stated to investigators, "During a climbing turn, Lt. [Charles S.] Wilkins' plane snap rolled. Lt. Hamlin immediately went into a roll. When he got on his back, it went into a dive and started spinning. He partially recovered at about 1,000 feet but then went into another spin and crashed." Investigators noted that the P-39 airplane has the tendency to go into a snap roll at slow speeds. The pilot apparently thought that the flight leader's unintentional snap roll was intentional and followed him into a roll. He apparently over corrected and stalled the airplane.

7-22-42B. Perrin Field, Texas. At 1155, a Vultee BT-13A stalled and spun into the ground six miles south of Perrin Field, Sherman, Texas, killing flight instructor 2Lt. Robert C. Gantz and student Aviation Cadet Warren M. Mozley. The aircraft struck the ground in a vertical attitude. The Accident Classification Committee could not determine why the airplane crashed.

7-22-42C. Cochran Field, Georgia. At 1525, a Vultee BT-13A on an instrument team flight stalled and spun into the ground eight miles east of Cochran Field, Macon, Georgia, killing RAF Flying Cadet Leonard A. Carter, London, England, and AAF Aviation Cadet Curtis N. Cooper, Atlanta, Georgia. Witnesses on the ground reported that the airplane spun at least three times from a low altitude. Investigation determined that the airplane was not spinning when it hit the ground but was in a shallow dive. The aircraft struck the ground nose first and immediately burst into flames upon impact.

7-23-42A. Bedford, Massachusetts. At 0900, a Curtiss P-40E returning from a six airplane formation training flight stalled and spun into the ground while in the landing pattern near the airport at Bedford, Massachusetts, killing pilot 2Lt. Clyde W. Kaffenberger, New Weston, Ohio. The pilot, leading the second element, entered a steep turn from base to final and lost flying speed, causing the airplane to stall and enter a spin at an altitude too low for recovery. The pilot, realizing that the aircraft was slipping off in the turn and entering a spin, gave the engine full throttle in an effort to correct the diminishing airspeed, loss of altitude and arrest the spin. The pilot momentarily stopped the spin but the aircraft failed to gain altitude and continued to slip sideways into a hill, impacting violently and coming to rest inverted. There was no fire but the aircraft was completely destroyed. The flaps and landing gear were in the extended position when the accident occurred.

7-23-42B. Camp Davis, North Carolina. At 1515, a Bell P-39D on simulated combat maneuvers struck and killed U.S. Army photographer 1Lt. W. W. Gates while deliberately buzzing a camera platform at Camp Davis, North Carolina. The pilot, 1Lt. Norbert C. Treacy, was uninjured and landed the aircraft safely at the Camp Davis airfield. "It is the opinion of the [Accident Classification] Committee that the main cause of the accident lies with the supervisory personnel in charge of the mission. The pilot, who flew four such attack missions previous to this one, had the understanding to fly as low as possible and simulate actual wartime attack. The mission was performed over varying terrain, resulting in turbulent air. The pilot is partly at fault, however, for flying too low and not realizing the condition of turbulent air. As he made his attack, his right wing dropped, hitting the photographer and injuring him fatally. The cause can also be laid to Lt. Gates who, while photographing this mission, neglected to take any safety precautions and neglected to jump from the platform on which he was standing."

7-24-42A. Lynchburg, Virginia. At 1035, a North American BC-1A flying in poor weather crashed on the Igloe Farm at Lynchburg, Virginia, killing pilot 2Lt. Russell W. Wetzel, Kewanee, Illinois, and passenger Capt. Harold C. Smith, 31, Norwood, Pennsylvania. The pilot was cleared to Lynchburg on an instrument flight plan. Weather at Lynchburg at the time was reported as visibility one and a half miles, light rain and a ceiling of 800 feet. Just prior to the time of the accident, the ceiling reportedly dropped to 400 feet. The airplane arrived over Lynchburg at an assigned altitude of 4,000 feet and requested the tower provide him an instrument approach procedure. The tower gave the pilot clearance even though the conditions were below the authorized minimums for a safe instrument let down. The aircraft let down through the overcast under control and was observed to make a 270 degree turn before the nose fell steeply. The pilot increased the throttle but was unable to maintain control of the airplane. The aircraft struck the ground in a vertical dive, exploding into flames upon impact. The force of the crash was so great that the aircraft engine was driven six feet into the ground. The pilot was ferrying the passenger from Bolling Field, Washington, D.C., to Lynchburg for a judicial matter involving a Bolling Field sergeant in a civilian criminal court.

7-24-42B. Waynesboro, Tennessee. At 1005, a North American O-47A crashed ten miles south of Waynesboro, Tennessee, killing three fliers. Killed in the crash were: 1Lt. Robert F. Turk, Wichita, Kansas, pilot; Maj. Gen. Frank C. Mahin, 54, Commander of the 33rd Infantry Division, Camp Forrest, Tennessee, passenger; TSgt. John L. Comerford, Alamo, Texas, observer. The aircraft took off from Tullahoma Army Air Field, Tennessee, to transport the general to Fort Sill, Oklahoma. The flight was scheduled to stop in Little Rock, Arkansas, en route to Fort Sill but apparently had engine trouble near Waynesboro. The airplane was observed by witnesses on the ground to make a 90-degree left turn to the south with the engine apparently malfunctioning. The airplane then turned to the east and apparently descended into the ground without power, careening for a short distance before coming to rest inverted. Investigators speculated that the pilot was attempting an emergency forced landing in one of the many small fields in a valley that is surrounded by heavily wooded terrain. Investigators confirmed that the landing gear was in the retracted position but were unable to determine if the flaps were retracted or the cause of the apparent engine failure. Weather was not considered a factor in the accident and it was noted that an exhaustive search failed to find the airplane flight report #1 and #1A. Maj. Gen. Mahin was on his way to Fort Sill to observe a ground troop training demonstration.

7-25-42A. Lake Dorr, Florida. At 0630 EWT crashed into Lake Dorr, Florida, killing 2lt. Delos J. Leavenworth, 25, Carleton, Michigan. The airplane was part of a flight of three P-40 airplanes on a bombing training mission at the AAF Bombing Range at Ocala, Florida. On the way back to the Army Air Base at Orlando, Florida, the formation was flying in a string formation when the flight leader dove on Lake Dorr and in the pull-up Lt. Leavenworth misjudged his position relative to the surface, failing to pull out of the dive in time to avoid impacting the water. The aircraft contacted the surface with its wings level and was seen to bounce off the water two times before smashing back into the lake, disintegrating and quickly sinking.

7-25-42B. Bennettsville, South Carolina. At 1540, two Stearman PT-17 airplanes collided in mid-air near the municipal airport at Bennettsville, South Carolina, killing civilian instructor Joseph A. Grork and student A/C Jackson E. White aboard PT-17 #41-8680. Civilian instructor Sidney J. Albert and student A/C William Wetmore were able to crash land PT-17 #41-25617 in a field. The Accident Classification Committee stated, "Witnesses to the accident claim Albert's ship was slightly cross tee, which let him drift to the left. At about 200 feet altitude, Albert sighted the other ship almost directly ahead and simultaneously collided. Albert's [airplane cut through] the other [airplane's] fuselage between the cockpit and the vertical stabilzer. The empennage of [Grork's] ship clung to Albert's right landing gear and wing. [Grork's] ship immediately dove into the ground. Albert was able to glide his ship ahead, landing in an open field, overturning with damage to the plane but none to personnel."

7-25-42C. Lavernia, Texas. At 1900, a North American BT-14 crashed into the ground while maneuvering six miles southwest of Lavernia, Texas, killing Aviation Cadet Kendall M. Stevenson. Aviation Cadet James H. Ballard witnessed the crash and later stated to investigators, "I observed a BT-14 do a slow roll at approximately 6,000 feet, finishing his maneuver and coming out in a right, wide radius, diving spiral, going into a vertical split-S maneuver at about 2,000 feet and from there into the ground. He slow rolled to the right. I did not see him make any attempt to recover."

7-25-42D. Olympia, Washington. At 1930, a Lockheed P-38F crashed after suffering the failure of its port engine just after take-off from the municipal airport at Olympia, Washington, killing pilot 2Lt. George Budraski, Jr. 1Lt. Wendell J. Kelley (AAF) witnessed the accident and stated to investigators, "The pilot retracted the wheels and was starting a normal climbing turn to the left when suddenly he steepened up his bank and immediately I knew he was in some trouble. The bank steepened to almost vertical and the ship lost altitude in the turn and crashed into some tall trees about [one-quarter of a] mile from the field. The ship had turned approximately 180 degrees from the direction of take-off before crashing. The trouble was due to left engine failure. The ship did not spin and was partially under control. The ship hit in a flat, partially stalled position and burst into flames."

7-26-42. Ajo, Arizona. At 1140, a North American AT-6A attempting an emergency forced landing stalled and spun into the ground five miles southeast of Ajo, Arizona, killing pilot SSgt. Francis G. Meyer and passenger Pvt. Lonnie B. West. Apparently the pilot ran one tank dry and the engine failed. The pilot lowered the wheels and attempted to extend his glide over a large tree on the approach to the area he chose for his landing. The aircraft stalled and dove into the ground at a steep angle from a low altitude just beyond the tree. The fuel selector switch was found in the "left" position and investigation revealed that this fuel tank contained no fuel. The starboard fuel tank contained adequate fuel to safely complete the flight.

7-27-42A. Okmulgee, Oklahoma. At 1030, an Aeronca L-3A practicing spins and spin recovery crashed into a cotton field 18 miles west of Okmulgee, Oklahoma, killing glider pilot student Cpl. John A. Russell, 22, Shidler, Oklahoma, and seriously injuring civilian instructor Denver E. Miner. Instructor Miner, a former pilot officer with the RAF Eagle Squadron, stated to investigators, "I had just demonstrated the spin recovery I wanted and I told the student to take me up and do another spin just like that one. Student pulled ship up immediately into a sharp left climbing turn, which appeared it might be a chandelle. Upon completing 90 degrees of maneuver, the airplane went into a violent spin to the left. Then I told the student that I had the airplane and started the recovery. I had stopped the spin and was on the pull-out when we hit the ground."

7-27-42B. Santa Maria, California. At 1315, a Stearman PT-13 suffering an apparent engine failure and attempting an emergency forced landing stalled and spun into a bean field six miles WNW of Santa Maria, California, killing civilian instructor Ted A. Kohler. The pilot apparently overshot the field he chose for his landing and stalled out of a turn in an attempt to avoid striking a house.

7-28-42A. Dorr Field, Florida. At 1450, a Stearman PT-17 stalled and spun into the ground while performing aerobatics five miles northwest of Dorr Field, Arcadia, Florida, killing Aviation Cadet Joseph T. Lane, Jr. Investigators speculated that the student had been performing slow rolls or Immelmanns at an altitude of less than 3,000 feet when the pilot stalled out of a maneuver and entered a spin at full power. The student effected a partial recovery but failed to reduce power resulting in an increasing spinning attitude of the aircraft. Witnesses in the air stated that the aircraft appeared to level out momentarily before it dove into the ground. The aircraft was seen to hit wheels first, collapsing the gear before nosing over and tumbling into trees 200 yards from where it first impacted the ground. The PT-17 did not burn but was totally destroyed.

7-28-42B. Corsicana, Texas. At 1630, a Fairchild PT-19A stalled and spun into the ground near Army Auxiliary Field #2 at Corsicana, Texas, killing A/C George F. Bradley. Investigators could not determine why the airplane entered a spin or why the student could not effect a recovery. The airplane spun in at a very flat angle.

7-29-42A. Williams Field, Arizona. At 0025, a Curtiss AT-9 attempting a landing at Williams Field, Chandler, Arizona, encountered a severe downdraft after inadvertently entering a thunderstorm and crashed two and one-half miles north of the field, killing Aviation Cadet Afton L. Smith and seriously injuring instructor 2Lt. Don R. Toland. The Accident Classification Committee stated, "The underlying cause of the accident was that the pilot flew a left hand traffic pattern after being directed to fly a right hand pattern to avoid the rainstorms [in the] area. He inadvertently flew into a severe rainstorm, which was accompanied by a strong wind. The pilot had no visual reference with the ground and went on instruments, and because of rough air conditions either lost control of his airplane or was unable to maintain altitude and as result struck the ground."

7-29-42B. Victorville, California. At 2100, two Beech AT-11 aircraft on a photographic mission collided in mid-air 10 miles southeast of Victorville, California, causing one to crash to the ground out of control, killing a sergeant pilot and a civilian photographer. Another civilian photographer was seriously injured when he miraculously fell from the aircraft and parachuted to safety. AT-11 #41-9579, piloted by 2Lt. Russell L. Bolin, landed at the Army Air Field at Victorville after passengers Capt. William R. Hulbush and 2Lt. Raymond W. Sullivan parachuted from the airplane. Lt. Sullivan was lightly injured in the bail out. Killed in the collision were pilot SSgt. Rodney M. Grim, 20, Ophiem, Montana, and photographer Clarence L. Block, 29, Los Angeles, California, aboard AT-11 #41-9581. Associated Press photographer Frank Filan, 37, was seriously injured in the collision and was thrown out of the airplane. He regained consciousness while falling through the air and immediately deployed his parachute. He was found wandering aimlessly in the desert in a dazed condition just before dawn the next day. Investigators determined that SSgt. Grim's aircraft ascended into the aircraft flown by Lt. Bolin because of an apparent over correction by SSgt. Grim while flying in turbulent air. Investigators also noted that the aircraft piloted by Lt. Bolin was carrying a load of bombs while SSgt. Grim's aircraft was not, speculating that this made Lt. Bolin's aircraft less maneuverable and might have hampered his ability to avoid the collision. Both pilots were faulted for failure to avoid the collision.

7-30-42A. Moscow, Arkansas. At 0728, a Fairchild PT-19A buzzing the home of the pilot's girlfriend crashed after striking a power line pole four miles south of Moscow, Arkansas, killing Aviation Cadet Gilbert F. Brendle and seriously injuring civilian flight instructor George E. Goodhead. Investigators stated, "Statements of witnesses and the condition of the wrecked airplane indicate that the airplane with full power on struck a REA power line pole about 30 feet above the ground with the left wing. The left wing and the landing gear were torn completely off and landed about 200 feet from the pole. The airplane continued on in a straight line for 100 yards, hitting the ground on the left side near the nose. The airplane then cartwheeled, tearing off the right wing and landing gear. Aviation Cadet Brendle, who was in the front cockpit, was killed instantly and Civilian Instructor Goodhead was seriously injured. From information supplied by witnesses, it is believed that Mr. Goodhead was well acquainted with a young woman who lived in a house at the scene of the accident. Mr. Goodhead had been to a party at this house the night before the accident. It was also stated by the witnesses that Mr. Goodhead had dived at the house on previous days. [The day of the accident] Instructor Goodhead dived within 100 feet of the ground near the home of Mr. H. P. Bryan. The airplane climbed, circled once, and again dived near the home of Mr. H. P. Bryan, hitting an electric power line pole with the left wing." Instructor Goodhead had been warned after the previous buzzing incidents that he would be disciplined if the buzzing continued.

7-30-42B. Houston, Texas. At 0930, a Cessna AT-17 suffered a structural failure of the port wing and crashed into a house at 2224 Riverside Drive, Houston, Texas, killing the civilian occupant Mrs. Ida Beatrice Baldwin, pilot A/C William R. Simpson, Jr., and co-pilot A/C Leland Scholl. Investigation revealed that the pilot was buzzing his girlfriend's apartment building. "According to the Accident [Classification] Committee, the airplane was flown in the vicinity of the Park Lane Apartments and put through a series of strenuous maneuvers, including rolls and vertical banks, at altitudes ranging from 50 to 1,500 feet. Indications are Cadet Simpson desired to attract the attention of a young lady friend residing at the Park Lane Apartments. It is believed the airplane was placed in a steep glide (just prior to the crash) at approximately 1,500 feet, (heading east over Herman Park toward the Park Lane Apartments) and during descent, two rolls to the right were executed. Sudden structural failure of the left wing appears to have occurred after the second roll, or during an almost vertical bank, which is believed to have been made for two 360-degree turns over the Park Lane Apartments. Airplane continued out of control (left wing low) for about four tenths of a mile distant where it crashed into the roof of a two story brick veneer house, fatally injuring Cadet Simpson (pilot), Cadet Scholl (co-pilot), and Mrs. Ida Beatrice Baldwin, civilian occupant of the house." Investigators noted that the port wing had been extensively damaged in an accident that occurred on 5-29-42 and speculated that the weakened condition of the wing might have contributed to the wing's structural failure under the severe stresses imposed by the maneuvers performed by the pilot.

7-30-42C. Minneapolis, Minnesota. At 1155, a Boeing C-73 (militarized version of the model 247D) crashed on take-off from Wold-Chamberlain Airport, Minneapolis, Minnesota, killing ten crewmembers. The aircraft and crew were on contract to the AAF from Northwest Airlines for ferrying personnel and training. The aircraft struck the ground in a vertical attitude and burst into flames immediately upon impact. The official accident report does not indicate the cause of the accident. Employees of Northwest Airlines killed in the accident were: John F. Fisher, Fremont, Ohio, pilot; Kenneth S. Olson, Minneapolis, co-pilot; Herbert Wiechman, Minneapolis; Leonard O. Larson, Minneapolis. Also killed were crewmembers from St. Paul, Minnesota: Winston A. Luchau, Robert A. Paulson, Wilbur C. Rose, James F. Painter, Lee D. Husby, and Roy Albert Josephson.

7-30-42D. Shaw Field, South Carolina. At 1630, two Vultee BT-13A airplanes collided in mid-air seven miles northeast of Shaw Field, Sumter, South Carolina, killing three fliers and seriously injuring a civilian flight instructor who parachuted to safety. Civilian instructor Noel E. Williams and student A/C Fray A. McMillan, flying BT-13A #41-10532, were practicing instrument flying at an altitude of approximately 3,000 feet. The instructor, acting as the safety pilot, commanded the student to enter a stall and spin under the hood and regain control. The student quickly complied and stalled the airplane, entering a two-turn spin to the right. He was beginning to recover when the aircraft descended into BT-13 #41-10641, which was piloted by instructor 2Lt. Hoyt Acker, Jr. and student A/C Curtis R. Risley. A/C Risley was flying under the hood with Lt. Acker as safety observer on an instrument flight. Lt. Acker attempted to parachute to safety but attempted his bail out too close to the ground to allow his parachute to fully deploy. Instructor Williams, although seriously injured and temporarily blinded, managed to escape the aircraft and parachute to safety. The cause of the accident was the failure of the two safety pilots to see and avoid the other aircraft.

7-30-42E. Mobile, Alabama. At 1855, a Lockheed A-29 suffering the failure of its port engine crashed into Mobile Bay seven miles southeast of Point Clear, Alabama, killing five crewmen. The aircraft was returning from a coastal patrol mission when it was observed flying over the bay, roll to the left through inverted and then dive into the bay. When the wreckage was recovered, investigation revealed that the aircraft was trimmed for single engine operation. Investigators could not determine the cause of the engine failure or why the pilot lost control of the aircraft. Killed in the crash were: 2Lt. James G. Dalrymple, pilot; Pvt. Morris L. Pike, bombardier; Pvt. Joseph M. Maglio, engineer; Pvt. Fred G. Paulsen, radio operator; Cpl. Antonio Rosato, gunner.

7-30-42F. Oxon, Maryland. At 2210, a Curtiss P-40F practicing night landings at Bolling Field, Washington, D.C., flew into the ground in a wooded residential area at Oxon Hill, Maryland, killing pilot 2Lt. Steven Bladykas. Investigation revealed that immediately after take-off a searchlight was trained on the aircraft by an aircraft-spotting unit. When the pilot attempted a right turn to maneuver the aircraft out of the light beam, the search light team continued to track the aircraft, blinding the pilot and causing him to fly the airplane into the ground. The aircraft exploded into flames upon impact and wreckage was scattered over a wide area. The careening aircraft knocked down an electric power pole and a telephone line. The propeller was found lodged in a front porch of a house about 500 feet from where the airplane initially struck the ground.

7-31-42A. Grider Field, Arkansas. At 0930, two Fairchild PT-19A aircraft collided in mid-air six miles south of Grider Field, Pine Bluff, Arkansas, killing A/C Harvey G. Thacker aboard PT-19A #41-14694. Student pilot 2Lt. William H. Bowick, 25, was uninjured and was able to guide heavily damaged PT-19A #41-2570 to an emergency landing in a field. The collision occurred at an altitude of approximately 500 feet agl. A/C Thacker's aircraft struck the ground inverted in a vertical attitude. Investigators speculated that A/C Thacker might have been attempting to jump but the aircraft hit the ground before he could get free.

7-31-42B. Bass Lake, California. At 1020, a Vultee L-1A crashed after striking power lines near Bass Lake Airport, 60 miles northwest of Fresno, California, killing pilot 2Lt. Meyers L. McCrory, Santa Ana, California, and passenger TSgt. John L. Coan, Des Moines, Iowa. The pilot had been attempting to assemble a formation of six Taylorcraft L-2 type aircraft for a navigation-training mission. The Accident Classification Committee stated, "Lt. McCrory took off from the Bass Lake Airport at approximately 1010 and climbed to 5,000 feet altitude. He then contacted Hammer Field tower and filed a flight plan by radio for himself and the six L-2A airplanes that were not equipped with radios. After doing this, he circled while the L-2As took off and after their take-off he proceeded to lose altitude and flew very low (below 100 feet) over the airport and lake and failed to clear power lines."

7-31-42C. Newark, New Jersey. At 2134, a North American BC-1A flying in poor weather crashed on Frelinghuysen Avenue in Newark, New Jersey, killing pilot Lt. Col. George R. Smith and passenger Maj. Jerome J. Simandl. The airplane, which was on an administrative flight, was flying on an instrument clearance to Bolling Field, Washington, D.C., but diverted to Mitchel Field, Long Island, New York, because of the deteriorating weather conditions, which included low fog. The pilot was cleared to let down for an instrument approach to Mitchel Field but for reasons unknown he flew the airplane into the ground at about a 60-degree angle, striking a parked automobile, cutting it in half and hurling pieces of it through the air and onto a parkway. The aircraft burst into flames upon impact with the ground and the automobile, scattering wreckage over a wide area. Investigators could not determine the cause of the accident and noted that the pilot was proficient in instrument flight and had made two similar instrument let-downs in the past several weeks.

August

8-1-42A. Yaqui Town, Arizona. At 0220, a North American AT-6B dove into the ground one mile south of Yaqui Town, Arizona, killing pilot 1Lt. Jack B. Childers, 27, Lubbock, Texas, and passenger Sgt. Wilson M. Smith, 26, Austin, Texas. The aircraft was one of a three-ship flight of AT-6 type aircraft on a night navigation training flight from Luke Field, Phoenix, Arizona, to Biggs Field, El Paso, Texas. 2Lt. Jack B. Kynard witnessed the accident and stated to investigators, "I was riding in the back seat of the [lead] airplane piloted by 1Lt. Jack D. Blanchard with 1Lt. Jack B. Childers' plane on the left and 1Lt. Stanley E. Gagon's plane on the right. [Twelve miles southeast of Sky Harbor Airport, Phoenix, Arizona] Lt. Childer's airplane broke formation and headed toward our tail, almost striking same, then suddenly [he was] going up about 100 to 150 feet above our airplane. He then winged over sharply, going straight to the ground, still under power as I could see the exhaust, which indicated that the engine was still on. The airplane exploded upon striking the ground."

8-1-42B. Lincolnton, North Carolina. At 1037, a Republic P-43A suffered a catastrophic failure of the starboard wing while maneuvering near Lincolnton, North Carolina, killing pilot 2Lt. William E. Grieve, Lombard, Illinois. The aircraft had taken off from Morris Field, Charlotte, North Carolina, on an individual training mission. The Accident Classification Committee stated that at an altitude estimated to be about 12,000 to 10,000 feet, the pilot put the aircraft into a vertical dive and at approximately 2,500 feet the starboard wing separated from the aircraft, being found three-quarters of a mile from where the main wreckage spun to the ground and exploded into flames.

8-1-42C. Midland, Georgia. At 1530, a Vultee BT-13 stalled and spun into the ground while maneuvering near Midland, Georgia, killing student pilot Pvt. J. G. Pemberton, Linwood, Kansas. Civilian witnesses on the ground reported that two BT-13 type airplanes were "dog fighting" and "chasing each other" when one of the aircraft suffered an apparent engine failure at low altitude. The pilot apparently attempted to turn toward a clearing for a forced landing. The aircraft was observed to bank steeply, rolling through to nearly inverted. The student attempted to return the aircraft to a level attitude but lost sufficient airspeed to cause the aircraft to stall and enter a spin at an altitude of approximately 500 feet. The BT-13 hit the ground at an angle of approximately 75 degrees, burying the engine, telescoping the fuselage and crushing the pilot.

8-1-42D. Mount Vernon, Washington. At 1730, a Lockheed P-38E on a vector-tracking mission crashed into Haystack Mountain, 13 miles east of Mount Vernon, Washington, killing pilot Lt. Col. Carl K. Bowen, 29, Milford, New Hampshire. The aircraft and pilot were overdue at McChord Field, Tacoma, Washington, and reported missing, eventually being spotted by an air search at about 1800 on 8-4-42. The pilot was eight miles east of the course that he had been assigned to fly. Investigators described the area where the crash occurred as "extremely rough and steep terrain." It was speculated that the pilot made a navigational error in the vicinity of Whidbey Island which, combined with wind drift, caused the airplane to collide with the mountain at an elevation of approximately 4,000 feet while flying in instrument conditions 35 miles north of Paine Field, Everett, Washington. Wreckage was spread over an area of 200 yards and it was speculated that the airplane might have cleared the mountain if it were devoid of the very tall trees that covered the top.

8-3-42A. Mojave, California. At 0630, a Bell P-39D practicing aerobatics near Muroc Dry Lake spun into the ground inverted 12 miles east of Mojave, California, killing pilot 2Lt. Robert E. Halprin. There were no witnesses to the crash and investigators could not determine its cause.

8-3-42B. Woodbury, New Jersey. At 1907, a Curtiss P-40F practicing aerobatics stalled and spun into the ground on the right-of-way of the Pennsylvania-Reading Seashore Railroad at Woodbury, New Jersey, killing pilot 2Lt. John S. Smith, Winterset, Iowa. The pilot was observed performing a slow roll at an altitude of approximately 5,000 feet when the aircraft stalled in an inverted position and spun to the ground in a tight spiral to the right. The aircraft exploded into flames upon impact.

8-4-42A. Columbia, South Carolina. At 1300, a North American B-25D returning from a local instrument flight crashed near the Army Air Base at Columbia, South Carolina, killing six fliers and seriously injuring one other. Investigators stated, "The pilot stalled the airplane while attempting to 'go around' after overshooting on a landing approach. The stall occurred at about 400 feet while the airplane was in a flat turn to the left. The wheels were nearly retracted at the stalling point. All eyewitnesses agreed that the engines were functioning normally as the airplane passed over the field. The plane went into a vertical dive and caught fire upon striking the ground."

Killed in the crash were: 2Lt. Neal Dow Curtis, pilot; 2Lt. Robert Roy Pattillo, co-pilot; Pvt. Floyd H. Shuman, radio operator; Pvt. Roy E. Graves, crewmember; Pvt. Benedict N. Streit, crewmember; Pvt. Alphie V. Tooke, crewmember. Pvt. Philip V. Downing was seriously injured.

8-4-42B. Romulus, Michigan. At 1329, a Bell P-39D on a transition training flight stalled and spun into the ground 24 miles northeast of the Army Air Base at Romulus, Michigan, killing pilot 1Lt. Theophilus H. Beville, Jr., 25, Gainesville, Florida. Investigators noted that the right hand cockpit door had been found some distance from where the airplane crashed, indicating that the pilot might have been contemplating a bail out attempt. Investigators determined that the airplane entered a flat spin at an altitude of approximately 4,000 feet but were unable to determine the events that led up to the crash or why the pilot did not exit the airplane. Investigators noted that a radio control box located on an after bulkhead station had come loose on other P-39 aircraft resulting in fouled control cables. It could not be determined if this had caused the spin or prevented the pilot from recovering the airplane.

8-4-42C. Spencer, Iowa. At 1420, a Taylorcraft L-2A dove into the ground three and one-half miles southwest of the airport at Spencer, Iowa, killing civilian flight instructor F.D. Evans and A/C Daniel W. Conklin. There were no witnesses to the accident.

8-4-42D. Warwick, Georgia. At 2345, a North American AT-6A on a weather check mission for Turner Field, Albany, Georgia, crashed to the ground three miles south of Warwick, Georgia, after being abandoned by its pilot. 2Lt. Bertram W. Thies, Baltimore, Maryland, was killed when he attempted to parachute to safety but was too low for his parachute to fully deploy. He was found a short distance from where the airplane hit the ground. The aircraft knocked down electric power lines when it crashed. Investigators could not determine what caused the pilot to abandon the aircraft.

8-5-42A. Clovis, New Mexico. At 0742, an Aeronca L-3B stalled and spun into the ground four miles north of Clovis, New Mexico, killing student glider pilot Sgt. Robert K. Houching, 20, Akron, Ohio. The aircraft was in the pattern for the airport at Clovis and attempting to turn to the base leg from the downwind leg when the pilot lost airspeed and stalled the airplane. The airplane was observed by witnesses to enter a four-turn spin to the left at an altitude of approximately 800 feet agl. At approximately 100 feet agl, too low to allow for a safe parachute jump, the pilot panicked and attempted to bail out, deploying his parachute while not completely clear of the airplane. He was found half out of the airplane with the partially deployed parachute lying beside him. Witnesses stated that the airplane partially recovered from the spin at about 100 feet above the ground and was observed to dive into the terrain at an angle of about 45 to 65 degrees.

8-5-42B. Gulf of Mexico, Florida. At 1905, a Martin B-26B dove into the Gulf of Mexico nine miles west of Egmont Key, Florida, killing the crew of three. Civilian observers on board a private vessel observed the aircraft to nose over abruptly and dive into the water with only a slight turn to the right on the way down. No effort by the pilots to recover from the dive was observed. The aircraft sank immediately and was never located. A parachute that was determined to be from the airplane was found floating in the sea a few days after the accident but the bodies of the crewmembers were apparently never recovered. Killed in the crash were: 2Lt. Martin R. Staum, pilot; 2Lt. Tom C. Young, co-pilot; Cpl. Wayne E. Johnson, engineer. Investigators could not determine why the airplane crashed.

8-6-42. Fletcher Field, Mississippi. At 1630, a Stearman PT-17 practicing elementary-eights stalled and dove into the ground seven miles west of Fletcher Field, Clarksdale, Mississippi, fatally injuring Aviation Cadet Jack J. McClennen, 22, Philadelphia, Pennsylvania. The aircraft did not burn and civilians working in the area quickly removed the cadet from the wreckage. He died about thirty minutes later.

8-7-42A. Camp Davis, North Carolina. At 0925, a Curtiss O-52 stalled and crashed to the ground just after take-off from Camp Davis, North Carolina, killing pilot 2Lt. Elmer A. Sommerfeld and passenger Pfc. Paul J. Bochicchio. Investigators stated, "The pilot, upon taking off, kept the plane in a stalling position with the nose unduly high for approximately 500 yards, never getting over 30 feet from the ground. The airplane stalled back to the ground just at the edge of the runway and [bounded into the air] with throttle still wide open, crashed into a large ditch, then turned over and burned."

8-7-42B. Keswick, Virginia. At approximately 1415, a Martin B-26A crashed after suffering an in-flight structural break up one and one-half miles east of Keswick, Virginia, killing two civilian ferry pilots. Pilot Walter R. Pinnell, 26, Nashville, Tennessee, and co-pilot Kenneth G. Clemens, 37, Nashville, Tennessee, took off from the Glenn Martin Aircraft Company Airfield, Baltimore, Maryland, at 1211 on a ferry flight to Nashville, Tennessee. The aircraft was heard flying northeast above the overcast. Witnesses on the ground then heard an explosion. The aircraft was then observed diving out of the clouds trailing smoke, fire and small bits of wreckage. Pieces of the aircraft, including the two engines, were observed falling away as the remainder of the aircraft tumbled and spun into a small field, exploding into flames. Wreckage was scattered over an area of over 500 yards. Investigators stated that conflicting witness statements and the premature removal of the wreckage by unauthorized and unqualified personnel hampered the investigation. Weather at the time of the accident was reported as overcast with a ceiling of approximately 6,000 to 8,000 feet with lower scattered to broken clouds at 2,000 to 3,000 feet with visibility at 4 to 6 miles with haze, smoke, light rain and occasional fog. The committee determined that the aircraft was off course but could not reach a conclusion as to what caused the accident or why the aircraft was off course.

8-7-42C. Colorado Springs, Colorado. At 1600, a Lockheed F-4 (photo-reconnaissance version of P-38 type airplane) dove into the ground two miles southeast of the Army Air Field at Colorado Springs, Colorado, killing pilot 2Lt. William H. Borgerding, 22, Belgrade, Minnesota. Witnesses observed the aircraft flying south five miles east of the field at an altitude of about 8,000 feet. The aircraft then started a slow roll to the right to an inverted position and attempted a diving split-S maneuver. At about 4,000 feet, the nose of the aircraft was seen to come up about 25 degrees before it fell off again and the airplane dove into an open field with tremendous speed. The aircraft burst into flames upon impact with the ground and was destroyed utterly. It is likely that the pilot put the airplane in a diving maneuver from which he could not recover because of the tendency of the P-38 type airplane to suffer from severe control surface buffeting caused by compressibility.

8-8-42A. Cherry Point, North Carolina. At 1045, a Douglas DB-7B crashed into a railroad tank car after suffering the failure of the starboard engine while in the traffic pattern for Cherry Point Marine Corps Air Station, North Carolina, killing pilot 2Lt. Harold E. Nankivell, 25, Denver, Colorado. A small fire erupted but was quickly extinguished by Cherry Point rescue personnel. The Accident Classification Committee stated, "The pilot was circling the field at 800 feet preparatory to landing. During some part of his approach the right engine failed. In attempting to reach the field he either turned to his right or permitted his right wing to lower. He was unable to raise it with full power on the left engine. [He selected] a field in which to land wheels up but in reaching that field he lost control, landed short, and struck a railroad spur and tank car."

8-8-42B. San Francisco Bay, California. At 0930, a Martin B-26B being ferried to Hickam Field, Oahu, Hawaii, crashed into San Francisco Bay, five miles southeast of Hamilton Field, California, killing the crew of four. Investigators stated, "The subject airplane dived through clouds, which were at an altitude of about 1,200 feet, according to weather report, and it assumed a vertical diving attitude which continued until the plane struck the water. The plane burst into flames and sank. Indications are that the pilot was returning to Hamilton Field for reasons unknown. Plane had been flying approximately 12 minutes and was not expected to be in the vicinity of the accident." Killed in the crash were: 1Lt. Robert B. Holland, Benton, Kentucky, pilot; 1Lt. Meinrad J. Endres, Jr., Munster, Texas, co-pilot; 2Lt. Richard R. Longnecker, Houston, Texas, navigator; Pvt. Roy M. Frette, radio operator, Badger, Iowa.

8-8-42C. Colorado Springs, Colorado. At 1130, a Lockheed F-4 (the photo-reconnaissance version of the P-38) crashed when its port engine failed just after take-off from the Army Air Field at Colorado Springs, Colorado, killing pilot 1Lt. Edward J. Peterson, 25, Alamosa, Colorado. The engine failed just after rotation and the aircraft reached an altitude of approximately 50 feet as it winged over to the left and into the ground east of the North/South runway. 1Lt. John E. Murray witnessed the accident and stated to investigators, "[Lt. Peterson] started his take-off run. As he rolled down the runway, I noticed a considerable amount of smoke pouring from his left supercharger. At about the time the plane left the ground, the smoke disappeared and then reappeared. As he put his gear up the plane rose in a climbing arc to the left, and completed the arc when the plane crashed about 50 feet from the runway to the east. The plane hit on the left motor and wing and immediately burst into flames." Lt. Peterson had been scheduled for a day navigation training flight. The Army Air Field at Colorado Springs, Colorado, was later named Peterson Field in memory of the pilot. Additionally, according to newspaper accounts reporting the crash, Lt. Peterson's dog was made a sergeant in the AAF.

8-8-42D. Craig Field, Alabama. At 1440, a North American AT-6A stalled and spun into the ground nine miles southwest of Craig Field, Selma, Alabama, killing Aviation Cadet Wiley M. Dunn. Investigators stated, "Cadet Dunn had mentioned several times previously that the method of catching up with the instructor's ship while circling is to retard the throttle and with an airspeed of approximately 80 or 85 mph and keeping the nose up with top rudder to pull the ship in and turn inside the instructor. It appears that such a method was put into practice and the ship apparently stalled and went into a spin from which it did not recover."

8-8-42E. Barksdale Field, Louisiana. At 1730, a Martin B-26B suffering the failure of the starboard engine crash landed 15 miles southwest of Barksdale Field, Shreveport, Louisiana, killing one crewmember and injuring two others, one seriously. Two other crewmembers escaped injury. The B-26 was on a navigation training flight and was flying at an altitude of about 2,500 feet approximately 25 miles southwest of Barksdale Field when the pilots noticed a drop in fuel pressure and diminishing performance of the starboard engine. The fuel pressure dropped to zero and the crew activated the fuel booster. Despite their efforts to keep it going, the engine quit and the crew feathered the propeller. The aircraft began to lose altitude and became hard to control. The pilot ordered the crew to prepare for a parachute jump and the bomb doors were opened. The aircraft continued to lose altitude and became even harder to control. The navigator opened the upper observation hatch and began exiting the aircraft. The pilot yelled for the crew to stop him because the aircraft was losing too much altitude too fast, making a parachute jump dangerous. Two members of the crew tried to grab the navigator and prevent him from exiting through the top hatch. The navigator, 2Lt. Edward V. Skahill, deployed his parachute before he was clear of the aircraft and was sucked out of the grasp of the crewmembers and out of the airplane. He struck the tail and his parachute fouled on the tail control surfaces, becoming entangled and interfering with the free movement of the elevator controls. Control of the airplane became almost impossible. The pilots were able to maintain enough control of the airplane to make a hard crash landing in a cotton field. The aircraft careened across the field for over 100 yards, breaking up and catching fire. The remaining crewmembers were able to escape the aircraft before it was enveloped in flames. Lt. Skahill was found dead in a cotton field about one-half mile from the scene of the crash landing. Pilot 2Lt. Milton Herman received minor injuries and co-pilot 2Lt. Richard H. Yeats was seriously injured. Crewmembers Cpl. Walter D. March and Sgt. Joseph P. Andrews were uninjured.

8-9-42A. Luke Field, Arizona. At 0110, a North American AT-6C flew into terrain 10 miles northwest Luke Field, Phoenix, Arizona, killing pilot 2Lt. William C. Dennis, 22, Dubuque, Iowa. The aircraft was one of a flight of three returning from a night navigation training flight. While flying at an altitude of 6,000 feet, the leader of the formation signaled to break up the formation. The left wingman dropped back and below while Lt. Dennis followed the leader and peeled off to the left. Lt. Dennis' airplane was observed by the left wingman to peel off to the left and continue to roll through inverted and back into a diving turn to the left until the airplane flew into the ground and exploded. Investigators speculated that the pilot lost visual reference to the horizon over the dark terrain while in the turn.

8-9-42B. Craig Field, Alabama. At 0323, a North American AT-6A stalled and spun into the ground four miles southwest of Craig Field, Selma, Alabama, killing RAF Flying Cadet Herbert Riding, Jr. The airplane had just taken off on a three-ship night formation flight. The cadet was attempting to maneuver into formation with the two other airplanes when he stalled the AT-6 and it spiraled to the ground, exploding into flames upon impact. There were no witnesses to the accident.

8-9-42C. Raleigh, North Carolina. At 0928, a Martin B-26B descended into trees at the end of the runway while attempting a take-off from the municipal airport at Raleigh, North Carolina, killing two crewmembers and seriously injuring five others. The pilot attempted the take-off on the 2,400-foot runway with no flaps, lowering one-quarter flaps when attaining 100 mph and becoming airborne grudgingly on the downward slope at the departure end. As the co-pilot retracted the gear, the aircraft descended into trees, the starboard wheel striking a six-inch diameter tree. The aircraft bounded into the air from the impact, traveling another 75 feet further before again descending into the terrain, breaking apart and catching on fire as it smashed through large trees. Cpl. Ross S. Ware was killed instantly and Pvt. John R. Rarrat died of his injuries at 1403. Seriously injured in the crash were: 1Lt. Frank E. Werneken, pilot; 2Lt. Michael A. Magnotta, co-pilot; 2Lt. Elmer J. Gedeon, navigator; TSgt. Lawrence D. Glover, engineer; Sgt. Maurice L. Clark, gunner.

8-10-42A. Gunter Field, Alabama. At 0815, a Vultee BT-13 stalled and spun into the ground while turning from the downwind let to the base leg in the traffic pattern one mile southeast of Gunter Field, Montgomery, Alabama, killing instructor pilot 2Lt. Torbert Slack Jr., Lake Charles, Louisiana, and RAF Flying Cadet Cyril E. Gray, Surrey, England. The airplane was seen to go into a chandelle type maneuver to the left, going to an almost inverted position and continuing in a low, wide spin to the left and into the ground. Despite numerous eyewitnesses and careful examination of the wreckage, investigators could not determine what caused the airplane to enter the maneuver and fly into the ground. It was noted by investigators that Lt. Slack was wearing moccasins instead of shoes on the flight.

8-10-42B. San Diego, California. At 1110, a Lockheed P-38F stalled and spun into San Diego Bay, California, while attempting a landing at North Island Naval Air Station, killing pilot 2Lt. Hatton W. Sumner. The aircraft was observed in the pattern on the downwind leg flying low and extremely slow. The aircraft then executed a slow, flat turn to the left to enter the base leg. The aircraft continued in the pattern with no flaps and with the landing gear in the retracted position. The pilot then attempted another slow, flat left turn from the base leg to final. Investigators speculated that the pilot lowered the landing gear while attempting to execute this left turn with low airspeed and possible failure of the port engine, stalling the airplane to the left and describing three quarters turn of a spin before the P-38 smashed into the bay.

8-11-42A. Mt. Meigs Field, Alabama. At 2245, a Vultee BT-13 on a local night flight flew into the ground while in the pattern for the airfield at Mt. Meigs, Alabama, killing RAF Flying Cadet Albert E. Ayling, Jr., Portsmouth, Hampshire, England. The BT-13 was observed making a steep right turn from the downwind leg to the base leg and continued in a descending turn to the right until it flew into the ground and exploded into flames. Investigators speculated that the pilot lost visual reference to the horizon while in the turn.

8-11-42B. Gray Field, Washington. At 2202, a North American O-47A collided with a revetment while attempting to take-off on a taxiway at Gray Field,

Fort Lewis, Washington, killing two crewmembers and seriously injuring another. The pilot made two landings at the field and then taxied up to a group of enlisted men who were operating the temporary field light system. The pilot made some comments about the position of several of the floodlights that were illuminating the field for the night landing exercises. The pilot then returned to his aircraft and taxied to the near taxiway rather than the one on the other side of the runway, which was being used for operations that night. The pilot turned onto the taxiway and began his take-off roll. The aircraft had become airborne and climbed about six inches off of the ground before it struck a gun revetment. Enlisted men who were in the revetment ran to safety when they realized that a collision was imminent. The landing gear was sheared off as the airplane plowed through the revetment. The airplane bounded back into the air and then smashed into the ground a short distance away, immediately bursting into flames. The enlisted men ran to the wreckage and attempted to rescue the fliers, pulling two of them out of the airplane before the flames became too intense. Investigation revealed that the pilots who had been practicing at the field had been landing between the smudge pots that lined the runway and taking off on the taxiway on the unobstructed portion of the field. Apparently the pilot thought that he was on the far taxiway when he attempted the take-off. Pilot 2Lt. James F. Faircloth and passenger Pvt. Robert L. Gilbert were killed and Sgt. Lendon C. Wright was seriously injured.

8-12-42A. Roswell, New Mexico. At 0119, a North American AT-6A flew into terrain one mile west of the municipal airport at Roswell, New Mexico, killing A/C Donald E. Angus, 24, Hayward, California. The student was attempting a landing and was told by the tower not to land and to go around. The student complied and gave the airplane full throttle for another approach. Investigators stated, "It is considered probable that due to the darkness of the night, the lack of any ground lights or nearby horizon lights, and the concentration of the student pilot in his landing gear, flap and stabilizer procedure, the student continued to lose altitude until he flew the airplane into the ground."

8-12-42B. Mobile, Alabama. At 0812, two Schweizer TG-2 type gliders collided in mid-air while being towed to altitude one quarter of a mile west of the municipal airport at Mobile, Alabama, killing one flier and injuring three others, one seriously. Civilian instructor John Seeley, 26, Elmira, New York (TG-2 #42-46634), was killed when he attempted to use his parachute at too low an altitude to allow it to fully deploy. His student, Sgt. Samuel G. Savory, was seriously injured in a crash landing of the two gliders, which remained stuck together after the collision. The tow airplane landed safely at its base. Civilian instructor George Rehill and student Sgt. Carl H. Sickau (TG-2 #42-53014) received minor injuries. The collision occurred at an altitude of approximately 600 feet above ground level. The Accident Classification Committee stated, "Instructor Seeley was on the short tow-rope on the right. Seeley allowed the glider to get on the left side of the [long] tow-rope of the other glider and the long tow-rope caught on top of the [starboard] wing of Seeley's glider and across the top of the fuselage. He then released and the glider started a climbing turn to the right, pulling the other glider into it. The other glider, piloted by Instructor Rehill, attempted to release. The gliders became interlocked in the air with Seeley's glider in an inverted position. Seeley jumped, but due to insufficient altitude his parachute failed to break his fall completely and he was killed instantly. Instructor Rehill still had a little rudder and elevator control and managed to guide [both gliders still locked together] into a tree, thereby, breaking the fall and possibly saving his own life and [the lives of] the two students."

8-12-42C. Luke Field, Arizona. At 0935, two North American AT-6B type airplanes collided in mid-air while maneuvering in a Lufbery circle 12 miles south of Luke Field, Phoenix, Arizona, killing passenger 1Lt. John D. Busiek, 26, Nevada, Missouri, and injuring pilot Lt. Col. Claude L. Brignall, who parachuted to safety from AT-6B #41-17363. 2Lt. Cyrus M. Birney, Jr. piloted damaged AT-6B #41-17230 to a safe landing. Passenger A/C John H. Allen, Jr. parachuted to safety. The airplanes were part of a six-ship flight that took off from Luke Field, Phoenix, Arizona, on an in-flight photographic mission. The flight consisted of two photo-taking airplanes and six subject airplanes, with the two photo-airplanes circling overhead of the formation. Investigators stated, "It is the opinion of this committee that the accident resulted when [Lt. Col. Brignall] attempted to join a Lufbery circle composed of six airplanes without his being scheduled to participate in the flight. [Lt. Col. Brignall] erred in judgment by deciding to join a formation for which he was not scheduled. A six-ship formation flight was being flown for the Photographic Department [Luke Field]. The formation went into a Lufbery circle to the left. The photographic ship [Lt. Col. Brignall] was outside and above the circle. [Lt. Col. Brignall] made an approach to the circle by dropping down on the outside and then up into position. He was trying to get between the sixth and lead ship. He miscalculated and came between the fifth and the lead. In doing so, his ship came up under the sixth plane and collided with it. His ship went out of control. The pilot jumped and the passenger went in with the ship. The other plane (the sixth ship) was able to return to its base. The cadet in the sixth airplane jumped under orders." Lt. Busiek was a flight instructor at Luke Field.

8-12-42D. Pond Field, California. At 1720, a Vultee BT-13 stalled and spun into the ground six miles west of Pond Auxiliary Field, California, killing instructor 2Lt. Stewart P. Smith and student A/C Herbert A. Jones. The Accident Classification Committee stated, "The Committee is unable to determine a definite cause for the accident. Judging from the path of the aircraft as described by the witness' statement and the attitude in which it struck the ground, it is believed that there is strong possibility the airplane became stalled, started to spin, and upon being pulled out abruptly, stalled into the ground in a comparatively level attitude and with a high rate of descent."

8-12-42E. Wasco Field, California. At 1730, a Vultee BT-13A stalled and spun into the ground while turning from base to final with power off three-quarters of a mile south of Wasco Auxiliary Field, California, killing A/C Paul C. Averill. Investigators stated, "[Cadet Averill] entered the traffic pattern in a normal manner. He was on his base leg and executed a 90-degree, power off turn into the field without a great degree of bank. He suddenly applied bottom rudder without proper aileron control. Ship entered a spin from which no recovery was made. Ship crashed into the ground and burst into flames."

8-13-42A. Nichols, Georgia. AT 0120, a North American AT-6B on a night navigation training flight flew into the ground 10 miles west of Nichols, Georgia, killing Aviation Cadet Carl G. Stridesburg. The aircraft struck power lines and exploded into flames when it smashed into the ground, the fire visible to students flying for miles around. Examination of the wreckage revealed no indication of mechanical or structural failure and enough fuel was on board to complete the navigation problem safely. Witnesses on the ground stated that the airplane's engine was operating and sounded to be running satisfactorily up until the time the AT-6 crashed. Investigators could not determine why the cadet flew the airplane into the ground.

8-13-42B. Barksdale Field, Louisiana. At 0924, a Martin B-26A stalled, entered a very tight spin and dove into the ground in a wooded area 10 miles northwest of Barksdale Field, Shreveport, Louisiana, killing the crew of four. Numerous civilians in the area saw the airplane dive straight down to the ground and explode into flames 100 yards south of Cooper Road, but much of their other testimony was conflicting. Investigators could not determine why the airplane dove into the ground but noted that the flaps were found to be down at 20 degrees. Killed in the crash were: Capt. Frank E. Locke, 24, Cleveland, Ohio, pilot; 2Lt. Alfred J. Woodard, 23, Roodhouse, Illinois, co-pilot; Cpl. Harry C. McCarthy, 23, Cross Creek, Pennsylvania, engineer; Pvt. Charles E. Harris, 22, Philadelphia, Pennsylvania, assistant engineer.

8-13-42C. Muroc Dry Lake, California. At 0820, a Bell P-39D on a gunnery practice mission crashed after colliding with the target at which it was shooting at Muroc Gunnery Range, California, killing pilot 2Lt. Thomas H. Vogel. Pvt. Jack Bailey witnessed the accident and stated to investigators, "The plane started coming in low and firing late for two trips. The third trip he was very low and firing late. He tried to miss the target but went through it instead. After he went through the target, he pulled the ship out of the dive and started up a range of mountains. He saw that he could not make it and then banked his ship to the south. When he did this, his motor shut off. He tried to land the plane, but instead hit the ground nose first, breaking off part of the prop. He skidded for about 25 yards, during which time the ship was burning. When the plane did stop, it was thrown over on its side, throwing the pilot out of the cockpit. By this time the ship was completely in flames."

8-13-42D. San Diego Bay, California. At 1155, a Lockheed A-29 on a target towing mission crashed into San Diego Bay, California, four miles southeast of North Island Naval Air Station, killing the crew of five. The Accident Classification Committee stated, "Because of lack of evidence the board finds that the cause of the accident is undetermined. The fact that the airplane crashed approximately five miles from the field after 11 minutes of flying time indicated that probably the failure of the instruments or the poor technique of the pilot caused the airplane to spin in. The low ceiling and 1,000 foot overcast is not considered a contributory factor except as it might bear on facts of the preceding statement." Killed in the crash were: 2Lt. Lloyd W. Tansing, pilot; Sgt. Elias Sauceda, engineer; Pvt. Winfield M. Bagley, radio operator; Cpl. Julian C. Mullinax, gunner; Pvt. William A. Cook, crewmember. Only Pvt. Bagley's body was immediately recovered. The official report indicates that the remaining crewmembers remain missing.

8-13-42E. Fort Benning, Georgia. At 1730, an Aeronca L-3B stalled and entered an inverted spin, crashing into the ground at Fort Benning, Georgia, and killing pilot 2Lt. J. N. Congdon and passenger Sgt. R. Loewit. Witnesses observed the airplane circling Fort Benning at an altitude of approximately 600 feet above ground level. The airplane was then seen to make an abrupt turn to the left. The airplane stalled and flipped over to an inverted attitude. The airplane then entered an inverted spin to the right, making about two full turns and crashing into the ground inverted in a vertical position.

8-13-42F. Sherman, Texas. At 1730, a Vultee BT-13 stalled and spun into the ground 17 miles southwest of Perrin Field, Sherman, Texas, killing instructor 2Lt. James W. Clements and student Aviation Cadet John R. Gray. The Accident Classification Committee stated, "While returning to Perrin Field from student instrument training rides Lt. Clements and Lt. Robert W. Hatch met in the vicinity of Gunter, Texas, and proceeded to fly formation. Both officers were flying from the front seat, each with a student in the rear seat. Lt. Hatch, the leader at the time of the accident, peeled off into a tight circle to the left. While following Lt. Hatch in the circle, the board assumes that Lt. Clements over controlled while in a steep bank, enough to cause the airplane to go into a violent spin. The pilot recovered from a spin to the right and immediately went into a spin to the left. Plane crashed while still in a spin. Due to poor technique and possibly the lack of altitude, he was unable to recover from the spin."

8-13-42G. Lakeland, Minnesota. At 1835, a Taylorcraft L-2A stalled and spun into the ground five miles southwest of Lakeland, Minnesota, killing civilian instructor H. Robert Christensen and student Pvt. Bruce Lee. Investigators stated, "Evidence points to a simulated forced landing and that coming up from the simulated forced landing, the airplane stalled out in a vertical turn close to the ground."

8-13-42H. Key Field, Mississippi. At 2213, a Douglas DB-7B crashed one mile southeast of the runway just after take-off from Key Field, Meridian, Mississippi, killing pilot 2Lt. Walter J. Miers, 23, Crowley, Louisiana. The Accident Classification Committee stated, "After comparing all witnesses statements, it is the opinion of the committee that approximately 300 feet in altitude after take-off into the south, the plane burst into flames, motors failed and the plane crashed into the ground at high speed, scattering wreckage for 300 yards. Motors were 100 yards further than any of the other wreckage. Motors were 300 yards from original point of impact."

8-13-42I. Key Field, Mississippi. At 2246, a Douglas DB-7B crashed one-half mile west of the runway while attempting a go-around at Key Field, Meridian, Mississippi, killing pilot 2Lt. John G. Brackett, 24, Arlington, Massachusetts. Investigators stated, "The pilot was landing to the west, he overshot the field. He increased his throttle and intended to go around for another approach. At approximately one-half mile west of the field the plane crashed into some trees and then burst into flames. The pilot apparently dumped his flaps too low to the ground and without sufficient speed causing him to drop into the trees which were approximately 150 feet higher than the runway."

8-14-42A. Victoria, Arkansas. At 0845, a Vultee BT-13A stalled and spun into the ground at Victoria, Arkansas, killing Aviation Cadet Herbert N. Perkins, Rich, Virginia. The aircraft was seen by witnesses on the ground to enter a spin at about 3,000 feet after the engine apparently stopped. Investigation revealed that the flaps were in the down position when the crash occurred and it was speculated that the student had forgotten to raise them when he took off from the airfield at Blytheville, Arkansas.

8-14-42B. Estherville, Iowa. At 1030, a Taylorcraft L-2A crashed while maneuvering three miles south of Estherville, Iowa, killing Aviation Cadet Purlmom H. Horne. "The Accident Classification Committee has come to the conclusion that the accident was caused by low flying on the part of the student, it is also concluded that the student, in the attempt to avoid hitting a large cottonwood tree suddenly appearing in front of him, over controlled in a sharp bank and went into a spin and did not have time nor altitude in which to recover."

8-14-42C. Bainbridge, Georgia. At 1320, a Vultee BT-13A crashed after suffering an engine failure on take-off from the AAF Basic Flying School at Bainbridge, Georgia, killing instructor 2Lt. Merrill C. Hinkson and student Aviation Cadet Glenn W. Connelly. The airplane was seen to take-off, gaining about 150 feet of altitude before the engine quit. The instructor attempted a left turn of about 100 degrees in an effort to get back into the field. The airplane stalled and nosed straight into the ground, bursting into flames upon impact. Investigators could not determine what caused the engine failure.

8-14-42D. Elfers, Florida. At 1740, a North American AT-6B flying in poor weather crashed to the ground out of control at Elfers, Florida, killing Aviation Cadet Paul J. Beeler. The aircraft had taken off from Spence Field, Moultrie, Georgia, and was to fly to Drew Field, Tampa, Florida, on a navigation training flight. When the pilot arrived in the Tampa area, the weather was deteriorating and he had trouble locating Drew Field. In his attempt to orient himself under the low hanging overcast, he struck and severed three power lines at Tarpon Springs, Florida. An aircraft spotter observed the airplane flying low over the New Port Richey observation tower, flying south then north before turning back to the south again and disappearing into the weather. A farmer working his field near Elfers observed the aircraft fall out of control from the overcast, plunging straight down into the ground and exploding into flames.

8-14-42E. Fort Morgan, Colorado. At 1605, a Taylorcraft L-2 stalled and spun into the ground while in the pattern for the auxiliary airfield six miles southeast of Fort Morgan, Colorado, killing civilian instructor Hubert H. Borden, 36, Morrell, Nebraska, and glider student Aviation Cadet Robert J. Drennan, 28, Marion, Kentucky. The aircraft was observed making a very steep downwind turn at an altitude of approximately 300 feet before it began slipping to the right. The aircraft continued to slip and struck the ground on its right wing, bounced into the air for a short distance before smashing violently into the ground. The aircraft tumbled for a short distance, breaking up and coming to rest inverted.

8-14-42F. Dowling, Michigan. At 2215, a Consolidated B-24D on a cross-country navigation flight crashed one and one-half miles north of Dowling, Michigan, killing the crew of nine. Killed in the crash were: 2Lt. Eugene C. King, Brawley, California, pilot; 2Lt. Louis H. Tally, Aledo, Texas, co-pilot; 2Lt. Maurice E. McCall, Estherville, Iowa, navigator; 2Lt. James J. Daly, Oakmont, Pennsylvania, bombardier; Cpl. George E. Strenge, Mountain Lake, Minnesota, engineer; Cpl. Ernest L. Prather, Beaver Falls,

Pennsylvania, assistant engineer; SSgt. John D. Hazen, Ipswich, Massachusetts, radio operator; Pvt. Steve Spataro, Wellsburg, West Virginia, assistant radio operator; Pvt. Raymond W. Carillion, Millersburg, Ohio, gunner. The Accident Classification Committee stated, "Lt. King was scheduled to make a flight, within the local flying area, from Willow Run to Saginaw to Kalamazoo to Willow Run [all in Michigan]. Take-off was made at 2107. It is believed that the pilot completed the first leg of his flight and was en route to Kalamazoo when he encountered bad weather. The pilot flew through the weather and was contact when he saw Hastings, Michigan, and mistook Hastings for Battle Creek. There was a fair in progress on the west side of town, bearing the same relation to Hastings as does Kellogg Field to Battle Creek. The pilot made a decision to land and circled to the left. Witnesses state the plane flew over Hastings from the east and tuned toward the southwest. Examination of the wreckage disclosed that the landing gear was down and locked. At the actual instant of the crash, it appears that the left wing tip first made contact with the ground, indicating a circle to the left and that the pilot might have been looking for the field and did not see the ground; probably the nose of the airplane dropped when made his turn and he did not take into account the slight difference in elevation of Battle Creek and that of Willow Run.

8-15-42A. Wade, Mississippi. At 1415, a Douglas DB-7B suffering a starboard engine fire crash-landed into a heavily wooded area at Wade, Mississippi, seriously injuring pilot 1Lt. Walter W. Ruoff and lightly injuring SSgt. Vernon W. Vogt. Bombardier 1Lt. Harold W. Prusa was killed in the crash landing. He had attempted to parachute to safety but apparently deployed his parachute while still in the aircraft, fouling the parachute and trapping him inside the airplane. Lt. Ruoff later stated to investigators, "I was flying en route from Hattiesburg, Mississippi, to Brookley Field [Mobile, Alabama] between 500 and 700 feet. I glanced down and noticed that the right oil temperature was about 85 degrees, which is a high temperature for this type of engine. I then opened the lower cowl and oil cooler flaps and a burst of flame came out of the engine. I pulled the release for the [fire extinguisher] on the right engine, but this did not have any effect in extinguishing the blaze. I then feathered the propeller, trimmed control tabs for single engine operation, and increased the left engine power. I lost about 300 feet, but with the increased power I was able to gain back about the same amount of altitude but could not seem to gain any more than about 300 feet. By this time, the whole right engine was a mass of flame, and I called Lt. Prusa and Sgt. Vogt and told them that the airplane was on fire and for them to jump or ride it down. Then I heard the trap door open in the front gunner's compartment and figured that Lt. Prusa had bailed out. I maintained this altitude for one to two minutes during which time I opened my hood and decided to bail out, but then decided I was too low for a safe jump and would have to make a crash landing. The flames at this time were completely enveloping the whole right side and tail of the airplane. I brought the airplane down to within about 100 feet of the ground and cut the left engine. I then eased it down close to the tree tops and attempted to hold it off as long as possible. I remember the right wing hitting some of the tree-tops, but do not know whether I jumped or was thrown from the airplane."

8-15-42B. Peru, Massachusetts. At 2132, a Douglas C-53 flying in poor weather collided with terrain near Garnet Peak at Peru, Massachusetts, killing 16 paratroopers and crew and seriously injuring three others. The aircraft was part of an eleven-airplane flight from Pope Field, North Carolina, to Hyannis, Massachusetts, with a refueling stop at Mitchel Field, New York. Weather at Mitchel was so poor that the flight leader contemplated spending the night. Orders came down from headquarters to take off. The clearance at Hyannis was cancelled because conditions were below minimums, so the flight was diverted to and cleared to Providence, Rhode Island. Once airborne, the formation could not stay together in the rapidly deteriorating conditions at the assigned altitude of 3,000 feet and was ordered to climb above the weather. The accident airplane flew on top of the weather with the formation before the clouds became increasingly thicker and the flight leader signaled for a slow instrument let down through the weather. The accident pilot's instrument card had been revoked after his last instrument check ride (7-26-42) and he was operating without a rated co-pilot and radio operator. The pilot began his let-down at about 300 feet per minute and flew the airplane into the ground under control with wings level and both engines functioning. The aircraft collided with large trees on a level ridge at an elevation of 2,005 above sea level, breaking apart and catching fire. The aircraft snapped off large trees in its path and came to rest 350 feet from the initial point of impact, scattering wreckage and dismembered bodies for many yards. Paratrooper Sgt. Robert Lee, Columbus, Ohio (the only passenger properly strapped in), miraculously escaped serious injury in the initial crash but was seriously burned while saving the lives of fellow paratroopers Pvt. James Fern, Abington, Virginia and Pvt. Alonzo L. Pearson, Windber, Pennsylvania, who were both severely burned. Sgt. Lee also pulled two men to safety not knowing that they were already dead. Sgt. Lee fired his pistol into the air to attract attention but no one noticed. Sgt. Lee then walked three miles to the state highway and notified State Police of the crash, guiding them back on foot to the crash site. The woods and underbrush were so dense that a trail had to be hacked through to allow rescue workers access to the wreck site. Weather at the scene of the crash was reported as solid overcast beginning at 2,500 feet with scattered clouds and occasional fog from 2,500 feet to ground level. Crewmembers killed in the crash were: 2Lt. Joseph J. Field, Jr., Atlanta, Georgia, pilot; SSgt. Charles M. Scott, York, Pennsylvania, co-pilot; SSgt. Robert W. Lamon, Shawnee, Oklahoma, engineer; Cpl. Austin E. Weeces, Craig, Nebraska, assistant engineer. Paratroopers (502nd Parachute Battalion, Fort Benning, Georgia) killed in the crash were: 1Lt. Gardner V. E. Plain, Ransomville, New York; Sgt. John H. Kelley, Titus, Alabama; Pvt. James F. Thompson, Los Angeles, California; Pfc. Stanley E. Racine, Pensacola, Florida; Pfc. Norman Sands, Chicago, Illinois; Pfc. Hyman R. Flinkman, Baltimore, Maryland; Pvt. Steve E. Schollin, Oak Haven, Illinois; T-5 Joseph Neurohr, Brooklyn, New York; Pfc. James D. Westbrooks, Anderson, South Carolina; Pvt. Jack E. Palmer, Antigo, Wisconsin; Cpl. Frank A. Lastosky, Sykesville, Pennsylvania; SSgt. Samuel Hathorn, Princess, Mississippi. Most of the men were sleeping on the floor at the time of the crash.

8-16-42A. Inadale, Texas. At 1400, a Consolidated B-24D flying in instrument conditions suffered the catastrophic failure of the port elevator and rudder and crashed three miles west of Inadale, Texas, killing three crewmembers. Pilot 2Lt. John W. McMahan, 30, Long Beach, California, co-pilot SSgt. Carl R. Bosse, 22, Goshen, Indiana, and engineer Pvt. Eugene E. Pence, 21, Godfrey, Illinois, were killed in the crash. The AAF accepted the B-24 on 8-14-42 and the crew had flown the aircraft to Tucson, Arizona, from Long Beach, California, on the first leg of a ferry flight to St. Paul, Minnesota. The airplane took off from Tucson without incident at 0939 MWT and the pilot reported to air traffic control at 1325 CWT, 15 miles west of Big Spring, Texas, requesting permission to let down to the assigned 7,000 feet from the 11,000 feet he was flying to stay on top. Permission was granted and the pilot reported that he would descend when he "found a hole" in the heavy cloud cover. At 1332, the pilot reported to be at 7,000 feet over Big Spring. At 1340, the aircraft was cleared for 9,000 feet on the airway to Ft. Worth, Texas. The pilot acknowledged the clearance and there was no further contact with the B-24. Witnesses on the ground heard and seen the B-24 plunge out of the overcast in a high-speed dive straight into the ground nine miles north of Loraine, Texas. The aircraft exploded into flames upon impact with the ground, scattering wreckage for over 100 yards. After a few moments, pieces of the tail section floated down to the ground one mile southeast of the main wreckage. Investigation revealed that the upper portion of the port vertical fin and rudder separated from the aircraft in flight causing the progressive failure of the port elevator and horizontal stabilizer and, due to the excess vibration, the starboard horizontal stabilizer, elevator, rudder and vertical fin. Investigators speculated that the pilot exceeded the structural limitations of the airframe while maneuvering the aircraft in instrument conditions. The sequence of events preceding the structural failure could not be determined. Weather in the area at the time was reported as overcast with ceilings at 2,000, visibility about two to four miles with light to moderate rain and mild turbulence at upper altitudes.

8-16-42B. Oldsmar, Florida. At 1545, a Bell P-39D stalled and spun into the ground at Oldsmar, Florida, killing pilot 2Lt. Harry I. Wilson. A witness on the ground reported seeing the airplane at an altitude of 1,500 feet in a tight, diving spin toward the ground. The starboard pilot's door was seen to leave the aircraft just prior to the airplane striking the ground nose first and exploding into flames. Investigators stated that they could not determine the cause of the spin or if an engine failure had occurred.

8-17-42A. Orlando, Florida. At 0831, a Curtiss P-40E taking off from the Army Air Base at Orlando, Florida, collided with a tractor-trailer fuel truck that attempted to cross the runway, killing pilot 1Lt. Edward J. Burke and fatally injuring the truck driver Pvt. Abraham Miller. Assistant truck driver Pfc. Eugene L. Hammer ran for safety when a collision seemed imminent, escaping injury. Another P-40E that was leading the take-off with Lt. Burke (who was echeloned to the right) barely missed hitting the truck, which was towing two fully loaded fuel tank trailers. The two P-40s were taking off to the north on the north-south runway and the truck was heading west on the southern east-west runway. Field procedures call for all ground traffic to get an affirmative (green) light signal from the tower to make a crossing of an active runway and all ground traffic is to halt and look to the tower for a green light signal prior to crossing. Tower personnel stated that the truck did not stop at the runway intersection before crossing and that the red signal light was illuminated in the tower the entire time that the flight was taxiing and on the take-off run. When the tower personnel perceived that an unsafe situation was developing, they radioed the departing flight that an incursion was taking place. No reply was received. The lead airplane was approximately 200 feet ahead of Lt. Burke's airplane and narrowly missed the tractor cab. Pfc. Hammer then ran from the truck when he saw Lt. Burke's P-40 heading directly toward them. Lt. Burke's P-40 was about four feet off the ground when it struck the fuel truck broadside at the forward end of the forward fuel trailer just behind the cab, both the truck and the airplane exploding violently into flames. The airplane knocked the truck over on its side and then bounded over the truck, coming to rest inverted about 30 feet beyond it. Pvt. Miller had either jumped out or was thrown out of the cab and was observed screaming while running down the runway with his clothes on fire. Pfc. Hammer grabbed Pvt. Miller and rolled him on the ground and rescue personnel extinguished his flaming clothing. Pvt. Miller, who died of his injuries at 2300, had been working at the base for about three days. The P-40s were taking off on a "scramble" mission.

8-17-42B. East Hartford, Connecticut. At 1106, a Curtiss P-40E stalled and spun into the ground after suffering an engine failure just after take-off from Rentschler Field, East Hartford, Connecticut, killing pilot 2Lt. Harry F. Sheraw. The aircraft gained about 200 feet of altitude when the engine quit. The pilot made a steep left turn to attempt an apparent emergency forced landing in a nearby field and stalled the airplane. The aircraft spun into the ground and burst into flames upon impact.

8-17-42C. Coleman, Texas. Aviation Cadet John A. Reynolds fell to his death when he failed to deploy his parachute after he was ejected from a Fairchild PT-19A at 1335, eight miles east of Coleman, Texas. His civilian instructor, O.H. Stansel, stated to investigators, "We had been in the air about 25 minute and had been practicing rectangular course at about 500 feet and I had told the student to find a road and do some S-turns. He had been having trouble holding altitude in turns. [He] started a turn, applying very much backpressure on the stick, making a steep, climbing turn. I corrected by putting forward pressure on the stick while in the turn and shoving the nose of the plane down. I noticed [that the] plane was out of control and glanced into the mirror and saw that the student had fallen out. I cannot say that his [safety] belt was on at the beginning of the flight. He did not open his parachute and was killed." Investigation revealed that the parachute was in correct working order and that the student did not strike any part of the airplane when he was ejected. Investigators speculated that the student must have panicked and neglected to deploy the parachute.

8-17-42D. Plains, Georgia. At 1500, a Stearman PT-17 crashed at the AAF Auxiliary Airfield at Plains, Georgia, killing civilian flight instructor Arthur K. Holloway, 23, Swainsboro, Georgia, and RAF Flying Cadet Douglas W. Flatau, 19, Somerset, England. Capt. Ernest F. Baldwin witnessed the accident from an aircraft that had just landed and rolled to a stop in preparation to beginning another take-off. Spotting another PT-17 close behind in the pattern, Capt. Baldwin halted his take-off. He stated to investigators, "I watched the airplane and suddenly observed that the pilot opened the throttle and pulled the nose up. He gained altitude from approximately 30 to 80 feet, and while doing so, he began a coordinated slow roll [to the left]. I was very surprised to see the airplane go into a roll at such a low altitude. The airplane rolled slightly past inverted whereupon the engine sputtered; the airplane lost flying speed; fell into a spin from inverted flight and completed about one-fourth to one half turn of an inverted spin. The airplane struck the ground about 50 yards from the airplane I was in, hitting the ground at an angle of about 70 degrees with right wing slightly first. The right wing crumbled; the nose hit; and the fuselage flopped to the ground. There were no other planes on the field at the time."

8-17-42E. Crookston, Minnesota. At 1551, a Taylorcraft L-2A practicing power off landings stalled and spun into the ground from an altitude of 100 feet at the AAF Auxiliary Airfield at Crookston, Minnesota, killing student pilot Pvt. Elmer M. Kulling, 20, Oil City, Pennsylvania.

8-17-42F. Okemah, Oklahoma. At 1700, a Taylorcraft L-2 and an Aeronca L-3B collided in midair three miles east of Okemah, Oklahoma, while practicing a three-airplane formation flight, killing civilian L-2 pilot Curtis Peter Hale, 32, Oklahoma City, Oklahoma, chief flight instructor for the Sooner Air Training Glider School. Civilian glider instructor James H. Taylor escaped injury and landed his damaged aircraft safely.

8-18-42A. Brooks Field, Texas. At 0850, a North American BC-1 landing at Brooks Field, San Antonio, Texas, collided with a BC-1I that had crossed its path on the landing roll out, killing the pilot of the BC-1I Aviation Cadet Harold F. Mastler. The pilot of the BC-1 that was performing the landing, Aviation Cadet Richard J. Downs, was seriously injured. Investigators stated that Cadet Mastler was instructed to taxi to the end of the field before attempting to cross the runway and had not.

8-18-42B. Brooks Field, Texas. At 1525, a North American BC-1I stalled and spun into the ground while attempting to enter the pattern two miles east of Brooks Field, San Antonio, Texas, killing Aviation Cadet Edward C. Brown. The Accident Classification Committee stated, "The pilot was flying in the rear position of a six-ship formation of two-ship elements. The formation was flying at about 1,500 feet indicated or 900 feet actual [above ground level] and making the first turn entering traffic. The leader of the last element states that the deceased pilot pulled up and above him and immediately afterward was seen diving into the ground. Some witnesses stated that the motor was running at all times. It is the opinion of the board that the pilot put his flaps down instead of his wheels, which gave the upward ballooning effect noticed by his element leader. Then the pilot noticed that he was overshooting his leader and thinking his wheels were down instead of his flaps, pulled back on the [throttle], stalling the ship and going into a dive from which he was unable to recover."

8-19-42A. West Point, Mississippi. At 0717, a Cessna AT-17 crashed into an 110,000 volt TVA power line after suffering an in-flight failure of the leading edge of its port wing two miles north of West Point, Mississippi, killing Aviation Cadet Charles P. King, Baltimore, Maryland, and Aviation Cadet Leon Lanfier, Long Island, New York. Witnesses on the ground observed the aircraft falling minus one wing from an altitude of approximately 6,000 to 7,000 feet. The airplane spun in and severed the power line before smashing into the ground and exploding into flames. Pieces of the port wing leading edge were seen floating to the ground and were found some distance from the main wreckage.

8-19-42B. Hemet, California. At 0945, a Ryan PT-22 crashed while performing aerobatics seven miles southwest of the airfield at Hemet, California, killing civilian instructor Harry V. Clugston and student Aviation Cadet Lloyd K. Burris. The Accident Classification Committee stated, "It is believed that at approximately 1,500 feet a half-roll was attempted. It is assumed that during the resulting pull out of the half-roll [at 1,000 feet] that the airplane was stalled by excessive backpressure, which resulted in a high wing loading. It further appears that this high wing loading stall resulted in the airplane again snap rolling or spinning from which recovery was not fully effected. The aircraft [dove in from about 500 feet and] struck the ground at approximately a 30 degree angle slightly left wing low, or as through it struck when turning to the left."

8-19-42C. Los Angeles, California. At 1525, a Lockheed P-38E dove into the ground at 135th Street and Inglewood-Redondo Road, Los Angeles County, California, killing pilot SSgt. William W. Conner. The aircraft was on an authorized aerobatic flight and was first observed in a vertical dive at an altitude of about 2,000 feet. Investigators stated, "The aircraft struck the ground in the middle of the macadam paved Inglewood-Redondo Blvd., one mile south of the southeast corner of the Los Angeles Airport. No automobile traffic was involved although the driver of a car, which had just passed the scene by approximately 300 feet, was burned slightly by the blast of the explosion, which followed impact. All vegetation in the vicinity was seared. The full supply of H. E. and other ammunition was discharged. Engine parts, propellers and armament were imbedded into the ground through the pavement a distance of 3-4 feet. Indications were that the aircraft had struck the ground in a slightly inverted attitude with the topside toward the southeast and bottom side toward the northwest. Wheels, landing gear parts and other wreckage was scattered to the northwest a distance of 150 feet. Pilot's remains were found to the southwest." It is likely that the pilot put the airplane in a diving maneuver from which he could not recover because of the tendency of the P-38 type airplane to suffer from severe control surface buffeting caused by compressibility.

8-19-42D. Ellington Field, Texas. At 2315, a Curtiss AT-9 experiencing engine trouble was abandoned by its crew and crashed three miles southwest of Ellington Field, Houston, Texas. A/C Peter W. Perrier, 21, Disney, Oklahoma, was killed instantly when he failed to deploy his parachute in time to avoid striking the ground. Co-pilot A/C Vincent L. Peters, Boyden, Iowa, parachuted to safety. The airplane, which was part of a six-ship flight, was returning to the field after successfully completing a formation training flight. The pilot, upon approaching the field, switched to his reserve fuel tank and both engines quit moments later. The pilot attempted to use the wobble pump to get fuel to the engines but failed to get them going again. The co-pilot left the airplane at about 1,000 feet above ground level. Cadet Perrier jumped soon after the co-pilot but delayed deployment of his parachute, pulling his ripcord just prior to striking the ground. Investigation found that the parachute was in perfect operating condition.

8-20-42A. Plant City, Florida. At 0935, a Martin B-26B crashed out of control three and one-half mile east of Plant City, Florida, killing the crew of six. Investigators stated, "The airplane fell, out of control, from an altitude of at least 1,200 feet, and possibly from as high as 7,000 feet, to about 100 or 150 feet from the ground, at which height the pilot was able to regain partial control and level it off. [The airplane was then seen to climb about 50 feet with black smoke emitting from the port engine] then went out of control again and struck the ground at an abrupt angle of descent, upside down, burying a portion of the airplane including the engines, and scattering the tail, wing tips, nose wheel and landing wheels beyond the first point of contact. It is possible that a failure of the left engine occurred when the pilot opened the throttles after leveling off, and the resulting yaw caused the airplane to roll over on its back, then plunge nose first into the ground. The cause of the accident is unknown." The aircraft struck the ground and buried itself deep in a swampy area and a large portion, including the engines, was never recovered. Army personnel dug out a large area of the swamp, but were unable to find the bodies of three of the crewmembers. The search was abandoned 8-22-42. Killed in the crash were: 1Lt. Joseph A. Nischan, 25, East Chicago, Indiana, pilot; 2Lt. Lawrence McCown, co-pilot and B-26 transition student; 2Lt. Franklin L. Pierce, B-26 transition student; SSgt. Howard Picton, engineer; Pfc. John W. Lunsford, assistant engineer; Pvt. Kenneth G. Burgdorfer, gunner.

8-20-42B. Aberdeen, South Dakota. At 1400 CWT, a Taylorcraft L-2 stalled and spun into the ground 19 miles southeast of Aberdeen, South Dakota, killing civilian flight instructor R.P. McClure and student TSgt. Thomas F. Quinlan. The aircraft was apparently practicing forced landings into a 20 mph wind from the southeast. Investigators speculated that the pilot either over-controlled or cross-controlled the airplane while trying to turn to make a field downwind.

8-21-42A. Monticello, Minnesota. At 1445, a Taylorcraft L-2A stalled and spun into the ground while in the pattern for the airfield at Monticello, Minnesota, killing student pilot Pvt. Benjamin L. Steim. The Accident Classification Committee stated, "The student tightened a low, steep turn in an attempt to prevent undershooting the field on the approach and spun through three fourths of a turn into the ground in a vertical position. The student died instantly."

8-21-42B. Zion City, Louisiana. At 1732, two Martin B-26B airplanes collided in mid-air and crashed near Zion City, Louisiana, killing nine fliers aboard both airplanes. Both airplanes fell into heavily wooded swamp. Killed in the crash of B-26 #41-17736 were: 2Lt. Marvin A. Zahn, pilot; 2Lt. William P. Mayhood, co-pilot; Pfc. Frederick R. Speiss, engineer; Pfc. Clyde E. Warden, gunner. Killed in the crash of B-26 #41-17739 were: 2Lt. William J. Dover, pilot; 2Lt. William R. Weber, co-pilot; Cpl. Frederick C. Edmonds, engineer; Pfc. Phillip J. Peterson, radio operator; Pfc. Vernon R. Davis, bombardier. The aircraft were flying at an altitude of approximately 1,000 feet in a six-ship formation training flight when Lt. Zahn, flying in the number-two position of the first element, dropped out of formation and then crossed under and climbed into the aircraft piloted by Lt. Dover, who was flying in the lead position of the first element. After Lt. Zahn suddenly left formation, he was observed flying under the leader with about eight feet of clearance between the two airplanes as he apparently tried to ease his way back into position. The airplane was then seen to suddenly climb into the lead airplane. The starboard propeller of Lt. Zahn's airplane contacted the fuselage of Lt. Dover's airplane at the forward end of the bomb bay. Lt. Zahn's airplane slid to the right and the airplanes became entangled momentarily as they slid against each other. Lt. Zahn's airplane fell away to the left and Lt. Dover's flipped over and fell away to the right, both spinning to the ground and exploding into flames. The other aircraft of the formation had to take evasive action to avoid colliding with the two falling airplanes.

8-22-42A. Myrtle Beach, South Carolina. At 0745, a Curtiss P-40F crashed into the ground after falling out of an Immelmann maneuver 10 miles northwest of Myrtle Beach, South Carolina, killing pilot 2Lt. Aldo J. Pieri. The pilot was on an authorized engineering mission that excluded loops and Immelmann turns. Civilian witnesses on the ground observed the aircraft begin to go into a loop. At the top of the maneuver, the pilot attempted to roll the airplane from inverted to a level, upright attitude. Before the airplane could be rolled to a level position, it entered a stall and was seen by witnesses to slip away on its starboard wing. The airplane then went into a tight, diving spiral to the right and slammed into the ground, exploding into flames in a swamp.

8-22-42B. Orlando, Florida. At 1015, a Curtiss P-40E on a maintenance test flight dove into the ground near the AAF Air Base at Orlando, Florida, killing pilot 1Lt. Wendell C. Miller. The Accident Classification Committee stated, "Lt. Miller, the [313th Fighter] Squadron Engineering Officer, was on a routine test hop testing an engine that had been noted running rough. Approximately 15 minutes after take-off, Lt. Miller came through a broken cloud layer (bottom of which was 3,000 feet from the ground) in a power dive, no spin, and crashed straight into the ground. The plane buried itself and was completely demolished, an explosion and fire followed the crash."

8-22-42C. Key Field, Mississippi. At 1635, a Douglas A-20A flying in a thunderstorm crashed two and one-half miles northwest of Key Field, Meridian, Mississippi, killing the crew of four. The aircraft had been reported missing after it failed to return from a local flight to the gunnery range 20 miles from the field. The A-20 was not found until 1900 on 8-24-42. A thunderstorm had moved into the area about the time that the airplane was due back at Key Field and investigators speculated that the pilot flew the airplane into the ground after entering instrument conditions. The airplane hit the ground at a steep angle in a small clearing. Killed in the crash were: 2Lt. Nolan A. Fanguy, 26, Houma, Louisiana, pilot; Sgt. Dominic A. Giannamore, 19, Stubenville, Ohio, gunner; Pfc. Claude J. Noel, 20, Richmond, Virginia, passenger; Pvt. William Noble, Jr., 20, Cleveland, Ohio, passenger.

8-22-42D. Bellevue, Washington. At 2215, a Lockheed P-38F dove into the side of a mountain five miles southeast of Bellevue, Washington, killing pilot 2Lt. William H. Wilson, 22, Chattanooga, Tennessee. The pilot was apparently blinded and disoriented by searchlights trained on his aircraft. The pilot had been instructed to fly the searchlight mission at an altitude of 11,000 feet and not to perform any extreme maneuvers to escape the lights. Three searchlights were tracking the aircraft when it was seen to go into a dive with no apparent effort by the pilot to recover. The aircraft continued in the dive until it collided with a mountain near Lake Washington, exploding into flames upon impact with the ground.

8-23-42A. Las Cruces, New Mexico. At 0215, a Boeing B-17E suffered a catastrophic structural failure and broke up in flight 16 miles west of Las Cruces, New Mexico, killing seven crewmembers. 2Lt. Quentin Hargrove, passenger and Engineering Officer 303rd BG, was lightly injured and radio operator SSgt. Walter E. Knox was seriously injured when they miraculously managed to parachute to safety after being ejected from the radio compartment as the bomber broke up. The B-17 had taken off from Biggs Field, El Paso, Texas, at 0145 on a practice navigation-bombing mission to Wendover, Utah, with a landing at Boise, Idaho. At an altitude of 13,500 feet the aircraft encountered thunderstorm activity accompanied by severe updrafts. The aircraft broke in half just forward of the radio compartment, the tail section and port wing tip being found intact a quarter of a mile from where the forward section slammed into the ground. The bomb load detonated upon impact and the remainder of the aircraft blown to bits by the resultant explosion. Wreckage and body parts were scattered over an area of about 2,000 yards. Two of the dead crewman had been thrown clear of the wreckage. Investigators speculated that the pilot over-controlled the aircraft while trying to recover from a severe updraft. Both survivors reported being hurled to the floor and held there by extreme force just prior to the aircraft breaking up. In his statement to investigators, Lt. Hargrove reported seeing the contents of the radio compartment "floating" out behind him as he was ejected from the airplane. Two other B-17s that had taken off from Biggs Field at about the same time turned back and returned to the field after encountering the thunderstorms and severe turbulence. Killed in the crash were: 2Lt. James E. Hudson, pilot; 2Lt. Eugene E. Rochester, co-pilot; 2Lt. Frank B. Johnson, navigator; 2Lt. Leon H. Shane, navigator; 2Lt. William R. Simno, bombardier; 2Lt. James Van George, assistant co-pilot; Sgt. Cato H. Myers, assistant engineer.

8-23-42B. Bogalusa, Louisiana. At 1030, a Martin B-26A on an instrument navigation training flight crashed five miles south of Bogalusa, Louisiana, killing six crewmembers and seriously injuring pilot 2Lt. Richard Farnsworth, 23, New London, Ohio. Lt. Farnsworth and co-pilot 2Lt. William L. Kelly, 25, Lamoni, Iowa, were thrown clear of the wreckage. Lt. Kelly died immediately after the accident and Lt. Farnsworth could remember no details of the accident or even taking off that day. The airplane had taken off from Harding Field, Baton Rouge, Louisiana, on a hooded instrument training flight. A civilian witness on the ground stated that he noticed the airplane gradually loosing altitude until it flew into the ground in a wooded area. Investigators noted that all crewmembers "had parachutes on and buckled" and that the ignition switches were found in the "off" position. Because of a lack of any real evidence, investigators could not determine the cause of the accident. Also killed were: 2Lt. Bernard H. Green, 23, Harlingen, Texas, navigator; Pvt. William R. McGehee, 25, El Paso, Texas, engineer; Pvt. Charles T. Kneipp, 22, Shreveport, Louisiana, gunner; Pvt. Edgar Stratton, 27, Globe, Arizona, bombardier; Sgt. J. C. Goodwin, 22, Clarksville, Tennessee, passenger.

8-23-42C. Paine Field, Washington. At 0910, a Lockheed P-38E suffering the failure of its port engine stalled and spun into the ground one-half mile southwest of Paine Field, Everett, Washington, killing pilot 2Lt. Jarrod T. Coleman. This was Lt. Jarrod's first solo flight in a P-38 type airplane. Lt. Coleman was approximately 700 feet above the field when he announced an emergency. Investigators reported that the pilot was instructed by the tower to raise his gear and make a shallow turn into the field after missing an initial opportunity to land. Lt. Coleman raised the landing gear at an altitude of about 300 feet at the middle of the field. When the airplane reached the west end of the field, the pilot turned into the bad engine and let the left wing drop in the turn. The airplane was observed to wing over to the left and fly into the ground in the opposite direction of the line of flight.

8-23-42D. Dale Mabry Field, Florida. At 1455, a Bell P-39D on a formation training flight crashed on take-off from Dale Mabry Field, Tallahassee, Florida, after suffering an engine failure, killing pilot 2Lt. Melvin F. Lacey. 2Lt. Edward C. Fletcher witnessed the accident and stated to investigators, "I took off as wingman in the second element led by Lt. Lacey of a five ship formation. I followed him off, breaking clear of the runway long before he did. I noticed when I was at about 50 feet that he was just leaving the ground. It appeared to me that he would have trouble clearing the trees at the extreme end of the East-West runway, and at this time he struck a large tree right at the center. Upon striking [the trees,] his left wing dropped as if in a gentle turn. The nose then dropped and I thought that he would make a crash landing straight ahead. As the ship went toward the ground it rolled into an inverted position to the left, striking the ground on the nose in this position." Investigation revealed that pieces of backfire screen were found in the cylinders, causing the engine failure. A Technical Order required the replacing of these particular type backfire screens because of a history of causing engine failure. This order had not been complied with even though the intake manifold had been marked as being in compliance.

8-24-42A. Atlantic Ocean, South Carolina. At 1015, a Curtiss P-40F practicing aerobatics crashed into the Atlantic Ocean eight miles northeast of the municipal airport at Myrtle Beach, South Carolina, killing pilot 2Lt. Cameron D. Martin. 2Lt. Harry J. McCullough witnessed the accident and stated to investigators, "I saw another P-40 type airplane performing acrobatics about one-half mile from me. We were both at approximately 14,000 feet. I completed a loop, followed by a slow roll. After recovery I estimate that I was at from 11,000 to 11,500 feet. I looked around and saw a P-40, presumably the same one, in a spin off to my right at least 1,000 feet above me. He was still about one-half mile away, but I could definitely distinguish the airplane as a P-40. I watched the airplane spin down and it seemed to recover from one spin and then go right into another. This happened two or three times, and finally the airplane disappeared into a small isolated cloud, which I estimated to be at about 1,500 feet. I expected to see the airplane come back into sight under control, for it appeared that the pilot had recovered again, just before I lost sight of the airplane. I saw nothing for a few seconds, and then I saw a tremendous splash as the airplane hit the water." The crash occurred about two miles off shore of the Ocean Forest Hotel.

8-24-42B. Key Field, Mississippi. At 2051, a Douglas A-20C crashed four miles west of Key Field, Meridian, Mississippi, just after take-off, killing pilot 2Lt. Joseph E. Hand, 21, Providence, Rhode Island. The aircraft was on a local night flight and scheduled to practice night landings and take-offs. The Accident

Classification Committee stated, "After questioning all witnesses, the plane appeared to be cruising normally and crashed into the ground and exploded and burned. Wreckage was scattered for 200 yards. The plane crashed in a wooded area and appeared to be in a vertical bank when it struck the ground. Because of darkness, no witnesses noticed the plane until it struck the ground and exploded. Witnesses heard the motors of several planes in flight and all appeared to be running satisfactorily. No particular plane was noticed to be in distress until flames were seen coming from the wooded area into which the plane crashed."

8-25-42A. Goodfellow Field, Texas. At 1320, a Vultee BT-13A stalled and spun into the ground at Goodfellow Field, San Angelo, Texas, killing flight instructor 2Lt. William R. Becker and student Aviation Cadet Sidney Goldberg. The Accident Classification Committee stated, "This accident occurred as a direct result of the airplane being stalled while in the first turn immediately after take-off. It was apparent from the testimony of witnesses that the pilot was climbing steeply. The airplane was under complete control until it started into [a left] turn after take-off, at which time the nose was observed to have been pulled up abruptly and without apparent reason, after which the airplane fell off on the left wing and headed toward the ground, making an approximate 90-degree left turn. The estimated altitude of the airplane at the time the stall occurred was between 350 feet and 400 feet above the ground. Partial control of the airplane had been regained before it struck the ground. At the moment of impact, the airplane's nose was about 45 degrees below the horizon and the right wing was slightly low."

8-25-42B. Blythe, California. At 1130, a Bell P-39D on a local transition flight spun into the ground 11 miles northeast of the Army Air Base at Blythe, California, killing pilot 1Lt. Rudolph E. Neubauer, 22, Winfield, Illinois. The aircraft had been reported missing when it failed to return to its base at Blythe AAB, being last seen by other army pilots at 1115 near Parker Dam in the vicinity of Parker, California, and the Colorado River, flying southwest at an altitude of approximately 9,000 feet. Despite a thorough aerial search, the airplane and pilot were not located until 10-01-42 when they were discovered on a remote desert mesa northeast of the base. The Accident Classification Committee stated, "It is the opinion of this committee that the pilot, Lt. Neubauer, was engaging in authorized aerobatic maneuvers from which the airplane went into a flat spin to the left and crashed into the ground in practically a horizontal position with little forward momentum. The absence of skid marks on the ground coupled with the positive shifting movement to the right of equipment in the airplane indicates the above deduction to be correct. The fact that all controls on the throttle quadrant were in the rear position, the pilot's safety belt had been unbuckled, the right escape door was open, and the pilot's feet were out the right door indicates that the airplane was out of control in a flat spin and that the pilot was attempting to abandon ship just prior to crashing. It must be borne in mind that the above deductions were arrived at six weeks after the accident occurred, since the wreckage had not been located until this time, and in view of the fact that there were no witnesses to the crash the committee classifies both the nature and the cause of the accident as undetermined."

8-25-42C. Fort Morgan, Colorado. At 2125, a Taylorcraft L-2A crashed while attempting a landing at an airfield six miles southeast of Fort Morgan, Colorado, killing Aviation Student George L. Fry, Jr. The aircraft came in high on the approach and the pilot executed a slight forward slip. He then began to violently S-turn the aircraft in an apparent attempt to lose altitude and not overshoot. As the aircraft began to stall, the pilot applied the throttle but could not recover the airplane, spinning in to the right from an altitude of approximately 250 feet above ground level.

8-26-42A. Hunter Field, Georgia. At 0837, a Vultee V-72 (similar to the A-31 type) crashed after suffering an engine failure just after take-off from Hunter Field, Savannah, Georgia, killing Pvt. Roy E. Gregg and seriously injuring pilot 2Lt. William R. Miller. Investigation revealed that the pilot had incorrectly set the controls for the two-stage blower, causing the engine to run extremely hot and ultimately leading to its failure.

8-26-42B. Hemet, California. At 0840, two Ryan PT-22 aircraft collided in mid-air one mile north of the airfield at Hemet, California, killing two students. A/C Joseph C. Mosner, flying PT-22 #41-15477, was in the traffic pattern preparing to land, flying at 500 feet above ground level and heading east on the downwind leg. Aviation Cadet Charles N. Roddy, flying PT-22 #41-15640, entered the traffic pattern flying southeast at a forty-five degree angle to the downwind leg, also at an altitude of 500 feet. Both pilots failed to see the other and the aircraft collided, locking together momentarily before falling away. Cadet Mosner's aircraft immediately dove toward the ground and he exited the aircraft in an attempt to parachute to safety but he struck the ground before he could pull the ripcord. Cadet Roddy's aircraft continued flying for several seconds before it too plunged to the ground.

8-26-42C. Youngsville, Louisiana. At 1130, a Douglas C-47 crashed six miles southeast of Youngsville, Louisiana, after it suffered a structural failure of its starboard elevator in flight, killing pilot 2Lt. Earle E. Hansen and co-pilot 2Lt. Robert L. Hollingsworth. Several civilian witnesses on the ground observed the aircraft flying east in a straight and level attitude at an altitude of approximately 2,000 feet when the starboard elevator was seen to suddenly fall from the aircraft. The aircraft immediately dove toward the ground. The pilots attempted to recover the aircraft from the dive and in an extreme pull-up, stressed the airframe to the point that the starboard wing was torn from the center section just outside of the engine nacelle. The aircraft continued in its dive to the ground and exploded into flames. The elevator was found about 150 yards northwest of the main wreckage. The starboard wing was found relatively intact about 100 yards from the wreckage. Investigators could not determine the cause of the structural failure. The C-47 was on a ferry flight from the Douglas factory at Long Beach, California to an unnamed airbase in the east.

8-26-42D. Lubbock, Texas. At 1245, two Curtiss AT-9 aircraft that were part of a fifteen-airplane formation collided in mid-air over the Lubbock Army Flying School, Lubbock, Texas, killing one cadet in each aircraft and three people on the ground. Aviation Cadet Jack J. Williamson, 22, Elkville, Illinois, was killed in the crash of AT-9 #41-12195 and co-pilot Aviation Cadet Owen D. Willard, 22, West Allis, Wisconsin, was seriously injured parachuting from the airplane. Student Aviation Cadet Charles C. Wylie, 22, Milwaukee, was lightly injured parachuting to safety and co-pilot Aviation Cadet Andrew R. Wood, 25, Canton, Massachusetts, was killed in the crash of AT-9 #41-12196. Aviation Cadet Clark I. Toombs, 22, Amarillo, Texas, Aviation Cadet Duke W. Harrison, Jr., 23, Greenville, Texas, and civilian Gilford W. Singleton, 36, Lubbock, Texas, were killed when Cadet Williamson's aircraft fell on the post tailor shop. The formation was flying over the field in an echelon type formation of twelve airplanes with the other three airplanes flying a V-formation behind the twelfth airplane of the echelon. During an attempted cross over maneuver to echelon, the last two aircraft of the V-flight hit the propeller wash of the echelon flight and collided.

8-26-42E. Lucas, Louisiana. At 1302, a Martin B-26A suffering the failure of its starboard engine crashed while attempting a forced landing in a field at Lucas, Louisiana, 10 miles southwest of Barksdale Field, killing four army fliers and lightly injuring a civilian tenant farmer. Leo Gullo, Lucas, Louisiana, received second-degree burns when the bomber careened through his house and burst into flames, destroying the house and its contents. Mr. Gullo had to be restrained to prevent him from entering the burning house to retrieve $800 in cash he had hidden away. Investigation revealed that the starboard propeller was almost completely feathered and the right mixture control was found in the "idle cut-off" position. The right feather control was in the "feather" position, leading investigators to determine that the starboard engine had probably failed and was not producing power. The pilot was attempting a wheels-up landing in a small farm field but was unable to make it, striking power lines and crashing into the farmer's house. Investigators could not determine the cause of the engine failure. Killed in the crash were: 2Lt. Louis F. Greis, 25, Syracuse, New York, pilot; 2Lt. Charles J. Preston, 21, Hollywood, California, co-pilot; 2Lt. William T. Stone, 26, Webb City, Missouri, navigator; Pvt. John M. Wood, 22, Bethesda, Maryland, engineer.

8-26-42F. Paine Field, Washington. At 1610, a Lockheed P-38F suffering the failure of its starboard engine crashed while attempting an emergency landing at Paine Field, Everett, Washington, fatally injuring pilot 2Lt. George C. Galbrath, who died of his injuries on 8-28-42. The aircraft was part of a flight of three P-38 type aircraft on a training mission when about eight miles from the field the pilot broke formation and declared an emergency. At about three or four miles south of the field the pilot lined up with runway 34 with wheels down. The aircraft attempted an approach but was too fast, the pilot pulled up the landing gear and went around, turning into the good engine. He again approached with excessive speed and, with landing gear down and no flaps, nosed the airplane down toward the runway. The pilot forced the aircraft down and finally made contact with the runway at mid-field, bouncing several times before going off the end of the runway, through a revetment and into a swamp. The aircraft turned over and burst into flames. After a great effort, enlisted men in the area were able to extricate Lt. Galbrath from the wreckage before he could drown or burn to death.

8-27-42A. Lubbock, Texas. At 1215, a civilian workman was struck and killed by a Laister-Kaufman TG-4A glider attempting a landing at the AAF Glider School at Lubbock, Texas. The pilot, SSgt. Ruby Harlan Dees, and his student, SSgt. Frank Gain, were not injured. The glider went out of control when it struck Phillip Otero with the port wing, killing him instantly and shattering the wing. Workers stated that Otero was sweeping off the runway and stepped into the path of the glider, seeing it only at the last moment. The pilot stated to investigators that he was aware of the workers and trucks on the west side of the north-south runway and attempted to make his landing (from north to south) favoring the east edge of the runway. Investigators stated that the accident was caused by the failure of the pilot to see and avoid the workman and the act of the workman walking into the path of the glider.

8-27-42B. Vernon, Texas. At 1450, a Fairchild PT-19A on student practice mission crashed five miles north of Vernon, Texas, killing Aviation Cadet Donald R. Vrooman. The pilot was on a scheduled flight to practice chandelles and lazy-eights. A farmer passing through the area in an automobile found the wrecked airplane in a field at approximately 1500. The cause of the accident could not be determined.

8-27-42C. Mulberry Grove, Florida. At 1630, a Martin B-26B suffering the failure of the port engine crashed into two houses and a garage at Mulberry Grove, Florida, killing six fliers and a civilian on the ground. The pilot, 2Lt. Claude R. Jackson (see 10-3-42F), suffered minor injuries and two civilians on the ground, Emma Lee Stevens and Charity Brown, received serious injuries. Killed in the crash were: 2Lt. Augustus A. Trulove, co-pilot; Pvt. Herman Ashner, engineer; Cpl. John C. Larkin, gunner; Pfc. Maurice B. Brady, gunner; Sgt. Kenneth D. Parker, passenger; Pvt. William Bickel, passenger. Also, Mattie Lee Brown, a civilian killed on the ground. Investigators stated, "Airplane was returning to Lakeland Army Air Base, Lakeland, Florida, after completing a two hour transition training mission. Pilot called the control tower for permission to "buzz" the field (low flying over open terrain is a part of the tactical training of the 320th [Bombardment] Group). After crossing the field at an altitude of approximately fifty feet, he pulled the airplane into a steep climb and at the same time making a turn to the left. While in this attitude at an altitude of about 800 feet, the left engine stopped completely. The pilot leveled off the airplane and started a gradual right turn noting at that time a drop in fuel pressure on the left engine. The throttle and prop controls were advanced so there was a manifold pressure of 40 inches and an RPM of 2,300 although the pilot believed the position of the throttles on the quadrant should have given a much higher manifold pressure reading. The airspeed indicated between 140 and 150 miles per hour. It was impossible to maintain altitude and airspeed and the pilot was of the opinion that he was not getting power out of either engine. When a forced landing became imminent, it was impossible to stretch the glide into a suitable field. All of the land surrounding the place where the accident occurred in is honeycombed with mounds and pits caused by mining operations of the International Mineral and Chemical Corporation. When the airplane was about fifteen feet above the ground the right wing dropped and struck the bank of a pit and skidded forward on its belly striking a garage containing a 1937 Ford Coupe destroying both and passing on through two frame houses setting them both on fire. The pilot, co-pilot and the engineer were thrown out of the airplane. The airplane and the private dwellings were destroyed by fire."

8-28-42A. Walden, New York. At 0956, a Vultee BT-13A spun to the ground at Walden, New York, fatally injuring instructor 2Lt. William P. Breunig. Student A/C R.H. Houser parachuted to safety and was uninjured. The instructor put the aircraft into a deliberate spin to the right at an altitude of about 6,000 feet. The instructor was having trouble recovering the airplane and ordered the student to bail out. As the student opened the canopy and rose in his seat and prepared to bail out, the pilot rescinded the order. The pilot still could not recover the airplane and then ordered the student to bail out again. The canopy tried to slam itself closed on the cadet as he rose to leave the aircraft the second time. After fully opening the canopy, he was able to successfully exit the airplane. The airplane spun into the ground in an open field about one-quarter of a mile away from where the cadet touched down in trees. Lt. Breunig died shortly after the accident.

8-28-42B. Muskogee, Oklahoma. At 1120, a Fairchild PT-19A crashed six miles south of Hatbox Field, Muskogee, Oklahoma, killing civilian flight instructor Charles Nelson, 31, Tulsa, Oklahoma, and seriously injuring student Aviation Cadet Kenneth E. Duewall, Ringgold, Texas. The aircraft had taken off from Hatbox Field at approximately 1115. A farmer driving by in an automobile spotted the wreckage of the aircraft in an open field at about 1200. There were no witnesses to the accident, but investigators speculated that the student was practicing a simulated power-off forced landing.

8-28-42C. Muroc, California. At 1130, a Bell P-39D on an individual combat mission with another P-39D spun to the ground six miles north of Muroc, California, killing pilot 2Lt. H. A. Minnich. Lt. Minnich's body was found outside of the aircraft and investigators speculated that he had exited the aircraft at low altitude in an attempt to parachute to safety, striking the ground before he could deploy his parachute. 2Lt. F.R. Scariano was flying with Lt. Minnich and witnessed the accident. He later stated to investigators, "We had been flying for approximately one-half an hour and Lt. Minnich was in front of me. We were flying at a relatively slow speed when his ship seemed to tumble end over end about three times, ending in a flat inverted spin. This was at approximately 12,000 feet. At approximately 7,000 feet, the plane flipped over and spun flat to the ground. From the air it seemed that the plane hit relatively flat and did not seem to be as damaged as much as one would expect from a crash beginning at such a high altitude."

8-28-42D. Ozark, Alabama. At 1400, two North American AT-6A type airplanes on a formation training flight collided in mid-air after entering a small cloud 10 miles east of Ozark, Alabama. A/C W.W. Bagby attempted to parachute to safety from AT-6A #41-16124, but was killed when he fell out of his improperly secured parachute harness. Witnesses on the ground observed the empty parachute float to the ground. A/C E.M. Braden parachuted to safety from AT-6A #41-16132 and was uninjured. Investigation revealed that the leader of the flight did not enter the cloud and that Cadet Bagby's aircraft over ran Cadet Braden's aircraft while flying in the cloud.

8-28-42E. Sardis, Georgia. At 1935, a Vultee BT-15 stalled and spun into the ground while maneuvering in the vicinity of Sardis, Georgia, killing student Pfc. Richard E. Funke. The aircraft was observed by witnesses on the ground to be performing snap rolls at a reasonable altitude. The aircraft then dove on the Kirshner Farm, pulling out at an altitude of about 50 to 100 feet. The aircraft then executed two steep turns around the Kirshner farmhouse. On the third turn around the house, the aircraft entered a stall at an altitude of approximately 100 feet, spinning to the ground and immediately bursting into flames upon impact.

8-29-42A. Quincy, Florida. At 0900, a Bell P-39D crashed while maneuvering five miles north of Quincy, Florida, killing the pilot. The airplane was part of a flight of several P-39 airplanes on a routine formation-training flight. The pilot, 2Lt. Robert A. Mariam, was killed when was he struck by the tail of the airplane while attempting to parachute to safety. His parachute did not open. The aircraft was seen by other pilots of the flight to execute a barrel roll at approximately 8,000 feet when it entered a nose down attitude and descended into clouds. Witnesses on the ground observed the aircraft emerge from the clouds and dive into the ground at a 45-degree angle. The pilot and the starboard escape door were seen to hit the ground about 1,500 feet from where the aircraft slammed into the ground.

8-29-42B. Glendale, California. At 1050, a Lockheed P-38F attempting an emergency landing after suffering the failure of its starboard engine stalled and spun into the ground one-quarter of a mile north of the air terminal at Glendale, California, killing pilot SSgt. Wayne M. Winks. The aircraft was observed by pilots to be approaching the field at an altitude of approximately 500 feet with the starboard propeller windmilling slowly. The pilot attempted a steep turn into the dead engine, resulting in a stall at an altitude too low to allow for recovery. The aircraft entered a spin to the right and smashed into the ground, exploding into flames upon impact. The Accident Classification Committee noted that pilots again be warned not to turn into the dead engine during single-engine operation of P-38 type aircraft. Investigation revealed pieces of neoprene gasket in the carburetor nozzle of the starboard engine.

8-30-42A. Shelby, Montana. At 1015, a Douglas A-20B collided with rising terrain 10 miles north of Shelby, Montana, killing pilot 2Lt. Robert Gustafson, 22, Islington, Massachusetts. The aircraft was part of a five-aircraft ferry flight originating at the Douglas Aircraft factory in Long Beach, California, and flying to Calgary, Canada. The aircraft was part of a flight that took off at 0855 from Great Falls, Montana. The aircraft was seen by the other pilots of the flight to leave formation while gradually losing altitude. The aircraft soon became separated from its flight and was not seen again by any of the other pilots. The pilot was not heard to radio that he was in any trouble. Witnesses on the ground reported seeing the aircraft flying below and well behind the formation prior to the crash. The A-20 collided with the crest of a low, smooth hill at a shallow angle at an estimated speed of 300 mph. After the initial impact with the hill, the aircraft bounded through the air for 400 yards before impacting the ground violently in a 30-degree dive with the starboard wing low, bursting into flames upon impact and scattering wreckage for over 900 yards. The pilot was thrown from the aircraft after the second impact and was found about 40 yards from the main path of the wreckage. Civilians that had arrived on the scene shortly after the crash reported that Lt. Gustafson had moved his head slightly just as they approached him but showed no signs of life afterwards.

8-30-42B. North Island, California. At 1040, a Lockheed P-38F engaging in simulated combat with another P-38 dove into the ground at North Island Naval Air Station, San Diego, California, killing pilot 2Lt. Clyde P. Baugh. The two P-38s were seen rolling and looping in simulated combat before separating and performing a head-on pass. The subject airplane was then seen to execute a split-S type maneuver after performing the head-on pass at the other airplane. At the bottom of the maneuver the subject airplane entered a dive from which the pilot was unable to recover. The day of the accident, the pilot had mentioned to other pilots that he was blacking out "easier" than he usually did. The pilot was also being treated for an ear infection but did not mention the enhanced tendency to black out to the flight surgeon. It was noted that the two P-38s had been operating at 20,000 feet prior to coming down to 10,000 feet for the simulated combat. Investigators could not determine if the pilot's health problems, oxygen starvation, or the flight characteristics of the P-38 type airplane were the cause of the accident.

8-31-42. Butler's Cove, Washington. At 1730, a Lockheed P-38E collided with the water of Puget Sound and eventually crashed to the ground in a wooded area near Butler's Cove, Washington, killing pilot 2Lt. Jose R. Noble. The P-38 was seen by other pilots to make a dive on the water from an altitude of approximately 800 feet at an angle of about 30 degrees. At the bottom of the dive, the P-38 struck the water with both propellers. The pilot recovered the airplane and then attempted to gain altitude but could not because both engines had stopped. The pilot then made a 90-degree turn to the east toward land and feathered both propellers. About three miles from where the P-38 struck the water, it descended into terrain and exploded into flames just short of a small clearing in the woods that the pilot had chosen for an emergency forced landing.

September

9-1-42A. Loachapoka, Alabama. At 0030, a Vultee BT-13 on a night navigation flight crashed three miles north of Loachapoka, Alabama, killing RAF Flying Cadet Thomas G. Hornsey. The BT-13 dove to the ground in a steep spiral with power and smashed through a number of pine trees before exploding into flames. There were no witnesses to the crash and investigators could not determine why the cadet flew into the ground.

9-1-42B. Craig Field, Alabama. At 1600, two North American AT-6A aircraft collided in mid-air 14 miles northwest of Craig Field, Selma, Alabama, killing RAF Flying Cadet Ernest R. J. Spooner, 20, Norwich, Norfolk, England, aboard AT-6A #41-417. RAF Flying Cadet Frederick Shippam, 20, Oxford, England, parachuted to safety from AT-6A #41-456. The aircraft were part of a three-ship formation maneuvering to land at the field. Cadet Shippam later stated to investigators, "At the time of the accident I was flying with my canopy open at 2,000 feet, with three ships in trail, my instructor leading. I was in number two position in trail when a signal for right echelon was given. I was moving straight ahead into formation when I heard an engine roaring. On checking my own ship, I found it was not my engine and immediately looked around and saw the other wing very slightly below on my right. At that moment, he struck my wing with the propeller about half way between the wing tip and the fuselage. My plane dived out of control and I released the safety belt and bailed out. I landed in an open field safely without injury."

9-1-42C. Angola, Louisiana. At 2100, a Martin B-26A suffering an in-flight fire in the starboard engine crashed in flames on a remote area of the state prison farm at Angola, Louisiana, killing seven fliers. Investigators could not determine the cause of the starboard engine fire that the pilot had reported to Harding Field, Louisiana, by radio. The aircraft crashed into heavily wooded sloping terrain soon after the radio call. Killed in the crash were: 2Lt. Louis J. Pigoni, Oak Park, Illinois, pilot; 2Lt. William E. Evans, Shreveport, Louisiana, co-pilot; 2Lt. Norman L. Vaughn, Logansport, Indiana, bombardier; SSgt. John P. Dye, Santa Barbara, California, radio operator; Pfc. Wade H. Fry, Ashville, Ohio, engineer; Pvt. Eugene L. Clayton, Memphis, Tennessee, student bombardier; Pvt. Louis H. Labertrandie, Jr., Yonkers, New York, gunner.

9-1-42D. Gardner Field, California. At 2145, a Vultee BT-13A on a night training flight collided with rising terrain 15 miles southeast of Gardner Field, Taft, California, killing Aviation Cadet Elmer D. Kimbrough. The Accident Classification Committee stated, "Cadet Kimbrough crashed into a mountain of approximately 2,000 feet [elevation]. Cadet Kimbrough had just completed five night landings at Kern Auxiliary Field and was practicing air work prior to making his sixth and final landing at the same field. He was assigned an altitude of 2,000 feet for his air work. The scene of the accident was between three and four miles south of his night flying area. The south boundary of the night flying area is two miles north of the edge of the foothills where Cadet Kimbrough crashed."

9-2-42A. Tampa Bay, Florida. At 0015, a Martin B-26B crashed into Tampa Bay seven miles south of MacDill Field, Tampa, Florida, killing three fliers. Civilian witnesses on board vessels in the bay observed the B-26 to be flying straight and level at an altitude of approximately 700 feet and then to dive directly into the bay, exploding into flames upon impact and scattering wreckage over 200 feet. One witness reported that a fire as large as "a city block" erupted on the surface of the water for a very brief time after the crash. The wreckage of the aircraft was found submerged in 25 feet of water. Investigators were unable to determine the cause of the crash. Killed in the crash were: 2Lt. William B. Rittenhouse, pilot; 2Lt. Stanley A. Hecker, co-pilot; Sgt. James E. Anton, engineer.

9-2-42B. Stamford, Texas. At 1000, a Fairchild PT-19A stalled and spun into the ground three miles southwest of Arledge Field, Stamford, Texas, killing civilian flight instructor Joe John Gault and student A/C Robert M. Warner. Investigators stated, "The crash resulted from an unintentional stall which occurred during, or immediately after a downwind gliding turn, without power, quite near to the ground."

9-2-42C. North Island, California. At 1545, a Lockheed P-38F collided in mid-air with a U.S. Navy North American SNJ-2 at North Island Naval Air Station, San Diego, California, killing army pilot 1Lt. James L. Hubbell, Jr. and U.S. Marine Corps pilot Capt. L.J. Nissen. Lt. Hubbell arrived at North Island with another P-38F at an altitude of approximately 1,500 feet after completing a gunnery mission. The leading P-38 peeled off and entered the traffic pattern followed by Lt. Hubbell. Lt. Hubbell made a left turn onto the downwind leg and descended into the SNJ that had apparently entered the downwind leg while Lt. Hubbell was in his turn. The P-38 exploded into flames and broke into pieces, falling into the sea. The SNJ entered a spiraling dive into the sea. Investigators faulted both pilots for failure to see and avoid each other in the pattern.

9-2-42D. Hunter Field, Georgia. At approximately 2300, a Vultee V-72 (similar to the A-31 type) crashed on a small island 10 miles south of Hunter Field, Savannah, Georgia, killing pilot 2Lt. Justin J. Haley and passenger Sgt. Roger F. Fish. The aircraft took off from Hunter Field at 2218 on a local flight and did not answer radio calls the tower made to it after 2300. The V-72 was declared overdue at 2400. An army A-17 located the aircraft at 0936 the next day. Investigators could not determine why the airplane crashed.

9-2-42E. Little Rock, Arkansas. At 2315, a Martin B-26A crashed seven miles northwest of Little Rock, Arkansas, killing the six fliers on board. The aircraft was seen flying to the west over the city and soon after apparently dove into the ground with both engines producing power. The B-26 exploded into flames upon impact and was destroyed utterly. Investigators could not determine the cause of the accident and witnesses' statements were vague and conflicting. Killed in the crash were: 1Lt. Kenneth E. Reddy, 22, Bowie, Texas, pilot; 2Lt. Charles S. Brachbill, 21, Williamsport, Pennsylvania, co-pilot; 2Lt. Phillip Williams, 23, Hinsdale,

Illinois, navigator; Sgt. Thomas T. Roberts, 22, Knoxville, Tennessee, engineer; Cpl. Dominic T. Moduno, 20, Brooklyn, New York, radio operator; Pvt. Thomas A. Naylon, 33, Pittsburgh, Pennsylvania, assistant radio operator. Lt. Reddy was a co-pilot on a B-25 (AAF serial no. 40-2249) that flew in Col. Jimmy Doolittle's Tokyo Raid on 4-18-42 and successfully parachuted with his crew into China after completing the mission.

9-3-42. Patterson Field, Ohio. At 1003, a Bell P-39K crashed while attempting an emergency forced landing at Patterson Field, Ohio, fatally injuring 2Lt. Joseph R. Jarvis. Just after take-off, at an altitude of about 300 feet, the aircraft had suffered an engine failure and the pilot attempted a 180-degree turn back into the field. The aircraft lost altitude and airspeed in the turn and the pilot attempted to stretch his glide, resulting in a stall and subsequent spin 50 yards short of the field. The aircraft struck the ground in a vertical bank to the left, exploding into flames immediately upon impact and scattering wreckage over a wide area. The pilot was thrown clear of the aircraft upon impact with the ground and was found alive and still strapped to his seat among the wreckage. According to published accounts of the accident, Lt. Jarvis' wife, the former Miss Jean Elizabeth Sullivan, was at his side when he died at the base hospital that same day. The pilot's mission was to fly the aircraft until the auxiliary fuel tank was empty of fuel. Investigators could not determine the cause of the engine failure but speculated that the engine might have failed because of air in the fuel line from the auxiliary fuel tank or the improper installation of the fuel tank feed lines. The auxiliary fuel tank had been jettisoned prior to the spin and was found 500 yards from the scene of the accident.

9-4-42A. Clovis, New Mexico. At 0840, an Aeronca L-3B stalled and spun into the ground seven and one-half miles WSW of Clovis, New Mexico, killing pilot Cpl. Rexford E. Beckett, 20, Longmont, Colorado. The aircraft was observed by witnesses on the ground to fly over a house. As the airplane flew over the house, it turned to come back over the house again. The airplane lost flying speed and stalled just as the pilot was completing the turn. The airplane turned 180 degrees as it spun in and struck the ground nose first. The pilot, a glider student, was killed instantly.

9-4-42B. Drew Field, Florida. At 1115, a Republic P-43A flew into the ground near Drew Field, Tampa, Florida, killing pilot 2Lt. Kingley Hawkins. The airplane was number three in a three-airplane formation when it was seen by witnesses on the ground to peel off to the left and fly directly into the ground. The other two P-43s continued to fly to the north and the pilots did not notice that the number-three airplane had crashed.

9-4-42C. Pearland, Texas. At approximately 1740, a Curtiss AT-9 flying in poor weather on an instrument training flight crashed out of control into a rice field five miles southwest of Pearland, Texas, killing flying students Sgt. Gerald F. Roberts, 22, Green Bay, Wisconsin, pilot, and Sgt. Nick R. Popa, 20, East Chicago, Indiana, co-pilot. Sgt. Roberts was scheduled to fly on instruments under the hood and Sgt. Popa was assigned as the safety pilot. Weather in the area at the time of the crash was reported as severe thunderstorms with heavy rain and a ceiling of approximately 1,000 feet. A witness on the ground observed the aircraft emerge from the clouds in a vertical dive with both engines providing power. The aircraft was seen to momentarily level out at a very low altitude before reentering the dive and smashing into the ground.

9-5-42A. Mitchel Field, New York. At 1023, a North American AT-6 crashed while attempting a take-off from Mitchel Field, Long Island, New York, killing pilot 1Lt. Jack R. Davies, 23, Florence, South Carolina, and passenger 1Lt. Allen M. Geissler, 38, Hot Springs, Arkansas. The Accident Classification Committee stated, "The cause of this accident is attributed to the fact the mixture control was pulled back to cut the engine out. Lt. Davies must have accidentally pulled the mixture control back to the rear instead of the propeller [pitch control] as so often has happened with this type of throttle quadrant. In doing so his engine stopped. Because of his extremely low altitude and the fact that he was over a congested area he could not land straight ahead. In making a turn to force land in a very small open area, he stalled the airplane." The airplane slammed into the ground and burst into flames one-half mile west of the end of the runway.

9-5-42B. Tampa Bay, Florida. At 1130, a Martin B-26B crashed into Tampa Bay, five miles northeast of St. Petersburg, Florida, killing the crew of five. The Accident Classification Committee report stated, "Airplane was traveling in a northerly direction. The airplane was seen going into a right climbing turn. The airplane approached the stalling point and fell off to the left. The airplane went into a spin to the left at an altitude of approximately 4,000 feet. It made about four turns of a spin, striking the water at an abrupt angle. Considering all of the spin data gathered from actual experience on the B-26B at this station, no pilot has made a recovery from a spin in less than 6,000 feet. Therefore, it is believed that the pilot did not have sufficient altitude to recover from this spin. The cause of the spin is unknown." Killed in the crash were: 2Lt. Harry N. Gahan, Ocala, Florida, B-26 instructor; 2Lt. Robert L. Martin, Long Beach, California, co-pilot/B-26 student; Sgt. Elmer Evans, Pittsburgh, Kentucky, engineer; 1Lt. John L. McKee, Charleston, West Virginia, B-26 student; 2Lt. William R. Addys, Oakland, California, B-26 student. The aircraft was on a transition training flight.

9-5-42C. Page Field, Florida. At 1320, a Martin B-26B suffered the failure of its starboard engine and crashed just after take-off from Page Field, Fort Meyers, Florida, killing the crew of six. The B-26 hit the ground approximately 400 feet from the runway. Witnesses on the ground reported that the aircraft lifted off the runway and reached an altitude of approximately 100 feet above ground level when the engine failed. The aircraft then half rolled to the right and flew into the ground inverted in a left wing down attitude, immediately exploding into flames upon impact with the ground. Investigators could not determine the cause of the engine failure. Killed in the crash were: 2Lt. Arthur W. Schmidt, Akron, Ohio, B-26 instructor pilot; 2Lt. Edward L. McBride, Williamsport, Pennsylvania, co-pilot and B-26 student; 2Lt. Gaston R. Boire, Berlin, New Hampshire, B-26 student; Cpl. John J. Dubois, Bucklin, Kansas, engineer; SSgt. James V. Trochesset, New Orleans, Louisiana, passenger; Cpl. Paul L. LaFleur, Haverhill, Massachusetts, passenger. The aircraft was on a transition training flight.

9-5-42D. New Augusta, Mississippi. At 1400, two Douglas A-20B type bombers collided in mid-air seven miles northwest of New Augusta, Mississippi, killing four crewmembers aboard A-20B #41-2982. The aircraft were part of a three-airplane formation training flight. Pilot 2Lt. Lawrence M. Bever, bombardier 2Lt. Oriel Mitcham and gunner Pvt. Council W. Roberts were uninjured and landed safely in A-20B #41-2965, which was flying in the number-two position, at Hattiesburg Army Air Base, Mississippi. Killed in the collision and crash of A-20B #41-2982 were pilot 2Lt. George R. Pritchard, bombardier 2Lt. Lethoe A. Groce, gunner Sgt. George W. Kaiser and radio operator SSgt. Floyd A. Lones. Lt. Bever's port wing struck flight leader Lt. Pritchard's starboard horizontal stabilizer and elevator, severing them and causing Lt. Pritchard's aircraft to spin to the ground. The aircraft exploded into flames upon impact with the ground, killing the crew instantly and scattering wreckage over a wide area.

9-5-42E. Wendover Field, Utah. At 2045, a Consolidated B-24D flew into the ground eight miles east of Wendover Field, Utah, killing the crew of six. Investigators could not determine why the airplane flew into the ground, striking on its port wing and flipping over to the right and onto its back before exploding into flames. Wreckage and bodies were scattered over a wide area. It was noted by investigators that there were a row of lights that were illuminating a road in the area and that a wheels down landing could have been made in any direction on the brilliant white salt flat upon which the airplane crashed. Examination of the wreckage did not uncover any conclusive evidence of mechanical failure. Weather at the time of the accident was reported as clear with ceiling and visibility unlimited and not considered a factor. Investigators speculated that the pilot might have attempted a thirty-degree banked turn to the left close to the ground, misjudging his position relative to the terrain and allowing the port wing to make contact with the ground. The B-24 had been practicing landings at Westover Field. Killed in the crash were: Capt. Ralph E. Dobrusky, San Antonio, Texas, pilot; 2Lt. Andrew Kulick, Washington, Pennsylvania, co-pilot; Sgt. Alfred E. Lutz, Philadelphia, Pennsylvania, engineer; Sgt. John L. McNeill, Spruce Pine, North Carolina, radio operator; Sgt. Clarence L. Hanson, Minneapolis, Minnesota, gunner; Cpl. Walter M. Lee, San Diego, California, gunner.

9-6-42. Hattiesburg, Mississippi. At 0830, a Douglas A-20B spun into the ground following the failure of its port engine two miles west of the Army Air Field at Hattiesburg, Mississippi, killing pilot 2Lt. Donald T. Baker. The aircraft was seen by witnesses on the ground to wing over into a spin to the left when the engine failed. The aircraft spun about four full turns before striking the ground and exploding into flames. Investigators speculated that the pilot failed to retard the starboard throttle and properly trim the aircraft for single-engine flight, allowing the differential power of the right engine to pull the aircraft over into a spin to the left.

9-7-42A. Lawson Field, Georgia. At 1400, Canadian Army Major D. H. Proctor (1st Canadian Parachute Battalion) was killed when Douglas C-47 #41-18503 struck him after he parachuted from C-47 #41-38607 during a paratrooper-training jump at Lawson Field, Fort Benning, Georgia. The aircraft received minor damage after striking the parachute and shroud lines with the starboard wing, causing Maj. Proctor to fall to his death. The collision occurred about 1,200 feet above the ground in intermittent light rain showers. The C-47 was piloted by 2Lt. Thomas P. Donigan and SSgt. Ralph P. Janes. Lt. Donigan stated to investigators, "At the point when the other ship was 300 or 400 yards in front of me, men started to jump. I saw several men leave the ship and the first canopy trail out and blossom in front of me. At this point I had the C-47 in a steep left turn with full power. I struck the shrouds high and well out on the right wing. I landed immediately, calling for [an] emergency landing when the tower directed I go around." The Accident Classification Committee faulted supervisory personnel for poor planning and execution, and the pilot and jumpmaster aboard 41-38607 for allowing the troops to exit the aircraft in such close proximity to Lt. Donigan's C-47.

9-7-42B. Lloyd, Florida. At 1710, a Bell P-39D spun to the ground after entering a solitary cumulus type cloud near Lloyd, Florida, killing pilot 2Lt. Carroll S. Appling. The airplane was part of a flight of two P-39 aircraft on an authorized aerobatic training flight. The pilot of the other P-39, 2Lt. Frank M. Cencak, stated to investigators, "I took off as wing man in a two-ship formation led by Lt. Appling. We climbed to about 8,000 feet, 17 miles east of [Dale Mabry Field, Tallahassee, Florida]. We did some maneuvers in trail and then decided to go up the side of a cumulus cloudbank. I noticed that I was losing airspeed rapidly and put the nose down but was slow in doing so. I stalled out and when I recovered was in a steep dive at about 4,000 feet, and at the level of the base of the cloud-bank. I closed the throttle, recovered and went into the left hand bank looking for the leading ship. At this time I saw his ship in a counter clockwise flat spin at 1,000 feet below me, and noticed that his nose went above and below the horizon in the spin. I was again losing airspeed, so I decreased the bank and used some throttle, at which time I lost sight of him. I could no longer locate him, and circled the spot until I noticed about eight cars leaving town and heading down a hard road. I returned to the field and notified the Group Surgeon and Squadron Operations Officer."

9-7-42C. Sarasota, Florida. At 1900, two Bell P-39D aircraft collided in mid-air four miles southwest of the Army Air Base at Sarasota, Florida, killing 2Lt. Harold R. Purdy (P-39D #41-38302) and 2Lt. Allyn F. B. Taylor (P-39D #41-7055). The aircraft were part of a formation training flight flying at an altitude of approximately 8,000 feet. Lt. Purdy had pulled up and out of a string formation of six airplanes while flying in the number-three position. He climbed and flew above the formation to allow all of the aircraft to pass before dropping back down into formation. When he attempted to drop back down in formation, he collided with the aircraft flown by Lt. Taylor, flying in the number-six position. Both aircraft broke up in flight and fell to the ground out of control. Lt. Taylor attempted to parachute to safety but his parachute only opened partially and he fell to his death, striking the water about 200 yards off shore.

9-7-42D. Greenville, Mississippi. At 2153, a Vultee BT-13A descended into the ground while flying in instrument conditions one mile southeast of the Army Air Field at Greenville, Mississippi, killing Aviation Cadet Heyward Harris, Jr. Weather in the area at the time of the crash was reported as scud type clouds at about 200 to 300 feet above ground level with higher broken clouds. Visibility was reported at four miles with light ground fog. The pilot had been called in to the field to land because of the deteriorating weather conditions.

9-8-42A. Roswell, New Mexico. At 1000, two Beech AT-11 aircraft collided in mid-air over target N-10 on the AAF Bombing Range ten miles northwest of Roswell, New Mexico, killing six fliers. Killed in the crash of AT-11 #41-27637 were: 2Lt. Oscar W. Predmore, 27, Franklin, Ohio, pilot; Aviation Cadet Gerald Ferdinand McKenna, 20, Pittsburgh, Pennsylvania, student bombardier; Pvt. Vernon T. McKelvey, 30, Ft. Cobb, Oklahoma, photographer. Killed in the crash of AT-11 #41-27591 were: 1Lt. David P. Griffiths, 24, Pittsburgh, Pennsylvania, pilot; 2Lt. William R. Newby, 22 San Jose, California, co-pilot; Aviation Cadet Allen William Hottois, 22, Lakewood, Ohio, student bombardier. Investigators stated, "Lt. Predmore was leading a three-ship formation which was engaged in 4,000 feet formation bombing on Target N-10. Lt. Griffiths was flying on the right (#2 position) and [2Lt. John L. Fields] on the left (#3 position). The formation had just completed dropping their bombs on their third run and the formation was making a gentle turn to the right. While on the turn, Lt. Griffiths was to move up and assume lead position. This was to be accomplished by Lt. Griffiths turning to the right and giving some throttle so as to gain in distance. Lt. Predmore was to drop down before turning, then place the ship in a gentle turn and join the formation on the right from under and behind. Lt. Fields was to continue in his turn and join on the new lead plane's left. Lt. Griffiths made a turn to the right as was customary. Lt. Predmore did not drop down at all, but started a gentle turn to the right and struck Lt. Griffiths' airplane just forward of the tail section, severing the tail. The two ships went down out of control and struck the ground at a steep angle about a half-mile apart. Both planes were completely demolished and the six occupants were killed. There was no evidence that any of the occupants of the two planes attempted to use parachutes."

9-8-42B. Newark, Ohio. At 1203, a North American B-25C flying in heavy rain showers and very poor visibility suffered a catastrophic structural failure of the port wing and crashed into a residence at 154 Hudson Avenue, Newark, Ohio, killing six fliers aboard the aircraft and two civilian women on the ground. Two civilian women were seriously injured and a seven-year-old boy received light injuries. The pilot, Col. Douglas M. Kilpatrick, 34, Houma, Louisiana, and co-pilot 2Lt. Lawrence S. Lawyer, 29, Wray, Colorado, both fell to their deaths in an unsuccessful attempt to parachute to safety. Col. Kilpatrick fell through the roof of the two story apartment/office building owned by Dr. Louis Mitchell at the corner of Clinton and Locust streets, passing through the attic and coming to rest in a kitchen. Lt. Lawyer fell to his death two blocks east of the crash site on a loading platform of the Baltimore and Ohio Railroad just outside the door of the freight office. Killed when the aircraft smashed into the residence were: 2Lt. Russell E. Newlin, 30, Indianapolis, Indiana; Cpl. R. A. Arens, 21, Dayton, Ohio; Pvt. Charles W. Watson, 31, Owensboro, Kentucky, photographer; Ovidio A. Picon, Dayton, Ohio, civilian crew chief (employed by the AAF at Wright Field, Dayton, Ohio). Mrs. Dolly Campbell, 73, Newark, was killed when a large piece of the wreckage smashed her body up against a building across the street from the crash site. Mrs. Jane Weston, 60, Newark, was killed by the exploding B-25 in the basement of 154 Hudson Avenue. Also in the residence at 154 Hudson were Mrs. Mary Welsh, 44, and Mrs. Paul Winsch, 28, both of Newark, who were both seriously injured by flying debris. Jerry Neighbor, 7, also of Newark, received minor injuries. The aircraft, bound for Mitchel Field, Long Island, New York, arrived over Newark during a severe local rainstorm an hour after taking off from

Wright Field, Dayton, Ohio. The aircraft circled the town several times at very low altitude, apparently trying to get its bearings. The aircraft then headed north and passed over a three-story office building, clearing it by about fifty feet. The aircraft again circled the area, losing more altitude and narrowly missed striking the Licking County Jail, climbing and banking steeply to the right to avoid striking it. The aircraft was then seen in a climbing turn to the right, leveling off at an altitude of about 1,000 feet. The port wing failed at the point half way between the landing light and the wing tip. The two pilots were soon seen to abandon ship. On the way down, the out of control aircraft struck a frame house with one wing and then struck another frame house at the corner of Hudson and Wyoming with the other wing before smashing into the three-story brick apartment house at 154 Hudson, completely demolishing it. Wreckage was scattered over a considerable area surrounding the crash site. It was noted in the crash report that the Newark, Ohio, Fire Department and Police force were inadequately equipped to deal with crash, having no foam to fight petroleum type fires. With the help of the heavy rain, the fire fighters were able to extinguish the fire. Hundreds of civilians crowded the streets in the crash area, hampering rescue personnel and becoming a burden for police. Investigators were puzzled at the presence of Pvt. Watson, a former civilian employee of Wright Field, aboard the aircraft. Investigation revealed that Pvt. Watson had no orders to New York and was absent without leave from Ft. Thomas, Kentucky, where he had re-enlisted in the army on 8-21-42. Pvt. Watson had been in and out of the army throughout the 1930s and his service record was somewhat checkered. Throughout 1942 Pvt. Watson was a civilian photographer at Wright Field. Investigators noted that throughout July and August, Pvt. Watson had been renting two Dayton apartments simultaneously and had been acting peculiar. He had tried unsuccessfully to secure a parachute and other flight equipment for himself at Wright Field. Immediately after re-enlisting at Fort Thomas, Pvt. Watson left Kentucky on his own motorcycle. He reportedly showed up at Wright Field "in uniform" with radio headphones although Fort Thomas quartermasters had not issued him any uniforms or equipment. He hung around Wright Field for several days and repeatedly told friends and former co-workers that he was leaving on a mission "east" in the near future. It was learned that Pvt. Watson gained passage on the aircraft by providing false leave orders that he secured from an unknown source. Despite Pvt. Watson's strange behavior, and after an extensive investigation of his past, there was no evidence that he was responsible for the airplane crash. Investigators speculated that the pilot stressed the aircraft beyond its design limitations in an attempt to re-establish level flight, causing the failure of the port wing.

9-8-42C. Hamilton Field, California. At 1630, a Lockheed P-38E dove into the ground 15 miles north of Hamilton Field, San Rafael, California, killing pilot 2Lt. Charley W. Stuart. Lt. Stuart and a wingman were flying in the vicinity of a cloud that began at 2,000 feet and topped out at 8,000 feet. It is believed that Lt. Stuart lost considerable airspeed while climbing in an attempt to get over the top of the cloud and he was last seen by the wingman at an altitude of about 7,000 feet. Investigators speculated that the pilot stalled the airplane and it fell off into the cloud and he was unable to make a transition to instrument flight, losing control of the P-38. Witnesses reported that the airplane dove toward the ground at an angle of approximately 45 degrees at very high speed. The P-38 failed to recover and dove into the ground where it exploded violently into flames upon impact. Investigators could not determine the precise cause of the crash but it is likely that the pilot put the airplane in a diving maneuver from which he could not recover because of the tendency of the P-38 type airplane to suffer from severe control surface buffeting caused by compressibility.

9-8-42D. Malheur County, Oregon. At 1955, Sgt. Donald J. Morton, while manning the tail gun position of a Boeing B-17E flying over the AAF Southeast Oregon Gunnery Range in Malheur County, Oregon, was fatally wounded by a fifty-caliber bullet that was accidentally fired by a faulty port waist machine gun that was in the stowed position. The flight was conducted in order to familiarize nine officers and non-commissioned officers with fifty-caliber aerial machine guns. At the beginning of the mission, the port waist machine gun was loaded and readied for fire by gunnery instructor Sgt. Norbert R. Rzeppa. The gun would not fire even though Sgt. Rzeppa charged the gun several times. Each time he charged the gun, a live round was ejected from the gun. Sgt. Rzeppa stated that he removed the ammunition belt from the port gun and pulled the charging handle several times to clear the gun of any live rounds. The instructor then readied the starboard waist gun for firing and that portion of the exercise continued. After several officers and sergeants fired the starboard gun, the charging handle broke. Sgt. Rzeppa decided to re-investigate the problematic port gun, which was in the stowed position. As Sgt. Rzeppa attempted to un-stow the gun, it went off. The round passed through a portion of an aft bulkhead and struck Sgt. Morton in the lower right back, passing out of his body through his chest. The round passed near an officer who was situated just behind Sgt. Morton observing him operate the machine gun and waiting his turn to fire. Several men testified that the gun was observed on "safe" and that Sgt. Rzeppa merely touched the gun when the accident occurred. Pilot Capt. Thomas E. Fenton and co-pilot 2Lt. William H. Scheil landed the B-17 safely at Gowen Field, Boise, Idaho, at approximately 2030. Sgt. Morton was immediately taken to the base hospital where he died of his wounds at 0520 on 9-9-42. Investigation by ordinance experts revealed that the gun was damaged and should have been taken out of service for repair. Investigation was hampered by the fact that the gun was cleaned and placed back in the armory after the incident. It was discovered that the live rounds that were ejected from the gun during the exercise exhibited a small dent on the primer. Investigators speculated that the live round had somehow "rode" the bolt during the instructor's attempt to clear the gun.

9-8-42E. MacDill Field, Florida. At 2215, a Martin B-26B crashed while attempting an emergency landing at MacDill Field, Tampa, Florida, killing three crewmen and seriously injuring a fourth. The number-two engine was not producing full power, prompting the pilot to call in to MacDill Field for an emergency landing. The pilot undershot the runway on the approach and descended into terrain about 1,500 feet from the end of the runway, the port wing striking a tree. The aircraft burst into flames as it smashed into the ground, scattering wreckage over an area of 600 feet. Killed in the crash were: 2Lt. Charles W. Armstrong, Kirkwood, Missouri, pilot; 2Lt. Jack W. Weston, Roanoke, West Virginia, co-pilot; Capt. James E. Glattly, Hot Springs, Arkansas, passenger. Engineer Pvt. Ervin Leshner, Philadelphia, Pennsylvania, was seriously injured. Investigation revealed that the wrong type of grease was being used in the magnetos, causing the points to burn badly. Both distributors were found to have clogged vent lines and the starboard distributor was found to have a broken spring and rotor contact brush.

9-9-42A. Morristown, New Jersey. At 0328, a Fairchild C-61 flying in poor weather flew into terrain west of Morristown, New Jersey, killing two fliers and seriously injuring two others. Killed in the crash were the pilot Capt. Paul J. Gorman, 28, Buffalo, New York, and 1Lt. Joseph Bernhardt, 28, Boston, Massachusetts. Passengers 2Lt. Richard E. Manches, Short Hills, New Jersey, and 1Lt. Leslie Meyers, Alexandria, Louisiana, were seriously injured. The aircraft had departed Dow Field, Bangor, Maine, for Orange, New Jersey, via Boston, Massachusetts, Hartford, Connecticut, and LaGuardia Airport, New York. The first leg of the trip to Boston was uneventful. The flight encountered instrument conditions on the leg to Brainard Field, Hartford. The fliers went into town at Hartford and ate dinner. The clearance to LaGuardia was cancelled because of poor weather and the flight was delayed for about two hours while they waited for the weather to clear. At 0138 the flight was finally cleared to LaGuardia. The flight soon entered instrument conditions but continued on to New York. When the airplane arrived over the New York/ New Jersey area, the weather prevented the crew from locating LaGuardia Airport or Newark Airport. The airplane then circled near Morristown for several minutes before making contact with tall trees. The aircraft bounded into the air for about 150 feet before it crashed into the ground, careening through large trees before flipping over and coming to rest inverted. The wreck and survivors were not found until about 1100. The purpose of the flight was to transport Lt. Manches to New Jersey to visit his seriously ill wife who was being treated in a hospital at Orange. It was noted by investigators that the aircraft was not equipped with a radio.

✓ **9-9-42B. Fletcher Field, Mississippi.** At 1307, two Stearman PT-17 aircraft collided in mid-air one mile southwest of Fletcher Field, Clarksdale, Mississippi, killing Aviation Cadet Seymour B. Meltzer, 21, Brooklyn, New York (PT-17 #41-8260), and Aviation Cadet Estin R. Teter, Jr., 22, Harman, West Virginia (PT-17 #41-25692). Both aircraft had just taken off from the field and were flying in the pattern and maneuvering to make touch-and-go landings when the collision occurred. Investigators faulted both pilots for their failure to see and avoid each other.

9-9-42C. Marion, Ohio. At 1415, a Piper L-4A collided with a hangar and crashed just after take-off from the municipal airport at Marion, Ohio, killing flying student SSgt. William E. Roberts and seriously injuring instructor 2Lt. John C. Townsend. Investigators stated, "Lt. Townsend took off in formation with two other L-4A aircraft from Marion Municipal Airport, his ship being on the left wing of the leader. The formation circled the field counter clockwise and on the return over the field the leader led the formation too low and too close to the hangar causing the ship on the left wing of the leader to strike the hangar with his left wing. After striking the hangar, the ship continued to climb to about 150 feet with the left wing low. At this point the ship fell off on the left wing, striking a 440-volt electric line and the right wing struck a tree, breaking the wing off. The ship fell to the right into a ditch 18 to 20 feet deep containing about 18 inches of water."

9-10-42A. Enterprise, Oregon. At 0430, a Boeing B-17F collided with rising terrain near Sheep Creek 25 miles southeast of Enterprise, Oregon, killing seven crewmembers. The B-17 had taken off at 0030 from Geiger Field, Spokane, Washington, on a night navigation mission. The airplane failed to return and was reported missing later in the day, finally being located 9-14-42 scattered along the 5,000-foot level of a wooded ridge. Weather at the time of the accident was reported as a low overcast with scattered light rain. It was determined that all four engines were operating at cruise rpm when the accident occurred. No evidence of structural failure was found. Investigation revealed that the B-17 was flying south when it struck a very tall tree 100 feet above the ground with the outer starboard wing panel, severing the panel and the top of the tree. The B-17 then rolled to the right to inverted and struck a stand of 100-foot timber, bursting into flames as it smashed through the trees and down a steep wooded ravine. Wreckage and bodies were scattered over an area of 800 yards. Killed in the crash were: 1Lt. Richard E. Pinneo, pilot; 2Lt. Anthony F. DiMenno, co-pilot;

2Lt. Peter Knudsen, Jr., B-17 student; 2Lt. Rane A. Fournier, navigator/bombardier; 2Lt. John W. Zurich, B-17 student; Pfc. George L. Schaffer, engineer; Cpl. Robert C. Raschka, radio operator.

9-10-42B. Gardner Field, California. At 0730, a Vultee BT-13A crashed six miles south of Gardner Field, Taft, California, killing instructor Theodore B. Ely, Jr. A/C Truman G. Kolls was seriously injured when he parachuted from the aircraft. The instructor was demonstrating various stall-spin exercises to the student and put the aircraft into a spin to the right. After the aircraft spun about five times, the instructor realized that he could not recover the aircraft and ordered the student to bail out. The spin was attempted at an altitude of 2,200 feet above ground level, too low to allow for a safe recovery.

9-10-42C. Bruce Field, Texas. At 1130, a Fairchild PT-19A stalled and spun into the ground four miles southwest of Bruce Field, Ballinger, Texas, killing Aviation Cadet Wilson P. Thedford. Witnesses on the ground observed the airplane spinning at an altitude of about 1,200 feet above ground level. A/C Thedford was killed when he attempted to parachute to safety at about 150 feet above the ground, too low to allow his parachute to fully deploy.

9-10-42D. Matagorda Island, Texas. At 1150, a North American AT-6C stalled and spun into the ground on Matagorda Island, Texas, killing Aviation Cadet Robert Earl. The aircraft was on a gunnery mission and was diving and firing on targets on the island. Apparently the student pulled up too sharply after one of his attacks and stalled the airplane. The aircraft whipped over in a half-roll, spun to the right and smashed into the ground in an inverted attitude.

9-10-42E. Clearwater, Florida. At 1430, two Curtiss P-40F aircraft engaging in simulated combat collided in mid-air and crashed at Clearwater, Florida, killing 2Lt. Carl L. Sabee (P-40F #41-13708) and 2Lt. Oliver J. Goldsmith (P-40F #41-13603). Witnesses on the ground observed the aircraft dog fighting and chasing each other. One aircraft reportedly made a climbing turn up into the bottom of a large bank of clouds at about 10,000 to 12,000 feet and momentarily was out of sight. The other P-40 was flying straight and level just below the bank of clouds heading in a northerly direction. Several seconds later, the P-40 that had climbed into the clouds descended out of the clouds in a slight dive and collided with the other head on. The two aircraft broke up into over a dozen pieces and smashed into the ground all over the vicinity of the Clearwater Causeway and the waters in the area. A large, flaming piece of wreckage fell onto the beach and into the waters north of the causeway. Another large piece of wreckage fell into shallow water near the drawbridge east of the causeway. The engine of one airplane landed behind the Wonder Bar and Grill. Lt. Sabee was killed in the initial collision and hurled from the wreckage. His parachute had burst open in the collision and his lifeless body was observed to float to earth in the fully deployed parachute, which was damaged and scorched. Two civilians in a motorboat recovered his body from about three feet of water south of the causeway. Witnesses on the ground reported that pieces of wreckage fell to the ground for many seconds after the collision.

9-10-42F. Douglas, Georgia. At 1445, a Stearman PT-17 crashed while attempting a forced landing in a cotton field at Douglas, Georgia, fatally injuring Aviation Cadet Joseph C. LaManna, Tuckahoe, New York. Investigators speculated that the student became lost and attempted to land in the field in an effort to determine his location. Civilian witnesses on the ground observed the aircraft circling the field at low altitude. Flying parallel to the highway and at an altitude of about 75 feet, the aircraft made a steep right hand turn to make an approach. The aircraft stalled and spun to the right, diving into the ground in between two tobacco barns. Civilians in the area immediately pulled Cadet LaManna from the wreckage and a passing motorist rushed him to Ocilla Hospital, Ocilla, Georgia, where he died a short time later.

9-10-42G. San Francisco Municipal Airport, California. At 1715, a Lockheed P-38F suffered a port engine failure just after take-off from San Francisco Municipal Airport, San Bruno, California, and crashed west of the runway, killing pilot 2Lt. Robert W. Chambers. Tower personnel reported that the engine failed at an altitude of 75 feet about two-thirds of the way down the runway. The pilot feathered the port propeller but apparently did not retard the throttle of the starboard engine. With landing gear still extended, the P-38 winged over to the left and flew into the ground, exploding into flames upon impact.

9-11-42A. Mt. Baldy, Arizona. At 0300, a Consolidated B-24D flying in poor weather collided with 11,590-foot Mt. Baldy at an elevation of 10,500 feet, 27 miles west of Alpine, Arizona, killing the crew of ten. The B-24 had taken off at 2037 CWT (9-10-42) on a night cross-country navigation flight from Omaha, Nebraska, to Davis-Monthan Field, Tucson, Arizona, having made the Tucson to Omaha trip earlier in the day. The airplane failed to return to Davis-Monthan Field and was reported missing, finally being located on 9-16-42 by two CAP pilots. The aircraft had cut a 700-foot path through the treetops before slamming into the terrain, smashing through tall trees and exploding into flames. The area of the crash was so remote and rugged that the bodies of the dead crewmen had to be taken down the mountain by pack animals. Investigators noted that the aircraft was 30 miles off of its intended course for Tucson. Investigators could not determine why the aircraft was off course and heading to the northwest, away from its intended destination. No evidence of structural or mechanical failure was found and all four engines were determined to be operating at the time of impact. Killed in the crash were: 1Lt. Earl E. Snell, Tucson, Arizona, pilot; Capt. A. W. Dwyer, Huntington, New York, co-pilot; 1Lt. J. L. Easton, Buffalo, North Dakota, assistant co-pilot; 2Lt. W. H. Prunty, Sanger, California, assistant co-pilot; 2Lt. B. B. Knudson, Los Angeles, California, assistant co-pilot; 2Lt. J. M. Hirsch, Denver, Colorado, navigator; 2Lt. J. W. Horton, Arkansas City, Kansas, navigator; 2Lt. D. L. Miller, Columbus, Ohio, bombardier; Sgt. J. R. Cerwenka, Tucson, Arizona, engineer; Sgt. Melvin H. Davis, Klingerstown, Pennsylvania, radio operator.

9-11-42B. Esler Field, Louisiana. At 1250, a Douglas DB-7B suffering the failure of its port engine just after take-off crashed one mile northeast of Esler Field, Alexandria, Louisiana. Killed in the accident were pilot, 2Lt. Walter H. Wyens, Hillsboro, Georgia, and gunners SSgt. Jack C. Farling, Frankfort, Indiana, and SSgt. George W. Lapointe, Claremont, New Hampshire. Witnesses on the ground reported that the DB-7 became airborne about two-thirds of the way down the runway and gained about 100 feet of altitude when the engine failed. The aircraft, with landing gear still extended, could not gain altitude and the pilot turned into the dead engine, losing airspeed and stalling the airplane. The aircraft slipped off to the left, barely missing trees adjacent to the airfield before smashing into the ground in a small clearing. The DB-7 did not catch fire and Esler Field rescue personnel removed the crewmembers from the wreckage immediately after the crash. Two crewmen were still alive when pulled from the wreckage, but they died shortly after the accident. Investigation revealed that the port propeller was in the feathered position. The port magneto breaker box was found full of dirty oil and the points badly burned, leading investigators to speculate that this condition caused the engine failure. Aircraft accident investigator Capt. C. E. Whitcomb accidentally started the wreckage on fire while investigating the cause of the crack-up. The aircraft was scheduled to fly a low-level navigation-training mission.

9-11-42C. Phoenix, Arizona. At 1215, a Stearman PT-17 stalled and spun into the ground 10 miles northwest of Thunderbird Field II, Phoenix, Arizona, killing A/C Frank A. Atkins. The aircraft smashed into the ground at a 45-degree angle and immediately burst into flames. A/C Atkins fell to his death when he abandoned the aircraft at very low altitude. His parachute burst open from impact with the ground and his body was found near the wreckage, having bounced 35 feet from where he initially hit the ground.

9-11-42D. Nashville, Georgia. At 1715, a Curtiss AT-9 crashed after flying into a thunderstorm three miles east of Nashville, Georgia, killing pilot Capt. Dallas L. Morris, Charlotte, North Carolina, and passenger 1Lt. Lawrence A. Gough, AAF Chaplain, New York, New York. The storms in the area were reported to contain severe turbulence, powerful wind with vertical drafts and heavy rain. Investigators could not determine why the airplane crashed but speculated that either the pilot lost control while flying in instrument conditions or the aircraft suffered a structural failure because of severe stress or over-control by the pilot.

9-11-42E. Foster Field, Texas. At 1620, a North American AT-6C stalled and spun while attempting to enter the pattern at Foster Field, Victoria, Texas, killing instructor pilot 2Lt. James A. Simpson, Uvalde, Texas, and student Aviation Cadet Edward S. Bissell, Harland, Kentucky. The aircraft was seen by witnesses on the ground and in the air attempting a turn into the pattern at an altitude of approximately 1,500 feet above ground level. The aircraft whipped over in a snap roll and spun about four times. The aircraft leveled out momentarily and then entered a spin in the opposite direction and smashed into the ground.

9-11-42F. Arabela, New Mexico. At 2210, a Cessna AT-17 flying in poor weather crashed 15 miles northeast of Arabela, New Mexico, killing instructor pilot 2Lt. Donald E. Ripsch, 24, Los Angeles, California, and student Aviation Cadet Thomas H. Cozens, 25, Encinitas, California. The aircraft was part of a three-airplane weather observation flight that had originated at Roswell Army Air Base, Roswell, New Mexico. The flight had inadvertently entered instrument conditions at the edge of a large storm front that was approaching the area. The flight leader ordered the flight to turn left to get out of the clouds and into the clear. It was speculated that Lt. Ripsch started his turn too late and turned into a part of the storm that contained very heavy rain and dangerously turbulent air. The other pilots in the flight observed Lt. Ripsch's airplane exit the clouds in a vertical dive. The aircraft dove into the ground and exploded into flames. The wreckage was not located until 0930 the next day.

9-12-42A. Kelly Field, Texas. At 0735, a Beech AT-10BH spun into the ground four miles south of Kelly Field, San Antonio, Texas, killing instructor pilot 2Lt. Alexander H. Peck. Aviation Cadet Daniel L. Bioscat parachuted to safety and was uninjured. The instructor was demonstrating various stall and spin maneuvers to the student at an altitude of approximately 3,300 feet above ground level. The instructor put the airplane in a stall and a deliberate spin. The instructor recovered the airplane and initiated another stall and spin immediately and at a significantly lower altitude. The airplane stalled and spun to the left and the instructor, unable to recover the airplane, ordered the student to bail out. The airplane spun into the ground and burst into flames, killing the instructor instantly. Investigators faulted the instructor for practicing spins at too low an altitude to allow for a safe recovery.

9-12-42B. Tuskegee, Alabama. At 1100, a Curtiss P-40E stalled and spun into the ground while maneuvering 10 miles northeast of Tuskegee, Alabama, killing 2Lt. Faythe A. McGinnis. Pilot 2Lt. Willie Ashley, Jr. witnessed the accident and later stated to investigators, "Lt. McGinnis was leading the second element of a six-ship formation which took off about 0945. I was fourth in the formation, flying Lt. McGinnis' wing. After taking off and gaining about 4,000 feet altitude, the formation was carried through a series of medium diving and climbing turns. The formation was loose, the airplanes about five ship lengths apart. We were at about 4,500 feet altitude when the leader did a slow roll, after which he dove to pick up speed and started into an Immelmann turn. As far as I could see, the leader and the second man completed the maneuver. Lt. McGinnis started into his turn and I was about to follow him, but after noticing the altitude, 5,000 feet, I decided to do a chandelle instead. Lt. McGinnis' plane stalled on its back after performing a half loop. The plane fell over into a normal spin, which appeared to include only one complete turn. Then it did a series of oscillations similar to that of a falling leaf. Recovering from these oscillations at about 200 feet, the nose fell. There was not enough altitude to pull out. The plane crashed and exploded." Both Lt. McGinnis and Lt. Ashley were original members of the 99th Fighter Squadron based at Tuskegee Army Air Field.

9-12-42C. Tallahassee, Florida. At 1310, a Bell P-39F slammed into the ground in a wooded area seven miles northwest of Tallahassee, Florida, killing 2Lt. William S. Richardson. Civilian witnesses on the ground observed the P-39 to go into a dive at an altitude of approximately 5,000 feet and continue the dive straight into the terrain. Investigation revealed that the airplane struck the ground in a dive at an angle of about 35 degrees and at an estimated speed of over 400 mph. The engine was producing power at the time the aircraft struck the ground. Investigators could not determine the series of events leading up to the accident or its cause but speculated that the pilot might have blacked out because there was no apparent attempt to recover from the dive or exit the airplane. The impact of the airplane with the ground was so terrific that wreckage was scattered over an area of three-quarters of a mile. The airplane was on a scheduled aerobatic flight.

9-13-42A. Columbia, South Carolina. At 0545, a North American B-25B returning from a night navigation training flight crashed two miles east of the runway while attempting a landing at Columbia Army Air Base, Columbia, South Carolina, killing the crew of seven. The aircraft had made three unsuccessful attempts to land prior to the fourth attempt that resulted in the crash. The pilot radioed the tower at 0510 for landing instructions. He was instructed to fly the pattern at 2,000 feet, to land to the west on the East/West runway and to call on the base leg. The pilot requested that the field floodlights be illuminated for his landing. On the pilot's first attempt to land, he came in too low and went around. On the second attempt he came in high and to the left. Again, he aborted the landing attempt and went around. On the third approach, the pilot came in high and on the right, lined up with the taxiway. The pilot again went around and radioed that he was having a hard time lining up with the runway. On the final attempt to land, the B-25 came in extremely low and had to pull up to avoid striking trees. The pilot attempted to gain altitude as the aircraft approached the end of the runway. The aircraft was observed to execute a very steep climbing turn to the right. The aircraft stalled and fell off to the right, smashing into the ground and exploding into flames. Killed in the crash were: 2Lt. Francis W. Sparks, 27, Columbia, South Carolina, pilot; 2Lt. Martin O. Vangsness, 22, Carpio, North Dakota, co-pilot; 2Lt. Donald V. Urquhart, 27, Ridgeway, Pennsylvania, assistant co-pilot; 2Lt. James F. Attridge, 23, Salem, Massachusetts, navigator; Sgt. Leon O. Gallis, 23, Northampton, Pennsylvania, engineer; Sgt. Charles O. Frame, 22, Richmond, Virginia, radio operator; Pvt. Roland G. Braun, 23, Beaver Dam, Wisconsin, assistant engineer.

9-13-42B. Big Spring, Texas. At 0925, a Vultee BT-13A attempting a forced landing crashed 20 miles southeast of Big Spring, Texas, killing Aviation Student Lewis J. Kastelic. The airplane was on an authorized navigation training flight. Civilian George E. Neely witnessed the crash and later stated to investigators, "I saw an airplane coming towards us from the east. The pilot seemed to be racing his motor and he was about 25 or 30 feet above the ground. The plane did not gain altitude and it went on west about one-fourth of a mile, circled to the left and came back again towards us from the east. This time it was about 20 feet high and the pilot signaled that he wanted to land, and I signaled him that it was okay. He then went on west about 200 yards, turned to the left and raced his motor several times but it never would gain altitude. He then came back about 200 yards east of me and banked his plane to land it. When he banked the airplane, it looked to me that he could not pull it out of it and it came down and hit the ground on the tip of the left wing and as soon as the plane hit the ground, it caught on fire."

9-13-42C. Brunswick River, South Carolina. At 1104, a Vultee V-72 (similar to the A-31 type) crashed into the middle of the Brunswick River where it meets Port Royal Sound, one mile east of Page Field, Parris Island, South Carolina, killing pilot 2Lt. Richard C. Burton, 22, Johnstown, New York, and gunner SSgt. Harry L. Massey, 21, Memphis, Tennessee. The aircraft was flying at 9,000 feet in a two-ship formation with another V-72, Lt. Burton flying the lead. Lt. Burton gave a signal to dive on a tanker in the Beaufort River. Lt. Burton and his wingman dropped their dive brakes and began their simulated bomb run. The wingman, 150 feet behind, observed that Lt. Burton's airplane began a slight turn to the right at about 3,500 feet and continued in the dive until it struck the water and exploded. The aircraft struck the water in a dive of about 80 degrees with no apparent attempt by the pilot to pull out or exit the aircraft. There was a fire on the water that covered about 50 yards and lasted about 45 seconds.

9-13-42D. Pacific Ocean, California. At 0956, a Lockheed P-38F flying in instrument weather conditions descended into the Pacific Ocean while trying to let down through overcast three miles northwest of La Jolla, California, killing pilot 1Lt. Wendell F. Seppich. The ceiling in the area at the time of accident was reported as 50 feet with visibility two miles. The pilot was attempting to find North Island Naval Air Station, San Diego, California, and witnesses on the ground observed the airplane descend through the clouds and into the ocean approximately 350 yards off shore.

9-13-42E. Dale Mabry Field, Florida. At 1555, a Bell P-39 stalled and spun into the ground while performing aerobatics seven miles south of Dale Mabry Field, Tallahassee, Florida, killing pilot SSgt. John J. Herles. A civilian witness on the ground stated that the aircraft stalled at the top of a loop and fell into a counter clockwise flat spin. The pilot made no attempt to leave the aircraft. Investigators noted that transition pilots were forbidden to perform aerobatics until they had ten hours on type. The pilot had very little time and no aerobatic training in P-39 type aircraft.

9-13-42F. Colorado Springs, Colorado. At 2012, a Lockheed F-4 (the photo-reconnaissance version of the P-38), apparently running low on fuel and attempting an emergency landing, crashed near the Army Air Field at Colorado Springs, Colorado, killing pilot 2Lt. Richard J. Copeland. The pilot was on an escorted navigation-training mission that had taken off from Memphis, Tennessee. The aircraft, having become separated from its flight, arrived at the base after dark and the pilot was instructed by the tower to proceed to Lowry Field, Denver, Colorado, because Colorado Springs had no runway lighting. The pilot radioed that he had to land immediately because he was low on fuel. The officer in the tower ordered that all the automobiles at the field be lined up to illuminate the field. The pilot, heading northwest, flew over the North/South runway with his landing lights on, but he intersected the runway at an angle and was unable to land. The P-38 then turned south and entered a standard pattern approach. The pilot attempted the turn from base to final and flew the airplane into the ground.

9-14-42A. Niagara Falls, New York. At 1050, a Bell P-39K spun to the ground one and a half miles northwest of the municipal airport at Niagara Falls, New York, killing civilian pilot Douglas T. Mitchell, 38, Johnsonville, Oklahoma. Mitchell, a test pilot for Bell Aircraft Corporation, was conducting the flight to test new fuel cells installed by Bell. The air traffic controller for Niagara Falls Airport, Richard C. Bonhurst, witnessed the accident and stated to investigators, "I first observed a P-39 west of the field about 4,000 feet to 6,000 feet doing repeated snap rolls. He was flying from north to south. He continued these rolls and fell out of a couple, losing 500 to 1,000 feet in altitude. He then reversed his direction, flying north and made approximately four or five rolls, losing considerable altitude. About 2,000 feet, the plane seemed to fall out of [a steep chandelle to the right] and fell towards the left, a quarter turn was made, and then it righted itself, the nose came up slightly. It then made a quarter turn to the right and seemed to be in a stalled position although it was nosing toward the ground. This gyration from left to right, and right to left, continued until the ship was approximately 500 to 800 feet off the ground. It then dove toward the ground without any visible change in attitude." Numerous qualified witnesses on the ground observed the accident and the maneuvers prior to the accident were captured on motion picture film, aiding the investigation. The actual crash was not recorded because trees obscured view of the airplane.

9-14-42B. Crookston, Minnesota. At 1015, a Taylorcraft L-2A stalled and spun into the ground seven and a half miles west of Crookston, Minnesota, killing civilian flight instructor Charles Schooley and student Cpl. Hugh C. Dodd. Statements collected from civilian witnesses, who were unfamiliar with aviation matters, were vague and conflicting. Investigation revealed that the airplane struck the ground in a diving spiral to the right.

9-14-42C. Gunter Field, Alabama. At 2230, a Vultee BT-13 crashed 14 miles southeast of Gunter Field, Montgomery, Alabama, killing RAF Flying Cadet Douglas A. Gell. The aircraft had just taken off from the field on a night navigation flight and was observed to gain about 200 feet when it began a descending turn to the right. Instructions were radioed to the student to straighten out and climb straight ahead. There was no acknowledgement by the student and the airplane flew into the ground in a diving turn to the right. Investigators speculated that the student lost his horizon or became distracted with turning off his landing lights and inadvertently flew the airplane into the ground.

9-15-42A. Burlington, New Jersey. At 0845, a Curtiss P-40F on an authorized aerobatic flight dove into the ground five miles east of Burlington, New Jersey, killing pilot 2Lt. John P. Heck. The aircraft was observed to enter a dive at an altitude of approximately 10,000 feet. The aircraft continued in the dive with power on at an angle of about 80 degrees until it crashed violently into the ground and exploded into flames. The airplane struck the ground at such a terrific speed that the wreckage was buried in the soft clay up to a depth of 11 feet. Only small portions of the pilot's remains were recovered.

9-15-42B. Lubbock, Texas. At 1355, a Taylorcraft L-2 on a pilot proficiency flight crashed just after take-off from South Plains Army Flying School,

Lubbock, Texas, fatally injuring student glider pilots SSgt. Edgell Smith and TSgt. Alex W. Uhl. The aircraft was observed to take-off and gain about 200 feet of altitude when the engine quit. The pilot stalled the airplane and it spun to the ground. Investigators noted that the pilot could have landed safely straight ahead.

9-15-42C. Hillsgrove, Rhode Island. At 1617, a Curtiss P-40F crashed just after take-off from State Airport, Hillsgrove, Rhode Island, killing 2Lt. Donald W. Hoefler. Civilian Air Traffic Controller James F. Jordan witnessed the accident and later stated to investigators, "A flight of three P-40s was given take-off instructions on Runway [6-24] (heading SW). [Lt. Hoefler] was in No. 2 position. The ship took off heading SW and banked to the left, a SE heading. Pilot had reached an altitude of approximately 500 feet, directly south of the field and between the SW and the SE runways, when he called for an emergency landing. He banked steeply to the left and was apparently going to try to come in on Runway No. 4. The ship then fell off on the left wing and dove into the ground. The last thing the pilot said just before the ship hit was 'emergency landing on...' and then he hit."

9-16-42A. Hermosa Field, Texas. At 0828, a Vultee BT-13A stalled and spun into the ground at Hermosa Auxiliary Army Airfield, 12 miles west of Pecos, Texas, killing Aviation Cadet Vernon M. Hoveland. The student was observed to execute a flat, nose high turn from base to final. The airplane stalled, flipped over and fell off into a one-turn spin from an altitude of approximately 600 feet. The wreckage caught fire many minutes after the crash.

9-16-42B. Birmingham, Alabama. At 1425, a Martin B-26B flying in instrument conditions collided with the wooded side of Double Oak Mountain 15 miles southeast of Birmingham, Alabama, killing the six crewmen on board. There were low hanging clouds and a heavy rain falling at the time of the crash. Visibility was reported as extremely poor. The aircraft was on a cross-country navigation flight from Lakeland, Florida, to Louisville, Kentucky. The aircraft exploded in flames as it careened through trees, leaving a trail of wreckage over 300 yards. Killed in the crash were: 2Lt. Frank W. Taylor, Houston, Texas, pilot; 2Lt. William F. Wade, Long Beach, California, copilot; 2Lt. John A. Johnson, Evanston, Illinois, bombardier; Sgt. William John Town, Detroit, Michigan, engineer; SSgt. Gordon B. Schlottman, Barrington, Illinois, radio operator; Sgt. Reynold J. Drott, Houma, Louisiana, passenger.

9-17-42A. Shaw Field, South Carolina. At 0230, a Vultee BT-15 practicing night landings crashed three miles northwest of Shaw Field, Sumter, South Carolina, killing Aviation Cadet Alfred C. Murphy, Jr. The student was contacted by the tower and instructed to return to his training zone because a flight of aircraft was going to land at the field. After the flight landed, Cadet Murphy did not return to the field and was reported missing. The airplane was found in a thickly wooded area after a brief search. Investigation revealed that the airplane crashed in an inverted position at a seventy-degree angle. Investigators speculated that the pilot lost the horizon while making a turn. Weather at the time of the crash was reported as clear with ceiling and visibility unlimited.

9-17-42B. Sarasota, Florida. At 0740, a Bell P-39D suffering a runaway propeller crashed one mile southwest of the Sarasota Army Air Base, Sarasota, Florida, killing 2Lt. Harry Winkler, Jr. This was Lt. Winkler's first solo flight in a P-39 type airplane. The Accident Classification Committee stated, "After 20 minutes duration on the first flight of the day, [at an altitude of 3,000 feet] and while within easily discernable distance of the Sarasota Air Base, the engine in the airplane suddenly emitted puffs of white smoke. A fraction of a second later, a sheet of flame was seen shooting from the left hand side of the engine. The airplane then went out of control and dove to the earth with the flames streaking from the power plant section. Prior to the time flame was seen, excessive high engine rpm was clearly audible, indicating that the propeller pitch changing mechanism had failed, either electrically or mechanically, advancing the propeller into full 'low pitch' position. It is believed that the extreme high rpm encountered resulted in sudden failure to some part of the engine; possibly partial disintegration of supercharger impeller, connecting rods, or some part of the fuel induction system. It is also noted that this engine was a special service test engine with copper-lead alloy main bearings. Due to the mangled condition of the engine, no examination for condition of the engine bearings was attempted."

9-17-42C. Drew Field, Florida. At 0930, a Republic P-43 attempting a landing at Drew Field, Tampa, Florida, stalled and was ground looped into another P-43 idling at the runway intersection, killing both pilots. Pilot 2Lt. Jock A.R. Stewart, attempting a landing in P-43 #41-6713, made his approach at a low airspeed, gliding in too slowly and causing a stall of the port wing. The wing stalled, dropped and began scraping the runway surface, possibly damaging the port landing gear. Lt. Stewart righted the aircraft and apparently saw that he was going to collide with P-43 #41-6719 at the intersection. The pilot tried to turn his airplane to the right in order to avoid striking P-43 #41-6719 and this time the starboard wing struck the ground. Lt. Stewart's airplane ground looped but could not avoid striking P-43 #41-6719 practically head on. Both aircraft exploded into flames. Lt. Stewart's airplane careened to the right and flipped inverted, coming to rest about 25 yards from the other P-43. The pilot of the other P-43, 2Lt. Robert W. Symonds, was able to exit his flaming aircraft and was seen running from the wreckage with his uniform totally engulfed in flames, collapsing a short distance from the wreck. He later died at the base hospital.

✓**9-17-42D. Taylor Field, Florida.** At 1520, a Stearman PT-17 attempting a landing at Taylor Field, Ocala, Florida, collided with an unoccupied PT-17 and a civilian on the flight line, killing the civilian and destroying both airplanes. Civilian linesman T.R. Butler, who was standing near the parked airplane, was struck and killed when PT-17 #41-25465 made a three-point landing through the flight line. Student pilot A/C Eugene Grenvicz was not injured in the accident. Investigators stated, "Student placed his base leg too close to the field, his glide was too fast and he overshot. Student failed to realize apparent danger since he was concentrating on a three-point landing and failed to notice surroundings. He made his landing beyond the half way mark on the field, held his plane straight, and from all accounts, did not see the parked airplane on the line until it was too late to avoid the crash, as he made no attempt to ground loop or to avoid parked ship. Student maintained a straight path with the wind tee. From the point of contact with the ground to where he crashed into parked airplane was a distance of some 350 feet. Student struck tail assembly of parked airplane with propeller and nosed his ship over on its back. Four lineman were working around the parked airplane when the crash occurred and one was killed." Civilian linesmen D. W. Howard, A. D. Miller and Thomas Knight received minor injuries.

9-18-42A. Gulf of Mexico, Florida. At 0735, a Bell P-39D was abandoned by its pilot and crashed into the Gulf of Mexico ten miles northwest of Sarasota, Florida. The aircraft was observed performing a series of steep banked turns prior to the accident. The aircraft then entered a steep dive and the pilot exited the aircraft. The pilot, 2Lt. Ernest Farrar, apparently parachuted safely from the aircraft and was seen to land in the water 200 yards from shore. His parachute and life jacket were found at the crash site and his body was recovered five days later 20 miles south of the point where he was seen to hit the water. Apparently the pilot drowned before he could be rescued from the sea.

9-18-42B. Columbia, South Carolina. At 0920, a North American B-25C suffered the failure of its starboard engine just after take-off and crashed at the Columbia Army Air Base, Columbia, South Carolina, killing the crew of four. AAF pilot 2Lt. William F. Butela witnessed the crash and later stated to investigators, "It was a normal take-off and as the plane was some distance in the air the flaps were raised. The plane started to mush and then I heard two pops. The B-25 immediately did a half snap roll to the right [at an altitude of approximately 300 to 400 feet]. An attempt was made to roll out and the plane then nosed up and fell to the ground apparently tail first. It went out of sight below the trees, its tail was straight down and a blast of flame arose immediately." Killed in the crash were: 2Lt. Arnold P. Libman, 24, Waukesha, Wisconsin, pilot; 2Lt. Lonnie C. Tucker, Jr., 24, Jacksonville, Texas, co-pilot; TSgt. Bert H. Ray, 20, Delta, Colorado, engineer; SSgt. Edward T. Davis, 19, Houston, Texas, gunner.

✓**9-18-42C. Sikeston, Missouri.** At 0915, a Fairchild PT-19A stalled and spun into the ground while practicing spins 12 miles east of the Harvey Parks Airport, Sikeston, Missouri, killing Aviation Cadet Charles R. Rhoads. The civilian instructor Howard H. Flentge received minor injuries when he parachuted to safety. He later stated to the Accident Classification Committee, "I was on a routine training flight with Cadet Rhoads, practicing high altitude air work. I instructed the student to do a spin to the right and to recover after completing two turns. Our altitude at this time was 3,500 feet. After two turns of the spin had been completed the student did not recover, so I called for a recovery immediately. He made no attempt to recover so I tried to take over. The student apparently froze on the controls, holding the right rudder on full and holding the stick all the way back and to the right. I attempted to break the rudder and get the stick forward but was unable to do so. Realizing that I could not get the controls, I ordered the student to jump. He appeared to stare straight ahead and showed no response to my instructions. I left the airplane between 600 and 700 feet. My parachute opened just before I reached the ground, and I did not see the airplane crash." The airplane exploded into flames upon impact with the ground, killing the student instantly.

9-18-42D. Ellsworth, Kansas. At 1130, a Lockheed F-4 (photo-reconnaissance version of the P-38) flew into terrain while flying in marginal weather 12 miles northeast of Ellsworth, Kansas, killing pilot 2Lt. Alfred W. Myers. A farmer working in his field observed the aircraft flying from northeast to southwest at a very low altitude in a shallow valley. The engines sounded normal. The P-38 climbed a little and went out of sight over a slight rise. Almost immediately he heard a crash and saw flames and pieces of the airplane exploding through the air. The ceiling at the time of the crash was reported to be approximately 250 to 500 feet agl and a light rain had just fallen prior to the accident. Investigators speculated that the airplane suffered a power-on stall while attempting a steep turn to avoid rising terrain and impacted the ground in a slipping, diving attitude to the right. It was noted by investigators that the airplane was more than 30 miles off course and the pilot had only flown only 11 hours in the previous 60 days.

9-18-42E. Leesburg, Georgia. At 1330, a Curtiss AT-9 damaged in a landing attempt crashed near the Auxiliary Airfield at Leesburg, Georgia, 12 miles north of Turner Field, Albany, Georgia, killing AAF instructor 2Lt. James H. Redding and seriously injuring RAF Flying Cadet Harry Eckersall. Instructor 2Lt. James S Boggs witnessed the accident and stated to investigators, "While taxiing back, southeast, for a take-off into the northwest, I saw an AT-9 making an approach to the end of the southeast-northwest runway.

From my position, the plane appeared to have down from one-half to three-quarters flaps and was making a normal approach to the end of the runway. The plane did not level off sufficiently, ran into the ground at a severe angle, and as a result, both landing gears gave way. The left gear gave way first then the right. From my vantage point, it appeared that both props were damaged with the left prop receiving the most severe impact with the runway. After making contact with the runway, the airplane bounced into the air and proceeded to climb. I gained radio contact with the pilot warning him that he damaged his landing gear and props. The plane continued to climb and about halfway down the runway it began to mush as if it were climbing with reduced power. At a position about 100 or 200 yards from the northwest end of the southeast-northwest runway, the airplane started a gentle turn to the left. The left wing dropped and the airplane went into a spin to the left from about 100 feet altitude completing half or three quarters of a spin before crashing and slowly catching fire."

9-18-42F. San Francisco, California. At 1055, a Lockheed P-38F crashed into San Francisco Bay, California, four miles north of San Francisco Airport, killing 2Lt. Colton R. Clark, Jr. The P-38 was observed approaching San Bruno Point at approximately 2,000 feet. The aircraft was then seen to begin a turn to the southwest toward San Francisco Airport. Without completing the turn and resuming level flight, the aircraft fell off on the port wing and into a dive. The aircraft was then seen to dive straight into the bay at a 60 degree angle in a wings level attitude 200 yards southeast of San Bruno Point with no apparent effort by the pilot to attempt a recovery. The impact of the airplane with the water produced a splash of approximately 100 feet. Investigators speculated that the pilot might have been switching fuel tanks and had become distracted, flying the airplane into the bay.

9-18-42G. Byron, Georgia. At 1525, a Vultee BT-13A suffered a catastrophic structural failure of the port wing and spun to the ground one mile east of Byron, Georgia, killing flight instructor 1Lt. Edgar S. Wardell, 22, Long Branch, New Jersey, and A/C A. N. Pendergrass, 22, Jefferson, Georgia. The aircraft was observed flying northwest when a loud report was heard. The aircraft was seen spinning to the ground minus its port wing, which fell to earth one-half mile from the scene of the crash. The aircraft exploded into flames upon impact with the ground. Investigation revealed that four lower outer wing panel attaching bolts on the port wing sheared off at the nuts, allowing the port wing to fold up and shear the upper wing panel attaching bolts, permitting the wing to separate.

9-18-42H. Huron Field, California. At 1615, a Vultee BT-13A stalled and spun into the ground while in the pattern at Huron Army Auxiliary Field, Huron, California, killing students Aviation Cadet John E. Hunter and Aviation Student Vernon M. Houlihan. The Accident Classification Committee stated, "The [pilots] had been up on a student team ride practicing instrument flying under the hood. They had completed the period and were returning to Huron Field. Two flights were operating from Huron Field, one flight using one-half of the field and one flight using the other half. Traffic was from south to north and A/S Houlihan was supposed to use the east half of the field. He apparently had entered traffic properly and had extended his base leg too far, which was noticed after he had started his gliding turn. In order to get back in his proper traffic lane he apparently steepened his bank to the right, holding backpressure on the stick and bottom rudder. Attempting to correct, the pilot applied opposite rudder still holding backpressure on the stick, which pulled the nose over high to the left and stalled. Apparently no throttle was used. The airplane hit the ground at about a 45 degree angle and burst into flames."

9-18-42I. Wendover Field, Utah. At 2000 PWT, student bombardiers A/C Charles C. Walker and A/C Harry W. Willits were burned to death and crew chief MSgt. Charles P. Chronowski was injured when a ground crew accidentally started a fire while attempting to fuel a Consolidated B-24D that was being prepared for a bombardier-training mission at Wendover Field, Wendover, Utah. Investigators stated, "The airplane was being refueled on completion of a mission and prior to take-off from a mobile service unit consisting of a tractor-truck and two 2,000-gallon trailers. [Note: This particular type of fueling trailer has two sets of hoses and reels on each side, one in the rear of the trailer and one forward of the wheels. The forward set of hoses and reels are referred to as the "side" reels.] Investigation reveals that the refueling crew habitually refueled from the left side of the trailers, which is always toward the nose of the airplane. This is done because the driver can more readily control the interval between the truck and the airplane. Because of this fact, the [right-hand] side [reel and hose] and [right-hand] rear reels and hoses are seldom used. In this particular instance, the nozzle on the right side reel of the [front] refueling trailer had been removed several weeks previous to the accident. The nozzle was defective and had been removed by a mechanic and turned in to the Sub-Depot for repair or replacement. Investigation shows that the crew chief demanded that 125 gallons of gasoline be placed in No. 4 tank and that 150 gallons be placed in No. 3 tank. Since the gasoline level in the forward trailer was higher than in the rear trailer, the refueling crew decided to gas both airplane tanks from the front trailer. In order to check the amount of fuel that was being pumped into the airplane tanks, it was necessary to use the pumps on both sides of the refueling trailer, there being only two meters on the refueling trailer, one on each side. This was contrary to normal procedure, [in that both front and back trailer units] are habitually used in order to determine from the [left] fuel flow gauges how much fuel is being pumped into the airplane. The refueling operator started his pump engine, and stated that he opened the valve that controls the right hose and reel. Investigation shows that the valve that controls the [front trailer right hand] side reel was open, as well as the valve that controls the right rear [reel]. The refueling crew pumped 125 to 150 gallons of gasoline into the airplane from each side of the [front] refueling trailer. Since the right hand side reel and hose had no nozzle attached and since the valve that controls it was open, 125 to 150 gallons of raw gasoline was pumped through the hose and onto the ground. This gasoline was ignited either through direct contact with the muffler of the refueling pump engine or through its vapors being blown to a smoke pot flare that was on a line with the prolongation of the fuselage and 40 feet away from the nose of the airplane. The gasoline was discharged from the open hose in such quantities that it rapidly spread out underneath the truck, trailer and airplane, setting [all] on fire. The two men who were in the nose of the airplane checking the installation of the bombsight were apparently trapped by the burning gasoline vapors.... After examination, evidence shows that valves on left side of trailer were closed, but the valves on right side of trailer were open on forward pump bellows, and the rear valve being opened accidentally caused the flow of gasoline through this forward right hand hose compartment, which having no nozzle attached, caused gasoline to flow on to the ground, from which it was ignited by a kerosene flare."

9-19-42A. Cherry Lake, Florida. At 1015, a Beech AT-10BH suffered a catastrophic structural failure of the tail section and crashed while attempting a peel-off maneuver one mile south of Cherry Lake, Florida, killing instructor 2Lt. Charles W. Ferguson, Jr., Tallahassee, Florida, and RAF Flying Cadet Victor Holman. The AT-10 was part of a three-airplane formation training flight. RAF Flying Cadet Thomas H. Hewett, flying the number-three airplane in the formation, witnessed the accident and later stated to investigators, "We followed the two other airplanes in a peel-off from an echelon formation, commencing at approximately 3,500 feet. We attained an airspeed of approximately 180 mph in the dive and kept rather higher than the other two airplanes during the descent. At a height of approximately 2,000 feet, I noticed small pieces of silver material emanating from the second airplane and gradually these pieces grew larger. I then realized that this airplane was coming apart, and finally it seemed to roll up into a ball with pieces still flying from it. Immediately when we realized what was happening to the second airplane, we pulled out of our dive. Thus we were well above the second airplane when it eventually plunged to the ground. After circling the scene for two or three times, noting that the pieces were spread over a large area of the field, we flew back to Rocky Ford Auxiliary Field and landed."

9-19-42B. San Francisco Bay, California. At 0715, a Consolidated B-24D suffering the failure of its number-two and three engines ditched into San Francisco Bay shortly after taking off from Hamilton Field, San Rafael, California. The crew was rescued and suffered various injuries but bombardier 2Lt. William R. Gunther, Paso Robles, California, was reported missing and presumed lost. Investigators stated, "Aircraft fully loaded with complete crew of eight men aboard, took off from Hamilton Field at 0705 PWT, in a southeasterly direction. The take-off was normal, radio contact was excellent and weather was clear with a slight haze. Aircraft after take-off headed in direction of Mt. Tamalpais where it was to meet with other aircraft and form for a flight to destination 'X' [Hickam Field, Oahu, Hawaii]. Before reaching Mt. Tamalpais, engines No. 2 and 3 missed and then completely cut out. The aircraft at this time was flying at about 1,500 to 2,000 feet. Pilot radioed flight leader that he was having engine trouble and that he was heading back. Aircraft then turned south and east. The aircraft lost altitude quickly and made a forced landing at about 140 mph in San Francisco Bay, midway between Angel Island and Treasure Island. The plane hit the water and bounded back into the air before nosing down and smashing back into the sea, damaging the front of the fuselage and also splitting the fuselage. Forced landing occurred at 0715 and plane sank at 0717. The number-two and number-three engines stopped because of a failure of the fuel system. It is probable that the fuel failure was caused by collapse of the fuel line from the bomb bay tanks. If take-off had been made on wing tanks as is proper and as is customary, this accident would not have occurred. It is believed that if the pilot had used maximum manifold and rpm on his remaining engines he probably could have sustained flight." Navigator 2Lt. William A. Walker, Rochester, New York, and radio operator TSgt. J. K. Nance, Moneta, Virginia, were seriously injured. Crewmembers receiving light injuries were: 1Lt. John P. Davis, pilot; 2Lt. Samuel D. Rose, co-pilot; SSgt. J. J. Durrett, engineer; TSgt. C. A. Dent, assistant engineer; Cpl. F. J. Mars, gunner. In March of 1943, Lt. Rose, as pilot in command, flew B-24D #41-24301, also known as the *Lady Be Good,* from Wichita, Kansas, to Soluch, Libya.

9-19-42C. Columbia, South Carolina. At 1027, a North American B-25D suffered the failure of the port engine and crashed while attempting a go-around at Columbia Army Air Base, Columbia, South Carolina, killing the crew of three. 2Lt. Robert E. Dales witnessed the accident and later stated to investigators, "The plane was approaching the field for a landing on the southeast runway. The plane leveled off about 50 feet off the ground. The pilot gave the plane the throttle in an effort to go around for another landing. The left motor did not catch and the plane rose in the air with the right motor pulling it around to the left. The plane climbed about 100 feet and dove at the field hitting on the left wing tip and then down on the nose. The plane then buckled in the middle just behind the wing section and the bomb bay. The right motor being thrown off clear of the plane." Investigators speculated that the port engine failed because the throttle had been retarded for an extended period on the approach. Killed in the crash were: 2Lt. Thomas M. Macteer, pilot; SSgt. Peter P. Conrad, co-pilot; SSgt. Alex W. Kaiser, engineer.

9-19-42D. Wrightwood, California. At 1015, a Vultee BT-15 crashed while maneuvering at low altitude at Wrightwood, California, killing pilot 2Lt. Alfo L. Baker, Wrightwood, California, and passenger Pfc. Howard B. Garlick. Investigators stated, "The plane passed over Wrightwood, where the pilot lived, at approximately 500 feet altitude. It turned and came back, losing altitude to approximately 150 feet. While making a [steep left] turn, the plane turned partially on its back, crashed and burned."

9-19-42E. Blairstown, New Jersey. At 1603, a Douglas C-47 flying in instrument conditions collided with Blue Mountain five miles west of Blairstown, New Jersey, killing seven fliers. The C-47 was on a ferry flight from Westover Field, Chicopee, Massachusetts, to the municipal airport at San Antonio, Texas. Civilian witnesses on the ground in the vicinity of the crash reported that it was very foggy at the time of the accident. The airplane was heard flying northwest at very low altitude just moments before smashing into 1,000-foot Kittatinny Ridge 150 feet from the top, exploding violently into flames upon impact. Investigators noted that the scene of the accident was 45 miles from the pilot's intended course. Newark, New Jersey, tower personnel reported that the pilot had called them and addressed them as "Philadelphia Tower." It was speculated that the pilot was lost and was attempting to establish his position when he flew the airplane into the ridge in a straight-and-level flying attitude. Further investigation revealed that the starboard engine might have failed prior to the accident. Two propeller blades on the starboard engine were found in the feathered position and were not twisted in the manner that would indicate that the engine was turning under power. The position rudder trim tab indicated that the starboard engine was not producing power. Investigators could not determine the nature of the engine failure or why the airplane was off course. Killed in the crash were: 2Lt. Leonard C. Cranford, Washington, D.C., pilot; 2Lt. William H. Smith, Bastrop, Louisiana, co-pilot; 2Lt. Harold, N. Seltzer, Long Island City, New York, navigator; Pfc. Gilbert Rosenberg, Louisville, Kentucky, engineer; Pvt. Harold W. Clegg, Pickford, Michigan, radio operator; Pvt. Francis A. Hallberg, Ogema, Wisconsin, assistant engineer; 1Lt. William H. Vail, East Orange, New Jersey, passenger.

9-19-42F. Red Bank, New Jersey. At 1740, a Lockheed P-38F flying in instrument conditions flew into the ground on the Henry L. Finch estate, Cooper Road at Headden's Corner, two miles north of Red Bank, New Jersey, killing pilot SSgt. Samuel P. Bradshaw, Nashville, Tennessee. The aircraft exploded violently into flames when it struck the ground, hurling wreckage over a large area of the estate. Visibility at the time of the accident was reported as poor. The pilot had been ferrying the P-38 from Long Beach, California, to Newark, New Jersey.

9-20-42A. Craig Field, Alabama. At 0345, a Cessna AT-17 on a night cross-country navigation flight crashed under unknown circumstances 24 miles south of Craig Field, Selma, Alabama, killing 2Lt. Harry F. Hawkins and 2Lt. Denny S. Durham. Weather at the time of the crash was reported as clear with ceiling and visibility unlimited. Heavy smoke and low scattered clouds were reported approaching from the west. The aircraft had been reported as missing and was not located until 9-22-42. There were no witnesses to the crash and the cause could not be determined.

9-20-42B. Dale Mabry Field, Florida. At 1010, a Bell P-39D that was part of a four-airplane formation crashed while maneuvering 12 miles south of Dale Mabry Field, Tallahassee, Florida, killing 2Lt. William M. Penick. After take-off at 0955, the formation climbed to 9,500 feet in a trail formation. Lt. Penick was piloting the number-three airplane. The formation leader began a series of lazy-8s and dived to an airspeed of approximately 270 mph and then began a pull-up. At the bottom of the dive, Lt. Penick's airplane stalled and was seen to flip backwards and then enter a normal type spin. At about 6,000 feet, the pilot released the port exit door. The airplane recovered from the spin and entered a steep dive. Condensation trails were seen forming on his wingtips as he pulled out. The aircraft then entered a flat, counter clockwise spin. The pilot exited the aircraft at an altitude of approximately 5,000 feet. He apparently deployed his parachute too early and it was fouled by the propeller. The airplane smashed into the ground and exploded, scattering wreckage over a wide area. The pilot struck the ground 20 feet from the wreckage.

9-20-42C. Greencastle, Indiana. At 1300, a Martin B-26B crashed into an abandoned golf course while maneuvering at low altitude two miles east of Greencastle, Indiana, killing the crew of three. The aircraft appeared over Greencastle at about 1245 and buzzed the town and the campus of DePauw University for fifteen minutes. The pilot attempted a vertically banked left turn at about 100 feet above ground level. The B-26 stalled, rolled through to an inverted attitude and crashed into the ground upside down, exploding into flames upon impact. Killed in the crash were: 2Lt. Thomas G. Murphy, 21, River Forest, Illinois, pilot; 2Lt. Dudley M. Steele, Stockton, California, co-pilot; Cpl. Norbert Charles Meyer, Green Bay, Wisconsin, engineer. Lt. Murphy and Cpl. Meyer were thrown clear of the exploding aircraft. It was discovered that Lt. Murphy was a graduate of DePauw and had been buzzing his former fraternity house.

9-20-42D. Monteagle, Tennessee. At 1210, a Vultee BT-13A flying in instrument conditions crashed in mountainous terrain near Monteagle, Tennessee, killing pilot 2Lt. Jack D. Sikes and passenger Pfc. Charles E. Hiltz. Apparently the pilot was lost and was flying low in an attempt to determine his position. Witnesses on the ground observed the aircraft flying up a valley and approaching rising terrain when the pilot attempted a steep right turn and flew into terrain.

9-20-42E. Plain Dealing, Louisiana. At 1400, a Martin B-26A crashed while practicing simulated engine failure procedures four miles west of Plain Dealing, Louisiana, killing three crewmen. Engineer Cpl. Norman O. Black, 24, Fort Worth, Texas, parachuted to safety. Killed in the crash were: 2Lt. Burton W. Basten, 23, Placentia, California, B-26 instructor pilot; 2Lt. Harold Kester, 24, Cory, Indiana, co-pilot and B-26 student (flying in left seat); SSgt. George W. Chandler, 23, Dallas, Texas, engineer. Cpl. Black reported that Lt. Basten had told him that he was going to test Lt. Kester with an unannounced simulated engine failure. Some minutes later, Lt. Basten cut the starboard engine. The airplane entered a steep spin to the right. The pilots struggled to gain control of the airplane but were not able to recover. Lt. Basten then ordered the men to bail out. Cpl. Black went to the bomb bay and waited for the doors to open. The airplane was pointing straight down and still spinning to the right. When the doors opened, he glanced back at the cockpit and observed the pilots still struggling with the controls in an attempt to recover the airplane. Lt. Basten was then seen by Cpl. Black attempting to leave the co-pilot's seat. The B-26 then entered a flat spin and Cpl. Black pushed himself away from the aircraft and parachuted to safety. The B-26 plunged to the ground and exploded into flames. Investigators speculated that neither Lt. Basten nor Lt. Kester was able to make the trim and power changes for single-engine operation in time to prevent loss of control of the airplane.

9-20-42F. Roscoe, California. At 1415, a Lockheed P-38G dove into the ground at the Hansen Reservoir, Roscoe, California, killing 2Lt. John A. Pfeifer. The airplane took off at 1330 from Glendale Airport with two other P-38s on a propeller feathering practice exercise. After the formation had reached an altitude of 10,000 feet, the two pilots and instructor feathered their propellers. After flying for 15 minutes on a single engine, the instructor gave the signal to the pilots to re-start their feathered engines. After complying, Lt. Pfeifer's P-38 was seen to enter a power dive. Witnesses on the ground reported that both engines seemed to be operating and at an altitude of about 1,000 feet the aircraft seemed to make a partial recovery from the steep power dive. The nose of the aircraft dropped abruptly and then the P-38 entered a vertical dive into the ground, exploding violently into flames upon impact. The aircraft was destroyed utterly with much of the wreckage being buried in the ground. Only small portions of the pilot's remains were recovered. Investigators speculated that the pilot might have blacked out in the partial pull-up or had inadvertently put the airplane in a diving maneuver from which he could not recover because of the tendency of the P-38 type airplane to suffer from severe control surface buffeting caused by compressibility. It was also speculated that the pilot might have been knocked unconscious by the cockpit canopy coming loose or separating during the dive, although no evidence of this could be found.

9-21-42A. Gowen Field, Idaho. At 0004, a Boeing B-17E suffering an in-flight fire in the number-four engine went out of control and crashed into an unoccupied B-17 parked on the ramp while attempting a landing at Gowen Field, Boise, Idaho, killing six fliers and seriously injuring four others. The B-17 was in line with the southeast runway on the approach at an altitude of 700 feet when the number-four engine caught on fire. The pilot cut the throttle to the engine and the co-pilot pulled the engine fire extinguisher handle but these efforts did not put out the flames. The aircraft continued to descend under control but still on fire. The control tower radioed the B-17 to advise it that it was on fire. At an altitude of approximately 150 feet, the B-17 went out of control and fell off to the right, barely missing buildings near the field and mushing into the flight line, colliding with a parked B-17E (41-2601). The two aircraft erupted in flames. Investigation revealed that the engine fire was caused by a fuel line becoming disconnected from the engine driven fuel pump, which in turn dumped raw fuel into the accessory section of the nacelle, being ignited by the heat of the supercharger or exhaust. It was speculated that the fire burned through the starboard aileron cables, causing the loss of control. It was noted that the pilot had not shut down the fuel pump and mixture control to the burning engine, causing fuel to continue to flow into the fire. Killed in the accident were: 2Lt. Martin J. Miller, New York, New York, navigator; 2Lt. Gilbert W. Planchon, Santa Ana, California, bombardier; Sgt. Edmond Buckanovitch, Norwood, Massachusetts, engineer; Cpl. Leroy G. Heintzleman, Allentown, Pennsylvania, assistant engineer; Pvt. Stephen O. Brown, Baltimore, Maryland, passenger; Pvt. Wilbert S. Hairiston, Winston-Salem, North Carolina, passenger. Seriously injured in the accident were: 2Lt. George H. Page, Buffalo, New York, pilot; 2Lt. James E. Smith, Los Angeles, California, co-pilot; Pvt. John A. Gore, hometown unknown, radio operator; Pvt. Postel A. Adams, Clinton, South Carolina, passenger. Privates Adams, Hairiston, and Brown were African-American airmen receiving maintenance training.

9-21-42B. Gulf of Mexico, Florida. At 1000, a Bell P-39D spun into the Gulf of Mexico four and one-half miles west of Tarpon Springs, Florida, killing 2Lt. Henry C. Garcia, New York. Lt. Garcia had been maneuvering in the area at about 9,000 feet with two other P-39s and a P-40 when he apparently stalled his airplane and spun in. None of the other pilots in the flight witnessed the accident or were even aware that it had taken place. Witnesses on the ground observed the aircraft in a violent spin. The aircraft had spun down a considerable distance when the tail was seen to separate from the airplane, hitting the water some moments after the airplane crashed into the water. A man and wife fishing in the area recovered the pilot's body immediately after the crash. Lt. Garcia stated to civilian and military dinner companions two days prior to the accident, "If a fellow just uses his head, he can always bring a P-39 out of a spin."

9-21-42C. Vinton, Louisiana. At 1400, a Martin B-26B suffered the catastrophic failure of its tail section and crashed into a shallow pond 14 miles southeast of Vinton, Louisiana, killing the seven fliers on board. The aircraft had taken off at 1315 from Barksdale Field, Shreveport, Louisiana, on a gunnery mission over the Gulf of Mexico. Civilian witnesses on the ground observed the airplane falling and spinning from several thousand feet minus its tail and with a considerable amount of debris trailing. Investigation revealed that one or both of the pilot's escape hatches had separated from the aircraft, impacting the tail and causing its failure. The aircraft went out of control, causing the progressive failure of various pieces including the main wheel doors and bomb bay doors, which were found among the tail section debris one-half mile from the main wreckage. Investigators could not locate the co-pilot escape hatch. The main wreckage smashed to ground in a small, swampy pond, exploding into flames and scattering wreckage over a large area. Witness statements were conflicting. Killed in the crash were: 1Lt. John E. Parson, Ardmore, Pennsylvania, pilot; 2Lt. Lewis H. Cohn, New York, New York, co-pilot; 2Lt. Guy O. Rowland, Alliance, Ohio, navigator; Cpl. Jaye F. Malone, engineer; Sgt. Charles A. Cannon, radio operator; Cpl. Wallace H. Gregory, gunner; Sgt. Paul R. Stocker, Boyertown, Pennsylvania, gunner. This accident was one of several fatal B-26 accidents in the United States and overseas in which the in-flight separation of the pilot's escape hatch and its impact with the tail have been thought to have caused the progressive failure of the tail section.

9-22-42A. Miami, Florida. At 0309, a Douglas B-18B taking off on an anti-submarine mission collided head-on with a Douglas C-49 that was taxiing toward it on the active runway at the 36th Street Airport, Miami, Florida, killing five fliers and injuring four others. The B-18 was at the west end of the East-West runway waiting for clearance from the tower to begin its take-off roll to the east. All aircraft taking off on tactical missions were under orders not to radio the tower for take off clearance but to make prior arrangements for visual signals and scheduling. The B-18 had been scheduled for a 0300 take-off and had been delayed clearance because an Eastern Airlines Douglas transport was taking off ahead of it. After the Douglas transport took off, the tower then cleared the C-49 (militarized version of the DC-3 and operated under contract for the AAF by Eastern Airlines) to taxi west on the active runway, turn around at the west end and await clearance for take off. As the C-49 taxied toward the end of the runway, the B-18 turned onto the runway and began its take-off, colliding head on with the C-49 approximately 150 yards from the west end of the runway. Both aircraft burst into flames and efforts to fight the fire were hampered by exploding machine gun bullets, bombs and depth charges. Killed aboard the B-18 were: 1Lt. Charles M. Green, 24, San Angelo, Texas, pilot; 2Lt. Newell B. Smith, 21, Tylertown, Mississippi, co-pilot; Sgt. Walter Boehm, 30, Methuen, Massachusetts, bombardier. Crewmembers injured aboard the B-18 were: 2Lt. Lester Weiss, 22, Yonkers, New York, navigator; Sgt. Gilbert H. Isenberg, 22, Corpus Christi, Texas, radio operator; Pvt. Wesley T. Johnson, 22, Conroe, Texas, assistant radio operator. The civilian crewmembers (all employees of Eastern Airlines) killed aboard the C-49 were Donald B. Johnston, 40, Coral Gables, Florida, pilot, and William F. Fortner, 22, Dothan, Alabama, co-pilot. Flight engineer John W. Bussey, 24, Deland, Florida, received serious injuries and burns. This crash prompted a congressional investigation.

9-22-42B. Perham, Maine. At 0745, a North American B-25C flying in instrument conditions suffered the catastrophic failure of its port wing and crashed five miles east of Perham, Maine, killing the crew of seven. The aircraft was part of a flight of eight B-25C aircraft en route to Gander, Newfoundland. The flight had landed at Presque Isle Army Air Base, Maine, to refuel and had planned on reassembling north of the field after take-off for the leg to Gander. Once the flight became airborne they encountered instrument conditions with ceilings as low as 600 feet and moderate fog. After flying around for some time with assembly becoming impossible due to the weather closing in on them, six of the B-25s returned to Presque Isle and landed safely. One aircraft departed the area and made it safely to Gander. The B-25 that crashed was found a short time later by a civilian who had witnessed the airplane fall out of the sky trailing debris. Investigators speculated that the pilot temporarily lost control of the aircraft in instrument conditions and stressed the port wing to the point of failure with an extreme pull-up maneuver in an effort to right his aircraft. When the wing failed and separated from the aircraft, it traveled from left to right and over the top of the fuselage, impacting the top turret and the starboard vertical fin and rudder assembly, causing them to separate from the aircraft. The aircraft struck the ground at a steep angle, exploding into flames and hurling wreckage over an area of 1,000 feet. Investigation revealed that the aircraft was over gross weight. Killed in the crash were: 2Lt. John F. Watson, pilot; 2Lt. John

W. Rives, Jr., co-pilot; SSgt. John S. Delano, bombardier; SSgt. James A. Kviz, engineer; SSgt. Eugene J. Crozier, gunner; SSgt. Richard K. Riddle, radio operator; Sgt. Frederick W. Rowbottom, passenger.

9-22-42C. Presque Isle, Maine. At 0946, a North American B-25C flying in instrument conditions suffered the catastrophic failure of its port wing and crashed seven miles NNE of Presque Isle Army Air Base, Maine, killing the crew of seven. The aircraft was part of a flight of eight that was headed for Gander, Newfoundland, and had stopped at Presque Isle to refuel. When assembly was attempted after take-off, instrument conditions were encountered and six of the airplanes returned to Presque Isle, all landing safely. One aircraft left the area and landed safely at Gander. One aircraft crashed after it suffered a similar structural failure shortly after take-off, killing its crew of seven (see 9-22-42B). At 0935, three of the six remaining B-25s began taking off in five-minute intervals. The accident aircraft took off second and soon encountered instrument conditions. Investigators speculated that the pilot temporarily lost control of the aircraft in instrument conditions and stressed the port wing to the point of failure with an extreme pull-up maneuver in an effort to right his aircraft. Investigators later stated, "The ship was heard in the clouds with the engines roaring and came out of the clouds at 200 feet above the field at an angle of about 35 degrees going north. The ship was then seen to make a sharp pull out. The left wing folded back, taking with it the empennage. The ship then made a half roll [to the left] and struck the ground upside down and exploded immediately on contact. The ship made contact with the ground at a sharp angle." Killed in the crash were: 2Lt. Ralph L Drogula, pilot; 2Lt. James Q. Crocker, co-pilot; SSgt. William H. Finch, bombardier; SSgt. Billy J. Hill, engineer; SSgt. George E. Simmons, radio operator; SSgt. Joseph Martino, gunner; SSgt. Lawrence A. Robinson, crew chief. When officers at the Presque Isle base learned of the accident, the three B-25s waiting to take-off were told to return to the line and their mission scrubbed. The other two B-25s that had taken off with the accident airplane made it safely to Gander. Investigation revealed that the airplane was over gross weight.

9-22-42D. Middletown, Pennsylvania. At 1241, a Lockheed P-38F stalled and spun into the ground after suffering the failure of its port engine just after take-off from the AAF Air Depot at Olmstead Field, Middletown, Pennsylvania, killing 2Lt. Herbert E. Smith. The pilot had been experiencing trouble with the port engine prior to take-off and he had returned to line to have the engine looked at. The spark plugs were changed and the engine run up and tested by the pilot. The pilot taxied to the runway, was cleared for departure and took off. The take-off appeared normal and the P-38 flew off with landing gear retracted. Witnesses on the ground reported that the aircraft never gained more than 300 feet of altitude after the take-off. A minute later the aircraft was seen coming back toward the field with the landing gear in the extended position. The pilot attempted a series of right turns over the town in an effort to get back to the field. The port wing was seen to drop and the aircraft winged over to an inverted attitude. The P-38 missed striking a school building by ten feet and just cleared a church steeple before plunging to the ground inverted and exploding into flames. The careening wreckage of the aircraft struck houses in the vicinity and burned down the home of Dr. O.H. Schwartz. The airplane was being ferried from Mines Field, California, to Newark, New Jersey.

9-22-42E. East Hartford, Connecticut. At 1515, a Curtiss P-40F crashed near Rentschler Field, East Hartford, Connecticut, killing pilot SSgt. Raymond R. Kroskiewicz. The pilot was coming in for a landing when he overshot the field and began to go around for another approach. He started to circle the field with his landing gear and flaps down, apparently near stalling speed. The pilot was keeping the nose up in a climbing attitude but he did not increase the power. He began a climbing left turn at an altitude of approximately 300 feet and stalled the airplane. The airplane rolled over, did one complete turn of a spin, and then plunged to the ground nose first. The pilot died a short time after the accident. He had only 1.5 hours in P-40 type airplanes.

9-22-42F. Decatur, Georgia. At 1520, a North American B-25C suffering the failure of the starboard engine crash landed in a wooded hollow between Ponce de Leon Avenue and Coventry Road, Decatur, Georgia, killing two fliers and injuring another. Four other crewmembers were uninjured. The aircraft was on a cross-country navigation training flight from Greenville, South Carolina, to Atlanta, Georgia, when the starboard engine quit due to a supposed fuel system problem. The B-25 instructor pilot (flying in the co-pilot seat) feathered the propeller and adjusted the power and trim for single-engine flight. A moment later, the port engine quit. The instructor re-trimmed the airplane for a non-powered glide and attempted a 180-degree left turn to make an emergency landing in an open area near a golf course. When he rolled out of the turn and realized that he could not make it, the instructor lowered the flaps and "held the nose up to stall the airplane in on the crash landing." The airplane stalled and smashed into the trees, shearing off both wings and erupting in flames. Killed in the crash landing were Cpl. Jack P. Lee, Macon, Georgia, passenger, and 2Lt. Robert S. Oberheiman, Norwood, Connecticut, passenger. Engineer SSgt. Claude E. Rieves received minor injuries. Crewmembers who escaped injury were: 2Lt. Harford P. Jenks, B-25 instructor pilot; 2Lt. Bud B. Binning, B-25 student; 2Lt. William R. Chandler, navigator; 2Lt. Michael A. Defeo, bombardier.

9-22-42G. Gulf of Mexico, Florida. At 1430 CWT, a North American AT-6A returning from an aborted target-tow mission stalled and spun into the Gulf of Mexico three miles southwest of Tyndall Field, Panama City, Florida, killing pilots Sgt. William Moore, 22, West Palm Beach, Florida, and Sgt. Frederick Heskes, 19, a Dutch sergeant pilot from Java. The aircraft was returning to Tyndall Field because the sleeve target it was towing had somehow become fouled on itself. Witnesses on the ground and in the air observed the AT-6 approaching the field at an altitude of approximately 500 feet and then enter a shallow and slow right hand spin, making two full turns. The aircraft appeared to level off momentarily before plunging into shallow water just off shore. Investigators could find no evidence that the tow target fouled the control surfaces of the aircraft and could not determine why the pilot was unable to maintain flying speed.

9-23-42A. Charlotte, North Carolina. At 0950, a Curtiss P-40F practicing low-level flying with another P-40F and a P-39D descended into the Catawba River ten miles southwest of the municipal airport at Charlotte, North Carolina, killing 2Lt. Frank C. Haggerty, Rockville Center, New York. The leader, flying in the P-39, led the P-40s in trail formation down to treetop level. The aircraft then flew at about 100 feet over the surface of the river and flew in a trail formation at this altitude for several minutes. The leader took the formation up to 500 feet to clear a bridge and power lines a mile ahead. After successfully clearing the bridge and wires, the leader then began another descent toward the river. Lt. Haggerty, apparently misjudging his position relative to the river, continued descending until his aircraft made contact with the water, bending the propellers and tearing away the lower engine fairing. The aircraft bounded into the air about 50 feet. The pilot tried to bank the aircraft toward shore but the aircraft slipped sideways to the left and struck the water in a left wing low, nose down attitude. Published accounts of the accident reported that Lt. Haggerty had gotten married on 9-19-42.

9-23-42B. Advance, Missouri. At 1758, a North American B-25D suffering the failure of its port engine crashed while attempting an emergency landing at the Army Air Field at Advance, Missouri, killing co-pilot SSgt. Melvin R. Woodward and seriously injuring pilot 1Lt. Walker L. Paine. The aircraft had taken off from Kansas City, Kansas, on a ferry flight to Memphis, Tennessee. When the B-25 was 25 miles north of Advance, the port engine failed. The pilot feathered the propeller and trimmed the aircraft for single-engine flight. The pilot arrived at the emergency airfield at Advance and circled the field. The pilot lowered the landing gear and made his approach, coming in too high. The pilot realized that he was going to overshoot and increased power for a go-around, climbing to approximately 75 feet when the starboard engine failed. The aircraft fell away to the left and smashed into the ground left wing low. Investigation revealed that the bottom cylinders of the starboard engine had blew out and it was speculated that the pilot might have mishandled his application of the flaps on the approach, causing him to overshoot the field.

9-24-42A. Newfoundland, New Jersey. At 0853, a Republic P-47B flying in instrument conditions crashed three miles east of Newfoundland, New Jersey, killing 2Lt. Edward J. Schreiber, Rochester, Minnesota. The pilot had been flying over Long Island for about an hour and headed into New Jersey, encountering deteriorating visibility and becoming lost. Witnesses on the ground observed Lt. Schreiber's P-47 under the overcast about a mile from where he crashed, flying east to west straight and level just over the treetops at which height he was breaking in and out of scud clouds, barely missing colliding with terrain and structures. The airplane attempted to execute a left turn but rolled over on its back and struck the ground inverted, bursting into flames and scattering wreckage over wide area near the Wanaque Reservoir.

9-24-42B. Florence, Arizona. At 0755 MWT, two Beech AT-11 aircraft collided in mid-air and crashed nine miles east of Florence, Arizona, killing eight fliers. Killed aboard AT-11 #41-27630 were: 2Lt. William B. Shea, 23, Kansas City, Missouri, pilot; 2Lt. Robert T. Ross, 22, Fort Huron, Michigan, co-pilot; Aviation Cadet Robert E. Cote, 19, Woonsocket, Rhode, Island, student bombardier; Aviation Cadet John H. Cwik, 27, Johnstown, Pennsylvania, student bombardier. Killed aboard AT-11 #41-27620 were: 2Lt. Donald J. Gibson, 24, Valley City, North Dakota, pilot; 2Lt. William P. Owen, 24, Magnolia, Arkansas, co-pilot; Aviation Cadet Wilbur C. Harter, 25, Delaware, Ohio, student bombardier; Aviation Cadet Matthew F. Farrell, 25, Lynn, Massachusetts, student bombardier. The investigators report stated, "While on a bombardier training mission, Lt. Shea and Lt. Gibson allowed their AT-11 aircraft to collide in mid-air at an altitude of 1,000 feet over bombing target E-6. Both planes were in flames as they fell earthward and contacted the ground about 90 degrees apart near the outer edge of the target. All eight crewmembers were killed. Lt. William P. Owen attempted to use his parachute but evidently pulled the ripcord before clearing the ship. His head was severed from his body by the stabilizer. From all indications his parachute was released from its pack. Apparently neither pilot saw the other in time to avoid the collision."

9-24-42C. Selfridge Field, Michigan. At 1522, a Curtiss P-40F on an engine modification test flight crashed near Selfridge Field, Mt. Clemens, Michigan, killing Packard Motor Car Company test pilot Edward A. Moore. The aircraft had been fitted with a Merlin engine. Investigators stated, "Pilot made a normal take-off using about one-half of the field. At 150 feet altitude, he put his wheels up. Between 350 and 400 feet, the engine suddenly quit. The pilot dropped his nose slightly and began a right turn back to the field. He then turned the plane back to the line of take-off. The ship lost flying speed and hit an ash tree and landed on its back in a road. It burst into flames immediately. As the fire destroyed almost all evidence, the cause of the accident cannot be determined. It is believed though that, as this was the first flight in a P-40 for this pilot, he might possibly have pulled the mixture control into idle cut-off position believing that he was pulling the prop control to reduce the prop rpm."

9-24-42D. Goodfellow Field, Texas. At 1700, a Vultee BT-13A dove into the ground 10 miles southeast of Goodfellow Field, San Angelo, Texas, killing Aviation Cadet Delmar M. Yoakum. The airplane struck the ground at angle of 70 degrees at extremely high speed. The airplane was observed by witnesses on the ground to dive straight into the ground from an altitude of 500 feet and with no apparent effort by the pilot to recover level flight. Investigators speculated that the dive probably began at a much higher altitude than that it was first witnessed.

9-24-42E. Petaluma, California. At 1550, a Bell P-39D stalled and spun into the top of a hill two miles southeast of Petaluma, California, killing 2Lt. Henry R. Jacobus. Lt. Jacobus was flying with three other P-39s in a Lufbery circle at 7,000 feet when the leader and the two other airplanes dove out of the circle and then executed a climbing left turn. Lt. Jacobus' P-39 fell out of this climbing turn in a stall and spin to the right, spinning down to about 4,000 feet where he momentarily recovered the airplane. The airplane then entered a spin to the right and spun into the ground, exploding into flames. The pilots of the other three airplanes were not aware that an accident had occurred.

9-24-42F. Waterford, California. At 1555, a student pilot abandoned a Vultee BT-13A after going into a spin nine miles east of Waterford, California. The student, Sgt. Eugene R. Glazier, Jr., 20, Bend, Oregon, was fatally injured when he was struck on the right leg by the airplane's rotating propeller. After parachuting to the ground still alive, the student was found by civilians in the area. Sgt. Glazier was conscious while he waited for an ambulance and made this statement to investigators one hour and forty-five minutes after the accident, "I went into a spin at 9,000 feet and jumped out of the airplane at 2,500 feet while it was still spinning. The propeller hit my leg." Sgt. Glazier was transported to Stanislaus County Hospital and died later that day.

9-24-42G. Gunter Field, Alabama. At 2000, a Vultee BT-13 suffering an apparent engine failure crashed just after take-off one and a half miles northwest of Gunter Field, Montgomery, Alabama, killing RAF Flying Cadet Charles N. Downs. Witnesses on the ground reported that airplane appeared to be trying to make it back to Gunter Field and was observed losing altitude in a wide turn back to the field. The aircraft stalled and dove into the ground, exploding into flames upon impact.

9-25-42A. Orlando, Florida. At 0925, a Douglas P-70 suffering the failure of its starboard engine crashed after overshooting the runway at the Army Air Base at Orlando, Florida, killing 1Lt. James H. Stewart and seriously injuring radar operator Pvt. J.S. Blankenship. The P-70 arrived over the field with the starboard propeller feathered with the airplane under control and flying satisfactorily. The pilot extended the landing gear and began his approach. The pilot did not retard the throttle enough, failed to lose airspeed and overshot the field, never getting lower than 25 feet. He retracted the landing gear and advanced the throttle in an attempt to go around again. He climbed to about 400 feet and began a turn into the bad engine. The P-70 lost altitude rapidly and fell off on the starboard wing, plunging to the ground and exploding into flames.

9-25-42B. Moody Field, Georgia. At 1015, a Beech AT-10BH crashed while attempting a take-off from Moody Field, Valdosta, Georgia, killing A/C Robert J. Hodgen and A/C Donald I. Ekstrom. Immediately after taking off the AT-10 began an extremely steep climb, losing airspeed. At an altitude of approximately 150 feet the aircraft stalled and fell off on its port wing, diving into the ground and exploding into flames. Investigation revealed that the pilots had neglected to unlock the flight controls prior to commencing the take-off.

9-26-42. Rimer, Ohio. At 1042, a Martin B-26B flying in instrument conditions suffered a catastrophic structural failure of the starboard elevator and horizontal stabilizer and crashed two miles north of Rimer, Ohio, killing the crew of seven. Investigators speculated that the pilots had lost control of the aircraft in the instrument conditions (low ceilings and light rain showers) and, in an effort to right the aircraft, had stressed the starboard elevator to the point of failure in an extreme pull-up maneuver. The aircraft was seen to emerge from a 2,000 foot overcast in a spin and then to dive straight down. The aircraft then entered another spin (to the left), oscillating between a flat spin and a nose down type. The starboard elevator was seen to fall from the aircraft and was found along the line of flight 200 feet from the point of initial impact. The aircraft dove into an open field with tremendous force and exploded into flames, scattering wreckage for over 300 yards. The navigator's astrodome had separated from the aircraft and was found over ten miles northwest of the crash scene on a farm near Grover Hill. It was determined that the astrodome had not struck any part of the aircraft after it had separated. Other pilots had reported in flight separations of the navigator's astrodome and engineers at Wright Field, Dayton, Ohio, soon solved the problem. The B-26 had taken off from Baer Field, Indiana, for Presque

Isle, Maine, on its way "to an active theatre of operations." Killed in the crash were: 2Lt. E.L. Newton, Kansas City, Missouri, pilot; 2Lt. Fred Bice, Tuscaloosa, Alabama, co-pilot; TSgt. A.J. Lamison, Three Springs, Pennsylvania; SSgt. P.J. Nelligan, Santa Rosa, California; Pvt. R.D. Risetter, Radcliff, Iowa; Pvt. O.R. Colestock, Hecia, South Dakota; Pvt. A.A. Wildt, Broadmead, Oregon.

9-27-42A. Clearwater Bay, Florida. At 1038, a Martin B-26B crashed into Clearwater Bay, Florida, one and a half miles west of Clearwater Beach Pier, killing five fliers and seriously injuring another. The crew of a U.S. Coast Guard patrol boat observed the B-26 flying from north to south just above the waves, which were estimated to be 6-7 feet. The airplane was then seen to climb to about 300 feet, momentarily level off and then dive straight into the bay at a shallow angle. Witnesses on the shore stated that the airplane appeared to be "standing still" before it plunged into the sea. The airplane disappeared under the water immediately. The Coast Guard boat located an oil slick that proved to be the crash site. The area was buoyed and dragged with the body of one of the pilots being retrieved. A special crash boat with divers arrived and attempted to locate additional bodies but rough water prevented any further attempt to retrieve the dead crewmen. Gunner Pvt. Charles E. Stuart, who was seriously injured, was rescued one mile north of the crash site by another Coast Guard patrol boat operating in the area. Investigators could not determine the cause of the crash and speculated that some kind of structural or material failure might have caused the accident. The B-26 was returning from a practice bombing and gunnery mission. Killed in the crash were: 2Lt. William T. Howell, pilot; 2Lt. Leland D. Yerkes, co-pilot; 2Lt. Marvin J. Smith, bombardier; SSgt. Edward W. Smith, radio operator; Pvt. Phillip B. Kreischer, engineer.

9-27-42B. Linthicum Heights, Maryland. At 1305, a North American BC-1A crashed while performing low altitude aerobatics at Linthicum Heights, Maryland, killing pilot 1Lt. Luther W. Cartwright, Jr., 22, Linthicum Heights, and passenger Sgt. Eli B. Williams, Tuskegee, Alabama, student mechanic from the experimental Tuskegee unit. Sgt. Williams was alive when he was pulled from the wreckage but died a short time later. Wreckage was scattered over an area of approximately 100 yards in a pasture a few miles from the home of the pilot's parents. Investigators stated, "Evidence indicates that in attempting a recovery from an inverted position at from 100 to 150 feet the ship lost altitude, the wing striking the ground and the airplane cartwheeling into a field, killing the pilot and passenger. No fire accompanied the crash."

9-27-42C. Atlantic Ocean, North Carolina. At 1705, a Lockheed A-29 on an anti-submarine coastal patrol mission crashed into the Atlantic Ocean 150 yards east of Cape Lookout, North Carolina, killing the crew of five. Witnesses on the ground observed the airplane flying east along the shore at an altitude of approximately 1,500 feet. The aircraft was then seen to dive straight into the sea. An oil slick, two empty life rafts, life vests, parachutes, and various debris were found floating on the surface but the wreckage was never positively located. Navy divers from Norfolk, Virginia, arrived on scene soon after but early the next day deemed the ocean conditions as too dangerous to conduct dives. The bodies of the pilot and the navigator were later recovered on the surface by Coast Guard patrol boats operating in the area. The search for the wreckage and the bodies of the three missing crewmembers was discontinued 10-2-42 because of deteriorating ocean and weather conditions. The aircraft was known to be carrying a full load of depth charges. Various witness statements were conflicting and investigators could not determine what caused the airplane to dive into the water. Weather was not considered a factor in the accident. Killed in the accident were: 2Lt. James G. Barr, pilot; 2Lt. Eugene J. Flood, navigator; Sgt. Ralph W. McDonnough, engineer; Pfc. Alan E. Duke, radio operator; Sgt. Charles G. McCleary, gunner.

9-28-42A. Plant City, Florida. At 1058, a Martin B-26B on a practice bombing and gunnery mission suffered a catastrophic structural failure of the tail section and the port wing, crashing six miles northeast of Plant City, Florida, and killing the crew of six. Witnesses on the ground observed the aircraft flying northeast at an altitude of about 2,500 feet when it entered a small bank of dark clouds. As the B-26 emerged from the clouds, numerous pieces of the tail section were seen separating and trailing behind the aircraft. The aircraft then entered a rolling dive, causing the failure of the port wing due to extreme loading. The rolling dive also caused the bomb doors and other pieces to separate from the aircraft prior to impact. The aircraft then stopped its rolling motion and entered a 45-degree dive toward the ground. The aircraft continued in the dive, striking a tree before slamming into the ground and exploding violently into flames. Wreckage was scattered over a wide area. Investigation revealed that the co-pilot's escape hatch had separated from the aircraft and struck the vertical fin, causing it to separate. The resultant vibration facilitated the progressive destruction of the remaining tail surfaces. The co-pilot's escape hatch and the vertical fin were never located. Weather was not considered a contributing factor in the accident. This is one of several fatal B-26 accidents in the United States and overseas in which one or both of the pilot's escape hatches separate in flight and strike the tail section, causing its progressive failure. Killed in the crash were: 2Lt. Francis Hallen, Flushing, Long Island, New York, pilot; 2Lt. Joseph W. O'Brien, Detroit, Michigan, co-pilot; 2Lt. Robert W. Carr, Homer, Indiana, bombardier; SSgt. Steve Sopp, Hammond, Indiana, radio operator; Pvt. Richard A. Griffith, Stony Creek Mills, Pennsylvania, engineer; Pvt. Chester H. Guzicki, South Bend, Indiana, gunner.

9-28-42B. Fort Morgan, Colorado. At 1310, a Taylorcraft L-2A stalled and spun into the ground seven miles east of Fort Morgan, Colorado, killing civilian flight instructor Walter Kitson, 36, and passenger Wesley E. Marsh, 16, both of Fort Morgan. It was later learned that Marsh was not authorized to be aboard the airplane. Investigators speculated that the pilot was unable to arrest the fatal spin because the passenger's left foot had become lodged in the rudder controls. Marsh had been discharged from the U.S. Marine Corps because of sickness a few weeks prior to the accident.

9-29-42A. Tiverton, Rhode Island. At 0825, a Curtiss P-40F collided in mid-air with a U.S. Navy Grumman JF-1 amphibian bi-plane while maneuvering near Tiverton, Rhode Island, killing the P-40 pilot and two passengers aboard the navy aircraft. The pilot of the JF-1, Lt. Cmdr. Clarence H. Hawkins, Wichita, Kansas, parachuted to safety but was seriously injured. The P-40 was flying southeast near Stone Bridge in the number-two position of a two-airplane tactical formation at 2,000 feet when the formation overtook the navy airplane flying at the same altitude. The lead P-40 began a gentle turn to the left in attempt to fly the formation clear of the navy airplane. The number-two P-40, piloted by 2Lt. Robert A. Marsh, was flying slightly below and about 100 yards to the right of the leader. The leader turned past the JF-1 but Lt. Marsh's P-40 collided with amphibian, striking it in the tail and the starboard wing. Both aircraft went out of control and plummeted into the Sakonnet River. Two of the three-man crew of the amphibian bailed out, but only Lt. Cmdr. Hawkins reached the surface alive, being rescued by a small fishing boat. The navy crewmen killed in the collision were Radio Mechanic 2nd class Homer D. McClendon, Providence, Rhode Island, and Aviation Machinist Mate 3rd class James H. Harris, Bridgeton, New Jersey.

9-29-42B. Spence Field, Georgia. At 1119, a North American AT-6A participating in a three-airplane Lufbery circle stalled and spun into the ground near Spence Field, Moultrie, Georgia, killing Aviation Student Benjamin Jenner. Investigators stated, "The student evidently applied excessive back pressure on the stick and his ship went into a spin at about 1,700 feet. The ship made two and one-half turns before it hit the ground. The ship exploded and burst into flames."

9-29-42C. Eglin Field, Florida. At 1145, a North American B-25D crashed into a beach after colliding with the sea 12 miles southwest of Eglin Field, Florida, killing five airmen and seriously injuring another. The aircraft was on a "low flying over water" training mission and had been executing simulated torpedo attacks prior to the accident. Pilot SSgt. B. W. Sherman was towing targets in the area and witnessed the accident. He later stated to investigators, "I observed [the B-25] about 5 miles out at a very low altitude approaching the shore at a 45 degree angle into the northwest. The prop wash left streaks on the water. I saw him hit the water 50 feet from the shoreline. [The B-25] came out of the water and staggered then skidded on the beach until it hit the sand dunes and then went end over end, then fell to pieces. I estimate that this aircraft was flying at one or two feet above the surface of the water." It was speculated that a downdraft or a large wave "tripped" the aircraft or that the pilot misjudged his position relative to the surface and flew the airplane into the water. Killed in the crash were: 2Lt. Leonard H. Woodworth, Ipswich, South Carolina, pilot; 2Lt. Ardith E. Crites, Charleston, West Virginia, co-pilot; TSgt. Floyd H. Fortenberry, El Paso, Texas, navigator; Cpl. Allen F. St. Claire, La Harpe, Illinois, engineer; SSgt. James W. Calvo, Mount Sterling, Illinois, gunner. Gunner Pvt. George A. Watkins, Woolrich, Pennsylvania, was seriously injured.

9-29-42D. Holtsville, Alabama. At 1622, a North American AT-6A on an instrument proficiency flight crashed into wooded terrain near Holtsville, Alabama, killing pilots 2Lt. William J. Wood and 2Lt. Frederick W. Crockett. The aircraft was seen circling the area at low altitude prior to the crash. There were no witnesses to the crash itself. The aircraft struck the terrain at an angle of 35 degrees at a high rate of speed, severing trees on the way down and exploding into flames upon impact.

9-30-42A. Hondo, Texas. At 1552, a Lockheed A-28A apparently suffering engine trouble just after take-off from the Hondo Army Air Field, Hondo, Texas, crashed three and a half miles ENE of the East/West runway, killing the crew of four. The aircraft was observed flying at about 200 feet above ground level with the engines sounding as if they were "spitting." The aircraft lost altitude as it continued its line of flight, coming very close to the ground when the engines appeared to "take hold again." The aircraft had climbed to an altitude of about 200 feet before entering a stall and falling off to the right and entering a spin. The airplane spun about three or four times before striking the ground and exploding into flames. Investigators could not determine the nature of the engine trouble. Killed in the crash were: Capt. John G. Rafferty, pilot; 1Lt. John P. Stewart, co-pilot; SSgt. Leland W. Bray, engineer; Cpl. Ervin E. Schield, passenger.

9-30-42B. Burbank, California. At 1531, a Lockheed P-38G on a maiden factory test flight crashed into parked aircraft while attempting a take-off from the Lockheed Air Terminal, Burbank, California, killing the pilot and a bystander on the ground. Two other bystanders were injured. Witnesses observed the P-38 lift off the runway and when the wheels were about two feet off the surface, the pilot retracted the landing gear. Before the landing gear retracted fully, the port wing was seen to dip slightly and the port propeller struck the runway, causing a loss of power. The P-38 became partially airborne (about 10 to 15 feet) before descending back into the runway left wing low. The P-38 smashed into the ground and then bounded back into the air with the port engine on fire. The P-38 struck the ground again then careened to the left, cartwheeling into four Ryan PT-22 airplanes (parked with eight others and waiting to be fueled) before crashing into and coming to rest in the hangar of Pacific Air Motive Company. The P-38 and three of the PT-22s exploded into flames, the fourth being smashed to pieces. Civilian flight instructor Edward C. Wike, 22, Hemet, California, was killed instantly while he was standing near the Pacific Air Motive Company hangar, being struck by the out of control P-38, which was piloted by Lockheed Aircraft Company pilot William M. McConnell, 46, San Fernando, California, who was also killed. John Waide, 30, Hemet, California, flight instructor, was lightly injured and Harold Keefe, 22, Hollywood, California, mechanic, was seriously injured.

9-30-42C. Columbus, Mississippi. At 2115, a Cessna AT-17 flew into terrain under control and with power one mile west of the Army Flying School at Columbus, Mississippi, killing Aviation Cadet Frank W. Peat and Aviation Cadet Arthur E. Reisman. The airplane had just taken off from the field on a scheduled night training flight. There were no witnesses to the crash, which was spotted from the air almost immediately. Investigation revealed that the airplane did not spin in but flew into the ground in a flying attitude, severing trees before bursting into flames upon impact.

9-30-42D. Garwood, Texas. At 2130, a Curtiss O-52 crashed 10 miles west of Garwood, Texas, killing pilot 2Lt. Sam M. Baker and passenger Lt. (j.g.) Leendert G. Coomans, U.S. Navy. The Accident Classification Committee stated, "The pilot and observer were on a night navigation training mission to proceed east out of Red Airway #32 to light #11 and return to Brooks Field [San Antonio, Texas]. The airplane crashed about one-quarter of a mile north of light #11 heading in a westerly direction. It struck the ground at about a 30-degree angle and at a high rate of speed while making a left turn. Both occupants were killed instantly. The airplane was exactly on course and on schedule at the time of the accident. Even though weather conditions were contact, other pilots reported that the atmosphere was hazy and the horizon indistinct. It is the opinion of the Committee that the accident was caused when the pilot started making his turn around light #11. Up to that time, he had been flying straight and level and had been orientating himself by means of lights on the ground. However, when he started his turn, he lost his horizon and failed to refer immediately to his instruments. He went into a spiral dive and struck the ground."

October

10-1-42A. San Antonio, Texas. At 1015, a Beech AT-10BH on a formation training flight dove into the ground one mile southeast of Auxiliary Field #3, 16 miles WSW of Kelly Field, San Antonio, Texas, killing A/C James A. Brown and A/C Dewey C. Brown, Jr. The AT-10 was flying in the number-three position of a three-ship formation that had peeled off in preparation to landing at the auxiliary field. The aircraft banked to the right in the peel-off, rolled to an inverted position and dove into ground, bursting into flames upon impact. Investigation revealed that that the starboard aileron fitting failed, preventing the free movement of the aileron. The material failure was a progressive type and subsequent inspection of other AT-10 airplanes at Kelly Field revealed the same type of failure evident on a number of the airplanes. It was noted by investigators that the aileron fitting was impossible to inspect without tearing away the aileron fabric. All AT-10 type aircraft were temporarily grounded after this accident.

10-1-42B. Selma, Alabama. At 1100, a Bell P-39D stalled and spun into the ground while maneuvering 15 miles northwest of Selma, Alabama, killing Aviation Cadet Harry H. Fuller. Investigators stated that the pilot had attempted a slow roll at an altitude of approximately 3,000 to 4,000 feet and stalled the airplane in an inverted position, causing it to enter a tumbling end over end spin along the pitch axis. The aircraft then entered a flat spin and crashed to the ground. The pilot was seen to leave the aircraft just above the treetop level and was found a short distance from the wreckage. His parachute had not been deployed. Examination of the wreckage revealed that the port door was open and still attached to the aircraft and the starboard door was secured and locked. By pulling an emergency release handle, either door on the P-39D type aircraft can be jettisoned to allow the pilot to leave the aircraft in an emergency. It was speculated that the pilot apparently struggled to exit the aircraft by opening the port door in the usual fashion rather than jettisoning the door with the emergency release handle.

10-1-42C. Antigo, Wisconsin. At 1230, a Piper L-4A stalled and spun into the ground six miles southwest of Antigo, Wisconsin, killing student pilot Pvt. C. W. Stanford. The aircraft entered a steep climb at low altitude after executing an apparent simulated forced landing. The L-4 stalled, fell off to the right and crashed to the ground in between a barn and a grain silo, barely missing colliding with both.

10-1-42D. Sequoia Field, California. At 1055, two Ryan PT-22 type airplanes collided in mid-air eight miles east of Sequoia Field, California, killing civilian flight instructor Edward J. Hedrick and student A/C James C. Schwindt aboard PT-22 #41-20658. A/C Mike Mumolo was killed in the crash of PT-22

#41-20661. The aircraft collided at an altitude of approximately 600 feet while both were turning. The aircraft collided with terrific force and locked together, falling to the ground on the Berry Foster Ranch. Investigators faulted both pilots for failing to see and avoid the other.

10-1-42E. Foster Field, Texas. At 2200, a North American AT-6A crashed just after take-off from Foster Field, Victoria, Texas, killing instructor 1Lt. Hugh C. Keith and student Aviation Cadet Jay R. Stevens. The airplane was seen to climb to about 500 feet after take-off and then enter a left turn. The aircraft then whipped over to the right and flew into the ground in a diving right turn. It was speculated that the pilot lost the horizon in the turn and flew the airplane into the ground. It was noted by investigators that the instructor was proficient in instrument flying. The ceiling was estimated to be about 500 to 600 feet with clear and unlimited visibility on top at 1,000 to 1,100 feet but the airplane was not seen to enter the clouds and was constantly in view of observers on the field from the time of take-off to the moment it crashed.

10-2-42A. Old Tampa Bay, Florida. At 1015, a Curtiss P-40F crashed into Old Tampa Bay while maneuvering eight miles northeast of Pinellas Field, St. Petersburg, Florida, killing pilot SSgt. Harold C. Kern. Witnesses on the ground observed the airplane recovering from a steep dive, leveling off at an altitude of approximately 5,000 feet and flying in an easterly direction. The airplane then entered a 45-degree dive toward the surface, turned over to inverted and dove straight down into the bay. The crash was so terrific that it produced a 150-foot splash. Investigators could find no evidence of mechanical or structural failure and could not determine why the pilot was unable to maintain control of his aircraft.

10-2-42B. Fairmont, Oklahoma. At 1510, a Vultee BT-15 practicing spins crashed five miles SSW of Fairmont, Oklahoma, killing instructor 2Lt. Gareth F. Gilson. Aviation Cadet Louis D. Berger parachuted to safety. Investigators stated, "[Lt. Gilson ordered] Cadet Berger [to] put the ship into a left-hand spin, using full back stick and full left rudder. To recover, he applied full right rudder and forward stick with the result that the spinning stopped. Upon attempting to neutralize the rudder, he found it jammed in the full right position. He immediately shook the stick as a signal for his instructor to take the controls and opened the [canopy]. By this time Lt. Gilson had gotten the ship into a medium turn to the right under increased power but was unable to maintain level flight. He attempted to contact the Enid [Oklahoma] control tower by radio, saying that the right rudder was jammed full on and that the occupants were about to jump. He then talked to the student, without switching to interphone, in an effort to produce concerted action against the rudder to free it. After three or four minutes he abandoned this attempt and ordered the student to jump. Berger jumped from the right side of the ship at approximately 1,500 feet altitude. The airplane entered a right-hand spin soon after Berger cleared it, continued to spin until it crashed into the ground and burned. Lt. Gilson cleared the airplane when it was below 500 feet and fell along side it, striking the ground about 120 feet west of the ship. When he struck the ground, his parachute canopy and half the length of the shroud lines were out of the pack."

10-2-42C. Cle Elum, Washington. At 1503, a Taylorcraft L-2A crashed near the airport at Cle Elum, Washington, killing pilot SSgt. Roger C. Larson and student Cpl. Joseph A. Morley. SSgt. Larson was taking off from the field on a local student flight. Investigators stated, "He got off the ground after using one-third of the runway. Halfway down the runway at an altitude of 75 to 100 feet he made a left hand 90-degree turn. He leveled the plane partially and started a second left hand 90-degree turn down wind at about the same altitude. About halfway through the second turn, Sgt. Larson was seen to be smiling. He banked the airplane into a very steep turn. The plane slipped to the left and hit the ground about 200 feet from the airport hangar. The engine was wide open. The airplane skidded about 20 feet then came to rest. In about 20 seconds the gasoline tanks ignited. The airplane burst into flames."

10-2-42D. Morris Field, North Carolina. At 1807, a Lockheed-Vega B-34 crashed after the failure of the port engine one and a half miles southwest of Morris Field, Charlotte, North Carolina, killing three crewmembers. The aircraft was on final approach for a landing at the field when the pilots had trouble getting the landing gear down. They radioed the tower that they were going to go around and try to get the landing gear straightened out. The airplane pulled up and began a turn to the left to re-enter the downwind leg. The port engine failed and the pilots feathered the propeller and notified the tower. The pilot was unable to continue his left turn back toward the field because he did not want to turn into the bad engine. Unable to maintain flying speed in level flight, the pilot apparently tried to mush the aircraft into a stand of 40-foot trees near the field. A heavily wooded area consisting of 75-80 foot trees surrounded the small stand. The airplane stalled, the starboard wing dropped and the airplane executed about one-quarter turn of a spin, striking the ground in a near vertical bank to the right, cartwheeling some distance before coming to rest against large trees and bursting into flames. Killed in the crash were: Joseph F. Cannon, Jr., Charlotte, North Carolina, civilian ferry pilot; J. R. Ibing, civilian ferry pilot (flying as co-pilot); 2Lt. Jenkins M. Robertson, Charlotte, North Carolina, AAF Air Traffic Control Officer, Love Field, Dallas, Texas, passenger. The aircraft was being ferried from the factory in California to Norfolk, Virginia, and had taken off from Love Field, Dallas, Texas, the day before. Lt. Robertson came aboard at Dallas to fly to Charlotte, North Carolina, where his wife and two children were living and the next day was to drive with them back to Dallas to relocate.

10-2-42E. Hondo, Texas. At 1858, a Lockheed A-28A taking off on a navigation training flight crashed into the ground two miles south of the AAF Navigation School at Hondo, Texas, killing eight fliers. The pilot was on an A-28 check-off flight with a rated A-28 check pilot on board. The Accident Classification Committee stated, "According to testimonies of several witnesses it appears that the airplane took off to the south on the 170 [degree] runway climbed for a short distance straight ahead and then made a gentle bank and turned to the left. It apparently continued in a wide turn through 90 [degrees], climbing slightly and then gradually rolling into a steeper bank and coming about in a turn of approximately 180 [degrees] from the direction of take-off. It appears that from this turn, gradually becoming steeper and tighter, the airplane rolled smoothly through the vertical, onto its back, and on over through a complete roll, continuing its rolling motion until the left wing started down a second time and then the airplane went into a tight, steep spin to the left and continued spinning, making one complete turn or more, until it struck the earth [bursting into flames]. It is not possible to determine the exact moment of occurrence, but apparently as the airplane went into the first roll there was a sudden and complete stoppage of both engines." Killed in the crash were: 1Lt. Milton A. Blaul, Jr., pilot; 1Lt. Roger Berg, co-pilot; Capt. Jack E. Sward, passenger; 2Lt. Donald E. Taylor, navigator instructor; 2Lt. Woodrow J. Vareika, student navigator; SSgt. Gerald E. Morrison, sergeant pilot and passenger; Sgt. Joseph G. Smithey, engineer; Alex Keresuk, civilian meteorology instructor.

10-2-42F. Union City, Oklahoma. At 2045, a Vultee BT-13A crashed into the ground in a diving turn to the right near Union City, Oklahoma, killing 1Lt. Robert A. Strauch. At 2020, the airplane had taken off from Oklahoma City with two other BT-13A aircraft for a ferry flight to Midland, Texas. Shortly after take-off, the other two pilots lost sight of Lt. Strauch and proceeded to Midland where they landed safely. It was not known why the pilot had left his flight or what caused him to fly the airplane into the ground.

10-2-42G. Sheridan, Texas. At 2059, a Curtiss AT-9 crashed five miles southwest of Sheridan, Texas, killing student pilots Pfc. Theodore G. Sneider, Hawarden, Iowa, and Pfc. Ernest B. Thomas, Jr., Dawson, Georgia. The airplane had been reported missing after it failed to return from a night navigation mission to Yoakum, Texas. A civilian woman and her son heard the airplane crash and mentioned it to a neighbor a couple of days later. He alerted officers at Kelly Field, Texas, and a search of the area was conducted. The airplane was found 10-6-42 by a Civil Air Patrol pilot. The AT-9 struck the ground at a steep angle and at high speed. Cadet Thomas was thrown clear of the wreckage. Cadet Sneider left the airplane at low altitude and was not able to deploy his parachute before striking the ground. Weather was not considered a factor in the accident and other airplanes on the same mission all landed safely after its completion.

10-2-42H. Greenville, Mississippi. At 2152, a Vultee BT-13A suffering an engine failure and attempting a forced landing crashed near the Army Flying School at Greenville, Mississippi, killing Aviation Cadet Harry E. Smith. The engine failed at an altitude of about 600 feet while the pilot was maneuvering in the vicinity of the field. The pilot alerted the tower of his engine failure but was unable to make the field for the emergency landing and collided with the east bank of Deer Creek, 15 yards from the west boundary of the field. Investigators speculated that the student might have inadvertently placed his fuel selector valve lever between the "left" and "off" positions or had experienced a carburetor problem of an unknown origin.

10-3-42A. Memphis, Tennessee. At 1000, a Vultee V-72 stalled and crashed while attempting an emergency landing at the municipal airport at Memphis, Tennessee, killing civilian pilot Trell M. Manley. The pilot had declared an emergency and entered the traffic pattern. He was observed making what appeared to be a normal approach but aborted the landing and attempted to go around. The airplane was then seen to enter a steep climbing attitude from which it stalled. The airplane fell off to the right from an altitude of about 50 feet. The pilot's head was out of the cockpit when the crash occurred and a light mist was seen emanating from the cockpit area by witnesses on the ground. Investigators speculated that the pilot was overcome by gasoline fumes that entered the cockpit due to raw gasoline pooling beneath the cockpit after a fuel line failure. Investigators noted that several V-72 type airplanes had been involved in similar incidents without loss of life.

10-3-42B. Rice, California. At 1020, a Bell P-39D stalled and crashed into the ground while turning from downwind to the base leg three miles south of Rice Army Air Field, California, killing pilot, 2Lt. Rex A. Shirk. The aircraft did not spin but dove straight into the ground. The aircraft caught fire soon after the crash and enlisted men who arrived at the crash scene could not rescue the pilot because of intense flames. The pilot had been on a local transition flight.

✓ **10-3-42C. Verden, Oklahoma.** At 1420, a Fairchild PT-19A crashed while performing a simulated forced landing five miles southwest of Verden, Oklahoma, killing civilian flight instructor Edward B. Haymaker, Caney, Kansas, and student A/C Horace L. Markland, Jr., 21, Waco, Texas. The airplane failed to clear a tree adjacent to the field it was shooting and struck the tree 10 feet from the top. The instructor, riding in the rear seat, attempted to regain control of the airplane but was unable to do so. The aircraft smashed into the ground about 150 feet from the tree it had struck.

10-3-42D. Will Rogers Field, Oklahoma. At 1431, a North American B-25C crashed while attempting a landing at Will Rogers Field, Oklahoma City, Oklahoma, killing three crewmen and injuring two others, one seriously. Investigators stated, "This airplane was flying approximately at 1,600 to 1,700 feet above the ground, below a 3,000 foot overcast with excellent visibility and was in the traffic [pattern] for a landing. There was evidently a propeller failure on the right motor causing the motor to turn at an excessive speed. This propeller was feathered and the left motor apparently also developed propeller trouble and indicated an excessive speed, with oil and smoke pouring out, and a lack of power. This caused the airplane to loose speed and altitude rapidly. The pilot decided that the airplane was getting beyond control and gave the signal to the crewmembers directly behind him and to the co-pilot to abandon ship. A few seconds later, the co-pilot and the pilot abandoned the ship through the top hatch. The airplane traveled a short distance and struck the ground almost in a vertical position, completely demolishing the ship. There was no resulting fire. The pilot and the co-pilot both successfully used their parachutes. There was one partially opened parachute suspended in the trees about 200 feet from the wreckage, and another about 75 feet from the wreckage. This committee was unable to determine whether these two parachutes and their wearers attempted to jump or were thrown clear of the wreckage. However, parts identified as belonging to three bodies were in the wreckage of the airplane." Killed in the crash were: TSgt. Samuel H. Bradford, Nashville, Tennessee, engineer; Pvt. Dale E. Fortune, Indianapolis, Indiana, gunner; Pvt. William Sittinger, Long Island, New York, passenger. Co-pilot 1Lt. George C. Kimsey, 28, Pilot Mound, Iowa, was seriously injured and pilot 2Lt. John K. Stevenson, 25, Indianapolis, Indiana, received minor injuries parachuting to safety.

10-3-42E. York, Alabama. At 1600, a North American AT-6A suffering an engine failure crashed while attempting a forced landing two miles south of York, Alabama, fatally injuring RAF Flying Cadet D. J. Calder. The AT-6 was seen to circle the area three times after the engine failure in an apparent attempt to find a place to land. The pilot found a clear area and attempted to land but aircraft smashed into the ground violently. The airplane did not burn. The pilot, who died later in the day, had been on a scheduled navigation training flight.

10-3-42F. Akron, Ohio. At 1856 EWT, a Martin B-26B suffering an engine failure on take-off crashed one-half mile west of the Municipal Airport at Akron, Ohio, killing seven fliers. Killed in the crash were: 2Lt. Claude R. Jackson, 22, Akron, Ohio, pilot (see 8-27-42C); 2Lt. Albert L. Spickers, co-pilot; 2Lt. William S. Holm, navigator; 2Lt. Ralph C. Shrigley, 23, Rootstown, Ohio, pilot-rated passenger; 2Lt. Thomas F. Schofield, Providence, Rhode Island, pilot-rated passenger; TSgt. John T. McDonald, engineer; SSgt. James J. Goluby, Allentown, Pennsylvania, radio operator. The B-26 had taken off from Baer Field, Fort Wayne, Indiana, on a navigation and transition flight, landing in Akron to refuel. A few of the crew were from the Akron area and had arranged beforehand for friends and relatives to meet them at the airport. Lt. Jackson's father, Capt. R. L. Jackson, Akron Fire Department, sister Joyce, and several other relatives and friends had bid the pilot farewell just minutes before witnessing the crash. Mr. and Mrs. Ralph J. Shrigley were there to visit with their son Lt. Shrigley and also witnessed the accident. Prior to take-off, the pilots had some difficulty starting the starboard engine. The pilot taxied out to the northeast end of the northeast/southwest runway and spent approximately five minutes running up the engines. The pilot was cleared for take-off and began his run to the southwest. After traveling about one-third of the runway, the starboard engine began malfunctioning. The pilot continued the take-off and finally pulled the airplane into the air at the end of the runway, immediately retracting the landing gear and beginning a gentle right turn to avoid rising terrain to the southwest. The aircraft began a climbing right turn to about 200 feet above ground level when the starboard engine failed. The starboard wing dropped and the B-26 fell off to the right and started to spin, striking the ground on its nose, left wing and engine, in a partially inverted attitude. The airplane struck the ground on the top of a 100-foot ridge and exploded into flames as the friends and family of the crew looked on in horror. An unidentified female relative of one of the crew had fainted and had to be carried away from the area. Some relatives and friends attempted to get to the accident scene but were kept away by local authorities. Investigation revealed that the starboard engine failure was caused by a gasket failure of the impeller shaft feed valve, "which permitted a surge of raw gasoline to by-pass and enter the blower section." There was evidence of a similar failure of the port impeller seal and reduced power in the port engine.

10-4-42A. Location Unknown. At 0124 EWT, the crew of a Martin B-26B radioed a distress call and announced that they were abandoning their aircraft. Radio operators in Raleigh, North Carolina, Spartanburg, South Carolina, and St. Louis, Missouri, heard the distress call. The crew did not report their position, hampering efforts to locate and rescue them. Despite a seven-day search effort, the B-26 and the crew were never seen or heard from again. The aircraft had taken off from MacDill Field, Tampa, Florida, at 1930 EWT on a cross-country navigation training flight to MacDill Field via Orlando, Jacksonville, Tallahassee, and Cross City. Missing and presumed lost at sea were:

2Lt. Arthur Young, Fall River, Massachusetts, pilot; 2Lt. Walter R. Pridemore, Fort Worth, Texas, co-pilot; 2Lt. James J. Degan, Philadelphia, Pennsylvania, navigator; Sgt. Carl L. Fountain, El Dorado, Arkansas, engineer; Cpl. Harvey H. Donahue, Crookston, Nebraska, radio operator.

10-4-42B. Wayne, Nebraska. At 0820, a Consolidated B-24D on an instrument calibration test suffered a triple engine failure and crash-landed two miles southeast of Wayne, Nebraska, killing two crewmembers that neglected to parachute to safety. The aircraft took off from the Sioux City Army Air Base, Iowa, at 0600, climbed to an altitude of 3,500 feet, and then circled the area of the air base until sunrise. The crew then began the calibration mission. The B-24, still flying at an approximate altitude of 3,500 feet, began experiencing roughness in the number-three engine. The pilots feathered the propeller and shut down the engine after it began to vibrate badly. After the engine shut down, the airplane dropped down to about 2,000 feet and then the number-two engine suddenly failed. The pilots feathered the propeller and after losing some altitude climbed a little bit to allow the crew to bail out. The number-one engine quit moments later. The pilot, satisfied that the crew had left the airplane, attempted a crash landing. The pilot picked out a field, made an approach and set the airplane down in a tail-low, gear-up attitude with no flaps at a speed of approximately 130 mph. The B-24 settled in and smashed itself to pieces as skidded along the ground, scattering wreckage over an area of 200 yards. Pilot 1Lt. Max J. Sanny and co-pilot 2Lt. William H. Jones received only minor injuries in the crash landing. Only after the bomber came to rest and the pilots extricated themselves from the wreckage were they aware that navigator 2Lt. James B. Sleigh and bombardier 2Lt. Howard C. Austin failed to bail out and were crushed to death when the bomber smashed into the ground. Receiving minor injuries parachuting to safety were: SSgt. Roy F. Beyler, radio operator; SSgt. Joe G. Herzog, engineer; SSgt. Jay S. Hansen, assistant engineer; Sgt. Alfred H. Williams, assistant radio operator; Sgt. Paul L. Berry, tail gunner. Examination of the wreckage by AAF investigators and engineers from Consolidated Aircraft Company revealed that severe carburetor icing caused the three engine failures.

10-4-42C. Brunswick, Georgia. At 1200, a Vultee V-72 crashed while attempting a take-off at Malcolm McKinnon Airport, Brunswick, Georgia, fatally injuring pilot 2Lt. Richard P. Hoffman and gunner Pvt. James V. Howlin, who died moments after the accident. The V-72 was unable to clear 100-foot trees at the end of the runway, striking them at about the 60 foot level. The airplane careened through the trees for 150 feet before coming to rest and catching fire. Pvt. Howlin was able to pull himself out of the wreck and made a short statement to rescuers before dying at the crash site. Lt. Hoffman's body was found half out of the airplane, suggesting that he attempted to crawl out of the airplane before he too died. Investigators deemed the field inadequate for use with V-72 type equipment and it soon became off limits to all high performance army airplanes.

10-4-42D. Point Reyes, California. At 1630, a Bell P-39D suffering an in-flight engine fire dove into a ridge two miles west of Point Reyes, California, killing pilot 2Lt. Ogden Gharing Wheeler. The airplane had been part of a six-airplane flight that had been maneuvering in the area at 2,000 feet (above sea level) on a formation-training mission. The airplane left the formation and other members of the flight reported seeing flames emitting from the starboard side of the engine compartment. One of the pilot's escape doors was seen to leave the airplane at an altitude of about 150 feet above ground level. The P-39 then entered a 45-degree dive toward the ground with no apparent effort by the pilot to level out or leave the airplane. The airplane smashed through trees before crashing into the ground and bursting into flames. Investigation of the wreckage revealed that an engine backfire caused pieces of backfire screen to enter the manifold and blower system causing the engine fire.

10-5-42A. Mather Field, California. At 0915, Aviation Cadet Hugh B. Campbell was killed when he was flung from a Ryan PT-22 while maneuvering at low altitude eight miles southwest of Mather Field, Sacramento, California. The cadet apparently fell out the airplane at less than 500 feet and managed to pull his ripcord but he struck the ground before his parachute could fully deploy. The airplane was seen to continue to fly normally for a mile before it dove into the ground.

10-5-42B. Bennettsville, South Carolina. At 1340, a Stearman PT-17 practicing "vertical coordination exercises" fell out of control and crashed to the ground near Bennettsville, South Carolina, killing student Aviation Cadet William R. Blair and seriously injuring his civilian instructor J. R. Hayworth. The airplane fell off to the left after encountering rough air while in a left vertical bank at low altitude. The airplane descended out of control and struck trees with the port wing, causing it to crash into the ground and burst into flames. The instructor, who could not remember the events preceding the crash, was rescued before the airplane became completely engulfed by fire.

10-5-42C. Church Point, Louisiana. At 1738, a Fairchild PT-19A suffering an apparent engine failure stalled and spun into the ground at Church Point, Louisiana, killing A/C William G. Browne, Baltimore, Maryland. The aircraft was seen to enter a spin to the right at about 3,500 feet. At about 300 feet, the airplane was seen to straighten out momentarily before it dove into the ground. The airplane did not catch fire.

10-6-42A. Greenville, Mississippi. At 0528, two Vultee BT-13A aircraft collided in mid-air one-half mile east of the Army Air Field at Greenville, Mississippi, killing student A/C Ernest J. Wiesmann, 21, Norwood, Massachusetts (BT-13A #41-11189), and student 2Lt. James H. Williams, Branford, Connecticut (BT-13A #41-11506). Investigators determined that Cadet Wiesmann overtook Lt. Williams and collided with his aircraft from the right rear and below at an altitude of 700 feet while in the pattern for the airfield at Greenville. The propeller of Cadet Wiesmann's aircraft destroyed the starboard elevator, starboard horizontal stabilizer, and the rudder of Lt. Williams' aircraft. Both airplanes plunged straight to the ground and burst into flames. The aircraft were on a local night training flight.

10-6-42B. Granger, Wyoming. At 1010, a Bell P-39L on a cross-country ferry flight suffered an apparent engine failure and crashed while attempting an emergency forced landing 18 miles SSE of Granger, Wyoming, killing pilot Capt. William Edwin Davis, Detroit, Michigan. It was speculated that the airplane stalled at low altitude while making an approach to a small field for an emergency landing, striking the ground right wing down with the landing gear retracted. The aircraft tumbled along the ground and scattered wreckage for almost 400 feet before the fuselage came to rest inverted and burst into flames 135 feet from the point of impact. The port wing was found 120 feet beyond the fuselage. Examination of the propeller indicated that it was not rotating under power at the time of impact. There were no known witnesses to the accident. The airplane had been reported as missing and was located 10-12-42 by search aircraft.

10-6-42C. Camp Stewart, Georgia. At 1610, a Lockheed-Vega B-34 on a target-towing mission crashed at Camp Stewart, Savannah, Georgia, killing five fliers and seriously injuring another. A sergeant pilot, with one hour of flying time on the B-34, was at the controls under the supervision of the squadron commander. There is no co-pilot station on the B-34, so the commander could not readily take over control of the airplane without changing seats with the student at the controls. While towing targets at an altitude of 1,400 feet for an anti-aircraft battery, the B-34 engines quit due to mismanagement of the fuel system by the pilot. The airplane descended rapidly and then began a left turn in an effort to crash land in a small clearing near the edge of the target area. The airplane lost flying speed and descended into heavily wooded terrain, crashing through the trees and exploding into flames. Investigators noted that the fuel system on the B-34 is somewhat complex and speculated that the pilot failed to switch to a full tank in time to restart the engines and prevent the airplane from crashing. It was also noted that the airplane had been flying too low to allow the commander to take over the controls and prevent the accident. Killed in the crash were: SSgt. William L. Ineson, 23, St. Petersburg, Florida, pilot and B-34 student; Capt. George M. White, 26, Edenton, North Carolina, Commanding Officer, 4th Tow Target Squadron, Hunter Field, Georgia, pilot observer; SSgt. Charles L. Giddens, 23, Tampa, Florida, sergeant pilot observer; SSgt. John N. Geotis, 24, Madison, New Jersey, engineer; SSgt. Adolph A. Budryk, 24, East St. Louis, Illinois, crewmember. The target reel operator, Pfc. Roland E. Anson, Reno, Nevada, was seriously injured.

10-6-42D. Valparaiso, Florida. At 1816 CWT, a North American AT-6A stalled and spun to the ground after suffering an engine failure near Auxiliary Airfield #5, ten miles northwest of Eglin Field, Valparaiso, Florida, killing pilot Capt. Anthony D. Piccolo and passenger 1Lt. Sam Futrovsky. The aircraft had taken off earlier in the day on a cross-country flight to Atlanta, Georgia, and had returned to land. As the aircraft circled the edge of the airfield, the pilot attempted a slow roll at an altitude of approximately 200 feet. When the aircraft became inverted, the engine quit. The pilot was unable to return the aircraft to a normal flying position and it dove into the ground in an inverted attitude, bursting into flames.

10-7-42A. Neptune, Tennessee. At 1055, a Douglas A-24A crashed while maneuvering near Neptune, Tennessee, killing pilot 2Lt. Jack Kennedy and passenger Sgt. Fred H. Holasek. Witnesses in the air observed the A-24 pull up from a dive in a steep, climbing turn to the left. The airplane then rolled to inverted. The pilot could not recover the aircraft and return to a normal flying position and the aircraft descended into the ground in a flat, inverted attitude from an altitude of approximately 1,500 feet.

10-7-42B. Selma, Alabama. At 1700, a Bell P-39D entered a tumbling spin along its pitch axis and crashed 12 miles northwest of Craig Field, Selma, Alabama, killing Aviation Cadet Christopher Lambert. The Accident Classification Committee stated, "It is believed that the pilot pulled the airplane into a stall at a safe altitude (10,000 feet). The airplane apparently stalled, slid backwards, and then went into a forward tumble. From this tumbling, the airplane then went into a flat spin [both inverted and normal]; and from the flat spin into a normal spin. In these events the airplane lost approximately 5,000 or 6,000 feet. Then from an altitude of about 4,000 feet, the airplane was spinning normally to the left and apparently the pilot over controlled and recovered from the left spin and then went into a right spin. The pilot, then seeing the futility of recovery, decided to abandon the aircraft. In so doing, he released the left door, which was to the outside of the spin and jumped, however, he was apparently struck by the tail of the aircraft or released his parachute too quickly through underestimating his altitude which was then approximately 2,000 feet. As a result, several panels were torn from the canopy of the parachute, which caused the parachute not to open.

It is believed that in the event that the pilot was not struck by the aircraft, had he delayed opening the parachute a few seconds longer, a safe jump could have been made." Cadet Lambert's body was found about 50 yards from where the aircraft struck the ground.

10-7-42C. Lubbock, Texas. At 2105, Pvt. Alvin F. Callahan, a ground crewman at South Plains Army Flying School, Lubbock, Texas, was killed when he walked onto the active runway and into the path of a Vultee BT-13A landing at the field. The BT-13 struck him with the port wing, fatally injuring him. Pilot Capt. Louis P. Egger and his passenger, who were on a glider tow mission, were not injured in the incident. The BT-13 had dropped the tow cable on the edge of the field prior to landing and in preparation of towing another glider. Pvt. Callahan and another enlisted man had been ordered to proceed to the other side of the field in order that they would be in position to attach the glider tow cable to the airplane when it taxied to a halt. Both enlisted men were warned to be alert for airplanes landing at the field. The other enlisted man crossed the runway successfully. Pvt. Callahan, who was lagging behind his partners, was last observed walking slowly with his hands in his pocket, his head down, looking neither left or right.

10-8-42A. Bastrop, Louisiana. At 1220, a Fairchild PT-19A stalled and spun to the ground while performing aerobatics near Bastrop, Louisiana, killing pilot 1Lt. Robert A. Broadie and passenger George D. Rhodes. The aircraft was maneuvering at about 1,900 feet with another PT-19A and the other pilot observed Lt. Broadie slumped over in the cockpit after pulling out of an extreme maneuver. The aircraft was then seen to enter an extreme nose up attitude, stall and flip to the inverted. The airplane then entered an inverted flat spin. After falling for a short time, the aircraft entered a normal spin to the right. The aircraft continued to spin slowly to the right until striking the ground. There was no apparent effort to recover the airplane or any attempt by either the pilot or the passenger to parachute to safety.

10-8-42B. Palm Springs, California. At 1530, a Douglas A-20B suffering the simultaneous failure of both engines crashed while attempting an emergency forced landing 10 miles north of Palm Springs, California, fatally injuring pilot Capt. Donald E. Brown, 25, Los Angeles, California. Capt. Brown, the son of the screen, stage and radio comedian Joe E. Brown, was on a cross-country ferry mission and had taken off from Palm Springs Airport with several A-20 type aircraft just minutes before the crash. The aircraft were cleared to Las Vegas, Nevada, and Ogden, Utah, to an undisclosed location east. After Capt. Brown and his wingman climbed to an altitude of approximately 1,500 feet above ground level, both of his engines failed and he attempted a wheels-up emergency landing on the desert. The aircraft executed a 180-degree turn and flew south for about 30 seconds before turning 90 degrees to the east. The A-20 was seen to descend with decent speed and then level off, flying at an altitude of about three feet above the ground for a short distance. The aircraft then contacted the terrain with excessive speed and slid across the desert for about 1,000 feet before striking a series of washes and some boulders, flipping over on its back and bursting into flames. A passing AAF Vultee BT-13 landed nearby after the crash to render assistance. Capt. Brown was found crawling about ten feet away from the airplane. The AAF men pulled Capt. Brown further away from the burning wreckage. Capt. Brown told the airmen that both engines quit and that he did not know why. He said that he was having trouble breathing and that he could not see. One of the airmen at the scene attempted artificial respiration but was unsuccessful in relieving Capt. Brown's discomfort and he died at approximately 1550 of a skull fracture and severe internal injuries. Investigators could not conclusively determine the cause of the double engine failure.

10-8-42C. Bruce Field, Texas. At 1742, a Fairchild PT-19A dove into the ground four and one-half miles SSE of Bruce Field, Ballinger, Texas, killing A/C Charles R. Foster. He fell to his death when he bailed out at an altitude of 100 feet, too low to allow his parachute to deploy. The PT-19 was seen in a steep dive at an altitude of about 1,800 feet. The airplane continued in this attitude until it struck the ground and exploded.

10-9-42A. Rapid City, South Dakota. At 0003, a Boeing B-17E on a landing approach collided with rising terrain three miles north of Rapid City Army Air Base, South Dakota, killing three crewmembers and injuring three others, one seriously. Two additional crewmembers escaped injury. The B-17 had turned from the downwind leg to the base leg far north of the usual turning point. The B-17 then turned from base to final. The pilot leveled from the turn too soon and was off on his approach to the North-South runway, making it necessary to make a slight left turn to line the airplane with the runway. The pilot stated later that while descending on the approach the runway and field lights disappeared from view but the town lights were still visible. The pilot, realizing he lost the field lights because of rising terrain, added full power and pulled back on the controls in an effort to gain altitude and clear the ridge in his flight path. The B-17, with landing gear extended, struck the ridge 20 feet from the top with the starboard wing and number-four engine, coming to rest on top of the ridge after smashing through trees, catching on fire and breaking in two at the after bomb bay section. Navigator 2Lt. Edwin W. Smith and radio operator Sgt. Harold F. LeFever were killed instantly in the crash. Gunner Sgt. James F. Hackney survived the crash but his leg was pinned under the fuselage wreckage and he was unable to move. Two crewmembers trained fire extinguishers on the approaching flames while other crew tried to free

Sgt. Hackney, who was standing on one leg. The crew was unable to free Sgt. Hackney's leg and he was burned to death. Investigation revealed that the pilots had never made a day landing at the air base and were not familiar with the terrain surrounding the field. As a result of this accident, lights were eventually installed on the ridge. Pilot 2Lt. Joe L. Rutan was seriously injured; co-pilot 2Lt. Roy S. Kline and assistant radio operator Sgt. Donald Stephens received minor injuries. Engineer SSgt. Leonard E. Scott and assistant engineer Sgt. Arthur Stern were uninjured.

10-9-42B. White Hall, South Carolina. At 1200, a North American B-25D crashed into a heavily wooded area near White Hall, South Carolina, killing the crew of three. The B-25 struck the ground at a steep angle, exploding violently into flames and scattering wreckage over a wide area. There were no witnesses to the crash and the aircraft was destroyed utterly. The cause of the accident could not be determined. The aircraft had taken off from Walterboro Army Air Field, Walterboro, South Carolina, (11 miles north of the crash site) on a local flight to locate the bombing practice range in the area. Weather in the vicinity at the time of the crash was reported as clear with unlimited visibility. Killed in the accident were pilot Capt. Charles T. Ceronsky, co-pilot 2Lt. Ronald A. McDonald, and engineer SSgt. Michael R. Poinson.

10-9-42C. San Francisco Bay, California. At 1059, a Bell P-39L dove into San Francisco Bay, California, while performing aerobatics, killing 2Lt. Lionel Melendez. Lt. Melendez had been flying a practice formation flight with other P-39s at an altitude of 6,000 feet when he and his wingman executed a peel off maneuver down to an altitude of approximately 2,000 feet. Lt. Melendez was seen to be climbing just after the maneuver. He was then observed to execute a half-roll and perform a split-S type maneuver, diving into clouds that were estimated to top out at about 1,500 feet. A civilian woman on the ground at Alameda reported that she had seen the aircraft dive into the bay 2,000 feet from shore at high speed with no apparent effort by the pilot to pull out or use his parachute.

10-9-42D. Ritter, South Carolina. At 1500, a North American B-25C on a practice bombing mission crashed near Ritter, South Carolina, killing the crew of four. Civilian witnesses on the ground reported that the airplane was flying in a nose-high attitude about 700 feet above ground level. The B-25 was then seen to stall and enter a spin, the airplane spinning about three times before diving into the ground nose first and exploding into flames. Investigators speculated that the pilot probably attempted a steep, climbing turn with low airspeed, causing the stall and spin at an altitude too low to allow recovery. Killed in the crash were: 2Lt. Albert E. R. Broyles, pilot; 2Lt. Don J. Butler, co-pilot; 2Lt. John Bugra, bombardier; Pfc. William L. Young, engineer.

10-9-42E. Godman Field, Kentucky. At 1710, North American P-51 performing low altitude aerobatics stalled and spun into the ground at Godman Field, Fort Knox, Kentucky, killing 1Lt. William M. Harmon. William H. Kunz, a civilian employee of North American Aviation Company, witnessed the accident and later stated to investigators that he observed Lt. Harmon dive at the field, level out and then perform a series of snap rolls. "[The] pilot once again snap rolled, but came out wavering. The ship then went into a spin. The pilot then recovered from the spin at about 500 feet, but was too low to bring the ship out of the dive. This was Lt. Harmon's first flight in a P-51 type aircraft. He was in the air approximately 20 minutes."

10-10-42A. Snowflake, Arizona. At 0808, a Vultee BT-15 on an administrative flight flew into the ground 20 miles east of Snowflake, Arizona, killing pilot 2Lt. Harold H. Burwell. "It is the opinion of the Aircraft Accident Classification Committee that Lt. Burwell caused the airplane to strike the ground with full flying speed. From observations made at the scene of the accident, the initial contact was made at an angle not exceeding ten degrees as there was no unusual gouging of earth at that point; however, it may easily be assumed that the airplane turned over end over end and proceeded through the brush and over the rough and stony ground for a distance of about 50 yards, the area scattered with fragments and broken parts."

10-10-42B. Valley Springs, Texas. At 1001, a Vultee BT-13A on a cross-country navigation training flight crashed six miles northwest of Valley Springs, Texas, killing student pilot Sgt. Odell N. Edmonds. Investigation revealed that Sgt. Edmonds was returning to his station at Curtis Field, Brady, Texas, on time and on course when the accident occurred. The aircraft struck the ground at a steep angle and at flying speed, exploding violently into flames upon impact and killing the pilot instantly. Investigators could find no evidence of material or structural failure and no witnesses to the accident could be found. Weather was not considered a factor in the accident. The cause of the accident could not be determined.

10-10-42C. Blythe, California. At 1345, a Douglas DB-7B on a gunnery-training mission flew into the ground 20 miles south of Blythe, California, killing pilot SSgt. Irving B. Kibler. The Accident Classification Committee stated, "The pilot, while practicing air-to-ground fixed forward firing on the gunnery course, was evidently attempting to align his sights on his assigned target and became so intent on his aim that he failed to realize how low to the ground he had come. The propeller of his port engine hit the ground along the line of targets causing the airplane to tumble towards the left. The airplane crashed through a tree in a wash directly behind the line of targets and first hit the ground about 75 yards to the left and rear of the target line. At this point the airplane completely disintegrated with parts of the airplane scattered for about 150 yards forward of this point. The body of the pilot was still strapped in the parachute harness and the parachute had broken open and was draped over one engine." Investigators speculated that the pilot might have used one eye in an attempt to line up with the target in the ring (gun) sight, losing depth perception and flying into the ground. It was recommended that all pilots be advised to use both eyes when lining up on a target using the ring sight.

10-10-42D. Ballou, Wisconsin. At 1616, a Taylorcraft L-2A stalled and spun into the ground while maneuvering nine miles northwest of Ballou, Wisconsin, killing student pilot Pvt. Gerard A. Dempsey.

10-11-42A. Topeka, Kansas. At 0230, a Consolidated B-24D crashed three miles west of the Army Air Base at Topeka, Kansas, killing the crew of eight. The B-24 took off from Topeka Army Air Base at 0204 on a bombing and gunnery training mission, and was to land at the Army Air Field at Rapid City, South Dakota. The airplane began to climb to altitude, reaching an altitude of approximately 7,000 feet when the number-one engine exploded and caught fire. The pilot shut down the engine and feathered the propeller and then contacted the tower at the air base for emergency landing instructions. The pilot received landing instructions and attempted a landing, which he overshot. The pilot aborted the landing and attempted to go around. The B-24 made a 180-degree left turn (into the bad engine) and attempted to gain some altitude. The airplane then began experiencing trouble with another engine (speculated to be number-two engine). Then the engine that had been shut down again erupted in flames. The pilot ordered the crew to clip on their parachutes and ready themselves to jump. It was speculated that the crew chief opened the bomb bay doors, creating additional drag on the airplane. The B-24 began to settle and reached an altitude too low for the crew to make a safe parachute jump. The B-24, with the number-one engine a mass of flames, descended into the ground on a road bank, shearing off the landing gear. The B-24 then bounded back into the air and smashed through some telephone poles, continuing for another 100 feet before smashing into the far bank of a small stream and exploding violently into flames. The airplane was carrying bombs and machine gun ammunition, which detonated in the resulting fire and hampered fire fighting and rescue personnel. Engineer SSgt. James C. Price, Trenton, Georgia, was found near the main wreckage badly burned but alive. He also had numerous broken bones, including a broken leg and internal injuries. Sgt Price was able to make a statement to army investigators before eventually dying of his injuries 10-12-42 at 0430. It was reported by ground crewmen that the pilot had trouble with the number-one engine prior to take-off. Pieces of engine cowling believed to be from the B-24 were found 2 to 3 miles from the crash site along the airplane's line of flight. Investigation revealed that the fire and failure of the number-one engine was caused by the failure of the number-two cylinder of the forward bank cylinders. Killed in the crash were: 2Lt. Ralph M. Dienst, 26, Pasadena, California, pilot; 2Lt. Charles W. Fischer, Jr., 22, Boston, Massachusetts, copilot; 2Lt. James H. Edwards, 24, Berkeley, California, navigator; 2Lt. James W. Holmes, 24, hometown unknown, California, bombardier; SSgt. Merlin O. Brager, 20, Madison, Wisconsin, assistant engineer; SSgt. Max Barron, 24, Pleasant Hills, Louisiana, radio operator; SSgt. Willis T. Barnes, 34, Columbus, Georgia, gunner; SSgt. Aubry H. Large, 23, Francis, Oklahoma, gunner.

10-11-42B. Clanton, Alabama. At 1930, a Vultee BT-13 crashed into a tree five miles east of Clanton, Alabama, killing Pvt. Jack M. Nolen, Montgomery, Alabama. Pvt. Nolen, who was not a rated pilot, took off from Gunter Field, Montgomery, Alabama, on an unauthorized flight at 1845 CWT. He told another mechanic who had seen him near the airplane that he was going on a "test hop" that evening. The mechanic assumed that the flight was to be conducted by a rated pilot and had no idea that Pvt. Nolen planned to fly the airplane himself. Witnesses on the ground observed Pvt. Nolen circling the airport at Clanton at an altitude of 200 feet in an apparent effort to land. The BT-13 was then seen flying low on an apparent approach, striking a tree and crashing into the ground. The aircraft erupted in flames upon impact.

10-12-42A. Astoria, New York. At 0905, two Republic P-47B aircraft collided in mid-air and crashed at Astoria, New York, killing one flier. P-47B #42-6047 smashed to the ground near 34th Street and 20th Avenue in Astoria, Long Island, New York, killing 2Lt. Donald W. Broady. 2Lt. Marvin V. Bledsoe was uninjured and was able to safely land severely damaged P-47B #42-6057 at La Guardia Field, New York. The aircraft were part of a four-airplane interception flight that had just taken off from La Guardia Field and were assembling in a string formation. Lt. Broady's P-47 was overtaking Lt. Bledsoe's P-47 from behind at high speed. Lt. Broady was unable to reduce his speed and avoid overshooting his position behind Lt. Bledsoe. Flying below Lt. Bledsoe's P-47, Lt. Broady ascended into him in a zooming pull-up, Lt. Bledsoe's propeller cutting into Lt. Broady's fuselage and severing the tail section just behind the turbo-supercharger. Lt. Broady's P-47 fell end over end from an altitude of 1,800 feet, crashing to the ground and exploding into flames, killing Lt. Broady instantly.

10-12-42B. New Canaan, Connecticut. At 0920, a Republic P-47B crashed into a colonial-era mansion at Valley Road, New Canaan, Connecticut, killing pilot 2Lt. Richard M. Craghill, 23, Corcoran, California. The airplane was engaging in unauthorized simulated combat with three other P-47 aircraft when it stalled and entered a spin, slamming into 197-year-old mansion and setting it on fire. The mansion was owned by Mrs. Ruth T. Weeks. Mrs. Weeks, who was not injured, was able to escape the home, which was completely destroyed by fire. A maid and three workmen who were in the home at the time of the accident were also able to escape and were uninjured. Lt. Craghill and his wingman had taken off from Bridgeport Municipal Airport, Bridgeport, Connecticut, on a blind flying mission and were soon intercepted by two P-47 type aircraft north of Norwalk, Connecticut, at an altitude of approximately 3,000 feet. Instead of continuing their blind flying mission, Lt. Craghill and his wingman engaged in simulated combat with the P-47s. The aircraft were trying to out turn each other in a Lufbery circle at an altitude of about 1,500 feet when Lt. Craghill tried to tighten his turn and stalled his P-47. He fell off into a spin to the left, completing two turns and then leveling off momentarily before entering a second spin to the right. The P-47 spun straight to the ground and into the home of Mrs. Weeks.

10-12-42C. Smyrna, Delaware. At 0920, a Curtiss P-40F dove into swampy ground near Smyrna, Delaware, killing pilot 2Lt. William J. Steele. The P-40 was one of a two-airplane flight that had taken off from Philadelphia Municipal Airport, Philadelphia, Pennsylvania, on a unit combat mission. The aircraft climbed to 16,000 feet and began a series of wingovers and split-S turns, leveling off at approximately 12,000 feet. At this point, the element leader lost track of Lt. Steele. Investigators could not determine why the airplane dove into the swamp at high speed and at a steep angle. The P-40 exploded violently upon impact, scattering wreckage up to 200 feet. The biggest piece of wreckage that could be found was a four-foot section of wing tip. The crash created a crater 20 feet wide by 10 feet long and about 15 feet deep. Only two-thirds of the pilot's remains were recovered. Recovery efforts were hampered by the crash site being located at an elevation close to sea level, causing the crater to persistently fill with water, which had to be constantly pumped out.

10-12-42D. Palacios, Texas. At 0955, a Douglas A-20C crashed on the right of way of the Southern Pacific Railroad three-quarters of a mile east of Camp Hulen, near Palacios, Texas, killing the crew of three. Soon after take-off from Camp Hulen the aircraft suffered a fire in the port engine. The pilot shut down the engine, feathered the propeller and trimmed the aircraft for single-engine flight. The pilot notified the field that he was in trouble. The engine and nacelle continued to burn and the pilot then began making a series of left turns (into the bad engine and with landing gear down) in an effort to get back to the Camp Hulen airfield. During the third left turn, the pilot put the aircraft in too steep of a bank and caused the aircraft to stall and enter a subsequent spin. The Camp Hulen control tower heard the pilot order the other two crewmembers to "bail out." Radio operator SSgt. Charles V. Cappadonna and gunner Pvt. Owen D. Douglas left the aircraft at an altitude of approximately 200 to 300 feet above ground level, too low to allow their parachutes to deploy and they fell to their deaths. Their bodies were found near where the aircraft struck the ground. Pilot SSgt. Ray Scott failed to bail out and was killed instantly when the A-20 spun to the ground and exploded into flames.

10-12-42E. Delaware River, New Jersey. At 1520, a Curtiss P-40F suffering an in-flight engine fire was abandoned and crashed into the Delaware River at Gibbstown, New Jersey. Pilot 2Lt. Charles E. Gibbs, 22, San Francisco, California, was able to successfully parachute from the airplane. He landed in the river and drowned before he could be rescued.

10-12-42F. Orlando, Florida. At 1540, a Curtiss P-40E, one of a two-airplane formation training flight, crashed while performing aerobatics ten miles northeast of Orlando, Florida, killing 1Lt. William B. Burge. The aircraft had performed a series of dives and turns and leveled out at an altitude of approximately 6,000 feet when Lt. Burge executed a slow roll to the left. At the completion of the roll the P-40 skidded violently to the left and snapped over into an inverted spin to the right, barely missing the second aircraft. The P-40 then entered a normal type spin, but the pilot failed to recover before the airplane spun into the ground and exploded into flames.

10-12-42G. Winder, Georgia. At 1600, two Piper L-4A type aircraft collided in mid-air near Winder, Georgia, killing pilot 2Lt. Raymond R. Van Sickle aboard L-4A #42-36524. The airplanes were part of a four-ship ferry flight to New Orleans, Louisiana. SSgt. William H. Willis parachuted to safety from L-4 #42-36525 after Lt. Van Sickles ascended into his airplane from directly underneath at an altitude of approximately 1,500 feet agl. Lt. Van Sickles abandoned his airplane at too low an altitude to allow his parachute to fully deploy. His body and the two airplanes fell on the farm of Henry Wages.

10-12-42H. Biggs Field, Texas. At 1402, a Consolidated B-24D suffering an engine failure and attempting an emergency landing crashed on the east side of Biggs Field, El Paso, Texas, killing gunner Sgt. Lansing Hibbard, Jr. Twelve other crewmembers received minor injuries. The aircraft had taken off from Biggs Field for a bombing and gunnery mission to Alamogordo, New Mexico, when its number-one engine failed. The pilot attempted the landing but was unable to keep the port wing from dropping to the point that it contacted the runway and caused the airplane to skid out of control. The B-24 broke up as it careened across the field, bursting into flames soon after coming to rest. The crew was able to exit the aircraft but could not rescue Sgt. Hibbard, who apparently had been rendered unconscious and was trapped in the wreckage. Receiving minor injuries in the accident were: 2Lt. Kenneth P. Iverson, pilot; 2Lt. Albert Greyhosky, co-pilot; 2Lt.

Ernest W. Fogel, Jr., B-24 instructor pilot; 2Lt. Walter G. Willis, navigator; 2Lt. Hubert R. Demeester, bombardier; SSgt. Claude Arnold, engineer; Sgt. Raymond M. McMahon, assistant engineer; Sgt. William H. Goode, gunner; Cpl. Albert E. Sprague, passenger; Pvt. Raymond E. Burd, radio operator; Pvt. Cosimo D. Papaleo, photographer; Aviation Cadet Frank S. Phillips, student bombardier.

10-12-42I. Moore Field, Texas. At 2025, a North American AT-6A on a night navigation training flight crashed 18 miles northwest of Moore Field, Mission, Texas, killing A/C Daniel R. Bates. The student was to fly to Laredo, Texas, to Kingsville, Texas, and then return to Moore Field. The airplane had been declared missing and was not located until 1710 on 10-13-42 when it was spotted from the air by search planes; investigators did not reach the wreckage until 10-14-42. Investigation revealed that the airplane had struck the ground in a 30-degree dive with 30 degrees of left bank with power on. It was determined that the airplane dove into the ground at a speed in excess of cruise speed.

10-13-42A. St. Petersburg, Florida. At 0845, a Bell P-39D stalled and spun into the ground five miles southwest of St. Petersburg, Florida, killing pilot 2Lt. Marvin M. Pearson. Capt. Andrew W. Salter, Jr. witnessed the accident and stated to investigators, "I was leading six P-39D-1 aircraft at approximately 11,000 feet. I signaled for string formation diving down to an indicated airspeed of 250 MPH, pulling up and slow rolling [to the left]. After the second roll, I made a turn to count the planes. I observed Lt. Pearson go into a [left] spin from level flight. My altitude at this time was 8,500 feet. Lt. Pearson, after approximately one and one-half turns, recovered and immediately went into another spin [to the right]. At this point I immediately gave orders over the radio [for Lt. Pearson] to bail out. Lt. Pearson recovered from the spin at least five times, going into a secondary stall and a subsequent spin [in the opposite direction]. The plane hit [the ground] and burst into flames. No parachute was observed."

10-13-42B. Tyndall Field, Florida. At 1015 CWT, a North American AT-6A (41-427) taxiing for take-off at Tyndall Field, Panama City, Florida, collided with an AT-6A (41-16635) that was parked on the line, fatally injuring Pvt. Carlton H. Davis who was standing guard near the airplane at the time. The pilot, SSgt. Marion C. Birmingham, was uninjured in the accident. The pilot stated that he had swerved his aircraft to avoid hitting an enlisted man who ran into his path. In doing so he collided with the tail section of the other AT-6, causing his AT-6 to careen to the right, striking Pvt. Davis with the port wing. Pvt. Davis died of his injuries on 10-19-42.

10-13-42C. Norwood, Massachusetts. At 1544, a Curtiss P-40F crashed out of control while performing aerobatics at low altitude near Norwood, Massachusetts, killing pilot SSgt. Watkins Mayo, 19, Tulia, Texas. The pilot had been on an assigned aerobatic practice flight and had been observed by witnesses on the ground diving on Boston Metropolitan Airport, making a pass over the field at about 50 feet. The pilot came back around and dove on the field again, pulling up to approximately 300 feet and into a double slow roll to the right. The P-40 stalled while inverted in the second roll, falling off to the right and spinning (to the right) about three-quarters of a turn. The pilot stopped the spin and the aircraft stalled and entered another spin in the opposite direction. The pilot was able to momentarily stop the spin to the left before the P-40 was then seen to dive into the ground from an altitude of less than 100 feet, exploding into flames upon impact.

10-13-42D. Marana, Arizona. At 1513, a Vultee BT-13A stalled and spun into the ground while on an instrument team mission 10 miles northwest of Marana, Arizona, killing A/C Harry H. Haldeman, Jr. and A/C Paul A. Dion. The aircraft was returning from a two-hour instrument flight and apparently entered a stall and spin at too low an altitude to allow the pilots to recover. The aircraft exploded into flames upon impact with the ground. Investigators speculated that the aircraft may have overshot an attempted simulated or actual forced landing and the pilot might have pulled up too sharply or entered too steep of a bank, allowing the BT-13 to enter a stall.

10-14-42A. Belen, New Mexico. At 0445, a Beech AT-11 flying on a simulated bombing mission dove into the ground after inadvertently entering instrument conditions near Belen, New Mexico, killing the pilot and two student bombardiers. Investigation revealed that the airplane had dove into the ground at high speed. The AT-11 was totally destroyed in the resulting explosion, scattering wreckage over 160 yards. Other pilots flying the same mission reported instrument conditions in the area at the time of the accident. Investigators stated that the deteriorating weather conditions and the relative inexperience of the pilot contributed to the crash. Pilot 2Lt. Boyd C. Knetsar and bombardier students A/C Earl W. Ferris and A/C John J. Fischer were killed in the crash.

10-14-42B. Alligator Point, Florida. At 0917, a Bell P-39D on a gunnery training mission collided with a fixed target at Alligator Point Gunnery Range, Florida, and crashed into shallow water in the Gulf of Mexico surf, killing pilot SSgt. Samuel W. Liggin. The pilot had apparently misjudged distance and dive angle on the approach for firing on the target. The pilot was seen by observers on the ground to change targets after his initial run. Investigators stated that this should not have affected his firing pass. The pilot executed his pass at very low altitude and abruptly pulled up but was unable to avoid striking the target with the port wing. The pilot lost control of the airplane and it crashed into the water about 600 feet from the target that had been struck.

10-14-42C. Drew Field, Florida. At 1115, a Bell P-39D crashed seven miles northeast of Drew Field, Tampa, Florida, killing pilot 2Lt. Stanley R. Sargent. The Accident Classification Committee stated, "Statements of witnesses indicate that the airplane was seen coming out of a spin at approximately 500 feet. This evidence was too indefinite to determine whether the ship was in a normal spin or in a tumble or flat spin. Evidence shows that the pilot left the airplane and it is believed he bailed out above 500 feet because the airplane apparently began to recover from the spin at that altitude. Examination of the pilot's body showed a gash on his head, which is believed to have been caused by his striking the tail of the airplane before he could pull the ripcord on his parachute. Pilot's body [with his right hand still clutching the D-ring] was found three-quarters of a mile from the crash and in the line of flight of the airplane."

10-14-42D. Richburg, South Carolina. At 2058, a North American B-25D on a night navigation-training mission crashed four miles east of Richburg, South Carolina, killing six crewmembers and a passenger. The airplane had been cleared for visual contact flying from Greenville Army Air Base, South Carolina, to Charlotte, North Carolina, to Columbia, South Carolina, and return to Greenville, but had encountered instrument conditions between Charlotte and Columbia. Witnesses on the ground observed the B-25 flying low and fast with both engines functioning normally. The pilot, apparently lost and trying to get his bearings, attempted a left turn close to the ground, striking trees with the port wing and crashing into the ground. The airplane exploded into flames upon contact, scattering wreckage over 400 yards. One witness observed the landing lights illuminate and extinguish just prior to the accident. At the time of the accident, the ceiling in the area of the crash was reported as almost down to ground level with very poor visibility. Killed in the crash were: 2Lt. Henry W. Curtis, pilot; SSgt. William S. Allen, co-pilot; 2Lt. Malcolm P. Getchell, navigator; SSgt. Woodrow W. Crowder, bombardier; Sgt. Edward C. Patrick, engineer; Sgt. George E. Stock, gunner; 2Lt. Arthur V. Pontius, passenger and rated army pilot.

10-14-42E. Oakwood, California. At 2133, a Vultee BT-15 flying in very poor weather and apparently lost, flew into the ground three miles north of Oakwood, Oklahoma, killing Aviation Cadet Wilmot D. Bolinger and Aviation Cadet Christopher T. McClellan. The Accident Classification Committee stated, "Aviation Cadet Bolinger and Aviation Cadet McClellan took off from Enid Army Flying School [Enid, Oklahoma] at approximately 1945 CWT to go to their assigned outlying field for night landing practice. They proceeded to the vicinity of the outlying field without identifying same and then started searching aimlessly in a large area to the northwest and west of Enid. They encountered hard rain in the vicinity of Oakwood, where they apparently attempted a landing but gave up due to roughness of the terrain and poor visibility, crashing soon thereafter into the ground in full flight. The ship struck in a wooded, rolling terrain and was completely demolished without burning. Both pilots were killed instantly."

10-15-42A. Cimarron, Oklahoma. At 1015, a Fairchild PT-19A stalled and spun into the ground four miles west of Cimarron Field, Oklahoma, killing Aviation Cadet Arthur W. Bagnai. The pilot was attempting to avoid a mid-air collision with another PT-19A at an altitude of approximately 500 feet, pulling the airplane up into a steep climbing turn to the right. The airplane went nearly inverted and lost flying speed, stalled and dove into the ground at a steep angle, bursting into flames upon impact.

10-15-42B. Summit, Illinois. At 1641, a Douglas C-49E flying in poor weather collided with a large tree and crashed in a cornfield at 6700 Harlem Avenue, Summit, Illinois, killing nine crewmembers and passengers. The C-49 was attempting to land at Chicago Municipal Airport and crashed two and one-half miles southwest of the runway. The C-49 had taken off from Sedalia, Missouri, earlier in the day and was on a cross-country navigation-training/personnel transport mission back to Chicago Municipal Airport, the crew's home station. Visibility in the area at the time of the accident was reported as very poor with low ceilings and light rain showers. Edward Salecki, Summit, witnessed the accident and later stated to investigators, "I was unloading straw from a truck when I noticed an airplane north of me not over half a mile away flying east and very low. Then it circled to the right quite steep and came directly over where I was standing, so low that it made me crouch down. While the airplane was still in the turn, the left wing struck the top of a cottonwood tree at about the same time the right wing struck the ground, or slightly before. The airplane then nosed over sharply toward the ground, bounced once, and exploded about 100 feet beyond, in the cornfield." Wreckage and bodies were strewn over a large area. Killed in the accident were: 1Lt. Walter Jackson, Warrensburg, Missouri, pilot; 2Lt. W. J. Tichy, Chicago, Illinois, co-pilot; Capt. C. A. Dafoe, Sedalia, Missouri, passenger; 2Lt. Walter L. Cravey, Warrensburg, Missouri, passenger; 2Lt. P. A. Stenz, Elm Grove, Wisconsin, student co-pilot; 2Lt. R. A. Roper, Chicago, Illinois, student co-pilot; 2Lt. C. J. Tompkins, Waskom, Texas, student co-pilot; SSgt. Max Bybee, Hammond, Indiana, engineer and crew chief; Sgt. Francis Kokenge, Seneca, Kansas, radio operator.

10-15-42C. Magdalena, New Mexico. At 2228, a Boeing B-17E on a night navigation training mission collided with the northwest slope of North Baldy Peak six miles southwest of Magdalena, New Mexico, killing the crew of nine. The B-17 collided with the mountain at an elevation of about 9,500 feet, approximately 75 feet from the peak. The tail section came to rest on the northwest slope while the wings, engines, center and forward section catapulted over the peak, coming to rest on the southeast slope. The flight was cleared for visual contact flying from Alamogordo Army Air Base, New Mexico, to Carrizozo, Vaughn, Albuquerque, Magdalena (all in New Mexico), and return to Alamogordo. The airplane apparently encountered instrument conditions near Magdalena. Witnesses on the ground reported that the B-17 had circled the town of Magdalena several times at an altitude of approximately 700 feet then departed the area on a southeast heading and climbing. Investigation revealed that the flight was on course and on time. Investigators stated that the pilot apparently misjudged his position and altitude relative to the surrounding terrain causing the B-17 to collide with the mountain. Killed in the accident were: 2Lt. John R. Pratt, pilot; 2Lt. Donald F. Jackson, co-pilot; 2Lt. Lawrence W. Van Tassel, navigator; 2Lt. Joseph L. Grant, bombardier; Sgt. Hensen E. Ortmyer, engineer; SSgt. Thomas C. Ferron, radio operator; Sgt. Robert C. Myers, assistant engineer; Sgt. William G. Walls, assistant radio operator; Sgt. Dale R. Rottier, gunner.

10-16-42A. Grapevine, Texas. At 1418, a North American B-25C flying in instrument conditions crashed after it collided with the 653-foot WFAA radio station transmitting tower at Grapevine, Texas, killing the crew of six. The B-25 had taken off from Will Rogers Field, Oklahoma City, Oklahoma, on an administrative flight that had originated at Rosecrans Field, St. Joseph, Missouri. The aircraft had missed three landing attempts at Love Field, Dallas, Texas, and was instructed by the tower to head west, climb to 4,000 feet and await instructions. Air traffic was beginning to stack up due to a Boeing B-17 that was also having trouble landing. The instruction to climb to 4,000 feet was acknowledged by the pilot but apparently not carried out. The pilot had indicated earlier that he might try Fort Worth as an alternate. Weather at Love Field was reported as ceilings below 1,000 feet, heavy overcast and visibility of less than a mile due to thunderstorms and rain. Weather at Fort Worth was similar but had been predicted to improve. After the B-17 finally landed at Love Field, tower personnel attempted to contact the B-25 to relay additional landing instructions but they were unable to do so. The B-25 had collided with the radio tower 502 feet above the ground with its starboard wing, severing the wing six feet from the tip. The wing panel and aileron pieces were found near the transmitting tower and pieces of aluminum were found in the tower itself, the top of which was obscured by clouds at the time of the collision. The tower was not toppled by the collision but sustained serious damage. The airplane rolled out of control and crashed to the ground in a steep left bank about 2,000 feet WSW of the tower, killing the crew instantly. The B-25 did not burn. Wreckage and body parts were strewn over a wide area. Ceilings in the area at the time of the crash were reported as 500 feet with rain showers and visibility of less than a mile. Initially, the pilot's request for a clearance from Will Rogers Field to Love Field was denied twice due to poor weather reported at Love Field. The pilot twice refused to produce an instrument card for the enlisted weather and clearance personnel at Will Rogers Field. The pilot, rated as a "Senior Pilot," cleared himself to Love Field after being refused clearance by the lower ranking personnel. Investigators learned that the pilot had successfully passed an instrument check ride the month before but were unable to explain why he would not produce his card for authorized personnel, who stated that the officer became "indignant" and appeared "insulted" after the request. The pilot's current instrument card could not be found among his personal effects. Killed in the crash were: Capt. James M. Treweek, pilot; Col. Edward S. Fee, co-pilot; SSgt. Wilfred E. Miller, engineer; Capt. Louis Rawlins, Jr., passenger; Cpl. Joseph L. Tyndall, passenger; Pvt. William Echko, passenger.

10-16-42B. Oxnard, California. At 1405, a Stearman PT-18 performing aerobatics at low altitude crashed into the ground at Oxnard, California, killing pilot 2Lt. Andrew E. Bowman. The Accident Classification Committee stated, "The plane was first observed making a low pass at a bean thrasher. The pilot pulled up to a steep wingover and came back over the thrasher at an extremely low altitude. He continued in a southerly direction across Fifth Street and made a low pass at a vehicle on the beach road. Pilot then pulled the airplane up into a steep wingover and was unable to recover to level flight. He attempted to make a split-S maneuver at a very low altitude and was fatally injured when his plane failed to recover and [dove into the ground] approximately two and one-half miles west of the home station [Ventura County Airport, California]."

10-16-42C. Fife, Texas. At 1630, a Vultee BT-13A on an instrument-training mission crashed while maneuvering two miles east of Fife, Texas, killing student pilots Cpl. Paul W. Brown and Pfc. Allen L. Johnson, who died a short time after the accident. The airplane did not burn and civilians in the area pulled the fliers from the wreckage just after the crash. Investigators stated, "Investigation at the scene of the accident revealed that the airplane dove from a considerable altitude, leveled off about 50 feet above the ground, and then proceeded to make a moderately steep bank to the right. After the airplane had completed about 90 degrees of the turn, the right wing struck a telephone line and tore through some mesquite trees, causing a loss of control and the crash."

10-17-42A. Des Moines, New Mexico. At 0615, a Boeing B-17E collided with the north slope of 8,732-foot Sierra Grande Mountain at an elevation of about 8,000 feet, seven miles southwest of Des Moines, New Mexico, killing the eleven fliers on board. The

B-17 had taken off from Rapid City, South Dakota, at 0311 MWT and was cleared to Duncan Field, San Antonio, Texas, on a navigation-training mission. Investigators speculated that after take-off the B-17 encountered strong easterly winds and the crew failed to compensate for 175 miles of drift, resulting in the course that led to the collision with the mountain. The crew attempted a position report for Garden City, Kansas, at 0527 but Garden City radio operators did not receive the transmission. The signal was received strongly by La Junta, Colorado, radio operators at 0527. The crew, apparently mistaking La Junta for Garden City, reported their position as 15 miles WSW of Garden City, flying at an altitude of 8,000 feet. A straight-line course from Rapid City to the accident site passes 15 miles west of La Junta. Investigation revealed that the B-17 was flying south in a straight and level attitude when it collided with the mountain and marks on the mountainside made by the propellers and engines indicated that all four were producing power. The aircraft exploded violently into flames in the collision and no readable instruments could be found, the flight deck being completely destroyed. A wristwatch that had stopped at 0615 was found among the wreckage that was strewn over an area of 100 yards by 700 yards. Civilians in the area reported that the weather at the time of the accident consisted of rain, light snow, and fog with poor visibility. Two sheepherders who had seen fire on the mountainside discovered the wreck on 10-19-42. Investigators speculated that the airplane was flying in instrument conditions and the pilot, unaware of his position relative to the terrain, flew the airplane into the mountain. Killed in the accident were: 1Lt. Paul D. Cunningham, pilot; 2Lt. Charles W. Mann, Jr., co-pilot; 2Lt. Robert S. Connor, Jr., navigator; 2Lt. Lawrence W. Hunter, bombardier; Aviation Cadet Edward W. Larson, student bombardier; SSgt. Joseph C. Lynch, engineer; SSgt. Emory F. Herzog, radio operator; SSgt. Ward Murie, assistant engineer; Cpl. Milton Lakin, assistant radio operator; Sgt. Jerome A. Krzmarzick, gunner; TSgt. William R. Patton, passenger.

10-17-42B. Shafter, California. At 1215, a Vultee BT-13A on an instrument training mission crashed 12 miles west of Shafter, California, killing pilot 1Lt. Harold N. Robinett and passenger Pvt. John H. Deremer. The Accident Classification Committee stated, "The pilot was cleared for an instrument training flight under the hood with an enlisted mechanic as safety observer. It is believed that the pilot was under the hood at the time the accident occurred. Eye witnesses stated that the airplane was in a power spin and upon pulling out of the spin at an excessive rate of speed, the wing panels and tail surfaces were seen to separate from the rest of the airplane at an altitude of 1,000 feet."

10-18-42. Chadron, Nebraska. At 1900, a Lockheed-Vega B-34 stalled and spun into the ground at Chadron Municipal Airport, Chadron, Nebraska, killing pilot 2Lt. Woodrow W. Syverson and passengers Pvt. Chester F. Obal and Pvt. Elmer W. Putzier. The aircraft was on a ferry flight with another B-34 when they approached Chadron Municipal Airport for a landing. Night had arrived when both aircraft circled the field three times in preparation for landing. The subject aircraft approached from the northwest with the intention of landing on the Northwest-Southeast runway. The approach was made at about 300 feet, approximately a quarter mile out and at an angle to the runway. The pilot corrected the glide path and, after passing over a row of trees about 500 yards from the end of the runway, he increased power, apparently fearing that he was about to undershoot the landing. The engines did not respond and the pilot pulled the aircraft up to extend the glide, stalled and entered a spin to the left, spinning one-quarter of a turn and into the ground. The B-34 exploded into flames upon impact.

10-19-42. Gulf of Mexico, Florida. At 0930 CWT, a North American B-25C on a simulated bombing mission crashed into the Gulf of Mexico two and one-half miles south of Destin, Florida, killing the crew of seven. Although no real evidence is present in the accident report and wreckage was recovered only in the area that the aircraft struck the surface, investigators speculated that the airplane might have exploded in mid-air before crashing into the sea. In fact, a witness stated that he did not see the aircraft explode before it dove into the water. Another witness stated that he observed "two black funnels of smoke" trailing the engines after hearing a loud report. Some of the witnesses thought that a "tanker" type merchant ship had blown up. All of the witnesses were non-flying civilians and their testimony vague. The cause of the accident remains "undetermined." Killed in the accident were: Lt. Col. George E. Pierce, pilot; 2Lt. John G. Dodd, co-pilot; SSgt. Billy B. Boggess, engineer; Pvt. Samuel A. Salli, radio operator; Pvt. Eugene A. Delisle, gunner; Pvt. Floyd Cheatham, passenger; Pvt. Joseph L. Daichman, passenger. The only remains recovered were the upper body of Pvt. Salli.

10-20-42A. Barksdale Field, Louisiana. At 1414, a Martin B-26A crashed while attempting a landing at Barksdale Field, Shreveport, Louisiana, killing four fliers. Engineer Sgt. Alvin M. Dunn, Pasadena, California, and assistant engineer SSgt. Arden R. Fox, Youngstown, Ohio, were killed instantly and 2Lt. Joseph B. Swann, Washington, Indiana, B-26 student and co-pilot, died of his injuries on 10-21-42 at 0726. B-26 instructor pilot 2Lt. Carl J. Wdowka, Barksdale Field, Louisiana, died of his injuries on 10-22-42. Investigators stated, "On a local transition flight instructor pilot Lt. Carl J. Wdowka was sitting in the pilot's seat and student pilot Lt. Joseph B. Swann was sitting in the co-pilot's seat. The airplane was approaching the field normally in a power-on glide with Lt. Swann at the controls, when Lt. Wdowka saw that the airspeed had dropped too low and called it to the attention of

Lt. Swann at least twice. At this point Lt. Swann lowered the nose to increase the airspeed. Lt. Wdowka evidently thinking the plane was very close to the ground, thought they were landing too short and pulled back on the control column. This caused the plane to stall and fall off on the left wing; the left propeller struck the ground, tearing the left engine from the plane and piercing a gas tank causing an immediate fire. The plane turned to the left and was burning fiercely before it came to a stop. Although the pilot and co-pilot succeeded in leaving the plane, they had received fatal burns. Although Lt. Swann allowed the airspeed to drop too low, it is the opinion of the board that the responsibility for the accident rests with Lt. Wdowka who evidently misjudged the distance to the ground and pulled the control column back when the airspeed was already indicating a stalling speed."

10-20-42B. Sikeston, Missouri. At 1550, a Fairchild PT-19A crashed seven miles east of Harvey Parks Airport, Sikeston, Missouri, after suffering a catastrophic failure of its port wing while attempting a spin recovery, killing civilian instructor Lowell C. Little. Mr. Little, who deployed his parachute too soon, became entangled in the nose-over strut while attempting to bail out of the airplane, which was out of control. He went down with the airplane. Aviation Cadet John Osorio received minor injuries parachuting to safety. Cadet Osorio's left hand and forearm became entangled with Mr. Little but he was freed when his parachute deployed and pulled him off.

10-21-42A. Lemoore, California. At 0719, a Vultee BT-13A, one of six ships attempting a formation take-off at the Army Flying School, Lemoore, California, ground looped on the take-off run and collided with another BT-13A, killing A/C Dino A. Berardo aboard BT-13A #41-21608. A/C Thomas A. E. Foster, flying BT-13A 11245, suffered minor injuries when Cadet Berardo's airplane ground looped to the left and into his BT-13 just as the tails of the airplanes began to rise. Cadet Berardo's airplane burst into flames in the collision. Investigation revealed that Cadet Berardo attempted his take-off with his rudder trimmed to the left 10 degrees, causing the ground loop and loss of control when the airplane began to get airborne. It was speculated that the student confused the 0 degree setting with the 10 degree setting of the rudder trim wheel when preparing the airplane for take-off.

10-21-42B. Norfolk, Virginia. At 1430, a Curtiss P-40F crashed into a house at 7438 Hampton Blvd., Norfolk, Virginia, killing pilot 2Lt. Burl J. Fink, who fell to his death attempting to parachute to safety. It was speculated that the pilot attempted his bail out at an altitude of approximately 50 feet, too low to allow the parachute to deploy. The aircraft was attempting a landing at the U.S. Navy Operating Base at Norfolk when the accident occurred. There were no witnesses to the actual crash. The P-40 and pilot were part of the 58th Fighter Squadron, 33rd Fighter Group, and were to be shipped aboard the aircraft carrier *USS Chenango* at Norfolk with 72 other P-40s, which were to be launched during Operation Torch. The rest of the group and their airplanes were loaded aboard the carrier and steamed for North Africa on 10-21-42. The group launched off of the aircraft carrier on 11-10-42 and established a base at the airport at Port Lyautey, French Morocco. The house belonged to J.M. Stokeley and it was completely destroyed by fire.

10-22-42A. Mindenmines, Missouri. At 1118, an Aeronca L-3B crashed into the ground and burst into flames after colliding with a L-3C two miles east of Mindenmines, Missouri, killing student pilot Sgt. Norman R. Smith. The pilot of the L-3C, Sgt. Howard Peck, landed his airplane after the collision and was uninjured. The collision occurred at an altitude of approximately 500 feet while both airplanes were preparing to land at the AAF Glider Training School near Pittsburgh, Kansas. The L-3B's tail section was damaged by the landing gear of the L-3C when the L-3C apparently descended into the L-3B. The L-3B dove straight into the ground and the pilot made no apparent effort to use his parachute. Sgt. Smith had a total of 6 hours 57 minutes flying time and was on his first supervised solo flight. Investigators faulted both students for failing to see and avoid each other and tower personnel for poor traffic management. The two airplanes were on identical legs of the pattern when the collision occurred.

10-22-42B. Sikeston, Missouri. At 1400, a Fairchild PT-19A crashed nine miles south of Harvey Parks Airport, Sikeston, Missouri, killing civilian instructor Oliver Mutrux and student Aviation Cadet Roy W. Erickson. Investigators speculated that the airplane was in a spin when it struck the ground. Civilian witnesses in the area did not see the airplane until just before it hit the ground. Investigators could not determine the cause of the spin.

10-22-42C. Wolfforth, Texas. At 1432, a Taylorcraft L-2H stalled and spun into the ground two miles southeast of Wolfforth, Texas, killing student pilots SSgt. James F. Rutherford and SSgt. John A. Beulter. Investigators speculated that the pilot had entered a spin at low altitude and partially recovered but was unable to prevent the airplane from diving into the ground.

10-22-42D. Pope Field, North Carolina. At 1700, a Piper L-4A crashed into a pine thicket four miles north of Pope Field, North Carolina, killing student pilots SSgt. Harold L. Mills and Sgt. Leo G. Herman. The airplane was part of a flight of four L-4 aircraft that had been ferrying personnel from Raleigh, North Carolina, to Pope Field. The aircraft was seen by other pilots in the flight in an extreme nose-high position before falling off on a wing and diving into the woods from an altitude of about 200 to 300 feet above ground level.

10-22-42E. Lawson Field, Georgia. At 1901, a Northrop A-17 collided with a dump truck while landing at Lawson Field, Fort Benning, Georgia, fatally burning pilot 1Lt. Thomas J. Burke and seriously injuring passenger 2Lt. Oswald K. Marquart, who also received serious burns. Civilian dump truck driver Joseph B. Simpson received minor injuries. The A-17, landing during dusk, struck the raised body of the dump truck with the port landing gear, which broke off and ruptured the fuel tanks, causing the airplane to erupt in flames as it slid down the runway. When the A-17 came to rest, Lt. Burke was able to help Lt. Marquart out of the burning airplane. Both were conscious and talking to rescue personnel in the ambulance on the way to the Fort Benning hospital. Investigation revealed that the truck driver did not get an affirmative green light signal from the tower that would have given him permission to enter an active runway with his truck, which was traveling down the center of the runway facing away from the approaching airplane. The pilot received no warning from the tower that there was a vehicle on the runway and was given permission to land. The truck, which had entered the runway unauthorized at least one other time before the accident, had dumped two loads of fill in a runway extension project taking place at the field.

10-22-42F. Hendricks Field, Florida. At 2106, a Boeing B-17B returning from a night transition flight crashed eight miles ENE of Hendricks Field, Sebring, Florida, killing five crewmembers. The B-17 was attempting to enter the pattern for the field when the crash occurred. The horizon was largely indistinct because of smoke from numerous grass fires in the vicinity. Straight down visibility and horizontal visibility (about 15 miles) were considered excellent with a full moon. Investigators speculated that the B-17 student (flying in the left seat) lost control of the airplane while trying to avoid a collision with a B-17E (41-9034) that was also in the pattern. The B-17B was seen to make a steep and abrupt climbing left turn and then enter an extreme nose-high attitude (approaching vertical), falling off into an inverted attitude and then into dive. Apparently the instructor, flying from the right seat, recovered the airplane momentarily before it was seen by other pilots flying in the pattern to descend rapidly into the ground in a flying attitude. The B-17B struck the ground at a 70-degree angle and exploded into flames upon impact, scattering wreckage and bodies over a wide area. Killed in the crash were: 1Lt. Jack D. Little, Los Angeles, California, B-17 student; Capt. Daniel E. Culver, Jr., Warwick, New York, B-17 instructor/co-pilot; 2Lt. Myron M. Malady, Syracuse, New York, B-17 student; 2Lt. Lyle F. McClelland, Western Springs, Illinois, B-17 student; Sgt. Howard J. Ginnavan, Baltimore, Maryland, engineer.

10-23-42A. Columbus, Mississippi. At 1050, a Lockheed A-29 crashed south of the Army Air Field at Columbus, Mississippi, killing instructor pilot 2Lt. Arthur G. Summerford, 22, Clio, South Carolina, and student Aviation Cadet Herman Adams, 23, Clifton, New Jersey. The Accident Classification Committee stated, "The pilot was making an approach at a time when traffic was exceptionally heavy and had turned in on his approach leg with wheels and flaps down. There were three planes also on the approach leg preceding him and apparently the pilot permitted the plane to approach its stalling speed before deciding to go around. At this point he opened the throttles, the left engine taking and the right engine cutting out, presumably due to loading up on slow approach, with the plane at approximately 800 feet altitude. The plane fell off to the right as through going into a right spin. The pilot was seen to push the nose down in an attempted recovery, but probably due to his close proximity to the tree tops at this point, he pulled the nose up before regaining sufficient flying speed and a secondary stall resulted in a spin into the ground from 200 feet [above ground level] from which he was unable to recover."

10-23-42B. Moore Field, Texas. At 1200, a North American AT-6A and an AT-6B collided in midair seven miles north of Moore Field, Mission, Texas, killing 2Lt. Samuel T. Jones aboard AT-6A #41-16520. 2Lt. Benjamin F. Marshall received minor injuries parachuting from AT-6B #41-17139. The two aircraft were flying in a four-ship diamond formation when the leader signaled for a change into string formation during which Lt. Jones mistakenly began maneuvering for a six-ship formation. Realizing his mistake, Lt. Jones attempted to regain his proper position in the four-ship string but descended into Lt. Marshall's airplane, which he apparently did not see. Lt. Marshall, upon finding his airplane uncontrollable, parachuted to safety. Lt. Jones was seen to leave his AT-6 at an altitude estimated to be less than fifty feet, too low to allow his parachute to deploy. Lt. Jones' body was found about ten feet from the wreckage, which had burst into flames upon impact with the ground. Investigators faulted both officers, who were non-rated student pilots undergoing flight training, for failing to see and avoid each other.

10-23-42C. Lemoore, California. At 1344, a Vultee BT-13A on an instrument training flight crashed 25 miles northwest of Lemoore, California, killing instructor pilot 2Lt. Don G. Crook and student Aviation Cadet Edgar O. Poleski. Investigators stated, "Instructor and [student] were flying under the hood. The cadet, sitting in the rear cockpit, was much larger than the instructor. It would appear that the plane entered a spin above 2,500 feet; that the pilot had failed to pull out or recover before crashing into the ground. It is believed that the weight of [Cadet Poleski] threw the [center of gravity] too far to the rear, which, combined with inexperienced pilots, resulted in an uncontrollable spin. The body of the instructor was found in the front cockpit. The cadet's body was found about 40 feet ahead of the wreckage. His parachute was opened and his cap was found 75 yards to the left and rear of the crash. The safety belt and seat showed no indications of unusual stress."

10-23-42D. Palm Springs, California. At 1715, a Lockheed-Vega B-34 and a Douglas DC-3 operating as American Airlines Flight #28 collided in midair over the San Gorgonio Pass near Palm Springs, California, killing 12 passengers and crew aboard the DC-3. The wreckage of the DC-3 smashed to earth on Chino Ridge near San Jancito Peak. The pilot of the B-34, 2Lt. William N. Wilson, 25, Long Beach, California, was able to land the damaged bomber safely at Palm Springs (at 1733) after the collision. Lt. Wilson and his passenger, SSgt. Robert R. Leicht, were uninjured. Killed in the crash of the airliner were: Charles F. Pedley, 43, Dallas, Texas, Captain of Flight #28 and American Airlines senior pilot (with over 20,000 flying hours); Louis F. Reppert, 26, Fort Worth, Texas, co-pilot, American Airlines; Estelle Regan, Dallas, Texas, Flight Attendant, American Airlines; Clinton Baker, Phoenix, Arizona; Frank Bird, Hollywood, California, Lockheed Aircraft Company; L.A. Hege, Winston-Salem, North Carolina; Cash Henderson, Phoenix, Arizona, Industrial Commissioner for Arizona; Ralph Rainger, Hollywood, California, composer; 2Lt. Joseph R. Rosser, Santa Ana, California, AAF; B.R. Vest, Jr., Indianapolis, Indiana, Allison Engine Company; E.H. Wallace, Las Vegas, Nevada; Charles M. Wert, Los Angeles, California. American Airlines Flight 28 originated at Burbank, California, as a trans-continental flight with the ultimate destination of New York City. Lt. Wilson was a pilot attached to the Ferrying Division of the Air Transport Command and was to ferry the B-34 from Long Beach, California, to Dallas, Texas. Co-pilot Reppert and Lt. Wilson were old friends who were flight instructors together at Lyle Flying Service in Santa Monica, California, before the war. After having dinner with mutual friends on the evening of October 22, 1942, Lt. Wilson and Mr. Reppert, and a civilian named Alvin Singer, had drinks together at the It Café, which was located at the Hollywood Plaza Hotel, Hollywood, California. The pilots agreed to meet in the air near Palm Springs the next day. Lt. Wilson took off from Long Beach at 1626 and Flight 28 departed Burbank at 1630. Lt. Wilson flew to the vicinity of Cal Aero Academy Airport, Ontario, and March Field, Riverside, where he circled at an altitude of 7,000 feet for several minutes, checking his engines and waiting for the DC-3. A rough running engine had delayed the scheduled ferry flight to Dallas a few days earlier. The DC-3 reported over Riverside at 1702 at an altitude of 9,000 feet. SSgt. Leicht spotted the DC-3 and informed Lt. Wilson who climbed the B-34 to the same altitude to catch up with and fly alongside his friend. Lt. Wilson wagged his wings to greet the DC-3. The aircraft flew along together at a relatively safe distance, the B-34 eventually getting about a mile ahead of the DC-3. When the two aircraft arrived over the San Gorgonio Pass, the B-34 executed a steep crossover turn to the right in front of and in close proximity to the airliner, in the approximate center of the pass. After the turn, the DC-3 was flying about a half mile to the left of the B-34 about 40 feet higher and about 50 feet behind. After a few minutes, Lt. Wilson executed another steep crossover turn to the left, also directly in front of the airliner. The DC-3 had apparently caught up with the B-34 and was at the same altitude as the B-34. Realizing he had come too close, Lt. Wilson began a climbing turn to the right in an attempt to avoid a collision with the DC-3, which had closed the distance and was pulling ahead. After climbing slightly, Lt. Wilson then lost track of the DC-3 and was looking for it in the few moments before the tail of the transport came up and in contact with the starboard wing and propeller of the B-34, which was apparently beginning to turn to the right. The DC-3's rudder was severed and its elevators seriously damaged. The DC-3 continued to climb to the right (out of control) and fell off up and over the starboard wing of the B-34, diving to the ground in a left spin and bursting into flames upon impact. Wreckage and body parts were scattered over a wide area. This accident sparked a congressional investigation (headed by U.S. Representative Jack Nichols, D-Oklahoma), which placed blame solely on the army pilot. Lt. Wilson was charged with 12 counts of involuntary manslaughter in a General Courts Martial and was eventually acquitted of all counts. Lt. Wilson was also acquitted in a civil trial. An enlisted man who witnessed the accident from the ground testified that the airliner "came up under the bomber." Lt. Wilson was scheduled to land at Palm Springs prior to resuming his flight to Dallas. It was noted that Lt. Wilson's younger brother was killed in a mid-air collision involving U.S. Navy aircraft just three months before this accident.

10-24-42A. Gilbert, South Carolina. At 1500, a North American B-25D suffered a catastrophic structural failure while attempting to recover from an inadvertent spin and crashed near Gilbert, South Carolina, killing the crew of five. The airplane was on a scheduled instrument training flight when the crash occurred 14 miles west of the Army Air Base at Columbia, South Carolina. Investigators stated, "It being impossible to accurately state the cause and nature of the accident, the following deductions and assumptions are presented: That the airplane was at an altitude probably in excess of 4,500 feet. It is assumed that the airplane was spinning or spiraling, and that upon emerging from the overcast the pilot (or the co-pilot) pulled back on the control column too rapidly causing the disintegration of the aircraft in the air to the extent that the wings (outboard sections), flaps (outboard), entire tail empennage, and miscellaneous parts of stressed skin were observed at distances of from 200 feet to one and a quarter miles from the point of impact. It appears that the airplane made contact in almost a 90-degree nose down vertical dive." Killed in the crash were: SSgt. Billy D. Pinkerton, 22, Farmington, Illinois, pilot;

SSgt. Morris O. Jones, 22, Minneapolis, Minnesota, co-pilot; SSgt. Morris Beital, 22, New York, New York, engineer; Pvt. Donald W. Goforth, 26, Guthrie, Oklahoma, gunner/radio operator; Pvt. Frank L. Good, Jr., 32, Philadelphia, Pennsylvania, gunner/radio operator.

10-24-42B. Paine Field, Washington. At 1250, a Bell P-39D stalled and spun into the ground 15 miles southeast of Paine Field, Everett, Washington, killing pilot 2Lt. James O. Woodruff. Lt. Woodruff fell to his death when he left the airplane at too low an altitude to allow his parachute to fully deploy. Witnesses on the ground observed the aircraft in a flat, counterclockwise spin until it struck the ground. The airplane did not burn but was destroyed utterly.

10-24-42C. Clarksville, Georgia. At 1706, a Curtiss P-40F flying in poor weather crashed to the ground two and one-half miles east of Clarkesville, Georgia, killing 1Lt. Harry T. Critz. The P-40 dove into the ground with power at an angle of approximately 45 degrees. Fog was reported at just above the treetops. Investigators stated, "[The] pilot was flying direct to Columbus, Mississippi, en route to Shreveport, Louisiana. He was letting down and upon breaking out of the low hanging clouds, approximately 200 to 300 feet [above ground level], executed a 180-degree turn, and either lost control of the ship or did not see the ground in time enough to prevent the crash. Location of the accident shows that the pilot was on course; although maps [were found] in the [P-40], no flight line was drawn. The weather conditions in this area were very bad."

10-24-42D. Blythe, California. At 1611, a Douglas DB-7B returning from a gunnery-training mission crashed while performing unauthorized aerobatics 20 miles north of Blythe, California, killing pilot SSgt. Cecil R. Lafollet and gunner Cpl. Robert P. Cady. The DB-7 was one of a flight of two DB-7B bombers that were observed flying at approximately 3,500 feet altitude after expending their ammunition at the gunnery range near Blythe. The two airplanes were seen executing slow rolls and loops by a 3rd Armored Regiment officer on the ground. Then while executing a steep peel-off to the right, one DB-7 stalled while inverted and fell off into a spin to the right. At approximately 500 feet above ground level and after about three turns of the spin, the bomber appeared to come out of the spin and go into a steep dive. The pilot seemed to have recovered control of the airplane momentarily. The DB-7 was then seen to dive into the ground and explode into flames upon impact.

10-24-42E. Bayou La Batre, Alabama. At 1939, a North American AT-6A flying in poor weather on a night navigation flight crashed under control with power into a swamp four miles northwest of Bayou La Batre, Alabama, killing pilot SSgt. Frank J. Borell, 26, Albany, New York, and passenger SSgt. Wilfred L. Barrios, 22, New Orleans, Louisiana. The airplane was on a scheduled training flight from Tyndall Field, Panama City, Florida, to New Orleans, Louisiana, when it flew into the ground at near flying speed, plowing through 150 yards of trees and underbrush before coming to rest. Thunderstorms were reported in the area at the time of the crash. The pilot had very few night-flying hours logged and very little instrument time. The airplane was not discovered until the morning of 10-25-42 and rescuers had to wade through two miles of swamp to reach the wreck.

10-25-42A. Atlantic Ocean, New York. At about 1100, two Republic P-47B aircraft collided in flight over Fire Island, Long Island, New York, and crashed into the Atlantic Ocean, killing 1Lt. Kenneth G. Presthus, 22, St. Paul, Minnesota (P-47B #41-6001), and 2Lt. Otto J. Zugelter, 23, Cincinnati, Ohio (P-47B #41-5999). The aircraft were part of a four-ship flight that had taken off from Bridgeport, Connecticut, on a gunnery-training mission. The flight flew at 2,500 feet in between overcast at 4,500 feet and a lower deck of clouds at approximately 1,000 feet above sea level. The flight sighted the target tow airplane at the three o'clock position and made a diving right turn in the general direction toward it, reaching a speed of 300 mph and losing about 1,000 feet of altitude. The flight leader, Lt. Presthus, and his wingman, Lt. Zugelter, descended into the lower cloud deck where they collided. The two other pilots, unaware that a collision had taken place, attempted to contact them by radio when they did not emerge from the clouds. The other two pilots circled the area for ten minutes before resuming their gunnery mission.

10-25-42B. Sweetwater, Texas. At 2005, a Vultee BT-13A flying poor weather collided with rising terrain 12 miles southeast of Sweetwater, Texas, killing pilot 2Lt. Vernon E. Dye, Pratt, Kansas, and passenger 1Lt. Thaddeus R. Hubbell, Watertown, Massachusetts. The airplane was on a navigation training flight and had taken off from Carlsbad, New Mexico, for Abilene, Texas, with another BT-13 at 1830. The two airplanes had been flying formation when they encountered low ceilings and instrument conditions. The other BT-13 turned away from the instrument conditions and at 2040 landed safely at Sweetwater, Texas, after circling the area for some time. The pilot had attempted to contact Lt. Dye by radio without success. Investigators speculated that Lt. Dye had attempted to fly beneath the overcast and collided with the terrain while descending. Investigation revealed that the airplane crashed violently into the ground at flying speed and bounded back into the air before descending back into the ground out of control, smashing itself to pieces and killing both fliers instantly. The wreckage and bodies were not located until 10-26-42.

10-26-42A. Stillwater, Minnesota. At 0837, a Taylorcraft L-2A stalled and crashed into the ground near Stillwater, Minnesota, killing student glider pilot Pvt. Michael Kuchynski and civilian flight instructor

Harry W. Stanke. The Accident Classification Committee stated, "It is the concensus of opinion that the accident was caused by the airplane stalling and falling into a spin at approximately 400 feet and striking the ground before recovery could be made. Upon examination of the airfoils immediately following the accident, they were found to be coated with frost. This is believed to be a contributing factor in that the stalling speed of the airplane was increased."

10-26-42B. Colorado Springs, Colorado. At 0837, an Aeronca L-3A attempting a take-off at the Army Air Base at Colorado Springs, Colorado, crashed out of control into a Lockheed F-4 (photo-reconnaissance version of the P-38 type) that was being worked on by four mechanics, killing three of them. The fourth mechanic, Pfc. Morris E. Krienski, received minor injuries. The L-3 pilot, 2Lt. Daniel W. Burrows, also received minor injuries. The mechanics killed on the ground were Pfc. Daniel J. Glacken, Cpl. Gerald F. Hogan and Pfc. Ernest E. Stanek. The pilot had taken off and climbed slightly before attempting a gentle left turn. During this turn, the aircraft apparently lost power. The pilot attempted to turn away from the ramp area but the airplane stalled and fell to the ground from an altitude of approximately 75 feet, striking the mechanics and the F-4 on the starboard wing and engine nacelle.

10-26-42C. Cross City, Florida. At 1025 EWT, a Bell P-39D attempting a landing at the Cross City Army Air Base, Florida, stalled and spun into the ground, killing pilot SSgt. Charles H. Wisler, Galion, Ohio. Investigators stated, "While on his approach, [SSgt. Wisler] was observed to overshoot the runway, raise the nose of the aircraft, and attempt a violent, nose-high turn. The airplane shuddered, seemed to recover, then, as though more violent pressure was applied, snapped over on its side, lost altitude, and went into the trees south of the field. The subsequent explosion demolished the airplane and cost the pilot his life. No material failure was indicated in this accident."

10-26-42D. Pendleton, Oregon. At 1358, a Lockheed-Vega B-34 crashed after being abandoned by its crew 50 miles west of Pendleton, Oregon, killing pilot 2Lt. John N. Boyd, who was unable to leave the airplane. The aircraft was on a tow target mission over the AAF Central Oregon Gunnery Range when the port engine failed. The pilot proceeded toward Walla Walla, Washington, but because of heavy haze, light rain and the failure of the radio, he became lost. The pilot flew toward Walla Walla for an hour and a half before he began circling the area of Pendleton. Believing the airplane to be low on fuel and unsure of his position, the pilot ordered the crew to abandon the airplane. The crew successfully parachuted to safety from an altitude of about 3,500 feet above ground level. A member of the crew who was descending in his parachute observed the B-34 to enter a medium banked turn to the right that turned into a tight spiral to the right before the airplane struck the ground and exploded into flames. Investigators speculated that the pilot failed to correctly trim the airplane before leaving his position and attempting to use his parachute, becoming trapped in the airplane by the forces of the spin. Engineer Pvt. Edward W. Winebrenner, assistant engineer Cpl. Stanley E. Pearson, and passenger Pvt. Robert L. Maisak were uninjured in their parachute jump.

10-26-42E. Luke Field, Arizona. At 1517, two Curtiss AT-9 aircraft collided in mid-air and crashed four and a half miles WSW of Luke Field, Phoenix, Arizona, killing two instructors and two students on board both airplanes. Both students were on scheduled dual transition flights. Investigation revealed that one AT-9 was flying south while the other was flying to the northeast when the collision occurred. Investigators faulted both crews for failure to see each other and avoid the collision. Civilian Flight instructor Kenneth R. Brooks, 30, San Gabriel, California, and student Aviation Cadet William H. Blake, 22, Tampa, Florida, were killed in the crash of AT-9 #41-5835; Instructor 2Lt. Robert P. Blair, Sacramento, California, and student Aviation Cadet Robert B. Longfellow, 21, Frankfort, Indiana, were killed in the crash of AT-9 #41-12031. The aircraft fell to earth on the Scott Libby Ranch near Waddell.

10-27-42A. Deadman Lake, California. At 0030, a Vultee BT-13A returning from a glider tow mission crashed near Deadman Lake, California, killing 1Lt. Corwin L. Stockton. The airplane had released its glider and apparently flew into the ground soon after. There were no witnesses to the crash and investigators speculated that the pilot might have lost the horizon and flew the airplane into the ground while making a turn in the pattern for the airfield at Deadman Lake.

10-27-42B. Blackshear, Georgia. At 1025, a Vultee V-72 (similar to the A-31 type) suffering an engine failure was abandoned and crashed at Blackshear, Georgia, killing 2Lt. Cecil G. Dulaney, who fell to his death when he attempted to use his parachute too low to the ground. His body was found near the wreckage with his parachute strung out behind him. Gunner SSgt. Wilburn E. Gillaspie received minor injuries parachuting to safety. The aircraft was part of a four-ship formation that was returning from a simulated bombing mission at the AAF bombing range 55 miles northeast of Waycross, Georgia.

10-27-42C. Bates Field, Alabama. At 1044, an Aeronca TG-5 glider crashed one mile southeast of Bates Field, Mobile, Alabama, killing civilian glider instructor Leon Kubinski, 26, Elmira, New York, and student Pvt. John H. Schmauss, 23, St. Paul, Minnesota. The tow ring on the tow airplane broke off at its attachment point on the fuselage, leaving two TG-5 gliders connected to each other by 600 feet of rope. Neither glider pilot was able to release the rope from their glider. After positioning the gliders to a point that forced the rope taught, the pilots attempted to break the rope with extreme maneuvers. The rope eventually broke and remained attached to Pvt. Schmauss' glider, causing considerable drag and making the glider extremely difficult to maneuver. Civilian glider instructor L. C. Schrader, who was piloting the other glider, reported seeing the rope draped over the port wing of the TG-5 and pieces of the glider falling away. The glider stalled and entered an extreme spin to the left until it struck the ground, killing both student and instructor instantly.

10-27-42D. King City, California. At 0908, a Vultee BT-15 crashed while attempting a take-off at the AAF airfield at King City, California, killing pilots 1Lt. Richard W. Mansfield and 2Lt. Norton S. Dale. The Accident Classification Committee stated, "Lt. Mansfield and Lt. Dale were taking off on a cross-country trip to Burbank, California. They had just taken off and started a right hand turn. Apparently they flew into a heavy fog bank and lost control of the ship before an accurate reading of the instruments could be made. The position of the ship, and the crash marks, indicate that it crashed out of a diving turn to the right. It is believed that an attempt was made to roll the ship out of the turn and to level off before it hit, because of the fact that it hit fairly level and traveled almost 100 yards after striking the ground."

10-27-42E. Lakeville, California. At 1215, a Bell P-39D on a scheduled aerobatic mission crashed out control into the ground three and one-half miles east of Lakeville, California, killing 2Lt. John J. Niosi, Hamilton Field, California. The Accident Classification Committee stated, "The committee is unable to determine the cause of the accident as no witnesses could be located who saw the airplane crash. The airplane evidently dove into the ground in an inverted position and at about a 60-degree angle. The crash occurred almost on the top of a 700-foot hill. The crash and resultant fire completely demolished the airplane. The bottom of the rudder, the bottom of the motor and the landing gear struts were exposed and on the top of the wreckage. The pilot had jumped or had been thrown out of the airplane before the crash and his body had landed about 25 feet to the rear of the tail assembly of the airplane, and the door for the right side of the cockpit was found 1,200 feet from the crash. The body of the pilot had landed in a face down position and in an area which was not burned over by the grass fire which resulted from the crash."

10-27-42F. Ingram, Texas. At approximately 1600, a Beech AT-10BH flying in instrument conditions flew into the ground two miles north of Ingram, Texas, killing pilot 2Lt. Roscoe L. Maris, El Paso, Texas, and passenger SSgt. Anthony J. Mirro, Yonkers, New York. The aircraft had taken off from Curtis Field, Brady, Texas, for Kelly Field, San Antonio, Texas, at 1440 CWT and soon encountered instrument conditions. The aircraft struck the top of a heavily wooded mountain in a diving vertical bank to the left, exploding into flames and scattering wreckage over a wide area. The ceiling in the area just after the crash was estimated at 300 feet agl. Investigators speculated that the pilot became disoriented in the overcast, which was reported by other pilots to have exceeded an altitude of 5,000 feet.

10-28-42A. Romulus, Michigan. At 1027, a Vultee BT-13A struck and killed a walking ground crewman while taxiing at Wayne County Airport, Romulus, Michigan. Sgt. Kenneth C. Pate was killed instantly when was struck on the body by the rotating propeller of the BT-13. Sgt. Pate walked onto the taxiway directly in the path of the airplane (with his back to the BT-13), which was taxiing to take-off on an instrument proficiency flight. Pilot 2Lt. Brooks L. McElroy and his civilian passenger T. B. Rapsis were not injured in the accident and the BT-13 suffered minor damage. Investigation revealed that Lt. McElroy was taxiing in a safe manner and could not avoid striking Sgt. Pate, who walked directly into the path of the airplane after exiting a hangar. Investigators speculated that Sgt. Pate could not hear the airplane directly behind him because of dozens of aircraft on the flight line with their engines running.

10-28-42B. Tehachapi, California. At 1140, a Douglas A-20C flying in instrument conditions dove into the east side of the north-south ridge of Double Mountain, seven and one-half miles southwest of Tehachapi, California, killing four fliers. The A-20 struck the terrain at an angle of about 90-degrees 100 feet from the top of the 7,950-foot peak, exploding into flames and hurling wreckage and bodies down the ridge. Investigators could not determine what caused the airplane to crash. Killed in the accident were: Maj. James M. Liddell, age and hometown unknown, pilot; Sgt. Harry W. Short, 20, Barnsdall, Oklahoma, radio operator; SSgt. Morris G. Deason, 24, Clinton, Arkansas, passenger; TSgt. Oliver W. Murphy, 35, March Field, California. Visibility in the area at the time of the crash was reported as extremely poor with scattered snowstorms in the higher elevations.

10-28-42C. Novato, California. At 1715, a Bell P-39D spun into the ground three miles north of Novato, California, killing 2Lt. Preston L. Day. The aircraft was observed in a flat spin to the left at an altitude of about 2,000 feet above ground level. The pilot appeared to recover the airplane momentarily but entered a secondary spin at about 1,000 feet. The P-39 hit the ground in a flat spin to the left with very little forward speed. The airplane did not burn and the pilot was killed instantly. The pilot was on an assigned high altitude mission. Investigators speculated that the pilot might have been suffering from oxygen starvation. The oxygen mask was found attached to the pilot's helmet but it could not be determined if it had been functioning properly.

10-28-42D. Venus, Florida. At 2028, a Martin B-26B on a local night transition flight dove into the ground near Venus, Florida, killing the crew of seven. Investigators could not determine why the airplane crashed. The B-26 struck the ground at high speed and at an angle of nearly 90 degrees, with pieces of wreckage buried up to eight feet underground. The aircraft exploded violently into flames, destroying the aircraft and crew utterly. Only small pieces of wreckage were found and the remains of the crew were unrecoverable with only "very small pieces of human anatomy [found] scattered throughout the area." Killed in the crash were: 2Lt. Thomas H. Dunbury, Jr., pilot; 2Lt. David Furman, co-pilot; Sgt. Emory D. Barton, engineer; Sgt. Lloyd U. Wyatt, assistant engineer; Sgt Douglas J. Johns, radio operator; Sgt. Vernal M. Skorison, gunner; Sgt. Charles Schwamberg, gunner.

10-28-42E. Abernathy, Texas. At 2110, a Curtiss AT-9 suffering the failure of its port engine crashed seven miles southwest of Abernathy, Texas, killing student pilot Aviation Cadet Charles C. Liscomb, who fell to his death after he left the airplane at too low an altitude to use his parachute. He struck the ground near the wreckage of the airplane. Aviation Cadet Samuel A. Lyons parachuted to safety and was not injured. The AT-9 was part of three-airplane formation training flight. Investigators speculated that a broken oil line caused the failure of the port engine.

10-28-42F. Helotes, Texas. At 2115, a Beech AT-10BH on a weather checking mission crashed one and a half miles northwest of Helotes, Texas, killing flight instructor 2Lt. Harold J. Gerow, 27, Middleburg, Vermont, and student Aviation Cadet Wilson W. Edwards, 23, hometown unknown. The pilot had radioed Kelly Field, Texas, about 15 miles northwest of the field at 4,500 feet agl and reported that he could see the field with no problem. He reported scattered clouds at approximately 3,500 feet east of San Antonio. The Kelly Field control officer ordered the pilot to return to Kelly Field. The pilot acknowledged the order and this was last he was heard of. The aircraft struck the ground in a diving vertical bank to the left and burst into flames, killing the pilots instantly. The cause of the accident could not be determined.

10-29-42A. Newburgh, New York. At 0945, a Vultee BT-13A crashed while attempting a landing at Stewart Field, Newburgh, New York, killing Aviation Cadet Ernest S. Barker, Jr., 21, New Jersey, and Aviation Cadet L. A. Curcuru, 23, Ferndale, Michigan. Investigators stated, "It is believed that the pilot of the airplane was concentrating on his method of approach to the traffic pattern at 700 feet and either he or his passenger suddenly saw another plane which they believed to be dangerously close. The plane was put into a very steep left vertical turn with probably too much pressure being applied to the left rudder and stick causing the plane to stall and spin down to the left. Apparently a partial recovery from the spin was made but the altitude above the earth was not sufficient for total recovery. The plane crashed into a wooded area."

10-29-42B. Chickasha, Oklahoma. At 1443, two Fairchild PT-19A aircraft collided in flight near the municipal airport at Chickasha, Oklahoma, killing three fliers and seriously injuring another. Aviation Cadet Merrill S. Leonard and civilian flight instructor Wesley A. Spikes, flying PT-19A #41-15075, were killed when they collided with PT-19 #41-15003, piloted by civilian flight instructor Clifford M. Wickhorst and Aviation Cadet Eugene B. Overhoff. Mr. Wickhorst was killed and Cadet Overhoff was seriously injured. Apparently Cadet Leonard's aircraft was entering the middle of the downwind leg of the traffic pattern at a 45-degree angle and Cadet Overhoff's aircraft was turning to the downwind leg shortly after taking off from the field. Extremely strong winds caused Cadet Overhoff's aircraft to travel slowly over the field and he had not flown far enough from the field to safely enter the pattern. Instructor Wickhorst, seeing that a collision was inevitable, grabbed the controls but was unable to avoid the other aircraft. Cadet Overhoff, also seeing that a collision was unavoidable, released his safety belt and left the airplane immediately after the collision. The aircraft were both in a turning attitude at approximately 500 feet agl when the collision occurred and each lost a wing. Both aircraft plunged to the ground and caught fire.

10-29-42C. Greenville, Texas. At 2058, a Vultee BT-13A on a night navigation flight dove into the ground near Greenville, Texas, killing Aviation Cadet Raymond R. Knott. The cause of the crash could not be determined but investigators speculated that the student might have lost his horizon or might have become "confused" while encountering scattered clouds and turbulent air and inadvertently flew the airplane into the ground. Investigation revealed that the airplane struck the ground at a speed of approximately 250 mph in a near vertical dive.

10-30-42A. Cochran Field, Georgia. At 0005, a Vultee BT-13A returning from a local night flight crashed two miles south of Cochran Field, Macon, Georgia, killing Aviation Cadet Elmslie B. Von Gemmingen. The pilot had entered the traffic pattern for Cochran Field from the west after being instructed to enter from the east. Another BT-13 was flying on the correct base leg and was spotted by the pilot who attempted to tighten his turn to avoid a mid-air collision. The BT-13 stalled and entered a spin from which the student could not recover. The aircraft smashed into the ground and burst into flames, killing the pilot instantly.

10-30-42B. Gila Bend, Arizona. At 1515, a North American AT-6A collided in flight with an AT-6B while both were trying to join an authorized Lufbery circle without permission six miles south of the AAF Gila Bend Gunnery Range, near Gila Bend, Arizona, killing SSgt. Ivan M. Coppess, DeKalb, Illinois, and injuring Virgil L. Jones, Gila Bend, Arizona, who parachuted to safety from the AT-6B. The pilot of the AT-6A, civilian Walter W. McLatchy, 25, formerly of Montana, parachuted to safety and received minor injuries. The Accident Classification Committee stated, "These airplanes had been on a mission to the ranges to show new pilots the range boundaries where they would be flying as [target] tow pilots. This was not a formation flight and the airplanes were to follow in trail at a safe distance from their leading airplanes, which were showing them the area. Their instructions had been to check the ranges and on completion, return and land at Gila Bend Airport. Two officers leading the group on this tour showed them the ranges and started back to Gila Bend where they landed without joining [an] authorized formation [of three airplanes flying in a Lufbery circle]. The remaining airplanes, however, joined this formation and after several turns, the mid-air collision occurred. As near as could be determined, they were at the end of the formation. The airplane flown by SSgt. Coppess apparently came from underneath and struck the airplane flown by Mr. McLatchy, breaking that aircraft in two. The leader of the formation apparently had started to bring the formation in to land on noticing additional airplanes attaching themselves to this formation." SSgt. Coppess was found on the ground near the wreck of the airplane with his parachute partially opened.

10-30-42C. Wickenburg, Arizona. At a time after 1600 MWT, an AAF training airplane crashed near Wickenburg, Arizona, killing A/C Melvin M. Smith, 22, Winters, Texas. The airplane had been missing for some time. No other information is available.

10-31-42A. St. Paul, Minnesota. At 1610, a Consolidated B-24D on a test flight crashed while attempting to take-off from the municipal airport at St. Paul, Minnesota, killing three civilian fliers. Pilot Clarence Bates, co-pilot Robert J. Raley and engineer/radio operator Robert D. Bunn were killed in the crash. All three were residents of St. Paul and employees of Northwest Airlines. The Accident Classification Committee stated, "A B-24D airplane took off from the St. Paul Airport, gained altitude normally until about 50 to 75 feet in height. At this point, there was a sharp change in the attitude of the airplane, causing the ship to climb at a very steep angle until it reached a height of approximately 300 feet. It then made a sharp turn to the left and falling rapidly it struck the ground at an angle of approximately 30 degrees, the left wing striking the ground first. The plane immediately burst into flames. The rudder and aileron surfaces of three other B-24D airplanes were damaged by the fire. The Committee examined the elevators in the wrecked plane and observed that these were locked in the 'up' position." Further investigation revealed that the rudders were locked in the neutral position. Bates, the pilot in command, had 8,175 hours of flight time.

10-31-42B. Langley Field, Virginia. At 2025 EWT, a Consolidated B-24D attempting a landing crashed three and one-half miles southeast of Langley Field, Hampton, Virginia, killing the crew of three. The B-24 had taken off from Patterson Field, Ohio, at 1739 EWT on a ferry flight to Langley Field and had approached the field sometime after 2000. The pilot had radioed the control tower that an engine had failed and requested emergency landing instructions. After obtaining landing instructions, the bomber exhausted all of its fuel and struck trees. The B-24 was completely destroyed as it smashed through the trees and crashed into the ground. A small fire broke out after the crash, which was quickly extinguished by Langley Field rescue personnel. Killed in the crash were pilot Capt. William J. Foley, co-pilot 1Lt. James H. Moore and engineer Pfc. Gilford Wright.

November

11-1-42. MacDill Field, Florida. At 1045 EWT, a Martin B-26B returning from a local transition flight crashed while attempting a landing at MacDill Field, Tampa, Florida, killing the crew of six. The Accident Classification Committee stated, "Because of trouble with the left engine, the pilot called the Control Tower and requested an emergency landing. In attempting to land, the pilot made a steep left turn very close to the field. He passed the north end of the North-South runway, going in a southerly direction and at an altitude of approximately 300 feet. Upon reaching the center of the field, the wheels were retracted and full power applied in an attempt to go around. The left engine failed and with full power on the right engine, the airplane went into a climbing turn to the left, rolled over [to inverted] and struck the ground in a near vertical position. Upon contacting the ground the airplane caught fire immediately. The pilot erred in that he overshot the field and attempted to go around knowing that he had partial failure of the left engine. The cause of the left engine failure is undetermined as the airplane was totally destroyed by the impact with the ground and by the fire." Killed in the crash were: 2Lt. William E. Kyle, pilot/B-26 student; 1Lt. William D. Malasky, co-pilot/B-26 instructor; Pfc. Charles E. Jacot, engineer; Pvt. Edmund Rohr, radio operator; Pvt. Harold W. Keebaugh, passenger; Pvt. Richard B. Rugger, passenger.

11-2-42A. Dorr Field, Florida. At 0840, a Stearman PT-17 crashed near Dorr Field, Arcadia, Florida, killing A/C Eugene A. McGee, Jr. The student was unable to recover from a deliberate spin. Witnesses stated that the airplane was spiraling down from a higher altitude to a lower altitude with partial power. The pilot apparently did not apply correct rudder inputs, which resulted in an increase in the steepness of the spin. The student was unable to recover the airplane in time to prevent it from striking the ground. The PT-17 burst into flames upon impact.

11-2-42B. Leesburg, Florida. At 0903, a Douglas DB-7B participating in a simulated interception mission spun to the ground four miles south of Leesburg, Florida, killing pilot 2Lt. Rudolph P. Weisal, Jr. Radio operator Pvt. William Thrasher parachuted to safety and was uninjured. The DB-7B was intercepted by four P-40 fighter aircraft at an altitude of approximately 11,000 feet and took evasive action. The fighters, flying slightly above and behind, peeled off and attacked from the left rear of the bomber. The DB-7 pilot attempted a steep climbing turn to the left. The airplane stalled and entered a flat spin to the left. The pilot ordered the radio operator to leave the aircraft at an altitude of approximately 5,000 feet. The pilot was unable to recover from the spin and the airplane crashed into the ground.

11-2-42C. Union, South Carolina. At 1130, a Vultee BT-13A on a navigation training flight crashed while attempting a deferred forced landing near Union, South Carolina, killing Aviation Cadet Lionel L. Lozier. The student was to fly a navigation problem between Laurens, South Carolina, and Maxton, North Carolina. The student was apparently attempting to land in a field in an effort to determine his location after becoming lost. The student, coming in with excessive speed, overshot the field. The accident occurred when he abruptly pulled the aircraft up in an effort to avoid striking trees and a power line at the edge of the field. The aircraft stalled at low altitude and crashed. Investigators could not locate the student's maps and speculated that he had lost them and became unsure of his position, prompting his attempted landing.

11-2-42D. Victory Field, Texas. At 1400, a Fairchild PT-19A suffered a catastrophic structural failure of the port wing after entering a deliberate spin three miles northwest of Victory Field, Vernon, Texas, killing civilian flight instructor Vaughn W. McCallister, Lovington, New Mexico. Aviation Cadet Ralph S. Travis received minor injuries parachuting to safety. Investigators stated, "After completing two spins, student climbed to 5,200 feet and entered a third spin. Plane stopped its rotation after two turns but failed to pull out of the ensuing dive. Student and instructor exerted full strength in an effort to pull stick back but failed. At 2,000 feet, left wing pulled off. At this point, student bailed out, after he experienced considerable difficulty in getting out of the plane. Instructor apparently attempted to jump but was struck by part of the airplane, possibly the tail group. Ripcord was still in the housing but [parachute] pack had burst with the force of the impact." The aircraft did not burn when it smashed into the ground.

11-2-42E. Muroc, California. At 1250, a Lockheed P-38 on a gunnery training mission collided with a target at the gunnery range near Muroc Army Air Base, Muroc, California, killing pilot 2Lt. Troy M. Bibb. Pvt. Lyle E. Rawson witnessed the crash and stated to investigators, "The plane [flew] through the target while in level flight, breaking away all but the right standard of the [target] frame. [The P-38] continued in level flight for a short distance, then turned slightly to the right, made a half roll and crashed in [an inverted] position. Upon contact with the ground, the plane burst into flames and bounced along for approximately one-quarter of a mile, disintegrating along the way and leaving an almost unbroken wall of flame. When I arrived at the place where the pilot lay, he was in a huddled position with his skull broken open and the right leg cut off just below the knee."

11-2-42F. Gardner, Florida. At 1645, a Martin B-26B on an instrument training flight crashed four miles ESE of Gardner, Florida, killing the crew of six. Killed in the crash were: 2Lt. Lewis O. Ryer, pilot; 2Lt. Billy Y. Elliott, co-pilot; 2Lt. Clayton A. Christensen, bombardier; Sgt. Robert W. Lewis, engineer; SSgt. Charles W. Simpson, radio operator; Sgt. Rudolfo Cortez, gunner. Aviation Cadet Fred Carman was flying in the area and witnessed the accident, later stating to investigators, "I was flying about 3,000 feet and saw the B-26 ahead of me go into a spiral which got tighter as the airplane went down. It turned seven or eight times and did not come out at any time. The ship exploded as it hit. I estimate that the ship was at cruising speed when the spiral started. The left wing was down [in] the spiral. The airplane hit [the ground] between two trees."

11-2-42G. Providence, Rhode Island. At 1710, two Curtiss P-40F aircraft collided in flight near Green Field, Providence, Rhode Island, killing 2Lt. William H. Pierson aboard P-40F #41-14135. SSgt. John W. Smallsreed parachuted to safety from P-40F #41-14183 and was uninjured. He later stated to investigators, "Lt. Pierson and myself were in formation climbing to participate in two ship combat. A third ship [a P-40E] made a few passes at us and we started after him. This was at approximately 7,000 feet northwest of Providence, six to eight miles from town. We broke formation and went into a Lufbery circle. The P-40E was in No. 1 position, I was No. 2, Lt. Pierson was No. 3. I felt a jar and saw Lt. Pierson break away to the right. That was the last I saw of him." Lt. Pierson's body was discovered 11-30-42 in Narragansett Bay.

11-2-42H. Hallettsville, Texas. At 2010, a North American AT-6C on a night navigation flight collided with terrain seven miles southeast of Hallettsville, Texas, killing A/C Nathan F. Sample. Investigators reported, "While on a night navigation flight, student pilot apparently flew into the ground when he attempted to dive out of a cloud formation rather than go on instruments. The angle of impact was beyond a 45 degree angle, engine was buried six to eight feet in the ground and the airplane was completely demolished." Other students that had taken off from Foster Field, Victoria, Texas, with Cadet Sample encountered the same cloud formations and successfully competed the navigation problem.

11-3-42A. St. Petersburg, Florida. At 1048, a Curtiss P-40F crashed one and a half miles southwest of Pinellas Field, St. Petersburg, Florida, killing pilot SSgt. James E. Campbell. The pilot had completed his turn from the downwind leg to the base leg and was seen to fall off in a stall to the right and dive into the ground. The aircraft burst into flames upon impact. Several witnesses on the ground reported seeing thick black smoke emitting from the engine area just prior to the crash.

11-3-42B. Maxwell Field, Alabama. At 1435, a North American AT-6A and an AT-6B collided in mid-air and crashed near Maxwell Field, Montgomery, Alabama, killing RAF Pilot Officer Lorenzo D. Coupland aboard the AT-6B. AT-6B pilot 1Lt. Russell O. Newburg parachuted to safety and was uninjured. RAF Pilot Officer Ernest C. Pye, flying the AT-6A, parachuted to safety and was uninjured. Investigators stated, "This accident happened during a four-ship formation training flight. The number-three [Newburg] and number-four [Pye] ships of the formation collided in a right climbing turn. Pilot Officer Pye, pilot of the number-four ship, was not in the proper position during this turn. Had he been, he would have been able to see the number-three ship on any maneuver. He states that he could not see the number-three ship after they went into the turn and at the same moment the two ships collided. The pilot in the rear seat of the number-three ship [Coupland] was not flying the airplane at the time of the accident and he failed to get out of airplane before it crashed. It is believed by this board that the tail assembly of the number-three ship hit the left wing of the number-four ship and from investigation, it was found that the propeller of the number-four ship cut the fuselage of the number-three ship in two just behind the rear seat."

11-3-42C. Greenville, Mississippi. At 1623, a Beech AT-10BH crashed while attempting to take-off from the AAF Flying School at Greenville, Mississippi, killing 1Lt. Phillip E. Thornton and 2Lt. S.T. Carter. "The Accident Classification Committee finds that this accident was the direct result of pilot error on the part of 1Lt. Thornton, in that he failed to unlock the controls prior to take-off. The airplane took off, climbed steeply to approximately 100 feet, stalled and dived into the ground. The pilots were killed and the airplane was completely destroyed in the crash and fire [that] followed. Examination of the wreckage disclosed that the control-locking pin was still in the locked position. Had the pilot or the co-pilot made a proper cockpit check prior to take-off in accordance with the pilot's check list in the cockpit, he would have unlocked the controls and the accident would not have occurred."

11-3-42D. Smyer, Texas. At 1950, a Cessna AT-17 crashed 10 miles south of Smyer, Texas, killing A/C Kenneth M. Reid and A/C William N. Phillippe. The AT-17 was part of a three-ship formation training flight that was forming up just after take-off. The other pilots in the formation noticed the airplane execute what appeared to be a cross-over turn, although the flight leader had given no signal for such a maneuver. It was also observed that the cockpit lights were on as the airplane passed over from left to right and fell out of view. The other pilots next noticed the airplane when it was exploding as it contacted the ground. The cause of the accident could not be determined but investigators speculated that the students might have been struggling with an engine failure.

11-3-42E. Enid, Oklahoma. At 2035, a Vultee BT-13A dove into the ground 12 miles west of the AAF Flying School, Enid, Oklahoma, killing civilian flight instructor Horton F. Gunn. Gunn was flying in the number-two position of a three-airplane flight on a night formation-training mission. The aircraft had been in the air about 30 minutes when they began to encounter clouds at the 4,000-foot level. The flight leader decided to descend through the rapidly thickening cloud layer, descending down to about 2,000 feet. Upon descending out of the clouds, Gunn fell well back of the formation and was observed flying erratically. The pilots in his flight lost sight of him temporarily and he was next seen in a dive at low altitude. After leveling off momentarily, the BT-13 dove back at the ground, striking with power on and at high speed. The pilot, who had logged over 3,576 flight hours, was killed instantly. There was no fire but the airplane was completely destroyed.

11-4-42A. Biggs Field, Texas. At 0303, a Boeing B-17E crashed just after take-off from Biggs Field, El Paso, Texas, killing three crewmembers and seriously injuring six others. The Accident Classification Committee stated, "The take-off was from west to east, on the West-East runway. The pilot stated that he attained a speed of 115 MPH and ran nearly to the end of the runway before he was able to get the plane into the air. [The] airplane crashed and was completely destroyed immediately following take-off on a local night flight. The airplane made a normal take-off but due to lack of power did not remain airborne and crashed about three-quarters of a mile from the point of take-off. The co-pilot, immediately after the plane became airborne and the wheels were retracted, and before the plane had attained sufficient altitude and flying speed, completely closed the turbo-superchargers. This caused the plane to lose the small amount of flying speed it had attained and [the B-17] settled until it contacted the ground and crashed." The B-17 burst into flames and Biggs Field rescue personnel were on the scene immediately. Killed in the crash were: 2Lt. Robert E. Brasfield, navigator; 2Lt. Lloyd G. Compton, bombardier; SSgt. Garfield A. Reusch, radio operator. Seriously injured in the crash were: 2Lt. Charles W. Byrd, pilot; 2Lt. William J. Denton, Jr., co-pilot; SSgt. Warren H. Clark, engineer; Sgt. Floyd R. Hudgens, assistant engineer; Sgt. Grant P. Donahue, assistant radio operator; Sgt. John C. Schleibaus, tail gunner.

11-4-42B. Palos Verde, Arizona. At 0313, a Curtiss AT-9 on a rendezvous mission crashed five miles north of Palos Verde, Arizona, killing 2Lt. William M. Bills and Aviation Cadet Wendell M. Baie. Investigators stated, "Pilot apparently entered a heavy bank of clouds and then called the Phoenix radio station for permission to fly under them. [The airplane] was heard flying low and first observed coming out of a low cloud with the engines wide open. The plane crashed into the ground with great violence, completely demolishing the ship and scattering wreckage over an area of 100 yards. Bodies were found about 75 to 150 yards from the plane. Parachutes were open but not used. It is believed that they were opened by impact with the ground. Both student and pilot were killed instantly. The cause of the accident could not be determined except that it is felt that the pilot became confused upon entering an area of cloud formation."

11-5-42A. Cross City, Florida. At 0840, a Bell P-39D crashed three miles northwest of the Army Air Base at Cross City, Florida, killing 2Lt. Ellis J. Ackerman, who was unable to recover the aircraft after entering a deliberate spin. The airplane was in a flat spin and struck the ground at high speed in a vertical attitude, exploding into flames upon impact. The pilot was assigned to conduct a center of gravity and spin test flight in response to a series of unusual spin accidents involving P-39 type aircraft.

11-5-42B. Myrtle Beach, South Carolina. At 1815, a Martin B-26B suffering a landing gear malfunction and a blown tire collided with a crash truck while attempting an emergency landing at the Army Air Base at Myrtle Beach, South Carolina, killing four civilian firemen aboard the crash truck. Three others aboard the truck were injured, two seriously. The B-26 blew its port main gear tire just as it became airborne on take-off, the tire and tube being shed from the rim. The pilot was ordered by tower personnel to circle the field so that other airplanes in the traffic pattern could be landed and the situation assessed before he attempted the emergency landing. The pilot radioed that he was getting an affirmative signal that the nose landing gear was down and locked while both main landing gear indicated they were in the retracted position, but did not show an affirmative signal for the locked position. Officers in the tower ordered the pilot to land on the north shoulder of the East-West runway with the nose gear down and the main wheels up. By this time, several crash trucks and ambulances were lined up on the edge of the field, ready to speed to the crash landing when it occurred. The pilot attempted the emergency landing and was able to keep the airplane on the shoulder for approximately 500 feet before it began veering to the right (south). Upon the shock of the landing, the port main gear dropped out of its well and gouged into the ground, ripping it from the airplane. The B-26 continued to veer to the right and lose speed. The line of crash trucks, anticipating the airplane coming to a rest, started onto the runway. The first crash truck in the line was unable to avoid being struck by the careening aircraft. There was no fire. Killed in the accident were: Don Allen, Malcolm A. Johnson, Newberry Sawyer and John D. Sessions, all residents of Conway, South Carolina. Civilian firefighter Joseph T. James, Loris, South Carolina was seriously injured. Capt. Richard T. Travis received minor injuries and Pvt. Wesley T. Collins was seriously injured aboard the crash truck (both AAF). Two members of the B-26 crew, bombardier 2Lt. William F. Carbonneau and engineer Sgt. William F. Crow received minor injuries. Pilot Straughan D. Kelsey, co-pilot 2Lt. Edward D. Sargent and navigator 2Lt. Melville A. Erickson were uninjured.

11-5-42C. Pearson, Texas. At 1800, a Vultee BT-15 crashed into the ground two miles west of Pearson, Texas, killing Brazilian Air Force student A/C Eduardo M. J. Demello. Witnesses on the ground reported that the aircraft had been performing aerobatics at low altitude. The aircraft had been seen diving on a freight train prior to the accident, clearing the cars by about 6 to 10 feet. The aircraft was seen to go into a vertical climb from which it stalled and fell over onto its back. The pilot was unable to recover the airplane and it crashed into the ground, bursting into flames and incinerating the pilot.

11-6-42A. Memphis, Tennessee. At 0449, a Douglas C-47A crashed just after take-off from the municipal airport at Memphis, Tennessee, killing pilot Flight Captain Raphael C. Clayton and co-pilot Pilot Officer Jesse Norman Rhew, both employees of Chicago & Southern Airlines. The Accident Classification Committee stated, "Pilot took off on the southwest runway in moderate rain. Upon reaching the southwest border he went into a very heavy rain. A half-mile from the border he crashed into trees approximately 75 feet high. Investigation of the control assemblies showed the elevator trim tab to be extremely nose heavy. The ship was not carrying any freight; [it had been serviced with 720 gallons of gasoline and carried no passengers]. Investigation of all witnesses proves that there was no engine failure. The committee believes that the combination of the ship's being trimmed nose heavy and its sudden contact with the heavy rain immediately following take-off caused the ship to make contact with the trees." The aircraft crashed and burst into flames approximately two miles southwest of the runway and had been cleared on an instrument flight plan for Jackson, Mississippi. The C-47 was to return to Memphis with an unspecified load of cargo.

11-6-42B. Spence Field, Georgia. At 0930, a North American AT-6B crashed after blowing a cylinder head while in the pattern for Spence Field, Moultrie, Georgia, killing instructor pilot 2Lt. Samuel L. Mitchell and student A/C Arthur J. Graf. Investigators stated, "It is the opinion of this board that a cylinder head broke and came off. In coming off, the ring cowl was broken allowing it to go forward into the propeller. The propeller caught the ring cowl and threw it into the canopy and tail section causing the plane to crash from 700 feet where the pilot was entering traffic. No information could be ascertained from the engine, which was completely destroyed, but the ring cowl, which was 400 yards from the accident, showed clearly where it had rubbed the propeller and where the propeller had caught and thrown it clear. The canopy had been severely smashed before the crash and the tail section showed signs of being severely damaged before the crash. The engine ring cowl had several small gashes which indicated that small pieces had been driven from the inside out."

11-6-42C. Denton, Texas. At 0927, a Taylorcraft L-2A smashed into the ground 12 miles northwest of Denton, Texas, killing civilian flight instructor Joseph W. Mosely and seriously injuring student A/S Albert P. Telles. Investigators speculated that the pilots became lost and low on fuel and were looking for a place to land. The pilot stalled the airplane at a low altitude and it spun to the ground before the he could safely recover from the maneuver.

11-6-42D. Muroc, California. At 1043, a Lockheed P-38F on a formation training flight suffered the failure of the starboard engine and crashed two miles northwest of the Army Air Base at Muroc, California, killing 2Lt. Wayne Hildreth, Jr. Investigators stated, "[Lt.] Hildreth was flying approximately 1,500 feet on the downwind leg of the traffic pattern when his right engine cut out, he had lowered his wheels and had made a right turn attempting to make a forced landing on the dry lake bed, putting his dead engine down and causing the airplane to enter into a right spin, making one and one-half turns before crashing into the ground on its nose; the airplane exploded and burned."

11-6-42E. Pinellas Field, Florida. At 1355, a Curtiss P-40F crashed near Pinellas Field, St. Petersburg, Florida, killing 2Lt. Virgil C. Ankron. The airplane smashed into the ground at high speed and at an angle of about 40 degrees. There were no witnesses to the actual crash but a civilian stated that he had seen the airplane perform two loops and other aerobatic maneuvers at low altitude just prior to the accident. The cause of the accident could not be determined.

11-6-42F. Lambert Field, Missouri. At 1640, a Martin B-26A crashed while attempting a take-off at Lambert Field, St. Louis, Missouri, killing engineer Cpl. C.L. Linn of Walnut, Iowa. Pilot 2Lt. E.F. Murphy, New York, and co-pilot 2Lt. C.J. Brown, California, were seriously injured. Investigators stated, "Pilot took off to the [northeast]. About half way down the runway the plane was pulled off the ground. The right engine appeared to lose power. Pilot set the plane down immediately about [540] feet from the [northeast] end. Brakes were applied 160 feet later. Tire marks show that the right brake grabbed and that the left brake was ineffective. This caused the plane to pull to the right until it finally left the runway 130 feet from the end. At this point, the [starboard main] tire blew out. The airplane hit the top of the far bank of a creek, the impact breaking the cockpit open enough so that the pilot and co-pilot escaped from the plane just before the fire started. The plane continued on for about 50 feet where the plane burst into flames. The third occupant of the ship was trapped and died in the fire." The B-26 was scheduled for a cross-country navigation training flight.

11-7-42A. Sumter, South Carolina. At 0830, a Vultee BT-13A stalled and spun into the ground eight miles south of Shaw Field, Sumter, South Carolina, killing Aviation Cadet Stephen J. Ofcarick. The airplane was one of a three-airplane flight that had been practicing formation flying. The formation was flying a loose trail type formation when the leader began a steep turn to the left. Cadet Ofcarick, flying in the number-three position, stalled the airplane and fell out of the turn at an altitude of 2,000 feet. The airplane entered a spin from which the pilot was unable to recover.

11-7-42B. Victoria, Texas. At 1000, a Republic P-43A crashed five miles south of Foster Field, Victoria, Texas, killing 2Lt. Robert L. Jones, 25, Western Springs, Illinois. Investigators speculated that the pilot endeavored to catch up with a flight of P-40 type aircraft that were flying nearby. The pilot put the airplane in a dive to pick up speed and attempted a steep chandelle maneuver from which he stalled out of at the top. The P-43 rolled inverted and began descending. The airplane was seen to be diving straight for the ground and the pilot apparently pulled the airplane through to a near flying attitude and almost recovered, striking the ground at an angle of about 15 to 20 degrees. The airplane skidded along the ground and burst into flames.

11-7-42C. Occoquan Bay, Virginia. At 1230, a Curtiss P-40F crashed into Occoquan Bay, Virginia, killing pilot 2Lt. Richard R. Barber, 22, Dallas, Texas. The aircraft was observed in a tight spin at an altitude of approximately 1,000 feet. The pilot, apparently trying to break the spin, pulled the airplane up into a half-loop and went inverted. The aircraft fell off in a vertical dive toward the surface and hit nose first and sank immediately. The tail of the airplane was seen to be moving side to side just before it impacted in about 8 to 12 feet of water. The airplane was completely destroyed and very little wreckage was recovered on the surface.

11-7-42D. Antigo, Wisconsin. At 1150, a Piper L-4A crashed 24 miles north of Antigo, Wisconsin, killing student pilot Pvt. William Wein. Investigators stated, "[The pilot] was seen to circle Post Lake at an altitude of approximately 300 feet then cut his motor and let down to about 50 feet from the water and approached a group of hunters who were standing on a pier that extended out into the lake. Upon approaching the bank, the plane banked sharply to the right and attempted to clear the embankment and trees. The embankment at this point was about 25 feet high and the trees were from 50 to 70 feet high. The plane failed to clear the trees and crashed and burned about 150 yards from the edge of the lake."

11-7-42E. Viola, Minnesota. At 1540, a Taylorcraft L-2A crashed near Viola, Minnesota, killing student pilot Pvt. C. R. Turner. Investigators stated, "Witnesses observed the airplane doing spins at a safe altitude. After the airplane had completely recovered from the spins, it glided down to an altitude above the level of the treetops. Witnesses stated that for a period of about fifteen minutes, the airplane was flown at a very low height and at times lower than the level of the treetops. At this low altitude, the airplane was flown in an area about two miles square that necessitated numerous turns. Just before the airplane crashed, witnesses observed the airplane start to turn at an altitude of about 50 feet when the nose of the plane was lowered and remained in this position until the airplane struck the ground."

11-7-42F. Jacksonboro, South Carolina. At 1730 EWT, a North American B-25C on an altimeter calibration flight crashed into a swampy area near Jacksonboro, South Carolina, killing the crew of four. Investigators stated, "The airplane crashed in a swamp from a considerable altitude according to civilian witnesses, which is estimated at 6,000 feet. Some witnesses stated an explosion was heard in the air before the aircraft crashed. It is possible that gasoline fumes had collected inside the wing or within the wing tanks in sufficient quantities to explode by static electricity or electric spark, or by spontaneous combustion caused by the latex sealant compound in the self-sealing fuel tanks. This would have blown off one engine and part of one wing. Parts and contents of the aircraft were scattered over a distance of two and one-half miles. A parachute found about one mile away was opened but had not been used by anyone in the crew as investigation of the rigging, seams, and creases proved. This parachute was thrown from the aircraft." Because of the lack of evidence the cause of the accident remains undetermined. Killed in the crash were: 2Lt. Douglas W. Cole, pilot; 2Lt. Forrest E. Carter, co-pilot; 2Lt. William C. Stewart, navigator; Sgt. Gene E. Lefler, engineer.

11-8-42A. Key Field, Mississippi. At 1133, a Douglas A-24A stalled and spun into the ground while attempting to land at Key Field, Meridian, Mississippi, killing pilot 2Lt. Victor Windus, 25, Allentown, Pennsylvania (and Denver, Colorado), and gunner Sgt. Jack L. Condo, 23, Monticello, Indiana. Pfc. C. McGarry witnessed the crash and stated to investigators, "While on duty as recorder in the tower, I was looking southwest from the tower. Over the North-South runway, a ship was seen in a spin about 100 feet above the ground. It crashed at about a 90 degree angle and burst into flames immediately."

11-8-42B. Battle Mountain Park, Oregon. At 1500, two Lockheed P-38F fighter aircraft flew into terrain while flying in formation five miles east of Battle Mountain Park, Oregon, killing both pilots. 1Lt. James A. Huffman and 2Lt. Robert A. Walstrom collided with Blue Mountain 25 miles SSW of Pendleton Field, Oregon, after losing contact with the other two P-38 aircraft that they were leading. The four-aircraft flight had taken off at 1345 from Burns, Oregon, for Pendleton Field an hour and a half after making the flight from Pendleton Field to Burns earlier in the day. Investigation revealed that the flight leader, Lt. Huffman, did not wait for the proper weather report before taking off. 1Lt. Gerald F. Leinweber and 2Lt. Grover E. Fanning were instructed by Lt. Huffman to make a 180-degree turn back for Burns if poor visibility was encountered. After dodging low-hanging clouds and rain showers, the flight encountered thick, low-hanging clouds just prior to arriving at Pendleton Field. Lt. Leinweber and Lt. Fanning executed a 180-degree turn, circled in a clear area, and soon found a clear way to Pendleton Field where they landed safely. Lt. Huffman and Lt. Walstrom were able to maintain formation while in the instrument conditions and soon collided with the mountain. Three hunters staying in a nearby cabin reported that they saw the fighters flying together before they disappeared into the clouds and collided with the mountain one-quarter of a mile distant.

11-8-42C. Battle Mountain Park, Oregon. At 1500, 2Lt. Robert A. Walstrom was killed when he flew into rising terrain with a Lockheed P-38F five miles east of Battle Mountain Park, Oregon. See 11-8-42B for details.

11-8-42D. Scott Field, Illinois. At 1930, a North American AT-6A flew into terrain while attempting an instrument approach to Scott Field, Belleville, Illinois, killing 1Lt. George C. Alchorn, 25, Harlingen AAB, Texas, and infantry officer passenger 2Lt. Robert F. Sullivan, 25, Lubbock, Texas. The AT-6, with the landing gear retracted, was letting down when the propeller contacted the ground. The airplane bounded back into the air and then contacted the ground at near flying speed 50 yards beyond the initial point of contact, bursting into flames and scattering wreckage for 250 yards.

11-8-42E. Waycross, Georgia. At 2110, a Vultee V-72 (similar to the A-31 type) returning from a night navigation flight crashed near the Army Air Field at Waycross, Georgia, killing pilot 2Lt. Rodney N. Sheain and seriously injuring gunner Sgt. J. W. Emminger. Sgt. Emminger was ordered by the pilot to leave the airplane and he was struck by the tail section while attempting to parachute to safety. He suffered a broken pelvis. Investigation revealed that Lt. Sheain left the airplane at an altitude too low to allow his parachute to fully deploy, causing him to fall to his death. His body was found near the wreckage with the parachute partially open. Evidently the airplane ran out of fuel while on the downwind leg for the field at Waycross.

11-9-42A. Dover, Delaware. At 1420, a Lockheed-Vega B-34 crashed while attempting a landing at the Dover Airdrome, Dover, Delaware, killing six fliers. The bomber made what appeared to be a normal approach for the runway with a 15 mph crosswind with gusting. The airplane lost an excessive amount of altitude and the pilot soon realized that he was going to undershoot the runway and collide with building materials that were stacked near a runway extension that was under construction. He added power and pulled up the nose of the aircraft. The B-34 climbed about thirty feet and stalled, falling off on the left wing. The airplane plunged to the ground in a diving left turn, exploding into flames upon impact. One crewmember was pulled from the wreckage alive but died soon after. Investigation revealed that the elevator trim tab was set to the nose high position. Killed in the accident were: SSgt. Gerald F. Nachreiner, pilot; Sgt. Richard B. Hens, engineer; Sgt. Leonard B. Cottongin, assistant engineer; Cpl. John H. Ramos, radio operator; Sgt. Ewald H. Bruns, crewmember; Pfc. Tony B. Latona, crewmember.

11-9-42B. Ridgeway, Illinois. At 2320, a North American B-25D ran out of fuel in flight and was abandoned by its crew, crashing to the ground near Ridgeway, Illinois, and killing Pvt. T. W. Owens, who neglected to parachute to safety and went down with the aircraft. Six crewmembers received minor injuries parachuting to safety. The aircraft was on a night navigation training flight from Columbia, South Carolina, to St. Louis, Missouri, and encountered instrument conditions in between St. Louis and Evansville, Indiana. The pilot decided to return to Evansville. He soon encountered severe and unfavorable winds and was unable to maintain a direct course for Evansville. The pilot, realizing the aircraft was extremely low on fuel, ordered the crew to get into their parachutes and to stand by for orders to bail out. The pilot gained a small amount of altitude before the starboard engine began to sputter because of fuel exhaustion. The pilot ordered the crew to leave the airplane. Investigators determined that Pvt. Owens' parachute was in working condition and it was not known why Pvt. Owens did not leave the airplane. Receiving minor injuries parachuting to safety were: 2Lt. Harold H. Wilkerson, pilot; 2Lt. Benjamin S. Howle, co-pilot; SSgt. Frederick S. Helms, engineer; SSgt. R. S. Berninger, radio operator; Sgt. George Goacher, gunner; Pvt. George G. Johnson, crewmember.

11-10-42A. Blytheville, Arkansas. At 0855, a Beech AT-10 crashed while attempting to take off from AAF Advanced Trainer Instructor School at Blytheville, Arkansas, killing student instructor pilots 2Lt. Harley D. Carrell and 2Lt. John E. Hayden. Immediately after take-off, the airplane was seen to go into a steep climb, stalling at approximately 200 feet above ground level. The AT-10 fell off on the port wing and plunged to the ground, exploding into flames upon impact. Investigators stated, "Upon investigation of the wreckage, it was found that the control lock was in position for locking and held there by the pin which passes through the lock yoke and aileron control chain. The lock, however, was not secured to the rudder bar. This allowed free movement of the rudder and backward movement of the control column, but aileron control and forward movement of the control column were restricted. It was possible to taxi out and take-off with such restricted control, but when forward movement of the column was required to keep ship from climbing too steeply, this movement was restricted and the ship stalled."

11-10-42B. Lockhart, South Carolina. At 1603, a Douglas C-47 suffered a progressive catastrophic structural failure after encountering severe turbulence, crashing six miles northwest of Lockhart, South Carolina, and killing the crew of three. The C-47 was part of flight of thirteen that had been cleared from Pope Field, North Carolina, to the AAF Air Depot at Mobile, Alabama. The flight encountered a severe front and broke up. Some airplanes went around the front while several others decided to punch through the front, including the accident airplane. Investigators stated, "The airplane was seen flying at a fairly high altitude with both engines running, and that parts of the airplane were seen falling to the ground. Airplane disintegrated in the air in a flight path ESE to WNW over an area of approximately two and one-half miles. Elevators, rudder, vertical and horizontal stabilizer, both wing tips, right aileron, sections of right wing to base of aileron well, cargo and passenger doors, door frame, left wing de-icer boot and large accumulation of unidentifiable scrap found in this area. Right engine fell out 100 yards beyond the crash, apparently while the ship was spiraling in." Killed in the crash were pilot 1Lt. George W. Peterson, co-pilot SSgt. Irwin C. Hutchinson and engineer TSgt. Donald Sweigart.

11-10-42C. Kelly Field, Texas. At 1545, a Beech AT-10BH crashed while attempting a take-off at Kelly Field, San Antonio, Texas, killing 1Lt. Burris W. Pacelia and 2Lt. James W. Bigbee. The Accident Classification Committee stated, "This accident occurred in the center of Kelly Field. The airplane started its take-off, gained flying speed on the ground, then suddenly climbed very steeply, almost vertically to an altitude of about fifty feet, did a wing-over to the left, and flew into the ground. Both pilot and co-pilot were killed in the crash. The airplane was completely demolished." Investigation revealed that the pilots attempted their take-off with the aileron and control column in the locked position with the locking pin in place. Free movement of the rudder and backward movement of the control column were possible with the lock in this position. Forward movement of the control column and aileron control would be restricted with the locking pin in place. Investigators faulted both pilots for not checking for the free movement of all control surfaces prior to attempting their take-off.

11-10-42D. Noble, Oklahoma. At 1710, a Curtiss P-40F crashed one and one-half miles southeast of Noble, Oklahoma, killing 2Lt. Levi A. Schaffer, 24, Johnston, Pennsylvania. Civilian witnesses on the ground stated that the P-40 was in a vertically banked turn to the right when it fell out of the turn and dove straight to the ground, exploding into flames upon impact. Lt. Schaffer had suffered a serious head injury 10-6-42 when he crash landed a P-40F and had recently returned to flying status following a brief stay in the station hospital at Will Rogers Field, Oklahoma City, Oklahoma.

11-10-42E. Tumwater, Washington. At 1705, a Lockheed P-38E on a high altitude test flight crashed while maneuvering near Olympia Airport, Tumwater, Washington, killing 1Lt. William Wirth. The P-38 was seen to dive into the ground and explode into flames. 1Lt. Charles J. Crocker, assistant operations officer, Olympia Airport, witnessed the accident and later stated to investigators, "I saw Lt. Wirth flying at approximately 30,000 feet. The plane was not visible but vapor trails were trailing the plane so its position was definite. As the plane approached the field, he lost altitude by a series of slow rolls. The plane dived toward the ground and pulled up as if to do a loop. When in a nose high position, it fell off on one wing and did two complete spirals down and I should judge at 15,000 feet it went into a vertical dive and continued that way until it hit the ground."

11-11-42A. Corona, California. At 1138, a Lockheed P-38E collided with high tension wires and crashed into Williams Canyon near Corona, California, killing 2Lt. Eugene W. Sheridan. The P-38 had taken off from the Army Air Field at Santa Ana, California, on a local transition flight and eventually had joined up with another P-38 that had been flying in the area. The two airplanes were seen maneuvering in the area just prior to the accident. The other pilot stated that Lt. Sheridan apparently noticed the four 115,000 volt power lines strung across the canyon and was seen maneuvering in an attempt to avoid them. The P-38 struck two of the lines, severing them and causing the lines to start a large brush fire. The airplane traveled a few hundred yards before plunging into the ground down the canyon and exploding into flames.

11-11-42B. Palestine, Texas. At 1545, a Vultee BT-13A crashed seven miles east of Palestine, Texas, killing pilot 2Lt. Gordon L. Whitmyre and passenger Sgt. Ellis Britton. The Accident Classification Committee stated, "It is the opinion of the committee that this crash was the result of the pilot losing control of his ship while doing aerobatics at a low altitude. We believe that the ship was rolled onto its back and from that position the nose dropped and the plane crashed before the pilot could roll it over. We believe that the plane hit the ground [while] still in a semi-inverted position."

11-11-42C. Spence Field, Georgia. At 2305, a North American AT-6C on a night navigation training flight slammed into the ground five miles northeast of Spence Field, Moultrie, Georgia, killing student pilot Cadet George M. MacMullin, Jr., U.S. Military Academy. The Accident Classification Committee stated, "It is the opinion of this board that the student lost his horizon while lowering his landing gear or checking some other detail in the cockpit and went into a spiral from which he did not recover. He was seen falling and was not spinning. In a thorough search of the airplane, no evidence of material failure could be found. It was evident that the ship was intact until it hit the ground. When the pilot was seen spiraling he had a landing light on and when the ship was found the landing gear was in the down position."

11-12-42A. Ames, Oklahoma. At 1025, a Vultee BT-15 smashed into the ground after failing to recover from an intentional spin one-half mile south of Ames, Oklahoma, killing A/C Warren Charles Exley. Investigators stated, "At approximately 1025 CWT [Cadet Exley] was observed to execute a series of snap rolls and then to enter a spin at approximately 5,000 feet. It is assumed from the evidence that the spin was executed intentionally for practice in recovery. Once entered, the spin continued until the airplane disappeared from the view of the witnesses. The position of the airplane on the ground indicated that it struck the ground with the right wing slightly low, nose down about 45 degrees with no forward speed and very little rotation but with great downward velocity. The pilot's body was discovered in the bottom of the remainder of the front cockpit with the safety belt still fastened, and without any indication of an attempt to use parachute." The airplane was on an authorized aerobatic practice flight.

11-12-42B. Selma, Alabama. At 1030, a Curtiss P-40 in a high speed dive came apart in the air and crashed 12 miles northwest of Selma, Alabama, killing 2Lt. Otto A. Breiner. Civilian farmer J.E. White witnessed the accident and later stated to investigators, "When I first saw the ship it seemed as though it was in a power dive. He seemed to be very high because the plane looked small. While he was still high, the wing seemed to snap off from the plane. The ship then began to whirl so fast that pieces began to fly off. Motor never stopped running until ship hit the ground. It never burst into flames. I never saw him jump out or try to jump out. I found his shoe and the name Otto Breiner was in his shoe. The wing is about one-fourth of a mile from the plane." It was speculated by investigators that the pilot might have stressed the P-40 while trying to pull out of the dive. Lt. Breiner had received his wings and his commission on 11-10-42.

11-12-42C. Stratford, Connecticut. At 1726, a Republic P-47C on a material test flight dove into the ground near the Bridgeport Municipal Airport, Stratford, Connecticut, killing pilot Capt. Robert K.

Noel, 23, Stratford. The pilot was under orders to test a new radio antenna mast that was installed on the P-47 and "was to dive at high speed in an attempt to tear the mast off." The pilot climbed the P-47 to about 15,000 feet where he executed two slow rolls before beginning his test dive. The Accident Classification Committee stated, "He apparently started the dive at about 12,000 to 15,000 feet. It was a nearly vertical dive and the airplane must have reached a very high speed. There seems to be no explanation for the accident except that the pilot did not allow himself sufficient altitude for the pull out." On the ground, pilots from Capt. Noel's unit (63rd Fighter Squadron, 56th Fighter Group) witnessed the accident and stated that the aircraft entered a vertical dive from a split-S maneuver just north of the field and dove into the ground at angle of about 80 degrees at an estimated speed of 600 mph, exploding into flames 300 yards east of the field. The P-47 and pilot were destroyed utterly. According to published accounts of the accident, Capt. Noel was to marry Miss Leila P. Redgate, Bridgeport, Connecticut, on 11-16-42.

11-13-42A. New Bern, North Carolina. At 0130, a Martin B-26B returning from a night training mission crashed three miles southwest of New Bern, North Carolina, after running out of fuel while attempting to land at a U.S. Marine Corps Auxiliary Airfield, killing the crew of six. A Marine sergeant on duty at the field reported that the airplane had circled the field and then made an approach but overshot. The airplane went around for another approach and crashed on the downwind leg when both engines stopped due to fuel starvation. The airplane crashed into a heavily wooded swamp, scattering wreckage over a distance of 500 feet. Killed in the crash were: 2Lt. Albert J. Trewitt, pilot; 2Lt. Clifton J. Duff, co-pilot; SSgt. Wilbert S. Calkins, engineer; Sgt. William Lockett, radio operator; SSgt. Orville D. Eberle, gunner; Cpl. Emil V. Landgren, bombardier.

11-13-42B. Berlin, Georgia. At 1413, a Bell P-39D crashed two and one-half miles north of Berlin, Georgia, killing Aviation Cadet John M. Kirk. The Accident Classification Committee stated, "It is the opinion of this board that the pilot tried to effect recovery from a spin too late. He was seen spinning at an altitude of 4,500 feet and continued into the ground. He tried to jump but went through the propeller. The left door was off and he jumped to the inside of the spin. The student was on an authorized training flight and was instructed to do maneuvers such as chandelles, lazy eights, steep turns, and one turn spins to the right and left."

11-13-42C. Corsicana, Texas. At 1328, two Fairchild PT-19A type aircraft collided in mid-air two miles northeast of Corsicana, Texas, killing Aviation Cadet Merritt L. Ring aboard PT-19A #42-2847. Civilian instructor LaVern Rosene and student Aviation Cadet Richard E. Gouzie received light injuries but were able to safely land PT-19A #42-2849 at Corsicana. Investigators stated, "The planes were flying level at about 600 feet. [Rosene] was flying due south. [Ring] was flying 10 degrees west of north. [Rosene] was looking back to find a plane that had been behind him. [Rosene] turned around and saw [Ring] flying towards him. It was too late to avoid a collision and [Rosene] banked his ship to miss the other. [Rosene's PT-19] sheared off a large portion of the right wing of [Cadet Ring's PT-19]. [Cadet Ring's PT-19] entered a spin from which it did not recover. No attempt was made to use parachute." Investigators faulted both pilots for failure to see each other and avoid the collision.

11-13-42D. Waycross, Georgia. At 1545, a Vultee V-72 (similar to the A-31 type) crashed near the Army Air Field at Waycross, Georgia, killing gunner SSgt. H. A. Mullins, who went in with the airplane. Pilot 2Lt. C.H. Eglehoff was seriously injured parachuting to safety. The airplane was returning to Waycross after completing a gunnery mission to Ossabaw Island gunnery range. The airplane was on the downwind leg of the pattern when a fuel line broke and began spraying gasoline into the cockpit. The pilot ordered the gunner to jump and then began to look for a place to execute a forced landing. The gasoline fumes intensified and then the pilot was sprayed directly in the face with gasoline, knocking off his sunglasses and blinding him. Hearing the engine begin to sputter, he exited the airplane and deployed his parachute. He reported that he hit the ground almost immediately after the canopy deployed. The pilot assumed that the gunner had left the airplane when ordered to do so. It was not known why the gunner did not leave the airplane. Several fatal and non-fatal accidents were attributed to a faulty fuel line fitting on the V-72 type airplane (See 10-3-42A).

11-15-42A. Will Rogers Field, Oklahoma. At 0905, a Douglas A-20C suffering an apparent engine failure stalled and spun into a field four miles south of Will Rogers Field, Oklahoma City, Oklahoma, killing pilot 2Lt. Jack C. Curry. The A-20 apparently spun to the ground, exploding into flames upon impact. Investigation revealed that the port engine ignition switch was found in the "off" position and the starboard engine ignition switch was found in the "on" position. One pilot flying in the area reported seeing "spurts of white smoke" emitting from the airplane prior to the crash. 2Lt. Norris J. Clark (AAF) witnessed the crash and later stated to investigators, "I noticed two airplanes south of Hangar No. 3, apparently in the traffic pattern of Will Rogers Field. Both planes were flying in an easterly direction, the first plane at 800 yards in the lead. The first plane nosed down as if to do an outside loop, and started a spin to the right after coming to an almost upside down position. Almost immediately the plane started to recover from the right spin and started into a left spin as it disappeared from my sight beyond the horizon."

11-15-42B. Conway, Arkansas. At 1645, a Vultee BT-13A performing aerobatics at low altitude stalled and spun into the ground at the municipal airport at Conway, Arkansas, killing pilot 2Lt. John W. Gillette, III, Birmingham, Michigan, and passenger Cpl. Donald H. Bucklin, Hale, New Hampshire. The airplane was observed to take off from the airport and then to make several steep turns around the airport. After circling the field, the pilot dove at the field to an altitude of about 300 feet. When at about the middle of the field the pilot began a normal climb and then executed a roll to the right. The BT-13 stalled after completing the maneuver, falling off to the right from an altitude of approximately 300 feet. The airplane dove into the ground and burst into flames in a field near the airport.

11-15-42C. Columbia, South Carolina. At 2115, a Martin B-26B crashed into the ground while maneuvering at low altitude seven miles southeast of the Army Air Base at Columbia, South Carolina, killing the crew of seven. The airplane was seen by civilian witnesses to be flying at an altitude of about 30 feet prior to the crash. The airplane began a turn to the right at low altitude, the starboard wing contacting the terrain and causing the B-26 to cartwheel into the ground. The airplane exploded into flames upon impact, scattering debris and body parts over several hundred yards. Killed in the crash were: 2Lt. Eugene L. Huntley, pilot; 2Lt. William A. Johnson, co-pilot: 2Lt. Charles M. Frissell, navigator; 2Lt. Harold W. Hope, bombardier; Cpl. Charles J. Miller, engineer; SSgt. Donald W. Evans, radio operator; Pvt. Vincent J. Meculskas, passenger.

11-15-42D. Delano, California. At 2150, two Vultee BT-13A aircraft collided in mid-air one mile east of Delano, California, killing student A/C Finis E. Roberts, Jr. aboard BT-13A #41-22382; instructor 2Lt. Frank Capel parachuted to safety and escaped injury. Cadet Roberts was killed when he went in with the aircraft and it was not known why he did not use his parachute. The pilot of BT-13A #41-1439, A/C Gerald Milbourn, was able to land his aircraft safely at Delano Auxiliary Airfield. Investigators stated, "The airplanes had been directed to assemble in formation as soon after take-off as possible. At an altitude of approximately 600 feet above the ground, the leading airplane [Lt. Capel], started a shallow turn to the right with the object of permitting [Cadet Milbourn] to take off, [who] was flying the No. 2 position, and to [allow him to] close up on the leader. While in the turn, [Lt. Capel] observed that [Cadet Milbourn] was over-running and immediately made a violent turn to the left in an effort to prevent a collision. The left wing of [Cadet Milbourn's airplane] collided with the right wing of [Lt. Capel's airplane]. [Cadet Milbourn] continued flight and landed successfully without further damage."

11-16-42A. Tina, Missouri. At 0730, a Cessna T-50 (similar to the AT-17 type) flying in poor weather dove into the ground 15 miles southwest of Tina, Missouri, killing pilot 1Lt. Charles E. McKinley and co-pilot 2Lt. Albert O. Dokken. The aircraft was on a ferry flight from St. Louis to St. Joseph, Missouri. Other aircraft on the flight reported penetrating a front and encountering turbulent air while en route. Investigators speculated that the aircraft suffered a catastrophic failure of the tail surfaces and structures as result of encountering severe turbulence. Pieces of wreckage were found scattered over a wide area suggesting an in-flight break up. This aircraft carried civilian registration N-34762 as well as AAF serial number 41-905.

11-16-42B. Marana, Arizona. At 0845, a Vultee BT-13A stalled and spun into the ground near Marana, Arizona, killing Chinese Air Force Flying Cadet Jui Kao. The BT-13 was observed spinning at an altitude of approximately 800 feet above ground level until it struck the ground and erupted in flames.

11-16-42C. Location Unknown. At 1430 CWT, a Consolidated B-24D and its crew were reported missing and are presumed to have crashed in the Gulf of Mexico in the vicinity of the Florida Keys while on a cross-country navigation training flight. The airplane was one of a flight of B-24s that took off from DeRidder Army Air Field, DeRidder, Louisiana, on a navigation flight across the Gulf of Mexico to a point near Key West, Florida. The flight was then to proceed to a practice-bombing mission at Waccasassa Bay, north of Tampa, Florida. At approximately 1430 CWT, when the flight was over Key West, the pilot radioed in to the flight that he was south of Key West. The flight circled the area of Key West for some time waiting for the airplane to catch up. When it did not appear, the flight proceeded with the remainder of its mission. The airplane was never heard from again. The B-24 was believed to have gone down in the vicinity of Key West. An extensive search of the area over several days revealed no clues, wreckage or bodies. Missing and presumed lost at sea were: 2Lt. Arden H. Hanes, pilot; 2Lt. Wilford A. Johnson, co-pilot; 2Lt. Stanley J. Trusiak, navigator; 2Lt. Zera W. Strait, bombardier; SSgt. Ernest M. Kennedy, engineer; SSgt. Richard C. Grubbs, radio operator; Sgt. Merlo R. Gray, gunner; SSgt. George I. Thompson, gunner.

11-16-42D. Geiger Field, Washington. At approximately 1230, two crewmembers were killed and two were injured, one seriously, when a Boeing B-17F suffered an oxygen equipment failure while returning to Geiger Field, Spokane, Washington, after completing a bombing-training mission at the AAF Boardman Bombing Range, Washington. Co-pilot 2Lt. John H. Noel was uninjured and was able to land the B-17 safely at the U.S. Naval Air Station at Pasco, Washington. After completing the practice bombing mission at 1100, the weather started closing in. The pilot, 1Lt. Robert H. Allyn, decided to climb over the clouds, which exceeded 25,000 feet. All hands were ordered to go on oxygen. Soon after beginning the climb, the radio ceased to function. The airplane leveled off at about 26,000 feet. Several minutes after flying at this altitude, the pilot passed out and the B-17 entered a violent spin, going inverted. Lt. Noel recovered the airplane and immediately began climbing again. The airplane leveled out at 31,000 feet where the radio began to function again and the pilot regained consciousness. He indicated that he was okay and then went to check on the crew who he found to be all right. The pilot took over the controls and after flying at 31,000 feet for some time, the no. 1 engine quit. He ordered Lt. Noel to go check on the crew. Lt. Noel reported to the pilot that A/C George F. Lenc, student bombardier, and engineer SSgt. Truman A. Bernard appeared to be dead. The pilot indicated that he was okay but then began to put the airplane in a dangerous attitude, prompting the co-pilot to again take over the controls. The pilot ordered Lt. Noel to again check on the crew but he refused to leave his station. The B-17 descended to 8,000 feet and the pilot's condition improved. The pilot and the co-pilot then landed the airplane. Student bombardier A/C William L. Yarborough received minor injuries. It was discovered that poorly fitted oxygen masks were the cause of the incident. The members of the crew were unfamiliar with the masks, which were of a brand new type. The co-pilot reported to investigators that he set his oxygen regulator on "manual" and set the flow on "full open." He also reported that the oxygen mask's elastic fasteners failed at altitude and that he had to hold the mask tight to his face or it would fall off.

11-16-42E. Muroc, California. At 1305, a Lockheed P-38E crashed while attempting to take-off from the Army Air Field at Muroc, California, killing pilot 2Lt. Clayton S. Lamb. Witnesses stated that the P-38 used up much of the runway on the take-off roll and was pulled into the air, the pilot retracting the landing gear immediately. The starboard engine appeared to be smoking slightly and the airplane appeared to enter a flat, skidding turn to the right. The P-38 was then observed to go into a steep, climbing turn to the right. At an altitude of about 300 feet above ground level, the airplane stalled, flipped over on its back, and fell off into a spin to the right, plunging to the ground and exploding into flames. The P-38 struck the ground southeast of the field. Investigators speculated that the starboard engine probably failed on take-off and that the pilot was unable to overcome the effect of the port engine in time to prevent the loss of control.

11-16-42F. Biggs Field, Texas. At 1600 MWT, a Vultee BT-15 collided with rising terrain west of Biggs Field, El Paso, Texas, killing pilot 1Lt. Edward T. McMillan and his passenger SSgt. Billy J. Hartsock. The Accident Classification Committee stated, "After clearing Biggs Field, Lt. McMillan, flying a BT-15 airplane, attempted to negotiate a climb over the mountain due west of the field by flying straight west after take-off. At the place he attempted to cross there were two ridges with a canyon in between. He was able to pass over the first ridge and it is the opinion of the committee that after passing over the first ridge, he encountered a down current of air in the canyon which made it impossible for him to climb over the second ridge out of the down current of air with the horsepower of the BT-15 airplane that he was flying. In his attempt to make a turn down the canyon, he crashed into the side of the mountain. The pilot and passenger were both killed and burned. The airplane and all equipment were completely destroyed."

11-16-42G. Gulf of Mexico, Florida. At 1905, a Martin B-26B crashed into the Gulf of Mexico about 40 miles southwest of the Army Air Field at Fort Myers, Florida, killing the crew of six. The B-26 had taken off from Fort Myers at 1810 EWT on a scheduled performance mission. The airplane radioed in the clear that the crew was abandoning ship over the Gulf of Mexico at 26 degrees 10 minutes North and 82 degrees and 20 minutes West. The nature of the problem was not stated. The crew and aircraft were not heard from again and was soon reported missing. The AAF and the U.S. Navy launched an extensive search that lasted several days. The bodies of pilot 2Lt. Donald E. Vail and co-pilot 2Lt. Fred Dees, Jr. were found on 11-20-42. Lt. Dees watch had stopped at 1905 EWT. The condition of the bodies suggested to investigators that the two pilots had parachuted into the sea and that they had drowned. It was speculated that the rest of the crew probably used their parachutes and drowned in the sea as well. Twenty-five miles south of Clewiston and in the vicinity of Fort Myers, the Civil Air Patrol spotted parachutes in the sea but a crash boat was unable to locate them. The Coast Guard found an oil slick southwest of Fort Myers but investigators could not positively determine if it was caused by the missing B-26. Missing and presumed lost at sea were: 2Lt. Louis R. Mikes, navigator; SSgt. Milton H. Newton, radio operator; SSgt. Richard Treat, engineer; SSgt. William G. Kittiko, gunner.

11-16-42H. Burbank, California. At 1609, a Lockheed P-38F attempting a landing at Grand Central Air Terminal, Glendale, California, crashed into an excavation behind the Burbank Municipal power plant at 174 W. Magnolia, Burbank, California, killing two Burbank municipal workers and seriously injuring three others. The pilot, 2Lt. James D. Koester, 21, Los Angeles, was also killed. The aircraft had peeled off third from its flight of four, entering the traffic pattern. The P-38 was then seen turning from base to final with landing gear down and no flaps. The aircraft attempted the turn to the right at an altitude of approximately 300 feet and was seen in a descending lateral skid to the right with the starboard wing down. The P-38 dropped to about 100 feet and then entered a nose down attitude, continuing in this attitude until it struck the ground behind the steam power plant. Several men escaped the path of the P-38 by sprinting from the work area. The P-38 struck a scaffold that was being used by two men, killing them instantly before smashing to the ground and exploding violently into flames. Ammunition from the aircraft began detonating in the resultant fire. Stanley T. Stone, 36, Burbank, and William D. Edens, 28, Los Angeles, were killed. William D. Amann, 59, North Hollywood, George Townsend, 52, Burbank, and Werner Zaiss, 37, Van Nuys, were seriously burned in the accident. All were city employees of Burbank.

11-17-42A. Taos, New Mexico. At 0405 MWT, a Consolidated B-24D on a celestial navigation flight collided with the top of domed shaped 12,441-foot Baldy Peak, 25 miles northeast of Taos, New Mexico, killing navigator 2Lt. Norman M. Godnick, New York, New York. Nine other crewmembers were injured, four seriously. The B-24 tore itself to pieces as it skidded for over 400 feet along the top of the peak after colliding with the mountain at near cruising speed. The B-24 did not burn and investigators described the crew's survival as "miraculous." The crew was marooned on top of the peak for four days, finally being brought down by rescuers on 11-20-42. While they waited to be carried out on horses or stretchers, the injured fliers were supplied with food, water and shelter by rescuers who first arrived by horseback on 11-18-42. The airplane had taken off at approximately 0230 CWT (over two hours later than originally planned) from Salina, Kansas, for the flight to Albuquerque, New Mexico, and Wink, Texas, and return to Salina. Because of a lack of oxygen supplies at Salina, the B-24 could not be fully serviced with oxygen and the crew would not be able to fly the entire mission on oxygen. The flight crew and operations officer concluded that the flight could be completed with only short periods of flight at altitudes exceeding 9,000 feet, eliminating the need for a full supply of oxygen aboard the airplane. The celestial cross-country navigation flight was cleared (in error) out of Salinas for Albuquerque for an altitude of 9,000 feet, the flight crew and operations officer apparently ignoring the fact that terrain exceeding 9,000 feet existed along the intended course. Just prior to the collision, the pilot climbed the aircraft to what he thought was 14,000 feet to escape extremely turbulent air. Investigation revealed that the altimeter was in error of approximately 1,290 feet. The pilot stated to investigators that he thought he was flying on a course south of Las Vegas, New Mexico. Investigation revealed that the B-24 was 70 nautical miles off course. Investigators cited the lack of experience on the part of the pilots, miscommunication between the flight crew and the operations officer, poor pre-flight planning, and a failure to correctly set the altimeter as contributing factors to the accident. Weather was not considered a factor in the accident. Seriously injured in the crash were: SSgt. Edward H. Paul, Teaneck, New Jersey, radio operator; SSgt. John J. Powers, Cresskill, New York, engineer; Pfc. Rudolph W. Matula, gunner; SSgt. Merlin G. Seeverts, assistant radio operator. Receiving minor injuries were: 2Lt. Anthol W. Riney, pilot; 2Lt. Arch Bishop, co-pilot; 2Lt. Thomas L. Mesenbourg, bombardier; SSgt. John B. Pursley, Chicago, Illinois, assistant engineer; SSgt. Peter G. Michel, Sigourney, Iowa, gunner.

11-17-42B. Mooresville, North Carolina. At 1545, a Spartan C-71 on a ferry flight crashed near Mooresville, North Carolina, killing Capt. James Gay Nall, 47, Charlotte, North Carolina. The airplane apparently developed some trouble in the air and the pilot attempted to bail out but was unsuccessful. The parachute's drogue chute was found over a quarter of a mile from the site of the crash. Investigators speculated that the pilot deployed his parachute before he was clear of the airplane and the parachute became fouled on the tail surfaces. The airplane plunged to the ground and burst into flames.

11-17-42C. Northern Field, Tennessee. At 1712, a Curtiss P-40F crashed on take-off from Northern Field, Tullahoma, Tennessee, killing 2Lt. William E. Charles. The pilot attempted the take-off with the fuel selector set on the auxiliary drop tank. After the take-off, the pilot attempted an extreme pull-up and the engine began to sputter because of fuel starvation. The pilot, remembering that he had the selector on the drop tank, attempted to switch the selector to the starboard wing tank. The engine eventually quit and the airplane crashed, bursting into flames upon impact with the ground. Investigators determined that the selector switch was set in between the main tank and the starboard wing tank, causing the engine to quit because of fuel starvation.

11-18-42A. Patterson Field, Ohio. At 0148, Douglas C-47 collided with terrain and crashed while attempting a landing at Patterson Field, Ohio, killing pilot Capt. Garton K. Wagar, New Rochelle, New York, and co-pilot First Officer Dyle C. Cole, Lansing, Michigan. Both were civilian employees of United Airlines and were operating under contract to the AAF. Investigators stated, "Several observers who had the plane in sight northwest of the field until impact stated that the ship appeared to be at normal altitude until the latter part of the downwind leg at which time he started losing altitude. His landing lights were turned on and the turn to the base leg [had been] made [and] the ship continued to lose altitude gradually. In the turn to the final approach it seemed to lose considerable altitude and at the last part of the turn the left wing struck a tree. There was a short flash of flame and the ship rotated counter clockwise, striking the ground upside down and bursting into flames shortly afterward. The impact of the left wing against the tree sheared the wing off just outside of the left engine nacelle, the wing falling 75 yards short of the wreckage. At the time of the accident the weather was intermittent rain, visibility two and one-half miles (in rain and smoke), ceiling about 1,200 feet, and it was observed that several thin wisps of fog appeared before the landing lights while the ship was on the base leg."

11-18-42B. Coker, Alabama. At 1015, a Vultee A-31 descended into terrain while letting down through instrument conditions near Coker, Alabama, killing pilot 2Lt. Waley H. Garrett and gunner SSgt. Kaz F. Kruzick. The aircraft burst into flames immediately upon striking the ground, killing both occupants instantly. Investigators stated, "Pilot was on a local routine training flight and went much further than he was supposed to. Apparently he became lost and confused and attempted to let down through a blanket of fog and smoke near Tuscaloosa, Alabama. All evidence points out that the plane was traveling at a tremendous rate of speed when it struck the ground."

11-18-42C. Foster Field, Texas. At 1015, a North American AT-6C dove into the ground 10 miles east of Foster Field, Victoria, Texas, killing Aviation Cadet Thomas J. Glynn, Jr. Investigators stated, "Cadet Glynn was flying individual combat with his instructor and had started at approximately 10,000 feet. The instructor would let Cadet Glynn on his tail and then would get on Cadet Glynn's tail through a series of passes. The flight worked down during this process to approximately 5,000 feet and at this point the instructor and the cadet were in a Lufbery [circle]. For no apparent reason, the cadet pulled his airplane over and out of the Lufbery, diving towards a large cumulus cloud that was approximately 3,000 feet below. The dive was approximately at a 70-degree angle and the instructor immediately tried to call him on the radio in an attempt to stop the dive from going into the cloud, however, he was unsuccessful, the airplane diving into the cloud. The instructor went down between the clouds and came out below this particular cloud and saw the crashed airplane. The instructor further stated that the base of the cloud was directly above him when he was flying 500 feet which would indicate that the cloud extended lower than those in the immediate vicinity."

11-18-42D. Douglas, Georgia. At 1334, a Stearman PT-17 stalled and spun into the ground near Douglas, Georgia, killing civilian instructor J.R. Puryear and student Aviation Cadet R. W. Post. The airplane had taken off from Douglas at 1321 on a routine training flight. The airplane was observed in a gliding descending turn when it entered a stall close to the ground. The instructor was unable to recover and the airplane smashed to the ground. The airplane did not burn but was destroyed utterly.

11-18-42E. El Cajon, California. At 1730, a Lockheed P-38F on an interception training mission dove into the ground at Leland Place, El Cajon, California, killing pilot 2Lt. Philip R. Sleadd. The aircraft exploded violently into flames upon impact and damaged at least five structures, including a chicken coop and a barn. Several chickens and a one-year-old horse were killed. The Accident Classification Committee stated, "Lt. Sleadd was engaged in a mission with five other squadron pilots which consisted of training in interception and combat between elements at 7,000 feet and 11,000 feet altitude. Lt. Sleadd is believed to have made a diving attack on 2Lt. John M. Kelly from which he did not recover. Several witnesses on the ground stated that the airplane crashed into the ground at an angle of 80 degrees at a terrific speed." Lt. Kelly, who dove out at a speed of 400 mph, estimated that Lt. Sleadd's speed exceeded 500 mph. It is likely that the pilot put the airplane in a diving maneuver from which he could not recover because of the tendency of the P-38 type airplane to suffer from severe control surface buffeting caused by compressibility.

11-18-42F. Kings Canyon, California. A Beech AT-7 collided with Mt. Darwin at Kings Canyon National Park, California, killing four army fliers. The aircraft and crew had been reported as missing and were not located until 9-24-47. No other information is available.

11-19-42A. Jackson, Tennessee. At 0800, a Fairchild PT-19A stalled and spun into the ground five miles southwest of Jackson, Tennessee, killing Aviation Cadet Howard Herman Miller, 23, Munster, Indiana. The Accident Classification Committee stated, "The student was climbing below the normal climbing speed at about 500 feet and when executing a climbing turn a stalled resulted. Student's apparent slow reaction time was such that he failed to pick up the wing and drop the nose and went into an accidental spin. The student recovered from the spin but as the stick went forward and the rotation ceased, the airplane struck the ground."

11-19-42B. Spence Field, Georgia. At 1007, a North American AT-6B collided head on with a Bell P-39D five miles northwest of Spence Field, Moultrie, Georgia, killing RAF instructor Pilot Officer Ronald MacKenzie, RAF Flying Cadet Marvin E. Lee, and the P-39 student pilot Aviation Cadet Edward F. Knotek, AAF. Cadet Knotek had just taken off from Spence Field and had gained about 2,000 feet in altitude and Cadet Lee was entering the pattern and lowering his landing gear. The airplanes collided head on and then fell to the ground in flames. Investigators faulted both pilots for failing to see the other and avoid the collision, speculating that both pilots had their eyes in the cockpit.

11-19-42C. Hammer, South Dakota. At 1400 CWT, two Piper L-4A type aircraft flying in a five-ship formation collided in mid-air one and a half miles northwest of Hammer, South Dakota, fatally injuring pilot 2Lt. Glen Campbell aboard L-4A #42-36694. He died at 2045 at Veblen Community Hospital, Veblen, South Dakota. Flight Officer Edward R. Robertson was uninjured and was able to land L-4A #42-36824 after the collision. Lt. Campbell's airplane struck F.O. Robertson's airplane from below, falling out of control and crashing into the ground. The propeller of

Robertson's L-4 was shattered. He was able to make a successful dead stick emergency landing.

11-19-42D. Gulf of Mexico, Florida. At 1615 EWT, a Bell P-39D crashed into the Gulf of Mexico eight miles west of Jena, Florida, killing pilot 2Lt. Vincent D. Pisauro. The airplane was in the number-four position of a four-ship flight that was on an aerial gunnery mission. After completing the gunnery practice, the flight leader noticed that the subject airplane was missing and then noticed a splash in the water some distance behind the formation. The P-39 was located from the air in about 20 feet of water by Capt. George Stillson. He stated later that he had observed the pilot's parachute "billowing up in the water from the cockpit." Capt. Stillson later spotted Lt. Pisauro's body floating in the water about 200 yards from the wreck. By the time that the crash boat arrived on the scene (after dark), the body could not be located. Despite an air and sea search that lasted several days, the body of Lt. Pisauro was never recovered.

11-19-42E. Bush Field, Georgia. At 1721, a Vultee BT-15 stalled and spun into the ground 20 miles southwest of Bush Field, Augusta, Georgia, killing Aviation Cadet Richard J. Newlon. The airplane had been observed by witnesses on the ground circling a field at an altitude of about 300 feet above ground level. The nose of the airplane dropped and the airplane was then seen to enter a spin and crash to the ground, bursting into flames upon impact.

11-19-42F. Tracy, California. At 2230, a Cessna AT-17B suffered a structural failure of the port wing and crashed eight miles southwest of Tracy, California, killing A/C Horace W. Cederstrom and A/C Robert O. Brunner. The airplane took off from Stockton Field, California, and was to assemble with several other AT-17 type airplanes at 4,000 feet for a night formation flight. After losing pieces of the wing, the tail group failed and the airplane "described an outside loop" and plunged to the ground, landing on its back and bursting into flames upon impact. A civilian found the airplane the next day at about 0800. Pieces of the port wing were found two miles from the main wreckage. Pieces of debris were found along the flight path from the spot were pieces of port wing was found and the crash site. The tail section was found about 100 yards from the main wreckage, intact. The engines were found 200 yards beyond the main wreckage, about one-quarter of a mile southeast of New Jerusalem Field.

11-20-42A. Morrison Field, Florida. At 0839, a Lockheed A-28A on a ferry mission crashed while attempting to take off from Morrison Field, West Palm Beach, Florida, killing the crew of six. Investigators stated, "This aircraft commenced its take-off at 0839 on the west end of the East-West runway and proceeded east on the runway. [The] runway is 4,430 feet in length, hard surface, the sky was clear, visibility unlimited, and practically no wind was prevalent. The aircraft had nearly reached the intersection of the East-West runway when take-off from the three-point position was made. The aircraft proceeded to climb at a steadily increasing angle until a power stall resulted at an angle in excess of 60 degrees and constantly approaching vertical, at an altitude greater than 200 feet and less than 300 feet. The aircraft then fell off on its left wing, apparently out of control, and fell almost vertical to the ground, near the east boundary of the airport, somewhat north of the East-West runway. No parts or other objects were seen to fall from the aircraft during the entire take-off run. By the time the aircraft was reached by persons on the ground, all occupants were dead, five of the six bodies having been thrown from the aircraft as a result of the crash. [The] elevator trim control was found to be in the full nose-up position." Killed in the crash were: 1Lt. James C. Cunningham, pilot; SSgt. Emmett A. Herbert, co-pilot; 2Lt. Elbert M. Garton, navigator; Cpl. James C. Beeson, engineer; Sgt. Francis H. Brady, radio operator; SSgt. F. B. Sutor, crewmember.

11-20-42B. Versailles, Ohio. At 1057 CWT, a Martin B-26B suffering the failure of its port engine crashed four miles northwest of Versailles, Ohio, killing the crew of seven. The airplane was part of a flight of six and was flying from Baer Field, Fort Wayne, Indiana, to Morrison Field, West Palm Beach, Florida, in preparation for a flight overseas. The accident occurred about 28 minutes after take-off. The B-26 was seen spinning and then to enter a dive toward the ground at angle of 45 degrees, exploding into flames upon impact. The Accident Classification Committee stated, "It is the opinion of the Committee that [the] B-26B crashed as a result of failure of the left engine; that it went into spin at 4,500 feet msl (3,500 feet above the terrain); that the pilot was unable to regain control of the airplane before it crashed to the ground. The cause of the failure of the left engine is undetermined due to the total destruction of the airplane and both engines as a result of the fire after the crash." Killed in the crash were: 1Lt. Homer R. Peck, pilot; 2Lt. Marcus L. Dean, co-pilot; 1Lt. Robert S. Clever, navigator; SSgt. George W. Barnes, engineer; SSgt. Rufus M. Roseberry, radio operator; SSgt. Delbert C. Armstrong, gunner; SSgt. Leroy Confer, gunner. Lt. Clever was a veteran of Lt. Col. James Doolittle's Tokyo bombing mission flown on 4-18-42.

11-20-42C. Castle Rock, Colorado. At 0858, Lockheed F-4 (photo-reconnaissance version of the P-38 type) crashed near Castle Rock, Colorado, killing pilot SSgt. Robey F. Beymer. Despite a thorough search by army airplanes and the Civil Air Patrol, the airplane wreckage and the pilot were not located until 11-28-42. Poor weather hampered the initial search efforts. Investigation revealed that the P-38 struck the ground in a vertical dive at high speed. The cause of the crash could not be determined, but it was speculated that the pilot put the airplane in a diving maneuver from which he could no recover because of the tendency of the

P-38 type airplane to suffer from severe control surface buffeting caused by compressibility.

11-20-42D. Pacific Ocean, Oregon. At 0915, a Lockheed A-29 on a search and patrol mission suffered the failure of its port engine and ditched in the Pacific Ocean west of Seaside, Oregon. The A-29 was returning from its patrol mission when about 50 miles from shore the aircraft suffered the failure of the starboard engine. After jettisoning the bomb load, the pilot began heading for Portland Army Air Base, Portland, Oregon. The pilot increased power on the port engine and trimmed the airplane for single-engine operation. After a short time of single-engine operation, the port engine began to run rough. Approximately three miles west of Seaside, the port engine failed. The pilot had given the crew the choice of bailing out or ditching. Two of the crew parachuted out and the pilot glided the aircraft to a ditching about 500 yards offshore near Seaside. Radio operator Sgt. Albert Povodnik drowned before he could be rescued. Pilot 1Lt. William S. Southern swam to shore and engineer SSgt. William R. Dart was rescued by a U.S. Navy Vought OS2U Kingfisher after spending a brief time in the water. Navigator 2Lt. Leroy Kline and gunner Sgt. Robert E. Pierce were missing after they elected to parachute into the sea three miles west of Seaside rather than ditch with the airplane. They were never found and presumed lost at sea.

11-20-42E. Cross City, Florida. At 1510 EWT, a Bell P-39D suffering an engine malfunction and attempting an emergency landing crashed 200 yards from the runway at Cross City Army Air Field, Cross City, Florida, killing pilot 2Lt. Robert H. Frear. Tower personnel observed the aircraft streaming white puffs of smoke and the pilot was told by the tower that the field was clear and he could attempt an emergency landing. The pilot missed his first approach and came around again with landing gear and flaps down, but too high and too fast. He missed the second approach and came around again. Black smoke was seen to billow from the aircraft on final and then the P-39 was seen to plunge to the ground 200 yards short of the field, exploding into flames upon impact. The pilot was killed instantly. Cross City Army Air Field rescue personnel were on the scene immediately and the fire extinguished after a short time. Investigators speculated that the coolant line had broken causing the white smoke. Because of the loss of coolant, the engine overheated, burned itself up and eventually failed.

11-20-42F. Malibu, California. A Lockheed P-38D crashed near Malibu, California, killing pilot 1Lt. D.F. Sharpe, Van Nuys, California.

11-21-42A. Salinas, California. At 0753, a Lockheed-Vega B-34 crashed while attempting a takeoff from at the Army Air Base at Salinas, California, killing five fliers and seriously injuring one other. The aircraft took off and climbed to about 200 feet and apparently suffered a loss of power in the port engine, causing the left wing to drop. The pilots seemed to get the aircraft under control and in a safe flying attitude. The pilot then put the B-34 into shallow bank to the left (into the dead engine) in an attempt to get back to the field. The aircraft fell off on the left wing in the turn and dove into the ground, bursting into flames upon impact. Investigation revealed that the port propeller was feathered and the starboard propeller distributor valve was stripped, causing a loss of pitch control of the propeller blades for the starboard propeller. The nature of the flight was a pilot check ride in type. Investigators speculated that the check pilot might have purposely cut the port engine although no real evidence suggests that this had occurred. Killed in the crash were: 2Lt. Francis Menante, Reno, Nevada, pilot; 2Lt. John M. McMurray, Charleston, South Carolina, co-pilot; Capt. Joseph W. Fanszick, Hamilton Field, California, passenger; Pfc. Herbert R. McCoy, Lancaster, Ohio, crewmember; Pvt. William D. Hyatte, hometown unknown, Kentucky, crewmember. Engineer SSgt. Herbert M. Walker, Los Angeles, California, was seriously injured.

11-21-42B. Wayne, Oklahoma. At 1630, a Douglas A-20C suffering a failure of the starboard engine crashed two and one-half miles south of Wayne, Oklahoma, killing three fliers. Gunner SSgt. Stanley E. Sloan parachuted to safety and received minor injuries. The A-20 was leading a three-airplane element on a local formation flight. The flight was peeling off to the left in preparation to land at Will Rogers Field, Oklahoma City, Oklahoma. During the steep bank in the peel-off, the airplane flipped over past inverted and the pilot recovered. The pilot then entered a steep diving turn to the right when the starboard engine failed. The pilot rolled the airplane to level but was unable to maintain control. The airplane rolled to the right several times, the pilot attempting to roll the A-20 to level flight each time. The pilot was unable to maintain control and the airplane entered a tight spin to the right and crashed to the ground, exploding into flames upon impact. Killed in the crash were: 1Lt. Thomas S. Rankin, pilot; SSgt. Stanley W. Murphy, pilot-rated passenger; Sgt. Grover J. Sparks, gunner.

11-21-42C. Greenville, South Carolina. At 2046, a North American B-25B suffering the failure of the port engine crashed while attempting an emergency landing at the Army Air Base at Greenville, South Carolina, killing the crew of three. Investigators stated, "[The B-25] took off at 2015 on a local night transition flight. At 2043, [the pilot] called to the [Greenville] control tower requesting an emergency landing clearance. The tower instructed him that all runways were clear for the landing. Control tower operator estimated his position to be seven miles northeast of the field at 800 feet above the terrain and lined up with the NE/SW runway. The pilot stated that he would land on the NE/SW runway to the southwest. According to the statements of reliable witnesses, the left engine was not running and throwing out white smoke. The plane was gradually descending up to a certain point where the descent increased probably due to the lowering of the landing gear. The ship crashed and burned at a point three miles northeast of the airdrome. An examination of the left engine showed that the propeller had not been feathered." Killed in the crash were: 2Lt. Charles T. Humphreys, pilot; 2Lt. Melvin R. Quast, co-pilot; SSgt. Otis K. Fryer, Jr., engineer.

11-22-42. Freda, California. At a time unknown, a Douglas A-24A dove into the ground 15 miles southwest of Freda, California, killing pilot 2Lt. Emery Gergely and gunner SSgt. Glen E. Wallace. The A-24 had taken off from Blythe, California, at about 1600 PWT on an assigned "high altitude oxygen flight." The mission was to circle the Blythe area at high altitude, terminating at 1800. When the airplane and crew failed to return they were reported as missing. A search effort was launched but was called off after ten days. The airplane and crew were not found until 2-17-43. Investigation revealed that the airplane dove straight into the ground and exploded upon impact, scattering wreckage over 100 yards.

11-23-42A. Roberta, Georgia. At 1315, a Douglas A-24A flying in poor weather crashed seven miles south of Roberta, Georgia, killing pilot 1Lt. Reynolds H. Middleton and passenger MSgt. David L. Dean. The airplane had taken off from the Army Air Base at Savannah, Georgia, at 1039 EWT on an instrument navigation training flight to New Orleans, Louisiana. Investigation revealed that the airplane was over sixty miles off course and an hour and fifteen minutes behind the flight schedule. The weather in the area at the time of the crash was reported as "ceiling estimated 500 to 600 feet overcast, visibility four miles with moderate rain and fog." Investigators speculated that the pilot was unsure of his position and lost control of the airplane while flying in instrument conditions. The airplane struck the ground in a vertical position, exploding into flames upon impact.

11-23-42B. Crockett, Texas. At 2035, a Beech AT-10 suffering the failure of the port engine was abandoned and crashed three and a half miles north of Crockett, Texas. A/C Sidney S. Lawrence parachuted to safety. A/C Everett A. Hilliard was killed when the aircraft struck him while he was attempting to bail out. His parachute deployed and his body floated to earth, being found by civilians the next day four miles north of Crockett. The aircraft, which was on a routine student night navigation flight, was at 3,500 feet above ground level and within sight of Crockett when the cockpit began to fill with smoke. The pilot (Cadet Lawrence) noticed that the port engine nacelle was covered with oil. He then noticed that the engine rpm went to zero. He shut down the engine, trimmed the aircraft for single-engine operation and began to look for a place to land. The aircraft began to lose altitude rapidly and the pilots decided to leave the airplane. Cadet Hilliard left the aircraft first at about 2,000 feet and Cadet Lawrence left the aircraft two miles north of town at an altitude of about 1,500 feet. The aircraft smashed into the ground and burst into flames.

11-23-42C. Perrin Field, Texas. At 2130 CWT, a Vultee BT-13A crashed into a parked BT-13A while attempting to take off from the Auxiliary Airfield 20 miles northeast of Perrin Field, Sherman, Texas, killing A/C Otto D. Sorrenson. Investigators stated, "The airplane and pilot in question were in the right hand (no. 2) position in a three ship V-formation. The formation took off and shortly after clearing the ground a few feet, the [no. 2] airplane struck an airplane, which was parked perpendicular to, and well to one side of, the intended line of take-off. During the take-off, the [no. 2] ship maintained its proper position in the formation, not swinging widely to the outside. At the time of the crash, a slight crosswind existed which would tend to drift the formation toward the parked airplanes. Also the air was hazy from dust raised by previous take-offs. The responsibility for the accident lays largely upon the leader of the formation who failed to orient himself properly before take-off, thereby taking off at an angle which did not give his wingman safe clearance."

11-24-42A. Centerville, Tennessee. At 1405 CWT, a Lockheed A-28A crashed 12 miles southwest of Centerville, Tennessee, killing pilot 2Lt. George A. Messing and co-pilot 2Lt.Robert E. Charley. The aircraft suffered an in-flight explosion and separation of the starboard wing after the pilot executed an extreme pull-up maneuver in an effort to right the A-28 after rapidly descending through overcast. The aircraft dove into the ground and exploded into flames upon impact. The starboard wing and engine were found 500 feet apart about 100 yards from the main wreckage. The starboard wing flap could not be located in the wreckage. The aircraft was being ferried from Dallas, Texas, to Nashville, Tennessee.

11-24-42B. Onalaska, Washington. At 1210, a Lockheed P-38D flying in poor weather on a cross-country navigation flight crashed near Onalaska, Washington, killing pilot 1Lt. Marcus M. Sur. The Accident Classification Committee stated, "The pilot was flying on a contact clearance between McChord Field, Tacoma, Washington, and Portland Army Air Base, Portland, Oregon. In the vicinity of the Toledo Range Station he ran into some bad flying weather, consisting of rain and hail, low ceilings, and poor visibility." Civilians on the ground heard the P-38 circling in the clouds above Onalaska just prior to the accident. It was then seen diving out of the overcast at tremendous speed, striking the ground at an angle of 60 degrees or greater, exploding into flames upon impact.

11-25-42A. Lake Murray, South Carolina. At 1350, a North American B-25C crashed into Lake Murray, Columbia, South Carolina, after losing power in both engines, killing four crewmembers and seriously injuring three others. Investigators stated, "Pilot was on a bombing mission at 1,000 feet over a target [located on an island] in Lake Murray. The interphone system was out so the co-pilot started to remove a [panel] which covered the cut [out] in the instrument panel for the installation of the automatic pilot to enable him to take visual signals from the bombardier up front. As he did so, both engines cut out simultaneously. Pilot checked switches, put booster pumps on and moved mixture control to full rich but engines would not restart. Pilot then attempted to reach shore but failed and made a crash landing on the water. Three members of the crew escaped, sustaining major injuries. [The crash was fatal] to the four remaining members of the crew. It is believed co-pilot cut the master switch when he removed the [panel] cover. When the engines cut out, the pilot checked the switches but his check was hurried due to the fact that he had very little altitude and so failed to detect switch in the off position. A diver was sent down to the submerged aircraft to check the master ignition switch to see if it was on or off. Switch was found all the way in the off position." Killed in the crash were: 2Lt. Carrol N. Peacock, Jr., Columbia, South Carolina, co-pilot; 2Lt. Charles E. Hoyt, Milwaukee, Wisconsin, bombardier; Pvt. Alfred E. Brandt, Peotone, Illinois, engineer. Missing and presumed lost was photographer Pvt. Howard P. Fonville, Cushing, Oklahoma. Pilot 2Lt. John W. Gilluly, bombardier 2Lt. Leonard V. Sheffield, and bombardier 2Lt. Edward R. Armstrong were seriously injured.

11-25-42B. Mt. Meigs, Alabama. At 1530, a Vultee BT-13A on an instrument team ride collided with trees and crashed six miles north of Mt. Meigs, Alabama, killing RAF Flying Cadet George G. Warner and seriously injuring RAF student Cpl. John Clark. After Cpl. Clark completed 55 minutes of flying on instruments under the hood, Cadet Warner took over control of the airplane. They were flying at an altitude of about 4,000 feet when Cadet Warner made two steep turns and then attempted a snap roll, ending up inverted. He pulled through in a half-loop and recovered the airplane. He then dove the airplane towards the ground and began to level off at 150 feet, coming within ten feet of the ground before completing the recovery. The pilot climbed the airplane up to 1,000 feet and then began a spiraling descent, failing to recover before colliding with trees. Cadet Warner was killed instantly when he was hurled into a tree after being ejected from the aircraft. A hunter who had witnessed the crash found Cpl. Clark wandering around the wreckage in a state of shock.

11-25-42C. Grand Ridge, Florida. At 2110 CWT, a Vultee BT-15 dove into the ground and exploded into flames near Grand Ridge, Florida, killing A/C Philip Weston, Jr. Investigators stated, "[The student] was dispatched on a night navigation training flight from Bainbridge, Georgia, to Quincy, Florida, to Marianne, Florida, to Quincy, Florida, and back to Bainbridge, Georgia. At approximately 2103 [CWT], Cadet Weston checked in with the control ship at Marianne, Florida. A short time later, he was seen circling at approximately 500 feet, about two miles south of Grand Ridge, Florida. After having circled two or three times, he was seen flying west at a very low altitude. At the same time, he flashed on his landing lights. His airplane then turned toward the south in a slight climb. Immediately thereafter, the nose dropped and the airplane dived into the ground. Cadet Weston was killed and the airplane burned. The Accident Classification Committee, after a thorough investigation, was unable to determine the cause of the accident."

11-26-42A. Freeport, Florida. At 0900 CWT, a Martin B-26B on a simulated torpedo run collided with a wooden battleship target and smashed into Scrub Hill near Freeport, Florida, killing four crewmembers and seriously injuring two others. The Accident Classification Committee stated, "Aircraft was making a simulated torpedo run on a battleship target, Range No. 54, off Hammock Point, Eglin Field Military Reservation. [The B-26] struck the battleship mast with the left engine, damaging the propeller, causing extreme vibration, and burning out the electrical system so that the propeller could not be feathered. The aircraft [crashed and burst into flames] on the top of Scrub Hill approximately fifteen miles NE of the battleship target in direct line of flight and burned. Pieces of the battleship target mast were removed from the engine; approximately thirteen feet of the mast was broken off." Killed in the crash were: 2Lt. William W. Chappell, pilot; 2Lt. James E. Calvert, co-pilot; SSgt. Richard O. Thompson, engineer; 2Lt. Merlin C. Vogelsang, bombardier/navigator. Radio operator SSgt. Roy F. Radford and tail gunner SSgt. Clarence R. Benson received serious injuries.

11-26-42B. Pensacola, Florida. At 1045 CWT, a Martin B-26B spun to the ground and exploded into flames 25 miles north of Pensacola, Florida, killing the crew of three. The Accident Classification Committee stated, "This aircraft was observed spinning after emerging from the overcast (approximately 1,000 feet ceiling). The horizontal stabilizer and vertical fin disintegrated and were found approximately [one-half] mile from the major portion of the wreckage. Aircraft was cleared on instruments at 12,000 feet and pilot was instructed by clearance officer that if turbulent air was encountered, he was to return to [Eglin Field]. It is the opinion of the Accident Classification Committee that this accident was caused by the pilot losing control of the aircraft in turbulent air." Other pilots flying over the area of the crash reported severe turbulence just after the time of the accident. Killed in the crash were pilot 1Lt. Duncan C. Murphy, co-pilot 2Lt. Albert J. Wieland and engineer Sgt. Herbert C. Wade.

11-26-42C. Fort Stockton, Texas. At 1145, a Stearman PT-17 spun to the ground eight miles northeast of Fort Stockton, Texas, killing Aviation Cadet

Leonard C. Nicholson. Cadet Nicholson plummeted to his death when he bailed out at too low an altitude to allow his parachute to fully deploy. Investigation revealed that the airplane struck the ground in a spin to the right at a speed in excess of 100 mph.

11-27-42A. Dayton, Ohio. At 0845 CWT, a Republic P-47C on an accelerated service test flight dove into the ground near a commercial radio station transmission tower at Dayton, Ohio, killing pilot 2Lt. Andrew C. Brenkus. The P-47 had climbed to 35,000 feet for a high altitude service test flight, which lasted about 30 minutes. The aircraft was then seen to enter a vertical dive from an altitude of about 35,000 feet, diving to the ground and exploding into flames. Witnesses on the ground stated that they saw no apparent effort by the pilot to leave the aircraft or pull the P-47 out of the dive. Investigators speculated that the pilot might have been suffering from oxygen starvation or was physically unable to recover from the dive because of a black out condition or the high-speed dive characteristics of the P-47 type airplane. The radio transmission tower broadcast for radio station WING.

11-27-42B. Columbia, South Carolina. At 1109, a North American B-25D crashed while attempting a single-engine emergency landing at Columbia Army Air Base, Columbia, South Carolina, killing two crewmembers and seriously injuring two others. The airplane was returning from a high-altitude performance flight when, at an altitude of 14,000 feet, the pilot decided to shut down the port engine to practice single-engine flight. He feathered the propeller and trimmed the airplane for single-engine flight. After flying in this configuration for a short time, the pilot found that he could not un-feather the port propeller. The pilot declared an emergency and the field was cleared for his landing attempt. Investigators stated, "The pilot applied excessive power to the good engine as he approached the end of the runway. It may have been that the pilot was trying to correct for drift caused by the strong prevailing winds on the only available runway or that his airspeed was allowed to drop excessively low. The plane, as it neared the runway, violently skidded to the right, climbed slightly, and then did an almost complete slow-roll to the left, striking the ground in an inverted position. It is possible that a row of culvert pipe lying approximately fifty feet to the left of the runway created a mental hazard to the pilot." Killed in the crash were the pilot 2Lt. Ross T. Morris and co-pilot 2Lt. Joseph F. Denning. Passenger and rated army pilot 2Lt. Orin T. Highfill and engineer SSgt. John D. Gray were seriously injured.

11-27-42C. Lake Charles, Louisiana. At 2340, a North American AT-6A on a five-airplane night-formation flight dove into the ground and exploded into flames 11 miles northwest of Lake Charles, Louisiana, killing Aviation Cadet Herbert A. Peterson. The student was flying in the number-three position of a five-ship flight that took off on a night formation flight. Other students observed the subject airplane leave the formation and change position from left to right in a gentle bank, losing altitude in the process. The tail and navigation lights were then seen descending rapidly until the airplane was seen to crash into the ground. Investigation revealed that the airplane struck the ground at an angle of about 70 degrees and that the student had unbuckled his safety belt prior to impact. The student was killed instantly upon impact.

11-28-42A. Bernalillo, New Mexico. At 0020, a Beech AT-11 on a night bombing training mission crashed 20 miles west of Bernalillo, New Mexico, killing the crew of four. The Accident Classification Committee stated, "Subject airplane was demolished after colliding with the ground in full flight. The airplane apparently mushed into the ground at a high rate of speed. The wings were level at the time of contact with the ground, since holes dug by the engines and propellers were equidistant and of approximately the same depth. After contacting the ground an explosion followed, scattering wreckage over an area of 250 yards radius. Due to the demolished condition of the airplane, no evidence of malfunction or sabotage could be determined. The cause of the crash is undetermined." Killed in the crash were: 2Lt. Merrell L. Cowan, pilot; 2Lt. Leo V. Cotter, bombardier instructor; Aviation Cadet James B. Case, student bombardier; Aviation Cadet Wells L. Cahoon, student bombardier.

11-28-42B. Santa Ana, California. At 1430, a Lockheed P-38F dove into the ground at the U.S. Navy Blimp Base at Santa Ana, California, killing 2Lt. Donald F. McAllister. The airplane was seen performing various maneuvers before it entered a dive at an altitude of approximately 3,000 feet. The pilot was unable to recover from the dive in time to avoid striking the ground. The P-38 struck the ground at an angle of about 80 degrees and exploded violently into flames upon impact, killing the pilot instantly. Portions of the starboard flaps were found one-third of a mile from the crash site. Investigators could find no definite cause for the accident but it is likely that the pilot put the airplane in a diving maneuver from which he could not recover because of the tendency of the P-38 type airplane to suffer from severe control surface buffeting caused by compressibility.

11-28-42C. Mt. Baker, Washington. At an unknown time, a Lockheed P-38E crashed into rising terrain at Mt. Baker, Washington, killing pilot 1Lt. Kenneth W. Ambrose, 24, Lincolnville, Kansas. The airplane had taken off from an Army Air Field in Alaska on a maintenance flight to Paine Field, Everett, Washington. The airplane failed to arrive at its intended destination and was declared missing. Two hikers found the airplane wreckage in September 1997. Published accounts noted that the pilot had played trumpet in the Count Basie Orchestra.

11-29-42A. Greer, South Carolina. At 1055, a North American B-25B flying in instrument conditions broke up in flight and crashed 11 miles north of Greer, South Carolina, killing the crew of four. The airplane had taken off from Greenville Army Air Base, Greenville, South Carolina, on a local instrument training flight. Weather was reported as ceiling of 1,800 feet with six miles visibility. Killed in the crash were: 2Lt. Herbert W. Barton, Jr., pilot; SSgt. James W. Britton, co-pilot; 2Lt. William T. Wright, bombardier; SSgt. Bertram J. Howard, engineer. "It is believed by the Accident Classification Committee that the accident was not caused by any internal explosion of aircraft, that [the] pilot, in letting down through the overcast while trying to regain visual contact with the ground, built up speed far in excess of limits permissible for B-25 type aircraft and, on coming out of the overcast and seeing how close he was to the ground, pulled back on the wheel too sharply and, in doing so, put too much strain on the wings, causing same to break. Sections of wing and tail group and sections of fuselage were scattered for a distance of [over] a mile from the place where the crash occurred."

11-29-42B. Mosquito Lagoon, Florida. At 1145, a Curtiss P-40E dove into Mosquito Lagoon near Titusville, Florida, killing 2Lt. Richard L. Byrne. Several witnesses on the ground observed the airplane dive into the lagoon and explode in flames. Investigators stated, "Lt. Byrne was discovered to be missing from a string formation of four airplanes, in which he was flying in the number-four position, after the formation had made a dive from 12,000 feet to 10,000 feet and pulled back up in a climbing turn to reform at 12,000 feet. During the dive, the leader and the number-two and three men in the formation executed an aileron roll to the left. It is not known whether the fourth airplane was flown through such a maneuver or not. The speed attained in the dive did not exceed 325 mph. It is further stated that the airplane piloted by Lt. Byrne never recovered from the dive and continued on down with power on until it crashed and exploded in the water. It is the opinion of the board that the cause of the accident is unknown."

11-29-42C. Anniston, Alabama. At 1604, a North American AT-6B flew into terrain while flying in instrument conditions at Anniston, Alabama, killing 2Lt. Rafford H. Gaulding. The AT-6 took off from Anniston Municipal airport for Maxwell Field, Montgomery, Alabama, at 1557 on an instrument flight plan. The pilot took off almost one-half hour later than planned. In that time the weather deteriorated to a 400-foot ceiling with one and one-half miles visibility in light rain, light fog and light smoke. Civilians on the ground at Anniston reported that they had heard the airplane circling in the clouds overhead before it plunged to the ground. The airplane did not burn but was completely destroyed.

11-29-42D. Danbury, Connecticut. At 1759, a Douglas C-39 lost power in both engines due to severe carburetor icing and was abandoned by its crew, crashing to earth at Danbury, Connecticut. The pilot, Capt. John F. Meehan, Jr., 43, Wyncote, Pennsylvania, and passenger Maj. Herman B. Leeth, 46, Indianapolis, Indiana, were killed in the crash. Five other crew and passengers parachuted to safety. Civilian Flight Engineer William Kuryle, Middletown, Pennsylvania, received serious injuries. Parachuting to safety and uninjured were: Col. George V. McPike, Hannibal, Missouri, Commanding Officer of the AAF Air Depot at Rome, New York; Maj. Robert V. Dunn, Marion, Indiana, co-pilot; Capt. Gerald Garrard, Cordele, Georgia, passenger; 1Lt. Ross DeLue, Chicago, Illinois, passenger. The C-39 had been cleared out of Westover Field, Chicopee, Massachusetts, for Grenier Field, Manchester, New Hampshire, on a supply and maintenance flight. The aircraft was equipped with carburetors that could produce dangerous and unrecoverable icing characteristics if mishandled by the flight crew. Investigators speculated that the severe carburetor icing was caused by the aircraft circling at low power settings while trying to establish radio contact with air traffic controllers. The C-39 lost power in both engines and the pilot was unable to maintain control of the airplane. The airplane fell off in a spin to the left and the fliers who did parachute successfully had a difficult time escaping the airplane. Capt. Meehan and Maj. Leeth were unable to escape the airplane before it struck the ground and exploded into flames.

11-29-42E. Merced, California. At 1605, a Vultee BT-13A spun into the ground ten miles northeast of Merced, California, killing Aviation Cadet Richard James White. Two cadets flying in the area observed the subject airplane enter a spin to the left at about 8,000 feet above sea level. The BT-13 spun for about 3,000 feet before coming out of the spin in a 70-degree nose down attitude. It remained in this position for a few seconds before entering another spin to the left that tightened until the airplane struck the ground. The pilot made no apparent effort to parachute from the airplane. The airplane did not burn but was destroyed utterly.

11-29-42F. Freeport, Florida. At 2000 CWT, a Curtiss P-40K flying in instrument conditions crashed five miles north of Freeport, Florida, killing Lt. Col. Boyd "Buzz" Wagner, 26, Johnstown, Pennsylvania. Lt. Col. Wagner, who achieved ace status as a lieutenant in the Philippines at the beginning of the war, had cleared himself from Eglin Field, Valparaiso, Florida, to Maxwell Field, Montgomery, Alabama, on an instrument flight plan. The airplane took off at 1949 CWT but never arrived at Maxwell Field and soon was declared missing. Despite an extensive search, Lt. Col. Wagner and the wreckage of the P-40 were not found until 1-6-43, several miles from Eglin Field. Investigation revealed that the P-40 spun to the ground and burned slightly after the crash.

11-29-42G. Sebastian, Texas. At 2315, a North American AT-6A crashed into a thick stand of mesquite trees five miles east of Sebastian, Texas, killing 1Lt.

Lloyd W. Rauch and 2Lt. Knox M. Fant. The airplane took off at 2210 from Austin, Texas, for a flight back to base at the AAF Gunnery School at Harlingen, Texas. The airplane had made a navigation training flight from Harlingen to Austin via Randolph Field, San Antonio, Texas, earlier in the day. Harlingen tower personnel reported that Lt. Rauch made a radio call at 2315 and was not able to complete the transmission. Harlingen tower personnel, despite repeated calls after 2315, were never able to re-establish radio contact with Lt. Rauch. The airplane was declared overdue at 2337 and was spotted from the air during the morning hour of 11-30-42. The AT-6 was destroyed utterly and the wreckage was scattered over a wide area. Both pilots were killed instantly and their bodies badly mutilated. Pieces of parachute were found in the mesquite trees but investigators determined by evidence found on the body parts that the parachutes opened because of severe impact with the terrain. Investigators could not determine what caused the pilot to fly the airplane into the ground.

11-30-42A. Sioux City, Iowa. At 0215, a Boeing B-17F flying in instrument conditions crashed one-third of a mile southwest of the Army Air Base at Sioux City, Iowa, killing the crew of ten. Weather at the time of the accident was reported as ceiling at 1,000 feet with visibility of three-quarters of a mile in light snow. Killed in the crash were: 2Lt. Henry Bunn, pilot; 2Lt. Edward E. Lowery, Jr., co-pilot; 2Lt. Harry J. Kaufman, navigator; 2Lt. Leon R. Christenson, bombardier; TSgt. James G. Kirkpatrick, engineer; SSgt. Wayne A. Sims, radio operator; Sgt. Robert E. Ables, gunner; Sgt. Alexander Armstrong, Jr., gunner; Sgt. Arlo H. Brown, gunner; 2Lt. Earl B. Reynolds, passenger. Shortly before the crash, the B-17 had passed over the field at approximately 600 feet agl. Moments later the tower officer observed "the whole sky illuminate by the explosive glow of the crash fire." Investigators stated, "There are no actual witnesses to the accident.... There is very little concrete evidence other than the actual telescoped wreckage to determine what happened and it is assumed that the pilot tried to instigate a contact approach to the field under instrument conditions and in doing so stalled the aircraft as soon as contact with the field was lost. The aircraft apparently stalled from a low altitude and crashed in a near vertical attitude before recovery could be accomplished. All other aircraft proceeded on instruments to alternate fields."

11-30-42B. Camden, Alabama. At 0908, a Fairchild PT-19A practicing simulated forced landings stalled and spun into the ground five miles north of Harrell Field, Camden, Alabama, killing civilian instructor William W. Duffy and student A/C Stephen H. Curtiss. Investigators speculated that the instructor was demonstrating forced landings and went into a steep right turn at an altitude of 250 feet. He apparently pulled sharply back on the stick, causing the airplane to snap over on its back. The airplane snapped back to the left and entered a spin to the left, striking the ground after completing about three-quarters of a turn of the spin.

11-30-42C. Miles, Texas. At 1215, a Vultee BT-13A on an instrument team ride dove into the ground ten miles northwest of Miles, Texas, killing students A/C Robert W. Holland and A/C Miller Hammons, Jr. Investigation revealed that the airplane dove into the ground with power on at an angle of 80 degrees and at a tremendous velocity. The students were killed instantly and the aircraft was destroyed utterly. Investigators could not determine why the students flew the airplane into the ground.

11-30-42D. Westfield, Massachusetts. At 1545, a Republic P-47B crashed into Higgins Swamp east of Barnes Municipal Airport, Westfield, Massachusetts, killing pilot 2Lt. Daniel V. Austin, 24, Dorchester, Massachusetts. Investigation revealed that the pilot was about an hour and a half overdue from his transition training flight. The P-47 was seen to fly over the airport to the southeast at an altitude of about 1,000 feet. The P-47 was seen to enter a turn to the left, descend, and then S-turn to the right and left at low altitude. The airplane was then seen to enter a steep left turn at low altitude. The pilot attempted to raise the nose and the airplane fell off to the left, spinning one-quarter of a turn before plunging to the ground and exploding in flames.

11-30-42E. Van Nuys, California. At 1335, a Lockheed P-38F suffering the failure of the port engine crashed while attempting an emergency landing at Metropolitan Airport, Van Nuys, California, killing pilot 2Lt. Ron R. Smith. 2Lt. Harry L. Smith, who was flying on the local formation training flight with Lt. Smith, witnessed the accident. He later stated to investigators, "Upon reaching the south extremity of the field, the aircraft turned 45 degrees to the left at what looked to be very low altitude, 100 feet, and then turned steeply to the left, diving into the ground and hitting the highway just west of the field. The ship immediately burst into reddish yellow flames, giving off a large volume of black smoke." Investigators stated, "The failure of the left engine, due to the loading up of the spark plugs with foreign material to such an extent that the plugs could not function will be credited with 50 percent of the accident cause."

11-30-42F. Moore Field, Texas. At an unknown time after 1546 CWT, a Republic YP-43 spun to the ground 23 miles NNE of Moore Field, Mission, Texas, killing A/C John P. Siew, San Antonio, Texas. The aircraft took off from Moore Field at approximately 1546 CWT and it was speculated that the crash occurred soon after. Investigators stated, "Upon examining the accident the committee has concluded that the plane crashed as a result of a right-hand spin. This was A/C Siew's first flight in this type of ship and was to be conducted in accordance with a pre-arranged transition course, which did not include voluntary spins on the first flight. Investigation has disclosed that A/C Siew had received a competent check-off in this type of aircraft prior to the flight."

11-30-42G. Anthony, Kansas. At 2012, a Vultee BT-15 crashed seven miles WSW of Anthony, Kansas, killing A/C John Henry Brettell. The student took off from Enid Army Air Field, Enid, Oklahoma, on a night navigation mission to Anthony, to Gage, Oklahoma, and return to Enid. The BT-15 landed safely at Anthony and took off soon after. Tower personnel at Anthony last observed the airplane flying southwest three miles west of the field at an altitude of about 1,500 feet agl. About five miles southwest of the field, witnesses on the ground observed the BT-15 turn left to a heading slightly east of south and then enter a power dive with the nose down approximately 30 degrees, striking the ground at high speed with the starboard wing slightly low. The airplane exploded violently into flames upon impact and wreckage was scattered over an area of approximately 500 yards. The student's body was found about 15 feet from the main wreckage with his partially deployed parachute lying near him. Investigators speculated that the student stood up in the cockpit and deployed the parachute just prior to the airplane striking the ground.

December

12-1-42A. Walnut Ridge, Arkansas. At 0020, a Vultee BT-13A crashed one mile northeast of the AAF Basic Flying School at Walnut Ridge, Arkansas, killing Aviation Cadet Charles Emil Thiede. At 0015, the airplane flew over the field at about 3,300 feet and the tower called the airplane in for a landing. The student acknowledged the landing instructions and began a descending turn to enter the pattern. The BT-13 failed to arrive for the landing and was not responding to repeated radio calls. The airplane was declared missing and was not located until 0825. Investigation revealed that the airplane struck the ground at a 20-degree angle at a speed estimated to be 120 mph, bursting into flames upon impact and killing the student instantly.

12-1-42B. Miles, Texas. At 1045, a Vultee BT-13A spun into the ground while practicing aerobatics four miles southwest of Miles, Texas, killing A/C Joseph Rangus. Investigators stated, "The student was on a solo acrobatic mission, and the witness first saw the airplane at an altitude of approximately 5,000 feet. The airplane did a series of six or seven slow rolls, leveling off after each one, and losing altitude down to approximately 2,000 feet. Upon completing the last roll, the airplane leveled off, but immediately fell into a spin. After several turns of the spin, the airplane recovered and was apparently in a straight and level position at a very low altitude, probably less than one hundred and fifty feet. However, the downward momentum and the increased speed resulting from the spin caused the airplane to crash into the ground." Witnesses reported that the airplane "pancaked' into the ground. The airplane did not burn but was totally demolished.

12-1-42C. Westhampton, Massachusetts. At 1540, a Republic P-47B flying in a snowstorm collided with rising terrain at Westhampton, Massachusetts, killing 2Lt. Charles C. May. The airplane was part of a three-ship formation training flight that had originated at Westover Field, Chicopee, Massachusetts. The flight leader inadvertently led the airplanes into instrument conditions and the formation soon broke up. It was speculated by investigators that the pilot had let down under a low ceiling and struck the side of a hill while trying to maintain visual contact with the ground. The ceiling was estimated to be about 1,500 feet with poor visibility in snow. The airplane crashed into a heavily wooded area and exploded violently into flames, scattering wreckage over a wide area. Investigation revealed that the pilot's only instrument training was performed on a Link Trainer.

12-1-42D. Leesburg, Virginia. At 1815, a Douglas DB-7B crashed ten miles northwest of Leesburg, Virginia, killing 1Lt. John Milburn. Passengers uninjured parachuting to safety and uninjured were 2Lt. Francis H. Nance, 2Lt. Robert L. Nattkemper and 2Lt. Stanley V. Molander. The aircraft became lost in instrument conditions while flying a cross-country navigation flight from Pope Field, North Carolina, to Greensboro, North Carolina. When it seemed that the aircraft was going to run out of fuel, the pilot ordered the passengers to bail out which they did successfully. Investigators speculated that Lt. Milburn did not attempt to use his parachute and was thrown from the aircraft when it smashed into the ground. It was not known why he was unable to leave the aircraft. Investigation revealed that the pilot took an incorrect heading soon after take-off and never made it to Greensboro. Apparently, the pilot circled a small town in Virginia at low altitude in an unsuccessful attempt to determine his position just prior to the crash.

12-2-42A. McKenna, Washington. At 0904, a Lockheed P-38F and a P-38E collided in mid-air near McKenna, Washington, killing P-38F pilot 2Lt. Jack N. Finley. 2Lt. Norman W. Jackson parachuted to safety from the P-38E. A four-airplane flight of P-38 type airplanes (led by 1Lt. Arthur A. Kitzman) from Olympia AAB, Tumwater, Washington, intercepted a four-airplane flight of P-38 airplanes from McChord Field, Washington, near McKenna at an altitude of approximately 11,000 feet. These flights made several passes at each other before withdrawing and reforming. It is believed that during this time Lt. Finley, who was the second element wingman, joined Lt. Kitzman's element and Lt. Jackson moved to Lt. Finley's position in the other element. A short time later, Lt. Finley fell out and lost the formation. He made a turn and tried to rejoin the formation, approaching Lt. Kitzman's element while climbing toward the direction of the sun. He was seen climbing almost vertically and coming head on and from below by Lt. Kitzman and the second element leader. The second element leader, Lt. Ralph H. Wandrey, realized that Lt. Jackson (flying on Lt. Wandrey's wing) was in the path of Lt. Finley's airplane. He and Lt. Jackson went into a vertical turn to the left in an attempt to avoid the collision but Lt. Finley's airplane collided with Lt. Jackson's P-38, striking the tail and severing it. Lt. Jackson, upon realizing that he had no control of the airplane, parachuted to safety and was uninjured. Lt. Finley's P-38 lost a wing, went out of control and spun to the ground in flames four miles northeast of where Lt. Jackson's airplane slammed to earth at McKenna.

12-2-42B. Robins Field, Georgia. At 1550, a Douglas A-20B collided with the ground just after taking off from Robins Field, Robins-Warner, Georgia, fatally injuring pilot Col. Francis M. Zeigler and civilian crew chief Arvil G. Copeland, Macon, Georgia. Investigators stated, "The pilot took off into the west on the east-west runway, making what appeared to be a normal take-off using a little more than half of the 5,000-foot runway. The plane, upon reaching an altitude of approximately 20 to 30 feet, leveled off, then nosed down in a flat dive. Continuing in its dive, the plane struck the shoulder of a road, which is located about 150 feet from the end of the runway. The impact caused the landing gear and the streamlined fairing of the right engine nacelle to be torn away from the airplane. The force of this impact caused the plane to bounce into the air again and it continued straight ahead, climbing to approximately 200 feet above the ground. The pilot then made a wide turn to left when it was noticed that the airplane was flying in an extremely tail low position and gradually losing altitude. Unable to maintain altitude, the pilot made an attempted landing in small field about two miles southwest of the runway. The plane struck the ground on its belly and flipped over on its back causing [its total destruction]. The pilot remained hanging in the airplane upside down in an unconscious state and a portion of the cockpit had to be cut away to remove him from the plane. The passenger was thrown clear of the plane and was found approximately ten feet from the plane in an unconscious state." They both died shortly after the accident at the station hospital. Investigators speculated that " ... the pilot, in attempting to raise the landing gear after take-off, looked down at the retracting controls and thus lost control of the airplane causing it to dive into the ground at the end of the runway."

12-2-42C. Foster Field, Texas. At 2045, a North American AT-6 crashed eight miles southeast of Foster Field, Victoria, Texas, killing pilot Maj. Elton Smith and passenger W/O George R. Perry. Investigators speculated that the airplane was in a dive toward the ground and the pilot was unable to recover. "[Major Smith] apparently saw how close he was [to the ground] and attempted to pull the airplane out with a violent backward motion of the stick, causing the airplane to start into a snap roll to the left. It is believed that the left wingtip struck the ground, throwing the airplane up on its nose and over on the right wing, thence on its back, smashing the vertical stabilizer, tearing the engine and firewall loose and catching on fire." Both occupants were killed instantly. The airplane had taken off from Foster Field on a routine night flight. W/O Perry was not scheduled to be on the flight, but Maj. Smith allowed him to come along at the last moment.

12-2-42D. Grace, Idaho. At 2010 MWT, a Boeing B-17F crashed out of control in mountainous terrain near Grace, Idaho, killing ten fliers. The pilot was attempting to remain in visual contact with the ground while flying under low overcast over mountainous terrain at night and was flying under the recommended safe altitude of 11,000 feet for this region. The airplane had taken off from Pocatello Army Air Base, Pocatello, Washington on a night navigation mission. The B-17 was flying from west to east when the pilot, apparently seeing that a collision with rising terrain was imminent, pulled up into a steep climb in an attempt to clear 9,500-foot Soda Peak, causing the B-17 to enter a violent stall and a vertical descent to the ground. The airplane cleared the peak but smashed to the ground on the other side in an area of heavy timber with trees reaching 70 to 80 feet in height. Only three trees were knocked down by the falling B-17 and only three others had branches stripped from top to bottom, indicating a vertical descent into the terrain. The B-17 exploded in flames upon impact, coming to rest at the 6,500-foot level on the eastern side of the mountain in Wilson Canyon. The pilot had been instructed to avoid flying in the mountains and to turn back if instrument conditions were encountered. The accident occurred approximately 30 minutes after take off from the Army Air Base at Pocatello, Idaho. Killed in the crash were: 2Lt. Harold K. Simpson, pilot; 2Lt. Wendell L. Sims, co-pilot; 2Lt. Thomas J. Wymond, navigator; 2Lt. Clarence G. Wenrich, bombardier; Sgt. Neal L. Slinker, engineer; SSgt. John N. Lazansky, radio operator; Cpl. Charles L. Walker, Jr., engineer; Pvt. Joseph R. Hughes, gunner; Pvt. Donald W. Seaney, gunner; Pfc. Nicholas Zompetti, assistant radio operator.

12-2-42E. Terrell, Texas. At 2225, a Vultee BT-13A and a Stearman PT-18 collided in mid-air four and one-half miles southeast of Terrell, Texas, killing BT-13 pilot 2Lt. William A. Reed and passenger Pvt. Wilton P. Toombs, Jr. The pilot of the PT-18, civilian flight instructor G. T. Derrick, was able to crash land the damaged airplane at Boykin Field, Texas, after the collision. The BT-13 was part of a 33-airplane weather-reporting mission and was to be flying at an altitude of 3,500 feet above sea level. For unknown reasons, the BT-13 descended to about 800 feet in the vicinity of

Boykin Field. The PT-18 had been practicing take-offs and landings at the field. The pilot of the PT-18 had just taken off and was climbing to 800 feet in the pattern when he saw the BT-13 heading straight at him. He pulled back on the stick in an attempt to avoid the collision but was unable to do so, the BT-13 striking the port landing gear of the PT-18 and bending it back. The BT-13 then described a chandelle type turn to the left, turning about 180 degrees before diving into the ground and bursting into flames.

12-3-42A. Livingston, Louisiana. At 0030, a Douglas DB-7B on a simulated night bombing mission dove into the ground at Livingston, Louisiana, killing pilot 2Lt. Charles Davis Smith and passenger Pvt. Thomas Jones. A farmer on the ground reported that both engines cut out and then restarted with a loud report that he described as a "backfire." He then observed the airplane dive into the ground in a heavily wooded area, exploding violently into flames upon impact.

12-3-42B. Willows, California. At 1544, a Vultee BT-13 and a BT-13A collided in mid-air and crashed to the ground two and one-half miles northwest of Willows, California, killing BT-13A pilot A/C Murry Gould. Investigators stated, "Cadet Gould lost sight of his formation leader when [the three-airplane] formation was put into a shallow climb. The formation leader, after gaining about 50 feet, started to level off. As he did, Cadet Gould's ship appeared directly in front of the lead ship and was still in a climb attitude. Lead ship's [propeller] cut leading edge of [Cadet Gould's] right horizontal stabilizer. The stabilizer then pressed against the lead ship's cowl and as Cadet Gould's ship was still rising, a portion of the stabilizer bent back or down. It is conjectured that this bent part of the stabilizer jammed the elevators sufficiently to prevent Cadet Gould from holding the ship in a level attitude. The ship dived from about 2,200 feet into the ground." Apparently, Cadet Gould stood up in the cockpit at an altitude of about 200 feet above ground level and pulled the ripcord of his parachute, being sucked out of the cockpit and plunging to his death with the partially deployed parachute trailing behind him. The pilot of the BT-13, 2Lt. Harry R. Bratt, was uninjured and landed his damaged airplane safely in an open field, suffering only bent propeller blades.

12-3-42C. Kingston, New York. At an unknown time after 2118 EWT, a Vultee BT-13A crashed into Big Indian Mountain 20 miles west of Kingston, New York, killing U.S. Military Academy Cadet Robert E. Horn. The airplane had taken off from Stewart Field, Newburgh, New York, at 1935 on a night cross-country navigation flight to Albany, New York, to Amsterdam, New York, to Albany, New York, and return to Stewart Field. The pilot checked in by radio at Albany at 2118 but was never heard from again after that. When the airplane failed to return to its base it was reported as missing. Investigators speculated that the airplane would be out of fuel by 0035. Despite an extensive seven-day air and sea search of the Eastern United States and the Atlantic Ocean, the cadet and the BT-13 remained missing. The pilot and airplane were found on 4-6-43 on Big Indian Mountain by a troop of Boy Scouts camping in the area.

12-3-42D. Gulf of Mexico, Florida. At 2155 EWT, a Martin B-26B on a simulated night bombing mission crashed into the Gulf of Mexico 100 yards off shore of Venice, Florida, killing the crew of seven. AAF officer Major Frank T. Hanby, a resident of Venice, reported that the B-26 was flying low and fast over Venice just prior to crashing into the water. Investigators could not determine why the B-26 crashed. Killed in the accident were: 2Lt. Parker J. Matthews, Kirkwood, Missouri, pilot; 2Lt. J. C. Workman, Hamilton, Texas, co-pilot; 2Lt. Daniel J. Murphy, Montreal, Ontario, navigator; 2Lt. Woodrow H. Hiebert, Newton, Kansas, bombardier; SSgt. John E. Chase, Cedar Grove, New York, engineer; SSgt. Henry Reicher, Newark, New Jersey, radio operator; Sgt. Marvin N. Paige, Burlington, Vermont, gunner.

12-3-42E. Biggs Field, Texas. At 2000 MWT, a Douglas O-46A on a searchlight tracking mission crashed 15 miles east of Biggs Field, El Paso, Texas, killing pilot 2Lt. Max H. Condron and radio operator SSgt. Herman D. Landis. Investigators speculated that the pilot became confused or disoriented because of several searchlight beams being trained on his aircraft. The O-46 was seen to go into a vertical dive and crash into the ground, killing both fliers instantly.

12-4-42A. Springfield, Ohio. At 1330 EWT, a Republic P-47C on an altitude test flight dove into the ground five miles southwest of Springfield, Ohio, killing pilot 2Lt. Edgar G. Greek, Farmingdale, Long Island, New York. The P-47 slammed into the ground in a vertical power dive from an altitude estimated to be 35,000 feet. The aircraft struck the ground with tremendous force and was destroyed utterly, exploding in flames and leaving a 12-foot deep crater. Investigators speculated that the pilot might have blacked out at altitude because of oxygen starvation or was physically unable to regain control of the aircraft. Investigation revealed that the pilot had borrowed the oxygen mask that he was using and that the aircraft was completely intact prior to hitting the ground. Investigators noted a similar accident occurring on 11-27-42 (see 11-27-42A).

12-4-42B. Ellaville, Georgia. At 1445 EWT, a Stearman PT-17 stalled and spun into the ground four miles east of Ellaville, Georgia, killing Aviation Cadet George C. Lew. Witnesses on the ground reported that the airplane was performing pylon-eights at an altitude of 500 feet above ground level. The airplane was then seen to go into a climbing attitude, stall, and then dive toward the ground, striking nose first in an open field. The cadet was killed instantly.

12-4-42C. Craig Field, Alabama. At 1425 CWT, a Curtiss P-40E stalled and spun into the ground while attempting a landing at Craig Field,

Selma, Alabama, killing Aviation Cadet William C. Dressler. "It is the opinion of the Aircraft Accident Classification Committee that the accident occurred as a result of the pilot's failing to first make a complete cockpit check and read his cockpit checklist prior to landing, thereby, failing to check his gasoline and to turn on a full tank; secondly, as he was overshooting [his landing] and his engine stopped, failed to lower the nose to keep gliding speed and attempted to turn back into the field, thereby stalling and spinning in from an altitude of about 40 or 50 feet. Had the pilot lowered [the] nose quickly when the engine stopped, and landed straight ahead, it is probable that he would have escaped serious injury."

12-4-42D. Artesia, New Mexico. At 1600, a Cessna AT-17B on an instrument team ride suffered the catastrophic structural failure of the starboard wing and crashed eight miles northeast of Artesia, New Mexico, killing Aviation Student Leslie G. Peterson and Aviation Student William P. O'Dell. Investigators stated, "Approximately one and one-half miles northeast of the crashed airplane, a section of the right wing was found. This section of fabric, plywood, and nose ribs had covered ribs #15 to #22 inclusive. From all indications the failure of this wing had begun as a failure of the leading edge, peeling the fabric and plywood from the bottom side, at the front of the main spar, forward and up and back along the topside of the wing to the trailing edge directly in front of the aileron. From the position of this large piece of fabric it appeared as if the plane then went into a flat spin traveling in a southwesterly direction, dropping small pieces of rib and plywood for a distance of one-half and three-quarters of a mile.... The flight path of the airplane from the first bits of wing wreckage was a gentle arc swinging from a southwesterly direction thru the west to northwest. While going in this northwesterly direction, the plane had evidently half-rolled and come straight down into the ground."

12-4-42E. Florence, Arizona. At about 1600 MWT, a Consolidated B-24D suffered a catastrophic structural failure and crashed 25 miles east of Florence, Arizona, killing seven crewmembers. Wreckage was scattered over a mile before the airplane struck the ground and exploded in flames. Killed in the crash were: 1Lt. Robert S. Adams, pilot; 2Lt. Arthur C. Edison, co-pilot; Sgt. James H. Akins, engineer; SSgt. Francis E. Santmyer, radio operator; Pvt. Edward Zientarski, assistant radio operator; Sgt. Armand Peltier, gunner; Pvt. Milton L. Kienholz, gunner. "It is the opinion of the Accident Classification Committee that the airplane encountered turbulent conditions while flying at a flight altitude above 9,000 feet above sea level; that the horizontal tail surfaces were subjected to severe upward forces; that at the time of subjugation of the horizontal tail surfaces to the severe upward forces, the right-hand tail group broke free of the airplane and was blown toward the east by strong westerly winds; that the flight control surfaces of the right-hand tail group was deformed or broken at the time the right-hand tail group was broken from the airplane; that the flight control surfaces were torn from the right-hand tail group during its descent; that at the time of subjugation of the horizontal tail surfaces to the severe upward forces, the plates on the lower surface of the left half of the horizontal stabilizer failed in tension; that sufficient of the structure of the left half of the horizontal stabilizer remained intact so that the left-hand tail group remained attached to the airplane until the airplane struck the ground and permitted partial control of the flight of the airplane; that sufficient control of the airplane was regained with remaining left-hand tail group to enable the pilot to recover from the spin and even though [yawing side to side] to a marked extent, fly away in a southeasterly direction after recovery from the spin; that the left-hand tail group was further stressed in recovering from the spin; and that the further cumulative weakening of the left half of the horizontal stabilizer caused the airplane to assume a stalling attitude and crash."

12-4-42F. Location Unknown. At an unknown time after 1930 EWT, a Martin B-26B disappeared with its crew of four and was presumed to have crashed into the Atlantic Ocean. The B-26 took off from the Army Air Field at Myrtle Beach, South Carolina, at 1930 EWT and failed to return to base at the completion of its two-hour night gunnery and bombing mission. Despite a massive search effort that lasted eleven days, which included help from airplanes and surface units of the U.S. Navy, Marines, Coast Guard, and the Civil Air Patrol, no trace of the bomber or crew could be found. Missing and presumed lost at sea were: 2Lt. Rosser C. Forbes, pilot; 2Lt. Norman H. Pearson, co-pilot; 2Lt. Billy B. Rocke, navigator; SSgt. Henry J. Haliday, engineer. 1Lt. George P. Gould, an AAF pilot stationed at Myrtle Beach Army Air Field, reported that he had observed a B-26 fly over the field at approximately 2355, heading due south. There is no evidence to suggest that this was the B-26 piloted by Lt. Forbes. The B-26 outfit stationed at Myrtle Beach Army Air Field suffered the loss of four B-26 airplanes that were ditched or abandoned in flight in separate non-fatal weather and navigation related accidents on 12-2-42.

12-5-42A. Mertzon, Texas. At 0020, a Beech AT-11 on a bombardment-training mission flew into the ground seven miles southwest of Mertzon, Texas, killing four fliers. Investigation revealed that the aircraft was returning to the Army Air Field at San Angelo, Texas, after completing its exercise. The airplane struck the ground in a wings level attitude in a slight dive, exploding into flames upon impact. Weather at the crash site was reported as a ceiling of 1,100 feet with unlimited visibility and in the target area (approximately 12 miles southwest of Mertzon) weather was reported as ceiling and visibility unlimited. There were no witnesses to the accident and investigators could not determine why the airplane crashed. Killed in the crash were: SSgt. Robert E. Martin, pilot; 2Lt. Murray Sandberg, bombardier instructor; Aviation Cadet Bennett F. Penn, student bombardier; Aviation Cadet Joseph L. Perkins, student bombardier.

12-5-42B. Fanlew, Florida. At 1758, a Bell P-39D flying in instrument conditions crashed three and one-half miles west of Fanlew, Florida, killing 2Lt. Woodrow L. Dick. The airplane struck the ground in a flying attitude, killing the pilot instantly and completely destroying the airplane. Investigators speculated that the pilot was letting down through the overcast and flew into the ground. The airplane had taken off from Cross City Army Air Field, Florida, at 1618 EWT on an administrative flight to Dale Mabry Field, Tallahassee, Florida. Cross City personnel tried to contact the pilot by radio to advise him of the deteriorating weather conditions but were unsuccessful.

12-5-42C. Maxwell Field, Alabama. At 2103 CWT, a Douglas C-47 attempting a landing crashed two miles south of Maxwell Field, Montgomery, Alabama, killing the crew of four and eight passengers. The Accident Classification Committee stated, "The subject airplane with the call letters 8494 cleared from Maxton Army Air Base, Maxton, North Carolina, for a non-stop flight to Brookley Field, Mobile, Alabama, [on a cargo and personnel transport mission] with an estimated time in route of 4 hours. The ship cleared Maxton Air Base at 1525 CWT and was due at Brookley Field at 1925 CWT. At 1840, Maxwell radio received instruction from Airways Traffic Control to try and contact Army 8494 to advise him to land at Maxwell Field, Alabama, due to weather conditions at Brookley Field. Maxwell Field radio was broadcasting blind and did not receive any response. At 2020 CWT, Maxwell Field radio again tried to contact 8494 but was unsuccessful. At 2045 CWT, Maxwell Field radio did hear 8494 calling but was unable to establish contact with him. At 2056, 8494 contacted Maxwell Field radio advising that he was circling Maxwell Field at 1,500 feet. Maxwell Field radio then heard him contact Maxwell Field control tower and receive landing instructions from the control tower. The control tower, having contacted 8494, asked if he was flying contact and the answer was 'Yes.' The pilot requested floodlights, and the tower operator immediately turned the Northeast bank of lights on. Maxwell Field tower then proceeded to give him landing instructions, advising him to land into the southwest on the lighted runway and to call again when he was on his base leg. The next call received by the control tower operator, the pilot of 8494 advised that he was making his approach. The control tower operator advised 8494 that he was cleared to land, and 8494 replied 'Roger.' A few minutes later, a call was received at the Operations Office that a cargo type airplane had crashed at the Reconsignment and Holding Depot which is located approximately 2 miles south of Maxwell Field. Upon investigation it was found that the crashed airplane was 8494. Floodlights are installed on power line poles at this depot, which at night give the impression from the air of a series of parallel runways. Upon investigation it was found that 8494 circled this depot several times with landing lights on.... It seems that the airplane was making his approach from the East for a landing into the West when approximately 40 feet above the ground the airplane struck an 11,000-volt power line. Four [power lines] were broken on each side of the pole, and it is the opinion of this board that only the landing gear struck the power line. A complete blackout of the area was caused by the failure of this power line. Witnesses state that after the airplane hit the power line, it continued descending until the pilot opened his throttles when the airplane was approximately 20 feet above the ground. The airplane then began to climb rapidly at a steep angle and was approximately 200 to 300 feet above the ground when it made a sharp turn to the left and began to lose altitude rapidly and struck the ground at approximately a 20 degree angle, tearing up railroad track and crashing into a power pole, exploded and burned. It is the opinion of the board that the pilot mistook the Reconsignment and Holding Depot for Maxwell Field and that he was descending for a landing when he struck the power pole. It is further the opinion of this board that if the pilot had continued to make a normal landing he would have been successful without undue damage to aircraft or injury to personnel, but it is believed that when the ship struck the power line and blacked out the entire area, the pilot realized that something was wrong and opened his throttles with the intention of pulling up and going around and that in pulling up, it is believed that he climbed at such a steep angle as to cause the ship to be in a stall. It is believed that the turn increased this stalling effect and in attempting to recover, the pilot dropped the nose but not having sufficient altitude crashed to the ground." Killed in the crash were: SSgt. Morton A. Stewart, pilot; SSgt. Garry F. Hinton, co-pilot; Sgt. Albert M. Killion, engineer; Cpl. William Bywaters, radio operator; 2Lt. James W. Leach, pilot-rated passenger; 2Lt. Robert E. Latas, pilot-rated passenger; Cpl. Niles A. Anderson, passenger; Sgt. Walter D. Gall, passenger; SSgt. Jesse J. Thomas, pilot-rated passenger; SSgt. William P. Graves, pilot-rated passenger; Sgt. Joe F. Rathgeber, passenger; Cpl. Russell E. Coleman, passenger.

12-6-42A. Stewarts Point, California. At 0718, a Consolidated B-24D crashed nine and one-half miles southeast of Stewarts Point, California, killing the crew of six. Investigators stated, "B-24D airplane #4262 took off from Hamilton Field [San Rafael, California, on a combat sea patrol mission] at 0615 PWT and was instructed to fly 120 degrees and climb on top of the overcast. [The pilot] reported having engine trouble at 0630. At 0635, 4262 stated he was on top at 9,500 feet and getting ice. He was advised to fly 300 degrees back to Hamilton Field at approximately 0700. Fifteen minutes later the tower was called by 4262 stating he was bailing out. The plane crashed about 60 miles Northwest of Hamilton Field with all personnel on board. Investigation at the scene of the accident shows no broken trees nearby and higher mountains to the west and north. The airplane, coming from the north, evidently passed over the mountains and crashed into the field. This indicates that the plane struck the ground at a very steep angle. From the scene of impact, the committee believes that the pilot dove into the valley and in pulling over the hill the plane partially stalled and when attempting to recover, the plane crashed [into a ridge and exploded into flames]. Witnesses in the area testified that the clouds were topping the mountains and there was an intermittent cold drizzle. It is believed that the pilot was either attempting to fly under the overcast to get his bearings, or possibly ice was forcing him down rapidly. The control tower log reveals that one engine was out, [the pilot] was encountering icing and was confused as to his position and the procedure to follow in this emergency." Killed in the crash were: 2Lt. Donald O. Pedrazzini, pilot; 1Lt. Herbert H. Knowlton, co-pilot; 2Lt. Charles F. Meyer, navigator; TSgt. Joseph D. English, engineer; SSgt. William C. Malone, radio operator; Pvt. Vern B. Smith, assistant radio operator. All went down with the airplane.

12-6-42B. Location Unknown. At an unknown time after 1001 PWT, a Consolidated B-24D disappeared and is presumed to have crashed into the Pacific Ocean. The crew of seven is presumed lost at sea. The B-24 had taken off from Hamilton Field, San Rafael, California, at 0640 PWT on a combat sea patrol mission. The crew had transmitted four routine position reports to Hamilton Field, the last one at 1000 PWT. The airplane reported its position as 37 degrees 30 minutes North and 131 degrees 15 minutes West and was never heard from again. Despite a search effort lasting two weeks, no trace of the airplane or crew was ever found. Missing and presumed lost at sea were: 2Lt. William W. Smith, Jr., pilot; 2Lt. Fay D. Villro, co-pilot; 1Lt. David L. Rosenberg, navigator; SSgt. Gilbert L. Simoneaux, engineer; Sgt. George H. Poot, assistant engineer; TSgt Richard F. Miskopf, radio operator; Sgt. Arthur Vigil, tail gunner.

12-6-42C. Wewahotee, Florida. At 2043, a Curtiss P-40E was abandoned and crashed at Wewahotee, Florida, killing pilot 2Lt. Alexander M. Wood, who was struck by the airplane while attempting to parachute to safety. His parachute successfully deployed and he floated the earth fatally injured. His body was found the next morning some distance from the crashed airplane. The cause of the accident could not be determined but investigators speculated that the airplane might have experienced engine trouble. The aircraft had taken off minutes earlier from Orlando Army Air Base, Orlando, Florida, on a searchlight and night pursuit mission.

12-6-42D. Fitzpatrick, Alabama. At about 2100, a Curtiss AT-9 dove into the ground near Fitzpatrick, Alabama, killing pilot Flight Officer Douglas A. Crother and passenger Pvt. Fred F. Rollins. The airplane took off at 2009 from Maxwell Field, Montgomery, Alabama, on an administrative cross-country flight to Turner Field, Albany, Georgia. Witnesses on the ground observed the airplane flying at low altitude toward the direction of Maxwell Field when it pitched up sharply to avoid rising terrain, stalled, then fell off on a wing and dove to the ground on top of a hill, bursting into flames upon impact. Investigators speculated that the pilot might have become confused (by losing the horizon) or lost and was attempting to find his way back to Maxwell Field. The ceiling in the area at the time of the crash was reported as 1,200 feet with light to moderate haze.

12-6-42E. Muroc, California. At 2110, a Lockheed P-38E on a night transition flight collided with terrain near the Army Air Base at Muroc, California, killing pilot 2Lt. John E. Garcia. The Accident Classification Committee stated, "After investigation of the crash it is the opinion of the board that the pilot was flying low over rolling terrain at night. Attempting to make a sharp left turn, the left wing struck the ground, causing the airplane to crash nose down. The plane then bounced into the air and exploded, throwing parts of the airplane for a thousand feet. There were no witnesses."

12-6-42F. Victorville, California. At 2210, a Beech AT-11 on a bombardier training mission collided with rising terrain and crashed near Target #5 on the AAF Bombing Range near Victorville, California, killing the crew of four. The Accident Classification Committee stated, "Plane flew into mountain ridge about 200 feet below crest. Elevation of crest slightly over 4,000 feet [above sea level]. Mission was being flown at approximately 7,000 feet. Immediately prior to the crash, other pilots on the same section of the range reported suddenly encountering snow, sleet, and freezing rain, where it had been perfectly clear a few minutes earlier. Clouds were also encountered at the altitude the planes were flying. As a result it was reported that carburetors were icing and that instrument flight was necessary." Killed in the crash were: 2Lt. Charles N. Wakefield, pilot; 2Lt. Richard D. Radmore, bombardier instructor; Aviation Cadet Edgar C. Bessels, student bombardier; Aviation Cadet William W. Bonner, student bombardier.

12-7-42A. Columbia, South Carolina. At 1310, a North American B-25C suffered a progressive catastrophic structural failure while flying in instrument conditions nine miles southeast of the Army Air Base at Columbia, South Carolina, killing the crew of seven. Wreckage was scattered over a wide area, some pieces becoming lodged in trees after falling to earth.

Investigators stated, "It is considered probable that the plane went into a very steep spiral from a right turn after passing the cone at CI range station and was pulled out violently when just below the overcast which at the time was approximately 600 feet. Due to the loadings imposed by the abrupt pull out, the plane disintegrated in the air. It crashed in an open field and fire followed the crash. Icing conditions encountered during the descent might have been a contributing factor." Killed in the crash were: 2Lt. John D. Trimmier, pilot; 2Lt. Robert E. Thomas, co-pilot; SSgt. Lawrence F. Ford, engineer; SSgt. Earl Fynan, engineer; Sgt. Julian C. Aycock, gunner; Sgt. Alton J. Jones, gunner; SSgt. William J. Gorman, gunner.

12-7-42B. Lancaster, California. At 1030, a Vultee BT-15 stalled and spun to the ground six miles west of Lancaster, California, while trying to enter the pattern for War Eagle Field, killing Aviation Cadet Walker L. Clapp and Aviation Cadet G. F. Bell. The airplane was returning from a scheduled instrument team ride and stalled out of a steep turn at an altitude of approximately 1,000 feet. The airplane went out of control and smashed into the ground, bursting into flames and killing both occupants instantly.

12-7-42C. Ajo, Arizona. At 1527 MWT, a North American AT-6A towing aerial gunnery targets suffered the failure of the port wing and crashed 12 miles WSW of Ajo, Arizona, killing civilian pilot Robert C. Moses and passenger Pvt. Frank C. Clemens. Investigators stated, "Pilot Moses was making a steep turn with the target out when his plane snap rolled and went into a spin, from which it never recovered. Aviation Cadet Fayette L. Weedin, who was making the passes at the target at the time the ship crashed, stated that it looked as it a wing flew off at approximately 1,500 feet above the ground. The left wing was found 300 yards from the point of impact. Scattered parts of canopy and the Form #1 [were] found nearly 400 yards from point of impact, in line with the detached wing. The wing evidently ripped from the bottom and peeled back, smashing the canopy before tearing away from the airplane. It is the opinion of the committee that the wing was torn off due to violent and irregular movement of the airplane and not in structural failure.... It is believed that the pilot was attempting to tighten a turn by holding bottom rudder and using excessive back pressure on the stick. It is also believed that the nose was in a nose high position at the time."

12-7-42D. Seguin Texas. At 2130, a North American AT-6A crashed six miles east of Seguin, Texas, killing pilot 2Lt. Merle F. Radtke and passenger Capt. Van B. Anderson. The airplane was on a night navigation flight from Brooks Field, San Antonio, Texas, to LaGrange, Texas, and return. The airplane struck the ground in a vertical position and at a high rate of speed. The cause of the accident could not be determined but investigators speculated that the pilot might have become disoriented and lost the horizon.

12-7-42E. Randolph Field, Texas. At 2200, two Vultee BT-13A type airplanes collided in mid-air two miles north of Randolph Field, San Antonio, Texas, killing A/C Stanley G. Chambers aboard BT-13A #41-21514. A/C Harry D. Junkin was killed in the crash of BT-13A #41-21447. The collision occurred at an altitude of about 600 feet above ground level about one-half mile south of the Austin Highway. The propeller of Cadet Chambers' airplane cut into the port wing of Cadet Junkin's airplane. Both airplanes fell to earth and exploded into flames. Neither cadet was able to use his parachute. Investigators faulted both pilots for failing to see each other and avoid the collision.

12-8-42A. Tampa Bay, Florida. At 1213 EWT, a Martin B-26B crashed into Tampa Bay just after take-off from MacDill Field, Tampa, Florida, killing six fliers. Investigators stated, "B-26, #41-17789 took off from MacDill Field on Runway 9 at 1212 EWT. Shortly after leaving the ground, the ship began a turn to the left in a climbing attitude. From here on, the ship was seen flying to the northeast in a [skid to the right (tracking 45 degrees to its heading) before going into a] violent yaw to the left with nose coming up higher all the time. The ship was seen to stall and rollover into a spin to the left, crashing into shallow water one mile northeast of MacDill Field. Upon examination of the [port] engine, the fuel pump drive shaft was found to be broken.... It is the opinion of the committee that the crash was caused by: 1. Loss of power on the left engine due to failure of the fuel pump shaft. 2. The pilot pulled the nose up until the ship stalled and spun." The B-26, scheduled for a formation flight and simulated bombing mission with another B-26, spun in from an altitude of approximately 300 feet. Killed in the crash were: 2Lt. John E. Williams, Los Angeles, California, pilot; 2Lt. Clarence O. Parsons, Electra, Texas, co-pilot; 2Lt. Norman J. Linne, Indianapolis, Indiana, bombardier; Sgt. John Mazzarina, New York, New York, engineer; Sgt. Samuel G. Lamond, Philadelphia, Pennsylvania, radio operator; SSgt. David H. Brown, Monongahela, Pennsylvania, gunner.

12-8-42B. King City, California. At 1645, two Ryan PT-22 type airplanes collided in flight near King City, California, killing Aviation Cadet William C. Pitt aboard PT-22 #41-20617. Aviation Cadet James L. Jacobson received minor injuries parachuting to safety from PT-22 #41-20818. Investigators stated, "Aviation Cadet Jacobson was flying rectangular patterns at 1,000 feet indicated [altitude]. He was recovering from a 90-degree turn, from a heading of north to a heading of east. As he rolled out he saw the other ship, traveling south, just as he hit it. He made an attempt to climb over but this try was not successful. He immediately bailed out and landed safely. Aviation Cadet Pitt seems to have made no attempt to leave his airplane. The responsibility rests equally upon the students as they were in command of the ships when the accident occurred. Stickers are being printed and are to be put into every airplane warning personnel to look around. All personnel are being warned that it is important to clear over the topside of a turn as well as into it." The setting sun was considered a contributing factor.

12-8-42C. Marana, Arizona. At 2130 MWT, a Vultee BT-13A dove into the ground near Marana, Arizona, killing Chinese Aviation Cadet Wen-Shen Pei. The airplane turned slightly as it struck the ground at high speed and at a steep angle, exploding into flames upon impact. The cause of the accident could not be determined.

12-9-42A. Mountain Home, Idaho. At approximately 1045 MWT, a North American AT-6A flying in poor weather conditions crashed into rising terrain 15 miles northwest of Mountain Home, Idaho, killing pilot 1Lt. Ray Irving Glasgow and passenger Maj. Harry J. Silver. The airplane had taken off from Luke Field, Phoenix, Arizona, on a cross-country flight to Portland, Oregon, on 12-8-42 at 1444 MWT, landing at Burley, Idaho, at 1840 because of weather. The fliers remained overnight and took off at 0953 MWT. They never arrived at Portland and were reported missing. Despite a massive search effort, the airplane and the dead fliers were not found until nearly seven weeks later at the 7,000-foot level of Bennett Mountain.

12-9-42B. Location Unknown. At a time between 0953 and 1200 PWT, a Bell P-39D and pilot 2Lt. Lewis M. Russell disappeared over the Western United States. The pilot was declared missing and presumed killed. The airplane took off from Paine Field, Everett, Washington, at 0953 on a local training mission. The airplane had about two hours worth of fuel. Weather was reported as ceiling at 1,000 feet with seven miles visibility, wind southwest at 10 mph. Despite a massive search effort, the airplane and pilot were never located. Poor weather conditions hampered the initial search effort, which was abandoned after 12-31-42. No other information is available.

12-9-42C. Glenview, Illinois. At 1345, a North American AT-6C suffering an engine failure crashed into a building while attempting an emergency forced landing on the recreation field at Great Lakes Naval Training Station, Glenview, Illinois. Seaman 2C Paul A. Landers, U.S. Navy, was struck by the airplane. He died of a skull fracture a short time later at the station hospital. Several other seamen avoided the airplane by sprinting from the area. Pilot 2Lt. Morton J. Dirnbyaur received a broken jaw and passenger SSgt. George E. Taylor was uninjured. The AT-6C had taken off from George Field, Lawrenceville, Illinois, on an instrument cross-country to Milwaukee, Wisconsin. The engine quit near the naval station and the pilot picked the recreation field for his emergency landing. The pilot was unable to stop the airplane once he landed on the recreation grounds, ground looping to the left and running into the southwest corner of Building 600. Investigators speculated that the engine failed due to carburetor icing.

12-9-42D. Scott, Louisiana. At 1630, a Fairchild PT-19A practicing simulated forced landings stalled and spun to the ground near Scott, Louisiana, killing civilian instructor Charles G. Mansfield and student Aviation Cadet Theodore R. Rivoir. Investigators stated, "The instructor allowed the student to glide too slow going downwind during a simulated forced landing. The airplane stalled at approximately 40 feet in the air. At this point, the instructor attempted to recover by use of the engine and allowing the nose to drop in a vain attempt to pick up flying speed. The airplane was at such a low altitude [that] it crashed before regaining enough speed to sustain flight." The aircraft did not burn but was completely destroyed.

12-9-42E. Boerne, Texas. At 2051 CWT, a Douglas B-18 flying in poor weather collided with rising terrain while attempting an instrument let-down seven miles southwest of Boerne, Texas, killing the crew of four. The B-18, based out of MacDill Field, Tampa, Florida, was flying to Texas to transport personnel back to MacDill Field. At the time of the accident there was a zero ceiling, zero visibility condition at the crash site (elevation approximately 1,450 feet). Killed in the crash were: Maj. Frank W. Potter, pilot; 2Lt. Otto E. Schroeter, passenger; Sgt. George W. Boykin, engineer; Sgt. Talmadge E. Hardy, radio operator. Investigators stated, "A B-18 airplane, with Major Potter as pilot, cleared on instruments from New Orleans Army Air Base [New Orleans, Louisiana] for Alamo Field, San Antonio, Texas. Time of departure was 1717, ETA 2057. At approximately 2030, aircraft passed over Alamo Field but was not visible. Airplane was next seen flying about 100 feet above terrain headed in a southwest direction six miles south of Boerne, Texas, at 2045. At 2050, airplane was again seen flying just above the treetops headed for hills of approximately 1,500 feet elevation in a southwesterly direction. About one minute later crash was heard and reported to the sheriff of Kendall County, Boerne, Texas. Sheriff instigated a searching party immediately which carried on all during the night and until the plane was located at 1030 December 10, 1942. Aerial search was instigated by Alamo Field and Kelly Field [San Antonio, Texas] at 1100 December 10, 1942 when ceiling lifted sufficiently for contact flight. Crash was located from air at 1140. Aircraft crashed in hilly country about seven miles southwest of Boerne, Texas. Indications are that the pilot was attempting an instrument let-down because he cleared a hill higher than the one first crashed into. Plane struck one hill, jumped a 50-foot canyon, struck the next hill, and stopped in another canyon 250 yards from the point of first impact. [The airplane] was headed in a southwesterly direction. Wreckage was strewn from the sight of the first impact to the final stopping point. Crash resulted in the complete [destruction] of the aircraft and the burning of the right wing. Bodies of the pilot and crew found burned beyond recognition on top of burned right wing about eight feet from the fuselage. Indications are that the right side of the fuselage was carried away by trees on first impact. Radio control panel, pencil, and a package of cigarettes were found on the first hill. It is the opinion of the board that equal responsibility for the accident can be charged to weather and the pilot."

12-9-42F. Ada, Oklahoma. At 2110, a Vultee BT-13A on a night navigation flight crashed four miles south of Ada, Oklahoma, killing pilot Capt. Delmer L. Biser and passenger 1Lt. Everett Caughey. The Accident Classification Committee stated, "Capt. Biser and Lt. Christenson [who was flying an AT-6 type airplane] were proceeding from the Army Air Base at Coffeyville, Kansas, to Perrin Field, [Sherman] Texas, on a routine night navigation flight. As the flight, with Capt. Biser leading, approached the vicinity of Ada, Oklahoma, Capt. Biser started to spiral down apparently to get under a broken cloud formation, which was developing under him. The spiral steepened until, as the airplane disappeared from Lt. Christneson's view, it was headed nearly straight down in a tight spiral. Investigation of the wreck shows that the airplane struck the ground in this position and at a considerable speed."

12-9-42G. Bowling Green Kentucky. At 2150, a North American B-25C on a navigation training flight crashed into the side of a hill while on an instrument let-down eight miles west of Bowling Green, Kentucky, killing the crew of six. The airplane took off with two other B-25 type airplanes from Anderson Field, Walterboro, South Carolina, for the formation flight to Nashville, Tennessee. The flight climbed to 5,000 feet just after take-off and soon encountered instrument conditions. The flight was able to stay together on top of the overcast and eventually diverted to Bowling Green. The accident airplane was leading the formation, initiated the descent and began his let-down first with the other two following individually in trail about 500 yards apart. The other two airplanes broke out at about 1,800 feet and landed safely at Bowling Green. The subject airplane descended into the ground at the base of a small hill, striking trees and exploding into flames upon contact with the ground. Wreckage and bodies were strewn all the way up the hillside and a 150-foot swath of trees were scorched or destroyed. Killed in the crash were: SSgt. Jack E. Blackledge, pilot (former RCAF pilot); SSgt. W. F. Hosey, co-pilot; SSgt. Jack Zelen, engineer; SSgt. Alfred T. Secor, Jr., radio operator; SSgt. Abe B. Seitz, assistant radio operator; SSgt. Forest C. Houck, gunner.

12-10-42A. Hernando, Mississippi. At 0130, a Vultee BT-13A on a night navigation flight crashed near Hernando, Mississippi, killing Aviation Cadet Carrol D. Whitacre. A civilian reported that the airplane had circled Hernando for a short time at an altitude of about 500 feet just prior to the crash, which he had heard. There were no eyewitnesses to the crash and investigators could not determine why the airplane flew into the ground with power on. Investigators noted that icing conditions existed in the area at the time of the crash. The airplane was one of three BT-13 airplanes from the Army Air Field at Blytheville, Arkansas (699th Training Squadron), involved in a fatal accident that morning (See 12-10-42B, C).

12-10-42B. Como, Mississippi. At 0330, a Vultee BT-13A on a night navigation flight crashed near Como, Mississippi, killing Aviation Student Jimmy O. Glasgow. Investigators speculated that the airplane might have picked up ice and spun to the ground, knocking down a power line and bursting into flames. The airplane was one of three BT-13 airplanes from the Army Air Field at Blytheville, Arkansas (702nd Training Squadron), involved in a fatal accident that morning (See 12-10-42A, C).

12-10-42C. Sardis, Mississippi. At 0330, a Vultee BT-13A on a night navigation training flight crashed at Sardis, Mississippi, killing student pilot 2Lt. Winlock O. Moore. The airplane apparently spiraled to the ground while flying in instrument conditions. Investigation of the airplanes maintenance records revealed that several instruments were inoperable and prevented the student from safely flying on instruments. Investigators noted that icing conditions existed but were probably not instrumental in causing the accident. The airplane was one of three BT-13 airplanes from the Army Air Field at Blytheville, Arkansas (702nd Training Squadron), involved in a fatal accident that morning (See 12-10-42A, B).

12-10-42D. Atlantic Ocean, Florida. At 0907, a Lockheed-Vega B-34 on an anti-submarine mission suffered a loss of power and ditched into the Atlantic Ocean three miles off shore of Key Largo, Florida, killing two crewmembers and injuring three others. Bombardier Pvt. William R. McCormick and radar operator Pvt. Clarence H. Laughlin were apparently injured or knocked unconscious during the ditching and were unable to free themselves and exit the airplane. Rescued by merchant seaman in a lifeboat after spending about two hours in the water were the pilot 1Lt. Jessie W. McFerrin, co-pilot 2Lt. Jordan M. Pennoyer, and radio operator Sgt. Warren G. Cook. Investigators stated, "The pilot was flying on his cabin tanks, and was circling a convoy under a low overcast at 850 feet altitude. The radar operator, Pvt. Clarence Laughlin, reported a target at 14 miles, thirty degrees right. The pilot turned from a left turn to a right turn and started for the target to investigate. After leveling out, the engines failed to catch, the pilot warned the crew to prepare for a water landing. The co-pilot, 2Lt. Jordan M. Pennoyer, had left his seat to check the gasoline in the cabin tanks, and the bombardier, Pvt. W. R. McCormick, passed by him on his way to the rear of the plane to a previously designated safety belt position. The co-pilot returned to his seat and attempted to help the pilot start the engines. He [the co-pilot] dropped the depth charges and closed the bomb bay doors. At this point, the plane was at an altitude of about 50 feet. The engines caught, and the plane climbed to about 100 feet, when they failed again. On the instructions of the pilot, both pilot and co-pilot feathered the props, and with the stick in its rearmost position, the plane landed in a wave trough, dragging the tail first. The plane came to a not particularly violent stop, and immediately the fuselage came up to the vertical. The pilot and co-pilot escaped from the pilot's hatch, and the radio operator escaped from the rear door. The airplane sank immediately, carrying with it the bombardier and the radar operator, both of whom had taken positions at the tail of the plane on the oxygen tank platform where safety belts had been installed. Upon being notified by the pilot to prepare for water landing, the radio operator left his seat to go the rear, then remembered his water landing procedure, and returned to send his SOS. He was in the act of locking down his key when the plane struck. He was thrown against his radio controls, but not seriously injured. He climbed out the cabin door, squeezing by the cabin [fuel] tanks, and over a pile of equipment, which had torn loose and fallen down toward the front of the aft cabin. He left by way of the rear door, noticing as he went the leg of the bombardier or radar operator amid the pile of debris near the doorway. After the co-pilot escaped, he swam to the rear of the plane and noticed that the bottom of the plane had been torn out, from the rear gun post to the bomb bay doors. The pilot picked up the cover from the de-icer reservoir, which is installed under the pilot's seat, which may indicate that a large area of the bottom of the plane was torn away. The trio remained in the water until picked up by a boat from the freighter they were investigating and they were later moved to shore by a coast guard boat that overtook the freighter for that purpose." Investigation speculated that the engine failure resulted from an apparent airlock in the fuel lines from the cabin [fuel] tanks. Investigators noted that the newly installed crash seats, to be used in case the airplane must make a forced water landing and located in the aft section of the aircraft, were unsafe and should be removed.

12-10-42E. Terre Haute, Indiana. At 1016, a Beech AT-10BH flying in poor weather spiraled to the ground and crashed into a backyard and garage at 1456 South 20th Street, Terre Haute, Indiana, killing A/C Milo G. Stevens, 26, White Plains, New York, and A/C William Sellers, 26, Ponchatoula, Louisiana. The airplane was on a cross-country navigation flight from George Field, Lawrenceville, Illinois, to Terre Haute, Indiana, to Evansville, Indiana, and return to George Field. Investigation revealed that this airplane was not equipped with an artificial horizon because of a shortage of these instruments at George Field. The airplane had encountered a sudden snowstorm while en route to Terre Haute, severely restricting visibility. The pilots lost control of the airplane and entered a steep spiral to the ground, apparently losing parts of the port wing on the way down. The airplane rolled wing over wing and then rolled inverted, knocking down a powerline pole before smashing into the backyard and garage of Lawrence Callahan. The airplane did not burn and no one on the ground was injured but several structures were damaged. A 1939 Chevrolet Sedan that was in the garage was completely destroyed. The port engine, breaking off of the wing after the airplane stuck the power line pole, was found in front of 1460 South 20th Street. Five airplanes from the training squadron at George Field were able to land safely after encountering the snowstorm.

12-11-42A. Colorado Springs, Colorado. At 0907, a Lockheed F-5A (photo-reconnaissance version of the P-38) suffering an apparent failure of the port engine crashed five miles ESE of the Army Air Field at Colorado Springs, Colorado, killing pilot SSgt. Walter E. Salsbury. The pilot had taken off from the base at 0857 and soon requested an emergency landing. The tower acknowledged and reported that the field was clear for an emergency landing. The pilot acknowledged the tower but never made it to the field. The airplane stalled and entered a spin, smashing into the ground and bursting into flames. Control tower personnel were trying to maintain radio contact with the airplane but then noticed a column of smoke rising in the southeast.

12-11-42B. Tucson, Arizona. At 1548, a Ryan PT-22 collided with a large saguaro cactus and crashed nine miles northwest of Ryan Field, Tucson, Arizona, killing civilian flight instructor Hugh B. Evans III and seriously injuring student A/C Charles H. Ingram. Prior to the accident the pilot had performed a spin and recovery (apparently at a safe altitude) followed by a series of low altitude snap rolls and vertically banked turns. Investigators stated, "The ship, piloted by the instructor, was engaged in low altitude flying over dangerous, mountainous terrain. The accident occurred as a result of this low flying when the right wing came in contact with the top of a tall saguaro cactus. This caused the ship to be thrown into a vertical right bank, and from point of contact with the cactus it traveled approximately 250 yards in an arc to the right, at which time its right wing came in contact with the ground, causing the ship to crash inverted. The distance from the spot where actual contact with the ground was made to the point where the ship actually ended up on the down slope of a hill in an upright position [the airplane facing opposite of the line of flight] was only 12 yards. Accident was directly caused due to disobedience of orders in deliberately flying at low and unsafe altitude over dangerous terrain. Pre-arranged meeting with persons on the ground for the purpose of exhibition of aerobatics was a contributory cause." Investigation revealed that Instructor Evans had apparently told friends in a bar the night before that he would buzz a geology field trip group, which included his girlfriend and other friends, the next day. A witness at the Santa Rita Bar reported that the girlfriend was very enthusiastic about the exhibition and had "egged" Evans on. Also, at the urging of the girlfriend, Evans agreed to have the cadet drop a pack of cigarettes to the group on the ground during the exhibition, which they did. The crash occurred one-quarter of a mile from the geology field trip group. The group did not witness the accident but had retrieved the cigarettes.

12-11-42C. Patterson Field, Ohio. At 1800 EWT, a North American A-36A on a fuel flow test flight crashed two miles northeast of Patterson Field, Ohio, killing pilot 1Lt. Albert E. Wright. The pilot had taken off from Patterson Field and climbed to 10,000 feet to conduct his tests. The weather deteriorated and the pilot was ordered to abort his tests and return to the field. Upon reaching the field, the pilot reported to the tower that he was not getting an indication on his instrument panel that his port main landing gear was down and locked. He flew past the tower and tower personnel advised him that the gear was down and appeared normal. The pilot stated that he was going to try to shake the landing gear down into the locked position by maneuvering. The airplane entered the overcast (ceiling about 5,000 feet) and moments later the pilot lost control of the airplane. Investigators stated, "From available reports, it is assumed that the pilot in his climb through the clouds stalled and went into a spin to the right. Upon reaching [visual] contact he recovered and due to over control, spun to the left. Upon recovering from the left spin, there was insufficient altitude for the pull out and he crashed at this point." Examination of the wreckage revealed that the locking pins for the main landing gear had engaged and were locked.

12-11-42D. Topton Airport, Mississippi. At 1705, three Vultee V-72 airplanes (similar to the A-31 type) collided in mid-air near Topton Airport, Mississippi, killing two fliers. Four other fliers were uninjured. The pilot of V-72 #875, 2Lt. William J. Wepfer, was uninjured parachuting to safety; gunner Sgt. Charles T. Mackey was killed when he went down with the airplane. Pilot 2Lt. Sterling W. Hancock was killed in an unsuccessful parachute jump from V-72 #910; gunner Sgt. John F. Mills was uninjured parachuting to safety. 2Lt. Edwin H. Crow, who was leading the three airplane flight in a left echelon formation in V-72 #542, and gunner Sgt. Robert S. Moore were able to land safely at the Topton Airport. Both were uninjured. The Accident Classification Committee stated, "It is the opinion of the committee investigating the accident that 2Lt. Wepfer, while flying in a three-ship formation, failed to bank his ship to the right during a shallow right turn, thereby causing a collision of the three airplanes while in flight."

12-11-42E. Foster Field, Texas. At 1717, a North American AT-6C landed on top of another AT-6C while both were attempting to land at Foster Field, Victoria, Texas, killing A/C Louis G. Bender. A/C Bender was flying in the rear cockpit of AT-6C #41-32269, which was piloted by A/C Clarence J. Deaton. A/C Deaton was uninjured. A/C John J. Mahoney was uninjured when he landed AT-6C #41-32359 on top of Bender's AT-6C, cutting into the rear cockpit with the propeller and killing the cadet instantly. The airplanes were part of two separate three-ship formations that were returning from gunnery practice missions at Matagorda Island. Investigators stated, "As the formation came over Foster Field at approximately 1715, the normal peel-off was accomplished and due to congested traffic conditions at that time the flight in which Cadet Deaton was No. 3 joined a very large traffic pattern and proceeded to the base leg prior to landing. As Cadet Deaton was on his base leg he heard the tower instruct the landing airplanes to shorten the base leg and come in closer to the field. He noticed another formation above that was forming a pattern inside of his traffic pattern. He was almost ready to turn on to his landing leg and continued in. Due to the large pattern the landing leg was made under power in a low approach and Cadet Deaton landed short on Runway 35R. Just prior to landing, the pilot stated that he saw an airplane pass over him and it continued on, landing between his airplane and the No. 2 airplane of his formation, which was already on the ground. The pilot further stated that he thought he had ample room in which to land and continued on. The airplane that passed over Cadet Deaton was piloted by 2Lt. Michael F. Robinson leading a formation in to Foster Field. Cadet Deaton had landed and rolled only a short distance when he heard his passenger, Cadet Bender, shout from the rear seat to 'Give it the gun and go around.' Almost simultaneously he heard the control tower tell the airplane that was landing to go around. Just as he applied throttle, an airplane landed on top of him, cutting his airplane nearly in two, the propeller striking Cadet Bender and killing him instantly. The airplane that landed on Cadet Deaton's airplane was piloted by Cadet Mahoney and was No. 2 in the formation led by Lt. Robinson. The airplanes rolled approximately 200 to 300 feet down the runway making 180 degree turn and neither airplane upset, coming to rest at the edge of the runway."

12-11-42F. Verdigris, Oklahoma. At 2229, a Lockheed-Vega B-34 on a navigation training flight crashed after it was abandoned by its crew after losing power in both engines near Verdigris, Oklahoma, killing three fliers and injuring three others, one seriously. The pilot, 1Lt. John B. French, was killed when his parachute fouled on the aircraft during his bail out attempt. 1Lt. William J. Burford, navigator instructor student, and engineer SSgt. J. B. Vance went in with the airplane and were killed. Also going in with the airplane was 1Lt. J. W. Hamilton, Jr., navigator instructor student, who suffered serious injuries. 1Lt. W. F. Preston, pilot rated passenger, and passenger Maj. M. B. Burnside both parachuted to safety and received minor injuries. Investigation revealed that the crew had mismanaged the fuel system and allowed the bomb bay fuel tank they were using to run dry, causing both engines to fail. Lt. French's body and his fouled parachute were found entangled in the wreckage of the tail section.

12-12-42A. Phoenix, Arizona. At 1140, a Cessna AT-17 on an instrument team ride crashed ten miles NNW of Sky Harbor Airport, Phoenix, Arizona, killing A/C Grady L. Davidson and A/C Frederick Fieldback. Investigators stated, "A/C Davidson as pilot and A/C Fieldback as safety pilot departed Williams Field [Chandler, Arizona] at 1345 on an instrument training mission on the east leg of the Phoenix Radio Range. Apparently the students passed over the cone of silence in ignorance of doing, and later (approximately 11 miles beyond) concluded that such was the case and perhaps attempted a steep turn to accomplish a return flight to the cone of silence. This maneuver may have caused the airplane to assume a highly abnormal attitude and a probable controversy arose between the pilot and safety pilot, which obstructed the corrective action in making a recovery to normal. In some details three witnesses differed in their testimony but all stated that from approximately 2,000 feet above the terrain the airplane was headed downward at about a 60 degree angle to the horizontal and struck the ground in that manner."

12-12-42B. Lewisville, Indiana. At 2200 CWT, a Douglas C-39 crashed two and one-half miles north of Lewisville, Indiana, killing eight crew and passengers. Investigation revealed that the airplane had taken off from Patterson Field, Ohio, at 2130 CWT on an instrument clearance for Stout Field, Indianapolis, Indiana, on a personnel and cargo flight. The airplane crashed into a cornfield about 70 miles from the point of take-off, on course and six minutes ahead of schedule. Investigators stated, "The plane was loaded approximately 2,762 lbs. over its gross weight at take-off and 2,432 lbs. at time of crash. The plane landed from west to east downwind, the left wingtip striking first, 90 feet farther the left engine, and 42 feet farther, and in a direct line, the right engine indicating that the airplane was in a steep left bank and cartwheeled from left to right." The cause of the crash could not be determined with any degree of certainty. Investigators speculated that overloading, carburetor icing, or the inadvertent shifting of the cargo as possible causes for the crash. It was noted by investigators that the subject C-39 was equipped with carburetors that could produce dangerous and unrecoverable carburetor icing characteristics if mishandled by the flight crew. Crewmembers killed in the accident were: Capt. John Nagel, 31, Louisville, Kentucky, pilot; 2Lt. Robert E. Goldenberg, 21, Cincinnati, Ohio, co-pilot; Cpl. Charles W. Bodenner, 23, Denver, Colorado, engineer; Cpl. Matthew Weisman, 24, Cambridge, Massachusetts, assistant engineer. Passengers killed in the crash were: 2Lt. Lloyd R. Pearson, 21, Ida Grove, Iowa; 2Lt. Robert B. Anthony, 21, Coral Gables, Florida; MSgt. Muriel O. Whittenburg, 43, Warrensburg, Missouri; Cpl. Lawrence F. Bogner, 25, Chicago, Illinois.

12-14-42. Eglin Field, Florida. At 1640 CWT, a North American A-36A suffered a catastrophic structural failure while maneuvering at AAF Gunnery Target Range #52, Eglin Field Military Reservation, Valparaiso, Florida, killing pilot 1Lt. Herman E. Watters. Prior to the accident, the A-36 had dropped its two bombs and had completed the first portion of its practice mission without incident. The accident occurred during the second portion of the mission, which included simulated dive-bombing. Investigators stated, "At 1633 CWT aircraft number 42-83666 was over range and was given instructions and a white [signal] panel to start his mission. At 1635, the plane made a dry run and his approach was from the east by the north. He dove on the target and pulled out of the dive heading southeast. This run appeared absolutely normal and he completed the pull-out at approximately 500 to 700 feet. At 1640, plane made its approach slightly north of due east. When plane went into the dive, it appeared that it started said dive from a position over and beyond its target. Plane nosed over into an inverted position headed towards said target. From 10,000 feet, at which altitude plane started its dive, it remained at this angle to approximately 2,000 to 3,000 feet, at which time plane started a half-roll to the right and completed it at between 500 and 1,000 feet. The position of the ship was almost vertical. Between 150 and 200 feet, the right wing of the plane buckled upward over the cabin cockpit and disappeared. During the course of the dive it could not be detected whether there was an attempt by the pilot to pull out of the dive. The plane crashed head on and buried itself about five feet into the ground and burned."

12-15-42A. Mitchel Field, New York. At 1125, a Republic P-47C crashed just after take-off from Mitchel Field, Long Island, New York, killing pilot Capt. John H. Porter. The pilot had taken off to the west on the East-West runway. The airplane gained about 1,000 feet of altitude and made a slight turn to the northwest when the engine failed. The pilot made a steep turn back towards the field in an effort to land to the southeast on the Northwest-Southeast runway. The airplane, with landing gear retracted and about 30 degrees of flaps, came down with excessive speed, estimated to be about 200 mph. Witnesses stated that the pilot was overshooting the landing. He forced the airplane down and the propeller struck the runway. The airplane plowed into the ground and started sliding to the right (to the south), careening for about 2,000 feet before striking a reinforced revetment head-on. The airplane bounded over the revetment and came to rest at the edge of the field. The pilot was killed when his head struck the gun-sight structure when the P-47 collided with the revetment.

12-15-42B. Berlin, Georgia. At 1600, a Bell P-39D suffering an in-flight fire crashed two miles west of Berlin, Georgia, killing pilot 2Lt. Glen W. Russell. A pilot flying two miles astern of the P-39 stated that he had seen the P-39 in a dive "a few degrees off of vertical," beginning at an altitude of about 5,000 feet. Witnesses on the ground stated that they had seen fire around the engine compartment while the airplane was at about 2,500 feet. Investigators could not determine the cause of the engine fire. The pilot attempted to bail out but was killed when he opened his parachute prematurely and became entangled in the tail section. The parachute tore into three pieces and the pilot fell to his death, his body being found about 150 yards from the mutilated parachute canopy, about one-quarter of a mile from the wreckage. Investigation revealed that the pilot had not been struck by any part of the airplane. Lt. Russell had received his wings and his commission on 12-13-42.

12-15-42C. Greenville, Mississippi. At 1535, a Vultee BT-13A on a navigation flight stalled and spun to the ground ten miles northwest of Greenville, Mississippi, killing Aviation Cadet Lawrence A. Tilley. The airplane entered an inadvertent spin at an altitude of approximately 4,000 feet. Investigators stated, "Cadet Tilley made several attempts to recover from the spin, but in each case apparently failed to use forward stick and succeeded only in reversing the direction of the spin by use of the rudder." The airplane burst into flames upon impact and the cadet killed instantly.

12-15-42D. Merced, California. At 1440, a Vultee BT-13A on a navigation and administrative flight spun into the ground two miles northwest of the AAF Basic Flying School at Merced, California, killing pilot Maj. Roland C. Ehret and passenger Capt. H.F. Phelan. The airplane, on an instrument clearance to Santa Ana, California, from Merced, was seen in a climbing left turn just after take-off from Merced, disappearing into overcast. The airplane was then seen to spin out of the overcast, striking the ground and bursting into flames upon impact. Weather was reported as instrument conditions with a ceiling of about 500 feet and visibility of about one and a half miles.

12-15-42E. St. Louis, Missouri. At 2147, a Consolidated B-24D crashed while attempting to take-off from Lambert Field, St. Louis, Missouri, killing seven fliers and injuring two others, one seriously. The airplane, which was on a navigation training flight, landed at Lambert Field at about 1500 because of deteriorating weather conditions. While the airplane was on the ground it was pelted by freezing rain, creating a layer of thin ice on the airplane. A light snow fell and covered the ice on the airplane. At about 1930, the weather seemed to be clearing up and the pilot decided to continue the flight. Several brooms were secured and the snow was swept off of the airplane. The thin coating of ice was not removed. The pilot was notified of this condition but elected to proceed with the flight anyways. Investigators stated, "The engines were run-up and the airplane checked. From available evidence the engines were functioning satisfactorily and the pilot called for take-off instructions at about 2130. He was given a choice of runway #6 or #4, the former being into the wind and longer, but having a grain elevator 100 feet high adjacent to the North end of the runway. The pilot selected #6 although the tower operator recommended #4, since it offered no obstacle to hinder the take-off and climb. At approximately 2145, the pilot took off. The airplane appeared to have difficulty in climbing and the pilot appeared to lift one wing to clear the grain elevator. The plane probably lost what little reserve speed it had in clearing this obstacle since it failed to gain altitude from that point to where it struck the tops of trees and the roof of a barn about two miles from the airport. The impact with the trees caused the airplane to crash into the ground and a fire immediately followed, demolishing the airplane and killing 7 of the 9 members of the crew. Gasoline sprayed over a dwelling adjacent to the crash and it was completely burned to the ground. A woman and her baby managed to escape from the dwelling and were uninjured." Killed in the crash were: 2Lt. William P. McDonald, pilot; 2Lt. Clarence B. Pyle, co-pilot; 2Lt. Warren E. A. Young, navigator; 2Lt. Stanley L. Rosen, bombardier; 2Lt. Alvin L. Sink, passenger; SSgt. Harry H. Stringer, engineer; SSgt. Sidney M. Schwartz, radio operator. Assistant radio operator Sgt. Robert T. Randol was seriously injured and tail gunner Sgt. Harwell J. Clements received minor injuries.

12-16-42A. Maxwell Field, Alabama. At 0325 CWT, a North American AT-6A flying in poor weather crashed 15 miles west of Maxwell Field, Montgomery, Alabama, killing flying student 2Lt. Albert S. Holbert. The airplane was part of a flight of 22 AT-6 type airplanes on a navigation-training mission. The flight had departed the municipal airport at Dothan, Alabama, at 0235 for the leg back to Maxwell Field. The flight soon encountered an unexpected front, which included extremely high winds, heavy rain showers and severe turbulence. It was speculated that the student lost control of his airplane soon after entering the front (See 12-16-42B, C, D).

12-16-42B. Maxwell Field, Alabama. At 0330 CWT, a North American AT-6A flying in poor weather crashed 18 miles southwest of Maxwell Field, Montgomery, Alabama, killing Aviation Cadet Arthur F. Messner, Jr. The airplane was part of a flight of 22 AT-6 type airplanes on a navigation-training mission. The flight had departed the municipal airport at Dothan, Alabama, at 0235 for the leg back to Maxwell Field. The flight soon encountered an unexpected front, which included extremely high winds, heavy rain showers and severe turbulence. It was speculated that the student lost control of his airplane soon after entering the front (See 12-16-42A, C, D).

12-16-42C. Maxwell Field, Alabama. At approximately 0330 CWT, a North American AT-6A flying in poor weather crashed 25 miles SSW of Maxwell Field, Montgomery, Alabama, killing Aviation Cadet Alan C. McDougall. The airplane was part of a flight of 22 AT-6 type airplanes on a navigation-training mission. The flight had departed the municipal airport at Dothan, Alabama, at 0235 for the leg back to Maxwell Field. The flight soon encountered an unexpected front, which included extremely high winds, heavy rain showers and severe turbulence. It was speculated that the student lost control of his airplane soon after entering the front (See 12-16-42A, B, D).

12-16-42D. Union Springs, Alabama. At approximately 0330 CWT, a North American AT-6A flying in poor weather crashed near Union Springs, Alabama, killing Aviation Cadet Thomas G. Mitchell. The airplane was part of a flight of 22 AT-6 type airplanes on a navigation-training mission. The flight had departed the municipal airport at Dothan, Alabama, at 0235 for the leg back to Maxwell Field, Montgomery, Alabama. The flight soon encountered an unexpected front, which included extremely high winds, heavy rain showers and severe turbulence. It was speculated that the student lost control of his airplane soon after entering the front (See 12-16-42A, B, C).

12-16-42E. Cottage Grove, Minnesota. At 1250, a Consolidated B-24D on a ferry mission suffered a catastrophic structural failure of the tail section and crashed out of control four miles east of Cottage Grove, Minnesota, killing the crew of three. Witnesses on the ground stated that just prior to the crash they had observed the airplane flying at low altitude, circling the vicinity of Cottage Grove, 16 miles southeast of the airport at St. Paul, Minnesota. Pieces of the tail section were seen to tear away from the airplane as it maneuvered at low altitude. The B-24 crashed and exploded into flames upon impact with a road embankment, scattering wreckage over a considerable area and knocking down power lines in the crash area. Investigators determined that the airplane had never entered the overcast and weather was not considered a factor in the accident. The airplane was being ferried from the AAF Modification Center at St. Paul to Memphis, Tennessee, after the completion of the scheduled modifications. Killed in the crash were the pilot Capt. Percival R. Davidson, Jr., co-pilot SSgt. Emile O. Leone, and engineer SSgt. John M. Lane.

12-16-42F. Port Angeles, Washington. At 1537, a Lockheed P-38E performing low altitude aerobatics crashed one mile east of the Army Air Field at Port Angeles, Washington, killing pilot 1Lt. Charles R. Hankins. It was reported that the airplane passed over the airfield from the west at low altitude and at a high speed. As the airplane passed over the end of the runway the pilot pulled up to about 300 feet. The airplane then rolled over on its back and then veered down to the right and into the ground. The airplane exploded into flames upon impact.

12-16-42G. Blythe, California. At 1946, a Douglas C-53 crashed near the Army Air Base at Blythe, California, after inadvertently towing aloft an unmanned Waco CG-4A glider, killing all eight crew and passengers aboard the C-53. The glider and passengers were being ferried to Victorville, California, from Tucson, Arizona. The glider cut loose and landed at Blythe after the glider pilot misread the blinker signal lights of the C-53. The pilot of the C-53 then landed. The glider pilot had been ordered earlier not to continue the ferry operation during night hours. The C-53 pilot informed the glider pilot that he (the C-53 pilot) had been ordered by his commanding officer to proceed directly to Victorville. There apparently was some confusion as to which orders were to be followed. The pilots decided to hook up the glider to the C-53 and proceed to Victorville. Investigators stated, "The glider crew then pulled the glider out to take-off position. The running lights were turned on because B-17s were operating off the runway. The co-pilot of the glider hooked the tow-rope to the C-53 while the pilot hooked up the glider. A mechanic from Blythe was sent forward with a flashlight with instructions on how to signal the tow plane pilot. As the glider crew [prepared to board], the [C-53] pilot noticed that the control [surface] locks had not been removed. He yelled to the [glider] co-pilot to get the right aileron while he and a mechanic removed the other blocks. It seems the mechanic with the flashlight signaled the tow airplane forward to take up the slack in the rope. The pilot of the tow plane evidently took this for the take-off signal so he took off. The pilot of the glider realized what was happening as the glider started to move, so he ran and tried to board the glider. He was unable to reach the door however, and was knocked down by the tail surfaces of the glider. As the tow plane started to move, the man with the flashlight gave a stop signal but evidently the pilot did not see it, as it was not heeded. The plane took off and shortly passed the end of the runway, made a 270-degree turn to the right, flew south a short distance, turned left and then made a 360-degree turn to the right. At this point he was over the south edge of the field at approximately 400 feet when the glider was cut loose. The tow plane was evidently in a complete stall when the glider was cut loose, as it immediately fell [to the ground] a short distance from the glider. As soon as the glider got away from its crew, they ran to the control tower to tell the ship to cut the glider loose and had just reached the foot of the tower when the glider was cut loose." The C-53 and the glider crashed about 40 seconds apart, the C-53 spinning to the ground first, exploding into flames upon impact. Glider pilot F/O Noel D. Addy and co-pilot TSgt. Lowell E. Nutting received minor injuries when the tail section of the glider knocked them down. Killed in the crash of the C-53 were: SSgt. Leroy J. Hopkins, 20, Hollis, New York, pilot; SSgt. Nathan J. Toretsky, 22, Long Island City, New York, co-pilot; SSgt. Gordon J. Jardina, 24, Dayton, Ohio, engineer; Sgt. Edward H. Nance, Jr., 21, Pear Valley, Texas, radio operator; 2Lt. J. H. Alabaster, Ft. Knox, Kentucky, passenger; Pvt. Donald R. Hinshaw, Beloit, Wisconsin, passenger; Pvt. James I. Smith, San Saba, Texas, passenger; Pvt. William C. Hart, 27, Orange, California, passenger. Pvt. Hart's father, William O. Hart, 57, Orange, California, was killed along with 16 other passengers and crew (two passengers survived) when the Western Airlines DC-3 they were aboard went down near Fairfield, Utah, on 12-15-42, just the day before.

✓ **12-17-42A. Greenville, Tennessee.** At 0955, two Fairchild PT-19A aircraft collided in mid-air five miles SSW Greenville, Tennessee, killing 2Lt. Hubert P. Sparks aboard PT-19A #42-34022. 2Lt. Michael T. O'Hare was able to land damaged PT-22A #42-34021 safely and was uninjured. The two subject airplanes were part of a five-ship ferry flight. Investigation revealed that Lt. Sparks was in the habit of flying behind a fellow flier and diving below him, then zooming up in front of the other flier in an effort to frighten him. Apparently, Lt. Sparks did this to Lt. O'Hare, colliding with Lt. O'Hare's ship at an altitude of about 5,000 feet and crashing to the ground out of control. The flight leader reported that Lt. Sparks had also performed an unauthorized loop just prior to causing the accident. The flight was cleared from Winston-Salem, North Carolina, to Knoxville, Tennessee, on a ferry flight from Hagerstown, Maryland, to Stamford, Texas.

12-17-42B. Moore Field, Texas. At 1345, a Republic P-43A stalled and spun into the ground while turning from base to final one-half mile west of Moore Field, Mission, Texas, killing pilot 2Lt. Joseph Bell, New York. Investigation revealed that the pilot, while in the pattern at an altitude of about 500 feet above ground level, had entered a steep right turn with insufficient airspeed, causing the airplane to stall, flip over on its back and enter a flat spin. The pilot righted the airplane after one turn of the spin but was unable to recover the airplane, spinning two more times before striking the ground and exploding into flames. Lt. Bell, an infantry officer before entering flight training, had received his wings on 12-13-42 and was undergoing transition training.

12-17-42C. Paine Field, Washington. At 1400, a Lockheed P-38D dove into the ground eight miles east of Paine Field, Everett, Washington, killing pilot 2Lt. Jack R. Kingery. The pilot had taken off from Paine Field at 1347 and was to rendezvous with another P-38 over the field at 20,000 feet for a high altitude oxygen flight. Lt. Kingery failed to meet the other airplane. A farmer observed the P-38 in a high-speed vertical dive toward the ground. The airplane exploded violently into flames upon impact, leaving a crater 30 feet across and 15 feet deep. The cause of the accident could not be determined with any certainty. The pilot might have suffered from oxygen starvation or had put the airplane in a diving maneuver from which he could not recover because of the tendency of the P-38 type airplane to suffer from severe control surface buffeting caused by compressibility.

12-17-42D. Coolidge, Texas. At 1710, a Vultee BT-13A on an instrument training flight crashed six miles west of Coolidge, Texas, killing A/C Timothy F. Keane and A/C Carl J. Marckmann. Investigation revealed that the airplane struck the ground in a vertical dive. Cadet Keane, in the rear cockpit on the hooded instrument flight, attempted to parachute to safety but was killed when the airplane struck him after he deployed his parachute prematurely. His body floated to earth in the fully deployed parachute, snagging on power lines and becoming suspending several feet above the ground. Cadet Marckmann, flying in the front cockpit as the safety pilot, went in with the airplane and was killed instantly. Investigators could not determine the cause of the accident.

12-18-42. Breckenridge Auxiliary Airfield, Oklahoma. At 2000, a Vultee BT-13A stalled and spun to the ground while turning from base to final at Breckenridge Auxiliary Airfield, eight and one-half miles northeast of the Army Flying School at Enid, Oklahoma, killing Aviation Cadet Bernard F. Stephens. The student had been scheduled to practice night landings at the auxiliary field. After two or three turns of spin, the airplane smashed into the ground in a nose down attitude and burst into flames.

12-19-42A. Kissimmee, Florida. At 1220, two Curtiss P-40E airplanes collided in flight near Kissimmee, Florida, killing 1Lt. Sidney O. Kane II (P-40E #41-24928) and Capt. Robert R. Brouk (P-40E #41-24956). The P-40s were part of a six-ship formation training flight that had originated at the Army Air Base at Orlando, Florida. The Accident Classification Committee stated, "It is found that Capt. Brouk was the leader of a six-ship flight which was acting as 'top cover' for two other flights that were making a simulated ground strafing attack on the Kissimmee, Florida, airfield. After the attack had been made by other flights, the leader of the 'top cover' flight gave orders by radio for his flight to execute a 'number one' attack on the airfield. This type of an attack is executed with individual airplanes of the flight flying in trail with approximately 100-200 yards between airplanes. It is further found that during the attack, the airplane flown by Lt. Kane in the number-two position of the formation, was entirely too close to the leader's airplane and that the number-two airplane struck the airplane flown by Capt. Brouk during a pull-up after the attack. The airplane flown by Capt. Brouk went out of control immediately after being struck and crashed. The other airplane continued on for a short distance gaining altitude, lost its propeller and crashed after an apparent attempt by the pilot to make a forced landing."

✓ **12-19-42B. Albany, Georgia.** At 1545, a Stearman PT-17 stalled and spun into the ground four miles north of Darr Aero Tech, Albany, Georgia, killing Aviation Cadet John R. Evans. The aircraft wreck was observed from the air by an instructor and reported to Darr Aero Tech. Supervisory personnel flew to the crash site and found the cadet to be fatally injured. The student had been scheduled to practice chandelles, lazy-eights, pylon-eights, and aerobatic maneuvers.

✓ **12-21-42A. Augustine Field, Mississippi.** At 0905, two Stearman PT-17 airplanes collided in mid-air one mile north of Augustine Field, Jackson, Mississippi, killing two fliers. Student pilot Cpl. Comyn F. Robinson was killed in the crash of PT-17 #40-1835; civilian instructor Carl R. Ziepprecht was killed in the crash of PT-17 #41-25770. Investigators stated, "The student was flying southeast at 700 feet and the instructor was flying west at the same altitude. The pilots apparently did not see each other until the last moment when they attempted to avoid each other but this was too late to [prevent] the collision. The left side and the top wing appeared to be damaged on the student's plane. The instructor's plane seemed to have both left wings sheared off near the fuselage. The instructor's plane, continuing in its original direction, spun into the ground, crashed and burned. The student's plane continued in its general direction towards the southeast for approximately 200 yards, going into a spin and [bursting into flames] upon hitting the ground. The accident occurred because neither pilot saw the other plane until too late to avoid the collision."

12-21-42B. Helena, Montana. At approximately 1700, a Vultee-Stinson L-9B flying in poor weather crashed three miles northwest of Fort William Henry Harrison, Helena, Montana, killing pilot 2Lt. Orville B. Verdery and passenger 2Lt. Leo B. Mansfield, an infantry officer from Ft. Harrison. Investigators were unable to determine the cause of the accident but speculated that the airplane collided with terrain in a flying attitude at full throttle while flying in instrument conditions. The aircraft was on a "reconnaissance snow flight." The aircraft and pilot were assigned to the First Special Service Force, Helena Municipal Airport.

12-21-42C. Waco, Texas. At 1910, a Beech AT-11 flew into the ground two and one-half miles southwest of the Army Flying School at Waco, Texas, killing pilot F/O Harold B. Jeffers, co-pilot Flight Officer William F. Venzke, and bombardier 2Lt. Harry D. Giles. Investigators noted that the pilot took off in deteriorating weather conditions. He possessed an instrument card and was cleared to the AAF station at Midland, Texas, from Waco, Texas. Investigators speculated that the pilot never intended to fly into the overcast and attempted to maintain visual contact under a low ceiling, estimated to be between 400 and 600 feet. The clearance officer reported that the pilot seemed particularly intent on flying to Midland that night, stating to him, "I have to be there."

12-22-42A. La Junta, Colorado. At 1015 MWT, two Consolidated B-24D airplanes on a bombing and gunnery training mission collided in mid-air approximately 12 miles southwest of La Junta, Colorado, killing seven fliers aboard B-24D #41-11826. Immediately following the collision, five crewmembers aboard B-24D #41-24041 parachuted to safety. Four others rode the airplane down to a safe emergency landing at Pueblo Army Air Base, Pueblo, Colorado, after it was discovered that two crewmembers were trapped in the nose and could not bail out. Killed in the crash of B-24D #41-11826 were: 1Lt. Gerald E. Smetts, Oakland, California, pilot; 2Lt. Charles T. Bowen, Jr., Flushing, New York, co-pilot; 2Lt. Adam C. Thompson, East Hampton, New York, navigator; 2Lt. Bliss M. Shannon, Brooklyn, New York, bombardier; Sgt. Victor H. McGregor, Casper, Wyoming, engineer; SSgt. Charles S. Meisenheimer, Jefferson, Texas, radio operator; Sgt. James H. Cooney, Baton Rouge, Louisiana, tail gunner. The pilot of B-24D #41-24041, 1Lt. Frank E. Slough, was able to land the airplane at Pueblo despite damage to both inboard propellers and the fact that pieces of the tail section of the other B-24 were lodged in the nose section, obstructing his view forward. Bombardier 2Lt. Henry W. Scott (minor injury) and navigator 2Lt. Lester J. Warner (minor injury) were trapped in the damaged nose section. Engineer SSgt. James E. Caillier (no injury) refused to bail out and, acting as co-pilot, landed with the pilot. Parachuting from the airplane were: 2Lt. Stewart P. Elliot, Jr., co-pilot (minor injury); Sgt. Elwood W. Harbison, assistant engineer (no injury); SSgt. Dan W. Kennon, radio operator (serious injury); Sgt. Herman J. Seigfeldt, assistant radio operator (minor injury); Sgt. Robert A. Griffin, tail gunner (no injury). The two airplanes were part of a three-ship formation that had made a bombing run on the AAF Bombing Range 60 miles southeast of Pueblo AAB, Colorado. The flight leader, Lt. Slough, discovered that a bomb rack circuit problem had caused the bombardier to inadvertently drop all of his bombs on the first run. Lt. Slough radioed the two wing B-24s that he was breaking up the formation and for them to conduct their remaining bombing runs in a two-ship formation and advised that he was going to fly to the Pueblo Gunnery Range (55 miles southeast of Pueblo AAB) to reconnoiter and wait for them. Lt. Slough assumed that his instructions were understood by both wingmen and broke away from the formation, flying toward the gunnery range at an altitude of 12,000 feet. Lt. Slough then began a gradual descent to about 7,500 feet to position his airplane for observing the gunnery range when the co-pilot, Lt. Elliot, reported that Lt. Smetts was still in the number-two position. Lt. Slough repeated his order to break off but Lt. Smetts did not reply. After a little while, Lt. Elliot could no longer see Lt. Smetts' airplane behind them and they assumed that he had broke off. A couple of minutes later, Lt. Smetts' B-24 appeared forward and below, rising in the path of Lt. Slough's airplane from underneath in a climbing turn to the left. Lt. Elliot yelled to warn Lt. Slough but not in time to avoid the collision. Lt. Slough's inboard propellers and nose section cut into Lt. Smetts' port vertical fin and rudder, and the horizontal stabilizer and elevators, severing them and causing Lt. Smetts' B-24 to go out of control and dive to the ground where it exploded into flames upon impact. No member of the crew was able to use his parachute. Pieces of Lt. Smetts' port vertical fin and rudder were lodged in the nose section of Lt. Slough's airplane. Lt. Slough's B-24 then began to lose altitude. At about 6,200 feet Lt. Slough was able to arrest the descent and regain control of the B-24. The pilot gave the order to bailout and five members of the crew bailed out immediately. It was soon discovered that the bombardier, Lt. Scott, and the navigator, Lt. Warner, were injured and trapped in the wreckage of the damaged nose section and could not bail out. The pilot and the engineer decided to stay with the airplane and save their crewmates. The engineer, Sgt. Caillier, acted as co-pilot for Lt. Slough during the emergency landing at Pueblo AAB. Sgt. Caillier had to lean out of the open co-pilot's window to help guide the landing because Lt. Slough's forward view was obstructed by pieces of Lt. Smetts' B-24 that were still lodged in the nose section. They were able to land the airplane safely without further injuring the bombardier and navigator. Investigation of the wreckage of B-24D #41-11826 revealed that it had also dropped its entire bomb load on the first run.

12-22-42B. Gordon, Texas. At 1145, a Douglas A-20C on a ferry mission crashed five miles northeast of Gordon, Texas, killing pilot 1Lt. William C. Martin. The airplane had been cleared from Big Spring, Texas, to Love Field, Dallas, Texas. A witness in the air stated that he had seen the A-20 perform a climbing turn to the left at an altitude of approximately 4,000 feet, turning 180 degrees from the line of flight. Just as the airplane reached the 180-degree position, the airplane was seen to enter a two-turn spin to the right. The pilot stopped the spin to the right and the airplane entered a spin to the left, spinning to the ground and bursting into flames upon impact.

12-22-42C. Saltfork, Oklahoma. At 1150, a Vultee BT-15 crashed two and one-half miles northeast of Saltfork, Oklahoma, killing pilot 2Lt. Henry R. Smith. Investigators stated, "2Lt. Henry R. Smith was ferrying an airplane from the Enid Army Flying School [Enid, Oklahoma] to Winfield, Kansas, and had reached the vicinity of Saltfork, Oklahoma. He was seen to descend to an extremely low altitude over the Salt Fork of the Arkansas River. He was seen to fly into a group of three high-tension wires at high speed. The wires broke and fell into the water after causing such damage that the airplane was unable to continue in flight. The right wing tip was seen to strike the water and then strike the riverbank to the right of the airplane. The airplane next struck the ground on its back a considerable distance from the water and was completely demolished without burning. The pilot was thrown clear and killed instantly."

12-22-42D. Otis Field, Massachusetts. At an unknown time between 1300 and 1330, a Curtiss P-40E crashed five miles north of Otis Field, Falmouth, Massachusetts, killing pilot 1Lt. Bart J. Judge, Jr. The pilot had been leading a three-airplane flight on a tactical training mission. The formation had made a high speed pass at a U.S. Navy Consolidated PBY and then broke off. The flight reformed and then participated in individual simulated combat. The flight had broken up during the maneuvering and when the two wingmen could not locate the Lt. Judge, they returned to their home station, Hillsgrove Army Air Field, Providence, Rhode Island, landing at approximately 1330. The wingmen had last seen Lt. Judge at approximately 1300. The airplane had been reported overdue and soon after declared missing. Despite a massive search effort, the airplane and the pilot's remains were not discovered until 3-22-43. Investigation revealed that Lt. Judge attempted to bail out at an altitude too low to allow his parachute to fully deploy. His body was found 15 feet from the main wreckage with the partially deployed parachute strung out behind him.

12-22-42E. Tucson, Arizona. At 1130 MWT, two Ryan PT-22 type airplanes collided in mid-air and crashed six miles west of Ryan Field, Tucson, Arizona, killing four fliers. Civilian instructor Edwin C. Melchior and student A/C Arthur B. Lewis were killed in the crash of PT-22 #41-15666. Civilian instructor Charles Stulls and student A/C Franklin C. Maville were killed in the crash of PT-22 #41-1903. There were no known witnesses to the actual collision, which was estimated to have occurred at approximately 2,500 feet agl. Several witnesses on the ground and in the air had seen the airplanes falling after the initial collision. The airplanes fell about one-quarter of a mile apart and a severed wingtip was found about 200 yards from the main wreckage area. There was no evidence that any of the fliers attempted to parachute to safety.

12-22-42F. Muskogee, Oklahoma. At 1745, a Fairchild PT-19A crashed three and one-half miles northwest of Muskogee, Oklahoma, killing Aviation Cadet Franklin R. Dennick. A witness on the ground stated that the airplane was maneuvering at an altitude of approximately 100 feet above the ground when it stalled and began to enter a spin. The pilot attempted a recovery by pointing the nose down but the airplane dove into the ground before a recovery could be effected.

12-22-42G. Fairfield, California. At 1558, a Douglas C-47 smashed into the ground after entering a spin while flying in overcast six miles west of Fairfield, California, killing the crew of five. The purpose of the flight was to swing the airplane's compass on the railroad right of way heading connecting Davis, California, and Elmira, California. The airplane was overdue at Elmira and declared missing. The wreck was located 12-27-42. Investigators stated, "After examination of the scene and the disposition of the wreckage, it is quite evident that the airplane struck the ground in a nearly vertical line of descent. There are distinct imprints of the leading edges of both wings, the two engines and the nose, and trees as close as 50 feet behind the wreck are untouched. The left wing imprint appears to be perpendicular to the apparent direction path, but the imprint of the right wing angles off in a forward direction approximately 20 to 25 degrees, which would indicate that the airplane was probably spinning when it struck. To augment this theory is the fact that the left engine, the left wing and the left propeller were within 30 feet of the point of impact while the right engine, right wing and right propeller were scattered across approximately level ground for a distance of about 200 feet. Further, it is believed that the speed of the airplane when it struck was not in excess of 120 or 130 miles per hour, judging from the imprint on the ground, the displacement of parts, and the general appearance of the wreck; which further backs the theory that the airplane was in a spin when it hit." Killed in the crash were: 2Lt. Ralph Martin Duffie, Bromley, Kentucky, pilot; 2Lt. Herman Loaffler, American Fork, Utah, co-pilot; 2Lt. James A. Howell, Middleton, Illinois, navigator; 2Lt. James M. Russell, Huntington, Indiana, navigator; MSgt. Reginald M. Tolar, Ennis, Texas, engineer.

12-23-42A. Grenier Field, New Hampshire. At 1340, a Curtiss P-40F suffering an engine failure crashed while taking off from Grenier Field, Manchester, New Hampshire, fatally injuring pilot 2Lt. Herbert Lawler, 21, Houston, Texas. A hunter who witnessed the accident rushed to the scene and pulled Lt. Lawler out of the flaming wreckage. Lt. Lawler, who was severely burned, died of his injuries on 1-4-43. The Accident Classification Committee stated, "Lt. Lawler took off as number-two man in a flight of four, led by 1Lt. Julian Adams. About the time Lawler had flying speed the engine began to miss and smoke and flame was seen coming from the exhaust stacks. He pulled it off the ground and cleared the trees at the end of the field and then crashed and burned in the trees beyond. The pilot was critically burned and the plane a complete wreck. It is the opinion of this committee that the accident resulted from engine failure, cause or causes unknown." The airplane was scheduled to participate in an aerial gunnery mission.

12-23-42B. Key Field, Mississippi. At 1430, a Vultee V-72 (similar to the A-31 type airplane) suffering an apparent engine failure crashed near Key Field, Meridian, Mississippi, killing pilot 1Lt. Frazier Curtis, who plunged to his death in an unsuccessful parachute jump. Passenger Cpl. William J. Tyler parachuted to safety and was uninjured. The airplane had taken off from Key Field and gained about 1,000 feet of altitude when the engine began to sputter. On the pilot's orders the crew chief bailed out, hearing the airplane crash soon after. Investigation revealed that Lt. Curtis attempted to use his parachute at too low an altitude to allow it to fully deploy, his body being found about 40 yards behind the wreckage.

12-23-42C. La Junta, Colorado. At 1330, a Cessna AT-17 on an instrument training flight suffered a structural failure of the starboard wing and crashed into North Blue Lake near La Junta, Colorado, killing A/C Donald E. Spencer, Seattle, Washington. A/C Vinton L. Stanfield was able to parachute to safety and was uninjured. Cadet Stanfield, flying as safety pilot, stated that he had noticed that the airplane was in a slight dive when Cadet Spencer, flying under the hood, pulled up on the controls abruptly. The starboard wing fabric tore away from the wing, leaving exposed ribs. The cadets were able to momentarily hold the airplane level with full left aileron but quickly decided to leave the airplane. The co-pilot, Cadet Stanfield, left the airplane first and thought that the pilot was right behind him. The airplane crashed into Blue Lake. Cadet Spencer's body was found in the wreckage. Cadet Stanfield was able to parachute to safety. He descended into the lake and swam to shore, suffering only mild exposure.

12-23-42D. Longtown, Mississippi. At 1550, two Vultee BT-13A airplanes collided in mid-air and crashed one mile east of Longtown, Mississippi, killing two fliers. 2Lt. Madison E. Roberts was killed in the crash of BT-13A #41-21383; 2Lt. John J. Conway was killed in the crash of BT-13A #41-22278. The airplanes were part of a nine-ship flight that had taken off from Blytheville, Arkansas, on a ferry mission to Greenwood, Mississippi. The tightly spaced formation was flying at about 2,000 feet above ground level with Lt. Roberts flying in the number-two position and Lt. Conway flying in the number-three position on the left side of a large V-formation. Investigation revealed that Lt. Roberts had momentarily lost his position and tried to rejoin the formation while it was commencing a shallow, diving turn to the right. He had dropped back about 100 feet above the other airplanes and in an attempt to rejoin the formation collided with Lt. Conway's airplane. Lt. Roberts, flying straight with wings level, dropped down into the formation while the formation was executing the shallow turn to the right. Lt. Conway's airplane overran and turned on top of Lt. Roberts' airplane. Lt. Conway's propeller struck the right side of Lt. Roberts' cockpit, cutting through the side of the airplane and probably killing Lt. Roberts instantly. Lt. Conway's wing hit Lt. Roberts' tail section, severing it. Both airplanes plummeted to the ground. Lt. Conway's BT-13 exploded into flames upon impact with the ground. Lt. Roberts' airplane fell to earth a quarter of a mile from the other but did not burn. Investigators faulted both pilots for failing to see each other and avoid the collision.

12-23-42E. Orlando, Florida. At 1735, two Curtiss P-40E airplanes collided in mid-air and crashed five miles east of the Army Air Base at Orlando, Florida, killing two fliers. 2Lt. Amherst W. Bingham was killed in the crash of P-40E #41-24993; 2Lt. Vance A. Schultz was killed in the crash of P-40E #41-36174. Investigators stated, "It is found that the airplane piloted by Lt. Bingham was being flown in the number-five position of a six-ship flight and that the flight was in a stepped-up right echelon formation just prior to the time of the collision. After getting the flight into position for a peel-off prior to landing, the leader gave the 'attention' signal, which is a flutter of the ailerons. As this signal was passed back by each individual pilot, the formation became disrupted to such an extent that the pilot of the number-five airplane in the formation skidded out of the formation, causing the number-six airplane to be forced to leave the formation entirely. On attempting to rejoin the formation, the pilot of the number-five airplane struck the number-four airplane piloted by Lt. Schultz. Both airplanes immediately went out of control and crashed. The P-40 flown by Lt. Bingham burned after the crash. The altitude at the time of the collision was approximately two thousand feet."

12-24-42A. Avon Park, Florida. At 0110 EWT, a Martin B-26B flying in thick fog crashed while attempting a landing at the AAF Bombing and Gunnery Range Airfield at Avon Park, Florida, killing two fliers and injuring three others, two of them seriously. The airplane had taken off from Memphis, Tennessee, at 2130 EWT on the return leg of a night navigation flight to MacDill Field, Tampa, Florida. Low ceilings were reported at MacDill and tower personnel advised the pilot to go to Hendricks Field, Sebring, Florida. The crew acknowledged the instructions and flew a course to Sebring. The airplane flew over the field at the Avon Park Bombing and Gunnery Range and the crew assumed that it was Hendricks Field, which in fact is nine miles east of the Avon Park range. They called Hendricks Field tower on the radio and requested landing instructions, being advised to use Runway 18. The lighted runway at Avon Park was 22, causing some confusion among the crew. The pilot called the tower at Hendricks Field again and they confirmed that the runway cleared for landing was 18. The pilot entered and flew the pattern for Runway 22 and then turned for his approach. The pilot lowered the landing gear, let out some flaps and began his descent, soon entering very thick fog. Visibility became extremely poor and the pilot flew the airplane into the ground 400 yards northeast of the runway. The landing gear collapsed and the airplane collided violently with a clump of palmettos and thick stumps, causing serious damage. The B-26 skidded across the ground until it came to rest in the drainage ditch surrounding the field. The airplane did not burn but was completely destroyed. Pilot 1Lt. Ralph E. Morrison and Pfc. Ralph E. Thomas were killed in the accident. Co-pilot 2Lt. Robert A. Broan and engineer SSgt. Sidney H. LaVance were seriously injured. Navigator 2Lt. Fred A. Meier received minor injuries.

12-24-42B. Karval, Colorado. At 1205, a Lockheed F-5A (photo-reconnaissance version of the P-38 type airplane) suffered a catastrophic structural failure of the tail booms and dove into the ground near Karval, Colorado, killing pilot 2Lt. Lee R. Crowe. Lt. Crowe took off from Louisville, Kentucky, at 1140 MWT on the return leg of a navigation flight to Peterson Field, Colorado Springs, Colorado. He was to climb to 30,000 feet and fly at that altitude until reaching the vicinity of Peterson Field. The airplane apparently entered a high-speed dive west of Karval, and when the pilot attempted to level out the tail section failed and separated. The airplane spun to the left several times and then dove into the ground, exploding into flames upon impact. Investigation revealed that the elevator trim tab push rod was broken and the broken ends had been pounding against each other. It was speculated that the vibration from this failure could have been severe enough to cause the failure of the tail section.

12-24-42C. Eagle Pass, Texas. At 1425, a Beech AT-7 on a navigation training flight crashed sixteen miles northeast of Eagle Pass, Texas, killing four fliers. At 1944, the airplane was reported overdue at its base at Hondo Army Air Field, Hondo, Texas, and was not located until 1300 on 12-26-42. Investigation revealed that the airplane flew into the ground under control with power in a flat area covered by stands of mesquite trees. The airplane broke apart and exploded into flames, killing the occupants instantly. Weather at the time of the accident was reported as ceiling and visibility unlimited. Investigators could not determine why the airplane flew into the ground. Killed in the crash were: 2Lt. Donald W. Pittsley, pilot; 2Lt. William L. Jones, navigator instructor; A/C Cecil A. Hovinen, student navigator; A/C James A. Hiller, student navigator.

12-24-42D. Drew Field, Florida. At 1645, a Curtiss P-40K on an authorized aerobatic flight spun into the ground near the Davis Causeway, ten miles west of Drew Field, Tampa, Florida, killing pilot 2Lt. Arthur M. Watson. Investigation revealed that the vertical fin, the upper half of the rudder, horizontal stabilizers and elevators had progressively failed in flight and separated from the airplane. The pilot had made no effort to leave the airplane, being found still strapped to his seat with the canopy closed.

12-24-42E. San Diego, California. At 1545, a Cessna AT-17 on a navigation training flight crashed 15 miles south of San Diego, California, killing pilot 1Lt. Jack M. Berg. Investigators stated, "Lt. Berg cleared for San Diego from Victorville via Indio and Plaster City [all in California]. This route was taken because it offered the best weather conditions. Lt. Berg reported his position when 15 miles west of Indio, but no subsequent reports were received. Upon receipt of a report from U.S. Navy personnel at San Diego of seeing what appeared to be parts of an airplane along the beach south of San Diego, a search party went out on January 4, 1943 and found numerous parts of an AT-17 along the beach about 15 miles south of San Diego. As no other AT-17 has been reported missing in this area, it was assumed to be Lt. Berg's plane. There is no evidence available to the Board to indicate how or why Lt. Berg crashed in the manner he did."

12-24-42F. Tucson, Arizona. A North American AT-6 type airplane crashed in the Rincon Mountains approximately 20 miles east of Tucson, Arizona, killing RAF Flying Cadet William James Watters, Glasgow, Scotland. The airplane had taken off from Falcon Field, Mesa, Arizona, on a routine training flight. See Note #22.

12-25-42A. Paine Field, Washington. At 1320, a Bell P-39D and a P-39L collided in mid-air and crashed two miles northwest of Paine Field, Everett, Washington, killing 2Lt. James R. Rudegeair (P-39D) and 2Lt. Charles G. Tison (P-39L). Investigators stated, "Lt. Rudegeair and Lt. Tison were participating in a routine formation training flight. The airplanes took off individually and had completed approximately 270 degrees of a turn around the field and were to fly a formation of elements of two. Lt. [first name unknown] Scott was leading and Lt. Rudegeair was assigned number two, Lt. Frederick E. Bullock assigned number three, Lt. Tison number four and Lt. George M. Lamb number five. In getting into position Lt. Tison apparently took up number-two position and Lt. Rudegeair pulled in too close behind and under Lt. Tison. Lt. Bullock and Lt. Lamb were trailing as Lt. Bullock was waiting for Lt. Tison to get into proper position. It is evident that Lt. Rudegeair came up under and collided with Lt. Tison's airplane and completely severed the empennage thus throwing both airplanes out of control. As this occurred at approximately 1,500 feet neither pilot was able to get out. The empennage of Lt. Tison's plane was found approximately 1,500 feet from the plane. Both airplanes were total losses but only Lt. Rudegeair's airplane burned. It is apparent that Lt. Tison did not follow instructions on assuming proper position in formation and other pilots in the flight had no doubt of the positions assigned."

12-25-42B. Twin Lakes, California. At 1800, a Cessna AT-17B flying in instrument conditions collided with rising terrain five miles southwest of Twin Lakes, California, killing four fliers. The airplane had been leading a flight of three AT-17 airplanes on a positioning flight cleared from Victorville, to Palmdale, to Bakersfield, California. The flight had been flying at 14,000 feet, proceeding toward Bakersfield and flying over the top of the overcast, which was estimated to be about 11,000 feet. Approximately twenty miles from the mountains in the vicinity of Tehachapi, the flight encountered extremely strong downdrafts and severe turbulence, knocking all of the airplanes into the overcast and breaking up the formation. The downdrafts were estimated to exceed 5,000 feet per minute. The pilots of the two other airplanes reported that they had picked up heavy ice (both rime and clear) shortly after entering the overcast and reported that they were unable to hold altitude. Eventually each broke out of the overcast at about 7,500 feet above sea level, about 500 feet above the terrain. The ice quickly melted after the airplanes entered clear air. One of the pilots returned to land at Victorville and the other landed safely at Bakersfield, having flown over the Pacific Ocean, approaching from the west. The accident airplane slammed into the western slope of the mountain forty feet from the peak in a flying attitude and was destroyed utterly. The purpose of the flight was to lead the two less experienced pilots to the destination. The airplane had been overdue at Bakersfield, reported as missing and was not found until 12-30-42. Killed in the crash were: 1Lt. Halbert Eugene Bing, Douglas, Arizona, pilot; 2Lt. Edgar T. Harder, Taft, California, passenger; Lt. Col. Charles M. Harootunian, executive officer, Douglas Army Air Base, Douglas, Arizona, passenger; SSgt. Robert L. Jacks, De Queen, Arkansas, passenger.

12-26-42A. Cliff, New Mexico. At an unknown time between 1000 and 1200 MWT, a Cessna AT-17 on a navigation training flight crashed 25 miles NNE of Cliff, New Mexico, killing pilot 2Lt. James R. Caldwell, Jr. and passenger SSgt. Maxwell W. Caldwell. The AT-17 took off from the Army Air Base at Roswell, New Mexico, at 0934 and was cleared to Williams Field, Chandler, Arizona. When the airplane failed to arrive at Williams Field it was reported as missing. Despite a massive search effort (including 94 airplanes from Roswell Army Air Base and 30 from Williams Field), the airplane remained missing. Two hunters discovered the airplane near Clear Creek on 11-19-43. Investigation revealed that the airplane hit the ground in a spin to the right. It was speculated that the pilots lost control of the airplane while they were flying in instrument conditions. Weather was reported as low ceilings and misting rain with light clouds covering the mountain tops.

12-26-42B. Monroe, Louisiana. At 2000, a North American B-25C flying in poor weather crashed after colliding with trees near Selman Field, Monroe, Louisiana, killing eight fliers. The airplane was on a flight from Barksdale Field, Shreveport, Louisiana, to Walterboro Army Air Field, South Carolina. The airplane was overdue at Walterboro and declared as missing. Two hunters discovered the wreckage of the B-25 and the remains of its crew in swampy land on 1-2-43. Investigators stated, "Pilot was advised against taking off from Barksdale Field due to a strong cold front moving in from the Texas panhandle. Pilot, in the face of existing weather conditions, ordered a take-off and ordered weather officer to sign his clearance. Take-off was made at 1919. [The pilot] was cleared to fly at 9,500 feet. The condition of the wreckage indicates that the pilot evidently attempted to stay underneath [the overcast] and remain partly [in visual] contact. In doing so, either the left wing or the tail assembly hit a tree. No attempt was made to use parachutes. Plane hit the ground at a shallow rate of descent." Killed in the crash were: Capt. Fred M. Hampton, pilot; 2Lt. Thomas A. Johnston, co-pilot; 2Lt. Henry R. Perchal, observer/navigator; 2Lt. Henry M. Barlow, passenger; TSgt. Robert P. Wiltshire, passenger; TSgt. Edwin J. Matthews, passenger; Sgt. Murray E. Scott, passenger; 2Lt. Charles F. Raymond, Jr., observer.

12-27-42. Haymarket, Virginia. At 1815, a Cessna UC-78B flying in instrument conditions crashed into the side of a hill near Haymarket, Virginia, killing the four fliers on board. The airplane, on a local instrument training flight that originated at Bolling Field, Washington, D.C., flew into the side of the hill while in a flying attitude at cruising speed. Weather at the scene of the accident was reported as low ceilings with light drizzle and visibility of two and one-half miles. Killed in the accident were: Maj. Horace J. Winn, Service Pilot; SSgt. Walter R. Best, engineer; Pvt. George E. Blanc, passenger; Pvt. Vernon M. Kragnes, passenger.

12-28-42A. Bruce Field, Texas. At approximately 0900, civilian flight instructor Robert H. Crawford was killed when he walked into the rotating propeller of a Fairchild PT-19A that was taxiing at Bruce Field, Ballinger, Texas. The PT-19, piloted by civilian instructor William H. Stewart and Aviation Cadet Alfred E. Iverson, was taxiing up to take-off position. Both pilots were uninjured. Mr. Crawford was walking diagonally toward the airplane and was hurled forward of the airplane upon being struck by the propeller. He was then struck by the left center section of the wing as the airplane moved forward. He was killed instantly. The airplane, moving at about 10 mph, rolled over Mr. Crawford and stopped about 15 feet from where the body of the flight instructor came to rest. The instructor and student had seen Mr. Crawford walking into the path of the moving airplane but were unable to stop the airplane in time to avoid striking him. There were several airplanes running at the time of the incident and it is possible that Mr. Crawford did not hear the airplane that hit him. Investigators speculated that Mr. Crawford was walking into an icy wind and that might have been a factor in the incident.

12-28-42B. Location Unknown. At an unknown time between 1039 and 2039 EWT, a Martin B-26B and seven fliers disappeared and are presumed to have crashed into the Gulf of Mexico while on a cross-country navigation flight from MacDill Field, Tampa, Florida, to Ellington Field, Houston, Texas. The B-26, carrying about ten hours worth of fuel, took off from MacDill Field at 1039 and was unreported ever since. The flight plan carried the airplane directly across the Gulf of Mexico. Despite a massive search effort, no trace of the airplane or its occupants was ever found. It was speculated that the airplane might have encountered poor weather. Missing and presumed lost at sea were: 2Lt. William A. Booth, Jr., pilot; Flight Officer Raymond J. Jewett, co-pilot; 2Lt. Edward E. Stevens, bombardier; Pvt. George A. Kennedy, engineer/gunner; Maj. Joseph C. C. Nate, passenger; 3rd Officer Eleanor C. Nate, Medical Corps, WAAC, passenger and wife of Maj. Nate; SSgt. Edwin H. Malone, passenger.

12-28-42C. Pine, Arizona. At an unknown time after 1210 MWT, a Curtiss AT-9 crashed 28 miles northeast of Pine, Arizona, killing Aviation Cadet Edward C. Hill and Aviation Cadet William A. Hough. The AT-9 had taken off from Rittenhouse Field at 1210 for Williams Field, Chandler, Arizona, a distance of eight miles. When the airplane failed to arrive at Williams Field it was declared missing. Despite a thorough search, the airplane and the remains of the pilots were not found until 6-4-43.

12-29-42A. Lake Park, Georgia. At 0035, a Beech AT-10BH on a night transition flight crashed three miles north of Lake Park, Georgia, killing A/C William H. Reid and A/C Leslie H. Carter. Investigators stated, "It was determined that the probable cause of the accident was a crash landing at night due to engine material failure, engine failure due to icing, or inexperience on the part of the pilot and co-pilot in flying without gyro flight indicator and gyro compass."

12-29-42B. Ludlow, California. At 0200, a Consolidated B-24D on a night navigation flight suffered a double engine failure and crashed near Ludlow, California, killing four crewmembers. Five crewmembers parachuted to safety, one receiving a minor leg injury. Navigator 2Lt. Rento Casini, assistant engineer Sgt. Layman L. Cash, and passenger Cpl. Sidney R. Vaughn were killed when they failed to bail out and went in with the airplane. Pilot 1Lt. Robert E. Johnson was killed when he fell out of his parachute harness while descending in his fully deployed parachute after successfully exiting the airplane. Parachuting to safety were 2Lt. Marshall E. O'Berg, co-pilot; 2Lt. Kenneth Duehring, bombardier; TSgt. James B. Callaway, engineer; TSgt. Thomas B. Peters, radio operator; Sgt. John L. Gannaway, assistant radio operator; Sgt. Earl E. Overstreet, tail gunner. Investigators stated, "The aircraft took off from Muroc [California] for a navigation flight to Albuquerque [New Mexico] and return at 1930. The flight was uneventful until the number-four motor stopped at 0145 on the return trip at about 10,000 feet altitude and about 30 minutes east of Muroc. When the engine stopped, the pilot Lt. O'Berg and the engineer, Sgt. Callaway, feathered the propeller and then Lt. Johnson took over as pilot as he was the more experienced pilot. In about 15 minutes, number-two engine started cutting in and out. At this time, the crew was ordered to be prepared to abandon ship and the pilot endeavored to keep the number-two motor running but the surging motor caused the airplane to yaw until it was out of control. The plane landed flat, apparently from a flat spin and Lt. Casini, Sgt. Cash and Cpl. Vaughn failed to abandon the airplane. Lt. Johnson had fallen out of his chute, and the rest of the crew bailed out successfully." It was speculated that the flight crew had mismanaged the fuel system or underestimated the amount of fuel being used, causing the engines to fail due to fuel exhaustion. It was noted that the airplane had not been fully serviced with gasoline prior to the flight.

12-29-42C. Coleman, Texas. At 0945, a Fairchild PT-19A practicing a simulated forced landing crashed four miles east of Coleman, Texas, killing A/C Christopher P. Jeansonne. Civilian flight instructor B.L. Bryan was uninjured. The instructor stated that the student froze on the controls with left stick and left rudder after the completion of a simulated forced landing attempt. The airplane hit a fence in a left bank, cartwheeling across a road and striking another fence before coming to rest. The cadet was killed in the crash. The instructor further stated that the student exhibited no abnormal or dangerous tendencies and was progressing in his training as well as other students with similar experience.

12-29-42D. MacDill Field, Florida. At 1200 EWT, a crewmember of a Martin B-26B was killed after he bailed out of the top escape hatch and struck the top turret, causing him to plummet to his death. The body of bombardier 2Lt. William A. Lasby was found by a U.S. Army truck driver near the Dale Mabry Road, in between Drew Field, Tampa, Florida, and MacDill Field, Tampa, Florida. Investigators stated, "Pilot [2Lt. Weldon L. Cherry] received contact weather clearance from Mobile [Alabama] to MacDill Field. A [weather] front passage was noticed en route although the weather was reported by pilot and passenger as being smooth. Upon arriving at destination instrument weather was encountered. Pilot attempted to fly under overcast but low ceilings forced him to go on instruments. Pilot attempted to climb through overcast and at an altitude of 4,000 feet, violent vertical currents were encountered resulting in stalling ship with spin following. Bombardier jumped from the ship on orders from the pilot. When body was recovered it was discovered that ripcord had never been pulled. Plane showed that Lt. Lasby struck the turret after jumping from the ship. Spin was recovered from once and then re-entered due to attempting too fast of a recovery. Final recovery was effected at approximately 700 feet [above ground level]. Lt. Col. John F. Batjer, passenger and rated pilot, took over controls from the pilot and landed the ship at Drew Field. Landing was of sufficient force to have caused severe damage." The pilot rescinded the order to abandon ship after regaining control of the airplane, preventing other crewmembers from bailing out. Uninjured in the emergency landing were: 2Lt. Weldon L. Cherry, pilot; Lt. Col. John F. Batjer, co-pilot; Sgt. Joseph D. Kolose, engineer; SSgt. David O. Tucker, gunner; 2Lt. Clyde W. Owen, pilot-rated passenger; 1Lt. Mark L. McAlpin, passenger; TSgt. Jerry D. Reed, passenger.

12-29-42E. Olympia Airport, Washington. At 1718, a Lockheed P-38F suffering the failure of the starboard engine crashed while attempting an emergency landing at Olympia Airport, Tumwater, Washington, killing pilot 2Lt. George S. Elliott. The Accident Classification Committee stated, "The pilot of this airplane was flying on a regular training mission and through mechanical failure, lost his right engine, which he feathered properly, and proceeded to make an emergency landing with one engine operating. His pattern and approach were excellent, and his airspeed about correct, but he failed to cut the throttle back at the proper time and he was overshooting slightly. Instead of landing the airplane on the runway and using his brakes to stay within the confines of the field, Lt. Elliott decided to go around for another landing and applied excessive power to this good left engine, causing the airplane to ski into the dead engine, and finally it fell off on the right wing and spun into the ground."

12-30-42A. Waco, Texas. At 0951, two Vultee BT-13A type airplanes collided in mid-air at the Army Flying School at Waco, Texas, killing three fliers. Instructor 2Lt. James A. Abney and A/C William H. Turner were killed in the crash of BT-13A #41-22734; A/C Paul G. Shudnick was killed in the crash of BT-13A #41-21744. The Accident Classification Committee stated, "A/C Paul G. Shudnick [was] on his approach for landing had overshot and was going around the traffic pattern again. 2Lt. Abney and A/C Turner had just taken off on the same runway. All students are instructed that in the event of an overshoot that the airplane in going around will uncover the runway to obtain better visibility. Apparently the ship going around [A/C Shudnick] overtook the ship that had just taken off and they collided on a prolongation of the runway when approximately 300 feet in the air just as both airplanes started their first turn in traffic." All three fliers were killed when the ships plunged to the ground and burst into flames.

12-30-42B. Roswell, New Mexico. At 0855, a Cessna AT-17 suffered a catastrophic structural failure of the port wing and smashed into the ground five miles southwest of the Roswell Auxiliary Airfield #1, Roswell, New Mexico, killing Aviation Cadet John L. Holbeach and Aviation Cadet Laverne E. Hogan. Investigation revealed that at an altitude of approximately 7,000 feet, the port wing leading edge and upper surface plywood and fabric separated from the wing in flight, causing the airplane to roll and spin to the ground, exploding into flames upon impact. Pieces of the wing that had peeled away were found about a mile from the crash scene. The airplane had been leveling out following a diving 360-degree turn.

12-30-42C. San Francisco Bay, California. At 1050, a Bell P-39D engaging in simulated combat with a U.S. Navy Grumman F4F-4 crashed onto a beach and shallow water in San Francisco Bay, three miles south of Newark, California, killing pilot 2Lt. Curtis D. Landrum. The P-39 was number two in a two-airplane combat formation that had executed several maneuvers with the F4F, which had initiated the combat. The flight leader put the formation in a steep wingover in an attempt to make a head-on pass on the F4F. Lt. Landrum's P-39 was seen to flip over on its back and dive into the beach. After the maneuver, the flight leader looked back and had seen the P-39 striking the beach and exploding into flames. Investigators speculated that the pilot was not able to pull the airplane out of the dive or flew through the propeller turbulence of the flight leader, losing control of the airplane.

12-30-42D. Eureka, Texas. At 1300, a Fairchild PT-19A dove into the ground at Eureka, Texas, killing Aviation Cadet Willard R. Hocker. Investigation revealed that the airplane dove into the ground in a near vertical position from an altitude of about 2,000 feet with power on. It was also noted that the airplane was not spinning when it hit the ground. The cadet was found in the rear cockpit with the safety belt unfastened.

12-30-42E. Musselshell, Montana. At 1430, a Boeing B-17F flying in icing conditions crashed to the ground out of control eight miles south of Musselshell, Montana, killing 12 crew and passengers. The B-17 had taken off from the Army Air Base at Great Falls, Montana, at 1300 MWT with two other B-17 airplanes on a navigation and formation training flight to Ainsworth, Nebraska, land, refuel, and return. The formation began its climb to altitude and entered the overcast at about 8,500 feet, breaking out on top at about 16,000 feet. The ascent took about 20 minutes at an indicated airspeed of 140 mph. All aircraft began to pick up rime ice about one inch thick on the underside of the leading edge. The subject airplane, which was also the lead B-17, also picked up some carburetor ice while climbing through the overcast. After about 20 minutes of flying on top, the lead B-17 was seen to descend into the clouds. There was no further visual or radio contact between the formation and the airplane after it began its descent. Witnesses on the ground reported seeing the B-17 spinning out of the overcast out of control. The airplane appeared to partially recover before entering an extreme nose-up position. The airplane then fell straight down and began to enter a flat spin, striking the ground with little forward motion and exploding into flames. Killed in the crash were: 1Lt. Edward T. Layfield, pilot; 2Lt. Gerald K. Beem, co-pilot; 2Lt. Regis J. Newland, navigator; 2Lt. Chester A. Knight, Jr., bombardier; SSgt. Hulon B. Dutten, engineer; SSgt. Charles T. Valys, radio operator; SSgt. Frederick T. Brown, gunner; Cpl. Fred E. Murray, gunner; Cpl. Hobart L. Hall, gunner; Maj. Orville A. Ralston, pilot rated passenger; Pfc. Jacob J. Reiss, passenger; TSgt. Wallace H. Hanson, passenger.

12-30-42F. Columbia, South Carolina. At 1706, a North American B-25D on an instrument calibration flight crashed just after take-off from the Army Air Base at Columbia, South Carolina, killing the crew of five. Witnesses reported that the airplane had taken off very quickly and was flying at near stalling speed. The airplane was seen to bank sharply to the right and then return to level flight. The airplane then was seen to bank sharply to the left and then regain level flight. Some witnesses reported that the starboard engine might not have been turning. The starboard wing was then seen to drop and the airplane then entered a spin and crashed to the ground from an altitude of about 800 feet, one and one-half miles from the runway. The B-25 struck the ground and exploded in a swampy gully, heading 180 degrees from its line of flight just prior to the crash. Killed in the crash were: Flight Officer Frank E. Mason, pilot; 2Lt. Lynn L. Beebe, co-pilot; 2Lt. Bradley C. Oliver, bombardier; MSgt. Milton J. Floan, engineer; Pvt. Cecil E. Rook, gunner.

12-30-42G. Hamilton Field, California. At 1500, a Bell P-39D crashed into the ground two miles north of Hamilton Field, San Rafael, California, killing pilot 2Lt. Lloyd G. Blythe. Investigators stated, "Pilot took off from Hamilton Field in a southeast direction. He made a climbing turn to the left over [San Pablo] bay. He reached a position about two miles northeast of the field and at an altitude of about 500 feet. The motor cut out at this point for some undetermined reason. He put the airplane into shallow dive as though he was planning on making a belly landing in a flat field straight ahead. He did not pull out of the dive but [dove] straight into a levee." The airplane exploded violently into flames upon impact with the ground and the pilot was killed instantly.

12-30-42H. San Francisco Bay, California. At 1805, a Bell P-39L crashed into San Francisco Bay, California, killing pilot 2Lt. Howard A. Medin. Investigators stated, "Lt. Medin took off with another plane on a local practice combat flight in a two-ship formation. They engaged in about a half-hour of practice combat maneuvers from 13,000 feet to 5,000 feet. At about 5,000 feet, Lt. Medin pulled his plane into a steep diving spiral with another plane on his tail. He pulled his plane into a vertical bank almost to the stalling point, and continued in this position gradually losing altitude at the same time. The other pilot following on his tail stated that Lt. Medin seemed well under control at all times, but did not realize he was so close to the water. Lt. Medin hit the water while in a tight vertical turn."

12-30-42I. Marionville, Missouri. At 2030, a Vultee BT-13A crashed into the ground at Marionville, Missouri, killing pilot 2Lt. William W. Nobles and passenger Cpl. John B. Randle. Investigators stated, "Lt. Nobles was a member of flight of six [BT-13] airplanes on a cross-country to St. Louis [Missouri] and return from Majors Field, Greenville, Texas. Because of unfavorable weather conditions en route, the flight landed at Springfield, Missouri, at approximately 1840. After being serviced with fuel and oil, the flight was cleared to Tulsa, Oklahoma, with the flight plan of 4,500 feet to Tulsa upon leaving Springfield (take-off approximately 2005). All pilots report that shortly after leaving Springfield at 4,500 feet, the formation had to lose altitude to remain below the clouds. The formation continued to lose altitude at the rate of about 500 feet per minute by a series of descents until the formation was down to about 2,000 [msl] near Marionville, Missouri. At this time, the formation turned back and landed at Springfield. Lt. Nobles was lost some time after the formation was at 4,000 feet [msl].... An officer supervising the flight noticed his absence and detached himself from the formation to look for him. The airplane was found some time later. It had crashed in a westerly heading in a medium dive (about 30 degrees), bounced 250 feet to a dike around a water tank and then tumbled another 250 feet, leaving pieces strewn along the way." It was speculated that the pilot might have failed to enrich his fuel during the descent and suffered an intermittent loss of power, losing flying speed and crashing as a result.

12-30-42J. Lovelady, Texas. At 2030, a Curtiss AT-9 on a night cross-country navigation flight suffered a starboard engine failure and crashed three miles north of Lovelady, Texas, killing Aviation Cadet Fred D. Hutchinson, 24, Oklahoma City, Oklahoma, and injuring Aviation Cadet William A. Kale, 26, Augusta, Georgia, who parachuted to safety. Investigation revealed that Cadet Hutchinson deployed his parachute while still in the cabin of the airplane. The parachute billowed out of the airplane and fouled in the tail section. The cadet was never able to leave the airplane and was found in the wreckage.

12-31-42A. Laredo, Texas. At 1500, a North American AT-6A and an AT-6C collided in mid-air 30 miles southeast of the Army Air Field at Laredo, Texas, killing SSgt. Arthur R. Hutcheson aboard the AT-6C. Pilot 2Lt. Gene M. Patton was able to parachute to safety from the AT-6C and was uninjured. The AT-6A, piloted by Flight Officer Lloyd E. Crimmins and SSgt. Carl B. Mathis, was landed safely at Laredo Army Air Field despite a badly damaged propeller. Both airplanes were attempting to join a formation led by a Douglas O-46. Lt. Patton was joining up from below and F/O Crimmins was joining the formation from above and the left. Crimmins' propeller severed the tail of Lt. Patton's airplane at an altitude of approximately 3,500 feet. Lt. Patton's airplane went over on its back and then spun to the ground, completing about six or seven turns before the pilot was able to abandon ship. Lt. Patton bailed out at the last possible moment, his parachute opening at about 100 feet above the ground. It is not known why Sgt. Hutcheson was unable to leave the airplane. His body was found in the badly mangled wreckage. It was noted that the angle of the sun might have inhibited the visibility of both of the pilots.

12-31-42B. Location Unknown. At an unknown time between 1317 and 1347 PWT, a Curtiss P-40E disappeared and is presumed to have crashed into the Pacific Ocean or in the mountainous areas near Portland, Oregon. Despite a massive search effort pilot 1Lt. James P. Shoemake remained missing and was presumed killed. The P-40 had taken off from Portland for Pendleton Field, Oregon, at 1317 on a mission to report clouds and weather in the area in between the two fields. It was noted that Lt. Shoemake did not report in at 1347 as instructed. At 1745, Lt. Shoemake and his airplane were declared missing. The search effort was terminated on 1-5-43.

12-31-42C. Craig Field, Alabama. At 1700, a Curtiss P-40G suffered the catastrophic structural failure of the port wing while maneuvering and crashed 14 miles east of Craig Field, Selma, Alabama, killing RAF Flight Officer Philip Jones. Investigators stated, "It is believed that Flight Officer Jones half-rolled his airplane in an endeavor to follow Pilot Officer Graham C. C. Pearson [RAF] and in so doing, centered his attention on the lead aircraft and failed to observe the airspeed of his own and, in attempting to recover from a power dive in which he attained a terrific airspeed, it is believed that he attempted a rather violent recovery and in so doing over stressed the wings of his aircraft, causing the left wing to give way; the airplane then fell into an uncontrollable spin and crashed."

12-31-42D. Muroc, California. At 1810, a Douglas A-20A flew into the ground while on a bombing and gunnery mission at Rogers Dry Lake near Muroc Army Air Base, Muroc, California, killing pilot 2Lt. Louis M. Williams. Investigators stated, "Pilot Williams was flying a low-level bombing mission on the south area of Rogers Dry Lake. His mission was to fire the front guns at a target 6 feet by 8 feet sitting on the lake then continue flight approximately one-half mile and drop a 100 pound practice bomb on another target, then circle to position to come in on the target and repeat the mission. He had just dropped his bomb and had continued about one mile from the target at a very low altitude with a speed of about 250 mph. His right propeller hit the ground first then the left propeller struck the ground, the propeller marks on the lake bed continued for about 200 feet then the airplane went into the air and traveled about one-quarter of a mile and landed on its belly. Both engines were torn out and [they] rolled for a thousand feet; the airplane traveled for another thousand feet in the air, crashing to the ground, the gas tanks exploded, burning whatever was left of the airplane. The pilot was still strapped to the seat, the top of his head was cut off and he was burned beyond recognition."

12-31-42E. Greenville, Mississippi. At 2100, a Vultee BT-13A acting as a weather ship for its student squadron crashed five miles northeast of the Army Flying School at Greenville, Mississippi, killing RAF Pilot Officer Michael T. Dickinson. The BT-13 had been circling the area of the Greenville Army Air Field, requesting landing instructions. The tower requested that the pilot stand by for landing instructions and he acknowledged. The airplane continued to circle for several minutes before smashing into the ground in a descending turn and bursting into flames.

1943

January

1-1-43A. Jackson, Tennessee. At 0930, a Fairchild PT-19A spun to the ground and crashed near Jackson, Tennessee, killing A/C Robert C. Swope. The student, while on an authorized spin practice flight, was unable to recover from an intentional spin entered at an altitude of 3,500 feet. Witnesses on the ground stated that the airplane spun in one direction all the way to the ground.

1-1-43B. Atlantic Ocean, Florida. At 1125, a Martin B-26B suffering a fire in flight crashed into the Atlantic Ocean approximately 150 miles ENE of West Palm Beach, Florida, killing three crewmembers, two being killed trying to parachute from the airplane. Three crewmembers parachuted to safety and were rescued from the sea by a U.S. Navy Consolidated PBY-5 operating in the area. The B-26 had taken off from the Army Air Field at Myrtle Beach, South Carolina, at 0910 EWT on a navigation-training mission to Nassau, Bahama Islands, and return. At about 1100, while flying at an altitude of approximately 5,000 feet, the pilot noticed that fuel pressure in the port engine was dropping. The engineer and the pilot turned on the booster pump and momentarily arrested the drop in fuel pressure. The engineer went to check the fuel feed lines in the bomb bay when he observed a flash followed by a fire in the bomb bay between the fuselage (port side) and the bomb bay fuel tank. The engineer returned to the flight deck and grabbed two CO2 fire extinguishers but was unable to extinguish the flames. The pilot jettisoned the bomb bay fuel tanks and ordered the crew to abandon ship. Because the port engine was losing power and the propeller windmilling, and the open bomb bay doors were causing additional drag, the pilot was barely able to hold the airplane steady and maintain flying speed. Engineer MSgt. Aaron R. Adams, gunner SSgt. Frederick C. Wooton and passenger Cpl. Ronald A. Creakbeam parachuted from the airplane and were picked up by the navy airplane after spending about three hours in the water. All were uninjured. After some delay, co-pilot 2Lt. Dan M. Erickson and radio operator Pvt. Walter Mann attempted to parachute to safety but were killed when they bailed out too low to allow their parachutes to fully deploy. Pilot 2Lt. John J. Cook was killed when he went in with the airplane. Investigators determined that a ruptured fuel line caused the drop in fuel pressure and the subsequent fire. Survivors stated that the radio operator had mistakenly attached his parachute to the seat straps instead of his parachute harness and became entangled. The engineer got him untangled and into a new parachute. The engineer then bailed out immediately. He witnessed the airplane dive into the sea moments later.

1-1-43C. Hicks Field, Texas. At 1345, a Fairchild PT-19 suffered a catastrophic structural failure and crashed to the ground six miles north of Hicks Field, Fort Worth, Texas, killing A/C Anthony R. Volpe. Civilian instructor Thomas F. Pemberton was able to parachute to safety. The Accident Classification Committee stated, "This accident occurred because of structural failure of main spar in the center section causing the right wing to come off at the junction where wing joins the fuselage. The failure resulted from an abrupt pull-up out of a dive after a spin recovery. The stress transmitted to the spar by the pull-up was not excessive and was no more than is to be expected in everyday student training. The student, due to extreme tenseness and fear, did not open his safety belt or make any other effort to get out. The instructor was thrown out when he attempted to reach the student and he made a successful descent by parachute."

1-1-43D. Stockton, California. At 1515, a Lockheed P-38G attempting an off-field landing crashed 12 miles west of Stockton, California, killing pilot 1Lt. Edwin P. Arneson. Investigators stated, "Lt. Arneson was one of a flight of four P-38 airplanes en route from Naval Air Station, San Diego, California, to Hamilton Field, California. At Van Nuys, California, they stopped over to refuel, then took off again for Hamilton Field via the San Joaquin Valley. Near Stockton, the left motor of one of the other planes failed and the pilot, Lt. O.B. Carter, made a successful single-engine landing on a highway. After landing, this pilot noticed Lt. Arneson circling him and dragging the nearby fields. Lt. Carter states that both [of Arneson's] motors seemed to be working properly and that he could see no reason [for him] to land except possibly to help Lt. Carter. Although no water showed on the ground, the first pilot landed on the road because the fields nearby looked muddy. After dragging one long field, Lt. Arneson lowered his landing gear, made a power approach to the field and attempted a landing. Two hundred yards after the wheels first hit the dirt, the plane nosed over on its back and immediately burst into flames. The pilot was killed and the plane entirely burned, except for the propellers, the rear half of the fuselage, and the empennage."

1-1-43E. Will Rogers Field, Oklahoma. At 2025, a Curtiss P-40F attempting a landing collided with an airplane tug vehicle that was traveling on the edge of the runway at Will Rogers Field, Oklahoma City, Oklahoma, killing the tug driver Sgt. Anthony Matonti. F/O Glenn F. Van Vilst, the pilot of the P-40, was uninjured but the airplane received major damage. The tug had been laying flares near the runway to assist in night landings that were taking place at the field.

1-2-43A. Vinita, Oklahoma. At 0200, a TG-1 type glider manufactured by the Cinema Company suffered a catastrophic structural failure and crashed at Vinita, Oklahoma, killing civilian glider instructor Stephen E. Brooks and student SSgt. Earl W. Ritchey. The glider was towed aloft and released over the municipal airport at an altitude of about 9,400 feet indicated altitude. Investigators stated that the glider "tore apart in the air," but they could not determine which part of the glider failed first. Wreckage was scattered over a two-mile area near the airport.

1-2-43B. Horseshoe Point, Florida. At 1050 EWT, a Bell P-39D collided with the ground and crashed at the AAF Gunnery Range at Horseshoe Point, Florida, killing pilot 2Lt. Roy D. Bristol. Investigators stated, "Lt. Bristol was flying on a ground gunnery mission. His first two passes on Target No. 2 were successful, but noticed to be excessively low. On the third approach to the target, he came so low that the airplane mushed into the ground on the pull-up about 30 yards beyond and slightly to the left of Target No. 2. A few pieces of the wing were found on this spot, and marks where the propeller had struck the ground. The ship bounced, and with what power it had left, was able to gain about 200 feet of altitude, while turning slightly to the right. Unable to maintain altitude or flying speed because of damage to the ship, it crashed and burned a mile and a half from the target."

1-2-43C. Elko, Nevada. At a time after 0853 PWT, a Boeing B-17F crashed into rising terrain in the Ruby Mountains 35 miles east of Elko, Nevada, killing the crew of ten. The airplane had taken off from Wendover Field, Utah, at 0810 PWT on a routine navigation flight to Elko, Nevada, to Burley, Idaho, and return to Wendover Field. The B-17 was originally to fly a navigation mission to Pocatello, Idaho, and Ogden, Utah, but this was cancelled in mid-air by controllers who advised that very unfavorable weather conditions existed along the route for that flight. The airplane acknowledged the change of flight and flew away to the west. The airplane called in a position report at 0853 and was never heard from again. The airplane remained missing until 6-24-43 when it was found to have flown into the side of a mountain that was undoubtedly shrouded in clouds at the time of the accident. Snow had been falling at the crash site at the time of the accident and it was speculated that the wreckage was covered with snow soon after the accident, hampering efforts to find the airplane. Killed in the accident were: 2Lt. Cyril J. Casey, pilot; 2Lt. Ray C. Houcheimer, co-pilot; 2Lt. Clifford Elliot, navigator; 2Lt. Arthur F. Kelly, bombardier; Sgt. Michael J. Disalvo, engineer; Sgt. Alexander C. Johnson, assistant engineer; Sgt. Seymour Sonkin, radio operator; Sgt. Byron E. Olson, assistant radio operator; Sgt. Lowell T. Stoner, gunner; Sgt. James A. Kornspan, gunner.

1-2-43D. Lubbock, Texas. At 1330, a Cessna AT-17 suffered the structural failure of the starboard wing and crashed 15 miles southwest of Lubbock, Texas, killing students A/C Joseph C. Smith and A/C Marvin E. Sutton. The airplane was part of a four-ship flight on a student formation training flight. The formation was flying at about 3,000 feet indicated altitude in an echelon to the right with the subject airplane flying in the number-four position. Each airplane peeled off of the formation to the left at seven-second intervals, simulating a peel-off for landing. The subject airplane began its peel-off when pieces of the starboard wing were seen to fall away. The airplane went into a sudden roll to the right, continuing to roll to the right and lose altitude until it smashed into the ground and exploded into flames. The starboard landing light, pieces of plywood and fabric, which later were identified to be almost the entire upper portion of the starboard wing, were found about one-half mile from the main wreckage.

1-2-43E. Collinsville, Illinois. At 1405, a Consolidated B-24D suffering a failure of the number-one engine and propeller was abandoned in flight and crashed seven miles north of Collinsville, Illinois, killing three fliers and injuring four others who parachuted to safety. Twelve other crewmembers and passengers parachuted to safety and were uninjured. Passengers Sgt. Earl A. Miller and 2Lt. Edward J. Hadraba fell to their deaths in unsuccessful parachute jumps. Passenger 2Lt. H. A. Huneywill was killed when he went in with the airplane. Co-pilot 2Lt. J. S. Preuss and passengers 1Lt. James B. Kendell, 2Lt. Vernon Miller and 2Lt. Frank Washburn received minor injuries parachuting to safety. Uninjured parachuting to safety were: 1Lt. John P. Davis, pilot; TSgt. Francis A. Loguercio, engineer; Cpl. Harry C. Kribbs, radio operator; 2Lt. Kensel E. Clutter, passenger; 2Lt. W. J. Burton, passenger; 2Lt. Herbert T. Kurz, passenger; SSgt. Roy H. Fuller, passenger; SSgt. Dennis S.

Lodge, passenger; Sgt. William E. Richards, passenger; Sgt. John E. Harmon, passenger; Sgt. Max G. MacFarlane, passenger; Sgt. Thomas J. Ray, passenger. The airplane had taken off from the Army Air Base at Alamogordo, New Mexico, on a ferry flight to Topeka, Kansas, with five other B-24s. The flight was several hours out of Alamogordo when the oil pressure in the number-one engine began dropping. The engineer speculated that the instrument was malfunctioning because the engine seemed to be performing satisfactorily and the airplane continued on the flight. When the B-24s arrived at Topeka, they were diverted to Scott Field, Illinois, because of poor weather. During the flight to Scott Field, the airplane dropped out of the formation and fell back about 20 miles as the flight crew tried to shut down the engine and feather the propeller. When the propeller shifted position and the engine began vibrating badly, the pilot ordered the 18 crew and passengers to put their parachutes on. The airplane continued to fly in this condition for about 30 minutes at an indicated altitude of 7,500 feet until the engine began smoking badly. The pilot ordered the passengers and crew to bail out. Investigators stated, "The five crewmembers on the flight deck left the ship by way of the forward bomb bay, three-fourths of which were filled with baggage. When the pilot and co-pilot bailed out, the airplane went into a spin and the 14 men in the rear had difficulty in using the bottom escape hatch due to congestion. Sixteen of the original 19 men parachuted safely to earth, two men bailed out so close to the ground that one parachute was partially opened and the other never opened. One man was still in the airplane when it crashed and burned. It is believed that he could not get out sooner due to congestion in the rear of the airplane. The airplane being in a spin set up terrific centrifugal force causing the men in the rear to have great difficulty in leaving the airplane."

1-2-43F. Gunter Field, Alabama. At 1415, a Vultee BT-13A attempting a take-off crashed at Gunter Field, Montgomery, Alabama, killing A/C Donald N. Walker. Investigators stated, "The student began his original run for take-off from the northwest corner of Gunter Field, heading in a southeasterly direction. As soon as the airplane had gained flying speed it immediately went up into a stall position, after which the left wing dropped and dug into the ground. This caused the airplane to ground loop approximately 90 degrees to the left, and from this point it seemed that additional throttle was given and an effort was made to take off in this new direction. The airplane took off in a stalled attitude and no recovery was made, resulting in the crash of the airplane."

1-2-43G. Columbus, Ohio. At 1902 EWT, a Douglas A-20B suffering an in-flight explosion and fire went out of control and crashed into a house at 1025 Seymour Avenue, Columbus, Ohio, killing pilot 1Lt. James S. Hansen. Investigation revealed that a broken bolt in one of the port wing fuel tank transfer pumps caused an explosion and fire in the port wing in between the fuselage and engine nacelle. The broken bolt allowed gasoline fumes from one of the numerous connections in the port engine nacelle to enter the transfer pump motor housing. A shorted terminal in the transfer pump motor housing ignited the gasoline fumes, causing the explosion. The airplane was one of 14 A-20s that had taken off from Nashville, Tennessee, on a ferry flight to Patterson Field, Ohio. The airplanes were flying on instruments individually on top of the overcast and upon arriving at Patterson Field found it to be socked in. The pilots had been ordered to "descend in the clear" and to return to Nashville if they could not. Only one airplane returned to Nashville. One landed at Lockbourne Army Air Field and another at Mount Sterling. The rest managed to let down in the vicinity of Patterson Field. Two of the airplanes were known to have exceeded the minimum ceiling in their let-downs. Investigators noted that none of the pilots had asked for weather conditions underneath when descending in the Patterson Field area. Apparently, Lt. Hansen diverted to Columbus and was attempting to land there when the airplane exploded in mid-air and crashed into a frame house at 1025 Seymour Avenue, the home of Mr. and Mrs. W.H. Cadden. They were not home at the time of the accident. Two other houses were damaged by the crash and subsequent fire.

1-2-43H. Shumans, Pennsylvania. At 1908 EWT, a Douglas O-46A flying in instrument conditions collided with rising terrain near Shumans, Pennsylvania, killing pilot 2Lt. Charles B. Koster. The airplane had taken off from Bolling Field, Washington, D.C. on a navigation training flight to Floyd Bennett Field, New York. Investigators noted that the airplane was more than 90 miles off course and that the pilot had not been cleared for instrument flight.

1-3-43A. St. Petersburg, Florida. At 1144 EWT, two Curtiss P-40F airplanes collided in mid-air and crashed at St. Petersburg, Florida, killing 2Lt. Frank W. Stevens aboard P-40F #41-13674. 2Lt. David L. Green was injured parachuting to safety from P-40F #41-14385. The airplanes had been part of a formation training flight that had originated at Pinellas Field, St. Petersburg, Florida. Investigators stated, "Lt. Green was leading a three-airplane flight in string formation. He rocked his wings as a signal to reform. F/O Roy N. Grafstrom immediately came up on the right, or the outside. Lt. Stevens came up fast and overshot Lt. Green. Lt. Green immediately came out of his turn and flew straight ahead. Lt. Stevens pulled out to the left, slowed down, and started into position. He came in too fast, and went underneath, and out of sight of Lt. Green. Lt. Green felt a hard jolt as Lt. Stevens' airplane struck his. F/O Grafstrom, who saw the accident, stated that Lt. Stevens went under and ahead of Lt. Green's airplane. Lt. Stevens then pulled up into Lt. Green's propeller and his tail section was cut off. Lt. Steven's airplane went into a spin and struck the ground. The pilot made no effort to bail out, as he was still strapped to the seat when he was found. Lt Green made a successful parachute jump. Both airplanes crashed in St. Petersburg, Florida, one striking a section of a house at 838 5th Avenue South, and the other landing in a lot behind a house at 935 8th Avenue South."

1-3-43B. Romulus, Michigan. At 1500 EWT, a Lockheed A-28A crashed while attempting a landing at Wayne County Airport, Romulus, Michigan, killing instructor pilot 2Lt. Vincent A. Barsdis. Pilot/A-28 student 2Lt. Jack V. McGuire and co-pilot/A-28 student 2Lt. Earl Pifer were seriously injured. Investigators stated, "Subject aircraft was on the final leg of its final approach for the field. It was carrying a heavy load of ice [on the airframe]. This excessive load of ice resulted in a high-speed stall followed by a horizontal roll during which all flying speed was lost and the aircraft struck the ground descending at approximately 30 degrees. The aircraft was inspected ten minutes after the crash. There was an excessive ice formation on the outer surface of the wing and approximately one-quarter inch of ice on the de-icer boots. The accident is attributed to two factors — a snow and rain squall moving in over and against the surface wind, plus the congestion of traffic caused by the sudden weather condition."

1-3-43C. Duncan Field, Texas. At 1442, a Douglas A-20B crashed while attempting a take-off from Duncan Field, San Antonio, Texas, killing pilot 2Lt. George R. Adams and engineer SSgt. James S. Eakes. The airplane was taking off on a ferry flight to Harding Field, Baton Rouge, Louisiana. The airplane had taken off behind several other A-20s on the same mission. The pilot retracted the landing gear and the A-20 gained about 25 to 50 feet of altitude and then crashed into the ground at high speed about 2,000 feet from the point where it became airborne. Investigators speculated that the airplane encountered the propeller turbulence of the preceding airplanes and the pilot lost control of the airplane, which crashed on railroad tracks near the field.

1-3-43D. Houston, Texas. At 1950, two Beech AT-10BH type airplanes collided in mid-air two miles south of Ellington Field, Houston, Texas, killing A/C Donald M. Cunningham, 21, Georgewest, Texas, who fell to his death in an unsuccessful parachute jump. Co-pilot A/C James E. Dolan, 21, Pontiac, Illinois, was injured parachuting to safety from AT-10 #41-9408. A/C Thomas E. Chandler and A/C Richard B. Cheatham were able to safely land AT-10 #41-26260 at Ellington Field. Both were uninjured. The two airplanes were part of a seven-ship formation that was coming in to land at Ellington Field because of a lowering ceiling. AT-10 #41-26260 was flying in the number-two position and AT-10 #41-9408 was flying in the number-four position, which was the lead ship of the second element. The number-two airplane drifted out of position and then into the path of the number-four airplane.

1-3-43E. Gunter Field, Alabama. At 2128, a Vultee BT-13A crashed 11 miles east of Gunter Field, Montgomery, Alabama, killing A/C Charles R. Bohnke, Jr. Investigators stated, "This student made a normal take-off from an auxiliary airfield 11 miles east of Gunter Field to begin a local night student training flight. From a height of approximately 400 feet he was seen making a turn to the left, which gradually increased until he went into a spiral and crashed. The committee finds that the student apparently lost his horizon and did not realize that he was increasing his bank to a dangerous degree. He continued to increase the turn until the airplane crashed."

1-4-43A. Walnut Ridge, Arkansas. At 0855, two Vultee BT-13A airplanes collided while taking off at the Army Air Field at Walnut Ridge, Arkansas, killing instructor 2Lt. Harry B. House and seriously injuring student A/C Norman A. Ulrich (41-21730). A/C Joseph P. Walker and A/C Hewet S. Hunsdon (41-23095) were uninjured. BT-13A #41-21730 was first in line to take-off on the Southwest runway and several BT-13s were behind waiting in line to take off. Before #41-21730 could achieve take-off speed it apparently lost power or the pilot closed the throttle and the airplane slowed down on the runway. BT-13A #41-21730 apparently was slowing to make a turn off of the runway back on to the taxiway. BT-13A #41-23095 was the next airplane in line to take off and the student began his take-off roll soon after BT-13A #41-21730. After traveling approximately 800 feet down the runway, BT-13A #41-23095 overtook and collided with BT-13A #41-21730. Just before the collision, the student in BT-13A #41-23095 apparently saw the airplane in his path and attempted to pull up over it. The second airplane climbed slightly before the propeller collided first with the rudder and vertical fin and then the cockpit area and fuselage of BT-13A #41-21730. The airplanes locked together and then turned to the right before separating. BT-13A #41-23095 came to rest about 50 feet beyond BT-13A #41-21730. There was no fire.

1-4-43B. Townsend, Georgia. At a time after 1400 EWT, a Douglas A-24A dove into the ground seven miles southwest of Townsend, Georgia, killing pilot 2Lt. Robert J. Chapman and gunner Sgt. Marion O. Gilmer. Investigators stated, "A-24A #42-6750 took off from Hunter Field in a formation of four ships for a bombing mission at Townsend, Georgia. The ship was last seen in a dive over the target range [while still] in the formation. After the final dive was accomplished, the ships were to return to their home station. When the ship did not return to its home station, a search was instituted, but the wreckage was not found until 48 hours later. The motor was buried six feet in the ground, and the airplane was scattered over an area of approximately ten acres. From the condition of the ship, it is impossible to determine the cause or nature of the accident."

1-4-43C. Bonham, Texas. At 1545, a Fairchild PT-19A spun to the ground six miles northwest of Jones Field, Bonham, Texas, killing A/C Emil J. Sipos. The student had entered a deliberate spin at an indicated altitude of 4,000 feet and was unable to recover. The student attempted to leave the airplane and it began to level off from the spin when he released the controls. The student left the airplane at an altitude of about 100 feet above ground level. His body was found about 25 feet from the airplane with the parachute partially unfolded.

1-4-43D. Whitesboro, Texas. At 1620, a Vultee BT-13A crashed six miles north of Whitesboro, Texas, killing student A/C Thomas G. Klix. The student was on his third solo flight and was told by his instructor to practice medium turns, coordinated exercises and stalls. The student apparently entered a deliberate spin at a safe altitude but was unable to recover the airplane. The airplane struck the ground in a spin at a high rate of speed. Apparently the student made no effort to leave the airplane.

1-4-43E. Fort Valley, Georgia. At 2243, a Vultee BT-13A crashed one and one-half miles southeast of Harris Field, Fort Valley, Georgia, killing student A/C James A. Creed. The airplane was observed on the base leg in the traffic pattern at the field but it failed to come in for a landing. It was discovered that the airplane struck the ground in a shallow power dive, bursting into flames and killing the student instantly. This was the student's first solo night landing.

1-5-43A. Fort Myers, Florida. At 0850 EWT, a Bell P-39D and a P-39F collided in mid-air eight miles west of Page Field, Fort Myers, Florida, killing P-39D pilot 1Lt. Edward F. Kernan. P-39F pilot 1Lt. James P. Holland was able to parachute to safety. The airplanes were part of a three-ship flight on an air-to-ground gunnery mission at Sanibel Island. The Accident Classification Committee stated, "The three airplanes took up echelon for the return trip to Page Field and instead of going direct, flew slightly northwest, necessitating a gentle right turn into the echelon toward the field. This turn momentarily put Lt. Holland [flying in the number-two position] directly into the sun but a cross-over was being effected in the turn when both Lt. Holland and 1Lt. Thomas R. Chantler [flying in the number-three position] noticed the lead airplane [Lt. Kernan] to pull up and saw Lt. Kernan looking back. Lt. Holland stated that immediately after, another more sudden pull-up was started and that because of Lt. Chantler directly behind him, he did not feel that he could duplicate the maneuver and tried to skid out of the echelon. Lt. Holland was not successful and the propeller of his airplane struck the wing of the lead airplane and both tail sections struck violently. Both airplanes went out of control and Lt. Kernan did not successfully use his parachute while Lt. Holland cleared the airplane and parachuted successfully. Both airplanes spun in and crashed."

1-5-43B. Shreveport, Louisiana. At 0930 CWT, a Martin B-26B crashed into Cross Lake six miles northwest of Barksdale Field, Shreveport, Louisiana, killing the crew of five. Killed in the crash were: 1Lt. Doyce D. Harden, Redfield, Iowa, B-26 instructor; F/O Malcolm J. Tahl, Cincinnati, Ohio, co-pilot/B-26 student; F/O James E. O'Connor, Wellsburg, Virginia, pilot-rated passenger; 2Lt. John P. Petraitis, Chicago, Illinois, bombardier; SSgt. John C. Aitchison, Sacramento, California, engineer. Investigators stated, "The Aircraft Accident Classification Committee met and the following was determined: Witnesses stated that they saw the airplane lose altitude rapidly and turned left as if in a spin. An examination of the wreckage showed that the airplane had struck at about a 60 degree angle and pieces of wreckage were strewn to the right of the aircraft, indicating that the airplane was in a spin to the left when impact occurred. Trim controls were in neutral position, indicating that both engines were functioning normal at the time of the accident. Probable cause was the instructor pilot stalled the airplane to a complete stall during a demonstration of stalling characteristics. In so doing, he lost complete control, and at this point the airplane started to spin and he had insufficient altitude to make the recovery. Due to the lack of complete information of the causal factor, the Board recommends that no assessment of pilot error be made." The airplane crashed into five feet of water in swampy terrain at the northwest edge of Cross Lake, about six miles northwest of Barksdale Field. This aircraft (41-17645) was the subject of a B-26 training film produced by the AAF in 1942. See Note #20.

1-5-43C. Columbia, South Carolina. At 1330, a North American B-25C on an instrument training flight crashed 15 miles north of the Army Air Base at Columbia, South Carolina, killing the crew of four. The airplane took off from the base at 1322 and was next sighted in a steep dive toward the ground. There was no apparent effort to recover the airplane and it dove into the ground at a 45-degree angle and exploded into flames. Investigators were not able to determine the cause of the accident. Killed in the crash were: 2Lt. Merritt M. Hopson, pilot; 2Lt. Walter R. Bone, co-pilot; Cpl. Hal W. Reed, engineer; SSgt. Vernon L. Arenas, gunner.

1-6-43A. Tallahassee, Florida. At 0006 EWT, a North American B-25C attempting a take-off crashed one and one-half miles northwest of Dale Mabry Field, Tallahassee, Florida, killing six crewmembers and seriously injuring navigator 2Lt. Ivor P. Evans. Killed in the crash were: 2Lt. William M. Putman, pilot; F/O Ralph R. Nored, co-pilot; 2Lt. Earl H. Brown, bombardier; SSgt. Oliver H. Brown, engineer; Sgt. Marvin E. Thompson, radio operator; Pfc. George F. Marinari. The airplane took off from the field on a night navigation mission and reached an altitude of about 60 feet when the pilot leveled off to pick up speed. The pilots apparently had their heads in the cockpit and let the airplane descend into trees in a flying attitude. Investigation revealed that the airplane descended into the terrain at an angle of about five degrees. The airplane burst into flames and was destroyed utterly. The navigator that had survived was in the radio compartment at the time of the crash.

1-6-43B. Marianna, Florida. At 0019 CWT, a North American AT-6C attempting a landing crashed near the Army Air Field at Marianna, Florida, killing A/C Kenneth O. Crowe. The airplane was coming in for a landing from a routine night flying exercise and was overshooting the runway badly. The control tower officer radioed the student and told him to go around. The student acknowledged and as he began to go around the tower officer told the student to turn off his landing lights, which he did. About 15 seconds later the airplane was observed to enter a dive of about 20 degrees at an altitude of approximately 400 feet, remaining in the dive until it struck the ground and exploded into flames. Wreckage was scattered over an area of 100 yards. The pilot was thrown about 75 feet from the main wreckage.

1-6-43C. Pacific Ocean, California. At about 1015, a Lockheed P-38F crashed into the Pacific Ocean about seven and one-half miles west of Oceanside, California, killing pilot 2Lt. James A. Bauman. The airplane was part of a four-airplane flight that had taken off from the Naval Air Station at San Diego, California, on a gunnery and bombing mission. The flight went to Clark Lake Bombing Range and each P-38 dropped two 100 lb. bombs. The flight then rendezvoused with an airplane towing a target at an altitude of 8,000 feet off shore of Oceanside. The flight made several passes at the target ship. Lt. Bauman was last seen in a pull-up after making a pass at the target ship. Lt. Bauman never rejoined the flight and failed to return to base and it was speculated that his airplane had crashed into the sea.

1-6-43D. Myrtle Beach, South Carolina. At 1448 EWT, a North American B-25D crashed in a wooded area near Gunnery Range V-18, approximately 15 miles southwest of the Army Air Base at Myrtle Beach, South Carolina, killing six fliers and seriously injuring gunner SSgt. T.B. Casten. Killed in the crash were: F/O Raymond G. Strover, pilot; F/O Warren L. Kendrick, co-pilot; Sgt. A.J. Cairns, gunner; Sgt. T.J. Connalonga, gunner; Sgt. G.L. Finney, gunner; Sgt. F.B. Dubois, gunner. The airplane had taken off from the Army Air Base at Columbia, South Carolina, on a turret gunnery mission. The airplane was flying at extremely low altitude over the gunnery range on several passes. After completing a firing pass the pilot attempted a steep left turn at low altitude. The airplane apparently slipped down in the steep turn and the port wing contacted treetops, sending the airplane out of control. The airplane smashed into the trees and exploded into flames. Gunnery range personnel immediately rescued Sgt. Casten from the burning airplane.

1-6-43E. Little Rock, Arkansas. At 1630, a Stearman PT-27 that was taxiing at Adams Field, Little Rock, Arkansas, struck and killed pilot 2Lt. Charles R. Whitwell, who was walking on the taxiway. The pilot of the airplane, 2Lt. Ralph C. Snow, was not injured. Investigation revealed that the Lt. Whitwell had walked into the path of the moving airplane on the port side, being struck by the rotating propeller. It was noted that at the time of the accident that many airplanes were in the immediate vicinity with their engines running. It was speculated that the pilot probably did not hear or see the airplane that struck him.

1-6-43F. White City, Kansas. At 1735 CWT, a Consolidated B-24D suffered a catastrophic structural failure of the tail section and crashed three miles west of White City, Kansas, killing 11 passengers and crew. Navigator-rated passenger 2Lt. Frank Maleckas parachuted to safety and was uninjured. The B-24 took off from the Army Air Base at Topeka, Kansas, at 1710 on an instrument training flight to Pueblo, Colorado. The airplane climbed to an altitude of 5,500 feet when it began picking up ice in the overcast. A crewmember in the rear of the airplane notified the pilots that the tail section was picking up a thick coating of ice. The pilot acknowledged the warning and said, "Do not worry." The airplane began losing altitude rapidly. The pilot ordered the crew and passengers to put on parachutes and to stand by to bail out. Lt. Maleckas stood by in the nose section and after noticing that the airplane was beginning to shudder violently he pulled the emergency hatch lever and bailed out. He stated that the pilot gave no order to abandon ship. Investigation of the wreckage indicated that the port horizontal stabilizer, elevator, vertical fin and rudder had separated from the airplane in one piece. The B-24 then dove straight ahead into the ground and exploded violently into flames. The severed port tail section was found over one-half mile from the scene of the main wreckage. Lt. Maleckas parachuted safely to the ground and ran to the wreckage immediately, but was kept away by exploding bombs, bullets and flares. The occupants had all been killed instantly. Investigators speculated that the structural failure was caused by an excessive amount of ice stressing the tail section and causing it to fail. Killed in the crash were: 1Lt. Robert Clyne, pilot; 2Lt. Norman A. Kendell, co-pilot; 2Lt. Joseph P. Hulder, navigator; 2Lt. Clifford A. Smallwood, bombardier; SSgt. Vernon A. Larson, engineer; SSgt. Thomas P. Merlino, radio operator; 2Lt. Alden A. Alleman, pilot-rated passenger; F/O James G. Masters, pilot-rated passenger; 2Lt. John Sedor, Jr., bombardier-rated passenger; SSgt. Leonard F. MacNabb, engineer-rated passenger; SSgt. Gustav Dzingoski, radio operator-rated passenger.

1-6-43G. Madill, Oklahoma. At 2030 CWT, a Consolidated B-24D suffered the catastrophic structural failure of the tail section and crashed 15 miles southwest of Madill, Oklahoma, killing the crew of ten. The airplane had taken off from the Army Air Base at Orlando, Florida, on a navigation training flight back to its home station at Salinas, Kansas, via Galveston, Texas. The B-24 was heard to pass over Perrin Field, Sherman, Texas, at high altitude several minutes before the crash. Apparently the airplane was not suffering any difficulty at that time. Investigation of the wreckage revealed that the entire tail section (horizontal stabilizers, elevators, vertical fins and rudders) had separated in one piece from the airplane prior to the crash. The severed tail section was found over 400 yards from the spot where the B-24 dove into the ground and exploded into flames. Investigators could not determine the cause of the tail section failure. It was speculated that "abnormal vibration" could be a factor in the accident. Killed in the crash were: 2Lt. Russell G. Bishop, Jr., Washington, D.C., pilot; 2Lt. John F. Howell, Asheville, North Carolina, co-pilot; 2Lt. Robert B. Secord, Cleveland, Ohio, navigator; 2Lt. James H. Dyer, Grand Rapids, Michigan, bombardier; 2Lt. Q.M. Lipsman, passenger; TSgt. Albert L. Baima, Chicago, Illinois, engineer; TSgt. Raymond E. Blazer, Greenfield, Ohio, radio operator; SSgt. Stanley J. Jaros, Chicago, Illinois, assistant radio operator; SSgt. Kelsey C. Horning, Hoisington, Kansas, gunner; Pvt. Jack Bagget, Dublin, Georgia, assistant engineer.

1-7-43A. Iron City, Georgia. At 0040 EWT, a Vultee BT-13A crashed at Iron City, Georgia, killing student A/C Conrad W. Christian. The student was on a routine night solo flight and was circling in his zone waiting to be cleared to come into land at Donalsonville Auxiliary Airfield. The student did not respond to radio calls and failed to return to his station at Bainbridge Army Air Field, Georgia. The airplane was found later that morning during daylight hours. The wreckage indicated that the airplane dove into the ground underneath its assigned zone. The student apparently lost his horizon and flew into the ground. The airplane did not burn but was destroyed utterly.

1-7-43B. Pecos, Texas. At 1244, a Vultee BT-13A crashed 15 miles southwest of the Army Air Field at Pecos, Texas, killing instructor 2Lt. Robert R. Coble and student A/C Frank W. Lewis. The airplane had taken off from the field at 1205 on a routine dual instruction flight. When the airplane failed to return to the field a search was instigated and the airplane was found at 1609. Investigation revealed that the airplane struck the ground with very little forward speed but with great vertical speed, indicating a stall and spin to the ground. The damage to the propeller indicated that it was not turning when the airplane struck the ground. Both occupants were killed instantly. The airplane did not burn but was destroyed utterly. It was speculated that the airplane had suffered an engine failure and the pilots were attempting an emergency forced landing when they entered a stall at an altitude too low to allow recovery.

1-7-43C. Lake Drummond, Virginia. At 1640, a Curtiss P-40F collided with the surface of Lake Drummond, Virginia, crashing into the water and killing pilot 2Lt. Robert Lloyd Eccles. The pilot was leading a flight of three P-40s in a V-formation. The flight was seen flying over the Lake Drummond at low altitude when Lt. Eccles' airplane struck the surface of the lake. The airplane bounded back into the air with serious damage. The airplane was unable to maintain flying speed and crashed into the center of the lake in about nine feet of water, coming to rest upside down. Lt. Eccles' body was recovered from the lake two days later. The P-40F flying to the right of Lt. Eccles, piloted by 2Lt. H. J. Morris, was damaged by flying wreckage and was forced to ditch in the lake. Lt. Morris suffered minor injuries.

1-7-43D. Woodville, California. At 1625, two Stearman PT-17 airplanes collided in mid-air and crashed three miles west of Woodville, California, killing two civilian instructors and two aviation cadets. Instructor Kenneth S. Coleman and student A/C Allan E. Peck were killed in the crash of PT-17 #41-8953. Instructor Albert F. Patrick and student A/C Olin S. Sellers were killed in the crash of PT-17 #41-8688. The Accident Classification Committee stated, "This accident occurred as the primary training airplanes were returning to the home field from the Tipton Auxiliary Airfield. Two airplanes, No. 41-8953 and No. 41-8688, were flying side-by-side 500 feet above the ground, at close interval, headed north with 41-8953 on the right of 41-8688. The two airplanes were overtaken by 41-7922, which approached 300 feet to the right of 41-8953 to check their field numbers and report them. Immediately after overtaking the other two airplanes, the middle airplane (41-8953) attempted to cross underneath the airplane on his left. He misjudged his distance and climbed directly into the flight path of 41-8688. The propeller of 41-8688 contacted the tail of 41-8953, severing the empennage from the airplane approximately two feet ahead of the vertical fin. The two airplanes then crashed out of control and burned. No attempt was made to use parachutes."

1-7-43E. Coffeyville, Kansas. At 2250 CWT, a Vultee BT-13A collided with trees and crashed three-quarters of a mile north of the Army Air Field at Coffeyville, Kansas, killing student A/C Raymond H. Mussack. The airplane was part of a three-ship flight that was returning to the field to land. The first two airplanes landed safely but the third failed to arrive. A search was commenced and the airplane was found a short time later. The airplane collided with trees while in the traffic pattern, crashing into a wooded area near the field. The BT-13 erupted into flames upon impact and the student was killed instantly. Investigators speculated that the student inadvertently flew the airplane into the ground.

1-7-43F. Gunter Field, Alabama. At 2330, a Vultee BT-13 crashed 11 miles southeast of Gunter Field, Montgomery, Alabama, killing student A/C Jeffery C. Ertz, Jr. Investigators stated, "This student took off from an auxiliary field on a local night training flight, from a runway outlined with smudge pots. The take-off was made into the east. At an altitude of approximately 500 feet he started a shallow turn to the right. After making a turn of approximately 90 degrees, he rolled out at a level attitude and was observed gradually losing altitude with power on. He continued losing altitude, in a straight line, until the airplane crashed and burned. From the testimony of witnesses, it has been determined that there was no power failure and that the pilot made no visible correction to level his airplane after it started its gradual descent."

1-8-43. Bishop, California. At 1630, a Vultee BT-13A collided with rising terrain and crashed 25 miles northwest of Bishop, California, killing student A/C Walter John Kania. The Accident Classification Committee stated, "The student took off on a routine solo practice flight in the local area at Bishop, California, at approximately 1530 and failed to return. The wreckage was located at approximately 1030, January 14, 1943. Due to the complete destruction of the aircraft, the Aircraft Accident Committee is unable to determine the cause of the accident. The damage to the aircraft and the marks left on the ground indicate that the airplane was in a nearly horizontal flight in approximately a 45-degree bank when it crashed. It is believed that the pilot was flying in or very low over the gorge. The elevation at the location of the crash being rather high (7,300')— and the location being on the leeward side of sloping terrain where prominent downdrafts prevail are believed to be contributory causes of the accident."

1-9-43A. Kissimmee, Florida. At 0040 EWT, a Douglas P-70 attempting a landing crashed near the Army Air Field at Kissimmee, Florida, killing pilot 2Lt. Miles J. Hemauer and radio operator Pvt. Joseph F. Hirl, Jr. The airplane was returning to the field following a night fighter training flight. Investigators speculated that the pilot inadvertently flew the airplane into the ground after entering patches of fog that were prevalent near the field. The airplane was in a steep bank to the left and in a moderate dive when it crashed into the ground, exploding violently into flames and killing the occupants instantly. Wreckage was scattered over an area of 100 yards.

1-9-43B. Denmark, Iowa. At 0100, a Curtiss AT-9 crashed two miles west of Denmark, Iowa, killing pilot 1Lt. Wayne S. Forsythe, Houston, Texas, and medical officer 1Lt. Lewis E. Barrick, Houston, Texas. The airplane had taken off from Kansas City, Missouri, on a navigation training flight to Chicago, Illinois. The airplane was cleared on an instrument flight plan and was assigned an altitude 3,000 feet. The airplane was on time and on course when it smashed into the ground out of control. The airplane exploded into flames upon impact and the occupants were killed instantly. Investigators could not determine why the pilot lost control of the airplane.

1-9-43C. Douglas, Georgia. At 1130 EWT, a Stearman PT-17 (41-25220) that was undergoing a pre-flight engine run-up at Douglas Municipal Airport, Douglas, Georgia, jumped the wheel chocks under power and struck and killed civilian crew chief Henry Thornton with the rotating propeller. The airplane continued out of control until it collided with PT-17 #41-25299, coming to rest a short distance away. Civilian mechanic Dudley Reed was seriously injured and mechanic Frank Scott, who was in the cockpit, was uninjured.

1-9-43D. Pacific Ocean, California. At 1230, a Bell P-39D was abandoned over the Pacific Ocean after suffering an engine failure three miles northwest of Point Reyes, California. Pilot 2Lt. Henry C. Dolde parachuted into the sea. He was struck by part of the airplane as he bailed out, suffering a compound fracture of the right leg. He died of exposure and shock before he could be rescued from the sea. The P-39 was one of a two-ship flight that took off from Hamilton Field, San Rafael, California, on an "Air Alert Mission."

1-9-43E. Gulf of Mexico, Florida. At 1600 EWT, a Curtiss P-40F crashed into the Gulf of Mexico ten miles northwest of New Port Richey, Florida, killing pilot 2Lt. John E. Rhoads. The airplane had taken off from Pinellas Army Air Base, St. Petersburg, Florida, on a gunnery mission. The Accident Classification Committee stated, "The pilot, after firing at the target in the water, pulled up, and at approximately 200 feet, the airplane rolled over on its back. The roll was to the left. The nose of the airplane dropped, and the roll continued until the airplane was again in a normal diving attitude. It crashed into the water at a 25-degree angle or less."

1-9-43F. Douglas, Georgia. At 2145 EWT, a North American AT-6A crashed ten miles west of Douglas, Georgia, killing student A/C Edmund T. Pardue. The airplane was on a navigation flight from Spence Field, Moultrie, Georgia, to Atlanta, Georgia, and return. The airplane took off from Atlanta for the return trip and checked in at Alma, Georgia, at 2121. When the airplane crashed it was on course and on time. Apparently the airplane flew into the ground under control, exploding into flames upon impact.

1-10-43A. Marianna, Florida. At 1325 CWT, a North American AT-6A stalled and spun to the ground at the AAF gunnery range five miles east of the Army Air Field at Marianna, Florida, killing instructor 2Lt. Paul B. Santee and student A/C Charles L. Rowe. The instructor and student were on a ground gunnery mission with the student flying the airplane from the front cockpit. The airplane arrived at the firing range where five other airplanes were circling in the firing pattern at an altitude of 700 feet. The subject airplane attempted to enter the pattern but aborted the attempt because the spacing was not right. Attempting to enter the pattern at a more favorable point, the airplane made a circling turn to the right. As the turn continued the pilot tightened the turn and was just about at the correct entry position when the airplane was seen to go into a snap roll to the right and then into a tight spin to the right. The airplane continued to spin until it struck the ground and burst into flames. The occupants were killed instantly.

1-10-43B. Savannah, Georgia. At 1545 EWT, a Vultee V-72 (similar to the A-31 type) collided with gasoline drums while taxiing at Hunter Field, Savannah, Georgia, killing gunner Sgt. Charles P. Weber and seriously injuring the pilot 1Lt. Norris M. Russell. Investigation revealed that the airplane was on a scheduled chemical bombing mission. The pilot was taxiing to the take-off position when the V-72 collided with an experimental rack of 55-gallon drums of aviation gasoline that was set up on the bomb-loading ramp. The airplane and the gasoline drum rack burst into flames upon impact. The gasoline drum rack was being used by the 84th Bombardment Group to train ground crews in "rapid refueling" techniques.

1-10-43C. Roswell, New Mexico. At 2115, a Cessna AT-17 and an AT-17B collided in mid-air and crashed ten miles north of the Army Air Field at Roswell, New Mexico, killing three crewmembers and injuring another who parachuted to safety. A/C William A. Garrett, 24, Decatur, Georgia, and A/C Charles A. Gettys, 22, San Louis Obispo, California, were killed in the crash of AT-17B #42-38829. A/C Harris L. Dippel, 24, Fillmore, California, was killed in the crash of AT-17 #42-412 and A/C Edward A. Dolton was injured parachuting to safety. Investigators stated, "Both airplanes were assigned to fly the traffic pattern at the municipal airport at 6,000 feet for one hour. There were six other AT-17s that were given similar instructions. They were to remain at 6,000 feet until called in by the control tower. Landings were being made on Runway 21 and traffic was right hand. Airplane No. 42-412 had made one trip around the pattern and was on its downwind leg of the second trip flying at 6,000 feet. The airplanes in the pattern were spaced quite evenly. Airplane No. 42-412 was about ready to start the turn onto the base leg when A/C Dolton noticed another airplane on the base leg, 100 to 200 feet below and almost ready to turn on its final approach. Then this airplane started a 180-degree turn to the left and A/C Dolton watched it until it disappeared under the left side of No. 42-412. A/C Dolton saw no other airplanes near his airplane and he prepared to make his turn onto the base leg. Suddenly A/C Dolton heard the engines of an approaching airplane and then saw it coming from the left front and below. A/C Dolton was unable to prevent the collision. After being thrown around considerably, A/C Dolton was able to free himself from the airplane and parachuted to the ground with minor injuries. A/C Dippel, co-pilot of airplane No. 42-412, remained in his seat and crashed with the airplane. A/C Garrett, pilot of airplane No. 42-38829, remained in his seat and crashed with his airplane. Airplane No. 42-38829 burned after crashing. A/C Gettys, co-pilot of airplane No. 42-38829, was thrown clear while the airplanes were still in the air. A/C Gettys had his parachute on and was still strapped to his seat when found about 130 yards from the airplane. The two airplanes crashed about 400 yards apart. From the scattered debris and wreckage it appears as if the right engine of airplane No. 42-38829 cut thru the left wing of No. 42-412, [and as airplane No. 42-38829 slid to the right] the left engine of No. 42-38829 hit the left engine of No. 42-412 and deflected on up thru the [left] side of the fuselage of airplane No. 42-412. It appeared as if the two ships were locked together during part of the descent and then crashed separately."

1-10-43D. Dewville, Texas. At 2340, a North American BT-14 crashed one and one-half miles north of Dewville, Texas, killing A/C Arthur M. Berry. The student was on a night cross-country navigation flight from Randolph Field, San Antonio, Texas, to Yoakum, Texas, to Austin, Texas, and return to Randolph Field. The airplane struck the ground in a steep bank to the right and burst into flames upon impact. Investigators speculated that the student lost the horizon and inadvertently flew the airplane into the ground.

1-11-43A. San Antonio, Texas. At 0020, a Curtiss O-52 crashed eight miles southwest of Brooks Field, San Antonio, Texas, killing pilot 2Lt. Oliver D. Briggs and passenger 1Lt. Kyle E. Ball. Investigation revealed that the airplane descended into the ground at flying speed and under control near the pattern area for Stinson Field.

1-11-43B. Oakland, California. At 0224, a Douglas C-53 being operated by a United Airlines flight crew for the AAF crashed at Oakland Municipal Airport, Oakland, California, fatally injuring civilian co-pilot Charles Sheppard Lamb. First Pilot John Adams Burks, United Airlines, was seriously injured and received serious burns. The airplane had taken off from Mines Field, Los Angeles, California, on an AAF cargo flight to McClelland Field, Sacramento, California. When the C-53 took off from Los Angeles, weather at McClelland Field was reported as zero ceiling and zero visibility in heavy fog. Fresno and San Francisco, California, were listed on the flight plan as alternate airports. When the C-53 was over Fresno, McClelland Field weather had not improved and the pilot diverted the flight to San Francisco. As the C-53 approached the San Francisco Bay area, San Francisco Municipal Airport closed down due to heavy fog, causing the airplane to divert to Oakland, which was still reporting clear. The pilot was able to pick up the Oakland radio beam and then he established visual contact with the field from an altitude of 4,000 feet. The pilot began a descent and then entered an approach for the field. The airplane made the approach end of the runway a little high and the pilot cut the throttles and waited for the airplane to settle. After traveling about halfway down the runway and still not touching down, the pilot realized that he would have to go around. He advanced the throttles and retracted the landing gear. Immediately after leaving the end of the runway the airplane encountered a thick bank of fog. The pilot immediately went on instruments and with flaps still down three-quarters began a turn to the left to pick up the pattern and make another approach. As the airspeed bled away in the turn, the pilot lowered the nose to pick up airspeed and inadvertently flew the airplane into the ground. The co-pilot died at Highland Hospital, Oakland, the day after the accident.

1-11-43C. Maxwell Field, Alabama. At 0815, a Beech AT-10BH crashed while attempting a take-off at Maxwell Field, Montgomery, Alabama, killing co-pilot 2Lt. C.E. Wheat and injuring pilot 1Lt. John K. Montmeat. The airplane crashed about 2,000 feet from the departure end of the runway. Investigation revealed that the airplane would not climb after take-off due to heavy frost on the airframe and wings. The pilot apparently had to pull the airplane into the air at the end of the 3,575-foot runway. The airplane would not climb and mushed into the trees. The airplane smashed itself to pieces as it careened through the trees, finally coming to rest in shallow water in the Alabama River.

1-11-43D. Colorado Springs, Colorado. At 1110, a North American B-25C crashed 20 miles northeast of Peterson Field, Colorado Springs, Colorado, killing the crew of six. Killed in the crash were: 2Lt. Richard P. Marx, B-25 instructor; F/O Earl L. Sumner, co-pilot/B-25 student; F/O Gordon F. Anderson, co-pilot/B-25 student; 2Lt. Robert C. Markway, navigator; Pfc. F.H. Hansfield, engineer; Pfc. N.L. Cadwell, radio operator. Pilot 2Lt. David W. Henderson witnessed the accident. He later stated to investigators, "I was flying directly north at 9,500 feet — airspeed 200 miles per hour — and Lt. Marx was flying in the same direction and the same airspeed about two or three miles ahead of me. He had been flying on the same course for about three or four minutes straight and level when he climbed about 100 feet straight ahead and did a complete and violent snap roll to the left without gaining or losing altitude. After the snap roll, he fell off on his left wing and did a one-turn spin to the left, again almost recovering. He again fell off to the left and did a one-half turn spin out of which he recovered, heading south. From our position, it appeared that he had pulled out and was almost straight and level when he hit and exploded immediately. To the best of my knowledge, he had both engines."

1-12-43A. Alamogordo, New Mexico. At 0100 MWT, a Consolidated B-24D crashed three miles west of the Army Air Base at Alamogordo, New Mexico, killing ten crewmembers. The airplane and crew were scheduled to fly a bombing training mission at 2000 (1-11-43) but it was postponed when the port landing light burned out after the pilots had started the number-one engine. There was also a report of low oil pressure in the number-one engine during the run-up. The airplane's paperwork also indicated that the Automatic Flying Control Equipment (AFCE) malfunctioned while the airplane turned to the right. At 2130, a new port landing light was installed and it promptly burned out and caught fire while the airplane was taxiing away from the front of the operations building where it was parked. The fire was immediately put out and there was no serious damage to the airplane. The crew chief sent for another landing light and the crew was told to get some rest. The pilots were sent to the link trainer and the bombardier uninstalled the bombsight and returned it to storage. The bombardier, 2Lt. William D. Pharmer, Jr., apparently thought that the mission had been called off. He went back to the officer's mess and then to his quarters to get some rest. He was not on board the B-24 when it crashed. At about 0030, the pilots decided that despite the inoperative landing light that they would go up for one pattern around the field and a landing prior to 0100, the time that their flying period was to conclude. At 0050, the airplane took off to the south and made a turn to the west and then began a right turn back toward the field. While flying over the extremely dark terrain, the pilots apparently turned on the AFCE at low altitude. The equipment malfunctioned, causing the B-24 to go into an uncontrolled dive. Before the pilots could recover the airplane it dove into the ground at a slight angle. The airplane smashed to pieces and exploded into flames, coming to rest about 250 yards from the initial point of impact. Investigators noted that there was no evidence of sabotage. Killed in the crash were: 2Lt. William F. Schulte, pilot; 2Lt. Elmer J. Molique, Jr., co-pilot; 2Lt. Billie D. Stults, co-pilot; 2Lt. Raymond W. Wherling, navigator; Pvt. Floyd J. Pekrul, engineer; Sgt. Paul D. Miller, assistant engineer; Sgt. Austin P. Young, radio operator; Sgt. Willie F. Rose, assistant radio operator; Sgt. Warren S. Leeper, gunner; Sgt. Robert C. McCormick, gunner.

1-12-43B. Winfield, Kansas. At 0905 CWT, a Vultee BT-13A attempting a take-off crashed at Strother Army Air Field, Winfield, Kansas, killing A/C Kenneth L. Merlis. The airplane took off to the north on the North-South runway. When the airplane reached an altitude of 25 feet, it entered an extreme nose-high attitude. The airplane then stalled at an altitude of about 50 feet and crashed approximately 150 yards west of the north end of the North-South runway. Just minutes before this accident occurred, a BT-13 attempting a landing ran through the fence at the end of the runway. Investigators speculated that the student might have been distracted by the mishap.

1-12-43C. Warner Robins, Georgia. At 1019 EWT, a North American A-36A suffering an in-flight fire crashed while attempting an emergency forced landing at Warner Robins Army Air Depot, Georgia, killing pilot 1Lt. Paul J. Fergen. Investigators stated, "The pilot took off into the west on the East-West runway. The take-off appeared to be normal in every respect, the airplane having obtained flying speed midway [down] the 5,000-foot runway from which point a gradual ascent was made, the engine sounding smooth during the entire take-off. Two or three minutes later the airplane was seen directly east of the field at an altitude of approximately 400 feet, the pilot having called the tower for emergency landing instructions. The crash alarm was sounded and at this time when the airplane, headed west on final approach, was seen to be leaving a trailing column of smoke while descending. When the airplane was approximately 150 feet above the ground, flames could be seen coming from the left side of the engine. The pilot made a normal descent toward the edge of the East-West runway until the flames, having grown in magnitude, enveloped the entire nose section at approximately 50 feet. From this altitude the airplane nosed down sharply and hit the runway, bounced into the air, winged over on its back and caused total damage to the airplane." Investigation revealed that the fire was caused by the failure of a fuel line from the port wing tank to the engine.

1-12-43D. Savannah River, Georgia. At 1035 EWT, a North American B-25C crashed into the Savannah River one mile north of Savannah, Georgia, killing the crew of six. The airplane had taken off with five other B-25 airplanes from Walterboro Army Air Field, South Carolina, on a navigation training flight. Near Savannah the flight went into a Lufbery circle for a couple minutes and then it broke up. The subject airplane, which was leading the flight, made one diving turn to the right and another to the left. When the pilot attempted an extreme pull-up at an indicated altitude of 3,000 feet, the starboard wing buckled and separated. The tail section then separated and the airplane plunged into the river and exploded into flames. Killed in the crash were: 1Lt. Arnold W. Kay, pilot; 2Lt. Floyd A.W. Hale, co-pilot; SSgt. Arnold A. Lassen, engineer; SSgt. Roland Medlin, gunner; TSgt. Roy S. Zufall, radio operator; Cpl. Georgia V. Silva, passenger.

1-12-43E. Lawson Field, Georgia. At 1430 EWT, a Aeronca L-3B crashed about 10 miles southeast of Lawson Field, Fort Benning, Georgia, killing pilot 1Lt. Edmond D. Humphrey and passenger TSgt. Thomas W. Kirkman. Investigation revealed that the airplane stalled out of a steep climbing turn to the right. When the airplane stalled it flipped over on its back. The pilot tried to roll through the maneuver, but before he could the engine quit because of fuel starvation caused by the inverted position. Due to the low altitude the pilot was unable to recover and the airplane smashed into the ground in an extreme nose-down position.

1-12-43F. Gulf of Mexico, Florida. At 1500 EWT, a Bell P-39D crashed into the Gulf of Mexico one mile offshore and due west of Perry Army Air Field, Perry, Florida, killing 2Lt. Harvey T. Davis. The airplane had taken off from Cross City Army Air Field, Florida, with five other P-39 airplanes on a gunnery mission. The formation had made several passes at the target when one P-39 broke off of the formation and flew away. The flight leader, flying in the firing pattern, noticed that the remaining four airplanes were flying in a string formation in a shallow turn to the right in an area to the right of the left-hand firing pattern that they were to be flying. The flight leader called the airplanes to return to the pattern and complete the firing mission. One airplane peeled off and flew back toward the pattern. The remaining three airplanes continued to fly in a shallow circle to the right. A moment later, Lt. Davis' airplane was seen to peel off to the left and then smash into the water.

1-12-43G. Baton Rouge, Louisiana. At 1500, a Piper L-4 airplane crashed while attempting a forced landing on a highway near Harding Field, Baton Rouge, Louisiana, killing pilot 2Lt. Robert G. Kibbe and engineer Cpl. William F. Doyle. The airplane struck telephone wires and a pole and flipped over on its back while attempting to land on the road. Investigators speculated that the engine lost power due to carburetor ice accumulation.

1-12-43H. King City, California. At 1515, a Ryan PT-22 collided with trees and a barn and crashed while attempting to land at the Army Air Field at King City, California, killing A/C William E. Trombley. The Accident Classification Committee stated, "A/C Trombley had made one unsuccessful attempt to land. On the second attempt he hit the ground in a stall with the left wing low. He immediately hit the throttle and took off in a steep, stalling turn to the left. The airplane traveled approximately 100 yards to the left of the runway. It clipped the top of some trees and hit a barn located at the bottom of the mesa upon which the airport is located. From [examination of the] evidence it seems that the left wing and engine were torn off by this impact, and the remainder of the airplane came to rest against another barn situated to the left of the first one."

1-13-43A. Sacramento, California. At 0010, a North American AT-6A attempting a take-off crashed at Mather Field, Sacramento, California, killing civilian pilot Sol Gold and seriously injuring civilian pilot Ray M. Grimshaw. Investigation revealed that Mr. Gold was piloting the airplane from the front cockpit on the night take-off. The airplane became airborne about halfway down the runway and at an altitude of about 40 feet the engine quit. The airplane continued to climb until it reached an altitude of about 100 feet when the pilot attempted a turn to the right back toward the field. Investigators stated that the turn was made in a nose-high attitude, causing the airplane to stall and fall off to the right. The airplane smashed into the ground nose first. The airplane did not burn. Investigation of the wreckage revealed that the pilot had taken off with the fuel selector valve in the "off" position, starving the engine of fuel and causing it to stop during the take-off.

1-13-43B. Myrtle Beach, South Carolina. At 0532, a Martin B-26B crashed five miles north of the AAF Bombing Range at Myrtle Beach, South Carolina, killing the crew of six. Killed in the crash were: 2Lt. Edward J. Zbyszewski, pilot; 2Lt. Frederick J. Hickman, bombardier; SSgt. Harold F. Boyer, engineer; SSgt. James C. Stewart, engineer; SSgt. James S. Perkins, gunner; Cpl. Robert R. McGee, radio operator. The airplane took off from the airdrome at the bombing range on a routine night training mission. The take-off was normal and the airplane flew away to the north. Two minutes after take-off the B-26 descended into wooded terrain in a bank to the left. The airplane immediately exploded into flames upon impact and the crew killed instantly.

1-13-43C. Cline, Texas. At 1145, a North American AT-6C crashed 15 miles northwest of Cline, Texas, killing A/C Cecil W. Crawford. The student was part of a six-ship student formation flight and was reported missing by the instructor/flight leader. A search was commenced and the student and the airplane were found at about 1330. Investigation revealed that the airplane struck the ground in a vertical dive while in an apparent spin.

1-13-43D. Medio, New Mexico. At 1125 MWT, a Boeing B-17F exploded in mid-air and crashed three miles east of Medio, New Mexico, killing the crew of seven. The airplane was returning to its base at Casper, Wyoming, after it had ferried an air-crew to Pyote, Texas, to ferry another B-17F back to Casper. The B-17 had checked in by radio while flying at 9,000 feet about ten miles north of Tucumcari, New Mexico, and had reported no problems. A short time later the airplane exploded in the bomb bay and radio room area, causing the airplane to break up in mid-air and fall to the ground. Pieces of the B-17 were scattered over a wide area but the main wreckage fell onto a remote mesa 35 miles north of Tucumcari. Five of the crewmembers were hurled from the wreckage at a high altitude and they apparently fell to their deaths. The two pilots were found strapped to their seats in the mangled wreckage. Investigators noted that the airplane was equipped with one 410-gallon bomb bay fuel tank that was filled to capacity. It was speculated that gasoline fumes from this fuel tank had accumulated in the bomb bay and radio room and were ignited by an unknown source. Killed in the crash were: 2Lt. Theodore L. Hinman, B-17 instructor pilot; 2Lt. Robert J. George, pilot; F/O Robert B. Goodman, co-pilot; 2Lt. Miller G. Ashcraft, navigator; TSgt. Maynard J. Marple, engineer; TSgt. Charles F. Zinn, radio operator; SSgt. Thomas L. Hamilton, gunner. Investigation revealed that instructor pilot Lt. Hinman had traded places with another instructor pilot for the flight to Pyote. Lt. Hinman wanted to make the flight to Pyote because it would give him the opportunity to buzz the home of his wife at Clovis, New Mexico, on the return trip. The wife of the instructor stated that a B-17 type airplane had passed over the house at low altitude the morning of the crash.

1-13-43E. Clarksville, Florida. At 1310 CWT, a Martin B-26B crashed three miles south of Clarksville, Florida, killing the crew of six. The airplane had taken off from MacDill Field, Tampa, Florida, on a "Heavy Load Take-Off and Fuel Consumption Test" flight. Investigators speculated that the airplane might have suffered an engine failure and the pilots were unable to make the power and trim adjustments for single-engine flight before losing control of the airplane. Killed in the crash were: 2Lt. Neil B. Snyder, pilot; 2Lt. Sanford O. Schulert, co-pilot; 2Lt. Abner R. Brooks, navigator; SSgt. Waldo W. Lindsey, bombardier; SSgt. Walter E. Chatham, engineer; Cpl. William Buckley, radio operator.

1-13-43F. Byron, Georgia. At 1440 EWT, two Vultee BT-13A type airplanes collided in mid-air one mile northwest of the auxiliary airfield at Byron, Georgia, killing RAF flight instructor Pilot Officer Anthony G. Mole, his student A/C Roy F. Scott, and solo student A/C William Sharlog. A/C Sharlog, in BT-13A #42-21168, was on a routine solo flight that had originated at Cochran Field, Macon, Georgia. P/O Mole and A/C Scott, in BT-13A #42-11257, were practicing landings at the auxiliary field. The airplanes were flying parallel to each other on the crosswind leg of the traffic pattern, with Cadet Sharlog on the outside. A/C Sharlog began a turn to the downwind leg when he collided with the airplane piloted by P/O Mole and A/C Scott.

1-13-43G. Mission, Texas. At 1500, a North American AT-6C crashed seven miles south of Moore Field, Mission, Texas, killing A/C L.Z. Smith, who fell to his death in an unsuccessful parachute jump. The student was on an authorized aerobatic training flight. An eyewitness stated that the airplane was in a spin when it hit the ground but she did not see the student leave the airplane. Examination of the wreckage revealed that the airplane hit the ground in a nose down attitude. The student's body was found some distance from the wreckage with the parachute partially deployed. Investigators speculated that the student put the airplane in a deliberate spin but was unable to recover.

1-13-43H. Willow Run, Michigan. At 1710 EWT, a North American AT-6C crashed eight miles north of Willow Run, Michigan, killing civilian pilots W.A. Pongratz and E.P. Skocdoloe. Both pilots were test pilots for the Ford Motor Company. Several witnesses on the ground observed the airplane falling out of control before it smashed to the ground and exploded into flames. Both pilots were killed instantly.

Investigators speculated that the pilot stalled the airplane at an altitude too low to allow recovery.

1-13-43I. King County, Washington. At 1900, a Lockheed P-38F flying in instrument conditions crashed into Coal Creek Mountain, King County, Washington, approximately four miles east of Mercer Island, killing 2Lt. Mason O. Thompson. The P-38 collided with the mountain at an elevation of about 1,000 feet. The airplane had taken off from McChord Field, Tacoma, Washington, on a cross-country formation flight to Olympia, Washington, and return. Lt. Thompson's airplane was delayed on the ground and he was not able to take off with his squadron. The flight took off without him but soon returned to the field when the weather began to deteriorate. Lt. Thompson ordered the fuel truck to stop refueling his airplane before it was fully serviced, stating that he had enough fuel to get to Olympia. Lt. Thompson, who apparently did not notice that his squadron had returned, took off on the flight to Olympia. His flight leader and the control tower attempted to contact the airplane and order it back to the field but without success. The airplane soon became lost and collided with the mountain.

1-13-43J. Fort Myers, Florida. At an unknown time, aerial gunner Sgt. William Wilmota, 22, Hobart, Indiana, was killed when he was strangled by equipment in a twin fifty-caliber turret he was operating on board a Consolidated B-24 that was participating in gunnery training near Fort Myers, Florida. No other information is available.

1-14-43A. Westover Field, Massachusetts. At 1120, two Republic P-47B airplanes collided in mid-air and crashed near Westover Field, Chicopee, Massachusetts, killing 1Lt. Joseph H. Freeman, Jr. 2Lt. Charles M. Allen was able to safely land his airplane and was uninjured. Investigators stated, "The flight of five airplanes took off from Bradley Field, Windsor Locks, Connecticut, at approximately 1100 EWT, for a local contact flight — Lt. Freeman leading in the #1 position and Lt. Allen flying on the right in the #2 position. Within approximately 20 minutes from the time of take-off, the flight entered the Westover Field local flying area at an altitude of approximately 3,000 feet. The flight thence descended to approximately 1,000 feet altitude over Westover Field, under what technically amounted to instrument conditions, at approximately 1120 EWT. Westover Field, being technically under instrument conditions, with a visibility of approximately one mile, due to fog and haze, had no other airplanes in the air at that time. At this point, Lt. Freeman, piloting P-47B #41-6055, was leading a tight five-ship formation, echelon down and to the right, at approximately 1,000 feet altitude — estimated airspeed 280/300 mph — when the tail of his ship was struck by the left wingtip of the #2 airplane in the formation, P-47B #41-6002, piloted by Lt. Allen. Apparently Lt. Allen at this point was so engrossed in this tight formation just descending into the haze, that he was not aware that the left wingtip of his airplane had struck the tail of Lt. Freeman's ship in front, thereby knocking the tail completely off the latter airplane, with fatal results to its pilot."

1-14-43B. Greenville, South Carolina. At 1410, a Laister-Kaufman TG-4A glider crashed three miles east of Greenville, South Carolina, killing glider student SSgt. Herman Lang and civilian glider instructor Fred W. Edelman. The glider was released from a BT-13A tow airplane at an altitude of 4,000 feet. The glider was seen in a spin to the right a short time later. The glider struck the ground at angle of 45 degrees and was completely demolished. Investigation revealed that the student's right boot was placed too high on the rudder pedal and became jammed in between the rudder pedal and the rudder pedal support structure. When full rudder was applied, the boot became jammed in the rudder pedals, preventing free movement of the rudder controls.

1-14-43C. Hahira, Georgia. At 1705 EWT, two North American AT-6C airplanes collided in mid-air and crashed two miles southwest of Hahira, Georgia, killing 2Lt. John M. Russell aboard AT-6C #41-32487. F/O James Summers and 2Lt Robert V. Rose were able to parachute to safety from AT-6C #41-32342, receiving minor injuries. The airplanes were on a simulated individual combat flight. Investigation revealed that Lt. Russell stalled his airplane above F/O Summers' airplane and fell into it at an altitude of about 9,000 feet, causing both airplanes to fall out of control. Lt. Russell was not able to parachute to safety and was found in the wreckage of his airplane.

1-15-43A. Gulf of Mexico, Florida. At 0915 EWT, a Douglas A-20B collided with an AAF crash boat and the surface of the Gulf of Mexico approximately 19 miles southwest of Page Field, Fort Myers, Florida, killing crewmen Sgt. Alvin J. Brewer and Sgt. Brooks Higginbotham. Pilot 2Lt. Robert H. Bell was injured. The pilot had taken off from Page Field at 0835 on a hooded instrument flight. After flying on instruments for a short period of time the pilot began flying over Fort Myers Beach at very low altitude. At 0910, the airplane was observed to buzz AAF Crash Boat J-2 at an altitude of about 500 feet. Investigators stated, "After Lt. Bell's first approach to the boat, he circled back and approached the boat at an altitude of about 20 or 25 feet. The pilot then made another circle over Sanibel Island and headed out to sea in the direction of the boat and in the general direction of the sun. In a diving attitude, the airplane came down to the boat, leveled off, and struck the water and the stern of the crash boat almost simultaneously, then crashed tail first. The propellers were knocked off first then the entire tail section back of the bomb bay and the left wing. The remainder of the airplane settled in the water and sank in about 12 feet of water." The occupants of AAF Crash Boat J-2, Capt. Walter Alderman, Cpl. William Sweeney and Pfc. Walter Humenck, received minor injuries. The stern of the boat had been knocked off and the boat began taking on water. Captain Alderman gave the boat full throttle and was able to get the boat to the shore. Capt. Alderman then commandeered a civilian rowboat and rescued Lt. Bell.

1-15-43B. Camp Upton, New York. At 1015, a Republic P-47C crashed near Camp Upton, Long Island, New York, killing 2Lt. Richard O. Simmons, Jr. The P-47 was seen in a steep climb and then it fell off in a spin to the right. The pilot recovered momentarily before entering a spin to the left. The pilot was unable to recover and the P-47 smashed to the ground and exploded into flames. The airplane was part of a four-ship flight that had taken off from Farmingdale, Long Island, on an interception mission.

1-15-43C. Whiteface, Texas. At 0955, a Cessna AT-17 suffered the catastrophic structural failure of the starboard wing and crashed ten miles north of Whiteface, Texas, killing A/C Charles R. Bishop and seriously injuring A/C Robert E. Bertelsen. The airplane was flying in the number-three position in a three-ship V-flight on a student formation flight that had originated at Lubbock Army Air Field, Texas. The formation was flying at an indicated altitude of 3,000 feet when the number-three airplane went out of control and crashed. Investigation revealed that the starboard wing failed at the leading edge from the engine nacelle outward. Pieces of the wing were found three miles from the scene of the main wreckage. The airplane broke up progressively as it fell and burst into flames upon impact. The students were killed instantly. Due to a similar accident that occurred 1-2-43, all AT-17 airplanes at Lubbock Army Air Field were temporarily grounded. See 1-2-43D.

1-15-43D. Headland, Alabama. At about 1000, a North American AT-6A and an AT-6C collided in mid-air ten miles north of Headland, Alabama, killing 2Lt. Robert L. Pedigo. RAF P/O Stanley McCreith and 2Lt. R.E. Parker were able to parachute to safety. P/O McCreith stated to investigators, "I took off at 0930 CWT, flying AT-6A [# 41-16402] with 2Lt. R.E. Parker in the rear seat. Following me in trail was 2Lt. R.L. Pedigo in AT-6C [#41-32392]. We were going on a combat mission and climbed to 10,000 feet to start combat. Combat was joined at approximately 0940 and the student made a poor showing. I then instructed him to follow me in long trail through aerobatics. I started aerobatic maneuvers at 9,000 feet and did slow rolls and barrel rolls. He did not seem to be able to hold altitude and stay with the lead ship. Lt. Pedigo always finished the maneuver at a lower altitude and further back than at the start. I led him through a series of steep turns, climbing and diving and also stall turns. These he accomplished satisfactorily. I then made a turn to the right to clear myself and dove to an airspeed of 200 mph. As I made the turn, I looked back and [Lt. Pedigo] was a good safe distance behind me. I pulled the ship up and did a normal Immelmann, rolling out at the top with approximately 90 mph. I then started a steady turn to the right, looking back to see if [Lt. Pedigo] had completed the maneuver satisfactorily. At that moment, I felt a sharp snap somewhat resembling the feeling of a high-speed stall. The ship went into an uncontrollable outside spin. This occurred at about 6,500 feet and I stayed with the ship for about 2,000 feet, trying to bring it out. The student bailed out and I followed him. [Lt. Pedigo] went past me in a flat spiral and crashed just below me." Lt. Pedigo received his commission and his army wings on 1-14-43.

1-15-43E. Kissimmee, Florida. At 1110 EWT, a Douglas P-70 crashed near the Army Air Base at Kissimmee, Florida, killing pilot 2Lt. James T. Jensen and passenger TSgt. Charles R. Badgett. Passenger TSgt. Ralph F. Wakerly was seriously injured. The Accident Classification Committee stated, "[The] pilot, while flying aircraft on a local test flight, lost the use of the left engine due to lack of fuel pressure from a fuel line leak. The left engine propeller was feathered and the pilot approached the field for a landing. After overshooting the field the pilot retracted the landing gear but erred in turning into his dead engine and attempting to climb in a left turn without sufficient airspeed to insure good control before attempting to gain altitude. As the airplane lost speed and started to settle, the pilot continued to pull up the nose until the airplane stalled and crashed."

1-15-43F. Jones Beach, New York. At 1500, a Republic P-47C crashed two miles east of Jones Beach, Long Island, New York, killing 2Lt. Richard W. Nellis. The airplane took off from Mitchel Field, Long Island, on an individual interceptor mission and was flying at high altitude when the engine caught on fire. The pilot was ordered to point the airplane toward the sea and then to bail out. The pilot bailed out and the airplane crashed into the ground and exploded near the shore. Winds carried Lt. Nellis over the water and he descended into the sea. He died of exposure before he could be rescued. His body was retrieved about an hour after he bailed out.

1-15-43G. Tallahassee, Florida. At 1528 EWT, a Republic P-47C crashed 15 miles northwest of Dale Mabry Field, Tallahassee, Florida, killing 2Lt. Andrew T. Sevok. Investigators stated, "This pilot was doing aerobatics at 7,000 feet. As he was recovering at the top of an Immelmann the engine section caught on fire. The pilot in an accompanying ship noticed fire coming from the waste gates, and the ship appeared to have on full power. The ship went into a spin. The pilot got out; apparently he pulled the ripcord too soon and the shroud lines of the parachute were cut by the horizontal stabilizer. The ship continued in a spin and crashed into the ground. The pilot's body was found 30 feet away."

1-15-43H. Yelm, Washington. At 1415, a Lockheed P-38G crashed two and one-half miles east of Yelm, Washington, killing 2Lt. Daniel M. Eby, who was fatally injured in an unsuccessful parachute attempt. Investigators stated, "Lt. Eby was flying in a formation of [six] P-38s on a high altitude mission when he suddenly left the formation and was seen to fly his airplane down through a hole in the overcast by another pilot in the flight. At that time his airplane was in no apparent trouble, although his instruments may have indicated engine trouble. The next time the airplane was seen was by ground observers, it was still above 10,000 feet but a streak of smoke was trailing it. The pilot then bailed out, his parachute opened, and he was carried nearly eight miles by the strong winds before approaching the ground. In the meantime his airplane plunged earthward, smoking badly. As he neared the ground, observers believed they saw Lt. Eby pulling on the shroud lines of his parachute in an effort to influence his direction, but the wind evidently dragged him through a wooded area, and dashed him against a tree or trees. A search was conducted for him and after about two hours he was found dead in a partially burned and logged hillside."

1-15-43I. Salina, Kansas. At 1923, a Consolidated B-24D attempting a take-off crashed at the Army Air Base at Salina, Kansas, killing five crewmembers and seriously injuring five others. The B-24 was taking off on a cross-country flight to DeRidder, Louisiana, en route to an overseas station. The airplane gained about 75 feet of altitude before going into a bank to the left and crashing into the ground and erupting in flames. Seriously injured in the accident were: Capt. Frank H. Beales, pilot; 2Lt. Norman R. MacLeod, co-pilot; MSgt. Arthur A. Schneider, engineer; SSgt. Willis A. Clark, assistant radio operator; Lt. Col. Harvey W. Regan, passenger. Killed in the accident were: 2Lt. Robert I. Van Horn, navigator; 2Lt. Edward G. Riley, bombardier; SSgt. Dale V. Brown, assistant engineer; SSgt. Robert L. Grace, radio operator; SSgt. Francis N. Ayers, gunner.

1-15-43J. Barksdale Field, Louisiana. At 1948, a Martin B-26B crashed while attempting a take-off at Barksdale Field, Shreveport, Louisiana, killing five crewmembers and seriously injuring two others. Investigators stated, "The pilot in making a night training flight was in the process of a routine take-off and pulled the airplane off the ground prior to the time that the airplane would become safely airborne. As a result of this action a violent stall ensued, whipping the airplane to the left. Marking of the propellers cutting the ground indicated that the pilot maintained full power on after a point of the stall." Killed in the crash were: 2Lt. William E. Mitchell, pilot; F/O Frank J. Krejoi, Jr., co-pilot; 2Lt. Weldon C. Randall, navigator; 2Lt. Melvin W. Kislingbury, bombardier; Thomas G. Pavlinich, engineer. Radio operator SSgt. Clyde S. Maddox and gunner Sgt. Phillip L. Brinson were injured in the accident.

1-15-43K. Tallahassee, Florida. At 2230 EWT, a Republic P-47C crashed two and one-half miles northwest of Dale Mabry Field, Tallahassee, Florida, killing 2Lt. Gerald D. Bullock. The airplane had taken off from Dale Mabry Field with 11 other P-47s on a routine night training flight. The weather was reported as clear until about 2200 when an overcast began moving into the area from the south. All airplanes from Dale Mabry Field were ordered back to the field in anticipation of the poor weather and limited visibility. The flight came into the field and had begun landing. Five of the P-47s had landed before a transient B-26 broke into the order of landings and the remainder of the P-47 flight had to wait until this ship landed. Lt. Bullock, flying the twelfth P-47, radioed the tower and stated that he was above the overcast and that he could not find the field. A senior pilot who was a member of the flight found Lt. Bullock and attempted to guide him down through the overcast and to the field. When the pilot broke out of the overcast at 400 feet, Lt. Bullock was missing. Lt. Bullock's airplane was heard passing over the field and was then heard to crash moments later. Examination of the wreckage indicated that the airplane struck the ground in a dive, exploding into flames upon impact and killing the pilot instantly.

1-15-43L. Gowen Field, Idaho. At 2323 MWT, two Boeing B-17E airplanes collided in mid-air ten miles southwest of Gowen Field, Boise, Idaho, killing ten fliers aboard B-17E #41-9181. Gunner Sgt. Joseph R. Moscicki managed to parachute to safety but was seriously injured. The twelve fliers aboard B-17E #41-2603 were uninjured and were able to land safely at Gowen Field. Investigators stated, "It appears that 41-9181, the airplane that crashed, was approaching 41-2603 at an angle of approximately 45 degrees to the left, head-on [at an altitude of 7,000 feet]. From further interrogation of the members of the crew of 41-2603, it appears that the passing light of 41-9181 was not [illuminated] at the time of the crash. From the interrogation of the pilot, co-pilot, navigator and bombardier on 41-2603 immediately after the crash, the passing light on 41-2603 was known to be on. The fuselage of 41-2603 was cut halfway through at the tail wheel station by the propellers of 41-9181, and the lower portion of the fuselage to the rear was torn away. One-third of the right elevator was torn away by the propellers of 41-9181. From these marks, the committee presumes that after the pilots [had observed B-17 #41-9181 coming at them from the left front quarter at angle of 45 degrees, pulled up] to avoid collision, the propeller marks on the tail section of this ship indicated that 41-9181 struck 41-2603 in an approximate direction of 45 degrees from the [port side and into the] left rear. Further marks on the tail of the fuselage of 41-2603 lead the committee to presume that 41-9181 had struck the tail of 41-2603 in such a manner as to cave in the upper section of the pilot's compartment of 41-9181, thereby temporarily stunning [or killing] the pilot and co-pilot of 41-9181 and allowing the ship to fall out of control." Killed in the crash of B-17E #41-9181 were: 2Lt. Paul C. Painter, pilot; 2Lt. William W. Reed, Jr., co-pilot; 2Lt. Raymond A. Hayes, navigator;

2Lt. Howard J. Norton, bombardier; SSgt. Rollen H. Eubank, engineer; Sgt. Pasqualino Patri, assistant engineer; SSgt. William T. Griffiths, radio operator; SSgt. Raymond A. Etherton, assistant radio operator; SSgt. Christian F. Wienhold, assistant radio operator; Sgt. Arthur P. Meyer, gunner. Uninjured in the crash landing of B-17E #41-2603 were: 2Lt. Charles M. Honaker, pilot; 2Lt. Charles D. Ferguson, co-pilot; 2Lt. Leroy Ford, navigator; 2Lt. Raymond E. Crowell, bombardier; TSgt. Clarence L. Campbell, engineer; SSgt. Claston D. Campbell, assistant engineer; SSgt. Isabel O. Garcia, assistant engineer; Sgt. Lemer C. Brunk, radio operator; SSgt. George L. Rose, assistant engineer; Sgt. James F. Jones, gunner; SSgt. Robert D. Tierney, gunner; Sgt. William D. Finkbiner, gunner.

1-16-43A. Petersburg, Virginia. At 1505, two Curtiss P-40F airplanes collided in mid-air five miles northeast of Petersburg, Virginia, killing F/O Clayton W. Huntsman aboard P-40F #41-14127. Capt. Roy B. Caviness was able to safely land heavily damaged P-40F #41-14131 at the Richmond Army Air Base, Virginia. Investigation revealed that Capt. Caviness was leading a three-airplane flight on a formation training flight. The formation was flying at an altitude of approximately 9,500 feet when Capt. Caviness gave the signal to echelon the flight to the right. Once proper spacing was achieved Capt. Caviness began a peel-off maneuver to the left, leveling off at about 6,000 feet. Capt. Caviness observed a dark object pass beneath his left wing. An instant later F/O Huntsman pulled up into the airplane of Capt. Caviness. Apparently F/O Huntsman overshot the lead airplane and pulled up in front of it. Capt. Caviness' propeller cut into the tail section of F/O Huntsman's airplane, damaging the empennage to such an extent to cause its progressive failure. The tail section fell away from the airplane in pieces, sending the P-40 out of control. The P-40 then spun to the ground and exploded into flames; F/O Huntsman was not able to leave the airplane and he was killed in the crash. Capt. Caviness was able to regain control of his damaged airplane and safely land at base.

1-16-43B. Big Spring, Texas. At 1705, a Vultee BT-13A crashed at Big Spring, Texas, killing pilot 2Lt. Souron S. Avedisan and passenger TSgt. Robert K. O'Bert. The airplane had taken off from Hensley Field, Texas, on a navigation flight to Pecos, Texas, and return. Apparently the airplane suffered an engine failure and the instructor pilot attempted an emergency forced landing on a gravel road. The airplane struck three power lines that were strung across the road. The airplane then smashed to the ground and erupted into flames, coming to rest about 500 feet from where the power lines were struck. A civilian who was driving an automobile east on the gravel road stated that the airplane was landing to the west and was coming straight at him, descending with the engine idling before striking the power lines and crashing near him.

1-16-43C. Tenaha, Texas. At 2030, a Martin B-26A suffering an in-flight fire was abandoned and crashed three and one-half miles southeast of Tenaha, Texas, killing engineer Cpl. Donald W. Katzenberger, 22, Mt. Carroll, Illinois, who fell to his death in an unsuccessful parachute jump. Crewmembers injured parachuting to safety were: 2Lt. Gus Weiser, 20, Monterrey, California, pilot; 2Lt. Rodney L. McCallister, 23, Wilmington, Delaware, co-pilot; 2Lt. William C. Sawyer, 21, Portland, Maine, navigator; SSgt. Raymond L. Downing, 22, Montpelier, Idaho, radio operator; Sgt. Ryan C. Ringler, 22, Hooversville, Pennsylvania, gunner. The airplane had taken off from Barksdale Field, Shreveport, Louisiana, on a navigation flight to Wichita Falls, Texas, to San Angelo, Texas, to San Antonio, Texas, and return. While flying at an altitude of about 6,000 feet msl and about one hour out of San Antonio the port engine caught fire so the pilots shut down the engine, feathered the propeller and trimmed the airplane for single-engine flight. The B-26 dropped to 3,000 feet before the pilots were able to put out the fire with the engine fire extinguisher. The pilot ordered the crew to put on their parachutes and to stand by for the bail out alarm. The pilot ordered the engineer to jettison the bomb bay fuel tank. The engineer dropped the tank but neglected to close the bomb doors, causing drag on the airplane. The airplane quickly fell to 2,000 feet and the pilot, unable to hold altitude, gave the bail out signal. The crew parachuted to the ground near a farmhouse. The B-26 smashed to earth and exploded about one-quarter mile away. Cpl. Katzenberger fell to his death when his parachute failed to open. His body was found the next day. Investigation revealed that Cpl. Katzenberger was struck in the head by part of the airplane as he attempted to exit.

1-16-43D. Elmdale, Texas. At a time between 2128 and 2148 CWT, a Beech AT-11 flying in instrument conditions collided with a tree and crashed one and one-half mile north of Elmdale, Texas, killing pilot 1Lt. Walter F. Huffman and student bombardiers A/C George F. Griffin and A/C Eldon C. Hemingway. Investigation revealed that the airplane was attempting to fly under a ceiling reported to be at 500 to 800 feet. The airplane struck a tree with the port wing and then cartwheeled into the ground. Wreckage and body parts were scattered over a considerable area. The pilot was found outside of the wreckage along the line of flight, indicating that he attempted to leave the airplane. The bodies of the cadets were found in the wreckage.

1-16-43E. Lake Charles, Louisiana. At 2318, a Consolidated B-24D suffering an in-flight fire was abandoned and crashed in the swamps 14 miles northeast of Lake Charles, Louisiana, killing pilot 2Lt. Benjamin E. Angell. Crewmembers injured parachuting to safety were: 2Lt. Wallace T. Moore, co-pilot; 2Lt. Edmond N. Haley, bombardier; 2Lt. Ted Greziak, navigator;

TSgt. Jack W. Butterbaugh, engineer; TSgt. Foster W. Wilson, radio operator; SSgt. Donald E. Lathrop, assistant radio operator; SSgt. Eugene E. Gaster, gunner. The airplane had flown into Tarrant Field, Fort Worth, Texas, the night before and was on the leg back to its station at DeRidder, Louisiana, when the engine instruments began giving fluctuating readings. It was noted that the airplane was delayed on the ground the night before because of fluctuating engine readings. Oil was found on the port side of the empennage and the line chief determined that it apparently came from the propeller domes. The gaskets were changed and instrument tubing and wiring were replaced and tested. The fluctuations in engine instrument readings seemed to be corrected and the pilots proceeded with the flight to DeRidder. Thirty minutes after take-off the engine instrument readings were fluctuating wildly. Oil pressure apparently began dropping dramatically in the number-one engine. An hour before the crash, the cylinder head temperature began rising and the pilots shut down the number-one engine and feathered the propeller. Cowl flaps were opened and the remaining engines seemed to cool off. A crewmember in the rear of the airplane reported that the number-four engine was on fire. The on board engine fire extinguisher was ineffective. The pilot and the engineer were unable to unfeather the number-one engine. Moments later, the number-two engine exploded into flames and the pilot ordered the crew to bail out. Surviving crew stated that airplane was trailing sparks and flames as it went down.

1-17-43A. Coalinga, California. At 0825, a Vultee BT-15 crashed six miles east of Coalinga, California, killing pilot 2Lt. Tillman B. Hendricks and seriously injuring pilot 1Lt. Oiva Maki. The pilots were flying the airplane from Lemoore, California, to Coalinga where they were to instruct students. Apparently the airplane was flying at low altitude and then attempted a steep 180-degree peel-off. Lt. Hendricks was unable to recover the airplane in time to avoid colliding with a small hill. Investigators noted that the air was "rough" near the scene of the accident.

1-17-43B. Moultrie, Georgia. At 1350 EWT, a Bell P-39D crashed five miles east of Spence Field, Moultrie, Georgia, killing 1Lt. Simon Rothstein, who fell to his death in an unsuccessful parachute attempt. Investigation revealed that the engine had failed and the pilot attempted to exit the airplane but was dragged to his death when his parachute opened prematurely and fouled on the airplane. It was noted that the pilot exited the airplane from the left side door and not from the right side, which is recommended.

1-17-43C. Mission, Texas. At 1445, a Martin B-26B crashed eight miles NNW of Moore Field, Mission, Texas, killing the crew of seven. Killed in the crash were: 2Lt. Luther N. Osborne, pilot; F/O G.S. Linder, co-pilot; 2Lt. P.T. Movelle, navigator; Sgt. A.D. Snyder, crewmember; Sgt. J.W. Lloyd, crewmember; Sgt. D.S. Bivens, crewmember; Cpl. L.L. Lewark, crewmember. The airplane had taken off from Moore Field on a navigation flight to its home station at Drane Field, Lakeland, Florida, via Duncan Field, San Antonio, Texas. A cowboy who was working in the area stated that he observed the airplane enter a dive toward the ground from an altitude of 200 feet. The airplane continued in this attitude until striking the ground, exploding into flames upon impact and scattering wreckage over a considerable area. Investigators could not determine the cause of the accident.

1-17-43D. Alamogordo, New Mexico. At 1500 MWT, a Consolidated B-24D crashed ten miles north of the Army Air Base at Alamogordo, New Mexico, killing 11 crewmembers and injuring radio operator SSgt. Richard Pugh, who parachuted to safety. The Accident Classification Committee stated, "Pilot was flying in the No. 4 position in a formation of four airplanes when the leader of the flight flew into a cloud. The pilot of the [subject] airplane became confused, attempted to slow the airplane down through throttling the engines, went into a spin or tight spiral from which he was unable to recover for some time; then in the recovery process the airplane had attained such a speed that abnormal stresses were [imposed], resulting in the loss of the left wing tip, rudder and elevator surfaces, resulting in total destruction of the aircraft." Killed in the crash were: 2Lt. James W. McClure, pilot; 2Lt. James D. Milligan, co-pilot; 2Lt. Daniel W. Phillips, navigator; 2Lt. Donald A. Irwin, bombardier; SSgt. William B. Dyson, engineer; Sgt. Chester C. Shoemaker, assistant engineer; Sgt. Robert T. Cole, assistant radio operator; Sgt. Melvin A. Plumber, gunner; Sgt. Leroy H. Wilkins, passenger; T/5 Steve Voytko, passenger; T/5 William J. Lewis, passenger.

1-17-43E. Gila Bend, Arizona. At 2045 MWT, a Cessna AT-17B crashed about 60 miles west of Gila Bend, Arizona, killing A/C Fred R. Thompson and injuring A/C Waldo I. Weeks, who parachuted to safety. Investigators stated, "Cadet Thompson and Cadet Weeks took off from Yuma Army Air Field, Yuma, Arizona, at approximately 2015 on a routine student navigation flight. About 60 or 70 miles out, they hit some clouds but instead of turning around and returning to Yuma Army Air Field they went on instruments and flew into the clouds. After they had been in the clouds for approximately five minutes they turned around and let down to an altitude of 5,000 feet, but were still in the clouds. They then tried to climb over the clouds, but at an altitude of approximately 10,000 feet they hit very turbulent air while still on instruments at which time they lost control of the airplane and Cadet Weeks was thrown clear of the ship when he pulled the emergency door release. [He was able to parachute to safety.] Cadet Thompson did not clear the airplane and went in with the wreckage."

1-18-43. Bassett Peak, Arizona. At 0327 MWT, a Consolidated B-24D crashed into Bassett Peak in the Galiuro Mountains 43 miles northeast of Tucson,

Arizona, killing the crew of eleven. The airplane had taken off from Davis-Monthan Field at 0037 MWT on a routine bombing training mission and never returned. The bodies of the crew and the wreckage of the B-24 were found 3-7-43 by a CAP airplane. Ground parties did not reach the airplane until 3-10-43. Examination of the wreckage indicated that the airplane collided with the peak while in full flight with all four engines operating at cruise speed. Killed in the crash were: 2Lt. John L. Wasson, pilot; 2Lt. Charles A. Beaucond, co-pilot; 2Lt. Frederick E. Neal, bombardier; 2Lt. Joseph Ferio, bombardier; Pvt. Henry C. Peterson, engineer; SSgt. Robert E. Story, assistant engineer; Pvt. Murty J. Huckaby, radio operator; Sgt. Alman L. Porter, assistant radio operator; Sgt. Stephen A. Safko, gunner; Sgt. Henry P. Pelletier, gunner; SSgt. Ernest M. Breier, gunner.

1-20-43A. Taft, California. At 0845, a Vultee BT-13A crashed 18 miles east of Gardner Field, Taft, California, killing student A/C Ambrose J. Walsh. Investigation of the wreckage indicated that the airplane had smashed into the ground in a spin to the left. Unknown civilians who had been traveling in an automobile had witnessed the crash and reported it to two telephone repairmen who were working in the area. The civilian motorists then drove away. AAF investigators could not locate the witnesses.

1-20-43B. Gulf of Mexico, Florida. At 1430, a Curtiss P-40K dove into the Gulf of Mexico near Lido Beach, Florida, killing 2Lt. Lawrence Gross. The airplane had taken off from the Army Air Base at Sarasota, Florida, on a local aerobatic flight. Witnesses observed the airplane in a vertical dive just prior to it striking the water. There was no spin. The cause of the accident was undetermined.

1-20-43C. Gunter Field, Alabama. At 2105, two Vultee BT-13 airplanes collided in mid-air two miles northwest of Gunter Field, Montgomery, Alabama, killing 2Lt. Michael A. Wood. A/C Richard R. Miller was uninjured parachuting to safety from his airplane. Investigators stated, "Lt. Wood, in BT-13 #40-1031, took off to fly a regular traffic pattern and practice night landings and take-offs. A/C Miller, in BT-13 #40-878, took off on the same mission as soon as Lt. Wood had cleared the field and had made his turn onto the crosswind leg. Both pilots proceeded as scheduled, except that Lt. Wood flew wide of the normal downwind leg. A/C Miller was turning on his downwind leg at the normal distance. Apparently Lt. Wood had already leveled off after turning onto the base leg of his approach for landing when A/C Miller started to turn onto his base leg. It was at this point that the airplanes collided. Normal altitude prescribed for the traffic pattern is 700 feet above the ground except after the last turn before landing and while gaining altitude after the take-off. A/C Miller immediately resorted to his parachute after the airplanes collided. Apparently Lt. Wood made no effort to use his parachute."

1-21-43A. Newport, Arkansas. At 0921, a Vultee BT-13A crashed 25 miles northeast of Newport, Arkansas, killing pilots 2Lt. William B. Campbell and 2Lt. Frederick G. Lake. Investigators stated, "It was apparent from the wreckage that the airplane had made a turn at a low altitude in preparation for a simulated forced landing, stalled in a steep turn, the stall being aided or caused by low airspeed and gusty wind. It appeared that approximately one-half of a turn of spin to the right was made prior to contact with the ground, and that the pilots had no opportunity to recover from a spin at that altitude. Examination showed all controls to be normal for a simulated forced landing—flaps at 20 degrees, propeller control in low pitch, gas tank on reserve."

1-21-43B. Muroc, California. At 0909, a Lockheed P-38E crashed 30 miles northeast of the Army Air Base at Muroc, California, killing 2Lt. Carl H. Drake. The airplane was part of a four-ship flight that had taken off from the base on a local formation flight. The formation was flying in trail when the leader began a shallow dive at 8,700 feet above ground level. The flight followed and when the dive approached the overcast, which topped out at about 8,000 feet, the leader pulled up and executed a slow roll. The second and third ships slow rolled. The fourth airplane, piloted by Lt. Drake, slow rolled and then fell into the overcast in a side-slipping attitude. Apparently Lt. Drake stalled the airplane after performing the slow roll and was unable to recover before striking the ground. The airplane exploded into flames upon impact.

1-21-43C. Circleville, Ohio. At 1415 EWT, a Bell P-39N crashed three miles northeast of Circleville, Ohio, killing 2Lt. Joseph L. Harrington. The pilot had taken off from Lockbourne Army Air Field, Columbus, Ohio, at 1350 to practice landings at the field. Immediately after take-off the pilot flew to an area south of the field and began to fly low over the farm of John Heffner. The airplane had made a low altitude pass at the farmhouse and pulled up steeply, stalling at the top and falling off to the left from an altitude of 500 feet. The airplane crashed into a grove of trees and burst into flames.

1-21-43D. Sweetwater, Texas. At 1630, a Fairchild PT-19A crashed 12 miles north of Avenger Field, Sweetwater, Texas, killing civilian instructor Felix H. McCain, 21, Sweetwater, and student A/C Harvey F. Cornwell, 21, Detroit, Michigan. The student and instructor were practicing simulated forced landings. The airplane flew over the east end of a large field in a westerly direction at an altitude of about 200 feet. The wind was from the west at 30 mph. After the airplane had crossed the western part of the field it apparently stalled. The airplane struck the ground in a nose-down attitude before the pilots could recover, killing them instantly.

1-22-43A. Kremlin, Oklahoma. At 1030, two Vultee BT-15 airplanes collided in mid-air and crashed two miles south of Kremlin, Oklahoma, killing two aviation cadets. A/C William O. Heim was killed in the crash of BT-15 #41-10129 and A/C John H. Green was killed while attempting to parachute from BT-15 #41-10101. Investigators stated, "Aviation Cadet John H. Green, while engaged in a routine formation training flight, attempted to execute a cross-under from a V-formation to left echelon unsuccessfully. In arriving at the number-three position in the left echelon, he evidently mistook the lead airplane for the other wingman and pulled into position on the left wing of the lead airplane just in front of the other wingman already there, damaging the leading edge of the left wing on the other airplane. After striking the leading edge of the left wing of the other airplane, he slowed his ship sufficiently so that his right wing rested on the on the left wing of the other airplane with the result that when he broke formation by turning to the left, the other airplane was forced to turn to the left also. After separating from the other airplane, he was seen to bail out at a sufficient height for his parachute to open properly. His parachute landing was observed to be normal, but the back of his skull was found to be crushed at the base, [an injury which] could be only have been received as he was leaving the airplane before opening his parachute. This fact is born out by the following circumstances: The parachute was opened normally and received no damage whatsoever, yet the parachute pack was splashed with drops of blood in such a manner as would be impossible after the pack was opened."

1-22-43B. Douglas, Arizona. At 1105, two Curtiss AT-9 airplanes collided in mid-air and crashed at the auxiliary airfield ten miles SSW of Douglas, Arizona, killing A/C Thomas E. Anderson and A/C Kenneth A. Anderson aboard AT-9 #41-11960. Instructor Jerome D. Kesterson and student A/C Robert J. Alsop were seriously injured in the crash of AT-9 #41-5786. Investigators stated, "Both airplanes proceeded to Auxiliary Field No. 2 for the second period of flying. The wind had shifted and was splitting the Northwest-Southeast and the Northeast-Southwest runways. Lt. Kesterton's airplane made an approach to land southwest on the Northeast-Southwest runway, and the student airplane made an approach to land southeast on the Northwest-Southeast runway. The two airplanes collided while approximately 30 feet in the air at the intersection of the Northeast-Southwest and the Northwest and Southeast runways."

1-22-43C. Union City, Tennessee. At 1345 CWT, a Stearman PT-17 on a routine training flight crashed one mile south of Embry Riddle Field, Union City, Tennessee, killing student A/C Clyde H. Harell and civilian instructor Robert Everett. Investigation revealed that the airplane stalled at an altitude of approximately 500 feet while attempting an emergency forced landing after suffering an apparent engine failure. Investigators were not able to determine the cause of the engine failure.

1-22-43D. Elmont, Virginia. At 1510, a Curtiss P-40F dove into the ground three miles west of Elmont, Virginia, killing 2Lt. William C. Gray. The subject airplane took off from the Richmond Army Air Base, Richmond, Virginia, at 1450 on a two-ship tactical formation flight. After practicing crossover turns at an altitude of approximately 5,000 feet, the flight leader led Lt. Gray into a steep dive. The flight leader pulled out of the dive at an altitude of 100 feet above ground level. Lt. Gray was not able to pull out of the dive before crashing into a wooded area at high speed in a near level attitude. The P-40 exploded violently into flames upon impact, scattering wreckage over a wide area; the pilot was killed instantly.

1-22-43E. Langley Field, Virginia. At 1620, a Republic P-47C attempting a landing collided with a Civil Air Patrol Culver Cadet airplane (NC-32470) on the active runway at Langley Field, Hampton, Virginia, killing the Culver pilot 1Lt. Clarence J. Parnell, AAF Material Command, Wright Field, Ohio. The pilot of the P-47, 2Lt. Leslie P. Cles, was not injured. The P-47 had arrived at Langley Field flying in the number-two position of a three-ship flight. The pilot of the subject P-47 had overshot on his initial landing attempt and had to go around, entering the pattern for his second landing attempt. At about this time the Culver airplane entered the traffic pattern about midway on the downwind leg. The P-47 quickly caught up with the Culver and both airplanes began a simultaneous final approach. Control tower personnel observed that an unsafe situation was developing and they signaled the P-47 to abort the landing with the red light signal and by radio. The P-47 pilot did not acknowledge either signal and continued with the landing attempt. The P-47 was rolling on the runway when the Culver landed in front of it. The P-47 overran the Culver, slicing through the airplane with the propeller and smashing it to bits. Investigators noted that Lt. Cles did not request landing instructions or proper permission to land while he was on his go-around. The P-47 received minor damage to the propeller, the starboard section of the engine cowling and the leading edge of the starboard wing. Investigators suggested that slower-type airplanes and high-performance airplanes use separate airfields.

1-23-43A. Midland, Texas. At 0355, a North American B-25D flying in adverse weather conditions crashed approximately 40 miles north of Midland, Texas, killing six crewmembers. Killed in the crash were: 2Lt. Edward W. Butler, Jr., McClelland, Arkansas, pilot; A/C Philip F. Deno, 21, Santa Paula, California, student bombardier-navigator; A/C Robert E. Paternoster, 25, Fairbury, Illinois, student bombardier-navigator; Pvt. J.V. Alexander, 25, Los Angeles, California, crewmember; Pvt. H.F. Trilschler, 36, Santa Barbara, California, crewmember; Pfc. John E. O'Connor, 23, Winthrop, Massachusetts, crewmember. The airplane had disappeared after taking off from Tinker Field, Oklahoma City, Oklahoma, on a student navigation flight to its home station at Roswell Army Air Field, Roswell, New Mexico. After a massive search effort, the airplane wreckage was found on 1-27-43. Investigators speculated that the pilot had lost control of the airplane while flying in instrument conditions at an altitude of about 6,000 feet. Examination of the wreckage revealed that the airplane had smashed into the ground at high speed at an angle of about 60 degrees, exploding violently into flames upon impact. Both engines and propellers were buried several feet into the ground. Weather in the area was reported as low ceilings, light rain, severe turbulence with light to moderate icing conditions.

1-23-43B. Dallas, Texas. At 1000, a North American AT-6C collided with trees and crashed at the intersection of Cedardale Drive and Floyd Branch Creek, ten miles south of Love Field, Dallas, Texas, killing pilot 2Lt. Willie C. Voight. Investigators stated, "The conclusions regarding the crash are based upon the sworn statements of witnesses, two of whom resided approximately one-half mile from the scene of the accident and four who were engaged in surveying a highway. According to these witnesses the airplane dived at the surveyors a total of six times, passing over their heads barley above the tree tops which were not more than 15 feet high at that point. Each time the airplane dived the men would scramble out of the road and seek cover. The second time the airplane dived at the group it struck the top branches of a tree beneath which one of the surveyors was seeking shelter. This apparently had no damaging effect upon the airplane as the pilot continued in wide, low circles and dived at the men four additional times before his airplane finally hit the trees at a terrific rate of speed, cutting a wide swath through them for over 200 yards before it finally came to a stop and burned. Wreckage was strewn along the path through the trees."

1-23-43C. Tahoka, Texas. At 1205, a Beech AT-11 crashed while attempting a forced landing six miles north of Tahoka, Texas, killing pilot 2Lt. Eugene F. Moriarty. Bombardier instructor 2Lt. Willard J. Hall, student bombardier A/C George D.A. Griffin and student bombardier A/C Edwin A. Stern parachuted to safety and were uninjured. The airplane had taken off from Big Spring, Texas, on a bombardier training flight when it became lost. Eventually the pilot ordered the crew to bail out and he attempted an emergency forced landing when the airplane ran out of fuel. Investigators stated, "It appears the pilot made a normal approach for [an emergency forced] landing but pulled the airplane up to avoid a telephone line, which, with the increased weight of the bombs, caused the airplane to stall in. It is believed the wheels were up, as the engines scooped out holes in the hard earth approximately one and one-half feet deep at the first point of impact. The airplane, after hitting [the ground], bounced forward about 75 feet without leaving a trace on the ground. It then hit the ground again, flipped over on to its back and came to rest about 75 yards beyond the initial point of impact."

1-23-43D. Will Rogers Field, Oklahoma. At 1215, a Douglas DB-7B and an A-20B collided in mid-air two miles southeast of Will Rogers Field, Oklahoma City, Oklahoma, killing three crewmembers aboard the DB-7B. Killed in the crash were pilot SSgt. Galileo J. Morganti, gunner Sgt. Robert R. Brock and gunner Sgt. Hewitt A. Maloy. The A-20B pilot, SSgt. Frank H. Giese, was able to safely land his damaged airplane and was uninjured. SSgt. Giese was flying the lead airplane of a three-ship V-flight with SSgt. Morganti flying in the number-three position. The formation lined up with the runway as it flew over the field at 3,000 feet in preparation to peeling off for a landing. SSgt. Giese signaled for a change to a right echelon formation. SSgt. Giese stated to investigators, "Suddenly I felt a severe jar and my left wing was forced up. I glanced to the left and there was a ship under and partly forward of my left wing. Immediately I pulled my ship up and to the left, and glanced back to my left to see what had happened to SSgt. Morganti's ship. It wasn't in sight. I figured that he must have gone down and to the right. I noticed my left wingtip was bent so rather than reform the formation I called the tower and told them to tell [Morganti] and [2Lt. Samuel T. Batson] to land. I was turning on my base leg about this time and let my wheels down and came in and made a normal landing. I turned off from the end of the runway and noticed another ship had just landed behind me. I turned up to the line and parked and noticed that the ship that had followed me in was [Lt. Batson]. I got out of my ship and looked at the wingtip. Major Dick and several other officers asked me what happened. I told them. They told me to come into the operations office with them. As I was walking away from my ship I asked where SSgt. Morganti was. Someone [asked] me, 'See that smoke? That's him.' That was the first I knew of SSgt. Morganti having cracked up."

1-23-43E. Great Falls, South Carolina. At 1415, a North American B-25C crashed two miles south of Great Falls, South Carolina, killing the crew of six. Killed in the crash were: 2Lt. Ralph F. Bladell, pilot; 2Lt. Virgil L. Blevins, co-pilot; 2Lt. Ralph L. Knepper, bombardier; SSgt. Percy J. McFaddin, engineer; SSgt. Jay G. Alma, radio operator; SSgt. Franklin McCue, gunner. The subject airplane was flying in the number-three position in a three-ship formation that had taken off at 1357 EWT from Columbia Army Air Base, Columbia, South Carolina, on a tactical formation training flight. Soon after take-off, the flight encountered instrument conditions. The flight leader ordered both airplanes to go on instruments and to proceed to Charlotte, North Carolina, where the airplanes would rendezvous. The flight leader ordered the number-two airplane to climb to 1,500 feet and the number-three airplane to climb to 2,000 feet. The other two airplanes landed safely at Charlotte, but the number-three airplane never arrived. Investigators stated, "The pilot evidently left the formation as instructed by the flight leader but instead of climbing to specified altitude decided to stay below the overcast. He was seen flying just above the treetops at various points along the valley and just as he evidently pulled up to clear the high tension wires before he crashed. It is assumed he again pulled up to clear the top of the hill on which he crashed and stalled the airplane. The airplane dove in at about a 60-degree angle to the ground as indicated by the angle at which two trees were sheared off at the point of crash. All parts of the airplane were at the spot of the crash."

✓**1-23-43F. Dutchtown, Missouri.** At 1410, a Fairchild PT-19A crashed one mile south of Dutchtown, Missouri, killing civilian instructor Carl J. Kallfelz and student A/C John R. Frankie. Investigation revealed that the PT-19A was performing S-turns along a road and was flying north on the west side of the road when it had maneuvered violently to avoid a mid-air collision with a Piper J-3 Cub that was operating in the area. Both airplanes were flying at about 500 feet above ground level. The J-3 was also flying on the west side of the road but was flying south. The PT-19 entered a steep bank to the right and passed beneath the J-3 before stalling and smashing into the ground. The falling airplane killed two cows when it struck the terrain. The J-3, piloted by civilian instructor Charles Boyd and civilian student Richard Taff, was able to land safely at Cape Girardeau, Missouri.

1-23-43G. Taft, California. At 1310, a Vultee BT-13A crashed one mile south of Gardner Field, Taft, California, killing instructor 2Lt. George L. Markham, Jr. and student A/C Joseph D. Almond. The airplane was flying in the number-two position of a three-ship V-formation that was preparing to land at Gardner Field. The Aircraft Accident Classification Committee stated, "Aviation Cadet Almond was on the controls of the airplane as it made the final turn of the right hand traffic pattern preparing to land. Cadet Almond banked his ship slightly and therefore had to use excessive rudder to hold the ship in the turn and consequently skidded. As the formation rolled out of the turn, Cadet Almond's airplane came very close to the lead airplane. Lt. Markham then took the controls and used them violently to avoid a collision. The airplane stalled and snapped over on its back. Lt. Markham attempted to recover by continuing in the roll. There was insufficient altitude for the resulting pull-out and the airplane crashed, striking the ground at approximately 30 degrees headed approximately 45 degrees to the right of the landing path. All of the air witnesses stated that they believed the airplane to be under control when it crashed and an inspection of the wreckage indicated that the airplane was not spinning at the instant of impact."

1-23-43H. Salina, Kansas. At 1530 CWT, a Consolidated B-24D suffering the failure of the number-three turbo-supercharger crashed while attempting an emergency landing at the Army Air Base at Salina, Kansas. Assistant radio operator Sgt. Alvin P. Adams was killed and bombardier 1Lt. Wendell H. Barnett and engineer Sgt. William A. Brown were seriously injured by pieces of the number-three propeller that broke off and entered the airplane. Pilot 2Lt. Robert A. Gearhart and co-pilot Roy E. Curtis received minor injuries. Engineer Pvt. James W. Alexander and radio operator SSgt. Verne D. Gardina received minor injuries parachuting to safety. The B-24 was on a high-altitude oxygen and bomb rack testing flight. The airplane had finished testing bomb racks at an altitude of 25,000 feet and then began climbing to reach the airplane's maximum ceiling. The B-24 was flying at an indicated altitude of 29,500 feet when the number-three turbo-supercharger failed and disintegrated, throwing pieces into the propeller and engine compartment, severing oil lines and seriously damaging the propeller. The pilots were unable to feather the number-three propeller and it shook itself free from the airplane. Pieces of the number-three propeller sliced through the side of the airplane, killing the assistant radio operator and injuring the pilot, co-pilot, engineer and bombardier. Two crewmembers bailed out on the pilot's command. The remainder of the crew were too seriously injured to use their parachutes and had to stay with the ship. The hydraulic system had been damaged and the landing gear had to be cranked down by hand. The pilot circled the field three times and then came in for a landing. The starboard landing gear collapsed on the landing roll and the airplane skidded to a halt.

1-23-43I. Ellington Field, Texas. At 1640, a Beech AT-10BH crashed while attempting a take-off at Ellington Field, Houston, Texas, killing pilot Capt. Stephen R. Batson, 25, Birmingham, Alabama, and infantry passenger 1Lt. Clarence R. Blume, 27, Hampton, Park, South Carolina. The airplane took off and entered an immediate steep climb. The control tower officer radioed the pilot to get the nose down. The airplane continued to climb and at an altitude of about 300 feet it stalled and fell off on the left wing. The airplane smashed into the ground and exploded into flames. Investigation revealed that the pilot attempted the take-off with the control column lock in place.

1-24-43A. Aquilla, Texas. At 0215, a Curtiss AT-9A and an AT-9 collided in mid-air three miles north of Aquilla, Texas, killing pilot 1Lt. Donald T. Chambers, Waco, Texas, and passenger Cpl. Benjamin Steimer, 19, Philadelphia, Pennsylvania, who were killed in the crash of the AT-9A. Flying students Pfc. Richard O. Foster and A/C Gordon W. Flint were able to safely land their damaged AT-9 at the Army Air Field at Waco, Texas. The Aircraft Accident Classification Committee stated, "1Lt. Donald Chambers and his passenger, Cpl. Benjamin Steimer, were leading a routine night formation flight of six AT-9 type airplanes at 7,000 feet altitude. Lt. Chambers signaled for an echelon of single ships to the right and confirmed it by radio. The student pilot [Pfc. Foster] in the number-three position of the leading element started down and under. As he was under the leader, the student took his eyes off the leader momentarily and looked back to see where the number-two airplane was. At that instant the student [Pfc. Foster] pulled up into the lead airplane, his left wing damaging the lead airplane's right elevator. No one in the formation saw the lead ship after that. The marks on the ground indicated that the airplane had come straight in, burying the engines about five feet in the ground. There was no apparent effort [by the occupants] to use their parachutes."

1-24-43B. Topeka, Kansas. At 1053 CWT, a Consolidated B-24D attempting a take-off crashed one mile south of the Army Air Base at Topeka, Kansas, killing the crew of eight. The airplane was taking off on a ferry flight to Salina, Kansas. Investigators stated, "This airplane took off at 1047, made a close circle of the field, was sighted with landing gear down west of the field, made a left 90-degree turn at about normal close in traffic pattern, went past the north-south runway (on which a ship was standing by preparatory to take-off), then upon seeing this ship clear the runway made a 180-degree turn, apparently to land, striking the ground while in a steep [left] bank and bursting into flames. It is believed the direct cause of the crash is due to error in pilot's judgment in making a steep [left] turn close to the ground — approximately 300-400 feet. No reason is known why the pilot returned for landing immediately after take-off, as he was not contacted by the tower and no evidence of trouble was observed except the close-in circle [of the field] and low altitude." Killed in the crash were: 1Lt. William L. Stephenson, B-24 instructor; 2Lt. Jack C. Mulligan, co-pilot; 2Lt. James B. Hinds, co-pilot; SSgt. Arthur J. Harvey, Jr., engineer; Sgt. Donald R. Groves, radio operator; SSgt. Louis A. Scheer, radio operator; SSgt. Blaine W. Huston, gunner; SSgt. George P. Richards, Jr., engineer.

1-24-43C. Alamogordo, New Mexico. At 1058 MWT, a Consolidated B-24D crashed in the White Sands National Monument 10 miles west of the Army Air Base at Alamogordo, Texas, killing 11 fliers. Killed in the crash were: 2Lt. William H. Ebert, Jr., pilot; 2Lt. Arthur M. Thomas, co-pilot; 2Lt. Paul Vistuba, navigator; 2Lt. Stanley G. Netz, bombardier; SSgt. Willard S. Shrader, passenger; SSgt. Glenn M. McKnight, engineer; Sgt. Robert A. Garrett, radio operator; Sgt. John T. Mikolich, gunner; Sgt. Joseph D. Rafac, radio operator; Sgt. Billie D. Francis, gunner; Sgt. Howard G. Lilley, gunner. Investigators stated, "Aircraft departed [Alamogordo Army Air Base] at 0745 on a Bombardment and Gunnery mission. Weather at the vicinity at the time showed broken alto-cumulus clouds 2,000 to 4,000 feet thick covering eight-tenths of the sky at estimated 5,000 feet above the terrain. Only one witness to the accident observed the aircraft emerge from the clouds in a moderate descent [and then] pull up into level flight or possible stall, then fall off into a spin which carried it to destruction. Other flights in the area at the time observed light to moderate turbulence and light rime icing in the clouds. Aircraft had just had its 100-hour inspection and was airworthy. Conclusion: Pilot lost control of the aircraft while in clouds and failed to effect recovery from resultant spin. Aircraft apparently was near recovery of level flight at moment of impact judging from distribution of wreckage."

1-24-43D. Mobile Bay, Alabama. At 1221, a Douglas A-20B crashed into Mobile Bay 12 miles ESE of Brookley Field, Alabama, killing passenger 1Lt. Richard T. Walsh. Pilot 2Lt. Sharon M. Johns, passenger Sgt. Anthony Russo and engineer Sgt. Sam Mamula received minor injuries. The airplane took off from Brookley Field on a flight to Tallahassee, Florida. Soon after take-off the airplane entered a thick bank of fog and descended into the bay at high speed, under control and with both engines operating. The airplane came to rest in water about four and one-half feet deep. Examination of the wreckage revealed that the engines and tail section had broken off of the airplane upon impact and that the nose section was completely demolished. It was noted by investigators that the fog bank reached from the surface of the bay to an altitude of about 1,500 feet. Visibility was zero.

1-24-43E. Gulf of Mexico, Louisiana. At 1420, a Douglas A-20C apparently crashed into the Gulf of Mexico approximately 35 miles south of Lake Charles, Louisiana, killing pilot 2Lt. Roger Crowe, gunner Sgt. William D. Cannon and gunner Sgt. James L. Gunter, Jr. The airplane had taken off from the Army Air Base at Lake Charles on an aerial gunnery mission over the Gulf of Mexico. The airplane arrived for its gunnery mission and rendezvoused with the target tow airplane. The two airplanes soon encountered instrument conditions. The subject airplane was never seen again after it entered the overcast. The airplane never returned to its base and was soon declared missing. Twenty-four hours after the airplane had taken off, the starboard main wheel and various pieces of wreckage washed ashore.

1-24-43F. Roswell, New Mexico. At 1430 MWT, a Vultee BT-15 crashed 15 miles west of the Army Air Field at Roswell, New Mexico, killing pilot 2Lt. Elmo Fannin and passenger Cpl. Frank O. Brimhall. Investigators stated, "Approximately 30 minutes prior to the time of the crash, Lt. Fannin and 1Lt. Perry W. Andrews taxied out in two BT-15s for take-off. The mission was to be local formation and general air work. The ships flew around in formation for a few minutes and climbed to an altitude of 6,500 feet indicated, where Lt. Fannin took the lead. He led the two-ship formation on up to an altitude of about 8,000 feet, leveled off, and flew straight for a short period.

The customary peel-off signal was given by Lt. Fannin and he peeled off to the right, and then rolled into a rather steep turn to the left. Lt. Andrews followed duplicating the maneuver. Two steep 360-degree turns were made to the left and Lt. Andrews states that with the turn becoming steeper, his ship stalled and he had to release the pressure on the stick to establish normal flying attitude. Lt. Fannin's ship stalled out of a steep right turn and whipped over on its back. The airplane did about one or one and a half turns of a spin to the right and stopped, then spun about three-fourths of a turn to the left and again stopped. It seemed the airplane faltered in a semi-stall position just prior to contacting the ground. Immediately after the crash, the airplane burst into flames." Investigators noted that this was Cpl. Brimhall's first flight in an airplane and that his seat belt was not fastened at the time of impact.

1-24-43G. Matagorda Peninsula, Texas. At 1540, a North American AT-6C and an AT-6B collided in mid-air and crashed 20 miles northeast of the Army Air Field at Matagorda Peninsula, Texas, killing three fliers. Pilot 2Lt. Ivor Kitterman and passenger Pvt. Phillip S. Todd were killed in the crash of AT-6C #41-32174 and A/C Grant Stidham was killed in the crash of AT-6B #41-17331. Lt. Kitterman was leading a six-ship formation in two, three-ship V-flights that were flying along the beach over the Matagorda Peninsula in a northeasterly direction at about 200 feet above sea level. A cadet flying in the second element radioed Lt. Kitterman that he had struck a bird and had apparently damaged his starboard wing, stating that he was pulling out of the formation. Lt. Kitterman radioed A/C Rayon Stokes, who was flying in the number-two position in the first element, to take over the flight while he dropped out to assist the other cadet. Apparently A/C Stidham misunderstood the instructions and when Lt. Kitterman left the formation in a climbing turn to the left he attempted to follow the maneuver. Lt. Kitterman's airplane lost airspeed while in the steep turn and A/C Stidham's airplane overran Lt. Kitterman's airplane from the left rear. A/C Stidham's starboard wing struck the tail section of Lt. Kitterman's airplane. Both airplanes fell out of control and crashed, immediately bursting into flames upon impact.

1-24-43H. Thompson, Texas. At 1700, two Beech AT-10 airplanes collided in mid-air and crashed near Thompson, Texas, killing A/C Albert B. Webb, Stillwater, Oklahoma, aboard AT-10 #41-1790. A/C Prescott G. Ward parachuted to safety and was uninjured. A/C Bernard F. Wilson and A/C Charles F. Zimmer parachuted to safety from AT-10 #41-1799 and were uninjured. The airplanes were part of a five-ship formation on a formation-training mission that originated at Ellington Field, Houston, Texas. The formation was in a gentle turn to the left when the collision occurred. AT-10 #41-1790 was flying in the lead position in the second element and AT-10 #41-1799 was flying in the number-three position in the second element. Apparently A/C Webb was gaining on the first element, causing him to throttle back to maintain spacing. Moments later, A/C Wilson's airplane overran A/C Webb's airplane from the left rear. Both airplanes fell out of control.

1-24-43I. Ruidoso, New Mexico. At 2005 MWT, a Consolidated B-24D crashed on the Mescalero Indian Reservation 15 miles south of Ruidoso, New Mexico, killing ten crewmembers. Navigator 2Lt. Richard L. Prather and navigator-rated passenger 2Lt. George W. Hepfner parachuted to safety and were uninjured. The B-24 was part of an eight-ship flight that had taken off from the Army Air Field at Topeka, Kansas, on a navigation flight to the Army Air Field at Alamogordo, New Mexico. Investigators stated, "From the evidence submitted, it is the opinion that the pilot flew from clear weather into clouds in the vicinity of the tips of the Sacramento Mountains at an elevation of approximately 12,500 feet. After flying in the clouds for approximately five minutes, he thought he was over Alamogordo and let down to 9,500 feet. He picked up ice, and undetermined amount, but evidently enough to make the airplane difficult to keep under control. Realizing the airplane was getting out of control, he started to give the signal to abandon ship. He crashed such a short time thereafter that only two occupants had a chance to jump. The airplane crashed out of control, exploded and burned. In the explosion, parts of the airplane were scattered to such an extent that it was impossible to tell the position of any controls." Killed in the crash were: 2Lt. Merle C. Brock, Wichita, Kansas, pilot; F/O Thomas N. Padon, Meadville, Texas, co-pilot; 2Lt. Walter C. Henning, Tucson, Arizona, bombardier; SSgt. Harry W. Ostrom, Chicago, Illinois, engineer; Sgt. Louis C. Fuller, Ironton, Ohio, radio operator; 2Lt. Reed J. Berrett, Ogden, Utah, pilot-rated passenger; F/O Robert O. Long, Kansas City, Missouri, pilot-rated passenger; 2Lt. David R. Fedder, Danville, New York, bombardier-rated passenger; SSgt. Guy L. Archambeau, Upland, California, radio operator/passenger; Sgt. Vernal O. Christensen, Fort Hall, Idaho, engineer/passenger.

1-24-43J. Ordway, Colorado. At 2305, a Boeing B-17F flying in poor weather collided with terrain and crashed near Ordway, Colorado, killing the crew of ten. Killed in the crash were: 2Lt. Edward E. Woodward, pilot; 2Lt. Lester A. Cattron, co-pilot; 2Lt. Norval W. Kelley, navigator; SSgt. Hascall J. Young, bombardier; SSgt. Carl E. Simpson, radio operator; SSgt. Charles J. Burns, gunner; Sgt. Charles D. McDonald, gunner; 2Lt. Martin J. Walden, passenger; 1Lt. Taylor D. Hunt, passenger; 2Lt. Vincent T. Larney, passenger. Investigation revealed that the airplane was trying to stay under a very low ceiling when it struck the ground with the port wing. The B-17 cartwheeled into the ground and exploded into flames. Wreckage and bodies were scattered over a considerable area. The airplane had taken off at 1654 MWT from the Pueblo Army Air Base, Colorado, on a local bombing mission and routine night flying.

1-25-43A. Page Field, Florida. At 0001 EWT, a Bell P-39D and a Douglas DB-7B collided on the runway while both attempted to land at Page Field, Fort Myers, Florida, killing P-39D pilot 2Lt. Raymond D. Cawyer. DB-7B pilot 2Lt. Edward W. Gamard received minor injuries. Investigation revealed that the P-39 had permission to land and had landed first. The DB-7B, which had an inoperative starboard landing light and was experiencing electrical problems, could not use its radio and did not have permission to land from the control tower. The DB-7B landed on top of the P-39 and collided with the fighter from the rear as it rolled down the runway. The P-39's landing gear collapsed and the fighter skidded to a halt and burst into flames. The DB-7B became temporarily airborne before falling off to the left and smashing to the ground. The DB-7B pilot had to be rescued from his smashed airplane.

1-25-43B. Matagorda Island, Texas. At 0901, a North American AT-6C crashed while on a firing exercise at the AAF Gunnery Range at Matagorda Island, Texas, killing A/C Harry E. Bradly. Investigation revealed that A/C Bradly was flying in the number-five position in a six-ship flight that had just taken off from the Army Air Field at Matagorda Island. The formation had to make two turns to get into position to use the gunnery range. Apparently A/C Bradly made too tight of a turn and entered a stalling attitude. The airplane whipped over on its back from a steep left turn and then smashed into the island nose-down in a semi-inverted banking attitude. The airplane burst into flames immediately and the student was killed instantly.

1-25-43C. New Albany, Ohio. At 1230 EWT, a Boeing B-17C flying in poor weather crashed two miles northwest of New Albany, Ohio, killing 12 crewmembers and passengers. Investigation revealed that the B-17 was attempting to let down through low clouds while trying to locate and land at Lockbourne Army Air Base, Columbus, Ohio. The B-17 collided with the ground in a flying attitude at high speed, exploding into flames and scattering wreckage, bodies and body parts over a considerable area. The airplane was returning to its station at Lockbourne after completing an administrative flight from Hendricks Field, Sebring, Florida. Killed in the crash were: Capt. Jack N. Pack, pilot; Capt. Robert L. Cooley, co-pilot; 1Lt. George H. Strobridge, co-pilot; SSgt. Chester B. Fudge, engineer; Cpl. Richard J. Klaetsch, radio operator; SSgt. John A. Sawatka, passenger; Sgt. Walter F. Baran, passenger; Sgt. James H. Foldesi, passenger; Pfc. Gordon O. Rutters, passengers; Pvt. Christian Mauthe, passenger; Pvt. George A. Box, passenger; Pvt. Salvatore B. Curcio, passenger.

1-25-43D. Dalhart, Texas. At 2049, a Waco CG-4A glider crashed into an unoccupied private residence while attempting an emergency forced landing after it was inadvertently released from a Douglas C-47A towing airplane, killing four glider students and a glider instructor. Glider students SSgt. Bernardo Fernandez and SSgt. Harold Kolom survived with serious injuries. Killed in the crash were glider instructor F/O Bradford K. Root and glider students SSgt. Claude C. Bruce, SSgt. Philo N. French, SSgt. Ernest J. Forbes, and SSgt. James Hyatt. The C-47 was piloted by 1Lt. Charles L. Harruff and 2Lt. William J. Metcalf. Both were uninjured and were able to land safely at the Army Air Field at Dalhart. Investigation revealed that the glider was towed aloft at 2040 to the northeast. At 2050, the C-47 radioed Dalhart that the glider had broke free and was lost. The pilots were told to release the tow-rope and attempt to locate the glider. Ground parties found the glider at about 0100 on 1-26-43. Investigation of the wreckage indicated that the glider pilot attempted to make an emergency landing and collided with the concrete structure. Investigators noted that CG-4A type gliders were not equipped with landing lights and should not be used by students at night until sealed beam headlights could be installed. It was discovered that the tow gear could be put in place with no positive indication that it is locked and safe to use. It was also noted that gliders had been inadvertently released on 15 other occasions since glider operations commenced at Dalhart on 11-20-42.

1-26-43A. Leesburg, Florida. At 1030, a Curtiss P-40F crashed while attempting an emergency landing at the Army Air Field at Leesburg, Florida, killing pilot 2Lt. Theora B. Burk. The pilot reported to the control tower that he was experiencing electrical problems that he was unsure if his landing gear was fully extended. The pilot was allowed to buzz the control tower and it was ascertained that the landing gear was extended. The control tower acknowledged that the landing gear was down and appeared in proper position. The pilot attempted a landing but he elected to go around when about ten feet above the runway. The pilot gave the airplane throttle for the go-around and gained about 100 feet of altitude. The engine quit and the pilot made a 45-degree turn to the right to make a crash landing in a swampy area. The airplane flipped over and smashed onto its back when the landing gear contacted the swamp. The airplane did not burn. The pilot was killed in the crash and his body was not retrieved from the airplane until 1115. Investigation revealed that the engine quit because the pilot had failed to switch to the fullest fuel tank while attempting his landing. It was not known why the pilot aborted the initial landing attempt, which looked to be satisfactory to personnel in the control tower.

1-26-43B. Flomaton, Alabama. At 1000, a Douglas C-67 (a B-23 type bomber modified into a high-speed transport) flying in poor weather suffered a catastrophic structural failure and disintegrated in mid-air, crashing to the ground 11 miles south of

Flomaton, Alabama, and killing the ten passengers and crew on board. The airplane had taken off from Drew Field, Tampa, Florida, on a personnel transport flight to Colorado Springs, Colorado, via Oklahoma City, Oklahoma. The Aircraft Accident Classification Committee stated, "This accident occurred at approximately [1000 CWT]. There was an active cold front with severe turbulence, heavy rain, with thunder and lightning in the vicinity of Flomaton, Alabama. Reports of other pilots indicate the intensity of this frontal activity. The airplane fell in a wooded area and came straight into the ground as the top on none of the trees were broken, indicating that the airplane did not glide in. Immediately upon impact with the ground the ship exploded and burned to an unrecognizable mass. It was difficult to identify the type of aircraft from the wreckage. One body was thrown clear and did not burn, however, the other nine were charred to such an extent identification was particularly difficult. Sections of both wings were found a distance of about two and one-half miles from the scene of the crash and a truckload of fragments was picked up over an area of about three and one-half miles from the scene. There was no indication or evidence that the airplane had come in contact with trees or the ground [prior to crashing]. The door to the fuselage and the life rafts were found about two miles from the scene. An inspection was made to determine whether or not the emergency release on the door had been pulled, however, it was found that the hinges on the door had been pulled out completely. The rivets on the parts of the wings were split loose. All of the bodies of the occupants had parachutes strapped on. From these facts, it is the conclusion of this board that either due to extreme turbulence, lightning, or an explosion due to low outside pressure, the wings gave way and the ship literally disintegrated in the air. This conclusion is based on the fact the ship and the bodies of the occupants therein, were found such a distance from the wings and other debris and the fact that the ship came in straight down. Apparently the ship was flying at an altitude high enough so that when the structural failure occurred, the engines and forward speed carried the fuselage some distance from the point the wings fell off. The parts of the airplane and wing which were picked up a distance of from one to three and one-half miles from the scene of the crash, ranged in size form small fragments to large sections of the wing. One of these fragments had markings thereon, with a pencil, indicating that some part of the airplane had been reinforced at one time or another. Whether the structural failure occurred at this point or elsewhere, this board is unable to state, but the conclusion seems to inevitably be that the wings and perhaps other parts of the airplane gave way in flight, resulting in the crash." Killed in the crash were: Maj. S.B. Gimble, Jr., Tampa, Florida, pilot; F/O Glenn A. Johnson, Oshkosh, Wisconsin, co-pilot; MSgt. Pius C. Witherow, Tampa, Florida, engineer; SSgt. J.C. Melder, Darlene, Idaho, radio operator. The passengers that were killed in the crash were all members of the staff of the commanding general of the Second Air Support Command, Colorado Springs, Colorado. They were: Brigadier General Carlyle H. Wash, Commanding General of the Second Air Support Command, Colorado Springs, Colorado; Lt. Col. James H. Cunningham, Colorado Springs, Colorado; Lt. Col. James A. Cain, Grynes Hill, Staten Island, New York; Maj. John R. Lee, Aide to General Wash, Riverside, California; 1Lt. John A. Simons, Albuquerque, New Mexico; 1Lt. Edward A. Drucia, Oak Park, Illinois. Gen. Wash was formerly the commanding general of the Third Air Force Command, Tampa, Florida (6-25-42 to 11-26-42).

1-26-43C. Donnelsville, Ohio. At 1330 EWT, a Consolidated B-24D flying in instrument conditions suffered a catastrophic structural failure of the port wing and crashed two miles west of Donnelsville, Ohio, killing pilot Capt. Harris F. Collier and co-pilot 2Lt. J.B. Westmoreland. Investigation revealed that the airplane had disintegrated at a very high altitude. It was speculated that the pilots were flying on instruments and had inadvertently put the airplane in an unusual attitude and upon recovering the airplane had subjected it to stresses and speeds that compromised the structural integrity of the wings and airframe. Investigation revealed that the port wing failed at a point 12 feet from the tip, separating and falling to earth about two miles from the main wreckage. The port elevator, stabilizer, vertical fin and rudder also failed and separated. The starboard wing failed and separated at a point just outside of the number-four engine nacelle. It is likely that extreme forces prevented the pilots from exiting the B-24 and parachuting to safety.

1-26-43D. Maxton, North Carolina. At 1520 EWT, a North American A-36A crashed 25 miles northwest of Laurinburg-Maxton Army Air Base, Maxton, North Carolina, killing 2Lt. Frank E. Turner. The airplane had been flying in the number-five position of a six-ship flight that was on a strafing mission. The flight leader had led the flight on a strafing pass in string formation and then performed a climbing turn to the left. During the maneuver, the first five airplanes inadvertently entered a cloud at an altitude of approximately 400 feet. Apparently Lt. Turner did not go on instruments soon enough and lost control of his airplane, causing it to crash to the ground.

1-26-43E. Yuma, Arizona. At 1730 MWT, a North American AT-6C suffered a catastrophic structural failure of the port wing and crashed ten miles southeast of Yuma, Arizona, killing A/C Ralph W. Bailey and injuring A/C William D. Altland, who was able to parachute to safety. Investigation revealed that the airplane was in an extreme dive, the port wing failing and separating when A/C Bailey tried to pull out of the maneuver. The students had been on a scheduled instrument team ride.

1-26-43F. Okmulgee, Oklahoma. At 2205, an Aeronca L-3B crashed 13 miles west of Okmulgee, Oklahoma, killing student pilot SSgt. Max D. Fensler, 21, Lincoln, Nebraska. The airplane crashed into a grove of trees while practicing landings at the Army Air Field at Okmulgee. Investigators stated, "The student was practicing 90-degree approaches making a left hand pattern and had made four successful solo landings. On the downwind leg of the fifth pattern, the airplane turned to the right approximately 45 degrees and struck a tree, which from all indications shattered the propeller, [the airplane] continued in the same general direction into the grove of trees."

✓ **1-27-43A. Sweetwater, Texas.** At 1130, a Fairchild PT-19A crashed near Avenger Field, Sweetwater, Texas, killing civilian instructor Allen W. Hanks. Investigation revealed that the instructor, who was taking an instructor refresher course, had put the airplane in a deliberate spin at an altitude of 3,000 feet, but was unable to recover the airplane before it struck the ground.

1-27-43B. Gladstone, North Dakota. At 1230 CWT, a Bell P-39M crashed three miles south of Gladstone, North Dakota, killing pilot 1Lt. Russell R. McManus. The pilot was on a ferry flight from Bismarck, North Dakota, to Great Falls, Montana. Witnesses on the ground stated that the airplane had flown over with the engine smoking and sputtering. The airplane was seen to climb and then to roll over to an inverted position before plunging to the ground and exploding into flames.

1-27-43C. Columbia, South Carolina. At 1355, a North American B-25D flying in poor weather suffered a catastrophic structural failure of the starboard wing and crashed five miles northeast of the Army Air Base at Columbia, South Carolina, killing the crew of seven. Killed in the crash were: Capt. Buell A. Bankston, 27, Houston, Texas, pilot; 2Lt. Lawrence W. Kiskaddon, 24, Seattle, Washington, co-pilot; 2Lt. Thomas M. Poole, 21, Muncie, Indiana, navigator; Capt. Arthur P. Vandergrift, 28, Memphis, Tennessee, passenger; SSgt. Ralph V. Metcalf, 22, Atlanta, Georgia, engineer; SSgt. Alfred Lefevre, 19, Weston, West Virginia, radio operator; SSgt. Lee J. Cobb, 19, Freedonia, New York, gunner. Investigators stated, "Capt. Bankston took off at 1330, [and] at 1340 reported to the tower at 5,400 feet, still climbing on instruments, and at 1350, again reported still on instruments at 8,000 feet. He gave instructions to clear all airplanes in his flight at 15-minute intervals [and ordered them to proceed to] 8,000 feet, and stated that he was proceeding to Atlanta [Georgia]. The pilot evidently lost control of the airplane and probably executed a violent recovery just below the overcast, literally tearing the airplane apart. The landing gear is believed to have been extended to slow down the airplane and possibly the flaps were also down." Pieces of the airplane were scattered over a mile along the flight path from where the fuselage smashed to earth. Investigation revealed that the starboard wing failed first and separated from the airplane, causing such stresses that the tail section progressively failed and also separated from the airplane. The remainder of the airplane crashed to the ground out of control and exploded violently into flames upon impact.

1-27-43D. Moore Field, Texas. At 1435, a Republic YP-43 crashed one mile west of Moore Field, Mission, Texas, killing 2Lt. James S. Woolfolk. Investigators stated, "2Lt. James S. Woolfolk was flying a P-43 airplane on an individual attack mission. He and 2Lt. John R. Murphy were flying near each other, one performing a maneuver and then the other doing the same thing. At 8,000 feet, Lt. Murphy was a good distance behind Lt. Woolfolk when the latter peeled off to the left in a steep dive. Lt. Murphy likewise peeled off to follow him, but at an indicated airspeed of 300 mph and at an altitude of 4,000 feet, Lt Murphy decided he had better pull up. At this time Lt. Woolfolk was approximately 1,000 feet lower, still going down. Lt. Murphy saw what appeared to be one wheel drop out of its well and Lt. Woolfolk's airplane make a violent dip to the right. Lt. Murphy pulled his ship up, and when he looked for Lt. Woolfolk again he saw the fire on the ground about 150 yards off the southwest edge of the field. 2Lt. Clifford Anderson, who was flying over the traffic pattern of the field, also saw Lt. Woolfolk's airplane diving at what appeared to him an almost vertical angle and an extremely high rate of speed. As he said, his first reaction was that the ship was going too fast and would never be able to pull out before hitting the ground. He noticed that the nose of the ship came up a few degrees as if Lt. Woolfolk were trying to pull it out, [but] then it went back into the vertical attitude; the ship seemed to be slowly coming out of its dive when it hit. Examination of the wreckage indicated that the ship struck the ground at about 60 degrees and apparently Lt. Woolfolk made no effort to get out, for he was found in the wreckage."

1-27-43E. Visalia, California. At 1620, two Ryan PT-22 airplanes collided in mid-air and crashed at Visalia, California, killing student A/C John T. McKinster and civilian instructor Howard G. Harvey, Jr. aboard PT-22 #41-20755. A/C Lafayette D. Petross was killed in the crash of PT-22 #41-20758. Investigation revealed that A/C Petross was in a diving and gliding turn to the left. The starboard wingtip of A/C McKinster's airplane, which was executing a slight climb, struck the port main wheel of A/C Petross' airplane. The impact swung A/C McKinster's airplane to the right and against the empennage of A/C Petross' airplane. The two airplanes broke apart in mid-air, falling to the ground out of control about one-half mile apart. There was no fire. None of the occupants were able to exit the airplanes.

1-27-43F. Pampa, Texas. At 2245, a Beech AT-10 attempting a take-off collided with a Cessna AT-17 that was stationary on the active runway at Laketon Auxiliary Airfield, Pampa, Texas, killing A/C William H. Kidd aboard the AT-17. A/C Walter H. Kelm was uninjured. A/C William H. Piles and A/C Ernest M. Quillen received minor injuries in the crash of the AT-10. Airplanes that were using the auxiliary airfield were to land on a designated landing strip and were to take off on a take-off strip that was situated to the left of the landing strip. Investigation revealed that A/C Kelm and A/C Kidd landed their airplane on the correct landing strip and then pulled off of the landing strip onto what they thought was a taxiway on the left, but was in fact the active take-off strip. All exiting traffic was to exit the landing strip to the right and all students had been briefed to the ground traffic pattern at the field. A/C Kelm and A/C Kidd smelled smoke in the cockpit and immediately shut down the engines and exited the airplane. A/C Kidd grabbed the on-board fire extinguisher but the smoke soon cleared. Both cadets believed that the fire hazard had disappeared and A/C Kidd began to stow the fire extinguisher in its original position. At that instant, A/C Kelm noticed that the AT-10 was taking off directly toward them. He yelled for A/C Kidd to exit the airplane. A/C Kidd, unaware of the danger, said, "Just a minute." Again A/C Kelm yelled for A/C Kidd to exit the airplane. A/C Kelm then jumped off of the wing and ran for cover. The AT-10, which just had achieved flying speed and was about to lift off, collided head-on with the AT-17 and came crashing to a halt. A/C Kelm rushed to the AT-10 and helped the pilots to safety and then ran to the AT-17 but was unable to free A/C Kidd, who was wedged in the wreckage. It was determined that he was killed in the impact.

1-27-43G. San Pablo Bay, California. At 2325, a Bell P-39D crashed into San Pablo Bay approximately 300 yards from the Hamilton Field Crash Boat House, San Rafael, California, killing pilot F/O Irvin E. Snell. The airplane, which was on a routine night training mission, took off on Runway 12 and was seen to make a gradual turn to the left. The airplane continued in the turn to the left until it smashed into the bay. The pilot had very little night flying experience, and it was noted by investigators that visibility was diminished because of light rain. It was speculated that the pilot had lost the horizon in the turn and flew the airplane into the bay.

1-28-43A. Uvalde, Texas. At 1140, a Fairchild PT-19A crashed five miles east of Garner Field, Uvalde, Texas, killing student A/C John M. Nyborg and seriously injuring civilian instructor Robert W. Anderson. The airplane, flying at an altitude of about 200 feet, stalled and spun out of a steep left turn and smashed into a grove of mesquite trees.

1-28-43B. Tonopah, Nevada. At 1340 PWT, a Bell P-39L stalled and spun into the ground on the AAF Bombing and Gunnery Range at Tonopah, Nevada, killing 2Lt. John V. Brown. The P-39 was observed in a climbing vertically banked turn when it stalled and snapped rolled to an inverted attitude at an altitude of 5,000 feet above ground level. The airplane entered a tumbling end-over-end spin and then fell into a normal type spin. The pilot recovered the airplane and then it fell into a spin in the opposite direction of the original spin. The airplane spun to the ground and burst into flames. Apparently Lt. Brown had released the starboard escape door just prior to the airplane striking the ground in a vertical attitude.

1-28-43C. Lookeba, Oklahoma. At 1705, a North American B-25C exploded in mid-air and crashed at Lookeba, Oklahoma, killing the crew of five. Killed in the crash were: F/O Clifford A. Briggs, Grand Ledge, Michigan, pilot; F/O William V. Capps, Talladega, Alabama, co-pilot; 2Lt. Hyman V. Braver, Kansas City, Missouri, navigator; SSgt. Edward A. Stanford, Jr., Spencer, Massachusetts, engineer; Pvt. Edward M. Lanum, Sullivan, Illinois, assistant engineer. The airplane had taken off from Colorado Springs, Colorado, on a navigation mission to Amarillo, Texas, to Oklahoma City, Oklahoma, and return. The airplane radioed in at 1616 CWT, reporting that it was 20 miles east of Amarillo. The airplane was next observed flying from the southeast at about 400 feet above ground level near Lookeba. The airplane circled the area twice then exploded in the bomb bay area. A second explosion was observed, severing the starboard vertical fin, rudder, stabilizer, elevator and a portion of the starboard wing. A third explosion rocked the airplane and it smashed to the ground where it exploded upon impact. Investigators could not determine the cause of the initial explosion.

1-28-43D. Pritchett, Colorado. At 2235, a Cessna AT-17B crashed ten miles west of Pritchett, Colorado, killing A/C Hugh P. Ness and A/C Lewis Newman. The airplane had taken off from the Army Air Field at La Junta, Colorado, on a night navigation flight to Borger, Texas, and return. Investigation revealed that the airplane had flown into the ground under control in a flying attitude at cruise speed. Wreckage was scattered over an area of 350 yards. Investigators could not determine what caused the students to fly into the ground.

1-29-43A. Phoenix, Arizona. At 1229, two Stearman PT-17 airplanes collided in mid-air and crashed three miles northwest of Thunderbird Field, Phoenix, Arizona, killing two students. A/C John W. Starr was killed aboard PT-17 #41-8045; A/C Robert V. Simone was killed in the crash of PT-17 #41-25373. Investigation revealed that both airplanes were attempting to enter the traffic pattern. Investigators stated, "The probable conclusions from viewing the wrecked airplanes and their positions is that A/C Starr's aircraft came in contact with A/C Simone's ship from above as the upper left wing tip of Simone's ship was cut off and Starr's ship had the tire, landing gear and one aileron cut by a propeller. The ships then stayed together for a short period, then separated and spun straight to the ground. Both student pilots were killed on contact with the ground."

1-29-43B. Patterson Field, Ohio. At 1518, a Consolidated B-24E attempting a landing crashed near Patterson Field, Ohio, killing co-pilot 1Lt. C.H. Parkey and seriously injuring pilot Maj. Roy W. Gustafson. Investigators stated, "This airplane was being delivered to Patterson Field for some project work. Landing traffic was using the sodded area of the field because of an accident on the runway. The pilot misjudged his distance and landed the airplane about 1,800 feet from the northeast boundary of the field. A drainage ditch, approximately 30 feet wide and 15 feet deep had been dug along the northeast boundary of the field and the airplane came to rest in this ditch with nose imbedded in the opposite bank. The tail broke off aft of the bomb bay. The pilot and co-pilot were imprisoned in their seats and could not be extricated until the front section was pulled free of the bank. This took considerable time because six wrecking trucks were not sufficient to move the wreckage. The large derrick was then called into use and succeeded in lifting the wreck sufficiently to get the pilots out. The co-pilot was badly crushed and died before reaching the hospital. After the pilots were removed the airplane caught fire and could not be extinguished."

1-29-43C. Placerville, California. At 1635, a North American B-25D crashed five miles north of Placerville, California, killing pilot Capt. L.T. Wagner and radio operator TSgt. Robert L. Morris. Five crewmembers were able to parachute to safety. Engineer SSgt. Harold Glarum and SSgt. Robert L. Huddleston received minor injuries parachuting to safety; co-pilot Alto F. Dolan, bombardier 2Lt. D.O. Sattler and passenger SSgt. George S. Ostrowski parachuted to safety and were uninjured. The airplane had taken off from Hammer Field, Fresno, California, on a navigation training flight to Medford, Oregon, and return. The airplane made the flight to Medford and was returning to Hammer Field when the accident occurred. Investigators stated, "The return trip was a series of ever-increasing difficulties until the airplane crashed. They lost the use of their radio, which was probably due to ice and snow. The extremely rough air made instrument flying difficult. The instruments seemed to give false readings, and the ice seemed to jam the controls so that eventually further flight became impossible. According to the testimony of two officers aboard, the airplane reached a speed of about 550 mph in an unavoidable dive, which the strength of both the pilot and co-pilot were needed to recover. It often took their combined efforts to break the ice from their rudders and elevators; yet, according to their testimony, the - de-icers never failed to work. They reached an altitude as low as 6,000 feet but icing continued. Since there were mountains in the area much higher than this and they were unable to determine their own position, they climbed back to 8,500 feet. By this time, the pilot decided further flight was impossible and gave the order to abandon ship. Five men jumped successfully, with two receiving minor injuries. The remains of Capt. Wagner and TSgt. Morris were found at the scene of the wreck, with nothing to indicate that they had attempted to jump."

1-29-43D. Camilla, Georgia. At 2115 EWT, a Vultee BT-13A flying in instrument conditions crashed six miles north of Camilla, Georgia, killing pilot 2Lt. Frank W. Roberts, Cartersville, Georgia, and passenger 2Lt. Arthur Krasne, Pittsburgh, Pennsylvania. The subject airplane and another BT-13A had taken off from Maxwell Field, Montgomery, Alabama, on a routine flight back to the home station at Walnut Ridge, Arkansas. About one hour after take-off the airplanes, which were flying at 3,000 feet, encountered instrument conditions. The pilot lost control of the airplane and it crashed into the ground. Investigation revealed that the subject airplane was not equipped for instrument flight.

1-29-43E. Dothan, Alabama. At 2240, a North American AT-6B crashed six miles north of Napier Field, Dothan, Alabama, killing A/C Otto H. Keller. The student had successfully completed a night navigation flight and was attempting to enter the pattern for the field. Investigation of the wreckage indicated that the airplane had impacted the ground in a flying position at flying speed with the landing gear and flaps in the retracted position.

1-29-43F. Roswell, New Mexico. At 2155 MWT, a Cessna AT-17B crashed eight miles southeast of the Army Air Field at Roswell, New Mexico, killing A/C John C. Thomas, 22, Mason City, Iowa, and A/C Alden L. Tingle, 21, Montpelier, Ohio. The airplane was returning to the field with two wing ships after successfully completing a local night formation flight. The flight approached the field in an echelon formation to the right. The leader began a gentle peel-off to position the formation for a landing at the field. The subject airplane peeled off and flew into the ground in a flying attitude. The airplane bounded back into the air for a short distance, stalled, and then fell off on the starboard wing. The AT-17 slammed back to the terrain on the starboard wing, cartwheeling into the ground where it exploded into flames. Investigators speculated that the pilots had realized their proximity to the ground at the last moment, pulling up just before the initial collision with the terrain.

1-30-43A. Biloxi, Mississippi. At 1515, a North American A-36A crashed at the AAF Hancock Bombing Range 17 miles southwest of Biloxi, Mississippi, killing pilot 2Lt. Laverne J. Wecker. Investigators stated, "Lt. Wecker appeared to approach his ground target in such a manner as to permit the propeller to contact the ground and part of the airplane to hit a small [tree] stump. It is the opinion of this board that Lt. Wecker reached forward in the cockpit and flipped the toggle switch to safety his gun when he had completed his

[pass]. There is evidence of the guns having been [put on safe] and such action probably took place when he passed the target flying very low at a speed of approximately 250 mph. The resultant contact with the ground knocked the propeller governor off and permitted the propeller to run away and later tear itself loose from the crankcase along with the reduction gear. After striking the stump, the ship ascended in a gradual climb and when it reached approximately 350 feet, two and one-half miles from the target, the propeller assembly tore loose and the airplane accomplished a fast wing-over to the left, plunging nose down into a cypress wooded swamp. It is believed that Lt. Wecker attempted to jump when the airplane assumed its final dive as his body was found clear of the cockpit and his parachute was partially unfurled."

1-30-43B. New Braunfels, Texas. At 2200, a Curtiss AT-9A crashed 12 miles northwest of New Braunfels, Texas, killing pilot 2Lt. Barney F. Bobo and co-pilot Capt. William E. Wilson, Jr. The Aircraft Accident Classification Committee stated, "Two student instructor officers were making a night navigation training flight from Sheppard Field, Wichita Falls, Texas, to Kelly Field, San Antonio, Texas. They were to fly from Sheppard Field to Austin, Texas, direct and then follow the light line to Kelly Field. After they had cleared Sheppard Field an overcast at about 1,800 feet built up south of Austin. Evidently these pilots tried to fly beneath this overcast and in doing so got too low. The airplane struck a hill and was totally demolished. Both of the student instructors were killed in the crash."

1-30-43C. Needles, California. At 2000, a North American A-36A was abandoned and crashed into the ground approximately 25 miles northwest of Needles, California, killing pilot 2Lt. Donald E. Ferguson. The airplane had been declared missing when it failed to return to its station at the Army Air Base at Rice, California, after taking off on a night navigation flight to Kingman, Arizona, to Ashfork, Arizona, and return. Lt. Ferguson was leading two other A-36 aircraft on the leg from Ashfork when the flight inadvertently entered a cloud. The flight broke up and the other two airplanes were able to go on instruments and eventually they broke into clear skies. The other pilots stated that they had not seen Lt. Ferguson emerge from the clouds. His body and the wreckage of his airplane were found on the side of a small mountain on 1-31-43. Apparently Lt. Ferguson attempted to parachute to safety but his parachute did not fully deploy and he was killed when he fell to earth. His parachute was found strung out behind him.

1-30-43D. Glendale, California. At 2008, a Vultee BT-13A flying in instrument conditions collided with high-tension wires and crashed in the hills near Chevy Chase Street between Pasadena and Glendale, California, killing pilot 2Lt. Earl W. Fenton and passenger Maj. Francis H. Partridge. Investigators stated, "The pilot had been cleared for a contact flight from Marana [Arizona] to March Field [Riverside, California]. It is understood from Lt. Col. W.S. Ford of Marana [Army Air Field] that the pilot had asked for a clearance to Burbank or Glendale, but because of the weather near Glendale his flight was cleared to March Field only. At about the time [the airplane] should have arrived at March Field, a BT-13 was observed circling San Bernardino Air Depot and the Control Tower at March Field attempted to contact him but received no answer. About 15 minutes later, a BT-13 was reported circling Ontario Observation Airdrome, but neither Ontario nor March Field could make radio contact with him. Cal-Aero Academy Flying Field at Chino, California, about 10 miles south of Ontario, was notified and lights were turned on. About 20 minutes later the aircraft crashed between Glendale and Pasadena, shortly after having been observed circling at low altitude over Pasadena. [The pilot was] apparently lost and attempting to fly contact under instrument conditions." Investigators noted that the ceiling in the area at the time of the crash was reported as being 400 feet above ground level.

1-30-43E. Tuskegee, Alabama. At 2120, a Curtiss P-40F attempting a take-off crashed at Tuskegee, Alabama, killing 2Lt. Richard Davis. The airplane had taken off on a local night training flight, gaining about 300 feet of altitude before going into a very shallow dive toward the ground. Investigators stated, "The airplane struck the ground at a very shallow angle, the propeller making six marks on the ground before it was sufficiently bent to cause it to tear itself loose from the engine, taking with it the reduction gears and part of the nose section. These pieces bounced and rolled for 800 feet from the point of the last propeller mark until they came to rest. After the first contact with the ground the airplane careened back into the air and traveled 1,260 feet before it touched the ground again. From the position of the wreckage, in respect to the last marks made on the ground, it is believed that the airplane hit its left wing first while in a skidding turn to the right then the nose of the airplane dug into the ground flipping the airplane over on its back, in which position it came to rest. The mark made by the wing, as it gouged into the ground, was about 32 feet long. Where the engine and nose struck it made a hole in the ground approximately two and one-half feet deep and 10 feet long. The airplane was completely destroyed by fire. There was no semblance of a fuselage left." Investigators could not determine why the airplane flew into the ground. The airplane and pilot were part of the 99th Fighter Squadron, 332nd Fighter Group.

1-31-43A. Castaic, California. At 1100, a Douglas C-53 that was towing a Waco CG-4A glider broke up in mid-air and crashed six miles northwest of Castaic, California, killing three fliers aboard the C-53. Three fliers were seriously injured parachuting to safety from the C-53. Two glider pilots received minor injuries parachuting to safety from the CG-4A. Pilot F/O Wilbur W. Boyd, pilot; co-pilot F/O Neil W. Rise and engineer TSgt. Charles E. Burnett were killed in the crash of the C-53; Radio operator Pvt. David P. O'Connor, and passengers Pfc. Stewart L. Nengel and SSgt. Robert Edelman were seriously injured parachuting to safety from the C-53. Glider pilots F/O Henry L. Greenfield and F/O James K. Rollins received minor injuries parachuting to safety from the CG-4A. The C-53 was ferrying the glider from Van Nuys, California, to Victorville, California. Investigators stated, "After being held at Van Nuys for two days on account of bad weather, F/O Boyd took off for Victorville towing a glider. Weather was clear at Van Nuys but low scattered clouds and ceilings of 2,500 to 3,500 feet were reported on the route through the mountains. Top of overcast was reported around 11,000 feet with icing in the clouds. After take-off the pilot climbed in an effort to get over the clouds. He was unable to clear them on a direct climb so he circled to gain altitude. While circling he drifted into clouds so he tried to climb through clear places. However, he was unable to avoid clouds entirely and finally was in a solid overcast. The airplane began to ice up immediately and encountered extremely turbulent air. It is the opinion of the Board that because of severe icing and turbulent air, both pilot and co-pilot together were unable to maintain control of the airplane with the result that it stalled, went into a spin, and disintegrated while spinning." The glider pilots stated that the C-53 had entered the clouds and soon disappeared from their view. After a short period of time the climb became erratic. Moments later they observed the towrope pointing straight down into the clouds. They immediately released from the C-53, but were unable to maintain control because of the heavy amount of ice that had accumulated on the glider. They jumped to safety.

1-31-43B. Richfield, Utah. At 1348 MWT, Vultee BT-15 crashed at Richfield, Utah, killing AAF instructor pilot 1Lt. Lewis E. White and civilian instructor Morris A. Fuellenbach. The airplane had taken off from Richfield Municipal Airport on what was described as a "local pleasure flight." The airplane took off normally and climbed to about 300 feet when the engine failed. The pilot began a turn to the right to make an emergency landing in an open field. The airplane stalled and entered a spin, smashing into the ground about one-quarter of a mile away from the airport. The airplane was destroyed utterly but did not burn. The pilots were killed instantly.

1-31-43C. Edinburg, Texas. At 1512, a North American AT-6C crashed seven miles northeast of Edinburg, Texas, killing instructor 2Lt. Glenn H. Howland and student A/C Benjamin F. Cotton. Investigation revealed that the port wing failed and separated while the airplane was performing a loop. Lt. Howland attempted to parachute to safety but bailed out at an altitude too low to deploy his parachute. He was found about 20 feet from the spot where the airplane slammed into the ground and exploded. His parachute was unopened. A/C Cotton was found in the charred wreckage.

1-31-43D. Waco, Texas. At 1710, two Vultee BT-13A airplanes collided in mid-air two miles northeast of Waco, Texas, killing pilot 1Lt. Nickerson Blood and passenger Pvt. Felix M. Rexach aboard BT-13A #41-22283. Pilot 2Lt. John A. Ullstrom and passenger Cpl. Francis McGruth were able to safely land BT-13A #41-22044 at the Waco Army Air Field. Investigators stated, "1Lt. Blood was leading a formation of three ships in a stepped up echelon to the right. Lt. Blood gave the signal to peel off and land. He started to peel off to the left, but in doing so pulled his ship up into the flight path of the number-two ship piloted by 2Lt. Ullstrom. Both the right aileron and right elevator of Lt. Blood's aircraft were damaged and it fell into a spin out of control. The peel off was attempted from an altitude of approximately 1,000 feet. Lt. Blood jumped, but from such a low altitude that the parachute never opened."

1-31-43E. Midland, Texas. At 2010, a Beech AT-7 on a navigation flight flew into the ground 18 miles east of the Army Air Field at Midland, Texas, killing five fliers. The airplane had taken off from Douglas, Arizona, on a student navigation flight to Midland, Texas. Investigation revealed that the airplane had flown into the ground in a flying attitude and at cruising speed. Investigators noted that the elevation of Douglas was 1,350 feet and the elevation at Midland was 2,856 feet. The ceiling in the area of the crash was reported to be 1,500 feet above ground level. Investigators speculated that the pilot, navigation instructor, and navigator cadets had failed to notice the difference in elevations of the departure airfield and the destination airfield, causing the airplane to fly into the ground and explode into flames. It was noted by investigators that this was the pilot's first night flight to Midland after being stationed at Williams Field, Chandler, Arizona. Killed in the crash were: 2Lt. Lindsey H. Barclay, pilot; 1Lt. Albert C. Dumont, navigation instructor; A/C R.A. Polhamus, student navigator; A/C W.E. Tretler, student navigator; A/C J.A. Powers, student navigator.

1-31-43F. Eagle Pass, Texas. At 2030, a North American AT-6C dove into the ground 13 miles northwest of the Army Air Field at Eagle Pass, Texas, killing student A/C Phillip C. Crowther, Geneva, New York. The airplane was flying in the number-two position of a three-ship night formation flight that had originated at Eagle Pass. The subject airplane left the formation soon after take-off and then dove into the ground where it exploded into flames upon impact.

1-31-43G. Dale Mabry Field, Florida. At 2215, a Republic P-47C attempting a take-off crashed at Dale Mabry Field, Tallahassee, Florida, killing pilot 1Lt. John B. Spence. Investigation revealed that the airplane had crashed just after take-off, 1,800 feet off the end of the runway. Investigators noted that there was no discernable horizon lighting in the area. It was speculated that the pilot had lost the horizon just after take-off and flew the airplane into the ground where it exploded into flames upon impact.

February

2-1-43A. Columbia, South Carolina. At 1155, a North American B-25D crashed five miles south of Columbia, South Carolina, killing five crewmembers and injuring two others. The airplane was part of a flight of ten that were taking off from the Army Air Base at Columbia on a flight to Aiken, South Carolina, and return. The subject airplane was the fifth to take off. The take-off appeared normal and the airplane reached an altitude of about 200 feet when it was observed to begin yawing to the left. The airplane lost altitude as it continued to yaw to the left. The airplane disappeared behind terrain for a short period before it was seen heading east at very low altitude. The pilot was apparently attempting to get back to the field when the port wingtip struck trees while the airplane was in a left turn. The B-25D smashed into the trees, broke up and burst into flames. Investigation revealed that the port engine began losing power just after take-off. Bombardier 2Lt. Homer H. Landes and crewmember TSgt. Lee T. Worsham were hurled from the wreckage and seriously injured. Killed in the crash were: 1Lt. Joseph R. Janovsky, pilot; 2Lt. Clair E. Lewis, co-pilot; SSgt. Howard A. Ferguson, engineer; Sgt. Edward A. Witt, radio operator; SSgt. Henry B. Evans, gunner.

2-1-43B. Mississippi Sound, Mississippi. At 1555, a North American A-36A crashed into Mississippi Sound five and one-half miles southeast of Keesler Field, Biloxi, Mississippi, killing pilot 2Lt. Alfred G. Edens, Jr. The pilot was leading a flight of three A-36A airplanes on a low altitude training flight when his propeller contacted the water. The airplane remained airborne for a moment before it crashed into the water. The pilot was killed when his head struck the interior of the airplane. Investigation revealed that the pilot's shoulder harness was improperly fastened.

2-1-43C. Hassayampa, Arizona. At 1845, a North American A-36A crashed six miles northwest of Hassayampa, Arizona, killing 2Lt. John L. Geidner. The airplane had taken off from the Army Air Base at Rice, California, with another A-36A on a night navigation flight to Phoenix, Arizona, via Blythe, California. The airplanes inadvertently entered a cloud at approximately 6,000 feet. About a minute later the other airplane emerged from the clouds, but the pilot did not see Lt. Geidner's airplane break out into the clear. Investigation of the wreckage revealed that the airplane struck the ground in a high-speed vertical dive, exploding into flames upon impact. It was speculated that when the airplane entered the clouds, Lt. Geidner was unable to go on instruments before losing control of his airplane.

2-1-43D. Waco, Texas. At 2049, a Vultee BT-13A dove into the ground ten miles southeast of the Army Air Field at Waco, Texas, killing student A/C Nyle Reed. Cadet Reed was flying on the right wing of the lead airplane in a two-ship night formation flight. After flying in formation for a short time, A/C Reed began falling behind. A/C Reed was then seen flying 180 degrees from his original flight path. The flight leader tried to contact the student by radio without success. The airplane failed to return to base and was declared missing, being found early the next day. Investigation of the wreckage indicated that the airplane had struck the ground in a high-speed vertical dive. The airplane did not burn but was destroyed utterly.

2-1-43E. Harper, Texas. At 2108, a Vultee BT-13A crashed 11 miles southeast of Harper, Texas, killing pilot 1Lt. F.C. Robertson and civilian passenger J.H. Draper. The airplane had taken off from Duncan Field, San Antonio, Texas, to return to its home station at Goodfellow Field, San Angelo, Texas, after completing an administrative flight. Investigation of the wreckage revealed that the airplane flew into the ground under control with power and in a flying attitude. The airplane had descended into trees, cutting a path about 100 yards long before crashing into the terrain. The airplane did not burn. Investigators speculated that the pilot had lost the horizon and inadvertently flew the airplane into the ground.

2-1-43F. Lancaster, California. At 2112, a Vultee BT-13A crashed six miles west of Lancaster, California, killing student A/C William C. Coffee. Investigators stated, "A/C Coffee was sent out on a night cross-country flight to Daggett [California], returned, and called the tower for landing instructions at 2108. The tower gave him the necessary instructions and the cadet let down in a southeasterly direction to enter traffic. He made his let down too far south of the field and as he swung around to enter the traffic pattern at an approximate traffic altitude, he flew into a hill. The airplane exploded upon hitting the side of the hill and burned causing death to the pilot and completely destroying the airplane. The hill mentioned above is about seven miles south of War Eagle Field, and rises to an altitude about 1,200 feet."

2-2-43A. Waco, Texas. At 0025, a Curtiss AT-9A crashed four and one-half miles northwest of Blackland Army Air Field, Waco, Texas, killing pilot 1Lt. Ralph L. Lightsey, Jr. and passenger Pfc. John D. Hanna, Jr. The airplane had taken off at 2210 from Blackland Army Air Field on a weather check flight. After flying for a short time, Lt. Lightsey, a squadron commander at the Blackland Army Flying School, called off student flying for the night. Lt. Lightsey apparently climbed through the overcast and flew above it until just shortly before the accident. Examination of the wreckage indicated that the airplane had struck the ground with its port wing while in a steep left bank.

The airplane cartwheeled from the point of impact and exploded into flames. Wreckage was scattered over hundreds of feet. Investigators were unable to determine the cause of the accident.

2-2-43B. Boerne, Texas. At 1100, a North American BT-14 flying in instrument conditions was abandoned and crashed 12 miles northwest of Boerne, Texas, killing pilot 2Lt. Harry D. Miller, who plummeted to his death in an unsuccessful parachute jump. Lt. Miller was not cleared for an instrument flight but had departed Randolph Field, San Antonio, Texas, into instrument conditions. The ceiling was reported to be at 600 feet and student flying had been called off. When the squadron commander discovered that Lt. Miller was out flying, he immediately attempted to contact him by radio from the squadron. Civilian controllers at San Antonio, Austin and San Angelo were also asked to try to contact the pilot and guide him to Dallas or Waco, which were reporting visual conditions. Radio contact with Lt. Miller was never established and soon calls were coming in that an airplane had crashed near Boerne. After a brief search, Lt Miller's body was discovered about 300 yards from the crashed airplane. His parachute was strung out behind him. Investigators stated that the pilot had attempted to use his parachute at an altitude too low to allow it to fully deploy.

2-2-43C. Coffeyville, Kansas. At 1330 CWT, a Vultee BT-13A flying in instrument conditions dove into the ground eight miles east of Coffeyville, Kansas, killing pilot 2Lt. Roy C. Lovell and student A/C Robert J. McWilliams. Investigators stated that there was an overcast at approximately 6,500 feet during the morning. At approximately noon, a lower level layer of cloud moved in underneath, forming an overcast of 4,000 feet. At 1310, the instructor and the student took off for an instrument training flight. They practiced instrument flying in between the layers. The student was in the rear cockpit, flying on instruments under the hood. Investigators speculated that the airplane somehow got out of control and that the instructor had allowed the student to attempt to recover the airplane. By the time the instructor took over the controls, he could not prevent the airplane from diving into the ground. It was estimated that the airplane was diving at a speed of 300 mph when it struck the ground. The airplane exploded upon impact and the occupants were killed instantly.

2-2-43D. Arnold, Kansas. At 1435 CWT, a Consolidated B-24D flying in poor weather was abandoned in flight and crashed eight miles north of Arnold, Kansas, killing navigator 2Lt. J.D. Graham, who fell to his death in an unsuccessful parachute jump. Co-pilot 2Lt. D.C. Carlson was injured parachuting to safety. Crewmembers parachuting to safety were: 1Lt. Roy M. Westley, pilot; 2Lt. W.G. Fullick, bombardier; SSgt. E.L. Olson, engineer; SSgt. R.L. Wells, radio operator. The B-24D had taken off from its station at the Army Air Base at Topeka, Kansas, on a bombing mission to the AAF bombing range at Pueblo, Colorado. The B-24D had completed its bombing mission and was flying in actual instrument conditions when the pilot lost control of the airplane and ordered the crew to bail out. Investigation revealed that the airplane had been flying above the overcast at about 9,500 feet msl when it encountered a huge front. The pilot attempted to climb out through the front, which topped out at about 20,000 feet. The airplane reached about 14,000 feet when the pilot saw a trough in the clouds and attempted to get back down through it. The terrain was spotted through breaks at about 6,500 feet. It was determined that it was raining heavily underneath the overcast. After a short time, the airplane began climbing through freezing rain. The pilot lost control of the airplane and it began to spin to the ground. The pilot ordered the crew to prepare to abandon ship. The co-pilot then took over the controls, regaining control of the airplane, re-establishing level flight and turning 180 degrees in order to get back to the clear area the airplane had descended through. The pilot then took over control of the airplane again and promptly entered a stalling attitude, entering another spin. The co-pilot attempted to re-establish level flight but the pilot fought him on the controls. The pilot ordered the crew to abandon the airplane. It was noted that the pilot twice had to order the co-pilot to leave his station and abandon ship. Investigators stated that Lt. Graham's parachute harness webbing and stitching failed at the "V-rings where the lifter webs of the parachute snap onto the harness."

2-2-43E. Bogue, Kansas. At 1720 CWT, a Boeing B-17F flying in poor weather crashed nine miles north of Bogue, Kansas, killing six crewmembers. Killed in the crash were: 1Lt. Paul M. Capaccio, pilot; F/O Latteimer Stewart, co-pilot; 1Lt. Jacob M. Madson, navigator; TSgt. T. Donlavage, engineer; SSgt. Frank Culver, radio operator; MSgt. H.M. Bloch, passenger. The B-17F had taken off at 1358 MWT from the Army Air Base at Casper, Wyoming, on a ferry flight to Tinker Field, Oklahoma City, Oklahoma. Investigators stated, "Airplane was observed about 1720 CWT, traveling northwest. It descended from a low overcast, ascended again into overcast, and briefly thereafter came out in a steep dive. Apparently there was some attempt at recovery, but airplane dived into the ground with sufficient impact to force the engines and nose three to five feet into the frozen earth. Airplane exploded on impact, with intense fire in impact area. Wreckage was scattered south of impact area in direction paralleling line of wings and to one side only, indicating that the airplane had a strong sideways movement at the time of impact. Airplane and bodies of personnel were extensively disintegrated. Controls and instruments were unrecognizable at time of inspection. There had been some thunderstorm activity in vicinity. Just prior to crash, the airplane was traveling in direction opposite to its proper course. It cannot be determined whether the pilot was deliberately returning toward Casper [Wyoming] to avoid thunderstorm areas or was not properly controlling his direction while under intense strain. In either event, it is evident that he did not have airplane under proper control to meet conditions [that] usually prevail at the base of a low overcast. The strong sideways movement of the airplane, in combination with the dive when it struck the ground, indicated a stall out of a turn at low altitude. Pilot apparently lost control at low altitude while alternating between contact and instrument conditions at the base of overcast."

2-2-43F. Avon Park, Florida. At 2048 EWT, a Martin B-26B attempting a take-off crashed three miles northeast of the Army Air Field at the Avon Park Bombing and Gunnery Range, Avon Park, Florida, killing the crew of six. Killed in the crash were: 2Lt. Charles A. Cox, pilot; 2Lt. Allen R. Meller, co-pilot; 2Lt. William D. Blevins, bombardier; Sgt. Royce A. Cantrell, engineer; Sgt. William S. Roberts, radio operator; Sgt. William D. Ensiminger, gunner. The airplane had just taken off from the Army Air Field at Avon Park when the port engine erupted into flames. The crash occurred when the pilot lost control of the airplane while attempting to position it for an attempted crash-landing. The airplane impacted the ground at a speed of about 140 to 170 miles per hour, exploding into flames upon impact. Investigation revealed that a hose clamp on a fuel line in the port engine accessory section came loose, causing a drop in fuel pressure and allowing gasoline to enter the accessory section. The gasoline was ignited, causing a massive fire in the port engine nacelle and wing, possibly burning the fabric control surfaces on the wing and tail section and subsequently causing the pilot to lose control of the airplane.

2-2-43G. Bedford, Virginia. At 2120, a North American B-25D crashed into Sharp Top Mountain 12 miles north of Bedford, Virginia, killing five crewmembers. The B-25D had taken off from its station at the Army Air Base at Columbia, South Carolina, on a night navigation flight to Florence, South Carolina, to Raleigh, North Carolina, to Lynchburg, Virginia, to Greensboro, North Carolina, and return to Columbia. The last contact with the airplane was at 2058 when it checked in on time at Raleigh. Investigation revealed that the airplane had passed over Bedford in a northwesterly direction before colliding with the mountain from the east at an elevation of about 3,000 feet above sea level. The airplane exploded violently into flames upon impact with the mountain. Investigators could not determine what caused the pilots to fly the airplane into the side of the mountain at the Peaks of Otter. Weather was reported as ceiling and visibility unlimited and was not considered a factor in the accident. Killed in the crash were: 2Lt. Paul M. Pitts, 21, Poteau, Oklahoma, pilot; 2Lt. William C. McClure, 22, Indianapolis, Indiana, co-pilot; 2Lt. George R. Beninga, 22, Marietta, Minnesota, bombardier/navigator; 2Lt. Hilary S. Blackwell, 22, Santa Monica, California, navigator; Cpl. Peter J. Biscan, 29, Chicago, Illinois, engineer.

2-2-43H. San Antonio, Texas. At 2035, a North American AT-6A crashed five miles SSE of Brooks Field, San Antonio, Texas, killing A/C Lucian F. Harris. Investigators stated, "The pilot was flying the left wing position in a three-ship formation at an altitude of approximately 1,000 feet at night. There was a thin overcast and the horizon indistinct. The pilot turned to the right and flew across the formation, causing the right wing man to dive in order to avoid [a mid-air collision]. [A/C Harris] apparently continued on to the right because the formation leader lost sight of him almost immediately and did not see him again. It is probable that [A/C Harris] spiraled or spun into the ground at this time. It is the opinion of the Committee that the pilot lost his horizon, developed a condition of vertigo, and either spiraled or spun into the ground. The airplane struck the ground going almost straight down."

2-3-43A. Fort Worth, Texas. At 1010, a Fairchild PT-19A crashed one and one-half miles northwest of Hicks Field, Fort Worth, Texas, killing A/C Raymond C. Hefton, who fell to his death in an unsuccessful parachute jump. Investigation revealed that the student failed to recover the airplane after it entered an inadvertent spin. The student did not unbuckle his safety belt and attempt to leave the airplane until the moment just before impact. He was hurled from the cockpit when the airplane hit the ground and his parachute broke open from the impact.

2-3-43B. Buckhorn Mountain, Virginia. At 1230, a Consolidated B-24D crashed into Buckhorn Mountain, Giles County, Virginia, killing three crewmembers. Killed in the crash were: 1Lt. James D. Houston, Omaha, Nebraska, pilot; 2Lt. Joseph J. Hays, El Paso, Texas, co-pilot; Pfc. Frank Rohozka, Chicago, Illinois, engineer. The airplane disappeared after taking off from the Army Air Field at Smyrna, Tennessee, on a ferry flight to Montreal, Quebec, Canada. The airplane remained missing until it was spotted from the air by the crew of a civilian airplane on 4-22-43. Poor weather hampered the initial search efforts. Investigation revealed that the weather in the area at the time of the crash was reported as instrument conditions with heavy snow. Investigation of the wreckage revealed that the airplane struck the west side of the mountain in a flying attitude at cruise speed. Because of the poor weather conditions, the exploding airplane was not observed by anyone on the ground. Four civilian woodcutters stated that it was snowing heavily on the day of the accident. They stated that they had heard a dull crash in the distance when they had stopped their truck due to poor visibility, but they did not associate it with an airplane crash.

2-3-43C. Helendale, California. At 2240, a Vultee BT-13A crashed three miles northeast of Helendale, California, killing student A/C Everett G. Elliot. The airplane had taken off from Lancaster, California, on a routine student cross-country flight to Daggett, California, and return. Investigation revealed that the airplane had dove into the ground and exploded into flames, killing the student instantly. Although the official accident report offers no real proof, some investigators speculated that the student might have lost control of the airplane while attempting to perform a slow roll or some other aerobatic maneuver at night. The airplane was on course and on time at the time of the accident.

✓ **2-4-43A. Sweetwater, Texas.** At 0850, a Fairchild PT-19A crashed eight miles northwest of Avenger Field, Sweetwater, Texas, killing civilian instructor Donovan H. D'Spain, 21, Sweetwater, and student A/C Paul M. Johnson, 27, Denver, Colorado. The airplane had taken off from Avenger Field at 0815 on a routine dual instruction flight. Investigation revealed that the airplane entered a deliberate spin to the right at an altitude of approximately 4,000 feet. Apparently the student and the instructor were unable to stop the spin and recover the airplane. Witnesses stated that the airplane described ten turns of spin before it smashed into the ground. There was no fire but the airplane was destroyed utterly.

2-4-43B. Patterson Field, Ohio. At 1210 EWT, a Martin A-30 attempting a take-off crashed into buildings at Patterson Field, Ohio, killing pilot F/O Billy B. Groce. Investigation revealed that the port engine began cutting out when the airplane had reached an altitude of about five feet on the take-off. The airplane climbed to about 75-100 feet when the engine quit completely. The airplane continued to climb slightly, turning to the left until entering an extreme left bank at an altitude of about 250 feet agl. The airplane rolled to an inverted attitude and then crashed into a garage and officers' quarters where it erupted into flames. The buildings were destroyed, but no one on the ground was injured or killed. Investigation revealed that the fuel selector valve lever for the port engine was positioned incorrectly.

2-4-43C. Marion, Massachusetts. At 1500, a Douglas A-20C flying in instrument conditions was abandoned and crashed at Marion, Massachusetts, killing observer Maj. Charles H. Cook, who went in with the aircraft. Radio operator Sgt. Albert D. Reposa and gunner TSgt. Donald T. Robinson were killed while attempting to parachute to safety. Pilot Maj. Arnold J. Bailey was able to parachute to safety and was uninjured. Investigators stated, "Major Bailey was cleared from Harrisburg [Pennsylvania] to perform a cooperative mission with the Third Armored Division in the local Harrisburg area. He was to fly to Lebanon, Pennsylvania, establish radio contact with the ground forces there and then return to land at Harrisburg. In the vicinity of Harrisburg the pilot pulled up into the overcast because of low forward visibility. After effecting a 180-degree turn on instruments, and while flying west in the direction of the airport, he again saw the ground but because of poor forward visibility climbed to the top of the overcast at 3,600 to 4,000 feet. According to [the pilot's] testimony, he was unable to return to the home station or land elsewhere. He claimed to have made two [unsuccessful] instrument [landing] approaches to the Harrisburg Airport [and] then proceeded southeast to Baltimore [Maryland] between layers. At Baltimore he found himself on instruments, (he possessed no current instrument certification) and flew to the vicinity of Cape Cod [Massachusetts] where he made a successful parachute jump when the fuel supply became low. The observer stayed with the airplane and was killed. The radio operator and gunner jumped, but their bodies have not been located after a five day search."

2-5-43A. Oklahoma City, Oklahoma. At 0507, a Douglas A-20B crashed two miles south of Will Rogers Field, Oklahoma City, Oklahoma, killing pilot 2Lt. Wallace O. Akins. The pilot was on an assigned navigation flight to Amarillo, Texas, and had taxied out to the North-South runway at 0500 where he performed his engine run-up. He told the control tower that he was going to stand by while he investigated a low magneto in one of the engines. After correcting the magneto problem by "burning out" the engine, the pilot was granted permission to take off. The airplane took off to the south at 0505 and climbed to an altitude of about 200 feet agl when the port engine began to lose power. The engine then quit and the airplane began to lose altitude. In an attempt to get back to the field, the airplane began a turn to the right, into the good engine, at an altitude of approximately 100 feet agl. The airplane was then seen to enter an extreme bank to the right and smash into the ground on the starboard wing. The A-20B cartwheeled into the ground and exploded into flames, killing the pilot instantly. Wreckage was scattered over an area of 680 yards. Investigators were not able to determine the exact cause of the port engine failure.

2-5-43B. Waco, Texas. At 1045, a Beech AT-10BH suffered a catastrophic structural failure and crashed 20 miles NNW of Blackland Army Air Field, Waco, Texas, killing instructor 1Lt. Joseph W. Jones and student A/C Salvatore J. Ciccio. Investigation revealed that the tail section had failed first, followed by the progressive failure of the fuselage and wings. The airplane slammed to the ground and exploded into flames. The tail section was found about 1,000 feet from the main wreckage. It was the testimony of Beech Aircraft Corporation engineer Paul E. Allen that five other AT-10 airplanes had crashed under similar circumstances when the pilots exceeded the cockpit placard speed limit of 175 mph with the cockpit canopy open. Investigators speculated that an adverse and dangerous vacuum condition was created in the aft fuselage when one of the cockpit canopy hatches separated from the airplane in flight, causing an explosive failure of the tail section. It was noted by investigators that the factory-installed cockpit placards (and those installed in accordance with an AAF Technical Order) limited the AT-10 type airplane to a speed of 175 mph with the canopy removed, but that new placards limited the airplane to a speed of 125 mph. It was not clear why the new placards had not been installed in the AT-10 type airplanes at the station. The fliers had made no apparent effort to leave the airplane and their bodies were found in the wreckage.

2-5-43C. Dalton, Georgia. At 1615 EWT, a Curtiss P-40E flying in instrument conditions crashed into the side of Stony Face Mountain five miles SSW of Dalton, Georgia, killing pilot 2Lt. Theodore B. Marxson, 24, Lafayette, Indiana. The airplane had taken off from Smart Field, Macon, Georgia, on a navigation training flight to Bowman Field, Louisville, Kentucky, and soon entered instrument conditions. The pilot was evidently trying to maintain visual contact with the ground when he collided with the mountain. The airplane exploded into flames upon impact and the pilot was killed instantly. Civilian witnesses reported that the mountain was shrouded in fog at the time of the accident. Ceiling was reported as low at the scene of the accident. It was noted by investigators that Lt. Marxson had been advised that adverse weather would be prevalent along his proposed route.

2-5-43D. Newberry, South Carolina. At 1730, two North American B-25C airplanes collided in mid-air and crashed 12 miles west of Newberry, South Carolina, killing 14 crewmembers aboard both bombers. Killed in the crash of B-25C #41-13281 were: 1Lt. John E. Wilson, pilot; 2Lt. Francis Bardell, co-pilot; 2Lt. Julius X. Zarchin, bombardier; SSgt. Harold B. Brown, engineer; SSgt. Franklin G. Morris, engineer; SSgt. Dean L. Loren, radio operator; Sgt. Walter W. Pratt, gunner. Killed in the crash of B-25C #41-13275 were: 2Lt. Kalerno K. Mannio, pilot; 2Lt. Donald D. Halley, co-pilot; 2Lt. Randolph V. Donaldson, navigator; TSgt. Lyle E. Vinson, engineer; Sgt. John C. McSalls, engineer; SSgt. Basil R. Sink, radio operator; SSgt. Allen M. Steen, gunner. The airplanes were part of a three-ship flight on a formation-training mission. The three-ship flight had been part of a seven-ship formation that had broken up. The formation had taken off from Morrison Field, West Palm Beach, Florida, and was flying back to the home station at the Army Air Base at Greenville, South Carolina. The formation was flying above the clouds at about 10,000 feet when the flight leader, Lt. Wilson flying in B-25C #41-13281, noticed a break in the overcast and began to descend through it. His flight followed him down and the other four airplanes remained above the overcast. The three-ship flight descended to about 1,500 feet agl near Lake Murray. After flying northwest at this altitude for a short period of time, flight leader Lt. Wilson made a sudden and steep turn to the right in an attempt to avoid low hanging clouds. Lt. Mannio, flying in B-25C #41-13275 in the number-two position, attempted to follow the maneuver and overran the leader. The two airplanes collided just as they were entering broken clouds. The airplanes fell from the overcast out of control and shedding pieces. The airplanes slammed to earth and exploded into flames. The tail section of Lt. Wilson's airplane was not located at the time of the investigation. The number-three airplane had made a turn to the left to avoid the instrument conditions and had avoided colliding with the other two airplanes.

2-5-43E. Montgomery, Alabama. At 1720, a Vultee BT-13A and a North American AT-6B collided in mid-air and crashed two miles south of Maxwell Field, Montgomery, Alabama, killing four fliers aboard both airplanes. 2Lt. Carlous J. Boeke, Jr. and 2Lt. Archibald J. Grefer were killed in the crash of the BT-13A; civilian instructor Joseph C. Cain and student Pvt. Herbert B. Gordon were killed in the crash of the AT-6B. Investigation revealed that both airplanes were attempting to enter the downwind leg of the traffic pattern at Maxwell Field. Witnesses stated that the AT-6B was already flying on the downwind leg and the BT-13A spiraled down from a higher altitude and dropped down into the pattern, apparently descending onto the AT-6B at an altitude of about 1,165 feet msl. Both airplanes fell to earth out of control and burst into flames upon impact. Investigation revealed that Lt. Boeke was piloting the BT-13 from the rear cockpit with Lt. Grefer as safety observer.

2-5-43F. Revielle Valley, Nevada. At 1730, a Bell P-39D on a ground gunnery mission crashed in the Revielle Valley near the AAF Bombing and Gunnery Range at Tonopah, Nevada, killing pilot 2Lt. Gordon W. Chapman. The airplane was part of a three-ship flight that had completed its gunnery exercise and was returning to the Army Air Field at Tonopah. The other two pilots did not see the airplane crash but had noticed it missing and had then spotted smoke on the ground. The flight leader flew over the smoke and observed scattered wreckage, confirming that the airplane had crashed. Investigation of the wreckage revealed that the airplane had mushed into the ground at high speed at an angle of 30 degrees. The pilot, still strapped to his seat, was hurled from the wreckage upon impact. He was found with the top of his head severed about 150 yards from where the airplane had initially struck the ground. The airplane, shedding pieces, bounded back into the air for about 50 yards before smashing back to earth and disintegrating. It was speculated that the pilot might have over-controlled his airplane while trying to follow the flight leader through a sudden maneuver. The airplane apparently stalled, rolled inverted, and crashed into the ground after the pilot had rolled through and was pulling up after diving through and nearly recovering.

2-5-43G. Kissimmee, Florida. At 2045 EWT, a Douglas XP-70 experimental night fighter (AAF serial #39-735) crashed while attempting a take-off at the Army Air Base at Kissimmee, Florida, killing pilot 2Lt. James H. Toal, Fort Worth, Texas. Radio operator Pvt. Jack Kirschbaum and crewmember Cpl. Philip D. Gorman received minor injuries. The airplane was taking off on a routine night flying mission when it flew into trees and crashed. Investigators stated, "It is found that shortly after take-off and just over the trees off the edge of the airdrome the pilot nosed the airplane down and struck trees. The radio operator noticed the pilot working the wobble pump just before they crashed. Up to the time of the crash both engines were functioning normally. The Aircraft Accident Classification Board feels that Lt. Toal, being used to flying the P-70 type aircraft with a 12-16 lb. fuel pressure system, may have possibly been concerned about the normal 5-7 lb. fuel pressure system of the XP-70 and was working the wobble pump for that reason. It is further offered that Lt. Toal may have had his head in the cockpit checking fuel pressure or working the wobble pump and allowed the airplane to nose down and strike the trees." The XP-70 was completely destroyed in the crash.

2-5-43H. Atlantic Ocean, Florida. At 2200 EWT, a Lockheed-Vega B-34 crashed into the Atlantic Ocean approximately 100 miles off shore of Melbourne, Florida, killing three crewmembers. Passenger Capt. Warren W. Werner was able to parachute to safety and was immediately rescued from the ocean by a civilian merchant vessel operating in the area. He sustained only minor injuries. Pilot Maj. Dellis E. Russell, observer Capt. Henry J. Langsenkamp and engineer TSgt. Joseph D. LeClair were apparently killed when the B-34 crashed into the sea. The pilot had been scheduled to fly another B-34 airplane on an over-water navigation flight from the 36th Street Airport, Miami, Florida, to San Antonio, Cuba, and return, but brake and magneto problems forced him to take the subject airplane instead. The airplane departed Miami and arrived at San Antonio, Cuba, at 1225 EWT with no apparent problems. The airplane took off for Miami at 1845 EWT. The airplane became lost when it was inadvertently flown past the southeast tip of Florida. At 2140 EWT, Miami air traffic controllers were alerted that the subject airplane was lost and out of fuel and was broadcasting a "Mayday" signal. At 2146, another mayday signal was received. Without an order to abandon ship, the crewmembers released the escape hatch. One crewmember kneeled down near the open hatch and then remarked, "I'll jump last." Capt. Werner immediately jumped through the open hatch, cleared the airplane and pulled the ripcord. As he descended in his parachute, he noticed an orange flash on the surface some distance away. Capt. Werner did not know if any crewmembers followed him through the hatch. He was rescued from the sea by crewmembers on board the *S.S. Cape Cod* shortly after entering the water. It was noted that the airplane was not equipped with navigation charts, radio facilities charts, flare cartridges, or adequate over-water safety equipment.

2-5-43I. Clinton, Illinois. At 2215, a Beech AT-7 attempting an emergency forced landing crashed six miles west of Clinton, Illinois, killing the crew of five. The airplane had taken off at 1853 CWT from Tinker Army Air Field, Oklahoma City, Oklahoma, on a night navigation flight to Scott Field, Belleville, Illinois. The airplane apparently became lost in very poor weather that consisted of heavy rain, lightning, high winds and poor visibility. The pilot circled the area near Clinton with his landing lights illuminated. The pilot was attempting to land in a clear area but apparently pulled up and stalled when he saw trees in his landing path. The AT-7 smashed into the ground, killing all on board instantly. The airplane immediately erupted into flames upon impact and was completely destroyed. Two other airplanes that had taken off on the same mission had arrived safely at Scott Field.

2-5-43J. Sacramento, California. At 2130, two North American AT-6A type airplanes collided in mid-air and crashed two miles northeast of Mather Field, Sacramento, California, killing four RAF fliers aboard both airplanes. RAF Pilot Officer John G. Moriarty and Pilot Officer James A. Paterson were killed in the crash of AT-6A #41-439. RAF Pilot Officer Fred Hodges and Pilot Officer John R. Latour-Eppy were killed in the crash of AT-6A #41-583. Investigation revealed that both airplanes were practicing take-offs and landings at Mather Field and were in the traffic pattern at an altitude of approximately 800 feet agl. AT-6A #41-583 was turning onto the base leg when the propeller of AT-6A #41-439 struck its fuselage from below and behind just aft of the rear cockpit, cutting through the fuselage and severing the tail section. Both airplanes then fell out of control. With its propeller damaged and the port horizontal stabilizer and elevator severed, AT-6A #41-439 rolled over to an inverted position and dove into the ground where it exploded into flames. AT-6A #41-583 immediately dove straight to the ground where it erupted into flames upon impact. All of the occupants were killed instantly upon impact.

2-6-43A. Sumter, South Carolina. At 1025, a Vultee BT-13A crashed into swampy terrain 15 miles east of Shaw Field, Sumter, South Carolina, killing student A/C James E. Guest. The student had taken off at 0945 for a one hour flight to Monaghan Field, an auxiliary airfield about 8 miles from Shaw Field. At about 1030, low-level clouds moved into the area and student flying was called off. All student airplanes were called in and it was soon discovered that A/C Guest was missing. The airplane wreckage was found in a remote wooded swamp on 2-10-43. Investigation of the wreckage revealed that the airplane had crashed into the swamp at a shallow angle and at a low airspeed.

The airplane did not burn but was completely destroyed.

2-6-43B. Tonopah, Nevada. At 0915 PWT, a Bell P-39L crashed six miles southeast of the AAF Bombing and Gunnery Range near Tonopah, Nevada, killing pilot Capt. Charles C. Johnson III. Investigation of the wreckage indicated that the airplane had struck the ground in an almost vertical angle. The airplane exploded into flames upon impact. It was speculated that the pilot was attempting to recover from an aerobatic maneuver but had insufficient altitude. The pilot apparently tried to parachute to safety but he left the airplane at an altitude too low to allow his parachute to fully deploy.

2-6-43C. Beech Grove, Arkansas. At 1155, a Vultee BT-13A suffered a catastrophic structural failure and crashed near Beech Grove, Arkansas, killing students A/C William M. Cozine, Jr. and A/C John L. Malm. The airplane had taken off from the Army Air Field at Walnut Ridge, Arkansas, at about 1130 on an instrument team ride with Cadet Malm under the hood in the rear cockpit and Cadet Cozine in the front cockpit as safety pilot. Examination of the wreckage indicated that both wings failed at a point 18 inches outside of the center section. One wing apparently struck the tail section causing its failure and separation. The airplane then spun to the ground and burst into flames upon impact. Investigators stated that the wings failed because of a sudden application of negative wing loading. It was not known what kind of maneuvers the students had put the airplane through prior to the wings failing and separating. Cadet Malm was killed when he had attempted to parachute to safety; he had left the airplane at an altitude too low to allow his parachute to fully deploy.

2-6-43D. Pendleton, Oregon. At 1759, a Lockheed P-38E crashed one and one-half miles southeast of Pendleton Field, Pendleton, Oregon, killing pilot 2Lt. Edwin Joseph Powell. The pilot had taken off with another P-38E on a local aerobatic flight. The P-38s climbed to an altitude of approximately 10,000 feet and began a "rat race." After performing a few maneuvers Lt. Powell did a complete loop, followed by a wingover maneuver into a steep dive. The airplane dove toward the ground at high speed with no apparent effort by the pilot to recover. The airplane then dove into the ground and exploded violently into flames, killing the pilot instantly. The cause of the accident was undetermined. It is likely that the pilot had put the airplane in a diving maneuver from which he could not recover because of the tendency of the P-38 type airplane to suffer from severe control surface buffeting caused by compressibility.

2-6-43E. San Angelo, Texas. At an unknown time after 2310, a Vultee BT-13A crashed four and one-half miles southeast of Goodfellow Field, San Angelo, Texas, killing A/C William R. Warner. The airplane had taken off from Goodfellow Field on a night cross-country flight to Sweetwater, Texas, to Harpersville, Texas, and return. The student had successfully completed the night cross-country and had returned to the field. The student requested landing instructions and had reported that his lights had been going on and off during the flight, but they were working at the time of the communication. The student was given landing instructions and told to land into the floodlights that were illuminating a portion of the field. The control officer did not receive a reply and soon became involved with another airplane. After a short period of time, the control officer noticed that Cadet Warner had not landed. He attempted to contact the student by radio but was unsuccessful. The student and his airplane were found the following morning. Examination of the wreckage indicated that the airplane had contacted the ground in a flying attitude and at cruise speed.

2-7-43A. Clarksdale, Mississippi. At 0815, a Stearman PT-17 crashed two miles southwest of Fletcher Field, Clarksdale, Mississippi, killing civilian instructor Joseph B. Patterson and seriously injuring A/C Abe J. Smith. Investigators stated, "While practicing steep turns at low altitude, approximately 500 feet (in violation of flying regulations), A/C Smith was being corrected for a mistake he made in a steep turn just previously attempted. He was told by Mr. Patterson that he had lost 100 feet. While the instructor was flying the airplane to demonstrate correct procedure, the airplane stalled and spun in."

2-7-43B. Fayetteville, North Carolina. At 1050, a Bell P-39D suffering an engine failure crashed five miles southeast of Fayetteville, North Carolina, killing 2Lt. Norman Henry Cisar. Investigators speculated that the engine had failed and the pilot was attempting to make an emergency forced landing in a small cornfield when the airplane stalled and spun to the ground. The airplane struck the ground at an angle of 45 degrees and burst into flames upon impact. The pilot was killed instantly. The airplane had taken off from Pope Field, Fort Bragg, North Carolina.

2-7-43C. Alvarado, Texas. At 1035, a North American BC-1 crashed near Alvarado, Texas, killing pilot F/O John H. Jackson and passenger Pvt. Raymond D. Shelton. The Aircraft Accident Classification Committee stated, "F/O John H. Jackson, reported to the Base Operations Office, Tarrant Field, Fort Worth, Texas, and requested an airplane to get in flying time for the month of February. Maj. J.W. Campbell, Assistant Base Operations Officer, qualified the credentials presented by F/O Jackson, and assigned him BC-1 airplane AAF #38-408. Maj. Campbell gave Jackson specific instructions that he would remain within 20 miles of Tarrant Field, that he would maintain radio contact with the control tower, and that he would not engage in aerobatics. At the flight line, Jackson picked up Pvt. Raymond D. Shelton as passenger and departed at 0920. Apparently they flew direct to Alvarado, Texas, which is 27 miles airline distance from [Tarrant Field]. Alvarado is the birthplace for both Jackson and Shelton. They went to grade school and high school together in this town. From shortly after 0900 until the time of the accident at approximately 1035, F/O Jackson flew his airplane with total disregard for his own safety and the safety of others on the ground by flying below treetops and over populated areas in violation of AAF regulations and contrary to all rules of good judgment. At approximately 1035, Jackson dove over the service station of J.E. Mahanay, and in doing so he became so low that he took off approximately eight feet of a tree alongside the station and tore out five wires carrying 2,300 volts. It is believed that at this time he lost control of his airplane since testimony shows that he gained some altitude after striking the wires. He then slipped off on his left wing over the Santa Fe Railroad Depot and the Western Union Telegraph lines along the depot, and struck a box car with the left wing while the airplane was in a vertical bank. The engine, fuselage and right wing carried past the top of the box car, tearing out sixteen wires of the Postal Telegraph line, the engine [tore loose], then the center section and finally the fuselage and wing came to rest about 150 feet from the box car." Numerous citizens of Alvarado observed the airplane flying at extremely low altitude over the town for some time before the accident.

2-7-43D. Eglin Field, Florida. At 1250 CWT, a North American AT-6A on a ground gunnery mission crashed at AAF Bombing and Gunnery Range 3A, near Eglin Field, Valparaiso, Florida, killing RAF Flying Cadet Felix Quinn. The airplane had made a normal pass at the target and was making a normal recovery when it was seen to roll to the right to an inverted attitude. The airplane continued to roll through until it slammed into the ground inverted in a banking attitude, striking initially on the starboard wing. The AT-6A exploded into flames upon impact and the student was killed instantly. The airplane and student were stationed at Craig Field, Selma, Alabama.

2-7-43E. Apache Junction, Arizona. At 1425 MWT, a Cessna AT-17 suffered the structural failure of the port wing and crashed three miles south of Apache Junction, Arizona, killing instructor 1Lt. James C. Shepard and Chinese Air Force student A/C Sun Yun Hsia. The Accident Classification Committee stated, "It is the opinion of this committee that the direct cause of this accident was the tearing away of the fabric attached to approximately one-half of the engine nacelle cowling from the upper surface of the left wing, causing a loss of lift and the disintegration of the wing with subsequent half-roll and dive into the ground. Witnesses observed the airplane in a long, shallow turn to the left at an altitude of about 1,000 feet above the terrain. The airplane then went slowly into a half-roll to the left and dove into the ground still rolling. A large section of fabric still attached to about one-half of the left engine nacelle was found approximately one mile from the wreckage of the airplane. The metal engine nacelle had ripped cleanly on a diagonal from the right rear to the left front. The door, having been released by use of the emergency release, was lying intact and undamaged about 500 feet from the wreckage of the airplane. It was evident that Lt. Shepard had tried to jump. It appeared that he was unable to do so because of the high seats installed on this airplane. The student's safety belt had not been unbuckled."

2-7-43F. Weatherford, Texas. At 1530, a Consolidated B-24D attempting an emergency forced landing crashed seven miles northwest of Weatherford, Texas, killing passengers Pvt. Thaddeus G. Searcy and Pfc. Charles D. Wilkins and injuring seven crewmembers and passengers. Injured in the crash landing were: 2Lt. James B. Turner, B-24 instructor; 2Lt. Noel L. Dahlander, co-pilot/B-24 student; 2Lt. Wilfrid A. Daly, co-pilot/B-24 student; 2Lt. Maurice L. Malone, co-pilot/B-24 student; 2Lt. Cecil Culbreth, passenger; SSgt. Philip D. Smiley, engineer; Pvt. Henry A. Wisz, passenger. The B-24D had taken off from Tarrant Field on a routine instrument training flight. Investigators stated, "During the first problem in instrument training, with Lt. Malone under the hood and Lt. Daly as co-pilot, the number-one engine was feathered and unfeathered using the normal procedure. This first problem was completed successfully and the airplane came down and contacted the control tower for landing instructions. They were instructed by the control tower to continue flying for a while since there was an 'Army Hour' [radio] program being broadcast from Tarrant Field. Lt. Turner had the airplane pull away from the field and placed Lt. Daly in the pilot's seat and Lt. Dahlander in the co-pilot's seat for a second [instrument training] problem. After flying for approximately ten minutes with Lt. Daly under the hood and Lt. Dahlander as co-pilot, the number-two engine was feathered. When it had been feathered approximately ten minutes, Lt. Turner instructed the co-pilot to unfeather and start the number-two engine. Simultaneously, with the unfeathering of the number-two propeller, a loud noise described as the slap of a hand against the ship, or a backfire, was heard by the crew. At this point the pilot under the hood began having trouble with the airplane, which began losing altitude and airspeed and began turning to the left, making it necessary to apply terrific right rudder. At approximately 1,000 to 1,500 feet, the instructor removed the hood and replaced Lt. Daly in the pilot's seat. From all testimony, it appears that as the number-two engine was being unfeathered, the number-one engine went out. With both engines out, sufficient altitude and airspeed could not be regained, and the airplane was crash-landed with the landing gear retracted, resulting in the death of [two] passengers [and] serious injury to two members of the crew and the complete destruction of the aircraft. The board believes that in the beginning of the operation of unfeathering of the number-two propeller, the switches on the number-one engine were cut off and the switches of the number-two engine never turned on. It is believed that the noise heard by the crew during the process of unfeathering the number-two engine, occurred in the number-one engine when the co-pilot unintentionally cut the switches on the number-one engine instead of turning on the number-two engine."

2-7-43G. Somerset, Kentucky. At 1715 CWT, assistant crew chief Pfc. Herman A. Tinder plunged to his death when he fell out of a Douglas C-47A that had been operating approximately twelve and one-half miles SSW of Somerset, Kentucky. The C-47A was piloted by 1Lt. Howard C. Vose, Jr. and 2Lt. Carl W. Campbell. Crew Chief Sgt. John F. Ewell and four airborne infantry passengers were uninjured. Investigators stated, "Army C-47 #41-18471 was cruising at 7,000 feet in the [vicinity] of Somerset, Kentucky. Co-pilot 2Lt. C.W. Campbell was flying the airplane at the time. The crew chief ordered Pfc. Tinder to remove the pilot's compartment door. Pfc. Tinder misunderstood the order and attempted to remove the main cargo door. Pfc. Tinder succeeded in opening the door and Sgt. Ewell, observing this action, immediately ran to help Pfc. Tinder close the cargo door. The two men had succeeded in nearly closing the cargo door when the slipstream tore the door loose and outwards. Pfc. Tinder was thrown clear of the airplane without a parachute and took a free fall from 7,000 feet above sea level. Sgt. Ewell fell backwards inside of the airplane. The exact spot of the accident cannot be determined because prior to this occurrence there had been engine trouble. The ship left Louisville [Kentucky] and returned on dead reckoning. However, the crew is in agreement that the accident occurred near Somerset, Kentucky, possibly 10 or 15 miles west or southwest of that town." Pfc. Tinder's body was not located at the time of the incident and it is not known if his body was ever found.

2-7-43H. Houston, Texas. At 1920, a North American AT-6A crashed into a private residence at 722 West 38th Street, Houston, Texas, killing instructor 2Lt. Joseph F. Folsom, 26, and seriously injuring student A/C Truman R. Theobald. The house, belonging to Mr. W.E. Dowdell, was nearly completely destroyed. The airplane had taken off from the Army Air Field on the Matagorda Peninsula on an instrument training flight. The student was supposed to practice on the Victoria, Texas, radio range. When the instructor could not tune in the Victoria range he ordered the student, who was flying from the rear cockpit, to fly to the Houston area. The instructor stated to the student that he lived in Houston and would fly over. He then took over control of the aircraft and flew to a point 15 miles WSW of Houston and began buzzing a farmhouse. The occupants of the farmhouse emerged and began waving at Lt. Folsom. Lt. Folsom returned the waves and then flew toward Houston. Once the airplane arrived at Houston, the instructor began to buzz the house of his sister at very low altitude. After several low-level passes the AT-6A collided with 50-foot trees. The airplane then went out of control and collided with the east wing of the frame cottage at 722 West 38th Street. The cottage was nearly completely destroyed. The airplane continued through the house and came to rest in the backyard of the house adjacent on the east. The airplane was a complete wreck and the surrounding landscaping was seriously damaged. Investigation revealed that the instructor had commented to another pilot that he would be flying for two hours. The student instrument flight was scheduled to last about 40 minutes only. Investigators concluded that the instructor had planned ahead of time to buzz the two locations. Mr. and Mrs. Dowdell and their two children (ages 2 and 4) were in the home at the time of the crash and miraculously escaped injury. The instructor and student were stationed at the Army Air Field at Eagle Pass, Texas.

2-7-43I. Merced, California. At 2030, a Vultee BT-13A crashed 14 miles south of the Army Air Field at Merced, California, killing instructor 1Lt. James Carl Jewell and student A/C Howard Gray Park, Jr. Investigation revealed that the airplane was flying at about 600 feet agl on the downwind leg of the traffic pattern at Potter Auxiliary Air Field when it entered a stall and a spin to the left, smashing into the ground and exploding into flames. The occupants were killed instantly. Investigators speculated that the airplane might have inadvertently entered a very small cloud that was at an altitude of 400 feet agl, causing the student to enter a stalling attitude.

2-7-43J. Longview, Texas. At 2325, a Vultee BT-13A crashed two miles north of Longview, Texas, killing student A/C Roscoe S. Stallworth, Jr. The airplane had taken off from the Army Air Field at Greenville, Texas, on a 38-ship night navigation flight to Parker Airport, to Wills Point, and return. The airplane became separated from the flight and the student became lost. The airplane entered a stall and spun to the ground while the student was trying to find a suitable place to land. Witnesses stated that the airplane had circled the Longview area several times before the crash. It was also reported that the student had been turning on the landing lights intermittently while he attempted to find a place to land.

2-8-43A. Hayward, California. At 0135, a Bell P-39D collided with rising terrain three miles east of Hayward, California, killing pilot 2Lt. Rex A. Fulghum. Investigators stated, "Lt. Fulghum and 2Lt. Howard B. Stivers took off of Mills Field at 0040 for a night formation flight. They entered a fog bank at 3,000 feet over Hayward. The flight leader made a 180-degree turn in order to clear the fog bank. The fog closed in during the turn. Lt. Fulghum crossed under the leader and that is the last time he was seen. Lt. Fulghum evidently thought the leader was going to let down in the direction of the [San Francisco Bay] area.

He began his descent and crashed on a mountain about 1,900 feet high. Lt. Fulghum crashed on the downward slope of a mountain just east of Hayward in Palmyra Canyon. The weather prior to take-off had been reported by telephone from the C.A.A. weather station at Mills Field [10 miles south of San Francisco] as visibility unlimited and scattered clouds at 4,500 feet."

2-8-43B. Sioux City, Iowa. At 2045 CWT, a Boeing B-17F crashed three miles southeast of the Army Air Base at Sioux City, Iowa, killing eight crewmembers and injuring two others. Killed in the crash were: 2Lt. Thomas V.M. Cryer, B-17 instructor; 2Lt. George W. Johnson, co-pilot; 2Lt. Robert N. Gribbin, navigator; 2Lt. John T. Quillin, crewmember; SSgt. Ben D. McCarson, engineer; SSgt. Preston W. Wheeler, radio operator; Sgt. Frederick R. Siegman, assistant radio operator; SSgt. William E. Reeves, gunner. Bombardier 2Lt. Robert S. Bartley and gunner Sgt. Thomas X. Pryal were seriously injured. Investigators stated, "[The B-17F] took off from Sioux City Army Air Base on a local training flight. The pilot contacted the control tower prior to take-off but no further radio contact was established. The airplane was fully serviced with gasoline and oil. The take-off was made to the southeast and almost immediately [the B-17] crashed approximately three miles southeast of the end of the Southeast-Northwest runway. Local weather was clear and unlimited with no moon and very dark. Investigation of the wreckage disclosed that the pilot's gyros were uncaged. According to the testimony of witnesses and investigation of the scene of the crash, the airplane was under control and was flying very low, but for some reason the pilot flew into the ground under full power in a flat position. [The B-17 bounded back into the air after the initial impact.] Then the airplane came apart in the air, exploded and crashed into a grove of trees approximately one-half mile from the initial contact with the ground. Examination of the initial contact marks on the ground indicated level flight (with the left wing slight lower than the right as the number-one propeller touched the ground first) and the airplane was apparently proceeding straight ahead under full power at a high rate of speed. The airplane was traveling due east. The wheels were up and witnesses who saw the crash stated there was no fire in the air and that the pilot did not turn on his landing lights or attempt a forced landing. It is presumed by the committee after examining the wreckage, marks on the ground and interviewing several witnesses to the actual crash that the pilot unknowingly or through lack of attention flew into the ground after take-off. Marks showed the propeller of the number-one engine touched the ground approximately eight yards in front of the barbed wire fence. The propeller and the left wing took out about 20 feet of the fence. The propeller of the number-two engine then struck the ground and in approximately 20 feet the propellers of the number-three and number-four engines made contact. The pilot evidently then attempted to pull up, striking the ball turret on the ground and causing the fuselage to break in two at this station. A section of the left wing and the number-one engine also flew off at this point of contact. The front part of the fuselage bounced or flew about one-quarter of a mile before striking the tops of some tall trees causing it to explode and burn. Tail section flew approximately the same distance to the right, not striking the trees and not being badly damaged. Wreckage was strewn all along the path form the first contact to where the main parts came to rest. None of the propellers were feathered. Surviving crewmembers stated that they were not aware of anything wrong in the air and it is presumed that the crash was caused by inattention on the part of the pilot and co-pilot. There was no evidence of sabotage and no available evidence of structural failure in the air. The pilot was a qualified instructor pilot with considerable experience and was at the controls at the time of the crash."

2-9-43A. Atlantic Ocean, Florida. At 0230 EWT, the crew of a Consolidated C-87 (the transport/cargo version of the B-24D) bailed out over the Atlantic Ocean approximately ten miles east of Hollywood, Florida, killing two crewmembers and injuring six others. The C-87 flew unoccupied for over 1,300 miles before crashing into El Viejo Mountain near Zaragoza, Nuevo Leon, Mexico. Assistant navigator 2Lt. Bernard A. Brannon and radio operator Pvt. Stanley Gutowsky were never found and were presumed lost at sea. Six other crewmembers suffered minor injuries after parachuting into the sea. They were: 1Lt. Robert C. Ulmer, pilot; 1Lt. Carl W. Hauth, co-pilot; 1Lt. Robert H. Digby, navigator; Pvt. Kenneth J. Riddle, engineer; 2Lt. Albert E. Lloyd, passenger; 2Lt. Gerald P. Knutson, passenger. Investigation revealed that the airplane had taken off at 0101 EWT from Morrison Field, West Palm Beach, Florida, on a ferry flight to an undisclosed overseas destination. The C-87 climbed normally after take-off and the pilots leveled out at an altitude of 9,000 feet, about 80 miles out to sea. The pilot reduced power and set the controls for cruise flight. Soon after leveling off and reducing power, the pilot noticed that the airplane was not gaining enough airspeed. Moments later the rudders and elevators began vibrating violently. The airplane began losing altitude at a rate of 300 feet per minute. The pilots increased power to maintain altitude but were barely able to sustain level flight. The pilot then turned the airplane back toward the mainland and ordered the crew and passengers to jettison all luggage and cargo. The pilot increased power and was able to maintain level flight, but the vibration in the tail controls became so violent that it was difficult for the pilots to keep their hold on the control wheels or their feet on the rudder pedals. The pilots struggled with the airplane until they reached the area near Miami, Florida, but the C-87 soon became uncontrollable and began to lose altitude rapidly. The pilot, fearing that the violent vibrations would tear the C-87 apart in mid-air, ordered all crewmembers and passengers to abandon ship. After all of the crewmembers had bailed out, the pilot set the automatic pilot and then exited the airplane. U.S. Coast Guard and civilian vessels operating in the area rescued six of the eight crew and passengers who had parachuted into the sea. The C-87, its control anomalies apparently self-corrected, flew unoccupied until it reached the area south of Zaragoza, Nuevo Leon, Mexico, at 0630 local Mexican time. The C-87 apparently circled the area for two hours before crashing into the south side of El Viejo Mountain approximately 400 feet below the peak at 0830 local Mexican time. It was noted by investigators that the subject airplane had had previous problems with tail surface buffeting.

2-9-43B. Dos Cabezos, Arizona. At 0914, a Curtiss AT-9 collided with rising terrain three miles north of Dos Cabezos, Arizona, killing instructor 2Lt. Martinjohn Knight and student A/C Marlin D. Olson. The airplane had taken off from the Army Air Field at Douglas, Arizona, at 0815 MWT on an instrument training flight to Cochise, Arizona, to Rodeo, Arizona, and return. The airplane never returned to base and was soon declared missing. The wreckage of the AT-9 was discovered from the air at 1525. Investigation of the wreckage revealed that the airplane had smashed into the east slope of the mountain at an elevation of approximately 6,500 feet while under control in a descending left turn. Investigators speculated that clouds had probably concealed the mountain at the time of the accident and that the student and instructor had probably miscalculated their position while making their descent, causing the airplane to collide with the mountain. It was further speculated that the airplane might have been descending and turning in actual instrument conditions to avoid icing conditions that had been reported by other pilots flying in the area. Investigators noted that this was the student's final instrument check ride for the course.

2-9-43C. Dalhart, Texas. At 1145, a Waco CG-4A glider crashed while attempting a landing at the Army Air Field at Dalhart, Texas, killing glider pilot F/O Richard B. Young and seriously injuring five other fliers. Students and crew seriously injured in the accident were: SSgt. John T. White, Jr., glider student; SSgt. William M. Whitney, IV, glider student; SSgt. Mason S. Wilt, glider student; SSgt. Robert L. Wonders, glider student; Pvt. James A. Roberts, crewmember. Investigators stated, "Glider No. 42-61879 entered the pattern [then] turned onto the base leg and started onto the approach leg. Extremely gusty winds fifty feet above the ground [and poor] visibility due to blowing dust caused the pilot to lose control of the ship. A gust of wind on the right wing followed by a lull in the gusts caused the ship to go into a [quarter-turn] spin to the left and the pilot lost control of the ship. The glider hit the ground nose and left wing first. Momentum carried the glider onto its back." The glider came to rest on the east end of the Northeast-Southwest runway.

2-9-43D. Westover Field, Massachusetts. At 1500, a Republic P-47B went out of control while attempting a take-off at Westover Field, Chicopee, Massachusetts, striking and killing two enlisted men on the ground and seriously injuring Pvt. Henry F. Morehead. Pvt. Jacob Adelsky and Pvt. Dewey A. O'Neal were killed instantly by the careening airplane; P-47B pilot 2Lt. Michael G. Kurpa was uninjured. Investigators stated, "The airplane started down the runway straight. At a point about one-third of the way down the runway, the tail began to lift, and the airplane bearing to the right. At about one-half way down Runway 01, the airplane struck a snow bank on the right, lost flying speed, and became out of control. The path of the airplane went down the right side of the runway, straddling a snow bank until it reached the intersection of Runway 01/19 and Runway 15/33, at which point the left landing gear was torn off of the ship. [The airplane] started immediately bearing to the left on the left wing tip and the right landing gear, crossed the runway in use, then proceeded down the right side of Runway 15 where a group of men were working on a snow clearing detail. [The P-47B] plunged through the men, killing two and injuring a third, and continued on for a distance of 50 feet where it came to rest, right side up and resting on the left wingtip and the right landing gear. At no time while the ship was out of control did the power seem to be cut. The group of men at work were well off the runway in use and were well out of danger for normal operations on that runway at the time of the accident."

2-9-43E. Newport, Arkansas. At 1420, Pvt. Clifford R. Hefner was killed when he ran into the rotating propeller of a Vultee BT-15 that was taxiing at the Army Air Field at Newport, Arkansas. The Accident Classification Committee stated, "The BT-15 airplane in which the student pilot, A/C Frank H. Pearman, and the pilot, 2Lt. Daniel S. Dickinson, were preparing to taxi out to take-off position and moving slowly out of the parking area with an enlisted man on each wing to prevent the airplane from hitting obstacles while taxiing. Suddenly the student pilot and the instructor pilot noted that the enlisted man on the [port] wing left his position [and] ran sharply in front of the airplane. The propeller knocked the enlisted man to the ground resulting in the amputation of [his] right arm by the propeller. The airplane immediately stopped and Lt. Dickinson went to [Pvt. Hefner's] assistance."

2-9-43F. Spofford, Texas. At 1500, two North American AT-6C airplanes collided in mid-air near Spofford, Texas, killing A/C Walter G. Rehm aboard AT-6C #42-3914. A/C Theodore A. Roamanek was able to parachute to safety from AT-6C #41-32595. He was uninjured. The two airplanes were part of two separate three-ship flights that had been split up after performing various maneuvers at 8,000 feet msl. After the flight leader split up the flights, he ordered A/C Rehm to lead the second three-ship flight in straight and level flight along a road. After performing various maneuvers with the first element, the flight leader sent A/C Roamanek to join the flight led by A/C Rehm. The student attached himself to A/C Rehm's flight and the two airplanes collided while A/C Rehm was leading the flight in a steep turn to the right. A/C Roamanek's airplane apparently overran A/C Rehm's airplane from behind. Both airplanes fell out of control and smashed to the ground where they exploded into flames upon impact. Apparently A/C Rehm attempted to parachute to safety but had pulled his ripcord too soon. When the parachute opened it fouled on the airplane, dragging the student along as the AT-6C spun toward the ground. The student was eventually flung free of the airplane at an altitude of 300 feet agl, but his parachute was seriously damaged and he fell to his death. His body was found near the wreckage of his airplane, about 30 miles north of his station at the Army Air Field at Eagle Pass, Texas.

2-9-43G. Mobile, Alabama. At 1515, a North American B-25C suffering engine and propeller problems crashed into trees while attempting an emergency forced landing 15 miles east of Brookley Field, Mobile, Alabama, killing passenger TSgt. Paul A. Compton and seriously injuring five crewmembers. Investigators stated, "Pilot after take-off headed on course and climbed to 1,700 to 2,000 feet and trimmed his airplane for cruising. Everything had worked normally on the run-up prior to departure. When approximately 15 miles east of Brookley Field, the right propeller began to flatten out with a loss of power, increased rpm and the usual sound resulting from a propeller runaway. Pilot and co-pilot attempted to reduce rpm but this had no effect. They then throttled, cut the engine, and worked the feathering switch on the right engine, however, right propeller would not feather. Although power was increased on the left engine the crew contends that they had a left engine failure and a crash was imminent. The pilot tried to make a small clearing, but hit trees approximately 100 yards from this clearing, cut the switches, and crashed to the ground." Investigation of the starboard engine revealed that the front and rear master rod bearings had both failed and seized, causing the propeller gears to become disengaged. Investigation of the port engine revealed no obvious problems and its apparent failure remained undetermined. Crewmembers seriously injured in the crash were: 1Lt. John B. Goode, pilot; 1Lt. John P. Lyle, co-pilot; SSgt. Clarence A. Thomas, engineer; W/O Sam Gacesa, passenger; Sgt. Richard A. Slatko, passenger.

2-10-43A. Monroe, Louisiana. At 0615, a Douglas A-20C crashed 50 miles southwest of Selman Field, Monroe, Louisiana, killing the crew of three. The airplane had taken off from Esler Field, Alexandria, Louisiana, on a routine cross-country navigation flight to Key Field, Meridian, Mississippi, via Monroe, Louisiana. Investigators stated that the airplane had begun to lose altitude and eventually collided with trees. The airplane then smashed into the ground and erupted into flames. Pilot 1Lt. Claude W. Vroma, navigator 2Lt. Charles R. Weaver and gunner SSgt. Theodore F. Ryder were killed instantly. Weather was reported as broken clouds at 2,500 feet to 4,000 feet with six miles visibility with light drizzle. Winds were from the southwest at 35 mph, gusting to 46 mph. Investigators could not determine why the airplane flew into the terrain.

2-10-43B. Salisbury, North Carolina. At 1030, a Vultee BT-13A suffering an in-flight fire crashed two miles south of Salisbury, North Carolina, killing pilot 2Lt. James A. Hooper, who fell to his death in an unsuccessful parachute jump. Investigators stated, "Lt. Hooper was a member of a flight ferrying BT-13 airplanes from Wilmington, Delaware, to various stations in Kansas. The flight landed at Greensboro, North Carolina, was serviced and took off at approximately 1000. From 10 to 20 minutes out of Greensboro, Lt. Hooper encountered fire in the engine section. He was seen to turn from the flight and circle in the vicinity of an open field with his nose section burning. Just before the crash, one witness noticed a burst of flames from the airplane and also saw the pilot attempt to leave the airplane from the left side. It is assumed that the pilot upon encountering fire picked out a suitable field and attempted to land but on his final approach found that the fire was so intense that he attempted to bail out at low altitude."

2-10-43C. Victoria, Texas. At 1030, two North American AT-6C airplanes collided in mid-air six miles north of Foster Field, Victoria, Texas, killing A/C Keith W. McCauley, who was killed attempting to parachute from AT-6C #42-3992. Instructor 2Lt. Charles M. d'Emery was uninjured and A/C Frederick J. McAnally received minor injuries in the collision, but they were able to safely land AT-6C #41-32358 at Wagner Auxiliary Airfield. Investigation revealed that the two airplanes were part of a six-ship flight on a formation-training mission. The formation was flying at 7,000 feet on top of the overcast. The leader began to let down through an opening at about 6,200 feet. As the flight began the let-down, A/C McCauley, flying in the number-six position, overran A/C McAnally, who was flying in the number-five position. The airplanes collided and A/C McCauley's airplane began an uncontrolled descent. A/C McCauley attempted to regain control of his airplane but was unsuccessful. He fell to his death after he bailed out at an altitude too low to allow his parachute to fully deploy.

2-10-43D. Sumter, South Carolina. At 1400, a Vultee BT-13A crashed 11 miles northeast of Shaw Field, Sumter, South Carolina, killing A/C John H. Maurer and seriously injuring instructor 2Lt. Gerald

C. Merchant. The student was performing a simulated forced landing exercise at an altitude of 2,000 feet agl when the airplane stalled out of a 180-degree turn to the left. The BT-13A rolled over to an inverted attitude and then fell into a spin to the left. The student was slow in effecting a recovery and the instructor immediately attempted to get control of the airplane. The student held the stick in the full aft position, preventing the instructor from regaining control of the airplane before it struck the ground. It was noted by the instructor that the student had failed to recover from a power-on spin the day before.

2-10-43E. Location Unknown. At a time after 1639 EWT, a North American O-47B flying in poor weather disappeared with its crew over Long Island Sound on a positioning flight from Bradley Field, Windsor Locks, Connecticut, to Mitchel Field, Long Island, New York. Pilot F/O Talmadge J. Simpson and passenger Cpl. Louis P. Vogt were declared missing and presumed lost at sea. Weather was reported as ceiling 4,000 to 6,000 feet with freezing rain. Investigators noted that the pilot had not flown on instruments in over six months. No trace of the airplane or crew was ever found.

2-10-43F. Dalhart, Texas. At a time between 1730 and 0030, a Douglas C-47 crashed 25 miles south of the Army Air Field at Dalhart, Texas, killing pilot 1Lt. Jules C. Burke, Memphis, Tennessee, and co-pilot 1Lt. Howard D. Bankston, Americus, Georgia. The C-47 had taken off from Dalhart and disappeared while on a routine instrument training flight. The C-47 was found smashed to bits in a dry riverbed on 2-20-43. Investigators could not determine why the C-47 had slammed into the ground out of control. The airplane had exploded violently into flames upon impact and wreckage was scattered over wide area.

2-11-43A. Providence, Rhode Island. At 1040, a Republic P-47B flying in instrument conditions crashed into a railroad switch yard in Providence, Rhode Island, killing pilot 2Lt. Robert F. Meyer. The subject airplane was part of a flight of four P-47B airplanes that took off from Green Field, Providence, Rhode Island, on a ferry flight to Trumbull Field, Groton, Connecticut. Conditions at Providence were reported as instrument with a 500-foot ceiling. The field had apparently been ordered closed shortly before the flight took off. Investigation revealed that the flight was cleared in error by Sgt. C.T. Simms without the signature of Capt. Francis G. Oksala, the Flight Operations Officer at Green Field. Sgt. Simms stated that he spoke with someone he thought was Capt. Oksala, who was in fact in a squadron meeting. Sgt. Simms, thinking he had authorization, cleared the flight. The flight, led by 1Lt. Gene F. Drake, took off at 1034 EWT and began to circle the field. After circling the field twice, only three airplanes were seen circling on the third pass. Witnesses had observed the subject airplane break off from the flight and begin a turn to the left. Witnesses reported that a thick fog had moved into the area at the time and that visibility was "bad." The airplane continued in a descending left turn until it crashed into the Auburn Freight Yard of the New York, New Haven & Hartford Railroad. The P-47 struck a boxcar with the port wing then smashed into the ground and exploded into flames, killing Lt. Meyer instantly. Investigators noted that the flight leader, Lt. Drake, did not have an instrument card on his person when filing his instrument flight plan. See 2-11-43B, C, D.

2-11-43B. Narragansett Bay, Rhode Island. At 1045, a Republic P-47B flying in instrument conditions crashed into Narragansett Bay, Rhode Island, killing pilot 2Lt. Raymond D. Burke. The airplane was part of a flight of four P-47B airplanes on a ferry flight from Green Field, Providence, Rhode Island, to Trumbull Field, Groton, Connecticut. The flight was led by 1Lt. Gene F. Drake and had been cleared in error by an enlisted man. The flight had taken off from Green Field at 1034 and began to circle the field. The flight circled the field twice, but on the third pass only three airplanes were seen circling. Lt. Burke crashed into the East Passage of Narragansett Bay, Rhode Island soon after the first crash. Lt. Burke's body was retrieved from the bay shortly after the accident. Witnesses reported that a thick fog had moved into the area at the time of the accident and that visibility was "bad." See 2-11-43A, C, D.

2-11-43C. Location Unknown. At about 1050 EWT, a Republic P-47B flying in instrument conditions disappeared soon after taking off from Green Field, Providence, Rhode Island, apparently crashing into the West passage of Narragansett Bay. Pilot 1Lt. Gene F. Drake was declared missing and presumed lost at sea. Lt. Drake was leading a four-ship flight of P-47B airplanes on a ferry flight to Trumbull Field, Groton, Connecticut. Lt. Drake did not have his instrument card on his person when he tried to file his instrument flight plan and the flight was cleared in error by an enlisted man. Lt. Drake referred the enlisted man to a Capt. Hovis at Trumbull Field. Capt. Hovis transferred the call to another officer and he allegedly verified that Lt. Drake did in fact possess a current instrument card. Investigation revealed that the flight was cleared in error by Sgt. C.L. Simms without the authorizations of Capt. Francis G. Oksala, the Flight Operations Officer at Green Field. Sgt. Simms stated that he spoke by telephone with someone that he thought was Capt. Oksala, who was in fact in a squadron meeting. Sgt. Simms, thinking that he had authorization, cleared the flight. Ceiling was estimated to be at 500 feet and the field had apparently been closed shortly before the take off at 1034 EWT. After the take-off the flight circled the field twice, but on the third pass only three airplanes were seen circling. Another P-47B crashed a short time later, followed by the apparent crash of Lt. Drake's P-47B at about 1050 EWT. Members of the

243rd Coast Artillery reported that they heard what they thought was an airplane crash in the West passage of Narragansett Bay near Fort Getty, Rhode Island. Witnesses reported that a thick fog had moved into the area and visibility was "bad." Lt. Drake's body was never recovered. See 2-11-43A, B, D.

2-11-43D. Location Unknown. At about 1050 EWT, a Republic P-47B flying in instrument conditions disappeared soon after taking off from Green Field, Providence, Rhode Island. Pilot 2Lt. John Pavlovic was declared missing and presumed lost at sea. It was speculated that the airplane had crashed into Narragansett Bay after taking off with three other P-47B airplanes on a ferry flight from Green Field to Trumbull Field, Groton, Connecticut. The other three airplanes had also crashed soon after take-off. Witnesses reported that a thick fog had moved into the area and visibility was "bad." Investigation revealed that the flight had been cleared in error by an enlisted man and without the authorizations of Capt. Francis G. Oksala, Flight Operations Officer at Green Field. The flight was led by 1Lt. Gene F. Drake. See 2-11-43A, B, C.

2-11-43E. Benson, Arizona. At 0930, a Curtiss AT-9 collided with a pole and crashed one-half mile east of Benson, Arizona, killing students A/C Albert I. Gray and A/C Weber A. Hattrem, Jr. Investigators stated, "Witnesses observed the airplane flying about 30 feet above the ground, following the road into Benson, Arizona, for about two miles. They then saw the airplane pull up abruptly, execute a half-roll to the left, and dive straight into the ground. According to witnesses, both engines evidently were functioning perfectly. A subsequent inspection of the area disclosed that the left wing of the airplane had struck a telephone pole, approximately 50 feet high, and that the pole had sheared off about four feet of the left wing, leaving that portion of the wing hanging on the pole. The road [that] the airplane had been following had no telephone or power line paralleling it. The pole [that] the airplane hit was the highest pole (the others being approximately 30 feet high) of a line [that] crossed the road diagonally."

2-11-43F. Walla Walla, Washington. At 1115, a Boeing B-17F flying in instrument conditions collided with rising terrain 17 miles east of Walla Walla, Washington, killing the crew of ten. Killed in the accident were: 2Lt. John T. Ray, pilot; 2Lt. Richard H. Reed, co-pilot; 2Lt. David T. Dunning, navigator; 2Lt. Hans N. Lehne, bombardier; TSgt. Alexander Dec, engineer; SSgt. Lloyd L. Ball, assistant engineer; SSgt. Milton D. Johnson, radio operator; SSgt. Claude W. Seif, Jr. assistant radio operator; SSgt. William C. Fankhanel, gunner; SSgt. Joseph F. Perkins, gunner. Investigators stated, "The pilot took off at 0615 for a local bombing and instrument practice. At 1050, near the end of his mission, he radioed in, requested and received clearance from Walla Walla CAA Radio Range Station to approach Walla Walla at 5,500 feet on the west leg. At that time he stated that his position was over Pasco [Naval Air Station], Washington. From that point Lt. Ray evidently proceeded to the Range Station, missed the cone of silence, and six minutes later at a distance of 17 airline miles east of Walla Walla, struck a ridge at approximately 6,000 feet msl near Table Top Mountain at 1115. The time of the accident being substantiated by various watches of the crewmembers that were all stopped at the time of impact." Investigators noted that a 30 mph tail wind might have propelled the airplane to the cone of silence faster than the pilot had anticipated. It was noted that Lt. Ray, an instrument instructor, was described as an "above average" pilot.

2-12-43A. Roswell, New Mexico. At 0526 MWT, a Consolidated B-24D crashed eight miles northwest of Roswell, New Mexico, killing eight crewmembers. Engineer SSgt. Ralph E. King, Onaway, Michigan, was able to parachute to safety and was uninjured. Killed in the crash were: 2Lt. Charles C. Wylie, Milwaukee, Wisconsin, pilot; 2Lt. Edward L. Games, Camden, New Jersey, co-pilot; 2Lt. Richard L. Morris, Philadelphia, Pennsylvania, bombardier; 2Lt. James E. McMullen, East Point, Georgia, navigator; Sgt. George Dickson, Barnsdall, Oklahoma, assistant engineer; SSgt. Robert I. Donaldson, Collingdale, Pennsylvania, radio operator; Sgt. Lesley W. Stinne, Jr., Louisville, Kentucky, assistant radio operator; Sgt. John J. Yauch, Holliday Cave, West Virginia, gunner. Investigators stated, "The mission started out as a four-airplane formation navigation and bombing mission, take-off at 0100. They were to fly from Biggs Field to Big Spring, Texas, to Amarillo, Texas, at 12,000 feet indicated en route to Alamogordo Bombing Range, New Mexico, drop ten bombs, and then return to Biggs Field, Texas. The four airplanes flew in loose formation to Big Spring, Texas; three airplanes then turned toward Amarillo; one airplane became separated and did not regain the formation (Liaison radio went out and he returned to Biggs Field). After reaching Amarillo, the formation turned toward the Alamogordo Bombing Range and started to climb at about 400 feet a minute. At 17,000 feet, the three airplanes executed an opening procedure placing the airplanes in an echelon to the right. The number-two airplane then dropped out of formation due to the failure of one of the crew's oxygen apparatus. The number-three airplane then started to close on the lead airplane, but encountered engine ice and lost altitude until he was about 1,500 feet below the leader. The pilot then eliminated the ice. His altitude at that time was about 20,000 feet. He then started to close up the formation and was going to take the number-two position. In doing this he began to over-run the leader, so he maneuvered his airplane so that he would stay in formation. At about 22,500 feet he noticed the lead airplane lose about 600 to 800 feet and then go down rapidly. The altitude of the leader just before he lost altitude was approximately

23,500 feet or 24,000 feet. The next time he looked at the leader's airplane he could see two red lights revolving (no green lights were visible). The airplane then went out of sight under his [port] wing. The pilot then started to turn to see where [the subject airplane] was going, instead he saw the sky light up and after turning farther he saw a fire on the ground." Investigation revealed that the airplane had entered a spin due to the engines quitting because of severe carburetor ice. When the airplane began spinning to the ground it suffered the catastrophic structural failure and separation of the tail section and the port wing. The bomb bay doors were found nearly five miles from the main wreckage. Most of the wreckage was scattered over a three-mile area. The airplane had exploded into flames upon impact. The engineer stated that he could not release his emergency hatch because of extreme g-forces. He also stated that he was either unconscious or extremely disoriented while the airplane continued to spin to the ground. When the airplane suddenly stopped spinning, he saw that the bomb doors were apparently open so he jumped out and deployed his parachute once clear of the aircraft. Investigation of the B-24D type airplane revealed that dangerous flight characteristics were exhibited when the airplane suddenly lost power on one or both port engines.

2-12-43B. Mirage Dry Lake, California. At 1145, two Waco CG-4A gliders collided in mid-air four miles north of Mirage Dry Lake, California, killing SSgt. Roy Kenneth Eden aboard CG-4A #42-78940. SSgt. Donald E. Englebrook was able to parachute to safety. SSgt. William J. Flanders and SSgt. James L. Faircloth were uninjured and were able to safely land CG-4A #42-61108 at the Army Air Field at Victorville, California. The gliders were being towed aloft for daily operations when the tow airplane pilots were ordered to assemble with two other airplane/glider tandems at about 3,000 feet agl for a photographic fly-by over the field. The four tow airplanes began to assemble in echelon formation but the glider tow-ropes were of different lengths. This caused the two gliders to be flying side-by-side during the formation assembly. There was moderate turbulence and the gliders had trouble holding position. Investigators stated, "Suddenly the pilot of number-two glider saw the number-one glider sliding toward him. He immediately pulled to the right but the wings of the two gliders collided, damaging the [starboard] wing tip and aileron of the number-one glider. The pilot of the number-one glider [SSgt. Englebrook] said he felt no jar and did not know when the collision occurred. However, the co-pilot was watching and when he saw the damaged wing he told the pilot to jump. The glider was cut loose and the pilot and co-pilot crawled back toward the door. The glider went into a steep spiral and the pilot had difficulty reaching the door. The co-pilot was unable to get out." The surviving pilot stated that the co-pilot had lost his grip and fell to the forward part of the glider as it gyrated to the ground.

2-12-43C. Knoxville, Tennessee. At 1505 CWT, two Lockheed P-38F airplanes collided in mid-air 12 miles north of Knoxville, Tennessee, killing both pilots. 1Lt. James W. Campbell was killed in the crash of P-38F #42-13167 and 1Lt. Raymond W. Burroughs was killed in the crash of P-38F #42-13176. Investigation revealed that the airplanes had taken off from Nashville, Tennessee, on a ferry flight to Newark, New Jersey. The airplanes apparently changed course and were observed flying together toward the northwest. The airplanes then changed course again and began flying toward the southwest. The airplanes were then seen flying to the south. Witnesses estimated that the airplanes were flying over the terrain at an altitude of approximately 200 to 400 feet. Investigation of the wreckage of the two airplanes indicated that the port propeller of Lt. Campbell's P-38 came in contact with the Lt. Burroughs' airplane. Lt. Burroughs attempted to parachute to safety but his parachute was fouled on the tail section and he fell to his death. His airplane smashed to earth and exploded at Raccoon Valley. Lt. Campbell's airplane, which was in a steep bank to the left, immediately dove into a hillside near Beaver Ridge and exploded. There were no witnesses to the actual collision.

2-12-43D. Reno, Nevada. At 1310, a Lockheed P-38F crashed near the Army Air Base at Reno, Nevada, killing pilot Maj. Edward E. Cobb, Jr. Investigation revealed that Maj. Cobb had flown past the reviewing stand at Reno Army Air Base and was flying parallel to the hangar line at an altitude of about 500 feet when he entered a climbing turn to the left. During the turn the airplane flipped over to an inverted attitude but Maj. Cobb was able to right the airplane. The P-38 flew straight for a moment before again flipping into an inverted attitude. The airplane became uncontrollable and the pilot was unable to recover before the P-38 slammed into the ground and burst into flames. Maj. Cobb was killed instantly.

2-12-43E. Lemoore, California. At 1330, a Vultee BT-13A crashed four miles southwest of the Army Air Field at Lemoore, California, killing pilot 2Lt. Jack D. Grove. Investigation revealed that the pilot had taken off from the Lemoore Army Air Field with another BT-13A on a ferry flight to Indian Auxiliary Airfield. When the two airplanes arrived in the area of the auxiliary airfield, Lt. Grover broke off of the two-ship flight and dove on the field. He then attempted to roll the airplane at low altitude. The pilot was able to roll through the maneuver but the airplane smashed into the ground right side up in a flying attitude. The airplane disintegrated and burst into flames as it careened across the terrain. Lt. Grove was killed instantly.

2-12-43F. Fort Worth, Texas. At about 1540, electrical technician Pvt. Harley K. Doerfler was killed when he walked into the rotating number-two propeller of a Consolidated B-24D that was being prepared for a mission at Tarrant Field, Fort Worth, Texas. Pilot Capt. James S. Michael ordered the ground crew that was conducting a pre-flight engine check to leave the engine running because the flight crew and the airplane were preparing to take off on a routine mission.

2-12-43G. Sarasota, Florida. At 1820 EWT, a Curtiss P-40K crashed near the Army Air Base at Sarasota, Florida, killing pilot Capt. Cloyd Woolley. The pilot was leading another P-40 in a simulated attack on the air base when he attempted a steep turn to the right at an extremely low altitude. The starboard wing struck the ground and the airplane cartwheeled into the ground and exploded. The pilot was killed instantly.

2-12-43H. Pueblo, Colorado. At 1813, a Boeing B-17F attempting a landing crashed 579 feet south of the south end of the North-South runway at the Army Air Base at Pueblo, Colorado, killing pilot 1Lt. Harold L. Etler and engineer TSgt. Herbert E. Sines. Six crewmembers were injured, two of them seriously. Crewmembers injured in the accident were: F/O Donald R. Fisher, co-pilot; Sgt. Leo R. Olsen, radio operator; Sgt. Albert L. Jensen, assistant radio operator; Sgt. Donald D. Russell, gunner. Navigator 2Lt. George J. Halaby and assistant engineer Sgt. William W. Hutchinson were seriously injured. Investigators stated, "Lt. Etler, while on a radius of action navigational mission, discovered that the hydraulic system had failed, causing a loss of pressure in the brake system. Pilot Etler, flying in a formation of five aircraft, called his commanding officer on the radio and informed him of the deficiency. The commanding officer instructed Lt. Etler to withdraw from the formation and make an individual landing, while the formation delayed its landing. Lt. Etler proceeded to the airfield and landed on runway 17 in a southerly direction. The airplane first touched the runway after approximately one-third of the runway had been traversed, rolled off the end of the runway and, in making a turn to the right, crashed into an arroyo, resulting in the death of the pilot and the chief engineer. Investigation of the crash indicated that the ship had approximately 600 pounds emergency hydraulic pressure at the time it landed and that the brakes were applied, slowing the ship up somewhat before reaching the end of the runway. However, the emergency pressure fell and the ship was not completely stopped before reaching the end of the runway and various witnesses estimated that it was traveling about 20 miles per hour at the time it left the end of the runway. The pilot, in an attempt to taxi out of the direct line of the runway, applied power to the number-one engine in an attempt to swing the ship to the right, but for reasons unknown did not see the arroyo and consequently failed to turn fast enough to avoid it."

2-13-43A. Stuttgart, Arkansas. At 0300, a Waco CG-4A crashed while attempting to land at the Army Air Field at Stuttgart, Arkansas, killing glider students SSgt. Herman E. Jordan and SSgt. Robert T. Kelnhofer. Investigators stated, "It is the belief of the committee that the two students on their last night of student flying were approaching to make a normal landing after dropping from tow. It was observed by the officer in charge of ground operations that the downwind leg was too close to the runway upon which the glider would land. It is believed that an excessively steep turn was attempted to remedy the error and that in the turn the rapid loss of altitude was not evident until too late, as the ship crashed from all evidence while still recovering from the turn."

2-13-43B. San Francisco Bay, California. At 0320, a Bell P-39N crashed into San Francisco Bay two miles southeast of Treasure Island, California, killing pilot 2Lt. Robert S. Rhea. The pilot had taken off from Oakland Municipal Airport at 0225 PWT on a local night flight. Investigators speculated that the P-39 began to suffer engine trouble and the pilot apparently attempted to make an emergency forced landing in the bay near Pier #42. The pilot's body was recovered from the bay a short time after the accident. The airplane wreckage was never recovered and investigators could not determine the cause of the accident.

2-13-43C. DeRidder, Louisiana. At 0840, a Piper L-4A crashed near the Army Air Base at DeRidder, Louisiana, killing observer Cpl. Wilbur L. Johnson and seriously injuring Liaison Pilot Cpl. Clair F. Chladek. Investigators stated, "Cpl. Cladek was on a tactical mission for the purpose of picking up messages from the 90th Division (U.S. Army), Message Center. Having contacted this Division and failing in the first attempt to pick up message from the ground, he circled the field and made another attempt. Coming in over the field approximately 20 feet high and 15 feet of towing line out, the [pick-up] bob evidently struck the earth, rebounded and became lodged somewhere underneath the tail. The pilot pulled up out of the field to an altitude of what he thought to be approximately 200 feet, turned around and asked his observer whether their attempt was successful. He was then informed that the line was caught. The same thing had occurred the previous day and he was successful in dislodging the line by rocking the ship. It was in this same manner, while attempting to dislodge this line that he struck a tree approximately 60 feet high with his right wing causing the fuselage to skid into another tree. The pilot stated that he had full control of the airplane at all times. [At no] time were the controls fouled. It is believed that Cpl. Johnson was leaning out of the door attempting free the line when the airplane struck the tree. Cpl. Johnson was thrown out of the ship, hitting the ground and causing his death."

2-13-43D. Muroc, California. At 0811, a Lockheed RP-38E suffering the failure of the starboard engine crashed while attempting an emergency forced landing near the Army Air Base at Muroc, California, killing pilot 2Lt. Bruce A. McKean. The Accident Classification Committee stated, "Pilot took off to test hop and slow time a new right engine. Immediately prior to leaving the ground the right engine began coughing and at an altitude of 200 to 300 feet it cut out completely. The airplane then fell to the right and straight down and upon striking the ground immediately burst into flames. The fire had completely enveloped the wreckage before anyone reached the scene of the accident."

2-13-43E. Poole, Virginia. At 1315, a Cessna UC-78B crashed near Poole, Virginia, killing pilot Lt. Col. H.H. Manchester, Jr. and passengers Cpl. Harold Loveland and Pfc. Robert W. Anderson. The airplane had taken off from Bolling Field, Washington, D.C., on a cross-country flight to Pope Field, Fayetteville, North Carolina, at 1140 EWT. The pilot checked in with Richmond, Virginia, Radio at 1228 reporting that he was at 2,000 feet above Richmond. Near Poole the airplane entered a heavy snowstorm and instead of turning back, the pilot pressed on toward Pope Field. A civilian witness on the ground stated that he heard an airplane flying as he came out of his dwelling. He observed the airplane coming straight at him from the southwest at very low altitude. The airplane was flying at treetop level and the pilot pulled up abruptly to avoid colliding with a tree, causing the airplane to stall and enter a half turn of spin. The airplane smashed to the ground, exploding into flames upon impact. It was speculated that the pilot was attempting to maintain visual contact with the ground. It was noted that the pilot had been ordered to maintain visual contact with the ground and to turn back for Bolling Field (or a suitable alternate field) if he encountered instrument conditions.

2-13-43F. New Braunfels, Texas. At 1230, a Vultee BT-13A crashed ten miles northwest of New Braunfels, Texas, killing pilots 2Lt. Chal N. Danie, Jr. and 2Lt. Calvin E. Giffin II. The airplane had taken off from Majors Field, Greenville, Texas, on a local flight to Randolph Field, San Antonio, Texas. The airplane fueled up and took off from Randolph Field at 1155 but failed to return Majors Field. After an extensive search, the airplane was located on 2-17-43. Examination of the wreckage indicated that the airplane was flown into the ground at a high rate of speed in a southeasterly direction in a banked attitude. The airplane cartwheeled into the ground with considerable force, exploding violently into flames and killing the occupants instantly. Wreckage was scattered over a wide area. Investigators could not determine why the airplane flew into the ground. It was noted that passenger Lt. Giffin was a pilot from Randolph Field.

2-13-43G. San Angelo, Texas. At 1345, a Vultee BT-13A spun to the ground six miles northeast of Goodfellow Field, San Angelo, Texas, killing student A/C Jack K. Gibson. The student was on an authorized aerobatic training flight. Witnesses on the ground observed the airplane in a spin to the left at an altitude of approximately 6,000 feet agl. The airplane continued to spin until it struck the ground and exploded into flames. Investigators speculated that the student had entered an unintentional spin while performing an aerobatic maneuver.

2-14-43A. Fort Myers, Florida. At 0925 EWT, a Bell P-39D crashed five miles northeast of Page Field, Fort Myers, Florida, killing pilot 2Lt. Fremont M. Holloway, 23, Anderson, Indiana. The pilot had taken off from Page Field to act as an observer on an instrument flight and had stalled and entered a spin out of a tight turn. Investigators stated, "During a rest period on an instrument flight, pilot of instrument ship did a split-S from 10,000 feet down to 5,000 feet with [Lt. Holloway] following. After this, he went into a steep turn to the right for approximately 270 degrees, and followed in a chandelle to the left. The pilot of the instrument ship saw his observer following in the steep turn but lost sight of his ship in the chandelle. He later observed the other ship burning on the ground, but did not see the ship go in."

2-14-43B. Wilmington, Delaware. At 0930, a civilian electrician was killed when he was struck by the port wing of a Vultee BT-15 that was taxiing at New Castle Army Air Base, Wilmington, Delaware. BT-15 pilots 1Lt. Moses C. Traver and 2Lt. William G. Thames were uninjured. Investigation revealed that George T. Roman, 43, Wilmington, was killed when he walked directly into the path of the airplane. The BT-15 was taxiing in a safe manner and making S-turns. The electrician apparently tried to cross in front of the moving airplanes that were operating near the edge of the runway.

2-14-43C. Rochester, Indiana. At 1017 CWT, a Martin B-26B flying in poor weather crashed two miles NNW of Rochester, Indiana, killing pilot 1Lt. Leonard C. Tall, co-pilot 2Lt. James A. Waldner and navigator 2Lt. Mayer Feingenbaum. The B-26 had taken off from Chanute Field, Rantoul, Illinois, on a ferry flight to Baer Field, Fort Wayne, Indiana. Investigation revealed that the airplane had flown into the ground while attempting to get under the ceiling in a heavy snowstorm. Witnesses on the ground observed the airplane circling at an altitude of about 200 feet in a figure-8 type maneuver until it crashed into the ground and exploded into flames. Investigators speculated that the pilot was attempting visual contact with the ground. It was further speculated that ship's radio compass was inoperative.

2-14-43D. La Mesa, California. At 1100, a Lockheed P-38F crashed at La Mesa, California, killing pilot 2Lt. William Edward Fox, Chicago, Illinois. The airplane had taken off from the U.S. Naval Air Station at San Diego, California, with four other P-38s on a routine training flight. The flight assembled and had climbed to about 33,000 feet. The formation had completed a turn to the left and was observed flying north, about 15 miles inland, paralleling the shore. The airplanes were flying stacked down to the left. The flight leader began another turn to the left, but Lt. Fox continued to fly straight ahead. The airplane was then seen to go into a dive that increased to vertical. There was no apparent effort by the pilot to recover and the airplane dove into the ground where it exploded violently into flames. It was later speculated that the pilot might have suffered a malfunction in the oxygen system.

2-14-43E. Jackson, Mississippi. At 1545, three crewmembers bailed out of a North American B-25D that went out of control while operating near the Army Air Base at Jackson, Mississippi. Radio operator SSgt. R.F. Deutsch was killed when he failed to deploy his parachute. Engineer SSgt. Mickey D. Pavkovich and gunner SSgt. Robert T. Henderson parachuted to safety and were uninjured. The pilots were able to regain control of the airplane and were able to safely land at the base. Pilot 1Lt. Lyle E. Anacker, co-pilot 2Lt. Dale Speicher and navigator 2Lt. John J. Howard were uninjured. Investigators stated, "The airplane was on a long range cruise with a maximum test load. The pilot encountered limited visibility [due] to instrument conditions. While in this weather the nose wheel door came off striking the right ring cowling, causing extensive damage. At that time the pilot lost control of the airplane and the co-pilot ordered the crew to abandon the airplane. The radio operator jumped first (fatal) followed by the engineer and the gunner. The latter two landed safely. The pilot, either during that interval or a short time later, regained control of the airplane after losing approximately 5,000 feet. He then orientated himself and landed at Jackson Army Air Base without further damage."

2-14-43F. Eldorado, Oklahoma. At 1630, a Douglas A-20C crashed eight miles northwest of Eldorado, Oklahoma, killing pilot F/O Charles P. Anderson, Jr., gunner Sgt. Howard J. Smith and gunner Sgt. Carlton Weall, Jr. Investigation revealed that the pilot was buzzing the farms of several relatives in the area and had buzzed the area on previous occasions. The airplane was observed in a steep left bank at about 150 feet when it stalled and fell off on the port wing and dove into the ground. The airplane exploded into flames upon impact and the occupants were killed instantly.

2-14-43G. Ilion, New York. At 1925, a North American B-25C collided with rising terrain at Cranes Corners, six miles south of Ilion, New York, killing pilot 1Lt. John R. Rodger, co-pilot 2Lt. Richard L. Vance and engineer TSgt. Arthur A. Sobczak. The airplane was on a ferry flight from the AAF Air Depot at Rome, New York, to Otis Field, Falmouth, Massachusetts, when the weather on the route deteriorated. Officers at Rome ordered the airplane to return to Rome. The pilot complied and soon entered instrument conditions that the control officers apparently did not know about. The pilot was flying on the Albany radio beam when he began descending. The B-25 collided with a ridge at an elevation of approximately 1,500 feet. It is probable that the pilots did not see the ridge due to instrument conditions and darkness. Investigators noted that the pilot might have completed the flight safely if he had been allowed to continue to Otis Field because conditions were much better there than the conditions he was ordered to fly into.

2-15-43A. Baston, Texas. At 0100, a Curtiss AT-9 crashed nine miles northwest of Batson, Texas, killing A/C Major F. Russell, 19, Nacogdoches, Texas, and A/C Thomas F. Rowland, Jr., 23, Pine Bluff, Arkansas. Investigation revealed that the airplane had taken off from Ellington Field, Houston, Texas, at 1438 CWT (2-14-43) on a day-night navigation flight to Austin, Texas, to Waco, Texas, to Hensley Field, Dallas, Texas, to Barksdale Field, Shreveport, Louisiana, and return. The airplane was scheduled to land at Hensley Field and Barksdale Field for servicing as needed. The AT-9 took off from Barksdale Field at approximately 0001 CWT but did not arrive at Ellington Field and was declared missing. Two civilians saw the AT-9 dive toward the ground and heard it crash an instant later, but they did not observe the actual impact. The wreckage was not located until 1240 CWT because it had crashed into a remote wooded area. Investigation of the wreckage indicated that the airplane had smashed into the ground in a vertical dive with both engines operating. The airplane exploded violently into flames upon impact and the students were killed instantly. It was speculated that the students had lost control of the airplane and were unable to recover in time before diving into the ground.

2-15-43B. Harrisburg, Missouri. At 0530, a North American B-25D flying in poor weather crashed after running out of fuel one mile west of Harrisburg, Missouri, killing pilot 2Lt. Ralph R. Boyvey. Five crewmembers parachuted to safety, injuring three. Instructor pilot Capt. Patrick O'Drago, co-pilot F/O Dennis R. McMonigle and navigator 2Lt. Ernest W. Vanina, Jr. received minor injuries; engineer Pvt. Archie M. Horn and radio operator Pvt. Otto C. Hoygaard were uninjured. The B-25 had taken off from Rosecrans Field, St. Joseph, Missouri, at 2050 CWT (2-14-43) on a long-range night celestial navigation flight to Maxwell Field, Montgomery, Alabama, and return. The navigator missed Maxwell Field and the airplane flew to a point near Atlanta, Georgia. The navigator then set a course for Rosecrans Field. When the airplane arrived in the vicinity of Columbia, Missouri, it encountered a severe snowstorm and poor visibility. The airplane lost the use of its radios because of interference and static caused by the snowstorm. Rather than turning around and heading back toward St. Louis, the pilots decided to continue to fly west, climbing to get "on top." Radio contact was made with Columbia and landing instructions were requested. The airplane was able to let down safely, find Columbia and enter the pattern. The port engine failed due to fuel starvation while the airplane was flying on the downwind leg of the pattern; the starboard engine quit about one minute later. The instructor had ordered the crew to stand by to bail out several minutes before the first engine failed. When the second engine quit he ordered the crew to bail out. The instructor, flying from the co-pilot seat, held the airplane under control while the pilot struggled to put on his parachute over the heavy-type flying suit he was wearing. The instructor stated to investigators, "Lt. Boyvey had his heavy sheep-lined flying suit on, and was having difficulty in getting his parachute on. Then the right engine quit and the ship was dropping fast. Lt. Boyvey was half standing in his seat attempting to pull on the straps of the parachute [harness] as I held the ship on an even keel. At 1,500 feet indicated he still did not have his parachute on; at 1,200 indicated, I steadied the wheel with one hand and slid my seat back with the other, then dove out head first through the navigator's escape hatch. My parachute opened and an instant later I hit the ground. At the practically the same time the ship hit the ground and caught fire approximately 100 to 150 yards away." Investigators stated that the pilots should have landed at St. Louis, Missouri, and refueled before resuming flight to Rosecrans Field. Investigators noted that the parachute harnesses issued to the crew were adjusted for use with regular flight clothing and not the heavy-type flying suit, which was not approved for use on the flight.

2-15-43C. San Angelo, Texas. At 1120, a Vultee BT-13A spun to the ground six miles northeast of Goodfellow Field, San Angelo, Texas, killing A/C Richard F. Gaffney. The student was on an authorized solo aerobatic flight. An aviation cadet flying in the area stated that he had observed the airplane in a spin to the right at an altitude of approximately 3,000 feet agl. The airplane partially recovered from the spin at about 2,000 feet but then went back into the spin until it hit the ground and burst into flames. Investigators speculated that the student let the airplane enter a spin while performing an aerobatic maneuver and was unable to recover before the airplane struck the ground.

2-15-43D. San Francisco Bay, California. At 1241, a Bell P-39N crashed into San Francisco Bay (near the San Mateo Bridge), California, killing pilot 2Lt. Lester Glickfeld, 22, Miami Beach, Florida. The airplane had taken off from Oakland Municipal Airport. Investigators stated, "Lt. Glickfeld was leading a flight of two airplanes on a local formation flight. At about 1,500 feet he went into a shallow dive, leveling off about ten feet above the water. He motioned his wingman to keep above him. He came down to a point where his propeller hit the water. The impact with the water bent the blades. This created severe vibration and reduced the engine output below that required for sustained flight. He pulled the airplane up to 100 feet and called the tower at Mills Field that he was going to come in for an emergency landing. The airplane went into a dive and hit the water in this attitude. Lt. Glickfeld's body was recovered several hours after the crash. His only external injury was a bruise on the chin. The medical officer stated that he died from drowning. His parachute harness was unfastened but the safety belt was fastened. The pilot and the seat were thrown clear of the airplane by the impact."

2-15-43E. Paragould, Arkansas. At 1500, a Vultee BT-13A crashed near Paragould, Arkansas, killing instructor 2Lt. Thomas E. Munger and student A/C William T. Hartman. Investigators stated, "Statements of witnesses indicate that Lt. Munger was executing a simulated forced landing on a large field about two miles southeast of Paragould, Arkansas. He came in low, downwind, about 200 feet above the ground and as he applied the throttle, the engine sputtered. Immediately, according to witnesses, the pilot started a steep turn, apparently in an effort to get back into the field. In so doing, the airplane stalled and then did a one-turn spin and crashed, nose down."

2-15-43F. New Albany, Indiana. At 1530 CWT, a Douglas RDB-7B crashed while attempting an emergency forced landing near New Albany, Indiana, killing pilot 2Lt. Thomas J. Kielminski. The pilot had taken off from Godman Field, Fort Knox, Kentucky, at 1252 CWT on a local flight. The pilot's flight plan indicated a large circle around the field to the left of the take-off direction. The pilot evidently deviated from the flight plan and at about 1530 CWT became lost, entering instrument conditions (with snow flurries underneath) somewhere near Louisville. Visibility in the area was described as "very limited." The pilot had made contact with Godman Field and was flying up and down the Ohio River. The pilot reported that his fuel was about to run out and that he was looking for a field to make an emergency forced landing. After selecting a suitable field, the pilot circled out over the river and turned back toward the field. The engines quit just as the airplane was banked in a 45-degree turn to the left as it crossed the tree line on the northwest bank of the river. The airplane smashed into the ground on the port wing and engine, immediately bursting into flames. The pilot was thrown clear and was killed instantly upon impact.

2-16-43A. Mesa, Arizona. At 1000 MWT, a Cessna AT-17 suffered the catastrophic failure of the port wing and crashed three miles east of Mesa, Arizona, killing Chinese Air Force students A/C Fu Chao Chu and A/C Sze-Chi Yuan. The airplane had taken off from Williams Field, Chandler, Arizona, on a routine instrument training flight. Investigation revealed that the port wing disintegrated in flight, scattering wreckage over an area of one and one-half miles along the flight path from where the wing failure occurred to the point where the airplane slammed into the ground and exploded.

2-16-43B. March Field, California. At 1420, navigator 2Lt. Warren J. Buckner, Jr. was killed when he walked into the rotating number-two propeller of a Consolidated B-24D that had just returned to March

Field, Riverside, California, after successfully completing a routine training mission. The pilot, Capt. Elliot T. Pardee, heard the propeller strike the navigator. The propeller was turning without power after the engine had been shut down.

2-16-43C. Albany, Georgia. At 1725 EWT, a Stearman PT-17 crashed seven miles south of Darr Aero Tech, Albany, Georgia, killing student A/C John W. Nivison. Investigation revealed that the airplane had struck the ground in a power-on spin from a considerable altitude. Investigators could not determine the events preceding the spin.

2-16-43D. Alexandria, Louisiana. At 1640, a Bell P-39N spun to the ground ten miles northeast of Esler Field, Alexandria, Louisiana, killing pilot F/O Roy G. Thomas. The airplane was apparently flying normally at 8,000 feet above ground level and was circling the area. The engine noise was barely audible from the ground. The pilot was maneuvering airplane and was seen to enter a spin that progressed into a flat spin. The airplane continued in a flat spin until it struck the ground. The pilot was heard to advance and retard the throttle several times but the airplane never stopped spinning. The P-39 slammed into soft ground in an upright, wings-level position while in a flat spin to the left. It was noted that the airplane was apparently turning slowly when it struck the ground. The airplane did not burn. The pilot was killed instantly.

2-17-43A. Gulf of Mexico, Florida. At 1245 EWT, a Curtiss P-40F crashed into the Gulf of Mexico surf at Madeira Beach, Florida, killing pilot F/O Ray N. Grafstrom, 23, Duluth, Minnesota. The airplane had taken off from Pinellas Army Air Base, St. Petersburg, Florida, on a target-towing mission. After two P-40 airplanes completed their gunnery mission at 15,000 feet, F/O Grafstrom returned to base, dropped his target at 1235 and then flew away from the field. A short time later he was observed buzzing Madeira Beach. Investigators stated, "Eyewitnesses state that this airplane was flying at approximately 100 to 200 feet along the beach. From this low altitude the airplane climbed sharply, dived at two women on the beach, leveled off just as it passed over them. The airplane continued straight for a distance of approximately 100 yards. It then started a turn to the left, the left wing struck the water and the crash resulted." Published accounts of the crash stated that the two women on the beach were collecting seashells when the airplane approached them at very low altitude. They then reportedly "threw themselves onto the sand" to avoid being struck by the oncoming fighter. Investigators noted that the pilot was unhappy that he had been held back from going overseas with his unit because of dental problems.

2-17-43B. Savannah, Georgia. At 1400 EWT, a Douglas A-24A crashed at Hunter Field, Savannah, Georgia, killing 1Lt. James E. Prechel and seriously injuring engineer-rated passenger Sgt. John D. Wentz. Investigators stated, "At about 1345 EWT, Lt. Prechel took off to fly in formation to DeRidder, Louisiana, for a permanent change of station. Immediately after take-off, Lt. Prechel contacted the control tower and stated that he was going to make an emergency landing. He had taken off on the NE/SW runway, and made a normal turn to land on the NW/SE runway. He never did retract his landing gear and did not state to the control tower why he was making an emergency landing. As he approached the end of the NW/SE runway, at approximately 150 feet altitude, his ship was seen by a number of people, including the control tower operators, to go into a spin, and it never recovered. As the pilot had not given any reason for his making an emergency landing, and due to the condition of the airplane, it is impossible to determine why the pilot returned to the field. It is assumed by the investigating committee that he lost flying speed, causing the ship to spin in."

2-17-43C. Siesta Key, Florida. At 1450 EWT, a Curtiss P-40K crashed into Siesta Key, Florida, killing 1Lt. Robert R. McNagny, 26, Columbia City, Indiana. The airplane had taken off from the Sarasota Army Air Base, Sarasota, Florida, at 1430 EWT on an authorized aerobatic mission. At approximately 1445, the airplane was seen flying at extremely low altitude in the vicinity of Siesta Key, Florida. The airplane collided with a tall pine tree, shearing four feet off of the starboard wing. The airplane went into an uncontrolled climbing roll, rolling three times to the right before diving and rolling into the ground. The P-40 exploded into flames upon impact about one-quarter of a mile from where it had struck the tree, killing the pilot instantly. Wreckage was scattered over an area of 400 feet long and 50 feet wide. The starboard wing contacted the tree about 10 feet from the top and the severed wingtip fell to earth about 200 feet further on.

2-17-43D. Huron, South Dakota. At 1530 CWT, a Boeing B-17F suffered a catastrophic structural failure and crashed near Huron, South Dakota, killing 12 AAF and civilian crewmembers. The B-17 had taken off from the Army Air Base at Rapid City, South Dakota, at 1349 CWT on a gunnery and instrument mission. Investigators stated, "After due consideration of all available information and the location of the several parts of the subject aircraft it is the assumption of the committee that the left wing panel came off first, immediately followed by the right wing panel. The airplane proceeded in a northeasterly direction for about three-fourths of a mile, losing the no. 1 engine. Upon losing the power on the left hand side by loss of the engine, the airplane was pulled around in a northwest direction losing altitude in a stalling manner until the crash. The right wing panel was seen to hit the ground a few seconds after the airplane crashed. The top covering of the left inner wing panel was found approximately one-fourth of a mile south of the crash and the bottom left center wing panel was found about one-half mile southwest of the airplane crash. These two pieces of covering are believed to have been pulled off due to the outer wing panels coming off first, the control cables to the ailerons holding the wing, caused the inner wing panels to be torn off. The rear door, which was found about 300 yards south of the airplane crash, could have been knocked off by the outer wing panel coming back and being held by the aileron control cables and hitting the fuselage. The right stabilizer was found about three-fourths of a mile southwest of the airplane crash, and the left stabilizer was found about one mile southwest of the crash. These two supporting surfaces are believed to have been knocked off by the outer wing panels after they came off." Killed in the crash were: 1Lt. Oliver L. Walker, pilot; 2Lt. Howard D. Stevens, Jr., co-pilot; 2Lt. Norman K. Mattice, navigator; 2Lt. George J. Morris, bombardier; TSgt. Gerald W. Baker, engineer; TSgt. Charles R. Hooberry, radio operator; SSgt. James D. White, assistant engineer; SSgt. Michael Mitzka, assistant radio operator; SSgt. Charles Missana, gunner; Vernon K. Etheridge, civilian technician; Kenneth Abbey, civilian technician; Roy L. Fischer, civilian technician.

2-17-43E. Prattville, Alabama. At 1610, a North American AT-6B spun to the ground four miles southwest of Prattville, Alabama, killing passenger Sgt. C.E. Bjork. Pilot 2Lt. Eugene W. Garges parachuted to safety and was uninjured. The airplane was part of a flight of three that were maneuvering on the orders of the flight leader. The flight leader had been flying in the number-three position for a short time. He then radioed the two other pilots that he was going to peel off and for them to follow him through the maneuver. He peeled off and then dove to pick up speed. The flight leader then pulled up into an Immelmann maneuver and the two airplanes followed. The subject airplane stalled at the top of the maneuver and began an inverted spin. After several turns, it was evident that the pilot was not able to recover the airplane and the flight leader radioed the pilot to bail out. The pilot bailed out at about 1,000 feet above the ground but the passenger went in with the airplane. The airplane spun to the ground in an inverted position. Investigators noted that the passenger in the subject airplane was not hooked up to the radio by headset and probably did not hear the order to bail out.

2-17-43F. Tucson, Arizona. At 1638 MWT, a Consolidated B-24D crashed while attempting a takeoff at Municipal Airport #2, Tucson, Arizona, killing six Consolidated Aircraft Company employees and injuring five others. Six Consolidated employees were uninjured. Consolidated employees killed in the crash were: Bernard W. Donner, 27, Ocean Beach, California; G.R. Goble, 44, Coronado, California; Edward Hill, 25, Pacific Beach, California, pilot-rated passenger; Paul Hill, 33, San Diego, California, pilot-rated passenger; M. Ross, 31, San Diego, California, engineer; F. Balch, employee. George S. Oberdorf, T. Lee, and W.T. Mims received minor injuries. F.M. Eachus and H.A. May were seriously injured. Consolidated employees uninjured in the crash were: Earl F. White, pilot; Hal Ellington, co-pilot; Hurst Gentry; H. Peck; M.C. Revy; W. Robinson. The airplane was taking off from the Consolidated Modification Center on a personnel transport flight to San Diego. The airplane, configured with 15 degrees of flaps, became airborne at 110 mph after rolling about 2,700 feet down the runway. The brakes were tapped to stop tire rotation and the pilot ordered the co-pilot to retract the landing gear. The co-pilot hit the switch and as he did the airplane settled into the runway on the retracting landing gear. The airplane skidded down the runway for over a thousand feet, coming apart and bursting into flames as it came to rest at the end of the runway. Investigation of the wreckage revealed no mechanical or structural failure.

2-18-43. Seattle, Washington. At 1226, a Boeing XB-29 suffering an in-flight fire crashed into the Frye & Company meat packing plant at 2203 Airport Way, Seattle, Washington, killing ten Boeing Aircraft Company crewmembers and 30 Frye employees on the ground. Twelve other Frye employees were seriously injured. A Seattle firefighter was killed and five others were seriously injured at the crash scene the day after the accident when they were poisoned by refrigerant gas that had escaped as they were battling a fire that had erupted in the smoldering ruins. This was the second B-29 prototype airplane (AAF serial #41-3) and it had taken off at 1209 from Boeing Field at Renton, Washington, on a flight to test engine performance and cooling, propeller governor performance, two-engine ceiling and general airplane performance. At 1217, a fire was noticed in the number-one engine and the crew radioed that the airplane was returning to the field. The flight crew activated the on-board fire extinguisher and it was thought that the fire was successfully extinguished. The airplane was flying north at an altitude of about 2,500 feet indicated when the number-one engine again burst into flames. The pilots feathered the number-one propeller and at 1222 a crewmember radioed the field that they were at 2,400 feet and descending with the number-one engine on fire. The crewmember requested that the fire equipment stand by for the emergency landing. The XB-29 was flying to the south and losing altitude rapidly, striking a power line before plunging into the top floor of the five-story meat packing plant where it exploded into flames at 1226. Investigation revealed that the fire was caused by the poor design of a wing leading edge fuel filler neck and cap. The fuel filler neck and fuel filler neck cover plates were located in the leading edge of the wing, and the low air pressure generated in these areas created a fuel leak that allowed fuel to enter the leading edge of the wing and the number-one engine nacelle. The raw fuel and fumes apparently accumulated in the leading edge wing structure and engine nacelle area and were probably ignited by contact with hot exhaust ducting or static electricity. The fire began to burn through the aircraft leading edge structures and the main inner wing areas and eventually to the bomb bay area. A magnesium de-icer valve casting burned fiercely, causing an explosion in the leading edge of the wing. This explosion ruptured the fuel filler neck and allowed raw fuel to escape, causing the port wing to erupt into massive flames. A crewmember was overheard on the radio/intercom saying to the pilot, "The wing spar is burning badly." The landing gear was seen to extend before the airplane slammed into the packing plant. The airplane exploded violently and the building erupted into flames. The crash killed several important members of the B-29 program's design and engineering teams and Boeing's chief test pilot. Boeing Aircraft Company employees killed in the crash were: Edmund T. Allen, Chief Boeing Test Pilot; Robert R. Dansfield, co-pilot; Fritz Mohn, engineer; Robert W. Maxfield, engineer; Thomas R. Lankford, engineer; Vincent W. North, engineer; Raymond L. Basel, engineer. Boeing engineers Edward I. Warseba, Claude E. Blaine and Harry W. Ralston attempted to parachute to safety but bailed out at an altitude too low to allow their parachutes to deploy. Frye & Company employees killed on the ground were: Domingo Amoroso, Harmon E. Burnison, Harry Fargin, Richard Faulkner, Albert J. Fowler, August Hoba, Frederick Hoba, Joseph A. Hoy, John G. Huddleston, Frank Kern, Phillippe LeBeau, Arthur Lindholm, Frank Lopez, Samson Oversoff, Archille Ricci, Mary Rosella, John Skorpan, Harry Tarasuk, Oscar Tuft. Nikifor Vaschenko, 71, Seattle, died of his injuries on 2-21-43 at Marine Hospital. Frye & Company employees seriously injured in the crash were: Culver L. Anderson, Harry W. Baird, Enok Greenkevich, Michael Greenkevich, Stanley Kovera, Joe Martinkus, Emil Mutti, George Schachtel, William Sutphin, Louis Thomas, Frank Vergillo, William Yanchenko. The fire at the packing plant had been put out that evening but the smoldering ruins erupted into flames the next morning, escalating into a three-alarm fire emergency. Seattle firefighter Luther D. Bonnar died of gas poisoning when he was overcome by refrigerant gas fumes that were released by burning refrigerator equipment. Five other firefighters were overcome by the gas fumes and were hospitalized in serious condition. The accident and resultant fire destroyed 107,000 pounds of pork and 80 live hogs, compounding a meat shortage in the Seattle area. Newspaper accounts of the accident reported that many Frye employees were on their lunch break at the time of the accident and were away from their workstations.

2-19-43A. Hutchinson, Kansas. At 0435 CWT, a Boeing B-17F crashed five and one-half miles northeast of Hutchinson, Kansas, killing the crew of nine. The airplane had taken off from the Army Air Base at Orlando, Florida, on a celestial navigation flight to its station at the Army Air Base at Kearney, Nebraska. There was no contact between the crew and ground stations after take off. The airplane apparently became lost and it was later observed circling the Hutchinson Naval Air Station, which was faintly visible to the crew through the broken clouds and fog. The B-17 crew believed that they were circling Smoky Hill Army Air Field at Salina, Kansas. They contacted the Smoky Hill control tower and requested landing instructions. They were given landing instructions and were cleared to land. The pilot announced that he had only 30 minutes of fuel left and it soon became evident that he was lost and unsure of his position. The Smoky Hill controller asked the pilot if he wanted to home in on the tower frequency. The pilot stated that the radio compass was inoperative. The tower then communicated the radio range frequency for Salina but received no reply from the B-17 crew. Weather at the accident sight was reported as ceiling zero with low-level fog prevalent. The pilot continued to circle Hutchinson and attempted to get under the fog. The B-17 collided with the ground under control in a flying attitude with power. The airplane skidded along the ground for 100 yards before bounding back into the air where it began coming apart and shedding pieces. The B-17 flew through the air about 400 yards, slamming back into the ground and erupting into flames before coming to rest. Killed in the crash were: 1Lt. Ned D. Knaphus, pilot; 2Lt. Bruce S. Upson, co-pilot; 2Lt. Carl T. Miller, navigator; 2Lt. Raymond L. Zeiter, bombardier; TSgt. Earl C. Woolenweber, engineer; TSgt. Bernard Budimirovich, radio operator; TSgt Max H. McArthur, gunner; SSgt. Walter E. Bybee, gunner; Pvt. James A. Farrell, gunner.

2-19-43B. Dona Ana County, New Mexico. At 1540 CWT, a Consolidated B-24D crashed while attempting an emergency forced landing at Dona Ana County, New Mexico, about 15 miles north of Biggs Field, El Paso, Texas, killing pilot 1Lt. Oscar Lee Cantrell, Jr. and navigator 2Lt. John J. Gessinger. Six crewmembers parachuted to safety and were uninjured. Crewmembers parachuting to safety were: 2Lt. Louis B. Yarbrough, co-pilot; 2Lt. Harry B. Meisel, bombardier; TSgt. Joseph J. Kasal, engineer; Pfc. Marvin M. Rounsaville, assistant engineer; SSgt. John T. Craig, gunner. Radio operator SSgt. Robert L. Hoendorf received minor injuries parachuting to safety. The airplane had taken off from Biggs Field on an instrument training flight at 1230 CWT. At 1530, co-pilot Yarbrough was flying on instruments under the hood in the pilot seat at 5,500 feet agl. The number-three engine then began vibrating and running away. The whole airplane began to vibrate badly. Smoke began to pour out of the number-three engine and then flames. The pilot pulled the number-three fire extinguisher and then attempted to feather the number-three propeller but was unsuccessful. The starboard wing kept falling and the pilot was having trouble keeping it up and maintaining level flight. Investigators speculated that the pilot apparently did not attempt to balance the power by retarding the port engines. The airplane continued to vibrate badly and the pilots could not read the instruments. The B-24 soon became almost uncontrollable so the pilot rang the bail-out bell and the crew abandoned ship. The airplane was observed on fire and flying to the south. It then pitched downward and dove into the ground where it exploded into flames. Wreckage was scattered over a wide area. Investigation revealed that the number-three propeller governor had failed, causing a runaway engine and propeller.

2-19-43C. Exeter, California. At 1720, a Ryan PT-22 crashed at Belmont and Firebaugh Avenues in Exeter, California, killing student A/C Jess A. Rosengren. Investigation revealed that the student had left the authorized practice area and began flying near Exeter at very low altitude. A short time later, the airplane was observed diving on a house on the outskirts of Exeter. The occupants exited the house and the student was seen to wave. The people on the ground waved back. The airplane was next seen to stall out of a steep left turn at an altitude of about 200 feet. The airplane snapped over to the right and dove into the ground. The student was killed instantly.

✓ **2-20-43. Helena, Arkansas.** At 1345, a Fairchild PT-19A crashed five miles south of Helena Aero Tech, Thompson-Robbins Field, Helena, Arkansas, killing civilian instructor B.W. Kirk and student A/C Robert C. Ahlers. Investigation revealed that the student stalled the airplane while in a steep right turn while performing simulated forced landings. The airplane entered a spin to the left at an altitude of about 200 feet agl.

2-21-43A. Amarillo, Texas. At 0905, a Lockheed P-38G crashed at the Army Air Field at Amarillo, Texas, killing pilot 1Lt. William H. Stuart. The pilot had taken off from English Field, Texas, on a ferry flight to Scott Field, Belleville, Illinois. The starboard engine apparently failed just after take-off. The pilot attempted an emergency forced landing two miles north of the field, but he failed to release the two 150-gallon auxiliary fuel tanks that were being carried on the under-wing pylons. When the pilot executed the forced landing with the fuel tanks in place, the fuel tanks ruptured and a fire erupted that engulfed the entire airplane. The blazing P-38 slid for 125 yards before coming to rest. The pilot made no apparent effort to escape the burning airplane. Intense heat, flames and heavy smoke kept rescuers back. An autopsy indicated that the pilot died of a skull fracture and was not burned to death. Because of the complete destruction of the airplane engines, investigators were unable to determine the cause of the starboard engine failure.

2-21-43B. Galveston, Texas. At 1130, a low-flying Lockheed-Vega B-34 collided with a 1935 Ford Tudor Sedan that was being operated on a beach seven miles west of the Army Air Base at Galveston, Texas, killing the five fliers aboard the B-34. Seven civilians in the automobile, a father and his six children, were injured, four of them seriously. Killed in the crash of the B-34 were: 1Lt. Edwin A. Cranmer, Pemberton, New Jersey, pilot; 2Lt. James I. Mitchell, Weldona, Colorado, bombardier; SSgt. Samuel Orenstein, Philadelphia, Pennsylvania, engineer; Pvt. Roger L. Tassin, Wallace, Louisiana, gunner; Pvt. Lemuel B. Tatum, Jr., Richmond, Virginia, gunner. Seriously injured in the accident were: Rosalind Gause, age 3; Helen Gause, age 6; Anna Marie Gause, age 12; Francis L. Gause, age 33, father. Boys Richard Gause, David Gause (ages unknown) and daughter Jean Gause, age 14, received minor injuries. The family had been residing in Galveston, but was originally from McAllen. Investigation revealed that the airplane had taken off from the Galveston Municipal Airport at 0946 CWT on a local flight to "swing the compass." At 1055, Lt. Cranmer joined another B-34 airplane for formation practice and to swing the compass. Spotty clouds and a low ceiling interfered with the formation practice and compass swinging so the two airplanes separated. Lt. Cranmer landed at Galveston at 1119 and then immediately took off again. The subject airplane was observed flying southwest at low altitude near Galveston Island and along the beach areas on the Gulf side the island. The airplane had made a left turn and had turned around over the Gulf of Mexico. A short time later the B-34 was observed flying to the northeast at extremely low altitude over the beach. Some witnesses estimated the airplane's altitude at about 15 to 20 feet agl. The airplane was seen pulling up to avoid colliding with a GMC 2½ ton truck that contained the range detachment crew from Ellington Field, Houston, Texas. Soon after avoiding the truck, the B-34 struck a 1935 Ford Tudor Sedan with the starboard wing. The airplane struck the automobile's radiator, hood, windshield and roof. The automobile's roof was torn off and the Ford was thrown 25 feet from the point of impact, landing right side up in the shallow water. The starboard wing was severed about two feet from the inboard edge of the aileron and it fell nearby. The airplane flew straight for a few seconds before entering a steep bank to the right and crashing into shallow water near the beach. Wreckage was scattered for about three-quarters of a mile along the beach from where the automobile was struck to the point where the airplane crashed into the surf.

2-21-43C. Cole, Oklahoma. At 1245, a Douglas A-20C crashed four miles south of Cole, Oklahoma, killing the crew of four. Killed in the crash were: F/O James D. Ferrante, pilot; Sgt. Frank M. Konoza, gunner; Pvt. Robert L. Cook, passenger; Pvt. Russell C. Cooney, passenger. The airplane had taken off at 1230 from Will Rogers Field, Oklahoma City, Oklahoma, with two other A-20s on a routine formation flight. The airplanes flew to the south and formed up in a right echelon formation, the subject airplane flying in the number-two position. The leader peeled off and the flight formed up in a chase formation at about 1,000 feet agl. The flight maneuvered for a short period of time, performing chandelles, some buzzing and tightly banked turns. The airplanes were in a steep turn to the right when the subject airplane stalled and entered a spin at an altitude too low to allow recovery.

2-21-43D. Calcasieu Lake, Louisiana. At 1435, a Martin B-26B suffering an engine failure crashed at Calcasieu Lake, Louisiana, killing engineer SSgt. Sidney H. LaVance and injuring five fliers, one seriously. Crewmembers injured in the accident were: 1Lt. Benjamin K. Ostlind, pilot; 2Lt. Jean H. Blackwelder, co-pilot; 2Lt. Deming D. Fairbanks, co-pilot; SSgt. Alfred F. Simard, radio operator. Gunner SSgt. Hancel E. Casto was seriously injured. Investigators stated, "The B-26B #41-18160, with a crew of six aboard, took off at approximately 1400 CWT for a skip bombing mission over Calcasieu Lake to test a new bomb sight. After reaching the region of the target, the front bomb bay doors were opened; the pilot made a left turn from approximately 500 feet altitude to the bombing approach leg, and started the run at approximately 30-foot altitude. As he started the run, he announced the fact to all the crew and warned the tail gunner to stand by to photograph the target. Pilot began pulling up to the usual 50 foot altitude with throttle and propeller settings giving him an airspeed of 205-210 mph, and as altimeter showed him to be passing a 40 foot altitude and at a distance of approximately three and one-quarter mile from the target, one engine cut out completely. Not being able to immediately tell which engine it was and being at such an altitude that instant action was necessary, the pilot and co-pilot 'got on' the rudders in order to hold a straight course and consequently were unable to say which side the airplane 'yawed' as their rapid efforts allowed very little such action nor could either one say to which rudder they applied the most effort, nor to the exact position of the rudder pedals. During this time the pilot heard the left propeller 'rev' down, then overspeed, and thus believing this to be the engine that had cut out, feathered the left propeller, and ordered all bombs dropped, as the safety pins had been removed. This necessitated opening the rear bomb bay doors by use of the bomb bay air bottle mechanism. The airplane lost speed most rapidly from 210 to 150 mph, and the pilot put on full right throttle and full low propeller pitch, but these apparently had no effect. The airplane could not be made to fly, particularly with both front and rear bomb bay doors open, and not having time to close these doors, the pilot warned the crew to stand by for a crash landing which was made on a mud flat in Calcasieu Lake, approximately one mile from the point where the engine failed."

2-22-43A. Albany, Georgia. At 0830 CWT, a Beech AT-10 crashed while attempting a take-off at Turner Field, Albany, Georgia, killing pilot 2Lt. James I. Platt and gunner-rated passenger Sgt. Manuel Perez. The Aircraft Accident Classification Committee stated, "The pilot started to make a normal take-off. However, as soon as the airplane gathered flying speed it started to climb at an ever increasingly steep angle until it stalled at approximately 100 feet, in about a 45-degree climb. It then dropped off to the left and made a half turn of a spin, crashing and exploding at approximately 100 feet to the left of the runway, and approximately 800 yards from the point of take-off. Subsequent examination of the pilot's control column and locking mechanism clearly indicated that the locking arm, which fastens the control column to the rudder bar, had not been completely unlocked and retracted so as to lie parallel to, and along the control column, but had been left in a horizontal (forward pointing) position which prevented the pilot from moving the control column forward of its neutral locked position. Deformation of the forward attaching end of this locking bar, and the attaching lugs on the rudder bar, clearly indicated that it was in contact with them at the time of the crash."

2-22-43B. Enterprise, Utah. At 1030 PWT, a Douglas C-47A crashed ten miles south of Enterprise, Utah, killing pilot 1Lt. Charles R. Butterwick and co-pilot 1Lt. C.R. Ives. The airplane was on a ferry flight from the Douglas factory to an undisclosed destination on the east coast. The airplane had taken off from Las Vegas, Nevada, for Hill Field, Ogden, Utah, at 0920 PWT, but failed to arrive at Hill Field. The airplane was last seen flying at 6,000 feet indicated altitude near Mormon Mesa, 68 miles from Las Vegas. Despite a massive search effort, the airplane wreckage was not found until 2-27-43 at about 1630. Examination of the wreckage indicated that the C-47 had apparently struck a ridge at an elevation of 7,100 feet at angle of 45 degrees with the starboard wing down. Evidence indicated that both engines were producing power at the time of impact.

2-22-43C. Alexandria, Louisiana. At 1455, a Bell P-39N spun to the ground one mile southwest of Esler Field, Alexandria, Louisiana, killing pilot Capt. Paul E. Logan. Investigation revealed that the P-39 was flying at approximately 6,500 feet indicated altitude at 200 mph when the pilot attempted a slow roll. The airplane stalled during the maneuver and entered a tumbling, end over end spin. After falling for some time the airplane entered a normal spin, and the pilot appeared to be trying to make a recovery at about 3,000 feet. The airplane continued to spin until it struck the ground and exploded. There was no apparent effort by the pilot to leave the airplane.

2-23-43A. Fort Jackson, South Carolina. At 0015, a Vultee BT-13A crashed in the north section of the Fort Jackson Military Reservation, South Carolina, killing student A/C Arthur J. O'Malley. The airplane had taken off from Shaw Field, Sumter, South Carolina, on a night navigation flight to Lumberton, North

Carolina, and return. The student arrived at Lane, South Carolina, as scheduled and then proceeded to Lumberton. The student apparently became lost on the leg from Lane to Lumberton. The airplane, more than 50 miles off course, arrived in the vicinity of Fort Jackson and circled at low altitude for about 15 minutes. The student selected a field, probably the Fort Jackson firing range, and attempted to perform an emergency forced landing. The student apparently undershot the intended field and collided with trees. Witnesses on the ground stated that the student had turned on the landing lights just prior to the airplane crashing into the trees in a flying attitude. The airplane was completely demolished and the student killed instantly. A BT-13A piloted by A/C Henry A. Post also became lost on the identical Shaw Field to Lumberton and return navigation mission. A/C Post abandoned his airplane in flight at 0150 EWT when it became low on fuel. He parachuted to safety but the BT-13A crashed into the northwest corner of a frame house belonging to Elizabeth E. McCall, Effingham, South Carolina. The four people sleeping in the house at the time of the accident were not injured. Weather was not thought to be a factor in either accident.

2-23-43B. Kolola Springs, Mississippi. At 0910, a Curtiss AT-9 crashed at Kolola Springs, Mississippi, killing instructor 1Lt. W.B. Wilson and student A/C K.G. Erickson. The airplane had taken off from the Army Air Field at Columbus, Mississippi, at 0850 with five other AT-9 airplanes on a navigation flight to Nashville, Tennessee, and return. The subject airplane dropped out of the flight and attempted to return to the field. Just as the ship approached the area of the field a large bank of low-level fog moved into the area, creating a localized zero-zero condition. The airplane descended into the ground in a flying attitude about one mile east of the field, bursting into flames upon impact. Investigators speculated that the instructor and student were apparently trying to get underneath the fog in an attempt to enter the traffic pattern for the field at Columbus.

2-23-43C. Muroc, California. At 0837, a Lockheed P-38F crashed near the Army Air Base at Muroc, California, killing pilot 2Lt. Marion C. Van Arsdell. The Aircraft Accident Classification Committee stated, "Pilot, Lt. Van Arsdell, took off at approximately 0755 on a local transition flight. This was his first flight in a P-38 type airplane. After he had flown approximately 40 minutes, his left engine started to emit white smoke and later [black] smoke, as witnessed by 1Lt. Roger M. Carter, who was flying nearby. Lt. Van Arsdell then became excited and dove the airplane to approximately 300 mph at 3,000 feet. He then tried to bail out. The condition of the parachute, as it was later inspected, indicates that he either pulled the ripcord while [he was] still in the cockpit or that he was not entirely free of the airplane and the shroud lines were severed by the horizontal stabilizer of the airplane. The airplane crashed [into the ground at a 60-degree angle] and disintegrated and the pilot was fatally injured." Investigators noted that another pilot who had been flying the P-38 could not feather the port propeller after suffering overheating problems with the port engine three days prior to the accident. The airplane had been landed with the port propeller windmilling and the starboard propeller running away. A burned out solenoid (in the port boom) was found and repaired but the cause of the failure was never determined. A bad oil temperature thermostatic valve on the port outboard oil cooler was replaced as well. Investigators speculated that on the fatal flight an electrical problem in the port boom prevented the oil cooler shutter doors, which are electrically operated, from functioning, causing the port engine to overheat.

2-23-43D. Marianna, Florida. At 1125 CWT, a North American AT-6C crashed near the Army Air Field at Marianna, Florida, killing Aviation Student David E. Stead. Investigators stated, "Aviation Student Stead was the fifth ship to take off in a flight of six scheduled for ground gunnery. The preceding aircraft had executed the standard 90-degree left turn with traffic and a 90-degree right turn out of traffic, at which time the aircraft, piloted by A/S Stead, was about 200 yards in the rear and about 100 to 150 feet below the preceding airplane. It is the opinion of the Aircraft Accident Classification Committee that in attempting to close this interval, A/S Stead executed too steep a turn at too slow an airspeed, resulting in a stall attitude and inadvertent spin to the right, and having only 500 feet altitude, he was [too low] to recover. Examination of the scene of the crash disclosed that the aircraft struck the ground in an attitude of approximately 40 degrees of left bank, at an angle of descent of about 70 degrees. It is believed that the student had stopped the spin but had insufficient altitude to recover to level flight."

2-23-43E. Husted, Colorado. At 1205, a Douglas C-49 flying in instrument conditions crashed out of control in mountainous terrain five miles west of Husted, Colorado, killing two civilian ferry pilots and an AAF pilot. The airplane, on an instrument clearance, had taken off from the Army Air Base at Pueblo, Colorado, on a cargo flight to Lowry Field, Denver, Colorado. The C-49 failed to arrive at Lowry Field. After an extensive search, the airplane was located from the air by Colorado Springs CAP on 2-28-43. Examination of the wreckage indicated that the airplane struck the mountain about 75 feet below the top. Investigators stated, "While flying in instrument conditions the airplane crashed into a heavily wooded mountainside approximately 13 miles northwest of Colorado Springs, Colorado, instantly killing all personnel on board. It appears as if a combination of icing conditions and engine failure caused the airplane to become uncontrollable." Civilian ferry pilot Baxter I. Ireland, 32, Denver, Colorado, co-pilot Francis Clark, 25, Arnett, Oklahoma, and AAF pilot-rated passenger

2Lt. Vencil C. Bohling, 24, Versailles, Missouri, were killed when the airplane struck the mountain while descending at a high rate of speed, exploding violently into flames upon impact. Wreckage was scattered over a wide area.

2-23-43F. Lake Murray, South Carolina. At 1420, a North American B-25C crashed into Lake Murray approximately 15 miles northwest of the Army Air Base at Columbia, South Carolina, killing five crewmembers and seriously injuring another. Pilot 1Lt. Doak A. Weston, 24, Aptos, California, miraculously survived when he was able to pull himself clear of the wreckage while underwater. Despite his serious injuries he was able to swim to shore (see 9-24-45A). Killed in the crash were: 2Lt. Marshall S. Hawke, 26, Muncie, Indiana, co-pilot; 1Lt. Clifford O. Sherry, 24, Chicago, Illinois, bombardier; 2Lt. John E. Hancock, 27, Carmel, Pennsylvania, bombardier; Sgt. George W. Rhine, 22, Englewood, California, engineer; 2Lt. Harold L. Feineuer, age unknown, Bay City, Michigan, passenger. The airplane was flying at extremely low altitude on a skip bombing training mission on the lake, which was described as "glassy smooth" by the surviving pilot. While attempting a left turn at low altitude the port wing tip stuck the surface of the lake, causing the airplane to cartwheel into the water. The B-25 sank in 125 feet of water. Investigators noted that the pilots had been instructed not fly at low altitude on the lake until they were on the actual bomb run.

2-23-43G. Meridian, Mississippi. At 1407, a Douglas A-20B crashed near Key Field, Meridian, Mississippi, killing pilot F/O Frank H. Giesse. Gunner Sgt. John T. Singland and gunner Sgt. Thomas L. Denning were able to parachute to safety and were uninjured. The airplane had taken off from Barksdale Field, Shreveport, Louisiana, on a ferry flight to Key Field. The airplane refueled at Barksdale after the pilot landed because of fuel concerns. After being refueled, the airplane took off for Key Field. When the airplane arrived at Key Field, tower personnel noticed a white mist and smoke emitting from the port wing. This was communicated to the pilot. The two gunners were seen to leave the airplane a few seconds later. Just moments after the second gunner cleared the ship the port engine nacelle and wing burst into flames. The airplane flew for a few seconds after the port engine exploded into flames and then rolled to the left and dove into the ground where it exploded into flames. Investigation revealed that a fuel line in the port engine nacelle came loose, allowing raw fuel to escape and flow to the port wing trailing edge. Investigators speculated that hot exhaust ducting or static electricity ignited this escaping fuel, causing the wing and engine to burst into flames.

2-23-43H. Littleton, Colorado. At 1455, a Vultee BT-13A was abandoned and crashed eight miles east of Littleton, Colorado, killing passenger 2Lt. Frank S. Larsell, Army Medical Corps, who was killed attempting to parachute to safety. Pilot Capt. Philip R. Anwyl was able to parachute to safety and was uninjured. The airplane and occupants were stationed at Marana, Arizona. The airplane had taken off from Lowry Field, Denver, Colorado, at 1447 MWT on an instrument navigation flight to Albuquerque, New Mexico. The airplane was flying at an indicated altitude of approximately 7,300 feet when the engine failed. The pilot turned the airplane back toward the east and attempted to restart the engine. He alerted Lowry Field that he was executing an emergency descent. When he was unable to restart the engine, the pilot ordered the passenger to bail out. Apparently Lt. Larsell waited too long before he pulled the ripcord and the parachute was unable to fully deploy before he struck the ground.

2-23-43I. San Antonio, Texas. At 1800, a Curtiss O-52 stalled and spun to the ground three miles north of San Antonio, Texas, killing pilot 1Lt. Robert C. Williams and observer SSgt. Jimmie R. Simpson. Investigators stated, "The pilot was making an approach to land on Runway 14 at Stinson Field. Traffic was to the right. His base leg was about a mile back from the end of the runway. He started the last 90-degree turn a little late and went past the line of the runway, and therefore, he continued his turn to the right in order to get back in line with the runway. At this point he had an altitude of about 200 feet, his motor popped 2 or 3 times and then quit, the right wing dropped and the airplane nosed down. The airplane turned more to the right and then straightened out. The motor started up with a roar and the airplane hit the ground in flat attitude. The airplane skidded in a straight line for about 125 feet, washed out the engine and lodged against a bank with the right wing broken off and folded over the left wing. The pilot was killed instantly and the passenger died a few hours later."

2-23-43J. Columbus, Mississippi. At 2045, a Beech AT-10 attempting a landing after crashed one mile north of the Army Air Field at Columbus, Mississippi, killing students A/C John J. Gallagher and A/C E.C. Gatchell. The airplane, which was returning from a night navigation flight, was cleared to land and was observed flying too low on final approach. Seconds later the airplane dove into the trees near the field and exploded into flames. Investigation revealed that both engines were producing power at the time of the accident. Investigators could not determine the cause of the accident.

2-23-43K. Gulf of Mexico, Florida. At 2200 EWT, a Martin B-26B crashed into the Gulf of Mexico approximately 300 yards off shore of Venice, Florida, killing the crew of six. Killed in the crash were: F/O Swain M. Higbee, pilot; 2Lt. Robert E. Jones, co-pilot; 2Lt. John A. Genty, bombardier; Sgt. Lawrence E. Broach, engineer; Sgt. Ernest H. Nichols, radio operator; Sgt. Gerald I. Crookston, gunner. The airplane had taken off from MacDill Field, Florida, on a night bombing mission. The airplane was observed circling over the beach area near Venice. The B-26 then turned toward the sea and gradually descended into the water. Investigators speculated that the airplane was attempting to make an emergency forced landing on the water.

2-24-43A. Matthews Field, Missouri. At 0750, a Fairchild PT-19A performing low altitude aerobatics crashed near Matthews Field, Missouri, killing civilian instructor James Smith. The pilot, a civilian flight school commander, took off from Matthews Field on a weather-checking mission. Matthews Field is an auxiliary airfield for the AAF Training Detachment at Sikeston, Missouri. Investigators stated, "Mr. Smith circled the field and started a shallow dive across the landing area from south to north into the wind. At the northern edge of the field he pulled the nose up and immediately started a roll to the left at approximately 100 feet above the ground. As the ship reached an inverted position the engine cut out and the nose yawed from north to northeast and fell off. The airplane sliced out of the last half of the roll with engine wide open, striking the ground just prior to completing the roll."

2-24-43B. Sumter, South Carolina. At 1340, a Vultee BT-15 crashed four miles northwest of Shaw Field, Sumter, South Carolina, killing student A/C Benjamin F. King. The student took off from Shaw Field on a routine student-training mission. Twenty minutes after take-off the airplane was seen at 2,000 feet agl in a spin. The airplane continued to spin until it hit the ground. There was no apparent effort by the student to recover from the spin or parachute from the airplane.

2-24-43C. Lake Charles, Louisiana. At 1636, a Lockheed-Vega B-34 performed an emergency forced landing on Lake Charles, Louisiana, killing pilot 2Lt. Roy H. Andes and gunner SSgt. Billy T. Davenport. Three other crewmembers were injured, one seriously. Co-pilot F/O Donald J. Bayne and engineer TSgt. Kenneth J. Bolson were injured and gunner Pfc. Thomas Minichillo received serious injuries. The airplane had taken off from the Army Air Base at Lake Charles on a target-towing mission. Soon after take-off, the port engine quit. The pilot switched fuel tanks but the engine would not restart. The pilots noticed that there was diminished fuel pressure in both engines. A moment later the starboard engine quit. The pilot configured the airplane for an emergency landing and ditched on Lake Charles. The pilot was able to perform a successful ditching and no one was injured. The pilot ordered the crew to abandon ship. As the crew exited the airplane it blew up violently and erupted into flames. The B-34 sank a short time later. The crewmembers that survived were rescued almost immediately. Pilot Lt. Andes was apparently fatally injured in the explosion. His body was recovered from the water soon after the accident. Gunner SSgt. Davenport was seriously injured but he drowned before he could be rescued. Co-pilot F/O Donald J. Bayne and engineer TSgt. Kenneth J. Bolson sustained minor injuries and gunner Pfc. Thomas Minichillo was seriously injured.

2-24-43D. Victoria, Texas. At 2130, a North American AT-6C crashed seven miles northwest of Aloe Army Air Field, Victoria, Texas, killing instructor pilot 1Lt. Paul C. Palmer, Jr. The instructor had taken off from Aloe Field at about 2100 on a weather-checking mission. A civilian reported that he had heard what he thought was an airplane crash at about 2130. Investigation of the wreckage indicated that the airplane struck the ground in a spin. The cause of the accident was undetermined.

2-25-43A. San Antonio, Texas. At 0015 a Vultee BT-13A suffered a structural failure and crashed five miles southwest of Kelly Field, San Antonio, Texas, killing pilot 2Lt. Charles W. Wilson and passenger SSgt. Thomas B. Robinson. The airplane had taken off from Kelly Field on a weather-checking flight and crashed soon after. Investigation of the wreckage revealed that the starboard wing had separated from the aircraft and struck the vertical fin and rudder, causing their failure and separation. The starboard wing and vertical fin were found about one-half mile from the scene of the main wreckage.

2-25-43B. Meridian, Mississippi. At 1125, a Douglas A-24A struck a farm building and crashed 15 miles northeast of Key Field, Meridian, Mississippi, killing pilot 2Lt. Edward F. Haynes and passenger Cpl. Robert H. Kotz. The airplane was flying in the number-three position of a three-ship V-formation on a formation training flight. Investigation revealed that the flight leader was leading the flight at altitudes reported to be as low as 15 feet. The number-three airplane struck a farm building with the port wing and crashed to the ground out of control. The airplane immediately burst into flames upon impact. It was speculated that the number-three airplane might have encountered propeller turbulence from the lead airplane and forced downward into the building. It was also speculated that the pilot of the subject airplane did not pull up in time to avoid colliding with the building.

2-25-43C. Elfrida, Arizona. At 2113 MWT, a Curtiss AT-9A and an AT-9 collided in mid-air three miles southwest of Elfrida, Arizona, killing instructor 1Lt. Edward A. Schwander aboard the AT-9A. His student, A/C Robert V. Saur, received minor injuries parachuting to safety. Aviation Student Thomas J. Ryan fell to his death attempting to parachute from the AT-9. A/C Robert Sandell parachuted to safety. The airplanes were part of a formation of four AT-9 type airplanes on a night formation training flight. The formation consisted of three student AT-9s flying in a V-formation at 8,000 feet indicated altitude (about 4,000 feet above ground level). Lt. Schwander, flying alternately behind and to the side of the formation in order to observe, attempted a crossover turn and collided with the AT-9 that was leading the formation. The instructor's AT-9A struck the lead student airplane from underneath at a 45-degree angle on the starboard wing trailing edge aft of the engine. The port wing of his airplane struck the bottom fuselage of the AT-9. The instructor's AT-9A continued upward and flipped over the top of the AT-9, falling off into a spin. It spun to the ground out of control. The AT-9 also went out of control, diving straight to the ground. Both airplanes exploded into flames upon impact. Cadet Saur, who was flying in the AT-9A pilot seat, reported that he had great difficulty escaping the cockpit of the spinning airplane. A/C Sandell, flying in the co-pilot seat of the AT-9, bailed out immediately after the collision, receiving minor injuries.

2-25-43D. Glenrock, Wyoming. At a time between 2131 and 2140, a Boeing B-17F crashed 25 miles north of Glenrock, Wyoming, killing the crew of ten. Killed in the crash were: 2Lt. Lewis E. Machala, pilot; 2Lt. Ned J. Devries, co-pilot; 2Lt. Harry N. Cannon, navigator; 2Lt. Joseph M. Hirshman, bombardier; SSgt. John Madden, engineer; SSgt. John F. Elson, assistant engineer; SSgt. Eugene D. Giles, Jr., radio operator; SSgt. Alfred W. Fuller, assistant radio operator; Sgt. James C. Fisher, gunner; SSgt. Robert C. Parrott, gunner. Investigators stated, "B-17F #42-5102 took off from Army Air Base, Casper, Wyoming, at 2110 MWT on a local bombing mission, to return at 2300 MWT. Pilot was cleared from Control Tower to high altitude bombing range, to bomb at 10,000 feet. Upon arriving at the range, the pilot reported clouds at the assigned altitude and requested permission to bomb at 8,500 feet. This new altitude was approved at 2130 MWT by the Control Tower and was the last radio contact with [the airplane]. Sometime between 2130 and 2140 MWT, B-17F #42-5102 crashed into the ground 25 miles north of Glenrock, about one-quarter mile west of Ross Road. The airplane was evidently traveling south when it crashed just below and to one side of the crest of a small hill. There was no evidence to show the airplane had touched at any other place than where it stopped, that is, it had not skipped. The #1 and #2 engines were uphill, and dug into the ground about 12 inches. The #3 and #4 engines were downhill and had dug into the ground about six inches deeper. The right wing tip was pleated with accordion-like folds — demonstrating that the right wing probably hit first. The tail and what was left of the fuselage ended up in the same place as the nose. Wreckage was strewn as far as 1,000 feet from the point of impact. The way pieces of flesh and the airplane were scattered indicated a terrific explosion at the time of impact. The burned area was about 400 feet long and 200 feet wide. The main wreckage was so tangled and burned that nothing could be determined. Propeller blades were in low pitch, and what would be the lower blade on each propeller was broken off. Propeller blades were comparatively unbent, considering the impact with which the airplane must have hit the ground. It is assumed that the pilot encountered instrument conditions and, inasmuch as he had a very limited amount of instrument experience and time, became confused. There were icing conditions in the clouds at the time of the accident and it is the Committee's opinion that the pitot tube might have iced up, adding to the pilot's confusion. A further assumption is that the pilot lost his bearings and fell out of the overcast out of control. Other pilots flying the same mission reported they encountered no trouble in flight, other than mild instrument weather and some icing."

2-26-43A. West Palm Beach, Florida. At 0330 EWT, a North American B-25D suffering a propeller failure crashed just after take-off one mile south of Morrison Field, West Palm Beach, Florida, killing the crew of five. Killed in the crash were: 1Lt. George E. Mudd, pilot; F/O Roy A. Young, co-pilot; 2Lt. Layton G. Zoubeck, navigator; Cpl. Samuel S. Powers, engineer; Cpl. John W. Walsh, radio operator. The airplane made what appeared to be a normal take-off to the southwest and then climbed to about 800 feet agl. The B-25 was in a left turn to the east when the port engine began to over-speed. Moments later the port engine erupted into flames and then quit, the propeller shearing off and separating from the engine. The airplane entered a steep bank to the left and began descending. The airplane stayed in this attitude until it struck the ground and exploded violently into flames. Investigation of the wreckage revealed that the port propeller pinion gear was sheared and the port propeller governor had failed. The port propeller had gone into extreme low pitch and allowed the port engine to over-speed beyond the maximum allowable rpm. The B-25 had been cleared to Borinquen Field, Puerto Rico, on a trans-Atlantic ferry flight to Oran, Algeria.

2-26-43B. El Paso, Texas. At 2119 MWT, a Boeing B-17F flew into the ground just after take-off from Biggs Field, El Paso, Texas, killing ten crewmembers and seriously injuring the pilot 2Lt. Edward M. Kelley. Killed in the crash were: F/O Charles G. Novak, co-pilot; 2Lt. James W. Bishop, navigator; 2Lt. Ernest F. Pasqua, bombardier; Sgt. V.A. Berrath, engineer; Sgt. A.J. Burke, radio operator; SSgt. Melvin P. Simms, Jr., assistant engineer; Sgt. Woodrow W. Erwin, assistant radio operator; SSgt. Joseph P. Bancom, gunner; Pvt. Robert M. Parcham, gunner; Sgt. Howard T. Speak, passenger. Investigators stated, "2Lt. Edward M. Kelley, flying B-17F No. 42-29528 on a night formation mission, began his take-off about thirty seconds after the B-17 ahead of him. His take-off was normal, but upon leaving the runway he glanced ahead to the lights on the B-17 [ahead of him], which began turning to the left and then [he, Lt. Kelley,] began flying on instruments. There was no distinguishable horizon and there were no lights ahead on the ground to judge his altitude above the terrain after he had passed the runway lights. The pilot stated that he followed the ship ahead, that he couldn't gain altitude, and that he was just beginning to fly 'good' when he hit the ground. Actually, his airplane hit the rough terrain several hundred feet to the right of where he thought he was flying, indicating that he may have been a victim of vertigo — meaning that due to the combination of flying by visual reference to the moving lights of the ship ahead and immediately changing to instrument flight, he actually was flying on a course over the ground different than he thought he was. The few seconds [that had] elapsed was not sufficient to correct, by reference to instruments, for the dangerous position of the aircraft. The pilot stated that he was successfully in correcting for instability caused by the [propeller turbulence] from the B-17 just ahead. If the landing lights had been used during this take-off, this accident might have been avoided."

2-26-43C. Gulf of Mexico, Florida. At 2222 EWT, a North American B-25D crashed into the Gulf of Mexico about 200 yards off shore of St. George Island in Apalachicola Bay, Florida, killing the crew of six. The airplane had successfully completed a trip from Tyndall Field, Panama City, Florida, to Atlanta, Georgia, on an important administrative and cargo flight. The airplane had taken off from Atlanta for the return trip on a visual clearance at 2107 EWT. The ETA for Tyndall Field was estimated to be 2237 EWT. Weather along the route was reported as ceiling and visibility unlimited and not considered a factor in the accident. When the airplane failed to arrive at Tyndall Field it was declared missing. The B-25D was found in three feet of water on 3-6-43 in Apalachicola Bay about 200 yards off shore of St. George Island, which is about 50 miles southeast of Tyndall Field. Investigation of the wreckage indicated that the airplane had flown into the water under control and at flying speed. A very small fire, with an area of about 20 inches, had erupted on the starboard wing but had been apparently doused by the sea. Pieces of the port wing had washed up on the bay side of St. George Island (the north shore) and showed no evidence of having contacted anything prior to the surface of the sea. Severe storms had passed through the area between the time of the crash and its discovery and it was thought that the wing had been washed ashore from the wreck site. Investigators speculated that the airplane had overshot Tyndall Field and had turned around over the Gulf of Mexico. The pilot apparently began descending and flew the airplane into the water. At 2222 EWT, a civilian bridge operator had seen a small flash fire on the surface of the sea on the night of the accident but did not associate this flash with an airplane crash. He had reported the information to the military police in the area and apparently the information was not acted upon. Killed in the accident were: F/O Edward A. Bagwell, pilot; 2Lt. Thomas F. Brown, co-pilot; F/O Albert D. Braswell, co-pilot; SSgt. Francis M. Morin, engineer; Sgt. James L. Smith, passenger; Cpl. Gene D. Hofherr, passenger.

2-26-43D. Orlando, Florida. At 2230 EWT, a Douglas P-70 crashed while attempting an emergency forced landing 8 miles east of the Army Air Base as Orlando, Florida, killing pilot F/O William C. Loeder. Radar operator 2Lt. Jasper Messmore III and crewmember Pvt. Robert A. Bittenbender parachuted to safety and were uninjured. Investigators stated, "F/O Loeder took off at approximately 1930 on a special Night Fighter training mission. After flying for about one hour [and failing to find the target vessel at sea] the pilot requested a [vector to return to Orlando Army Air Base]. The pilot was given homing vectors for about 45 minutes, which were 180 degrees off, putting the pilot further away from the base. The [Oscar] Homing Station finally gave up their efforts to get F/O Loeder back to his base and turned him over to [Lisbon] control, which gave F/O Loeder [a correct vector] back to his home base. The pilot ran out of gas while approximately 10 miles from the base and the radio operator and passenger bailed out safely and the pilot received fatal injuries in the crash. It is the opinion of the Aircraft Accident Classification Committee that the personnel charged with the operation of the Homing Station were responsible for the crash that resulted in the death of the pilot because of the wrong vectors given to the pilot in their effort to bring him back to the base." Investigators stated that the homing station personnel also gave the pilot erroneous estimates of the distance the airplane had flown and how far it had to go, further confusing the pilot. It was speculated that a more experienced controller could have issued vectors to an airfield that was closer to the airplane, allowing the pilot to safely complete the flight. Other pilots reported that the Oscar Homing Station had also given them incorrect vectors. Investigators noted that the pilot had flown for seven hours and twenty minutes earlier in the day and might have been fatigued.

2-27-43A. Kansas City, Missouri. At 1205 CWT, a Beech AT-10BH attempting a landing at Kansas City Municipal Airport, Kansas City, Missouri, crashed into an army mess hall at the AAF station, killing civilian waitress Leona Childers, Kansas City, Missouri. Three civilian employees of the mess hall were injured. Lillian Lightner, Gloria Baker, both of Kansas City, and Beulah Trucker, Independence, Missouri, all received minor injuries. Pilot Capt. Leonard J. Lang, Jr. was uninjured and passenger Sgt. Samuel Melman received minor injuries. The airplane had taken off at 1010 CWT from St. Louis, Missouri, on a cross-country flight to Kansas City. The airplane had initially taken off from its base at Moody Field, Valdosta, Georgia, at 0030 EWT and had landed Atlanta, Georgia, and at Smyrna, Tennessee, where it took on fuel before flying to St. Louis. The Aircraft Accident Classification Committee stated, "When the airplane reached a point 10 miles east of Kansas City Municipal Airport the pilot contacted the tower and received landing instructions. The tower reported to the pilot in code, wind 20-25 mph with gusts. The pilot made a normal approach and landed the aircraft in a three-point attitude. After rolling along the runway few feet the right wing came up and the airplane started a slow turn to the left, the left wing tip dragging in the dirt off of the runway. The airplane continued along the ground in a slow turn until it had turned approximately 90 degrees from the runway. About this time the airplane was rolling on three wheels but the pilot, fearing that he would run into some frame buildings, decided to take off. He applied full power, left the ground, struck a small frame building, shearing off the left wing tip, continued 50 feet further and crashed into a frame mess hall, killing a negro cook's helper and destroying the aircraft." Investigators stated that the pilot should have ground looped the airplane instead of attempting to take off, which was downwind. The airplane apparently was stalled just after becoming airborne. The airplane was out of control and was 150 degrees from its original heading on the landing when it crashed into the buildings.

2-27-43B. Atlantic Ocean, Georgia. At about 1600 EWT, a Lockheed-Vega B-34 was abandoned and crashed into the Atlantic Ocean about 60 miles off shore of St. Catherines Island, Georgia. Six crewmembers parachuted from the airplane. The crew of the civilian freighter *Alcoa Prospector*, which was operating in the area, was able to rescue three crewmembers. Three crewmembers were not rescued and declared missing. The B-34 had taken off from the Army Air Field at Charleston, South Carolina, at 1146 EWT on an anti-submarine patrol mission. Investigation revealed that the airplane had flown too far out to sea because of a malfunctioning compass. After becoming lost and low on fuel, the radar operator was able to locate a surface vessel at 5 degrees to the right of the airplane's course. The pilot flew toward the vessel and tried to contact it by radio without success. The airplane crew attempted to contact the ship's crew with a light-signal lamp, but this effort was also unsuccessful. After circling around the ship at 2,000 feet for several minutes, the pilot ordered the crew to abandon the airplane. The surviving crewmembers stated that all of the parachutes were observed to have deployed. The three crewmembers that survived stated that a lifeboat from the freighter soon appeared and rescued them. After searching for the remaining crewmembers for a few hours, the freighter set a course for Charleston, arriving there at 0200 on 2-28-43. An AAF B-34 and two U.S. Navy Vought-Sikorsky OS2U type airplanes searched for the missing crewmembers for many hours without success. Radio operator Sgt. Arthur F. O'Hare, radar operator TSgt. Earl E. Ragsdale and engineer TSgt. Roy R. Jones were declared missing and presumed lost at sea. Pilot Capt. Raymond A. Fortin, co-pilot 2Lt. Ralph O. Griffin and navigator-bombardier 2Lt. Philip E. Atkinson were able to parachute to safety and were uninjured.

2-27-43C. Vicksburg, Mississippi. At 1505, a Beech AT-10 suffering an engine failure entered a spin and crashed while attempting an emergency landing at the Vicksburg Municipal Airport, Vicksburg, Mississippi, killing pilot 2Lt. Perry W. Blain. Investigators stated, "Pilot, Lt. Blain, was en route from Waco, Texas, to Jackson [Mississippi] with a flight of three AT-10s. When approximately 130 miles west of Jackson the oil pressure of his left engine began to drop, becoming steadily lower until the left engine could not be operated. Upon reaching Vicksburg he requested the Flight Leader to land and circled the field and then came in for a normal landing with the wheels and flaps down. The wind was south at 5 mph. Lt. Blain made a high approach attempting to land into the north (downwind). After using up two-thirds of the field he decided that it would be impossible to land and attempted to go around. The field from that point on is up grade with trees at the end. He cleared the trees but in doing so he lost flying speed, stalled out and started to spin as he hit the ground. The airplane burst into flames immediately upon striking the ground."

2-27-43D. Santa Maria, California. At 1530, a Vultee BT-15 crashed 10 miles ESE of Santa Maria, California, killing pilot Capt. Wallace M. Johnson and passenger 1Lt. Jay H. Caldwell. The airplane had taken off from Hancock Field, California, on a routine training flight. The airplane was observed to fly through the saddle of two peaks from the east at an altitude of approximately 6,000 feet msl. The engine was heard to sputter and then the airplane was seen to go into a spin. Investigators stated, "From witnesses and from the observation of the wreck, the airplane hit with tremendous impact on the side of a hill at about a 60-degree dive. The fuselage was thrown down the side of the hill approximately 50 yards from the first point of impact. The right wing, stabilizer and elevator, and vertical fin were crumpled, but intact. The remainder of the airplane was strewn over the side of the hill and completely destroyed."

2-27-43E. Encino, Texas. At 2110, a North American AT-6C crashed 14 miles southeast of Encino, Texas, killing student A/C William L. Crow. The airplane had taken off from Moore Field, Mission, Texas, at 1939 on a night navigation flight to Rio Grande City, Texas, to Hebbronville, Texas, to Falfurrias, Texas, and return to Moore Field. No landings were scheduled for the flight. The student checked in by radio at Hebbronville and Falfurrias on time and on course. The airplane failed to return to Moore Field and was soon declared missing. The AT-6C was discovered from the air the next day by a search plane. Examination of the wreckage indicated that the airplane had hit the ground at great speed in a diving turn to the right. The student was killed instantly and wreckage was scattered over an area of 350 yards. It was later speculated that the student had lost the horizon and flew the airplane into the ground. Weather was reported as ceiling and visibility unlimited and not considered a factor in the accident. See 2-27-43G.

2-27-43F. Cuero, Texas. At 2140, a North American AT-6C crashed at Cuero, Texas, killing students A/C Marquis B. Taylor and A/C Warren Pickett.

The Aircraft Accident Classification Committee stated, "The airplane took off from Brooks Field [San Antonio, Texas] at 2058 on a night training mission and crashed on the Brayton Flying Field near Cuero at approximately [2140 CWT]. The airplane was on course and on schedule. The airplane was seen to go into a spin at approximately 3,000 feet altitude. It stopped turning at about 500 feet and went straight into the ground [with] power on. It is believed that the student lost his horizon, developed a condition of vertigo and went into a spin. He was effecting his recovery and had stopped turning when he hit the ground."

2-27-43G. Guerra, Texas. At 2217, a North American AT-6A crashed eight miles east of Guerra, Texas, killing student A/C William E. Hildreth. The airplane had taken off from Moore Field, Mission, Texas, at 1930 on a night navigation flight to Falfurrias, Texas, to Hebbronville, Texas, to Rio Grande City, Texas, and return to Moore Field. No landings were scheduled for the flight. The student had checked in by radio at Falfurrias on time and on course. The student failed to check in at Hebbronville and he failed to return to Moore Field and was soon declared missing. The airplane was discovered from the air the next day by a search plane. The wreckage was found to be on course for Rio Grande City from Hebbronville. Examination of the wreckage indicated that the airplane had flown into the ground under control and at flying speed. It was later speculated that the student had lost the horizon and flew the airplane into the ground. Weather was reported as ceiling and visibility unlimited and not considered a factor in the accident. See 2-27-43E.

2-28-43A. Pleasant Hill, Louisiana. At 0100, a Beech AT-7 crashed seven miles southwest of Pleasant Hill, Louisiana, killing five fliers. The airplane had taken off from Monroe, Louisiana, on a night navigation flight to the home station at Hondo, Texas. The AT-7 had made the flight from Hondo to Monroe earlier in the day. Investigators could not determine what caused the airplane to smash into the ground at high speed. Killed in the crash were: F/O Roy A. Lafser, pilot; 2Lt. Frank J. Remein, navigator instructor; student navigators A/C Harry P. Bestor, A/C Frederick C. Bigler, A/C John B. Binger.

2-28-43B. Avon Park, Florida. At 1205 EWT, two Martin B-26B airplanes collided in mid-air at the AAF Bombing and Gunnery Range at Avon Park, Florida, killing ten fliers. The airplanes were part of a four-ship of flight of B-26 airplanes in two, two-ship flights. The formation was flying over the northern end of Arbuckle Lake at about 1,100 feet agl. B-26B #41-17619 was leading the second element of two airplanes. The formation went into a shallow turn to the right. B-26B #41-17625, the wing ship, lost about 50 feet in the turn. The wing ship then climbed and pulled up almost abreast of the lead ship. The wing ship dropped down and began drifting to the left. The lead ship had suddenly gained about 50 feet and then began to settle back down into position. At the same time, the wing ship began to climb and to move back over to position. The wing ship ascended into the lead airplane, its number-two propeller contacting the fuselage of the lead ship. Both airplanes then fell out of control and slammed into the ground where they exploded into flames upon impact. Killed in the crash of B-26B #41-17625 were: 1Lt. Ramon G. Buela, pilot; F/O Harry G. Smith, co-pilot; 2Lt. Dane Hansen, bombardier-navigator; Sgt. Eldon H. Kinn, engineer; Sgt. Elroy P. Gainer, radio operator. Killed in the crash of B-26B #41-17619 were: 1Lt. Frank L. Wassell, Jr., pilot; 2Lt. Robert J. Brude, co-pilot; 2Lt. Arthur R. Kuntz, bombardier-navigator; Sgt. Richard A. Sofinski, engineer; SSgt. Raymond Dykes, radio operator; SSgt. Rex E. Lee, gunner.

2-28-43C. Austin, Texas. At 1147, a Douglas C-49F crashed six miles northeast of Austin, Texas, killing the crew of five. The airplane had taken off from Del Valle Army Air Base, Texas, on a routine training mission at 0830 and entered the vicinity of Austin Municipal Airport, Austin, Texas, at 1130. The airplane made a landing and performed a running take-off. The airplane re-entered the traffic pattern and traveled a bit farther than usual before turning on the downwind leg. Another pilot flying in the vicinity observed the C-49 at an altitude of about 1,000 feet above ground level while flying in the pattern. The airplane began a left turn onto the base leg with the landing gear extended. The turn was observed to be "flat" and then the degree of bank was seen to increase dramatically. The airplane entered a spin and smashed into the ground, exploding into flames upon impact and killing the occupants instantly. Killed in the crash were: 2Lt. Fran H. Horton-Billard, Jr., C-47 instructor pilot; TSgt. Russell E. Sampson, engineer; C-47 students/co-pilots 2Lt. Chester F. Steward, 2Lt. Jack E. McNeil, 2Lt. William E. Ernest.

2-28-43D. Mississippi River, Missouri. At 1630 CWT, a North American AT-6A crashed into the Mississippi River 12 miles ENE of Hayti, Missouri, killing A/C James E. Morrison, Jr. The airplane had taken off from Blytheville, Arkansas, with two other AT-6 type airplanes. The students had agreed to meet in the air at Hayti and agreed to watch each other perform air work and various maneuvers. The pilots followed each other in a wide circle for a short period. A/C Morrison then broke away from the two other ships and flew down to an altitude of about five feet above the surface of the Mississippi River. The other two AT-6 airplanes followed at an altitude of approximately 1,100 feet above ground level. After flying at extremely low altitude above the river for several minutes, A/C Morrison's propeller struck the surface of the river. The airplane bounded back into the air for a short distance and then crashed into the river. The airplane and the pilot disappeared into the river and were not recovered.

2-28-43E. Savannah, Georgia. At 2142 EWT, a Martin B-26B crashed while attempting a take-off at Hunter Field, Savannah, Georgia, killing the crew of six. The Accident Classification Committee stated, "B-26B airplane serial number 41-17997 called the control tower for take-off instructions and was given taxi and take-off clearance for a southwest take-off on the left-hand southwest runway. It was observed to make a normal take-off run and take-off to the southwest at 2141 EWT. Approximately one minute after take-off, the aircraft crashed and burst into flames two miles southwest of the field. There was no contact made with the aircraft after take-off. Due to the fact that the airplane was completely destroyed by fire and both motors and one magneto were completely demolished, it is impossible for this committee to determine the cause of the accident." Killed in the crash were: 1Lt. Harold P. Johnston, pilot; 1Lt. Bernard A. Anderson, co-pilot; 2Lt. Edward J. Visser, navigator; TSgt. John W. Smith, engineer; MSgt. Peter Bader, passenger; TSgt. Merivin E. Chagnon, passenger.

2-28-43F. Nixon, Texas. At 2245, a North American AT-6A crashed two miles north of Nixon, Texas, killing instructor pilots 2Lt. Tracey F. Wilson and 2Lt. Ernest E. Clough. The airplane was part of a flight of ten AT-6 airplanes that had taken off from Stinson Field, San Antonio, Texas, at 2200 on a "Blinker Mission" to Yoakum, Texas, to Pawnee, Texas, and return to Stinson Field. Lt. Wilson's AT-6A failed to return to Stinson Field and it was soon found to have crashed. Examination of the wreckage indicated that the airplane had spun into the ground with full power. The airplane exploded into flames upon impact. It was speculated that the pilot had lost the horizon and then lost control of the airplane. Two other AT-6A airplanes that were part of this Blinker Mission were involved in fatal accidents. The airplanes were stationed at Brooks Field, San Antonio. See 2-28-43G, H.

2-28-43G. Cuero, Texas. At 2300, a North American AT-6A crashed 20 miles northwest of Cuero, Texas, killing instructor pilots 2Lt. Michael Stazo and 1Lt. Jack M. Christ. The airplane was part of a flight of ten AT-6 airplanes that had taken off from Stinson Field, San Antonio, Texas, on a "Blinker Mission" to Yoakum, Texas, to Pawnee, Texas, and return to Stinson Field. Lt. Stazo's AT-6A failed to return to Stinson Field and it was soon found to have crashed. Examination of the wreckage indicated that the airplane had spun into the ground with full power and exploded into flames upon impact. It was speculated that the pilot had lost the horizon. Two other AT-6A airplanes that were part of this Blinker Mission were involved in fatal accidents. The airplanes were stationed at Brooks Field, San Antonio. See 2-28-43F, H.

2-28-43H. Yoakum, Texas. At 2340, a North American AT-6A crashed five miles north of Yoakum, Texas, killing instructor pilot 2Lt. Royce M. Clemens and passenger SSgt. John Lazan. The airplane was part of a flight of ten AT-6 airplanes that had taken off from Stinson Field, Texas, on a "Blinker Mission" to Yoakum, Texas, to Pawnee, Texas, and return to Stinson Field. Lt. Clemen's AT-6A failed to return to Stinson Field and it was soon found to have crashed. Examination of the wreckage indicated that the airplane had spun into the ground with full power. The airplane exploded into flames upon impact. It was speculated that the pilot had lost the horizon and then lost control of the airplane. Two other AT-6A airplanes that were part of this Blinker Mission were involved in fatal accidents. The airplanes were stationed at Brooks Field, San Antonio. See 2-28-43F, G.

2-28-43I. Poso Field, California. At 2145, a Vultee BT-13A crashed one mile northeast of Poso Field, California, killing student A/C David W. Wilds. The student had been flying in his assigned zone when he was called in for a landing. He successfully made a landing and a running take-off and then he was ordered to return to his zone. After flying for a short time, the student was called in for another landing. The student overshot his landing and was ordered to return to his assigned zone and altitude. After flying around for a short time, the student was again called in for a landing. The student was observed to turn toward the field and he was ordered to illuminate his landing lights to identify his position. The student was observed to illuminate his landing lights at an altitude of about 600 feet agl. He then flew into the ground at a steep angle in a steep bank to the right with the landing lights still illuminated. The student and airplane were stationed at Minter Field, Bakersfield, California.

March

3-1-43A. Boston Harbor, Massachusetts. At 0832, a Republic P-47C dove into Boston Harbor one-half mile off shore of Squantum Naval Air Station, Massachusetts, killing pilot 2Lt. Stanley Jankowski. The airplane was one a two-ship flight of P-47s that had taken off from the Army Air Field at Bedford, Massachusetts, at 0730 on a scheduled individual combat training mission. The two airplanes twice climbed to 11,000 feet indicated altitude and engaged in individual combat down to approximately 8,000 feet. The two airplanes were climbing up for a third encounter when a lone P-47 approached and lined up for an attack. The two wingmen broke up while attempting to engage the lone P-47 and apparently had lost each other. After failing to locate Lt. Jankowski and unable to contact him by radio, the other pilot returned to Bedford alone. At 0830, Lt. Jankowski encountered a flight of four P-47s at about 8,000 feet and he engaged the second element. The two airplanes out maneuvered Lt. Jankowski and got on his tail. Lt. Jankowski then attempted a split-S maneuver, rolling the P-47 inverted and then pulling the airplane into a dive. Lt. Jankowski was unable to pull out and the airplane dove straight into Boston Harbor in a high-speed vertical dive. Investigators speculated that the pilot had entered the maneuver at a speed exceeding 200 knots and had blacked out during the dive.

3-1-43B. Baltimore, Maryland. At 1145, a Republic P-47C crashed into Chesapeake Bay near the mouth of the Middle River at Baltimore, Maryland, killing pilot 2Lt. Meyer Rothblatt. A fisherman witnessed the P-47C dive into the bay, but after a thorough search investigators were unable to find the wrecked airplane. On 3-11-43, an A-4 type seat cushion and part of the trim-tab control panel were found to have washed ashore near the area where the airplane was reported to have crashed. The pilot's body was never recovered.

3-1-43C. Avon Park, Florida. At 1444 EWT, a Martin B-26B attempting an emergency forced landing crashed three and one-half miles from the airfield at the AAF Bombing Range at Avon Park, Florida, killing bombardier-navigator 2Lt. Stanley Kirshman and radio operator Sgt. Frank L. Szarfiniski. Engineer SSgt. Herbert E. Buck, Jr. was seriously injured. Pilot 1Lt. Walter E. Caples, co-pilot 2Lt. William A. Nolte, Jr. and gunner Sgt. Arnold Elliot received light injuries. The airplane had completed its bombing mission and was about to turn for its station at MacDill Field, Florida, when the starboard engine began to malfunction. The pilot feathered the starboard propeller and trimmed the airplane for single-engine flight. The B-26 was flying at about 700 feet agl when the pilot turned back toward the airfield at the bombing range, but upon completion of the turn the airplane had dropped to about 200 feet agl and was about one mile off to the west of the airfield. The pilot, unwilling to attempt a turn at that low altitude, kept flying straight ahead until he crash landed in an open area about three and one-half miles west of the airfield.

3-1-43D. Davenport Field, Texas. At 1510, a North American BT-9C crashed one-half mile west of Davenport Field, Texas, killing A/C Albert H. Brousseau. The airplane had taken off from Randolph Field, San Antonio, Texas, on a routine training flight. Investigators stated, "This student was practicing accuracy landings at Davenport Field. He had made four perfect landings and was coming in for his fifth. Witnesses to the accident stated that the airplane was in a normal glide on the base leg, but when he got to the last turn (at approximately 500 feet) it looked as though he had gone too far past his point of entry into the line of approach and pulled back on the stick to make the turn [steeper]. When he did, the airplane slowly peeled off into a spin from which he had almost recovered when it went into a secondary spin."

3-1-43E. Blanco, Texas. At 2210, a Vultee BT-13A crashed 12 miles southeast of Blanco, Texas, killing pilot 2Lt. William E. Walter. The airplane had taken off from Kelly Field, San Antonio, Texas, on a student night cross-country navigation flight to Junction, Texas, to Austin, Texas, and return to Kelly Field. Investigators noted that the airplane was well ahead of schedule when it crashed on the Tantermeuhl Ranch while on the Austin to San Antonio leg of the flight. It was speculated that the pilot might have made a navigational error and had picked up the Austin to San Antonio light line in an effort to get back to Kelly Field. Examination of the wreckage revealed that the airplane crashed into the ground in a steep diving attitude at high speed. The airplane was completely destroyed and the pilot was killed instantly.

3-1-43F. New Braunfels, Texas. At 2330, a Vultee BT-13A crashed 20 miles north of New Braunfels, Texas, killing pilot 2Lt. Frederick W. Williams. The airplane had taken off from Kelly Field, San Antonio, Texas, on a student night cross-country navigation flight to Junction, Texas, to Austin, Texas, and return to Kelly Field. Investigators noted that the pilot had checked in at Austin on schedule at 2305. Examination of the wreckage revealed that the airplane had crashed into the ground in a steep diving attitude at high speed. The airplane was completely destroyed and the pilot was killed instantly.

3-2-43A. Avon Park, Florida. At 0005 EWT, a Martin B-26B crashed while attempting a take-off from the airfield at the AAF Bombing Range at Avon Park, Florida, killing five crewmembers and seriously injuring gunner Sgt. Robert H. Foor. Killed in the crash were: 2Lt. Eugene Laurier, pilot; 2Lt. William H. Wahl, co-pilot; 2Lt. Jeffery J. Fitzpatrick, bombardier; Sgt. James J. Gorman, engineer; SSgt. Thomas M. Brown, radio operator. The Aircraft Accident Classification Committee stated, "B-26B airplane serial no. 41-17715 took off at 0002 EWT into the southwest. Shortly after leaving the ground, fire was visible to the witnesses [observing] on the right side of the airplane. The airplane continued to climb for a moment, then lost altitude quickly, striking the ground, rebounding into the air, and moving forward approximately one-half mile in an almost straight line with the runway. At or immediately following the initial contact with the ground, the right engine was torn from the airplane. There was no sign of fire at the point of contact with the ground either on the airplane parts lost on impact, or in the dry grass where impact occurred. Upon striking the ground the second time, the aircraft rebounded again, swerved to the left for a distance of one hundred yards and came to a stop, burning. It is believed that the rear part of the fuselage, from the rear bomb bay back broke off when the airplane swerved to the left." Sgt. Foor, who had apparently been thrown clear of the wreckage, was found "walking around" in shock about 75 feet from the crash site.

3-2-43B. Lockbourne Army Air Field, Ohio. At 0715 CWT, ground crewmember Pfc. John J. O'Sullivan was fatally injured when he walked into the rotating number-one propeller of a Boeing B-17D (40-3082) that was being readied for a training mission at Lockbourne Army Air Field, Columbus, Ohio. Pfc. O'-Sullivan died at 0815 at the base hospital. Investigation revealed that the number-one, number-three and number-four engines were running. Ground crewmembers were attempting to start the number-two engine when a small fire broke out in the number-two engine nacelle. Pfc. O'Sullivan, who was standing fireguard at the number-one engine, walked into the rotating number-one propeller while attempting to train a fire extinguisher on the fire in the number-two engine nacelle.

3-2-43C. Gulf of Mexico, Florida. At 1045 EWT, a North American AT-6A crashed into the Gulf of Mexico while on a target-towing mission near the AAF Gunnery Range at Apalachicola, Florida. Pilot 1Lt. Charles H. Williams was missing and presumed lost at sea. Target tow operator Pfc. Charles A. Riola was killed. The airplane had been participating in a gunnery mission with 16 other airplanes. The weather had been described as "variable" with a ceiling estimated at 5,000 feet above sea level with unlimited visibility underneath. Investigators stated that the weather had suddenly closed in on the area, producing a zero-zero condition in a matter of minutes. Controllers were unable to contact the subject airplane by radio. The airplane never returned to its base at Tyndall Field, Panama City, Florida, and was soon declared missing. Pfc. Riola's body had washed ashore on the Gulf of Mexico side of St. Georges Island on 4-1-43. His body, with his parachute still attached, was found strapped to a rear seat from an AT-6A type airplane. Lt. Williams' body was never recovered.

3-2-43D. Lucasville, Ohio. At 1415 CWT, a Waco CG-4A suffered a structural failure and crashed two miles northwest of Lucasville, Ohio, killing glider pilot F/O Leon T. Workman, 24, Kansas, and injuring glider pilot F/O Albert Cohen, age unknown, New York. Examination of the wreckage indicated that the "Gibson fitting" had failed and broke loose from the elevator. Still attached to its cable, the Gibson fitting flailed in the slipstream and whipped against the port elevator for over thirty minutes, causing the elevator to eventually fail. The CG-4A then went into a steep climb. When the glider became uncontrollable, the pilots cut loose from the tow airplane. The pilots left their seats and attempted to exit the CG-4A but had experienced great difficulty because the glider had entered a series of loops and stalls as it fell out of control. Unable to parachute to safety, the pilots attempted to get back to the controls and attempt to recover the glider. The co-pilot, F/O Cohen, was able to get to the area behind the pilot's seat but the glider crashed to the ground on its right side before he could regain control. He was seriously injured. The pilot, F/O Workman, who was fatally injured, was crouched behind the co-pilot's seat at the time of the accident. Examination of the CG-4A gliders at Laurinburg-Maxton Army Air Base, Maxton, North Carolina, the subject glider's base, revealed that several of them had cracked Gibson fittings and "unsatisfactory" control surface fabric.

3-2-43E. Oklahoma City, Oklahoma. At 1620, a Bell P-39D crashed while attempting a take-off at Will Rogers Field, Oklahoma City, Oklahoma, killing pilot 2Lt. William E. Glass. Investigators stated, "When the airplane reached the runway it was given a red light from the tower to allow a A-20 on its final approach to land. When the runway was clear the green light was given to the P-39 and it started to take off. During the take-off run the airplane seemed to gather speed normally; however, several witnesses remarked on the loud noise of the motor about the time it gained flying speed. The climb seemed normal. At about fifty feet above the ground the engine sputtered a little but took hold again and ran smoothly until about 400 feet when it started to miss badly. At this point the motor quit (subsequent investigation showed ignition switches on and other controls in apparent normal operation settings). The moment power was lost the airplane assumed a gliding angle and then the left door came off (it was found some distance away from the crash). The airplane continued its glide for a short distance then the left wing dropped, the airplane made a [half-turn spin] and crashed nose first. When the wing dropped [the airplane's] approximate altitude was 100 feet. Upon investigation at the scene of the accident it was observed that the left wing and nose made contact with the ground first as evidenced by marks in the soil. It is certain that the pilot was in the airplane at the time of impact; however, he was found some ten or fifteen feet from the wreckage. It is the opinion of the Aircraft Accident Classification Committee that, due to an overload in the propeller electrical system, the thermal overload switch kicked out and made the propeller electrical systems inoperative. The pilot did not put it back to the 'on' position to continue normal operation. Thus, with the propeller fixed in low pitch, the engine and propeller overspeeded."

3-2-43F. Pacific Ocean, California. At 1708, a Bell P-39N crashed into the Pacific Ocean 12 miles west of the Golden Gate Bridge, San Francisco, California, killing pilot 2Lt. Robert E. Mahood, Jr. The airplane had taken off from Mills Field, California, on an altitude mission. Investigators speculated that the pilot suffered a failure of his oxygen system, causing him to lose consciousness. The P-39N was seen to crash into the sea in an almost vertical dive. The pilot's body was never recovered.

3-2-43G. Alamogordo, New Mexico. At 2341 MWT, a Consolidated B-24E suffering an in-flight fire crashed three miles northwest of Alamogordo, New Mexico, killing eight crewmembers. Radio operator SSgt. Edmund F. Williamson parachuted to safety and was uninjured. Killed in the crash were: 2Lt. Thomas N. La Pia, pilot; 2Lt. Richard S. Browne, co-pilot; 2Lt. William F. Hogmann, Jr., navigator; 2Lt. Leonard L.

Taix, bombardier; SSgt. Frederick S. Jennings, engineer; SSgt. William A. Lonnevik, assistant engineer; Sgt. William E. Luchowski, assistant radio operator; SSgt. Louis J. Fontana, gunner. The Accident Classification Committee stated, "The airplane departed Alamogordo Army Air Base at 2317 [and] proceeded to climb in the direction of the Alamogordo Bombing Range. At sometime shortly thereafter but prior to 2337, a fire started in the no. 2 engine. From the evidence of witnesses on the ground, the pilot apparently turned in the direction of the Army Air Base and dove his airplane. When the aircraft was first observed by witnesses on the ground, it was south of the field, approximately one mile, heading in a northerly direction, with sparks streaming from the left side of the airplane. As [the airplane flew] across the airdrome, continuing to lose altitude, the fire was spreading rapidly until prior to the moment of impact [when] the airplane was completely engulfed in flames. Then the airplane fell off, and hit the ground with great force, exploded and was completely demolished by fire. From the statement of witnesses, the origin of the fire could not be determined, nor could it be determined whether the fire originated immediately adjacent to the no. 2 engine and in front of the firewall, or immediately behind the firewall in the inner wing panel. It is also evident that the fire was the immediate cause of this accident, inasmuch as at the time of [the airplane] passing over the airdrome, the airplane was apparently under control. There was an apparent lapse of approximately 4 to 5 minutes between the time that the fire was noticed in the no. 2 engine and the time of the actual crash. During this period of time the fire did not gain very much headway until reaching the vicinity of the Alamogordo Army Air Base, or possibly the latter 30 seconds to one minute of this time, and it appeared that the pilot was not unduly concerned, inasmuch as he was attempting to land on the airdrome and had given no signal for the crew to abandon the airplane. Upon examination of the wreckage it was determined that the no. 2 engine, which was the original source [of the fire], did not have its propeller feathered upon impact. There appears to be no evidence of structural failure."

3-3-43A. Springerville, Arizona. At a time after 1230 MWT, a Boeing B-17F collided with rising terrain and crashed 20 miles southwest of Springerville, Arizona, killing nine crewmembers. The B-17 had taken off from Portland, Oregon, and disappeared while on a leg of a cross-country navigation flight from Lockbourne Army Air Base, Columbus, Ohio, to Portland, to Las Vegas, Nevada, to Tucson, Arizona, and return to Columbus. The airplane, discovered on 7-12-43, was found to have crashed approximately 180 miles off course in the Burro Mountains in the White Mountain Range. Investigators stated, "It is assumed the accident was caused by the ship becoming lost during flight and being forced down due to bad weather. The pilots, not knowing they were over high mountains and being on instruments, may have attempted to let down in order to resume [visual] contact flying. The aircraft hit the trees and the side of the mountain apparently in level flight and at a normal cruising speed. When contact was made with the trees, [the airplane] was headed NNE. The ship apparently started breaking up on first contact with the trees and continued hitting lower on the trees as the slope of the hill gradually rose, the wings tearing off and gas tanks exploding and burning on contact. The nose section of the ship also burned on contact. The ship was totally demolished. The bombsight was recovered and turned in to the proper authorities; the wreck [was] salvaged by the 87th Sub-Depot." Killed in the accident were: Capt. Keith N. Busch, B-17 instructor; Capt. Rex K. Latham, co-pilot; Capt. Lynn L. Motley, co-pilot; 1Lt. Darwin H. Pfaender, co-pilot; 2Lt. William H. Startzer, co-pilot; TSgt. Clyde L. Miller, engineer; Sgt. William P. Appleby, assistant engineer; Pfc. Charles E. Arthur, radio operator; Pvt. Geronimo Medina, passenger. U.S. Senator Harry S. Truman (D-Missouri) made an inquiry on the search efforts on behalf of the family of Pfc. Arthur.

3-3-43B. San Antonio, Texas. At 1420, a Fairchild PT-19A suffering an apparent engine failure crashed 400 yards north of Kelly Field, San Antonio, Texas, killing civilian instructor William S. Luton. Witnesses on the ground stated that the engine failed at an altitude of approximately 200 feet above ground level. The pilot apparently stalled the airplane when he attempted to make a turn back toward Kelly Field. The airplane fell off into a spin and smashed into the ground. Investigators noted that the pilot could have made a safe emergency landing straight ahead rather than attempting a turn back to the field after the engine failure.

3-3-43C. Shoemaker, New Mexico. At 1445 MWT, two Lockheed F-4 type airplanes (the photo-reconnaissance version of the P-38 type) collided in mid-air eight miles northeast of Shoemaker, New Mexico, killing 2Lt. Nemo R. Leeman, the pilot of F-4 #41-2181. 2Lt. George N. Payne, the pilot of F-4 #41-2193, parachuted to safety an instant before his airplane slammed to earth and exploded. He was uninjured. The two airplanes had taken off from the Army Air Base at Colorado Springs, Colorado, on a navigation-training flight to Roswell, New Mexico, to Amarillo, Texas, and return to Colorado Springs. Investigators speculated that F-4 #41-2181 collided with F-4 #41-2193 from underneath and Lt. Leeman was killed in the collision. Both airplanes fell out of control, smashing to the ground and exploding violently into flames. Lt. Payne landed in his parachute about 200 yards from where the airplanes fell to earth.

3-4-43A. Columbia, Alabama. At 0820, a North American AT-6A crashed at Auxiliary Airfield #5, Columbia, Alabama, killing RAF instructor F/O Joshua Johnston and seriously injuring aviation student Sgt. P.J. DiCicco. The airplane was buzzing the area of the field north of the North-South runway at an altitude of 200 feet. The airplane executed a slow roll at 200 feet and then entered the traffic pattern. The airplane flew over the field at low altitude and then attempted another slow roll at the north end of the North-South runway. The pilot was unable to recover the airplane in time to avoid colliding with the ground.

3-4-43B. Pueblo, Colorado. At 0832, a Lockheed F-5A (the photo-reconnaissance version of the P-38 type) attempting an emergency forced landing crashed near the Army Air Base at Pueblo, Colorado, killing 2Lt. James N. Schlachter. Investigators stated, "Lt. Schlachter took off at 0752 on a local training mission. The clock was recovered from the wreckage indicated that he crashed at 0832. It appears from the condition of the left wing and the pitot tube that the aircraft hit the ground in an inverted position. The drop tanks could not be found; this fact indicates that the pilot had released them because he was attempting one-engine operation. The right propeller indicated it was either windmilling or feathered when the aircraft hit the ground, probably the latter. Investigation showed that the coolant radiator in this engine was completely empty of coolant fluid. It seems unlikely in this case that the coolant was lost after the aircraft hit the ground. Thus it seems likely that the engine failed from overheating due to coolant leak or improper servicing of coolant. What further made it impossible for the pilot to maintain control in single-engine operation remains undetermined."

3-4-43C. Roswell, New Mexico. At 1001 MWT, a Cessna AT-17B crashed while attempting a take-off at the Army Air Field at Roswell, New Mexico, killing the crew of three. Investigators stated, "It was evident from the time the airplane started rolling for the take-off that the pilot was having a great deal of trouble keeping the airplane straight. After traveling approximately 100 yards the left wing of the ship dropped decidedly, actually pulling the right wheel clear of the ground; the pilot regained control, straightened the airplane out and continued the take-off run. The airplane again swerved and the pilot then pulled the ship off of the ground at a low flying speed, evidently in an attempt to get the ship in flight and fully under control. After leaving the ground the left wing dropped again, as if a strong gust of wind had lifted the right wing. The left wing continued to drop then it appeared as if the pilot had opened the throttles wide and pulled the stick all the way back. The airplane climbed abruptly to an altitude of approximately 80 feet with the left wing down and a slight turn to the left. The airplane appeared to be almost stationary as it stalled to the left and fell over in the beginning of a spin. The airplane seemed to hit the landing mat on the left wing and engine first, bursting into flames immediately and then the airplane cartwheeled on over the right engine and wing and the fuselage broke over toward the front and left. The fire was under control and almost completely extinguished within ten minutes but the occupants remained in the airplane and were either killed in the crash or were so badly hurt that they were burned to death. From the time flying started at 0730 until 1001 when the accident occurred, traffic direction had been changed five times, each time 180 degrees, due to shifting winds. Traffic at the time of the accident was on Runway 12 and out on the mat the airplanes were landing with a considerable tail wind. All [pilots] landing and taking off on Runway 12 were having considerable difficulty controlling airplanes on the ground under the stiff cross and tail wind." Killed in the accident were: 2Lt. Kingdon C. Gish, 25, University City, Missouri, instructor pilot; A/C Harry K. Nave, 21, Elizabethton, Tennessee, student; A/C James W. Nicoson, 22, San Francisco, California, student.

3-4-43D. Platte City, Missouri. At 1105 CWT, a Fairchild PT-19A collided with a tree and crashed on the Wilson Farm one mile east of Platte City, Missouri, killing pilot 2Lt. John E. McGinn and pilot-rated passenger 2Lt. John E. Alber. The airplane was seen flying at low altitude, twice circling the farmhouse occupied by Betty Kitchen. The third time around the pilot waved at the civilian occupants of the house who had emerged to observe the airplane. The airplane then entered a steep bank to the left and passed between two tall trees as it flew away to the north. The airplane then collided with a tree on a small rise across the highway after passing between the two trees. The airplane then entered a shallow dive toward the ground. The airplane smashed into the ground and then bounded back into the air, shedding pieces for about 80 yards before smashing back into the ground and colliding with a telephone pole. The remnants of the airplane came to rest wrapped around the pole. Both pilots were killed instantly. The occupants of the farmhouse denied knowing either pilot.

3-4-43E. Vernon, Texas. At 1230, a Fairchild PT-19A crashed ten miles southeast of Victory Field, Vernon, Texas, killing student A/C Jerome J. Donovan and seriously injuring civilian instructor Ray L. Henry. Investigation revealed that the airplane stalled while performing a simulated forced landing. The instructor stated that the student had apparently "froze" on the controls after overshooting the intended field and attempting a turn to the left at low altitude in a gliding turn.

3-4-43F. Garden City, Kansas. At 1730 MWT, a Vultee BT-13A crashed 13 miles northeast of Garden City, Kansas, killing students A/C William A. Keeler and A/C Carlton R. Waehler. The airplane had taken off from the Army Air Field at Garden City at 1540 on a local instrument training flight. When the airplane failed to return to base it was declared missing. A search effort was immediately launched but was hampered by a blizzard the next day. The airplane was eventually located by a search plane on 3-6-43. Investigators were unable to determine why the airplane smashed into the ground in a flying attitude. It was speculated that the students might have been "hedge hopping."

3-4-43G. Needles, California. At 1700, a Taylorcraft L-2A crashed 16 miles south of Needles, California, killing pilot 2Lt. Verne J. Hawes and observer Capt. Howard B. Larlee. Investigators speculated that the airplane had suffered an apparent engine failure and was attempting an emergency forced landing. The pilot pulled the nose up abruptly when he realized that he was going to hit telephone lines near Power Line Road, causing the airplane to stall. The airplane entered a spin to the right, executing one-half turn of spin before it smashed to the ground about one-eighth of a mile east of California Highway 95. The airplane had taken off from the Army Air Field at Desert Center, California, while on a joint operation with the command of the 4th Armored Division. Only one propeller blade had been broken off while one remained essentially intact, indicating a stopped engine.

3-4-43H. Guymon, Oklahoma. At 2059 MWT, a Vultee BT-13A crashed 15 miles WNW of Guymon, Oklahoma, killing A/C William S. McAllister. Investigators stated, "The student pilot took off from the base airfield at about 1945 MWT on a triangular night navigation training flight from Garden City Army Air Base to Ulysses [Kansas] to Liberal [Kansas] to Garden City, with no landings except at Garden City. The student checked in by radio at Ulysses on schedule at about 2020, and proceeded toward Liberal on the second leg of the flight. This was the last radio contact made with the student. He failed to check in at Liberal, and he failed to return to Garden City as scheduled at 2120. A search was made by several ships in that area, but no contact was made. Several patrol ships reported the weather at the approximate time and place of the accident as being excellent for student night navigation flying. The following day [3-5-43] a blizzard prevented a further aerial search. The next day (3-6-43), however, the wreck was located first by a ground searching party, and immediately after by a ground searching airplane."

3-4-43I. Crandall, Texas. At 2230, a Vultee BT-13A crashed four miles SSE of Crandall, Texas, killing A/C John E. Kane. The airplane had taken off from Majors Field, Greenville, Texas, on a night navigation flight to Parker Airfield, Texas, to Wills Point Airfield, Texas, and return to Majors Field. The only scheduled landing was at Majors Field. Clearance was filed for 44 BT-13A airplanes to be dispatched at three-minute intervals for the identical navigation mission. Only 31 airplanes actually took off on the mission, the last departing at approximately 2130. Twelve of the airplanes landed at Parker, Texas, and remained overnight due to lowering ceilings. Three airplanes landed and remained overnight at Wills Point, two airplanes remained overnight after landing at Terrell, Texas, and two others landed and remained overnight at McKinney, Texas, all due to deteriorating weather. Two of the student airplanes crashed at about 2230, killing both aviation cadets. A/C Kane checked in at Parker on time and on course and he elected to continue to Wills Point. The controller cautioned the student to turn back for Parker if he encountered ceilings below 2,000 feet agl. Investigation of the wreckage indicated that the airplane had crashed into the ground in a steep dive, killing A/C Kane instantly. There were no witnesses. Investigators speculated that the student had probably lost control of the airplane after inadvertently entering instrument conditions. BT-13A #41-21899 crashed 12 miles northwest of Ennis, Texas, at 2230, killing student pilot A/C Frank Lauth. Cadet Lauth had departed Wills Point at 2135, reporting ceilings below 2,000 feet agl. He radioed that he was returning to Parker but he never arrived. His wrecked airplane was found the next day. Investigation of the wreckage revealed that the airplane had smashed into the ground while in a steep dive, killing the student instantly. Investigators speculated that the student had probably lost control of the airplane after inadvertently entering instrument conditions. There were no witnesses. See 3-4-43J.

3-4-43J. Ennis, Texas. At 2230, a Vultee BT-13A crashed 12 miles northwest of Ennis, Texas, killing student A/C Frank Lauth. The airplane was part of a student flight of 31 BT-13A airplanes that had taken off from Majors Field, Greenville, Texas, on a night navigation flight to Parker Airfield, Texas, to Wills Point Airfield, Texas, and return to Majors Field. The only scheduled landing was at Majors Field. The student had checked in at Parker Airfield on time and had also arrived in the vicinity of Wills Point on time, reporting a ceiling below 2,000 feet. The student radioed that he was returning to land at Parker. The airplane never landed at Parker or returned to Majors Field. His wrecked airplane was found the next day. Examination of the wreckage indicated that the airplane had crashed into the ground in a steep dive, killing the student instantly. There were no witnesses to the crash. Investigators speculated that the A/C Lauth had lost control of the airplane after inadvertently entering instrument conditions. A/C John E. Kane, who was a member of the same 31-ship navigation flight, was killed when his airplane crashed near Crandall, Texas. See 3-4-43I for details.

3-4-43K. Union Springs, Alabama. At 2315, a North American AT-6A crashed ten miles southwest of Union Springs, Alabama, killing A/C Paul W. Frank. The student had become lost while on a student night navigation flight from Napier Field, Dothan, Alabama, to Maxwell Field, Montgomery, Alabama, to Opelika, Alabama, and return to Napier Field. The student apparently attempted an emergency forced landing at night near Alabama Highway 6. Investigation of the wreckage revealed that the student had about one hour worth of fuel remaining and could have safely completed the flight. Investigators noted that had the student flown about 20 minutes north or west, he would have intersected a light line and would have been able to follow it to an illuminated airport. Investigators noted that all students had been instructed to use their parachute rather than to attempt an off-airport emergency forced landing at night.

3-5-43A. Flagstaff, Arizona. At 0145 MWT, a Boeing B-17F flying in instrument conditions crashed into Mount Elden five miles east of Flagstaff, Arizona, killing the crew of five. The airplane had taken off from the Army Air Base at Blythe, California, on a night instrument flight. The pilot had apparently become lost after losing the Blythe radio beam. The airplane apparently encountered poor weather that consisted of snow, freezing rain, low clouds and icing conditions. In an attempt to orient himself, the pilot apparently began to let down in a descending right turn when the B-17 collided with the east side of Mount Elden at an elevation of approximately 8,000 feet msl. The airplane exploded violently into flames upon impact and the crew was killed instantly. Killed in the crash were: F/O Keitt O. Martin, Jr., B-17 instructor; 2Lt. Ralph E. Majors, co-pilot/B-17 student; 2Lt. Frederick C. Ruhl, co-pilot/B-17 student; TSgt. Walter H. Ehlers, engineer; TSgt. Robert E. Sanford, radio operator.

3-5-43B. Alamogordo, New Mexico. At 0305, a Consolidated B-24E crashed near the Army Air Base at Alamogordo, New Mexico, killing nine crewmembers. Navigator 2Lt. Charles B. Tyler parachuted to safety and sustained only minor injuries. Killed in the crash were: 2Lt. John G. Striplin, pilot; 2Lt. Joseph L. Shumsky, co-pilot; 2Lt. Royce G. Matthews, bombardier; TSgt. Patrick McKay, engineer; SSgt. William F. George, radio operator; Sgt. Marvin Litz, assistant engineer; Sgt. John S. Swank, assistant radio operator; Sgt. Lawson R. Armstrong, gunner; Sgt. James R. Early, gunner. The surviving navigator stated that the airplane was flying at high altitude and had entered what he thought was a dive or a spin. He stated that he lost consciousness after he was violently forced into the floor at about 8,000 feet msl, or about 4,000 feet above ground level. The navigator stated that he did not know how he left the airplane, stating that he came to on the ground with his parachute deployed. Investigation of the wreckage indicated that an explosion occurred in the starboard wing in between the number-three engine and the fuselage. An altimeter that had been blown clear indicated a reading frozen at 7,800 feet. The starboard wing was blown off and it impacted the tail section, causing its failure and total separation. The navigator probably was thrown free as the airplane came apart in the air. Investigators could not determine what caused the airplane to enter an extreme dive or spin or what had caused the explosion in the starboard wing.

3-5-43C. Warwick, Georgia. At 1000 CWT, a Beech AT-10BH collided with a tree and crashed one-half mile south of Warwick, Georgia, killing students A/C Frank D. Borghetta and A/C Paul W.B. Berge. Investigators stated, "Crash of this airplane occurred in the vicinity of Leesburg Auxiliary Airfield No. 1 [auxiliary to Turner Field]. This airplane had taken off from the field approximately five minutes before the crash. [All witnesses interviewed by the committee stated that immediately before the crash] the airplane had flown very low over one field which was being cultivated, zoomed over some trees and then flew very low over the next field. At the end of this field the tail section struck a tree as the airplane attempted to climb over them. After this, the airplane rolled over in an inverted position and came down in a wooded section. This airplane was completely demolished and both occupants killed instantly. This flight was scheduled as an instrument team ride, therefore, this low flying was unauthorized."

3-5-43D. Fisher Island Sound, Connecticut. At 1130, a Republic P-47B suffering an apparent engine failure ditched into Fisher Island Sound near Trumbull Field, Groton, Connecticut, killing 2Lt. Carl P. Bock. The pilot had completed his gunnery exercise with his flight, which was returning to land at Trumbull Field. The pilot overshot the runway on his landing attempt so he retracted the landing gear, but not the flaps, and gave the airplane power for a take-off and a go-around. Witnesses stated that the airplane did not gain altitude as it flew over the gunnery range. The P-47 was seen to lose altitude steadily and the pilot was seen to open the canopy. A short time later, he either stalled the airplane or had hooked the port wing in the water, causing the airplane to cartwheel into the sea. The pilot had apparently loosened his safety belt and collided with the interior of the airplane. He was thrown clear of the wreckage and was found face down in the water. An autopsy determined that the pilot had suffered a head injury but had actually died of drowning. Investigators noted that the pilot might have survived the ditching if he had kept his safety belt and shoulder harness fastened. Investigators noted that all pilots should be warned that when performing an off-field emergency forced landing that they should not loosen or unfasten their safety belt or shoulder harness until the airplane comes to a complete stop. The airplane was never recovered from the sea and the cause of the engine failure was never determined.

3-5-43E. San Bernardino, California. At 0858, a Consolidated B-24D suffered a catastrophic structural failure and crashed in Cajon Pass near Devore Station, eight miles northwest of San Bernardino, California, killing the crew of five. The airplane had taken off from March Field, Riverside, California, at 0817 PWT on a routine transition flight. Investigation revealed that wreckage was scattered over a distance of two miles from the point where the B-24D slammed into the side of a hill in a steep dive. The airplane exploded violently into flames upon impact. The tail section had broken off in several pieces and was found scattered over the terrain. The port wing had separated near the number-one engine. Investigators speculated that the pilots probably put the airplane in a maneuver that exceeded the structural limits of the airframe, causing the failure of the tail section and the port wing. Killed in the crash were: Capt. Marvin J. Frum, pilot; 1Lt. Marshall E. O'Berg, 23, co-pilot/B-24 student; TSgt. Merle F. Draper, 23, Westmoreland, Kansas, engineer; SSgt. Fred Bedford, radio operator; SSgt. Russell P. Boyd, 21, Milwaukee, Wisconsin, gunner.

3-5-43F. Richmond, Virginia. At 1430, a Curtiss P-40F crashed into a wooded area two miles south of the Army Air Base at Richmond, Virginia, killing 2Lt. Jack R. Lewis. Investigators could not determine why the airplane crashed into the woods in a medium dive at full speed. The airplane exploded into flames upon impact and the pilot killed instantly. It was noted that this was the pilot's second solo P-40 transition flight.

3-5-43G. Eagle Pass, Texas. At 2230, a North American AT-6C crashed five miles east of the Army Air Field at Eagle Pass, Texas, killing pilot 2Lt. Herbert I. Sullivan and seriously injuring pilot-rated passenger F/O Charles Phippen. The airplane was returning to the field after completing a routine night flight when the pilot flew the airplane into the ground with the landing gear extended while on a straight-in approach to the active runway. The pilot-rated passenger in the rear seat noticed the dangerously low altitude on the approach and as he reached for the microphone to warn the pilot, the airplane flew into the ground.

3-6-43A. Colorado Springs, Colorado. At 0948 MWT, a Lockheed F-4 (the photo-reconnaissance version of the P-38 type) suffering an apparent engine failure crashed while attempting an emergency forced landing north of the Army Air Base at Colorado Springs, Colorado, killing 2Lt. James R. Brooks, Jr. It was determined that the pilot was unable to maintain altitude while returning to the field after suffering a failure of the starboard engine. The airplane smashed into rising terrain during the emergency landing. It was speculated that if the airplane had 25 more feet of altitude he might have been able to crash land on the crest of the hill that he crashed into. The airplane did not burst into flames upon impact, indicating that the pilot probably had cut the switches prior to impact. The pilot was thrown clear of the airplane and was killed when he collided with the ground. It was later speculated that the pilot might have survived had he not loosened his safety belt and shoulder harness before impact.

3-6-43B. King City, California. At 1417 PWT, a Ryan PT-22 crashed after entering a spin near King City, California, killing A/C James A. Newstrom. A cadet flying in the area reported that he had observed the airplane in a spin at an altitude of 3,000 feet above the terrain. The airplane described about six or seven turns of spin before smashing into the ground. The student made no obvious attempt to recover the airplane or to parachute to safety. Investigators speculated that the student froze on the controls after entering an unintentional spin.

3-6-43C. Newton, Texas. At 1752, a Martin B-26B crashed ten miles ENE of Newton, Texas, killing four crewmembers. Radio operator SSgt. James Thompson was able to parachute to safety and sustained only minor injuries. Sgt. Thompson reported that he was pushed from the airplane by gunner SSgt. Charles Arthur Colson, who was killed in an unsuccessful attempt to parachute to safety. Pilot 1Lt. Paul Elmer Smith, bombardier 2Lt. Richard F. Edwards and engineer SSgt. R. Mac Corbett went in with the airplane and were killed instantly. Investigation revealed that the airplane had taken off from the Army Air Base at Lake Charles, Louisiana, on a routine training flight. The pilot apparently attempted a slow roll at an altitude of 11,000 feet msl. The airplane stalled during the maneuver and fell into a flat spin to the left. The pilot partially recovered but the airplane entered a conventional spin to the left. The B-26 remained in the spin to the left until it struck the ground and exploded into flames. A moment after the B-26 entered the second spin the pilot rang the bailout bell and opened the bomb bay doors with the emergency switch. Sgt. Thompson left the airplane at an altitude of about 7,000 feet msl. He landed about two miles from the point where the airplane crashed. Investigation revealed that Sgt. Colson had deployed his parachute before he cleared the airplane. His parachute fouled on the some part of the B-26 and he was dragged along as the airplane went down. He was flung free of the airplane just before impact. His body, still attached to the damaged parachute, was found suspended in trees about 50 yards from the main wreckage. Investigators noted that performing aerobatic maneuvers in B-26 type aircraft was prohibited.

3-7-43A. Sylvester, Georgia. At 0500 CWT, a North American AT-6B crashed 12 miles southeast of Sylvester, Georgia, killing A/C Harry Freedman. The airplane had taken off at 0322 CWT from Spence Field, Moultrie, Georgia, on a night navigation flight to Butler, Georgia, and return to Spence Field. The airplane checked in at Butler at 0425 CWT on time and on course but never arrived at Spence Field. Despite a massive search, the airplane remained missing until a farmer discovered the wreck on 4-4-43 at about 0900 CWT. Investigation revealed that while on the leg from Butler to Spence Field the airplane crashed into a heavily wooded area in a diving left turn at an angle of approximately 80 degrees, exploding violently into flames upon impact. The impact was so great that two large trees were knocked over and a third cut in half. It was estimated that the airplane, heading due north, dove into the ground at a speed in excess of 200 mph. Investigators could not determine why the airplane crashed. It was later speculated that the student had lost the horizon and inadvertently flew into the ground.

3-7-43B. Donaldsonville, Louisiana. At 1500, a North American O-47A crashed at Donaldsonville, Louisiana, killing pilot 2Lt. Eugene D. Kuhre, and passengers Capt. Charles S. Anderson and Sgt. Edward E. Hardison. The airplane had taken off from the Army Air Base at New Orleans, Louisiana, on a routine training flight. The airplane was observed circling the vicinity of Donaldsonville for a short time before it flew into the ground in a descending turn. The airplane exploded into flames upon impact and the occupants were killed instantly. Weather was not thought to be a factor in the crash.

✓ **3-7-43C. Quitman, Texas.** At 1730, a Fairchild PT-19A crashed three miles south of Quitman, Texas, killing 2Lt. Billie E. Morehead. The airplane was flying in the number-six position of a six-ship formation on a ferry flight from the Fairchild Aircraft Company factory at Hagerstown, Maryland, to the Army Air Field at Coleman, Texas. The flight departed Shreveport, Louisiana, at 1630 CWT en route to Dallas, Texas. Witnesses on the ground observed the first three airplanes fly over at about 2,500 feet above ground level. Two other PT-19s passed over at about the same altitude about two minutes later. Moments later, Lt. Morehead's airplane was seen to pass over at an altitude of about 15 feet agl. The pilot pulled up to avoid striking a three-wire power line but was unsuccessful and collided with it. The landing gear fouled on the wires and the airplane dragged the power lines for over 150 yards before it clipped a telephone pole with the starboard wing. The airplane veered to the right and then hit the highway hard on its wheels. The airplane bounded back into the air about 40 feet before smashing into the highway about 75 yards further on, striking on the port wing and engine. The engine and propeller tore loose from the airplane and came to rest some distance from the main wreckage.

✓ **3-7-43D. Houston, Texas.** At 1745, a Fairchild PT-19A crashed six miles south of Houston, Texas, killing Women's Air Force Service Pilot student Margaret S. Oldenburg and civilian instructor Norris G. Morgan. Witnesses on the ground observed the airplane in a spin at an altitude of about 800 feet. The airplane had stopped spinning an instant before it struck the ground. The student and instructor were killed instantly.

3-7-43E. Midland, Texas. At 2257, a Beech AT-11 crashed just after take-off from the Army Air Field at Midland, Texas, killing the crew of four. Killed in the crash were: F/O Leland W. Slingerland, pilot; 2Lt. John H. Ferrell, bombardier instructor; A/C Lewis J. Lochner, student bombardier; A/C Milton Lipsky, student bombardier. The airplane was taking off on a routine bombardier-training mission. The take-off appeared normal until the airplane reached an altitude of about 400 feet agl when it went into a steep dive into the ground. The AT-11 exploded into flames upon impact. Investigators speculated that the autopilot might have been inadvertently activated with the elevator control in the "down" position, causing the flight gyro unit to put the airplane into a dive toward the ground. The airplane was at such a low altitude that the pilot was not able to overcome the autopilot before the airplane struck the ground.

3-8-43A. Lake Lawtonka, Oklahoma. At 1130, a Taylorcraft L-2B crashed one-half mile east of Lake Lawtonka, Oklahoma, killing Liaison Pilot 1Lt. Robert P. Stallings and seriously injuring Liaison Pilot 2Lt. Norris Littlepage. Investigation revealed that the airplane had stalled out of a steep left turn performed at an altitude of approximately 200 feet. Investigators speculated that the pilot had probably attempted the turn with insufficient power, excessive backpressure and too much bottom rudder.

3-8-43B. Fort Myers, Florida. At 1235 EWT, a Bell P-39D crashed near Page Field, Fort Myers, Florida, killing 2Lt. Earl L. Noble. Investigators stated, "Pilot took off on Runway 4 and climbed to approximately 200 feet. He then started turning to the left and attempted to turn back into the field on Runway 12. As he neared the runway he continued to [increase] the bank, and tighten the turn until the airplane snapped over on its back and dived into the ground. This occurred from an altitude of approximately 100 feet. The airplane hit on the extension of Runway 12 approximately 200 feet from the end of the runway. The airplane burst into flames and was totally destroyed."

3-8-43C. Donalsonville, Georgia. At 2030, a Vultee BT-13A crashed just after take-off from the Auxiliary Airfield at Donalsonville, Georgia, killing instructor 2Lt. Arnold F. Haspert and student A/C J.H. Gooding, Jr. Investigation revealed that the airplane made what appeared to be a normal take-off and initial climb. The airplane leveled out at 200 feet agl momentarily to pick up speed before continuing with normal climb. At about this time the engine apparently failed. The pilot set up a glide for an emergency forced landing but it was crosswind. The pilot put the airplane into a turn to the right as he attempted to position the airplane into the wind. The glide was too steep and the airplane crashed into the ground while in a steep bank to the right. The occupants were killed instantly. The airplane was practicing night take-offs and landings at the field when the accident occurred.

3-8-43D. Warrensburg, Missouri. At 2300, a Waco CG-4A glider crashed while attempting a landing at Sedalia Army Air Field, Warrensburg, Missouri, killing engineer Pfc. Harold J. Monoham. Glider pilot F/O Remus C. Hooten and glider co-pilot Elmer C. Johnson were seriously injured. Glider pilot F/O Robert L. Duffy received minor injuries. Investigation revealed that the glider had undershot the landing area and crashed into trees adjacent to the landing area. Apparently the pilot tried to stretch the glide.

3-9-43A. Albany, Georgia. At 0725 CWT, a Beech AT-10 crashed while attempting a take-off at Turner Field, Albany, Georgia, killing students A/C Mance D. Stark, Merryville, Louisiana, and A/C Robert J. Swain, Shreveport, Louisiana. Investigators stated that the airplane appeared to be losing power while on the take-off run. The airplane became airborne but had difficulty maintaining flying speed. After reaching an altitude of about 75 feet, the airplane turned slightly to the right then turned slightly to the left. The port wing dropped and the airplane crashed into the trees where it exploded into flames upon impact. Investigators could not determine what caused the diminished performance of the engines while the airplane was on the take-off run.

3-9-43B. Muroc, California. At 0820, a Lockheed RP-38E crashed near the Army Air Base at Muroc, California, killing pilot 2Lt. Alvah J. Miller. The airplane was on a ground gunnery mission. The Aircraft Accident Classification Committee stated that the pilot was making a low level pass at the target. He came in too low and had to abruptly pull up to avoid colliding with the target. The pilot pulled up abruptly and the airplane entered a high-speed stall, snapping over on its back. The pilot attempted to right the airplane but he over controlled and struck the ground with the port wing and propeller. The airplane cartwheeled into the ground and exploded violently into flames.

3-9-43C. Highland, New York. At 1220, a Vultee BT-13A crashed four miles west of Highland, New York, killing Cadet Harold A. Young, Jr. and seriously injuring Cadet Richard C. Snyder. Both were students at the U.S. Military Academy at West Point. The students had taken off from Stewart Field, Newburgh, New York, at about 1000 EWT on an instrument team ride. Investigation revealed that the students had been observed flying in the vicinity of Highland at low altitude. A short time later the airplane collided with trees and crashed into terrain.

3-9-43D. Cheshire, Massachusetts. At 1615 EWT, a Republic RP-47B crashed while attempting a forced landing near Cheshire, Massachusetts, killing 2Lt. Sommers D. Levermore. The pilot had taken off from Westover Field, Chicopee, Massachusetts, on a transition flight. The pilot became lost and began circling the area near Cheshire, apparently trying to get his bearings. The airplane began to get low on fuel and the pilot elected to perform a wheels-up landing in a field. The pilot undershot the field and he attempted to stretch the glide. The airplane stalled, entered a spin and crashed into trees adjacent to the field the pilot had chosen for his landing.

3-9-43E. Muroc, California. At 1357, two Lockheed RP-38E type airplanes collided in mid-air at Muroc, California, killing 2Lt. Robert E. Karch (41-2278). 1Lt. Hawley P. Mills was uninjured and was able to safely land his damaged airplane (41-2106) at the Army Air Base at Muroc. The airplanes were part of a flight of five P-38s that had taken off from Muroc Army Air Base on a local formation training flight. The formation climbed to an indicated altitude of approximately 7,500 feet. Lt. Mills was leading the second element and Lt. Karch was flying in the wing position on the right. Lt. Karch fell slightly out of position and in an attempt to regain his position he overtook the element leader, Lt. Mills. Lt. Karch banked to the right but was unable to avoid a collision. Lt. Karch's port wing collided with Lt. Mills' port boom from underneath. Lt. Karch's port wing continued to slide forward until it contacted the port propeller of Lt. Mills' airplane. Lt. Karch's port engine burst into flames and quit and his P-38 fell out of control and into a spin. The airplane continued to fall out of control until it smashed into the ground where it exploded into flames upon impact, killing the pilot instantly. Lt. Mills' port engine quit so he feathered the port propeller and trimmed the airplane for single-engine flight. He was able to safely make an emergency wheels-up landing at the base at approximately 1400. Lt. Mills, who was uninjured, stated that he never saw the airplane that collided with him.

3-9-43F. San Diego, California. At 1430, two Lockheed P-38G type airplanes collided in mid-air over the U.S. Naval Air Station at San Diego, California, killing ace pilot Capt. Virgil W. Lusk aboard P-38G #43-2307. Capt. Mark J. Mourne was uninjured parachuting to safety from P-38G #43-2253. The airplanes were part of a four-ship flight that had taken off at 1415 from North Field, San Diego Naval Air Station to demonstrate fighter tactics they had used in the African Campaign to members of the 330th Fighter Squadron, which provided two P-38s to simulate an attack on the four-ship flight. The four-ship flight consisted of Capt. Lusk, Capt. Mourne, Capt. W.J. Hoelle and Capt. William Paul Moore. The two-ship flight consisted of 1Lt. Olin B. Carter and 1Lt. John M. Kelly. The flight was at an altitude of about 6,500 feet msl when the attacking P-38s made their first pass. The two airplanes flew approximately 400 yards above the four-ship flight as they made their pass. The four-ship flight was flying in the standard four-ship combat formation of two, two-ship elements. The wing element was flying to the left of the lead element. After the attacking P-38s pulled away in a climbing turn to the left, the leader of the four-ship flight made a turn to the right and then made a turn to the left. It was at the conclusion of the right turn and just before the left turn when the collision occurred. Capt. Lusk was crossing over from left to right and had not dropped down far enough. It was speculated that he might have still been watching the attacking P-38s as they banked to the left and climbed away. The tail section of Capt. Lusk's P-38 was seen to separate and the airplane was on fire. Capt. Lusk was seen to fall from the airplane, his parachute opening immediately. It was speculated that he had been either killed in the collision and thrown clear, his parachute burst open by the collision, or he had opened his parachute before he was completely clear of the airplane and had apparently been struck by some part of the aircraft. Capt. Lusk was found dead in San Diego Bay by a U.S. Navy crash boat. His fully deployed parachute was still attached to his body. His P-38 slammed to earth and exploded on the parade ground at the U.S. Marine Corps basic training base at Camp Pendleton. Published accounts of the accident reported that marine recruits and drill instructors had to run for cover as the airplane crashed to the ground. Capt. Mourne's airplane also burst into flames and went out of control immediately, entering a flat spin to the right. After about four turns of a flat spin, the P-38 entered a conventional spin to the right. Capt. Mourne was unable to regain control of the airplane and he parachuted to safety with great difficulty. He had trouble releasing his canopy but managed to release it with his fingers instead of the release lever, which would not work. He got onto the port wing and was being forced into the still rotating port propeller by the spin. He grabbed hold of a vent and braced his right foot on the port engine nacelle and was able to prevent his body from being forced into the port propeller by the rotation of the spin. He was able to climb back onto the port wing and then was able to roll down and off of the port trailing edge. He opened his parachute at the last possible moment. His airplane also slammed to earth at Camp Pendleton. The wing airplane of the second element nearly collided with debris after the collision. Published accounts of the collision reported that Capt. Lusk had shot down five enemy airplanes in the Mediterranean War Theater.

3-9-43G. Eagle Pass, Texas. At 1640, a North American AT-6C crashed 14 miles east of the Army Air Field at Eagle Pass, Texas, killing instructor 1Lt. Luciana Marsh and student A/C Malvern L. Hillard. The airplane had taken off from Eagle Pass on an authorized aerobatic flight. The airplane had smashed into the ground in a high-speed vertical dive. The airplane had exploded into flames upon impact and was destroyed utterly. There were no known witnesses to the accident and investigators could not determine its cause.

3-9-43H. Mullet Key, Florida. At 1855 EWT, a Douglas A-24A suffering an engine failure was successfully crashed landed on an unnamed island at Mullet Key, Florida. The pilot, 2Lt. Theodore P. Welgoss, was uninjured. Gunner Sgt. Max A. Tevebaugh drowned after successfully parachuting to safety. Soon after take-off from Drew Field, Tampa, Florida, on a scheduled bombing training mission, the A-24's engine began to sputter and smoke. The cockpit filled with smoke and the pilot gave the gunner the option of bailing out or crash landing. The gunner elected to bail out and drowned before he could be rescued from the sea. The pilot was able to make a successful wheels-up landing on the beach. Investigation revealed that a broken breather seal on the propeller mechanism caused the engine failure.

3-9-43I. Drew Field, Tampa, Florida. At 2210 EWT, a Martin B-26B suffering an engine failure crashed while attempting an emergency forced landing at Drew Field, Tampa, Florida, killing five crewmembers and injuring two others. Investigators stated, "B-26B No. 41-18137 en route from Eglin Field [Valparaiso, Florida] to Avon Park [Bombing and Gunnery Range, Avon Park, Florida] had an engine failure four minutes north of Drew Field. Pilot feathered right [propeller] and came in downwind on Runway 27 at an exceedingly high speed clearing a bus at the east end of the field by inches. The airplane touched the runway on the last [third], [the pilot] applied brakes hard and [the ship] skidded ahead for 100 yards, struck a five-foot [deep drainage] ditch [the landing gear collapsed and the B-26 broke up as it] slid another 100 yards in flames. Two crewmembers got out but could not help others in the airplane due to the heat of the fire." Drew Field rescue personnel were on the scene immediately but could not save the remaining crewmembers. It was noted by investigators that the airplane had not contacted the tower at Drew Field prior to attempting the emergency landing and it had been given a red light by tower personnel as it approached. The surviving engineer stated that the starboard engine oil pressure began to fluctuate then the pressure dropped very low, but the cylinder head temperature was okay. Moments later the starboard engine burst into flames and then quit, causing the pilot to attempt the emergency landing at Drew Field. It was noted that the starboard vacuum pump and hose had been replaced the day before. Killed in the crash were: 2Lt. Elmer O. Bischmann, pilot; 2Lt. Victor R. Tharp, co-pilot; 2Lt. James J. Boyle, navigator; SSgt. Lawrence E. Ferguson, engineer; Pvt. George M. Walker, radio operator. Engineer SSgt. John Prystach and gunner Cpl. Byron G. Dubois suffered only minor injuries escaping the burning airplane.

3-9-43J. Mission, Texas. At 2230, a North American AT-6C crashed three miles south of Moore Field, Mission, Texas, killing RAF instructor P/O Jack R. Powell and AAF student A/C John A. Sanborn. The airplane had taken off from Moore Field on a routine dual night transition flight and had been circling in its zone waiting for its turn to come in for a practice landing. The airplane was called in and it was seen "peeling off" to the right to enter the pattern. Tower personnel observed the airplane in a steep descending turn to the right and attempted to warn the pilots as the airplane flew into the ground. The AT-6C exploded violently into flames and the occupants were killed instantly. It is likely that the student and instructor had lost the horizon and inadvertently flew the airplane into the ground.

3-9-43K. Stockton, California. At 2215, a Cessna AT-17B suffered a catastrophic structural failure of the port wing and crashed two miles south of the Army Air Field at Stockton, California, killing students A/C Harold C. Bridges and A/C Noble Brewster. The airplane was flying at an altitude of about 1,000 feet agl and attempting to enter the traffic pattern at Stockton when the port wing failure occurred. The students lost control of the airplane and it smashed into the ground where it exploded into flames. Pieces of the port wing were found along the airplane's line of flight. The airplane had been on a student night training flight from Stockton Army Air Field to New Jerusalem Auxiliary Airfield and return.

3-10-43A. Bascon, Florida. At 0405 CWT, a North American AT-6C crashed one-half mile northwest of Bascon, Florida, killing student A/C Stephen Faniglula. The airplane had taken off from the Army Air Field at Marianna, Florida, and had flown to Auxiliary Airfield #3 where it practiced night take-offs and landings for an hour and a half. The student was then ordered to return to Marianna Army Air Field, which was 11 miles away and visible from above the auxiliary airfield. After 30 minutes the airplane had still not arrived and was declared overdue. The airplane was found at about 0600 CWT. Investigation of the wreckage indicated that the airplane had dove into the ground.

3-10-43B. Fort Myers, Florida. At 1135 EWT, a Bell P-39N crashed five miles north of Page Field, Fort Myers, Florida, killing 2Lt. Leslie F. Jarvis. The airplane failed to return from a routine training flight and was not found until 3-14-43. Investigation of the wreckage indicated that the airplane had struck the ground in a spin to the right. The pilot was apparently attempted to bail out. The starboard escape hatch was found partially opened. The parachute had apparently deployed in the cockpit and had fouled on the door cowl. The starboard door had apparently closed on the deployed canopy, allowing the airplane to drag the pilot to his death. Examination of the crash scene revealed that the parachute shroud lines were strung out for their entire length with the pilot still attached in his harness.

3-10-43C. Sumter, South Carolina. At 1500 EWT, a Vultee BT-13A crashed 14 miles northeast of Shaw Field, Sumter, South Carolina, killing A/C Benjamin Crawford. The airplane had taken off from Shaw Field at 1420 on a routine solo flight. The airplane was next observed in a spin, describing at least 12 turns of spin before smashing into the ground and killing the student instantly. Investigation of the wreckage indicated that the student had attempted to leave the airplane but had failed to do so. His safety belt and harness were found unfastened and he had been thrown clear of the cockpit. A witness stated that the airplane looked as if it were beginning a recovery just before it hit. Investigators described the accident as a "spin from considerable altitude."

3-10-43D. Phoenix, Arizona. At 1350 MWT, a North American A-36A and a Stearman PT-17 collided in mid-air and crashed six miles north of Phoenix, Arizona, killing PT-17 student A/C Donald Harnack, 22, Remsen, Iowa. A-36A pilot 2Lt. Arthur D. Durnbaugh parachuted to safety and was uninjured. The PT-17 had taken off from Thunderbird Field, Glendale, Arizona, on a routine training flight. The A-36A was part of a flight of four that was part of a larger formation that had taken off from Rice Field, California, on a simulated bombing mission on Cutter Field, Arizona. While the A-36 formation was returning to Rice Field at an altitude of 5,500 feet agl, one A-36 pilot in the trailing four-ship flight radioed that he had some troubles and was going to land at Luke Field. The malfunctioning A-36 left the formation and turned toward Luke Field. The leader of the trailing four-ship flight spotted a lone PT-17 flying east at about 1,500 feet agl. The flight leader broke off to make a simulated attack on the trainer from the east and then after the pass he was going to land at Luke Field. Lt. Durnbaugh, flying as wingman to the flight leader, followed the flight leader down. The flight leader stated that he about was 300 feet above the trainer as he made his pass at 300 mph. Investigators stated that "Lt. Durnbaugh was definitely flying his leader's wing" very tight. Investigation revealed that the flight leader passed very close to the trainer on the simulated pass, causing Lt. Durnbaugh to collide with the trainer in a head-on dive. Lt. Durnbaugh, his eyes glued to his leader, stated that he never saw the PT-17 while on the diving pass and thought that his airplane had simply blew up in mid-air. The A-36A struck the PT-17 squarely with the lower wing surface, killing the student instantly and causing the PT-17 to erupt in flames and break up completely in mid-air. A/C Harnack had been thrown clear of the PT-17 and his badly mutilated body was found east of Central Avenue north of Phoenix. Investigators stated that the PT-17 was "completely demolished in mid-air [and] fell in pieces among which no one was larger than one man could carry." The A-36 lost its port wing and spiraled to the ground out of control where it exploded into flames. Lt. Durnbaugh received minor injuries parachuting to safety.

3-10-43E. Del Rio, Texas. At 1540 CWT, a Martin B-26 suffering an apparent engine failure crashed while attempting an emergency forced landing 12 miles north of Laughlin Field, Del Rio, Texas, killing the crew of four. The airplane had taken off from Laughlin Field on a routine training flight when it apparently began suffering engine problems. Investigators speculated that the pilot was attempting an emergency forced landing when the he overshot his intended field. The pilot pulled the nose up abruptly in an effort to clear a ravine in front of him. The airplane stalled and fell into the ravine at a steep angle, exploding violently into flames upon impact. The crew was killed instantly. Killed in the crash were: 2Lt. Benjamin L. Littlejohn, pilot; 1Lt. Walter F. Griffin, co-pilot; SSgt. John R. McGuire, engineer; SSgt. Robert H. Steeds, engineer.

3-10-43F. Selma, Alabama. At 1730, a TG-8 type glider (AAF serial #43-3139) being towed on a ferry flight suffered a catastrophic structural failure of the port wing and spun to the ground 32 miles southeast of Selma, Alabama, killing glider student SSgt. Rex Collin Newman. The glider was being towed from Mobile, Alabama, to Maxwell Field, Montgomery, Alabama, by a Vultee BT-13A type airplane at an altitude of about 1,500 feet agl when the port wing failed and separated from the glider. The glider broke the towline and immediately spun to the ground, the glider student making no apparent effort to leave the glider. He was killed instantly upon impact. The tow-plane circled the area of the crash several times and then flew away.

3-11-43A. Walhalla, South Carolina. At a time between 0200 and 0215 EWT, a North American B-25C crashed into rising terrain 22 miles north of Walhalla, South Carolina, killing the crew of five. Killed in the crash were: F/O Richard S. Brock, pilot; 2Lt. Bart S. Monroe, co-pilot; 2Lt. Philip J. Graziano, bombardier/navigator; SSgt. Harvey Capellan, engineer; Sgt. Michael Sekel, radio operator. The Aircraft Accident Classification Committee stated, "Pilot F/O Brock, in B-25C airplane, No. 41-12740, departed at 2010 EWT, March 10, 1943, on a night navigation flight from Greenville Army Air Base [Greenville, South Carolina] to [Key Field] Meridian, Mississippi, with no stop en route contemplated. Aircraft was cleared under Contact Flight Rules at 3,500 feet and the route was direct both ways. At 2215 EWT, the pilot contacted Meridian and reported his position as 20 miles East of Meridian at 3,500 feet and requested the weather at Greenville. No further contacts were reported. The estimated time en route was four hours and the aircraft carried enough fuel for five and one-half hours of flight. When sufficient time elapsed to insure the complete exhaustion of the fuel supply the aircraft was listed as missing.... The ensuing search was hampered by inclement weather. Aircraft from [Greenville Army Air Base] searched the mountainous area north and west of Greenville but no trace of the missing aircraft was sighted, it having crashed in a thickly wooded section 22 miles north of Walhalla, South Carolina, near South Carolina State Highway 107. An engine of the aircraft was discovered in a ravine a distance of over 1,000 feet from the scene of the wreck by a young boy named Sean Crane, who notified Mr. Ben Rogers of his find. Mr. Rogers scoured the area, found the wreck [on 3-23-43] and notified [the authorities]. It is the opinion of the committee that the pilot was temporarily lost and off course due to a southeast wind. The pilot, not knowing or realizing the altitude of the mountains in the vicinity due to the poor visibility, flew the ship into a tree on the highest point losing the left wing. The ship crashed farther on in the timber. It is also the opinion of the committee that the accident was not caused by material failure or due to exhaustion of fuel. At the time of the crash the aircraft was headed in an easterly direction on the west leg of the Greenville beam."

3-11-43B. Perry, Oklahoma. At 0356, a Douglas RDB-7B flying in poor weather crashed five miles north of Perry, Oklahoma, killing pilot F/O Joseph H. Gillepsie and gunner Pvt. Kenneth R. Isom. "It is the opinion of the Aircraft Accident Classification Committee that the pilot, immediately upon take-off, either purposely or erroneously took a course other than that assigned him and started to fly toward Wichita, Kansas. He ran into a severe local snow flurry around Perry and for reason or reasons unknown, circled some lighted areas at a very low altitude (oil wells being drilled and an airport under construction). Evidently he was having engine trouble at that time and the combination of a bad engine plus instrument conditions may have been more than the pilot was able to cope with and he flew into the ground. Before the flight the pilot drew the course on his maps and was 'briefed' on the mission." Investigators noted that Pvt. Isom was an unauthorized passenger.

3-11-43C. Lutz, Florida. At 1030 EWT, a Martin B-26B suffered a catastrophic structural failure of the tail section and crashed at Lutz, Florida, killing the crew of five. The airplane took off from MacDill Field, Tampa, Florida, on a gunnery-training mission. Killed in the crash were: 2Lt. Victor H. Kolb, pilot; 2Lt. Robert Dawson, co-pilot; 2Lt. John R. Nelson, bombardier/navigator; SSgt. Warren H. Allen, engineer; SSgt. Mitchell J. Luczak, radio operator. Investigators stated, "The airplane was observed in a steep dive from an altitude of 3,000 to 4,000 feet at one time, and again just after completing a [steep left] turn at low altitude. The ship was pulled out of the second dive and the tail was seen to disintegrate just prior to the time the ship hit the ground. The horizontal stabilizers were found approximately three-quarters of a mile from the scene of the accident. The vertical fin was found approximately 150 feet closer to the crash than the horizontal stabilizers. The fact the tail assembly disintegrated, according to civilian witnesses, would cause the airplane to crash. The cause for the disintegration of the tail assembly is undetermined since the airplane burned."

3-11-43D. Savannah, Georgia. At 1621 EWT, a Douglas A-20C collided with the control tower and crashed at Hunter Field, Savannah, Georgia, killing pilot 2Lt. Ernest A. Edens, gunner Sgt. Thomas W. Thomas and engineer Sgt. Charles E. Taylor. The airplane took off at 1615 EWT from Chatham Field, Savannah, on a cross-country flight to the Army Air Base at Homestead, Florida. Investigators stated, "Just prior to Lt. Edens' take-off, a formation of three [A-20 airplanes] had taken off.... After leaving Chatham Field, Lt. Edens joined this flight, and all four ships flew low over Hunter Field. Lt. Edens flew directly across the field from the northwest to the southeast, while the other three ships passed directly over the building area. Lt. Edens' ship was observed to crash into the 65-foot control tower located in the center of the east parking area. The left wing was sheared off, and the ship burst into flames and crashed on the southeast edge of the field. An unsuccessful attempt was made by the control tower to contact the ship when it was first observed heading toward the field from the northwest." Investigators noted that the control tower, which was demolished, was newly built and was not yet in operation.

3-11-43E. Indian Springs, Nevada. At 1335 PWT, a North American AT-6A crashed near the municipal airport at Indian Springs, Nevada, killing pilot F/O Jack B. O'Hagan and passenger Pvt. Jack W. Chappell. The airplane had taken off from the Army Air Field at Las Vegas, Nevada, on a "Range Estimation" mission. Qualified witnesses on the range observed the airplane flying away from them at an altitude of about 500 feet agl. When the airplane reached the spot where it should have turned around, the pilot pulled up into steep right turn. After the airplane turned approximately 90 degrees it started to slip downward on the starboard wing. The airplane then entered a spin to the right and smashed into the ground and burst into flames. The occupants were killed instantly.

3-11-43F. Deerland, Florida. At 1600 CWT, a North American AT-6C suffered a catastrophic structural failure of the port wing and crashed four miles north of Deerland, Florida, killing pilot 2Lt. Simon D. Lynde and passenger SSgt. John R. Lawrence. Investigators stated, "The aircraft was flown from Auxiliary Airfield No. 3, Eglin Field [Valparaiso, Florida]. Statements of witnesses indicated that the aircraft was being dived and pulled up into vertical snap rolls. The second time this maneuver was attempted, the [port] wing was seen to tear from the aircraft and float away with other parts of the wing. The aircraft crashed in a river bed."

3-11-43G. Mechanicsburg, Pennsylvania. At 1722, a Douglas C-47A flying in instrument conditions collided with trees and executed a successful emergency forced landing at Mechanicsburg, Pennsylvania, killing a government employee on the ground. The four C-47 crewmembers and eight passengers were uninjured. The airplane had taken off at 1445 EWT from Pope Field, Fort Bragg, North Carolina, on a transport flight to Olmsted Field, AAF Air Depot, Middletown, Pennsylvania. The airplane was on an instrument clearance and when it arrived at a point about 20 minutes from Olmsted Field instrument conditions were encountered. As the airplane neared Harrisburg, Pennsylvania, it let down from 5,000 feet indicated altitude to about 3,000 feet. The pilots requested weather information and Harrisburg Radio reported a ceiling of 1,500 feet, three-quarters of mile visibility with no wind. Investigators stated, "The aircraft proceeded toward the Harrisburg Cone and was instructed by Harrisburg Radio to proceed out on the West leg of the beam for four minutes at 3,000 feet. The aircraft proceeded as instructed and started a turn around procedure, letting down during the turn (Note: This is contrary to correct procedure for Harrisburg). At 1706, the airplane was instructed by Harrisburg Radio that they were number one to approach Middletown and to report leaving 3,000 feet, also to report when over the Harrisburg Range, and when in contact. The pilot let down to an indicated altitude of 1,400 feet — not breaking out into contact conditions, nor was the ground visible. The aircraft then struck some trees and damaged the leading edge of both wings and the right motor, which caused the aircraft to vibrate (Note: It is believed that the C-47 struck a ridge which is parallel and approximately four miles to the right of the West leg of the Harrisburg Range. The mountain is approximately 1,000 to 1,500 feet in height). The pilot succeeded in flying a short time but was unable to gain sufficient altitude to permit abandoning the aircraft. The right motor then lost power and the aircraft began losing altitude rapidly and to vibrate more violently. Breaking through the overcast when only several hundred feet above the ground and encountering intermittent banks of scud clouds, the pilot selected a field and made an emergency landing with the wheels up. The landing was carried out skillfully. The aircraft broke through telegraph wires, which established the time of landing at 1722. The airplane skidded across the rolling field, through a fence, and into a 12-foot excavation. Just before reaching the excavation, Mr. Albert C. Kraft, a Civil Service employee of the U.S. Bureau of Yards and Docks, was struck by the aircraft and killed instantly. It was impossible for the pilot to avoid hitting this person." Pilot Capt. Terry G. Hutton, co-pilot F/O Robert L. Raley, radio operator Sgt. Louis R.E. Savela and engineer SSgt. Robert D. Sims were uninjured as were the eight U.S. Army passengers.

3-11-43H. Fort Sumner, New Mexico. At 1630, a Taylorcraft L-2A suffering engine trouble crashed just after take-off from Fort Sumner, New Mexico, killing Glider Pilot F/O James E. Murphy. The airplane took off from the main airfield at Fort Sumner and climbed to an altitude of about 300 feet agl when the engine lost power. Investigators stated, "It is believed that engine trouble was due to carburetor ice. Conditions were very favorable to carburetor icing and several other airplanes of similar design experienced severe carburetor icing. Eyewitnesses report that as the engine lost power, the pilot held the nose of the airplane up, losing airspeed and maintaining flight in a semi-stalling position. This attitude was held for approximately 300 yards when the airplane finally stalled completely and spun from an altitude of approximately 300 feet. The airplane made one-half turn [of spin] and crashed, still spinning, demolishing the aircraft and instantly killing the pilot."

3-11-43I. Garden City, Kansas. At 1646 MWT, a Vultee BT-13A spun to the ground nine miles NNE of Garden City, Kansas, killing A/C George O. Washburn. Investigators stated, "This student pilot took off from the Garden City Army Air Base at 1600 on a local solo transition training flight. It was his third solo flight and he had been declared proficient in spin recovery by his instructor. However, his instructor had told him not to practice spins solo. The student was not heard from after he took off. The crash was reported by a student who said he first saw the airplane spinning rapidly at about 700 feet. He saw it spin on in and crash."

3-11-43J. Ontario, California. At 1637, two Stearman PT-18 type airplanes collided in mid-air one mile south of Cal Aero Academy, Ontario, California, killing A/C Zandel Boucher aboard PT-18 #40-2013. A/C John J. Forrester was able to safely land PT-18 #40-1996 at Cal Aero Academy. A/C J.F. McCullough witnessed the accident. He later stated to investigators, "I was on the approach leg that intersects the downward leg which runs parallel to South Cal-Aero when I first became aware that two airplanes were flying dangerously close together. The airplanes were to the right and in front of me. [PT-18 #40-1996] was just banking out of the approach leg and beginning to level off on the downwind leg. [PT-18 #40-2013] was on the downwind leg. They appeared to be at the same height. As [40-1996] saw a collision was imminent, he made an attempt to avoid the collision by zooming above [40-2013]. As he did so, his propeller hit [40-2013's] tail assembly. The tail assembly then flew into the air and [40-2013] plunged earthward. The falling airplane appeared about a 90-degree spin before imbedding its nose into the ground. There was no fire." Investigators noted that the pilot of PT-18 #40-1996 was eliminated from Primary Training.

3-11-43K. Tucson, Arizona. At 1750 MWT, a Consolidated B-24D crashed three miles south of Davis-Monthan Field, Tucson, Arizona, killing eight crewmembers and injuring two others. Navigator 2Lt. Bayard J. Squire and gunner SSgt. Edwin S. Park survived the accident with serious injuries. Killed in the crash were: 2Lt. Clinton L. Lee, pilot; 2Lt. Jay R. Thompson, co-pilot; 2Lt. Walter T. Chase, bombardier; SSgt. Charles E. Bennewitz, engineer; SSgt. Fredie C. Eifert, assistant engineer; SSgt. Frank D. Snell, radio operator; SSgt. Theodore A. Rusnack, assistant radio operator; SSgt. Robert D. Kamp, gunner. The airplane had taken off from Davis-Monthan Field on a bombing and gunnery mission. The airplane climbed to about 7,500 feet and proceeded to the bombing range. Clouds obscured the bombing range so the airplane returned to the area of Tucson and made several simulated bombing runs. The airplane soon lost power on the number-one and number-two engines. The pilot attempted to glide the airplane to Davis-Monthan but was unable to maintain control. The B-24 stalled and then fell off on the left wing and plunged to the ground. Investigation revealed that the engine failure was caused by a short circuit of the main ignition switches of the number-one and number-two engines. It was found that the number-one ignition switch had been turned past its stop, causing the insulator to break in several pieces and to short out the number-two ignition switch. Investigators stated, "The apparent cause of the accident was the killing of the #1 and #2 engines ignition switches by the co-pilot accidentally while in flight. The engine ignition switches on B-24D aircraft are located on the right front side of the co-pilot's position near an oxygen regulator. While in flight, the co-pilot placed his right (heavy flying) boot on the knob of the oxygen regulator thus striking the #1 engine ignition switch, breaking the bakelight stop and [associated] metal parts which in turn shorted the #2 engine ignition switch. [The] investigation [of] the B-24Ds here at [Davis-Monthan], reveals that the co-pilots place their right foot on the knob of the regulator, which is located near the #1 engine ignition switch. This is evidenced by the wear on top of the oxygen regulator knobs [on all B-24Ds on the base]. With the #1 and #2 ignition switches shorted out, the propellers on the #1 and #2 engines begin to windmill; thus the loss of engine power and increased drag on the left side of the airplane. It is believed that the pilot did not feather the #1 and #2 propellers due to the fact that he expected his engines to operate again. Apparently the airplane, trying to make the base, lost flying speed, stalled, and crashed from loss of these two engines." Investigators noted that the crash occurred over terrain where a normal emergency landing could have been made. Investigators suggested that guards be installed in all B-24D aircraft to prevent the co-pilot's boot from breaking the ignition switches or inadvertently switching them off.

3-11-43L. Langley Field, Virginia. At 2015, a Consolidated B-24D crashed into the Langley View section of Langley Field, Hampton, Virginia, killing five crewmembers and injuring engineer Cpl. Joseph A. Wood. Killed in the crash were: 1Lt. Paul A. Kirk, pilot; Maj. Paul M. Person, co-pilot/B-24 instructor; 1Lt. Milton R. Montgomery, co-pilot; Sgt. Edward W. May; Cpl. Joseph M. Keyes, radio operator. The airplane had been practicing night take-offs and landings at Langley Field when the accident occurred. The airplane had made its second take-off and soon after leaving the ground the airplane "swerved." The pilot apparently lowered his head in the cockpit to adjust the trim tabs when the co-pilot inadvertently switched off the turbo-superchargers as he adjusted the throttles, which had crept back. The airplane was unable to maintain altitude and descended into the ground where it smashed itself to pieces and erupted into flames.

3-12-43A. Orlando, Florida. At 1210 EWT, a Republic P-47C crashed into an unoccupied house one mile east of the Army Air Base at Orlando, Florida, killing pilot 2Lt. James L. Carroll and injuring a civilian woman on the ground. The airplane was approaching Orlando AAB from the northeast in preparation to landing to the southwest on the Northeast-Southwest runway when it apparently stalled while turning from the base leg to the final approach. The airplane fell off into a spin to the left and smashed into the house. The propeller was embedded into the ground and the fuselage and engine had smashed into the house, which erupted into flames. About 40 seconds after the crash the P-47 exploded violently, causing the house to erupt into a raging fire. Ammunition was exploding in the burning wreckage. The pilot was killed upon impact. The airplane was part of an eight-ship flight (four P-47s and four P-40s) that was returning to the base following a routine flight. Civilian Marie Huffman, occupying an adjacent dwelling, suffered a broken rib, contusions and "hysteria." She was taken to Orange General Hospital, Orlando.

3-12-43B. Meridian, Mississippi. At 1230, a North American A-36 flew into the ground three miles south of Key Field, Meridian, Mississippi, killing 2Lt. Richard R. Shaw. Investigators stated, "Lt. Shaw had told fellow officers on different occasions that he had a tendency to black out while doing steep turns in the A-36 type airplane. According to one witness, a turn was made before the fatal crash. It is our opinion that Lt. Shaw blacked out while making a steep turn and crashed into the ground before recovering. The airplane struck the earth at about a 90-degree angle and was buried approximately fifteen feet."

3-12-43C. Montgomery, Alabama. At 1310 CWT, a Vultee BT-13 crashed two miles northwest of Gunter Field, Montgomery, Alabama, killing pilot 2Lt. Sanford Friedman, Greensboro, North Carolina, and passenger Pvt. James A. Rolfe, Orlando, Florida. The airplane had taken off from Gunter Field on a 50-hour inspection flight. The airplane made a normal take-off and was seen to make a 90-degree turn to the right. The airplane then made a 90-degree turn to the left to leave the traffic pattern. The airplane was then seen to enter a steep climb until it stalled and entered a spin to the left. The pilot was able to stop the rotation of the spin but the airplane dove into the ground in a vertical position. Both occupants were killed instantly.

3-12-43D. Sheridan, Arkansas. At 1545 CWT, a Boeing B-17F flying in poor weather crashed six miles north of Sheridan, Arkansas, killing the crew of nine. The airplane had taken off from Smoky Hill Army Air Field, Salina, Kansas, on a flight to the European Theater of Operations via Morrison Field, West Palm Beach, Florida. The pilots were issued an instrument clearance and were assigned to fly at 10,000 feet indicated altitude. At about 1544, the B-17 was observed circling the area north of Sheridan, Arkansas, at low altitude. The pilot was apparently trying to get under the overcast. The B-17 circled for a short time before entering a stall and smashing into the ground where it exploded into flames in an area of heavy timber near Arkansas Highway 35. The weather was reported to be instrument conditions with a ceiling reported to be about 400 feet agl with heavy rain, lightning, and fog. Killed in the crash were: 2Lt. George H. Davis, Dubuque, Iowa, pilot; 2Lt. Robert Y. Turchetto, Nutley, New York, co-pilot; 2Lt. Leo E. Dolan, St. Louis, Missouri, navigator; 2Lt. Philip E. Niewolak, Dunkirk, New York, bombardier; TSgt. Dewitt H. Tyler, Porterville, California, engineer; SSgt. David F. Secorsky, Detroit, Michigan, assistant engineer; TSgt. Peter K. Ivanovich, Bisbee, Arizona, radio operator; SSgt. Kenneth D. Cain, hometown unknown, assistant radio operator; SSgt. Arthur N. Potter, Sutton, Massachusetts, gunner.

3-12-43E. Glendale, California. At 1810, a Curtiss AT-9 crashed onto Grand View Avenue while attempting a take-off at the municipal airport at Glendale, California, killing three fliers and seriously injuring passenger TSgt. Joseph Correll. Pilot Capt. Robert H. Kacy, co-pilot SSgt. Merle F. Robison and passenger Pfc. George E. Herman were killed in the accident. The airplane was attempting to take-off on an administrative flight to Muroc Army Air Field, Muroc, California, when the accident occurred. The airplane had a short take-off run and apparently took off in a stalling turn to the left. The airplane barely missed striking the control tower as it fell off to the left. The airplane began mushing toward the ground in a nose high attitude. East of the Southern Pacific Railroad tracks the tail section struck a power-line pole and separated from the airplane, dropping to the ground near the pole. The engines were heard to either throttle back or lose power as the airplane rolled to the left to an inverted attitude. The AT-9 missed several other power-lines and then smashed into Grandview Avenue near San Fernando Road in an inverted attitude. The airplane did not burn but was destroyed utterly.

3-12-43F. Lubbock, Texas. At 2143, a Lockheed C-60A crashed one-half mile south of South Plains Army Air Field, Lubbock, Texas, killing the crew of three. The airplane had taken off at 2115 from South Plains Field on a glider-towing mission. The airplane-glider tandem circled the area southeast of the field and the pilot reported that heavy clouds were moving into his zone. The Control Officer advised the airplane to stay under the clouds. The airplane-glider tandem disappeared into the overcast as they approached the field. The ceiling was estimated to be 200 feet agl. Moments later the airplane was seen to fall from the overcast and smash into the ground where it exploded into flames. The glider was able to release and make a safe landing at South Plains Army Air Field.

3-13-43A. Williams Field, Chandler, Arizona. At 0900 MWT, mechanic Pvt. James E. Thomason was killed when a Cessna AT-17 collapsed on him at Williams Field, Chandler, Arizona. Investigation revealed that Pvt. Thomason had inadvertently removed the bolts and nuts from the landing gear retracting screw bracket without positioning jacks or other supports, causing the landing gear to retract and the airplane to settle to the ground. As the airplane settled to the ground, Pvt. Thomason's head was caught between the engine mount brace and the landing gear cradle, breaking his neck and killing him instantly. The airplane was not damaged.

3-13-43B. Tampa Bay, Florida. At 1455 EWT, a Martin RB-26A crashed into Tampa Bay, Florida, killing the crew of five. Killed in the crash were: 1Lt. Alan M. Okell, Cranford, New Jersey, B-26 instructor; 1Lt. Francis P. Martin, MacDill Field, Florida, pilot; 2Lt. Roy A. Skow, Omaha, Nebraska, co-pilot; SSgt. Archie E. Dively, MacDill Field, Florida, engineer; SSgt. Reginald R. Reynolds, Lueders, Texas, student engineer. The airplane had taken off from MacDill Field, Tampa, Florida, on a local transition flight. Investigators stated, "The airplane was seen spinning from an altitude of 4,000 to 5,000 feet. When first observed by pilots in the air the airplane was spinning to the right [completing about three or four turns of spin] and as it approached the water, the spin flattened out and the airplane struck the water in a horizontal position." It was speculated that the airplane might have suffered from a runaway starboard propeller. Witnesses stated that they observed no apparent effort by the crew to leave the airplane.

3-13-43C. Dunsmuir, California. At 1320, a Lockheed-Vega R-37 (similar to the B-34 type) crashed into the west slope of Hirz Mountain, approximately 23 miles south of Dunsmuir, California, killing five crewmembers. Co-pilot 1Lt. Robert M. Smith was able to parachute to safety and was uninjured. A/C G. Martin, a student pilot from the AAF school at Santa Ana, California, was killed attempting to parachute to safety. He had jumped first and apparently had been struck by some part of the airplane. His body was found some distance from the wreckage. Killed in the crash were: 2Lt. Joe K. Hanna, pilot; Pvt. Boyd P. Lawhorn, engineer; Cpl. C.D. Hazelton, passenger; Sgt. H. Morris, passenger. The airplane had taken off from Red Bluff, California, on a ferry flight to Medford, Oregon, at about 1300 PWT. The airplane was on a visual clearance and was assigned to fly at 11,000 feet msl. The airplane penetrated a light overcast soon after leaving Red Bluff and continued to climb through light layers until it reached the vicinity of Redding, California, when instrument conditions were encountered. The pilot elected to continue climbing on instruments and the airplane soon began picking up ice. The pilots activated the de-icing equipment but after a short period of time extreme icing conditions were encountered. The pilot then made a 180-degree turn and ordered the crew to put on their parachutes. The airplane became very heavy with ice and almost uncontrollable so the pilot ordered the crew to abandon ship. The co-pilot and student left the airplane immediately and the airplane soon went out of control. Lt. Smith parachuted to the ground in a remote area and he walked for about four hours before coming to California Highway 99.

3-13-43D. Amber, Oklahoma. At 1720, a Fairchild PT-19A crashed one and a half miles south of Amber, Oklahoma, killing A/C George N. Sabella. Investigation revealed that the airplane suffered an engine failure while flying at 500 feet on the downwind leg of the pattern at Auxiliary Airfield #2 (auxiliary to the AAF station at Chickasha Municipal Airport, Chickasha, Oklahoma). The PT-19A entered a spin to the left and smashed into the ground, killing the student instantly. There was no fire but the airplane was completely destroyed. Investigation of the engine revealed that the oil pump shaft had failed.

3-13-43E. Laurel, Mississippi. At 1747, a North American O-47B struck a tree and crashed 14 miles east of Laurel, Mississippi, killing three fliers. Pilot 2Lt. John K. White, crew chief Sgt. Jack Formby and passenger Cpl. Lloyd I. Berkoff were killed in the accident. The airplane had taken off from the Army Air Field at Laurel on a triangular navigation mission. The Aircraft Accident Classification Committee stated, "Evidence has shown that the aircraft struck a small white oak tree, approximately 35 feet high, cutting off [the top of the tree] eight feet from the top. This oak tree sheared off 13 feet of the right wing from the wingtip to approximately 12 inches beyond the aileron. This portion of the wing broke off and traveled 60 yards on the line of flight and landed intact. The shearing action and the loss of the right wing turned the airplane over on its back, striking the ground [inverted] 200 yards from the oak tree. The airplane made a hole approximately 25 feet long, eight feet wide and three feet deep, demolishing the complete canopy and the top of the fuselage, shearing off two top cylinders of [the starboard] engine, engine mounts, fire wall, oil tanks and the right wing root. The airplane then bounced and continued on for 40 yards. The engine was torn out and thrown clear approximately 50 yards from the fuselage. No evidence of fire was shown as the battery and electrical equipment was completely demolished and scattered over a wide area. Reviewing the statements of witnesses, it is [believed] that this airplane was flying very low and at a fast rate of speed when it struck the tree."

3-13-43F. Tallahassee, Florida. At 2210 EWT, a Douglas A-24A suffering an apparent engine failure was abandoned and crashed near Dale Mabry Field, Tallahassee, Florida, killing one flyer. Passenger T-3 Donald G.J. Hartig was killed when he was struck by the aircraft while attempting to parachute to safety. Pilot 1Lt. Giles C. Wright parachuted to safety and was uninjured. The airplane had taken off from Key Field, Meridian, Mississippi, at 1950 on a cross-country navigation flight to Dale Mabry Field. The pilot entered the traffic pattern for Dale Mabry Field and extended the landing gear on the downwind leg and switched to the fullest fuel tank when the engine began sputtering. The airplane was flying at about 1,200 feet agl so the pilot ordered the passenger to abandon the aircraft. As the pilot abandoned the airplane on the left side, he noticed that the passenger was standing on the starboard wing and holding on to the airplane. The passenger attempted to use his parachute at an altitude too low to allow the parachute to fully deploy and he fell to his death. The pilot landed about 60 feet from the spot where the airplane smashed into the ground and exploded into flames.

3-13-43G. Alamogordo, New Mexico. At 2310 MWT, waist gunner SSgt. Eugene C. Yandle died of oxygen starvation aboard a Consolidated B-24D flying on a bombing mission at the AAF Bombing and Gunnery Range near Alamogordo, New Mexico. The B-24D was flying at approximately 30,000 feet msl when SSgt. Yandle disconnected his oxygen line, pulled his oxygen mask down past his chin, left his post at the port waist gun position and went to the tail gun position. The tail gunner noticed SSgt. Yandle and motioned him to put his mask on and get back to his station. Investigators noted that there was only one oxygen connection at the tail gun position. SSgt. Yandle acknowledged the tail gunner and started back toward his station. Moments later SSgt. Yandle was seen lying face down on the waist section deck. The tail gunner notified the pilot who dove the airplane down to a lower altitude. A crewmember performed artificial respiration on the downed gunner but was unsuccessful in reviving him. Investigators noted that SSgt. Yandle had been instructed on the proper use of oxygen equipment aboard B-24D airplanes. The oxygen equipment was found to be in working condition. It was later speculated that the gunner, who was using the A-8B type oxygen mask (with balloon-type re-breather bag), might have had a frozen oxygen mask. The early A-8B type oxygen mask was prone to freezing from breath condensation and was eventually replaced by the A-9 and A-14 type masks. Seven crewmembers were uninjured. The B-24D was piloted by 2Lt. Roy E. Braly and co-pilot Capt. Doyle Hicks.

3-14-43A. Scottsdale, Arizona. At 1115 MWT, two Curtiss AT-9 type airplanes collided in mid-air two miles east of Scottsdale, Arizona, killing A/C Fred G. Thayer, who was attempting to parachute from AT-9 #41-12111. A/C William B. Robbins was able to parachute to safety from AT-9 #41-12111, receiving minor injuries. A/C Leland R. Shuyler and A/C Lawrence C. Vogt were uninjured and were able to safely land AT-9 #41-11951 at Williams Field, Chandler, Arizona. Investigators stated, "[AT-9 #41-12111] and AT-9 #41-11951 were flying closing courses varying only about three degrees in the same direction. [AT-9 #41-12111] overran AT-9 #41-11951 approaching from the right. The pilots of both airplanes pushed the sticks forward, but the tail assembly of [AT-9 #41-12111] struck the top of the left engine of AT-9 #41-11951. The left propeller of AT-9 #41-11951 sheared off the elevator of [AT-9 #41-1211], which crashed after both occupants bailed out, one unsuccessfully. Evidence indicated that A/C Thayer left the airplane somewhat prematurely, pulled the ripcord and caused his parachute to become fouled on the airplane." Examination of the body and marks on the ground indicated that the cadet was pulled to his death by the airplane. The airplanes had taken off from Williams Field on a routine student formation training flight.

3-14-43B. Ardmore, Oklahoma. At 1400, an Aeronca L-3C crashed seven miles southwest of Ardmore, Oklahoma, killing glider pilot F/O Emil M. Horkavi, Gary, Indiana, and glider pilot F/O Frank M. Dimond, Wilmington, California. The airplane smashed into a wooded area in a flying attitude. The airplane did not burn but was destroyed utterly. There were no witnesses and investigators could not determine the cause of the accident.

3-14-43C. Whitesboro, Texas. At 1435, a Taylorcraft L-2A crashed three miles west of Whitesboro, Texas, killing civilian instructor Henry J. Frances and seriously injuring student pilot 2Lt. Ervin L. Meyer. The airplane had taken off from the Army Air Field at Denton, Texas, on a cross-country navigation flight to Gainesville, Texas, to Sherman, Texas, and return to Denton. The airplane was seen flying to the east toward Sherman at extremely low altitude along Texas Highway 82. The airplane was flying below treetop level when the pilot attempted to pull up to avoid striking a tree. The port wing struck a tree and the airplane went out of control and smashed into another tree. The airplane did not burn but was completely destroyed.

3-14-43D. Puget Sound, Washington. At 1640, a Lockheed P-38F and a U.S. Navy Lockheed-Vega PV-1 collided in mid-air over Puget Sound near Everett, Washington, killing P-38F pilot 1Lt. Harrell H. Rule and the six navy crewmembers aboard the PV-1 (BuNo 29733). Killed in the collision were: Lt. (j.g.) George R. Millikan, San Francisco, California, pilot; Ensign Walter R. Baman, co-pilot; Ensign Gerald S. Bennett, Freeport, Maine; Aviation Radioman 2nd Class James B. Krenciprock, Niles, Ohio; Aviation Machinist's Mate 3rd Class Frank A. Plaia, Sacramento, California; Aviation Radioman 2nd Class, Niles, Ohio; Aviation Metalsmith 1st Class Thomas J. Jeffries, Burlington, North Carolina. The P-38 had taken off from Paine Field, Everett, on a local training mission. The PV-1 was on an instrument-training mission. The PV-1 was flying in actual instrument conditions and the P-38 was flying in an area near the clouds. The PV-1 suddenly emerged from a cloud and smashed into the P-38. Both airplanes burst into flames, went out of control and broke up as they fell into Puget Sound. Only the body of Jeffries was recovered from the water.

3-14-43E. St. Johns, New Mexico. At a time unknown, a Beech AT-11 crashed into a mountain near St. Johns, New Mexico, killing the crew of four. The airplane had taken off from Kirtland Army Air Field, Albuquerque, New Mexico, on a routine training mission. No other details are known. Killed in the crash were: 1Lt. Dorbert A. Garrett, Minneapolis, Minnesota, pilot; A/C Robert A. Callahan, Portland, Oregon, student bombardier; Sgt. Joseph J. Gibbs, Franklin, New Carolina, crewmember; Pvt. Smooth E. Tolman, Bell, California, crewmember.

3-15-43A. Taft, California. At 1010, a Vultee BT-13A crashed near Gardner Field, Taft, California, killing passenger/civilian mechanic Howard W. Cantrell. Pilot F/O Robert B. Yeomans was able to parachute to safety and was uninjured. The airplane was being test flown after being repaired at an area repair depot. F/O Yeomans was flying the airplane at about 6,500 feet agl when he decided to enter a spin. After some difficulty, the airplane grudgingly entered a spin to the left. After about two turns of spin, the pilot popped the stick forward. It jammed momentarily and then moved freely and without tension. The pilot was unable to affect the pitch or bank attitude and was unable to recover from the spin. The airplane went out of control and the pilot ordered the passenger to bail out several times. The pilot stated that the passenger only looked over the side and failed to comply. F/O Yeomans exited the airplane at about 1,200 feet agl and parachuted to safety. After the pilot bailed out the airplane banked sharply to the left and entered a vertical dive, smashing into the ground and exploding into flames. Investigation revealed that the flight controls were improperly rigged.

3-15-43B. Anderson, South Carolina. At 1520, a Lockheed P-38G attempting an emergency landing crashed two miles west of Anderson, South Carolina, killing Service Pilot 1Lt. Lee Dwell Byron. The pilot was apparently attempting a wheels-down landing in a rye field but the airplane undershot the field and smashed into a grove of trees where it erupted into flames. Wreckage was strewn over an area of 1,000 feet. The P-38 had taken off from Atlanta, Georgia, on a ferry flight to Greensboro, North Carolina.

3-15-43C. Gulf of Mexico, Florida. At 1615 EWT, a North American AT-6A crashed into the Gulf of Mexico 15 miles south of Fort Myers, Florida, killing 2Lt. Burwell L. Bradfield and gunner Pvt. Ralph L. McCoy. The airplane was flying at an altitude at about 500 feet msl when it suddenly went out of control and dove into the sea. Investigators speculated that the rear gunner, while firing his weapon, had apparently fouled the control cables with his right foot, causing the pilot to lose control of the airplane. Investigators suggested a guard for the gunner's foot.

3-15-43D. Camp Rucker, Alabama. At 1720, a Douglas DB-7B crashed three miles southeast of Ozark Army Air Field, Camp Rucker, Alabama, killing the crew of three. The airplane had taken off with two other DB-7B airplanes on a mission in conjunction with ground units at Camp Rucker. The airplane was the second to take off and it soon entered a climb to join up with the lead airplane. Investigators stated, "The airplane was observed to be joining with the lead airplane. After making a 180-degree left turn, and upon assuming the No. 2 position on the right wing of the lead airplane, the DB-7B pulled up into a very steep left chandelle and wing-over, passed over the tail of the lead airplane, fell into a dive followed by a vertical spin to the left, and narrowly missed colliding with the third airplane, [then] crashed and burned. The third airplane was approaching from the rear to join the formation in the No. 3 position on the lead airplane's left wing. The accident occurred 3 to 4 minutes after take-off and the wreckage was located in a [heavily wooded area] approximately three miles southeast of the airfield, completely destroying and burning [the airplane]."

3-15-43E. Location Unknown. At approximately 1845 PWT, a Lockheed P-38G disappeared and is presumed to have crashed into the Pacific Ocean offshore of San Mateo Point, California. Pilot 1Lt. Timothy A. Pedley III, age 23, was missing and presumed lost at sea. The airplane had taken off from Orange County Airport, Orange County, California, on a coastal patrol mission with another P-38. Lt. Pedley was flying at about 8,000 feet over the sea when he began having trouble in the port engine. The pilot feathered the port propeller, trimmed the airplane for single-engine flight and turned back for the mainland. The P-38 gradually lost altitude until it descended into the clouds at about 1,800 feet msl. The P-38 disappeared after entering the overcast, which was estimated to be about 600 feet thick. The next day, a U.S. Navy vessel recovered a piece of fuselage and the pilot's headset from the sea. The pilot's body was never recovered. It was later speculated that the pilot lost control of the airplane while flying in the clouds and was unable to recover before smashing into the sea. Investigators noted that the pilot was experienced in single-engine flight and proficient at instrument flying. Lt. Pedley was the son of Col. Timothy A. Pedley II, Fort Howze, Texas.

3-16-43A. Emhouse, Texas. At 1105, a Fairchild PT-19A suffered a catastrophic structural failure of the port wing and crashed two and one-half miles west of Emhouse, Texas, killing civilian instructor Elbert E. Baldwin, who fell to his death attempting to parachute to safety. Student A/C Robert L. Webb parachuted to safety and suffered only minor injuries. This flight was to be the student's final check ride in Primary Flight Training at the AAF Primary Flight School at Corsicana, Texas, prior to graduating to Basic Flight Training. The student stated that they had climbed to approximately 6,500 feet msl and had completed several maneuvers when the instructor announced that he wanted to perform an Immelmann turn. The instructor, flying in the rear cockpit, took over and dove the airplane to pick up speed. He then pulled up sharply to climb into the Immelmann when the port wing snapped off. The airplane fell out of control for a couple of thousand feet before the instructor gave the student the signal to bail out. Both pilots left the airplane immediately. Instructor Baldwin was found about 300 yards from the airplane wreckage with his parachute canopy free of the pack but still partially folded and with the shroud lines in the pack still secured by the elastic straps. He had been killed instantly upon impact. Apparently the instructor pulled the ripcord too late to allow the parachute to fully deploy.

3-16-43B. Galena Park, Texas. At 1155, a Beech AT-10 suffered a catastrophic structural failure of the starboard wing and crashed near Galena Park, Texas, killing students A/C Allen K. Lamb, 21, Franklin, Kentucky, and A/C Francis X. Lavallee, 24, Lowell, Massachusetts. Witness on the ground stated that they heard the sound of an airplane running at high RPM but could not see the airplane because it was in the clouds. Moments later the airplane was seen spinning out of control from the 1,200-foot overcast minus the starboard wing, crashing into the ground and erupting into flames upon impact. Witnesses reported that pieces of the airplane rained down for two minutes after the airplane crashed into the ground in a vertical dive. A major portion of the wing was found over a mile from the spot where the airplane crashed. Investigation revealed that the leading edge of the wing had failed and the plywood upper surface tore away. The airplane had taken off from Ellington Field, Houston, Texas, at 1100 CWT on a routine transition-training mission.

3-16-43C. Utica, Mississippi. At 1540, a Beech AT-11 suffered a catastrophic structural failure of the elevators and crashed three miles southwest of Utica, Mississippi, killing pilot 1Lt. William A. Moores. Co-pilot 2Lt. Warwick V. Smith, passenger Maj. Millard N. Tutwiler and civilian passenger John M. Fountoulskis parachuted to safety. Maj. Tutwiler suffered minor injuries. The airplane had taken off from the Army Air Field at DeRidder, Louisiana, on an important administrative flight to the Army Air Base at Jackson, Mississippi. The pilot had encountered deteriorating weather conditions and elected to turn back toward DeRidder. At a point approximately five miles west of Utica, the pilot changed course slightly to avoid a major portion of a huge front that was moving into the area. The airplane was flying about 1,000 feet below the bottom of a large formation of cumulus clouds when severe drafts caused the elevators to fail and break off. The airplane then fell out of control and the pilot ordered the passengers and co-pilot to bail out which they did immediately. Investigation revealed that the pilot had bailed out at an altitude of about 150 feet above ground level and was pulled to his death when he deployed his parachute before he had cleared the ship, causing it to be caught on the horizontal stabilizer.

3-16-43D. Walnut Ridge, Arkansas. At 1545, two Vultee BT-13A type airplanes collided in mid-air near Walnut Ridge, Arkansas, killing A/C Robert S. Hawes aboard BT-13A #41-11007. Instructor 2Lt. Louis R. Semrau and student Wyman E. Osness were uninjured and able to make a successful emergency landing in BT-13A #41-22333 at the Army Air Field at Walnut Ridge. Both airplanes were in the traffic pattern for Walnut Ridge. Investigators stated, "According to the statement of witnesses, the pilot of airplane #41-11007 was flying on a course of approximately 270 degrees and about 700 feet above the ground when he collided with the undercarriage of BT-13A #41-22333, which was in a bank of 20 degrees to the right and on a course of approximately 180 degrees. Examination of wreckage disclosed that the right landing gear assembly of airplane #41-22333 struck the front cockpit of airplane #41-11007, the propeller [of BT-13A #41-22333] demolishing the rear cockpit. The left landing gear struck the vertical fin, tearing it off in the impact. BT-13A #41-11007 went into an uncontrolled dive and crashed into a cultivated field." The landing gear of BT-13A #41-22333 was badly damaged, but the instructor was able to safely crash land the airplane after giving the student the option of bailing out or staying with the ship. He stayed with the airplane. Investigation revealed that BT-13A #41-22333 had flown a little too far out from the regular traffic pattern. The instructor ordered the student to turn the airplane to the right and get closer to the center of the airfield. The instructor's airplane was attempting a 45-degree entry but did not enter the pattern at the correct point, entering in the last portion of the downwind leg. It was during this time that the collision occurred. Investigators faulted both pilots failing to see and avoid the other aircraft.

3-16-43E. Bay City, Texas. At 2145, a North American AT-6C suffered a catastrophic structural failure of the port wing and crashed at Bay City, Texas, killing A/C Billy G. Elliot. Investigators stated, "A/C Elliott took off at 2028 on a scheduled cross-country flight from Aloe Field [Victoria, Texas] to Bay City, Texas, to Houston, Texas, to Yoakum, Texas, and to return and land at Aloe Field, with no landings being scheduled en route. Shortly after 2100 CWT he was heard over Bay City at an altitude to which the townspeople are accustomed and was then heard diving on the southern portion of the city. He pulled out and leveled off over the business section just above the tops of the buildings and a number of witnesses stated that they ran out of the theater, garages, hotel lobby and restaurants, thinking the airplane was going to hit the buildings. The noise was described as 'terrific' and 'deafening.' The airplane proceeded north, the noise lessening, and turned west about two blocks beyond the business district, turned south four blocks coming around into the east heading back over the business district and diving down. Witnesses described these turns as being made at great speed and in steep banks, the green lights being directly above the red lights. Upon leveling out to the east on the last turn the noise rose again to the extent that several witnesses living in the vicinity said that they and their families instinctively ducked. One witness, who was in his yard, stated that he saw a bright flash as the airplane started over the field into which it crashed. Both wings were badly wrinkled from exceeding the limits of the airplane, the left wing having opened up on the leading edge near the center section in such a way as to allow the gun camera to fall out as the airplane [passed over] the field. This opening turned the airplane over on its back and the bright flash described by a witness was the left wing falling off, landing flat on the ground on the right side of the path of the airplane. The airplane continued upside down for about 200 feet, taking off the rudder, some of the rear canopy and the nose plunged into the ground making a scoop shaped hole 12 feet long and 30 inches deep, the right wing tip cutting the ground 50 feet before the impact." The wreckage of the airplane and body parts were scattered over an area of approximately 300 yards.

3-17-43A. Clovis, New Mexico. At 0410 MWT, a Consolidated B-24D crashed five miles northeast of the Army Air Base at Clovis, New Mexico, killing five crewmembers and injuring six others. Killed in the accident were: 2Lt. Jack R. Moss, pilot; 2Lt. Stephen P. Mezei, navigator; 2Lt. Merle K. Clark, bombardier; Sgt. Bruce H. Davis, Jr., assistant radio operator; Sgt. Maxwell Wolfe, gunner. Crewmembers injured in the crash were: 1Lt. David E. Council, B-24 instructor; 2Lt. William D. Harrison, co-pilot; SSgt. Francis J. Eckery, engineer; Sgt. Walter J. Kalczynski, radio operator; Sgt. William J. Gabonay, assistant engineer; SSgt. Leonard Powers, gunner. The airplane was returning to Clovis after successfully completing a routine navigation training flight. The B-24 instructor was in the co-pilot's seat. The pilot received landing instructions from the tower and then entered the traffic pattern. The airplane, flying at 5,200 feet msl, overshot the turn from the base leg to final approach and the pilot had to S-turn to get the airplane in line with the runway. The instructor was calling off the airspeed but apparently no one was minding the altitude. Just as the instructor realized that the airplane was descending toward the ground, the airplane flew into the terrain and began disintegrating. Investigation of the wreckage revealed that the flaps were in the retracted position and the landing gear was in the down and locked position. Investigators noted that the base leg that was being flown was abnormally far from the field.

3-17-43B. Siloam Springs, Arkansas. At 1020, a Fairchild PT-19 suffered a catastrophic structural failure of the starboard wing and crashed 15 miles west of Siloam Springs, Arkansas, killing civilian instructor John H. Cramer. Student A/C Neil D. Naiden suffered minor injuries parachuting to safety. The airplane had taken off from Tulsa Municipal Airport, Tulsa, Oklahoma, on a dual cross-country navigation flight to Siloam Springs and return. The airplane performed a scheduled landing at Siloam Springs Municipal Airport, and then took off at 1012 CWT on an ordered flight to Auxiliary Airfield #7. At 1015, the instructor took over the controls and performed a snap roll. The cadet then took control of the airplane again and executed a slow roll. The instructor took over again and performed a "90-degree inverted turn." After completing the maneuver, the airplane was righted and began climbing back to its assigned altitude of 4,500 feet msl. After achieving the assigned altitude, the instructor dove the airplane at the ground at a 50-degree angle. As the airspeed approached 160 mph, the instructor applied excessive left rudder and backpressure on the stick in an apparent effort to perform a single or double snap roll. The starboard wing snapped off about one and a half feet from the wing root. The instructor ordered the student to bail out and he complied immediately. The airplane continued flying for a short distance before it dove out of control into the ground. Investigators speculated that the instructor had made no apparent effort to leave the airplane because his safety belt and shoulder harness were still fastened when he was found in the wreckage.

3-17-43C. Camden, Arkansas. At 1816, a Stearman PT-17 crashed six miles east of Harrell Field, Camden, Arkansas, killing A/C Clarence G. Church. The airplane had taken off from Harrell Field at 1727 on a routine solo training flight but did not return. The airplane was found at about 1930. Investigation of the wreckage revealed that the airplane had dove into the ground at an angle of approximately 75 degrees. Investigators could not determine the chain of events that led to the airplane diving into the ground.

3-17-43D. Walnut Creek, California. At 1815, two Bell P-39N type airplanes collided in mid-air five miles SSW of Walnut Creek, California, killing 2Lt. Jack F. Scriven, who was attempting to parachute from P-39N #42-8885. The pilot of P-39N #42-8852, F/O Robert H. Smidt, was able to make a successful wheels-up landing at Hamilton Field, San Rafael, California. He received minor injuries. The Accident Classification Committee stated, "The accident happened while the airplanes were flying formation on top of broken clouds, F/O Smidt leading, with Lt. Scriven in trail. After completing a dive, F/O Smidt turned right for a short interval, then started a chandelle to the left when the collision occurred. Lt. Scriven's left wing collided with F/O Smidt's right wing, tearing the aileron off of Smidt's right wing, plus part of his right wing, leaving the wing tip and leading edge undamaged. F/O Smidt did not see the other airplane at the time of impact or thereafter. It is believed that Lt. Scriven's airplane went into a spin after the impact. He crashed two miles southeast of Oakland. The pilot apparently attempted to bail out of the airplane, but jumped too late, and did not have time to pull the rip cord before striking the ground, killing him."

3-17-43E. Blythe, California. At 1927, a Boeing B-17F suffering an in-flight fire crashed five miles south of Blythe, California, killing the crew of seven.

Killed in the crash were: F/O Donald R. Krumm, pilot; 2Lt. Joseph L. Blunck, co-pilot; SSgt. Donald W. Lahman, engineer; Pvt. Oren Monceaux, radio operator; SSgt. Oleus J. Bobichaux, assistant radio operator; 2Lt. Benjamin W. Jump, pilot-rated passenger; 2Lt. Joseph P. Barron, pilot-rated passenger. The Accident Classification Committee stated, "[At about 1925 PWT] B-17F #42-5450 was observed flying at a very low altitude about five miles south of Blythe, California. The airplane was observed flying in a southwesterly direction at about 200-300 feet above the ground with its No. 4 engine on fire. The airplane made a left turn heading back toward the northeast. Shortly after making this turn the airplane crashed and burned. The pilot made no contact with any ground radio station indicating that he was in trouble. It is the opinion of this board that the accident was due to fire in the No. 4 engine, probable loss of the No. 4 engine, possible loss of the No. 3 engine, and the inexperience of the pilot."

3-18-43A. Winnsboro, South Carolina. At 1520, a North American B-25D flying in instrument conditions crashed two miles southeast of Winnsboro, South Carolina, killing the crew of eight. Killed in the crash were: 1Lt. Stephen G. Link, pilot; 2Lt. William B. Aubert, co-pilot; 1Lt. Jack D. Ledbetter, pilot-rated passenger; SSgt. Dennis A. Wihl, engineer; Sgt. Edward J. Dubicki, radio operator; SSgt. Gerald W. DeBord, radio operator/gunner; SSgt. Aaron F. White, passenger/radio operator; SSgt. Ernest J. Storm, passenger/radio operator. The Aircraft Accident Classification Committee stated, "The ship which crashed was flying as the right wingman in a three-ship formation en route from Greenville, South Carolina, to Myrtle Beach, South Carolina. The formation was flying at about 500 feet above the terrain when it became necessary for the leader to make a slight turn to the left to avoid lower ceilings and smoke. The formation leader did not see the right wingman again. It is the opinion of the [Committee] that the pilot flying in the number-two position in the formation, Lt. Link, went on the pre-arranged instrument formation procedure upon finding himself in instrument conditions. This procedure called for his turning 90 degrees to the right from the course of the leader and descending to 500 feet below the leader's altitude. It is possible that Lt. Link became confused in carrying out this procedure, did not check his altitude carefully enough, watch his rate of descent carefully enough, thus causing him to fly into the ground."

3-18-43B. Centerville, Alabama. At 1500, a Douglas A-20C crashed eight miles southeast of Centerville, Alabama, killing pilot F/O Harry T. Rockwell and engineer Pvt. Harold D. Snider. The airplane was on a ferry flight from Memphis, Tennessee, to Tallahassee, Florida. The Aircraft Accident Classification Committee stated, "Witnesses stated that the aircraft's engines stopped about two miles from the scene of the crash and that the pilot was apparently attempting to make a crash landing. The terrain is quite rough and from the position of the wreckage it could be construed that the left wing of the aircraft struck the side of a hill and threw the nose down into the earth. The nose section was apparently torn loose from the fuselage and did not burn. The baggage carried was evidently stowed in the nose section and was so scattered by the force of the crash that it was not consumed by fire."

3-18-43C. Eagle Pass, Texas. At 1525, a Curtiss P-36C (#38-191) crashed five miles east of the Army Air Field at Eagle Pass, Texas, killing student A/C Raymond H. Allen, Kensington, Pennsylvania. The airplane was taking off to the southwest on Runway 21 on a routine training flight. Witnesses stated that the airplane seemed to be having engine trouble on the take-off roll. The airplane, with the landing gear still extended, was seen in a turn to the left and trying to gain altitude. The airplane was then seen to enter a dive toward the ground. The airplane crashed into the ground and erupted into flames, killing the student instantly. Investigation of the engine revealed that the entire supply of engine oil was blown out through the oil breather vent, causing the engine to seize. There was no engine oil found in the crankcase or reduction gear. Another P-36C (#38-35) was involved in an accident the day before at 1550 CWT, suffering an identical engine failure about five miles north of the Army Air Field at Eagle Pass. The pilot, A/C Harry L. Burdette, was able to parachute to safety, suffering only minor injuries.

3-19-43A. Cudjoe Key, Florida. At an unknown time between 0837 and 1237 EWT, a Lockheed-Vega RB-34 crashed in the Florida Keys ten miles north of Cudjoe Key, Florida, killing the crew of five. Killed in the crash were: 2Lt. Meredith M. Aylor, pilot; F/O Robert E. Langdon, co-pilot; Cpl. Thomas J. Cruise, radio operator; SSgt. Arthur C. Wetterman, engineer; Pfc. Charles J. Zeleski, passenger. The airplane had taken off from Boca Chica Army Air Base, Key West, Florida, at 0837 EWT on a local anti-submarine patrol mission. The airplane was scheduled to return to base at approximately 1137 EWT but failed to return. The airplane was located the next day. Wreckage was scattered over an area of 400 yards 25 miles northeast of the base. Investigation revealed that the starboard propeller was in the full-feathered position.

3-19-43B. Baton Rouge, Louisiana. At 1035, a Curtiss P-40L crashed one mile east of Harding Field, Baton Rouge, Louisiana, killing 2Lt. Ralph W. Thornburg. The airplane was the fifth airplane to take off in a five-ship flight. The first four airplanes made a 180-degree left turn. The number-five airplane almost overran the number-four airplane while in the turn. The pilot, in order to join the formation and not over-run the number-four ship, performed a steep left turn and stalled the airplane. The P-40L entered a spin and smashed to the ground, killing the pilot instantly.

3-19-43C. Honey Grove, Texas. At 1700, a Vultee BT-13A crashed 12 miles north of Honey Grove, Texas, killing A/C John K. Nowling. The airplane was found in a small clearing surrounded by very tall trees, none of which were damaged, indicating that the airplane crashed to the ground in a vertical position. Investigators could not determine why the airplane crashed. The airplane had taken off from Perrin Field, Sherman, Texas, on a routine student training flight.

3-19-43D. Tucson, Arizona. At 1627 MWT, a Ryan PT-22 crashed five miles west of Ryan Field, Tucson, Arizona, killing student A/C Earl R. Kaweck. There were no witnesses to the accident. Investigators speculated that the student had stalled out of an unknown maneuver at low altitude and was unable to recover before striking the ground. Examination of the wreckage indicated that the airplane dove into the ground at a 45-degree angle about 12 minutes after take-off. The propeller indicated that it was turning at low rpm at the time of impact.

3-19-43E. La Junta, Colorado. At 2325, two Cessna AT-17 type airplanes collided in mid-air and crashed near the Army Air Field at La Junta, Colorado, killing three fliers and injuring one. A/C Lawrence M. Ganes was killed in the crash of AT-17 #42-38. A/C James E. Grant was able to parachute to safety from AT-17 #42-38, suffering minor injuries. A/C George T. Dupuis and A/C Albert J. Ellwood were killed in the crash of AT-17D #42-13995. The airplanes were part of a seven-ship flight that had taken off from La Junta Army Air Field. The flight, led by instructor 2Lt. R.E. Roberts, was flying in a stepped-down V-formation at about 2,000 feet above the ground. Investigators stated, "After flying a short time with this seven-ship formation, Lt. Roberts gave the signal that his ship was taking the lead [after having given up the lead for a short period of time]. This was accomplished by the leading ship and the three ships on the right dropping back, and Lt. Roberts and the two ships flying on his rear moving up one position. The formation after [Lt. Roberts retook the lead] was a loose, step-down V-formation, four ships to the left and four ships to the right. Immediately after assuming this formation, Lt. Roberts gave the signal for echelon to the right in preparation for coming in for a landing. The altitude of the formation at this time was 6,100 feet above sea level, about 2,000 feet above the ground. In executing this maneuver, aircraft #42-38, which was the first ship on Lt. Roberts' left wing, [moved] back and to the right and pulled up between the last two aircraft on the right side of the V.... When aircraft #42-38 pulled up in between the last two aircraft, the fuselage was cut off just aft of the wing section by the last ship in the formation. This ship, AT-17D #42-13995, caught fire immediately and crashed.... The two airplanes crashed about 200 yards apart. The tail section of AT-17 #42-38 was found 250 yards west of the main fuselage.... After the accident occurred, the remainder of the formation returned [to La Junta Army Air Field] and landed without mishap."

3-19-43F. Marana, Arizona. At 2346 MWT, two Vultee BT-13A type aircraft collided in mid-air while in the traffic pattern for the Army Air Field at Marana, Arizona, killing A/C Ralph L. DePauw aboard BT-13A #41-22526. A/C Harold B. Murray was able to make an emergency landing in BT-13A #41-21340. He suffered only minor injuries. Investigation revealed that A/C Murray was flying a base leg that was a little farther out than usual. A/C DePauw was flying an extended straight-in approach. The collision occurred when A/C Murray turned from the base leg onto the approach leg. A/C DePauw's airplane spun to the ground, killing him instantly.

3-20-43. Santa Rosa, California. At 1315, a Bell P-39N crashed at Santa Rosa, California, killing 2Lt. Paul A. Leydens. The Aircraft Accident Classification Committee stated, "Lt. Leydens was flying a wing position at approximately 10,000 feet in a two-ship formation. The leader of the flight performed a slow roll to the left, leveled out and then performed a slow roll to the right. The last time Lt. Leydens was seen, he was flying straight and level. It is believed that Lt. Leydens fell into a spin from which he could not recover while trying to do a slow roll. The possibilities were that the pilot might have tried to recover with throttle on, as witnesses heard the roar of an engine. [Investigation] at the scene of the crash [indicated] there is a possibility that the right elevator was disconnected or sheared off before the crash."

3-21-43A. Mitchel Field, New York. At 1140, two Republic P-47C type airplanes collided in mid-air near Mitchel Field, Long Island, New York, killing 2Lt. James M. Egan, who fell to his death attempting to parachute from P-47C #42-6553. The pilot of P-47C #42-6569, 2Lt. Robert H. Farley, parachuted to safety and was uninjured. Lt. Farley later stated to investigators, "We had just taken off on the alert flight and I was flying last position. The leader of the second element was below and in trail of the first element. I joined on the outside of the element away from the turn. As I did so the No. 3 man started to play the inside of the turn, I crossed under and was in position with our element on the approximate level of the first element. As I was crossing under for this position the No. 3 man changed his mind and motioned to play the turn on the other side. In crossing under, we were pretty well on the same level as the first element. When we had reached trail position I started to cross under. I had already reached my position on the outside of the turn and I started to turn my ship and checked my instruments. When I looked up the nose of the No. 3 ship was opposite and above my wing and I immediately felt my tail go down and my nose went up and the ship went out of control. I tried pushing the stick forward but got no response, so I jumped. I do not know what part of No. 3 ship hit my ship but he was above me."

3-21-43B. Pekin, New York. At 1215, a Bell P-39N suffered a structural failure and crashed at Pekin, New York, killing civilian pilot Clarence P. Sengpiel. Investigators stated, "In the opinion of this Board this accident was caused by structural failure in the empennage group. It is impossible to determine by investigation whether or not failure occurred in normal flight or as the result of aerobatic maneuvers. Pilot's attempt to abandon ship occurred at an altitude too low for parachute to be effective, resulting in the death of the pilot. A thorough investigation of elevator and rudder quadrants developed the fact that flaws exist in some elevator quadrant castings. Immediate steps are being taken by Bell Aircraft Corporation to redesign and replace all quadrants in service and in production." The airplane had taken off from the municipal airport at Niagara Falls, New York, on a factory test flight.

3-21-43C. Southmayd, Texas. At 1315, a Vultee BT-13A crashed three miles southwest of Southmayd, Texas, killing student A/C John A. Faw. The airplane had taken off from Perrin Field, Sherman, Texas, on a routine solo training flight. Investigators speculated that the airplane had probably spun into the ground after stalling out of an unknown maneuver. No other information is available.

3-21-43D. Harvey, Illinois. At 1440, a Piper L-4 crashed while attempting a take-off at Rubinkam Airport, Harvey, Illinois, killing pilot 1Lt. Walter G. Yarbrough, Jr. and injuring 2Lt. Gerritt W. Fitzgerald. The wind was variable so take-offs could be attempted on either runway. The L-4 began taking off on the north runway and encountered a civilian Porterfield type airplane that had been taking off about 800 feet behind on the northwest runway. The L-4 was in a climbing turn to the left when it turned in front of the civilian airplane. Just as a collision seemed imminent, the civilian ship pulled up to the left and Lt. Yarbrough also made a violent climbing turn to the left. The L-4 lost airspeed and stalled, falling off on a wing and smashing into the ground on the nose and the starboard wing. The civilian pilot, Vance LaBolle, was able to make a successful take-off. The L-4 was stationed at Chicago Municipal Airport, Illinois.

3-21-43E. Merkel, Texas. At 1530, two Vultee BT-13A type airplanes collided in mid-air and crashed ten miles south of Merkel, Texas, killing WASP pilot Cornelia Fort, Nashville, Tennessee, aboard BT-13A #42-42432. AAF pilot F/O Frank E. Stamme, Jr. was able to land BT-13A #42-42450 at Abilene, Texas. He was uninjured. The two airplanes were part of a seven-ship fight that had taken off from Midland, Texas, on a ferry mission from Long Beach, California, to Dallas, Texas. Investigators stated, "It appears that the landing gear leg on F/O Stamme's aircraft broke off the left wing tip of Pilot Fort's aircraft and peeled approximately six feet of the leading edge of the wing toward the center of the aircraft. At this point the aircraft apparently pulled free of each other. An examination disclosed that the left wing tip of Pilot Fort's aircraft was wood, and the right wing tip was metal (This fact was subsequently verified by the factory). This would account for the ease with which the left wing tip broke off at the time of contact with the other aircraft. F/O Stamme's aircraft did not go out of control. Pilot Fort's aircraft went out of control, and after executing a series of rolls it went into an inverted dive, slowly rotating to the left. There was apparently no attempt to recover or to use the parachute. The emergency latch on the hatch release was found to be locked. The aircraft apparently hit vertically, and did not move from the point of contact. The marks outlining the wings were very distinct on the ground, and the engine was buried approximately two feet in the ground, and showed no movement after impact." F/O Stamme circled the accident site for about ten minutes before flying to Abilene. The remaining aircraft in the flight were able to successfully complete the ferry flight to Dallas. Pilot Fort was one of two female flight instructors working on Oahu, Hawaii, on 12-07-41. During the Japanese attack she and a student were on a routine dual instruction flight when they encountered a huge formation of Japanese Navy airplanes en route to Pearl Harbor, Oahu, Hawaii.

3-21-43F. Rumsey, California. At 1558, a Bell P-39N crashed in a remote area near Rumsey, California, killing pilot 2Lt. Lionel M. Herman, 21, St. Louis, Missouri. The airplane was part of a formation of four P-39 aircraft that had taken off from Hamilton Field, San Rafael, California, on a formation navigation flight. The formation was flying on top of an overcast at 10,000 feet msl. About four miles south of Clear Lake, California, the flight leader looked back and noticed that Lt. Herman was missing. The wreckage of his airplane was not found until 5-5-43. Investigation revealed that the airplane had smashed into a mountainside after the pilot attempted to parachute to safety. Investigation revealed that he had bailed out at an altitude too low to allow his parachute to deploy. It was speculated that the pilot might have been letting down through instrument conditions after suffering some unknown mechanical failure. A P-39N from the same squadron on the same navigation mission, but with a separate three-ship flight, crashed at Middletown, California, killing the pilot (See 3-21-43G).

3-21-43G. Middletown, California. At 1600, a Bell P-39N crashed one and one-half miles northeast of Middletown, California, killing pilot 2Lt. Robert E. Hayen, who fell to his death attempting to parachute to safety. The Aircraft Accident Classification Committee stated, "Lt. Hayen was number-three in a three-ship formation on a cross-country flight when his leader decided to go through a hole in the overcast. The leader turned to miss a cloud and lost Lt. Hayen. Lt. Hayen later was seen flying low over Middletown in the rain. He was gunning his engine and witnesses at the scene of the accident said they saw the airplane and heard a loud report and [saw] the parachute of the pilot and part of the airplane coming down separate from the airplane itself. The airplane was found on a ridge southeast of Middletown and the pilot and the tail section were found on another ridge 600 feet east of the airplane. The cause of the accident is believed to be due to the fact that the pilot got lost in bad weather. It is believed that when the pilot [bailed] out of the ship it rolled over and he caught himself on the left antenna and struck the left horizontal stabilizer. The impact of the body and sudden opening of the parachute caused the tail section to snap off." A P-39N from the same squadron on the same navigation mission, but with a separate three-ship flight, crashed at Rumsey, California, killing the pilot (See 3-21-43F).

3-22-43A. Gulf of Mexico, Florida. At 0830 CWT, a North American AT-6B and an AT-6C collided in mid-air over the Gulf of Mexico at Gunnery Range 34, near Eglin Field Military Reservation, Valparaiso, Florida, killing AT-6C pilot A/C Warren C. Hussey. AT-6B pilot Robert C. Cottrell was seriously injured in a crash landing. Both airplanes were flying on a fixed gunnery mission off shore of the Eglin Field Military Reservation. The AT-6C was in the number-one position and the AT-6B was flying in the number-two position and both were taking turns firing on a towed target. While making a pass on the target tow ship, A/C Cottrell collided with the tail section of the airplane piloted by A/C Hussey. A/C Hussey's airplane dove into the sea and his body was never recovered. A/C Cottrell was seriously injured after he crash-landed on a beach after the collision. Pieces of the AT-6C tail section were found embedded in the engine of the AT-6B. The search for A/C Hussey was called off on 3-30-43.

3-22-43B. Wrightsville, Arkansas. At 1015, a Taylorcraft L-2A crashed at Wrightsville, Arkansas, killing student liaison pilots SSgt. John W. Coleman and SSgt. Joseph J. Cunzich. Investigation revealed that the airplane entered a stall and a spin at an altitude of about 300 feet above ground level while performing a simulated forced landing. The student apparently made a steep turn with low airspeed, causing the stall.

3-22-43C. Harlingen, Texas. At 1240 CWT, a Lockheed AT-18 crashed while in the traffic pattern for the Army Air Field at Harlingen, Texas, killing ten fliers. The airplane had attempted a landing at the field, but it had landed in a tail-high attitude and bounced back into the air after first contact with the runway. The pilot elected to go around and he applied throttle for take-off. The airplane began turning to the left with the port wing down. The airplane reached an altitude of about 100 feet agl about 45 degrees to the left of the runway heading with the port wing down. The pilot applied full right aileron to straighten the airplane out but it stalled and rolled violently to the left to an inverted position, diving into the ground upside down where it exploded into flames. All on board were killed instantly. Killed in the crash were: 2Lt. Coburn B. Snook, Fayetteville, New York, pilot; F/O Wayne E. Hill, Edinburgh, Indiana, co-pilot; Sgt. William A. Hillier, Beverly Hills, California, gunnery instructor; SSgt. Jack B. Schiraga, Philadelphia, Pennsylvania, gunnery instructor. Student gunners killed in the crash were: Pvt. Walter P. Baliski, Detroit, Michigan; Pvt. Charles H. Rahe, Cincinnati, Ohio; Pvt. Joseph R. Martelli, Campbell, Ohio; Pvt. Oval M. Howe, Indianapolis, Indiana; Pvt. James E. Tate, hometown unknown; Pvt. John W. Dickey, hometown unknown.

3-22-43D. Hugo, Colorado. At 1616, a Lockheed F-4 (the photo-reconnaissance version of the P-38 type) suffering a fire in flight crashed near Hugo, Colorado, killing 2Lt. Harry C. Echols. The airplane had taken off from Peterson Field, Colorado Springs, Colorado, at 1450 on a routine two-hour training flight. At approximately 1600, the airplane was seen flying at about 20,000 feet. At 1610, the airplane was observed spiraling down to a lower altitude trailing smoke. Investigation revealed that the port engine had apparently caught on fire and the pilot had feathered the propeller. The port engine became a mass of flames and the pilot elected to bail out. Investigators stated, "The pilot apparently bailed out on the right side of the cockpit and in so doing the rip cord caught on some protruding object and the parachute opened before the pilot could clear the airplane. Both the pilot and the parachute became entangled in the tail surfaces, evidenced by the fact that the main canopy of the parachute was found approximately one-half mile from the body. The shroud lines had cut against the tail surfaces and had become detached from the main harness. Apparently the ship exploded in mid-air about 200 feet above the ground. The pilot and parts of the tail were found along with the left wing, while the main portion of the airplane was found in an inverted position about three-quarters of a mile from the tail, wing and other small parts. Only the main portion of the ship burned. No evidence of fire was found on any of the scattered parts or the left wing. Parts of the airplane were found over an area of approximately 500 yards wide by one-half mile long. No parts were found between this one area and the spot where the ship crashed and burned about three-quarters of a mile away."

3-22-43E. Sacramento River, California. At 1931, a Vultee BT-13A struck a telephone line cable and crashed into the Sacramento River four miles southeast of Red Bluff, California, killing instructor pilot 2Lt. Frederick G. Wool and passenger Cpl. Orville J. Rose. The Aircraft Accident Classification Committee stated, "The pilot was flying at an extremely low altitude en route from Chico Army Flying School, Chico, California, to Red Bluff Airport, Red Bluff, California [to act as a control ship for a student night cross-country flight]. The aircraft struck and broke a heavy steel cable supporting telephone lines crossing the Sacramento River resulting in the airplane plunging into the river."

3-22-43F. Midland, Texas. At 2355, a Beech AT-11 on a bombardier-training mission suffered a structural failure of the elevators and crashed near Midland, Texas, killing four fliers. Killed in the crash were: 2Lt. Rees H. Davies, pilot; 2Lt. Frank E. O'Connor, bombardier instructor; A/C Paul Browning, student bombardier; A/C George P. Hallook II, student bombardier. Investigators stated, "This crash occurred at 2355 as indicated by two watches recovered from the wreck. Immediately prior to the crash, it is believed that the pilot and crew were in the process of conducting a normal bombing mission, during which they were flying at 11,000 feet above the ground and had dropped nine of the ten bombs carried. Oxygen regulators and masks recovered showed that the crewmembers were using oxygen with settings of from 18,000 to 20,000 feet on the regulators. This pilot had flown one previous mission on the same targets as indicated by entries in the Form-1. Another pilot who was flying the same type of mission on these targets at the same time reported that no unusual conditions were encountered, and that visibility was good with a few scattered clouds approximately 400 to 1,000 feet above the mission altitude. This pilot completed his mission and left the target area approximately 10 minutes before the accident occurred. At that time 2Lt. R.H. Davies was flying in the target pattern ahead of him and was flying in an entirely normal manner. The airplane itself struck the ground in an inverted attitude at approximately 45 degrees to the vertical. On striking the ground the airplane did not move forward but fell back in the up side down attitude so that it came to rest in that attitude. Upon inspection of the wreckage it was found that the entire elevator assembly was missing. Search disclosed the fact that the elevator assembly had separated from the airplane while at a considerable altitude as the elevator assembly broke into three pieces. The right half of the elevator hit the ground approximately one mile from the airplane. The outer approximately 18 inches of the left elevator assembly containing the counter balance hit the ground approximately one and one-half miles from the airplane. The remaining inner portion of the left elevator hit the ground approximately two miles from the airplane. These pieces were in a direct line with the airplane itself. From the position of the airplane and its attitude upon impact with the ground, it is believed that the pilot was turning onto or had completed his turn on to the course to the next target. It is believed that the elevator assembly failed during this turn or immediately after this turn and that the airplane, upon loss of elevator control either went into a dive which increase beyond the vertical or began a roll to the upside down position. It is the opinion of the Aircraft Accident Classification Committee that this accident was due only to structural failure of the elevator assembly hinges during routine flying of the airplane. No evidence can be found showing that the pilot was flying the airplane in any but in a normal manner. Immediately following this accident, all AT-11 type airplanes at this station were grounded and an inspection was made of the elevator assemblies. Numerous elevator hinges were found to bowed slightly and were found to have what appeared to be small cracks along the bottom surface of the hinges indicating metal fatigue. It is recommended by this command that all AT-7 and AT-11 type airplanes be grounded immediately and thorough inspection made of the elevator assemblies. It is further recommended that the hinges on the elevator assemblies on AT-7 and AT-11 airplanes be strengthened so that they can properly stand loads that may be placed upon the assembly during routine operation of the airplane."

3-23-43A. Hempstead, Long Island, New York. At 0520, a Republic P-47C crashed into a science lab at Hofstra College, Hempstead, Long Island, New York, killing pilot 1Lt. Earl D. Hayward. The P-47C (41-6292) had been attempting a take-off at Mitchel Field when it collided on the runway with a P-47C (41-6557) piloted by Capt. Rush M. Shortley, who was uninjured. Investigators stated, "It is believed that Lt. Hayward became confused while taxiing out for take-off. Apparently upon reaching the intersection of the ramp and the North-South Runway he paused long enough to let his flight leader get out of sight and then assumed that the flight leader had taken off, and he followed. As a result of this mistake Lt. Hayward took off down wind with a wind of approximately 15 miles per hour and used only one-half to two-thirds of a 3,800-foot runway that had 60-foot lighted obstruction at the end. During the course of this take-off he struck the flight leader's airplane [taking off and reaching an altitude of about 75 feet agl] and then flew thru some electric light wires subsequently crashing on the grounds of Hofstra College. Further investigation shows that this pilot had no time at night in the past six months. Also the flight leader, Capt. Shortley, had not flown a P-47 type airplane at night — yet these two pilots were the only two in the squadron considered competent for handling alert operations at night."

3-23-43B. Coleman, Texas. At 0520, a Fairchild PT-19A crashed six miles northwest of Coleman, Texas, killing student A/C Richard Law. Civilian instructor O.H. Stansel was injured parachuting to safety. Investigation revealed that while demonstrating maneuvers to the student, the instructor was thrown out of the aircraft at an altitude of about 600 feet agl. He landed on the tail and was hanging onto the vertical fin. His weight on the vertical fin was forcing the airplane into an extreme nose-high attitude. The instructor let go and deployed his parachute. The last time the airplane was observed it was in a slight dive. Apparently the student was not able to recover from the stall and spin. The airplane crashed into the ground nose first, killing the student instantly. Despite the instructors insistence that his safety belt had been fastened, investigators were able to determine that the instructor had not fastened his safety belt. The safety belt was found to be adjusted too short to fit around the instructor.

3-23-43C. Marana, Arizona. At 1030 MWT, a Vultee BT-13A crashed four miles NNW of the Army Air Field at Marana, Arizona, killing student A/C George S. Frias. The student had taken off from Marana on a routine student solo flight. Investigation of the wreckage indicated that the airplane had struck the ground at a 45-degree angle, killing the student instantly. Investigators could not determine the events leading up to the accident.

3-23-43D. Bennettsville, South Carolina. At 1440, a Stearman PT-17 crashed at Bennettsville, South Carolina, killing student A/C Robert G. Welborn. The airplane had taken off from Bennettsville on a routine student solo flight. Witnesses observed the airplane in a power-on vertical dive at about 2,000 feet agl. The airplane dove to about 1,000 feet and then pulled up about ten degrees before falling back into a vertical dive and smashing into the ground. The student was killed instantly.

3-23-43E. Avalon Auxiliary Airfield, Mississippi. At 1520, a Vultee BT-13A crashed near the Avalon Auxiliary Airfield, Mississippi, killing 2Lt. James A. Potter and student A/C Henry J. Kolata. The airplane had landed at Avalon Airfield while on an instrument training flight that had originated at Shaw Field, Sumter, South Carolina. The airplane took off and circled the field. The airplane then made a high-speed low-altitude pass over the airfield. The airplane cleared a training airplane holding at the end of the runway by inches. The pilot, thought to be the instructor, then pulled up into a steep chandelle type maneuver. The airplane stalled and fell off on wing and then dove into the ground.

3-23-43F. Shaw's Station, North Carolina. At 1630, a Piper L-4A crashed near Shaw's Station, North Carolina, killing Liaison Pilot SSgt. John R. Mancuso and passenger Pvt. Mitchel F. Rotundo. Investigators could not determine what caused the airplane to fly into the ground at a 45-degree angle.

3-23-43G. Bakersfield, California. At 1757, a Vultee BT-13A crashed seven miles south of Minter Field, Bakersfield, California, killing student A/C David S. Lefkowitz, who was fell to his death attempting to parachute to safety. Investigators stated, "The student cleared for one hour of solo flight to practice slow rolls and snap rolls. The airplane was first observed in right hand spin by [another pilot] at an altitude of approximately 7,000 feet. The spin appeared to be slightly slower and flatter than normal, the horizontal axis being inclined at approximately 45 degrees to the vertical. The spin continued without interruption to the ground and no attempt at recovery was observed. The position of the body approximately six feet outside the cockpit, together with lack of evidence of injury sustained by the pilot inside the cockpit indicated that he probably cleared the aircraft just before the crash. The safety belt was found unbuckled and the parachute ripcord had not been pulled."

3-23-43H. Clifton, Tennessee. At 2045 CWT, a Vultee BT-13A crashed one and one-half miles from Clifton, Tennessee, killing A/C John A. Langley, Jr. The airplane had taken off from the Army Air Field at Courtland, Alabama, on a night navigation flight to Goodrich, Tennessee, to Corinth, Mississippi, and return. The student had checked in on time and on course at Goodrich. Witnesses observed the airplane flying at 2,500 feet agl over Goodrich with the engine sputtering but flying straight and level. The airplane was then seen to go into a spin to the left. The airplane continued to spin to the left until it struck the ground and erupted into flames. The student made no effort to leave the airplane and was killed instantly upon impact.

3-23-43I. Rock Island, Texas. At 2059, a North American AT-6C crashed five miles south of Rock Island, Texas, killing pilots 1Lt. Lyman S. Sylvester, Madison, Wisconsin, and 1Lt. William G. Bright, Chicago, Illinois. The airplane had taken off from Foster Field, Victoria, Texas, on a night navigation flight to Barksdale Field, Shreveport, Louisiana, and return. The airplane had made it to Barksdale Field and took off for Victoria at 1826 CWT. The airplane never arrived at Victoria and was declared missing. Witnesses reported that they had heard the airplane flying over the area at low altitude with the engine sputtering. They heard the airplane crash a few minutes later. They reported the accident but investigators were unable to find the airplane in the area of the reported crash. Investigation revealed that the airplane had dove into the ground at a steep angle at approximately 250 mph. The airplane was completely demolished and the occupants "mangled beyond recognition."

3-24-43A. Orlando Army Air Base, Florida. At 0828 EWT, a Lockheed F-5A (the photo-reconnaissance version of the P-38 type) crashed northwest of the Army Air Base at Orlando, Florida, killing pilot 2Lt. Charles G. Harris. The airplane, equipped with two external 150-gallon auxiliary fuel tanks, took off to the north from the Western North-South Runway at Orlando AAB on a photo-reconnaissance mission. Soon after take-off the port engine failed. The pilot feathered the propeller, trimmed the airplane for single-engine flight and then called in for an emergency landing. The pilot turned to the left and entered the pattern for a landing to the southeast on the Northwest-Southeast Runway. The airplane was flying at 500 feet agl and just above stalling speed when the pilot extended the landing gear and made his turn from the base leg to final approach into the dead engine. The airplane lost airspeed rapidly, stalled, flipped over and entered a spin to the right. The airplane smashed into the ground among a group of large army tents near the base, exploding into flames upon impact. The pilot was killed instantly. A light-duty cargo trailer was destroyed by fire.

3-24-43B. Eufaula, Alabama. At 1300 CWT, a Vultee BT-13 crashed while attempting a forced landing 15 miles southwest of Eufaula, Alabama, killing student A/C Alexander Horner, Jr. The student had taken off from Gunter Field, Montgomery, Alabama, on a routine student navigation flight. The student became lost and when the airplane ran out of fuel he attempted a forced landing. The student chose a field for his landing and made his approach. The airplane stalled and fell off into a spin to the left, making approximately one and one-half turns of spin before smashing into the ground nose first. It was later speculated that the student apparently attempted to extend his glide and caused the airplane to stall.

3-24-43C. Lake Martin, Alabama. At 1452 CWT, a Curtiss P-40F crashed at Lake Martin, Alabama, killing pilot 2Lt. Earl E. King. The airplane had taken off from the Army Flying School at Tuskegee, Alabama, on a routine navigation flight. The airplane was seen flying near a large dam in the area and then flying at low altitude over Lake Martin. A civilian witness stated that the airplane smashed into the lake with the engine running. Investigation revealed that the pilot had inadvertently flew the airplane into the lake while making an authorized simulated low-level attack on a small, uninhabited island on Lake Martin. The airplane flipped over on its back and immediately sank to the bottom of the lake. The propeller, propeller spinner, and radiator were the only wreckage recovered just after the accident. Apparently they had separated upon impact and had been hurled through the air a great distance, smashing into shallow water near the island. Examination of the propeller revealed that the engine was running at full power at the time of impact. The pilot's body and the airplane were not recovered until 12-21-43, after the lake level had lowered over 40 feet. The airplane and pilot were part of the 99th Fighter Squadron.

3-24-43D. Rock Hall, Maryland. At 1630, two Republic P-47C type airplanes collided in mid-air south of Rock Hall, Maryland, killing 2Lt. Roy J. Reighard aboard P-47C #41-6500. F/O James F. Bowen was able to parachute to safety from P-47C #41-6501. He was uninjured. The airplanes were part of a four-ship flight that had taken off from Baltimore Army Air Base, Maryland, on an interception mission. The flight reached an altitude of about 7,000 feet indicated altitude when the flight leader made an abrupt 180-degree turn to the right. In an effort to maintain position and not overrun the leader, F/O Bowen, flying on the leader's left wing, had to drop back in the turn and cross under to the inside of the turn. Lt. Reighard's flight leader made a shallow turn. The first element was far enough ahead and to the left to permit the second element to begin to approach it at a 90-degree angle. Lt. Reighard, who was supposed to be flying on the second element leader's right wing, was actually flying above and behind his flight leader during the turn. The second element leader apparently tightened the turn and made up the distance but still approached the first element leader at an awkward angle. During Lt. Reighard's turn, F/O Bowen had apparently dropped back too far during the crossover and pulled up and into Lt. Reighard's P-47, shearing off the tail section at the supercharger. Both airplanes immediately went out of control and crashed into shallow water at Rock Hall.

3-24-43E. Venice, Florida. At 1703 EWT, a Douglas A-24A dove into the ground at the AAF Bombing and Gunnery Range at Venice, Florida, killing pilot 2Lt. Joseph V. Michl and passenger 2Lt. Myron T.S. Dirickson. The airplane had taken off from Drew Field, Tampa, Florida, on a dive-bombing mission. The Aircraft Accident Classification Committee stated, "[The] airplane dove into the bombing range area. Apparently the pilot failed to start recovering from the dive soon enough. Indications were that the airplane hit the ground in a [wings] level position with the nose low. This was determined by the fact that the imprint of both wing panels was clearly shown plus a hole where the nose of the airplane hit. Evidently the airplane exploded upon impact, as there were no complete parts of the airplane found." Wreckage was scattered over an area of about 700 feet.

3-24-43F. Cross City, Florida. At 1810 EWT, a Republic P-47C crashed near the Army Air Field at Cross City, Florida, killing pilot 2Lt. Richard F. Belsan. The pilot had taken off from Cross City Army Air Field with several other P-47 aircraft on an "altitude aerial gunnery" mission. Investigators stated, "After an hour and a half and just before the formation returned to the field, Lt. Belsan left his formation at 1,500 feet and headed for the field. His flight leader called to him several times but was unable to get an answer. Upon reaching the base he made the standard approach for a landing into the northeast without contacting the tower for landing instructions. He came over the field and made a gentle climbing turn to the left preparatory to landing. At the top of this turn it appeared that he changed his mind and decided to land southeast. In trying to make this 270- degree turn he tightened it up too much and consequently stalled the airplane at an altitude of 800 feet, which was too low to recover. The airplane made a half-turn to the right, a half-turn to the left and went into the ground at a 60-degree angle just off the end of the Northwest-Southeast Runway."

3-24-43G. Seguin, Texas. At 1745, a North American AT-6C crashed three miles west of Seguin, Texas, killing pilot 2Lt. Willis L. Jones and RAF Pilot Officer Ronald A. Dixon. Student airplanes from Randolph Field, San Antonio, Texas, were practicing at the airfield at Seguin and flying in the vicinity. When a weather alert came in, Capt. William J. O'Donnell, who was in charge of the training activities at Seguin, was able to contact all student airplanes and get them to return to Randolph Field ahead of the weather. Capt. O'Donnell spotted Lt. Jones about a mile and half ahead of him. As he flew up to Lt. Jones to follow him to Randolph Field, Lt. Jones entered a gentle right turn until he got on Capt. O'Donnell's tail. Flying at about 2,300 feet agl, Captain O'Donnell and Lt. Jones entered a Lufbery turn, which reversed direction about three times. During the last turn, Lt. Jones dropped about 15 degrees of flaps in an effort to tighten his turn. The airplane stalled and flipped over on its back before going into a tight spiral toward the ground. The pilot failed to recover and the AT-6C crashed into a plowed field where it exploded into flames.

3-24-43H. Hope Valley, Rhode Island. At 1900, a Republic RP-47B crashed at Hope Valley, Rhode Island, killing pilot F/O Oscar C. Kline. The airplane had taken off with another P-47 from Westover Field, Chicopee, Massachusetts, at 1715 on a routine formation training flight. The flight was led by 1Lt. Richard A. Cline. Darkness arrived and the two airplanes became lost and low on fuel. Lt. Cline elected to attempt a landing on a highway near Hope Valley. He made seven approaches before he was able to put the airplane down on the highway. He struck a post with the port wing, causing only minor damage to the wing and the port landing gear. He taxied to a clear area and parked. After observing Lt. Cline's successful landing, F/O Kline attempted a similar highway landing about two miles away. Just as the airplane touched down and began rolling down the highway F/O Kline spotted an automobile coming toward the airplane. He gave the airplane throttle for take-off and pulled the airplane into the air. He successfully avoided the automobile but lost flying speed and fell off in a stall back down to the highway. The port wing struck a telephone pole, causing the airplane to veer to the left and cartwheel into the ground. The airplane smashed into an area of heavy timber about 75 feet from the edge of the highway.

3-24-43I. Grenier Field, New Hampshire. At 1933, passenger 1Lt. Alan J. Bamberger (Quartermaster Officer), New York, New York, was killed when he walked into the rotating starboard propeller of a North American B-25C preparing to depart Grenier Field, Manchester, New Hampshire. Investigators stated, "Lt. Bamberger walked west out of the Base Operations [building] to the airplane which was parked headed south with the engines running at approximately 1,000 RPM. He then walked past the airplane in front of the port engine at which point he was motioned by the pilot to climb in from the rear, however Lt. Bamberger continued on past the nose and then out back towards the rear of the airplane on a direct line to the hatch which had opened, this path led directly thru the propeller arc, and although the co-pilot yelled a warning, Lt. Bamberger continued on until struck by the propeller." Lt. Bamberger was killed instantly and his parachute pack ripped open by the propeller. The airplane was piloted by 1Lt. George J. Nied and co-pilot F/O Eugene V. Burgess.

3-24-43J. Seville, California. At 1720, a Ryan PT-22 spun to the ground one-quarter mile southwest of Seville, California, killing student Owen Milton Opsal. Investigation revealed that the airplane had struck the ground in a steep spin to the right, striking the ground nose first. The airplane had taken off from the Army Air Field at Visalia, California, on a routine student solo flight.

3-24-43K. Location Unknown. At an unknown time after 2048 EWT, a North American B-25D disappeared and is presumed to have crashed into the Atlantic Ocean with its crew of six while on a navigation training flight. The airplane had taken off from the Army Air Base at Columbia, South Carolina, on a night navigation-training mission to Charleston, South Carolina, to a point 50 miles at sea, to Savannah, Georgia, to Charleston, and return to Columbia. At 2048, the crew radioed Charleston that the airplane was heading for Savannah after it had reached its navigation point at sea. One airplane radio operator reported hearing what he thought was the missing airplane making an incomplete radio transmission to the mainland at about 2200 EWT but this could not be verified by ground operators or radio logs. The airplane was never heard from again. Despite a massive search effort, no trace of the airplane or crew was ever found. Missing and presumed lost at sea were: 2Lt. Rufus S. Humphries, Jr., pilot; 2Lt. Haig P. Wylie, co-pilot; 2Lt. Frank J. Moore, navigator; SSgt. Horace C. Holt, engineer; SSgt Joseph E. Messina, radio operator; Sgt. Stanley F. Warshak, gunner.

3-25-43A. Gulf of Mexico, Florida. At 1427 EWT, a Lockheed-Vega R-37 (similar to the B-34 type) suffering an engine failure ditched into the Gulf of Mexico about 110 miles southwest of Boca Chica Army Air Base, Key West, Florida. Special Equipment Operators Cpl. David D Reinhart and Cpl. Robert B. Townsend were missing and presumed lost at sea. Co-pilot/navigator 2Lt. William E. Shackleton and radio operator Pfc. Daniel H. Coutu suffered minor injuries. Pilot 1Lt. Anthony B. Mitchell and radio operator Sgt. Dominick Carboni were uninjured. The airplane had taken off from Boca Chica AAB on an anti-submarine patrol mission. Investigators stated, "The right engine failed intermittently at 700 feet over open water. The pilot feathered the right engine and applied power to the left engine, which did not take power immediately. The left engine was cut back momentarily then power was again applied. Bombs were jettisoned and all instruments checked to ascertain the probable cause of the failure of the right engine. When the bombs were jettisoned and the bomb door handle [placed] in the closed position, the airplane was at 150 feet, 120 to 130 mph and losing altitude. The pilot decided that a water landing was inevitable, therefore feathered the left engine and attempted to land into the wind at the crest of a wave. The airplane submerged on contact with the third wave. The responsibility of the accident is divided between failure of the right engine from causes unknown and the pilot's hesitancy to apply normal rated power on the left engine after it had momentarily failed when power was first increased." The surviving crewmembers floated in a life raft for about 45 minutes until a civilian vessel operating in the area rescued them from the sea. There was no sign of the two Special Equipment operators and they are presumed to have drowned soon after the airplane ditched and broke apart.

3-25-43B. Waycross, Georgia. At 1442 EWT, a North American A-36 suffered a catastrophic structural failure of the port wing and crashed on the AAF Bombing Range at Waycross, Georgia, killing 2Lt. James B. Tubb. Investigators stated, "The accident occurred during a practice glide bombing mission. The pilot attempted an abrupt recovery from his bombing run after releasing his first bomb. The sharpness of the recovery put too great a stress on the airplane causing the left wing to pull off. The wing struck the horizontal stabilizers, tearing them off the airplane. The airplane continued on into the ground approximately 1,000 feet beyond the target. The result was instantly fatal to the pilot and the airplane was a total loss. Investigation revealed that the left wing had failed adjacent to the dive brake."

3-25-43C. Lake Elsinore, California. At 1545, a Lockheed P-38E crashed one mile north of Lake Elsinore, California, killing F/O Claire A. Skewes. Investigators stated, "Flight Officer Skewes was assigned an oxygen mission and took off from Ontario Army Air Field [California] at 1452. An aircraft spotter observed this airplane at a very high altitude at 1545 in a vertical dive. Other witnesses observed the airplane at a lower altitude in a spin with parts floating down separately. From examination of the wreckage, the following is concluded: That, for an unknown reason, the pilot went into a vertical power dive at a high altitude and was unable to regain control of his airplane. He apparently shut off his engines, attempted to lower the landing gear and flaps and then attempted to bail out. The shroud lines to his parachute were severed and it is believed that he or his parachute struck the tail surfaces of the ship, causing the ship to snap out of the dive and immediately disintegrate. [The port wing and the tail section separated from the airplane before it struck the ground.] All pilots have been constantly warned against allowing this type of aircraft to attain extremely high speeds. In this particular instance, the pilot might have become unconscious at a high altitude and lost control of the airplane." It was later speculated that the pilot had put the airplane in a diving maneuver from which he could not recover because to the tendency of the P-38 type airplane to suffer from severe control surface buffeting caused by compressibility.

3-25-43D. Trinidad, Colorado. At 2100, a Consolidated B-24D crashed into the side of a mountain 45 miles west of Trinidad, Colorado, killing the crew of eight. The airplane had taken off from the Army Air Base at Clovis, New Mexico, at 1949 MWT on a navigation training flight to Pueblo, Colorado, and return. Apparently the B-24 was flying in instrument conditions when it struck the mountain in a straight and level attitude at cruising speed. The airplane exploded violently into flames upon impact with the mountain. Killed in the crash were: 2Lt. George W. Wiggins, pilot; 2Lt. William R. Bowerman, co-pilot; 2Lt. John T. Crowley, navigator; 2Lt. Hoyt A. Phelps, bombardier; SSgt. Walter S. Ciesielski, engineer; Sgt. Walter G. Dymond, radio operator; SSgt. Franklin C. Bynum, assistant engineer; Sgt. Edward T. Foley, assistant radio operator.

3-25-43E. Alamogordo, New Mexico. At 2210, a Consolidated B-24E flew into the ground just after take-off from the Army Air Base at Alamogordo, New Mexico, killing seven crewmembers and injuring five others. Investigation revealed that the number-two engine caught on fire immediately after the B-24 became airborne. The pilot shut down the engine and feathered the propeller. The pilot then entered a gentle left turn in an effort to turn back for an emergency landing at the base. The airplane was flown into the ground while in this turn. The airplane cartwheeled into the ground and burst into flames as it broke up. Killed in the crash were: 2Lt. Frank W. Gardner, pilot; 2Lt. Will D. Willis, navigator; SSgt. Price A. Keyes, assistant engineer; SSgt. Ralph W. Clark, gunner; SSgt. Lawrence W. Supernand, gunner; Sgt. John A. Jarvis, passenger; Pfc. Francis X. Joyce, passenger. Crewmembers seriously injured in the crash were: 2Lt. Edward G. Taylor, bombardier; SSgt. Floyd E. Veasey, engineer; SSgt. Robert W. Davis, radio operator; Sgt. John L. Rice, assistant radio operator. Co-pilot 2Lt. Harold R. McMillan suffered only minor injuries.

3-25-43F. Yuma, Arizona. At 2300 MWT, a North American AT-6C crashed ten miles southwest of Yuma, Arizona, killing student A/C Dwight E. Linkey. The airplane had taken off from Yuma Army Air Field on a night navigation mission to Tonopah, Nevada, to Blythe, California, and return. The airplane successfully completed the mission and returned to Yuma, flying over the field at an altitude of about 3,500 feet agl. The tower gave the pilot permission to land and ordered him to enter the pattern. The airplane flew into the ground moments later. Investigation revealed that the airplane struck the ground in a bank to the right. The airplane burst into flames upon impact and the student was killed instantly.

3-26-43A. Kasbeer, Illinois. At 1145, a Bell P-39N crashed at Kasbeer, Illinois, killing pilot 2Lt. Elwood L. Highhouse. The airplane was part of flight of three that had taken off from Baer Field, Fort Wayne, Indiana, on a ferry flight to Des Moines, Iowa. The airplanes were flying at 4,000 feet agl near Ottawa, Illinois, when the two wingmen became separated from the flight leader in deteriorating weather conditions. The two wingmen then descended separately through the clouds in an attempt to make visual contact. When the airplanes broke out underneath they were heading straight for the ground. Lt. Elwood pulled up sharply but entered a stall at an altitude too low to allow recovery. The airplane smashed into the ground and exploded into flames. The pilot was killed instantly. Seconds later the other wingman, 2Lt. Steve S. Macut, was killed when he crashed into the ground about five-eighths of a mile from Lt. Highhouse's crashed airplane. Investigators noted that the pilot had less than ten hours of flying time on P-39 type airplanes. The flight leader was able to land safely at Des Moines. See 3-26-43B.

3-26-43B. Kasbeer, Illinois. At 1145, a Bell P-39N crashed at Kasbeer, Illinois, killing pilot 2Lt. Steve S. Macut. The airplane was part of a flight of three that had taken off from Baer Field, Fort Wayne, Indiana, on a ferry flight to Des Moines, Iowa. The airplanes were flying at 4,000 agl near Ottawa, Illinois, when the two wingmen became separated from the flight leader in deteriorating weather conditions. The two wingmen then descended separately through the clouds in an attempt to make visual contact. When the airplanes broke out underneath they were heading straight for the ground. Lt. Macut pulled up sharply but entered a stall at an altitude too low to allow recovery. The airplane smashed into the ground and exploded into flames. The pilot was killed instantly. A few seconds earlier, the other wingman, 2Lt. Elwood L. Highhouse, was killed when he smashed into the ground five-eighths of a mile away from Lt. Macut's crashed airplane. Investigators noted that the pilot had less than ten hours flying time on P-39 type airplanes. The flight leader was able to land safely at Des Moines. See 3-26-43A.

3-26-43C. Port Dover, Ontario. At 1406 EWT, a Bell P-39N crashed three miles west of Port Dover, Ontario, Canada, killing Service Pilot 1Lt. John R. Anderson. The airplane was the number-three airplane in a six-ship formation that had taken off from Niagara Falls, New York, at 1340 EWT on a ferry flight to Romulus Army Air Field, Michigan. At 1406 Lt. Anderson dropped out of the formation, which was at 1,500 feet agl and began to descend. Another airplane followed Lt. Anderson down to about 1,100 feet when he observed Lt. Anderson to jettison his auxiliary fuel tank. The subject airplane continued to lose altitude. The pilot was killed when he bailed out at an altitude too low to allow his parachute to deploy.

3-26-43D. Plainview, Texas. At 1635, a Piper L-4A crashed near the municipal airfield at Plainview, Texas, killing pilot 1Lt. Henry O. McNerney and seriously injuring civilian passenger Gaylord W. Elbert. Investigation revealed that the airplane stalled while attempting a simulated forced landing. The airplane fell off on a wing at an altitude of about 200 feet agl and smashed into the ground nose first. Lt. McNerney died a short time after the accident.

3-26-43E. Mount Hood, Oregon. At an unknown time after 1620 PWT, a Lockheed RP-38E collided with rising terrain near Mount Hood, Oregon, killing pilot 2Lt. Alan C. Strader, Spokane, Washington. The airplane had been reported missing after failing to return to Portland Army Air Base, Oregon, after a routine training flight. The airplane and the remains of the pilot were not found until 7-9-62. Investigators stated at the time of the airplane's disappearance, "The formation of two P-38s with 1Lt. Robert J. Meyer leading and Lt. Strader in the number-two position was flying at about 8,000 feet altitude amid heavy broken clouds. At the time, there was a great deal of flying activity among the fighter planes [that were in the immediate vicinity]. Evidently while the formation was flying in and around the broken clouds, Lt. Strader found himself in the overcast unexpectedly, and shortly thereafter, lost. Lt. Strader reported to Lt. Meyer first, that he was at 10,000 feet on instruments, and shortly afterward, at 11,000 feet on instruments [approximately seven miles east of Portland AAB]. Lt. Meyer directed Lt. Strader to fly on a course of 300 degrees and to descend to 6,000 feet. Lt. Meyer does not know whether this maneuver was completed. The [Portland AAB] tower was not in contact with Lt. Strader." Twenty-seven airplanes and over 100 transient airplanes were involved in the search for Lt. Strader, which was discontinued on 4-6-43.

3-26-43F. Morton, Mississippi. At 2015, a Douglas A-24A crashed seven miles northwest of Morton, Mississippi, killing pilot 2Lt. Donald W. Goerke and passenger Sgt. Theodore R. Radomski. The airplane had taken off from Key Field, Meridian, Mississippi, on a local night navigation flight. The pilot encountered instrument conditions and apparently lost control of the airplane, allowing it to fly into the ground.

3-27-43A. Gardner Field, California. At 0825, a Vultee BT-13A crashed near Gardner Field, Taft, California, killing student A/C William L. Sherin. Investigation revealed that the student had made two successful landings during a 90-degree power-off landing stage. While turning from the base leg to the final approach on his third landing, the student stalled the airplane. The airplane oscillated to the left and right as the student struggled with the controls. The airplane snapped over to an inverted position and smashed into the ground upside down at an angle of 65 degrees. The airplane careened along the terrain for a short distance before flipping back to an upright position as it disintegrated. The engine separated and came to rest some distance from the main wreckage. The student was killed instantly.

3-27-43B. Desert Center, California. At 0850, a Bell P-39F spun to the ground one mile south of the Army Air Field at Desert Center, California, killing pilot Capt. Ralph W. Soverel, Jr. Investigation revealed that the airplane entered a flat spin at an unknown altitude. The pilot apparently could not recover from the spin and attempted to parachute to safety. He fell to his death when he left the airplane at an altitude of about 300 feet agl, too low to allow his parachute to fully deploy.

3-27-43C. Tonopah, Nevada. At 0940 PWT, a North American AT-6 suffering an engine failure and attempting an emergency landing on a highway collided with a 1941 Ford Tudor Sedan 16 miles west of Tonopah, Nevada, killing AT-6 passenger AAF Flight Surgeon 1Lt. Ralph Hunt Sullivan, Jr. and seriously injuring pilot 2Lt. Davis True Perron. The driver of the automobile, Mrs. C.E. Hill, Reno, Nevada, and her two young children suffered minor injuries in the accident. Investigators stated, "Lt. D.T. Perron took off from [Tonopah Army Air Field] at about 0839. After climbing toward the north he turned toward the west to get over the large valley west of Tonopah. He started to do some lazy-eights, but his passenger, Lt. Sullivan, told him that he did not like lazy-eights. Lt. Perron then put the airplane in a diving turn, and while he was in the dive his engine stopped running. When he realized the engine was not running, he told Lt. Sullivan to jump, but Lt. Sullivan either was unable to get his canopy open or decided to stay with the airplane since the pilot was going to try to make a landing. Lt. Perron made the decision that he would attempt a wheels-down landing on the highway [to] prevent damaging the airplane. He made a gliding turn to the right and lined up with the highway; then he lowered his landing gear and flaps. He saw a car parked by the highway, which seemed to be making no attempt to get out of the way. He turned his passing and landing lights on and off in hopes that the car driver would move off of the road. The car remained on the side of the road and he was unable to pull up over it, causing his left wing to hit the top of the car. His airplane turned over while still in the air and started cartwheeling along the side of the road. The airplane was completely demolished." Mrs. Hill could see that the airplane was coming straight at the car so she gathered her children and they ducked down as far as they could on the front floor of the car, avoiding serious injury when the AT-6 hit the car and smashed in the roof on the driver's side.

3-27-43D. Vichy, Missouri. At 1145, a Republic P-43D crashed near the Army Air Field at Vichy, Missouri, killing pilot 2Lt. Morton D. Barnes. The pilot was on a routine transition flight and had made one successful landing. The airplane took off again, entered the pattern and extended the landing gear in preparation for another landing. The airplane began a gentle turn to the left and began to lose altitude. The P-43D continued to descend in a gentle turn to the left until it smashed into trees three-quarters of a mile from the field. The horizontal stabilizer was sheared off and the airplane bounded back into the air for another half mile, crashing into treetops before smashing to the ground in a small field. The airplane tumbled several times, breaking up and bursting into flames. The pilot was thrown out of the cockpit as the airplane tumbled. His mangled body was badly burned in the resultant fire.

3-27-43E. Oliver Springs, Tennessee. At 1235 CWT, a Douglas C-47 flying in instrument conditions suffered a catastrophic structural failure and crashed at Oliver Springs, Tennessee, killing the four-man crew and eight passengers. The Aircraft Accident Classification Committee stated, "Pilot took off on an instrument flight from Pope Field [North Carolina] to Tulsa [Oklahoma] to cruise at 8,500 feet. At 1223 CWT, pilot called Knoxville, Tennessee, Airways Radio, stating he was at 11,500 feet and asked for instructions. Knoxville radio replied 'no instructions.' Pilot broke out of the overcast at about 3,000 feet in a descending attitude 32 miles west of Knoxville. Descent continued for about 2.2 miles holding a generally straight heading. The ship then assumed an abrupt left climbing turn, at which time a section of the metal skin over the left tanks and a small piece of left stabilizer left the aircraft. Immediately thereafter an explosion occurred which completely disintegrated all the forward portion of the aircraft from the main entrance door, also the tail surfaces, scattering portions over about one and one-half square miles. One propeller was feathered, the other almost completely feathered, we believe, because pilot had detected excessive vibration or flutter. It is the belief of this Accident Committee that this accident was caused by one or more of the following: Disintegration of the fabric covered control surfaces due to excess speed. Structural failure of left wing and stabilizer in sudden pull-up, which might have been caused by either pilot making this maneuver on sight of the ground, or shifting of passengers into the tail for possible crash landing after realizing loss of control surfaces by shredding, or failure of elevator hinge bracket by a resonate vibration set up in using 1,200 to 1,600 rpms. The descent from original altitude was probably due to conditions beyond the pilot's control, such as icing of one or more of the following: engines, outside instrument vents on ship, mechanical failure of either or both engines, failure of controls due to resonant frequency, or controls becoming jammed. The explosion was probably caused by the auxiliary venting through the filler cap of the fuel tanks installed in the cabin or the fracture of the gas lines leading to the wing tanks from these auxiliary cabin tanks at the time of the crumpling action in the left wing." Crewmembers killed in the accident were: F/O Charles E. Fontaine, pilot; 2Lt. J.C. Wingate, co-pilot; SSgt. Joseph A. Stahlman, engineer; Cpl. Walter C.

Oughton, radio operator. Passengers killed in the accident were: Pvt. George P. Bacca, Pvt. James E. Berry, Pvt. Claude A. Brasier, Pfc. William C. Hoover, Cpl. Willis F. McElroy, SSgt. William G. Snell, and 2Lt. Allen C. Deatherage.

3-27-43F. Shreveport, Louisiana. At 1404, a Martin B-26B crashed ten miles northwest of Barksdale Field, Shreveport, Louisiana, killing the crew of six. The Aircraft Accident Classification Committee stated, "The aircraft cleared Barksdale Field at 1359 and left traffic in a normal manner, the right engine almost immediately catching on fire and [the airplane] crashing at 1404, the time being determined by a watch found in the wreckage. Evidence of jettisoned bomb bay tanks and bombs indicates the pilot was preparing for a crash landing. Loss of control in the later stages is evident from the fact that the airplane turned to the right and dove into the ground." Killed in the crash were: 2Lt. Alexander D. Neilson, pilot; 2Lt. William A. Mihalovich, co-pilot; 2Lt. Howard W. Niemer, bombardier; SSgt. Robert R. Peer, engineer; SSgt. Armand J. Martini, radio operator; Sgt. Marion Anderson, gunner.

3-27-43G. Marion, Texas. At 1400, a Beech AT-10BH crashed one mile east of Marion, Texas, killing pilot 1Lt. John E. Pettaway and co-pilot 2Lt. Ollie L. Reynolds. Investigation revealed that the airplane had just taken off from the auxiliary airfield at Marion. A farmer and his wife had observed the airplane in a spin and they stated that it had straightened out momentarily just before it hit the ground and exploded into flames. Investigators speculated that the airplane was in the second turn of traffic at the time it entered the spin.

3-27-43H. Phoenix, Arizona. At 1530 MWT, a Curtiss P-40F crashed nine miles ENE of Luke Field, Phoenix, Arizona, killing Chinese Air Force pilot Sub-Lt. Han-Ju Chen. Investigation revealed that the airplane had smashed into the ground at high speed while apparently performing aerobatics at low altitude. The airplane had taken off from Luke Field on an engineering test flight.

3-28-43A. El Paso, Texas. At 0308 MWT, a Beech AT-10 crashed 35 miles ESE of Biggs Field, El Paso, Texas, fatally injuring pilot 2Lt. Leonard C. Jago. Investigation revealed that both engines had failed while the airplane was flying at 8,000 feet indicated altitude. The airplane lost altitude as the pilot tried to restart the engines. The airplane smashed into the crest of a hill at an elevation of about 6,000 feet and skidded for about 50 yards before coming to a rest in a canyon and bursting into flames. When AAF personnel and investigators arrived at the scene of the crash, they could not find the body of Lt. Jago. Lt. Jago had been found alive by soldiers in a jeep about ten miles from the scene of the accident. In an effort to get aid, he had apparently walked and crawled over very rough terrain with serious burns on his head, face, hands, arms and legs to the mid point of the thighs. The pilot gave a statement as he was driven to a hospital. Lt. Jago, suffering second and third degree burns over 60 percent of his body, died at 2237 MWT at the station hospital at Biggs Field.

√**3-28-43B. Ballinger, Texas.** At 0900, a Fairchild PT-19A crashed four miles east of Bruce Field, Ballinger, Texas, killing student A/C James C. Lloyd. The student was practicing solo landings at Johnson Auxiliary Airfield. The student had made three successful landings but had slightly overshot on each of them, landing on the last two-thirds of the runway each time. The student overcorrected on the fourth attempt and undershot the field. Instead of using full throttle and maintaining a constant altitude, the student used partial throttle and tried to "drag it in" from an altitude of about 100 feet. Seeing that the airplane was going to collide with a tree, the student pulled the stick back abruptly but could not avoid colliding with the tree. The airplane had struck the tree with the starboard wing and the right side of the lower fuselage and climbed straight up before flipping over to an inverted position at an altitude of about 50 feet. The airplane then smashed into the ground upside down, killing the student instantly.

3-28-43C. Savannah, Georgia. At 1113 EWT, a Martin B-26B crashed five miles southwest of Chatham Field, Savannah, Georgia, killing the crew of four. The airplane had taken off from Hunter Field, Savannah, on a local test flight. The airplane was observed by control tower personnel at Chatham Field (about ten miles Northwest of Hunter Field) flying in a southwesterly direction at about 1,000 feet above ground level. The airplane then began a gentle turn to the right and descended into the ground in a flying attitude, erupting into flames when it came to rest. Investigation of the wreckage indicated that both engines were producing power at the time of the accident. The propellers gouged into the ground for 100 yards before the fuselage smashed into the terrain. The airplane skidded along the ground for another 100 yards before it smashed into a wooded area and burst into flames. Investigators could not determine why the airplane flew into the ground. Killed in the crash were: F/O Floyd R. Thompson, pilot; F/O Joseph R. Whitehead, co-pilot; SSgt. William Fill, engineer; Sgt. Edgar L. Jackson, radio operator.

3-28-43D. Headland, Alabama. At 1115, a North American AT-6C and an AT-6A collided in mid-air five miles northeast of Headland, Alabama, killing AT-6C pilot A/C Edward J. Conley. AT-6A pilot 2Lt. Lawrence E. Allison was able to parachute to safety and was uninjured. Investigators stated, "While on a tactical formation mission, the instructor placed his formation in trail with approximately 200 yards distanced between airplanes. When performing aerobatic maneuvers while in trail, each student had been instructed to do them in turn waiting until the airplane ahead had completed its maneuver before starting his and to execute them from cruising speed. A/C Conley, flying in the No. 3 position, lost approximately 500 feet executing the slow roll. The pilot of the No. 4 airplane [Lt. Allison] lost sight of A/C Conley and pulled up approximately 200 feet. After completing the slow roll, A/C Conley had much excess speed and pulled up attempting to rejoin the other airplanes. A/C Conley's airplane collided with [Lt. Allison's] airplane, causing both airplanes to crash."

3-28-43E. Winslow, Arizona. At 1205 MWT, two Vultee BT-13A airplanes collided in mid-air 30 miles west of Winslow, Arizona, killing F/O Keith W. Travis and civilian instructor George R. Leech aboard BT-13A #41-1586. Glider pilot-rated passenger F/O Norman H. Weber was killed attempting to parachute to safety from BT-13A #41-1602. Pilot 2Lt. Joseph W. McClellan was able to parachute to safety and was uninjured. Investigators stated, "Capt. R.E. Brett, leading a flight of seven BT-13As, was returning to Twenty-Nine Palms, California, from Albuquerque, New Mexico. They had towed six gliders to Albuquerque. A stop was made at Winslow, Arizona, for gasoline. The airplanes took off from Winslow at approximately 1145 MWT, and [climbed] to 10,000 feet [in] a loose V-formation. [2Lt. Victor W.] Healy was in the No. 3 position, Lt. McClellan was in the No. 5 position and Mr. G.R. Leech was in the No. 7 position. Lt. Healy noticed a C-47 approaching from the west and as he was not sure of its altitude, broke the formation to the left. His passenger, Glider Pilot [F/O James J.] Kirsch, watched the C-47 go by, and in doing so, saw airplane #41-1586, piloted by Mr. Leech, pull to the right and up into airplane #41-1602, piloted by Lt. McClellan, and saw fragments begin to fly from both ships. Lt. McClellan bailed out immediately after the collision and landed safely. Passenger F/O Weber [bailed out at an altitude too low to allow his parachute to deploy]. His parachute apparently opened just as he reached the ground. No reason was uncovered to explain his delay in jumping and an examination of his parachute showed no evidence of malfunctioning. The passenger in airplane #41-1586, F/O Travis, made no attempt to jump and from the testimony given it is the opinion of the Committee that he was fatally injured in the collision."

3-28-43F. Augusta, Georgia. At 1425 EWT, a Vultee BT-13A suffered a catastrophic structural failure and crashed ten miles east of Bush Field, Augusta, Georgia, killing students A/C Vincent C. Dill and A/C John Carl Brent. The airplane had taken off from Bush Field on an instrument cross-country flight. The airplane apparently entered a power-spin at an undetermined altitude, describing several turns of spin before recovering with full power still applied. The port wing failed just after the recovery and the airplane went out of control. The airplane smashed to the ground and erupted into flames, killing the students instantly. Investigators noted that the students had crashed outside the authorized local flying area. Investigators determined that the students had exceeded the design limitations of the airplane, causing the port wing to fail at the point where the outer wing panel joined the wing center section.

3-28-43G. Merced, California. At approximately 1430, a Vultee BT-13A crashed 13 miles northeast of Merced, California, killing student A/C Earl A. Greving. The student had taken off from the Army Basic Flying School at Merced at about 1400 on a routine training flight. When the student did not return by 1600 he was declared missing. The airplane wreckage was found in the bottom of a small gully at 0830 the next day. The airplane had struck the ground in a vertical position at a speed estimated to be 200 mph. The airplane did not burn but was destroyed utterly.

3-28-43H. Baton Rouge, Louisiana. At 2051, a Curtiss P-40M flew into the ground one mile east of Harding Field, Baton Rouge, Louisiana, killing pilot 1Lt. Willis J. Dance, Jr. The airplane had taken off from Harding Field on a routine night transition flight. The Aircraft Accident Classification Committee stated, "Shortly after taking off, [at an altitude of about 300 feet agl] the airplane went into a turn to the right and down from which it never recovered. Upon impact with the ground, [the airplane] burst into flames and was completely destroyed. The pilot was thrown a few feet away from the main wreckage and was fatally injured. In the opinion of the [Committee], the pilot was looking down into the cockpit to make some adjustment such as retracting the landing gear, and was not aware of the airplane's altitude until too late."

3-28-43I. Roswell, New Mexico. At 2100, a Cessna AT-17 crashed five miles west of the Army Air Field at Roswell, New Mexico, killing student A/C Klyde H. Young and seriously injuring A/C Carl A. Lorenzo. The airplane had been flying in its zone and was called in to land by the control officer at Roswell. A short time later, the airplane apparently flew into the ground on the entry leg for Runway 17 with power and under control. The airplane shed pieces as it careened across the terrain for 100 yards before coming to rest. There was no fire but the airplane was completely destroyed.

3-29-43A. Ellensburg, Washington. At 0245 PWT, a Boeing B-17F collided with rising terrain 16 miles east of Ellensburg, Washington, killing the crew of nine. Killed in the crash were: 2Lt. Derek Bovingdon, pilot; 2Lt. Andrew L. Bowman, co-pilot; 2Lt. Ennis D. Hake, navigator; 2Lt. Eryle W. Boettcher, bombardier; SSgt. Max E. Miller, engineer; SSgt. William Miller, radio operator; SSgt. Edward R. Miller, assistant engineer; SSgt. Ralph C. Wagner, gunner; Sgt. Ray W. Stitzel, gunner. The B-17F had taken off from the Army Air Base at Ephrata, Washington, at 0037 PWT on a training mission to practice night landings, instrument flight, and dry bomb runs. Lt.

Bovingdon made two landings and then at 0200 requested clearance to proceed to the bombing range. Clearance was granted and Lt. Bovingdon and his crew were instructed to make a position report every half-hour. At 0240, the B-17 crew had made a contact request, which was acknowledged by Ephrata AAB. The B-17 radio operator gave the signal indicator and began a routine position report, which, according to the ground operator, suddenly cut off in mid-transmission at 0245. No further radio contact was made with the ship. At 0310, an explosion was noted about five miles northeast of Ephrata AAB. Control Officers then called in all of the B-17s that were operating from Ephrata AAB, but they were unable to contact the subject airplane, B-17F #42-29500, or B-17F #42-5313. At first it was thought that these two aircraft had collided in mid-air and crashed together but investigation of the explosion revealed that only the wreckage of B-17F #42-5313 was present at the crash site (See 3-29-43B). A Civil Air Patrol search aircraft located the wreckage of B-17F #42-29500 near Whiskey Dick Mountain on 4-1-43. The Aircraft Accident Classification Committee stated, "Pilot evidently flew into a mountain ridge as furrows indicate propellers dug into the ground on near top of side of ridge and wreckage was discovered on top and down the far side of the ridge.... Investigation at the scene of the accident revealed the fact that the airplane had been flying approximately due west when it hit the ridge about 40 feet from the top, the propellers cutting furrows for a distance of approximately 250 feet. The airplane then bounced over the ridge and the main portion of the wreckage was strewn from the top of the ridge down the far side and covered an area of approximately 150 yards in width and 500 yards in length. The elevation of the terrain where this airplane crashed is 3,520 feet above sea level. After the crash, fire evidently broke out in one wing panel and in the wrecked control cabin. There was no evidence of other fire. It is the opinion of this Accident Committee that the accident occurred while the pilot was practicing instrument flying, that the pilot did not know his actual position with relation to his distance from the airfield and that the pilot was flying the Ephrata Range although he had not been cleared to do so." The ceiling at the crash site at the time of the accident was reported to be about 3,300 to 5,000 feet and was thought to be a contributing factor in the accident.

3-29-43B. Ephrata, Washington. At 0310 PWT, a Boeing B-17F flew into the ground five miles northeast of the Army Air Base at Ephrata, Washington, killing 11 fliers. Killed in the crash were: 2Lt. Marvin R. Furch, pilot; 2Lt. Joseph J. Whalen, co-pilot; 2Lt. James A. Calnon, observer-pilot; TSgt. Harold L. Mason, engineer; TSgt. William T. Freeland, radio operator; SSgt. Joseph E. Brielman, assistant radio operator; SSgt. Harold W. Bunce, gunner; SSgt. John G. Bowen, assistant engineer; TSgt. Herman E. Stephens assistant engineer; Sgt. Abraham Kaden, passenger; Sgt. John D. Naus, passenger. Investigation revealed that B-17F #42-5313 took off from Ephrata Army Air Base at 0254 PWT on a one hour and thirty minute local training mission to practice night landings and night instrument flight on the Ephrata Range. There was no radio contact with this B-17 after take-off. At 0310, an explosion was noted five miles northeast of Ephrata AAB. Control Officers then called in all of the B-17s that were operating from Ephrata AAB, but they were unable to contact the subject airplane, B-17F #42-5313, and B-17F #42-29500. At first it was thought that these two aircraft had collided in mid-air and crashed together but investigation of the explosion revealed that only the wreckage of B-17F #42-5313 was present at the crash site. Investigation of the wreckage revealed that B-17 #42-5313 had dove into the ground and exploded violently, scattering wreckage and body parts over an area of one-half mile long by 200 yards wide. It was speculated that the pilot probably did not transition from contact to instrument flight, or from instrument to contact flight, soon enough to prevent the airplane from flying into the ground. Civil Air Patrol search aircraft located the wreckage of B-17F #42-29500 on a ridge near Whiskey Dick Mountain, 16 miles east of Ellensburg, Washington, on 4-1-43 (See 3-29-43A).

3-29-43C. Kershaw County, South Carolina. At 0834, four paratroopers collided with a Douglas C-47A that was participating in a paratroop drop at Kershaw County, South Carolina, killing three paratroopers and injuring another. The C-47A was so damaged that it had to make an emergency forced landing four miles southwest of Camden, South Carolina. The pilot received minor injuries and the remaining four crewmembers were uninjured. Sixteen paratroopers were able to parachute safely from the subject C-47. Investigation revealed that the airplane was flying in the number-three position of a nine-ship V-formation of C-47s that was participating in a simulated parachute invasion of Kershaw County. The formation had been flying at 200 feet above the terrain and then climbed to an altitude of 600 feet agl, slowing to about 100 mph for the parachute drop. Just as the operation began, the number-three C-47 encountered severe propeller turbulence and began descending to the right. Just as the pilot began to regain control of the airplane, a para-pack container that had been released by the lead C-47 struck the starboard engine, damaging the propeller and rendering the engine inoperable. In quick succession three paratroopers collided with the C-47. A falling paratrooper struck the starboard wing and his parachute became snagged on the wing. He remained attached to his parachute and was dragged along by the aircraft. Another falling paratrooper smashed into the fuselage just above the pilot's windscreen, caving in the cockpit ceiling. He was killed instantly upon impact. Another falling paratrooper collided with the port engine, causing his parachute to become fouled on the port propeller. The port propeller severed his parachute shroud lines and he fell to his death without a parachute. The port engine became inoperable and the pilot could not maintain flying speed. The pilot elected to make a wheels-up emergency forced landing four miles southwest of Camden. The paratrooper who was still connected to his parachute that was fouled on the starboard wing was apparently dragged to his death during the crash landing. Paratrooper Pvt. Reed D. Hughes struck some part of the airplane, or collided with the damaged para-pack or a body, sustaining serious injuries. Paratroopers Sgt. H.R. Callahan, Cpl. C.A. Hedlins and Pfc. A.N. Elliot were killed in the accident. Pilot 2Lt. Earl C. Huffington received minor injuries when the paratrooper struck the cockpit. Copilot W.R. Ennis, engineer Sgt. C.J. Backstrom, radio operator Cpl. L.D. Barshal and engineer-rated passenger Pfc. J.W. Crise were uninjured in the crash landing.

3-29-43D. Benhaden, Florida. At 0900 EWT, a Curtiss P-40K suffered a catastrophic structural failure and crashed near Benhaden, Florida, killing pilot 2Lt. Everett R. Edwards. The airplane was flying with another P-40 at 12,000 feet when the flight leader started a series of aerobatic maneuvers. When the flight leader looked back after the maneuvers, the wingman was not there. Investigation of the wreckage revealed that both wings had separated from the airplane before contact with the ground. Wreckage was scattered over a great area. Investigation revealed that the pilot had been smashed through the cockpit canopy with great force after the wings left the airplane and the fuselage structure collapsed and buckled.

3-29-43E. Shaw Field, South Carolina. At an unknown time after 1130 EWT, a Vultee BT-13A crashed into a remote swampy area 16 miles southwest of Shaw Field, Sumter, South Carolina, killing A/C Warren G. Nichols, 22, Brooklyn, New York, and A/C Keith Wayne Jessup, 21, Gary, Indiana. The Aircraft Accident Classification Committee stated, "A/C Nichols and A/C Jessup were dispatched at 1040 on an instrument team ride with A/C Nichols under the hood in the rear cockpit and A/C Jessup as observer in the front cockpit. The mission was a dead reckoning flight at 3,500 feet from Shaw Field to Bishopville, South Carolina, to Manning, South Carolina, to St. Matthews, South Carolina, [and return] to Shaw Field. No further report on the ship was obtained and it was classified as missing at 1530. A search by air was begun and continued until dark. The search continued the following day and at approximately 1530 the airplane was sighted in the Santee River swamp near the junction of the Wateree and Congaree Rivers. The point of the crash was approximately on the course from St. Matthews to Shaw Field. Judging from the marks on the vegetation in the swamp, the airplane had not spun or dived in, but had come in straight with considerable horizontal velocity and low vertical velocity. The airplane was headed west at the time of the crash. The course from St. Matthews to Shaw Field is 37 degrees. Surface winds were generally southwest at the estimated time of the crash. From the above facts it appears likely that a forced landing was being attempted. No other reason can be seen for turning off course when the ship was within ten minutes of completion of the mission. The airplane is completely wrecked and partially submerged in a dense swamp. Although an emergency forced landing appears probable, it cannot be definitely classified as such, since the approach was made toward the middle of the swamp rather than toward the edge. The swamp is approximately five miles across, and the airplane crashed approximately two miles from the eastern edge while it was headed west."

3-29-43F. Hopkinsville, Kentucky. At 1115 CWT, a Bell P-39D crashed one and one-half miles south of Hopkinsville, Kentucky, killing 2Lt. Walter H. Bell, 22, Elkton, Kentucky. The pilot had been buzzing the home of his uncle, Thurston Layne, at Lafayette Road just before the accident, making diving and climbing passes at treetop level. The pilot attempted a steep turn at about 300 feet when the airplane flipped to an inverted attitude. The nose fell toward the ground and the airplane smashed into the ground in an inverted position on the right of way of the Tennessee Central Railroad, exploding into flames and scattering wreckage for several hundred feet. After the initial impact, the pilot had been hurled through the air for over 150 feet, clearing a coal car before slamming into the ground dismembered and decapitated. The pilot had taken off from Godman Field, Kentucky, on a local training flight.

3-29-43G. Eureka, Texas. At 1130 CWT, a Fairchild PT-19A suffered a catastrophic structural failure of the port wing and crashed four miles north of Eureka, Texas, killing civilian instructor Thomas H. Coombe and student A/C Julius Frankel. The airplane had taken off from the Army Air Field at Corsicana, Texas, on a routine training flight. Investigation revealed that the airplane had leveled out at about 1,000 feet agl after completing a diving turn. The port side of the center wing section broke off near the fuselage joint, causing the airplane to go out of control and crash to the ground. The student and instructor were killed instantly. The port wing, the port center section and port landing gear were found a few hundred feet from the main wreckage.

3-29-43H. Marfa, Texas. At 1715, a Cessna AT-17 collided with an antenna pole on an airfield radio truck and crashed at the South Auxiliary Airfield near Marfa, Texas, killing instructor 2Lt. William F. Hoover, Los Angeles, California, and A/C George Hugo Hallin. 2Lt. Howard B. Yeager, a pilot-rated bystander, was seriously injured attempting to rescue the two flyers. Investigation revealed that the airplane had taken off on Runway 17. The airplane then made a 180-degree turn and flew parallel with the runway at low altitude and high speed. The airplane flew toward the radio truck and pulled up abruptly at the last moment, clearing the truck but striking the top three inches of the 25-foot radio antenna pole being used by the truck with the port wing. Pieces of the port wing began separating from the airplane and it went into an uncontrolled roll to the left, striking the ground upright ten yards south of the north end of Runway 17. The airplane bounded back into the air over a small drainage ditch, slamming back to earth 25 yards further on. The airplane burst into flames as it skidded through a fence and came to rest about 85 yards from where it made the second contact. Pieces of the airplane were scattered over a wide area. The engines came to rest 90 yards and 30 yards ahead of the main wreckage. Lt. Yeager pulled both fliers out of the airplane alive. Instructor Lt. Hoover died at approximately 2200 the day of the accident. Cadet Hallin died at the Marfa Army Air Field station hospital on 4-1-43 with his wife present.

3-29-43I. San Antonio, Texas. At 1831, a Fairchild PT-19A crashed three miles southeast of Kelly Field, San Antonio, Texas, killing pilot 2Lt. George C. Duvall. Pilot-rated passenger, 2Lt. Rueben H. Gross, received minor injuries. The airplane took off from Kelly Field and climbed to an altitude of 1,200 feet. The pilot then dove to extremely low altitude and attempted a right turn in close proximity to the ground. The starboard wing struck the terrain and the airplane cartwheeled into the ground.

3-29-43J. Garden City, Kansas. At 1840 MWT, a Vultee BT-13A crashed 22 miles northeast of the Army Air Field at Garden City, Kansas, killing instructor 2Lt. Hansford L. Simmons and student A/C Mack W. Young, Jr. The airplane had taken off from Garden City at 1720 on a routine training flight and failed to return, being declared missing when the fuel was estimated to have run out. The airplane was found the next day. Investigation revealed that the airplane was completely destroyed when it smashed to the ground in a vertical manner. It was speculated that the airplane had entered a spin at an altitude too low to allow for recovery. There were no witnesses to the accident.

3-29-43K. Georgiana, Alabama. At 2030, a North American AT-6C crashed 12 miles west of Georgiana, Alabama, killing student A/C Richard J. O'Conner. The airplane had taken off from the Army Air Field at Marianna, Florida, on a night cross-country mission. A civilian witness on the ground observed the airplane circle a farm field three times at low altitude with landing lights illuminated. The airplane was apparently making an approach for an emergency forced landing when the student realized that he was going to collide with a farmhouse. He pulled up abruptly to avoid the house and stalled the airplane. The airplane fell off in a spin and smashed to the ground about 150 feet beyond the house. The airplane did not burn but was destroyed utterly. It was speculated that the student might have been experiencing engine trouble.

3-29-43L. Columbus, Mississippi. At 2103, a Beech AT-10 crashed while attempting a landing at the Auxiliary Airfield #1, Columbus, Mississippi, killing A/C H.E. Walker and A/C R.K. Smith. Investigation revealed that the airplane was turning in the traffic pattern at an altitude of 800 feet when it stalled and entered a spin to the ground. It was noted that it was a dark, moonless night.

3-30-43A. Malvern, Alabama. At 0820, a North American AT-6C suffered a catastrophic structural failure and crashed near Malvern, Alabama, killing A/C Jack N. John. A/C Lowell Schurr was able to parachute to safety. The airplane had taken off from Napier Field, Dothan, Alabama, on an instrument training flight. The airplane was flying at an indicated altitude of 6,700 feet with A/C Schurr flying under the hood. After flying on instruments for about 40 minutes, A/C Schurr turned the aircraft over to A/C John. A/C Schurr stated that A/C John immediately put the airplane in a steep peel-off maneuver. The port wing panel and part of the port center section sheared off and struck the fuselage as the airplane dove down. A/C Schurr reported that something crashed into the left side of the fuselage. He stated that he had opened the canopy and unfastened his safety belt, but does not remember falling free as the airplane broke up and spun to the ground. The next thing he remembers is waking up in a farmhouse surrounded by people. A/C John spun in with the airplane and was killed upon impact. Investigators stated that A/C John had probably exceeded the design limits of the airplane during the maneuver, causing the port wing to fail and separate.

3-30-43B. Laredo, Texas. At 1250, a North American AT-6C crashed 20 miles NNE of the Army Air Field at Laredo, Texas, killing pilot 2Lt. Clyde W. Smith, 23, Dallas, Texas, and student gunner Pvt. Casper Agnew, 22, New Albany, Indiana. Investigators stated, "2Lt. Clyde William Smith took off on a routine gunnery training flight at 1229 and proceeded to the Aerial Gunnery Range. After the gunner had fired his first 100 rounds Lt. Smith pulled out of firing position to give the next gun ship the position to fire from. When pulling up in the rear of the formation, Lt. Smith fell off into a spin to the right, seemed to recover for an instant, and then went into a spin to the left. The airplane hit the ground in a hard skid completely demolishing the airplane and killing Lt. Smith and Pvt. Casper Agnew instantly."

3-30-43C. Byran, Texas. At 1315, a Vultee BT-13A collided with a road grader operator while attempting a take-off at Byran Field, Byran, Texas, killing the civilian operator Jesse D. Clay. Pilot Capt. Robert J. Seabolt and pilot-rated passenger 1Lt. R.G. Smith were uninjured. The Accident Classification Committee stated, "At approximately 1310, instructor Capt. Seabolt and [instrument] student Lt. Smith taxied out to the North end of the North-South Runway to take-off on the instrument training flight. The take-off was to be an instrument take-off. As the airplane took off and gained flying speed, it was being flown by the student who was under the hood. As the airplane left the ground and started drifting toward the west side of the runway, the instructor took over the controls and attempted to control the airplane so that it would continue in the general direction of the runway. He was unable to do so probably due to the fact that the wind was gusty and slightly from the east. A road grader pulled by a tractor was working about 3 feet from the west side of the runway the right wing struck the operator of the road grader, killing him instantly. The pilot recovered control of the airplane and landed on the Northwest side of the field. The accident was an unavoidable accident, and no one was at fault. The work that was being done by the civilian employees incident to the installation of a badly needed temporary lighting system for the field, which was essential to the removal of the hazard presented by the open ditch along the runway. The basic underlying cause of the accident was the exigency of war and the pressing urgency which required that training begin before the field was entirely completed."

✓ **3-30-43D. Arcadia, Florida.** At 1430 EWT, a Stearman PT-17 crashed five miles east of Dorr Field, Arcadia, Florida, killing civilian instructor Kenneth E. Faxon and injuring student A/C William T. Wissing. Investigators stated, "It is believed that the instructor and student were performing stalls without having commenced at the minimum altitude of 2,500 feet, continued a series of stalls without clearing the area beneath themselves, and without checking the altitude. It is further believed that the instructor, after having his student perform these stalls, again without clearing the area or checking the altitude, had the student enter a spin, and without realizing the low altitude failed to commence recovery until too late."

3-30-43E. Penelope, Texas. At 1530, a Vultee BT-13A crashed three miles southeast of Penelope, Texas, killing student A/C Willis A. Shaffer, Jr. Investigation revealed that the airplane had entered a spin to the right at an altitude too low to allow recovery. The student was killed when he attempted to parachute to safety just before the airplane smashed into the ground. The student was found near the tail section with his parachute partially open. The ripcord D-handle was found a short distance from the body.

3-30-43F. Greenville, Texas. At 1615, a Vultee BT-13A crashed ten miles south of Majors Field, Greenville, Texas, killing student pilot 1Lt. Douglas M. Boudeman and seriously injuring instructor 2Lt. Harold H. Fox. The airplane had apparently spun to the ground. There were no witnesses and the surviving instructor could not remember the events leading up to the accident. Investigators noted that the instructor was convinced that he had crashed the airplane while attempting a landing at Majors Field.

3-30-43G. Challis, Idaho. At 2010 PWT, a Boeing B-17F was abandoned in flight and at approximately 2110 PWT crashed into a mountain at an elevation of 8,000 feet, 22 miles southeast of Challis, Idaho. Eight crewmembers were able to parachute to safety. Engineer SSgt. Henry G. Van Slager was declared missing and was presumed killed in an unsuccessful parachute jump. His body was not found. Five members of the crew were able to find shelter at a remote forest ranger's mountain cabin about 50 miles northwest of Challis. They were able to eventually telephone for help and were rescued after spending two nights in the cabin. Crewmembers rescued at the cabin were: F/O Harold E. Thompson, Springfield, Oregon, co-pilot; 2Lt. Austin, Broken Bow, Oklahoma, navigator; SSgt. Morris Becker, Ozone Park, Long Island, New York, radio operator; SSgt. Harvey R. Wiegard, Walla Walla, Washington, assistant radio operator; SSgt. Erwin W. Grundman, Compton, California, gunner. Pilot 2Lt. Joseph R. Brensinger, bombardier 2Lt. George W. Smith and assistant engineer SSgt. Howard A. Pope were found scattered through out the area. They were not injured. Investigators stated, "At approximately 1300 PWT, B-17F #42-29514 took off on a calibration mission from Walla Walla Army Air Base, Washington. The airplane was fully serviced with gas and oil according to the Form-1A. After completing the calibration mission, the airplane returned and landed at Walla Walla and let the instructor-navigator out. After the instructor-navigator got out, they taxied right out and took off at approximately 1530 PWT to do some local flying and practice instrument flying. After running into some bad weather and flying on instruments they became lost. Due to static conditions, they could not make radio contact. When the gasoline got low in the main and wing tanks, an effort was made to transfer gasoline from the bomb bay tanks, which were presumed to be full. Upon examination, the bomb bay tanks were found to be empty. The pilot then ordered everyone to bail out, he himself being the last to leave the ship after setting the A.F.C.E. [automatic flight control equipment]. Before jumping at 2030 PWT, the pilot made sure everyone was out of the ship. The airplane then continued to circle and crashed into the side of a mountain approximately 8,000 feet [elevation] at approximately 2210 MWT. In the search that followed, eight members of the crew were located alive in a primitive area approximately 50 miles northwest of Challis, Idaho. Aerial search for Sgt. Van Slager was abandoned after 18 days, during which no trace of him could be found." An AAF Beech C-43 sent to search for the missing crew vanished on 4-5-43. The C-43 had crashed into the side of a mountain 50 miles west of Challis, Idaho, killing two AAF officers and a forest ranger. The C-43 wreckage was found on 4-13-43. See 4-5-43E.

3-30-43H. Kerrville, Texas. At 2345, a Beech AT-7 flew into rising terrain 40 miles west of Kerrville, Texas, killing pilot 2Lt. Ernest B. Paxton and crewmember Sgt. Clarence L. Crandall. Navigator 2Lt. Alfred E. Conradt was seriously injured. Navigator 1Lt. Cecil L. Tolleson and Sgt. Arvel D. Wright received minor injuries. Investigators stated, "The airplane was returning on a night navigation mission from Hensley Field, Texas, to the AAF Navigation School, Hondo Army Air Field, Hondo, Texas. At approximately 2300, the air became quite rough and celestial observations became impossible. At this time the airplane was flying at an altitude of approximately 8,000 feet. 1Lt. Tolleson determined from radio fix positions that the aircraft was to the right of course. The pilot, 2Lt. Paxton, believed the aircraft was left of course. At this time, due to the roughness of the air, the pilot was advised to return to Hondo by means of radio. The pilot started his descent. Sometime between 2330 and 2345, 1Lt. Tolleson observed the altimeter reading at 2,500 feet. The altimeter setting on the navigator's altimeter was correct according to the reading at Hondo at the time of the crash. The altitude at the scene of the crash was 2,495 feet. Immediately following [checking the altitude] the pilot started an ascent. It was at this time that the crash occurred."

3-31-43A. Las Vegas, Nevada. At 0730 PWT, two North American AT-6A type airplanes collided in mid-air 30 miles NNW of the AAF Gunnery School at Las Vegas, Nevada, killing F/O Edward J. Laird aboard AT-6A #41-534. Student gunner Pvt. C. Ulery, Jr. received minor injuries parachuting to safety. Pilot 2Lt. Lyle D. Crawford and student gunner Pvt. L.J. Parmo were uninjured parachuting to safety from AT-6A #41-16312. The two airplanes were part of a five-ship flight on a gunnery-training mission. The formation was flying in trail at 8,500 feet msl in the area where they were to rendezvous with the target tow ship. When the formation met with the target tow ship, the leader (AT-6A #41-534) signaled that he was leaving the formation to fire on the target ship. The lead airplane then made a turn to the right to get behind the target ship, which was flying ahead and to the right of the formation. The second airplane followed the lead airplane. When the lead airplane leveled out, the second airplane did not and collided with the tail of the lead airplane. Both airplanes went out of control and crashed.

3-31-43B. La Grange, Georgia. At 0915 CWT, a Vultee BT-13A crashed one mile northeast of La Grange, Georgia, killing A/C William T. Holliday. The student had taken off from Gunter Field, Montgomery, Alabama, on a routine cross-country. The airplane approached the La Grange Municipal Airport at 3,000 feet and requested landing instructions. The student was advised that the field was operating with a traffic pattern to the left and was instructed to land to the south. The student, flying at an altitude of 700 feet, entered a traffic pattern to the right. When the student turned on the base leg, the control ship ordered him to fly a left hand pattern. The student gave the airplane throttle and pulled up abruptly. The airplane stalled and entered a spin at an altitude too low to allow recovery. The airplane smashed to the ground about one mile north of the airport, killing the student instantly.

3-31-43C. Houston, Texas. At 0945, two Beech AT-10 type airplanes collided in mid-air at Ellington Field, Houston, Texas, killing instructor 2Lt. Jack A. Wheeler and student A/C Carlton P. Wilkins aboard AT-10 #41-26291. A/C Sam N. Webb and A/C Warren Webbeking were able to land AT-10 #41-26469 at Ellington Field and were uninjured. Both airplanes were attempting to land on Runway 13 and were on the approach leg at an altitude of about 400 feet agl. Investigation revealed that control tower personnel noticed that the airplanes were too close to each other to make a safe landing. The tower ordered both airplanes to retract the landing gear and go around. AT-10 #41-26291 then pulled up into AT-10 #41-26469. The starboard wing of AT-10 #41-26291 was severed and fell away. AT-10 #41-26291 climbed slightly then fell away and smashed into the ground where it erupted into flames upon impact. AT-10 #41-26469, with its starboard wing severely damaged, was thrown into a vertical bank to the right. The students recovered at extremely low altitude and with great difficulty they were able to land the airplane.

3-31-43D. Big Spring, Texas. At 1345, a North American AT-6C crashed at Big Spring, Texas, killing pilot 1Lt. James H. Brown and seriously injuring 1Lt. Wesley W. Harris. Investigation revealed that the airplane was suffering engine trouble and the pilot was attempting to make an emergency forced landing when he flew into the ground. The airplane skidded at a right angle across the furrows of a plowed field for a short time before flipping over to an inverted position. Investigators noted that the pilot-rated passenger could not remember the accident or the events leading up to it.

3-31-43E. Canaan, Connecticut. At an unknown time between 1534 and 1604 EWT, a Curtiss P-40E flying in instrument conditions crashed into a remote ridge one-quarter mile east of the Childs' Estate near Canaan, Connecticut, killing pilot 1Lt. Daniel H. Thorson. The airplane had taken off from Mitchel Field, Long Island, New York, and disappeared while on a ferry flight to Bradley Field, Windsor Locks, Connecticut. The airplane was discovered on 4-24-43. Investigators speculated that the pilot had gotten lost and flew into the mountain while attempting to get below the ceiling.

3-31-43F. Location Unknown. At a time after 1705 EWT, a North American BT-14 disappeared while on an administrative flight from Farmingdale, Long Island, New York, to Trumbull Field, Connecticut. The pilot, Capt. William J. Hennon, was presumed lost at sea. The pilot had been in contact with Republic Tower, New York, at 1705 and was never heard from again. Investigators speculated that the airplane had probably suffered engine failure and crashed into the Atlantic Ocean. It was also discovered that the pilot's proposed route was reported as instrument conditions at the time of the disappearance. The search for Capt. Hennon was discontinued on 4-16-43.

3-31-43G. Coffeyville, Kansas. At 1625 CWT, a Vultee BT-13A crashed 14 miles ENE of Coffeyville, Kansas, killing instructor 2Lt. William M. Miener and student A/C Larkin G. Henry. Investigators stated, "The student and instructor were engaged in low altitude air work, probably elementary-eights at 200 feet altitude. During this maneuver the instructor either gave the student a simulated forced landing or was demonstrating that maneuver. It is also possible that he had an actual forced landing. Then, during the rapid climb and turn out of the elementary-eight into the turn for position to land into the wind, he stalled the airplane. His recovery was not made rapidly enough to regain flying speed before the complete loss of altitude, and he crashed into the ground."

3-31-43H. Yuma, Arizona. At 1615 MWT, a North American AT-6C crashed 20 miles northeast of Yuma, Arizona, killing students A/C Robert L. Copeland and A/C Earle H. Cunningham. Investigators stated, "While on a routine instrument team ride A/C Cunningham and A/C Copeland came below the specified minimum altitude of 5,000 feet and in the opinion of the board were deliberately participating in extremely low flying. Statements of Officer personnel who witnessed the crash show that the cadets made an extremely low pass at a group of men below and came up in an extremely low 180-degree wingover. The airplane clipped the top of a low hill, hit the top of a second hill and crashed into a third hill, resulting in the death of the Aviation Cadets." The airplane was completely destroyed and wreckage was scattered over a wide area.

3-31-43I. Bakersfield, California. At 1825, two Vultee BT-13A type airplanes collided in mid-air one and one-half miles east of Minter Field, Bakersfield, California, killing instructor 2Lt. Alvah L. Fisher and student A/C Farion M. Williams aboard BT-13A #41-23122. Student A/C George K. Mount was able to safely land BT-13A #41-10939 at Minter Field. He was uninjured. Investigators stated, "Subject airplane #41-10939 was being flown at an indicated altitude of 1,000 feet on the 45 degree approach to the downwind leg of the traffic pattern, flying with flaps down and propeller in low pitch when it was overtaken from the right rear by airplane #41-23122 flown by a student and instructor. The left wing tip of [BT-13A #41-23122] made contact with the right landing gear strut of [BT-13A #41-10939], tearing off five feet of the left wing of [BT-13A #41-23122]. [BT-13A #41-23122] crashed and burned with fatal results to personnel. [BT-13A #41-10939] continued flight and landed without further damage."

April

4-1-43A. Murrieta, California. At 1615, a Lockheed RP-38E dove into the ground at Murrieta, California, killing pilot 2Lt. Everett P. McBeth. The P-38 had taken off at 1515 PWT from March Field, Riverside, California, on a transition flight. Investigation revealed that the airplane had dove into the ground at high speed at a steep angle from an undetermined altitude, exploding violently into flames upon impact. Investigators could not determine why the airplane crashed. It was speculated that the pilot had put the airplane in a diving maneuver from which he could not recover because of the tendency of the P-38 type airplane to suffer from severe control surface buffeting caused by compressibility. Investigators noted that the pilot had very little P-38 flying time.

4-1-43B. Camp Polk, Louisiana. At 2158, a Douglas A-24A crashed 14 miles southeast of Camp Polk, Louisiana, killing pilot 1Lt. Elmer R. Cawthon, Jr. The airplane had taken off from DeRidder Army Air Base, DeRidder, Louisiana, on routine training flight. The airplane was observed flying over the Camp Polk area about 50 minutes after take-off. Several witnesses stated that the airplane's engine had quit momentarily and then had restarted before the airplane dove into the ground at a 70-degree angle. The A-24A exploded violently into flames upon impact, killing the pilot instantly.

4-1-43C. Clovis, New Mexico. At 2121 MWT, a Consolidated B-24D crashed near Clovis, New Mexico, killing the crew of nine. Killed in the crash were: 1Lt. Ray G. Hanson, pilot; 2Lt. J.W. Foster, co-pilot; 2Lt. Fred Scagnetti, navigator; 2Lt. H.J. White, bombardier; TSgt. P.H. Pricer, engineer; TSgt. R.H. Blume, radio operator; SSgt. V.D. Berreth, radio operator; SSgt. J.M. Day, gunner; SSgt. W.H. Block, gunner. The B-24D had taken off from Clovis Army Air Base on a night navigation mission. Immediately after take-off the airplane suffered a failure of the number-one engine. The pilot feathered the propeller, trimmed the airplane and attempted to re-enter the traffic pattern. Attempting to enter the downwind leg of the traffic pattern, the pilot made a steep left turn at low altitude and apparently flew into the ground south-southwest of the field. The B-24 exploded into flames upon impact, scattering wreckage and bodies over a wide area.

4-2-43A. Sebring, Florida. At 1000 EWT, a Boeing RB-17B crashed 19 miles south of Hendricks Field, Sebring, Florida, killing the crew of five. Killed in the crash were: Capt. Joseph B. Chrishoph, pilot; 2Lt. James H. Miller, co-pilot; 2Lt. Ralph S. Massey, co-pilot; Capt. John C. Yeuell, co-pilot; Cpl. Tommy F. Cherry, engineer. The airplane had taken off from Hendricks Field on a routine transition flight. Investigators stated, "The pilot evidently realized that he was going to crash as the radio hatch and the pilot's compartment hatch were released and all throttles were cut. Apparently the pilot held the airplane above the treetops as long as possible, resulting in a gradual decrease of flying speed until the airplane again stalled and dove into the ground at approximately a 45 degree angle from a very low altitude as the wreckage indicated very little speed at the time of impact. It [was] thought that the floats in the carburetors were stuck when the airplane fell off into a spin or some other malfunction of the fuel system existed, which in turn resulted in a dead engine or engines and the inability of the pilot to maintain altitude after a recovery had been accomplished." The B-17 burst into flames upon impact and was completely destroyed.

4-2-43B. Seville, California. At 1047, a Ryan PT-22 crashed four miles east of Seville, California, killing student A/C Foster N. Anderson. It was speculated that the student was attempting a forced landing or a simulated forced landing in a large wheat field when the airplane stalled and entered a spin to the right at an altitude of 50 feet.

4-2-43C. Alexandria, Louisiana. At 1855, a Piper L-4A suffering engine problems crashed into the Red River six miles southeast of Alexandria, Louisiana, killing observer Cpl. B.F. Hawkins and injuring Liaison Pilot Cpl. William A. Halonan.

4-2-43D. Lake George, Florida. At 2130 EWT, a Douglas A-24A flying in poor weather crashed near Lake George, Florida, killing pilot 2Lt. Roy F. Dunn and gunner Sgt. Phillip J. Torella. The airplane had taken off at approximately 2030 EWT from Drew Field, Tampa, Florida, on a night navigation flight to Daytona Beach, Florida, to Gainesville, Florida, and return to Drew Field. The pilot encountered instrument conditions and apparently lost control of the airplane.

4-2-43E. Tullahoma, Tennessee. At 2125 CWT, a Douglas A-20C crashed two miles northeast of Northern Field, Tullahoma, Tennessee, killing pilot 2Lt. Clifford W. Hanson, Gary, Indiana. The airplane was on a scheduled night navigation mission. Investigation revealed that the pilot attempted the take-off with the elevator trim control set in the full nose-down position and the aileron trim tab set in the full starboard wing down position. The airplane began the take-off roll and the pilot had great difficulty in getting airborne. The pilot forced the airplane into the air and it flew for a short time before smashing into a wooded hill and exploding into flames.

4-3-43A. Fort Myers Beach, Florida. At 1020 EWT, a Curtiss P-40K crashed into the Gulf of Mexico surf at Fort Myers Beach, Florida, killing pilot 2Lt. Carl A. Seibel. The airplane had taken off from the Army Air Field at Sarasota, Florida, on an instrument training flight. The Aircraft Accident Classification Committee stated, "The pilot peeled off from approximately 2,000 feet and descended to the water level. The pilot proceeded down the beach, just off shore, pulled up over a pier, which extended from the beach out into the [Gulf of Mexico], and then dropped back down to water level. Immediately after this, the pilot started to make a turn away from the beach, but straightened out again. As the pilot did this, the airplane struck the water. The airplane caught fire and was completely destroyed. The pilot was killed."

4-3-43B. Rittenhouse Field, Arizona. At 0830, a Lockheed RP-322 crashed three miles northeast of Rittenhouse Field, Arizona, killing student A/C Benjamin G. Conrath, Jr. The student had taken off from Williams Field, Chandler, Arizona, on a routine training flight. The Aircraft Accident Classification Committee stated, "Subject airplane was first observed by the co-pilot of an AT-9 airplane which was several miles northwest of it at about 3,000 feet. He stated that the subject aircraft appeared to be in a vertical dive at high speed when first observed at an altitude of about 3,000 feet. He stated that he saw the top of the subject aircraft and watched its unwavering descent until it hit the ground, there [being] a cloud of dust and a burst of flame [visible]. A civilian ground witness claimed to have seen the aircraft coming down at about a 40-degree angle with the landing gear extended. Since the left propeller in the wreckage was found to be feathered, it is the opinion of the committee that engine trouble was experienced and the pilot, contemplating a forced landing in the desert, lowered the landing gear and while manipulating controls with his head in the cockpit, inadvertently dove into the ground." It was also speculated that the student had extended the landing gear in an unsuccessful attempt to arrest a dive that got out of control.

4-3-43C. Shreveport, Louisiana. At 2030, a Martin B-26 attempting a take-off crashed two miles east of Barksdale Field, Shreveport, Louisiana, killing the crew of four. Killed in the crash were: 1Lt. Roy F. Wanamaker, B-26 instructor; 2Lt. Dwight W. Corrier, co-pilot/B-26 student; SSgt. Robert H. Fry, engineer; SSgt. Edward E. Long, gunner. "The Aircraft Accident Classification Committee met and the following was determined: The aircraft was taking off on a night flight with a student pilot [flying in the pilot's seat] for the first time at night and the instructor flying in the co-pilot's seat. The airplane crashed in swampy ground, the trees completely demolishing the aircraft and forming a mat, preventing total submerging and permitting the aircraft to burn beyond possible study." Investigators could not determine the cause of the crash.

4-4-43A. New Bedford, Massachusetts. At 1205, a Republic P-47C crashed while attempting a dead-stick landing at the Army Air Base at New Bedford, Massachusetts, killing pilot 2Lt. Fred V. Bright, Jr. Investigators stated, "Pilot Lt. Bright took off from Green Field, Providence, Rhode Island, at 1106 in a P-47C-5 #41-6562 scheduled for a local aerobatic training mission. At approximately 1156 he called the

Providence Control Tower, said that he was in trouble and that he was going to attempt an emergency landing at a field that he saw under him. Circling the field at New Bedford, Massachusetts, and gradually losing altitude, the pilot made a steep stalling turn [to the left] with wheels and flaps down at an estimated altitude of 300 feet [agl], dropped out of the turn, and crashed approximately 100 yards from the edge of the runway."

4-4-43B. Corvallis, Oregon. At 1600 PWT, an Interstate L-6 crashed at Corvallis, Oregon, killing Liaison Pilot Pvt. Alvin L. Baer and passenger Pvt. Warren E. Buchholz. The airplane was observed flying at extremely low altitude prior to the accident. The airplane made another low-level pass and was pulled up in a steep climbing turn to the left. The airplane fell off to the left and dove into the ground, killing the occupants instantly.

4-4-43C. Avon Park, Florida. At 2050 EWT, a Martin B-26B crashed at the AAF Bombing Range at Avon Park, Florida, killing the crew of six. Killed in the accident were: 2Lt. Stanton W. Johnston, Evanston, Illinois, pilot; 2Lt. William A. McCarty, Jacksonville, Florida, co-pilot; 2Lt. John P. Shatrwka, Kenosha, Wisconsin, navigator; Sgt. Wallace E. Erler, Ridgewood, New York, engineer; SSgt. Edgar V. Cournoyer, West Warwick, Rhode Island, radio operator; Sgt. Wayne C. Martinek, Elk Rapids, Michigan, gunner. The Aircraft Accident Classification Committee stated, "B-26B #41-17930 crashed from flight during a practice bombing mission. The estimated altitude of the airplane just prior to the accident was 750 feet. Having completed the ninth bombing run across the target from the south to the north, the airplane was being flown in a turn to the east in preparation for a bombing run from east to west. The point of the airplane's contact with the ground is one and three-fourths of a mile from the north tower of the bombing range and approximately 75 degrees true from the tower. The airplane had reached a heading of some 120 degrees at this point. The airplane struck the ground in an inverted position at an angle of 40 to 55 degrees to the horizon. All evidence indicates that the aircraft was stalled in the turn, causing it to roll into a position from which the pilot was unable to recover. The cause of the stall cannot be definitely fixed. The two possibilities are a stall due to the pilot allowing the airplane to lose flying speed or stall due to engine failure. Both engines were damaged by the crash and subsequent fire beyond possibility of determining whether or not there was actually a power failure."

4-5-43A. Enid, Oklahoma. At 0930, a Vultee BT-15 crashed seven and one-half miles west of the Army Air Field at Enid, Oklahoma, killing pilots 2Lt. John E. Britton and 2Lt. Omar E. Bryant. The student instructors were apparently performing a simulated forced landing when the accident occurred. The airplane was seen "wobbling" from side to side at an altitude of 200 agl before climbing slightly then falling to the ground and bursting into flames. Investigation of the wreckage indicated that the student instructors had apparently stalled the airplane and entered a spin to the left at an altitude too low to allow recovery. Both pilots were killed instantly.

4-5-43B. Hamilton Field, California. At 1045, a Bell P-39N crashed two miles north of Hamilton Field, San Rafael, California, killing pilot 2Lt. Edward K. Forsythe. The airplane was returning to Hamilton Field after successfully completing a ground gunnery mission when the engine seized due to the loss of all its oil. The propeller continued to turn against the seized engine, snapping the drive shaft in two and causing the propeller torque tube to twist and distort the fuselage. The starboard armament panel blew off in the slipstream. The airplane then entered a spin, crashing to the ground and bursting into flames. The starboard door was found some distance from the wreckage, indicating that the pilot had contemplated leaving the airplane.

4-5-43C. Pinellas Field, Florida. At 1350 EWT, a Curtiss P-40F crashed at Pinellas Field, St. Petersburg, Florida, killing pilot 2Lt. Chester W. Baumgartner. Investigators stated, "As the pilot was taking off, and as the tail of the airplane came up, the tow target came loose from the airplane. As far as could be determined the engine was running smoothly at this time. Investigation of the tow-rope and connections show no reason for the rope dropping loose. It is impossible to determine whether the tow-rope dropped loose or the pilot pulled the release. The airplane took off and climbed to about 300 to 450 feet. The engine failed and from witnesses' statements, the pilot must have attempted to make a 180-degree turn to return to the field. In a diving turn, the airplane lost altitude and did not break the descent before hitting the ground."

4-5-43D. Bakersfield, California. At 1110, a Vultee BT-13A smashed into the ground 23 miles northwest of Minter Field, Bakersfield, California, killing student A/C William C. McDonald. The Accident Classification Committee stated, "The airplane was first observed at approximately 4,500 feet indicated altitude in a right hand spin, which continued without interruption until the airplane crashed. The pilot was found about 15 feet from the wreck with parachute released, indicating that he probably cleared the airplane and pulled the ripcord very near the time he made contact with the ground. The safety belt in the front cockpit was found unbuckled."

4-5-43E. Challis, Idaho. At 1447 MWT, a Beech C-43B collided with rising terrain near Roughneck Peak 50 miles due west of Challis, Idaho, killing two AAF flyers and a U.S. Forest Department Ranger. The airplane had come from Hill Field to assist in the search for the crew of a B-17 that had parachuted into the wilderness area approximately 50 miles west of Challis (See 3-30-43G). The airplane had taken off at 1447 MWT from the AAF emergency airfield at Challis to search for the B-17 crew. When the C-43B failed to return to Challis, it was declared missing. On 4-13-43, the airplane was found to have crashed into the side of a mountain very near the top at an elevation of approximately 8,500 feet msl. Pilot Capt. William Kelly, co-pilot 2Lt. Arthur A. Crofts, Jr. and forest ranger Charles J. Langer were killed in the accident.

4-5-43F. Sapelo Island, Georgia. At 1700 EWT, a North American B-25C crashed at Sapelo Island, Georgia, killing the crew of seven. The airplane had taken off from the Columbia Army Air Base, South Carolina, on a navigation-training mission. The Aircraft Accident Classification Committee stated, "The investigating officer talked to James Miller, a Coast Guard [enlisted man] who was on duty in a watch tower, and he stated that he sighted the airplane flying at approximately 300 feet, coming straight toward the tower. All at once there was a terrific roar from the motors and the airplane went absolutely straight up for approximately 1,000 feet, did a complete loop, and dived straight down into the beach." The airplane exploded into flames upon impact. The wreckage was submerged by high tide. Killed in the crash were: 2Lt. James W. Peterson, pilot; 2Lt. Henry H. Hoffman, co-pilot; 2Lt. Robert M. Reser, bombardier-navigator; 2Lt. Thomas H. Fail, bombardier-navigator; SSgt. William W. Colby, engineer; SSgt. Joseph T. Dervan, radio operator; Sgt. James S. Welch, gunner.

4-5-43G. Robins Field, Georgia. At 1915 EWT, a Vultee BT-15 crashed near Robins Field, Warner-Robins, Georgia, killing F/O Lloyd H. Morrell and SSgt. Adrian H. Jacobs. The airplane had taken off from Bush Field, Augusta, Georgia, on a ferry flight to Robins Field. The airplane was seen to enter a snap roll at about 2,000 feet agl. The pilot apparently failed to recover and the airplane dove into the ground. Investigators noted that a snap roll could be successfully performed in a BT-15 type airplane at an altitude of 2,000 feet agl.

4-6-43A. Vernon, Texas. At 1010, a Fairchild PT-19A suffered a catastrophic structural failure of the port wing and crashed three miles east of Victory Field, Vernon, Texas, killing civilian instructor George A. Beardshear. Student A/C Elmer R. Queen parachuted to safety and was uninjured. The student was receiving dual instruction in two-turn spins and after making a poor recovery the instructor took over and performed a spin, recovering in one and a half turns. As the instructor recovered from the spin, the port wing separated from the airplane. The instructor ordered the student to bail out, which he did. The instructor then exited the airplane but did not deploy his parachute in time to avoid striking the ground.

4-6-43B. Haines City, Florida. At 1115 EWT, a Douglas P-70 crashed into an unoccupied civilian residence at 816 Moss Street, Haines City, Florida, killing pilot 2Lt. James E. Giellum and radar operator 1Lt. Benjamin M. Morehouse. Nobody on the ground was injured. Investigators stated, "P-70 #39-752 [was] on an interception mission using a B-24D as a target. At an estimated altitude of 3,000 feet over Haines City the P-70 was seen to go into a flat spin from which it did not recover. The airplane crashed within the city limits, both pilot and observer being killed. From statements of eyewitnesses it is believed that the P-70 was brought in too close to the B-24 and that the pilot cut back his throttle to avoid over-running the target ship. The P-70 evidently hit the slip stream of the target ship which threw it into a flat spin from which the pilot could not recover."

4-6-43C. Winfield, Kansas. At 1042 CWT, a Vultee BT-13A crashed six miles southeast of Strother Army Air Field, Winfield, Kansas, killing instructor 1Lt. Glenn E. Peterson and student A/C Leslie F. Littlefield. The airplane was observed to go into a vertical dive at an altitude of about 1,500 feet agl and then to smash into the ground, bursting into flames upon impact. Investigation of the wreckage revealed that the airplane was pulled up slightly before it struck the ground. Investigators could not determine the events prior to the airplane going into the dive to the ground.

4-6-43D. Key Field, Mississippi. At 1323, a Douglas A-24A crashed while attempting a take-off at Key Field, Meridian, Mississippi, killing pilot 2Lt. Robert R. Ebling, Warren, Ohio, and gunner Sgt. Robert M. Bethany, Coleman, Texas. The A-24A was taking off on a navigation flight from Key Field to Biloxi, Mississippi, and return. The Aircraft Accident Classification Committee stated, "The pilot was taking off into the northeast on the Northeast-Southwest Runway. The pilot ran off the left side of the runway about half way down the runway. The airplane was skimming along just off of the ground. The airplane continued on the grass and across the East-West Runway and across the taxi ramp. The airplane was a few feet off the ground when the right landing gear hit a tractor roller. After hitting the roller, the airplane crashed into the railroad tracks about 1,500 feet off the end of the Northeast-Southwest Runway. The airplane was a total wreck and both the pilot and the gunner were killed. Examination of the control-locking device revealed that the controls were locked."

4-6-43E. Stuttgart, Arkansas. At 1600, a Laister-Kaufman TG-4A crashed at the Army Air Field at Stuttgart, Arkansas, killing pilot 2Lt. Edward K. Dwyer and seriously injuring glider pilot 1Lt. Francis T. McHenry. Lt. Dwyer entered the traffic pattern at about 600 feet agl. He attempted a turn at about stalling speed with the nose up, losing airspeed and altitude and causing the glider to enter a spin. The pilot was able to stop the spin but he was unable to recover control before the glider dove into the ground.

4-6-43F. Tujunga, California. At 1400, a Vultee BT-13A crashed into Mt. Lukens ten miles northwest of Tujunga, California, killing Maj. William T.

Nuna. The airplane disappeared on a routine positioning flight from Ontario Army Air Field, California, to Merced Army Air Field, California. Despite a massive search effort the airplane was not found until 12-12-43. The BT-13A was found to have crashed into the north slope of Mt. Lukens at an elevation of approximately 4,000 feet msl, about 400 feet from the peak.

4-6-43G. Apalachicola, Florida. At 1740 EWT, two North American AT-6A type airplanes collided in mid-air at the Apalachicola Gunnery Range, Florida, killing F/O John K. McClellan the pilot of AT-6A #41-218. Student gunner Pvt. Ernest S. Harding parachuted to safety and was uninjured. Pilot F/O James C. Morris and student gunner were able to safely land AT-6A #41-16515 at Apalachicola Field, Florida. They were uninjured. The airplanes, stationed at Tyndall Field, Panama City, Florida, were on an air-to-air gunnery mission. Investigators stated, "F/O Morris stated that the tow ship had reached the east end of the towing range and had turned 180 degrees to a westerly heading. F/O Morris flew into the turn under the tow cable and a little ahead of the target and was completing his turn preparatory to lining up on the target for his gunner to resume firing. At this moment F/O Morris states that there was an impact of sufficient force to throw him forward in his seat and cause his sunglasses and headset to be thrown off. He further states that he, at no time, saw the airplane, which collided with him, either before or after the impact. F/O Morris was able, although his tail group was severely damaged, to return to the Apalachicola Field and accomplish a safe landing."

4-6-43H. Fort Myers, Florida. At 1750 EWT, two North American AT-6A type airplanes collided in mid-air near Buckingham Field, Fort Myers, Florida, killing the two pilots and the two student gunners. Pilot F/O Ivan G. Fulwider, South Bend, Indiana, and student gunner Cpl. C.G. Gillispie, Ivydale, West Virginia, were killed in the crash of AT-6A #41-15841. Pilot 2Lt. James W. Hall, Philadelphia, Pennsylvania, and student gunner Pvt. Robert D. Heggem, Denver, Colorado, were killed in the crash of AT-6A #41-251. Investigators stated, "[The students] had completed a routine gunnery mission and were returning to the field to land. Lt. Hall entered traffic on a wide pattern and [flew] a deep base leg. He turned onto the approach and commenced a power let-down. F/O Fulwider in the meantime had flown a normal pattern for an AT-6 type airplane at 1,000 feet. When he turned onto the approach leg, the airplane piloted by Lt. Hall was several hundred feet beneath and a little in front of him. Apparently neither pilot was aware of the presence of the other airplane. Both airplanes continued in for a normal landing. The operators in the control tower did not observe the danger. F/O Fulwider, with more altitude to lose, let down more rapidly and collided with the tail section of Lt. Hall's airplane at an approximate altitude of 300 feet. Both airplanes spun to the ground out of control and crashed about 100 yards short of the runway."

4-6-43I. Tonopah, Nevada. At 1645 PWT, a Bell P-39D dove into the ground at the AAF Bombing and Gunnery Range 42 miles southeast of Tonopah, Nevada, killing pilot 2Lt. Rudolf H. Bisterfeldt. The airplane was part of a five-ship flight on a high-altitude formation-training mission. Investigators stated, "The formation had been up to 18,000 feet msl and were letting down in trail formation. At about 15,000 feet the formation had to fly around a small cumulus cloud, which was about 300 yards in width. This cloud was described by Lt. Becker as being similar to a fair-weather cumulus cloud, which the rest of the formation had gone around. [The pilot entered the cloud and evidently lost control, causing the airplane to go out of control and smash to the ground.] The force of the airplane hitting the ground shook houses in a mining camp about a mile from the scene of the accident. The four wing guns were imbedded in the ground where the airplane hit, and they indicated that the airplane must have struck the ground upright at about a 70-degree angle. The airplane and engine completely disintegrated. The pilot apparently did not attempt to jump because the safety belt latch was found locked."

4-6-43J. Randolph Field, Texas. At 2055, two Curtiss AT-9A airplanes collided on the runway at Randolph Field, San Antonio, Texas, killing four pilots. Pilot 2Lt. Albert A. Cullen and co-pilot 2Lt. Milton A. Elliot were killed aboard AT-9A #42-56966. Pilot 2Lt. Donald S. Graves and co-pilot 2Lt. Howard B. Johnson were killed aboard AT-9A #42-56923. The Aircraft Accident Classification Committee stated, "On the night of this accident two lighted runways were being used on the west side of the field for landings. The space between [the two runways] (533 yards) was being used for take-offs by the airplanes landing on both runways — the airplanes from the right-hand runway [were taking off on the grass parallel with] the lights on his runway, and the airplanes from the left-hand runway [were taking off on the grass parallel with] the lights on his runway. Lt. Cullen and Lt. Elliot were flying from the runway nearest to the hangar line. Lt. Graves and Lt. Johnson were flying from the runway on the outside of the field. They were taking off to the south. Lt. Graves took off slightly ahead and slightly above the airplane piloted by Lt. Cullen, who was taking off in the lane on Lt. Graves' left. Witnesses stated that Lt. Graves took off in a straight line and that Lt. Cullen's airplane veered to the right on take-off. The two airplanes collided when they were approximately 50 feet above the ground. Both airplanes fell to the ground and burst into flames."

4-7-43A. Las Animas, Colorado. At 0005 MWT, a Vultee BT-13A crashed one-quarter mile northeast of Las Animas Auxiliary Airfield, Colorado, killing 2Lt. Herbert L. Jesser. Investigation revealed that Lt. Jesser relieved the officer in command of the control ship at Las Animas Auxiliary Airfield at 2220. The last student left the airfield at 2350. The crash truck crew and the lighting crew observed Lt. Jesser take off at about 0005. The crash truck and the lighting truck left the field immediately after the take-off. When Lt. Jesser failed to arrive at La Junta Army Air Field, La Junta, Colorado, he was declared missing. The airplane was found during daylight hours later that morning one-quarter mile northeast and to the left of the runway at Las Animas. Investigators speculated that during his first turn after taking off the pilot lost the horizon and flew into the ground.

4-7-43B. Leakey, Texas. At 0245, a Beech AT-7 flying in instrument conditions crashed into rising terrain 12 miles north of Leakey, Texas, killing five fliers. The airplane had taken off from the Army Air Base at Jackson, Mississippi, at 2241 on a navigation-training mission to its base at the AAF navigation school at Hondo Army Air Field, Hondo, Texas. Investigation revealed that the airplane was attempting to stay underneath the overcast, which was estimated to be about 1,000 agl at the time the airplane collided with terrain at an elevation of approximately 2,200 feet. The airplane exploded violently into flames upon impact. Killed in the crash were: 2Lt. Robert L. McReynolds, pilot; 2Lt. John H. Edyvean, navigator instructor; A/C Rollie H. Lesher, student navigator; A/C Donald E. Lantz, student navigator; A/C Daniel E. McGurren, student navigator.

4-7-43C. Fort Sumner, New Mexico. At 0820 MWT, a Taylorcraft L-2A crashed near Auxiliary Airfield #3, six miles southeast of Fort Sumner, New Mexico, killing glider pilot F/O Jim E. Webb. The pilot had been practicing landings at the auxiliary airfield and was observed to overshoot on two previous landings. The airplane was seen making the turn from the base leg to final approach when it stalled and fell off on a wing and dove into the ground 100 yards west of the of the field.

4-7-43D. Collinsville, Texas. At 1520, a Vultee BT-13A crashed six miles east of Collinsville, Texas, killing A/C Lonnie V. Crumbley and seriously injuring A/C Frank L. Danner. Investigators stated, "[The students] were dispatched on a Student Instrument Team Ride. Cadet Crumbley was Observer in the front cockpit and Cadet Danner, who was to fly the instruments, [was] in the rear cockpit. At about 1520 CWT, their ship crashed six miles east of Collinsville, Texas, killing Cadet Crumbley and seriously injuring Cadet Danner. We are unable to obtain a statement from Cadet Danner, as he apparently has suffered a lapse of memory. He states that he believes that he was in the front cockpit and last thing he remembers is that they were at an altitude of 5,000 feet over the Red River at a point northwest of Perrin Field [Sherman, Texas]. However, statements of W.R. Childs and O.E. Faulkner indicate that the ship was flying at an extremely low altitude. They stated that the ship, while hedge-hopping, pulled over a clump of trees and then dropped back down towards the ground, whereupon the right wing hooked into the ground, cartwheeling the airplane [into the ground]."

4-7-43E. Mission, Texas. At 2120, a North American AT-6C crashed 38 miles northeast of Moore Field, Mission, Texas, killing A/C Chester F. Robin. The airplane had taken off from Moore Field, Mission, Texas, on a night navigation flight to Brownsville, Texas, to Alice, Texas, and return to Moore Field. The airplane failed to return to Moore Field and was declared missing. The airplane wreckage was discovered on 4-9-43. Investigation of the wreckage indicated that the airplane crashed into the ground in a nose-down position. It was speculated that the student had lost control of the airplane after inadvertently entering scattered clouds, which were reported to be at layers ranging from 1,500 to 2,000 feet agl. Another student from Moore Field was killed while flying the identical night navigation mission but in reverse (See 4-7-43F).

4-7-43F. Raymondville, Texas. At approximately 2200, a North American AT-6C crashed 13 miles northeast of Raymondville, Texas, killing A/C Joseph P. O'Brien. The airplane had taken off at 2015 CWT from Moore Field, Mission, Texas, on a night cross-country navigation mission to Alice, Texas, to Brownsville, Texas, and return to Moore Field. The airplane checked in at Alice and was never heard from again. The airplane wreckage, with the dead student in the cockpit, was located on 5-4-44. Investigation of the wreckage indicated that the airplane crashed into the ground in a nose-down position. It was speculated that the student had lost control of the airplane after inadvertently entering instrument conditions. Another student from Moore Field was killed while flying the identical night navigation mission but in reverse (See 4-7-43E).

4-7-43G. Taylor Field, Alabama. At 2235, a Vultee BT-13A crashed one mile southwest of Taylor Auxiliary Airfield, Alabama, killing student A/C Paul B. Manouvrier. Investigators stated, "The student was making his initial take-off for a series of night local training flights at Taylor Auxiliary Airfield. He took off normally into the south and at an estimated altitude of 500 feet above the ground began a 90-degree right turn toward his zone. After approximate completion of the turn the airplane was not observed for the moment until attention was attracted to it by a burst of fire on the ground about one mile southwest of the control tower. The airplane struck the ground in an uncultivated field in a steep right bank, nose-down, and power-on, with sufficient force to scatter parts over a distance of about 100 yards. The student was killed instantly and the wreckage caught fire. It is believed that the student was deceived as to his attitude due to a momentary loss of orientation."

4-7-43H. Douglas, Arizona. At a time between 2145 and 2300 MWT, a Cessna AT-17 crashed four miles northeast of Douglas, Arizona, killing A/C John

D. Patterson and A/C Carlton A. Pedley. The airplane had taken off from the Army Air Field at Douglas on a local night navigation flight. The airplane failed to return and was found to have crashed. Investigation of the wreckage indicated that the airplane flew into the ground while under control at cruise speed with landing gear and flaps retracted. Both students were killed instantly and wreckage was scattered over a great distance. It was speculated that the students had misread the altimeter and allowed the airplane to fly into the ground.

4-8-43A. Baer Field, Indiana. At 0042 EWT, a Boeing B-17E attempting a take-off crashed three-quarters of a mile southwest of Baer Field, Fort Wayne, Indiana, killing three crewmembers and seriously injuring two others. Killed in the crash were: 1Lt. Leroy L. Ronnfeldt, Columbus, Ohio, co-pilot; Pvt. Devere Lambert, Mishawaka, Indiana, passenger; Pvt. Bennie J. Grabiec, Chicopee, Massachusetts, engineer. Pilot 1Lt. Ernest Pokral, Chicago, Illinois, and radio operator Pfc. Leo J. Kadlubowski, Syracuse, New York, were seriously injured. The airplane, which was on a B-17 instructor training flight, had flown from the airplane's station at Lockbourne Army Air Base, Ohio, earlier in the day and was taking off to the southwest for the return trip. Soon after take-off the B-17's port wing collided with three large trees at an elevation of about 35 feet above the ground, clipping off the tops. The B-17 continued in the air for another 225 yards before smashing into the ground and bursting into flames. Wreckage was scattered over an area of 490 yards. Baer Field rescue personnel were on the scene immediately. Investigators could not determine why the airplane collided with the trees. It was later speculated that the pilots had inadvertently flew the airplane into the ground while being preoccupied in the cockpit.

4-8-43B. Smithville, Georgia. At 1215 CWT, a Curtiss AT-9 crashed five miles west of Smithville, Georgia, killing A/C Jack C. Salisbury, Nicholson, Pennsylvania, and A/C John L. Singletary, 26, Cairo, Georgia. The airplane took off from Turner Field, Albany, Georgia, at 1145 CWT and assembled in the number-two position of three-ship formation training flight. The formation flew to Smithville Auxiliary Airfield in a V-formation. As the formation approached the field, the leader signaled for an echelon to the right. The airplanes made the formation change and then the leader peeled off to the left to land at the field. The subject airplane was observed to peel off in a steep bank and then spiraled into the ground. Both engines were apparently operating normally at the time of the accident. The airplane was completely destroyed and the students killed instantly.

4-8-43C. Craig Field, Alabama. At 1330, a Curtiss P-40F crashed five miles south of Craig Field, Selma, Alabama, killing A/C William H. Benjamin. "It is the opinion of the Aircraft Accident Classification Committee that this accident was the result of the student losing control of his aircraft, for reasons unknown, and crashing into the ground. From an examination of the wreckage it appeared that the student had attempted to use his parachute but was unsuccessful in clearing the aircraft before it struck the ground. A witness stated that he had seen 'something white' string out behind the aircraft just before it struck the ground but it did not detach itself before the airplane crashed. The force of the crash was such that the wreckage was badly scattered making it very difficult to construe a logical sequence of events. Another student, who happened to be flying at an altitude of 3,700 feet over the scene of the crash, noticed the airplane spinning down below him. He lost sight of it momentarily and then saw a burst of flame at the moment it struck the ground. He estimated the other aircraft was about 1,500 to 2,000 feet below him at the time he first saw it."

4-8-43D. Dublin, California. At 1410, a Bell P-39N crashed into rising terrain five miles southwest of Dublin, California, killing pilot 2Lt. James A. Jago. The airplane was part of a four-ship formation that had taken off at 1320 from Oakland Municipal Airport, Oakland, California, on a formation training flight. The P-39s were flying a string formation at an indicated altitude of 9,000 feet, above a layer of broken clouds at 5,000 feet. The subject airplane left the formation and crashed into a hill where it exploded into flames, killing the pilot instantly. The formation leader stated that he had not seen Lt. Jago leave the formation. Investigators speculated that the pilot might have run a fuel tank dry and was unable to switch tanks and restart the engine in time to avoid entering the overcast and crashing into the terrain immediately after breaking out underneath. It was speculated that the pilot might have recovered satisfactorily underneath the overcast but had probably collided with the terrain because he could not maneuver in the narrow canyon.

4-8-43E. Inez, Texas. At 2250, a North American AT-6A crashed 15 miles north of Inez, Texas, killing student A/C Richard W. Hocker. The airplane had vanished after having taken off from Foster Field, Victoria, Texas, at 2029 on a night navigation mission to Temple, Texas, and return. The airplane was found on 4-18-43. Investigators speculated that the student had gotten lost and became low on fuel. The student, with landing gear retracted, landing lights illuminated and with full flaps, attempted an emergency forced landing in a farm field. The student apparently misjudged the wind direction and his height above the ground, causing the airplane to smash into the ground hard with the starboard wing. The airplane cartwheeled into the ground and was completely demolished. The student was killed instantly. It was noted that overcast conditions had moved into the area at about the time of the crash. Another student from Foster Field who was flying on the same night navigation mission was involved in a fatal accident near Yoakum, Texas (See 4-8-43F).

4-8-43F. Yoakum, Texas. At 2308, a North American AT-6C crashed one-half mile northeast of Yoakum, Texas, killing student A/C Charles R. McRoberts. The airplane had taken off from Foster Field, Victoria, Texas, at 2002 on a night navigation mission to Temple, Texas, and return. Investigators stated that the student encountered poor weather conditions when he arrived in the Yoakum area. The student managed to stay under the 300-foot ceiling and was able to find the Yoakum airport. The student began circling the airport. He entered the traffic pattern and then extended the landing gear and flaps and illuminated the landing lights. The student began a left turn from the base leg to the final approach when he apparently flew into the ground. The port wing struck the terrain and the airplane cartwheeled into the ground, bursting into flames upon impact. Another student from Foster Field who was flying on the same night navigation mission was involved in a fatal accident near Inez, Texas (See 4-8-43E).

4-9-43A. Mayer, Arizona. At 0830 MWT, a Vultee BT-13A was abandoned and crashed five miles east of Mayer, Arizona, killing civilian instructor Max K. Lehman, who fell to his death in an unsuccessful parachute jump. The airplane had taken off from Wickenburg, Arizona, on a flight to tow a Taylorcraft TG-6 glider to Winslow, Arizona. The pilot had inadvertently entered a cloud formation and visibility went to zero. The glider pilot released the glider from the towline. The BT-13 pilot apparently became disoriented and bailed out. The pilot apparently deployed his parachute too soon and it fouled on the towline that was still attached to the airplane. The pilot fell to earth about one-half mile from where the airplane smashed to the ground.

4-9-43B. Thomasville, Alabama. At 1225, a North American A-36A flying in poor weather crashed eight miles southwest of Thomasville, Alabama, killing pilot 2Lt. Howard N. Teague, Jr. Investigators stated, "The pilot cleared from Meridian, Mississippi, to Waycross, Georgia, on an instrument flight. At the time the flight was made, Craig Army Air Field was reporting a severe thunderstorm. The pilot in all probability flew into the thunderstorm and either tried to fly through it, or realizing his mistake attempted to return to Meridian. Upon examination of the wreckage and the immediate surround terrain, it appears that the pilot lost his horizon in flight or an overload of wing ice combined with partial or complete engine failure due to ice threw the ship partially or completely out of control and in attempting to regain control of his aircraft, through instrument flight, caused the aircraft to go into a steep spiral. The aircraft struck the ground with one wing down and excavated a furrow about 50 feet in length before striking the base of a tree which caused it to explode, scattering a parts over a half acre of ground. The accident occurred during a violent thunderstorm and the aircraft crashed about one and one-half miles from the nearest house. Occupants of this house stated they heard an explosion about 1225, but attributed the noise to a clap of thunder. The aircraft and the body of the pilot was found a day later by a hunter in a heavily wooded area. It is the opinion of the Aircraft Accident Classification Committee that this accident was the result of the pilot's losing control of the aircraft and crashing into the ground through causes unknown. The pilot suffered fatal injuries in the crash and the aircraft was completely demolished. There was no evidence to show that the pilot attempted to use his parachute."

4-9-43C. Augusta, Georgia. At 1740 EWT, a Vultee BT-15 crashed 15 miles southwest of Bush Field, Augusta, Georgia, killing A/C Luther A. Harrell, Jr. and A/C Robert J. Schweiger. The airplane took off from Bush Field on an instrument team ride. The airplane was seen in a steep diving turn but the pilot failed to effect a recovery before the airplane struck the ground. Investigation of the wreckage indicated that the airplane had struck the ground in a level attitude at high speed, smashing itself to pieces before coming to rest in a small stand of trees 125 yards later. The airplane did not burn but was completely demolished. Wreckage was scattered over an area of approximately 400 feet.

4-10-43A. Wetmore, Texas. At 0900, a Fairchild PT-19A crashed one mile northwest of Wetmore, Texas, killing civilian instructor trainee Rodney E. Fundis. Investigators stated, "According to the description of one witness, the pilot made a roll in the airplane at approximately 1,000 feet altitude from which it went into a spin, doing approximately three turns of spin before striking the ground without any apparent recovery. When [the airplane] struck the ground it burst into flames."

4-10-43B. Easthampton, Massachusetts. At 1140, a Republic RP-47C crashed five miles west of Easthampton, Massachusetts, killing pilot 2Lt. John F. Reed, who fell to his death in an unsuccessful parachute attempt. The airplane was part of a four-ship flight that had taken off from Westover Field, Chicopee, Massachusetts, on a formation-training mission. The formation was flying at about 10,000 feet when it was noticed that Lt. Reed, who leading the second two-ship element, was leaking fuel badly from the underside. Lt. Reed peeled off over Westover Field and descended in a circle, losing altitude rapidly. At an altitude of about 100 feet agl, the pilot left the airplane and it dove into a wooded area where it exploded into flames. The pilot was found to have slammed into the ground about 50 feet from the wreckage. The pilot had pulled the ripcord and the pilot chute had popped out, but the main parachute had not deployed.

4-10-43C. Tampa Bay, Florida. At 1420 EWT, a Martin RB-26A crashed into Tampa Bay four miles WNW of MacDill Field, Tampa, Florida, killing the crew of five. The airplane had taken off from MacDill

Field on a routine training flight and was flying at 2,500 feet when it went into a spin to the right. The airplane was seen spinning to the right with its nose pointing straight down, both propellers windmilling, bomb bay doors closed and the landing gear in the retracted position. The spin seemed to flatten out as the B-26 neared the water but continued to spin until it struck shallow water five miles north of the western end of the Gandy Bridge, about one mile off shore. Investigators were unable to determine what caused the airplane to enter the spin. Killed in the crash were: 1Lt. Joseph W. Shipley, Bristol, Virginia, B-26 instructor; Capt. John H. Duncan, Jr., Lindale, Texas, B-26 student; 1Lt. Robert N. Rouhier, Louisville, Ohio, B-26 student; SSgt. Ocie H. Marlow, Thalia, Texas, engineer; Capt. Samuel A. Barkoff M.D., New Orleans, Louisiana, passenger.

4-10-43D. Enid, Oklahoma. At 1520, a Vultee BT-15 crashed while attempting a go-around at Enid Army Air Field, Enid, Oklahoma, killing A/C Charles F. Stelmokas. The airplane was returning to the field following a routine student solo flight. The Aircraft Accident Classification Committee stated, "On final approach Cadet Stelmokas allowed the airplane to drift to the left edge of the runway. He was landing towards the southeast into an 18 mph wind from the south-southeast. Upon making contact with the left side of the runway, he applied some power in an effort to keep the airplane rolling straight on the runway; whereupon his right wing [lifted] so that his right wheel lifted off of the runway and the airplane swerved to the left on the left wheel, leaving the runway and turning approximately 45 degrees to the left so that the airplane was heading straight towards the hangar line. When approximately 40 yards [away] from the runway, the pilot applied full power and depressed the nose so that both wheels struck the ground heavily, causing a bounce which resulted in an abnormally nose-high climbing condition of flight. When approximately 25 feet off of the ground, the airplane stalled, breaking to the left and striking the ground with sufficient force to demolish the airplane, with fatal results to the student."

4-10-43E. Philadelphia, Pennsylvania. At 1640, a Republic RP-47C exploded in mid-air and crashed near Philadelphia Municipal Airport, Philadelphia, Pennsylvania, killing pilot 1Lt. Otto A. Loesch. The Aircraft Accident Classification Committee stated, "RP-47C #41-6521 exploded in mid-air over the Philadelphia Municipal Airport at an altitude of approximately 500 feet, and as a result of the explosion, went out of control and crashed into a residential district one-half mile north of the airport. Lt. Loesch had taken off from Philadelphia Municipal Airport at 1620 for the purpose of testing and slow timing a new engine [that] had been installed in the airplane. Approximately 15 minutes after take-off, he contacted the Control Tower and requested emergency landing clearance and received landing clearance from the tower via radio and green signal light. While over the field with landing gear in the extended position preparatory to making approach for landing, an explosion was heard and results of the explosion observed by several persons who were on or in the immediate vicinity of the airport. Immediately following the explosion parts of the airplane were observed to fall to the ground. The airplane immediately went into a diving turn and crashed, burning immediately. The cause of the explosion of the airplane in mid-air is unknown."

4-10-43F. Ardmore, Oklahoma. At 2215 CWT, a Vultee BT-13A spun to the ground near Lake Murray, 15 miles southeast of Ardmore, Oklahoma, killing A/C Ralph E. Gunnels. The airplane had taken off from Perrin Field, Sherman, Texas, on a solo night navigation flight to Pauls Valley, Oklahoma, to Gainesville, Texas, and return to Perrin Field. A park ranger at Lake Murray reported that the airplane had circled the area four times before crashing into the ground and exploding into flames. Investigation of the wreckage revealed that the airplane had struck the ground in a spin with power on. The Aircraft Accident Classification Committee stated, "[A/C Gunnels'] roommate had prepared his maps and tried to coach him considerably in regards to this particular navigation flight. It is interesting to note that his roommate made a bet of one dollar with Aviation Cadet Gunnels that he would become lost on this navigation flight."

4-11-43A. Beowawe, Nevada. At an unknown time between 0300 and 0330 PWT, a North American B-25C crashed six miles southwest of Beowawe, Nevada, killing the crew of three. Pilot 1Lt. Hugh O. Hoffman, co-pilot 2Lt. Leonard K. Smith and engineer Sgt. John P. Conway were killed in the crash. The airplane had taken off from Mather Field, Sacramento, California, on a night navigation flight to Elko, Nevada, to Boise, Idaho, and return to Mather Field. Five other B-25 airplanes had taken off from Mather Field on the same mission. Weather reports at the time of the flight forecast broken clouds to overcast clouds from 8,000 to 10,500 feet msl between Mather Field and Humboldt, Nevada, with the clouds topping out at approximately 12,000 feet msl. Visibility on top of the overcast was reported as unlimited. The pilots of two other B-25 airplanes that had taken off after Lt. Hoffman noticed a bright flash of light shining up through the overcast that they were flying over. Both pilots noted the position on their maps and reported the fact when they returned to Mather Field. When Lt. Hoffman failed to return to Mather Field, a search was commenced. The wreckage was spotted from the air in the area where the pilots had reported seeing the flash of light. Investigation of the wreckage indicated that the B-25C had dove into the ground at a 30-degree angle at high speed, exploding violently into flames upon impact and scattering wreckage and body parts over a wide area. Investigators could not determine what caused the pilots to fly the airplane into the ground. No evidence of structural or mechanical failure could be found.

4-11-43B. Alligator Point, Florida. At 1000 EWT, a Curtiss P-40K crashed at Alligator Point, Florida, killing pilot 2Lt. Samuel P. Sumner. The Aircraft Accident Classification Committee stated, "Eyewitnesses saw this aircraft complete a dive on the temporary bombing range without dropping his bomb and pulled up steeply. At the top of the climb [at] approximately 1,000 to 1,500 feet [agl] he fell off on the right wing and plunged directly into the bay. The aircraft exploded upon hitting the water and only fragments of the aircraft have been recovered. The aircraft had not appeared to be on fire or to have had any material failure as far as ground eyewitnesses could see. Committee believes that this pilot may have blacked out in too sharp a pull out."

4-11-43C. Griffin, Georgia. At 1020 EWT, a Vultee BT-13A crashed near Victory Airport, Griffin, Georgia, killing instructor pilot 2Lt. Jesse G. McMillian, Mt. Airy, North Carolina, and passenger Sgt. Eugene E. Breon, Columbus, Ohio. The airplane had taken off at 0845 EWT from Cochran Field, Macon, Georgia, with two other BT-13A type airplanes, both piloted by students, on a formation training flight. The flight landed at Victory Airport. Investigators stated, "After remaining on the ground for about ten minutes, Lt. McMillian ordered his students to take off, climb to 1,000 feet above the airport and wait for him. Shortly after the students left the ground, Lt. McMillian took off in a steep chandelle over the hangar. He continued around the airport and then made a straight pass at extremely low altitude along the full length of the airport. As he reached the boundary marker, he pulled up to an altitude of approximately 100 feet and executed a slow roll to the left. The airplane 'washed out' at the bottom of the roll, slid out to the right, turned approximately 90 degrees in a flat skid and struck the ground. The airplane immediately burst into flames and both pilot and passenger were killed."

4-11-43D. Santa Ana, California. At 0815, a Lockheed P-38G crashed near Orange County Army Air Field, Santa Ana, California, killing pilot 2Lt. Clyde K. Sellers, 25. The Accident Classification Committee stated, "Lt. Sellers took off from Orange County Airport as fourth man in a four-ship patrol flight, and returned to land after two hours with the rest of the flight at 0815. Inasmuch as the direction of landing was to the north (Runway 3) the formation came over the runway from the south, peeled up to the right and landed individually in a right hand pattern. Because Lt. Sellers was the last man his approach was long. According to eyewitness reports, as the airplane reached about one-half mile from the edge of the runway, the airplane suddenly turned to the right and spiraled down, or spun down, until it hit the ground at about a 25-degree angle in a soft plowed field. The airplane did not burn, but was [completely] destroyed and the pilot killed." It was later speculated that the P-38 had suffered a failure of the starboard engine because of fuel starvation, causing the pilot to lose control of the airplane.

4-11-43E. Dawson, Texas. At 1430, a Piper L-4B crashed one mile east of Dawson, Texas, killing pilot 1Lt. Harry C. Victery and medical officer passenger Capt. Walter D. Campbell. The airplane took off from Waco Army Air Field, Waco, Texas, on a local flight. Investigators stated, "Evidence at the accident showed conclusively that the accident was caused by the airplane hitting a wire of a rural power line that crosses [Texas Highway 31] at this point. The wire [that was] struck, a number-six steel wire [that] was 30 feet above the ground, did not break from the impact but was torn loose from the insulator on one of the nearest supporting poles. Evidently the airplane went out of control after striking the wire, hit the pavement nose first, and slid about 75 feet before coming to a stop. It is believed that friction sparks caused by sliding along the pavement were the cause of the fire that resulted. Just prior to the accident this airplane was observed flying low over the town of Hubbard, Texas, which is located on State Highway 31, four miles southwest of where the crash took place. The airplane was flying in a northwesterly direction. It was next observed following State Highway 31 at a low altitude. Witnesses stated that there was no indication that the pilot saw the power line before hitting it."

4-11-43F. Pendleton Field, Oregon. At 1640, an Aeronca L-3C crashed five miles southeast of Pendleton Field, Pendleton, Oregon, killing 1Lt. Harold C. Pfau. The airplane had taken off from Pendleton Field with two other L-3 type airplanes on a local training flight. Soon after take-off the airplanes separated. A few minutes later one of the pilots noticed a puff of smoke on the ground. Investigators could not figure out why the pilot flew into the ground. The airplane skidded for about 30 feet before flipping over. The airplane was completely demolished and the pilot killed instantly.

4-12-43A. Sandersville, Georgia. At 0001 EWT, a Vultee BT-13A crashed at Sandersville, Georgia, killing student A/C Harry E. Morton. The student had taken off from Cochran Field, Macon, Georgia, on a cross-country navigation flight to Alma, Georgia, and return. The student had checked in at Alma on the return leg but there was no further radio contact after that. The airplane was observed circling the Sandersville area before it dove into the ground at a steep angle. It was speculated that the pilot had lost the horizon and entered a spin or had inadvertently flew the airplane into the ground.

4-12-43B. Tucson, Arizona. At 0606 MWT, a Consolidated B-24D crashed three and one-half miles southeast of Davis-Monthan Field, Tucson, Arizona, killing the crew of ten. Investigation revealed that the B-24 had taken off to the southeast on Runway 12 at 0603 on a gunnery exercise and simulated bombing mission. Apparently, the B-24 flew into the ground about two and one-half miles ESE of the southeast end of Runway 12 shortly after take-off, exploding into flames upon impact. Investigators could not determine why the airplane apparently flew into the ground. Killed in the crash were: 2Lt. William A. Furner, pilot; 2Lt. Frank Berry, Jr., co-pilot; 2Lt. Robert H. McCelvey, navigator; SSgt. Albert G. Conto, radio operator; SSgt. Henry F. Warner, assistant radio operator; SSgt. Kenneth J. Oestreich, engineer; Sgt. Robert E. Funk, engineer; SSgt. Valmore Beausoleil, gunner; Pfc. Comer L. Jenkins, gunner; Cpl. Robert J. Norris, gunner.

4-12-43C. Ridgeland, South Carolina. At 0840, a North American B-25D crashed ten miles southeast of Ridgeland, South Carolina, killing seven fliers. Killed in the crash were: 1Lt. Leslie W. Shepard, pilot; 2Lt. Thomas C. Galbraith, co-pilot; 2Lt. Warren E. Smith, navigator; TSgt. Weldon E. Boone, engineer; TSgt. Bill G. Comerford, radio operator; Cpl. Harold F. Lemieux, gunner; Sgt. Walter J. Szychulski, passenger. Investigators stated, "The investigation officer talked to Capt. Charles M. Hagest, who was leading the flight of three airplanes from Walterboro, South Carolina, to Savannah, Georgia, at 0840 EWT. He states that Lt. Shepard was flying on his left wing, and as the weather was extremely rough, they were flying in a very loose formation. Everything was proceeding as usual when Capt. Hagest noticed that Lt. Shepard started to make a steep left hand turn. As it appeared to him, the nose of Lt. Sheppard's airplane seemed to be rather high, and at first he thought of the motor going out, but could not ascertain if that was the case from the distance that he was viewing it from. After he had turned approximately 180 degrees, Lt. Sheppard's airplane fell off into a spin to the left, crashed, and immediately burst into flames. Capt. Hagest circled the scene of the crash and then proceeded to Hunter Field [Savannah, Georgia] to report it. The investigating officer drove from Savannah to the scene of the accident, a distance of approximately 45 miles, with the ambulance, fire truck, and crash equipment. The airplane was completely demolished and was still burning when he arrived there. Evidently the ship was spinning when it [contacted] the ground as it came straight down through the treetops. The condition of the airplane was such that it is impossible to determine the cause of the accident."

4-12-43D. Apponaug, Rhode Island. At 0845, two Republic P-47C type airplanes collided in mid-air four miles south of Apponaug, Rhode Island, killing 2Lt. Gordon M. Kimpel (P-47C #41-6634) and 2Lt. Eldred G. Howard (P-47C #41-6174). Investigators stated, "At approximately 0830, 2Lt. Kimpel took off on a formation training mission. The flight of four airplanes was led by 2Lt. Howard, with Lt. Kimpel flying the number-two position. While flying in an extended string formation at 8,000 feet, Lt Howard, the leader, pulled up vertically and in trying to follow the maneuver, Lt. Kimpel evidently lost him in the sun and hit the lead ship with the wing of his airplane. The airplane piloted by Lt. Kimpel went down and to the right and a few seconds later he was seen opening his parachute. The airplane crashed in a swampy wooded area near Apponaug, Rhode Island, with no damage to private property. Lt. Kimpel evidently was struck by the tail of his airplane because he was found dead in a small pond off of Greenwich Bay. The airplane piloted by Lt. Howard went straight down and crashed into Narragansett Bay. The pilot made no attempt to use his parachute."

4-12-43E. Brown Field, Texas. At 1055, two Fairchild PT-19A type airplanes collided at Brown Auxiliary Air Field, Texas (Auxiliary to Jones Field, Bonham, Texas), killing student pilot 2Lt. W. N. Scabad aboard PT-19A #41-20206. Civilian instructor David Pruet and A/C Arthur E. Ruona received minor injuries in the crash of PT-19A #41-15059. Investigators stated, "Student pilot 2Lt. W.N. Scabad, in airplane No. 41-20206, in haste to take off on his first unsupervised solo flight into the practice area taxied to take-off position, stopped momentarily 45 degrees to the direction of take-off, glanced briefly in the direction of the airplanes approaching for landing, turned and proceeded to take off. Airplane No. 41-15059, piloted by Civilian Instructor David R. Pruet with Aviation Cadet Arthur E. Ruona was on the approach leg for landing. Neither pilot observed the other airplane; as a result the solo student took off and proceeded to climb while the dual airplane continued to glide until the collision occurred approximately 50 feet above the ground. The instructor in the landing airplane observed the airplane below him simultaneously with the collision and was unable to alter his course to avert the collision. Both airplanes fell to the ground in the inverted position and the position of the body of the fatally injured student, plus the injuries [that] he received, indicate that he either jumped free or fell from his airplane as it fell to the ground. The statement of the cadet that cranked the solo student's airplane indicated that his safety belt was fastened at the time of take-off."

4-12-43F. Paradise Valley, Arizona. At 1635 MWT, a Stearman PT-17 crashed at Paradise Valley, Arizona, killing student A/C Francis V. Bellows. The Accident Classification Committee stated, "[A/C Bellows] was following two other airplanes at low altitude outside the local flying area. He centered his attention upon the other two airplanes [that] he was following and did not notice the rise in the irregular terrain. When he noticed how close to the ground he was, he attempted a right turn and pulled the stick back so far that he caused a high-speed stall in an attempt to get the nose above the horizon. The airplane then mushed into the ground. The ship broke in two directly behind the rear cockpit, the gas tank exploded, and the ship burned. The pilot crawled about 30 feet away from the ship. His clothing was completely burned off. A motorist in the vicinity of the airplane saw it disappear behind a small hill and then saw smoke and flames coming from that vicinity. The pilots of the other two airplanes did not see the crash occur and at that time were headed back for the home field."

4-12-43G. Plain Dealing, Louisiana. At an unknown time, a twin-engine bomber crashed near Plain Dealing, Louisiana, killing four unidentified fliers. Published accounts of the crash reported that witnesses had observed the airplane in flames before it collided with trees while attempting to make an apparent emergency forced landing in a clearing. One crewmember fell to his death when he attempted to parachute to safety but was too near the ground to allow his parachute to deploy. The airplane slammed into the ground and exploded violently into flames. Officers at Barksdale Field, Shreveport, Louisiana, stated that the airplane was not from that station. No other information is available.

4-13-43A. Gowen Field, Idaho. At 0816 MWT, a Boeing B-17F (42-5751) attempting an instrument take-off collided with two parked B-17F airplanes at Gowen Field, Boise, Idaho, killing four crewmembers aboard B-17F #42-5751 and an engineer on the ground. Two B-17 crewmembers were injured. Five crewmembers aboard B-17F #42-22987 received minor injuries and three ground crewmen were injured while they were working near B-17F #42-29548. Killed in the crash of B-17F #42-5751 were: 1Lt. Richard Pease, instructor pilot; 2Lt. William D. Workman, B-17 student; 2Lt. Elbert J. Williams, B-17 student; SSgt. Clifford J. Hawthorne, radio operator. B-17 student 2Lt. Kenneth W. Athens received minor injures and engineer SSgt. Lloyd O. Hipps sustained serious injuries. Crewmembers injured aboard B-17F #42-22987 were pilot 2Lt. Edward B. Apperson and crewmembers SSgt. Wayne M. Brown, SSgt. John A. Gaglio, SSgt. Hubert C. Galloway and Sgt. Dush Kekich. Sgt. Archie D. Heifner, Cpl. Carl E. Lintores and Pfc. Fred S. Dickerson were injured on the ground. Aerial engineer Sgt. Thomas E. Lovelace was killed when he was struck by the subject airplane, B-17F #42-5751. The airplane was on an instrument training flight and was lined up on the southeast runway for an instrument take-off. Investigators stated, "2Lt. Workman was under the hood practicing an instrument take-off. 1Lt. Pease, an instructor pilot of considerable experience (1,667 hours of military time [and] 663 hours of B-17 time), was sitting in the right seat. The airplane commenced a normal take-off and ran 1,550 feet down the runway. At this time the airplane commenced to swerve slightly to the right. The airplane turned off the runway under full power. It was straightened out momentarily while still under full power and swerved further to the right. Finally, after proceeding in a wide arc for approximately three-tenths of a mile from the runway, [the starboard wing] hit the left wing of an airplane parked on the north side of the ramp (B-17 #42-29548). It swung further to the right, still under full power at an estimated speed of 70 or 80 miles per hour and proceeded south across the ramp and smashed head-on into another parked airplane (B-17F #42-22987). As the airplane came onto the ramp, witnesses stated that it was in a nose-down attitude under full power. The propellers were observed to strike the ground and the left tire blew out just before the airplane struck the wing of the first parked airplane. Two witnesses saw fire or smoke coming from the hub of the left wheel. Approximately ten witnesses agreed that the airplane was still traveling under full power when it crashed into the airplane on the ramp. At the time of the accident the airplane was traveling at an angle of approximately 45 degrees or more from the original take-off direction, and was swerving violently to the right. When 42-5751 collided head-on with 42-22987, both airplanes were badly damaged and fire started, without explosion, which completely destroyed both airplanes. Four men in the airplane were killed as a result of the crash and their bodies were removed by Army personnel before the fire reached them. One man who was on the ramp was struck and killed by the airplane. It will be necessary to replace the complete left wing on B-17F #42-29548. From a testimony of witnesses, including two survivors of the airplane who were in the cockpit prior to the crash, the following sequence of events seems most probable: The pilot under the hood lost partial control of the airplane on take-off and ran off the runway. Lt. Pease, the instructor pilot, took over the controls just as the ship left the runway and attempted to straighten the airplane out. He succeeded for a short stretch, but the airplane continued to swerve to the right until shortly before the crash. Apparently the brakes were used during this period, especially the left brake, which was observed burning when the airplane crossed the ramp. It is probable that this brake was used excessively in an effort to straighten out the airplane. Just before the airplane went on the ramp, it was observed to be exceptionally tail-high, so much so that the propellers struck the ground, indicating a possibility of exceptional brake use just before the crash. During this entire time, the airplane was traveling under take-off power and there was no evidence of engine or propeller failure. There were many witnesses to this accident, including the two surviving members of the crew who were in the cockpit. The consensus of opinion was that there was no mechanical failure in the airplane. The two survivors of the crash testified that the controls were not locked, and neither of them were aware of any engine failure or other mechanical failure. It appears that the airplane started to turn to the right; the instructor pilot took over and attempted to straighten it out; did so momentarily, and the airplane swerved again and crashed. The presumption of the committee is that Lt. Pease, a very experienced and well-qualified instructor pilot, took over the controls as the airplane [swerved off of] the runway, feeling sure that he could take it off safely. Without reducing power, he fought the controls in an effort to straighten the path of the airplane. There is the possibility that the two pilots were both on the controls. It is evident that the brakes were used excessively during the latter part of the airplane's path on the ground and that full power was never reduced. Possibly one pilot was attempting the take-off and the other attempting to stop it. The engineer of the airplane, who was standing behind the two pilots, stated that after Lt. Pease said 'I've got it,' that the man under the hood dropped his hands and did not touch the controls further. Lt. Althen, who claims that he was standing between the [pilots], said he observed Lt. Pease manipulating the controls in an effort to right the ship, and it appeared to him that Lt. Workman may also have been working the controls at the same time." It was reported by several witness that the number-one and number-three engines were on fire when the airplane smashed into B-17 #42-22987. The field carpentry shop was also destroyed.

4-13-43B. Muroc, California. At 0901, a Lockheed P-38F crashed near the Army Air Base at Muroc, California, killing pilot 2Lt. Joe Sara. The airplane was in the number-four position of a four-ship flight that had taken off from Muroc Army Air Base at 0800 on a routine formation flight. After an hour of flying and maneuvering, the formation returned to Muroc to land. The pilot, still flying in the number-four position, approached the runway with landing gear and flaps extended. The airplane encountered the propeller turbulence of the ships in front. The pilot apparently over controlled the airplane and entered a stall at the end of the runway. The airplane dove into the ground and burst into flames, killing the pilot instantly.

4-13-43C. Tullahoma, Tennessee. At 1130 CWT, a Bell P-39D crashed six miles southwest of Northern Field, Tullahoma, Tennessee, killing pilot 2Lt. Rufus E. Mitchell. The airplane took off at 0950 on a local training flight, which was to terminate at 1115. Before the airplane returned to the field, a cold front moved into the area and the ceiling had dropped. The Control Officer ordered all airplanes be called in before the weather closed the field. The tower was only able to contact a Curtiss P-40, which promptly came in to land. The tower was unable to contact the three P-39s that were out. One by one the P-39s came in to land. The subject airplane came in last. The ceiling in the vicinity of the field had dropped to about 600 feet agl with scattered rain showers. The pilot inadvertently entered a rain cloud and lost control of the airplane. The airplane was seen to dive out of the overcast at an angle of about 60 degrees. The airplane dove into the ground and exploded into flames.

4-13-43D. Augusta, Georgia. At 1723 EWT, a Martin B-26B crashed near Daniel Field, Augusta, Georgia, killing the crew of seven. The airplane had taken off from the Army Air Base at Greenville, South Carolina, on a routine navigation flight to its station at the Avon Park Bombing Range, Avon Park, Florida. Investigators stated, "[The B-26] was first observed by the Operations Dispatcher about 1720 EWT, five or six miles distant in a southwesterly direction. The airplane was flying at approximately 200 feet altitude and proceeding in a northwesterly direction on a straight course for Daniel Field. The airplane altered its course twice during the next three minutes of flight, turning first to the right and [then] to the left shortly before hitting the treetops a mile and one-fifth from the center of Daniel Field. The airplane lost altitude during the above maneuvers and due to the elevation of the southern end of the northeast-southwest runway the airplane was obscured from view at the time of impact with the ground. A path 200 yards long was cut through a patch of woods by the airplane before crashing against the bank of a small creek. Due to the impact with this creek the left motor was torn loose and the airplane swerved around 180 degrees and caught fire immediately." Killed in the crash were: 1Lt. Robert L. Floyd, pilot; 2Lt. Richard G. Larsen, co-pilot; SSgt. Carl Justin, engineer; Sgt. Charles S. Elder, engineer; Sgt. Lewis F. Halmi, gunner; Cpl. Woodrow O. Rosell, bombardier; Pfc. James S. Nelson, radio operator.

4-14-43A. Marianna, Florida. At 0745 CWT, a North American AT-6C crashed 15 miles SSE of Marianna, Florida, killing A/C Lorenzo H. Knight. The airplane had taken off from the Marianna Army Air Field with another AT-6 on an authorized individual combat training flight. The airplanes were in a Lufbery circle when the subject airplane was pulled up into a steep climbing turn at an indicated altitude of approximately 3,000 feet. The airplane fell off into a power stall and entered a violent spin. The student was able to recover the airplane just above the treetops. The airplane flew along the tops of the trees for a few moments before falling off on the port wing and smashing into a wooded area. The student was killed instantly.

4-14-43B. Sanibel Island, Florida. At 0900 EWT, a Bell P-39D crashed into the surf at the AAF Ground Gunnery Range at Sanibel Island, Florida, killing 2Lt. Warren W. Kranz. The airplane had taken off with another P-39 from Page Field, Fort Myers, Florida, on a ground gunnery and dive-bombing mission. The airplanes had dropped their bombs on the bomb range and then began the gunnery exercise. The pilot had made two machine gun passes at the target and then made an extremely long burst on the third run. The airplane barely cleared the target but pilot was not able to pull the airplane up in time to avoid crashing into the Gulf of Mexico 50 yards off shore.

4-14-43C. Westerly, Rhode Island. At 0915, a Republic P-47B suffering an engine failure crashed while attempting an emergency forced landing one mile east of Westerly, Rhode Island, killing pilot 2Lt. Robert W. Coleman. The airplane had taken off from Trumbull Field, Groton, Connecticut on a ground gunnery mission. The airplane suffered an engine failure and the pilot attempted to make it to the Naval Air Station at Westerly for an emergency landing. The pilot realized that he was losing too much altitude and that he was not going to make the field. The pilot was then forced to make a wheels-up emergency forced landing in a field that was littered with huge boulders. The pilot made a perfect landing but the P-47 collided with several boulders, causing it to break up and burst into flames.

4-14-43D. Coffeyville, Kansas. At 1120 CWT, a Vultee BT-13A crashed 16 miles east of SSE of Coffeyville, Kansas, killing A/C John H. Dumas. Investigators stated, "2Lt. Leon M. Phillips and two solo students were flying a V-formation at about 2,800 feet indicated [altitude]. Cadet Dumas was on the left wing of the instructor and Cadet Carson O. Thomas on the right. Lt. Phillips, after cautioning the students to stay above him, peeled off gently to the left and went down to about 1,400 feet indicated. The two cadets were to follow in trail and then fall in their former positions. This maneuver was being accomplished to practice 'joining a formation after take-off.' About six seconds after the instructor peeled off to the left, Cadet Dumas followed him in a gentle peel-off, according to Cadet Thomas. The latter continued to fly straight ahead and did not see Cadet Dumas' airplane again. The instructor did not see Cadet Dumas' airplane between the time he (the instructor) peeled off and the time the cadet crashed. It is the opinion of the [Aircraft Accident Classification Committee] that Cadet Dumas continued his peel-off, losing altitude all the while, and crashed, never having seen the ground."

4-14-43E. Brooks Field, Texas. At 1237, a Curtiss O-52 crashed at Brooks Field, San Antonio, Texas, killing pilot 2Lt. Robert E. Kuhn and passenger 2Lt. Robert O. Fechtner. The airplane was returning to Brooks Field after successfully completing a routine navigation flight. The pilot entered the pattern and began a turn for the approach leg at a reduced airspeed. The airplane stalled and fell off into a spin, describing one turn of spin before smashing into the ground, killing the occupants instantly.

4-15-43A. Long Island Sound, New York. At 0730, a Republic P-47C crashed into Long Island Sound about one-half mile off shore of Long Island, New York, killing pilot 1Lt. Don G. Shofner. The airplane had taken off from Mitchel Field, Long Island, on a routine formation training flight when it suffered a mid-air explosion just aft of the engine at approximately 4,000 feet indicated altitude. The entire engine and propeller were blown off of the airframe and it was speculated that the pilot was probably killed or seriously injured in the explosion. The airplane fell out of control into the sound about one mile west of U.S. Coast Guard Station 94. Despite a thorough search, the wreckage of the airplane and the pilot's body were never recovered. It was speculated that because of a leak between the fuel pump and the carburetor, gasoline had gathered in the accessory section behind the engine and was ignited by the hot exhaust manifold.

4-15-43B. Shrewsbury, Massachusetts. At 0905, a Republic P-47C crashed at Shrewsbury, Massachusetts, killing pilot 2Lt. James F. Lyons. The airplane was part of a four-ship flight that had taken off from Bedford Army Air Field, Bedford, Massachusetts, on a high-altitude formation-training mission. The subject airplane was flying in the number-four position at about 28,000 feet msl. After successfully completing the mission the formation broke up. The subject airplane was seen to go into a split-S maneuver at about 27,000 feet and then continue in a dive straight into the ground. The airplane exploded violently upon impact and the pilot killed instantly.

4-15-43C. Whitesboro, Texas. At 0815, a Vultee BT-13A crashed four miles southeast of Whitesboro, Texas, killing A/C Donald J. Jones. Two cadets who were flying in the area reported seeing the airplane spinning at an altitude of approximately 6,000 feet msl. The airplane continued in a flat spin until it smashed into the ground where it burst into flames. The cadet made no observable effort to leave the airplane.

4-15-43D. Goodnight, Texas. At 1255, a Douglas A-20C crashed four miles east of Goodnight, Texas, killing pilot 2Lt. Eugene S. Hulette. The pilot was leading a three-ship flight on a formation-training mission. The three airplanes entered a Lufbery circle and the subject airplane stalled when the pilot tightened his turn too much. The airplane fell out in a spin to the left. The pilot recovered momentarily but the airplane entered another spin to the left. The pilot was unable to recover the airplane and it spun to the ground where it exploded into flames upon impact.

4-15-43E. Fort Myers, Florida. At 1640 EWT, a Lockheed AT-18 crashed while attempting a landing at Buckingham Army Air Field, Fort Myers, Florida, killing 11 fliers. Investigators stated, "[The pilot] was returning from a routine gunnery mission and attempted to land into the north on the 5,000-foot North-South Runway. He made a normal approach, attempted a three-point landing with full flaps and bounced once. A witness from the control tower stated that the pilot then apparently moved his control column forward and bounced a second time, approximately 30 feet [into the air]. It was at this point that the pilot decided to go around again. He applied power and retracted his landing gear. It was observed that the [airspeed] was low and the airplane was in an excessive climb attitude. The airplane was observed to mush and yaw. Inspection of the wreckage indicated that the flaps [were] in the full up position when the crash occurred. At an altitude of approximately 100 feet the airplane stalled off on the left wing and fell into a spin to the left, bursting into flames immediately upon crashing. The airplane had turned approximately 270 degrees to its original heading when it crashed. The airplane crashed on the outer edge of the ramp, approximately 1,000 feet short of the north end of the North-South Runway." Killed in the crash were: 2Lt. Ivy N. Gardner, pilot; SSgt. Herman L. Eversole, engineer; Pvt. Edward S. Gajzik, student gunner; Pvt. Edmon W. Pressley, student gunner; Pvt. Jack M. Qualls, student gunner; Pvt. Chester I. Seratta, student gunner; Pvt. George Rabey, student gunner; Pfc. Robert L. Lemper, gunnery instructor; Pfc. Raymond W. Brande, gunnery instructor; Pvt. Robert M. Clarkson, gunnery instructor; Pvt. Jack W. Naugher, passenger.

4-15-43F. Muskogee, Oklahoma. At 1629, a Douglas A-20C crashed near the Army Air Base at Muskogee, Oklahoma, killing pilot 2Lt. Tad Blackston. The pilot had taken off from the base at about 1430 for his last two hours of transition training. The pilot had made two satisfactory landings and was "doing very well." While he was coming in for his final landing, personnel in the control tower noticed that the pilot was crabbing the airplane in, correcting for a 15 to 25 mph crosswind. The approach was to the southwest and looked satisfactory, and the airspeed appeared to be adequate. When the pilot broke the glide and released the crab in an attempt to land, the airplane began to drift to the left. As the airplane drifted to the left it appeared to begin to turn to the left. Witnesses reported that the port propeller appeared to be turning slower than the right propeller. The airplane was heading due south at about 15 feet above the southwest runway. The Control Officer radioed the pilot to give the airplane throttle and go around. There was no response from the pilot. The A-20 was observed to enter a steep bank to the left, continuing to roll to the left until the airplane rolled through inverted and the starboard wing struck the ground. The airplane smashed into the ground and immediately exploded into flames. The pilot was killed instantly.

4-16-43A. Peterson Field, Colorado. At 1230, a Lockheed F-4 (photo-reconnaissance version of the P-38) suffering an engine failure crashed while attempting an emergency landing at Peterson Field, Colorado Springs, Colorado, killing pilot 2Lt. Joseph G. Willette. Investigation revealed that this was the pilot's first solo flight in a F-4 type airplane and it was speculated that the pilot was "nervous." The pilot had asked for taxi instructions but did not request take-off clearance before he took off. An officer in the control tower noticed that the port propeller was apparently in the manual position. The tower tried to contact the pilot and warn him of the propeller condition but received no response. It was noted that all F-4 and F-5 airplanes are to keep in radio contact with the field at all times. Just after take-off, the port propeller apparently ran away and the pilot feathered the propeller. Investigators stated that the pilot did not use the correct emergency procedures and did not trim the airplane for single-engine flight when he attempted to return to the field. The airplane was then seen to go into a spin to the left. The pilot was unable to recover and the airplane spun to the ground, exploding into flames upon impact. Investigators noted that the pilot had very little study of emergency procedures for the P-38/F-4 type airplanes.

4-16-43B. MacDill Field, Florida. At 1832 EWT, a Martin RB-26A crashed while attempting an emergency forced landing at MacDill Field, Tampa, Florida, killing three fliers. The airplane had taken off from MacDill Field on a routine instrument training flight. As the airplane flew over MacDill Field the port engine began "cutting out." Investigators stated, "The pilot called the tower from an altitude of 2,500 feet over the field, requesting an emergency landing as his left engine was missing. He was cleared to land and then he proceeded to make a right-hand traffic pattern. While [turning onto] the final approach leg at an altitude of about 1,500 feet, wheels down, the left engine burst into flames. At this point, power was still on the bad engine. At an altitude of about 200 feet the left wing dropped and the ship [rolled to the left and] into the ground and completely burned just off of Runway 13, killing all persons on board." Pilot 2Lt. Homer A. Wood, Sherman, Texas, co-pilot 2Lt. James J. Fahy, Philadelphia, Pennsylvania, and engineer Pfc. Arthur Latter, New York, New York, were killed in the crash.

4-16-43C. Warrensburg, Missouri. At 2017, a Douglas C-47 and a C-47A collided in mid-air near Sedalia Army Air Field, Warrensburg, Missouri, killing three fliers in the crash of the C-47. The C-47A was able to crash land safely at Sedalia Field, and the six crewmen and passengers were uninjured. Pilot 2Lt. Henry J. Trawinski, co-pilot Ross L. Livasy and engineer Pfc. Philip H. Bacon were killed in the crash of C-47 #41-18383. Crewmembers uninjured in the crash landing of C-47A #42-23367 were: 2Lt. Fitzhugh S. Bostwick, pilot; 2Lt. Samuel W. Fairchild, co-pilot; Pfc. William P. Thyne, engineer; Cpl. Albert J. Stein, passenger; Pvt. Adelard A. Granger, passenger; Pvt. Roland E. Klawitter, passenger. Investigators stated, "At 1959, pilot Lt. Trawinski [in C-47 #41-18383] called the control tower and gave his position as four miles southwest of the field requesting permission to enter the traffic pattern. Meanwhile Lt. Bostwick in Airplane #42-23367 was accomplishing follow through landings and take-offs with two other aircraft in the pattern. At 2009, 41-18383 called on the base leg for landing clearance. Approximately 30 seconds later [the pilot of] 42-23367 called giving [his] position [as being on the] base leg, requesting landing clearance. The tower operator stated, 'Ships were positioned satisfactorily on the base leg.'

Evidently neither pilot knew the position of the other aircraft. Tower did not assign landing sequence for either airplane, only stated, 'Cleared to land' to each pilot at each pilot's radio contact. The next time the tower looked at the two approaching aircraft they appeared to be very close [to each other, with] one trailing the other [that was] slightly lower. The operator called to the ship in the #2 position to 'Pull up, Pull up.' Airplane #41-18383 [in the number-one position] pulled up immediately into the propellers of Airplane #42-23367, which struck the fuel tanks; Airplane #41-18383 exploded and plummeted to earth.... The crew of Airplane #41-18383 was killed and the aircraft was demolished by [the] crash and fire.... The pilot of Airplane #42-23367 stated that as he approached the runway at one-quarter mile from the end of the runway he felt and heard a terrific crash and the aircraft wobbled and stalled out of control. The pilot then saw a flash from the explosion of the other aircraft, and then saw all but the empennage rise in front of his airplane, and then it fell away. Lt. Bostwick and Lt. Fairchild, both using all of their strength, forced the control column forward and the nose went down, engines and switches were cut, gasoline off, flaps in one-half position down. Lt. Bostwick landed straight ahead off of the runway. The [port] elevator was burned off and the lower surface of the right elevator burned. The right propeller blades were bent, left propeller was torn off at the [time of the collision] in full flight. The nose section of the left engine was torn away with the propeller."

4-17-43A. Sea Bay, New Jersey. At 0900, a Republic P-47D crashed into Sea Bay about three miles north of Seaside Heights, New Jersey, killing pilot 2Lt. Morgan J. Barton. Investigators stated, [Lt. Barton] was seen in a spin over Sea Bay, New Jersey. When the airplane was approximately 3,000 feet [above] the ground, it came out of the spin and went into a steep dive in the opposite direction and crashed into the bay. The pilot was not seen to jump from the airplane but was found about 50 yards from the crash floating face down. The pilot's parachute had not been [deployed]."

4-17-43B. Starke, Florida. At 0945 EWT, a Douglas DB-7B and an A-20C collided in mid-air at Starke, Florida, killing gunner Sgt. Albert J. Posey aboard the DB-7B. DB-7B pilot 2Lt. Alliston Slade and observer 1Lt. Paul R. Hable were able to parachute to safety. Gunner Sgt. Stanley Ruszkiewicz and SSgt. George Lewis parachuted to safety from the heavily damaged A-20C. The pilot, 2Lt. John R. Crandall, was able to make a successful emergency wheels-up landing in an open field just outside of Starke. SSgt. Lewis received serious injuries parachuting to safety. The Aircraft Accident Classification Committee stated, "Lt. Crandall, flying A-20C #41-19178, was leading a two-plane formation. The other ship, DB-7B #AL-353, was being piloted by Lt. Slade. The formation had been executing a series of diving and climbing turns. A climbing turn to the right had been completed when Lt. Slade, flying the left wing, experienced some trouble with his right engine. He pulled forward to the left and below Lt. Crandall's airplane. Lt. Crandall failed to observe this maneuver. Lt. Slade then checked his instruments and while this check was being made the airplanes collided [with] Lt. Slade's airplane apparently coming from below and from the left, his propeller cutting through the nose section of the A-20C. AL-353 then went out of control. Lt. Slade and Lt. Hable bailed out successfully. Sgt. Posey remained with the airplane. Sgt. Ruszkiewicz and Sgt. Lewis jumped successfully from A-20C #41-19178. Lt. Crandall then proceeded to make a crash landing."

4-18-43A. Homestead, Florida. At a time between 1100 and 1115 EWT, a North American B-25D suffered a catastrophic structural failure and crashed 15 miles SSW of the Army Air Field at Homestead, Florida, killing the crew of six. Killed in the crash were: 1Lt. Ralph B. Wildenhaus, pilot; 2Lt. James B. Jellison, co-pilot; TSgt. Andrew J. Moll, engineer; SSgt. Jay H. Fisher, passenger; Seaman 2nd Class Ralph I. Shriner, USN passenger; Seamen 2nd Class Robert M. Lillard, USN passenger. Investigation of the wreckage revealed that the port wing leading edge and the tail section failed in flight and caused the airplane to go into an uncontrolled flat spin. The airplane smashed to the ground in a remote area and exploded into flames. It was later speculated that the port wing leading edge had failed, separated from the wing and collided with the tail section, causing its progressive failure. The port wing leading edge and tail section, broken in two pieces, were found about one mile southwest of the main wreckage. The airplane had taken off from Boca Chica, Florida, on an anti-submarine training flight.

4-18-43B. Alamogordo, New Mexico. At 1535, a Consolidated B-24E broke up in flight and crashed approximately 30 miles WSW of the Army Air Base at Alamogordo, New Mexico, killing seven crewmembers and injuring two others. Killed in the crash were: 2Lt. Howard G. Homberger, navigator; 2Lt. Vytant J. Bakas, bombardier; SSgt. Carl A. Merkling, engineer; SSgt. James M. Eltzroth, radio operator; SSgt. Jess Sympson, assistant engineer; SSgt. Anthony W. Hafer, assistant radio operator; SSgt. Ralph V. Pelletier, gunner. Pilot F/O Harry H. Herrmann was seriously injured and co-pilot 2Lt. Earl B. Gray suffered minor injuries parachuting to safety. Investigators stated, "Airplane No. 42-7025 took off at approximately 1453 MWT, climbed to 18,000 feet for a local bombing mission. During the climb, the Automatic Flight Control Equipment was used for a period of 4 to 5 minutes, after which time it was turned off and the airplane continued in flight, the co-pilot flying, the pilot checking check lists, equipment, etc. At approximately 18,300 feet the co-pilot leveled off and assumed a compass course, as directed by the navigator. During the course of this flight the co-pilot saw clouds ahead and apparently called this fact to the attention of the pilot, who told the co-pilot to fly around the clouds, not realizing the clouds were as close as they were. Just before entering the clouds the co-pilot turned the controls over to the pilot, who immediately attempted to turn and get away from the cloud formation. However, it was so close that he was unable to turn away from the clouds and did go into them. At this point he was losing altitude rapidly and attempted to nose the airplane up to reduce speed. The co-pilot, believing that the pilot needed help on the controls, applied pressure to bring it out quicker, and the airplane went into a tail stall condition and flipped over on its back. In addition, excessive turbulence was encountered which helped to throw the airplane into an unnatural position, stunning the pilot and co-pilot and thereafter the airplane started tearing apart. Some ice and hail were encountered in the cloud, which tends to show that there was excessive turbulence in this cloud. There is no evidence that any explosion occurred until such time as the fuselage and what was left of the wings, with gas cells, struck the ground, at which time it exploded and burned. There was excessive tearing apart of the airplane before reaching the ground, inasmuch as two wing tips, broken off just outside of the outer engine cells, the right wing stub, including the inboard engine and the outboard engine mount and one wheel, tail surface and other minor parts of the airplane, were scattered over some mile and a half to two miles from the burned part of the wreckage. The nose section of the airplane was very close to the wreckage and was broken off at about the forward part of the bomb bay. The tear of the two wing tips indicate that the airplane was upside down at the time the two wing tips broke off, inasmuch as the tear was from the top toward the bottom, indicating that the stress on the wing tip was on the top side of the wing, rather than on the bottom. It is possible that the airplane started [coming] apart at about 12,000 to 15,000 feet and that the [disintegration] continued until [the airplane] reached the ground. Weather reports indicate that some turbulent conditions existed in the cumulus clouds over the mountains. The clouds encountered by this airplane were built up over the San Andreas Range and the wreckage, when found, was on the side of the foothills building into this mountain range. At the point of breaking off of the nose section, the pilot and co-pilot were thrown free and descended in parachutes. The pilot's parachute showed evidence of having been either torn during the [aircraft breaking up] or was damaged from falling debris."

4-19-43A. Fort Myers, Florida. At 0925 EWT, a Bell P-39F dove into the ground while performing aerobatics near Fort Myers, Florida, killing pilot 2Lt. Leon E. Forsyth. The airplane had taken off from Page Field, Fort Myers, Florida, on a routine transition flight. The Aircraft Accident Classification Committee stated, "The pilot was on his first flight in this type of aircraft. The pilot started a loop from approximately 3,000 feet and completed the same at approximately 600 feet. It is believed that the pilot cut the throttle as he completed the loop in order not to collide with the ground. This conjecture is derived from witness' reports that the engine was sputtering. The aircraft completed the loop in a high-speed stall. This is indicated by the fact that witnesses said that the aircraft was wobbling. As the pilot opened the throttle again the nose of the aircraft dropped, and the aircraft dove into the ground. The accident was fatal to the pilot [and] the airplane completely destroyed. It is the opinion of the committee that the pilot exercised poor judgment in executing this maneuver at so low an altitude."

4-19-43B. El Reno, Oklahoma. At 0920, a Fairchild PT-19A crashed one and three-quarters miles southeast of El Reno, Oklahoma, killing civilian instructor H.W. Parker and student A/C Allen T. Reasoner. The airplane had taken off from El Reno Army Air Field on a "Civilian progress check flight" at about 0855. Investigators could not determine why the airplane crashed.

4-19-43C. San Marcos, Texas. At 0950, a Lockheed AT-18A crashed two miles north of San Marcos, Texas, killing the crew of three. The airplane had just taken off from the Army Air Field at San Marcos on a local transition flight when the port engine apparently failed at an altitude of approximately 300 feet above ground level. The pilot attempted to regain power and maneuver the airplane back toward the field. The port wing was seen to drop but the pilot was able to right the airplane. The port wing dropped again and the pilot lost control. The airplane rolled over to the left to an inverted attitude and then entered a spin, describing one turn of spin before smashing into the ground and exploding into flames. Two crewmembers were thrown clear upon impact but were killed instantly. Pilot 1Lt. Bradford K. Herman, co-pilot 2Lt. Earl P. Williams and pilot-rated passenger F/O Roger B. Denman were killed in the crash.

4-19-43D. Pittsburgh, Texas. At 1030, a Vultee BT-13A crashed four miles northwest of Pittsburgh, Texas, killing A/C Joseph P. Scully, Jr. and A/C William J. Schroeder. The airplane had taken off from Majors Field, Greenville, Texas, at 0930 CWT on an instrument team ride to Mineola, Texas, to Pittsburgh, Texas, and return to Greenville. The airplane was seen at an altitude later estimated to be about 3,000 feet agl when it went into a medium diving 180-degree turn until it was flying south. The airplane leveled out and then began another diving turn until it was flying at low altitude to the west. The airplane then dove on a farmer who was plowing a field. The airplane pulled up abruptly at a low altitude and mushed into trees in a wooded creek bottom near the plowed field. The airplane was estimated to be traveling at about 240 mph when it collided with the trees. Wreckage was scattered over an area of 150 yards and both students were killed instantly. Investigators noted that this was the last scheduled flight for the students prior to graduating from Basic Flight Training.

4-19-43E. Barksdale Field, Louisiana. At 1052, a Martin B-26B crashed while attempting a landing at Barksdale Field, Shreveport, Louisiana, killing three crewmembers and injuring two others. The airplane was returning from a routine training flight. Investigators stated, "The airplane made a normal approach and definitely had an unsafe slow speed and was landing on Runway 32R. At approximately 100 feet altitude, the airplane was seen to rock as though it had either struck [propeller turbulence] or had reached the stalling point. The right wing then dropped and the airplane struck the ground about 800 feet to the right of the runway with the right wing first, the right engine and then the nose, and then [the B-26 cartwheeled into the ground]. The pilot stated that he was not obtaining full power on his right engine and substantiated by examination of the wreckage, which indicated that he had 13 degrees of rudder control. The pilot did not give the Control Tower the impression that he was shooting an emergency landing." Killed in the crash were: SSgt. Lloyd M. Barnes, engineer; SSgt. William F. Hinds, engineer; SSgt. John M. Randall, bombardier. Pilot F/O James O. Stutler and co-pilot 2Lt. George W. Manning were seriously injured.

4-19-43F. Fritch, Texas. At 1310, a Douglas A-20A collided with telephone wires and crashed five miles north of Fritch, Texas, killing pilot 2Lt. George M. Mickie. The airplane had taken off from the Army Air Field at Amarillo, Texas, on a local navigation-training mission. The Aircraft Accident Classification Committee stated, "A rancher in the vicinity reports subject airplane flying at low altitude along the canyon formed by the Canadian River bed north of Fritch, Texas. Suddenly he heard a crash, which he estimated to have been at a point where he knew a telephone [line] supporting cable to be suspended across the riverbed. This was a three-eighths inch cable across the river, which is approximately five-eighths of a mile wide at this point. This rancher was the only witness to the accident. The witness reports that he did not see the airplane hit the cable, but that he heard another crash a few seconds after the first one. Investigation found the wreckage of the airplane imbedded into the side of the cliff forming the south wall of the river canyon. The condition of the wreckage indicated that the airplane landed upside down while traveling at a high rate of speed. The location of the crash was approximately five-eighths of a mile distant from the severed telephone [line] supporting cables. The pilot's cap, part of the engine top cowlings, parts of the cowl flaps, and parts of the vertical tail fin were found at the approximate spot where the airplane is believed to have hit the telephone [line] supporting cables. The bulk of the airplane was found in the side of the cliff along the riverbank a short distance away. The height of the cables were estimated to have been approximately 30 feet above the river bed where the airplane struck."

4-19-43G. Phoenix, Arizona. At 1410 MWT, a Curtiss P-40F crashed one and one-half mile northeast of Auxiliary Airfield #1 (located five miles north of Luke Field) Phoenix, Arizona, killing pilot 1Lt. Barry P. Fisher. The airplane made a normal take-off and reached an altitude of about 250 feet when the engine apparently failed. Instead of making an emergency landing straight ahead, the pilot turned the airplane to the left, straightening it out after about 90 degrees of turn. The starboard wing suddenly dropped and the airplane entered a spin. The airplane struck the ground with the starboard wing and burst into flames immediately upon impact. Investigation revealed that the pilot had stood up in the cockpit just before impact and was killed instantly.

4-20-43A. Titusville, Florida. At 1037 EWT, a Douglas P-70 crashed 12 miles southwest of Titusville, Florida, killing pilot 2Lt. Benton C. Gant. The airplane had taken off from Orlando Army Air Base, Orlando, Florida, on a "supervised single-engine operation flight." The airplane was flying with the port engine shut down and the port propeller feathered at about 7,000 feet msl when the pilot lost control. The airplane entered a spin to the left and the pilot could not recover. The instructor radioed for the pilot to break the spin by retarding the throttle on the starboard engine and using the rudder. The instructor repeated his instructions but the airplane continued to spin. When the airplane reached an altitude of 4,500 feet, the instructor ordered the pilot to abandon ship. The P-70 continued to spin and smashed to the ground, killing the pilot instantly.

4-20-43B. Richland, Texas. At 1450, a Fairchild PT-19A crashed at Richland, Texas, killing civilian instructor Harry F. Hernandez and seriously injuring student A/C Vincent W. McGinnis, Tulsa, Oklahoma. The airplane had taken off from Corsicana Army Air Field, Texas, on a routine dual instruction flight. From the statements of witnesses, investigators speculated that the instructor and student were performing a simulated emergency landing when the crash occurred. As of 4-29-43, A/C McGinnis still had not regained consciousness.

4-21-43A. Luverne, Alabama. At 1000, a North American AT-6C crashed five miles north of Luverne, Alabama, killing student pilot 1Lt. Harold R. Maher, Medford, Massachusetts. The airplane had taken off from Napier Field, Dothan, Alabama, on a low-altitude navigation mission. Investigation revealed that the airplane had been buzzing a field prior to the accident. Two farmers who were plowing a field stated that the airplane had flown toward them at a very low altitude from the direction of Luverne. The airplane then pulled up steeply and entered a steep turn, the student waiving at the two men as he flew by. The airplane circled the area and dove on the two farmers again. The pilot was not able to pull out of the dive and the airplane smashed into the ground where it erupted into flames. The student was killed instantly. Civilian Moses Balford received minor injures trying to rescue the student.

4-21-43B. Fort Myers, Florida. At 1620 EWT, a Bell P-39D crashed near Page Field, Fort Myers, Florida, killing pilot 2Lt. Jesse Moore, Jr. Investigators stated, "The pilot was the number-three man in a three-ship string formation practicing turns, lazy-eights, and chandelles [at approximately 12,500 feet indicated altitude]. At the completion of a steep chandelle [to the left at an indicated altitude of approximately 10,500 feet] the pilot stalled the aircraft and it snapped off into a spin. The pilot made three recoveries from the spin but snapped back into one each time. The aircraft crashed with the pilot [on board]. It is the opinion of the Aircraft Accident Classification Committee that the pilot [put] the aircraft into a spin as a result of reefing the aircraft in the chandelle until it snapped, and that he failed to recover."

4-21-43C. Location Unknown. At a time after 2052 EWT, a Consolidated B-24D with a crew of ten was presumed to have crashed into the Atlantic Ocean approximately 210 miles east of Virginia Beach, Virginia. The airplane had failed to return to its base at Langley Field, Hampton, Virginia, from an anti-submarine patrol and was declared missing. Investigators stated, "A distress (SOS) signal was intercepted at 2420Z [Greenwich Mean Time]. There was no associated message. After sending MO's interspersed with distress signals until 2452Z, all communication ceased. Two aircraft and a merchant ship in the vicinity were notified immediately and a Navy destroyer was also dispatched to search and render assistance. Eight airplanes of the 18th Anti-Submarine Squadron took-off at approximately 0900Z [0500 EWT], April 22, 1943 to conduct a search in the vicinity of the position last determined by radio direction finders. No personnel or material has been located." Crewmembers missing and presumed lost at sea were: 1Lt. Fred M. Speake, Jr., pilot; F/O Charles L. Barbour, co-pilot; 1Lt. Charles H. Dingus, navigator; 2Lt. James W. Millikan, bombardier; TSgt. Williard G. Garrett, engineer; SSgt. Titus E. Hearn, assistant engineer; SSgt. Harold G. Danley, radio operator; Pvt. Sheldon J. Wonderdahl, assistant radio operator; Cpl. Eugene T. Bragg, gunner; Cpl. Howard F. Prestige, gunner.

4-21-43D. Evansville, Indiana. At 2326 CWT, a Lockheed C-60A crashed while attempting a landing at the Evansville Municipal Airport, Evansville, Indiana, killing the ten crewmen and passengers on board. The Aircraft Accident Classification Committee stated, "Pilot was [on a personnel transport flight] from Cleveland, Ohio, to Selman Field [Monroe, Louisiana]. It is the opinion of the board that half way en route the pilot encountered weather, which caused him to decide to return to Cleveland or to some alternate station. After checking the weather along the route, it is the opinion of the board that the weather at Memphis, Tennessee, was such that due to the limited amount of time acquired in this model airplane, it caused the pilot to hesitate to proceed to his destination. The board believes the radio was out, based on the fact that Evansville Tower could hear the airplane, and gave him several long calls, but he apparently never received the tower. Upon failing to contact Evansville Tower, the pilot circled the landing field twice, received a green light from the control tower and in turn acknowledged the green light by blinking his navigation lights. He then proceeded to land from the south into the north, which was into the wind. Evidently the pilot realized that he was high on his approach and dropped full flaps and complete power off. This condition would require extreme tail-heavy setting of the elevator trim tab. While approximately 50 feet in the air, the pilot apparently decided to go around. He then pushed both throttles forward to go around and evidently failed to re-trim the elevator control tabs. Due to the sensitivity of the control tab on this type of airplane, it is believed by the Board that the pilot could not push forward hard enough on the control column to prevent the airplane from zooming [up] violently, therefore causing the airplane to assume an unusual angle of climb, consequently stalling. The left wing dropped viciously and the airplane went into the ground, practically nose first and immediately exploded and burned. The trim tabs [that remained intact] on the elevator were both in the full tail heavy position." Crewmembers killed in the crash were: Capt. William V. Litvin, pilot; 1Lt. Carlton B. Overton, Jr., co-pilot; 2Lt. Charles W. McLaughlin, navigator. Passengers killed in the crash were: 1Lt. George F. Dennison, pilot; 2Lt. Howard D. Baer, pilot; Capt. Paul D. Jones, navigator; 2Lt. Michael W. Schaffer, navigator; 2Lt. John W. Kerrigan, navigator; SSgt. Arthur N. Vandeli, engineer; SSgt. James P. Taylor, engineer.

4-22-43A. Farmingdale, New York. At 1020, a Republic P-47C dove into the ground at Farmingdale, Long Island, New York, killing pilot 2Lt. Raymond G. Barnes. The airplane had taken off from the Army Air Base at Farmingdale on a high-altitude training mission. It was speculated that the oxygen system failed and the pilot lost consciousness, causing the airplane to enter a steep dive toward the ground from 28,000 feet msl. The pilot apparently regained consciousness while the airplane was diving in instrument conditions. As the airplane dove out of the overcast at 2,000 feet agl the pilot pulled up sharply and the airplane climbed vertically into the overcast. The airplane emerged from the clouds seconds later in a power dive toward the ground. The airplane smashed into the ground, exploding violently into flames upon impact. The airplane was destroyed utterly. Investigators noted that the remaining small bits of the airplane were so "demolished that most of the parts were unrecognizable."

No pieces of the oxygen system could be found to verify an oxygen failure and it was noted that the pilot's body was so badly mangled and dismembered that conclusive evidence was impossible.

4-22-43B. Fitzgerald, Georgia. At 1500 EWT, a Vultee BT-15 crashed three miles west of Fitzgerald, Georgia, killing students A/C William H. Young and A/C Raymond J. Murphy. The airplane took off from Cochran Field, Macon, Georgia, at 1330 EWT on an instrument team ride to Manchester, Georgia, to Fitzgerald, Georgia, and return to Cochran Field. A/C Murphy was the instrument pilot and A/C Young was the safety pilot in the front cockpit. The airplane was observed by witnesses on the ground flying at an altitude of approximately 2,000 feet. The airplane was seen to enter a steep dive toward the ground. The airplane climbed steeply then performed a slow roll at about 2,000 feet. The airplane was then observed flying at very low altitude. The airplane performed a turn at low altitude and the port wing struck the terrain, causing the airplane to smash into the ground. The students were killed instantly. There was no fire but the airplane was completely destroyed.

4-22-43C. Matagorda Peninsula, Texas. At about 1400, armorer Pvt. Robert Benjamin was killed when he was shot in the head by a thirty-caliber machine gun installed in the starboard wing of a North American AT-6C, which had accidentally discharged while being serviced at the Army Air Field at Matagorda Peninsula, Texas. Pvt. Benjamin was granted permission to ride on the gunnery mission that was to be flown by 2Lt. Norman G. Bedgood. Pvt. Benjamin was assisting Sgt. Charles W. Carter in synchronizing the cowl machine gun. Pvt. Benjamin stepped back after turning the propeller while assisting Sgt. Carter and stood directly in the line of the starboard wing machine gun. Sgt. Carter alerted Lt. Bedgood, who was sitting in the front cockpit filling out the Form-1, that the cowl machine gun was loaded and ready. Sgt. Carter, not knowing that the wing machine gun was loaded, reached into the cockpit, turned on the power and then inadvertently fired the guns. Pvt. Benjamin was struck in the head by four thirty-caliber rounds and killed instantly.

4-22-43D. Hilliard, Florida. At 1950 EWT, a North American A-36A crashed five miles southeast of Hilliard, Florida, killing pilot 2Lt. Carl D. Horn. The Aircraft Accident Classification Committee stated, "The subject pilot disappeared from his [three-ship] formation while on a supervised night cross-country flight from Tampa, Florida, to [his base at] Waycross, Georgia. Neither of the other men in the flight noticed his absence for several minutes since they were in the act of changing positions in the formation. The pilot gave no indication of trouble in spite of the fact that he had just held [inter-airplane] communication with the other ships in the flight. For this reason, the search for the missing ship was only localized within a 25-mile area and considerably prolonged because of it. The pilot, deceased, and the completely demolished airplane were found April 26. Due to the condition of the wreckage nothing could be determined as to the cause of the accident from it. The pilot's parachute had been opened at an extremely low altitude; it was found opened to the full length, unfolded but [the canopy] not open. The radio antenna wire was wrapped around the pilot's right ankle. It is probable that he delayed opening the parachute in an attempt to free himself from the wire. It is possible that the pilot was suffering from some degree of vertigo. It was at his radioed request that he was taking over the lead ship position and this might be taken as some explanation of his failure to parachute to safety and to call the other ships as to his emergency condition. It is believed that his airplane must have been in an unusual attitude to cause his entanglement in the antenna wire. It appears that the only way that the pilot could have become entangled in the antenna would be to bail out in an inverted or near inverted position."

4-23-43A. Fredericksburg, Texas. At 0905, a Beech AT-7 flying in instrument conditions collided with terrain 22 miles southeast of Fredericksburg, Texas, killing three crewmembers and seriously injuring another. Student navigators A/C Frederick Swift, A/C Charles A. Taylor, Jr. and A/C John R. Zaic, Jr. were killed in the crash. Pilot 2Lt. Robert A. Gibson was seriously injured. The airplane was part of a flight of ten that had taken off from the Army Air Field at San Marcos, Texas, on a routine student navigation mission to Fredericksburg, to Cross Plains, Texas, to Clifton, Texas, and return to San Marcos. When the flight encountered poor weather near Blanco, Texas, the flight leader ordered all of the airplanes to make a 180-degree turn and return to the field at San Marcos. Evidently the subject airplane did not get the message and attempted to remain under the overcast, which apparently extended to ground level. Examination of the wreckage indicated that the airplane, while flying due north, had apparently descended into the terrain on the downward slope of a hill at an elevation of 1,900 feet, about 400 feet from the crest. The airplane cut a path downhill through the trees for a distance of about 600 yards before crashing to a stop and bursting into flames. The pilot and two student navigators were thrown clear of the wreckage. Investigators noted that the airplane was about three miles off of its intended course. The other nine airplanes were able to land safely at San Marcos Army Air Field.

4-23-43B. Hyman Field, Alabama. At 1015, a North American AT-6C suffered a catastrophic structural failure and crashed five miles north of Hyman Field, Alabama, killing instructor 2Lt. Elmer G. Raymond and A/C Jack Howell. The subject airplane was the lead airplane of a three-ship flight that had taken off from Napier Field, Dothan, Alabama, on a routine formation training flight. The formation was flying in trail at about 2,000 feet agl when the lead airplane peeled off and entered a split-S into a steep dive. The trailing ships followed and the leader pulled up at about 500 feet at an estimated speed of 200 mph. The airplane snapped into a roll and after one complete roll the number-three pilot observed the starboard wing separate from the airplane. The airplane continued to roll to the right and then went out of control, smashing into a stand of trees. The airplane did not burn but was completely destroyed.

4-23-43C. Paris, Texas. At 1030, a Vultee L-1 suffered a structural failure and crashed one mile south of Cox Field, Paris, Texas, killing pilot 1Lt. James E. Wurdack and engineer-rated passenger SSgt. William B. Valle. The airplane was on a scheduled test flight. Investigators stated, "Approximately two minutes before the airplane crashed, it flew over the field and dropped a message as follows: '[I am] calling on [frequency] 4495. Will received on 4495.' Shortly after this, the airplane was seen going into a shallow, diving turn south of the field at between 1,000 feet and 1,500 feet. At about 700 feet the dive and bank to the right became much steeper and the ship turned over on its back and nosed straight down. The airplane buried itself in the ground about three to four feet deep and the occupants were killed instantly. The two bodies were buried beneath the wreckage and could not be seen at all until part of the wreckage had been removed. There was no fire. At 1850, April 27, 1943, Sgt. Kenneth F. Linson found another message a few feet off the end of the southeast runway. This message was evidently dropped several minutes before the second message and was not seen by anyone on the field. The first message [read:] 'Right [wing] [lift] strut broken, right wing bending. Cannot slow down or will spin. Stand by to pick us up. Calling on 4495. Receiving on 4495. Wurdack.'"

4-23-43D. Tyndall Field, Florida. At 1122 CWT, a Lockheed-Vega AT-18 crashed while attempting a take-off at Tyndall Field, Panama City, Florida, fatally injuring pilot 1Lt. Raymond I. Pipal. Crewmember Sgt. H.E. Mills and engineer Sgt. D.A. Alm were injured. Six gunnery students and three passengers were injured, one seriously. Gunnery students injured in the accident were: Pvt. J. Alton, Pvt. F.D. Anderson, Pvt. R.C. Anderson, Pvt. R.S. Diamond, Pvt. T.J. Maderak, Pvt. A.L. Thompson. Two pilot-rated passengers, 1Lt. Dudley W. Gilmore and 1Lt. Don L. McDowell, sustained minor injuries and passenger Pvt. E.H. Morgan was seriously injured. The Aircraft Accident Classification Committee stated, "According to the statement of Lt. Gilmore, the aircraft made a normal take-off, climbed a few feet in the air and the left wing dropped slightly, was straightened up by the pilot and then the left wing dropped again. The aircraft struck the ground in an attitude of approximately 35 degrees nose down. The left wing of the aircraft contacted the ground and the remainder of the aircraft struck the ground, both engines bursting into flames. From the discussions with the crew of the aircraft and observations made shortly after the aircraft crashed, Lt. Pipal abandoned the aircraft via the pilot emergency escape hatch and received bad burns, which resulted in his death approximately 12 hours later.... It is the opinion of the committee ... that the aircraft left the ground before safe flying speed had been obtained. It is believed that a wingtip stall aggravated by the use of the ailerons resulted in the crash of the aircraft. The aircraft was taking off on a scheduled gunnery-training mission.

4-23-43E. Venice, Florida. At 1317 EWT, two Martin B-26C type airplanes collided in mid-air near the Army Air Base at Venice, Florida, killing 13 fliers. Killed in the crash of B-26C #41-34818 were: 1Lt. Paul R. Sterpole, pilot; 2Lt. Thomas F. Allen, Jr., co-pilot; 2Lt. Walter S. Koze, bombardier; 2Lt. Edward R. Luoma, observer; SSgt. Frank E. Smith, TSgt. Marvin R. Van Natta, radio operator; Sgt. Clyde F. Trainer, gunner. Killed in the crash of B-26C #41-34790 were: 2Lt. Emil J. Hausch, pilot; F/O Robert J. Baskin, co-pilot; 2Lt. Randolph E. Bailey, bombardier; Pvt. Louis C. Beal, engineer; SSgt. J.H. Jones, gunner; Pfc. Frank P. Miller, gunner. The airplanes were part of a six-ship flight that had taken off from Drane Field, Lakeland, Florida, on a bombing training mission. The formation was flying along a railroad track at an altitude of approximately 100 feet above ground level and a few miles from the bombing target. B-26C #41-34818 was the lead airplane of the second element of three. B-26C #41-34790 was flying in the number-three position of the second element. The flight leader ordered the number-two ship and the number-three ship (41-34790) of the second element to change positions. The airplanes were changing over from right to left echelon when the collision occurred. Investigation revealed that B-26C #41-34790 flew into lead aircraft. The port wing of B-26C #41-34790 was severed in the collision and the aircraft fell to the ground out of control. B-26C #41-34818, which lost the starboard wing in the collision, crashed to the ground with considerable forward momentum, exploding violently into flames upon impact. The port engine and propeller of B-26C #41-34790 were found in the wreckage of B-26C #41-34818.

4-23-43F. Lacoochee, Florida. At 2115 EWT, a Beech AT-11 flying in poor weather crashed east of Lacoochee, Florida, killing the crew of four. The airplane had taken off from the Army Air Base at Orlando, Florida, on a routine searchlight mission and had apparently entered an area of heavy rain. The airplane failed to return to Orlando and was declared missing. After an extensive search, the AT-11 was found on 4-25-43. Investigation of the wreckage indicated that the airplane had struck the ground in a spinning attitude. Killed in the crash were: 2Lt. Earl E. Dosier, pilot; 2Lt. James R. Hillebrandt, co-pilot; Cpl. Vincent Costain, passenger; Pvt. Bennie R. Sodczak.

4-23-43G. Avon Park, Florida. At 2155 EWT, a Martin B-26B crashed while attempting a landing at the AAF Bombing Range Air Field at Avon Park, Florida, killing three crewmembers and seriously injuring two others. The pilot had completed a routine three-hour transition flight and was returning to base when the accident occurred. The airplane was about 10 feet above the runway and about to touch down when the pilot decided to go around. The airplane gained about 150 feet of altitude before it sank into the terrain, skidding for over 600 yards before crashing to a stop and bursting into flames. Navigator-bombardier 2Lt. Nicolas J. Bonnell, passenger SSgt. Billy T. Cartwright and SSgt. Leo F. Rohlfs were killed in the crash. Pilot 2Lt. Norman A. Mauer and co-pilot 2Lt. Sherman G. Knowles were badly burned.

4-24-43A. San Antonio, Texas. At 0900, a Bell P-39F crashed one mile south of Kelly Field, San Antonio, Texas, killing pilot Lt. Col. Alexander G. Greig. Investigators stated, "Lt. Col. A.G. Greig took off at approximately 0854 CWT toward the south on Runway 18 at Kelly Field, Texas. Upon reaching an altitude of approximately 300 feet, at a distance of about one-fourth of a mile south of the field, black smoke appeared to come pouring from the exhaust stacks. The pilot then turned approximately 15 degrees to the right as if to attempt to return to the field. Then [he] immediately turned back to the left to a course approximately parallel to the line of take-off. At a distance of approximately one-half mile south of the field, a large puff of black smoke was observed to come from a point near the rear of the engine. The airplane then lost altitude rapidly and crashed in an open field approximately one mile south of Kelly Field." Pieces of the engine cowling were found scattered along the line of flight. Evidence indicated that the airplane was on fire before it smashed into the ground. Investigators speculated that the accident was caused by the pilot "over-revving of the engine on take-off, resulting in backfire which caused an explosion in the blower section resulting in a fire, [the] flames of which reached the cockpit and prevented a safe crash landing."

4-24-43B. Hemet, California. At 1110, a Ryan PT-22 crashed ten miles southwest of Hemet, California, killing student A/C Irvin F. Offill. Investigation revealed that the airplane had entered a tight spin to the right at an altitude of approximately 3,000 feet agl from which it did not recover. The student had taken off from the AAF station at Hemet on a routine student solo flight.

4-24-43C. Great Salt Lake, Utah. At 1719 PWT, a North American B-25D crashed into the Great Salt Lake approximately five miles south of Locomotive Springs, Utah, killing the crew of five. The airplane had taken off from Mather Field, Sacramento, California, and had failed to return to base following a routine navigation-training mission. The airplane was found on 4-26-43. The airplane had been seen flying at low altitude at various times along its route. Investigation revealed that the pilot had been flying over the Great Salt Lake at very low altitude when he inadvertently flew the airplane into the lake. The propellers had struck the surface and the pilot pulled up sharply, causing the tail section to strike the water and apparently compromising its structural integrity. The starboard wing contacted the surface and the airplane smashed into the lake and broke up. The fuselage snapped in two pieces just aft of the upper turret and the nose separated from the fuselage. The main wreckage was found submerged in about 15 feet of water approximately 1,000 yards off shore. Investigators experienced great difficulty salvaging the wreckage because the airplane broke through the salt bed and was mired in thick mud. Investigation revealed that the crew had apparently survived the initial crash and had drowned or "suffocated" in the salty water. Killed in the accident were: 1Lt. Louis E. Johnson, pilot; 2Lt. James T. Jensen, co-pilot; 2Lt. Frederick P. Rollston, pilot-rated passenger; Cpl. Harold V. Avansino, engineer; Pvt. Dorsey U. Flaherty, passenger. Investigators noted, "The bodies of Lt. Johnson and Lt. Rollston were found 4-27-43, on the east shore of the lake, approximately ten miles southeast of the wreck. The bodies of Lt. Jensen and Cpl. Avansino were found 4-28-43 on the east shore of the lake approximately seven miles southeast of the wreck. Parts of clothing of each of the above named men were removed, i.e., shoes and shirt, indicating that personnel prepared themselves for swimming to shore before leaving the wrecked airplane."

4-24-43D. Tullahoma, Tennessee. At 2230 CWT, a Douglas A-20B crashed one mile NNW of Northern Field, Tullahoma, Tennessee, killing the crew of three. Investigation revealed that the airplane was taking off on a routine night transition flight. The take-off appeared to be normal. When the airplane reached an altitude of approximately 150 feet it suddenly burst into flames. The flames died down as the airplane turned about 45 degrees to the left of the take-off heading and then dove into a wooded area at an angle of approximately 30 degrees, striking the ground left wing first. The airplane exploded violently upon impact, scattering wreckage and body parts over a wide area. Pilot 2Lt. Theodore J. Schwink, observer 1Lt. Harley W. Jefferson and gunner SSgt. Edward E. Eames were killed instantly.

4-24-43E. Pocatello, Idaho. At an unknown time, a Boeing B-17F crashed near the Army Air Base at Pocatello, Idaho, killing three unidentified officers and injuring six crewmembers. No other information is available. Crewmembers seriously injured in the crash were: 2Lt. Thomas C. Casady, Montgomery, Alabama; Sgt. Marshall Groce, Peytonsburg, Kentucky; Sgt. Clifford C. Province, Cincinnati, Ohio; Sgt. Richard C. Schick, Philadelphia, Pennsylvania; Sgt. Woodrow W. Shoemaker, Fabens, Texas; Sgt. Stanleight E. Turner, Conway, Missouri.

4-25-43A. Gowen Field, Idaho. At 0129 MWT, a Boeing B-17F collided with rising terrain seven miles northeast of Gowen Field, Boise, Idaho, killing four crewmembers and seriously injuring five others. Killed in the crash were: F/O Horace V. Carpenter, pilot; SSgt. Walter F. Stocklas, radio operator; SSgt. Allen B. Vincent, gunner; SSgt. Robert H. Kingsbury, gunner. Crewmembers seriously injured in the accident were: 2Lt. Craig W. Harrington, co-pilot; 2Lt. Raphael R. Strauss, navigator; 2Lt. Herbert H. Smith, bombardier; SSgt. Clarence H. Nelson, engineer; SSgt. George J. D'Amato, assistant engineer. The airplane had taken off from Gowen Field at 0125 MWT on an instrument training flight, which was also to include some dry bombing runs in the area west of the field at an altitude of 7,500 feet. The airplane made an exceptionally long take-off run to the southeast, lifting off and flying off to the east for about four minutes. The airplane then turned to the north and collided with a shallow slope below Lucky Peak at an elevation of 5,800 feet where it burst into flames upon impact. Investigation revealed that all four engines were operating at the time of the collision and weather was not considered a factor in the accident. Investigators could not explain the unusually long take-off run. The surviving crewmembers were able to walk to a house where an ambulance picked them up later that morning.

4-25-43B. San Pablo Bay, California. At 1629, a Bell P-39N attempting an emergency ditching crashed into San Pablo Bay, California, about 200 yards south of the Hamilton Field Boat House, killing pilot 2Lt. Stanley F. Pennes. Investigators stated, "Pilot took off on a formation flight. About 12 minutes after take-off [the] pilot called the tower [stating] that he was out of gas and requested emergency landing instructions. Pilot was cleared to come straight in for an emergency landing. About one mile from the runway the motor apparently stopped and the airplane [with the landing gear in the extended position] nosed into the bay and went over on its back. The pilot was dead when he was recovered." Investigation revealed that the fuel selector valve was in the auxiliary fuel tank position. The airplane was not carrying an auxiliary fuel tank.

4-25-43C. Hallettsville, Texas. At 2025, a North American AT-6C crashed 12 miles southeast of Hallettsville, Texas, killing A/C Warren C. Price and A/C Harry J. Durr. The airplane had taken off from Brooks Field, San Antonio, Texas, and failed to return from a routine night navigation flight. Investigation revealed that the airplane smashed into the ground in a diving turn with power on. The airplane burst into flames immediately upon impact and the students were killed instantly. Investigators speculated that the students had probably lost the horizon in hazy conditions and inadvertently flew into the ground.

4-25-43D. La Junta, Colorado. At 2110, two Cessna AT-17D airplanes collided in mid-air near the Army Air Field at La Junta, Colorado, killing students A/C Roger M. Daly and A/C Earl G. Davidson in AT-17D #42-14015. 2Lt. Leland L. Jamison, A/C A.C. Davis and A/C D.A. Dane were uninjured and were able to successfully land AT-17D #42-13959 at La Junta. Investigators stated, "2Lt. Jamison was instructed to lead a formation of three AT-17s for two hours of night formation. The three airplanes took off at approximately 2040. At an altitude of 6,500 feet and some three miles north of Las Animas Auxiliary Airfield they were in a three-ship V-formation led by Lt. Jamison in aircraft #42-13959. On the right wing was airplane #42-13912 and on the left wing of the lead ship was aircraft #42-14015, piloted by A/C Davidson. Statements made by the occupants of the lead ship and no. 2 ship reveals that in making a shallow right turn the ship piloted by A/C Davidson lost position while making the turn and was far to the left and rear of the formation when the formation straightened out. By radio Lt. Jamison corrected A/C Davidson and instructed him to rejoin the formation. Evidence shows that A/C Davidson, in rejoining the formation, over controlled and for some unexplained reason continued in a right skid, hitting the lead ship [and] damaging the outer half of the right horizontal stabilizer and elevator. The damage caused to the aircraft piloted by A/C Davidson must have been of such an extent that it caused the airplane to immediately go into an uncontrolled vertical dive. No attempt had been made by either [student] to jump and the airplane crashed." See 8-19-43H.

4-26-43A. Location Unknown. At an unknown time between 0245 and 0330 EWT, a Douglas A-24A disappeared on a flight from Montbrook, Florida, to Drew Field, Tampa, Florida. Pilot 2Lt. Marvin L. Huddle and passenger Sgt Frederick M. Washburn were missing and presumed to be lost at sea. The airplane had taken off from Drew Field on a night navigation flight to Ocala, Florida, to St. Augustine, Florida, and return to Drew Field. The pilot had gotten lost and landed at Montbrook, 40 miles off course, to get his bearings. The pilot took off at 0245 EWT and was due at Drew Field at 0330 EWT. The airplane was never heard from again.

4-26-43B. Fort Campbell, Kentucky. At 0545 CWT, SSgt. F. Clinton Metz walked into the rotating starboard propeller of a North American B-25C that was undergoing a pre-flight run-up prior to taking off from the Army Air Field at Fort Campbell, Kentucky, on a routine training flight. Pilot 1Lt. Thomas R. Hatfield, co-pilot 2Lt. Vernon H. Olson and pilot-rated passenger 1Lt. Robert Lawley were uninjured. The Aircraft Accident Classification Committee stated, "At approximately 0545, Lt. Hatfield and Lt. Olson were starting the engines on a B-25C airplane. The right engine was started and upon receiving the clear signal from [a ground crewman], the left engine was engaged. The engine threw out some flame, which was extinguished as the engine took hold. The flame was normal and a small amount of gasoline on the ground caught fire, [and the fire] was put out immediately by the alert crew. The alert crew then signaled the pilot to cut the engines. The pilot then saw SSgt. Metz, the crew chief, lying under the right propeller. SSgt. Metz had just previously walked around the right engine to get into the navigator's compartment and was in the navigator's compartment when the left engine was started. It is the belief [of the committee] that due to the fact that it was dark and the excess amount of flames that came out of the engine must have frightened SSgt. Metz, causing him to run from the airplane between the fuselage and the right engine. SSgt. Metz was struck on the right side of the head and was fatally injured."

4-26-43C. Montgomery, Alabama. At 0750 CWT, a Vultee BT-13A crashed eight miles south of Gunter Field, Montgomery, Alabama, killing student A/C Earl S. Dickson. Investigators stated, "Cadet Dickson was on a routine training flight with specific instructions to practice chandelles, lazy-eights, and maximum performance climbing turns. His airplane was observed spinning at an altitude of 5,000 feet by 2Lt. George A. Holmes, an instructor in another airplane at that altitude. The spin continued uninterrupted to a low altitude, estimated to be about 500 feet. At that point the rotation stopped but the airplane dove into the ground almost vertically. Examination of the wreckage indicated an estimated airspeed of about 200 miles per hour upon impact and that the rotation had definitely stopped. The pilot's body was found about three feet in front and to the right of the telescoped fuselage, indicating that he was not in the cockpit at the time of impact. The logical deduction from the evidence is that the student was exerting back-pressure on the stick during his attempts to recover from the spin and when he found himself unable to recover from the spin and at a low altitude, he got out of the cockpit intending to jump. When he released the controls the airplane stopped spinning and remained in the recovery dive. The airplane either hit the ground before the pilot got clear, or the pilot upon noticing that recovery had taken place, tried unsuccessfully to get back into the cockpit at too low an altitude to jump."

4-26-43D. Sarasota, Florida. At 1050 EWT, a Curtiss P-40K crashed 15 miles northeast of the Army Air Field at Sarasota, Florida, killing pilot 2Lt. Ralph D. Searl. The airplane had taken off from the field at Sarasota with two other P-40 airplanes on an authorized low-altitude training flight. The pilot was flying in the number-three position of a three-airplane V-formation. Investigators stated, "The flight leader made a 90-degree left climbing turn upon completion of the mission, thereby turning into Lt. Searl. Upon further interrogation of the flight leader and the right wing man, it is the belief of this board that Lt. Searl overran the leader and lost sight of him in the turn and [while in a vertically banked turn to the left] spun in from an altitude of between 500 and 700 feet."

4-26-43E. La Paz County, Arizona. At 0750 PWT, a Boeing B-17F crashed while attempting an emergency forced landing in hilly terrain in La Paz County, Arizona, approximately 20 miles southeast of Blythe, California, killing four crewmembers and seriously injuring four others. Crewmembers killed in the crash were: 2Lt. Edward C. Rickel, bombardier; SSgt. Willard R. Pierce, engineer; Sgt. Leonard R. Rostankowski, tail gunner; SSgt. Richard C. Henderson, gunner. Crewmembers seriously injured in the accident were: 2Lt. John D. Abrell, pilot; 2Lt. Eugene R. Johnston, co-pilot; SSgt. Carl J. Sisco, radio operator; Sgt. Olaf K. Sagaas, assistant engineer. The Aircraft Accident Classification Committee stated, "B-17F No. 42-3126 was engaged in firing on the air to ground gunnery range at about 0750. The co-pilot was flying the airplane with the bombardier sitting in the co-pilot's seat. The pilot was firing from the upper turret. At the end of a pass over the gunnery range at an altitude of approximately 400 feet above the ground, the airplane was banked to the right in order to bring the top turret to bear on the last target. The co-pilot straightened out from the turn and attempted to climb, increasing his power and having the bombardier increase the rpm. The co-pilot stated that at this point he became aware of a feeling of loss of power and could not maintain altitude. The pilot, leaving the upper turret, was also aware of a loss of power and jumped into the co-pilot's seat, vacated by the bombardier. The pilot stated that he made a hurried check of all switches and instruments and found everything normal. Due to the low altitude and apparent lack of power the pilot had to make a crash landing. After [crash] landing the airplane caught fire and burned. It is the opinion of this board that the accident was apparently caused by a partial stall in a turn and failure to apply corrective action. There was no concrete evidence of power plant failure."

4-26-43F. Muroc, California. At 0750, a Lockheed P-38G crashed at the AAF Ground Gunnery Range near the Army Air Base at Muroc, California, killing pilot 2Lt. Donald T. Tilzey. The airplane, which was stationed at Ontario, California, had taken off from Muroc on a ground gunnery mission. The airplane dove on Target #2 with too great a speed and the pilot pulled up abruptly, causing the airplane to mush into the ground about 20 yards beyond the target. The airplane burst into flames as it smashed into the ground and the pilot was killed instantly upon impact.

4-26-43G. Location Unknown. At about 0800 PWT, a Lockheed P-38G disappeared and is presumed to have crashed into the Pacific Ocean while flying between Santa Catalina Island and Newport Beach, California. The pilot, 2Lt. John Cleary, Jr., 23, was missing and presumed lost at sea. The airplane was part of a four-ship patrol flight that had taken off at 0600 PWT from Orange County Airport, Santa Ana, California. The pilot requested to fly in the number-four position. The flight leader noticed that the airplane was missing as the flight approached Newport Beach, California. Despite a massive search effort, no trace of the airplane or pilot could be found.

4-26-43H. Lee Pope, Georgia. At 1110 EWT, a Vultee BT-13A crashed near Lee Pope, Georgia, killing A/C John H. Cossey, Iuma, Mississippi. The airplane had taken off from Cochran Field, Macon, Georgia, at 1040 EWT on a routine student solo flight. The airplane was seen to go into a spin to the right at an estimated altitude of 1,500 feet agl. The airplane described three turns of spin without any apparent effort by the student to recover. The airplane smashed into the ground and burst into flames upon impact. The student was killed instantly.

4-26-43I. Ballinger, Texas. At 1132, a Consolidated B-24D flying in instrument conditions broke up in flight and crashed nine miles northeast of Ballinger, Texas, killing the crew of ten. Killed in the crash were: 1Lt. John W. Chambers, pilot; 2Lt. Lester L. Braman, co-pilot; 2Lt. John E. Hull, navigator; 2Lt. Mark E. Jacobs, bombardier; TSgt. James Taylor, engineer; SSgt. Russel S. Frantz, radio operator; SSgt. John E. Hammond, assistant engineer; SSgt. Ernest C. Lewis, assistant radio operator; SSgt. Hobson A. Pearson, gunner; SSgt. Francis S. Hughes, gunner. The B-24 was part of a six-ship formation that took off from Clovis Army Air Base, New Mexico, on a flight to Kelly Field, San Antonio, Texas. Investigators stated, "Airplane was in No. 2 position of the leading element of a six-ship formation. The mission was scheduled for 25,000 feet. At that altitude, the formation was flying above an up-sloping overcast. At 25,000 feet it was found necessary to climb or enter instrument conditions. The flight commander, unable to do the former, broke [up] the formation [and] instructed all ships to climb and reassemble on top. Four airplanes broke away, climbed and were on top at 29,000 to 31,000 feet. No turbulence or icing was encountered. The fifth airplane turned 180 degrees and descended straight ahead. No heavy turbulence was encountered though there was indication of some carburetor ice, but no wing ice. [The subject airplane] was last seen turning 90 degrees to the original heading." Investigators speculated that the airplane suffered a catastrophic structural failure at high altitude, causing the airplane to break up and fall to the ground out of control. Investigators could not determine the cause of the in-flight break-up or what part of the airframe failed first.

4-26-43J. Lee Field, Missouri. At 1400, a Fairchild PT-23 crashed while performing aerobatics at Lee Field, Missouri, killing civilian instructor Dominick Cassera and student A/C David B. Frederick. The Aircraft Accident Classification Committee stated, "[The instructor] attempted a slow roll to the left at approximately 500 feet. The nose was apparently forced too high and the ship stalled out completely after it finished the roll. Apparently considerable altitude was lost in the actual roll, and the unusually high position of the nose caused an immediate and severe whip stall. The ship struck the ground at an obtuse angle about half way between the inverted and the vertical. The ship burst into flames when it struck the ground. The engine was reported to be wide open through the entire [maneuver]."

4-26-43K. Harding Field, Louisiana. At 1442, two North American A-36A type airplanes collided in mid-air at Harding Field, Baton Rouge, Louisiana, killing 2Lt. Charles C. Cunningham who crashed in A-36A #42-83752. 2Lt. Leslie W. Sokay received minor injuries but was able to safely land A-36A #42-84031 at Harding Field. The airplanes were part of a three-ship flight that had taken off from Harding Field on a routine transition flight. The airplanes were circling the field in a left turn at 3,000 feet msl when Lt. Cunningham rapidly overtook the airplane piloted by Lt. Sokay from underneath and to the left. Lt. Cunningham then pulled up and into the path of the airplane piloted by Lt. Sokay. The tail section of Lt. Cunningham's airplane contacted the propeller of Lt. Sokay's airplane. Lt. Cunningham's airplane rolled over and then spun to the ground out of control, bursting into flames upon impact. Lt. Sokay was able to maintain control of his airplane and was able to make a successful crash landing on Harding Field.

4-26-43L. Barksdale Field, Louisiana. At 1515, a Martin RB-26A crashed while attempting an emergency forced landing one mile south of Barksdale Field, Shreveport, Louisiana, killing co-pilot 2Lt. John P. McKinney. Pilot 2Lt. Monroe M. Hunsicker, engineer Sgt. Richard W. Diekman and radio operator SSgt. Coy S. Brown received serious injuries; bombardier-navigator 2Lt. Edward J. Collins suffered minor injuries. Investigators stated, "The pilot made a normal take-off and went approximately 100 feet in the air. The right engine cut out, caught, and cut out again. The pilot reduced power on the good engine in order to maintain control while trimming the aircraft. Loss of altitude was such that he could not trim the airplane in order to use additional power on the good engine or go into single-engine operation. The pilot proceeded with a wheels-up landing in the available field, but in doing so struck a drainage ditch about 70 degrees to the [horizontal]. This extreme angle permitted the nose of the airplane to strike the far bank, completely wrecking the aircraft. Since this accident, investigation has shown that air lock will form between the fuel pump and the carburetor in the fuel system, contributing to the loss of power and probable cutting out of the engine, and this is believed to be the causal factor in the accident."

4-26-43M. Belen, New Mexico. At 1945 MWT, two Beech AT-11 airplanes collided in mid-air 15 miles west of Belen, New Mexico, killing four fliers. Pilot 2Lt. Eugene W. McEntee, bombardier instructor 2Lt. Howard L. Bondy and student bombardier A/C Henry P. Bradshaw were killed in the crash of AT-11

#41-9505. Bombardier student A/C Raymond E. Breckman was killed attempting to parachute to safety. AT-11 #41-9504 was able to land safely at Kirtland Army Air Field, Albuquerque, New Mexico, after the collision and its crew uninjured. They were: 1Lt. Thomas L. Gammill, pilot; SSgt. James J. Picucci, bombardier instructor; A/C Phillip V. Cancelleri, student bombardier; A/C Lee L. Ferris, student bombardier. The airplanes had taken off at approximately 1835 MWT from Kirtland Army Air Field, New Mexico, on a bombardier-training mission. Investigators stated, "Statements of the personnel [on board AT-11 #41-9504] reveal that they had just dropped a bomb and had spotted it [on the ground] when the photographer noticed another bomb hit almost immediately after their bomb. The cadet who had just dropped the bomb [from AT-11 #41-9504] then noticed [AT-11 #41-9505] coming from their left rear and before he could warn the pilot the collision occurred.... Lt. McEntee was apparently flying too close behind and slightly below the ship piloted by Lt. Gammill. After releasing the bombs Lt. McEntee probably looked back over his shoulder to spot the bomb through the bomb bay doors and in doing so pulled up and into [AT-11 #41-9504].

4-26-43N. Potter Field, California. At 2230, a Vultee BT-13A attempting a landing at Potter Field, California, collided with a parked BT-13A, killing A/C Thomas J. O'Neil. The airplane had taken off from Merced Army Air Field, Merced, California, on a routine night training flight. The Aircraft Accident Classification Committee stated, "A/C O'Neil, pilot of A.C. #41-10754, attempted a cross tee night landing at 2230 at Potter Field (auxiliary to Merced Army Air Field, 18 miles south) and crashed while attempting to take-off again. The attempted landing was made toward several parked airplanes and floodlights located on the south half of the field. It is believed that the reason for the landing was that the pilot became confused and thought that he was making a normal landing in the correct direction until his wheels touched the ground. At this point, he apparently realized his mistake and opened the throttle in an attempt to pull up over the parked airplanes. Then, it is believed that he was blinded by the floodlights and lost the horizon. When the airplane banked to the left, his [port] wing tip struck a parked airplane's canopy [BT-13A #41-11441] and he crashed on the field."

4-27-43A. Cape Charles, Virginia. At 1515, a Consolidated B-24D suffering an in-flight fire crashed near Capeville Road, Cape Charles, Virginia, killing pilot 1Lt. David H. Cominole. Ten crewmembers were able to parachute to safety. The airplane had taken off from Langley Field, Hampton, Virginia, on a gunnery training flight. Investigators stated, "1Lt. Addison L. Brown, navigator, while firing the left waist gun, had the gun jam and while trying to clear and charge the gun, it ran away, sending a burst of shots into the wing just outboard and behind the number-one engine. The shells burst upon striking the wing, setting fire to the airplane. This all took place over [the Atlantic Ocean] and the pilot flew the ship over land and ordered the crew to bail out. The flames spread with increasing intensity in the meantime and before the pilot could bail out, the ship partly disintegrated and crashed with the pilot still at the controls." Crewmembers parachuting to safety were: F/O Grafton W. Stull, co-pilot; 1Lt. Addsion L. Brown, navigator; 2Lt. Salvatore P. Casarella, Jr., bombardier; Sgt. John M. Patton, engineer; Cpl. Donald DeShazo, assistant engineer; TSgt. James G. Kehoe, radio operator; Cpl. Elmer M. Link, assistant radio operator; SSgt. Carl S. Boomgaard, gunner; Cpl. Joseph J. Cagney, crewmember; F/O Kenneth A. Houp, pilot-rated passenger.

4-27-43B. Location Unknown. At an unknown time after 1616 EWT, a Republic P-47C disappeared after it took off from Mitchel Field, Long Island, New York, on a scheduled aerobatic flight. The pilot, 2Lt. Jerald W. Powell, was declared missing and presumed lost at sea.

4-28-43A. Parkersburg, West Virginia. At 0830, a Fairchild PT-23 collided with a tree and crashed eight miles north of Parkersburg, West Virginia, killing instructor F/O Richard R. Gould and civilian flight instructor Joseph O'Shea. Investigators stated, "Subject airplane took off [from Stewart Airport] with its pilot and a local airplane instructor for a local flight while on a ferrying mission. After engaging in aerobatic maneuvers for a few moments, it was seen to fly in the direction of [Wood County Airport]. Observers at the other field watched the airplane approach at approximately 600 feet, make a 180-degree turn, gradually ascend to 800 feet, and when about a mile away from this field, go into a steep dive and right turn. [The airplane] passed over a ridge during [this] maneuver and was seen to be in a steep dive and a steep right turn at the time it struck the tree located at the foot of the ridge, the ship crashing in a field adjacent to the trees." Investigators noted that Instructor O'Shea was not authorized to be on board the aircraft.

4-28-43B. Vichy, Missouri. At 0930, a Douglas A-20A crashed near the Army Air Field at Vichy, Missouri, killing pilot 2Lt. Ernest Hanson. The airplane had taken off from Vichy Army Air Field at 0830 on a local training flight. Investigation revealed that the airplane was observed flying at approximately 2,000 feet above ground level when black smoke began pouring from the port engine. The airplane began circling to the left as the pilot looked for a place to make an emergency forced landing, losing altitude until it entered a stall and spin to the left at an altitude of approximately 300 feet. The airplane "turned over" and spun to the ground. The pilot died moments after he was found in the smashed cockpit area, which had tore loose from the fuselage. Investigators noted that the port propeller was not feathered.

4-28-43C. Coyote Wells, California. At 1100, a Lockheed P-38G crashed ten miles southwest of Coyote Wells, California, killing pilot 1Lt. William P. Ackerman, Jr. Investigators stated, "Lt. Ackerman was leading a three-ship flight on a scheduled simulated strafing mission near Coyote Wells, California. On about his third pass at a medium tank, Lt. Ackerman mushed into the ground about fifty feet beyond the tank where he lost both propellers, reduction gears, and parts of the engine nacelles. After his first contact with the ground the airplane was thrown into the air again with the left engine burning, to a height of about 300 to 500 feet when it made a climbing turn to the left [winging over to an inverted attitude] and then crashed. It is the opinion of this investigating board that Lt. Ackerman was probably knocked unconscious on the first impact, and had no control of the airplane after it was thrown back into the air." The flight had originated at North Island Naval Air Station, San Diego, California.

4-28-43D. Exeter, California. At 1325, a North American B-25D crashed in the Yokohl Valley 13 miles southeast of Exeter, California, killing the crew of four. The airplane had taken off from Long Beach, California, on a ferry flight to Hamilton Field, San Rafael, California. The airplane reported to ground stations at 1233 while flying at 11,000 feet over Bakersfield, California, the pilot stating that they were going to "swing their compass." Several witnesses, including an official airplane spotter at Porterville, California, had heard the airplane flying above the overcast. The airplane was then seen to dive out of the overcast before smashing into rising terrain. Investigators stated, "Evidence indicates that the airplane contacted the ground while traveling in a northeasterly direction at a steep angle of descent, with the right wing and engine hitting first, cartwheeling to the left. A terrific explosion occurred, scattering parts of the airplane in a 200 to 300 yard radius, mostly in a southwesterly to northwesterly direction. The airplane came to rest on its back and was completely destroyed by fire." Investigators noted that the airplane was about 25 miles off course and that the nose landing gear door had separated from the airplane and struck the starboard propeller when the pilot attempted an abrupt pull-out just before striking the ground on the Gill Ranch. Killed in the crash were: 1Lt. Grant B. Schley, pilot; 2Lt. Tillman F. Oehler, co-pilot; 2Lt. Morgan J. O'Brien, navigator; Cpl. Hansard A. Thompson, radio operator.

4-28-43E. Clovis, New Mexico. At 1651 MWT, navigator 2Lt. Ray B. Lorimer was killed when he inadvertently bailed out of a Consolidated B-24D 20 miles southwest of Clovis, New Mexico. The Aircraft Accident Classification Committee stated, "Aircraft was ... flying on a routine high-altitude mission ... at 31,000 feet when turbo-supercharger on No. 3 engine disintegrated and damaged the aircraft's fuselage.... Excessive manifold pressure was used ... causing overspeeding ... resulting in disintegration of the turbo-supercharger. Navigator bailed out at 31,000 feet, on preliminary bail out warning, without oxygen." Pilot 1Lt. Calvin W. Parker and co-pilot 2Lt.Ray T. DeMerritt were uninjured and were able to safely land the B-24D at Clovis Army Air Base. Seven other crewmembers were uninjured.

4-28-43F. Waycross, Georgia. At 2015 EWT, a Martin B-26C was mistakenly abandoned in flight and crashed two miles south of the Army Air Field at Waycross, Georgia, killing two crewmembers. Four other crewmembers parachuted to safety and were uninjured. The airplane was part of a 16-ship flight that had taken off from the Army Air Field at Lakeland, Florida, on a simulated bombing and strafing mission on the Army Air Field at Waycross, Georgia. While the attack was taking place, a B-26 suffered an engine fire and was successfully landed on the field. The B-26 flight leader noticed that the airplane on the field was on fire and had transmitted a warning over the radio, ordering the crew to "get out." The subject airplane was then seen climbing steeply from low altitude to about 2,000 feet agl about two miles west of the field with airmen bailing out. The airplane stalled, fell off to the left and spun to the ground. Radio operator/top turret gunner Cpl. Claude W. Mack and tail gunner Pvt. James F. Vondra, Jr. did not bail out and went in with the airplane. The surviving pilot stated that about "three minutes" had passed between the time the bail out alarm was given and the time that the pilot had left the airplane. It was not known why the two gunners did not leave the airplane. Investigators speculated that Cpl. Mack "had not gotten clear of the turret" at the time of the crash and was not wearing a parachute. It was later speculated that Cpl. Mack had inadvertently left his parachute forward in the radio compartment and was not able to access it when he needed it. Pvt. Vondra was found wearing a parachute. Investigators stated that the crew mistakenly abandoned the airplane as a result of a verbal message to another aircraft that was burning on the runway at Waycross Army Air Field. Crewmembers parachuting to safety were: 1Lt. James E. Roberts, pilot; 1Lt. John F. Neyenhouse, co-pilot; 2Lt. John F. McCarty, navigator; Pvt. Frank Brazes, engineer.

4-28-43G. Rembert Auxiliary Airfield, South Carolina. At 2130, a Vultee BT-13A crashed near Rembert Auxiliary Airfield, South Carolina, killing instructor 2Lt. William A. Gumuka and student A/C Stanley W. Fish. The airplane was practicing night landings at the field when it stalled and entered a spin while turning to the right from the base leg to final approach. It was speculated that the turn was wide and the pilots attempted to tighten the turn, causing the airplane to snap roll to the right and spin to the ground.

4-29-43A. Cross City, Florida. At 0901 EWT, a Republic P-47C crashed at the Cross City Army Air Base, Cross City, Florida, killing pilot 2Lt. John H. Thompson. The Aircraft Accident Classification Committee stated, "2Lt. Thompson was taking off on a scheduled formation mission. He taxied out to the end of the runway and took his position on the right wing of the leader for a formation take-off. The aileron signal was passed and acknowledged signifying the readiness of both ships. As they moved down the runway Lt. Thompson began to trail behind. Suddenly he crossed the runway to the left and behind the lead ship, fighting [propeller turbulence] as he went. At this time the tower operator told him to cut the throttle and engine. Lt. Thompson continued out across the field and forced the airplane into the air. With wheels down he hit the trees, turned on his back and somersaulted to an upright position where the airplane burned." Investigation revealed that the tail wheel was locked in a turned position.

4-29-43B. Pecos, Texas. At 1015, a Vultee BT-13A crashed one mile north of Toyah Auxiliary Airfield, Pecos, Texas, killing student A/C Ross H. Hibbard, Jr. The airplane had taken off from the Pecos Army Air Field on a routine training flight. Investigators stated, "[The airplane] crashed while making the last turn into the field for landing. [The student] was returning to Toyah Auxiliary Airfield after having practiced air work. Traffic was to the right and the last turn into the field was made a little late. In order not to land on the [dusty] area in the center of the field, the turn appeared to be tightened by possible backpressure on the stick. The airplane stalled and broke to the left, recovered and broke to the right, and continued dropping with each wing alternately going down until the airplane struck the ground and immediately burst into flames."

4-29-43C. Sacramento, California. At 0853, a North American B-25C attempting a take-off crashed one-quarter mile south of the Sacramento Municipal Airport, Sacramento, California, killing B-25 student 2Lt. Leland G. Larkin. B-25 instructor 1Lt. George H. Custard and B-25 student 2Lt. Ray P. McLaughlin received minor injuries. Three B-25s from Mather Field, Sacramento, California, had been practicing night take-offs and landings at the field in the dark early morning hours. At 0315 visibility dropped below minimums and all flying was temporarily halted. The pilots waited until daylight hours to attempt a flight back to base at Mather Field. Because of a persistent bank of fog over the area of the municipal airport, the three airplanes were required to file instrument flight plans for the flight back to Mather Field, which was reporting visibility of seven miles with ceiling unlimited. The tower operator ordered all three airplanes to take off individually on tower command and to report when on top of the fog layer. When the first airplane broke out on top of the fog layer, the tower gave permission for the second airplane to take-off. The airplane took off but settled back into the ground with landing gear retracted about one-quarter mile south of the airport. The airplane flew into the ground and smashed itself to pieces, bursting into flames soon after coming to rest. 2Lt. Larking was trapped in the wreckage and burned to death.

4-29-43D. Harlingen, Texas. At 1115, a Lockheed AT-18 crashed one mile northwest of the Army Air Field at Harlingen, Texas, killing nine crewmembers and student gunners. Investigation revealed that the airplane was returning to the field following a routine gunnery training flight. The airplane apparently entered a stall while turning from the base leg to the final approach at an altitude of about 800 feet agl. Soon after completing the turn the port wing was seen to suddenly drop. The pilot brought the wing up, but the airplane snapped over on its back and entered a spin. The airplane spun about one and a half turns before striking the ground and exploding into flames. Pilot SSgt. Meldon A. Noah, engineer Sgt. Harold J. Gorman and gunnery instructor SSgt. Arthur L. Markley were killed. The six gunnery students killed in the accident were: Pfc. Laverne Barkley, Jr., Pfc. Keith M. Hill, Pvt. Clifford B. Jackson, Pfc. LeRoy E. Johnson, Pvt. John J. O'Neill and Pvt. Morse D. Pigg.

4-29-43E. Red Bluff, California. At 0919, a Bell P-39N crashed at Red Bluff, California, killing pilot 2Lt. Dan J. Haas. The airplane had taken off from Santa Rosa, California, on a routine training flight. The airplane had buzzed the municipal airport twice, performing a steep pull-up each time. When the pilot buzzed the airport a third time his pull-up was extremely steep and the airplane entered a high-speed stall. The airplane snapped over on its back and then dove into the ground where it exploded into flames upon impact.

4-29-43F. Camp Campbell, Kentucky. At 1445 CWT, a Piper L-4 crashed at Camp Campbell Military Reservation, Kentucky, killing observer 1Lt. John C. Braswell and seriously injuring pilot Sgt. George O. Cline. The airplane had been picking up and dropping messages, and was seen to make an exceptionally long glide during the last message pick-up. After the airplane had picked up the message, the pilot attempted to gain altitude but the airplane would not respond. The pilot performed an emergency forced landing but the airplane struck a fence pole. The airplane went out of control and smashed into the ground, bursting into flames after coming to rest. The occupants were able to escape the airplane with their clothes on fire. The observer was overcome by flames and fell into the burning terrain where he was burned to death before he could be rescued.

4-29-43G. Garden City, Kansas. At 2250 MWT, a Vultee BT-13A crashed near the Army Air Field at Garden City, Kansas, killing student A/C Richard F. Malchew. The airplane had taken off from Garden City on a routine night flight. The student was practicing night take-offs and landings with five other BT-13s in the traffic pattern. The airplane failed to return to the field when night flying was terminated and was found the next day at about 0645 MWT about one-half mile from the departure end of the runway. Investigation revealed that the airplane had hit the ground very hard in a perfectly flat wheat field. The airplane bounced twice, nosing in at a steep angle on the second bounce and then smashing into the ground inverted about 600 feet from the initial contact. The wheels and the engine had separated from the airplane. It was speculated that the engine had failed just after take-off and the pilot was attempting a straight ahead forced landing.

4-30-43. Famoso Auxiliary Airfield, California. At 1425, a Vultee BT-13A crashed at Famoso Auxiliary Airfield, California, killing A/C Philip J. Nelson. Investigators stated, "Student was practicing 90-degree power-off stages at Famoso Auxiliary Airfield. Upon starting take-off, the airplane swerved to the left as the tail left the ground. The airplane continued to swerve to the left to an angle of about 30 degrees from the original direction of take-off. The airplane changed direction slightly to the right, was pulled off the ground sharply directly in front of a wire fence, caught the top strands of wire on the tail wheel, continued to climb sharply, finally falling off, at the same time rolling to the left through telephone wires, and struck the ground in an inverted position on the highway approximately 50 yards from the west edge and even with the center of the field. The engine caught fire but was extinguished immediately before spreading to the rest of the aircraft."

May

5-1-43A. Waco, Texas. At 0900, a Vultee BT-13A crashed ten miles north of the Army Air Field at Waco, Texas, killing instructor 2Lt. Howard Hertz and student A/C Valentine F. Schott, Jr. Investigation revealed that the instructor was demonstrating forced landing technique to the student. The airplane made a successful approach as it shot a simulated forced landing on a farm field. Full throttle was then applied in an attempt to climb the airplane back to a safe altitude. The engine did not respond and the airplane stalled, falling off on the right wing and crashing into the ground. The occupants were killed upon impact. Investigators speculated that the instructor had failed to "clear" the engine while in a prolonged glide.

5-1-43B. San Bernardino, California. At 1605, a Douglas A-20G suffering an in-flight fire crashed while attempting an emergency forced landing at the Army Air Field at San Bernardino, California, killing pilot 1Lt. Andrew G. Luchsinger and engineer Cpl. Leo J. Neuman. Passenger Pfc. Ervin Hess was able to parachute to safety. He suffered minor injuries. The airplane was being ferried from the Douglas Aircraft Company factory at Long Beach, California, to Gore Field, Great Falls, Montana. The airplane had taken off from San Bernardino Army Air Field at 1545 PWT, cleared for Hill Field, Ogden, Utah. At approximately 1600, the airplane returned to the area of San Bernardino Army Air Field with the starboard engine nacelle on fire. The passenger bailed out at low altitude. The pilot attempted an approach for Runway 2, but the engine nacelle fire had compromised the structural integrity of the starboard wing. The wing failed and folded up. The airplane fell off from an altitude of 75 feet agl and smashed into the ground where it exploded into flames.

5-2-43. Gulf of Mexico, Florida. At an unknown time after 0915 CWT, a Lockheed P-38G crashed into the Gulf of Mexico on Range #35, Eglin Field Military Reservation, three miles south of Tower Beach, Fort Walton Beach, Florida, killing pilot 2Lt. Clyde M. Adams. Investigators stated, "This aircraft took off from Auxiliary Air Field #6, Eglin Field Military Reservation, on a simulated minimum altitude ground strafing mission. The aircraft had been diving on the water and firing machine guns and cannon. No eyewitnesses have been located and the actual time of the accident and cause are unknown. When the aircraft was overdue, a search was made over the entire area and an oil slick was sighted on the Gulf of Mexico on the range over which the aircraft was scheduled to fly. Divers who were sent down located pieces of the wreckage and only a small portion of the body of the pilot. It is the opinion of the Aircraft Accident Classification Committee that this accident was caused by the pilot's misjudging the distance of the aircraft above the surface of the water in a simulated ground strafing attack while performing a scheduled mission."

5-3-43A. Mountain Home, Idaho. At 0307, a Boeing B-17F suffered an in-flight explosion and crashed one mile north of the Army Air Base at Mountain Home, Idaho, killing eight crewmembers. The airplane had taken off at 0110 from Gowen Field, Boise, Idaho, on an instrument navigation flight. The airplane was also scheduled to make a dry bombing run in the area east of Gowen Field at an altitude of 10,000 feet. At 0255 MWT, the airplane made a routine position report. The airplane was flying three miles northeast of the base when the port wing exploded, causing a large portion of the port wing to fail and separate. The navigator's compartment was also destroyed in the explosion. The airplane went out of control and broke up, the fuselage breaking just aft of the radio compartment. The main wreckage slammed into the ground and exploded violently into flames. Wreckage and bodies were scattered over a large area. Some wreckage had fallen onto structures in the area, causing minor damage. Investigators speculated that gasoline fumes had caused the wing explosion. Killed in the crash were: 2Lt. Rector Arney, pilot; 2Lt. Thomas M. Homer, co-pilot; 2Lt. Luther Chase, navigator; 2Lt. Lewis A. Lillard, bombardier; SSgt. Richard M. Campbell, engineer; Sgt. John W. Strickland, assistant engineer; Sgt. Otis L. Blackwood, radio operator; Sgt. Melvin Kaizer, gunner.

5-3-43B. Del Rio, Texas. At 0943, a Martin B-26B crashed 25 miles northwest of Laughlin Field, Del Rio, Texas, killing four crewmembers. Killed in the crash were: 2Lt. Francis A. Thomas, B-26 instructor; 2Lt. Russel R. Sanders, B-26 student; Cpl. Henry H. Wtulich, engineer; Pfc. Roy S. Smith, assistant engineer. Investigators stated, "It is believed by this board that Lt. Thomas was instructing a student, that in so instructing, he was teaching power-off stalls which is part of the course given here. It is further believed that in doing this power-off stall, the aileron stalled before the wing root, causing the airplane to spin in and be completely demolished on impact and burned. This stall occurred in a B-26B-10MA. The remaining airplanes of this type at this station [Laughlin Field] have a spoiler in the wing root approximately one foot long. It is not definitely known that this particular airplane had a spoiler of this length, but it can be inferred that it was similar to the other models received. There is no spoiler under the de-icer boot, similar to the B-26B-4 model, which is approximately two feet long. It is believed that if this model airplane had a spoiler similar to the spoiler on the B-26B-4, that the pilot would feel a buffeting of the controls prior to getting a wing tip stall."

5-3-43C. Fendell Grove, Virginia. At 1115, engineer SSgt. Douglas B. Gertsen was killed while attempting to parachute to safety from an out of control Consolidated B-24D that was operating over Fendell Grove, Virginia. The engineer had waited too long to pull the ripcord and the parachute did not fully deploy before he smashed into a dead tree, killing him instantly. B-24 student 2Lt. Frederick W. Wiese, assistant engineer SSgt. John Dillinger and radio operator Sgt. William M. Holmes parachuted to safety and were uninjured. The pilots, B-24 instructor Capt. Harold E. Gooding and B-24 student 1Lt. Robert F. Valentine, were able to regain control of the airplane and were able to land safely at Langley Field, Hampton, Virginia. The airplane entered a turn to the left, which developed into an uncontrolled skidding turn. The instructor took control of the airplane, but despite adding full power and full right rudder the airplane continued to turn to the left and lose altitude. The instructor ordered the crew to bail out. Lt. Valentine stayed behind to help the instructor. The instructor shut down the number-one engine, feathered the propeller, retarded the remaining throttles and was able to regain control of the airplane. The instructor was able to climb to 9,000 feet indicated altitude. The instructor flew on three engines for several minutes and then restarted the number-one engine. The engine started and appeared to perform normally. The pilots flew the B-24D to Langley Field where they were able to land safely.

5-3-43D. Muroc, California. At 0920, a Lockheed RP-38E crashed while attempting a take-off at the Army Air Base at Muroc, California, killing pilot 2Lt. Raymond H. Hanson. Investigation revealed that the port propeller ran away just as the P-38 became airborne. Investigators stated that the pilot used excessive power on the starboard engine, causing the airplane to roll to the left and enter a stall. The airplane entered a spin and smashed into the ground where it exploded into flames. The pilot was killed instantly.

5-3-43E. San Francisco Bay, California. At 1032, a Bell P-39N crashed into San Francisco Bay, California, killing 2Lt. Jonathan Hofto. Investigators stated, "Lt. Hofto cleared from San Francisco Army Airdrome for an intelligence flight to the vicinity of Watsonville [California] and return. He departed San Francisco at 0925. At approximately 1015 he crashed in the bay about three miles west of Moss Landing. Two fisherman report that they saw the airplane skimming over the water at about 100 feet and then dive into the bay about 300 yards from where they were fishing. The engine appeared to be running normally at the time of the crash. They reported that the visibility was so poor that they did not see the airplane until it was about a half-mile away. A search was made of this area to locate the airplane and pilot but all that was found was the first-aid kit and few small remnants of airplane. The cause of the accident was undetermined. From what evidence is obtainable it is possible that the pilot was flying in very poor weather and was trying to stay below the fog bank and flew into the water without knowing it."

5-3-43F. La Junta, Colorado. At 1145, a North American AT-6A crashed near La Junta, Colorado, killing pilot 1Lt. Joe A. Herbert and pilot-rated passenger 1Lt. Ernest M. Elkin, Jr. The airplane was observed flying at low altitude prior to the accident. Investigators speculated that the pilot had stalled the airplane at the top of a steep pull-up maneuver. The airplane spun to the ground and burst into flames.

5-3-43G. Zephyrhills, Florida. At 1543 EWT, a Curtiss P-40E crashed while attempting an emergency landing at Zephyrhills Army Air Base, Zephyrhills, Florida, killing pilot Capt. Earl R. Boyd. The pilot had taken off with another P-40E at 1532 EWT on a routine interception training flight. When the other pilot reached 6,500 feet, he noticed that Capt. Boyd was not with him. Investigators stated, "At approximately 1540, Capt. Boyd called Zephyrhills Tower for an emergency landing and the field was cleared. About that time an airplane appeared at about 2,500 feet just east of the field. It passed over the field at approximately 2,000 feet flying west and it was noticed that the propeller was not windmilling. He passed over the edge of the field and he started a steep 180-degree turn, but continued to lose speed and altitude very rapidly. He had almost completed his turn and was in a position to land northeast when the airplane went into spin at approximately 250 feet. It crashed and burned about 300 yards short of the runway. The pilot suffered major injuries and never recovered consciousness. Investigation of the crash showed two rods thrown through the crankcase, causing the failure of the engine. Pilot overshot while attempting a forced landing and spun in."

5-3-43H. Barksdale Field, Louisiana. At 1500, a Martin B-26C crashed while attempting a take-off at Barksdale Field, Shreveport, Louisiana, killing engineer SSgt. Thomas J. Grier and bombardier/navigator 2Lt. John R. Beattie. Crewmembers seriously injured in the crash were: 2Lt. Ervine J. Button, pilot; 2Lt. Merlin E. Johnson, co-pilot; Sgt. Roger J. Thompson, radio operator; Sgt. Jordan L. Bushnell, gunner. Investigators stated, "The airplane was taking off on a 330 degree heading on the grass. The airplane apparently struck a bump and was thrown into the air. The left wing was down and the airplane was skidding. The airplane was not at flying speed when contact was again made with the ground; the pilot attempted to hold the airplane on the ground. The skidding effect obtained from the bump and accentuated in the air, is believed to have sheared both the nose landing gear and [both] main landing gear when contact was made with the ground after the bump. The shearing of the right landing gear broke the gasoline lines [in the starboard engine nacelle], causing the airplane to burn. Fire spread through the entire airplane with such rapidity that in a few seconds the fire was uncontrollable, exploding ammunition further preventing the approach of fire fighting equipment."

5-3-43I. Britton, Oklahoma. At 1615, a Vultee BT-13A crashed one mile east of Britton, Oklahoma, killing student A/C Clyde R. DeLautre. Investigators stated, "A/C DeLautre was dispatched on a student cross-country flight from Strother Army Air Field, Winfield, Kansas, to Bartlesville, Oklahoma, to Stroud, Oklahoma, and return to Strother Field. He was instructed to fly at an altitude of 3,500 feet indicated on all legs of the course. He completed the cross-country as far as Stroud satisfactorily. He was dispatched from Stroud by the control officer there to Strother Army Air Field at approximately 1555. Instead of returning to Strother Field he proceeded to the vicinity of Britton, Oklahoma, which is 55 miles southwest of Stroud. He was 100 degrees off course. As he approached Britton from the northeast he made three coordinated turns of approximately 45 degrees each at an estimated altitude of 1,200 feet above the ground. He then flew straight and level until he entered a medium dive, which he held to an altitude of approximately 500 feet before he pulled up into approximately a 35-degree climb. He held this climb until he had regained approximately his original altitude at which point he entered a second dive, which was steeper than the first. He continued to dive to an altitude of approximately 300 feet at a point approximately over his grandmother's house. His pull-out from this dive appeared to be the beginning of a loop, but instead of looping he suddenly did a half-roll at an estimated altitude of 1,200 feet. He then attempted a straight pull-out but just as he was passing the vertical position the ship stalled and did one and three-quarters turns of a spin and crashed into the ground. The ship was at a 45-degree angle with the ground when it hit and after it hit, the fuselage fell back on the ground so that the ship was lying right side up with the nose headed west. Flames immediately started in the engine and quickly spread to the fuselage and wings. The airplane was completely wrecked and A/C DeLautre killed."

5-3-43J. Baton Rouge, Louisiana. At 2200, a North American P-51A crashed near Harding Field, Baton Rouge, Louisiana, killing pilot 2Lt. Earl Kielman. The pilot had made two landings at Harding Field and then went back to his assigned zone to wait for permission to make a third landing. The airplane then crashed into the ground and burst into flames in a diving turn after the pilot had apparently lost the horizon.

5-3-43K. Location Unknown. At an unknown time after 2329 EWT, a North American AT-6A disappeared after taking off from MacDill Field, Tampa, Florida, on a routine cross-country navigation flight. The pilot, 2Lt. Donald Alfred Gray, was declared missing and is presumed lost at sea.

5-3-43L. Mobile Bay, Alabama. At 2324, a Douglas C-53 crashed into Mobile Bay, Alabama, killing three fliers and injuring two. Pilot F/O Alan A. Hayden, co-pilot 2Lt. R.E. Fennelly and passenger 1Lt. H.W. Tikkanen were killed in the crash. Engineer TSgt. R.D. Beeman and crewmember Pfc. A.L. Ritter survived with minor injuries. They were rescued from the bay soon after the crash. The airplane had taken off from Brookley Army Air Field, Mobile, Alabama, on a flight to Lawson Field, Fort Benning, Georgia. Investigators stated, "Indications are that the pilot made a turn to the left over the bay while still at low altitude, and flew the ship into the water under power while still in the turn. A night take-off to the south at Brookley Field requires instrument flying almost immediately after leaving the runway lights and a turn to the left over Mobile Bay necessitates a smooth normal instrument turn. This airplane could not have reached the point where it crashed without making a very steep turn as was determined the following day by taking off down the South Runway in the same type of ship and turning to the point of the crash. It is probable that the pilot either attempted a steep turn to pick up the lights of Mobile or attempted a steep instrument turn at low altitude (500 feet or less) and struck the water before he became aware that he was losing altitude. The movement of TSgt. Beeman and Pfc. Ritter from the forward section of the cabin to the rear door just at the time of take-off may have made the ship tail heavy and caused it to climb more steeply than it would have otherwise. This was a possible contributing factor in the accident as it may have caused the ship to approach a stall position just as the pilot was trying to go from contact flying from the runway lights to instrument flying over the dark area south of [Brookley Field]."

5-3-43M. George Field, Illinois. At 2330, a Beech AT-10BH crashed two miles south of George Field, Lawrenceville, Illinois, killing A/C Hal Bigger Feldman and A/C Howard W. Forestall. The airplane was returning from a routine night training mission when it stalled and spun to the ground while turning from the base leg to final approach.

5-4-43A. Casper, Wyoming. At 0126 MWT, a Boeing B-17F crashed just after take-off from the Army Air Base at Casper, Wyoming, killing eight fliers and seriously injuring pilot 2Lt. Robert C. Brown. Killed in the accident were: 2Lt. Oran A. Beardslee, co-pilot; 2Lt. Thomas E. McCarty, navigator; 2Lt. Floyd N. Jackson, bombardier; SSgt. William F. Bauman, engineer; Sgt. Albert T. Reese, assistant engineer; Sgt. Robert W. Williams, radio operator; Sgt. Lewis M. Mahaffey, assistant radio operator; Sgt. Robert M. Orton, gunner. The airplane took off to the west on Runway 25 on a local night flight. The airplane made a normal take-off and climbed to an altitude of about 150 feet when it began to descend rapidly until it smashed into the ground about one-half mile off the end of the runway. The B-17 exploded into flames upon impact. The pilot does not remember how he exited the ship. Investigators speculated that the pilot had lost the horizon just after taking off and inadvertently flew into the ground.

5-4-43B. Cross City, Florida. At 1143 EWT, a Republic P-47C attempting a take-off crashed near the Army Air Field at Cross City, Florida, killing pilot Capt. Maurice L. Stephens. Investigation revealed that the airplane took off normally and climbed to about 100 feet when the engine apparently failed. Heavy black smoke was seen pouring from the engine area just after take-off until the airplane crashed. The P-47, banking to the left and losing altitude rapidly, smashed into a wooded area where it burst into flames immediately upon impact. The pilot was killed instantly. The airplane was on a scheduled two-ship high-altitude training flight.

5-4-43C. Ledyard, Connecticut. At 1312, a Republic P-47C crashed at Old Indian Town Road, Ledyard, Connecticut, killing pilot 2Lt. Charles H. Brown. The airplane was part of a flight of three P-47s that had taken off from Groton Field, Connecticut, on a local training mission. The airplane was seen in a tight turn to the left at about 4,500 feet when it fell "out of control." The airplane then entered a 45-degree dive until it smashed into the ground where it exploded into flames immediately upon impact. The pilot was killed instantly. Examination of the propeller blades indicated that the propeller was not turning under power at the time of impact. It was later speculated that the engine had apparently failed while the P-47 was in a tight turn to the left, causing the airplane to stall and fall out of control. The P-47 smashed into the ground before the pilot could regain control of the airplane.

5-4-43D. Ocala, Florida. At 1400 EWT, a Stearman PT-17 crashed near Taylor Field, Ocala, Florida, killing student A/C Adolph Wiskosky. The airplane was at an altitude of about 3,500 feet when it went into an intentional spin and the student failed to recover. The student attempted to parachute to safety but had left the airplane at too low an altitude to allow his parachute to fully deploy. The airplane came out of the spin when the student left the airplane.

5-4-43E. Gunter Field, Alabama. At 1340, a Vultee BT-13A crashed 12 miles south of Gunter Field, Montgomery, Alabama, killing A/C Louis J. Charanza, Caldwell, Texas. Instructor 2Lt. August A. Ellwood parachuted to safety and was uninjured. Investigation revealed that the airplane was flying at 7,000 feet indicated altitude when the instructor began demonstrating spins and recovery technique. The instructor put the airplane into an intentional spin to the right. The airplane had described two and one-half turns of spin when the instructor attempted to break the spin. The instructor stated that the airplane did not respond to recovery inputs. The instructor neutralized the controls and began another attempt to recover. When this attempt failed, the instructor ordered the student to jump. The student bailed out of the right side of the aircraft and was thrown into the rotating propeller by the force of the spin.

5-4-43F. Pennsville, New Jersey. At 1500, a Republic RP-47C exploded in flight and crashed near Pennsville, New Jersey, killing pilot 2Lt. Richard H. Harms. The airplane had taken off at 1315 EWT from Philadelphia Municipal Airport, Philadelphia, Pennsylvania, with another RP-47C on a high-altitude formation flight. The airplanes were descending on the inbound leg back to base when Lt. Harms' engine exploded at an indicated altitude of 5,000 feet. Debris was seen falling from the airplane as it exploded. The airplane flew straight for a few seconds and then fell off to the left into a steep dive to the ground, trailing heavy black smoke and debris. The airplane struck the ground at a steep angle and exploded violently into flames upon impact, killing the pilot instantly.

5-5-43A. Salome, Arizona. At 0330 MWT, a North American AT-6C crashed three and one-half miles ENE of Salome, Arizona, killing A/C John S. Clark, Jr. The student was on a night navigation mission and had successfully checked in at Salome. The student was heading back to base at Luke Field, Phoenix, Arizona, when he apparently lost the horizon and inadvertently flew the airplane into the ground.

5-5-43B. Gordonsville, Tennessee. At 0709 CWT, a Bell P-39N crashed while maneuvering nine miles west of Gordonsville, Tennessee, killing pilot 2Lt. Roy C. Davidson. The airplane had taken off from Chattanooga, Tennessee, on a "visual reconnaissance mission for the Red Army of the Tennessee Maneuvers." Investigation revealed that the airplane stalled out of steep turn at low altitude while engaging in simulated combat with an opposing force fighter airplane. The pilot was unable to recover and the airplane smashed to the ground and exploded into flames.

5-5-43C. Leesburg, Florida. At 0844 EWT, a Curtiss P-40E crashed two and one-half miles northeast of the Army Air Base at Leesburg, Florida, killing pilot 2Lt. Francis H. Delano. The airplane had taken off from Leesburg Army Air Base. Investigators stated, "Lt. Delano was on a routine transition training flight when his airplane was seen approaching the field at approximately 1,000 feet. When about a mile north of the field, the engine stopped completely and the pilot made a very steep right turn in an attempt to land at Leesburg Air Base. The airplane stalled and crashed out of control, apparently catching fire before it hit the ground."

5-5-43D. Independence, Kansas. At 1020 CWT, a North American BT-14 crashed while attempting to take-off from the Army Air Field at Independence, Kansas, killing instructor 2Lt. Robert W. McCarthy and student A/C Don I. Francis. The airplane was taking off to the south on an instrument training flight. The instructor was flying the airplane from the front cockpit. The airplane took off normally and reached an altitude of approximately 400 feet agl when witnesses observed smoke pouring from the engine area. The airplane climbed to 600 feet and then turned to the left instead of following normal traffic to the right. The airplane established level flight to the east for about 30 seconds. The smoke ceased and the airplane entered a controlled glide toward the ground. The airplane began a 90-degree turn to the left for a forced landing when it stalled. The airplane entered a spin to the left, describing three-quarters of a spin before smashing into the ground.

5-5-43E. Savanna, Illinois. At 1050, a Cessna UC-78B suffered the catastrophic structural failure of the port wing and crashed onto the grounds of an Army Ordinance Depot at Savanna, Illinois, killing 2Lt. Raymond G. Peterson and 2Lt. Reino A. Ranta. The airplane had taken off from Chicago Municipal Airport, Illinois, on a "strafing mission with a unit on maneuvers from Fort Sheridan [Illinois]." Witnesses on the ground reported hearing a loud report before the airplane entered a rolling dive toward the ground. The airplane smashed into the ground in an inverted attitude, scattering wreckage over a wide area.

5-5-43F. Location Unknown. At a time after 0926 PWT, a Bell P-39N was thought to have disappeared approximately ten miles west of Half Moon Bay, California. The pilot, 2Lt. Russell E. Peters, was missing and presumed lost at sea. The airplane had taken off from Mills Field on an "Air Alert" formation training flight. Investigators stated, "Lt. Peters was flying in a two-ship formation on an air alert mission. He was returning to shore, flying as wingman in the formation. The formation leader observed him at 8,000 feet at about 10 miles west of Montara Point. At this time flying appeared to be normal. About thirty seconds later the formation leader looked back and Lt. Peters' airplane was no longer in sight. The formation leader turned around to go back and try to find him but could not see any signs of the airplane or pilot. As the formation leader was getting short of gasoline he discontinued the search and returned to Mills Field."

5-5-43G. Delmar, California. At 1036, a North American AT-6A crashed three miles south of Delmar, California, killing pilot 2Lt. John F. Phelan and engineer-rated passenger Sgt. Richard J. Kopp. The airplane had taken off from the Naval Air Station at San Diego, California, on a local flight. Investigation revealed that the airplane was seen to make several passes over Camp Callan at low altitude. The airplane performed a steep pull-up maneuver, stalled, fell off on wing and crashed to the ground north of Camp Callan. It was noted that Lt. Phelan had been scheduled to fly a P-38 but found it out of commission upon his arrival. The pilot took the AT-6A because it was the only aircraft readily available for flight.

5-5-43H. Ryan Field, Tucson, Arizona. At 1210 MWT, a Ryan PT-22 collided with a civilian operated truck while on the landing roll at Ryan Field, Tucson, Arizona, fatally injuring civilian truck driver Frank Valasco. Civilian Flight Instructor Leo Gardner and student A/C Willard L. Gluth were uninjured. The airplane was returning to the field following the successful completion of a routine training mission. The student made a poor approach while attempting a landing to the north on the main north-south runway and the instructor allowed the landing to continue. The airplane began to roll off of the right edge of the runway and collided with the truck, which was pulling an implement and operating about 30 feet to the right of the active runway. Mr. Valasco died about 14 hours after the accident.

5-5-43I. Barnsboro, New Jersey. At 1630, a North American AT-6A crashed near Barnsboro, New Jersey, killing pilot 1Lt. Robert V. Runyon and pilot-rated passenger 2Lt. Edward P. Botsford. The airplane had taken off from Philadelphia Municipal Airport, Philadelphia, Pennsylvania, on a "test hop." The airplane was observed making low passes in the vicinity of the crash. The airplane was observed to pull out of a diving maneuver at low altitude. The airplane made a steep left turn at low altitude, the port wing contacted the ground and the airplane cartwheeled into the terrain and burst into flames. Burning wreckage was scattered over a wide area.

5-5-43J. Youngsville, Louisiana. At 1720, a Fairchild PT-19A crashed near Youngsville, Louisiana, killing student A/C Thomas L. Dwyer, Jr. Investigation revealed that the engine had failed and the pilot was attempting to shoot an emergency forced landing when he entered a stall at an altitude of about 40 feet. The student was unable to recover and the airplane smashed into the ground in a steep nose down attitude. Investigators speculated that the student was attempting to stretch his glide after undershooting the landing.

5-5-43K. Muroc, California. At 1919, a Lockheed RP-38E crashed just after take-off from the Army Air Base at Muroc, California, killing pilot 2Lt. Everette W. MacKenzie. The airplane was part of a three-ship flight taking off on a routine formation mission. The subject airplane was the third airplane to take off. Soon after take-off, the engine began smoking badly. The airplane made 90-degree turn and the pilot feathered the propeller and trimmed the airplane for single-engine flight. The airplane had reached an altitude of about 1,500 feet when it was seen to fall off to the right and then spiral to the ground where it exploded into flames approximately four miles northwest of Muroc Army Air Base. Investigation revealed that the starboard engine had caught fire just after the landing gear had been retracted. The pressure relief valve had apparently "unscrewed" itself and allowed all of the engine oil to be expelled from the crankcase, causing the engine to seize.

5-5-43L. Soda Springs, Idaho. At 2233 MWT, a Boeing B-17F crashed while attempting an emergency forced landing two miles southeast of Soda Springs, Idaho, killing seven fliers and injuring three others, one seriously. The airplane had taken off at 1930 PWT from Wendover Field, Utah, on a routine training flight to Pocatello, Idaho, and return. The airplane was flying on top of instrument conditions and was to pick up the Pocatello beam. The crew picked up the beam but lost radio contact about 15 minutes later. The pilot turned around, thinking that they had flown past Pocatello. The pilot let down through a hole in the clouds and flew toward the lights of Soda Springs. The pilot did not know his position and continued to circle the town. Several automobiles went to the edge of town and lined up along a clear area and signaled the pilot to land. The pilot buzzed the field and then turned around to make an approach on the field. The airplane collided with the ground on the approach. The port wing had contacted the ground and the starboard wing collided with a power pole and high-tension lines. The B-17 smashed into the ground and burst into flames, skidding out of control and scattering wreckage for approximately 300 yards. Killed in the crash were: 2Lt. Donald T. Arnold, navigator; SSgt. Richard L. Atkinson, engineer; SSgt. Floyd F. Journeay, engineer; SSgt. Adolph Zuelly, crewmember; SSgt. Thomas A. Smith, radio operator; SSgt. Elmer C. Barsig, gunner; SSgt. George E. Clausius, Jr., gunner. Pilot 2Lt. Melvin H. Williams and bombardier 2Lt. Dixie E. Reese received minor injuries and co-pilot 2Lt. Audley B. Pratt was seriously injured.

5-6-43A. Bowie, Texas. At 1515, a Piper L-4F crashed four miles north of Bowie, Texas, killing pilot 2Lt. James D. Freeman. Investigation revealed that the airplane was observed flying at extremely low altitude while following a highway. The airplane collided with power lines that were strung across the highway at an elevation of approximately 35 feet as the highway passed over a low hill. The airplane smashed into the ground and careened 165 feet, exploding into flames and killing the pilot instantly.

5-6-43B. Lawson Field, Georgia. At 2007 EWT, a Douglas A-20C crashed near Lawson Field, Fort Benning, Georgia, killing the crew of four. The airplane had taken off from Warner-Robins, Georgia, on an administrative flight to Lawson Field. Killed in the crash were: 1Lt. John M. Baker, pilot; TSgt. Joseph Dzuris, passenger; Sgt. John L. Pobeshein, passenger; Pvt. Edward J. Annucci, passenger. The pilot made a normal approach to land to the south at Lawson Field. Just as the airplane approached the runway, the port engine failed and the airplane began to drift to the left. The airplane had drifted so far to the left that it was in danger of landing on a row of Douglas C-47 airplanes parked east of the runway. The pilot gave the good engine full power for a go-around, feathering the port propeller and attempting to trim the airplane for single-engine flight. The airplane, still drifting to the left, was seen to pass over buildings at Lawson Field at an altitude of about 100 feet before it stalled and spun to the ground on the southeast edge of the field. The airplane burst into flames upon impact.

5-6-43C. Whitney, Texas. At 2130, a Cessna UC-78B flying in poor weather struck a power line and crashed six miles WSW of Whitney, Texas, killing student A/C Walter E. Scott and A/C Richard M. Schumacker. The airplane had taken off at approximately 2050 CWT from Blackland Army Air Field, Waco, Texas, with 11 other AT-17/UC-78 type aircraft on a night navigation flight back to base at Altus, Oklahoma. The airplane crashed approximately 25 miles NNW of Blackland Army Air Field. Investigation of the wreckage indicated that the airplane smashed into the ground in a diving, vertically banked left turn. Investigators speculated that the students had entered a severe thunderstorm and apparently lost control of the aircraft before transitioning to instrument flight. Weather at the scene of the crash was reported as severe thunderstorms with heavy rain, gusty winds and lightning.

5-7-43A. Warner-Robins, Georgia. At 0920 EWT, two Vultee BT-13A type airplanes collided in mid-air one and one-half miles west of Warner-Robins Field, Warner-Robins, Georgia. Instructor pilot 2Lt. Randolph D. White and student A/C Robert F. Outlaw were killed in the crash of BT-13A #41-1424, which had taken off from Cochran Field, Macon, Georgia, on an instrument training flight. Instructor 2Lt. Norton D. Eastment and student A/C Robert L. Datz were uninjured and able to execute an emergency forced landing in BT-13A #41-22036. BT-13A #41-22036 was part of a three-ship formation that had taken off from Cochran Field at 0915 on a routine training flight. Investigators stated, "The lead airplane of this [three-ship] flight was piloted by 2Lt. Robert Forrester. Following a normal take-off, the formation broke traffic and continued climbing. At an altitude of approximately 1,500 feet, and approximately five miles south of the field, Lt. Forrester noticed an airplane near the formation. This airplane later proved to be BT-13A #41-1424, piloted by 2Lt. White and A/C Outlaw, who were engaged in a routine instrument flight. When first seen, this airplane was approximately 500 feet above and 1,000 feet ahead of the formation, flying on a parallel course. [BT-13A #41-1424] then executed a split-S to the right and dove directly on the formation from about eleven o'clock. Lt. Forrester, due to lack of time, was unable to maneuver the formation out of the path of Lt. White's airplane. 2Lt. Eastment and A/C Datz, flying the left wing position of the formation in BT-13A #41-22036, did not see the approaching airplane. The left wing of Lt. White's airplane struck the airplane piloted by 2Lt. Eastment, tore off the left horizontal stabilizer and elevator, and jammed the rudder. The left wing of Lt. White's airplane was broken off at the center section and the airplane crashed, killing both occupants. Lt. Eastment maintained control of his damaged airplane and effected a safe landing."

5-7-43B. Bowling Green, Kentucky. At 0945 CWT, a Bell P-39D crashed one and one-half miles south of the municipal airport at Bowling Green, Kentucky, killing pilot Capt. James H. Kaden, who fell to his death in an unsuccessful parachute attempt. The airplane had taken off from Godman Field, Kentucky, on a mission to escort a flight of North American B-25 bombers that were engaged in a simulated attack on Bowling Green Municipal Airport. The subject airplane passed over the airport at an altitude of about 1,000 feet agl and about 200 yards behind his flight. The airplane then executed a steep turn to the left and the pilot released the port escape door. The airplane entered a spin and the pilot exited the aircraft at an altitude of about 100 feet. Lt. Kaden's parachute did not fully deploy and he was killed when he hit the ground about 180 feet from where the airplane smashed to earth and exploded violently into flames.

5-7-43C. Marana, Arizona. At 0950 MWT, a Vultee BT-13A crashed eight miles southwest of the Army Air Field at Marana, Arizona, killing instructor pilot 2Lt. Avery N. Segall and student A/C James R. Smart. The airplane had taken off from Marana on a routine instrument training flight. The instructor was also scheduled to demonstrate a five-turn spin. An AAF officer flying the control ship observed the subject airplane at an altitude of about 7,000 feet in a normal spin. The airplane continued to spin until reaching about 2,000 feet when the rate of spin slowed considerably. The airplane then went back into a fast spin and spiraled into the ground. The airplane did not burn but was destroyed utterly. The instructor had made no apparent effort to leave the airplane. The student's seat belt was found to be unfastened and hanging at the sides of the seat. The student had been hurled from his seat and was found at the bottom of the rear cockpit.

5-7-43D. Del Rio, Texas. At 1145 CWT, a Martin B-26B crashed 35 miles east of Laughlin Field, Del Rio, Texas, killing four fliers. Killed in the crash were: 1Lt. Harold E. Dyer, Jr., pilot; F/O Carlos W. Moss, co-pilot/B-26 student; F/O Dwight H. Orahood, co-pilot/B-26 student; SSgt. John P. Frigerio, engineer. The airplane had taken off from Laughlin Field on a routine training mission at 0800 CWT, but had returned to the field when student engineer Sgt. E.D. Weaver became airsick and could not continue. The airplane took off at 0950 and continued on its mission to practice single-engine operation. The airplane apparently stalled and entered a steep, diving spin to the ground. Witnesses on the ground reported seeing the airplane in a spin at about 3,000 feet agl with both engines "roaring." The airplane smashed into the ground in a vertical, nose-down attitude, exploding violently into flames upon impact. Investigators noted that the airplane was equipped with a "short spoiler at the wing root." Investigators speculated, "It is possible that the installation of a longer spoiler would give the pilot a buffeting at tail surfaces before a wing-tip stall could occur."

5-7-43E. Hamilton City, California. At 1030, a Vultee BT-13 crashed near Hamilton City, California, killing instructor 1Lt. Richard L. Knautz and student A/C Herman Gilbert, Jr. Investigation revealed that the instructor and student were on a routine "check" flight that had originated at Chico Army Air Field, California. The airplane was seen to enter an inverted dive at an altitude of 4,500 feet at a high rate of speed. The airplane appeared to attempt to roll through the inverted dive, causing the port wing to fail and separate from the airplane. The airplane smashed into the ground in a spin but did not burn. The instructor and student were killed instantly. Investigators could not determine why the airplane had entered an inverted dive.

5-7-43F. Oscoda, Michigan. At 1345, a Curtiss P-40F crashed while attempting an emergency forced landing at the Army Air Base at Oscoda, Michigan, killing pilot 2Lt. Jerome Thompson Edwards. The airplane had taken off from Oscoda AAB on an aerial gunnery mission. Investigators stated, "The pilot was flying in formation across the field at about 1,500 feet, going west. His engine cut out on him and he executed a steep turn to the left, putting his wheels down in an attempt to land on the East-West Runway, landing east he tightened the turn and did not complete it. He hit about 100 feet before the runway. The airplane had not leveled off when it hit the ground in a shallow dive. The wheels were knocked off and the airplane bounced up and fell again to the ground. All pilots have been repeatedly told about forced landings being shot with wheels up."

5-7-43G. Baltimore, Maryland. At 1515, a Republic P-47D crashed into the Patapsco River near the Army Air Base at Baltimore, Maryland, killing pilot

2Lt. James E. Tolan. The airplane had taken off from Baltimore Army Air Base as the number-two airplane of a four-ship flight on an altitude formation mission. While flying at 22,000 feet indicated altitude, the flight leader noticed a thick trail smoke emitting from the subject airplane. The flight leader ordered Lt. Tolan to return to the field immediately. The airplane circled the base to the south with the propeller windmilling. The field was immediately cleared in anticipation of an emergency landing. Lt. Tolan made a 90-degree pattern around the field and turned onto final approach for Runway 32. The airplane was too far from the field to land on the runway and the pilot attempted to stretch the glide, causing the airplane to stall and enter a spin to the right. The airplane described about three-quarters of a turn of spin before smashing to earth and exploding into flames 300 yards short of the runway in about four feet of water.

5-7-43H. Dalhart, Texas. At 2208, a Boeing B-17F crashed four miles west of the Army Air Base at Dalhart, Texas, killing eleven crewmembers and seriously injuring tail gunner SSgt. Richard I. Stoney. Killed in the crash were: 2Lt. Lloyd E. LeRoy, pilot; 2Lt. George H. Wuelfing, co-pilot; 2Lt. Charles C. Enegbrepson, navigator; 2Lt. Willis M. Rodgers, bombardier; SSgt. Ernst A. Siros, engineer; SSgt. Francis J. Wethington, radio operator; Sgt. Alton E. Barrineau, gunner; Sgt. Leon Skiha, gunner; SSgt. Paul H. Snider, gunner; TSgt. Joseph T. King, engineer instructor; Cpl. Rex L. Clifford, photographer. The airplane was returning to Dalhart Army Air Base after completing a routine training mission. The airplane, with landing gear down and flaps at about one-third, had made a steep turn to the left from the downwind leg to the base leg. The airplane continued to slip to the left until it contacted the ground on the port wing. The airplane cartwheeled into the ground and burst into flames. The tail section had separated from the main wreckage and did not burn.

5-8-43A. Cash, Texas. At 1015, a Vultee BT-13A suffered a catastrophic structural failure of the starboard wing and crashed one and one-half miles northeast of Cash, Texas, killing instructor 2Lt. Norbert V. Delanoy, Jr. and student A/C Ernest L. Leonard. The airplane had taken off from Majors Field, Greenville, Texas, on a routine training mission. The airplane was scheduled to practice chandelles, stalls, elementary-eights and other basic maneuvers. Examination of the wreckage indicated that four nuts were stripped off of their respective bolts on the lower side of the forward part of the front spar, allowing the outboard starboard wing to separate at the point where it joins the forward part of the center section. Investigators speculated that the wing failed after an extreme maneuver. Witnesses on the ground believed that the accident was a result of a mid-air collision. No evidence of contact with another airplane could be found. The outboard starboard wing was found approximately 75 yards from where the airplane smashed into the ground and burst into flames. The occupants made no apparent effort to bail out. Investigators noted, "That no pilot in the air at the time of the accident claims to have seen or know anything about this accident."

5-8-43B. San Juan County, Utah. At 1115 MWT, a Consolidated B-24E crashed in San Juan County, Utah, approximately 35 miles northwest of Shiprock, New Mexico, killing four fliers. The airplane was being ferried from Kirtland Army Air Field, Albuquerque, New Mexico, to the Army Air Base at Pocatello, Idaho. The airplane was seen in a flat spin when it broke through a 500-foot overcast. The tail section failed and separated before the B-24E slammed into the ground in an inverted attitude and exploded into flames. The tail section smashed to earth about one-half mile from where the airplane struck the ground. The pilot, 1Lt. Luther A. Ponath, was killed when he attempted to use his parachute at too low an altitude to allow his parachute to deploy. Co-pilot 2Lt. Attilio Dogliotti, engineer MSgt. Louis M. Baumgartner and radio operator TSgt. Jack H. Keller went in with the airplane and were killed instantly.

5-8-43C. Shorter Auxiliary Airfield, Alabama. At 1445 CWT, a Stearman PT-17 crashed near Shorter Auxiliary Airfield, Alabama, killing instructor 2Lt. James E. Brothers and passenger Pvt. Ross C. Stewart. The airplane had taken off from Tuskegee Army Air Field, Tuskegee, Alabama, on an individual proficiency flight. An instructor flying back to the Tuskegee Army Air Field observed the airplane performing aerobatic maneuvers at a low altitude. The airplane was also seen by other witnesses performing split-S maneuvers and half-rolls at an altitude below 2,000 feet agl. The pilot apparently failed to pull out of one of the maneuvers before striking the ground. The occupants were killed instantly. The instructor that had initially witnessed the low altitude maneuvers was reporting the incident in the operations room when he and others were informed that the airplane had crashed.

5-8-43D. Oklahoma City, Oklahoma. At 2142 CWT, a Douglas C-47A flying in instrument conditions crashed two miles north of Tinker Field, Oklahoma City, Oklahoma, killing five crewmembers and ten passengers. The airplane had taken off from Tinker Field at 2132 on a cargo/personnel ferry flight to Alliance Army Air Base, Nebraska, and crashed into a plowed field ten minutes later. The Accident Classification Committee stated, "Investigation revealed that the airplane struck the ground while in a left turn, the left wing striking the ground first. Approximately 200 yards past the first contact with the ground, the airplane struck with such force that major parts of the airplane, including the engines, were thrown off the airplane and all occupants were thrown out. There was evidence that the airplane was burning after this contact with the ground. The airplane then bounced and finally came to rest approximately 300 yards further on where it continued to burn." Investigators speculated that the pilot became confused while flying on instruments and allowed the airplane to enter a diving attitude toward the ground. The airplane was seen to emerge from the 700-foot overcast in a diving attitude. The pilot was unable to recover in time to prevent the airplane from striking the ground. Crewmembers killed in the crash were: 1Lt. Robert H. Pair, pilot; 2Lt. Norman C. Keister, co-pilot; F/O Joseph N. Phelps, co-pilot; Sgt. Robert E. Bennett, engineer; Cpl. Stephen R. Mendry, radio operator. Passengers killed in the accident were: 2Lt. Irving S. Greenburg, Cpl. Frank C. Sutterby, Cpl. Douglas N. Grover, Cpl. Grayson C. Naden, Pfc. Jack D. Rooker, Pfc. Albert C. Tilley, Pfc. Cecil C. Motley, civilian passenger G.W. Bradshaw, civilian passenger R.E. Fitzgerald, civilian passenger J.C. Coleman.

5-9-43A. Lake Huron, Michigan. At 1157 EWT, pilot 2Lt. Wilmeth Sidat-Singh disappeared after he parachuted from a Curtiss P-40F over Lake Huron, Michigan, about three miles east of Tawas Coast Guard Station. The airplane was seen flying at an altitude of about 3,500 feet above the surface of the lake with the engine smoking. The pilot bailed out and was seen to parachute into the lake, but disappeared before a crash boat could rescue him. Despite a massive search, no trace of the pilot could be found. The pilot was declared missing and was presumed lost. The airplane had taken off from Oscoda Army Air Field, Oscoda, Michigan, on local aerobatic training flight.

5-9-43B. Hempstead, New York. At 1300, a Republic P-47D crashed at Hempstead, Long Island, New York, killing pilot 2Lt. Robert O. Jamison. The P-47 was the number-three airplane in a three-ship flight that had taken off from Mitchel Field, Long Island, on an aerial gunnery mission. The airplane was apparently suffering engine trouble when the flight had reached 20,000 feet. The pilot returned to Mitchel Field and entered the traffic pattern in an attempt to land. The airplane was turning from the base leg to final approach when it entered a spin to the left at an altitude of 400 feet agl. The pilot was unable to recover and the airplane smashed into the ground and exploded into flames.

5-10-43A. Fort Stockton, Texas. At 0930, a Fairchild PT-19A crashed at Fort Stockton, Texas, killing A/C Marvin E. Bales. Investigation revealed that the airplane had taken off from Gibbs Field, Fort Stockton, at 0759 CWT on a routine training flight. The airplane had landed and returned to the line at 0853, taking off again at 0900. At 0930, the airplane was seen to enter a deliberate spin at approximately 6,500 feet. The airplane spun about four or five times and then stopped spinning momentarily. The airplane then entered a spin in the opposite direction. The student was not able to recover and the airplane spun into the ground, killing him instantly. The airplane did not burn but was destroyed utterly.

5-10-43B. Tuskegee, Alabama. At 1000, a Curtiss P-40E crashed five and one-half miles northwest of Tuskegee, Alabama, killing pilot 2Lt. Sidney J. Mosley. The airplane had taken off from Tuskegee Army Air Field on a routine training flight. Investigation of the wreckage indicated that the airplane had struck the ground in an inverted spin. Investigators speculated that the pilot might have attempted a loop or an Immelmann maneuver and apparently stalled out at the top, entering an inverted spin. There were no witnesses to the crash.

5-10-43C. Midland, Texas. At 1019, a North American AT-6C crashed at Midland, Texas, killing pilot 2Lt. Robert J. McGinn. The airplane was seen flying at an altitude of about 1,500 feet when it went into a diving turn to the right. Witnesses then observed the airplane in a steep dive toward the ground. The pilot never recovered and the airplane smashed into the ground at high speed, exploding into flames and killing the pilot instantly.

5-10-43D. Lubbock, Texas. At 1200, a Cessna AT-17B and a Cessna UC-78B collided in mid-air 15 miles southwest of Lubbock, Texas, killing four fliers. A/C Nicholas G. Kirincich and A/C Earl W. Kenny were killed in the crash of UC-78 #42-39178. A/C Hector D. Lane, Jr. was killed in the crash of AT-17 #42-38793 and A/C Edward I. Lang was killed while attempting to parachute to safety. The airplanes were part of a three-ship flight that had taken off from Lubbock Army Air Field on a routine formation training flight. The airplanes were flying in an echelon formation at about 2,100 feet. The student formation leader gave the signal to return to a V-formation. The airplanes collided while attempting to execute the change from echelon to V-formation. There were no witnesses to the actual collision. A farmer observed both airplanes falling to the ground. Investigation revealed that A/C Lang deployed his parachute before he was clear of the AT-17 and it was fouled by the aircraft structure, causing him to fall to his death.

5-10-43E. Chandler, Arizona. At 1127 MWT, a Lockheed RP-322 suffering an in-flight fire crashed five miles southeast of Williams Field, Chandler, Arizona, killing student A/C Herbert Brenner. Investigators stated, "Pilot called [Williams Field] control tower to report his right engine on fire. He was then about to enter traffic. Shortly thereafter the engine fell from the aircraft, the aircraft half-rolled, dove into the ground and burned. Investigation showed the connecting rod, #4 cylinder, right bank, [had] been driven thru the crankcase. It is assumed that the fire then started within the engine, spread to the magnesium parts and burned the engine mounts, allowing the engine to part with the aircraft."

5-10-43F. Waller-Franklin Airfield, California. At 1040, a Stearman PT-13A and a PT-13B collided in mid-air two miles west of Waller-Franklin Airfield, California, killing A/C Frank Unger aboard

PT-13A #37-245. A/C Numa E. Snider, Jr. was able to safely land PT-13B #40-1729 at the field. He was uninjured. Both airplanes had taken off from Hancock Field, California, on routine training flights. Investigators stated, "Cadet Snider was just breaking his glide at approximately 700 feet in order to enter the traffic pattern at Waller-Franklin Field. At this point, another airplane, piloted by Cadet Unger, came in fast behind and a little below Cadet Snider. Apparently Cadet Unger was so intent on getting the airplane in that he had his eyes fastened on the field and just as he noticed Cadet Snider's airplane, he hit [Cadet Snider's] lower left wing with his right upper wing. Cadet Unger pulled his airplane up and to the left. The airplane snapped into a spin, made two complete turns and crashed into the ground, killing [Cadet Unger] instantly. Cadet Snider straightened his airplane out of a right bank into which he had been thrown and then landed on Waller-Franklin Field."

5-10-43G. Sprague, Washington. At 1225, a North American B-25D crashed three and one-half miles northwest of Sprague, Washington, killing the crew of five. Killed in the crash were: 2Lt. John F. Kirbey, pilot; 2Lt. Homer B. Glover, co-pilot; TSgt. Clarence A. Murray, engineer; SSgt. Edward C. Hartley, radio operator; SSgt. Lewis J. Asmer, radio operator. The airplane was part of a three-ship flight that had taken off from Geiger Field, Spokane, Washington, on an instrument flight to Portland Army Air Base, Portland, Oregon. The flight encountered instrument conditions and the formation broke up. The airplane, flying in instrument conditions, entered an area of extreme turbulence at about 11,000 feet indicated altitude when the pilots apparently lost control. The pilots were unable to regain control of the airplane and it slammed to the ground where it exploded into flames upon impact.

5-10-43H. Cocoa, Florida. At 1610 EWT, a Douglas P-70 flying at low altitude collided with a palm tree and crashed two miles west of Cocoa, Florida, killing pilot 2Lt. George R. Ewing. The airplane had taken off from Orlando Army Air Base, Orlando, Florida, on a routine training flight. The airplane had been seen flying at very low altitude in the vicinity prior to the crash. Investigators stated, "On the last pass the airplane, flying very low, apparently pulled up, at which time the airplane mushed slightly. At the same time, the pilot attempted a bank thus causing the right wing to hit a palm tree. As the airplane hit the tree it was forced to spin around 180 degrees, hitting the ground approximately 50 feet from this tree. The airplane skidded along the ground for about 200 yards during which time three palm trees were knocked down. When the airplane hit the fourth tree it burst into flames at which time the complete path [that the airplane traveled] burst into flames apparently due to bursting gasoline tanks spreading gasoline over the area."

5-10-43I. San Diego, California. At 1333 PWT, a Consolidated XB-32 crashed into a U.S. Marine Corps Base while attempting to take-off from Lindbergh Field, San Diego, California, killing the Consolidated test pilot and three U.S. Marines on the ground. Six XB-32 crewmembers were seriously injured. Thirty-one marines were seriously injured and hospitalized and 48 marines received minor injuries when the XB-32 smashed through five base buildings and 19 enlisted barracks. Killed in the crash were: Consolidated Aircraft Company test pilot Richard Addison McMakin, 31, Coronado, California, pilot; Pvt. George C. Thomas, Detroit, Michigan, USMC; Pvt. Oral Meranda, Scottsburg, Indiana, USMC; Pvt. Jerry M. Myers, Pinckneyville, Illinois, USMC. Pvt. Meranda died at 1645 and Pvt. Myers died at 1700. Consolidated Aircraft Company crewmembers seriously injured in the crash were: Don H. Scott, Mission Beach, California, co-pilot; William Ponder Easley, San Diego, California, engineer; George Markian Stetsuk, Chicago, Illinois, engineer; Jasper Franklin Woodall, Chicago, Illinois, engineer; John Frederick Scott, San Diego, California, observer; William Frank Chana, Berwyn, Illinois, observer. Investigators stated, "Subject airplane was attempting to take-off on a routine test flight, but failed to get off the ground before reaching the end of the runway. Under full power, it struck the wire fence separating Lindbergh Field from the USMC Base, then crashed into a number of buildings. Finally coming to rest in the Recruit Barracks area, about 1,000 feet from the fence. The airplane and all the buildings it struck were completely demolished. Fire broke out charring the wreckage and all nearby buildings. The pilot was burned to death; the other crewmembers escaped with major injuries. Three Marines were killed and approximately 80 injured. After evaluation of all existing evidence by [the Aircraft Accident Classification Committee], the probable cause of this accident appears to be [that the] flaps were not set at the proper degree for take-off as specified in the flight plan for this flight." Twenty-four buildings were destroyed in the crash.

5-10-43J. Okolona, Kentucky. At 1728 CWT, an experimental Curtiss C-76 transport suffered a catastrophic structural failure and crashed at Okolona, Kentucky, killing three civilian crewmembers. The airplane had taken off from Louisville, Kentucky, on a test flight. Investigation revealed that the airplane was flying at an altitude of approximately 2,000 feet agl when it came apart in mid-air. Wreckage was scattered over a wide area. Investigators speculated that extreme vibration caused the structural failure. Curtiss-Wright Corporation employees killed in the crash were: Ed Schubiner, pilot; Dick Trowbridge, co-pilot; Robert Scudder, engineer.

5-10-43K. Edgemont, California. At 1830, a Lockheed P-38F suffering an in-flight fire crashed at Edgemont, California, killing pilot 2Lt. Richard Thomas. Investigators stated, "This accident resulted from an engine failure which resulted in fire. It appears that when the left engine failed the pilot feathered the propeller and headed for a landing at March Field

[Riverside, California], but just before reaching the field the airplane caught on fire and the pilot attempted to leave the ship but his parachute was caught on the stabilizer and he was carried down with the crash. The airplane was so completely demolished that no positive information could be obtained." The airplane had taken off from Ontario Army Air Field, California, on a routine training flight.

5-11-43A. Brady, Texas. At 0910, a Fairchild PT-19A crashed 12 miles west of Curtis Field, Brady, Texas, killing civilian instructor Jessie E. Medley, 32, Santa Rosa, New Mexico, and student A/C William H. Wardell, 26, Alameda, California. Investigation revealed that the airplane had entered a spin at an altitude too low to allow recovery. The airplane described two turns of spin before it struck the ground. The airplane did not burn but was completely demolished and the occupants killed instantly.

5-11-43B. Glendale, Arizona. At 0810 MWT, a Stearman PT-17 spun to the ground five miles east of Thunderbird Field, Glendale, Arizona, killing student A/C William C. Schrouder, Jr. Investigators speculated that the airplane had entered a spin at an altitude too low to allow recovery. There were no witnesses to the crash. The PT-17 had taken off from Thunderbird Field on a routine solo flight.

5-11-43C. Ashland, Virginia. At 1345, a Republic P-47D crashed five miles north of Ashland, Virginia, killing pilot 2Lt. Herman L. Krouse, South Bend, Indiana. Investigators stated, "Pilot was flying as wingman in a two-ship formation climbing to 20,000 feet. Leader of the formation states that at 17,000 feet subject pilot pulled abreast of him and gestured in such a way that the leader understood wingman was having difficulty with oxygen and was going to leave the formation. Leader motioned approval to descend and the wingman made a violent peel-off into a steep dive in which the leader lost sight of him. Hearing the noise of the descent, witnesses on the ground looked up to see the airplane just below the clouds and evidently recover from a dive into a stall, recover from the stall just over the trees, fly along for several seconds and then plunge into ground. It crashed on the Timberton Dairy Farm, exploded and killed the pilot. Evidence at the scene of the accident indicated that the pilot had his left hand on the trim tabs at the time of the crash." Investigators noted that the pilot could not secure an A-10 type oxygen mask prior to the flight and was unable to fly at high altitude for any prolonged period of time.

5-11-43D. Victoria, Texas. At 1620, a North American AT-6A spun to the ground while attempting to land at Foster Field, Victoria, Texas, killing student A/C Carl M. Clark. Investigators stated, "A/C Clark was observed to make his approach at the proper altitude of 1,500 feet over the runway on which he intended to land and then make an excessively steep peel-off to the base leg, such peel-off being a turn that placed his ship in a position quite a bit beyond the vertical bank. This in itself attracted some attention, enough so that he was then observed to make his base leg approach at 500 feet with his final turn on to the approach very near the beginning of the runway. A/C Clark rolled into his approach turn which he made very steep and at a speed estimated to be lower than safety would demand and in rolling out of this turn his ship was observed in a stall which resulted immediately in a spin and [it] dove straight into the ground. His ship crashed very nearly beneath the spot A/C Clark made his turn and upon crashing burned."

5-11-43E. Shreveport, Louisiana. At 1730, a Martin B-26C suffering the failure of the starboard engine crashed five miles WSW of Barksdale Field, Shreveport, Louisiana, killing pilot 2Lt. Leonard J. Cook, Hayden, Colorado. Six fliers were able to parachute to safety and were uninjured. Investigators stated, "The crew had returned from a Navigation flight and were going to complete the mission on the Bombing Range. They had dropped three practice bombs when the engineer discovered they had lost an excessive amount of hydraulic fluid and recommended they return to the field immediately. The pilot pulled up and entered traffic in a normal manner. He was making a turn to the downwind leg when the right engine failed. The pilot went into single-engine operation and then made a left turn, away from traffic. He was unable to maintain altitude on one engine. He decided to lighten the airplane by jettisoning the remainder of practice bombs and bomb bay tanks with the emergency system. They were unable to open the bomb bay doors to jettison the load. The pilot then decided to have the crew abandon the airplane through the nose wheel well. There was an appreciable decrease in the rate of descent after five crewmembers had left the airplane. The engineer again tried to open the bomb bay doors from the bombardier's position, but was unable to accomplish this. While the engineer was returning to the pilot's compartment, he felt the airplane lurch to the left and [he] immediately abandoned the airplane. The airplane was then seen to go into a steep dive and turning to the left as it crashed. The Committee believes that with the failure to jettison the load, the additional drag of the wheels being lowered to enable the crew to jump, necessitated that the pilot use maximum power from his left engine to maintain his airspeed. This Committee also believes that this caused the left engine to fail, which would account for the diving turn to the left." Crewmembers parachuting to safety were: 2Lt. Robert R. Murray, co-pilot; 1Lt. Joseph M. Coyle, navigator; 2Lt. Daniel T. Sufana, bombardier; SSgt. Earl C. Pittman, engineer; SSgt. Orville Hahn, radio operator/gunner; SSgt. Michael J. Sikora, gunner.

5-11-43F. Ocala, Florida. At 1840, two North American A-36A type airplanes collided in mid-air near Ocala, Florida, killing the pilot of A-36A #42-83797, 2Lt. Charles D. Pacatte. 2Lt. Ferne G. Lovell received minor injuries parachuting to safety from A-36A #42-83787. The airplanes were part of a four-ship flight that had taken off from Alachua Army Air Field, Gainesville, Florida, on a local formation flight. The flight practiced formation flying for about 30 minutes. The flight leader then had to maneuver the formation to avoid a thunderstorm that was approaching the area. The airplanes flew to Ocala and the flight leader noticed that there was no flying activity at the flying school. The flight peeled up and then buzzed the field individually, reforming after three passes and climbing to 2,000 feet agl. The airplanes then formed a Lufbery circle over the field. The leader then reformed the flight and flew to the south. The two airplanes collided moments later. Lt. Pacatte was flying as wingman of the first element leader and Lt. Lovell was flying as wingman of the second element leader.

5-11-43G. Peterson Field, Colorado. At 1708, a Lockheed F-4A crashed while attempting a forced landing near Peterson Field, Colorado Springs, Colorado, killing pilot 2Lt. Robert F. Granz. Investigators stated, "It is believed that the pilot was attempting a forced landing after having had a complete power failure, the cause of the which is undetermined. This was the first flight in this type of aircraft by the pilot. The landing was attempted downwind and downhill in a plowed wheat field with the landing gear down. The pilot failed to drop his [auxiliary] drop tanks before attempting the landing. The surface wind was approximately 30 mph. No flaps were used in this attempted landing. Impact with the ground was hard enough to [collapse] the landing gear and cause the airplane to explode. The impact and the explosion resulted in the death of the pilot."

5-11-43H. Indianapolis, Indiana. At 2335 CWT, a Vultee BT-13A crashed 30 miles west of Stout Field, Indianapolis, Indiana, killing Service Pilot 2Lt. Gilbert L. Worley, Jr. and pilot-rated passenger 2Lt. Robert H. Martin. The airplane had taken off from Stout Field on a flight back to its base at Garden City, Kansas, when it encountered instrument conditions. The ceiling was reported as 100 feet agl. The pilot flew in instrument conditions for several minutes before deciding to return to Stout Field. The pilot attempted a 180-degree turn and flew into the ground at an angle of about 30 degrees. The aircraft burst into flames upon impact and the pilots were killed instantly.

5-12-43A. Mullet Key, Florida. At 0730 EWT, a Douglas A-24A spun into Tampa Bay at the AAF Bombing Range at Mullet Key, Florida, killing pilot 2Lt. Theodore H. Work and gunner Sgt. Charles L. Simpson. The airplane was the number-two airplane in a three-ship flight that had taken off from Drew Field, Tampa, Florida, on a dive-bombing mission. The formation had made two dive-bombing passes at the target range and the subject airplane was pulling up to rejoin the lead airplane after the second pass. The subject airplane was flying at a 90-degree angle to the flight path of the lead airplane. The pilot put the airplane into a vertically banked turn to the right in order to parallel the flight leader's flight path. The airplane fell into a spin, describing three or four turns of spin before smashing into the bay. The airplane exploded upon impact and sank immediately. The occupants killed instantly.

5-12-43B. Vinegar Bend, Alabama. At 1115, a North American B-25A suffered a structural failure and crashed near Vinegar Bend, Alabama, killing five crewmembers and injuring pilot Lt. Col. Adam J. Heintz, who was able to parachute to safety. Killed in the crash were: 2Lt. H.A. Thomas, co-pilot; SSgt. Milford Hershkowitz, radio operator; Maj. J.W. Dodson, passenger; Maj. Cheney Cowles, passenger; SSgt. H.E. Bracken, passenger. Investigation revealed that the number-one engine seized while at cruise power, causing severe vibration and stresses that threw the engine and propeller into the fuselage structure. The pilot reported that he was flying at 3,500 feet indicated altitude and felt a sudden shock and the sound of metal tearing. In the next instant the top of the cockpit canopy began to crush down around the pilot's head and shoulders. The cockpit windscreen and windows shattered and the cockpit began to disintegrate around the pilot. The pilot unfastened his safety belt and soon he was falling through the air clear of the airplane. He immediately deployed his parachute and floated safely to earth.

5-12-43C. Pope Field, North Carolina. At 1415, an Aeronca L-3B crashed five miles southeast of Pope Field, Fayetteville, North Carolina, killing rated glider pilot F/O Charles L. Wiess and observer passenger 1Lt. John G. Graney. The airplane had made several passes over a column of vehicles, dropping sandbags. During the final pass the airplane stuck the top of a large tree, severing the starboard wing. The airplane smashed to the ground approximately 200 yards from the tree.

5-12-43D. Tampa, Florida. At 1630 EWT, a North American BT-14 crashed 15 miles northwest of Tampa, Florida, killing passenger TSgt. Artie U. Mitchell and pilot 1Lt. Claude M. Orr, who fell to his death in an unsuccessful parachute attempt. The airplane had taken off from Drew Field, Tampa, Florida, on a local test flight. The airplane was observed in a spin at an altitude of about 8,000 feet. The airplane continued to spin until it struck the ground in a wooded area. A parachute was seen trailing the airplane. Apparently the pilot deployed the parachute while still standing in the cockpit. The parachute was found in a tree near the wreck.

5-12-43E. Roswell, New Mexico. At 1554 MWT, a Vultee BT-15 crashed 28 miles northwest of the Army Air Field at Roswell, New Mexico, killing pilot 2Lt. Mervin E. Krohn and seriously injuring bombardier-rated passenger 1Lt. John C. Fehrs, Jr. The airplane had failed to return to base and was declared missing. The airplane was found on 5-14-43. Investigation revealed that the airplane had been flying at low altitude and collided with a small hill that it had failed to clear. The pilot was found dead in the cockpit. The passenger was not at the wreck and he was found on 5-14-43 wandering in the desert with very serious injuries. The survivor could not remember any details of the flight or the crash. The pilot had received his wings and his commission on 5-12-43.

5-12-43F. Olympic Mountains, Washington. At approximately 1530 PWT, a Lockheed P-38E crashed into rising terrain in the Olympic Mountains 40 miles southwest of Port Townsend, Washington, killing pilot 2Lt. Dean T. Mangold. Lt. Mangold was leading a four-ship flight of P-38 airplanes on an aerial gunnery mission over the Straight of Juan de Fuca. Investigators stated, "On completion of the mission the P-38s started the return flight to McChord Field [Tacoma, Washington]. [The pilot] flying the tow target ship, [a Lockheed] RB-34, sighted questionable weather ahead and called the P-38 flight leader to close up the flight on the RB-34. The flight was in the process of closing up in echelon on the right wing of the RB-34 when they struck a snow squall about three miles southeast of Blyn, Washington. Instructions were then given to the P-38s to climb ahead and cautioned them about the mountains on the right of the flight path. The break out on top was at about 10,000 feet. [The RB-34 pilot] then contacted the flight and found two P-38s in the clear and en route McChord Field and one P-38 on his wing. [The RB-34 pilot] was unable to contact Lt. Mangold." On 11-1-46, civilian Glenn Harbough found the airplane wreckage smashed into a wooded slope in the southern Olympic Mountains.

5-12-43G. Pampa, Texas. At 1815, a Curtiss AT-9A and a Cessna UC-78B collided in mid-air two miles west of Pampa, Texas, killing the instructor aboard the AT-9A and seriously injuring his student. Instructor 2Lt. William A. Gibbons was killed in the crash and student A/C Melvin J. Hamann was seriously injured in a barely successful parachute jump. Student pilots 2Lt. Paul D. Pattinson and 2Lt. William R. Nielson were able to parachute to safety from the UC-78B, receiving minor injuries. The airplanes were flying in a three-ship V-formation under the direction of Lt. Gibbons. The instructor ordered the UC-78, which was leading the formation, to drop back to the number-three position. The UC-78, flying slightly below, dropped back and then climbed into the propellers of the instructor's AT-9, which had been flying on the right wing of the lead AT-9 and was attempting to drop back. The collision severed the tail of the UC-78, sending it spinning to earth. The airplane broke up after three or four spins, hurling the students into space at an altitude of about 200 feet. They immediately deployed their parachutes. The AT-9 went out of control in a spin to the right and the instructor immediately ordered the student to bail out. The student rolled out onto the port wing and apparently deployed the parachute before he was clear of the airplane. The student's parachute apparently fouled and it was seen to "collapse" and "string out" as the student descended. Both airplanes slammed to earth within 200 yards of each other.

5-13-43A. Chesapeake Bay, Maryland. At 0757, a Republic P-47D crashed into Chesapeake Bay near Galesville, Maryland, killing pilot 2Lt. Dean E. Scroggie. The subject airplane was the lead airplane of the second two-ship element of a four-ship flight that had taken off from Baltimore, Maryland, on a high-altitude training flight. The flight, which had been flying at 29,000 feet msl, had been cleared to return to base at Baltimore. The airplanes began descending and at an altitude of 25,000 feet the lead airplane began a diving left turn. The subject airplane followed but was not able to pull out of the dive. The airplane dove into four feet of water about 150 feet off shore, striking at high speed and exploding violently into flames upon impact.

5-13-43B. Zephyrhills, Florida. At 0855 EWT, a Douglas A-24A stalled and spun to the ground three miles northeast of the Army Air Field at Zephyrhills, Florida, killing pilot 2Lt. Harold E. Kievit and pilot-rated passenger 2Lt. Clare W. Brown. Investigation revealed that the engine had apparently failed and the pilot was attempting to execute an emergency forced landing. The airplane stalled at an altitude of approximately 60 feet agl and entered a spin to the left. The pilot was unable to recover the airplane and it smashed to the ground at a steep angle in a nose-down attitude. The airplane had taken off from Drew Field, Tampa, Florida, on a cross-country navigation flight to Daytona, Florida, and return.

5-13-43C. Gulf of Mexico, Texas. At 0810, two North American AT-6A type airplanes collided in mid-air 23 miles northeast of the Army Air Field at Harlingen, Texas, and crashed into the Gulf of Mexico, killing three fliers. Student gunner Pfc. Frank M. Loconsolo was able to parachute to safety from AT-6A #41-704. Pilot 2Lt. Albert S. Collins was killed in the crash. Pilot 2Lt. Robert W. Stahley, Jr. and student gunner Pfc. Lindel M. Adamson were killed in the crash of AT-6A #41-16104. AT-6A #41-16104 was being flown as the target tow ship and AT-6A #41-704 was the gunnery ship on a gunnery-training mission. Lt. Stahley extended his target reel and made a 180-degree turn to establish a flight path parallel with an island for the gunnery run. Apparently Lt. Stahley was unaware that Lt. Collins had pulled his airplane into a firing position when he began maneuvering. Witnesses observed AT-6A #41-16104 (Lt. Stahley) make an abrupt turn into AT-6A #41-704 (Lt. Collins).

5-13-43D. Camp MacKall, North Carolina. At 0925, a Vultee L-5 crashed near Camp MacKall, North Carolina, fatally injuring Liaison Pilot SSgt.

John T. Hart. He died of his injuries 6-10-43. Observer 1Lt. Earl H. Bridgewater was seriously injured. Witnesses stated that the airplane was in a steep bank at an altitude of approximately 100 feet above the ground. The nose was seen to drop and the airplane mushed into the ground. Investigation of the wreckage revealed that the engine was producing power at the time of impact. The airplane had taken off from Camp MacKall on a reconnaissance flight.

5-13-43E. Old Tampa Bay, Florida. At 1055 EWT, a Curtiss P-40L crashed into Old Tampa Bay, Florida, killing pilot 2Lt. Joseph S. Svienty. The airplane had taken off from Pinellas Field, St. Petersburg, Florida, on a local transition flight. Witnesses on the ground reported hearing a loud explosion and then seeing the airplane, minus one wing, in a spinning dive toward the bay. The airplane disintegrated in the air and the pieces fell into the bay. Only a very small amount of human remains were recovered from the water. Investigators could not determine what caused the airplane to break up in mid-air.

5-13-43F. Fort Collins, Colorado. At 1110 MWT, a Boeing B-17F flying in instrument conditions crashed into rising terrain 14 miles west of Fort Collins, Colorado, killing pilot 1Lt. William T. Averitt and engineer Cpl. Stanley Luxenberg. The airplane had taken off from Denver Municipal Airport, Colorado, on a ferry flight to Great Falls, Montana. The pilot became lost after he inadvertently flew toward Cheyenne, Wyoming, when he wanted to fly toward Laramie, Wyoming. Investigators stated, "The pilot apparently flew toward Cheyenne and became lost, finally turning up the Buckhorn Creek Valley to the northwest in an effort to reach Laramie. In trying to fly contact the pilot stayed in the Buckhorn Creek Valley until he reached a point approximately 14 miles west of Fort Collins, Colorado, or eight miles northwest of Masonville, Colorado. At this point the elevation of the mountains decidedly increased and, as the pilot was just under the clouds, he evidently was forced to climb into them. Visibility at this time and at this point was restricted to about 30 feet and the pilot barely cleared a small ridge. After crossing this ridge, the pilot evidently attempted to turn out of the mountains. In doing so, he turned into a box canyon, striking a small, jutting knoll with the port wing, which turned [the airplane] into the mountain approximately 300 feet further on. The right wing and fuselage [collided with the mountain]." The airplane burst into flames upon impact. Investigators noted that the flight could have been completed safely had the pilot headed toward Laramie, which was reporting visual conditions. Investigators also noted that the flight should not have been cleared without a complete flight crew.

5-13-43G. North Stonington, Connecticut. At 1438, a Republic P-47D crashed near North Stonington, Connecticut, killing pilot 2Lt. Harry J. McElroy, Jr. The pilot was leading a flight of five P-47s on a high-altitude oxygen mission from Bradley Field, Connecticut, to Groton Field, Groton, Connecticut, via New Haven. The flight had reached an altitude of 31,000 feet and was flying west. Lt. McElroy began a descending left turn and the flight followed. The subject airplane continued in a dive until it struck the ground at high speed, exploding violently into flames upon impact. Evidence suggests that the airplane disintegrated when it reached the point of compressibility in the dive. The remainder of the flight attempted to pull out of the dive at about 23,000 feet. The other pilots reported that their airplanes came out of the dive with much difficulty and physical exertion at altitudes ranging from 4,700 and 8,000 feet. One of the pilots attempted to establish radio contact with Lt. McElroy without success. The airplanes circled the area for about 15 minutes and then returned to Groton Field where they landed safely.

5-13-43H. Candia, New Hampshire. At 1710, a Republic P-47D crashed near Candia, New Hampshire, killing pilot 2Lt. Charles R. Ralph. The airplane had taken off from Grenier Field, Manchester, New Hampshire, with two other P-47D aircraft on a scheduled aerobatic training flight. Ground witnesses reported that the subject airplane had been performing loops and other aerobatic maneuvers just prior to the crash. Investigation revealed that the airplane had struck the ground in a spinning attitude where it exploded violently into flames upon impact, killing the pilot instantly. Investigators could not determine what caused the airplane to go into a spin or why the pilot was unable to recover.

5-13-43I. Hereford, Arizona. At 1600 MWT, a Cessna AT-17B suffered the catastrophic structural failure of the starboard wing three miles southeast of Hereford, Arizona, killing A/C Roger W. Peterson and A/C Henry R. Phinney. The airplane had taken off from the Army Air Field at Douglas, Arizona, on a routine formation training flight. The airplane had rendezvoused with two other AT-17 type airplanes and was waiting for a fourth to catch up. The flight leader peeled off to circle Hereford while waiting. The other two airplanes followed. The subject airplane, flying in the number-three position, suffered the catastrophic structural failure of the starboard wing at an altitude of approximately 3,000 feet agl. The airplane entered a tight spin to the right and smashed into the ground where it exploded into flames.

5-13-43J. Windsor Locks, Connecticut. At 1930, a Republic P-47D crashed near Bradley Field, Windsor Locks, Connecticut, killing pilot F/O Claude G. Baker. The airplane had taken off from Bradley Field with two other P-47D type airplanes on a high-altitude formation flight. The flight had been flying at about 27,000 feet when the subject airplane entered a split-S maneuver. The airplane gained too much speed on the way down and the tail surfaces failed when the pilot attempted to pull out of the dive. The airplane went out of control and dove into the ground, exploding violently into flames upon impact. The pilot made no apparent effort to leave the airplane and was killed instantly upon impact. Investigators speculated that the airplane reached a point of compressibility in the dive and the pilot was unable to recover.

5-13-43K. El Paso, Texas. At 2318 MWT, two Consolidated B-24E airplanes collided in mid-air four miles northeast of Biggs Field, El Paso, Texas, killing fifteen fliers and seriously injuring co-pilot 1Lt. Dean E. Woodley, who had miraculously parachuted to safety from B-24E #42-7000. Killed in the crash of B-24E #42-7000 were: 1Lt. Avalee E. Huie, pilot; 1Lt. Alvin Glickman, navigator; 1Lt. Richard H. Kemp, bombardier; SSgt. Jack W. Dempsey, engineer; SSgt. Harry H. Crowe, Jr., engineer; SSgt. Clair C. McFeathers, Jr., radio operator. Killed in the crash of B-24E #42-6990 were: Lt. Col. Samuel C. Mitchell, pilot; Capt. Robert S. Von Schrilitz, co-pilot; SSgt. Alfred R. Longo, radio operator; Sgt. John W. Hoerr, engineer; Capt. Ira B. Benger, pilot-rated passenger; 1Lt. Frederick M. Edler, pilot-rated passenger; Capt. Jack S. McCarty, passenger; Pvt. E.M. Radicot, passenger; Sgt. Carl Sitek, passenger. Investigators stated, "Lt. Col. Mitchell, flying B-24E #42-6990, had taken off from March Field, [Riverside] California, at 1851 PWT. The flight plan listed a 'Contact Flight Rules' flight with an estimated four hours of elapsed flying time direct to Biggs Field. At 2318 MWT this aircraft, while flying in a [climbing] left turn at about 6,000 feet above the level ground four miles north of the El Paso Range Station, collided head on into the left wing and left [fuselage] side of B-24E #42-7000, which was headed south. The statement made by the only survivor, co-pilot of B-24E #42-7000, 2Lt. Woodley, who was thrown clear during the resulting explosion, established the fact that neither his pilot nor he had seen B-24E #42-6990 in time to avoid the collision. Apparently neither the Colonel nor his co-pilot saw the other aircraft in time to change their course. 1Lt. Huie had taken off at 2240 CWT from Biggs Field in B-24E #42-7000 [on a] scheduled bombing mission. The bombing was not carried out, but the pilot and co-pilot decided to change the mission to navigation because their bombsight was not operating. The pilot then began the navigation mission soon after take-off, which meant flying on the established airway and working the El Paso Range. While flying in a wide left circle toward the south, B-24E #42-6990 collided with [the] left wing [of B-24E #42-7000]."

5-14-43A. Topsfield, Massachusetts. At 0637, a Republic P-47C crashed at Topsfield, Massachusetts, killing pilot 2Lt. Ernest W. Hagar, 20, Sweetwater, Texas, who fell to his death in an unsuccessful parachute jump. The airplane had taken off at 0604 EWT from Bedford Army Air Field, Bedford, Massachusetts, in the number-three position of a four-ship flight on a formation-training mission. After climbing to 4,000 feet agl, the flight assembled into a string formation. The flight passed through a very thin layer of clouds at 5,000 feet and continued to climb, leveling off at an altitude of 6,500 feet. At this time a thick column of white smoke was seen trailing the number-three airplane. The flight leader ordered Lt. Hagar to leave the formation and find an airport to land at. The subject airplane dropped back though the thin layer, attempting to get to Bedford, which was directly behind, or East Boston, which was to the right. The P-47D then lost all engine power and began to lose altitude rapidly. The pilot was unable to make it to an airport, but waited too long to bail out. He left the airplane at an altitude too low to allow the parachute to fully deploy and he fell to his death.

5-14-43B. Tarpon Springs, Florida. At 0848 EWT, a Curtiss P-40L suffered a catastrophic structural failure of the starboard wing and crashed one-half mile north of Tarpon Springs, Florida, killing pilot 2Lt. James H. Toomer. Witnesses on the ground heard a loud report and observed the airplane falling in a spin minus the wing. The airplane slammed to the ground in an area of scattered pine trees. Investigation revealed that the starboard wing struck the tail section as it separated. Small bits of wreckage were scattered over a three-quarter mile area. The starboard wing was found about one-half mile from the spot where the airplane slammed to the ground and exploded into flames. The airplane had taken off from Pinellas Army Air Field, Pinellas, Florida, on a routine transition flight.

5-14-43C. Sopchoppy, Florida. At 1145 EWT, a Curtiss P-40K crashed into a swamp five miles north of Sopchoppy, Florida, killing pilot 2Lt. Earl A. McKinnon. The airplane had taken off from Dale Mabry Field, Tallahassee, Florida, on an aerial gunnery mission. The airplane apparently developed engine trouble and the pilot chose a field and attempted an emergency forced landing. The airplane overshot the field and the pilot pulled up, stalling the P-40 and entering a spin at low altitude. The airplane smashed into a swamp, killing the pilot instantly.

5-14-43D. Marfa, Texas. At 1500, a Cessna AT-17B suffered a catastrophic structural failure of the port wing and crashed 17 miles southwest of Marfa, Texas, killing instructor pilot 2Lt. Daniel P. Strickland. The airplane had taken off from Marfa Army Air Field with two other AT-17 type airplanes (flown by cadets) on a routine formation-training mission. The formation was flying in string at 9,000 feet agl when the cadets observed pieces falling away from Lt. Strickland's port wing. The wing failed and the airplane entered a spin to the left, smashing to the ground and exploding into flames. A civilian husband and wife reported that the airplanes were chasing each other and rolling before the airplane crashed on their ranch. These civilians, who were unfamiliar with aviation matters, would not sign a statement, fearing that to do so would jeopardize their chances to collect government benefits that they were due.

5-14-43E. Amarillo, Texas. At 1929, a North American B-25G crashed near the Army Air Field at Amarillo, Texas, killing pilot 2Lt. Manfred A. Ekstrom, co-pilot 2Lt. James E. Lambeth and engineer Cpl. Leif M. Halverson. The ship had taken off from Kirtland Army Air Field, Albuquerque, New Mexico, with another B-25G on a ferry flight from Long Beach Army Air Field, Long Beach, California, to Fairfax Field, Kansas City, Kansas. As the two airplanes approached the Amarillo area, it became evident that low overcast covered the area and extended to the east where the destination airfield was located. The airplanes attempted to get underneath the ceiling, but the clouds seemed to slant toward the ground. Rather than entering instrument conditions, the other B-25G turned around to return to Albuquerque. Lt. Ekstrom continued toward Amarillo Army Air Field. Ceiling in the area was reported as low as 200 feet agl. The pilot, against orders, entered instrument conditions and eventually arrived in the area of the field. The Amarillo Army Air Field traffic control personnel issued a weather report to the subject airplane. The pilot radioed the Amarillo control tower that he would advise them if he elected to return to Albuquerque. The airplane was then seen breaking out of the overcast, which was estimated to be about 100 feet agl, descending steeply at high speed. The B-25 then suddenly banked to the left and smashed into the ground where it exploded violently into flames, scattering wreckage over a wide area. The crew was killed instantly. Investigators noted that all ferry pilots were instructed not enter instrument conditions and not to fly below ceilings of 1,500 feet agl. The pilot had begun his AAF flying career as one of the very first sergeant pilots.

5-15-43A. Karnack, Texas. At 0530, a Boeing B-17F crashed three miles southeast of Karnack, Texas, killing nine crewmembers. Killed in the crash were: 2Lt. Kyriakos K. Passepulo, pilot/B-17 student; Capt. John R. Heckman, B-17 instructor; 2Lt. Edward F. Smith, B-17 student; 2Lt. Samuel J. Cleaver, B-17 student; F/O George C. Harsa, B-17 student; Cpl. Willard M. Nowell, engineer; Cpl. Roy E. Cox, radio operator; Pvt. Russell R. Werdon, assistant engineer; Pvt. Aubrey C. Terry, passenger. The B-17F had taken off from Lockbourne Army Air Field, Columbus, Ohio, on a cross-country navigation mission to Hobbs, New Mexico, and return. Just prior to the crash, the airplane was first seen flying west and was then seen to turn east and head in the direction of Shreveport at an altitude of 3,000 feet agl. The engines were advanced to high rpm and the airplane was seen to climb then to "rock in the air," indicating that the airplane was apparently entering a stalling condition. Some observers stated that the airplane was on fire before impact. Investigation of the wreckage indicated that the airplane had been trimmed to maximum right rudder and aileron, indicating a loss of power on the port engines. Magnesium pieces on the port engines were melted and the port propellers were not feathered. Wreckage indicated that the airplane was on fire and out of control at the time of impact. The pilots were apparently attempting an emergency forced landing in a small clearing on the grounds of the Longhorn Ordinance Works. The clearing was surrounded by wooded terrain. Investigators speculated that the pilot attempted to stretch his glide toward the clear area, permitting the airplane to stall and enter a spin at an altitude too low to allow recovery. The B-17 smashed into the ground and exploded, killing all on board instantly. The position of the bodies in the waist section indicated that the airmen were attempting to leave the airplane. Investigators noted that it was likely that the pilots did not know that the clearing — the grounds of the Longhorn Ordinance Works — was an explosion hazard. There were no injuries or damage on the ground.

5-15-43B. San Angelo, Texas. At 1230, a Vultee BT-13A and a North American BT-9 collided in mid-air six miles east of Goodfellow Field, San Angelo, Texas, killing two fliers and injuring a third who managed to parachute to safety. A/C Gerald E. Johnson was killed in the crash of the BT-9, which was returning to Goodfellow Field from a routine solo flight. Instructor 2Lt. Barney L. Rouse was killed and A/C Donald A. Regner was seriously injured in the crash of the BT-13A, which also had taken off from Goodfellow Field on a routine dual training flight. The airplanes were at an altitude of approximately 1,500 feet agl. The BT-9 was heading southwest and the BT-13A was heading southeast. The engine and propeller of the BT-13 struck the starboard wing and fuselage of the BT-9. The BT-9 went out of control and began to disintegrate in the air; the engine and tail section separated and smashed to the ground. The BT-13A lost its engine and propeller in the collision. The BT-13A then went out of control and "pancaked" into the ground, killing the instructor instantly.

5-15-43C. Potts Mountain, Craig County, Virginia. At 1344, a Cessna UC-78B flying in instrument conditions crashed into Potts Mountain, Craig County, approximately 28 miles northwest of Roanoke, Virginia, killing the crew of three. The airplane had taken off from Shaw Field, Sumter, South Carolina, on a cross-country navigation flight to Roanoke, Virginia, and return. The airplane checked in at Woodrum Field, Virginia, at 1330 on the inbound leg, reporting at 5,000 feet indicated altitude on instruments. The airplane was heading northwest and collided with the Potts Mountain at an elevation of 4,010 feet, first striking a tree 75 feet from the top of the mountain and then smashing into the terrain. The starboard engine was hurled over the top of the mountain and came to rest on the West Virginia side of Potts Mountain. The airplane was destroyed utterly and wreckage was scattered over a considerable area. Two crewmembers were hurled from the airplane at the time of impact and one remained in the wreckage. Pilot Capt. Jere A. Wells,

Jr., navigator-rated passenger Maj. Robert M. Bell and passenger Maj. James P. Citta were killed in the crash.

5-15-43D. Sherman, Texas. At 1315, a Vultee BT-13A crashed one mile south of Perrin Field, Sherman, Texas, killing instructor 2Lt. Cyril O. Hartley and student A/C Leon E. Daubert. The airplane was flying the number-two position in a two-ship flight that had taken off from Perrin Field. The subject airplane had fallen behind the lead ship. Some minutes later the subject airplane was able to catch up. The student attempted to rejoin the formation and started to slip toward the lead airplane. The instructor took over the controls and "applied corrective measures so abruptly that the ship stalled and snapped into an inverted position." Lt. Hartley, in an attempt to recover, completed the roll and pointed the airplane down. The airplane approached the ground rapidly and the instructor pulled up abruptly, causing the airplane to enter a secondary stall close to the ground. The instructor was unable to recover and the airplane entered a spin, smashing into the ground on the starboard wing and bursting into flames upon impact. Both occupants were killed instantly.

5-15-43E. Wichita, Kansas. At 1455 CWT, a Cessna UC-78B crashed 19 miles north of Wichita, Kansas, killing pilot 1Lt. Herbert C. Stucke. The airplane, which had taken off from Cessna Field, Wichita, on a factory test flight, was seen engaging in a series of climbing turns and shallow dives just prior to the crash. The airplane failed to recover from the third dive, striking the ground in a vertical, nose-down attitude. The airplane did not burn, but was destroyed utterly. The pilot was killed upon impact.

5-15-43F. Ginger Cake Mountain, North Carolina. At 1700, a Douglas C-47 flying in instrument conditions collided with Ginger Cake Mountain, approximately 25 miles NNW of Morgantown, North Carolina, killing four crewmembers and fourteen passengers. Crewmembers killed in the crash were: F/O W.B. McBroom, 20, Coryell, Texas, pilot; 2Lt. Robert C. Mack, 25, Birmingham, Alabama, co-pilot; TSgt. Gerald J. Sutherland, 22, Duluth, Minnesota, engineer; SSgt. Lester J. Kuchar, 23, Chicago, Illinois, radio operator. The airplane had taken off at 1500 EWT from Laurinburg-Maxton Army Air Base, Maxton, North Carolina, on a personnel transport flight to Lunken Field, Cincinnati, Ohio. The airplane encountered instrument conditions near Charlotte, North Carolina. The pilot made two 360-degree turns while deciding whether or not to turn back for Maxton. He elected to continue on to Cincinnati and regained the course for Lunken Field. The airplane, while flying on instruments, smashed into the 4,350-foot mountain at an elevation of about 4,000 feet msl. The airplane exploded violently into flames upon impact, scattering wreckage and bodies for over 500 yards. Investigators noted that the airplane should have been flying at 8,000 feet msl while passing over the mountainous terrain. Newspaper accounts of the crash described the visibility at the scene of the crash as "pea soup" fog. Passengers killed in the crash were: 2Lt. Donald D. Lewin, 22, Huntington Park, California, pilot-rated passenger; 2Lt. James H. Valentine, 21, Cincinnati, Ohio, pilot-rated passenger; 2Lt. James E.R. Lauderdale, 25, Artesia, New Mexico, pilot-rated passenger; 2Lt. Joseph E. Plaza, 25, Elko, Nevada, pilot-rated passenger; F/O James C. Armstrong, Jr., 25, Knightstown, Indiana, glider pilot; F/O Orville M. Buchheit, 23, Greencastle, Indiana, glider pilot; 2Lt. Thomas B. Oksendahl, 31, Barrington, Illinois, glider pilot; TSgt. Edward T. Russell, 23, Bellville, Ohio, engineer-rated passenger; Sgt. Melvin F. Mullinax, 22, Decatur, Illinois, radio operator passenger; SSgt. William B. Goddard, 24, Cincinnati, Ohio, radio operator passenger; 1Lt. Morton M. Katz, 24, Cincinnati, Ohio, passenger; SSgt. Francis P. McMahon, 26, Portsmouth, Ohio, passenger; SSgt. Guy A. Partridge, Jr., 22, Marion, Indiana, passenger; Pvt. William T. Templeman, 20, Chicago, Illinois, passenger.

5-15-43G. Oklahoma City, Oklahoma. At 1844, a Douglas A-20G crashed two miles southwest of Will Rogers Field, Oklahoma City, Oklahoma, killing pilot 2Lt. Jerome H. DeLisle. The airplane was in the number-two position of a three-ship formation. The leader gave the signal to peel off for landing and he then peeled off of the formation. The subject airplane peeled off and rolled past vertical, continuing in a dive. The pilot pulled back on the column and the airplane continued to dive and roll through 270 degrees of turn. The pilot was unable to regain control of the airplane, causing it to stall and enter a spin. The airplane smashed into the terrain and exploded into flames.

5-15-43H. Harding Field, Louisiana. At 1850, a North American P-51A suffered a catastrophic structural failure and crashed at the AAF Bombing Range near Harding Field, Baton Rouge, Louisiana, killing pilot 2Lt. John J. O'Brien. Investigators stated, "While on a local dive bombing mission the pilot came around on a normal approach to the range and at the end of the dive, after having released his bombs and at the beginning of the pull-out, the right wing failed under stress. This wing failure opened up the skin and at this speed the air pressure [that] built up inside the ship caused the airplane to disintegrate. The pilot was thrown clear in mid-air and his [parachute] opened but at this speed it failed, splitting along the seams of the canopy. The pilot was fatally injured."

5-16-43A. Ephrata, Washington. At 0055, two Boeing B-17F airplanes collided at a runway intersection while attempting to take-off at Ephrata Army Air Base, Ephrata, Washington, killing ten fliers. Killed in the crash of B-17F #42-5345 were: 1Lt. Lloyd H. Nygard, pilot; 2Lt. Leland W. Crumbach, co-pilot; 2Lt. Daniel H. Stone, Jr., pilot-rated passenger; 2Lt. Billy J. Evans, bombardier; Sgt. George Seukarian, engineer;

Sgt. Kenneth L. Johnson, radio operator; SSgt. Ray Cook, engineer; Pfc. Mark D. Flynn, radio operator; Pfc. Edward F. Pych, gunner. Bombardier 2Lt. Donald Wright was killed aboard B-17F #42-29501. Crewmembers uninjured aboard B-17F #42-29501 were: Capt. Edward A. Steedman, B-17 instructor; 2Lt. Forrest Poore, B-17 student; 2Lt. John E. Ward, B-17 student; Pvt. Francis A. Talbot, bombardier; Pfc. James F. Crews, radio operator. The two B-17 airplanes were preparing to take-off on a routine night training mission. The Accident Classification Committee stated, "While the pilot of airplane #42-29501 was off the extreme southeast end of Southeast/Northwest runway checking engines, pilot of airplane #42-5345 called the tower for taxi instructions. Tower told #42-5345 to stay on the ramp. When #42-29501, with Capt. Steedman as instructor in the co-pilot seat, called for take-off clearance, the tower called all airplanes to see if [they were] clear of the runway. Airplane #42-29501 was cleared for take-off. Airplane #42-29501 [taking off to the northwest on the Southeast/Northwest runway] crashed into the right side of and wing of airplane #42-5345 [which was taking off to the north on the South/North runway] at the intersection of the South/North and Southeast/Northwest runways during the take-off run, at which place #42-5345 burned with its crew. On seeing green running lights of #42-5345, pilot of #42-29501 tried in every way possible to avoid the accident. Right wing tip and about 15 feet of the right wing from #42-5345 ripped into and stuck in the nose and along the leading edge of the left wing of #42-29501, causing bombardier of #42-29501 to be thrown from the airplane [resulting] in fatal injuries. Number-one engine of #42-29501 was knocked off of the airplane, causing fire to break out along the left wing. Fire was blown out by [the propeller blast] of [the number-two] engine. Other crewmembers of #42-29501 escaped without injury."

5-16-43B. Buffalo, New York. At 1513 EWT, a Curtiss P-40N performing low-altitude aerobatics crashed near Buffalo Municipal Airport, Buffalo, New York, killing Curtiss-Wright Corporation test pilot Hugh Van Alstyne. The airplane was on a "Demonstration flight at a Bond Sales Dedication." Investigators stated, "The pilot crossed the field going north at about 250 MPH while at an altitude of approximately 300 feet. At mid-field the nose of the airplane pulled up slightly and the roll started. From eyewitness accounts gathered from pilots with experience in similar types and models [of this airplane] it appears that the [pilot] encountered an accelerated stall condition while pulling out of the last half of an improperly executed slow roll. The aircraft hit the ground with one wing, having stalled after recovery to level flight at about 10 feet of altitude, then cartwheeling [into the ground], completely destroying it and killing the pilot instantly. The aircraft would possibly have struck the ground regardless of the stall."

5-16-43C. Lake Charles, Louisiana. At 1542, a Martin B-26B crashed while attempting a take-off at the Army Air Field at Lake Charles, Louisiana, killing two crewmembers and injuring four others, one seriously. Pilot 1Lt. Howard G. Mealey and co-pilot 2Lt. James F. Gambill were killed in the crash. Bombardier 2Lt. Thomas H. Sharp, engineer SSgt. Arthur Fernandez and radio operator TSgt. George S. Robison received minor injuries; navigator 1Lt. Richard R. Chapman was seriously injured. The airplane was taking off on a cross-country navigation flight to Avon Park, Florida, and return. The airplane was attempting to take off on Runway 17, straight into a 25 to 30 mph wind. Investigation revealed that the airplane never got off of the ground on the take-off. The airplane was seen to slow down about two-thirds of the way down the 4,610-foot runway. The airplane rolled off of the end of the runway and skidded across a sodded area 200 feet wide, striking a five-foot deep ditch with the nose wheel. The airplane smashed into the ditch and flipped over on its back. Investigation revealed the port main tire blew out while on the take-off run and the pilots terminated the take-off, but were not successful in stopping the airplane and avoiding a crash.

5-16-43D. Waycross, Georgia. At 1715 EWT, a North American A-36A crashed six miles north of the Army Air Field at Waycross, Georgia, killing 2Lt. Ronald P. Heasley. Investigators stated, "Lt. Heasley was acting as an observer on a team instrument flight at 10,000 feet when his airplane went into a dive which was continued into the ground. The only reasonable conclusion is that the pilot lost consciousness due to causes unknown at the time of the accident." It was later speculated that the pilot might have been suffering from oxygen starvation.

5-16-43E. Waxahachie, Texas. At 1615, a Vultee BT-13A crashed three miles northeast of Waxahachie, Texas, killing pilot 1Lt. William S. Farish, Jr. and passenger Sgt. Jasper J. DeMaria, Jr. The airplane had taken off from Waco Army Air Field, Waco, Texas, with eight other airplanes on a formation training flight to Waxahachie. The flight arrived at Waxahachie Auxiliary Airfield and landed at 1430. The airplanes of the flight took off individually at 1545. Lt. Farish was the second from last to take off. While circling the field at 2,000 feet agl prior to assembling the formation for the flight back to Waco, Lt. Farish attempted to perform a slow roll to the right. The airplane began skidding to the left as the pilot attempted to recover from the inverted position. The airplane recovered in a dive toward the ground. The pilot was unable to recover the airplane before it smashed into the ground. The airplane burst into flames upon impact and the occupants were killed instantly.

5-16-43F. Meridian, Mississippi. At 2030, a Curtiss AT-9 suffering an apparent engine failure crashed 16 miles north of Key Field, Meridian, Mississippi, killing instructor 2Lt. L.V. Gruver. A/C Glenn

F. Nesselbeck was able to parachute to safety. The Accident Classification Committee stated, "The instructor and student were cleared from Columbus [Army Air Field, Columbus, Mississippi] on a night formation and transition flight with Meridian as a check point. Approximately 15 miles northwest of Meridian the left engine quit and the cadet was ordered to bail out. He bailed out successfully and landed in a swamp, but the instructor's jump was fatal due to the lack of altitude when he bailed out. The ripcord had been pulled and the parachute was strung out behind the body, but had not checked the fall at all."

5-17-43A. Dyersburg, Tennessee. At 0358 CWT, a Boeing B-17F crashed two miles north of the Army Air Base at Dyersburg, Tennessee, killing seven crewmembers and injuring three others. The airplane was on a scheduled night training mission and was practicing night take-offs and landings. The airplane was coming in for a landing when the number-three engine began losing power. The pilots mistakenly feathered the number-four propeller. They increased power on the port engines but were unable to maintain flying speed because of the drag of the failing number-three engine. The airplane stalled and fell off to the right, smashing into the ground on the starboard wing at a steep angle. The B-17 cartwheeled into the ground and began to break up. The starboard engines were torn free and the fuselage broke up, fire breaking out on some pieces of the scattered wreckage. The airplane was completely destroyed. Killed in the crash were: 2Lt. James W. Koller, pilot/B-17 student; 2Lt. Paul L. Hamman, co-pilot/B-17 student; 2Lt. Gilbert T. Smith, B-17 instructor; 2Lt. Cloyd D. Zimmerman, navigator; 2Lt. Albert Algosi, bombardier; SSgt. Robert L. McCune, engineer; Sgt. Henry A. Wells, assistant radio operator. Radio operator Sgt. Anthony J. Matthei and assistant engineer SSgt. Bernard J. Grieco received minor injuries and gunner Sgt. Joseph V. Farley was seriously injured.

5-17-43B. Seymour, Indiana. At 0951 CWT, a Beech AT-10BH crashed near Freeman Army Air Field, Seymour, Indiana, killing student pilot 2Lt. Francis M. Costello and injuring student pilot Capt. Green R. Davis, Jr. The airplane was taking off on a routine student training flight. Just after becoming airborne the airplane entered a steep climb, stalling out and falling off into a spin. The airplane struck the ground violently, bursting into flames several seconds after impact. Capt. Davis was able to escape the wreckage, but after several attempts was unable to free Lt. Costello, who was very seriously injured. Lt. Costello was burned to death before he could be freed from the wrecked airplane. There was no evidence that the control column was locked and the student stated that the column was unlocked as far as he could remember. Investigators could not find any structural failure or apparent mechanical failure that would cause the airplane to go out of control. The cause of the accident was undetermined.

5-17-43C. Patterson Field, Ohio. At 1045 CWT, a Curtiss A-25A crashed two miles north of Patterson Field, Ohio, killing pilot 1Lt. Melvin G. Bertzyk. The airplane was on an Accelerated Service Test Flight. When the pilot attempted to switch fuel tanks, the selector valve would not operate. As the pilot attempted to troubleshoot the fuel valve, the airplane began to lose altitude. When the pilot noticed that the airplane was flying toward the ground, he pulled up abruptly and stalled the airplane. The airplane fell off on the starboard wing and into a spin at an altitude too low to allow recovery. The airplane smashed to the ground near Bellefontaine Road on the Wahlay Farm. The wreckage careened over a small creek, coming to rest on the opposite bank and the terrain beyond. The pilot was killed instantly. Investigators noted that a similar fuel valve selector malfunction had occurred with another A-25A type airplane.

5-17-43D. Douglas, Arizona. At 1620 MWT, a Cessna AT-17B crashed eight miles NNW of the Army Air Field at Douglas, Arizona, killing pilot 2Lt. Edwin M. Gabrielson and passengers SSgt. Donald G. Mossey and Cpl. Robert H. Appel. The airplane had just returned to its station after undergoing wing strengthening and brake repair at an AAF Maintenance Depot. The airplane was on an engineering test flight and a 50-hour test check flight. Witnesses on the ground observed the airplane descending rapidly in a steep spin. The airplane stopped spinning and then smashed to the ground in a 45-degree dive. The airplane burst into flames upon impact and the occupants killed instantly. Investigators noted that there was no evidence of wing failure.

5-17-43E. Lake Mead, Nevada. A Sikorsky S-43 twin-engine amphibian aircraft crashed during test operations on Lake Mead, Nevada, killing CAA inspector William M. Cline. Pilot Howard Hughes, the head of the Hughes Aircraft Company, and Hughes engineer Gene Blanford received minor injuries, and were rescued immediately following the crash by civilian craft that were operating in the area. Co-pilot Charles W. Rosenberg and Hughes engineer Richard Felt were seriously injured and were immediately picked up by a second boat operating on the lake. According to published accounts of the crash, which was reported to have happened late in the day, the experimental amphibian aircraft violently nosed into the water while operating on the surface of Lake Mead near Bolder Dam. The airplane sank after the accident. Hughes Aircraft Company spokesmen stated in a published account of the crash that the aircraft would be raised from the bottom of the lake. Hughes was testing the airplane after the installation of more powerful engines.

5-18-43A. Corinth, Mississippi. At 0025, a Vultee BT-13A and a Vultee BT-15 collided head-on while attempting to take off at the municipal airport at Corinth, Mississippi, killing two fliers and injuring two others. 2Lt. Robert L. Wilson and passenger Pvt.

Kenneth H. Shorter were killed aboard BT-13 #42-42627. 2Lt. Kenneth M. Levy and SSgt. Raymond Cassille received serious injuries aboard BT-15 #42-41663. The BT-13A was taking off to the north on the main North/South runway and the BT-15 was taking off to the south on the same runway. The airplanes collided at about mid-field just prior to reaching flying speed. Pvt. Shorter was killed instantly and Lt. Wilson died minutes after being pulled from the wreckage. The BT-13A had burst into flames after the collision. Lt. Levy and SSgt. Cassille were unconscious when they were pulled from the wreckage of their airplane, which did not burn but was completely destroyed. Both airplanes came to rest facing each other opposite of the original take-off direction, effectively switching positions. Investigation revealed that the direction of take-off was from north to south at the time of the accident. Both airplanes were taking off on scheduled night navigation missions.

5-18-43B. Charleston, South Carolina. At 0540, a North American B-25D crashed while attempting a take-off from the Army Air Base at Charleston, South Carolina, killing the crew of seven. Killed in the crash were: 1Lt. John S. Dreher, pilot; F/O Joseph W. Lewis, co-pilot; 1Lt. Irving T. Katz, bombardier/navigator; Cpl. Salvatore J. Sciremammo, gunner; SSgt. Frederick Renter, Jr., engineer; SSgt. Ernest E. Fritz, engineer; Sgt. Joseph A. Krzeski, gunner. The airplane was taking off on Runway 21 (7,000 feet) on an anti-submarine patrol mission over the Atlantic Ocean. The airplane crashed into a wooded area about 1,000 yards to the right and 90 degrees to the right of the take-off heading, indicating a 90-degree turn immediately following take-off. The airplane collided with trees in a 30-degree bank to the right, smashing down several trees before slamming into the ground and bursting into flames. Investigation revealed that the starboard engine had failed just after take off and the pilot could not maintain flying speed. He apparently turned into the bad engine and stalled the airplane at an altitude too low to allow recovery.

5-18-43C. Gulf of Mexico, Florida. At 0855 EWT, a North American AT-6A crashed into the Gulf of Mexico about 25 miles southeast of Dale Mabry Field, Tallahassee, Florida, killing pilot 2Lt. Thomas A. Ward and pilot-rated passenger 2Lt. Richard L. Smith. Investigators speculated that the airplane stalled out of an aerobatic maneuver.

5-18-43D. Boaz, Alabama. At 0930, a Vultee BT-13A collided with trees and crashed ten miles southwest of Boaz, Alabama, killing A/C Stanley J. Smith and fatally injuring A/C Olin A. Stafford. A/C Stafford died at about 0800 on 5-19-43. The airplane had taken off from Courtland Army Air Field, Courtland, Alabama, on a dual instrument navigation flight to Gasden, Alabama, and return. The airplane was seen to descend rapidly in a diving turn. The airplane then collided with a tall pine tree with the starboard wing, severing the wing and causing the airplane to go out of control. The airplane crashed into the ground about 300 yards beyond the tree. The airplane was reported to be flying at very low altitude in the area prior to the crash.

5-18-43E. Marianna, Florida. At 1000 CWT, two North American AT-6C airplanes collided in mid-air 14 miles NNE of the Army Air Field at Marianna, Florida, killing A/C Lloyd R. Coffman in the crash of AT-6C #42-48846. A/C Harold C. Cobb was seriously injured parachuting to safety from AT-6C #42-43970. The airplanes were part of a three-ship flight that had taken off from Marianna on a routine formation training flight. The airplanes landed individually at Auxiliary Airfield #1 and then a few minutes later took off individually. The flight leader and A/C Coffman had taken off and positioned their aircraft in formation at about 800 feet agl, A/C Coffman flying on the instructor's port wing. A/C Cobb took off and attempted to join the formation. A/C Cobb approached the formation at about a 90-degree angle and then made a steep right turn to join up. A/C Cobb was going too fast and pulled up, mushing the AT-6C into the number-two aircraft, piloted by A/C Coffman. A/C Coffman's airplane went out of control and immediately spun to the ground where it exploded into flames upon impact. The cadet made no obvious effort to leave the airplane.

5-18-43F. Tuscaloosa, Alabama. At 1410, two Stearman PT-17 airplanes collided in mid-air near Tuscaloosa, Alabama, killing three fliers. Student pilot Capt. William Alexander Crossland, Jr. was killed in the crash of PT-17 #41-8261; A/C Harold E. Miller was killed in the crash of PT-17 #41-963 and civilian instructor John W. Smith fell to his death in an unsuccessful parachute attempt. Investigators stated, "This accident occurred while the student [Capt. Crossland] was practicing low-altitude maneuvers. It is believed he was working on pylon-8s and was so intent on watching the movement of the pylon that he failed to see another ship, which apparently was practicing S-turns across the road. According to witnesses, there is no doubt that each pilot of the respective airplanes was unaware of the other's presence until they collided (or interlocked) wings from a head-on position. There was no evidence of any carelessness whatsoever concerning this incident. After colliding, [PT-17 #41- 8261] plummeted straight to the ground with wings trailing behind. The collision occurred at an approximate altitude of 500 feet. Apparently Capt. Crossland was stunned because he made no attempt to use his parachute. After colliding, [PT-17 #41-963] with the student and instructor plummeted head-on to the ground. Evidently the instructor attempted to use his parachute and must have pulled the ripcord while standing in the front seat. The parachute became entangled with the tail group. The instructor was found lying underneath the rudder with the parachute entangled in the tail of the airplane. Upon contact with the ground, the airplane smoldered for a few minutes and then burst into flames."

5-18-43G. Long Island Sound, New York. At 1513, two Republic P-47C airplanes collided in mid-air over Long Island Sound, New York, killing 2Lt. Edward E. Woolery, who fell to his death in an unsuccessful parachute jump from P-47C #42-6561. 2Lt. Alexander S. Koczak received minor injuries parachuting from P-47C #42-6636. The airplanes had successfully completed an instrument-training mission and were returning to Groton Field, Groton, Connecticut. The airplanes began engaging in simulated individual combat and apparently Lt. Woolery's P-47 collided with the tail section of Lt. Koczak's P-47. Witnesses on the ground reported that the airplanes were flying west at an altitude of about 4,000 feet agl when Lt. Woolery was seen to make a steep left turn into Lt. Koczak's airplane, severing the tail of Lt. Koczak's airplane and sending it out of control. One airplane smashed into the sound and exploded about 2,000 yards out from the west end of Great Gull Island and the other slammed into the water and exploded about 1,000 yards out from the northeast end of Great Gull Island. Lt. Koczak was rescued from the sea east of the island.

5-18-43H. San Antonio, Texas. At 1615, a Bell P-39K suffered a catastrophic structural failure and crashed seven miles south of Kelly Field, San Antonio, Texas, killing pilot Capt. John L. Robinson. Investigation revealed that the starboard stabilizer and elevator had separated from the aircraft due to extreme stress imposed by the execution of an unknown maneuver. The airplane smashed into the ground in a flat spin to the right. Witnesses reported seeing the airplane in an inverted position and attempting to roll through just prior to spinning to the ground.

5-18-43I. Pacific Ocean, California. At 1452, a Bell P-39N crashed into the Pacific Ocean approximately five miles west of the mouth of the Russian River on the west coast of California. The pilot, 1Lt. Donald K. Hopkins, was seen to parachute to safety but apparently drowned before he could be rescued. The airplane was part of a three-ship flight that had taken off from the Army Air Field at Santa Rosa, California, on a routine gunnery mission. While returning to Santa Rosa after the gunnery mission, other pilots in the flight reported seeing a large amount of white smoke issuing from the starboard engine compartment of the subject airplane. Lt. Hopkins radioed that he was bailing out and was seen to safely parachute into the sea from 4,000 feet into what was reported as "high seas." He apparently drowned a short time later.

5-18-43J. Studio City, California. At 1913, a Bell P-39D attempting a forced landing crashed into a residential section of Studio City, California, killing pilot Capt. Varian K. White. Witnesses on the ground reported seeing the airplane in a wide left turn at very low altitude with smoke trailing from the engine compartment with the engine "missing." The engine "started" again and the airplane climbed slightly, barely clearing trees and a gasoline service station building at the southeast corner of Whitsett Avenue and Moorpark Street. The airplane, still trailing smoke, then veered northeasterly across Whitsett Avenue and struck tall trees at 12427 Moorpark Street with both wings. When the airplane struck the trees, the engine was heard to shut down. The airplane smashed to earth in a field west of Whitsett Avenue and north of Moorpark Street, exploding into flames and scattering wreckage over a wide area. Several houses were damaged, one seriously, when pieces of wreckage smashed into the structures. The pilot was killed instantly. The airplane had taken off from Burbank, California, for a flight to the AAF Gunnery Range at Tonopah, Nevada.

5-19-43A. Lebanon, Tennessee. At 0835 CWT, a Martin B-26C collided with trees and crashed eight miles east of Lebanon, Tennessee, killing three crewmembers and injuring three others. The airplane had taken off from Godman Field, Kentucky, on a mission in cooperation with Second Army. The airplane made a simulated strafing attack on a forty-millimeter anti-aircraft gun emplacement, which was in place on a small hilltop. The airplane came in so low that the pilot had to bank sharply to the right to avoid striking the gun, barely missing an army truck parked to the right of the gun emplacement with the low starboard wing. In its bank to the right, the airplane headed for a group of tall trees. The airplane then banked sharply to the left to raise the starboard wing to avoid striking a tall tree. The starboard wing struck the tree, causing the airplane to go out of control and smash to the ground in a field about 250 yards beyond the tree. The airplane struck the ground on the port wing, flipping over on its back and bursting into flames. Investigators noted that the elevation of the tree was lower than the elevation of the 40-millimeter gun emplacement. Pilot 2Lt. Emil Nelson, Jr. and gunners SSgt. O.E. Bennett and SSgt. L.E. Hoag were killed in the crash. Navigator Roy F. Humes and engineer SSgt. Fred M. Yeager received minor injuries and co-pilot 2Lt. Ray G. Burkett was seriously injured.

5-19-43B. Grenier Field, New Hampshire. At 1228, two Republic P-47D airplanes collided on the runway while attempting to land at Grenier Field, New Hampshire, killing 2Lt. Russell C. Wilson aboard P-47D #42-8024. 2Lt. Gilbert L. Jamison was able to escape P-47D #42-22344 without injury. The airplanes were part of four-ship flight that was landing on Runway 35 at the field. The first two airplanes landed without incident and turned off of the runway. The number-three airplane, piloted by Lt. Wilson, landed well down the runway, slowing down to taxi speed about two-thirds down the runway. The number-four airplane, piloted by Lt. Jamison, came in fast and floated down to the middle of the runway. Lt. Jamison touched down and, still traveling fast, rolled onto the slower moving P-47 piloted by Lt. Wilson. The collision occurred about 50 yards from the intersection of Runway 35 and Runway 24. The two airplanes careened off of the right side of the runway. Lt. Wilson's P-47, which was still entangled with Lt. Jamison's P-47, was burning near the tail section just after coming to rest. The engine of Lt. Wilson's P-47 was running at full throttle after coming to rest. Rescue personnel and other pilots were unable to access any part of the smashed cockpit to rescue the pilot or shut down the engine, which continued to run as flames erupted underneath the airplane. The propeller blast hampered efforts to extinguish the flames at the back and underside of the bottom airplane. The flames soon spread to the cockpit. A pilot who had been attempting to rescue the pilot and shut down the engine was forced to jump off of the airplane after a small explosion occurred in the cockpit. The engine stopped running only after the airplane burst into large flames about fifteen minutes after the accident. The rescue crew extinguished the fire only after great difficulty. Investigation revealed that the rescue crews did not have adequate equipment on hand to cut into the cockpit to rescue the pilot and shut down the engine. Lt. Wilson's airplane again burst into flames when a large wrecker type truck pulled the two airplanes apart.

5-19-43C. Gulfport, Mississippi. At 1650, a Douglas A-24A collided with a tree and crashed about 15 miles north of Gulfport, Mississippi, killing pilot 2Lt. James J. Ryan and passenger Pvt. Richard S. Hartman. The subject airplane had taken off with three other A-24 airplanes from Key Field, Meridian, Mississippi, on a simulated bombing and strafing mission and a navigation mission to Hattiesburg, Mississippi. The flight leader led the formation in a peel off to the left. The subject airplane, which was flying in the number-three position, did not bank enough and began to overrun the number-two airplane. Investigators stated that the subject airplane "winged over and lost altitude to avoid hitting another airplane in the formation and when opposite controls were applied, a violent roll was the effect; [the A-24] went over on its back and the [starboard] wing struck a tree, causing the airplane to crash." The airplane dove into the ground and burst into flames upon impact, killing the occupants instantly and burning their bodies beyond recognition.

5-19-43D. Mojave, California. A Northrop N9M-1 Flying Wing aircraft crashed near Mojave, California, killing Northrop test pilot Max Constant. It was reported that the pilot was conducting stall tests when the aircraft suffered a "flight control reversement [sic] at slow speed."

5-20-43A. Cumberland Springs, Tennessee. At 0300 CWT, a Douglas A-20C attempting an emergency forced landing crashed two miles west of Cumberland Springs, Tennessee, killing observer 2Lt. Fred J. Habian. Pilot 2Lt. Charles E. Dearing was seriously injured; gunner SSgt. Paul A. Rupert escaped injury. The airplane had taken off at 2355 CWT (5-19-43) from William Northern Field, Tennessee, on a night navigation mission to Birmingham, Alabama, to Chattanooga, Tennessee, and return. The airplane returned to Northern Field and entered the traffic pattern. The pilot extended the landing gear and moved what he thought were the propeller controls and both engines stopped. The pilot attempted an emergency forced landing and initially executed a "good" off-field landing (during darkness) with the landing gear extended. The airplane rolled through a wheat field, ran over a fence, crossed a ditch about ten feet across and two feet deep, smashed through a small stand of trees, careened across a gravel road, smashed through another fence, plowed through a corn and bean patch, and smashed through another fence before coming to rest inverted in a peach orchard. The airplane did not burn. The observer was killed in the crash. The gunner was able to escape the wreckage, but the pilot was unable to extricate himself from the smashed cockpit. He was finally cut free of the cockpit two and a half hours after the accident. Investigators could not determine the cause of the double engine failure. It was later speculated that the pilot had inadvertently manipulated the mixture controls, moving them to "idle cut-off."

5-20-43B. Columbia, Alabama. At 0930, a North American AT-6C crashed while performing low-altitude aerobatics at Hyman Field, Columbia, Alabama, killing pilot 2Lt. John T. Linney. Investigation revealed that the airplane had taken off from Napier Field, Dothan, Alabama, on an instrument team ride. Pilot 2Lt. R.O. Watson had performed an instrument take-off and flew under the hood to Hyman Field with Lt. Linney as the safety pilot. The two pilots landed and took a cigarette break. Lt. Linney then took off solo, buzzing the field at very low altitude whereupon he "zoomed" up to about 150 feet, turning around and buzzing the field again. Lt. Linney climbed the airplane to about 100 feet and began to perform a slow roll. When the airplane became inverted the engine cut out and the airplane stalled. The airplane staggered as it rolled through and the engine restarted just as the airplane dove into the ground. The airplane burst into flames upon impact and the pilot killed instantly.

5-20-43C. Ralston, Oklahoma. At 1000, a Vultee BT-13A crashed into the Arkansas River two miles north of Ralston, Oklahoma, killing passenger Cpl. Leroy K. Murrill and seriously injuring Service Pilot 2Lt. Glen E. Erickson. The subject airplane was the first of six BT-13 airplanes to take off from Tulsa, Oklahoma, on individual cross-country navigation missions to Garden City, Kansas, and return. In the vicinity of Ralston, the subject airplane separated from the other airplanes and descended to an altitude of about 75 feet above the Arkansas River, presumably to observe the heavy flooding that was present at the time. The pilot spotted the Santa Fe Railroad Bridge in the flight path and attempted to pull up, but before he could the airplane struck three power lines that were strung across the river. The airplane then dove into the river at a 20-degree angle, flipping over on its back. The pilot, after experiencing great difficulty with the strong current, was able to swim to a pile of debris that included a large tree branch where he able to hang on. Three civilians in a small boat rescued him a short time later and brought him to shore about two miles down river from the crash site. The river was about one-half mile wide where the airplane went in. The body of Cpl. Murrill, still strapped to his seat, was found on 5-24-43.

5-20-43D. Garden City, Kansas. At 1015 MWT, a Vultee BT-13A crashed seven miles northeast of Garden City, Kansas, killing A/C Don F. Taber. The airplane had taken off from Garden City Army Air Field on a routine solo flight. The student was scheduled to practice climbing turns, gliding turns, stalls, and spins. A witness on the ground observed the airplane in a spin at an altitude of approximately 1,000 feet above ground level. The airplane continued to spin until it struck the ground. The cadet was killed instantly.

5-20-43E. Chicago, Illinois. At 1140 CWT, a Consolidated B-24E flying in instrument conditions crashed into the side of a 20,000,000 cubic foot natural gas holding tank at 73rd Street and Central Park Avenue, Chicago, Illinois, killing 12 fliers. The airplane had taken off at 0650 CWT from Tarrant Field, Fort Worth, Texas, on a cross-country navigation flight to Chicago Municipal Airport, Chicago, Illinois, and return. The airplane had flown on top of the overcast or on instruments for a majority of the time on the inbound flight to Chicago. The airplane encountered real instrument conditions when it arrived in the vicinity of Chicago. Weather conditions were reported as a ceiling below 500 feet with visibility of three-quarters of a mile in light rain, light fog, and light smoke. The B-24 instructor pilot had contacted Chicago air traffic controllers and was cleared to make an instrument approach. The pilot was cleared to land to the northeast on Runway 4 Center. The airplane passed directly over the field at 1136 but the pilot was unable to get below the ceiling and make visual contact with the field, forcing him to execute a missed approach. The airplane flew to the northeast and the pilot began a right turn until the airplane was heading southwest. The pilot was having trouble communicating with the Chicago control tower, possibly due to static or perhaps confusion over the missed approach. As the pilot attempted to get under the ceiling the control tower again cleared the B-24E to land to the northeast on Runway 4 Center, transmitting this instruction twice. The pilot did not respond to the instruction because the airplane apparently collided with the gas holding tank at that instant. The 20,000,000 cubic foot gas holding tank, which was located approximately two miles southeast of the airport, was described as a 285-foot wide 28-sided polygon, "381 feet from ground level to the eaves; 404 feet high to the highest point of the crown; 416 feet high to the top of the roof ventilator; 486 feet high to the top of the beacon tower; and 490 feet high to the top of the 1,000 watt light." Investigation revealed that the B-24E, flying from the northeast, collided with the gas holding tank at approximately the 250-foot level, causing the tank to explode violently into flames. The flames, which reached hundreds of feet into the air, were reported to be visible through the fog. The terrific explosion was heard and felt for miles. All but one flier, who had been hurled clear of the wreckage when the tail was severed in the blast, were incinerated in the massive blaze. Only a small amount of human remains were recovered. No one on the ground was killed or injured. The gas holding tank was owned and operated by The Peoples Gas Light and Coke Company, Chicago. It was noted that Springfield, Illinois, air traffic controllers alerted the B-24 instructor pilot that Chicago Municipal Airport was reported to be closed down. The pilot stated that he would "try to make it anyway." Heavy fog was reported in the area of the crash and a witness on the ground reported that only the first 50 feet of the tank was visible in the fog. Investigators noted that the flight was cleared in error at Tarrant Field. Killed in the crash were: Capt. James R. Gilcrease, Houston, Texas, B-24 instructor; 2Lt. David S. Alter, Pittsburgh, Pennsylvania, pilot/B-24 student; 2Lt. Frederick L. Dutt, Wadsworth, Ohio, B-24 student; 2Lt. John C. Wallace, Luling, Texas, B-24 student; 1Lt. H.B. Messick, Indianapolis, Indiana, navigator; SSgt. Norman W. Yutzy, Canton, Ohio, engineer; TSgt. Ben F. Zumwalt, Ingram, Texas, engineer; Pfc. Nick Lonchar, Weirton, West Virginia, student engineer; Capt. A.W. Lent, Hamilton Field, California, passenger; Sgt. Arthur A. Huber, New York, New York, passenger; Capt. John M. Wallace, Camp Barkeley, Texas, passenger; 2Lt. A.L. Gentry, hometown unknown, passenger.

5-20-43F. El Reno, Oklahoma. At 1415, a Fairchild PT-19A crashed near Mustang Field, El Reno, Oklahoma, killing civilian instructor John F. Harrison and student pilot Capt. Edward R. Halbach. Investigation revealed that the airplane stalled and entered a spin at an altitude too low to allow recovery. Both occupants were killed instantly.

5-20-43G. Fort Myers, Florida. At 1530 EWT, a North American AT-6C crashed five miles east of Page Field, Fort Myers, Florida, killing pilot 2Lt. Owen H. Anderson and student gunner Pfc. William D. Howell. The airplane was part of a flight of four AT-6 type aircraft that had taken off from Page Field at 1455 on a routine flexible gunnery mission. Investigators stated, "The flight proceeded out to the range and commenced to fire. Lt. Anderson, piloting one of the firing airplanes, finished firing [with] his student gunner first and left the flight in order to return to the field. This was the last [time] the airplane [was seen by anyone in the flight]. Approximately one hour later, a check of the field indicated the airplane was missing. A search was instituted and the crashed airplane was located three miles from the gunnery range and approximately adjacent to the position where it had been observed to have turned off the range. Inspection of the wreckage indicated that he airplane had spun or dived in. The airplane crashed into tall grass and was undisturbed. The complete destruction of the airplane indicated that the crash [had] been violent. The tail section had broken off completely and was thrown clear of the wreckage. Inspection of the propeller indicated that the engine was dead when the crash occurred. One blade was bent to a 90-degree angle; the second blade was only slightly damaged. The airplane did not catch fire."

5-20-43H. El Reno, Oklahoma. At 1515, a Fairchild PT-19A crashed near Mustang Field, El Reno, Oklahoma, killing civilian instructor William E. Calderwook and seriously injuring student A/C Amos A. Zoellers. Investigation revealed that the airplane suffered an apparent engine failure and was attempting an emergency forced landing. The airplane stalled when the pilot attempted a turn to the right at low altitude and low airspeed. The airplane struck the ground with the starboard wing. The airplane did not burn but was destroyed utterly. The surviving cadet could not remember any details of the accident or the events that led up to it.

5-20-43I. Lancaster, California. At 1535, a Vultee BT-13A crashed six miles west of Lancaster, California, killing student A/C Leo L. Beck, Jr. The airplane had taken off from War Eagle Field, California, on a routine solo training flight. The airplane failed to return and was found to have crashed. Investigation revealed that the airplane had struck the ground at high speed in an extreme nose-down position. There were no witnesses and investigators could find no apparent cause for the accident. Investigators speculated that the student had entered a deliberate spin at a safe altitude and was unable to recover before striking the ground.

5-20-43J. Norene, Tennessee. At 1905 CWT, a Vultee L-5 crashed two and a half miles north of Norene, Tennessee, killing pilot Sgt. Robert L. Dickinson and observer 1Lt. Paul E. Dean. The airplane had taken off from Lebanon, Tennessee, on a message-dropping mission with the Fifth Armored Division. The observer was engaged in controlling ground traffic of armored vehicles and truck convoys participating in ground maneuvers. The airplane made a low level pass over a column of vehicles moving over a road in the maneuver area and the observer dropped a message to the column commander. After dropping the message, the pilot attempted a left turn and smashed into trees with the port wing. The airplane then plunged out of control into a wooded area and burst into flames. Both occupants were killed instantly.

5-20-43K. Westfield, Texas. At 1935, a North American B-25C suffered a catastrophic structural failure of the starboard wing and crashed three-quarter of a mile southeast of Westfield, Texas, killing the crew of seven. The airplane had taken off at about 1900 CWT from the Army Air Field at DeRidder, Louisiana, on a cross-country navigation flight to Ellington Field, Houston, Texas, and return. Four other airplanes that had taken off at intervals on the identical navigation mission reported severe thunderstorms were encountered near Beaumont, Texas, and a course change was made to avoid them. The subject airplane was observed flying in a southeasterly direction at an altitude of about 4,000 feet agl when it was seen to enter the thunderstorm at an airspeed exceeding 200 mph. Witnesses on the ground reported that the thunderstorm was severe with heavy rain and abundant lighting. It was speculated that the airplane encountered severe turbulence that was present in the thunderstorm. The airplane was subjected to extreme stresses and the outer starboard wing failed at a point just outboard of the engine nacelle. The wing separated and struck the starboard tail structure, causing its failure and separation. The airplane went out of control and entered a spin to the right, continuing to spin until it struck the ground and exploded into flames. The other four airplanes were able to safely complete the identical navigation mission. Killed in the crash were: 2Lt. Cecil C. Walkey, Jr., pilot; 2Lt. George H. Rich, co-pilot; 2Lt. Benjamin Gazevitz, navigator; 1Lt. Robert E. Ward, bombardier; SSgt. George B. Clarke, radio operator; Pvt. Herbert F. Barnes engineer; Sgt. Billy D. Hulbert, gunner.

5-20-43L. Rapid City, South Dakota. At 2225 MWT, a Boeing B-17F suffering an in-flight fire crashed two miles west of the Army Air Base at Rapid City, South Dakota, killing five crewmembers. Four other crewmembers parachuted to safety and were uninjured. The airplane had taken off from Rapid City Army Air Base at 2011 MWT on a routine bombing mission. The airplane climbed to 20,000 feet msl in preparation to using the AAF Bombing Range approximately 45 miles from Rapid City. At about 2205, the pilot noticed that the oil pressure in the number-two engine was falling. The pilots attempted to feather the number-two propeller, but it would not feather because of a lack of oil in the system. The propeller continued to turn at high rpm and it soon ran away. A fire soon broke out in the number-two engine, eventually burning through the firewall and into the wing structure. The pilot first elected to land at the base with a straight-in approach, but the fire became so intense that he ordered everyone to bail out. The pilot stated that at an indicated altitude of 14,000 feet he ordered the engineer to order the rest of the crew in the aft part of the airplane to bail out. The crewmembers in the forward part of the airplane then parachuted to safety. The pilot, satisfied that the entire crew had bailed out, stated that he flew the airplane to about 1,500 feet above ground level before he left the airplane and parachuted to safety. The airplane crashed into the ground, bursting into flames as it broke up. The airplane struck the ground at a relatively flat angle, causing one investigator to speculate that one of the enlisted men might have been attempting to crash land the B-17. No crewmember's body was found in the cockpit wreckage. Despite the pilot's testimony that he ordered the engineer to order the enlisted men to abandon ship, it was not known why all of the enlisted men at the rear of the airplane did not bail out when the pilot allegedly gave the command. The co-pilot testified that the pilot had activated the bail out alarm bell and gave the crew a verbal command to bail out. Two enlisted crew were found wearing parachutes. One enlisted crewmember was found in the wreckage wearing a parachute that apparently had deployed upon impact. It was later speculated that the enlisted men might have assumed that the pilot was going to fly the airplane into the base and elected to remain aboard against orders. It was also speculated that an enlisted crewmember's parachute might have inadvertently deployed in the aircraft, preventing the remainder of the crew from exiting the airplane. Two enlisted men were killed upon impact and the other three died a short time later. One enlisted man, who was pulled from the wreckage alive, was reported to have said while in delirium, "Get out of the way, I want to jump." He died a short time later. Crewmembers uninjured parachuting to safety were: 1Lt. Frederick F. White, pilot; 2Lt. Jack E. Ryan, co-pilot; 2Lt. Mario W. Del Re, Jr., navigator; 2Lt. Wayne S. Rader, bombardier. Killed in the crash were: SSgt. Paul E. Doerr, engineer; SSgt. Stanley Cushman, assistant engineer; SSgt. Alvin W. Boock, radio operator; SSgt. Harold Hall, gunner; SSgt. Marshall A. Foster, gunner. It was noted that the crew's regular radio operator was not aboard the aircraft for this flight.

5-21-43. United States. *No official U.S. Army Air Forces aircraft accident reports of any kind could be found for this date. See note #2.*

5-22-43A. Drew Field, Tampa, Florida. At 0845 EWT, two Douglas A-24A type airplanes collided while taxiing at Drew Field, Tampa, Florida, killing gunner Sgt. Joseph Tyo. The pilot of A-24A #42-60795, 2Lt. Ben H. Nicholls, escaped injury. The pilot and gunner of A-24A #42-60869, 2Lt. Earl E. Jackson, Jr. and Sgt. Angelo Gambino, also escaped injury. The subject airplane was the number-two airplane in a four-ship flight that was returning to the field after completing a dive-bombing practice mission. The lead airplane landed on Runway 14 and taxied to the end of the runway. As the lead airplane exited the runway, the subject airplane, which had just landed, taxied to the end of the runway. The subject airplane came to a halt at the end of the runway and Sgt. Tyo got out of his cockpit, taking up a position on the starboard wing to help the pilot taxi back to the parking area. The number-three airplane landed and began taxiing toward the end of the runway, colliding with the subject airplane, which was stopped and still had not cleared the runway. The number-three airplane struck the number-two airplane from the right rear, striking Sgt. Tyo with the rotating propeller, severing his right leg and killing him instantly.

5-22-43B. Atlantic Ocean, South Carolina. At 1010, a Bell P-39D crashed into the Atlantic Ocean ten miles southwest of Myrtle Beach, South Carolina, killing pilot 2Lt. Philip R. Moore. The airplane was strafing water targets on the surface at the AAF Bombing and Gunnery Range southwest of Myrtle Beach. The airplane, flying at extremely low altitude, collided with the target. The P-39D was damaged to such a degree that it crashed into the sea about two miles beyond the target, which was located at Gunnery Range T-20. The airplane had climbed to about 40 feet and was in a bank to the right when it struck the water. Pieces of wreckage and wood debris (from the target frame) were seen falling away from the airplane as it flew away. Witnesses stated that the pilot had made his previous eight firing passes at very low altitude, clearing the target by inches each time.

5-22-43C. Bartow, Florida. At 1515 EWT, two North American P-51A type airplanes collided in mid-air seven miles southeast of the Army Air Base at Bartow, Florida, killing 2Lt. Robert P. Dudley aboard P-51A #43-6064. 2Lt. Harry R. Fisk was uninjured and was able to make a successful wheels-up emergency landing at Bartow Army Air Base. The airplanes were part of a four-ship flight that had just taken off from Bartow AAB on a formation training flight. The first three airplanes had lined up in correct position in an echelon formation to the right. Lt. Fisk, flying in the number-four airplane, attempted to join the formation, overrunning and colliding with Lt. Dudley's airplane, which was flying in the number-three position. Lt. Dudley's airplane immediately went out of control and crashed to the ground where it exploded into flames, killing him instantly. Lt. Fisk's airplane was heavily damaged and it was only with great difficulty that he was able to make a safe emergency landing.

5-22-43D. Holyoke, Massachusetts. At 1820, a Republic P-47C and a Republic RP-47B collided in mid-air at Holyoke, Massachusetts, killing P-47C pilot 2Lt. Charnelle Larsen. Capt. Edward F. LaClare was uninjured parachuting to safety from the RP-47B. The airplanes were part of a two-ship flight that had taken off from Westover Field, Chicopee, Massachusetts, at 1818 EWT on a routine formation training flight. Capt. LaClare was leading with Lt. Larsen flying his port wing. The wingman was attempting a cross-over from the left to the right when he collided with the leader from underneath. Both airplanes went out of control. Lt. Larsen's airplane smashed to the ground in an alley where it burst into flames. The airplane wreckage struck several structures, destroying a garage and a 1937 Ford automobile. Capt. LaClare's starboard wing was severed in the collision and he was able to parachute to safety at an altitude of 800 feet agl. Capt. LaClare's airplane crashed into an unoccupied house, destroying the structure. There were no fatalities or injuries on the ground. The official report did not indicate the address or street where the airplanes smashed to the ground.

5-22-43E. Marysville, California. At 1710, a Bell P-39N crashed five miles north of Marysville, California, killing 1Lt. James T. Olmstead. The airplane was leading a flight of seven P-39 type aircraft on a flight from Oakland Municipal Airport, Oakland, California, to Oroville, California, via Sacramento, where the formation had stopped to refuel. Investigation revealed that the subject airplane performed a slow roll at an indicated altitude of 2,500 feet. The airplane lost about 1,000 feet during the roll. The airplane climbed back to the original altitude and performed a second slow roll, again losing about 1,000 feet. The airplane again climbed back to 2,500 feet before performing a third slow roll. The airplane stalled out during the third roll, entering a spin. The pilot was unable to recover the airplane and attempted to parachute to safety. The pilot fell to his death when he bailed out of the airplane at an altitude of 100 feet, too low to allow his parachute to deploy.

5-22-43F. Coffeyville, Kansas. At 2200 CWT, a Vultee BT-13A crashed about six miles southeast of Coffeyville, Kansas, killing A/C John I. Ziegenbusch. The airplane was returning to the Army Air Field at Coffeyville, following a routine night training flight. The student alerted the tower that he was over the field at 3,000 feet. The tower gave the student landing instructions and then he attempted to enter the traffic pattern. The student apparently lost the horizon while descending into the traffic pattern, allowing the airplane to descend into the ground. The airplane stuck the ground hard but remained flyable, bounding back into the air apparently under control but on fire. Fuel apparently was leaking and it caught on fire as the student applied full throttle and pulled up steeply. The airplane then stalled and entered a spin at an altitude too low to allow recovery. The airplane smashed into the ground and exploded into flames, killing the student instantly.

5-22-43G. Rapid City, South Dakota. At 2230 MWT, a Boeing B-17F crashed five miles southeast of the Army Air Base at Rapid City, South Dakota, killing 11 fliers. The airplane had taken off at 2112 MWT from Rapid City AAB on a cross-country navigation and bombing mission. The aircraft had arrived at the area of the North Bombing Range but was unable to contact the bombing range operator. The bombing range tower operator observed the B-17 circling the range and tried to contact the bomber but was unsuccessful. The range operator lost sight of the airplane and two or three minutes later observed a large explosion near the range. Investigation revealed that the airplane had smashed into the ground at an 80-degree angle, exploding violently into flames. The rear entrance door was found intact and undamaged by fire approximately 50 feet from the main wreckage. Investigators speculated that the door's emergency release was pulled an instant before the B-17 slammed into the ground. Investigators could not determine what caused the airplane to dive into the ground at a steep angle. It was later speculated that the pilots might have lost the horizon, causing the airplane to enter steep dive toward the ground. Killed in the crash were: 2Lt. Emmett C. Christian, pilot; 2Lt. John J. Doyle, co-pilot; 2Lt. Carl E. Kissling, navigator; 2Lt. Marshall H. Ruess, bombardier; SSgt. Roger C. Kellog, engineer; Sgt. Stanley H. Gramza, assistant engineer; SSgt. Jack L. Tricker, radio operator; SSgt. Mather P. Milligan, assistant radio operator; Pvt. Charles G. Brancamp, gunner; Sgt. Lawrence Burbank, gunner; 2Lt. Leeman L. Gains, passenger.

5-22-43H. Peterson Field, Colorado. At 2247 MWT, a Boeing B-17F crashed two miles east of Peterson Field, Colorado Springs, Colorado, killing three crewmembers and three female Army Medical Corps passengers. The airplane had taken off from Peterson Field on a flight to check off the co-pilot for night flying in B-17 type airplanes. The airplane descended into the ground under control while attempting to enter the traffic pattern for Peterson Field, exploding violently into flames upon impact. It was later speculated that the airplane had encountered a thunderstorm that had passed near the field and flew into the ground while trying to stay under the ceiling. Crewmembers killed in the crash were: 2Lt. Leroy S. Walton, pilot; 1Lt. James Ross Wilson, co-pilot; Cpl. Cecil L. Blanke, photographer/engineer. Medical Corps passengers killed in the crash were 2Lt. Evelyn M. Purdy, 2Lt. Jeanne A. Tusow and 2Lt. Mafalda E. Arney. The official AAF aircraft accident report did not indicate if the three female Medical Corps passengers, who were assigned to the station hospital at Peterson Field, were authorized to be on board the B-17.

5-23-43A. Yellowstone National Park, Montana. At 0039 MWT, a Boeing B-17F crashed on the grounds of Yellowstone National Park four miles south of West Yellowstone, Montana, killing ten crewmembers. Bombardier 2Lt. William F. McDonald was able to parachute to safety and sustained only minor injuries. Killed in the crash were: 2Lt. Roy E. Thompson, pilot; 2Lt. James J. Highley, co-pilot; 2Lt. George A. Brash, navigator; SSgt. Leo E. Thorn, engineer; SSgt. Donald W. Rice, radio operator; SSgt. Lawrence W. Medlin, gunner; Sgt. Alexander G. Jurkowski, gunner; SSgt. Gilbert E. Underwood, assistant engineer; SSgt. Willard McCune, crewmember; 2Lt. Robert K. Edwards, passenger. The airplane had taken off from Marysville, California, on a flight back to its home station at Lewiston, Montana. As the airplane neared

Reno, Nevada, the pilot climbed to about 15,000 feet indicated altitude to escape turbulent air and icing conditions. Soon after reaching this altitude, the airplane again encountered very turbulent air and icing conditions. As the airplane neared Yellowstone National Park the pilot ordered the crew to prepare to bail out. The co-pilot shouted an order to "Get out now, get out now" at the same instant. The bombardier immediately bailed out of the nose hatch and the B-17 crashed seconds later. Investigators speculated that the airplane had picked up a thick coating of ice, causing the pilots to lose control of the B-17. Investigation of the wreckage revealed that the airplane had slammed into the ground in a nose-down attitude, exploding into flames upon impact.

5-23-43B. McBee, South Carolina. At 0618, a Douglas A-24A crashed at McBee, South Carolina, killing pilot 1Lt. John M. Martin, Jr., who fell to his death in an unsuccessful parachute jump. The airplane had taken off from Laurinburg-Maxton Army Air Base, Maxton, North Carolina, on a mission in cooperation with ground units. The airplane was last seen flying in the correct position in its formation when the flight leader began a 360-degree turn at an altitude of approximately 450 feet agl. After completing the turn, the flight leader noticed that the airplane was missing. A ground party located the airplane wreckage at about 1200 EWT. Investigation revealed that the airplane had struck the ground out of control at an angle of about 40 degrees. The airplane did not burn, but was destroyed utterly. Lt. Martin's body was not found in the wreckage. The body was located a few hours later, approximately 250 yards from the main wreckage. It was speculated that Lt. Martin had bailed out at an altitude too low to allow his parachute to deploy.

5-23-43C. Huntersville, Tennessee. At 0720 CWT, a Stearman PT-27 crashed one mile east of Huntersville, Tennessee, killing civilian instructor Vance G. Burke and seriously injuring student A/C John P. Hartz. Investigators stated, "Mr. Burke and Cadet Hartz took off around 0715 with the intention of practicing acrobatics. Estimated ceiling 2,000 feet. Cadet Hartz took off and climbed to 1,000 feet indicated. At that altitude Mr. Burke tapped his head to signify that he was taking over the controls. He dived the aircraft to 110 mph indicated and attempted what seemed to be the start of a slow roll. As the aircraft approached the inverted position, Cadet Hartz became apprehensive because the aircraft seemed so low to the ground. He counteracted Mr. Burke's control pressure and attempted to roll the aircraft back the same way [it] entered [the roll and back] to level flight. Before the wings could be [leveled], the aircraft struck the ground inverted."

5-23-43D. Silver Lake, California. At 0910, two Beech AT-11 airplanes collided in mid-air 12 miles southwest of Silver Lake, California, killing four fliers. Crewmembers killed aboard AT-11 #41-9480 were: 2Lt. William S. Barnes, pilot; 2Lt. Harold R. Dobson, bombardier; Sgt. John A. Lowry, passenger; Sgt. William E. Thomas, passenger. AT-11 #41-27480 was landed safely with only minor injuries to passenger Sgt. Albert F. Masci. Pilot Capt. Corwin J. Papik, co-pilot 2Lt. Raymond J. Daugherty and passenger SSgt. Paul P. Stock were uninjured. The airplanes were part of a three-ship flight that had taken off from the Army Air Field at Victorville, California, on a routine formation/navigation training flight to the Army Air Field at Las Vegas, Nevada. Lt. Barnes, flying in the number-three position, was attempting to change position when he passed below the number-two airplane, which was piloted by Capt. Papik. Lt. Barnes, apparently unaware that he was directly beneath the number-two airplane, climbed abruptly. The cockpit area of AT-11 #41-9480 collided with the port propeller of AT-11 #41-27480. AT-11 #41-9480 fell away out of control and crashed to earth where it exploded into flames. Capt. Papik was able to regain control of his badly damaged airplane and make a successful emergency landing at Silver Lake Airport.

5-23-43E. Lees Summit, Missouri. At 1240, a Curtiss AT-9A crashed at Lees Summit, Missouri, killing pilot Capt. William J. Burke and passenger Maj. John H. Waldo. The airplane was part of a flight of six that had taken off from Kansas City, Missouri, on a navigation flight to Wichita, Kansas, and return. The airplanes were flying in two elements of three. The subject airplane was flying in the number-three position of the first element. The formation encountered an area of poor visibility. The subject airplane fell out of formation and crashed into the ground in an apparent spin. The remainder of the formation was able to safely complete the flight to Wichita and the pilots were unaware that the crash had occurred. Investigators speculated that the pilot might have encountered severe carburetor icing. It was also speculated that the pilot might have lost control of the airplane in instrument conditions. Investigators noted that the weather conditions were "just above minimums." The cause of the crash remained undetermined.

5-23-43F. Parker, Arizona. At approximately 1900 PWT, a Douglas A-24A crashed ten miles southwest of Parker, Arizona, killing pilot 2Lt. James N. Goe and passenger 1Lt. James P. Matthews. The airplane had taken off at 1826 PWT from the Army Air Field at Rice, California, on a local training flight and never returned. The airplane wreckage was found in a remote area 500 feet from the east bank of the Colorado River on 5-25-43 at about 1420 PWT. The tops of three trees located about 500 yards from the scene of the crash had their tops sheared off. It was speculated that the airplane had collided with these tress before crashing.

5-24-43A. Niagara Falls, New York. At 0914 EWT, a Bell P-39Q suffered a catastrophic structural failure of the port wing and crashed at Niagara Falls,

New York, killing pilot 1Lt. Bennett B. Battaile. Investigation revealed that the airplane had entered a high-speed dive and was pulled up abruptly, causing the port outer wing panel to fail and separate. The airplane went out of control and smashed to the ground where it burst into flames upon impact. The pilot was killed instantly.

5-24-43B. Miami, Florida. At 1007 EWT, a Martin B-26C crashed while attempting an emergency landing at Navy Master Airport, Miami, Florida, injuring three crewmembers and killing radio operator SSgt. Edward V. Peterson. The airplane had taken off at 0826 EWT from Morrison Field, West Palm Beach, Florida, on a ferry flight to an active theater of war via Borinquen Field, Puerto Rico. The airplane circled the field once and climbed before turning on course for Borinquen Field. After flying for about 30 minutes the airplane suffered an electrical failure and the pilot began experiencing difficulty in keeping the propellers synchronized. The pilot regained control of the propellers but soon discovered that all "electrical and autosyn instruments were out and the radio was dead." The pilot checked the circuit breakers but could not get the instruments back up. The pilot attempted to get back to Morrison Field but soon began experiencing trouble with the port engine, electing to attempt an emergency landing at Miami. The pilot circled the Navy Master Airport once, lowered the landing gear and entered the traffic pattern for an approach for the southeast runway. The nose landing gear would not lower. The co-pilot lifted the cockpit hatch and hydraulic fluid immediately sprayed into the cockpit area. After the hydraulic fluid stopped spraying, a pool of hydraulic fluid was found in the area directly underneath the nose landing gear. The pilot went around the pattern while the nose landing gear was pumped down manually. Visibility in cockpit was hampered because hydraulic fluid had sprayed on the windscreen. The pilot then attempted an emergency landing on the northeast runway. The pilot stated that the only way he could see out was by looking out of the open pilot's side window. The airplane passed over 6,000 feet of runway before touching down. The airplane had touched down approximately 1,500 feet from then end of the runway, running off the end of the runway and becoming airborne again for about 200 feet before hitting a pile of "coral fill," tearing off the nose landing gear. The airplane bounded back into the air and then smashed back to the ground in a slight ravine where both engines and both main landing gears sheared off. The airplane smashed into and damaged a U.S. Navy Ford Pick Up truck and then careened through a wire fence, the fuselage breaking in two pieces just aft of the wings before coming to rest about 2,500 feet from the end of the runway. The airplane did not burn but was destroyed utterly. Pilot 2Lt. Elston C. Johnson, co-pilot 2Lt. James K. Blanton, Jr., and navigator 2Lt. Ernest R. Bradley sustained minor injuries in the crash landing.

5-24-43C. Ajo, Arizona. At 1245 MWT, a North American AT-6B crashed at the AAF Gunnery Range near Ajo, Arizona, killing A/C Clyde R. Daniel. Investigators stated, "The student smashed into the ground following an attack on a ground gunnery target. The crash occurred when the student attempted an abrupt pull-out after performing a high-speed dive on the ground gunnery target. The airplane entered a high-speed stall and struck the ground 24 feet behind the target. The airplane struck the ground tail first, bounding back into the air and striking the ground again 225 yards from the initial point of impact. The exact reason for the student carrying out his attack as low as he did is not known. According to witnesses, he made a high approach at approximately a 45-degree angle, pushing the nose down to keep the sights on the target. The airplane kept diving steeper, picking up speed until very low on the target. The abrupt pull-out caused the ship to stall."

5-24-43D. Dyersburg, Tennessee. At 2245 CWT, a Boeing B-17F crashed while attempting to take off from the Army Air Base at Dyersburg, Tennessee, killing four crewmembers and injuring four others, two seriously. The B-17 was taking off on a routine training flight to practice night take-offs and landings. The B-17 student, flying in the left seat, performed a good visual take-off, but he failed to go on instruments soon enough, allowing the airplane to strike trees approximately 1,500 to 2,000 feet beyond the end of the runway. The airplane smashed into a wooded area and burst into flames. Dyersburg AAB rescue personnel were on the scene immediately and were able to rescue four crewmembers. Killed in the crash were: 2Lt. John L. Toth, bombardier; Sgt. Anthony C. Semola, radio operator; Sgt. Nathan S. Higgins, assistant radio operator; Sgt. Delaparte Howald, gunner. Pilot 2Lt. Gillette V. Patterson and assistant engineer Pvt. Thurston H. Caswell sustained serious injuries. B-17 instructor pilot 1Lt. Kendrick R. Bragg and engineer Sgt. Joseph Millich, Jr. received only minor injuries.

5-25-43A. San Pablo Bay, California. At 0417, a North American B-25D crashed into San Pablo Bay two miles east of Hamilton Field, San Rafael, California, killing four fliers. The airplane had taken off with another B-25D from Hamilton Field at 0414 on a ferry flight to an active war theater via Hickam Field, Oahu, Hawaii. Investigators stated, "Thin broken clouds at 900 feet with 12 miles visibility existed at the time of the accident. The airplane made a normal take-off to the southeast and proceeded straight ahead for approximately one minute. At that time he started a normal climbing turn to the left and disappeared into the broken clouds. Approximately two minutes later the airplane appeared very suddenly below the clouds heading northwest and in a steep dive. It then hit the water and exploded." Investigators could not determine what caused the airplane to dive into the bay. Killed in the crash were: 1Lt. F.R. Chapman, Jr., pilot;

2Lt. C.J. O'Brien, Jr., co-pilot; 2Lt. F.E. Effrige, navigator; Pvt. H.L. Karsh, radio operator.

5-25-43B. Leesburg, Florida. At 1030 EWT, a Douglas P-70 crashed into a small lake 10 miles south of the Army Air Base at Leesburg, Florida, killing three fliers. Pilot 2Lt. Arthur C. Tilton, radar operator Sgt. Walter H. Johnson and radio operator Pvt. Albert J. Thonen were killed in the crash. Investigation revealed that the P-70 had taken off from Kissimmee Army Air Field, Florida, with another P-70 on an interception mission. The P-70s had completed their interception mission and soon were jumped by two Curtiss P-40 airplanes. The airplanes briefly engaged in simulated combat. The subject airplane, in an effort to turn into one of the P-40s and make a head-on pass, entered a vertically banked turn at an altitude of 9,500 feet. The subject airplane stalled, rolled over and entered a flat spin, describing 13 or 14 turns of spin before crashing into a small lake and bursting into flames. The occupants made no apparent effort to parachute to safety.

5-25-43C. Camp Maxey, Texas. At 1042, a North American O-47A collided with terrain and crashed three miles west of Camp Maxey, Texas, killing the crew of three. Investigators stated, "Pilot and crew were on a simulated strafing mission with units from Camp Maxey, Texas, which were on maneuvers on the Camp Maxey training area. In attempting a simulated strafing of some troops on a road in the area, the pilot made a 90-degree approach to the road, crossed the road, and then attempted a steep turn. The airplane was too low and due to either a wing tip stall, mushing, or slipping, it hit first on the left wing tip, then hit some saplings, which tore off the left elevator, causing the airplane to turn over on its back. At this point the airplane was approximately 30 to 40 feet in the air. It kept losing altitude then hit on the nose, skidded along on its nose upside down until it hit a small hill where it turned right side up again. The airplane then burned." Pilot 1Lt. Max Price, 25, Plymouth, Indiana, crewmember Pfc. Ray A. Cowan, and crewmember Pfc. George L. Dillahunty were killed in the crash.

5-25-43D. Huron Auxiliary Airfield, California. At 0950, two Vultee BT-15 airplanes collided in mid-air one mile southeast of Huron Auxiliary Airfield, California, killing four fliers. Instructor pilot Capt. Ernest C. McCorkle, 23, Oldfield, Virginia, and student A/C Jack W. Young, 24, Bernardsville, New Jersey, were killed in the crash of BT-15 #41-10198. Instructor pilot 2Lt. Elliot R. Barber, 23, Melrose, Mississippi, and student A/C Clifford A. Pilcher, 24, Lawton, Iowa, were killed in the crash of BT-15 #42-41314. The airplanes had taken off separately from Lemoore Army Air Field, Lemoore, California, on scheduled dual instruction flights. Investigation revealed that the collision occurred at an altitude of about 2,000 feet agl. Both airplanes fell to the ground out of control and burst into flames, killing the occupants instantly.

5-25-43E. Smyrna, Tennessee. At 1305 CWT, a Consolidated B-24D crashed two and one-half miles north of the Army Air Field at Smyrna, Tennessee, killing the crew of seven. The airplane had taken off from Smyrna Army Air Field, leading a three-ship flight of B-24 aircraft on a student formation mission. Investigators stated, "Pilot after being informed by other ships in the formation that his No. 2 [fuel tanks] were discharging gasoline, was returning to the field for an emergency landing. [Number-one propeller] was feathered. Airplane spun in and crashed on base leg [while turning to the] approach to the field. Examination of the wrecked airplane ... revealed that the airplane had hit the ground in a vertical position out of a spinning turn to the left, because the airplane landed in a moderately, thickly wooded sector of high trees and only those trees in direct vertical flight were damaged by the airplane; that the No. 1 propeller had been feathered but that from what remained of the instruments, it appeared that the other three engines were functioning normally, that the landing gear was down at the time of the crash, that the clock in the cockpit had stopped with the hour hand broken off and the minute hand at five minutes, indicating the time of the crash at 1305, that the flaps appeared to have been lowered approximately 10 degrees. The fuel booster pumps were off, all fuel selector valves were ON and fuel tank caps were on and tight. The Aircraft Accident Classification Committee finds that the accident occurred after failure of the No. 1 engine in flight, and as a result of the loss of flying speed and control, the airplane went into a spin to the left while the pilot was making his approach to the home field for an emergency landing. It is believed that the pilot retarded the throttle of the No. 2 engine preliminary to feathering the propeller of the No. 2 engine because gasoline was spraying from that vicinity, and at this time found that the No. 1 engine was dead and merely windmilling or wasn't putting out adequate power, so again advanced the throttle on the No. 2 engine and feathered the propeller of the No. 1 engine. It is also possible that the No. 2 engine wasn't putting out adequate power either because of mechanical difficulty or the loss of gasoline from the No. 2 fuel cell." It was also speculated that the pilot might have feathered the wrong engine. Killed in the crash were: Capt. Stuart K. Oliver, 24, St. Paul, Minnesota, B-24 instructor; 2Lt. Charles R. Harris, 21, Franklinton, North Carolina, pilot/B-24 student; 2Lt. Samuel E. Fennegan, 21, Fairmont, North Carolina, co-pilot/B-24 student; Cpl. William N. Clayton, 19, Memphis, Tennessee, engineer; Pvt. Harold J. Linane, 31, Matick, Massachusetts, crewmember; Pfc. Thomas F. Monahan, Jr., 20, Minto, New Brunswick, Canada, crewmember; Pfc. James R. DeRusha, 20, Roxbury, Massachusetts, passenger.

5-25-43F. Maplesville, Alabama. At 1315, a Vultee BT-15 crashed three miles west of Maplesville, Alabama, killing pilot 2Lt. Robert Delos Van Tassei.

The airplane had taken off from Gunter Field, Montgomery, Alabama, on a ferry flight to the Army Air Field at Columbus, Mississippi. The airplane failed to arrive and it was later found to have crashed. Searching parties located the wreckage at approximately 1830. There were no witnesses to the crash. Examination of the wreckage and crash scene led investigators to speculate that the airplane might have collided with a small tree limb, estimated to be about three inches in diameter. A three-foot section of the starboard wing was found about one-quarter of a mile southeast of the area of the crash. The airplane had smashed into the ground upright with the port wing low. The nose of the BT-15 smashed into the ground and the airplane began to flip end over end, breaking the fuselage in two pieces. The pilot was killed instantly upon impact.

5-25-43G. Matagorda Peninsula, Texas. At 1440, ground crewman Pvt. Graham S. Rife was fatally injured when he was run over by a North American AT-6C that was taxiing at the Army Air Field at the AAF Bombing and Gunnery Range at Matagorda Peninsula, Texas. The AT-6C had retuned from a gunnery mission and was taxiing toward the parking area when the port wing struck Pvt. Rife, who had walked into the path of the moving aircraft. The ground crewmember was knocked to the ground and hit his head on the pavement, knocking him unconscious. Pvt. Rife never regained consciousness and died several hours later. The airplane was piloted by 2Lt. Harold H. Clopton.

5-25-43H. Cross City, Florida. At 2210 EWT, a Beech AT-10BH was abandoned in flight and crashed 15 miles northeast of the Army Air Field at Cross City, Florida, killing A/C Martin B. Enos, Paulsboro, New Jersey, who fell to his death in an unsuccessful parachute jump. A/C Stephen DeLura was able to parachute to safety and was uninjured. The airplane had taken off from Moody Field, Valdosta, Georgia, on a night navigation flight to Dunnellon, Florida, and return. Investigation revealed that the students had lost control of the airplane while flying on instruments in turbulent air. The airplane entered a spin at an indicated altitude of 1,500 feet. Investigators stated that the students had tried to let down below the ceiling after entering instrument conditions. Investigators stated that the students should have made a 180-degree turn rather try to let down at night. It was noted that several students were able to safely fly through the instrument conditions and complete the flight. The surviving student, who had landed in a remote swampland, struggled over the terrain all night until finding an abandoned railway at 0600. A civilian truck rescued the student at about 1100.

5-25-43I. Roxie, Mississippi. At 2155, a North American A-36A crashed near Roxie, Mississippi, killing pilot 2Lt. John G. Hall. The airplane had taken off from Harding Field, Baton Rouge, Louisiana, on a flight to Vicksburg, Mississippi, to Jackson, Mississippi, and return. Investigation revealed that the airplane had struck the ground with power at a 50-degree angle with the starboard wing low. The airplane burst into flames upon impact and the pilot was killed instantly. Investigators could not determine what caused the airplane to dive into the ground.

5-26-43A. Romulus, Michigan. At 0706 EWT, a Bell P-39Q crashed one-quarter of a mile south of the Army Air Field at Romulus, Michigan, killing pilot 2Lt. Ronald E. Ruelle. The airplane took off at 0658 from Runway 22 on a local transition flight. The airplane was observed south of the field at an altitude of about 5,000 feet. The airplane was then seen descending rapidly. As the airplane approached an altitude of about 2,000 feet, it was seen to enter a steep bank to the right. The airplane snapped over to the left and entered a flat spin to the left. The airplane continued to spin to the left until it hit the ground and exploded.

5-26-43B. Mather Field, California. At 1115, a Lockheed RA-28A crashed ten miles SSE of Mather Field, Sacramento, California, killing four fliers and injuring two others. Passenger Cpl. Louis F. Vermillion escaped the wreck uninjured. Killed in the crash were: 1Lt. Marvin A. Dillwood, pilot; 2Lt. Edward L. Szymanski, co-pilot; MSgt. Abel M. Lineberger, engineer; SSgt. James C. Shaffstall, passenger. Navigator 2Lt. Milton W. Monson and passenger TSgt. T.P.L. Davis were seriously injured. The airplane had taken off from Mather Field on a scheduled training flight and was returning to the field when the port engine began to lose power. The pilot radioed the tower for landing instructions as the airplane approached the area of the field at an altitude of about 1,000 feet agl. The co-pilot, who was flying the airplane, lowered the landing gear and prepared to enter the traffic pattern when the port engine began to fail or lose power. Lt. Dillwood, who was a more experienced pilot, took over the controls and attempted to fly the airplane in. Lt. Dillwood was unable to maintain altitude and he ordered the crew to prepare for a forced landing. The pilot was forced to land downwind on rolling terrain. He was able to choose a flat area surrounded by hills for his emergency landing. Because of the tailwind, the pilot overshot the smooth lowland and smashed into the slope of the hill beyond. The pilot had retracted the landing gear just before the airplane smashed into the hill. The airplane burst into flames upon impact and a secondary explosion occurred while the survivors were escaping the flaming wreck.

5-26-43C. Mullet Key, Florida. At 1535 EWT, a Douglas A-24A dove into the sea at the AAF Bombing and Gunnery Range at Mullet Key, Florida, killing pilot 2Lt. Cranston B. Lott and passenger Pvt. Embry S. Thurman. The airplane had taken off from Drew Field, Tampa, Florida, on a dive-bombing mission. The airplane had performed four successful dive-bombing runs, but the pilot was unable to pull out of the fifth dive. The A-24A began its fifth dive-bombing run at an altitude of about 8,000 feet msl and was observed diving vertically toward the surface at very high speed. The pilot was unable to pull out of the dive and the airplane continued toward the surface. The pilot, making one last attempt to pull up prior to striking the water, was able to get the nose up but the A-24 mushed into the water and bounded back into the air on fire and shedding pieces. The airplane smashed back into the sea an instant later and burst into flames. Both occupants were killed upon impact. Investigation revealed that when Lt. Lott came to the flight line to take the airplane on the bombing mission, he asked Pvt. Thurman, who was standing nearby, if he wanted to go along. Pvt. Thurman agreed to go on the mission, drawing a parachute and boarding the airplane.

5-26-43D. Davon, Pennsylvania. At 1600 EWT, a Republic P-47D crashed at Davon, Pennsylvania, killing pilot 2Lt. Vernon H. Langston. The airplane was part of a four-ship flight that had taken off from the Municipal Airport at Philadelphia, Pennsylvania, on a high-altitude formation mission. The subject airplane was flying in the lead position of the second two-ship element. The leader entered instrument conditions on the climb. When the flight broke out on top, the subject airplane began to suffer engine problems. The pilot radioed that he was leaving the formation and returning to the field. The subject airplane was next seen flying in a severe rainstorm at low altitude. The engine was heard "cutting out" and the pilot was apparently attempting an emergency forced landing. The airplane was circling at an altitude of about 100 feet agl when the port wing was seen to drop and the airplane spun into a wooded area where it burst into flames. Investigators speculated that the pilot was attempting to "stretch his glide" while trying to set down in an open field.

✓ **5-26-43E. Rochelle, Texas.** At 1500 CWT, a Fairchild PT-19A crashed one and one-half mile southeast of Rochelle, Texas, killing civilian instructor Marion E. McLaurin and student A/C Everal H. Lemley. The airplane had taken off from Curtis Field, Brady, Texas, on a routine training flight. Investigation revealed that the airplane was performing "ground track maneuvers" when it stalled and entered a spin at an altitude of 500 feet agl. The instructor broke the spin but there was not enough altitude to pull out of the dive. The airplane smashed into the ground at an angle of 60 degrees.

✓ **5-27-43A. Pine Bluff, Arkansas.** At 0845, a Fairchild PT-19A crashed 23 miles southeast of Grider Field, Pine Bluff, Arkansas, killing student A/C Harold J. Parkes. The student took off at 0820 from Mathews Auxiliary Airfield, Arkansas, on a routine solo training flight, proceeding to his practice area in the vicinity of Varner, Arkansas. The pilot entered a deliberate spin to the left at an altitude of 2,500 feet indicated altitude and was unable to recover before striking the ground. The airplane was completely destroyed and the student killed instantly.

5-27-43B. White River, Illinois. At 1020, a Beech AT-10BH crashed into the White River, 20 miles southeast of George Field, Lawrenceville, Illinois, killing pilot 2Lt. Grover L. Justice and passenger Pvt. Gordon Bennett Smith. The airplane had taken off from George Field with another AT-10BH on a local formation flight. The subject airplane flew in the number-two position and then the airplanes switched positions. Investigators stated, "[The] possible cause of the crash was that the pilot lowered his flaps while slowing down for the other ship to pull into formation and in doing so, reduced his speed and stalled the ship. The ship was thrown almost on its back. Lt. Justice recovered to a steep spiral and had the ship almost in level flight when he came to a row of trees on the bank of the White River. He banked steeply and went about half way through them when his left wing caught and threw him nose-first into the river. The ship sank in about 15 feet of water. The airplane was completely demolished. Both pilot and passenger were killed instantly."

5-27-43C. Tonopah, Nevada. At 0920 PWT, two Bell P-39D airplanes collided in mid-air and crashed at the AAF Bombing and Gunnery Range at Tonopah, Nevada, killing one flier and injuring another. 2Lt. Bryce R. VanCott was killed in the crash of P-39D #41-28267. 1Lt. Glendon V. Davis was able to parachute to safety from P-39D #41-28285, sustaining only minor injuries. The Accident Classification Committee stated, "Lt. Davis' [three-ship] flight took off [from Tonopah Army Air Field] for a dive-bombing mission and aerial gunnery mission. After dropping their bombs, the flight proceeded to the aerial gunnery range and joined their tow ship. After the flight made a few passes at the target using the parallel approach pattern, Lt. Davis noticed that Lt. VanCott was not carrying his attack past the target as he was supposed to do. He tried to contact him and motioned with hands the proper pattern to use. On the next pass Lt. VanCott corrected his error. Lt. Davis made another pass and was crossing back 2,000 feet above the tow-plane when another P-39 struck the bottom of his airplane. The force of the collision temporarily stunned him. Upon recovering he remembers seeing Lt. VanCott's airplane doing a roll with part of the left wing missing. Since Lt. Davis was unable to control his airplane at indicated airspeeds of less than 200 mph, and because of the violent vibration — even with the motor off— he decided to jump. At the time he jumped his altimeter read 12,200 feet, or over 7,000 feet above the ground. The tow-plane and the remaining airplane returned to the field and notified the tower of the accident." Investigators stated that Lt. VanCott pulled up into Lt. Davis in a steep climbing turn. The airplanes crashed onto the range six miles apart.

5-27-43D. Drasco, Arkansas. At 1129, a Vultee BT-13A crashed four miles north of Drasco, Arkansas, killing student A/C Thomas E. Arnesen. The student had taken off from the Army Air Field at Newport, Arkansas, on a solo navigation flight. Investigators stated that the student had apparently become lost and had attempted an emergency forced landing in rough terrain after encountering lowering ceilings and heavy rain. The airplane apparently was stalled after the student cut the ignition while attempting to stretch his glide. The airplane entered a spin at an altitude too low to allow recovery, describing one-eighth of a turn of spin before smashing into the ground.

5-27-43E. Dale Mabry Field, Florida. At 1440 EWT, two Republic P-47D airplanes collided in mid-air ten miles south of Dale Mabry Field, Tallahassee, Florida, killing one flier and injuring another. 2Lt. James O. Sheek was killed in the collision and was blown clear of P-47D #42-22386. 2Lt. Alexander M. Ritsch parachuted to safety from P-47D #42-22326, suffering only minor injuries. The airplanes were part of two separate two-ship flights that had taken off from Dale Mabry Field. The airplanes were engaged in a Lufbery circle when the number-two airplane, piloted by 2Lt. Sheek, straightened out and then entered a wingover maneuver. The number-two airplane, in an apparent attempt to rejoin the circle, climbed back into the circle and into the tail section of the number-three airplane, piloted by 2Lt. Ritsch. The collision severed the tail of the number-three airplane and caused the number-two airplane to explode, sending both airplanes out of control. Lt. Ritsch bailed out immediately. Lt. Sheek was blown out of the wreckage by the explosion.

5-27-43F. San Pablo Bay, California. At 1540, a Bell P-39N crashed into San Pablo Bay near San Pablo Point, California, killing pilot 2Lt. Donald W. Barger. Investigators stated, "Pilot was flying on an ordered acrobatic mission at 12,000 feet. He completed one slow roll successfully [and] then started another one. As he came out of this slow roll he made what appeared to be a sloppy recovery and the airplane went into a skid to the left and to the right, then the nose dropped and the airplane went into a tumble for abut three turns. Something fell off of the airplane; it is believed that this was the door that the pilot released in an attempt to get out of the airplane. The airplane straightened out and again went into an irregular tumble, then straightened out at about 9,000 feet and dove almost vertically into the bay." The official AAF accident report indicates that the body of the pilot was not recovered.

5-27-43G. Smyrna, Tennessee. At 2140 CWT, a Consolidated B-24D flying in poor weather collided with a tree and crashed near the Army Air Field at Smyrna, Tennessee, killing six crewmembers and injuring two others. The airplane had taken off from Smyrna Army Air Field on a triangular navigation flight to Chattanooga, Tennessee, to Knoxville, Tennessee, and return to Smyrna. Investigators stated, "Student pilot, on return from a night triangular navigation flight, joined traffic after normal pattern and radio landing calls, then flew over the runway, on advice from the instructor, instead of landing. Forward visibility was restricted by rain on windshield but side visibility was approximately six miles in light rain. Student pilot pulled up but ran into scud at about 1,400 feet indicated (field elevation 512 feet; normal night traffic pattern 2,000 feet indicated) and endeavored to hold 1,200-1,400 feet indicated. Close-in approach pattern as taught was not flown accurately on downwind leg due to student's difficulty in holding altitude. The difficulty was caused by the airplane being fully loaded with gasoline, power not being applied until after speed was lost and nose held high, and possibly from intercooler shutters being closed. Instructor took over on the base leg (from right seat) and overshot turn into final approach. In angling back into runway in right turn, right wing tip struck a tree. Application of power in a semi-stalled, nose-high position is known to increase spin tendency on heavy type B-24D airplanes and the right wing continued crashing through trees until the airplane hit telephone cable and flat rocks just south of the railroad tracks [surrounding] the field, 3,800 feet southwest of the Northeast-Southwest Runway. Airplane struck tree with right wing tip in banking back into the approach for the runway, continued through the trees, crashing 200 yards further and burning up. Accident was caused by instructor and student crew losing too much altitude on approach; and by not flying a precision instrument pattern when forward visibility was restricted by rain." Killed in the crash were: 2Lt. Thomas N. Arrington, B-24 instructor pilot; 2Lt. Donald E. Cochran, pilot/B-24 student; Sgt. J.R. York, engineer; Pvt. James K. Blanchard, crewmember; 1Lt. B.F. Fizzono, pilot-rated passenger; Pilot Officer G. Lamie, RAF, pilot-rated passenger. B-24 student 2Lt. A.J. Demetrakopulos and crewmember Pvt. D.M. Lawson were pulled from the wreckage alive, suffering very serious injuries.

5-27-43H. Monroe, Louisiana. At 2150, a Martin B-26B crashed while attempting a take-off at Selman Field, Monroe, Louisiana, killing five crewmembers and seriously injuring another. The airplane was taking off on a cross-country navigation flight back to its home station at Barksdale Field, Shreveport, Louisiana. The Accident Classification Committee stated, "Pilot started the motors on the ramp, taxied to the northeast end of the Northwest-Southeast runway for take-off into the southeast. Proceeded to take off into the southeast on the 5,000-foot runway. According to the statement of ... [a] member of the Field Night Lighting Crew, the pilot never did run his engines up prior to take-off. This may have been a contributing factor to the accident. One or both motors apparently cut out or lost power about the time that the airplane reached the end of the runway. In the opinion of the Board the pilot got to the end of the runway, pulled his airplane off the ground, pulled his wheels up and did not have sufficient flying speed to stay in the air. About 1,000 feet after leaving the end of the runway, the airplane mushed back into the ground. Both motors were knocked out. The airplane then continued approximately another 1,000 feet, hit a telephone line, which runs along the railroad right-of-way, crossed the railroad right-of-way and landed on the highway where it burned." Killed in the crash were: 1Lt. Glyn E. Stephens, pilot; 2Lt. Norman F. Brown, co-pilot; 2Lt. Raymond L. Knott, bombardier; SSgt. Percy J. Hamilton, engineer; SSgt. Gerald F. George, gunner. Radio operator SSgt. Richard F. Benders received serious injuries.

5-28-43A. Conifer, Colorado. At 0010, a Curtiss C-46A flew into the side of a mountain three miles northwest of Conifer, Colorado, killing pilot Capt. Robert E McIntyre and co-pilot 2Lt. William H. Eisman. The airplane took off from Lincoln, Nebraska, on a ferry flight to Lowry Field, Denver, Colorado. Investigators stated, "According to statements of witnesses, this airplane was first observed passing between two peaks in the foothills of the mountain range approximately ten miles southwest of Littleton, Colorado. The airplane maintained a constant altitude in flight until it struck the mountain known as Black Mountain or Shadow Mountain, at a point about 8,500 feet above sea level. The night was clear with perfect visibility. The airplane appeared to be functioning perfectly. Witnesses could see the airplane from a distance of from three to four miles." It was speculated that the pilots might have fallen asleep with the autopilot on. It was noted that the pilots had begun flying very early the previous day.

5-28-43B. Lancaster, California. At 0045, a Vultee BT-13A crashed one and one-half mile southwest of War Eagle Field, Lancaster, California, killing A/C George R. Bohac. The airplane flew into the ground while attempting to let down from the upper traffic pattern. Investigators stated that the airplane struck the ground in a "very flat dive."

5-28-43C. Manor, Georgia. At 1115 EWT, two North American P-51A aircraft collided in mid-air and crashed six miles south of Manor, Georgia, killing both pilots. 2Lt. Walter J. Herman, Jr. was killed in the crash of P-51A #43-6297; F/O Richard V. Norris was killed in the crash of P-51A #43-6278. The airplanes were part of a six-ship flight (three, two-ship elements) that had taken off from Waycross Army Air Field, Waycross, Georgia, on a ground gunnery mission. The flight had separated while making their passes at the gunnery range. F/O Norris was leading the third two-ship element back to base when Lt. Herman approached and attempted to join up with the two airplanes. Lt. Herman dropped down and flew into the tail of F/O Norris, severing the tail section. The airplanes went out of control and slammed to earth where they both exploded into flames. Neither pilot was able to bail out.

5-28-43D. Pecos, Texas. At 1440, a Vultee BT-13A crashed nine miles east of the Army Air Field at Pecos, Texas, killing student A/C James G. Newton. The airplane took off from Pecos at 1415 on a routine student solo flight. The BT-13 was observed to enter a spin to the right at approximately 3,000 feet agl, continuing in the spin until it struck the ground and burst into flames. The student was killed instantly.

5-28-43E. Midland, Texas. At 1539, a Lockheed-Vega RB-34A suffered a catastrophic structural failure and crashed 40 miles north of the Army Air Field at Midland, Texas, killing five fliers. The airplane took off from Midland at 1501 on a routine training flight. A civilian observed the airplane flying at approximately 7,000 feet agl. He stated that the engines were not making the usual noise. The airplane rolled over to an inverted attitude and then entered a spin. The pilot attempted to recover and apparently stressed the airplane beyond its design limits. The starboard rudder and vertical fin failed and separated. The airplane, shedding pieces, smashed into the ground and exploded upon impact. Investigation revealed that the port propeller was feathered. Investigators discovered that the pilots had discussed practicing single-engine stalls and recoveries before take-off. Investigators noted that the RB-34A type airplane had "vicious stall characteristics, particularly with uneven power" and would "whip over to inverted" when stalled with one engine out. Killed in the crash were: F/O Robert L. Williams, B-34 instructor; Capt. Samuel S. Revis, pilot; Samuel W. Thomas, co-pilot; Pvt. Donald P. Beaulieu, passenger; Pvt. Wayne H. Maki, passenger.

5-28-43F. Melvin, Texas. At 1641, a Fairchild PT-19A crashed two miles west of Melvin, Texas, killing pilot 1Lt. Jackman M. Shattuck. Investigators stated, "Lt. Jackman M. Shattuck attempted to take off from a field two miles west of Melvin, Texas, in which he had ferried a pilot to fly out a plane that had a forced landing. In taking off, the left wing of the plane came in contact with the limb of a tree causing the airplane to go out of control ... and crash."

5-28-43G. Morris Field, North Carolina. At 1900, a Piper L-4A crashed ten miles ENE of Morris Field, Charlotte, North Carolina, killing Liaison Pilots Cpl. James F. Roach and Cpl. Edwin P. Bortz. The airplane was seen to make three low-altitude passes near a farmer working in his field. The airplane, which was flying at about 30 feet agl, was in danger of colliding with a power-line so the pilot pulled up in a steep chandelle-type maneuver. The airplane stalled out of the maneuver, the pilot lost control and the airplane dove straight into the ground. The pilots were killed instantly.

5-28-43H. Alliance, Nebraska. At 1720 MWT, an Aeronca L-3B crashed at Alliance Municipal Airport, Alliance, Nebraska, fatally injuring pilot F/O

Charles B. Edwards. The airplane took off from Alliance Army Air Base on a routine training flight. The pilot flew to the municipal airport to practice landings, entering the left-hand traffic pattern. Another L-3 was practicing landings at the airport and that airplane went over to a right-hand pattern and lost sight of the subject airplane. Investigation revealed that the subject airplane was in a glide at about 400 feet agl when it began to stall and mush toward the ground. The pilot applied power and raised the nose, but the airplane stalled and then dove into the ground. The pilot died a short time after the accident.

5-29-43A. MacDill Field, Florida. At 0944 EWT, a Martin B-26B crashed at MacDill Field, Tampa, Florida, killing six fliers. The airplane was part of a flight of three that was returning to MacDill Field after flying a "review" flight. The formation was to land at MacDill and refuel for a flight to the home station at Myrtle Beach Army Air Field, Myrtle Beach, South Carolina. The flight leader radioed the tower while the formation was flying on the downwind leg of the traffic pattern. The formation, which was echelon right, did not break up until the pilots were turning on the base leg. All three ships extended their landing gear at the same time while turning onto the final approach. The first ship was cleared to land and the second ship, which was about 200 feet above the level of the other two airplanes, was ordered to go around by the control tower operator. The number-two ship retracted the landing gear and climbed slightly as it flew away. The number-three airplane, flying at about 200 feet agl, apparently encountered severe propeller turbulence from the lead airplane and was seen to enter a vertical bank to the left. The pilot leveled out, but the airplane began to yaw to the left and then to the right. The subject airplane then entered a half-roll to the right and smashed into the ground in an inverted attitude, exploding violently into flames upon impact. Killed in the crash were: 2Lt. Elmer D. Martin, pilot; 2Lt. Edgar R. Richardson, co-pilot; 2Lt. Edward B. Pogonowski, bombardier/navigator; SSgt. Samuel R. Fasone, engineer; SSgt. Fred G. Kimmerele, radio operator; SSgt. Frank Canavit, gunner.

5-29-43B. Poplar Bluff, Missouri. At 1205, a Vultee BT-13A crashed ten miles east of Poplar Bluff, Missouri, killing instructor 2Lt. Seymour E. Auborn and student A/C Charles F. Conn, Jr. The airplane took off at 1130 from Walnut Ridge Army Air Field, Arkansas, on a navigation flight to Poplar Bluff and return. The airplane arrived in the vicinity of Poplar Bluff at about 1200 and began to engage another BT-13 in simulated combat. The airplanes were maneuvering at about 3,500 feet indicated altitude. The other BT-13 broke away at about 2,200 feet indicated altitude (1,850 feet agl), but the subject airplane attempted to re-engage in simulated combat. The subject airplane dove on the other BT-13 from the front quarter, passing beneath it. The subject airplane did not have enough altitude to complete the maneuver and it smashed into the ground at a shallow angle at high speed. The airplane burst into flames and the occupants were killed instantly upon impact.

5-29-43C. Providence, Rhode Island. At 1530, a Republic P-47C crashed near Green Field, Providence, Rhode Island, killing pilot 2Lt. Millard F. Parsley, Jr. The pilot radioed the control tower and requested an emergency landing. The pilot did not state what the problem was. A Douglas DC-3 operated by American Airlines, which was on final approach, was ordered to go around to make room for the emergency landing. The subject airplane entered the pattern at about 500 feet agl. The subject airplane stalled at about 300 feet agl while turning from the downwind leg to the base leg, describing one turn of spin before smashing into the ground nose-first and exploding.

5-29-43D. Niagara Falls, New York. At 1555, a Bell P-39M crashed at Niagara Falls, New York, killing Bell Aircraft Company test pilot Richard N. Kraschel. Investigators stated, "The crash occurred while conducting flight tests on spin chutes attached to the vertical fin and to the wings. The Board also believes that the pilot was unable to release the chute from the airplane, resulting in a loss of control. After the loss of control, the ship went into a spin or spiral. An unsuccessful jump was attempted by the pilot at 500 [feet agl]. The Board also believes that the parachute ripcord was never pulled because pilot was unconscious, probably having struck part of the empennage in exit from the airplane."

5-29-43E. Location Unknown. At a an unknown time after 1518 PWT, a Consolidated B-24D disappeared and is presumed to have crashed in the Pacific Ocean while on a routine navigation training flight. The crew was missing and presumed lost at sea. The B-24 was on individual over-water navigation flight and was to rendezvous with nine other B-24s at a point 140 nautical miles out to sea. The airplane never arrived at the rendezvous point and was never seen again. Investigators noted that the airplane had nine hours worth of fuel on board when it took off from Marysville Army Air Field, California. The B-24 and crew were stationed at Pocatello Army Air Field, Pocatello, Idaho. Missing and presumed lost at sea were: 2Lt. Paul V. Pavel, pilot; 2Lt. Forrest H. Branson, co-pilot; 2Lt. Paul R. Chlhault, navigator; 2Lt. Raymond R. Ramey, bombardier; SSgt. Edwin A. Johnson, engineer; Sgt. Carleton H. Lewis, radio operator; SSgt. Edwin L. Grundstrom, assistant engineer; Sgt. Henry J. Marra, assistant radio operator; SSgt. Harry E. Williford, gunner; SSgt. Edgar H. Schmidt, gunner.

5-29-43F. Cross City, Florida. At 1835 EWT, a North American BT-14 crashed seven miles southwest of the Army Air Field at Cross City, Florida, killing pilot 2Lt. Wilford E. Hushaw and pilot-rated passenger 2Lt. Ronald S. Brown. The airplane took off from Cross City on a routine instrument training flight. Lt. Hushaw as flying as the instrument pilot and Lt. Brown was flying as the safety pilot. The airplane failed to return to Cross City and a search effort was launched. The airplane wreckage was found by a state forest ranger about an hour after the search commenced. The airplane struck the ground at a steep angle, breaking up and bursting into flames upon impact. Examination of the wreckage indicated that the flaps were in the down position. Investigators speculated that the pilots were attempting an emergency forced landing in a clearing near the crash site.

5-30-43A. Salt Flat, Texas. At 0330 CWT, a Boeing B-17F collided with rising terrain 32 miles southeast of Salt Flat, Texas, killing 11 fliers. Investigators stated, "Aircraft took off from Army Air Base, Pyote, Texas, 2100, May 29, 1943, for a six-hour local bombing and night flying mission. Aircraft last reported in to Group Ground Radio Station at 0220 May 30, 1943, giving position over Odessa, Texas. Nothing more was heard from the aircraft. The aircraft was sighted from the air at 1600 May 30, 1943 on mountain 32 miles southeast of Salt Flat, Texas. The elevation of the mountain is approximately 4,990 feet. The aircraft was flying eastward, hit the west side of the mountain approximately 175 feet from the top and bounced over the top of the mountain, coming down on the other side. The tail section broke off aft of the side entrance and fell to the ground at the summit of the mountain. The rest of the aircraft, excluding the engines, was approximately 200 feet down the slope. The engines rolled about 600 feet down the slope. The airplane burned destroying all evidence. Engines and accessories were battered to such an extent that no evidence or indication of engine failure could be determined." Weather was reported as ceiling and visibility unlimited with high, thin scattered cirrus with a little lower altocumulus to altostratus east of Salt Flat and scattered light thunderstorms west of Salt Flat. Weather was not considered a factor in the accident. Killed in the crash were: 2Lt. William J. Sauer, pilot; 2Lt. Charles G. Ryan, co-pilot; 2Lt. Joe A. Chamberlain, navigator; 2Lt. Frank N. Jones, bombardier; Cpl. Julian B. Wilkerson, engineer; SSgt. Martin J. Patten, engineer; Sgt. George W. Byrnes, radio operator; Sgt. Edward J. O'Hearn, assistant radio operator; Sgt. John J. Hefferan, gunner; Sgt. Edward D. Reed, gunner; TSgt. William B. Clark, radio operator instructor.

5-30-43B. Long Island Sound, Connecticut. At 0935, a Republic RP-47B crashed into Long Island Sound three miles southeast of New Haven, Connecticut, killing pilot 2Lt. Neil C. Donovan. The pilot was leading a three-ship flight that took off from Westover Field, Chicopee, Massachusetts, and was scheduled to fly a high-altitude formation flight at 20,000 feet. After flying at high altitude for a short time, the pilot led the formation down to low altitude and was flying to the north very low over Long Island Sound one-quarter mile off of the southern Connecticut shore. He apparently misjudged his altitude above the surface and struck the water with his propeller. The pilot pulled up in a steep chandelle to the left. After about 90 degrees of turn, the airplane stalled and fell off on the left wing, plunging straight down for the sea. The pilot leveled out momentarily, but the airplane smashed into the water at an angle of about 30 degrees about 300 yards off shore, heading west. The pilot was killed instantly upon impact.

5-30-43C. Laughlin Field, Texas. At 1300 CWT, a Martin B-26B crashed while attempting a landing at Laughlin Field, Del Rio, Texas, killing three fliers. The airplane had made a flight the previous day from Laughlin Field to Tarrant Field, Fort Worth, Texas. The airplane was delayed on the ground overnight due to weather conditions. The next day the airplane took off from Tarrant Field for the return flight to Laughlin Field. The airplane arrived at Laughlin Field, entered the traffic pattern and made a normal approach. Investigators stated, "The pilot delayed the pullout from the approach glide too long. Then when the pullout was attempted, it is probable that too rapid application of control pressures resulted in a high-speed stall at low altitude." The airplane plunged to the ground at a steep angle and exploded violently into flames upon impact. Pilot F/O Robert D. Jones, co-pilot F/O Milburn B. Henry and engineer Sgt. A.W. Kanzler were killed in the crash.

5-30-43D. Covelo, California. At an unknown time after 1130 PWT, a Boeing B-17F flying in instrument conditions collided with rising terrain near Covelo, California, killing the crew of six. The airplane was part of a 12-ship formation that had taken off from the Army Air Field at Casper, Wyoming, on May 28, 1943 and flew to the Army Air Field at Marysville, California. On the day of the accident, the airplane took off from Marysville Army Air Field on a routine over-water navigation-training mission. The airplane was to fly to a point at sea 150 miles west of Marysville, then to a point at sea 150 miles west of Eugene, Oregon, then to Eugene, and then back west 200 miles, then south to a point at sea west of Clear Lake, California, then east to Clear Lake for an interception problem, and then return to Marysville. The airplane made nine position reports, the last coming at 0923 PWT. The airplane reported on time and on course about 180 miles out at sea. The airplane failed to return to rendezvous for the interception problem and failed to return to base. The airplane was declared missing and a large search effort conducted. The airplane was not found until 8-10-43, when a civilian farmer found the wreck on the side of a mountain near Covelo. Investigation revealed that the airplane was flying east when it collided with the west side of the mountain. Had the airplane passed over the peak, it would have surely collided with a taller mountain a short distance to the east. The airplane exploded into flames upon impact.

The mountain was undoubtedly shrouded in clouds at the time of the accident. Local civilians who were interviewed after the wreck was found stated that they did not hear or see anything unusual the day of the accident. Local weather at the time of the accident was reported as low ceilings with steady rain. Killed in the crash were: 2Lt. James O. Westbury, pilot; 2Lt. William Blalock, co-pilot; 2Lt. Joseph J. Weitzel, navigator; 2Lt. Ralph S. Johnson, Jr., bombardier; SSgt. Ralph J. Sicking, radio operator; SSgt. Paul P. Scheffer, engineer.

5-30-43E. Oakland, California. At 1749, a Vultee BT-13A on an unauthorized flight crashed one mile northeast of Oakland Army Airdrome, Oakland, California, killing non-rated pilot Cpl. Arthur G. Anderson. Investigators stated, "Cpl. Anderson started this transient airplane and proceeded to take off without [permission].... Two enlisted men in the vicinity tried to prevent Cpl. Anderson from taking off in the airplane but without success. He took off with the ignition switch on the left magneto and with the pitot cover on. After about 15 minutes of very erratic flying, consisting of aerobatics and low flying, Cpl. Anderson pulled the airplane up into a steep climb. The airplane stalled and fell to the ground, killing [Cpl. Anderson] and totally demolishing the airplane." Investigation revealed that Cpl. Anderson had about 90 hours of logged flight time, but had been eliminated from an AAF Basic Flying School.

5-30-43F. Williams Field, Arizona. At 2200 MWT, a Curtiss AT-9A crashed five miles east of Williams Field, Chandler, Arizona, killing students A/C William L. McCollough and A/C Robert A. Smith. Investigators stated, "Subject aircraft was flying over the traffic pattern at night without instructions to enter traffic. The aircraft was not observed again until the fire of the crash was observed. The appearance of the wreckage showed the aircraft to have hit with wings level and traveling from north to south. The crash occurred on the open desert five miles northeast of Williams Field. Ensuing fire completed the destruction of the aircraft. It is thought that the pilot misread or [incorrectly set] his altimeter, flew too low and accidentally collided with the ground."

5-31-43A. Coffeyville, Kansas. At 1150 CWT, a Vultee BT-13A crashed ten miles south of the Army Air Field at Coffeyville, Kansas, killing student A/C James W. Brandt. The Accident Classification Committee stated, "The Board is of the opinion that the accident was caused by an inadvertent spin which was the result of a partial snap roll at a low altitude. After the ship executed three-quarters of a snap roll, the student found himself in a vertical bank. The student ... [used] the controls in such a manner that instead of righting the airplane, he went into a spin from which he was unable to recover."

5-31-43B. Harlingen, Texas. At 1157, student gunner Cpl. William D. Irwin, Jr. was killed when the walked into the rotating propeller of a North American AT-6A operating at the Army Air Field at Harlingen, Texas. Pilot 2Lt. George J. Bokun and student gunner Pfc. Rueben W. Nyquist were uninjured. Investigators stated, "Lt. Bokun ... had completed a regularly scheduled gunnery run at approximately 1157 CWT. At the time he taxied his ship from the taxiway into the main ramp he noticed a gunner, later identified as Cpl. William D. Irwin, Jr., carrying an ammunition box on his right shoulder, moving in a path that would coincide with the path being taken by the airplane. [The pilot] made an S-turn to the right to avoid this man, and then became aware of a tug [tractor] moving on the ramp directly in front of him. His attention was evidently became centered on the tug, which he finally ascertained had moved from his path, and he did not see the gunner who was evidently directly in front of his ship. He was not further aware of this man until [the man] was struck by the propeller."

5-31-43C. Pradentown, Florida. At 1400 EWT, a Martin RB-26A flying in poor weather crashed at Pradentown, Florida, killing five fliers. The airplane took off from MacDill Field, Tampa, Florida, on an instrument training flight. The pilot and co-pilot were to have equal time flying under the hood; while one pilot flew under the hood as the instrument pilot, the other acted as the safety pilot. There were no witnesses to the crash. Examination of the wreckage indicated that the airplane had struck the ground in a diving spiral to the right. The B-26 exploded violently into flames upon impact. Investigators speculated that the pilots had inadvertently entered a thunderstorm and lost control of the airplane. Killed in the crash were: 2Lt. George J. Ganow, pilot; 2Lt. Joseph O. Esser, co-pilot; 2Lt. Charles H. Carroll, navigator; 2Lt. William M. Williamson, bombardier; SSgt. Harry J. Farrell, Jr., engineer.

5-31-43D. Autaugaville, Alabama. At 1320, a Curtiss P-40F performing low-altitude aerobatics crashed eight miles southwest of Autaugaville, Alabama, killing student A/C James S. Savage. The P-40 took off from Craig Field, Selma, Alabama, on a routine training flight. The airplane was observed doing various aerobatic maneuvers before the crash. The airplane was seen to enter a dive and then pull up into a loop. The airplane came straight down with the port wing down and in a slip to the left. The airplane leveled off and performed two rolls. The airplane then entered a spiral to the left and smashed into the ground where it exploded into flames.

5-31-44E. Lake Kerr, Florida. At 1555 EWT, a Republic P-47G crashed ten miles south of Lake Kerr, Florida, killing 2Lt. Robert E. Sawers. The airplane took off from the Army Air Field at Cross City, Florida, on a routine navigation flight to East Palatka, Florida, to Lacoochee, Florida, and return. The subject pilot's commander engaged the subject airplane in an authorized interception near Lake Kerr. The subject airplane dove down and performed a steep pullout, causing streamers to form from the wingtips. The airplane stalled at an indicated altitude of 3,500 feet. The airplane snapped over to inverted and then entered a vertical dive toward the ground. The airplane exploded violently into flames upon impact and the pilot was killed instantly.

5-31-43F. Toyah Auxiliary Airfield, Texas. At 1553, a Vultee BT-13A crashed at Toyah Auxiliary Airfield, 15 miles southeast of the Army Air Field at Pecos, Texas, killing student pilots Maj. Walter J. Fitzgerald, Jr. and A/C Raymond G. Beck. The airplane was on a routine instrument team ride that had originated at Pecos Army Air Field. Student pilot Maj. Fitzgerald was in the front cockpit and A/C Beck was in the rear cockpit. The students successfully completed their instrument work and then flew to Toyah Auxiliary Airfield to perform landings. The airplane, flying at about 500 feet agl, stalled on the turn from the base leg to final approach. The airplane rolled to an inverted attitude and then entered a spin, describing one-half turn of spin before smashing into the ground and exploding into flames.

5-31-43G. Starke, Florida. At 1820 EWT, a Douglas A-20B collided with a fence and crashed four miles north of Starke, Florida, killing pilot 2Lt. Wirt E. Billingslea. The airplane had taken off from Alachua Army Air Field, Gainesville, Florida, on a routine training flight. The airplane arrived in the area of Starke and began buzzing a house at very low altitude. The airplane made two circling passes at very low altitude. The airplane came back for a third pass and entered a vertically banked turn to the left, striking a fence with the port wing. The airplane went out of control and cartwheeled into the ground, exploding into flames upon impact. Investigation revealed that the airplane had buzzed the same area at 0830 that morning and that a similar airplane had buzzed the area on several occasions in the weeks leading up to the accident.

5-31-43H. Smith Creek Valley, Nevada. At approximately 1702 PWT, a Douglas C-67 (a modified B-23) flying in instrument conditions collided with rising terrain at Smith Creek Valley, 36 miles west of Austin, Nevada, killing eleven crewmembers and passengers. The airplane took off from Peterson Field, Colorado Springs, Colorado, on a personnel transport flight to Reno Army Air Base, Reno, Nevada. Investigation revealed that the C-67, while flying in instrument conditions and at cruise speed, collided with Desatoya Peak at an elevation of approximately 10,000 feet msl, exploding into flames upon impact. The airplane, flying southwest, hit about 150 feet from the top of the peak. Investigators speculated that the pilots were in a very shallow descent in an effort to establish visual contact with the ground, causing the airplane to collide with mountainous terrain. The Superintendent of the Nevada State Fish Hatchery observed a low flying airplane pass over the hatchery at low altitude and then disappear into the clouds. He stated that he had heard a "strange" noise a short time later, followed by silence. The superintendent reported the incident to authorities on 6-3-43 when he made a trip into Austin. AAF authorities located the airplane on 6-4-43 and reached the wreck the next day. Killed in the crash were: 1Lt. George Adams, pilot; 1Lt. E.V. Anderson, co-pilot; Cpl. W.L. Corey, engineer; Cpl. Albert R. Schneider, radio operator. Passengers killed in the crash were: Col. P.R. Love, TSgt. Stanley Krysztowick, Cpl. Frank Fugazza, Cpl. Freddie E. Jones, Pvt. Ralph M. Godsell, Cpl. Don C. Brill, Cpl. Hugo Knight Vissher.

5-31-43I. Avon Park, Florida. At 2335 EWT, a Martin B-26C crashed while attempting a landing at the Army Air Field at Avon Park Bombing Range, Avon Park, Florida, killing six fliers. The airplane had taken off from Avon Park Army Air Field on a local familiarization flight. Investigators stated, "At approximately 2325 ... B-26C #41-34756 called Avon Park Tower for landing instructions. The tower told him he was number-two in traffic to land on Runway 31 (NW). On turning on final approach, the tower told 41-34756 to go around, as he was too close to the preceding ship to land. B-26 #41-34756 went around and called again on his downwind leg and again requested landing instructions. Avon Park Tower cleared him to land on Runway 31. No other contact was made with 41-34756. A few moments after last contact a flash of light was seen three or four miles from the tower." The pilots apparently lost control of the airplane and it smashed to the ground in a diving spiral. Killed in the crash were: F/O William R. Norman, pilot; 2Lt. Milton O. Edwards, co-pilot; SSgt. Clarence R. Hughes, engineer; Sgt. Richard M. Condit, passenger; Sgt. Leroy J. Bording, passenger; Sgt. John W. Collins, gunner.

5-31-43J. Oklahoma City, Oklahoma. At 2330, a Douglas A-20B suffering an apparent engine failure crashed two miles north of Will Rogers Field, Oklahoma City, Oklahoma, killing three fliers. The airplane took off from Will Rogers Field at 2320 on a routine night navigation mission. The airplane made a normal take-off and climbed to about 3,000 feet indicated altitude when the port engine quit. The pilot radioed the control tower of the fact and requested an emergency landing. The tower cleared the pattern and the runway for the airplane to make the emergency landing. The subject airplane entered the traffic pattern on the downwind leg. The starboard engine quit just after the airplane turned on the base leg. The airplane stalled and the pilot lost control, entering a diving spiral toward the ground. The airplane burst into flames upon impact and all on board were killed instantly. There were no witnesses to the accident. Investigators could not determine the cause of the engine failures. Pilot F/O Edward J. Carroll, gunner Sgt. Charles W. Gilliam and SSgt. David H. King were killed in the crash.

June

6-1-43A. Cleburne, Texas. At 0030, a Beech AT-10BH dove into the ground six miles northwest of Cleburne, Texas, killing A/C Glen J. Ogden and A/C John R. Pertmer. The students took off from Blackland Army Air Field, Waco, Texas, on a routine navigation mission to Abilene, Texas, and return. Investigators stated, "The accident occurred ... 55 miles north of Waco, Texas. The students were on the light line and just four lights away from their home field. The committee believes the students had been lost as they failed to check in at Abilene, they had been in the air one hour longer than was necessary to make the flight, and they were over 50 miles off of their correct course. However, in view of the fact that they were headed in the right direction down the light line, we believe they were not lost at the time of the accident. The airplane was so completely demolished that no evidence could be found to indicate the real cause of the crash. The marks on the ground indicated that the right wing struck the ground first and the airplane cartwheeled straight ahead, scattering parts over an area of 100 yards."

6-1-43B. Pampa, Texas. At 0045, Cessna AT-17 dove into the ground one-quarter of a mile east of Pampa Army Air Field, Pampa, Texas, killing A/C James A. Minor and A/C Frank R. Monroe. Investigators stated, "The [cadets] had been engaged in local night formation and crashed in the traffic pattern after peeling off of formation from over the field. Evidence reveals that the cadets were the last in the landing sequence. They were advised by the tower to call on the approach leg. Immediately they reported that they had been cut out and had to go around. Their absence was not confirmed until approximately 0300 when student flying was over. Search was immediately started and the wreckage was discovered at 0630, one-quarter of a mile east of the field. Both [cadets] were found dead in the wreckage.... It is the opinion of the committee that the airplane was flown inadvertently into the ground, left wing striking first.... It is felt by the committee that the cadets had lost the horizon and failed to resort to their instruments."

6-1-43C. Fort Valley, Georgia. At 0245 EWT, a Vultee BT-13A dove into the ground two and one-half miles northeast of Harris Field, Fort Valley, Georgia, killing pilot 1Lt. Herbert R. Messenger, Jr. and passenger Pfc. Peter Lekakis. The airplane was observed flying in a southwesterly direction at about 4,000 feet indicated altitude with landing lights illuminated. The airplane was seen to "peel off" to the right, executing a dive to about 2,000 feet and then performing a steep pull-up, falling off to the left and then diving into the ground and exploding into flames. The airplane had been acting as a control ship for the student airplanes flying out of Cochran Field, Macon, Georgia.

6-1-43D. Tifton, Georgia. At 0850 EWT, a Curtiss P-40F crashed eight miles east of Tifton, Georgia, killing A/C Warren W. West. Investigation revealed that the engine was operating until the airplane struck the ground and it was noted that this was the student's first flight in a P-40 type airplane. Investigators speculated that the student might have fell into an unrecoverable spin while performing unauthorized aerobatics on what was to be a basic familiarization flight that did not include aerobatic maneuvers.

6-1-43E. Pinellas Field, Florida. At 1315 EWT, Crew Chief Pvt. Didley T. Sweany was killed when he was struck by the rotating propeller of a Curtiss P-40L being taxied to a parking area at Pinellas Army Air Field, St. Petersburg, Florida. Pilot 2Lt. Charles B. Hardy was uninjured. Investigators stated, "The pilot started taxiing out with a crew member at each wing tip. When the airplane had cleared all the parked airplanes, the wing men left the airplane and the man on the right wing gave the pilot an all clear signal. The pilot angled across the ramp into the 'A' flight parking area and the propeller hit the crew chief, who was waiting for his airplane to return to the ramp."

6-1-43F. Tifton, Georgia. At 1800 EWT, a Curtiss P-40F on a "test flight" spun to the ground one-quarter mile east of Tifton, Georgia, killing pilot 1Lt. Jack C. Phelps. Investigators stated, "[Lt. Phelps] was seen following a [North American] P-51 in trail, circling the Tifton Airport. The P-51 peeled off and came down [to an altitude of 250 feet above ground level] just off the edge of the field. Lt. Phelps followed the P-51 during the peel-off and dive. The P-51 executed a steep turn [at about 200 feet above ground level] and Lt. Phelps attempted to follow him. Lt. Phelps' airplane affected a high speed stall from which it never recovered."

6-1-43G. Avon Park Bombing Range, Florida. At 1938 EWT, a Martin B-26B crashed while attempting an emergency landing at Avon Park Bombing Range, Florida, injuring seven crewmembers and killing gunner SSgt. William F. Whitener. Investigators stated, "Pilot was giving single-engine operation and practice feathering [instruction to B-26 students] at 6,000 feet. The right propeller was feathered once and functioned properly. After another hour the right propeller was feathered again and failed to unfeather. The pilot came over the field at 4,000 feet and made his pattern at 2,000 feet. His wheels touched down about three-quarters of the way down the runway. Due to his excessive speed, he bounced off the runway and overshot. The [pilot lost control of the airplane and] crashed about one-half mile beyond the runway and caught fire." Receiving minor injuries in the crash were: 2Lt. Clarence J. Potter, pilot and B-26 instructor; 2Lt. Howard F. Stateman, co-pilot and B-26 student; 2Lt. Frank E. Teltschick, co-pilot and B-26 student; 2Lt. Joseph H. Vorrier, bombardier; SSgt. Arthur E. Botsford, engineer; SSgt. Charles Schnabel, radio operator; SSgt. William J. Hughes, radio operator.

6-2-43A. Hill Field, Utah. At 1308, a Douglas A-20G attempting a landing at Hill Field, Ogden, Utah, suffered a failure of the port engine and crashed into a hangar, killing civilian mechanic student Wendell R. Powell, St. Anthony, Idaho. Five civilian mechanics and helpers were seriously injured and 23 other civilians received minor injuries. Pilot Capt. Woodrow M. Bell also received minor injuries. The pilot took off on a ferry mission from McCarran Field, Las Vegas, Nevada, at 1000 MWT. After an uneventful flight to Hill Field, air traffic control personnel ordered the pilot to hold south of the field while existing traffic cleared the area. After a short time the A-20 was cleared to land on Runway 31. On final approach, about one-half mile from the end of the runway at an altitude of 200 feet, the A-20 was flying at about 120 mph with landing gear and flaps extended when the port engine suddenly quit. The airplane immediately began swerving to the left. The pilot increased power on the starboard engine (rather than cutting power and landing straight ahead), causing the airplane to continue banking to the left. The pilot lost control and the A-20 continued in a turn to the left, colliding with a repair hangar about 1,000 yards to the left of Runway 31. The airplane struck the roof with the port wing and crashed through the top. There was no fire. The five Ogden, Utah, civilians seriously injured in the accident were: Foreman Arlie Benton, Mechanic Helper Mildred DeNovellis, Worker Zea Hepler, Student Mechanic Della Pacheza, Worker Harold Sadler.

6-2-43B. Hemet, California. At 1415, a Ryan PT-22 stalled and spun to the ground after attempting a steep right turn immediately after take-off from Auxiliary Airfield #A-2, ten miles west of the main Army Air Field at Hemet, California, killing civilian instructor Harold J. Zamora and fatally injuring A/C Harold E. Imel. The airplane was seen to take off normally, reaching an altitude of 300 feet before entering a turn to the right, against the traffic pattern. The airplane was seen to enter a violent spin to the left after passing through about 10 degrees of turn, describing about one and one-half turns of spin before striking the ground. The airplane stuck the ground on its port wing and nose but did not catch on fire. A/C Imel died of his injuries at 1445 on 6-3-43.

6-2-43C. Tarriffville, Connecticut. At 1720, a Republic P-47D dove into the ground from 23,000 feet seven miles east of Tarriffville, Connecticut, killing pilot 2Lt. Frederick P. Strickland. Lt. Strickland and 1Lt. Donald K. Dilling flew to Lake Candlewood at 23,000 feet and engaged in simulated combat for 15 minutes. After the exercise, Lt. Dilling noticed Lt. Strickland beginning to fly erratically. Lt. Strickland began to climb and turn 180 degrees. Lt. Dilling caught up with Lt. Strickland but was unable to get close enough to the airplane to observe the pilot. Lt. Dilling turned toward Westover Field and attempted to contact Lt. Strickland by radio without success. Lt. Dilling turned his P-47 around and was able to spot Lt. Strickland's airplane. Lt. Strickland's P-47 was then seen to enter a dive toward the ground with no apparent effort by the pilot to pull out or use his parachute. Investigators speculated that the pilot was incapacitated by oxygen starvation when he inadvertently dove the airplane into the ground on the bank of the Farrington River four miles southwest of Bradley Field, Connecticut.

6-2-43D. Richmond, Virginia. At 2007, a Republic P-47D and P-47B collided in mid-air while in the pattern at Richmond Army Air Base, Richmond, Virginia, killing 2Lt. Clifton B. Norman (P-47B) and injuring 1Lt. David L. Gross (P-47D). Investigators stated that Lt. Norman had executed an unusually wide pattern and a long approach directly into the sun. Lt. Gross had executed a normal pattern and turned from base to final at an altitude of 500 feet, apparently into the path of Lt. Norman's airplane, which was also flying at 500 feet. Both airplanes had been cleared to land individually with Lt. Norman cleared to land first. Realizing that a dangerous situation was developing, control tower personnel radioed Lt. Gross and ordered him to go around. A defective radio receiver prevented Lt. Gross from understanding the warning from the tower. Both airplanes (with landing gear extended) descended on a collision course during the final approach and personnel in a mobile radio unit near the field, anticipating the collision, repeated the control tower's warning and order. Lt. Gross got the message and gave his airplane the throttle in an attempt to maneuver out of the trouble. Lt. Norman's P-47 descended into view just as Lt. Gross (who was overtaking Lt. Norman) turned to try to avoid the collision, the airplanes colliding at about 300 feet agl. The port wing of Lt. Gross' P-47 impacted the tail section of Lt. Norman's airplane. Lt. Norman was killed instantly when his airplane crashed onto the runway and exploded violently into flames. Lt. Gross received minor injuries when his P-47 crashed out of control into a stand of trees in the middle of the airbase. Investigators faulted both pilots for failing to see and avoid each other in the traffic pattern. The airplanes had been on an interception mission.

6-2-43E. Florence, South Carolina. At 2250, a North American B-25D crashed while attempting a landing at Florence Army Air Field, Florence, South Carolina, killing three crewmembers and seriously injuring engineer Sgt. Elmo E. Steele. The Accident Classification Committee stated, "[The pilot] entered the traffic pattern in a normal manner; on the base leg, he called the tower for landing instruction. He was instructed not to land because numerous planes were waiting for take-off instructions. As he turned onto his final approach, he acknowledged the instructions given by the control tower. He continued on the final approach, losing altitude and crashing into trees about two and one-half miles from the end of the NE/SW runway. The plane cut a path in the trees 500 yards long; the bodies were found 350 yards down the cleared path from the point of impact. The plane was so badly demolished that inspection revealed nothing factual concerning the cause of the accident." Killed in the accident were: 2Lt. George D. Morgan, pilot; 2Lt. Glenn H. Tucker, Jr., co-pilot; Sgt. Milton E. McConkey, engineer. The airplane had been on a local night transition flight.

6-3-43A. Cross City, Florida. At 1200 EWT, a Republic P-47C crashed-landed four miles northeast of Cross City, Florida, killing pilot 2Lt. Robert McCann. The airplane was on a scheduled "Altitude Aerial Gunnery Mission" with another P-47 when the propeller ran away. The pilot called his flight leader and was advised to use manual controls to hold the propeller at a constant rpm. The pilot attempted this but failed to get control of the propeller. The leader radioed that they would return to Cross City and land immediately. Unable to make it to the field, Lt. McCann attempted a wheels-up forced landing in an open field. Lt. McCann successfully crash-landed the airplane but suffered a depressed skull fracture and a broken right arm and died of his injuries 6-7-43.

6-3-43B. Pinellas Field, Florida. At 1735, a Curtiss P-40L crashed while attempting a landing at Pinellas Army Air Field, St. Petersburg, Florida, killing pilot 2Lt. Harold E. Cecil, Jr. Maj. Bruce F. Parcell witnessed the accident and stated to investigators, "An echelon of airplanes of the 440th Fighter Squadron flew across the field at approximately 1,500 feet, and peeled off for landing. I saw the second airplane approximately two seconds after it peeled off. My attention was called to it by an enlisted man who raised his arms and pointed upward. He was apparently so surprised that he could not say anything. When I saw the airplane, it was out of control, nose down, and had flipped over on its right side. It was apparent that the airplane had partially stalled. The airplane, nose still down, rolled abruptly to the left, then to the right, as the pilot evidently fought to regain control. At approximately 600 feet, the airplane stopped turning, but was descending at about a 70-75 degree angle. The pilot pulled 'streamers' [wingtip condensation trails] from this point until the airplane crashed in the center of the field. Upon impact, the airplane disintegrated. There was no fire nor were there any explosion connected with the crash."

6-4-43A. Castroville, Texas. At 1000, a Lockheed-Vega R-37 (similar to the B-34 type) suffered a catastrophic structural failure and crashed at Castroville, Texas, killing pilot 1Lt. Robert J. Hanish, co-pilot 2Lt. Charles M. Lortie and engineer TSgt. Leon J. Cortese. The airplane was cleared on an instrument flight from Victoria, Texas, to Hondo, Texas, taking off at 0840. Investigation of the wreckage revealed that the port wing structure had failed, causing the wing to separate from the airplane.

6-4-43B. Farmingdale, New York. At 1720, a North American AT-6A and a Republic P-47D collided in mid-air at Farmingdale, Long Island, New York, killing AAF pilot 1Lt. Joseph V. Chapman aboard the AT-6A and fatally injuring Charles H. McLevedge, civilian test pilot for Republic Aviation Corporation. Republic Aviation officials interviewed McLevedge before he died, stating later, "McLevedge took off in P-47D-5 #42-8462 from Republic Field using #7 runway. Both pilot and ground observers noticed that an excessive amount of smoke was coming from the left side of the engine cowling. The Radio Control Tower contacted McLevedge and advised him of this fact. McLevedge circled the field and flew over the tower and asked whether or not smoke was still coming from his engine. He was advised that it was, so he decided to land. At that time he was headed north and he lowered his gear and made a turn to the left. Just as he was coming out of his bank on his east leg, an AT-6A appeared immediately in front of him and at the same altitude, crossing his path towards the field at angle of approximately 30 degrees. McLevedge immediately attempted to pull up over the trainer but was unable to clear it. After the airplanes had collided, McLevedge's airplane seemed to fly satisfactorily and he thought that he could bring the ship in for a safe landing but smoke and flames came into the cockpit from the floorboard. McLevedge states that he rolled the airplane and after considerable difficulty in releasing his safety belt due to the fact that the adjustment was such that the belt latch was several inches off-center, he left the airplane with it in an attitude with the wings approximately vertical and the nose down. As he left the airplane he passed through flames. McLevedge landed very close to the airplane and went through the flames caused by the explosion of the ship when it hit the ground." McLevedge died of severe burns 6-7-43 at 0127. Lt. Chapman failed to bail out and was killed in the crash of the AT-6. He was returning to the field following a cross-country flight to Bridgeport, Connecticut.

6-4-43C. Mather Field, California. At 2100, an aerial engineer was killed when he walked into the arc of the rotating port propeller of North American B-25D that was preparing to take off from Mather Field, Sacramento, California. 2Lt. James E. Jones, pilot and B-25 student, 2Lt. Paul L. Bradley, co-pilot/B-25 student, and engineer Pfc. John R. Johnson were preparing to take off on a night transition flight, running up the engines in the run-up area prior to turning onto the runway when the starboard engine caught on fire. Smoke emanating from the starboard engine began entering the cockpit. When the pilots determined that the starboard engine was on fire, Lt. Bradley deployed the emergency fire extinguisher for the blazing engine while Lt. Jones cut the switches. Lt. Bradley then ordered Pfc. Johnson to get out of the airplane and stand by with a fire extinguisher. The engineer climbed out of the airplane through the rear hatch and proceeded forward in a crouched position. He then passed under the port wing on the outside of the port engine where he was struck in the head by the port propeller, which was still rotating. Pfc. Johnson was killed instantly.

6-5-43A. Pecos, Texas. At 0747, two Vultee BT-13A aircraft collided in mid-air three miles southeast of Pecos Army Air Field, Pecos, Texas, killing 2Lt. Robert F. Heinrich aboard BT-13A #41-22929. 2Lt. John P. Trowbridge, flying BT-13A #41-9868, was uninjured and able to land his damaged airplane at Pecos. Investigators stated, "Both officers took off from Pecos Army Air Field at 0740 in a group of seven airplanes to conduct a search within the local flying area for a student that was lost on a night cross-country flight. The airplanes were to conduct the search in echelon to the right and spaced 500 yards apart. After take-off a turn was made to the left followed by a turn to the right until approximately 4,300 feet indicated. Lt. Trowbridge leveled off to take up his position for the search. The two airplanes collided just as Lt. Trowbridge leveled off with Lt. Heinrich's airplane in a climbing position, the tail section of this airplane striking the propeller and nose section of Lt. Trowbridge's airplane. Half of the vertical stabilizer and rudder of Lt. Heinrich's airplane were sheared off and the airplane plunged vertically toward the ground without spinning until it crashed."

6-5-43B. Turner Field, Georgia. At 0930 CWT, a Beech AT-10 crashed five miles northeast of Turner Field, Albany, Georgia, killing pilot 2Lt. Clark Espenship, Highland Park, New Jersey, and co-pilot 1Lt. John L. Mood, Fort Worth, Texas. The airplane had just taken off from Turner Field on a routine training mission when the starboard engine failed and caught fire. An improperly installed starboard engine gauge unit capillary tube became entangled in the starboard landing gear causing the starboard engine to fail and catch on fire when the landing gear was retracted. Investigators stated, "It is the opinion of the committee that this landing gear drive shaft became entangled with the extra length of capillary tube which in turn pulled the entire group of engine controls into the revolving drive shaft and at the same time pulled the gasoline line off, allowing gasoline to run out of this section of the wing. This gasoline was ignited by the exhaust pipe, which is directly under this position. This, also, caused the engine to stop. This all occurred at the time the landing gear drive shaft was turning when the landing gear was being raised."

6-5-43C. Charlotte Harbor, Florida. At 1045 EWT, a Curtiss P-40L crashed at Charlotte Harbor, Florida, killing pilot 2Lt. Homer D. Herring. The Accident Classification Committee stated, "Witnesses stated that the pilot was leading the flight on one leg of a low altitude cross-country [navigation training flight], when he went down to buzz a boat in Charlotte Harbor, against the orders of the flight leader. The pilot pulled up over the boat and crashed into the water on the other side. The plane bounced back into the air, did two rolls, caught fire and crashed nose-first into the water."

6-5-43D. Atlantic Ocean, South Carolina. At about 1100, a North American B-25D on an upper turret gunnery mission crashed into the Atlantic Ocean five miles off shore of Myrtle Beach, South Carolina, killing eight fliers. The airplane had taken off from Myrtle Beach Army Air Base at 1015 EWT. The bodies of pilot 2Lt. William C. Fallon and navigator 2Lt. James R. Carney were recovered soon after the accident. Missing and presumed lost at sea were: 2Lt. Chester L. Cook, co-pilot; SSgt. Richard E. Muller, engineer; SSgt. Walter W. Buege, radio operator; SSgt. Robert J. Burns, gunner; SSgt. Carl Hurin, gunnery instructor; 2Lt. Marvin R. Moore, pilot-rated passenger. Although a significant amount of wreckage was recovered from the sea, investigators could not determine the cause of the accident.

6-5-43E. Mather Field, California. At 1000, a North American B-25D crashed six miles south of Mather Field, Sacramento, California, killing pilot Maj. Roy Gilbert, co-pilot 2Lt. Major H. Marsh and engineer Sgt. Martin J. Lysaught. Investigators stated, "Witnesses testified that the left engine was turning over very slowly. Just prior to the accident, the right engine seemed to be operating normally. When at an altitude of between 500 and 1,000 feet, the airplane was seen to nose up and then spin to the right. It made several turns before striking the ground, where it crashed and burned. No attempt was made to parachute to safety. Investigation of the wreckage indicates that the airplane had recovered from the spin sufficiently to stop turning just prior to striking the ground. Both ignition switches were in the 'on' position. The airplane was apparently trimmed for single-engine procedure. Neither propeller was feathered. It is the opinion of this board that the instructor had given the student a practice single-engine condition by pulling back on the throttle on the left engine." The airplane had taken off from Mather Field on a routine training flight.

6-5-43F. Orlando, Florida. At 1409 EWT, a Lockheed P-38G crashed into two houses west of Orlando Army Air Base, Orlando, Florida, after suffering an apparent engine failure on take off, killing four civilians and seriously injuring two civilians. AAF pilot 2Lt. Ellis I. DeGroff was seriously injured. The P-38 had been scheduled for a cross-country navigation flight but that was canceled so the P-38 could take part in a photographic mission with a Bell P-39, a Curtiss P-40, a Republic P-47, a North American P-51, and a Douglas P-70. The pilot of the P-38 had experienced some overheating in the starboard engine prior to take-off. Personnel had attempted to prevent Lt. DeGroff's take-off because of an apparent fueling error. Despite several attempts, tower and operational personnel were unable to contact the pilot by radio or with light signals before he began his take-off run. The pilot thought that he had heard the tower transmit "Cleared to take off" and he began his take-off on the southernmost East-West runway. The starboard engine failed to produce sufficient power and about halfway down the 5,300-foot runway the pilot throttled back on the port engine. He then pushed both throttles to full and attempted to pull the P-38 into the air. Unable to take off, the pilot then cut both throttles and attempted to use his brakes, blowing out his port main tire. Investigators stated, "The airplane crashed through the boundary fence at the western end of the runway, crossed a road, and crashed into a house about 100 feet from the fence, and after running completely through the first house, ran through the wall of a second house and burst into flames. Here the pilot escaped from the burning wreckage. Fire completely destroyed the two houses and partially destroyed a third." Civilians killed on the ground were: Betty Lou Giddens, age 14, Orlando; Elizabeth Campbell, age 54, Orlando; William Fred Smothers, age 52, Orlando; Rueben D. Foreman, age 35, Winter Garden. Seriously injured were Carol Margie Giddens, age 12, Orlando, and Mr. Dewey Pickett, age unknown, Orlando. The Smothers' house (140 Crystal Lake Drive) was struck first, the roof totally collapsing, and Mr. Smothers was observed being struck by the P-38 in the front yard as he tried to escape the path of the careening fighter. Investigators stated that although the fuel pumped into the P-38 prior to the flight was an experimental high-performance grade, it was not thought to have caused the failure of the starboard engine. The cause of the engine failure was not determined.

6-5-43G. Jersey, Georgia. At 1545 EWT, a Martin B-26B crashed while maneuvering at low altitude near Jersey, Georgia, killing the crew of six. The airplane had been cleared on a navigation-training mission from Myrtle Beach, South Carolina, to Atlanta, Georgia. The airplane was observed to pass beneath high-voltage wires near Bostwick, Georgia. Investigators stated, "The pilot had been flying at a low altitude, estimated by various observers at from 20 to 50 feet for at least eight miles prior to the point of the crash. Immediately prior to the accident the ship was flying up a slight draw and passed over a house one-eighth of a mile behind which was hill. The tail and both propellers of the plane struck the hill 150 feet from the top. The ship bounced into the air and the left engine stopped. Parts of the airplane were strewn from this point to the point [where the airplane came to rest]. The plane hit the ground again 800 feet from the first point of impact and right engine stopped. Ship bounced again and skidded along the ground for 770 feet and the fuselage broke just forward of the upper gun turret, 25 feet before the point of rest. Both engines apparently left the ship at this point and the wings and the [remaining] fuselage then made a 210-foot turn to the right and came to rest, on fire. Ship caught fire after first bounce and gasoline exploded when it hit the second time. Witnesses state that there was a change in the attitude of the airplane just prior to the accident, indicating that the pilot tried to gain altitude suddenly, which, connected with the first marks on the ground when the ship hit, leads to the belief that the ship mushed into the hill." Killed in the crash were: 2Lt. George R. Barkhurst, Dallas, Texas, pilot; 2Lt. Phillip J. Props, Lafayette, Indiana, co-pilot; 1Lt. Benjamin L. Holmes, San Francisco, California, navigator/bombardier; SSgt. William G. Maki, Atlantic Mine, Michigan, engineer; SSgt. William J. Maurer, Spencer, Iowa, radio operator; Cpl. Leibel I. Engel, Indianapolis, Indiana, crewmember. A carcass of a large German Shepard type dog was found near the wreckage and was thought to have been on board the B-26 at the time of the crash. One crewmember was thrown clear of the wreck.

6-5-43H. Havana, Florida. At 1715 EWT, a Republic P-47D dove into the ground and exploded violently into flames four miles northwest of Havana, Florida, killing pilot 2Lt. John F. Dooley. The airplane was in the number-four position number of a four-ship flight that had taken off from Dale Mabry Field, Tallahassee, Florida, on a high-altitude formation mission. The formation had leveled out at 30,000 feet when the subject airplane was observed to suddenly leave the formation. The flight leader called the pilot in but received no reply. The P-47 then entered a steep dive and crashed into the ground at high speed, destroying the airplane utterly and killing the pilot, who made no effort to leave the airplane. Investigators speculated that the pilot was suffering from oxygen starvation and inadvertently dove the airplane into the ground.

6-5-43I. St. Anthony, Idaho. At 1600 MWT, a Douglas A-20G crashed 12 miles north of St. Anthony, Idaho, killing pilot 2Lt. Hubert P. Hampton, Ulysses, Kansas. The A-20 was on a ferry flight from Daggett, California, to Great Falls, Montana, via Pocatello, Idaho. The airplane was observed by civilians in the area to descend from the overcast in a spin. The A-20 was then seen to apparently recover from the initial spin, climb steeply, and then to enter another spin from which the pilot was unable to recover. The airplane did not burn when it struck the ground.

6-5-43J. Terrell, Texas. At 2130, a Vultee BT-13A on a night navigation flight dove into the ground one and one-half miles southwest of Terrell, Texas, killing pilot 2Lt. Truman S. Crabtree and passenger Sgt. Thomas H. Patterson. The airplane was part of a seven-ship flight that had taken off from Beaumont, Texas, on a day navigation flight to Majors Field, Greenville, Texas. Investigators speculated that the airplane had probably encountered unfavorable weather conditions, reported as thunderstorms with heavy rains, high winds and poor visibility. Four airplanes in the flight landed at Love Field, Dallas, Texas, one landed at Tarrant Field, Fort Worth, Texas, and one pilot managed to get back to Majors Field. Investigation revealed that the airplane dove into the ground at a steep angle. RAF flying cadets based at Terrell attempted to find the wreck that evening but they were driven back by the intense downpour.

6-5-43K. Texline, Texas. At 2345, a Boeing B-17F suffering top turret fire was abandoned and crashed ten miles south of Texline, Texas, killing five crewmembers. Four crewmembers were able to parachute to safety; one was seriously injured while two others received light injuries and another was uninjured. The airplane, which was on a simulated bombing mission, was flying at 14,000 feet when a fire broke out on the port side of the base of the upper turret. The fire became so intense that the flames ignited the powder charges in the practice bombs that the airplane was carrying. The B-17 was seen by witnesses on the ground to descend into the terrain trailing about forty feet of flames. This fire was the first of several fires aboard B-17F type aircraft in the United States and overseas in which the wiring in the base of the upper turret becomes chafed, causing short circuits in the junction box of the top turret electrical system. The electrical arcs that were produced compromised the oxygen line that passed through the same part of the turret system junction box at the base of the turret, causing oxygen to escape from the line. The electrical arcing combined with the oxygen leak in the base of the top turret caused the intense fire. Surviving crewmembers reported hearing three explosions prior to exiting through the rear door. It is likely that the pilots were overcome by fumes and smoke and were unable to control the airplane or parachute to safety. Killed in the crash were: 2Lt. Edgar C. Hanks, pilot; 2Lt. Phillip R. Sallee, co-pilot; 2Lt. Harlan S. Severson, bombardier; SSgt. Richard L. Waters, engineer; SSgt. Edward M. Profozich, gunner. Assistant engineer Sgt. Frederick Babb was seriously injured; Radio operators SSgt. Oscar A. Nelson and Sgt. Charles R. Patterson received light injuries and gunner Stephen A. Wedge was uninjured.

6-6-43A. Pyote, Texas. At 0156, a Boeing B-17F suffering a fire in the number-three engine crashed five miles south of Pyote Army Air Base, Pyote, Texas, killing eight crewmembers. Tail gunner Sgt. Kennard C. Stevens miraculously survived but was seriously injured. All airplanes in the area were called in because of deteriorating weather conditions and at 0155 the subject airplane was given landing instructions after it had called Pyote AAB control tower. A moment later the number-three engine burst into flames. Sgt. Stevens stated that he had seen that the number-three engine was on fire as the B-17 entered the pattern for the airbase. The pilot ordered the crew to secure their parachutes and to stand by. As the crew was securing their parachutes, the airplane dove into the ground at a steep angle. Witnesses on the ground observed the starboard engine and wing ablaze as the B-17 went in, starboard wing down. Wreckage and bodies were strewn over a wide area. Sgt. Stevens came to lying next to the tail cone and waited for rescuers to come to his aid. Killed in the crash were: 2Lt. Joseph C. Folsom, Jr., pilot; 2Lt. Norman P. Beck, co-pilot; 2Lt. Edwin E. Curran, bombardier; Sgt. Allen Y. Mumford, engineer; Pvt. Edward W. Trapp, assistant engineer; Sgt. David Lachter, assistant radio operator; Sgt. Robert L. Been, assistant radio operator; Alphonse M. Kwedar, ball turret gunner.

6-6-43B. Marfa, Texas. At 1030, two Cessna UC-78B airplanes collided in mid-air near Marfa, Texas, killing A/C John W. Yeoman and A/C John R. Ray aboard UC-78 #544. A/C Charles F. Youmans and A/C Paul J. Stahle, aboard UC-78 #773, received minor injuries when they made a wheels-up landing near Marfa Army Air Field. After making four landings at Marfa, A/C Stahle took over from A/C Youmans and performed a take-off. They climbed to 800 feet and made their first turn in the pattern when airplane #544 descended on top of them, the starboard engine nacelle of #544 contacting the upper fuselage of #773. The windscreen and part of the nose of #773 had been ripped away in the collision and A/C Stahle was knocked out momentarily. Youmans gained control of the airplane and was able to crash land two miles southeast of the field. Aircraft #544 plunged to the ground and exploded one and one-half miles southeast of the field.

6-6-43C. Hollywood, California. At 1120, a Lockheed P-38G crashed into a house at 4443 Strohm Avenue, Hollywood, California, killing pilot 2Lt. Leon D. Mitchell and civilian resident Mrs. Myrtle Baker. At 1115, Lt. Mitchell was observed making low altitude passes from east to west and west to east near Riverside Drive and Cahuenga Boulevard. Published accounts of the crash stated that the P-38 had circled the area three times and had dove on "mythical targets" on three occasions, pulling up just above the treetops each time. The airplane was reported to have flown so low that it "shook the buildings in the vicinity." On the fourth diving maneuver, the pilot apparently misjudged his altitude or was unable to pull out of the dive in time to avoid colliding with a row of eucalyptus trees. The airplane then collided with the houses of Arthur Horn and A.C. Simels, shearing off the roof corners of both houses before striking the Simels' backyard barbeque pit. The P-38 then careened through the Baker garage and exploded violently into flames. The force of the collision and explosion hurled the Baker automobile against the back of the Baker home, killing Mrs. Baker and igniting a huge fire. Mrs. Baker, who apparently was standing on the service porch watching the low flying, was partially cremated in the resultant fire. In all, eight houses were damaged or destroyed.

6-6-43D. Gulf of Mexico, Florida. At 1615 EWT, a Bell P-39D crashed into the Gulf of Mexico near Captiva Pass, Florida, killing pilot 2Lt. Edward C. Leach. 2Lt. Richard J. Love witnessed the crash and later stated to investigators, "Lt. Leach, in the lead ship, had made two skip bombing runs and was on his third. He passed over the target toward the Gulf at an approximate altitude of 75-100 feet, dropped his bomb and started to chandelle up to the right. His bomb did hit the water, and while he was chandelling to the right from a bank of about 70 degrees and an altitude of about 250 feet, it was noted that the plane snapped or was knocked about 45 degrees to the right, almost on his back. Lt. Leach pointed the nose down slightly and leveled his wings and then the right wing dropped again. The plane hit the water with the right wing first in what appeared small slip to the right. The plane also appeared to burst into flames a brief second before striking the water. The plane sunk immediately and did not float."

6-7-43A. Maxton, North Carolina. At 0400, a Douglas C-47A flying in poor weather crashed four miles northeast of Laurinburg-Maxton Army Air Base, Maxton, North Carolina, killing four crewmembers and 16 glider pilots. The four crewmembers killed in the crash were: 2Lt. Lendrum A. Foster, pilot; 2Lt. Leslie J. Stephens, co-pilot; Pfc. Anton J. Hirschman, engineer; Pfc. Donald W. Neal, radio operator. Sixteen rated glider pilots killed in the crash were: 1Lt. Joseph L. Duckworth, F/O Everett Bennett, F/O Robert W. Burns, F/O William M. Douglass, F/O Albert C. Hunter, F/O Thomas J. Jones, F/O Christ Kalamatas, F/O Victor C. Lebsock, F/O Martin Levine, F/O Edgar E. Llewellyn, F/O Lawrence B. Oberg, F/O William E. Odell, F/O Douglas C. Oland, F/O Edward W. Orton, F/O Zail E. Roberts, F/O James L. Williamson. The airplane had taken off from Pope Field, Fort Bragg, North Carolina, at 0339 EWT on a flight to transport the glider pilots to Laurinburg-Maxton AAB and encountered a powerful line of thunderstorms upon arrival. The storms included heavy rain and "terrific" lightning. The C-47, trying to find the air base in the severe thunderstorms, was flying under the ceiling about 200 feet above ground level. The pilot apparently attempted a right turn at very low altitude, causing the starboard wing to strike the terrain. The airplane smashed in on its starboard wing and engine, cartwheeling across the ground and exploding into flames. Investigators speculated that the pilot was unaware of his proximity to the ground when he attempted the turn. An airplane that had departed Pope Field five minutes after the accident airplane returned to Pope Field after encountering the storms.

6-7-43B. Paris, Texas. At 1100, a Vultee BT-13A crashed six miles northwest of Paris, Texas, killing A/C Joseph C. MacCabe, Greenville, Texas, and seriously injuring A/C Kenneth MacKenzie, Portland, Oregon. The airplane was on a scheduled instrument team ride from Majors Field, Greenville, Texas, to Mt. Vernon and back to Majors Field. Instead of turning at Paris for a course to Mt. Vernon, the cadets began buzzing a small lake. After a series of dives and steep pull-outs, A/C MacCabe attempted a steep left turn at low altitude, the port wing striking the terrain and causing the airplane to cartwheel across the ground and explode violently into flames. A/C MacKenzie was pulled from the burning wreckage by local youths who had been fishing at the lake.

6-7-43C. Garden City, Kansas. At 1408 MWT, a Vultee BT-13A crashed 25 miles north of Garden City, Kansas, killing A/C Albert E. Butcher and A/C Charles W. Collier. The two students were on an instrument team ride from Garden City Army Air Field to Lake McKinney, to Friend, Kansas, and return to Garden City. Witnesses on the ground stated that the airplane was flying at low altitude, making steep turns and buzzing an 85-foot oil derrick just before the accident. The pilot attempted a steep right turn at low altitude, the starboard wing striking the terrain and causing the airplane to cartwheel across the ground and explode violently into flames.

6-7-43D. Gulf of Mexico, Florida. At 1730 CWT, a North American AT-6A on a gunnery-training mission crashed into the Gulf of Mexico, 12 miles south of Apalachicola, Florida, near St. George Island, killing pilot 1Lt. William A. Finney and passenger Pfc. Joseph P. Yurich. A pilot flying nearby reported that the airplane had been flying straight and level at an altitude of 800 feet when the nose suddenly rose and the airplane began a steep climb. The airplane then fell off into a spin and crashed into about six feet of water near the north shore of St. George Island in St. George Sound.

6-7-43E. Gardner Field, California. At 1550, a Vultee BT-13A crashed ten miles northeast of Gardner Field, Taft, California, killing instructor 2Lt. Erik M. Olleson and student A/C George R. Shirey. Investigators stated, "At 1550 (established by the stopped watch of the dead pilot) this airplane crashed, killing both occupants. The airplane struck the ground in a vertical dive, apparently not spinning, for there were no marks on the ground to indicate it. The condition of the wreckage indicates that the airplane was not moving at a terrific rate of speed at the time of impact."

6-7-43F. Avenger Field, Texas. At 2310, a Vultee BT-15 on a night cross country navigation training flight flew into the ground 40 miles west of Avenger Field, Sweetwater, Texas, killing civilian instructor Henry S. Aubrey and WASP Flying Cadet Jane Champlin. Investigators stated, "The student was making a turn around the town and allowed the turn to get too steep. Both the student and the instructor then lost the horizon and allowed the ship to get into a spiral. The instructor failed to go on instruments and prevent the crash."

6-8-43A. March Field, California. At 0045, a North American B-25D crashed two miles west of March Field, Riverside, California, killing four crewmembers. The airplane had arrived at March Field with another B-25 at 2350 on 6-7-43, flying in from Mather Field, Sacramento, California, on a night navigation training mission. The subject airplane took off first for the return trip to Mather Field and soon after take-off encountered fog and low-level clouds, which were about 1,000 feet thick, at about 100 feet above ground level. Investigators speculated that the pilot apparently became disoriented and flew the airplane into the ground while in a descending left turn. Killed in the crash were: 2Lt. Leroy E. Webb, Sweetwater, Texas, B-25 instructor pilot; 2Lt. Robert L. Alexandar, Pecos,

Texas, co-pilot/B-25 student; SSgt. Aulton A. Collins, engineer; SSgt. Jim T. Cauley, radio operator. SSgt. Cauley was originally scheduled to fly aboard the other B-25 (41-13207), which also encountered the fog but was able to fly through and land safely at Mather Field at 0245. Investigators were not certain whether the instructor or the student was flying the airplane at the time of the accident.

6-8-43B. Ochlocknee, Georgia. At 0815 EWT, a Bell P-39N crashed at Ochlocknee, Georgia, killing 2Lt. Ralph L. Nutter. The pilot took off on a transition flight at 0800 and climbed to 10,000 feet. The P-39 was observed to execute a lazy-eight, the pilot attempting to split-S out of the maneuver. The airplane stalled out of the maneuver and entered a series of high-speed stalls that changed from left to right. At about 5,000 feet, the airplane entered a very slow turning flat spin from which it did not recover.

6-8-43C. Camp Gordon Johnston, Florida. At 0915 EWT, a North American P-51A crashed to the ground and exploded violently into flames 100 yards from the motor pool at Camp Gordon Johnston, Florida, killing pilot 2Lt. Russell Love. The airplane was observed performing barrel rolls and loops just prior to the accident. Witnesses reported that a light trail of "grayish" smoke was trailing from the airplane's starboard side just prior to the crash. Investigators stated, "From the statements of witnesses, it seems that the pilot was doing aerobatics at low altitude, lost control of the airplane and crashed. The plane was so badly damaged that a complete investigation of the engine and airframe could not be made. Pilot was scheduled for an instrument mission and should not have been over the gulf doing aerobatics."

6-8-43D. Post, Texas. At 0920, a Cessna UC-78B on a cross-country navigation flight crashed 18 miles east of Post, Texas, fatally injuring A/C Arthur Frank DeLooff and seriously injuring A/C Leroy J. Decuir. Investigation revealed that the airplane had been flying below 100 feet for 10 miles prior to colliding with a stand of mesquite trees about 10 feet tall. The airplane smashed into the ground about 150 yards further on, bounced into the air about 40 yards before striking the ground sideways, collapsing the starboard wing and rolling inverted, skidding along the ground for another 100 yards.

6-8-43E. Camp Campbell, Kentucky. At 1215 CWT, a Bell P-39N crashed while attempting a landing at Campbell Army Air Field, Camp Campbell, Kentucky, killing pilot 2Lt. Norman Hermansen. The airplane was part of a flight of P-39 airplanes that had been escorting B-25 aircraft on a simulated bombing mission. Investigators stated, "Lt. Hermansen was cleared to land by the control tower. When he turned onto his approach leg he apparently was not squared away with the runway, but rather than go around, due to the fact that there were two C-47s towing gliders in the pattern, Lt. Hermansen, according to his own statement, decided to tighten his turn and land; the ship mushed into the ground, caught fire and burned." The pilot died of third degree burns in the post hospital soon after making a statement to investigators.

6-8-43F. Eureka, Texas. At 1415, a Fairchild PT-19A crashed three miles south of Eureka, Texas, killing A/C Edward Kovitz. The airplane had been flying at an altitude of about 3,000 feet above ground level when it entered a nose down attitude followed by a roll to the right. The airplane then entered a full spin to the right. The airplane continued in a vertical spin to the ground, the pilot stopping the spin an instant before the airplane struck the ground in a vertical attitude.

6-8-43G. Amado, Arizona. At 1345 MWT, a Ryan PT-22 on a day navigation training flight collided with terrain while maneuvering seven miles west of Amado, Arizona, killing civilian flight instructor Donald Shaw, Tucson, Arizona, and seriously injuring A/C Eugene V. Marshall. The Accident Classification Committee stated, "It is evident from the witness statements and the actual examination of all evidence that the instructor was engaged in hedge-hopping and other low altitude maneuvers prior to and at the time of the accident. Witness statements indicate that the plane had made at least one previous pass at the field and at the time of the accident the pilot misjudged his altitude, the wheels coming in contact with the soft-plowed ground, causing the ship to nose in 48 yards from the point of impact. From this spot the ship evidently somersaulted several times through heavy mesquite growth at the end of the field, coming to rest 123 yards from the first point of contact with the ground."

6-8-43H. Sherman, Texas. At 1615, a Vultee BT-13A crashed while attempting a take-off at Gaskin Field, 12 miles southwest of Sherman, Texas, killing A/C Grant W. Fiege and seriously injuring instructor 1Lt. Howard W. Bobo. Investigators stated, "Cadet Fiege started to take off on the North-South runway to the south. The Cadet failed to have the propeller in high pitch so Lt. Bobo cut the throttle and had the cadet taxi back and start again. Lt. Bobo states that on the second attempt the airplane took off and had climbed approximately 15 to 20 feet above the ground. At this point the left wing suddenly dropped and the airplane started to swerve to the left.... The airplane continued to swerve to the left until the left wing tip hooked into the ground and the airplane continued over on its back. It is believed that Cadet Fiege would have survived this crash except for the fact that his seat, which was of the new wooden construction, broke loose from its fittings throwing him forward into the wreckage of the motor."

6-8-43I. Merced, California. At 1525, a Vultee BT-13A crashed nine miles northeast of Merced Army Air Field, Merced, California, killing A/C Ralph H. Smith. Witnesses in the air stated that they had seen the BT-13 in a spin at approximately 7,000 feet. The airplane continued in a slow spin to the left until it hit the ground. There were no obvious attempts by the pilot to recover from the spin.

6-8-43J. Jourdanton, Texas. At 1800, a Cessna AT-17B and a Cessna UC-78B collided in mid-air five miles south of Jourdanton, Texas, killing A/C James W. Pate aboard the AT-17B. A/C John L. Murtha was uninjured and safely landed his damaged UC-78B at Brooks Field, San Antonio, Texas. Investigators stated, "A five ship formation, led by an instructor, had divided, three ships flying cloud formation and two ships [Pate and Murtha] flying ordinary formation in a stand-by area. The instructor, having finished the cloud formation flying, called for the two ships to join his formation and return to Brooks Field. A/C Murtha started to comply and made a gentle turn, with A/C Pate on his right wing. A/C Pate flew a very tight formation and collided with A/C Murtha's ship's elevator surfaces, damaging [them] slightly. The impact evidently damaged A/C Pate's ship sufficiently to throw it into a spin. A/C Pate's ship spun down several thousand feet then dove into the ground. Evidence is available that A/C Pate had been repeatedly warned against flying too tight a formation in the AT-17. [It is recommended that] cadets are warned against flying too tight a formation in AT-17 airplanes."

6-8-43K. San Pablo Bay, California. At 2057, a North American B-25D crashed into San Pablo Bay, five miles southeast of Hamilton Field, San Rafael, California, killing four crewmembers. The airplane had just taken off from Hamilton Field on a ferry flight to Hickam Field, Hawaii. The airplane was seen to enter a steep climb just after take-off, disappearing into low-level clouds. Personnel in the control tower at Hamilton Field stated that moments later they observed a huge explosion on the bay about five miles southeast of the end of the runway. Killed in the crash were: 1Lt. Harmon W. Honsinger, service pilot; 2Lt. William B. Monk, co-pilot; 2Lt. Homer L. Byrd, navigator; Pfc. Ernest P. Colburn, radio operator.

6-9-43A. Lambert Field, Missouri. At 1420, a Boeing B-17F crashed while attempting a landing at Lambert Field, St. Louis, Missouri, seriously injuring seven crew and passengers and fatally injuring radio operator SSgt. Raymond C. Hedrick, who died at 0745 on 6-11-43. The B-17 had been cleared from Lowry Field, Colorado, to Bolling Field, Washington, D.C., on a "Flight in connection with AAF photo matters." The pilot had been flying at 9,000 feet on instruments when the he noticed excessive fuel consumption during the first two hours of flight. The pilot elected to land at Lambert Field to have the airplane serviced even though Lambert Field was officially closed to "Army traffic except on official business by command of Lt. General H.H. Arnold." Upon arriving at the field the pilot overshot the landing attempt, coming in too high and too fast. In making contact with the runway, the B-17 bounded through the air several times, using up a considerable portion of the runway. Once on the ground, the pilot attempted to use the brakes but was unable to prevent the B-17 from running off the end of the runway and colliding with a ditch. The airplane did not burn but was completely destroyed. Seriously injured in the crash were: Col. Paul T. Cullen, pilot; 2Lt. Maurice D. Lee, co-pilot; 1Lt. Manner O. Clements, navigator; TSgt. John F. Buecker, engineer; 2Lt. H.E. Oberg, passenger; 2Lt. R.J. Borben, passenger; Sgt. A.B. Jenemore, passenger.

6-9-43B. Broome Auxiliary Airfield, Texas. At 1440, a Vultee BT-13A stalled and spun to the ground while turning from base to final at Broome Auxiliary Airfield, Texas, killing instructor 2Lt. Warden H. King, Jr. and student A/C Joseph W. Karklus. Investigators stated, "The airplane stalled at an altitude of approximately 450 feet, and the wings were in a vertical position with the right wing low. The airplane fell out of this position, slipping to the right, with the nose down. The wings were level, and the nose was slightly down, but the airplane crashed into the ground in a 'mushy' dive." The student and instructor were based at Goodfellow Field, San Angelo, Texas, six miles north of the accident site.

6-9-43C. Gardner Field, California. At 1400, a Vultee BT-13A spun to the ground 12 miles ENE of Gardner Field, Taft, California, killing A/C George E. Bialas, Chicago, Illinois. The Accident Classification Committee stated, "A/C Bialas had been sent out for a local solo practice flight to work on climbing turns, stalls, and gliding turns. No one observed his air work and no one observed the start of his spin. However, one solo cadet working at 3,500 feet in the same area where the crash occurred did see an airplane spin past him at approximately 1400. The airplane was so close that he interrupted his own work to get out of the way and then tried to locate the spinning airplane again. He could not see it and so dismissed it from his mind and went about his own work. Two witnesses on the ground about one and one-half miles from the spot where the airplane crashed noticed airplane 41-10907 a few hundred feet above the ground, saw it make two turns of a spin and strike the ground. It is the opinion of this committee that A/C Bialas fell into a spin from his stall practice, became frightened and excited, and was unable to effect a recovery. All control cables were intact and control surfaces still mounted and moved freely after the crash. No material failure is indicated."

6-9-43D. Venice, Florida. At 1745, a Bell P-39N on an instrument training flight suffered a catastrophic structural failure and crashed near Venice Army Air Base, Venice, Florida, killing pilot 2Lt. John W. Ballas. Investigators stated, "Lt. Ballas entered the thunderhead cloud formation at approximately 10,000 feet altitude. The committee is of the opinion that violent turbulence was encountered. The excessive speed of the plane in a dive plus the turbulent air caused structural failure of the tail. Noting the loss of control of his ship, Lt. Ballas decided to jump. Lt. Ballas left the plane and his parachute canopy caught on the ragged skin of the tail section. The parachute tore loose and Lt. Ballas descended to the ground with his parachute open. The pilot's back was broken causing instant death. The tail section of the airplane tore off and fell intact approximately one-quarter of a mile from the main wreckage."

6-9-43E. Graton, California. At 1900, a Bell P-39N entered a flat spin and crashed eight miles west of Graton, California, killing pilot 2Lt. Virgil E. Wyss. The pilot attempted to parachute to safety but left the airplane at too low an altitude to allow his parachute to deploy. The airplane was seen to oscillate violently along its pitch axis as the pilot struggled to break the spin. At an altitude of approximately 300 feet, the escape door was seen to leave the airplane followed by the pilot, who struck the ground near where the airplane crashed.

6-9-43F. Villaldma, Mexico. At a time unknown, a multi-engine AAF airplane was abandoned by its crew near Villaldma, Mexico, after becoming lost and out of fuel, killing pilot 2Lt. Ernest R. Blakely, Woodbridge, California, and crewmember Sgt. Charles V. Kane, Sterling, Illinois. Two unnamed crewmembers parachuted to safety. The airplane, which had taken off from Esler Field, Alexandria, Louisiana, was scheduled to terminate its mission in the U.S. and its flight plan did not include flight over Mexico.

6-10-43A. Weed, California. At 0240, a Consolidated B-24E on a night navigation mission crashed at Weed, California, killing six fliers. Four crewmembers parachuted to safety and were uninjured. Killed in the crash were: F/O George Churchwell, co-pilot; 2Lt. George H. Clarke, navigator; SSgt. Willet F. Draker, engineer; SSgt. William W. Powell, assistant engineer; SSgt. Alonzia E. Johnston, gunner; SSgt. John F. Kennedy, gunner. Parachuting to safety were: 1Lt. Douglas Thornburg, B-24 instructor (see 7-5-43D); 2Lt. Alfred J. Demeusy, pilot; 2Lt. Charles L. Weist, bombardier; SSgt. Alvin Dschaak, assistant radio operator. After a considerable delay on the ground due to leaking hydraulics and radio problems, the B-24 took off from Salinas Army Air Base, California, at 2350 (6-9-43) on a flight to Sacramento, Scotia, Redding, Stockton, Bakersfield and return to Salinas with Fresno as an alternate. There were two B-24 student pilots and a B-24 instructor on the flight deck. The flight was normal until the airplane reached an area north of Redding when the pilots noticed a drop in rpm. Thinking that they were picking up carburetor ice, they shut the inter-cooler doors. The rpm increased but a few minutes later the pilots noticed an oscillation of the number-two engine oil pressure gauge. The pressure dropped to zero and then the engine quit. The pilots attempted to feather the engine without success. The pilots noticed blue sparks behind the feather buttons when they were engaged. The number-two engine was finally feathered when the pilots noticed that the number-one propeller feathered itself. The pilots attempted to unfeather the engine but the blades turned "backwards." The instructor ordered the crew to secure their parachutes and prepare to bail out. Soon after this order was given, the airplane lost all electrical power and instruments. Sgt. Dschaak used a flashlight to signal the crewmembers in the after part of the ship to bail out. The co-pilot, F/O Churchwell, was piloting the airplane and agreed to hold the airplane while the crew bailed out. Thinking that the rest of the crew had bailed out, Lt. Demeusy and Lt. Thornburg left the ship. The airplane crashed seconds later when the co-pilot apparently attempted to leave the ship. Investigation revealed that SSgt. Kennedy had deployed his parachute prematurely and it became fouled in the escape hatch, preventing him from leaving the airplane. It was speculated that the other crewmembers were trying to free SSgt. Kennedy when the airplane slammed into the ground. Lt. Weist and SSgt. Dschaak landed near the wreckage and, with the help of civilians who were on the scene, pulled SSgt. Kennedy's body free of the burning wreckage.

6-10-43B. Moody Field, Georgia. At 1100 EWT, a Beech AT-10BH crashed one and one-half miles southeast of Moody Field, Valdosta, Georgia, killing instructor 2Lt. Paul J. Durbin and student A/C Albert J. Baehr. The airplane had been practicing single-engine landings when the pilot overshot the landing. He increased the throttle for a go-around, retracted the landing gear and was observed to climb straight ahead. The airplane was then seen to make a left turn at an altitude of about 400 feet above ground level. The turn continued beyond 90 degrees and the port wing was seen to drop. Witnesses reported that the port wing dropped and then the starboard wing dropped. The airplane was near stalling speed and the pilot was observed using the rudder to lift the wings as they alternately dropped. Just before striking the ground in a wooded area, the engines were heard to "rev up" and then to stop altogether just before impact. The airplane exploded into flames. Investigators stated, "It is the opinion of the committee that while attempting to switch from single engine operation to both engines, the pilot turned the engine selector [fuel] valve to the 'off' position and noticed his mistake as the engines stopped. There was not sufficient altitude to allow the engines to start again in time to continue flight."

6-10-43C. Camp Beale, California. At 0835, a Bell P-39N on a low-level formation strafing mission collided with a hill and crashed one-half mile east of Camp Beale, Marysville, California, killing 2Lt. Eugene A. Seagren. The formation was scheduled to strafe tanks and half-tracks at the Camp Beale range. Lt. Seagren was observed flying lower than his fight leader and had just strafed one tank at extremely low level. He began a slight climbing turn to the right in order to make another attack, striking the side of the top of the hill with his starboard wing. The airplane bounded into the air and was observed to climb in what appeared to be a "slow roll," rolling onto its back and stalling at 200 feet above ground level before plunging to the ground and exploding into flames.

6-10-43D. Trenton, Florida. At 1500 EWT, a Republic P-47G on an altitude navigation flight crashed one and one-half miles west of Trenton, Florida, killing pilot 2Lt. Robert S. Coleman. It was speculated by investigators that the pilot became disoriented after entering a thick rain cloud. The airplane was observed diving out of the cloud base, estimated to be approximately 1,500 to 2,000 feet, at an angle of about 80 degrees. The pilot attempted to pull the airplane out of the dive, but the altitude was insufficient and the P-47 continued into the ground at high speed at an angle of approximately 30 degrees, exploding violently into flames. Investigators noted that the airplane wreckage was smashed into a hole approximately 18 feet deep and 30 feet wide.

6-10-43E. Madill, Oklahoma. At 1615, a Vultee BT-13A crashed 16 miles southwest of Madill, Oklahoma, killing A/C Randolph Glance. It was speculated that the airplane hit the ground in a flat spin. Investigators could not determine if the spin was a power-off or power-on type.

6-10-43F. Luray, South Carolina. At 1915, a North American B-25D crashed one-half mile southeast of Luray, South Carolina, killing six fliers. The airplane was on a ferry flight from Harris Neck, Georgia, to Charlotte, North Carolina, when it encountered severe thunderstorms. Apparently portions of the port wing separated in flight while the airplane was flying in the storm, either through over-control or flying at high speed through rough air conditions. The B-25 was shedding pieces as it came out of the clouds at an altitude estimated to be about 200 feet above ground level. The airplane smashed into the ground and exploded violently into flames seconds later. Killed in the crash were: 1Lt. Melvin E. Levin, pilot; 2Lt. James P. McCarty, co-pilot; 2Lt. Emmett McCann, observer; TSgt. James J. Barry, engineer; Sgt. Joseph O. V. Benard, radio operator; SSgt. J. L. Pelton, engineer/gunner.

6-10-43G. Salton Sea, California. At 1848, a Curtiss O-52 suffering an apparent engine problem made an emergency forced landing in the Salton Sea, 500 feet off shore of Salton Beach, California. Pilot 2Lt. Fred E. Warner was able to swim to safety but passenger Pfc. Robert E. Somers drowned while trying to reach the shore. The airplane was flying at an altitude of 600 feet above the surface of the lake when the engine began to lose power. The pilot attempted to remedy the engine problem without success. The airplane was losing altitude and it became unsafe to use parachutes so the pilot ditched parallel to the shore at Salton Beach. The airplane sank immediately. The two fliers attempted to swim to shore but Pfc. Somers was unable to continue. Lt. Warner left Pfc. Somers his parachute and swam to shore to get help. The pilot stripped off his clothes and secured a board to help rescue Pfc. Somers. As Lt. Warner reached the halfway point between Pfc. Somers and the shore, he observed Pfc. Somers let go of the floating parachutes and go under.

6-11-43A. Temple Terrace, Florida. At 1000 EWT, a Curtiss P-40K crashed at Temple Terrace, Florida, killing pilot 2Lt. Roger H. Massie. The aircraft had taken off from Hillsborough County Airfield and was on a scheduled aerobatic mission. Investigators speculated that the pilot was performing aerobatics at a low altitude and dove into wooded terrain at a 35-degree angle. Witnesses stated that the airplane had been flying low in the area just prior to the accident.

6-11-43B. Westover Field, Massachusetts. At 1050, a Republic P-47B stalled and spun into the ground one mile south of Westover Field, Chicopee, Massachusetts, killing pilot 2Lt. Bruce Cowan. The pilot was in the pattern for a formation landing with his flight leader and stalled out of the turn from base to final at an altitude of about 1,300 feet above ground level. The airplane dove into the ground and exploded into flames in a wooded area between Westover Field and Chicopee Falls.

6-11-43C. Auburn, New Jersey. At 1127, a Republic P-47D crashed one mile south of Auburn, New Jersey, killing pilot 2Lt. Frederick W. Moeller. The airplane was number four in a four-airplane "controlled alert mission." The airplanes became separated from the flight leader when the flight encountered broken clouds at about 1,200 feet above ground level. The three remaining airplanes were observed by witnesses on the ground to emerge from the clouds and then to enter a climbing turn to the right. The third airplane, piloted by Lt. Moeller, broke off from the formation and dove into the ground and exploded into flames.

6-11-43D. Marianna, Florida. At 1045 CWT, a North American P-51A performing an aerobatic demonstration flight crashed at Marianna Army Air Field, Marianna, Florida, killing pilot 2Lt. Paul C. March. The flight had originally consisted of four airplanes but one P-51 aborted after it lost its canopy in flight. The flight leader dove the three-airplane flight at the field and leveled off at about 1,000 feet above ground level, the number-two airplane flying at about 1,300 feet and Lt. March's airplane flying at about 600 feet. The flight leader then led the flight into a slow roll to the left. Lt. March completed about half of the roll when his airplane suddenly snapped downward, plunging into the ground and exploding into flames. Investigators speculated that the airplane was caught in the propeller turbulence of the preceding airplanes, causing the pilot to lose control and fly into the ground.

6-11-43E. Keystone Heights, Florida. At 1150 EWT, a Bell P-39M stalled and spun into the ground two miles southwest of Keystone Army Air Field, Keystone Heights, Florida, killing pilot 2Lt. Bernard K. Timm, Jr. Investigators stated, "Lt. Timm was circling the Keystone Army Air Field after returning from an escort mission. Two North American P-51 type airplanes were also circling the field and Lt. Timm was apparently trying to follow one of them. The P-51 made a sharp turn, which Lt. Timm attempted to duplicate. While in the turn he stalled and spun and was unable to regain control before hitting the ground. Witnesses stated the plane was almost under control just prior to the crash. Apparently the airplane stalled a second time, just over the treetops."

6-11-43F. Mt. Pleasant, Texas. At 1145, a Lockheed P-38F attempting an emergency forced landing crashed and burst into flames at Mt. Pleasant, Texas, killing pilot 1Lt. Clyde D. Campbell, Jr. The airplane was on a ferry mission from Kelly Field, San Antonio, Texas, to Nashville, Tennessee. Evidence indicates that the port engine threw the number-six connecting rod, causing the port engine to fail. The pilot tried to jettison the two fully serviced drop tanks attached to the P-38 but was unsuccessful. The inability to discard the drop tanks caused the starboard engine to overheat and diminish in performance. The pilot was unable to hold altitude and was attempting to make an emergency landing. The airplane struck the treetops and the side of a hill, careening approximately 200 feet and turning 180 degrees before coming to rest and bursting into flames. The pilot was unable to free himself from the wreckage and was burned to death.

6-11-43G. Long Boat Key, Florida. At 1310 EWT, a Vultee BT-13A stalled and spun to the ground at Long Boat Key Gunnery Range, Florida, killing pilot 2Lt. Richard W. Stone and passenger SSgt. Arthur L. Jensen. The pair had been sent to the gunnery range to relieve the crew on station. Investigators stated, "The pilot flew low over the gunnery range ... from the east ... then pulled up into a climbing turn to the right ... from treetop level. At the top of the climb the airplane lost considerable speed and stalled, the nose dropped and the airplane went into a dive and struck the ground at approximately a 30-degree angle. There was insufficient altitude for recovery."

6-11-43H. Eagle Pass, Texas. At 1600, a North American AT-6C flew into the ground two miles east of Spofford Road, ten miles north of Eagle Pass Army Air Field, Eagle Pass, Texas, killing A/C Verl Chamberlain and A/C Frederick M. Carrington. Investigation revealed that the airplane collided with the terrain in a bank to the right exceeding 45 degrees and at high speed. There were no witnesses to the accident.

6-11-43I. Bloomfield, Kentucky. At 1615 CWT, a Martin B-26B suffering the failure of the port engine crashed while attempting a forced landing five miles south of Bloomfield, Kentucky, killing co-pilot 1Lt. Paul J. Reidleberger and engineer SSgt. Rex E. Dickey, who was fatally injured in his bail out attempt. Pilot Maj. Ernest N. Ljunggren was injured seriously when he and the co-pilot went in with the airplane. Passenger Capt. Charles B. Hodges was uninjured parachuting to safety. Also parachuting to safety and receiving minor injuries were: 2Lt. William W. Mosser, navigator/bombardier; TSgt. John Perry, radio operator; 2Lt. Kenneth C. Fuller, passenger. Tail gunner SSgt. John J. Hardy received serious injuries parachuting to safety. The B-26 had been cleared from Myrtle Beach, South Carolina, to Godman Field, Kentucky, on a personnel transport mission. After passing Asheville, North Carolina, the oil pressure gauge for the port engine began fluctuating. Experiencing no apparent drop in performance, the pilots decided to proceed to Godman Field. A short time later the port engine failed. The pilot adjusted the trim and power for single-engine flight and ordered that the bomb bay fuel tank be jettisoned. The crew was unable to jettison the fuel tank and the airplane began to lose altitude. The pilot ordered the passengers and crew to bail out and held the airplane up until a large hill was cleared. The crew and passengers bailed out. The airplane was sinking in a nose-up attitude and stalled just above the treetops, striking the terrain tail first and then pancaking in and bursting into flames.

6-11-43J. Desert Center, California. At 1524, a Bell P-39N collided with rising terrain in the Orocopia Range, 15 miles southwest of Desert Center Army Air Base, Desert Center, California, killing pilot 2Lt. Arthur B. Chandler. The airplane impacted the mountain and exploded violently into flames 30 feet below the peak. Weather was not considered a factor in the accident.

6-11-43K. Healdsburg, California. At 1600, a Bell P-39Q spun to the ground and crashed into a wooded area two and one-half miles east of Healdsburg, California, killing pilot 2Lt. Gail E. Palmer. Investigation revealed that the airplane entered a tumbling, end-over-end type spin along its pitch axis before entering a flat spin from which the pilot was unable to recover. The pilot had jettisoned one of the escape doors but was unable to parachute to safety. Investigators noted the unfavorable spin characteristics of P-39 type airplanes as a contributing factor in the accident.

6-11-43L. Pocatello, Idaho. At 2141 MWT, a Consolidated B-24D on a bombing and gunnery training mission crashed 16 miles northwest of Pocatello Army Air Base, Pocatello, Idaho, killing the crew of nine. At 2120, the Pocatello AAB Control Tower Officer radioed all airplanes and ordered them back to base because of extremely high winds forecasted to arrive in the area. A short time later, all flying was called off and the base was closed. The tower advised the B-24 to fly to Boise, Idaho, or Wendover Field, Utah, or Ogden, Utah, as alternates because winds were blowing at 60 mph at Pocatello. The airplane had been flying east and attempted a turn south toward the field when it rolled onto its back, describing approximately two and one-half rolls before striking the terrain in an inverted position and exploding violently into flames. Investigators stated, "The accident occurred because of sudden and unpredictable strong winds arising and causing the pilot to lose control of his ship." Killed in the crash were: F/O Lenval H. Cash, pilot; 2Lt. William T. Allard, co-pilot; 2Lt. Edward E. Wilson, navigator; 2Lt. Frank E. Kellog, bombardier; Sgt. Donald H. Lindsay, engineer; Sgt. Fred W. Dilworth, radio operator; Sgt. George A. Basnight, assistant radio operator; Sgt. Warren O. Bacon, gunner; Sgt. Chester M. Wingate, gunner.

6-12-43A. Atlantic Ocean, Staten Island, New York. At 1430, a Curtiss O-52 on an artillery spotting mission crashed into the Atlantic Ocean at Lower Bay, 400 yards off shore of Fort Wadsworth, Staten Island, New York, killing pilot 2Lt. Marvin Winburne and observer 1Lt. S. Ellsworth Duff. Investigators speculated that the pilot lost control of the airplane when he inadvertently entered a large bank of coastal fog (observed only over the water) and suffered possible spatial disorientation. Personnel with the 245th Coastal Artillery Brigade stated that visibility was about 50 yards or less.

6-12-43B. West Point, Virginia. At 1430, a Republic P-47D flew into the ground five miles SSW of West Point, Virginia, killing 2Lt. Henry J. McCaffery. Investigators stated, "A flight of [three] P-47Ds from Langley Field [Virginia] were on a high-altitude training mission at 25,000 feet. Practice dives were made in formation with airspeeds indicating from 250 to 300 mph. At 23,000 feet the flight went into a dive with [2Lt. James R. Nall (see 6-12-43C) and 2Lt. Henry J. McCaffery] in wing positions. The Flight Leader [2Lt. Milton E. Soward] started recovery from the dive at 15,000 feet and completed it at 11,000 feet with an indicated airspeed of 400 mph. Observing that the two wing planes were not in position, he made efforts to locate them but was unsuccessful. He then proceeded to a rendezvous point and when the two planes failed to rejoin him he returned to Langley Field. The missing aircraft were later located five miles southwest of West Point, Virginia. Unable to recover from the dive, they had crashed at high speed in a wooded area and were approximately two miles apart."

6-12-43C. West Point, Virginia. At 1430, a Republic P-47D flew into the ground five miles SSW of West Point, Virginia, killing 2Lt. James R. Nall. Investigators stated, "A flight of three P-47Ds from Langley Field [Virginia] were on a high-altitude training mission at 25,000 feet. Practice dives were made in formation with airspeeds indicating from 250 to 300 mph. At 23,000 feet the flight went into a dive with [2Lt. James R. Nall and 2Lt. Henry J. McCaffery (see 6-12-43B)] in wing positions. The Flight Leader [2Lt. Milton E. Soward] started a recover from the dive at 15,000 feet and completed it at 11,000 feet with an indicated airspeed of 400 mph. Observing that the two wing planes were not in position, he made efforts to locate them but was unsuccessful. He then proceeded to a rendezvous point and when the two planes failed to rejoin him he returned to Langley Field. The missing aircraft were located five miles southwest of West Point, Virginia. Unable to recover from the dive, they had crashed at high speed in a wooded area and were approximately two miles apart."

6-12-43D. Dodge City Army Air Field, Kansas. At 1425 CWT, a Martin B-26C ran off the end of the runway while trying to take off from Dodge City Army Air Field, Dodge City, Kansas, killing assistant engineer Sgt. Lawrence N. Shuger and radio operator Pvt. Mitchell S. Foreman. Pilot and B-26 student 2Lt. Joseph G. Berkely and engineer Pvt. Glen N. Elliot received minor injuries. B-26 instructor and co-pilot 1Lt. Robert E. Stewart was seriously injured. The Accident Classification Committee stated, "The aircraft failed to gain sufficient speed to become airborne which resulted in continuation of the end of the runway, coming to rest at a point approximately 2,000 feet. After striking the ground in several places, both engines were torn loose from the aircraft at a point approximately 1,500 feet from the end of the runway. At this point, fire immediately developed, probably in one or both engine nacelles. All personnel escaped from the burning aircraft after coming to a stop, but the extent of injuries resulted in fatalities to two of the crewmembers." Despite detailed examination of the engines and statements made by the surviving pilots, investigators were not able to determine the cause of the loss of power of the engines on the take-off run.

6-12-43E. Olfin, Texas. At 1730, a Fairchild PT-19A on an individual solo flight crashed two miles south of Olfin, Texas, killing A/C Gerald L. Campbell, St. Joseph, Missouri. Investigators stated, "There was a thorough search for witnesses but none were reported. From all evidence found, the examining board reports that the airplane hit the ground at an angle near vertical and had not skidded across the ground. The engine made a hole about three feet in depth in hard ground. The pitot tube stuck into the ground at an almost a vertical angle. It was indicated that no parts came off before the airplane hit the ground. All parts were in the immediate vicinity of the airplane. The aircraft did not burn. The cause of the accident is unknown."

6-12-43F. Soledad, California. At 1700, a Consolidated B-24E crashed while attempting an emergency forced landing two miles west of Soledad, California, killing five crewmembers and injuring three others, one seriously. Killed in the accident were: Capt. Jerral W. Derryberry, B-24 instructor (see 1-8-42C); 1Lt. George M. Lake, B-24 student and co-pilot; 2Lt. Joseph H. Comer, Jr., B-24 student and co-pilot; 2Lt. Phillip W. Graham, B-24 student and co-pilot; Pfc. John N. Prindeze, radio operator. Engineer TSgt. Joe H. Sutton and assistant engineer TSgt. Wade H. Loos, Jr. received minor injuries and tail gunner Sgt. Leroy L. Klug was seriously injured. Witnesses stated that the B-24 passed beneath a row of power lines and then collided with a second row of power lines before crashing to the ground in an open field one-quarter a mile away. Investigators stated, "While on a routine training flight in the vicinity of Soledad, California, B-24E #42-7131 made an emergency landing about 20 minutes after take-off [from Salinas Army Air Base, California]. The airplane was observed flying at a very low altitude (200-300 feet) for about a mile before it crashed. Very little information as to probable cause of the accident could be obtained from the survivors or from witnesses near the scene of the accident. It was known that Capt. Derryberry, an experienced B-24 pilot, was giving instruction in two and three-engine operation to Lt. Comer, Lt. Graham and Lt. Lake. It is the opinion of this committee that some unknown difficulty was encountered in the course of this instruction, which resulted in the accident."

6-12-43G. San Francisco Bay, California. At 1808, a Bell P-39N crashed into San Francisco Bay off shore of Coyote Point, California, killing 2Lt. Gordon C. Miller. Investigators stated, "Lt. Miller had taken off at 1804 and made a wide 180 degree turn, when his ship was seen to go in a spin from about 2,000 to 3,000 feet. He broke the spin and attempted a recovery but as he was pulling up the plane struck the water with its wing. Smoke was seen coming from the plane on its way down. The cause of the accident is undetermined."

6-12-43H. Marianna, Florida. At 2230 CWT, a North American AT-6C on a night navigation flight crashed seven miles west of Marianna Army Air Field, Marianna, Florida, killing A/C William N. Puckett. The Accident Classification Committee stated, "Cadet Puckett ... was approximately ten miles from [Marianna AAF] when he entered a thunderstorm, accompanied by severe lighting that had moved in from the south. According to a witness who spotted the airplane at a somewhat low altitude, there was a severe bolt of lighting. At that time, the airplane appeared as "a big ball of fire." This appearance subsided and the navigation lights remained lighted. Immediately, the plane made a sharp turn, circled the home of the witness, and crashed into the ground."

6-13-43. Fedora, South Dakota. At 1545, two Boeing B-17F type airplanes collided in mid-air near Fedora, South Dakota, killing eleven fliers aboard B-17F #750, which crashed to the ground two miles southwest of Fedora. B-17F #790 crash-landed nine miles west of Canova, South Dakota, injuring ten crewmembers, three of them seriously. One crewmember was uninjured. The Accident Classification Committee stated, "Three B-17s #42-5790 [2Lt. Billy F. Fuller] leading, airplane #42-5750 [2Lt. Owen B. Groves] flying in number-two position and airplane #42-5446 [2Lt. Frederick R. Glazier] flying in number-three position were on an authorized formation training flight [at an altitude of approximately 4,500 feet agl]. A mid-air collision occurred during three-ship formation when aircraft #750 crossed under lead ship (#790), came up under it and struck it, resulting in damage to both aircraft to such an extent that aircraft #750 fell out of control and crashed immediately and aircraft #790 made an immediate forced landing. Lt. Glazier in airplane #446 flying number three position had crossed over to the right side of the formation, forming a right echelon, then Lt. Groves flying in airplane #750 started his cross over to number two position. Evidently Lt. Groves did not drop down enough to clear [lead] airplane's propeller wash before crossing over." 2Lt. Edward Knowles, Jr., aboard B-17 #446, stated to investigators, "B-17 #446 dropped back and down and echeloned to the right (stepped down) and #750, in attempting to cross under #790, hit #790 [in the] ball turret with ... its tail assembly, [which] broke loose at the main entrance door and flew up and hit the right stabilizer of #790 and broke it off. B-17 #750 went into a vertical dive and B-17 #790 peeled off to the right and went into a gentle glide and made a crash landing. B-17 #446 circled the area once and headed for Sioux Falls, South Dakota." Killed in the crash of B-17 #750 were: 2Lt. Owen B. Groves, pilot; 2Lt. Michael R. Hric, co-pilot; 2Lt. Roy B. Lever, instrument navigator; 2Lt. James J. O'Connell, navigator; 2Lt. Robert T. Cassidy, bombardier; SSgt. Marvin J. Curtis, engineer; SSgt. Dean M. Garland, assistant engineer; SSgt. Sidney S. Packer, radio operator; SSgt. Murry V. Whitehead, assistant radio operator; SSgt. Bruno E. Twardy, gunner; SSgt. Charles R. Bowers, gunner. Injured in the crash landing of B-17 #790 were: 2Lt. Billy F. Fuller, pilot; 2Lt. James S. Law, B-17 instructor pilot; 2Lt. Willis D. Johnston, co-pilot; 2Lt. William F. Mattulke, navigator; 2Lt. Robert L. Hagen, bombardier; SSgt. Carl O. Blake, radio operator; SSgt. Lawrence H. Dressen, assistant radio operator. Engineer SSgt. George D. Evans, assistant engineer SSgt. Loyal H. Clark and gunner SSgt. Edward J. Hillyer received serious injuries. Gunner Sgt. Russell D. Gunyan parachuted to safety and was uninjured.

6-14-43A. Edmond, Oklahoma. At 0925, a Douglas A-20G crashed one and one-half miles northeast of Edmond, Oklahoma, killing pilot 2Lt. Robert W. Hampson. The airplane had taken off from Tinker Army Air Base, Oklahoma, at 0907 on a "routine check flight." The airplane was seen at an altitude of 1,200 feet with the port engine nacelle in flames. The airplane soon crashed into a wooded ravine and exploded into flames. The port wing was found 100 yards from the main wreckage, indicating to investigators that the wing had separated from the airplane before it struck the ground. Wreckage was scattered over an area of 500 yards.

6-14-43B. Hammond, Louisiana. At 1020, a Beech AT-7 inadvertently entered a thunderstorm and crashed at Hammond, Louisiana, killing all five fliers on board. The aircraft was one of a flight of 12 AT-7 aircraft flying with a flight of four Lockheed C-60 aircraft on a navigation training flight. The flight encountered scattered thunderstorms and maneuvered to avoid flying through them. The AT-7 became separated from the flight and crashed soon after. The remaining airplanes were able to avoid the storms and land safely after completing their mission. Killed in the crash were: 2Lt. Alvin R. Grossman, pilot; 2Lt. Stanley P. Dabrowski, navigator instructor; A/C Francis G. Conway, student navigator; A/C Rafel L. Cooper, student navigator; A/C Robert E. Cooper, student navigator.

6-14-43C. Sarasota, Florida. At 1135, a Curtiss P-40N dove into the ground two and one-quarter miles north of Sarasota, Florida, killing pilot 1Lt. Cecil H. O'Dell. The airplane was part of formation of four P-40 airplanes on a scheduled high altitude aerobatic flight. Lt. O'Dell led the flight through a series of aerobatic maneuvers at an altitude of 20,000 feet. The flight leveled out at an altitude of 15,000 feet. After a few moments of level flight, Lt. O'Dell executed a half-roll, executing a split-S maneuver into a steep dive. The flight followed the leader until reaching an indicated airspeed of 480 mph, pulling out of the dive at an altitude of 8,000 feet. The other pilots observed Lt. O'Dell in a vertical dive at a speed estimated to be 500 mph or better, disappearing from sight as he passed through a layer of clouds at about 3,000 feet. The airplane struck the ground in a vertical dive and exploded into flames.

6-14-43D. Pevely, Missouri. At 1115, a Vultee BT-13A crashed one mile west of U.S. Highway 61, near Pevely, Missouri, killing A/C James J. Nolan and A/C James G. Overby. The airplane was on a scheduled "cross-country instrument team ride" from Malden Army Air Field, Missouri, to Lucas Airport, Crystal City, Missouri, and return. The airplane did not check in with the control airplane at Crystal City. Investigators stated, "According to witnesses, the airplane circled at a low altitude two or three times, diving, zooming and making a loud noise. The airplane was then seen to dive out of sight on the south side of a hill, and then pull up in a steep climbing turn to the right. As the airplane pulled up, the right wing struck the top of a tree, cutting off a number of branches and demolishing the wing. The airplane struck the ground approximately 30 yards north of the tree with such terrific force that the wreckage was strewn up the side of the hill for distance of about 100 yards." Investigators speculated that the cadets were buzzing a large house on top of the hill near where they crashed.

6-14-43E. Atlantic Ocean, Long Island, New York. At 1645, a Republic P-47D on an aerial gunnery mission dove into the Atlantic Ocean five miles southwest of Bay Shore, Long Island, New York, killing pilot 2Lt. Thomas J. Rainey. The airplane had taken off from Mitchel Field, New York, at 1520 and engaged in the aerial gunnery exercise at 20,000 feet. Investigators were unable to determine what caused the airplane to dive into the sea from a high altitude.

6-14-43F. Lubbock, Texas. At 1705, a Cessna AT-17 crashed when it suffered a catastrophic structural failure of the port wing four miles south of Lubbock Army Air Field, Lubbock, Texas, killing instructor pilot 2Lt. Robert C. Morrison and student A/C Robert D. Hodges, Jr. Investigation revealed that the plywood leading edge of the port wing separated from the wing at an altitude of approximately 1,000 feet above ground level while the airplane was climbing out following take-off.

6-15-43A. Ruidoso, New Mexico. At 0650, two Consolidated B-24E aircraft collided in mid-air ten miles north of Ruidoso, New Mexico, killing 19 crewmembers aboard both airplanes. SSgt. Julius G. Robinson miraculously parachuted to safety from B-24E #42-7071. Killed in the crash of B-24E #42-7071 were: 2Lt. Donald R. Lembright, pilot; 2Lt. Harold E. Simmons, co-pilot; 2Lt. Vernon L. Laschanzky, navigator; 2Lt. Michael Iwaszuk, bombardier; Sgt. Raymond B. Post, assistant engineer; SSgt. Leo F. Sheehan, radio operator; Sgt. Richard W. Gray, assistant radio operator; Sgt. Harold Collin, gunner; Sgt. Joseph J. Huddy, gunner. Killed in the crash of B-24E #42-7146 were: 1Lt. Donald W. Kaspervik, pilot; 2Lt. Ray E. Ward, co-pilot; 2Lt. John V. Andrew, navigator; 2Lt. Bruce G. Ellis, bombardier; SSgt. Eldon F. Rossler, engineer; SSgt. Arthur P. Wikle, assistant engineer; Sgt. Allen Chamovitz, radio operator; SSgt. George L. Scott, gunner; SSgt. Andrew J. Alexander, gunner; 2Lt. Edward V. Handy, passenger. The aircraft were part of a formation of eight B-24 airplanes flying in three elements of three in a javelin type formation on a bombing and gunnery training mission. The formation was flying north and had leveled out at about 13,000 feet (4,000 feet above ground level) when B-24 #7071, flying in the number-three position of the first element, began sliding over to the right, eventually flying above and behind B-24 #7146, which was flying in the lead position of the first element. The port propellers of B-24 #7071 made slight contact with the port vertical fin and rudder of B-24 #7146. B-24 #7071 backed off for an instant before colliding with the tail of B-24 #7146 again, severing the upper third of the port vertical fin and rudder with the inboard port propeller. The starboard propellers chewed into the starboard vertical fin as B-24 #7071 slid to the right while the port propellers severed the remainder of the tail section. B-24 #7146 fell to earth and exploded into flames. According to the surviving engineer, Sgt. Robinson, B-24 #7071 was under control after the accident and he stated that the pilot stalled the airplane and it entered a spin to the ground. Being held in the airplane by the forces of the spin, Sgt. Robinson was able to pull himself to a hole in the fuselage and was flung to safety. Wreckage of the airplanes fell five miles west of Ft. Stanton and ten miles north of Ruidoso, approximately 50 miles north of the bomber base at Alamogordo. It was noted by investigators that the rising sun was a contributing factor in the accident.

6-15-43B. Hondo, Texas. At 1318, a Taylorcraft L-2B collided with a thicket of mesquite trees and crashed eight miles south of Hondo, Texas, killing 2Lt. Richard S. McNamara and passenger Pvt. Donald G. Dickensheets. There were no witnesses to the crash and the wreckage was found later in the day.

6-15-43C. Tucson, Arizona. At 1350 MWT, a Ryan PT-22 crashed four miles northeast of Ryan Field, Tucson, Arizona, killing student A/C Theron D. Wood. The student put the airplane in a deliberate spin at an altitude of 3,000 feet agl. The student was unable to recover and the airplane described ten complete rotations before striking the ground at angle of 80 degrees. Since there was little rotational effect visible at the crash site, investigators speculated that a recovery might have been in progress. The student was hurled through the port front part of the fuselage upon impact. His safety belt was found intact and unhooked, indicating that the student might have been trying to leave the airplane just prior to impact and had unhooked it.

6-15-43D. Delta, Utah. At 1450 MWT, a Douglas A-20G suffering the failure of the starboard engine crashed while attempting a single-engine emergency landing at the municipal field at Delta, Utah, killing pilot 1Lt. Dellie O. Ullrich. The A-20 was on a ferry flight from Daggett, California, to Hill Field, Ogden, Utah, and had taken off from McCarran Field, Las Vegas, Nevada, at 0325 PWT. At 1429 MWT, the pilot called Delta air traffic control and requested emergency landing instructions and was told that the field was clear and that he could land on Runway 30 to the northwest. Contrary to the tower instruction, the A-20 began a landing attempt to the southeast (without flaps extended), picking up a quartering tailwind. The approach was described as "steep" and with excessive speed. After "skimming" the runway for a considerable distance, the airplane touched down 1,000 feet from the southeast end of the runway. The airplane was unable to stop in time to avoid overrunning the end of the runway, rolling 1,700 feet past the end of the runway before the nose wheel dug in and the airplane flipped over onto its back, killing the pilot instantly.

6-15-43E. Portland, Oregon. At 1411, a Bell P-39N stalled and spun to the ground while turning from base to final at Portland Army Air Base, Oregon, killing pilot 2Lt. Jim E. Duncan. The pilot was attempting a left turn with landing gear and flaps extended when the airplane stalled and entered a "partial spin," plunging to the ground in a nose down attitude and exploding violently into flames.

6-16-43A. Smyrna, Tennessee. At 0721 CWT, a Consolidated B-24D overran the end of the runway and crashed while attempting a take-off from Smyrna Army Air Field, Tennessee, killing five fliers. Investigation revealed that the flight crew did not set the flaps at 20 degrees for take-off. Witnesses reported that the airplane was "pulled" off the runway and but did not become fully airborne. Investigators determined that the pilots forced the B-24 into the air at a point on the runway where the aircraft would normally become airborne with the proper use of flaps. The B-24 mushed back to the ground and continued off the end of the runway, colliding with a large ditch and careening over a concrete road 2,000 feet south of the south end of the North-South runway. The airplane nosed over and crushed the flight deck and the tail section tilted up 70 degrees. The fliers were crushed and trapped in the wreckage, which had erupted into flames. One flier was alive in the wreck but flames enveloped the wreckage before he could be rescued. Killed in the crash were: Capt. Clifford J. Gormley, B-24 instructor pilot; Capt. Dorsey A. Pulliam, co-pilot and B-24 instructor student; Capt. Daniel T. Pope, co-pilot and B-24 instructor student; Pfc. Vito W. Bandenelli, engineer; Pvt. William P. King, engineer. Investigators stated that the cause of the accident was the improper use of the take-off checklist and the flight crew's failure to set the flaps at 20 degrees for take-off. Each pilot on board had at least 1,300 hours of military flying experience.

6-16-43B. Lake Huron, Michigan. At 0842, a Vultee BT-13A on a weather checking flight crashed into Lake Huron two miles off shore of Oscoda, Michigan, killing pilot 2Lt. Nathaniel M. Hill and observer 2Lt. Luther Blakeney. The airplane was seen to enter a fog bank immediately after take-off. The airplane crashed six minutes later. Lt. Hill's body was not recovered.

6-16-43C. Pinellas Field, Florida. At 1045, two Curtiss P-40L airplanes collided in mid-air five miles northwest of Pinellas Field, St. Petersburg, Florida, killing 2Lt. Arthur M. Easterlin (P-40L #42-11062) and 2Lt. Clemons C. Miley (P-40L #42-11108). Investigators stated, "Lt. Easterlin was on an authorized instrument flight during which 2Lt. Donald E. Claycomb was acting as observer. After a period of approximately 30 minutes on instruments, Lt. Claycomb had Lt. Easterlin come from under the hood for a rest. Lt. Claycomb then led the flight and attempted to join a four-ship flight of similar airplanes [that] were in a string formation. After joining behind the last ship, Lt. Claycomb decided to cut across the circle and take his position behind the number-two ship in the other formation, as the distance between ships was too great. Upon trying to close in to assume this position, Lt. Claycomb found that he was not in a position to do so at which time he pulled away from the formation. Lt. Easterlin, his wingman, had evidently not followed his leader in such a manner as to keep the other ships of the formation in view and in trying to stay with the formation allowed his airplane to collide with [Lt. Miley's airplane]."

6-16-43D. Abilene, Texas. At 1000, a Vultee L-5 crashed ten miles southwest of Abilene Army Air Field, Abilene, Texas, killing pilot SSgt. John D. Montmorency and passenger Pvt. Carl Glassman. Investigators stated, "During the flight the pilot made several loops in the airplane. From all reports by witnesses, the pilot tried two loops in succession, the second of which was too low for a complete recovery. It is possible that the excessive speed attained in completing the second loop placed so great a pressure on the control surfaces that the pilot was unable to control the airplane."

6-16-43E. Garden City, Kansas. At 1120 MWT, a Vultee BT-13A stalled and spun to the ground eight and one-half miles east of Garden City, Kansas, killing instructor 2Lt. Robert K. Richardson, 20, Oxnard, California, and student A/C Charles W. Simpson, 21, Brooklyn, New York. Maj. L.D. Van Mullum, Assistant Director of Training, Garden City Army Airfield, witnessed the accident and stated to investigators, "I noticed an airplane in a steep left turn close to the ground. I assumed that the pilots were doing elementary-eights, since the airplane was directly over a road that was crosswind and at about the altitude elementary-eights are practiced. As I watched the ship, it seemed to do a snap roll to the left and almost completed a full turn before hitting the ground at a steep angle. The airplane immediately burst into flames upon striking the ground."

6-16-43F. Bowman Field, Kentucky. At 1511 CWT, a Bell P-39D crashed one-half mile from Bowman Field, Kentucky, killing pilot 2Lt. Vincent A. DeJulius. The airplane was part of a three-airplane formation navigation flight to Hopkinsville, Nashville and return to Godman Field, Fort Knox, Kentucky. The formation encountered poor weather and elected to return to the vicinity of Godman Field. The flight participated in simulated combat for about an hour before entering a Lufbery Circle. Lt. DeJulius entered a "center section stall out of a very tight turn," the airplane falling in a flat spin and striking the ground in between two houses. The nose section damaged one house. No one on the ground was injured. The airplane did not burn but was completely destroyed.

6-16-43G. Luke Field, Arizona. At 1615 MWT, a Curtiss P-40E on a routine training flight crashed 35 miles northwest of Luke Field, Phoenix, Arizona, killing A/C Donald J. Ashpole. There were no actual witnesses to the accident but investigation revealed that the airplane had struck the ground in a nose down attitude at high speed.

6-16-43H. Hopkins Wells, California. At 1940, a Bell P-39Q crashed three miles southwest of Hopkins Wells, California, killing pilot 1Lt. William W. Rogers. The Accident Classification Committee stated, "From viewing the wreckage the airplane apparently crashed out of a flat spin. The right door was found approximately 100 yards east of the wreck. The body was found about three feet from the right wingtip with parachute almost completely opened. A large dent in the top of the wing was spattered with blood. This probably indicates that the pilot was on the wing when the airplane struck the ground. All aerobatics have been prohibited in this type of airplane."

6-17-43A. Williams Field, Arizona. At 0010, ground crewman Pfc. Roy E. Peterson was killed instantly when he walked into the rotating propeller of a Lockheed RP-322 at Williams Field, Chandler, Arizona. A/C Robert A. Williamson, pilot of the RP-322, was not injured. The student had started both engines and gave Pfc. Peterson the signal to remove the wheel chocks. Pfc. Peterson, standing in front of the aircraft on the right side, walked across the front of the RP-322 and into the rotating starboard propeller.

6-17-43B. Camp Carson Military Reservation, Colorado. At 1050, a Lockheed F-5 on a high altitude training flight crashed onto the Camp Carson Military Reservation, Colorado, killing the pilot 2Lt. Forest E. Farr. The airplane fell to earth at high speed in a vertical attitude. Investigators speculated that the pilot, suffering from oxygen starvation, was unable to prevent the airplane from crashing to the ground.

6-17-43C. Rapid City, South Dakota. At 1217, a Boeing B-17F crashed while attempting to take off from Rapid City Army Air Base, South Dakota, killing six fliers and seriously injuring four others. The Accident Classification Committee stated, "Aircraft #42-30448 took off toward the west on the East-West runway at 1217 MWT as the eighth ship of a ten-ship formation. The take-off of the first seven ships was normal in all respects, as observed by eyewitnesses. [The seven B-17s] passed over the west end of the runway at low altitude. [The pilot, believing the B-17 was not getting off the ground, closed the throttles as he approached the end of the runway, thinking that he could run off the end of the runway onto a grass field and stop. The B-17 ran off an earthen bank and into the air, crossing a small pond and crashing into the opposite bank.] A careful observation of the tracks made on the runway at the west end indicated that the airplane did at some point leave the runway and came back onto it at a point about 300 feet from the west end. The main landing wheels touched the runway at that point and when the aircraft passed off the west end of the runway all three wheels were on the ground. The aircraft rolled a distance of approximately 330 feet, at which point there is a nearly vertical bank on the east side of a small pond which is about 25 feet below the top of this bank. The aircraft took off from the edge of this bank and went through the air approximately 300 feet, over the pond, and crashed against the upslope on the other side of the pond, apparently in a nose high position, as the tail hit in the water at the edge of the pond, the remainder of the landing gear hitting up on the slope." Pilot 2Lt. Lloyd T. Clark, navigator 2Lt. Emery M. Mittuch, bombardier 2Lt. John S. Pannell and engineer SSgt. Daniel Antolick were seriously injured. Killed in the crash were: 2Lt. Richard R. Billman, co-pilot; SSgt. William Playford, assistant engineer; SSgt. Samuel Compton, radio operator; SSgt. James L. Piccolo, tail gunner; Sgt. Charles Williams, gunner; Cpl. Howard Estep, gunner.

6-17-43D. Shaw Field, South Carolina. At 1430, a Vultee BT-13A spun to the ground five miles north of Shaw Field, Sumter, South Carolina, killing A/C Frank H. Quertermous. The airplane had entered a spin to the left at an unknown altitude from which the student could not recover. Investigation revealed that the student attempted to parachute to safety and had been standing on the wing when the airplane struck the ground. He was found lying across the port horizontal stabilizer. The airplane did not burn but was destroyed utterly.

6-17-43E. Shaw Field, South Carolina. At 1430, a Vultee BT-13A crashed eight miles north of Shaw Field, Sumter, South Carolina, killing A/C Robert H. Stinson. The student had been practicing spins and failed to make recovery. He was able to escape from the airplane but was killed instantly when he was struck in the head by the rotating propeller. Cadet Stinson's body floated to earth in the fully deployed parachute.

6-17-43F. Shaw Field, South Carolina. At 1515, a Vultee BT-13A crashed 15 miles northeast of Shaw Field, Sumter, South Carolina, killing A/C Malcom D. Carnahan. The student was practicing spins at an altitude of approximately 6,000 feet. The Accident Classification Committee stated, "This accident was caused by the failure of the pilot to use the correct spin recovery procedure in attempting to recover from a three turn spin. It is believed that the pilot, in attempting to recover from the original spin to the right, inadvertently went into a spin to the left from which he did not recover."

6-17-43G. Elk City, Kansas. At 1445 CWT, a North American BT-14 crashed seven miles east of Elk City, Kansas, killing 2Lt. Frederick J. Dowcett and A/C Henry E. Wick. Investigators stated, "Indications at the scene of the wreckage of the airplane showed that the airplane made contact with the ground nose first in a nearly vertical attitude, with the left wing slightly low. It is believed that ... at a low altitude of approximately 300 to 500 feet, this airplane stalled and spun into the ground in a spin to the left, causing the resultant damage and fatalities."

6-17-43H. Laredo, Texas. At 1745, a Lockheed AT-18 returning from a target tow mission encountered poor weather and crashed near Laredo Army Air Field, Laredo, Texas, killing eight fliers. The airplane was approaching the area where it was to drop its sleeve type tow target when it entered a thunderstorm and crashed to the ground nose first in an almost vertical attitude, exploding violently into flames upon impact with the ground. Killed in the crash were: 2Lt. Frank W. McLin, pilot; 2Lt. Norbert F. Woss, co-pilot; Sgt. Harold J. Ree, gunnery instructor; Pvt. Bobby E. McGee, engineer and tow reel operator; Cpl. George E. Fortado, student gunner; Cpl. James W. Foust, student gunner; Pfc. Julius Foreman, student gunner; Pfc. Edward A. Forrest, student gunner.

6-17-43I. Newton, New Jersey. At 1945, a North American B-25D flying in poor weather crashed six miles north of Newton, New Jersey, killing the crew of five. The airplane had taken off from Bolling Field, Washington, D.C. on a flight to Stewart Field, New York, at 1858, climbed to 5,000 feet for cruise flight and soon encountered severe thunderstorms. Investigators stated that the pilot had not reduced speed and lowered the landing gear for flying in severe turbulence and as a result the aircraft suffered a structural failure of the port wing and possibly parts of the tail section. The aircraft was seen by witnesses on the ground to emerge from the overcast at an altitude of 1,000 feet, shedding pieces as it dove into the ground, exploding violently into flames upon impact. Killed in the crash were: 1Lt. Leonard Rand, pilot; 2Lt. Samuel B. Hannah, co-pilot; SSgt. Floyd B. Edmonds, engineer; Cpl. Murray Schwartz, radio operator; TSgt. Steve Sandusky, passenger.

6-18-43A. Newport, Arkansas. At 0915, a Vultee BT-13A spun to the ground three miles northwest of Newport, Arkansas, killing A/C Clark N. Bevin and A/C Joseph M. Holmes. Investigators stated, "Student pilots took off on a routine navigation instrument training flight, and shortly after clearing airport traffic pattern at an altitude of approximately 1,500 to 2,000 feet were observed by eyewitnesses to enter a spin from which recovery was not effected. The aircraft struck the ground nose first in the spin and immediately burst into flames."

6-18-43B. Romulus, Michigan. At 1110, a Bell P-39Q crashed while attempting a landing at Romulus Army Air Field, Romulus, Michigan, killing pilot 2Lt. Mark K. O'Toole. The airplane was observed making a long and low approach to runway 31. The pilot maintained this low approach on final until he reached a point approximately one-quarter of a mile east of the field when he began to stretch his glide. Qualified witnesses on the ground stated that the pilot would have had to added power to climb up and establish a proper glide. The tower advised the pilot to add power and "fly it in." There was no response from the pilot. Witnesses reported that the airplane was flying at an altitude of 50 to 60 feet with flaps and landing gear extended and appeared near stalling speed. Twice the P-39 almost fell into a spin. Realizing he was not going to make the field, the pilot rapidly applied power and caused the airplane to roll to the left and fly into the ground inverted about 200 yards east of the runway. The P-39 struck the ground 30 degrees from the horizontal, flipping back upright as it came to rest and burst into flames.

6-18-43C. Grider Field, Arkansas. At 1015, a Fairchild PT-19A crashed five and one-half miles northeast of Grider Field, Pine Bluff, Arkansas, killing civilian instructor Donald P. Safris and student A/C Harold B. Dohner. The airplane entered a spin to the right at low altitude, making it impossible for the pilots to recover the airplane before it struck the ground. The occupants were killed instantly.

6-18-43D. Casper, Wyoming. At 1630, a Consolidated B-24E suffering an engine failure crashed while attempting to make an emergency forced landing 28 miles northwest of Casper Army Air Base, Wyoming, killing bombardier 2Lt. William A. Burton and injuring six crewmembers, two seriously. Three crewmembers were uninjured. The B-24 took off from Casper Army Air Base on a gunnery mission at 1600 MWT, making two dry runs over the gunnery range and returning for a third pass. The airplane had been flying at an altitude of 6,700 feet and was about to commence firing on its gunnery mission when the fuel pressure dropped to zero on the number-three engine and the pilot was not able to maintain altitude with the remaining three engines despite adding full power. The number-three engine failed and the pilot feathered the propeller. The pilot states that he did not have complete control of the ship due to a tremendous yawing of the airplane and that his altitude could not be maintained. The pilots were not able to lower the flaps because the flap hydraulics operates off of the number-three engine. The airplane crash-landed downwind at a speed higher than normal landing speed. The airplane skidded along for a few hundred feet until it struck a small ditch, the nose and flight deck twisted in the impact, killing the bombardier and trapping the radio operator. Pilot 2Lt. James McMullin, assistant engineer Sgt. Mervin F. Barkey and gunner Sgt. Robert J. McVeigh were uninjured. Engineer SSgt. Maurice Cutright and assistant radio operator Sgt. Jesse J. Froho were seriously injured. Receiving minor injuries in the accident were: 2Lt. Clyde Peters, co-pilot; 2Lt. John A. Worhach, navigator; Sgt. Edmond A. Chaloux, radio operator; Sgt. Jesse J. Colvin, passenger.

6-18-43E. Van Nuys, California. At 2121, a Lockheed P-38G crashed one and one-half miles northwest of the Army Air Field at Van Nuys, California, killing 2Lt. Eugene B. Sullivan. The pilot had radioed Van Nuys and reported that the engines were "missing" and that he was going to abandon the airplane. Investigation revealed that the pilot pulled his ripcord too soon and the parachute became fouled on the lower counter-balance located on the elevator. The pilot was dragged down by the airplane and was killed when the airplane slammed to the ground. His body was found near the port engine nacelle and wheel well with the fouled parachute lying near him. The pilot's body was dragged from the crash area when the wreckage erupted into flames. Investigators were not able to determine the cause of the apparent engine failure.

6-19-43A. Carlsbad, New Mexico. At 0100, a Beech AT-11 crashed while attempting a landing at Carlsbad Army Air Field, New Mexico, killing two fliers and seriously injuring two others. "After a thorough investigation of the accident and questioning of all witnesses, it is the opinion of the Accident Classification Committee that the pilot, having completed his mission, proceeded to the field and made a normal approach for a landing. The wheels struck the ground once or possibly twice and then the pilot opened his throttles to either go around or to make a new landing approach. However, he evidently failed to check his stabilizer setting and when the airplane started abruptly up, lost his horizon and control of the airplane, which resulted in the complete stall, falling off on the left wing and then straight into the ground, making a 180-degree turn during the descent. The tail of the airplane then dropped down so that the airplane was headed in the opposite direction of the approach when it came to rest." Pilot 2Lt. Richard E. Buckley and engineer Pfc. Joseph M. Miller were killed and navigators 2Lt. Harold R. Levant and 2Lt. Louis F. Klayman received serious injuries.

6-19-43B. Mount Calm, Texas. At 0850, a Vultee BT-13A crashed four miles south of Mount Calm, Texas, killing A/C Henry L. Profe. Investigators stated, "The cadet was on a solo training mission. According to available evidence and witnesses, the cadet was last seen to be in a spin at an altitude of approximately 6,000 feet, from which he never recovered. Prior to the crash of the airplane, the airplane partially recovered from the spin, stalled again, and dived approximately straight down. The airplane crashed, being completely demolished and the pilot killed."

6-19-43C. Lake Murray, South Carolina. At 1045 EWT, a North American B-25C crashed into Lake Murray, South Carolina, killing the crew of four. The airplane was on a low-altitude bombing mission on Target #3 with another B-25, flying at altitude of about 200 feet on the bomb run. After passing the target, the lead B-25 climbed to 400 feet in a climbing turn but the accident airplane made his turn at low altitude, losing altitude until the starboard wing struck the water. The airplane cartwheeled into the lake and burst into flames north of the target. Killed in the crash were: 1Lt. Floyd C. Grazier, Jr., pilot; 2Lt. John G. Schuten, co-pilot; 2Lt. Howard H. Dailey, bombardier; 2Lt. William L. Price, bombardier.

6-19-43D. Thompson-Robbins Field, Arkansas. At 1015, a Fairchild PT-23 crashed five miles southwest of Thompson-Robbins Field, Helena, Arkansas, killing civilian instructor John B. Sisk and student A/C Joseph F. Bauer, Jr. The PT-23 was apparently shooting a simulated forced landing at an altitude of about 25 feet when the pilot became aware of a set of telephone wires in the flight path. He pulled up abruptly and added power rapidly, gaining about 50 feet, clearing the wires and entering a snap roll to the right. The pilot was unable to maintain control and the PT-23 crashed into the ground on the nose and starboard wing, killing the occupants instantly.

6-19-43E. La Junta, Colorado. At 1059, a North American B-25D suffering a failure of the port engine crashed while attempting an emergency landing at La Junta Army Air Field, La Junta, Colorado, killing co-pilot 2Lt. Robert J. Grissell. Pilot 2Lt. Herbert G. Taylor, engineer TSgt. Richard H. Terry and engineer Pvt. James R. Stevenson escaped injury in the crash. Investigators stated, "[The pilots] were returning from a local instrument flight. As they were on their downwind leg preparatory to landing, wheels down, the engineer noticed a sudden drop in oil pressure on the left engine. The pilot feathered the left propeller, retracted the landing gear, and called the tower that he was coming in for single-engine landing. The tower acknowledged the fact and the pilot started a normal single-engine approach. At approximately 200 feet above the ground and some distance from the runway, the tower called the aircraft stating that his wheels were up and for the airplane to go around. The co-pilot was just about to lower the wheels at the time the tower called to go around. The indicated airspeed was approximately 125 mph. Nevertheless, the pilot followed the tower's instruction. Full throttle and rpm were applied. The airplane was able to mush along until high-tension power lines appeared at the end of the field. In pulling over these the aircraft lost flying speed and shortly thereafter stalled from approximately 30 feet above the terrain. On coming in contact with the ground the plane burst into flames. The top escape hatch was released and the aerial engineer, assistant aerial engineer, and the pilot were able to escape. The co-pilot, for some reason, was unable to clear himself from the airplane before flames enveloped the cockpit."

6-19-43F. Miami, Florida. At 1304, a North American B-25D crashed while attempting a take-off from the 36th Street Airport, Miami, Florida, killing pilot Capt. Charles R. Dimke. Co-pilot 1Lt. William G. Simpson, gunner TSgt. Phillip Wells and gunner SSgt. William B. Norman received serious injuries. Investigators stated, "After rolling approximately 2,000 feet on the take-off, the landing gear was raised before the aircraft had attained flying speed and it is believed that the flaps were also raised due to the fact that the flap valve was in the 'UP' position when the aircraft had come to rest. According to eyewitnesses, the landing gear was raised before the aircraft left the ground and the pilot dragged the tail in an attempt to stall the aircraft off of the ground. After the landing gear was raised, the aircraft settled and the left propeller dug 3-inch holes in the runway and the right propeller dug one-inch holes at the intersection of the E-W runway. The pilot then succeeded in getting the aircraft into the air and it started into a left bank with one engine apparently running away (due to a propeller blade being broken off of the left propeller or due to damage to the propeller governor) and the aircraft settled to the ground, hitting on the left wing and cartwheeling, destroying the airplane."

6-19-43G. Florence, South Carolina. At 1415, a Taylorcraft L-2A collided with a high-tension wire at Florence Army Air Field, Florence, South Carolina, killing pilot 2Lt. Marshall J. Burlow and seriously injuring passenger 2Lt. Edward W. Cordes. The pilot had flown low over the field and returned for a lower pass, hitting the wires and crashing to the ground out of control.

6-19-43H. Hoover Field, Texas. At 1430, a Cessna UC-78B and a UC-78C collided in mid-air over Hoover Auxiliary Airfield, Pampa, Texas, killing all four fliers aboard both airplanes. Students 1Lt. William G. Meggs and 2Lt. William L. MacCoy were killed in the crash of UC-78 #577. Students A/C Theodore B. Allen and A/C Stanley P. Anderson were killed in the crash of UC-78 #989. The aircraft were in the traffic pattern for the field and were standing by for orders from the tower to start practicing 90-degree spot landings. Investigation revealed that UC-78 #989 had lost about 300 feet of altitude and had dove to pick up airspeed and, in an attempt to climb back up to regain his position behind the leader (#577), had climbed into #577, severing its tail. The collision occurred at an altitude of approximately 4,000 feet above sea level. Both aircraft plunged to the ground and exploded into flames within the boundary of Hoover Field.

6-19-43I. Great Falls, Montana. At 1343, a Boeing B-17F suffering a double engine failure crashed while attempting to make an emergency landing at Great Falls Army Air Base, Montana, killing seven crewmembers and seriously injuring four others. The airplane had taken off from Great Falls at 1332 MWT on a navigation flight to Pendleton Army Air Base, Oregon. Soon after take-off, the number-two and number-four superchargers ran away. The pilot feathered the propellers on both of the engines and radioed Great Falls that he was coming in for an emergency landing. The airplane entered the traffic pattern and made its approach, the pilot holding back on the flaps until he was sure he was not going to undershoot the field. He gave the order to drop flaps but they would not come down. The pilot ordered the engineer to crank the flaps down manually. During the lapse of time needed to lower the flaps manually, the airplane was overshooting the field. The B-17 only had about one-third flaps at an altitude of about 40 feet at about mid-way down the runway. It became apparent to the pilot that he would not be able to land safely so he gave the remaining two engines full power for the go-around, neglecting to rescind the order to the engineer to drop flaps. The engineer continued to lower the flaps as the airplane struggled to gain altitude. The engines were not able to produce enough power to keep the B-17 airborne and it descended into the ground, striking a gully, breaking up and bursting into flames. Killed in the crash were: 2Lt. Daniel B. Kline, pilot; 2Lt. Raymond L. Duncanson, navigator; 2Lt. Stanley A. Smith, bombardier; SSgt. Albert Walsh, radio operator; Sgt. William G. Neerlag, gunner; Sgt. George N. Scanzoni, gunner; Cpl. Everett C. Hastings, assistant engineer.

Co-pilot 2Lt. John H. Ulrich, engineer SSgt. Henry C. Dangil, assistant engineer SSgt. Peter G. Bertram and assistant radio operator SSgt. Melvin A. Schreyer were seriously injured.

6-19-43J. Zuehl Field, Texas. At 1500, a Vultee BT-13A crashed four miles southeast of Zuehl Field, Texas, killing 2Lt. Gerrard J. Hunt and 2Lt. John E. Glasrud. Witnesses reported observing the airplane flying straight and level at an altitude of approximately 1,500 feet agl. The airplane then went into a gliding turn to the right. The airplane made two turns, losing altitude and apparently entering a spiral. The spiral became steep and speed increased until the airplane struck the ground.

6-20-43A. Biscayne Bay, Florida. At 0850, a Douglas C-49K crashed into Biscayne Bay 100 yards off shore of Miami, Florida, killing three fliers. The airplane had taken off from the 36th Street Airport, Miami, at 0806 on a local instrument training flight. Witnesses on the ground observed the airplane flying with apparent engine trouble, diving toward the bay at a 45-degree angle until it contacted the surface and exploded. Investigation revealed that the port engine was not producing power, the port engine throttle was closed and the port propeller was in the full low position. Investigators speculated that the crash was caused by "A simulated or actual engine failure and a stall at the cruising altitude of 2,000 feet and the inability to recover flying speed before hitting the water." Killed in the crash were civilian twin-engine instructor William A. McAfee, student 2Lt. Harry L. Gebler and student 2Lt. Arthur E. George.

6-20-43B. Gulf of Mexico, Florida. At 1050 EWT, a Curtiss P-40L crashed into the Gulf of Mexico seven miles southwest of Sarasota Point, Florida, killing pilot F/O Raymond C. Chambers. Investigators stated, "[The pilot] was a member of a four-ship flight on a scheduled aerial gunnery mission. F/O Chambers was flying the third airplane in the flight. The flight of four proceeded to the gunnery range, making their normal passes at the tow target, which was being towed [at 5,500 feet] by 2Lt. George W. Stovall in a P-40 type aircraft. On the third pass at the tow target, F/O Chambers, in making his pass, apparently shot the target off. He was in such a position and traveling at such a speed that he was unable to avoid colliding with the target. The airplane fell off into a spin, and the pilot, unable to recover, crashed into the Gulf."

6-20-43C. Lake Istokpoga, Florida. At 1320, a North American P-51A crashed into Lake Istokpoga, four miles east of Hendricks Field, Sebring, Florida, killing pilot 2Lt. Paul J. Hollobaugh. The airplane was cleared on a local flight from Page Field, Fort Myers, Florida. The airplane was observed to fly over the lake at extremely low altitude, circling around the lake following the shoreline. The P-51 was in a 70-degree bank when the starboard wingtip struck the water. The airplane cartwheeled into the shoreline, coming to rest upright in shallow water. The pilot was killed instantly.

6-20-43D. Rapid City, South Dakota. At approximately 1230 MWT, a Boeing B-17F broke up in flight and crashed at the AAF gunnery range, 25 miles south of Rapid City Army Air Base, South Dakota, killing nine crewmembers. The airplane had taken off from Rapid City AAB at 1140 MWT on an aerial gunnery mission. The airplane was standing by in the gunnery range pattern waiting its turn to fire at a towed target when it was seen on fire and breaking up in flight by witnesses on the ground. The B-17 fuselage broke in half at the radio room compartment near the starboard wing. Pieces of the starboard wing, various fire blackened wreckage and bodies were found strewn over an area of one-quarter mile wide by one mile long on the edge of an area called Red Shirt Table. The body of the bombardier was found one and one-half miles SSE of the main wreckage some 56 hours after the accident. Apparently the bombardier was attempting to parachute to safety; he died of a severe head injury and was seen floating to earth in his fully deployed parachute. The body of the ball turret gunner was found in the smashed wreckage of the ball turret. Investigation of the wreckage revealed that the airplane suffered a catastrophic internal explosion of "gasoline and air" in the starboard wing at an altitude of 3,500 feet above ground level. The damage and subsequent fire caused the airplane to break up and crash. Investigators noted that the B-17 was the "latest overseas type of B-17F" that had been modified with extra wing fuel tanks at the modification center at Cheyenne, Wyoming. Killed in the crash were: 1Lt. Robert R. Owens, pilot; 2Lt. Frederick W. Buckley, Jr., co-pilot; 2Lt. Yale Kweskin, navigator; 2Lt. Martin Gelbert, bombardier; SSgt. Francis P. Owens, engineer; SSgt. Thomas E. Strauser, radio operator; Sgt. John C. Kozlowski, assistant engineer; Sgt. Raymond Spencer, assistant radio operator; SSgt. Robert C. Bailey, gunner.

6-20-43E. Hebbronville, Texas. At a time unknown, a North American AT-6C dove into the ground seven miles south of Hebbronville, Texas, killing A/C James W. Brown, Youngstown, Ohio, and A/C Howard R. Humphreys, Omaha, Nebraska. The AT-6 had taken off at 1545 CWT from Moore Field, Mission, Texas, on an instrument cross-country from Moore Field to Hebbronville, to Falfurrias, and return to Moore Field. A search effort was commenced when the airplane became overdue. The only witness to the accident, a civilian unfamiliar with aviation matters, told investigators that the airplane dove into the ground and exploded.

6-20-43F. Westover Field, Massachusetts. At 1820, a Republic P-47D and a P-47B collided on the runway while landing at Westover Field, Chicopee, Massachusetts, fatally injuring 2Lt. Richard M. Burdick and seriously injuring 2Lt. Warren F. Donovan. Investigators stated, "Lt. Burdick [P-47D] made a normal landing after completion of an altitude formation flight. He had slowed down and was almost ready to turn off the runway when Lt. Donovan [P-47B] overtook him after making a very 'hot' landing without flaps and bouncing several times. Both Lt. Burdick and Lt. Donovan were on the right side of the runway. Lt. Burdick tried to get out of the way by taxiing off the left side of the runway but it was too late. Lt. Donovan's airplane ran completely over that of Lt. Burdick. The propeller cut through his tail and the left landing gear passed through the fuselage just forward of the tail assembly, cutting it completely off. The left wing struck the cockpit, leaving the pitot tube wedged in the armor plate behind Lt. Burdick. Both pilots were injured; Lt. Burdick died the following day."

6-20-43G. Albany, Texas. At approximately 2300, a Martin B-26 crashed while attempting an emergency forced landing 18 miles north of Albany, Texas, killing the crew of five. The airplane was cleared on a flight from Laughlin Field, Del Rio, Texas, to Abilene, to Wichita Falls, to Fort Worth, and return to Del Rio. The B-26 was flying at an altitude of 3,000 feet agl when the starboard engine failed. The pilot trimmed the airplane for single-engine operation and elected to attempt an emergency landing in a cleared area. He attempted a wheels-up landing, hitting the ground tail-first on the last third of the field, which was estimated to be about one mile long. The airplane was unable to stop in time to avoid smashing into trees at the end of the field, causing the B-26 to break up and burst into flames. Killed in the crash were: 1Lt. Robert T. Lee, Jr., pilot; 2Lt. Marion P. Messingham, co-pilot; SSgt. Bernard K. Pollock, engineer; Pfc. Robert T. Groome, crewmember; SSgt. Herbert C. Arnold, passenger. Civilian ranchers found the wreckage at 1100 the next day.

6-21-43A. Moses Lake, Washington. At 1029, a Boeing B-17F crashed 30 miles southwest of Moses Lake, Washington, killing the crew of eight. The airplane had taken off from Moses Lake Army Air Base on a local bombing and training mission at 0745. The airplane had completed its bombing mission and was later observed by witnesses in the air falling in a vertical dive at about 8,000 feet (6,500 feet above ground level). The airplane began rolling slowly as it dove toward the ground. At an altitude of approximately 5,000 feet, the tail assembly was seen to separate from the airplane followed by the starboard wing. The port wing soon left the airplane and the fuselage progressively broke apart, the wreckage falling to the ground over a wide area. No fire was observed before or after the airplane began breaking up. Weather was not considered a factor in the accident. Several B-17 crews reported the ceiling at about 12,000 feet with generally good flying conditions. Investigation revealed that the number-four propeller was in the feathered position. Killed in the crash were: 2Lt. Lyle E. Graham, pilot; 2Lt. Kenneth F. Westover, co-pilot; 2Lt. Russell H. Wolff, Jr., bombardier; Sgt. Charles H. Land, engineer; Sgt. Guy H. Reeves, radio operator; Sgt. John J. Babb, radio operator; Sgt Jackson B. Woodley, gunner; Sgt. Louis B. Sacerino.

6-21-43B. Marana, Arizona. At 1218 MWT, a Vultee BT-13A stalled and spun while turning from base to final at Marana Army Air Field, Marana, Arizona, killing A/C John Ed Hester, Jr. Investigators stated, "[The student] pulled his airplane into too steep a turn for his final approach into the home airfield. The airplane went into a snap spin, making a turn and a half before striking the ground. The airplane struck in a position 180 degrees from the original flight direction."

6-21-43C. Columbus, Georgia. At 1455 EWT, a Bell P-39Q crashed five miles southeast of Columbus, Georgia, killing pilot 1Lt. Joseph S. Bartlett, Asheville, North Carolina. The pilot was practicing chandelles at a safe altitude when the airplane fell into a flat spin to the left from which the pilot was unable to recover. The pilot bailed out at an altitude of about 1,000 feet above ground level but was struck by part of the airplane and was unable to pull his ripcord, causing him to plunge to his death.

6-21-43D. Hartford, Connecticut. At 1548, two Republic P-47B aircraft collided in mid-air ten miles south of Hartford, Connecticut, killing 2Lt. William C. Ives aboard RP-47B #41-6052 and injuring 2Lt. Elmer Buss, who parachuted to safety from P-47B #41-6035. The aircraft were part of a four-airplane formation flight that had become lost in poor visibility. The flight leader led the flight west to intersect the Connecticut River with a view to turning north to follow the river to Springfield. When the leader spotted the river, he initiated the right turn to the north. Lt. Ives and Lt. Buss collided at an altitude of 2,200 feet during this turn. Lt. Buss parachuted to safety from his spinning airplane. Lt. Ives was unable to parachute to safety and went in with his airplane.

6-21-43E. Santa Ana, California. At 1346, a Lockheed P-38G crashed while attempting a landing at Orange County Airport, Santa Ana, California, killing pilot 2Lt. Jack C. Albright. The airplane was observed on its final approach, crossing the end of the runway at about 300 feet. The P-38 was seen to glide to about 100 feet when the pilot apparently elected to go around. The pilot applied throttle and the airplane began skidding to the right. The airplane (with landing gear and flaps extended) then rolled to the right and into the ground, exploding violently into flames. Investigators speculated that the starboard engine might have failed, causing the airplane to roll to the right when power was applied.

6-21-43F. Parker Peak, California. At 2125, a Consolidated B-24E collided with Parker Peak eight miles southwest of Mono Lake, California, killing seven fliers. The B-24 had taken off from Tonopah, Nevada, at 2015 and flew a bombing mission at the Tonopah

Bombing Range but failed to return to its base at Salinas, California. Weather was forecasted as broken clouds over the mountains with high winds and occasional severe turbulence. Eight miles southwest of Lee Vining, the B-24 collided with Parker Peak at an elevation of 12,800 feet, about 100 feet below the summit, killing all on board instantly. The B-24 was not found until 6-5-45. Killed in the crash were: 2Lt. William E. Hunt, pilot; 2Lt. Keith E. Delp, co-pilot; 2Lt. Dwight G. Bacon, navigator; 2Lt. Clair Hartzell, bombardier; TSgt. San Meyerhoffer, engineer; SSgt. Elmer Swaving, radio operator; Pfc. Maurice A. Smith, photographer.

6-22-43A. Alpena, Michigan. At 1030, an Aeronca L-3C collided with a power line and crashed at Alpena, Michigan, killing pilot SSgt. Ralph L. Diven and passenger MSgt. Johnnie J. Boyle. Investigators stated, "The pilot was attempting to practice a forced landing and ran into the power line. The wheels were caught in the power line and the aircraft was pulled into ground. The aircraft hit the ground and flipped over on its back."

6-22-43B. Oneco, Florida. At 1115, a Curtiss P-40K crashed two miles east of Oneco, Florida, killing 2Lt. Edward C. Martin. The airplane was observed in a shallow dive. The witness stated that the airplane did not appear to be in trouble and that the engine was running up until the time the airplane impacted the ground and exploded violently into flames. The pilot made no apparent attempt to parachute to safety. Investigators noted that the parachute was not connected to the pilot's harness and was found 250 feet from the airplane. Investigators could not determine the cause of the accident.

6-22-43C. Floresville, Texas. At 1145, a Bell P-39D stalled and plunged to the ground five miles northwest of Floresville, Texas, killing pilot F/O Norman L. Boswell. Investigation revealed that the airplane stalled out of a split-S maneuver while engaged in simulated combat with another P-39 type airplane. Investigators noted that the pilot had only five hours in P-39 type equipment and that aerobatics are permitted only if the pilot has 15 hours on type.

6-22-43D. Adel, Georgia. At 1530 EWT, a Beech AT-10BH collided with a civilian Waco UPF-7 biplane five miles south of Adel, Georgia, killing AAF instructor 1Lt. Robert B. Benfield and A/C Joseph R. Leonardo aboard the AT-10. Civilian instructor Lamon Williams and civilian student Dorsey Lee Sproles were uninjured when they parachuted to safety from the Waco. The collision occurred at an altitude of about 2,300 feet. The AT-10 was part of a two-ship flight. The pilot of the other AT-10 stated that the Waco was maneuvering close to the AT-10 formation in an unsafe manner and that flew his formation in close proximity to the Waco to "get his number." Instructor Williams stated to investigators, "I was flying with my student between 2,000 and 2,500 feet preparing to do a series of stalls when two AT-10s dived down from about 300 feet above me on the left, and I fell in behind the two airplanes in a tight turn to the right. One [AT-10] was in view ahead, above and to the left of me. The next thing I knew I felt a big jar and looked out and my right wing was gone. I tried the airplane and it was out of control so we bailed out. I never saw the airplane after it hit me."

6-22-43E. Davis, Oklahoma. At 2315, a Vultee BT-13A crashed at Davis, Oklahoma, killing A/C Charles W. Ritter. The student had taken off from Perrin Field, Sherman, Texas, at 2200 on a night navigation flight to Ada, Oklahoma, to Gainesville, Texas, and return to Perrin Field. Investigators stated, "At approximately 2315, Cadet Ritter crashed near Davis, Oklahoma, apparently on course from Ada, Oklahoma, to Gainesville, Texas. A/C Ritter had circled the town and then cut his motor and attempted to land. The position and condition of the plane indicated that it was very likely that A/C Ritter had attempted to land. The plane, although damaged beyond repair, was not entirely demolished and was in an upright position. It appeared that he attempted to land in a ravine and had run head on into a steep bank, the plane coming to rest approximately one-half way up the slope of this bank. It is the opinion of this board that if the front seat had been a metal one instead of a wooden type, this crash might not have been fatal."

6-23-43A. Reverie, Tennessee. At 0650 CWT, a Boeing B-17F crashed into a swamp adjacent to the Mississippi River on Island No. 35 at Reverie, Tennessee, killing five fliers. For 30 minutes prior to the accident, the B-17 had been observed buzzing the steam barge *J.D. Ayres*, which was underway on the river. The B-17 had made several passes at the barge and on the last pass had cleared the barge by only 15 feet. Witnesses stated that the airplane made a steep right turn at low altitude and collided with trees with the starboard wingtip, crashing into the swamp at the edge of the riverbank at Mile 193 and exploding into flames. The B-17 had taken off from the Army Air Base at Dyersburg, Tennessee, on an instrument-training mission that did not include low altitude flight. Witnesses on the ground stated that they could see crewmembers waving at them as the airplane flew over. It was also reported that the barge would blow its horn as the B-17 flew over it. Killed in the crash were: 2Lt. Alexander Jasman, pilot; 2Lt. Gerrit Swier, co-pilot; 2Lt. James M. Odegaard, co-pilot; Sgt. Joseph B. Audio, engineer; Sgt. Joseph S. Ciriono, radio operator.

6-23-43B. Columbiana, Alabama. At 0830, a Vultee BT-13 crashed two miles south of Columbiana, Alabama, killing students 2Lt. William E. Cadle and A/C Donald A. Callender. The students had taken off from Gunter Field, Montgomery, Alabama, on an instrument navigation team ride. Investigators were not able to determine why the airplane crashed into the ground and exploded.

6-23-43C. Melvin, Texas. At 0847, a Fairchild PT-19A crashed five miles east of Melvin, Texas, killing civilian instructor J. Park Kemp and student A/C Anthony Ruffino. Investigators stated, "J. Park Kemp and A/C Anthony Ruffino were cleared from Bruce Field [Ballinger, Texas] at 0805 CWT for a short dual cross country to Melvin, Eden, and return to Bruce Field, proposed landing only at Bruce Field. According to statements from witnesses the airplane was engaged in low aerobatics for about 20 minutes, recovering as low as 50 feet above the ground, and was flying at times just above the ground level. Immediately after a roll, the airplane stalled and hit the ground at approximately a 60-degree angle to the horizontal. The airplane hit the ground and bounced about 20 feet along the same direction as original flight where it came to rest. It was indicated that no parts came off of the airplane before it hit the ground."

6-23-43D. Nesmith, South Carolina. At 1145, a North American B-25G crashed three miles northeast of Nesmith, South Carolina, killing the crew of seven. The airplane had taken off from the Army Air Base at Myrtle Beach, South Carolina, on a bombing and gunnery mission and then was to return to base at Greenville, South Carolina. The airplane was seen flying at an altitude of 2,500 feet when it began skidding and weaving from side to side. The airplane went into a steep turn to the left and then crashed to the ground and exploded violently in flames. Investigation revealed that the nose landing gear door inadvertently came open in flight causing unfavorable flying characteristics that the pilot was unable to control. The door was seen to separate from the airplane and was found some distance from the main wreckage. Killed in the crash were: 2Lt. John L. McCann, pilot; 2Lt. Harold R. Anderson, co-pilot; 1Lt. Ralph D. Keeney, co-pilot; Sgt. Edwin C. McNeil, engineer; Sgt. Bernard Lermen, radio operator; Sgt. Robert Kik, gunner; Sgt. Ellis G. Lee, gunner.

6-23-43E. Love Field, Texas. At 1211, a Lockheed P-38H crashed at Love Field, Dallas, Texas, killing pilot 2Lt. Joseph F. Grucza. The airplane took off from Love Field at 1200 on a ferry flight to Atlanta, Georgia. At 1209, tower personnel noticed the airplane two miles south of the field with the starboard propeller feathered. The field was cleared for an emergency landing. The airplane entered the pattern and then made a turn onto final approach, flying at about 500 feet. The approach appeared normal. The pilot lowered the landing gear and then the flaps. The airplane crossed the runway at about 150 feet, continuing at this altitude for about one-third of the runway. The pilot applied power for a go-around and retracted the landing gear. The airplane rolled to the right, then to the left, then to the right again, mushing into the ground on its starboard wing and exploding into flames upon impact. The airplane was carrying two fully serviced external fuel tanks.

6-23-43F. Fort Smith, Arkansas. At 1640, two Vultee BT-13A aircraft collided in mid-air one mile northwest of the municipal airport at Fort Smith, Arkansas, killing 2Lt. Clifford Zweck and Pvt. Archie H. Hazard aboard BT-13A #41-10921. 2Lt. Vincent J. Farkas, Jr. was uninjured and able to safely land heavily damaged BT-13A #41-22373 at the airport. Investigators stated, "At 1425, a three-ship formation led by Lt. Frank W. McElwee, with Lt. Farkas and Lt. Zweck as wingmen, was cleared [from Perrin Field, Texas] for Tulsa, Oklahoma, via Fort Smith, Arkansas. The formation arrived at Fort Smith at 1635. As they were over the field at 1,500 feet, Lt. McElwee signaled for a peel-off. Lt. Farkas dropped back, spacing himself for the landing. Lt. Zweck was directly over Lt. Farkas' ship. As Lt. Zweck was losing altitude, his tail section came in contact with the left wing and propeller of Lt. Farkas' ship. The damage caused Lt. Zweck's airplane to go into a vertical dive, which resulted in a fatal crash."

6-23-43G. Phoenix, Arizona. At 1650 MWT, a Curtiss P-40L crashed one-half mile north of Luke Field Auxiliary Airfield #4, Phoenix, Arizona, killing Chinese Air Force pilot Sub-Lt. Ping-Jen Wu. The airplane was part of a three-airplane flight (two Chinese students and an AAF instructor) on a ground gunnery mission. The instructor had led the flight through several passes and then signaled a return to Luke Field Auxiliary Airfield #1. The flight leader executed a steep aileron turn to the left and then signaled for formation. Leveling off at approximately 4,000 feet, the leader noticed that the number-three airplane was on the ground and burning. There were no witnesses to the accident and investigators speculated that the student stalled out of the turn.

6-24-43A. Westborough, Massachusetts. At 1158, two Republic P-47D airplanes collided in mid-air at Westborough, Massachusetts, killing 2Lt. Douglas Smith, Eads, Tennessee, (P-47D #42-8186). 2Lt. Donald L. Murrie (P-47D #41-8208) parachuted to safety and was uninjured. The airplanes were part of a four-ship flight on a local interception mission. The formation was echeloned to the right and executed a 45-degree turn to the right in order to come out line abreast. During the turn, Lt. Murrie (the number-three airplane) lost sight of Lt. Smith (the number-two airplane), overtaking him and causing the number-four airplane to alter his turn to avoid an unsafe situation and keep Lt. Murrie in sight. The number-four airplane maneuvered out of the way just as the two airplanes collided. Both airplanes plummeted to the ground, crashing near each other and exploding into flames. Lt. Murrie landed in his parachute nearby. The wreckage of Lt. Smith's airplane came to rest on a railroad track and was struck by a passing train before the engineer could be warned.

6-24-43B. Goshen, Kentucky. At 1300 CWT, an Aeronca L-3C crashed on the Belknap Farm at Goshen, Kentucky, killing F/O Charles J. Kosinski and

F/O John W. Gardner, both rated glider pilots. Investigators stated that the airplane was attempting a simulated forced landing and stalled out of a turn, spinning to the ground.

6-24-43C. Greenwood, Mississippi. At 1447, a Vultee BT-13A spun to the ground seven miles southwest of Greenwood, Mississippi, killing A/C William B. Brubaker. A witness in the air observed the airplane enter a spin at an altitude exceeding 6,000 feet. The airplane described six turns of a spin to the left and started to straighten out, re-entering the spin and continuing until it reached a low altitude. At this point the airplane apparently came out of the spin and then dove into the ground straight ahead. The cadet was seen to leave the airplane at an altitude of approximately 100 feet, too low to allow his parachute to fully deploy and causing him to fall to his death. His body was found 20 feet from the airplane with the parachute partially opened.

6-24-43D. Bernie, Missouri. At 1730 CWT, a Vultee BT-15 crashed ten miles northwest of Bernie, Missouri, killing A/C Arnold D. Krause. The airplane was on a scheduled local transition flight and was observed by witnesses on the ground to circle a farmhouse several times. The airplane was seen to perform a roll at an altitude of about 50 feet above ground level. The airplane stalled out of the roll and crashed to ground.

6-24-43E. Hobbs, New Mexico. At 1636 MWT, a Vultee BT-15 crashed while attempting a takeoff from Hobbs Army Air Field, Hobbs, New Mexico, killing pilot 1Lt. Gene S. Cullumber and seriously injuring civilian passenger John M. Wolgast, an employee of Wright Aeronautical Corporation. The airplane took off and reached an altitude of 100 feet agl when the engine failed. The pilot made a turn to the left in an effort to land on the wide shoulder of Runway 26. The airplane struck the ground hard at a 30-degree angle and skidded for about 100 yards, shearing off the landing gear. The airplane came to rest and then burst into flames. Mr. Wolgast attempted to free the pilot from the cockpit but the canopy was jammed closed.

6-25-43A. Sylvester, Georgia. At 0100 EWT, a North American AT-6B crashed 14 miles southeast of Sylvester, Georgia, killing 1Lt. Willard H. Keeling. The airplane took off from Spence Field, Moultrie, Georgia, at 2305 (6-24-43) on a "weather patrol night flight" to Sylvester and return. The pilot returned to Spence Field but was told to return to his assigned area to complete the mission. The pilot acknowledged and returned to the area of Sylvester. The airplane smashed into the ground at a very steep angle in a vertical left bank at an estimated speed of 225 mph. Rain and thunderstorms were reported in the area of Sylvester. Investigators could not determine the cause of the accident.

√ **6-25-43B. Albany, Georgia.** At 1030 EWT, a Stearman PT-17 performing low altitude aerobatics crashed 19 miles SSW of Albany, Georgia, killing A/C Robert L. Celko. Investigators stated, "[The witness stated] that the airplane made a power-off dive in his direction while, he, the witness, was plowing an open field with a tractor. The airplane came low enough to cause subject witness to hurriedly dismount from his tractor to avoid being hit by the airplane. Apparently the student then made a power-on recovery from the dive and then attempted to execute a slow roll to the left in the ensuing pull-up. At approximately 200 feet, the cadet executed half of the slow roll and was then unable to recover from the inverted flight position. As he saw the ground beginning to get closer and closer, due to the fact that he began to lose altitude, his reaction was to pull the stick back, which caused the airplane to do a split-S from the inverted position and the airplane struck the ground head-on in a slightly inverted angular position."

6-25-43C. Tonopah, Nevada. At 1005, a Bell P-39N crashed two miles northeast of the runway of the AAF Bombing and Gunnery Range at Tonopah, Nevada, killing pilot 2Lt. Albert J. St. Germain. Investigators stated, "After Lt. St. Germain took off on his first transition flight in a P-39, he started turning to the right with his wheels still extended. The degree of the plane's bank continued to increase until suddenly the plane snapped over on its back and dove into the ground while still inverted."

6-25-43D. Luke Field, Arizona. At 1245 MWT, a Curtiss P-40E stalled and spun to the ground 15 miles southwest of Luke Field, Phoenix, Arizona, killing Chinese Air Force Sub-Lt. Yo Chen. The airplane was part of a flight of three P-40s on a formation and aerobatics flight. The flight flew in formation for about 45 minutes when they peeled off and began a series of aerobatic maneuvers, eventually ending up in a Lufbery circle. The flight climbed to about 11,000 feet and began performing Immelmann turns. Lt. Chen stalled out of an Immelmann and spun to the ground, exploding into flames upon impact. There was no obvious effort of the pilot to parachute to safety.

6-25-43E. Newton, Texas. At 1625, a Douglas A-20G and an A-20C collided in mid-air ten miles southeast of Newton, Texas, killing six fliers aboard both bombers. Killed aboard A-20G #790 were: 2Lt. Howard J. Hildebrandt, pilot; Sgt. Lawrence E. Zelley, gunner; 2Lt. Vincent J. Jasinski, passenger. Killed in the crash of A-20C #149 were: 2Lt. Paul F. Curdy, pilot; SSgt. Richard L. Wentling, gunner; Pvt. Thomas F. Riley, passenger. The Accident Classification Committee stated, "This mission consisted of a two airplane formation on a simulated low altitude bombing attack on a tank column operation in the area southeast of Newton, Texas. The airplanes were first sighted by the members of medium tank crews of Company G, 41st Armored Regiment, 11th Armored Division, which was proceeding in column, southerly along a road in that area. When first seen, they were approaching the line of march from the southeast. They were flying in close formation with A-20 #790 to the left of and slightly to the rear of A-20 #149. The pilot of #149 started a turn to the right as if to assume position over the tank column. The pilot of #790 continued momentarily on the line of flight on which it had approached the column and then started a right turn. It is apparent he turned more steeply than #149. The horizontal interval between the airplanes had, for that brief instant, increased but on the execution of the steeper turn by #790, that interval was closed up. Witnesses agree that the first point of contact of the two airplanes was the left wing tip of #149 and the right wing tip of #790, while both were in a banked turn to the right. One of the propellers of airplane #790 cut through the left wing of airplane #149 just ahead of the aileron. The left engine fairing was cut off cleanly. The fuselage of #149 was cut in two places, just in front of the tail group and again at the rear gun turret. Both airplanes crashed into trees approximately 200 yards ahead and approximately 50 yards east of the tank column. Just before #149 struck the trees, one man, later identified as Pvt. Riley, was seen to leave the plane. From the condition of his injuries, and from the fact that he came from the rear part of the wreckage, it is believed that he probably received fatal injuries in the collision of the two airplanes and fell from the airplane rather than jumped. On impact with the ground, both airplanes burst into flames."

6-26-43A. Presque Isle, Maine. At 0815, a Martin B-26B crashed into a horse drawn wagon after striking the ground with its starboard wingtip nine miles NNE of Presque Isle, Maine, killing five crewmembers and four civilian farm hands working in a potato field near the wagon. The aircraft had taken off from the Army Air Base at Presque Isle on a flight to the European war zone. The pilot's father-in-law, Carl Rasmussen, owned a farm near the airfield and the pilot had told friends and relatives that he was going to "fly over" after he took off from the base on his flight overseas. He buzzed the Rasmussen house twice at extremely low altitude. On the second pass, the pilot also buzzed two horse drawn wagons and a group of farm workers in a field nearby. The pilot attempted a climbing turn to the right. The starboard wing struck the ground 200 feet in front of the team of horses, wagons and five workers. The B-26 cartwheeled into the ground and exploded into flames, cutting down the horses, wagon and workers in its path. Wreckage of the B-26, the wagon and dismembered bodies of victims and horses were strewn over potato fields on the farms of Carl Rasmussen and O.B. Griffith. Army fliers killed in the crash were: 1Lt. Bertrand H. Robertson, pilot; 2Lt. Herbert F. Meyers, co-pilot; 1Lt. Edwin M. Hankinson, navigator; SSgt. William H. Jochim, engineer; TSgt. John M. Kuser, radio operator. Civilians killed were: Alfred Winter, 36; Alfred Winter, Jr., 9; Ann Therialt, 23; Eloise Newton, 18; all were from Caribou, Maine. Freeman Hitchcock, 48, Caribou, Maine, escaped with minor injuries when he lashed and drove a team of horses and a wagon into nearby woods when a collision seemed imminent.

6-26-43B. Fulton, California. At 0721, a North American AT-6A suffered a catastrophic structural failure of the port wing and crashed three miles southeast of Fulton, California, killing pilots 2Lt. Joseph M. Osborn and 2Lt. Fonemor Saxton. The airplane had taken off from Santa Rosa Army Air Field, California, on a scheduled instrument navigation training flight. Witnesses on the ground stated that they had heard a loud report and looked up to see the airplane spinning to the ground in a nose down attitude minus a wing.

6-26-43C. Cameron, South Carolina. At 1040, a Stearman PT-17 crashed one mile north of Cameron, South Carolina, killing A/C James L. Cobb. The student was on an authorized aerobatic flight. The airplane was seen flying south approximately 30 feet over Highway 31, passing over an automobile. The driver of the automobile reported that the airplane's bottom wing was at nearly the same level as the automobile's roof and that the pilot waved with his left hand as he flew over. Still flying at an extremely low level, the pilot began to rock his wings and then attempted a climbing 180-degree turn to the right. As the airplane approached a north heading, it hung on its nose for a moment before it dove into the ground on the edge of a cotton field at an angle of 60 degrees. The airplane burst into flames within three minutes of hitting the ground.

6-26-43D. Tyndall Field, Florida. At 1000 CWT, a Taylorcraft L-2B crashed 12 miles northeast of Tyndall Field, Panama City, Florida, killing passenger Cpl. Verne L. Sharp and seriously injuring pilot 2Lt. Herman B. Bray. The airplane wreckage was not found until 1100 on 6-28-43. Lt. Bray was found at about 1600, one mile northeast of the wreckage of the airplane, "wandering about in a dazed condition."

6-26-43E. Perry Field, Florida. At 1620 EWT, two Republic P-47D type airplanes collided in midair 20 miles south of the Army Air Field at Perry, Florida, killing 2Lt. Norman R. Messier aboard P-47D #42-22380. 1Lt. Weston M. Lennox escaped injury and was able to land heavily damaged P-47D #42-22385 at the field. The airplanes were on a scheduled ground strafing training mission. Lt. Lennox stated to investigators, "After having flown straight and level about 1,500 feet and approximately south for about 15 miles I started diving on a pier. As soon as I started my dive, I called Lt. Messier on the radio and told him that we would simulate an attack on the dock. After making a pass on the dock, I pulled up turning to the left and away from my wingman. While I was still in a steep bank and at about 1,500 feet, I looked back and to the left to check on Lt. Messier's position. I saw his ship directly under mine. He was closing in so rapidly that I could not react before his ship hit my low wing. Lt. Messier's ship was in a shallow turn to the left. He came from behind and inside of my turn, passed directly under my ship, and swept ahead of my ship and to the outside of my turn. I rolled out of my bank, glancing at my damaged left wing and checking aileron control; then I looked to the right to see what had happened to Lt. Messier's plane. He was still in a shallow turn to the left and then straightened out in slight dive. By this time, I was slightly above, to the left and almost parallel to Lt. Messier's ship. There was a large tear in the top of the fuselage of his ship just aft of the intercooler doors and the torn metal skin was hanging at the left side. I called for him to pull up several times; it appeared that he had no elevator control. I started to tell him to bail out but he was very low. The ship dove into a wooded area."

6-26-43F. War Eagle Field, California. At 2335, a Vultee BT-13A dove into the ground seven miles southeast of War Eagle Field, Lancaster, California, killing civilian instructor Robert E. Gilchrist and student A/C Harry C. Mullens. Investigation revealed that the airplane dove into the ground from an almost vertical position.

6-27-43A. Tonopah, Nevada. At 0738, a Bell P-39N crashed at Tonopah Airport, Tonopah, Nevada, killing pilot 2Lt. Milan Larry Jasper. Lt. Jasper peeled off of his formation and was in the pattern for the airport, attempting a left turn from base to final with landing gear extended at about 500 feet. As the turn continued, the degree of bank increased. The pilot tried to straighten out but stalled out of the turn and entered a spin. The airplane spun one complete turn and smashed into the ground 200 yards southeast of the end of runway 33, 360 degrees from the line of flight.

6-27-43B. Adrian, Georgia. At 1230 EWT, a Martin B-26C crashed while attempting an emergency forced landing in a cotton field four miles east of Adrian, Georgia, killing radio operator SSgt. Peter G. Dorsch. Pilot 2Lt. Arthur C. Nelson and co-pilot F/O Frank J. Ryan were seriously injured. Investigators stated, "[The airplane] took off from Hunter Field [Georgia] at 1030 EWT on a routine fuel consumption test. The pilot took off on an instrument clearance for Columbus, Georgia, with Atlanta, Georgia, as his alternate airport. He flew to Atlanta, called in, and then proceeded on course back to Hunter Field. About halfway back, the pilot sated that he decided to transfer fuel. He transferred the left-hand auxiliary to the right-hand main without difficulty, but in transferring the right-hand auxiliary to the left-hand main, the fuel could not be transferred. The emergency system was also tried but it would not work. Suspecting that he would have to go to single-engine operation, the pilot decided to get below the overcast, which was at approximately 800 feet. The pilot flew at 700 feet for about five minutes before the left engine cut out because of lack of fuel. The left [propeller] was then feathered, and the ship continued to fly on course for approximately five minutes, even though the airspeed began to drop. It then became necessary to sacrifice altitude for airspeed and at 500 feet the pilot decided to make a wheels-up landing in the nearest available field. Three attempts were made to land, and finally a cotton field and peanut patch was chosen, and an emergency landing was made. In choosing the field for landing, the pilot did not see three terrace rows, which were at the edge of the field. His airplane struck the terraces, crashed and caught fire."

6-27-43C. Lakeville, California. At 1050, a Bell P-39Q crashed 15 miles north of Hamilton Field in the vicinity of Lakeville, California, killing pilot 2Lt. Andrew Melek, Jr. Investigators stated, "Lt. Melek was on a combat training flight. In his attempt to evade his attacker, he stalled [the] airplane and went into a spin to the right. On his recovery he over controlled his airplane and went into a spin to the left and crashed. From investigation at the scene of the accident it is believed he was recovering from the left hand spin because the point of impact showed the airplane hit in a flat attitude."

6-27-43D. Kingsport, Tennessee. At 1320 CWT, a Vultee BT-13A attempting an emergency forced landing crashed at Kingsport, Tennessee, killing pilot 1Lt. Joseph W. Madden and fatally injuring passenger W/O Joseph A. Cepko. The airplane was on an individual navigation flight from Pittsburgh, Pennsylvania, to Knoxville, Tennessee, when the pilot became lost. The pilot flew around for about two and a half hours trying to figure out where he was with out success. Running low on fuel, the pilot endeavored to make an emergency landing. He flew over the golf course at Kingsport but did not attempt a landing there. The pilot chose a field and attempted his landing. The starboard wing struck an earthen mound while the airplane skidded across the ground. The airplane swung around and crashed into the mound, bursting into flames. The passenger was pulled from the wreckage alive but he died at approximately 2300 that evening. The pilot, passenger and airplane were stationed at Gunter Field, Montgomery, Alabama.

6-27-43E. Randolph, Vermont. At 1535, a Boeing B-17F suffering the failure of all four engines and attempting an emergency forced landing, crashed at Randolph, Vermont, after striking trees on the approach, killing pilot 2Lt. Leonard L. Bolon, co-pilot 2Lt. Harry A. Turner and gunner SSgt. Donald E. Burgan. Seven crewmembers were able to parachute to safety from about 8,000 feet, suffering only minor injuries. Investigators stated, "The inboard engines were feathered due to a drop in oil pressure. Immediately after this, oil pressure on the outboard engines dropped to zero and the engines began smoking and vibrating. The pilot ordered the crew to jump and seven members of the crew complied with the order. The tail gunner, co-pilot and pilot stayed with the ship in an effort to make a crash landing. On the approach to an open field, a large tree in their gliding path disrupted their flight and caused them to crash. The airplane was demolished and the three crewmembers were killed instantly." Investigation revealed that the co-pilot had inadvertently set the inter-cooler controls to "hot" instead of the normal position. This caused a thinning of the engine oil and, combined with over-leaning the fuel mixture, caused detonation in the engine cylinders. The increased load on the outboard engines caused their failure. The engineer had noticed the incorrectly set inter-cooler levers and corrected them, but not in time to prevent the failure of the outboard engines. Suffering minor injuries parachuting to safety were: 2Lt. John F. Cordell, navigator; 2Lt. John R. Sampson, bombardier; TSgt. Jack B. Morrison, radio operator; TSgt. Jessie V. Pace, engineer; SSgt. Oscar F. Krummel, assistant engineer; SSgt. Charles D. O'-Connell, assistant radio operator; Pvt. Cleo F. Stevens, gunner. The B-17 had taken off at 0630 CWT from Grand Island, Nebraska, for a flight to the European Theatre of Operations via Bangor, Maine.

6-27-43F. Louisville, Kentucky. At 1530 CWT, three Aeronca L-3B airplanes encountering a severe local windstorm crashed at the 2000 and 2200 blocks of Douglass Avenue in Louisville, Kentucky, killing Pfc. John E. Stilley, 19, Kansas City, Missouri, aboard L-3B #43-26833. The L-3B had taken off and climbed to an altitude of about 500 feet and remained in the pattern for the airport at Louisville. A severe updraft forced the L-3 up to about 1,500 feet before the pilot was able to regain control. The airplane was then forced to the ground by a severe downdraft, striking the roof an apartment house at 2053 Douglass and then colliding with another apartment house at 2057 Douglass before crashing to the ground in the driveway. Three female civilians received minor injuries when the airplane crashed into the apartment building. They were: Mrs. Ernest Finch, Sheboygan, Wisconsin, Mrs. R.E. Smith and Miss Virginia Smith, 2057 Douglass, Louisville. The pilot, F/O John A. Werner, Medford, Wisconsin, received very serious injuries. F/O Roy C. Lewis, Los Angeles, California, suffered serious injuries and passenger Pfc. Franklin Wolf received minor injuries when their L-3B (#43-26834) crashed into a tree in the backyard at 2064 Douglass. Pilot F/O William J. Hargis and passenger Pfc. Marvin Woods received minor injuries when their L-3B (#43-26839) crashed into the backyard at 2240 Douglass. Mrs. W.P. Oldham, 74, suffered a heart attack at the scene.

6-27-43G. Huntington Beach, California. At 1525, a Lockheed P-38G suffering engine trouble was abandoned by its pilot over the Pacific Ocean and crashed onto the shore at Huntington Beach, California, killing four children playing on the beach. Six other civilians were injured, three of them seriously. The pilot, F/O Gene R. Fair, parachuted to safety and was uninjured. Investigators stated, "F/O Fair was airborne at 1422 PWT for a local training mission in number-two position in a flight of four P-38 airplanes. After practicing combat maneuvers at an altitude of 10,000 to 12,000 feet for approximately one hour he noticed the right coolant warning light flickering on and off. He then opened both coolant shutters and shortly thereafter the right engine lost power. The pilot feathered the right propeller and proceeded to trim the airplane for single engine operation and started a gentle left turn. At this time his position was approximately 8,000 to 10,000 feet altitude over Huntington Beach, California. Before the left turn was completed, F/O Fair noticed a loss of manifold pressure on the left engine and abandoned the airplane. After the pilot bailed out, the aircraft made a sweeping 300 degree turn to the left, crashing on the beach." Witnesses on the ground reported that the airplane was pointed out to sea before it turned around and crashed onto the beach in a nose down attitude. The airplane exploded into flames upon impact. The pilot parachuted from the airplane at an altitude of approximately 3,000 feet, descending into the Pacific Ocean about four miles from the scene of the accident. Killed in the crash were: Roselva Barrego, 5, killed instantly; Frances Silva, 4, died at 0145 on 6-28-43; Faith Barrego, 8, died at 0345 on 6-28-43; Mary Silva, 13, died at 0700 on 6-28-43. Seriously injured were: Frank Barrego, age 6; Ruben Silva, age 6; Rudolph Silva, age 9. All resided at 17th and Verrano Streets, Huntington Beach. Mr. Harold S. Keeler, age 50, Florentine Keeler, age 48, both of Arcadia, California, and Genevieve Keeler, 39, of Garvey, California, received minor injuries.

6-27-43H. Castle Rock, Colorado. At 2030, a Lockheed P-38H collided with terrain three miles east of Castle Rock, Colorado, killing pilot 2Lt. Roy F. Frew. The pilot had taken off from Peterson Field, Colorado Springs, Colorado, on a flight to Chugwater, Wyoming. Near Castle Rock, the pilot encountered severe thunderstorms with a low ceiling. The pilot soon became lost and apparently attempted to make a forced landing. The airplane was observed flying very low over Castle Rock. Moments later the P-38 collided with a ridge at an elevation of approximately 6,800 feet, about 500 feet above the elevation of Castle Rock. The airplane exploded violently into flames and the pilot was killed instantly. Local civilians reported that the weather at the time of the accident was "very poor" with very low clouds, thunderstorms and heavy rain.

6-27-43I. Sandersville, Georgia. At approximately 2300 EWT, a North American B-25C collided with rising terrain 18 miles northeast of Sandersville, Georgia, killing the crew of six. The airplane had been on a night navigation training flight from Columbia Army Air Base, Columbia, South, Carolina, to Dadeville, Alabama, and return. The airplane did not return to Columbia and was reported missing. The B-25 was eventually located and identified on 8-31-43. Investigation revealed that the airplane collided with a heavily wooded knoll while flying in instrument conditions.

The airplane struck the knoll in a level attitude at about cruise speed, exploding into flames and killing the crew instantly. Investigators speculated that the pilot had attempted to maintain visual contact with the ground by flying under the ceiling, estimated to be less than 1,000 feet. Killed in the crash were: 2Lt. Gilbert P. Johnson, pilot; 2Lt. Mourse A. Nielson, co-pilot; 2Lt. Michael R. Rusnak, bombardier/navigator; SSgt. Daniel Hall, engineer; Sgt. Joseph E. Redmen, radio operator; Sgt. Horace S. Bowen, Jr., gunner.

6-27-43J. Augusta, Georgia. At 2320 EWT, a North American B-25C crashed 20 miles NNE of Augusta, Georgia, killing the crew of seven. Investigators stated, "[The airplane] was returning from a routine navigational training flight [from Columbia Army Air Base, Columbia, South Carolina, to Dadeville, Alabama] and about 20 miles NNE of Augusta, Georgia, crashed into a large tree. The airplane struck in a clearing, bounced over some smaller trees and stopped about 500 feet beyond. Parts were scattered along the path and fire followed the crash. It is believed that the pilot was flying low to maintain [visual] contact with occasional reference to instruments due to weather conditions. If he were on instruments he might have failed to check elevation of terrain along the route or failed to allow for altimeter changes." Killed in the crash were: 2Lt. Eugene W. Feldott, pilot; 2Lt. Matt A. Pansky, co-pilot; 2Lt. Rodney F. Russell, bombardier/navigator; 2Lt. Donald L. Knapp, bombardier/navigator; SSgt. Ralph Powell, engineer; SSgt. Donald W. Austin, radio operator; Sgt. Henry E. Brown, gunner.

6-27-43K. Moore Field, Texas. At 2325, a North American AT-6C crashed at Auxiliary Airfield #1, seven and a half miles east of Moore Field, Mission, Texas, killing 1Lt. Richard T. Ellison. The pilot had returned to Auxiliary Airfield #1 following the completion of an assigned weather checking mission when, at an altitude of about 5,000 feet and with landing lights illuminated, he began to execute a series of slow rolls and snap rolls directly over the field. The AT-6C was seen by qualified witnesses on the ground to enter a tight spiral to the right following a series of rolls. At about 2,000 feet, the pilot appeared to regain control of the airplane, pulling up the nose and leveling the wings. The AT-6C immediately went into steep spiral to the left, crashing into the ground and exploding into flames. Investigators noted that the airplane struck the ground with such velocity that the airplane was completely destroyed and the pilot severely mangled and dismembered.

6-28-43A. Thomasville, Georgia. At 0900 EWT, a Bell P-39N spun to the ground four miles southeast of Thomasville, Georgia, killing pilot 1Lt. James P. Emerson. Investigators speculated that the airplane went into a flat spin and the pilot was unable to recover before spinning to the ground. Investigation revealed that the pilot had released the starboard door prior to impact with the ground. The pilot's body was found about five yards from the airplane.

6-28-43B. Ponder, Texas. At 1007, a Bell P-39Q crashed two miles west of Ponder, Texas, killing pilot 2Lt. James H. Edwards. Witnesses on the ground reported that when first noticed, the airplane was in a spin at an altitude of about 6,000 or 7,000 feet. The airplane went in and out of a flat spin and the direction of the spin changed on at least two occasions. The pilot stopped the spin momentarily at about 1,000 feet. The airplane then dove into the ground at angle of about 45 degrees with engine wide open.

6-28-43C. La Junta, Colorado. At 1000, a Vultee BT-13A crashed at Mesa DeMayo near La Junta, Colorado, killing pilot Capt. Frederick R. Ireland and seriously injuring passenger 2Lt. Robert D. Blackman. The airplane had departed Lowry Field, Colorado, for Goodfellow Field, Texas, at 0820 and encountered instrument conditions. The pilot apparently was attempting to stay below the ceiling and collided with the mesa 400 feet from the top. Two cowboys found Lt. Blackman wandering around in shock about four and a half miles from the scene of the accident. Investigation revealed that the flaps were down 20 degrees at the time of impact and the ceiling was estimated to be about 500 feet above ground level.

6-28-43D. Winchester, Tennessee. At 1120 CWT, a Bell P-39F crashed nine miles west of Winchester, Tennessee, killing pilot 2Lt. George Shawkins. Investigators stated, "Lt. Shawkins was given a combat formation mission after which he was to proceed on a single ship strafing mission against ground troops with a minimum authorized altitude of 100 feet above the terrain. At completion of the formation mission he continued the strafing mission alone. While he was in the process of strafing ground troops, he made one run then swung up into a steep turn to the right and started back for another run at a second position close to the first. The airplane hit the top of a tree about 50 or 60 feet high with the right wing, then proceeded to drop where it hit the trunk of a second tree about 100 feet from the first. This tree trunk was about 24 inches in diameter where the right wing broke it off. The right wing of the airplane was torn off and remained lodged beneath the fallen tree. The rest of the airplane continued approximately another 50 feet where it tore out the corner of the Harmony School building. The airplane disintegrated, leaving the propeller in a path between the right wing and the corner of the schoolhouse. There was no hole in the ground to indicate a point of impact. This would indicate that the airplane might have dropped close to the ground after hitting the first tree and into the school building. The pilot was killed and his body was found intact inline with the main part of the wreckage about 500 feet beyond. Parts of the airplane were scattered for distance of approximately 600 feet beyond the building."

6-28-43E. Liberty Field, Georgia. At 1425 EWT, a Lockheed-Vega R-37 (similar to the B-34 type) suffering a double engine failure crashed while attempting an emergency landing near Liberty Field, Camp Stewart, Georgia, killing pilot 1Lt. Norman E. Schewe and injuring four crewmembers. Investigators stated, "The pilot was in the process of changing fuel tanks when both engines started cutting out. He was forced down in a heavily wooded swamp area with both engines dead. The aircraft was on a low (1,500 feet) towing mission. After 40 minutes of cruising, the fuel tanks in use were changed to the auxiliary fuel tank. The engine failure resulted from the loss of fuel pressure. All efforts to restart engines by changing fuel tanks [and the] throttle and mixture settings were unsuccessful. The aircraft was losing altitude all this time and the pilot attempted to reach a clearing ahead for a crash landing. But realizing the distance was just a little more than his glide, he was obliged to pancake the ship into a forest adjacent to the clearing, about 200 feet away. The plane came in tail low, clipping the treetops for 100 feet before it plunged down, cutting through more trees and finally coming to a stop against the trunk of a tree having a diameter of four feet. The fuselage lay parallel with the ground, both wings crushed approximately midway from the wing roots and both engines and propellers damaged. Empennage was completely smashed. The nose section received the most damage, being crushed back as far as the radio compartment." Injured in the crash were: SSgt. Bernard F. Wojcinski, engineer; Cpl. Oliver C. Hornick, crewmember; Pfc. Richard E. Barrie, radio operator; Cpl Charles Q. Racer, passenger.

6-28-43F. Pittsburgh, Kansas. At 1515, an Aeronca L-3B crashed three miles west of Pittsburgh, Kansas, killing civilian instructor Max W. Larson and student 2Lt. Royal E. Forrester. Investigators stated, "This student and instructor had been doing a series of lazy-eights and had executed two simulated forced landings. Just before to the accident, the airplane was seen gliding to the east at an estimated altitude of 50 feet. The airplane started a turn to the left and suddenly dropped out of sight. The airplane struck the ground at approximately a 60 degree angle."

6-28-43G. Visalia, California. At 1730, a Ryan PT-22 crashed near Visalia, California, killing A/C William D. Rencher, Jr. A civilian witness stated that he observed the airplane spinning at an altitude of about 300 feet and that the airplane made one complete turn of spin before impacting the ground on its nose.

6-29-43A. Ten Sleep, Wyoming. At a time after 0001 PWT, a Boeing B-17F collided with rising terrain in the Big Horn Mountains near Cloud Peak, 18 miles NNE of Ten Sleep, Wyoming, killing the crew of ten. The B-17 took off from Pendleton Field, Oregon, at 2052 PWT (6-28-43) and disappeared on a flight to Grand Island, Nebraska. The B-17 was found on 8-11-45. The airplane was last heard from when it transmitted a routine position report at 0000 PWT, reporting its position as approximately 60 miles due south of where it collided with a ridge at an elevation of 12,800, about 50 feet from the peak. Investigators speculated that the pilots sighted the terrain at the last moment and attempted to pull the airplane up and over the ridge. The airplane smashed into the west side of the mountain in a tail-low attitude, breaking up as it careened over the peak. Wreckage and bodies were scattered over one-third of a mile across the peak and both the western and eastern slopes. Killed in the crash were: 2Lt. William R. Ronaghan, pilot; 2Lt. A. J. Tilotta, co-pilot; 2Lt. Leonard H. Phillips, navigator; 2Lt. Charles H. Suppes, bombardier; TSgt. James A. Hinds, engineer; SSgt. Ferguson T. Bell, radio operator; SSgt. Lee V. Miller, assistant engineer; SSgt. Charles E. Newburn, assistant radio operator; SSgt. Jake F. Penik, gunner; SSgt. Lewis M. Shepard, gunner.

6-29-43B. Saluda, Virginia. At 1740, a Republic P-47C collided with terrain five miles northwest of Saluda, Virginia, killing 1Lt. Raymond C. Zettel. The P-47 collided with trees while trying to fly under a 300-foot ceiling at an estimated speed of 200 mph. The P-47 had taken off from Middleton, Pennsylvania, on a ferry flight to Langley Field, Hampton, Virginia.

6-29-43C. Alamogordo, New Mexico. At 1618, a Consolidated B-24E crashed while attempting a landing at Alamogordo Army Air Base, Alamogordo, New Mexico, killing ten crewmembers and seriously injuring another. A thunderstorm was approaching the field when the airplane asked for landing instructions for Runway 3. The pilot overshot his first attempt for Runway 3 and went around. As the airplane flew the pattern, the wind changed in favor of Runway 7. As the storm approached the field, winds increased from 7 knots to 35 knots, crosswind for Runway 3. The tower radioed the airplane in order to advise changing to Runway 7 but was unable to make contact and on the second attempt the airplane again failed to land. Moments later the control tower lost power. The airplane attempted a third landing and again came in high and failed to line up because of the severe crosswind. On the last go-around, a severe rainsquall moved into the local flying area and the tower lost visual contact with the B-24 as the visibility deteriorated. The surviving crewmember, assistant engineer SSgt. Jimmie V. Staffiera, who suffered serious injuries, reported that visibility was very poor and severe turbulence violently rocked the airplane persistently in the pattern. The airplane was flying in the pattern at an altitude of about 500 feet with landing gear and flaps extended when the pilot lost control of the airplane in the rainstorm and crashed approximately one mile northwest of the airbase. Killed in the crash were: 2Lt. Leander M. Sherman, pilot; 2Lt. John E. Hennessey, co-pilot; 2Lt. Leonard M. Sundahl, navigator; 2Lt. Warren L. Krausse, bombardier; TSgt. Franklin C. Houseman, engineer; Sgt. James H. Supinsky, radio operator; Sgt. Wilbur R. Zimmerman, assistant radio operator; Sgt.

John L. Peterson, gunner; Sgt. Roy J. C. Payne, gunner; Cpl. Fred D. Mathews, passenger.

6-29-43D. Blackland Army Air Field, Texas. At 2045, civilian crash truck crewman W.R. Elrod was struck and fatally injured by the rotating propeller of a Beech AT-10BH at Blackland Army Air Field, Waco, Texas. The airplane had gotten stuck while it was attempting to taxi to the parking area at the field. After a couple of attempts to free the aircraft by themselves, instructor 2Lt. Robert A. Servais and student A/C Manuel A. Pidcock were joined by the crash truck crew, who were ordered to the airplane by the officer in the control tower. As Lt. Servais was attempting to clear clumps of dirt from in front of the port main wheel, Mr. Elrod walked into the path of the rotating port propeller, being struck in the head and left arm. Mr. Elrod died at 0945 on 6-30-43.

6-30-43A. Newhall Pass, California. At 1100, a Lockheed P-38G crashed at Newhall Pass, California, killing pilot 2Lt. Loncie L. Tucker, 28. The airplane had taken off from Glendale, California, with three other P-38 airplanes on a formation cross-country flight. The airplane was observed to drop back from the formation and feather the port propeller. Investigators speculated that the pilot might have been endeavoring to make a single-engine forced landing when the airplane entered a flat spin to the left. Witnesses reported that the main landing gear, but not the nose landing gear, were extended just prior to the airplane entering the spin. The pilot attempted to parachute to safety but apparently deployed his parachute prematurely, becoming caught on the tail surfaces and being dragged down with the airplane, which landed upright and relatively intact. The starboard engine erupted into flames and burned for a short time.

6-30-43B. Amagon Field, Arkansas. At 1347, a Vultee BT-13A crashed one-half mile northwest of Amagon Auxiliary Airfield, Arkansas, killing A/C Albert L. Grewe, 19, Des Plaines, Illinois. Investigators stated, "The [airplane] was observed in a right spin at an estimated altitude of 4,000 feet. At approximately 1,500 feet the rate of spin slowed down, but increased again. The pilot cleared the cockpit on the right side, evidently planning to make a parachute jump, but he apparently jumped from the airplane just prior to ground contact, was thrown through the propeller, instantly killed and the airplane completely wrecked. The parachute was never opened. From examination of the wreckage, the airplane was in a flat spin attitude when it crashed into the ground."

6-30-43C. Saxe, Virginia. At 2210, a Beech UC-45A crashed at Saxe, Virginia, killing six crew and passengers. The airplane had taken off from Charlotte, North Carolina, on personnel flight to Bolling Field, Washington, D.C. Witnesses on the ground reported that the airplane had circled the town twice before entering a 45-degree dive toward the ground and that the engines were "raced" just before impact. The Accident Classification Committee stated, "The committee is of the opinion that there was a failure of some part of the engines or airplane which necessitated a forced landing, that the approach was made downwind, that the pilot thought that he was undershooting so he stretched his glide, applied power too late to prevent a stall and the airplane dove into ground. An explosion occurred on impact which completely disintegrated the airplane and engines." Killed in the crash were: Maj. William M. Marks, pilot; SSgt. William Evans, Jr., engineer; Col. L. Richard Baughman, passenger; Col. Kamiel Maertens, passenger; Lt. Col. Milton H. Pressley, Jr., passenger; Lt. Col. Felix A. Todd, Jr., passenger.